A Half-Century of Physical Asymptotics and Other Diversions

Selected Works by **Michael Berry**

reproduced by permission of Chrystal Cherniwchan

A Half-Century of Physical Asymptotics and Other Diversions

Selected Works by **Michael Berry**

Michael Berry
University of Bristol, UK

World Scientific

NEW JERSEY • LONDON • SINGAPORE • BEIJING • SHANGHAI • HONG KONG • TAIPEI • CHENNAI • TOKYO

Published by

World Scientific Publishing Co. Pte. Ltd.
5 Toh Tuck Link, Singapore 596224
USA office: 27 Warren Street, Suite 401-402, Hackensack, NJ 07601
UK office: 57 Shelton Street, Covent Garden, London WC2H 9HE

Library of Congress Cataloging-in-Publication Data
Names: Berry, Michael V., author.
Title: A half-century of physical asymptotics and other diversions :
selected works by Michael Berry / Michael Berry, Bristol University, UK.
Description: Singapore ; Hackensack, NJ : World Scientific, [2017] |
Includes bibliographical references.
Identifiers: LCCN 2017016848| ISBN 9789813221192 (hardcover ; alk. paper) |
ISBN 9813221194 (hardcover ; alk. paper) | ISBN 9789813221208 (pbk. ; alk. paper) |
ISBN 9813221208 (pbk. ; alk. paper)
Subjects: LCSH: Mathematical physics. | Geometric quantum phases. | Caustics (Optics) |
Quantum theory. | Quantum chaos.
Classification: LCC QC19.6 .B47 2017 | DDC 530.15/6362--dc23
LC record available at https://lccn.loc.gov/2017016848

British Library Cataloguing-in-Publication Data
A catalogue record for this book is available from the British Library.

Preface

When Dr K K Phua invited me to publish a collection of my papers, selecting them and ordering them seemed easy. But it wasn't. Should I choose my favourite papers, or my most-cited papers, or select on some other basis? And the simplest ordering — chronological — would be incoherent, because a lifetime's work is multiply connected: over a variety of areas of physics, and over time. I recalled the 1976 retirement speech of my late colleague Sir Charles Frank: "Physics is not just Concerning the Nature of Things, but Concerning the Interconnectedness of all the Natures of Things". Here, 'things' applies not only to objects, but also to the conceptual structures that have concerned me as a theorist.

In the end, I compromised. The book is organised around research themes — areas where I have made significant contributions. Each theme is the subject of a chapter, comprising a number of my works and a descriptive introduction emphasising connections between the themes. A life in science consists of more than published research papers, and so, to give a more nuanced picture, I also included some tributes to other scientists, a selection of book reviews, travel reports, unpublished speeches and other musings.

Many of my papers — more than is customary nowadays — are single-author, and the introductions to the chapters mention few people — essentially, only those I have collaborated with. But I am not under the illusion that mathematical science is a solitary activity. Quite the contrary, I have been inspired and influenced by the discoveries and scientific styles of a number of physicists and mathematicians. Tributes to some of them constitute Chapter 7. I have enjoyed collaborations with my graduate students and other scientists; their contributions have been indispensable at all levels from conceptual to technical. Chance conversations have led to a number of 'claritons' (the elementary particles of sudden understanding); these are indicated by acknowledgements in the resulting published papers. And of course, the lists of references in all my papers implicitly recognise the work of many others.

Since my undergraduate studies at Exeter (1959–1962) and my Ph.D at St Andrews (1962–1965), my whole career has been at Bristol. For more than 50 years, I have benefited enormously from the friendly and intellectually supportive atmosphere of the Bristol physics department, from successive Bristol University administrations — largely hands-off but helpful when needed — and from the Royal Society, which supported me as a research professor for several decades. I have been offered positions at other prestigious universities in the UK and elsewhere. But after serious consideration, I always decided to remain in Bristol (to the disappointment of my late mother, even when I told her that the prestige comes from being invited — you don't get more if you actually go).

I thank my children for surviving a father who was often travelling, and lost in abstraction even when physically present. Above all, I thank my wives Eve, Lesley and (for the longest time) Monica, for their tolerance and their loving support during all these years of my obsession with physics.

Bristol 2017

Contents

Note on citations

In the introductions to the chapters, papers reprinted in this book are referred to by chapter and section numbers (e.g. [1.1]). Papers that are not reprinted here are referred to by their number in my full publications list (e.g. [B34]) in Chapter 11. Unpublished works are labelled U (e.g. [U1]).

Acknowledgements

I thank my co-authors, and the publishers listed below, for graciously giving permission to reproduce the indicated items. When citing these papers, please give the full original publication details, in addition to references to this book.

American Institute of Physics: [1.9], [6.2]
American Mathematical Society: [7.6], [7.10]
Current Science: [7.2]
Editions Belin: [9.16]
Elsevier: [2.8], [6.1]
Nature: [1.7], [2.5], [7.1], [9.4]
Oxford University Press: [6.4]
Princeton University Press: [4.5]
Royal Society of Edinburgh: [7.8]
Royal Society of London: [1.1], [1.5], [1.6], [1.8], [3.2], [3.3], [4.1], [4.3], [4.6], [4.7], [9.6], [9.7]
SIAM Review: [3.6]
SPIE: [1.3], [1.4]
Springer: [3.4], [6.3]
Swiss Physical Society: [7.11]
Taylor and Francis: [2.2], [2.12]
Times Higher Education: [9.5]
UK Institute of Physics: [1.2], [2.1], [2.3], [2.4], [2.6], [2.7], [2.9], [2.10], [2.11], [2.13], [3.1], [3.5], [4.4], [5.2], [5.3], [7.5], [9.1], [9.2], [9.3], [9.13]

Full acknowledgement of each item reproduced is printed on the first page of that item.

Chapter 1
Phases

From the beginning, I was fascinated by waves, especially their characteristic property, namely phase. My first paper [B1], published in 1965, when I was still a graduate student in St Andrews, concerned the phase difference between two waves. It had been claimed that this was different for moving observers. We showed that this is wrong: the phase difference is invariant under Lorentz (also Galilean) transformations.

That was essentially a pedagogical exercise concerning phase. My first original contribution in this area, with John Nye [1.1], reported the discovery of the singularities of phase as ubiquitous features of all kinds of waves, e.g. light and sound, as well as the more exotic waves in superfluids and superconductors. Around such a singularity, the phase changes by a multiple of 2π. The waves swirl around the singularities, and the phase gradient is an irrotational field, so phase singularities are wave vortices. They are places of zero intensity, so they are also nodes. Geometrically, these nodal sets are points in the plane, and lines in space; they are stable under perturbation, generalising the dark fringes of classical interferometry, which are surfaces in space, and not stable. We originally used the term 'wavefront dislocations', because phase singularities are analogous to defects in crystals. Close to a singularity, the phase varies on scales smaller than the wavelength, so phase singularities exemplify superoscillations (Chapter 5) — though it was several decades before I recognised the connection.

The collaboration arose from Nye's research in glaciology. We had been discussing the application of radio waves to explore the Antarctic ice sheet and the underlying terrain. Nye devised a student laboratory project, in which ultrasound replaced radio waves and crinkly kitchen foil replaced Antarctica's bottom. He noticed that, as the transmitter/receiver was moved, the number of wave crests in the echo changed; we eventually realised that this indicated that the reflected wave contains phase singularities, the general theory of which we developed.

The 1974 research with John Nye is my only significant work with a senior colleague. Nye was and is a master of clear thinking and clear writing as well as an exemplar of decent scientific behaviour.

An early application revealed phase singularities in the eponymous Aharonov-Bohm (AB) quantum effect, discovered in Bristol in 1959 and confirmed experimentally by our late Bristol colleague Robert Chambers (at that time, I was a schoolboy applying to universities, and Bristol rejected me). In 1980, in an undergraduate project, we found theoretically, and explored experimentally [1.2], a classical analogue of the AB wavefunction, with ripples on the surface of water (instead of electron waves) encountering a bathub vortex (instead of a magnetic flux line). We included Chambers as an author of the paper after he showed us his unpublished sketches from 1960, showing wavefronts emerging from the phase singularity at the magnetic flux line.

The AB arrangement can be regarded as a kind of interferometer, in which a shift of fringes measures the phase change generated by magnetic flux. Much later [B397], this led to the understanding that every interferometer is threaded by lines of phase singularity, whose number is an integer approximation to the (usually non-integer) phase being measured. I have explored many physical and mathematical aspects of the AB wavefunction [B97, B157, B309, B310, B487], and remain fascinated by it.

The initial response to the paper with Nye was muted, but following optical experiments in the Ukraine in the early 1990s and connections with the orbital angular momentum of light, the subject of optical vortices has exploded into a substantial area of research worldwide. Here, I include two conference reviews, [1.3] and [1.4], with the latter including polarization singularities (discovered by Nye and Hajnal in the 1980s) and also the original phase singularity, discovered by William Whewell in the 1830s in the wave patterns of the tides (more history is in [B105] and [B342]).

In three dimensions, optical vortices are lines which can have non-trivial topology: they can be linked and knotted. This was first shown in a collaboration ([1.5] and [B333], see also [B328]) with Mark Dennis, who has greatly extended and deepened this area of wave topology and knot theory generally. (For knotted LEGO, see the end picture and [B280].) Also with Dennis came the germination ([B364], see also [B404]) of an idea I had seeded several decades before, that optical vortices possess quantum cores: the dark light of a phase singularity can be regarded as a window, opening to our view the faint glimmering of the quantum vacuum (see also Chapter 6). The analogous acoustic vortices — threads of silence — also possess cores, disturbed by the whispering of Brownian motion.

In white light, interference produces characteristic colour patterns associated with phase singularities ([B346], [B347]). This inspired Edward Cowie to compose 'Colours of dark light' for string quartet and oboe.

Rather different was my 1984 paper [1.6] describing the geometric phase accumulated by the quantum states of systems that are slowly cycled. This has turned out to a clarifying and unifying concept that has influenced many areas of physics; it has been reviewed many times, including by me in [B187] and [B198]. I include here an account [1.7] of how I found the phase, and anticipations of aspects of it by other people (see also [B212]). The counterpart for non-chaotic systems in classical mechanics, namely John Hannay's 'geometric angle' (see [B132]) has turned out to be influential in several areas of dynamics.

The application of geometric phase ideas that pleased me most, but alas, has not enjoyed the wider appreciation we hoped for, was to illuminate the spin-statistics connection (SS) that is fundamental in quantum physics. SS relates the phase (zero or π) associated with the exchange or permutation of identical quantum particles to their spin (integer or half-integer). In papers [1.8] and [1.9], resulting from a collaboration with Jonathan Robbins, SS is associated with the topology of a Hilbert space incorporating identicalness.

Proc. R. Soc. Lond. A. **336**, 165–190 (1974)
Printed in Great Britain

Dislocations in wave trains

By J. F. Nye and M. V. Berry
H. H. Wills Physics Laboratory, University of Bristol

(*Communicated by F. C. Frank, F.R.S. – Received* 17 *January* 1973)

When an ultrasonic pulse, containing, say, ten quasi-sinusoidal oscillations, is reflected in air from a rough surface, it is observed experimentally that the scattered wave train contains dislocations, which are closely analogous to those found in imperfect crystals. We show theoretically that such dislocations are to be expected whenever limited trains of waves, ultimately derived from the same oscillator, travel in different directions and interfere – for example in a scattering problem. Dispersion is not involved. Equations are given showing the detailed structure of edge, screw and mixed edge–screw dislocations, and also of parallel sets of such dislocations. Edge dislocations can glide relative to the wave train at any velocity; they can also climb, and screw dislocations can glide. Wavefront dislocations may be curved, and they may intersect; they may collide and rebound; they may annihilate each other or be created as loops or pairs. With dislocations in wave trains, unlike crystal dislocations, there is no breakdown of linearity near the centre. Mathematically they are lines along which the phase is indeterminate; this implies that the wave amplitude is zero.

1. Observation of dislocations

In this paper we introduce a new concept into wave theory: using elementary arguments, we show that wavefronts – that is, surfaces of constant phase – can contain dislocation lines, closely analogous to those found in crystals. The work originated in attempts to understand radio echoes from the bottom of the Antarctic ice sheet; the spatial fine structure, or 'fading pattern', of the echo may be used for precise determination of position (Nye, Kyte & Threlfall 1972; Walford 1972; Nye, Berry & Walford 1972). A laboratory analogue experiment was carried out, using ultrasound instead of radio waves. The relatively low frequency of ultrasound enabled the detailed phase structure of the echo to be studied, and wavefront dislocations were observed. We suspect that dislocations may often have been observed in phase sensitive experiments without their significance being appreciated (see, for example, figure 5 of a paper by Findlay (1951) showing dislocations in a radio wave field reflected from the ionosphere, and the remarks in §8 of the present paper about amphidromic points). It is possible that dislocations may find application in remote sensing as 'markers' in a wave field, because they are definite features recognizable even in the presence of noise. In this paper, however, our purpose is simply to demonstrate that wavefront dislocations exist, to deduce their detailed structure and to examine some of their properties.

We begin by describing the original observation. Ultrasonic pulses from a small source were incident, in air, on a rough surface, and the scattered pulses were received by a small moveable microphone and displayed on an oscilloscope. Each pulse from the source consisted of about 10 sinusoidal waves (frequency 100 kHz) within a smoothly varying (approximately Gaussian) envelope (figure 1*a*). The echo from the rough surface was of longer duration (figure 1*b*), consisting of 50 or more approximately sinusoidal waves of fluctuating amplitude. By moving the microphone along a line it was possible to find points in space where the envelope of the echo had zero strength for some particular time delay. Figure 2*c* shows a

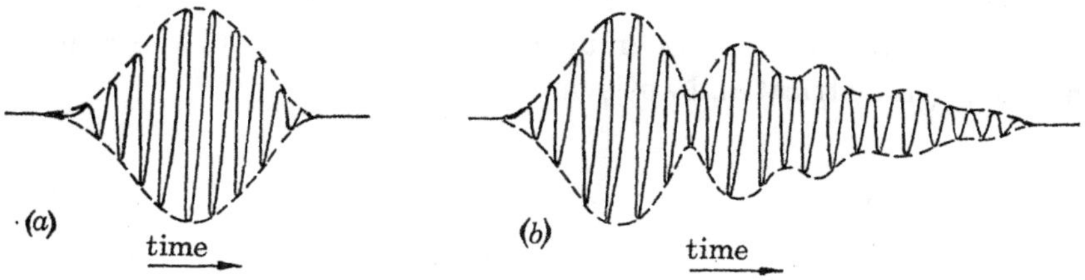

FIGURE 1. (*a*) Quasi-monochromatic pulse from a source. (*b*) The echo from a rough surface.

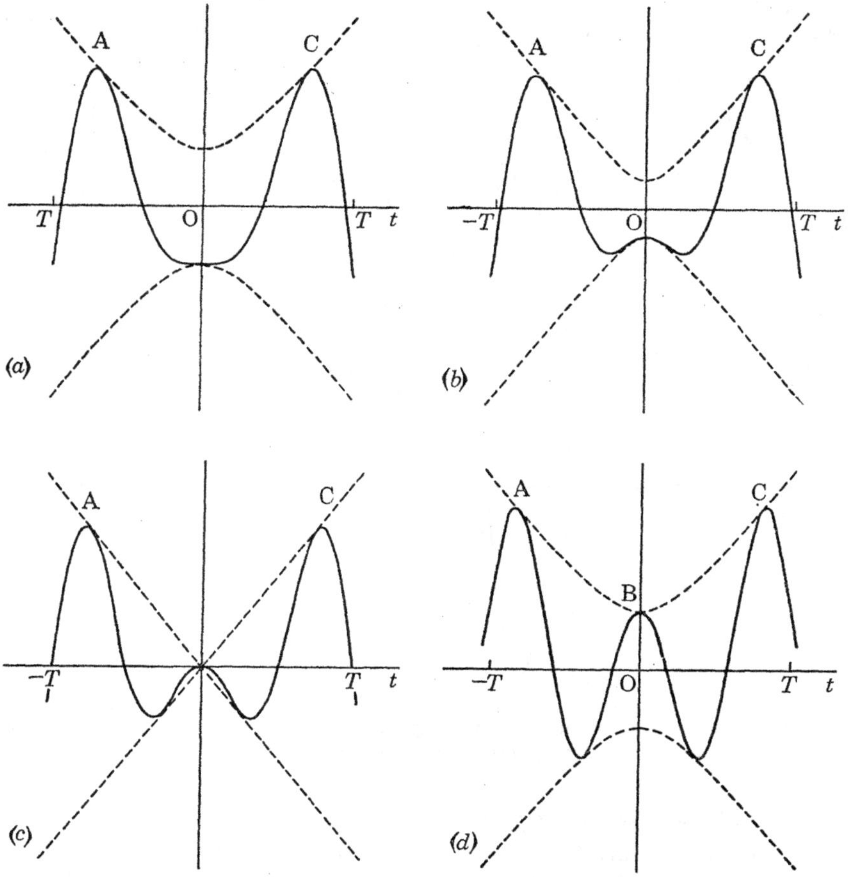

FIGURE 2. An edge dislocation – symmetrical case. Signal versus time t for four different positions of a receiver: (*a*) $x = -2\beta_e^*/k$, (*b*) $x = -\beta_e^*/k$, (*c*) $x = 0$, (*d*) $x = 2\beta_e^*/k$. $T = 2\pi/\omega$ = period.

symmetrical example of the time variation of the signal at such a point; the broken line is the envelope. By moving the microphone first to one side (figures 2*a*, *b*) and then the other (figure 2*d*) one could follow the crests A and C continuously through the transition and see that a new crest B was created between them. Figures 3*a*, *b*, *c*, *d* show an antisymmetrical example of the same thing; again a new crest B appears between crests A and C. In general, the phase of the carrier wave relative to the envelope is such that one observes an unsymmetrical curve which is a linear combination of the symmetrical and antisymmetrical cases.

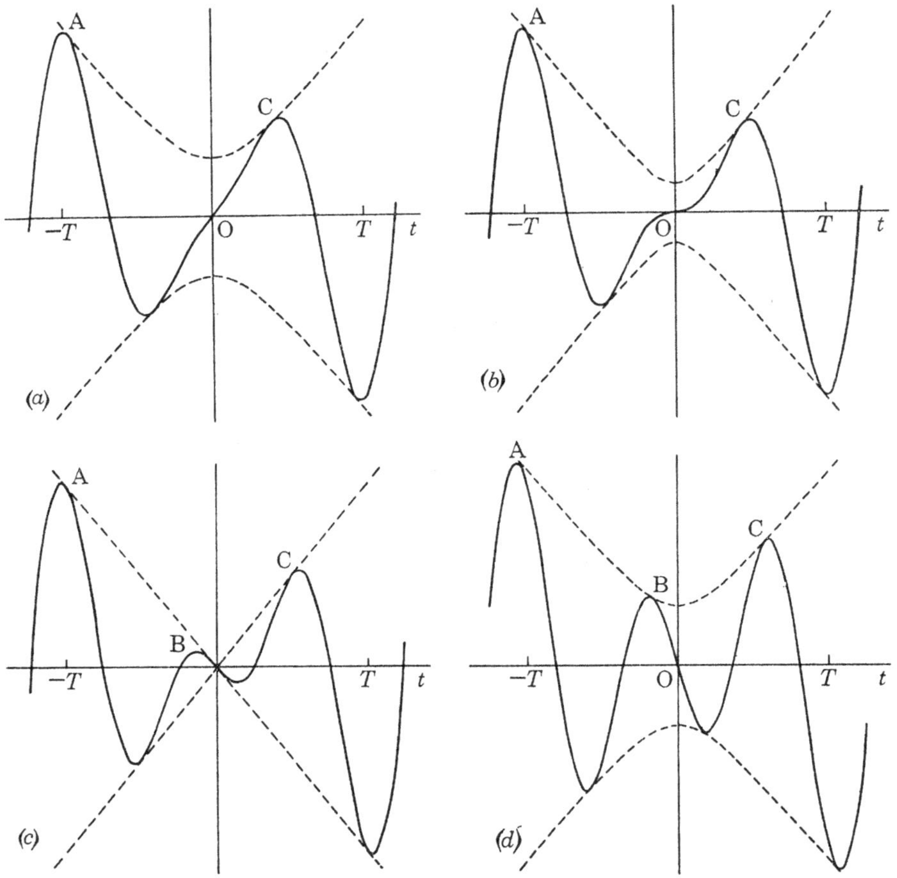

FIGURE 3. As figure 2 – antisymmetrical case. Figures 2 and 3 are calculated by taking the real and imaginary parts respectively of equation (13).

A possible spatial variation at a given time of the echo corresponding to figures 2*a* to *d* is shown in figure 4. As later examples will show, the wavefront dislocation may move backwards, forwards or sideways independently of the motion of the wavefronts; but for simplicity in this introductory section we consider the case in which the dislocation travels with the wavefronts, in a direction normal to their mean plane. As the pattern of crests sweeps upwards a receiver at P notes the passage of crests A and C, and as the receiver is moved continuously to R a new crest B is found to appear as the receiver passes point Q. The spatial pattern has a crest wavefront that ends at N. (The precise location of N depends on what one defines as a crest; we

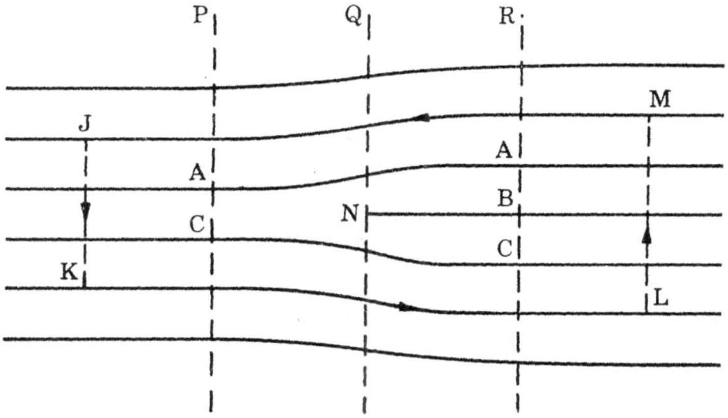

FIGURE 4. Spatial pattern of wave-crests at a fixed time. There is an edge dislocation at N.

shall deal with this presently.) In three dimensions the wavefront ends in a line. We shall call this linear structure an edge dislocation by analogy with the edge dislocations found in crystals (Read 1953; Nabarro 1967). In practice, since the amplitude falls to zero at N, the details of the transition may be submerged in the noise, but, by following a 'Burgers circuit' such as JKLM which is everywhere in places of strong signal and counting the number of wavefronts crossed, one can be quite sure that a dislocation has been enclosed. The analogy with crystal dislocations goes further: we shall see that it is possible to have both screw and mixed edge-screw dislocations, that wavefront dislocations are capable of movement by glide and climb, that they can intersect, that they can collide and rebound, or annihilate each other or be created as loops or pairs.

In general, pulses rather than monochromatic waves are essential for dislocations. In a monochromatic wave the time variation at a point must be strictly sinusoidal. This is clearly impossible at Q in figure 4, for the passage of N past Q is a unique event. At the same time there must be some periodicity if we are to identify a dislocation. Thus the dislocations are structures that disappear at both the monochromatic and the white noise limit; they need both the localization property of a pulse and the oscillation property of a continuous wave. Dispersion is not involved.

However, in certain degenerate cases dislocations can be produced by continuous waves. One is the non-localized interference fringe FF shown in figure 5, as produced, for example, by a Young's two-slit experiment. Monochromatic waves are moving upwards and their amplitude is zero on FF; there is a phase change of π across FF. This could be described as a row of edge dislocations of alternating sign or as a single infinitely extended dislocation. Because all the crests behave identically the time variation at any point is still strictly sinusoidal. Other degenerate cases are the stationary pure screw dislocations and the localized interference fringe which we shall describe later.

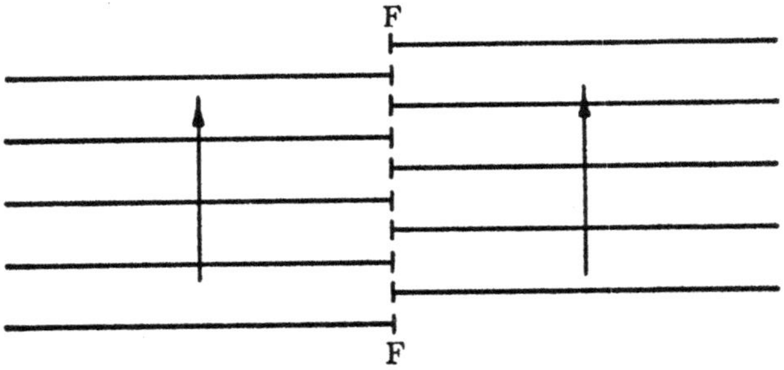

FIGURE 5. A non-localized interference fringe FF. The waves travel upwards.

We have introduced wavefront dislocations as structures observed when a pulse is diffracted from a rough surface and we shall deal more fully with this method of producing them in a separate paper. Here we show that they are a general feature to be expected whenever there is interference between pulses derived from a common oscillator; diffraction by a rough surface is but one way of making them.

2. THE FUNDAMENTAL MATHEMATICAL PROPERTY OF A DISLOCATION

What distinguishes a dislocation from other features of a wave field? To answer this, we must first define precisely the *amplitude* and *phase* of a wave. Let there be an oscillator with fixed angular frequency ω. After an arbitrary phase change the signal from it is amplitude-modulated and made to drive a source of scalar waves, the wave amplitude being proportional to the driving signal. We call such a source *quasi-monochromatic*. At the same time the original signal from the oscillator is changed in phase and amplitude-modulated in other fixed ways and made to drive other sources in a similar fashion. All the wave trains so produced travel together in an undispersive medium and their combined effect is observed at a certain point P. If the original oscillation is proportional to $\cos(-\omega t)$ we denote the resulting signal at P by $\psi_c(t)$. The phase of the original carrier wave is now changed by $\frac{1}{2}\pi$, while everything else is kept fixed. Thus, let the original oscillation now be proportional to $\sin(-\omega t)$, while the envelopes representing the various amplitude modulations and all the phase changes are held fixed, and denote the resulting signal at P by $\psi_s(t)$. We can say that, if the original oscillation were proportional to $e^{-i\omega t}$, the signal at P would be the complex wave function

$$\psi(t) = \psi_c(t) + i\psi_s(t) = \rho(t)\, e^{i\chi(t)}, \quad \text{say,} \tag{1}$$

(χ being real and ρ being real and positive) on the understanding that either the real or the imaginary parts represent the physical quantities. Then

$$\rho^2(t) = \psi_c^2(t) + \psi_s^2(t) \quad (\rho(t) \geqslant 0), \tag{2}$$

and

$$\tan \chi(t) = \psi_s(t)/\psi_c(t). \tag{3}$$

It is natural to define $\rho(t)$ and $\chi(t)$ as the amplitude and phase of the wave; they are both time-varying quantities deducible from the observed functions $\psi_c(t)$ and $\psi_s(t)$. It is important to notice that $\rho(t)$, for example, is not strictly deducible by observing, say, $\psi_c(t)$ alone; it is essential to change the phase of the carrier wave to obtain complete information about the amplitude and phase. With this definition $\rho(t)$ is the envelope of the observed oscillation as the phase of the original oscillation is varied. This is the strict meaning of the broken line in figures 2 and 3.

We now examine the behaviour of $\rho(t)$ and $\chi(t)$ in the neighbourhood of a dislocation. The argument applies to any dislocation, moving in any way; in particular it applies to figure 4, which represents a pure edge dislocation moving with the wave. The changing complex wave function at a fixed position P in space can be represented by a moving point on an Argand diagram whose axes are ψ_c and ψ_s. Because of the quasi-monochromaticity of the source, we expect the point to encircle the origin quasi-periodically, the average time for a circuit being $2\pi/\omega$. Figure 6*a* shows (schematically) the single loop traversed by the point on the Argand diagram between two successive crests A and C of the component ψ_c; this corresponds to the signal sketched in figure 2*a*. At an observation position R, which lies on the other side of the path of dislocation, an extra crest B has appeared between A and C, so that (figure 6*d*) *two* loops now lie between A and C on the Argand diagram, corresponding to the signal of figure 2*d*. (If time is plotted perpendicular to the diagrams of figure 6 the changing wavefunction at a fixed point in space is represented by a helix with varying cross-section and pitch. Viewed in this way the crests A and C remain identifiable, as distinct turns of the helix, throughout the transition.)

As the observation position moves from P to R, the curve representing $\psi(t)$ must change continuously between that of figure 6*a* and that of figure 6*d*, which has different topology. This can only happen if, for some intermediate observation position Q, *the curve passes through the origin*, where ψ_c and ψ_s, and hence the amplitude ρ, are zero. One possible form for the curve at Q is shown in figure 6*c*, while the curve for a position between P and Q is shown in figure 6*b*. We note, in passing, that in the configurations (not drawn) intermediate between figures 6*b* and *c* the phase χ would be retrograde for part of the curve. It is easily verified that the behaviour of $\psi_c(t)$ for the four curves in figure 6 reproduces the symmetrical signal versus time curves of figure 2, while the behaviour of $\psi_s(t)$ reproduces the antisymmetrical curves of figure 3.

Precisely where the dislocation lies is a matter of definition, and we choose the position where the amplitude ρ vanishes. This is the most fundamental definition from a theoretical point of view (see § 6); it is preferable to choosing the cusp in figure 6*b* where a new loop is just about to appear; if we choose the position where the pair of new zeros first appears, we should have a dislocation whose position depends on which 'projection' of the complex wave function represents the physical signal (that is, on the phase of the carrier wave). Since ρ is zero at a dislocation, it follows from the properties of polar coordinates in the Argand diagram that the phase χ is indeterminate. However, it is possible to have zeros of ρ, with indeterminate χ, that do not represent dislocations – the curve in the Argand diagram may

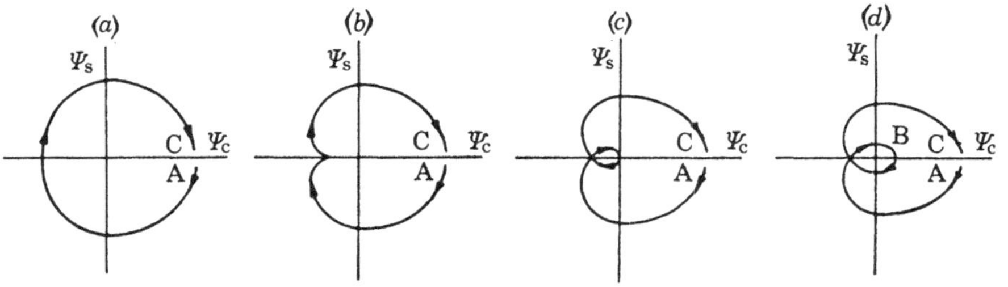

FIGURE 6. Curve in Argand diagram representing the complex wave function as a function of time for four positions near a dislocation. (c) is the configuration at a point through which the dislocation passes.

move up to the origin, and then move away again, without having swept through it; an example is equation (13) of §4 with $(kz-\omega t)$ replaced by $(kz-\omega t)^2$. The essential property of a dislocation is that χ changes by a multiple of 2π on a closed circuit around it. Thus, although the condition $\rho = 0$ is a useful indication of where to look for a dislocation, it is necessary to check that the structure located in this way does indeed have the topology that characterizes a dislocation. We shall examine this topology in a little more detail in §6.

The dislocations in a given wave field are situated at positions where $\psi_C(t) = 0$ and $\psi_S(t) = 0$. At a given instant t these two equations define surfaces in three-dimensional physical space, which intersect in a family of lines (except in some degenerate cases). As time proceeds, the surfaces move and hence the lines – the dislocations – also move. As they move the dislocation lines sweep out surfaces in space. Thus a receiver that displays the signal at a function of time need only explore a line in order to find a dislocation, which will be at the point where the line intersects the surface (with Dr M. E. R. Walford we have mapped some of these surfaces for the special case of the ultrasonic wave field described in §1).

3. Theoretical construction of dislocations: qualitative treatment

We now describe a simple way of constructing dislocations theoretically. Consider identical pulses of scalar waves emitted simultaneously from two point sources S_1 and S_2, and let the signal be observed at a point P. Figure 7*a* shows one of the pulses, as observed at P, a monochromatic wave modulated in amplitude by an envelope, and the convention in figure 7*b* will be used to denote the pulse envelope and the positions of the crests and troughs within it. We use the words crests and troughs rather loosely here, and our argument is only qualitative; a strict treatment will follow later. The signal observed at P will depend on the path difference $S_1P - S_2P$. In figure 7*c* the position of P is such that the path difference is two wavelengths, and so the carrier waves reinforce one another to give the resultant shown

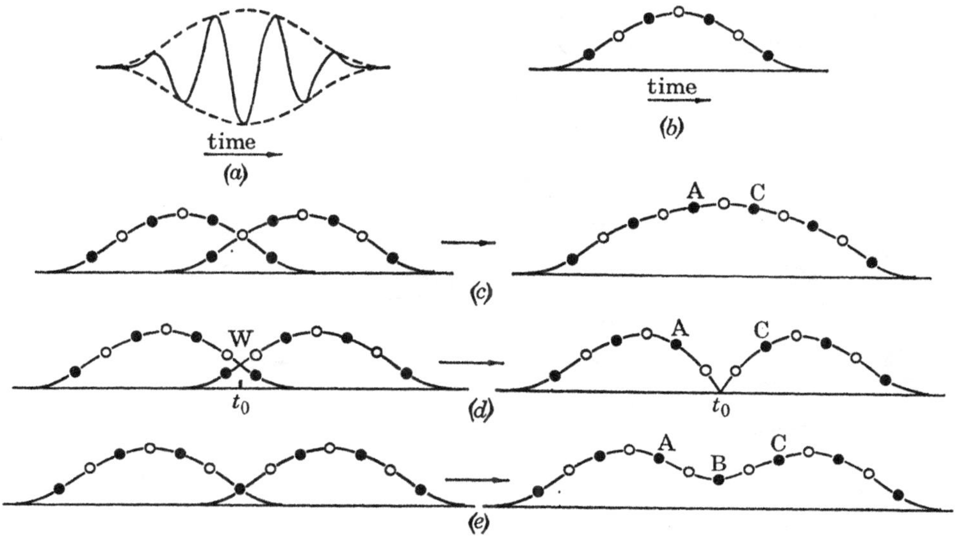

FIGURE 7. The construction of edge dislocations by the interference of pulses. Signal versus time. (*a*) the pulse form; (*b*) schematic representation of pulse: ●, crests, ○, troughs; (*c*) path difference is two wavelengths; (*d*) path difference is $2\frac{1}{2}$ wavelengths, giving destructive interference at time *t*; (*e*) path difference is three wavelengths, and a new crest B has appeared.

on the right (we ignore the fact that the two pulses will be of slightly different strengths). In figure 7*d* P has moved so that the path difference is $2\frac{1}{2}$ wavelengths. There is destructive interference, but it is only complete at one time t_0. In figure 7*e* the path difference is three wavelengths and the carrier waves reinforce again. There are six crests in figure 7*c*, a new one is on the point of appearing in figure 7*d*, and there are seven in figure 7*e*. Clearly, when the path difference is zero there will be four crests, and when the path difference is large there will be two separate pulses with eight crests in all; crests appear as the path difference increases. Each time one appears there is a dislocation. The essential point is that, when the path difference is such that the carrier waves interfere destructively, the interference will only be complete at one time, namely the time when the two envelopes intersect; at this time the amplitude ρ of the combined wave is zero. This principle still holds when the two pulses are of unequal strength and duration, except that then the envelopes may intersect, for a given path difference, at more than one time. The spatial pattern of crests at fixed time shown in figure 4 corresponds with figure 7.

To proceed analytically we retain only those features of the theoretical model just described which are essential for the production of an edge dislocation. The sources are placed at infinity so that there are now two pulses of plane waves crossing at a small angle, and instead of an approximately Gaussian envelope we choose a linear variation of amplitude through the pulse, one pulse rising as the other falls. This amounts to approximating the envelopes near the intersection point W in figure 7 *d* by their tangents. In this way we may calculate in detail the structure in time and space near a dislocation, but the results will not have significance outside the range where the envelopes may be approximated by their tangents.

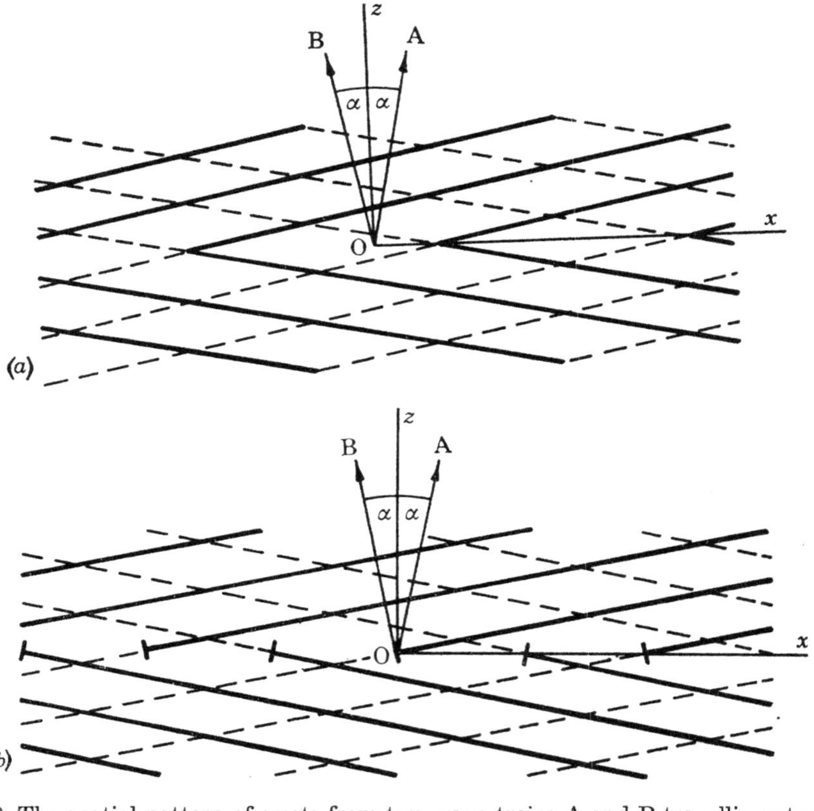

FIGURE 8. The spatial pattern of crests from two wave-trains A and B travelling at an angle to one another. A predominates in the lower half-space and B predominates in the upper half-space. (*a*) symmetrical case; (*b*) alternating case.

The general nature of the result to be expected is seen in figure 8 *a*, which shows the spatial pattern of wave crests at fixed time. The wave train A that travels upwards and to the right has an amplitude that rises towards the rear of the train; the wave train B that travels upwards and to the left has an amplitude that rises towards the front of the train. The amplitudes are arranged to be equal on Ox, so that in the upper half-space train B predominates and in the lower half-space A predominates. Thus, as we move along Ox, new wave crests appear in pairs. In crystal dislocation language Ox is a tilt boundary. Note that figure 8*a* has been drawn so that the origin, where the two waves have equal amplitude, is a zero (approximately midway between a crest and a trough) for both waves; this is the same relationship as in figure 7*d*. Figure 8*b*, on the other hand, shows the pattern of crests that results if the phases are arranged so that at the origin, where the waves have equal amplitude, a crest of B falls on a trough of A. Edge dislocations of alternate character appear along Ox. Figures 8*a* and *b* merely show which of the two wave trains is dominant at any point; the true pattern of resultant wave crests will be similar but with local readjustments. The exact analysis which follows essentially calculates these local readjustments.

4. Exact analysis for edge dislocations

We shall consider complex wave functions ψ satisfying the scalar wave equation

$$c^2\nabla^2\psi = \frac{\partial^2\psi}{\partial t^2}, \tag{4}$$

the wave velocity c being constant.

Let the wave trains A and B, as in figures 8*a* and *b*, be the following solutions of (4):

$$\left.\begin{aligned} \psi_A &= a_0\{1-\beta_e(k_1x+k_3z-\omega t)\}\exp[i(k_1x+k_3z-\omega t-\tfrac{1}{2}\pi)] \\ \text{and}\quad \psi_B &= a_0\{1+\beta_e(-k_1x+k_3z-\omega t)\}\exp[i(-k_1x+k_3z-\omega t+\tfrac{1}{2}\pi)], \end{aligned}\right\} \tag{5}$$

with

$$\omega/k = c, \quad k_1 = k\sin\alpha, \quad k_3 = k\cos\alpha, \tag{6}$$

where a_0 and β_e are constants, ω is the fixed angular frequency at the original oscillator, k is the wave number corresponding to ω, and α is the angle between the wave normals and the z-axis. Put

$$\xi = k_1x, \quad \zeta = k_3z-\omega t. \tag{7}$$

The resultant complex disturbance ψ is given by

$$\psi = \psi_A+\psi_B \equiv \rho\, e^{i\chi}. \tag{8}$$

That is

$$\rho\, e^{i\chi} = 2a_0\{(1-\beta_e\xi)\sin\xi + i\beta_e\zeta\cos\xi\}\, e^{i\zeta} \tag{9}$$

Hence

$$\rho^2 = 4a_0^2\{(1-\beta_e\xi)^2\sin^2\xi+(\beta_e\zeta)^2\cos^2\xi\}, \tag{10}$$

and

$$\chi = \arctan\left\{\frac{\beta_e\zeta}{(1-\beta_e\xi)\tan\xi}\right\}+\zeta+2n\pi, \tag{11}$$

the ambiguity of an odd multiple of π in the value of arctan being resolved by making sure that the sine of the angle has the same sign as $\beta_e\zeta\cos\xi$. If the integer n is chosen so that χ is between $-\pi$ and π, χ becomes what we shall call the reduced phase χ_0; otherwise n is zero and χ is a continuous, but multivalued, function. It is sufficient for our purpose to keep β_e small and restrict attention to a limited region of (ξ, ζ) space around the origin, so that the envelope expressions in (5) do not go negative.

It is easily seen from (9) or (10) that the dislocations, which occur where ρ is zero, lie along the lines parallel to the y-axis defined by $\zeta = 0$, $k_1 x = m\pi$, where m is any integer or zero, and that they alternate in character. Figures 9 *a* and *b* show lines of constant reduced phase χ_0 and constant amplitude ρ at time $t = 0$ for $\tan\alpha = 0.1$ and $\beta_e = 0.1$. If one traverses a closed circuit anticlockwise, χ increases by 2π for every dislocation enclosed. The lines of constant χ_0 may be read in different ways depending on the phase of the carrier wave relative to the pulse envelope in the

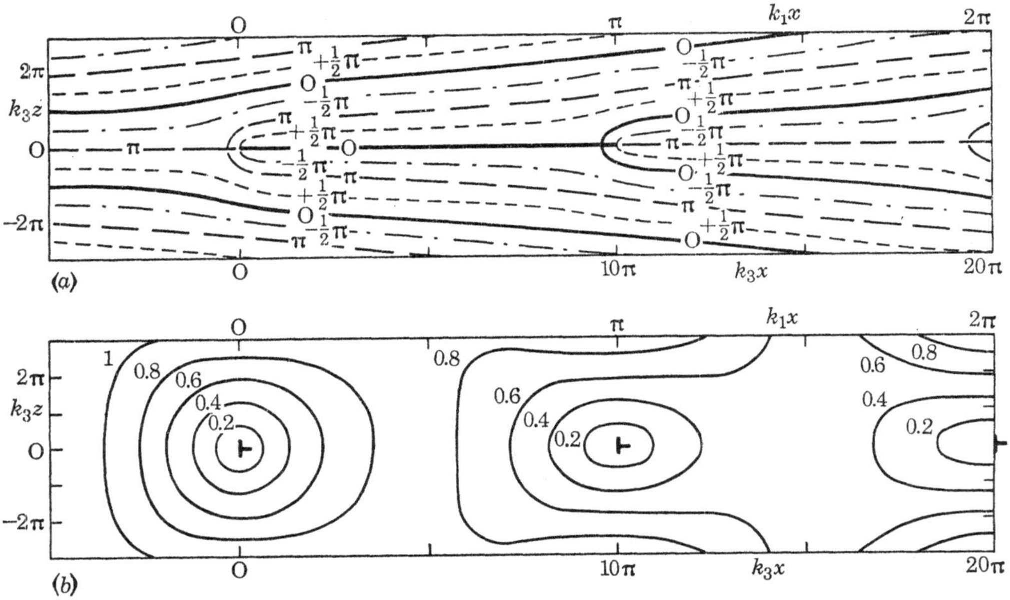

FIGURE 9. (*a*) A row of edge dislocations (tilt boundary) formed by the interference of two pulses of plane waves, each modulated linearly, $\tan\alpha = 0.1$ and $\beta_e = 0{\cdot}1$. The scales are arranged so that the pattern is a true-to-scale map in the (x, z) plane at $t = 0$. Lines of constant reduced phase χ_0 are shown; the dislocations are singularities of χ_0. Any arbitrary value of χ_0 may be chosen to represent crests. The whole pattern moves upwards with velocity $c \sec\alpha$, unchanged in form. (*b*) As (*a*) but showing contours of wave amplitude ρ. The numbers on the curves are values of $\rho/2a_0$. The dislocations are at places where $\rho = 0$.

original oscillation. For example, if the original oscillation is $\cos(-\omega t)$, the resultant is $\rho(x, y, z, t)\cos\chi(x, y, z, t)$ and the full lines where $\chi_0 = 0$ are to be read as crests; this corresponds to figure 8 *a*; at even values of m a new crest appears, while at odd values a crest splits into two. If, on the other hand, the original oscillation is $\sin(-\omega t)$ the crests are the lines where $\chi_0 = \frac{1}{2}\pi$. It can be seen that these run into the dislocations alternately from above and below, corresponding to figure 8 *b*. Alternatively, the lines of $\chi_0 = 0$ may be read as troughs and $\chi_0 = \pi$ as crests, and so on. Crests and troughs defined in this way are of course not necessarily exactly maxima and minima of the disturbance, but they are usually very close to them.

The disturbance described by (9) is not quite periodic in ξ, because both the constituent wave trains grow stronger towards the left. The time dependence of the disturbance is contained entirely in ζ, and thus the whole pattern simply sweeps upwards parallel to Oz with velocity $\omega/k_3 = c\sec\alpha$. Thus the wavefronts move upwards at a speed greater than c, carrying the dislocations with them.

The fact that the dislocations are equally spaced is of course simply a consequence of using plane wave trains, and at the same time the details of the wave pattern far from the origin are not of primary interest because they depend on the unrealistic and artificial assumption of a linear envelope. We are really interested in the structure of a single dislocation. In order to study this structure undistorted by interaction with neighbouring dislocations in the row we use a limiting process in which the angle 2α between the two primary wave trains is decreased to zero, thus moving the dislocations apart. First define new constants A_0 and β_e^*:

$$A_0 = 2a_0 \sin\alpha \quad \text{and} \quad \beta_e^* = \beta_e \cot\alpha. \tag{12}$$

Then write equation (9) in terms of A_0, β_e^*, k, α, x and z, and let $\alpha \to 0$, keeping A_0, β_e^* and k fixed. In the limit

$$\psi = A_0\{kx + \mathrm{i}\beta_e^*(kz-\omega t)\}\exp\mathrm{i}(kz-\omega t), \tag{13}$$

so that

$$\left.\begin{aligned} \rho^2 &= A_0^2\{k^2x^2 + \beta_e^{*2}(kz-\omega t)^2\} \quad (\rho \geqslant 0), \\ \text{and} \qquad \chi &= \arctan\frac{\beta_e^*(kz-\omega t)}{kx} + kz - \omega t + 2n\pi. \end{aligned}\right\} \tag{14}$$

These equations describe a single edge dislocation parallel to Oy, passing through O at $t = 0$, and moving parallel to Oz at a speed $\omega/k = c$. Figure 10 shows ρ and reduced phase χ_0 at $t = 0$. The meaning of the parameter β_e^* (which determines the x scale) may be seen by noting that before the limiting process the change in amplitude of one of the primary wave trains over one period was $2\pi\beta_e \boldsymbol{a}_0 = \pi\beta_e^* A_0$. Thus β_e^* can be regarded as a measure of the non-monochromaticity of the pulse, or of the bandwidth. In the monochromatic limit, $\beta_e^* = 0$, the dislocation becomes infinitely extended along Oz and the phase map turns into figure 5. The larger β_e^* the more compact is the dislocation in the z direction (in crystal dislocation language β_e^* is an inverse measure of the width of the dislocation; it determines the core structure). When $\beta_e^* = 1$ the ρ contours for $t = 0$ are circles and the inverse tangent in (14) is simply the polar coordinate angle θ measured from the x-axis. Thus, near the origin $\chi = \theta$.

At time $t = 0$, or with respect to coordinates carried along with the wave, we have, for $\beta_e^* = 1$,

$$\begin{aligned} \operatorname{grad}\chi &= \operatorname{grad}\theta + k\boldsymbol{n}_z, \\ &= \boldsymbol{n}_\theta/r + k\boldsymbol{n}_z, \end{aligned} \tag{15}$$

where $\boldsymbol{n}_\theta$ and $\boldsymbol{n}_z$ are unit vectors. Thus: if χ is regarded as a potential, we find that the field $\operatorname{grad}\chi$ is the sum of a uniform field along the z axis and a vortex (we were led to this interpretation by a suggestion of Professor F. C. Frank, F.R.S.). Since the uniform field is the undistorted wave, the disturbance produced by the dislocation is the vortex. In this sense the dislocation may be pictured as a vortex that is carried along with the wave, rather as a vortex in a river is carried along by the main flow. When $\beta_e^* \neq 1$ the general result is the same but the vortex is distorted. There is a stagnation point to the left of the origin where the backward grad χ due to the vortex just cancels the grad χ due to the forward flow. It is readily found that this occurs at

$$x = -\beta_e^*/k. \tag{16}$$

This is the point S in figures 10 and 11, where two contours of $\chi_0 = \pi$ cross at right angles; it is a saddle-point for χ. Figure 11 shows the trajectories of grad χ at $t = 0$ when $\beta_e^* = 1$. Their equation is

$$(kx)^2 + (kz)^2 = Q_0 \exp\{-2(1+kx)\},$$

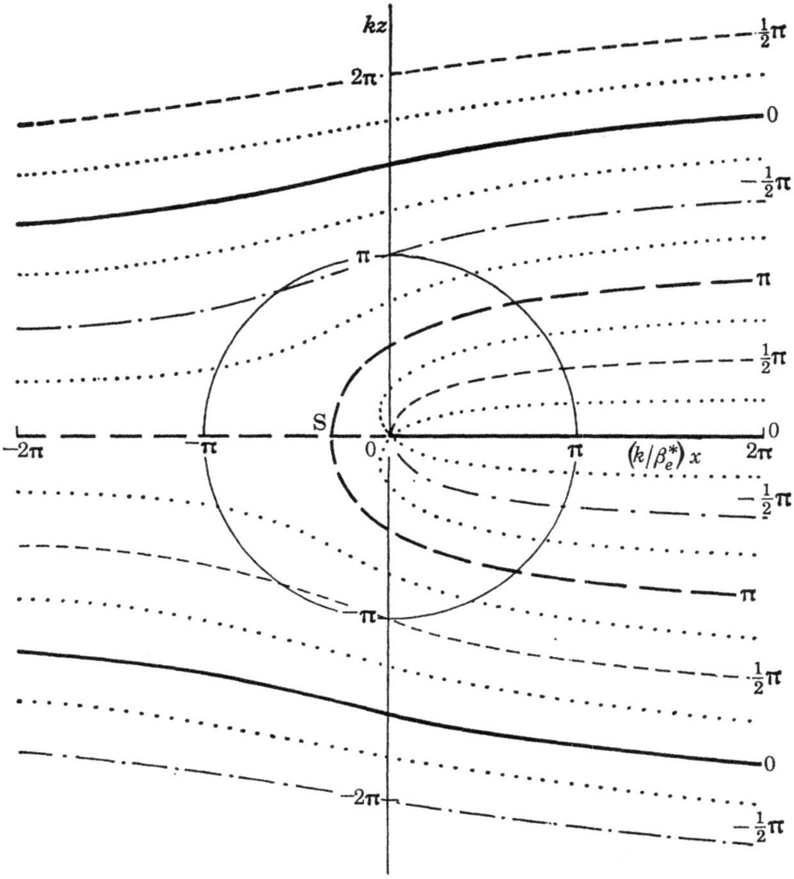

FIGURE 10. A single edge dislocation at the origin showing the lines of reduced phase χ_0 at $t = 0$. Any arbitrary value of χ_0 may be chosen to represent crests. The figure is a true-to-scale map when $\beta_e^* = 1$. ρ is proportional to the radial distance from the origin; the circle shows the contour for $\rho = \pi A_0 \beta_e^*$. S is the stagnation point.

where the constant Q_0 labels the different curves. We have tried without success to find a physical interpretation of this vortex motion (for example, as momentum or energy flux) valid for all types of nondispersive wave; in quantum mechanics $\rho^2 \operatorname{grad} \chi$ would represent a local expectation value of momentum.

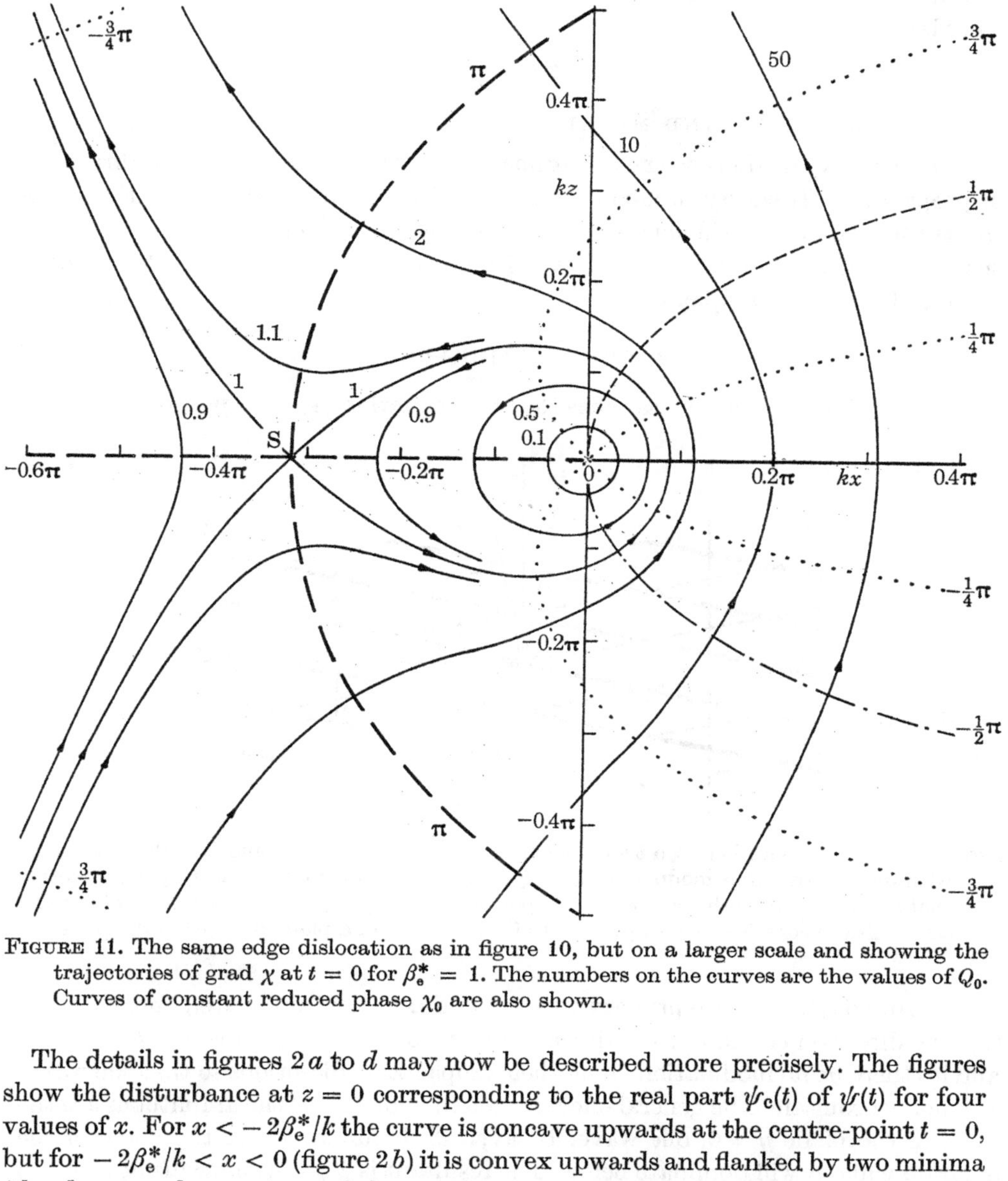

FIGURE 11. The same edge dislocation as in figure 10, but on a larger scale and showing the trajectories of grad χ at $t = 0$ for $\beta_e^* = 1$. The numbers on the curves are the values of Q_0. Curves of constant reduced phase χ_0 are also shown.

The details in figures 2*a* to *d* may now be described more precisely. The figures show the disturbance at $z = 0$ corresponding to the real part $\psi_c(t)$ of $\psi(t)$ for four values of x. For $x < -2\beta_e^*/k$ the curve is concave upwards at the centre-point $t = 0$, but for $-2\beta_e^*/k < x < 0$ (figure 2*b*) it is convex upwards and flanked by two minima (the downward curvature of the envelope outweighing the upward curvature of the cosine). Two new zeros appear precisely at the centre of the dislocation (figure 2*c*), but they are preceded by the appearance of the new maximum slightly below the zero level. In the antisymmetrical structure of figure 3, which corresponds to the imaginary part $\psi_s(t)$ of $\psi(t)$, the two new zeros appear at $x = -\beta_e^*/k$. As discussed in § 2, the unique property of the dislocation line $x = 0$, $\zeta = 0$ is that the disturbance is zero for *all* values of the phase of the carrier wave relative to the pulse envelope, or, in other words, for all linear combinations of the real and imaginary parts of the complex disturbance ψ.

5. Screw and mixed edge-screw dislocations

So far we have modulated the two primary wave trains linearly in their directions of propagation. If we try to modulate a train of plane waves parallel to the wavefronts the modulation will normally spread out by diffraction, but it may be verified that if the modulation is merely linear with position the wave propagates without change. Thus, for example, the wave

$$\psi = (a + bx)\exp[\mathrm{i}(kz - \omega t)], \tag{17}$$

where a and b are constants, satisfies the wave equation (4) (cf. equation (13)).

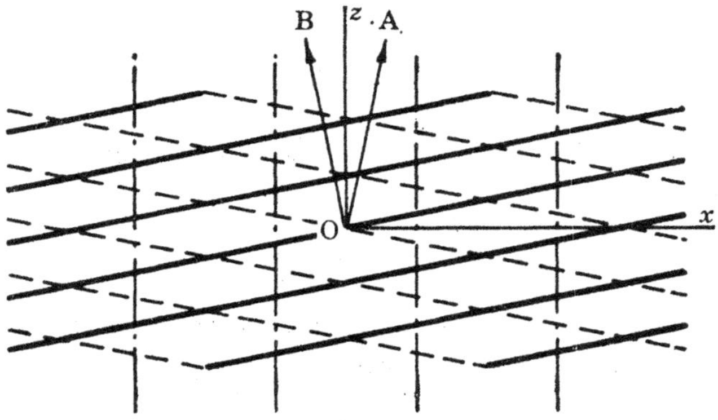

Figure 12. Interference between two monochromatic plane waves A and B making an angle with one another and modulated linearly in opposite directions along Oy, the direction that is common to both wavefronts. Oy points into the page. The result is a set of parallel screw dislocations (—·—·—) parallel to Oz and lying in a plane parallel to the page.

Accordingly, let the two primary waves A and B be now modulated linearly along Oy, the direction common to both wavefronts, so that as one rises the other falls, and let there be no modulation along their propagation directions so that the waves are monochromatic. The general effect is seen in figure 12. If the amplitudes are the same in the plane $y = 0$, one wave, B say, will dominate above the plane of the diagram while A will dominate below. The resultant is a set of screw dislocations in the plane $y = 0$, all parallel to Oz and equally spaced along Ox. It can be seen from figure 12 that they are right-handed. In crystallographic terms this is a twist boundary.

For mixed screw-edge dislocations we simply add modulation along the propagation directions. The analysis for screw, edge and mixed screw-edge dislocations is conveniently done all together. Let the two primary waves now be

$$\left.\begin{aligned}\psi_A &= a_0\{1-\beta_e(k_1x+k_3z-\omega t)+\beta_s k_1 y\}\exp[i(k_1x+k_3z-\omega t-\tfrac{1}{2}\pi)],\\ \psi_B &= a_0\{1+\beta_e(-k_1x+k_3z-\omega t)-\beta_s ky\}\exp[i(-k_1x+k_3z-\omega t+\tfrac{1}{2}\pi)],\end{aligned}\right\} \quad (18)$$

where β_s is a constant. Writing $k_1 y = \eta$, we find that the resultant disturbance $\psi = \psi_A + \psi_B$ is

$$\psi = 2a_0\{(1-\beta_e\xi)\sin\xi + i(\beta_e\zeta - \beta_s\eta)\cos\xi\}\exp(i\zeta) = \rho\exp(i\chi). \quad (19)$$

Hence $$\rho^2 = 4a_0^2\{(1-\beta_e\xi)^2\sin^2\xi + (\beta_e\zeta-\beta_s\eta)^2\cos^2\xi\} \quad (\rho \geqslant 0), \quad (20)$$

and $$\chi = \arctan\left\{\frac{\beta_e\zeta-\beta_s\eta}{(1-\beta_e\xi)\tan\xi}\right\} + \zeta + 2n\pi, \quad (21)$$

where the conventions about the π are the same as before.

As always, the dislocations lie along the lines where $\psi_c = \psi_s = 0$; from (19), they have the equations

$$\left.\begin{aligned}\xi &= m\pi,\\ \beta_e\zeta-\beta_s\eta &= 0,\end{aligned}\right\} \quad (22a)$$

or, at $t = 0$,

$$\beta_e^* z - \beta_s y = 0 \quad (22b)$$

(ignoring the solution $\beta_e\xi = 1$ as being outside the field of interest). Define constants β and δ so that

$$\left.\begin{aligned}\beta_e^* &\equiv \beta\cos\delta,\\ \beta_s &\equiv \beta\sin\delta.\end{aligned}\right\} \quad (23)$$

δ is then the inclination of the dislocation lines to Oy. $\delta = 0$ corresponds to $\beta_s = 0$ and pure edge dislocations. $\delta = \frac{1}{2}\pi$ corresponds to $\beta_e^* = \beta_e = 0$ and pure screw dislocations.

Proceeding to the limit $\alpha \to 0$ as before, we find

$$\psi = A_0[kx + i\{\beta_e^*(kz-\omega t) - \beta_s ky\}]\exp[i(kz-\omega t)], \quad (24)$$

so that

$$\left.\begin{aligned}\rho^2 &= A_0^2[k^2x^2 + \{\beta_e^*(kz-\omega t)-\beta_s ky\}^2] \quad (\rho \geqslant 0),\\ \chi &= \arctan\frac{\beta_e^*(kz-\omega t)-\beta_s ky}{kx} + kz - \omega t + 2n\pi.\end{aligned}\right\} \quad (25)$$

This describes a single mixed dislocation lying in the yz plane at an angle δ to Oy, passing through O at $t = 0$ and moving parallel to Oz at velocity c. To obtain a pure screw dislocation put $\beta_e^* = 0$. Then

$$\chi = -\arctan\frac{\beta_s y}{x} + kz - \omega t + 2n\pi. \quad (26)$$

If one encircles the z-axis at constant z and fixed t, χ changes by 2π for each revolution. If $\beta_s = 1$ the surfaces of constant phase at given time are helicoids and we obtain the simple helicoidal wave

$$\psi = A_0 kr \exp[\mathrm{i}(kz - \omega t - \phi)], \tag{27}$$

where r, ϕ, z are cylindrical polar coordinates. $\beta_e^* = 0$ represents the monochromatic limit; the screw dislocation is one of the special cases where a dislocation can exist in a pure monochromatic wave.

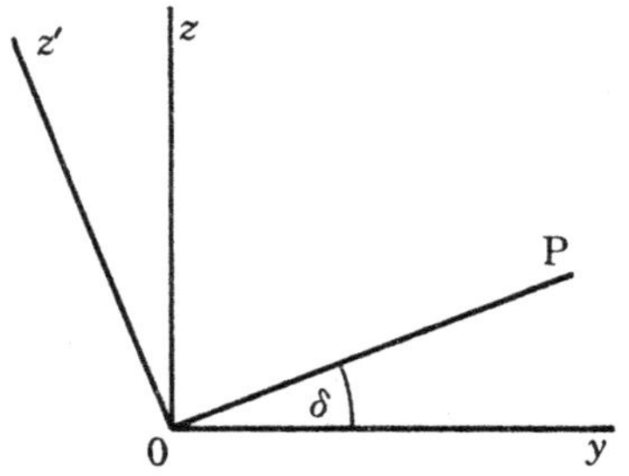

FIGURE 13. Axes for a mixed edge-screw dislocation. OP is the dislocation line; Ox points out of the plane of the paper.

For the mixed dislocation take a new axis Oz' in the yz plane and perpendicular to the line of the dislocation (figure 13); thus

$$z' = z\cos\delta - y\sin\delta. \tag{28}$$

Then equations (25) become, at $t = 0$,

$$\rho^2 = A_0^2 k^2 (x^2 + \beta^2 z'^2), \quad (\rho \geqslant 0), \tag{29a}$$

$$\chi = \arctan(\beta z'/x) + kz + 2n\pi. \tag{29b}$$

This may be pictured as a vortex circulating about the dislocation line in a uniform flow parallel to Oz. If the inclination δ (which determines the edge-screw character) is varied while β is held fixed, equation (29a) shows that the dislocation always has the same field of ρ when it is viewed in its own coordinate system. β is a measure of the nonmonochromaticity of the pulses, by extension of the argument already given for a pure edge dislocation; β is also an inverse measure of how far the dislocation spreads out in the yz plane perpendicular to its length.

Now that we have derived equations (13) and (24) for isolated dislocations, by mixing two plane wave trains and using a limiting process, we can be wise after the event and see that these equations (but not those for rows of dislocations) could have been guessed *a priori*, by using the principle that a single plane wave train can be amplitude modulated linearly in any direction in its wavefront and the fact that this still holds if the coefficients are complex. From this point onwards, our method of constructing dislocations will be more direct. We shall simply present complex wave functions ψ which satisfy the wave equation (4) and shall then interpret them as representing various kinds of dislocations. ρ and χ, as always, are the amplitude and phase of ψ.

One way of realizing such dislocations physically would be to place on the plane $z = z_0$, where z_0 is large and negative, that distribution of sources which will produce the correct boundary condition $\psi = \psi(x, y, z_0, t)$. In each case it will be found that the required sources are quasi-monochromatic in the sense defined in §2, being derived by (complex) amplitude modulation from the primary oscillation $e^{-i\omega t}$.

Equation (27) is the simplest example from a class of helicoidal waves describing *multiple* pure screw dislocations, of 'strength' s; a more general expression which can easily be shown to satisfy the wave equation (4) is

$$\psi = (Ar^s + [B/r^s]) \exp [i(kz - \omega t \pm s\phi)], \tag{30}$$

where A and B are constants and s is a positive integer; the choice of sign in the exponent determines the 'handedness' of the dislocation. The solutions associated with the coefficient B are singular at the origin. Equation (30) is still not the most general expression for an isolated screw dislocation; for example, the wave reflected from a surface consisting of one turn of a helicoid near $z = 0$, whose pitch is s half-wavelengths, has the asymptotic form

$$\psi \underset{\substack{r \to \infty \\ z \to \infty}}{\longrightarrow} A \exp [i(kz - \omega t \pm s\phi)], \tag{31}$$

which is simpler than (30), but it can be shown that the core structure (r small) is very complicated.

Thus multiple screw dislocations exist. What about multiple edge and mixed dislocations? We conjecture that these cannot exist, although we are unable to give a general proof. The pattern of figure 8*a* is suggestive on this point. It shows 'extra half-planes' coming together in pairs as they would at edge dislocations of double strength. But when the local readjustments have taken place (figure 9*a*) each double dislocation has split into a pair of single-strength dislocations with phase-structures differing by π. A further argument is that the slightest perturbation splits a double pure screw dislocation into two single-strength mixed dislocations. To show this, we proceed by analogy with equation (24), which gives a mixed dislocation as the sum of a pure single-strength screw and a linearly modulated plane wave; thus we consider the wave

$$\psi = A_0\{r^2 e^{-2i\phi} + i\beta_e^*(kz - \omega t)\} \exp [i(kz - \omega t)]. \tag{32}$$

The dislocations are the lines where ψ_c and ψ_s are simultaneously zero, that is where

$$\left.\begin{aligned} r^2 \cos 2\phi &= 0, \\ r^2 \sin 2\phi &= \beta_e^* \zeta, \end{aligned}\right\} \tag{33}$$

ζ now denoting $kz - \omega t$. These describe two parabolas lying in orthogonal vertical planes, namely

$$\left.\begin{aligned} \zeta &= r^2/\beta_e^*, & \phi &= \tfrac{1}{4}\pi, -\tfrac{3}{4}\pi, \\ \zeta &= -r^2/\beta_e^*, & \phi &= -\tfrac{1}{4}\pi, \tfrac{3}{4}\pi. \end{aligned}\right\} \tag{34}$$

The dislocations are of single strength, and change their character with ζ, being pure screw when $|\zeta| = \infty$ and pure edge when $\zeta = 0$. In the limit when $\beta_e^* = 0$ the parabolas degenerate into the original double pure screw along the z-axis, and in the limit $|\beta_e^*| = \infty$ they degenerate into two single-strength edge dislocations intersecting orthogonally in the plane $\zeta = 0$.

As well as suggesting that multiple dislocations do not exist unless they are pure screw, the wave (32) introduces two new properties of dislocation lines: they may be curved, and two edges may intersect. A simple model showing the intersection of a screw and an edge is

$$\psi = A_0\{kx + \mathrm{i}\beta y(kz - \omega t)\} \exp[\mathrm{i}(kz - \omega t)], \tag{35}$$

corresponding to dislocations where

$$x = 0 \quad \text{and} \quad y\zeta = 0, \tag{36}$$

that is, to a screw along Oz intersecting an edge parallel to Oy.

All the patterns of straight, curved, edge, screw and mixed dislocations discussed so far have arisen from waves which contain z and t only in the combination ζ; that is, they are of the form

$$\psi = \psi(x, y, \zeta). \tag{37}$$

It would be possible to generate a great variety of dislocation patterns, using the fact, derivable from (4), that waves of this type satisfy the two-dimensional Laplace equation

$$\frac{\partial^2\psi}{\partial x^2} + \frac{\partial^2\psi}{\partial y^2} = 0. \tag{38}$$

However, all such dislocations would be rigidly attached to the wavefronts, and in order to investigate the ways in which dislocations can move relative to the wavefronts, and to one another, it is necessary to widen the class of waves considered, to include solutions where z and t do not appear solely in the combination ζ. Some waves of this type will be examined in §7. In order to describe the motion of the resulting dislocations we first introduce some concepts from crystal dislocation theory.

6. Burgers circuit, glide and climb

The theorem, well-known for crystal dislocations, that the Burgers vector is conserved along the length of a single dislocation applies equally well for dislocations in wave trains. A closed circuit (e.g. JKLM in figure 4) is traversed at fixed time in the dislocated wave and the value of

$$(\lambda/2\pi)\oint \mathrm{d}\chi = N\lambda, \quad \text{say}, \tag{39}$$

where λ is the wavelength $2\pi/k$, is the analogue of the Burgers vector. N is necessarily an integer or zero and equals the total strength of the dislocation lines encircled (with due regard to their signs). If the shape and position of the circuit is altered the value of the integral remains the same provided the circuit continues to encircle the same dislocation lines. In crystal dislocation language the Burgers vector of each dislocation is $\lambda \boldsymbol{n}$, where $\boldsymbol{n}$ is a unit vector along the wave normal in an undislocated reference wave. The difference is that in the wave, as distinct from the crystal, components of the Burgers vector parallel to the reference wavefront are of no significance. Only the magnitude and sense of the Burgers vector is significant and these are given by $N\lambda$ (positive or negative).

The Burgers circuit just described was traversed at fixed time. If we now allow the circuit to travel with the wave it follows by continuity that the Burgers vector for the circuit remains the same provided it is not crossed by a dislocation line; the magnitude of the Burgers vector for the moving circuit remains equal to the total strength of the dislocation lines encircled multiplied by the wavelength.

Wave trains, unlike crystals, are oscillatory in time as well as space, and therefore one may consider a more general kind of 'Burgers circuit' where the line element can be wholly or partly time-like as well as space-like. If the wave disturbance is now thought of as varying in the four dimensions (x, y, z, t), the line integral of χ round any closed circuit drawn in this space must necessarily be a multiple of 2π; thus

$$\oint \mathrm{d}\chi = 2N\pi, \tag{40}$$

where N is an integer or zero. If the circuit lies in a plane for which t = constant, we have (39); on the other hand, the method of observing a dislocation described in § 1 and the argument about the Argand diagram in § 2 can be regarded as employing a circuit in the (x, t) plane. By shrinking the circuit around a dislocation continuously to zero one can see from (40) that χ must be singular on the dislocation itself – and it then follows from the properties of the Argand diagram, together with the physical requirements that the wave function must be continuous and single-valued, that ρ must be zero. Thus, by *defining* a dislocation by means of a Burgers circuit one is led automatically to the conclusion that the precise location of the line is the place where $\rho = 0$ (rather than, for example, various other alternative possibilities that might be suggested by figure 6).

With crystal dislocations a Burgers circuit always has to be chosen so as to avoid passing through 'bad crystal' where the lattice is so badly distorted that the crystallographic directions cannot be unambiguously identified (Frank 1951). It is interesting to notice that this particular complication does not appear with wave dislocations; χ can be well defined everywhere (provided the waves are ultimately derived from a source with a carrier frequency ω) and a dislocation has no disordered core. There is no breakdown of linearity in the wave equation near the singularities. Thus, however close together the dislocations may be, they do not lose their identities; they are always recognizable as singularities of χ where ρ is zero. With crystals, on the other hand, the dislocation concept itself has to be abandoned when the dislocations are very close together. Of course, in practice it may be impossible to resolve individual wave dislocations when they are very close together because of the presence of noise, but that is another matter.

A dislocation may move either by *glide*, defined as motion in the plane containing the Burgers vector and the direction of the dislocation line (the *glide plane*), by *climb*, defined as motion perpendicular to this plane, or by a combination of glide and climb. Thus a pure edge dislocation, which is normal to its Burgers vector, glides by moving perpendicular to the wavefronts (more strictly, perpendicular to the reference wavefronts), and climbs by moving parallel to the wavefronts. For a pure screw dislocation, which is parallel to its Burgers vector, all vertical planes are glide planes, and climb has no meaning.

7. Motion, birth, annihilation and collision of dislocations

First we consider edge dislocations parallel to the y-axis. As we have already seen, to obtain moving dislocations, it is necessary that z and t should not appear only in the combination ζ. The simplest monochromatic wave of this new type involves z linearly, and can be seen by inspection of the wave equation to be

$$\psi_1 = (k^2x^2 + \mathrm{i}kz)\exp[\mathrm{i}(kz - \omega t)]. \tag{41}$$

The inevitable appearance of quadratic modulation in x means that moving dislocations obtained using ψ_1 will occur in pairs.

To obtain *glide*, we add ψ_1 to a multiple of the wave function (13) describing a single edge dislocation; this gives

$$\psi = \{Akx + k^2x^2 + \mathrm{i}B(kz - \omega t) + \mathrm{i}kz\}\exp[\mathrm{i}(kz - \omega t)]. \tag{42}$$

The dislocation lines satisfy

$$\left.\begin{aligned} Akx + k^2x^2 &= 0, \\ B(kz - \omega t) + kz &= 0, \end{aligned}\right\}$$

or

$$\left.\begin{aligned} x &= 0 \quad \text{or} \quad -A/k, \\ z &= \frac{Bc}{B+1}t. \end{aligned}\right\}$$

There are thus two dislocations, which glide parallel to Oz, and have velocity v_{g} given by

$$v_{\mathrm{g}} = \frac{B}{B+1}c, \tag{43}$$

while the undistorted wavefronts move with velocity c. Thus in a frame of reference moving with the dislocations the lines of constant χ sweep through in sequence, backwards or forwards; glide is equivalent to a steady change of phase of the dislocation.

v_{g} can have all values, positive or negative, according to the value of B. For $-1 < B < 0$, v_{g} is negative: the dislocations move backwards. As $B \to -1$, the dislocations become infinitely spread out in the z direction and $|v_{\mathrm{g}}| \to \infty$, rather as the point of coincidence of a vernier can be made to travel very fast even though the scales themselves are only moving slowly. If $|B|\to\infty$, $v_g\to c$ and the disturbance consists of two non-gliding edge dislocations. Finally, if B is zero, v_g is zero; this is the monochromatic limit, and the two dislocations are stationary features of the wave, in fact two localized interference fringes. From this point of view our moving dislocations may be regarded as moving interference fringes.

It is also possible to obtain the 'glide' wave function (42) by a limiting process analogous to that used in § 4 to obtain the single edge dislocation: two primary plane wave trains, linearly modulated in their directions of propagation, are also linearly modulated parallel to their wavefronts and perpendicular to their common direction.

By replacing the linear factor $(kz-\omega t)$ in (42) by a quadratic factor, which corresponds to taking two interfering wave trains whose pulse envelope is not linear but quadratic, pairs of edge dislocations can be made to annihilate by glide, or to appear spontaneously. The wave function is

$$\psi = \{Akx + k^2x^2 + \mathrm{i}B(kz-\omega t)^2 + \mathrm{i}kz\}\exp[\mathrm{i}(kz-\omega t)], \tag{44}$$

whose dislocations lie at

$$\left.\begin{aligned} x &= 0 \quad \text{or} \quad -A/k, \\ z &= ct - \frac{1}{2Bk}\{1 \pm \sqrt{(1-4B\omega t)}\}. \end{aligned}\right\} \tag{45}$$

If $B > 0$ there are two pairs of approaching dislocations of opposite sign which glide together at $t = (4B\omega)^{-1}$ and mutually annihilate, while if $B < 0$ the pairs of dislocations suddenly appear at $t = -(4|B|\omega)^{-1}$ and then glide apart.

To obtain *climb* of edge dislocations we simply set $A = 0$ in (42), and replace B by $B-\mathrm{i}C$; thus the resulting dislocations satisfy

$$x^2 = \frac{Cz}{Bk}, \quad z = \frac{Bct}{B+1}. \tag{46}$$

There are two dislocations of opposite sign: both have the same z value which need not change with speed c, so that the dislocations can glide. In addition they may separate or approach one another, that is, they may climb. If $C/(B+1) > 0$, the two dislocations are born at $t = 0$ and separate by climb, initially with infinite speed; this would manifest itself as the sudden tearing of a wavefront (when $B+1 > 0$) or as the appearance of an expanding strip of wavefront (when $B+1 < 0$). (When a pulse is reflected from a scattering object the waves that have travelled the shortest path, and therefore arrive first, are usually travelling in the direction of the average reflected wave, whereas later arrivals tend to be oblique. It may be shown that this results *on the average* in the disappearance of wavefronts. Therefore, in this situation, we expect the appearance of a new piece of wavefront to be a comparatively rare event.) If $C/(B+1) < 0$, the two dislocations climb towards each other and annihilate at $t = 0$; this would manifest itself as the sudden disappearance of a strip of wavefront (when $B+1 < 0$), or, rarely, as the spontaneous healing of a tear (when $B+1 > 0$). In both cases the surface swept out by the moving dislocations is a parabolic cylinder whose axis is parallel to Oy. These solutions may also be obtained by adding a third wave train, travelling along Oz, to the two considered previously (§4 and figure 8), and taking the limit $\alpha \to 0$.

To obtain a *circular climbing edge dislocation loop* (climbing prismatic dislocation) it is only necessary to notice that if x^2 is replaced by $\frac{1}{2}r^2$ in (41), where r is the radial coordinate of cylindrical coordinates, ψ_1 will satisfy the wave equation. Thus the wave

$$\psi = \{\tfrac{1}{2}k^2r^2 + \mathrm{i}kz + \mathrm{i}(B - \mathrm{i}C)(kz - \omega t)\} \exp [\mathrm{i}(kz - \omega t)] \tag{47}$$

describes a circular edge dislocation loop sweeping out a paraboloid of revolution whose axis lies along Oz. If $C/(B+1) > 0$ the loop is born at $t = 0$ (appearing as the sudden puncture of a wavefront, or, rarely, as the birth of an expanding circular island of wavefront), while if $C/(B+1) < 0$ the loop vanishes at $t = 0$ (disappearance of an island of wavefront, or, rarely, the spontaneous healing of a puncture). The solution (47) may also be obtained by a limiting process involving the interference of a plane wave with a toroidal wave (for example, from an annular source).

It is possible for edge dislocations to *collide and rebound* without annihilation. To construct this situation we require a slightly more complicated wave than (41), namely

$$\psi_2 = (\tfrac{1}{3}k^3x^3 + \mathrm{i}k^2xz) \exp [\mathrm{i}(kz - \omega t)]. \tag{48}$$

Combining ψ_2 with a plane wave with linear pulse envelope, we obtain

$$\psi = \{\tfrac{1}{3}k^3x^3 + B(kz - \omega t) + \mathrm{i}k^2xz\} \exp [\mathrm{i}(kz - \omega t)], \tag{49}$$

which has two edge dislocations of opposite sign, parallel to Oy, satisfying

$$\left.\begin{aligned} xz &= 0, \\ \tfrac{1}{3}k^3x^2 + B(kz - \omega t) &= 0. \end{aligned}\right\} \tag{50}$$

The dislocation for which $x = 0$ moves along Oz with speed c (i.e. it neither climbs nor glides), while the dislocation for which $z = 0$ moves along Ox with a speed which varies, becoming infinite at $x = 0$ (that is, it climbs along Ox while gliding backwards at speed c). At time $t = 0$ the two dislocations collide. Afterwards one moves off along Ox and the other along Oz. If we identify the dislocations by their signs it may be shown that each has been deflected through a right angle.

Finally we consider the *motion of pure screw dislocations*. In their simplest form, given by equation (27), these are monochromatic, and it is clear that the addition of another monochromatic wave such as ψ_1 (equation 41) cannot produce motion. But there is a 'complementary' solution to ψ_1, where t appears linearly instead of z; it is easily verified that this solution is

$$\psi_3 = (\omega t - \mathrm{i}k^2x^2) \exp [\mathrm{i}(kz - \omega t)]. \tag{51}$$

Adding ψ_3 to a real multiple of the 'screw' wave function (27), we obtain

$$\psi = \{\omega t - \mathrm{i}k^2x^2 + Ak(x - \mathrm{i}y)\} \exp [\mathrm{i}(kz - \omega t)], \tag{52}$$

which has a single dislocation, satisfying

$$\left.\begin{aligned} x &= -\omega t/Ak, \\ y &= -kx^2/A. \end{aligned}\right\} \tag{53}$$

Since it is parallel to Oz this is a screw dislocation; it glides along a parabola in the xy plane. Replacing A by $\mathrm{i}B$, we find that the resulting wave has two screw dislocations of opposite sign, satisfying

$$\left.\begin{aligned} x = 0, \quad x &= B/k; \\ y &= -ct/B. \end{aligned}\right\} \tag{54}$$

These glide parallel to the y-axis while maintaining a constant separation. For *annihilation and creation of screw dislocations* we replace A in (52) by $A+\mathrm{i}B$, obtaining the dislocation line equations

$$y = -\frac{ct+Ax}{B} = \frac{-kx^2+xB}{A}. \tag{55}$$

These describe two screw dislocations moving along a parabola in the xy plane; they are created and glide apart if $AB > 0$, and they glide together and annihilate if $AB < 0$.

8. Concluding remarks

We have shown that wavefront dislocation lines, along which the amplitude is zero, are perfectly analogous to crystal dislocations so far as their kinematic properties are concerned. On the other hand, there is nothing analogous to the forces between crystal dislocations or to their line tensions. We may also note that, while their topology is virtually identical, the superposition properties of the two sorts of dislocations are different. Crystal dislocations are lines where, mathematically, the stress and strain become infinite; so when two elastic strain-fields, both containing dislocations, are added together, the dislocations remain in place. In wave trains, on the other hand, $\rho = 0$ on the dislocations, and so when two dislocated wave fields are added each one tends to destroy the dislocations of the other, although new dislocations may be created elsewhere by interference. (If the second field is added gradually, the original dislocations move continuously, but they may be annihilated and new ones may be created.) Another difference is that, in crystals, climb tends to be a slow process compared with glide because it requires diffusion of lattice vacancies, whereas with wavefront dislocations there is no corresponding restriction on the speed of climb.

It is interesting to compare wavefront dislocations with caustic surfaces and focal lines. These are envelopes of the rays of geometrical optics, that is, loci of points where the density of rays is infinite. At caustics and foci the wave amplitude is generally large, becoming infinite in the limit when the wavelength becomes vanishingly small (in comparison with the radii of curvature of rays and the dimensions of diffracting objects). Thus, when geometrical optics is a valid approximation, caustics and foci are the dominant features in the wave field. However, in this limit dislocations are unobservable, because they can only be unambiguously identified by measuring the phase change around a Burgers circuit, and phase is a quantity which varies infinitely fast when the wavelength is zero. In long waves, on the other hand, dislocations are readily observable, but caustics and foci lose their prominence because the amplitude is no longer especially large. Thus dislocations are entities complementary, in a certain sense, to foci and caustics.

For simplicity, and also to emphasize that dispersion is not necessary for the production of dislocations, we have restricted ourselves to the scalar wave equation (4). But dispersive wave trains can also contain dislocations; in particular, the time-dependent Schrödinger equation of quantum mechanics may be analysed by the methods of this paper, leading to the conclusion that matter waves contain a tangled 'cobweb' of dislocation lines on which the electron density is zero. These cobwebs are not superposable, even though the wave equation is linear.

Wavefront dislocations can also exist in two dimensions (for example in water waves), and take the form of points instead of lines. Only edge dislocation points may occur. Examples of the monochromatic case, where the dislocation points are localized interference fringes (see the paragraph following equation (43)) occur in a paper by Braunbek & Laukien (1952): their figure 2 shows the lines of constant phase for the field of a plane wave diffracted by a half-plane; several dislocation points may be seen. Other examples of this special case are the 'amphidromic points' for tides (Whewell 1833; Defant 1961) around which the lines of constant phase travel like the spokes of a wheel (the phase structure is identical with that very near the origins of figures 10 and 11). There are two amphidromic points in the North Sea.

It is clear that the use of other combinations of allowed modulation and the addition of further wave trains will produce more complex arrangements of dislocations moving in more complex ways. Moreover, the waves carrying the dislocations need not be plane. Reflexion of a pulse from a rough surface is one way of producing a large number of interacting pulses and may therefore be expected to result in a complex field of dislocations. Scattering of a pulse from any spatial array of scattering centres will produce a similar effect. These are subjects for further study, and some progress has been made on them. In this paper we have simply shown that dislocations exist and have tried to explore some of their elementary properties.

We are grateful to Dr M. E. R. Walford for his valuable help and advice. The ultrasonic apparatus used for the observations of § 1 was made by R. G. Kyte and D. C. Threlfall, to whom we also express our thanks, and an improved version was designed and made by Dr Walford. The experimental work was partly supported by a grant from the Natural Environment Research Council.

References

Braunbek, W. & Laukien, G. 1952 *Optik* **9**, 174–9.

Defant, A. 1961 *Physical oceanography*, volume 2. London: Pergamon Press.

Findlay, J. W. 1951 *J. Atmos. Terrest. Phys.* **1**, 353–366.

Frank, F. C. 1951 *Phil. Mag.* Series 7, **42**, 809–819.

Nabarro, F. R. N. 1967 *Theory of crystal dislocations*. Oxford: Clarendon Press.

Nye, J. F., Kyte, R. G. & Threlfall, D. C. 1972 *J. Glaciol.* **11**, 319–325.

Nye, J. F., Berry, M. V. & Walford, M. E. R. 1972 *Nature, Phys. Sci.* **240**, 7–9.

Read, W. T. 1953 *Dislocations in crystals*. New York: McGraw-Hill.

Walford, M. E. R. 1972 *Nature, Lond.* **239**, 93–95.

Whewell, W. 1833 *Phil. Trans. R. Soc. Lond.* **123**, 147–236.

Eur. J. Phys. 1 (1980) 154–162. Printed in Northern Ireland
154

Wavefront dislocations in the Aharonov–Bohm effect and its water wave analogue

M V Berry, R G Chambers, M D Large, C Upstill and J C Walmsley

H H Wills Physics Laboratory, University of Bristol, Tyndall Avenue, Bristol BS8 1TL, England

Received 20 June 1980

Abstract We study the wavefronts (i.e. the surfaces of constant phase) of the wave discussed by Aharonov and Bohm, representing a beam of particles with charge q scattered by an impenetrable cylinder of radius R containing magnetic flux Φ. Defining the quantum flux parameter by $\alpha = q\Phi/h$, we show that for the case $R = 0$ the wave ψ_{AB} possesses a wavefront dislocation on the flux line, whose strength (i.e. the number of wave crests ending on the dislocation) equals the nearest integer to α. When α passes through half-integer values, the strength changes, by wavefronts unlinking and reconnecting along a nodal surface. In quantum mechanics this phase structure is unobservable, but we devise an analogue where surface waves on water encounter an irrotational 'bathtub' vortex; in this case α depends on the frequency of the waves and the circulation of the vortex. Experiments show dislocation structures agreeing with those predicted. ψ_{AB} is an unusual function, in which incident and scattered waves cannot be clearly separated in all asymptotic directions; we discuss its properties using a new asymptotic method.

Résumé L'article est consacré aux surfaces d'onde (surfaces de phase constante) de l'onde, introduite par Aharonov et Bohm, qui décrit un faisceau de particules de charge q diffusé par un cylindre impénétrable, de rayon R, traversé par un flux magnétique Φ. Soit $\alpha = q\Phi/h$ le paramètre adapté à une description quantique du flux. On montre que, pour $R = 0$, l'onde ψ_{AB} de Aharonov et Bohm présente une dislocation de la surface d'onde sur la ligne de flux; la 'force' de cette dislocation, c'est-à-dire le nombre de maxima de vibration se terminant sur la dislocation, est égale au nombre entier le plus proche de α. Quand α passe par une valeur demi-entière, la 'force' change, les surfaces d'onde se réarrangeant le long d'une surface nodale. Cette structure des surfaces d'onde n'est pas observable en Mécanique Quantique, mais l'on peut proposer une analogie hydrodynamique, où des ondes de surface sur l'eau rencontrent un tourbillon irrotationnel (du type 'vidange de baignoire'); dans ce cas, α dépend de la fréquence des ondes et de la circulation du tourbillon. L'expérience met en évidence des structures de dislocations en accord avec les prédictions théoriques. ψ_{AB} est une fonction de comportement inhabituel, pour laquelle les ondes incidente et diffusée ne peuvent être clairement distinguées dans toutes les directions asymptotiques; ses propriétés sont analysées à l'aide d'une méthode asymptotique originale.

0143-0807/80/030154+10$01.50 © The Institute of Physics

1 Introduction

In an influential paper, Aharonov and Bohm (1959) studied the quantum mechanics of a beam of particles (with charge q and mass m) incident normally on a long thin cylinder containing a magnetic field $\boldsymbol{B}(\boldsymbol{r})$ parallel to its length. They supposed that the electrons could not penetrate into the cylinder and that the magnetic field could not leak out. This mutual inaccessibility of particles and field ensures that in classical mechanics the scattering pattern of particles beyond the cylinder cannot depend on the field inside. But in Schrödinger's equation it is the magnetic vector potential $\mathbf{A}(\boldsymbol{r})$, and not the field, that determines the wavefunction, and $\mathbf{A}(\boldsymbol{r})$ outside the cylinder contains the imprint of the field inside via the magnetic flux Φ, given by the Stokes relation

$$\Phi = \oint \mathbf{A}(\boldsymbol{r}) \cdot \mathrm{d}\boldsymbol{r} = \oiint \boldsymbol{B}(\boldsymbol{r}) \cdot \mathrm{d}\boldsymbol{S}, \qquad (1)$$

where the integration path encloses the cylinder. Aharonov and Bohm (1959) showed that the quantum mechanical scattering pattern does indeed depend on Φ, in a manner confirmed experimentally by Chambers (1960) and Möllenstedt and Bayh (1962). This result, and its implication that in quantum mechanics the vector potential has a direct physical significance (as opposed to classical mechanics where it is a mathematical device), have become known as the Aharonov–Bohm (AB) effect.

The AB effect has given rise to controversy, to which we shall devote a few remarks at the end of the article. Our main purpose, however, is to describe and illustrate some little-known properties of the AB wavefunction. In §2 we explain that, far from the cylinder, the incident and scattered waves can be distinguished everywhere except near the forward direction, where they are inextricably connected; without this connection both incident and scattered waves would be multivalued. The results are most transparently obtained by some unusual asymptotics described in the Appendix.

Section 3 is the heart of the paper, and concerns the wavefronts, i.e. the lines of constant phase of the wavefunction. Their behaviour depends on the quantum flux parameter α, defined by

$$\alpha = q\Phi/h, \quad (2)$$

where h is Planck's constant. We show that a number of wave crests, equal to the nearest integer to α, end on the cylinder containing the flux, which is therefore a 'wavefront dislocation' in the sense of Nye and Berry (1974). In quantum mechanics these phase singularities are unobservable, since as shown by Wu and Yang (1975) all observables (such as the AB scattering cross section) depend not directly on α but on $\exp(2\pi i\alpha)$, and so $\alpha+1$ and α are indistinguishable.

The presence of interesting but unobservable topology in the wavefunction prompted us to devise an analogue system in which the wavefronts can be seen directly. We show in §4 that surface waves on water crossing an irrotational (bathtub) vortex constitute such a system. Qualitative and quantitative experimental confirmation of the predicted properties of the wavefront dislocations are presented in §5.

2 Aharonov–Bohm wavefunction

Let the incident particles be represented by a plane wave with wavenumber k incident from $x=+\infty$, $\theta=0$ (figure 1) on an impenetrable cylinder of radius R centred on the z axis and containing the flux Φ. A suitable vector potential yielding zero field outside the cylinder and satisfying (1) is

$$\mathbf{A}(\mathbf{r}) = (\Phi/2\pi r)\hat{\boldsymbol{\theta}}, \quad (3)$$

where $\hat{\boldsymbol{\theta}}$ is the azimuthal unit vector. Of course infinitely many other vector potentials are possible, related to (3) by gauge transformations; a careful discussion of these is given by Ingraham (1972).

The wavefunction $\psi(\mathbf{r})$ must satisfy the following conditions:

(i) Schrödinger's equation:

$$\frac{1}{2m}(-i\hbar\nabla - q\mathbf{A}(\mathbf{r}))^2\psi(\mathbf{r}) = \frac{\hbar^2k^2}{2m}\psi(\mathbf{r}), \quad (4)$$

with $\mathbf{A}$ given by (3).

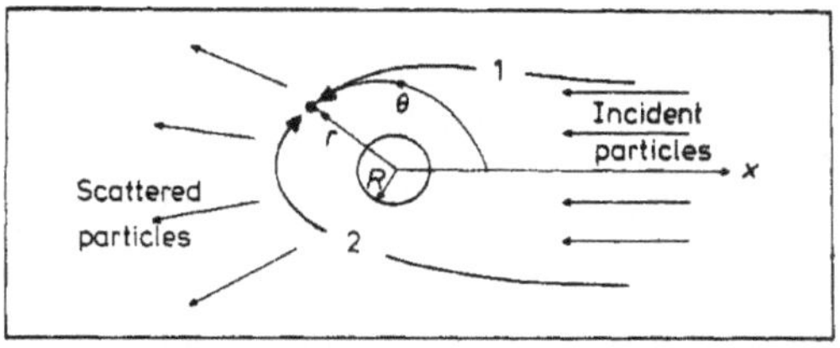

Figure 1 Scattering geometry and coordinates for the Aharonov–Bohm effect, showing two paths reaching the same point by topologically different routes.

(ii) Single-valuedness (for detailed discussions of this requirement see Merzbacher (1962), Tassie and Peshkin (1961) and Kretschmar (1965a)).

(iii) Impenetrability:

$$\psi = 0 \qquad \text{when } r = R. \quad (5)$$

(iv) Asymptotics: as $r\to\infty$, ψ must be the sum of the incident wave plus a purely outgoing wave; the mathematical expression of this condition will be discussed later.

To see how subtle the problem is we first describe a tempting but unprofitable approach. Consider the partial-wave expansion of the wave $\psi_0(\mathbf{r})$ corresponding to scattering by the cylinder containing zero flux, namely

$$\psi_0(\mathbf{r}) = \sum_{l=-\infty}^{\infty} \frac{(-i)^{|l|}\exp(il\theta)}{H_{|l|}^{(1)}(kR)} \times\{J_{|l|}(kr)H_{|l|}^{(1)}(kR) - J_{|l|}(kR)H_{|l|}^{(1)}(kr)\}, \quad (6)$$

where J and H denote the usual Bessel functions (Abramowitz and Stegun 1964). This satisfies conditions (i)–(iv) with $\Phi=0$. When $\Phi\neq 0$, it is easily shown by direct substitution that Schrödinger's equation (4) is satisfied by

$$\psi(\mathbf{r}) = \psi_0(\mathbf{r})\exp\left\{iq\int^{\mathbf{r}}\mathbf{A}(\mathbf{r}')\cdot d\mathbf{r}'/\hbar\right\} = \psi_0(\mathbf{r})\exp(i\alpha\theta), \quad (7)$$

where α is the flux parameter (equation (2)). This also satisfies conditions (iii) and (iv) but violates (ii) because it is not single-valued unless α is an integer. Therefore it is not the correct solution.

The failure of single-valuedness should not be interpreted as a requirement that α must be quantised (as in the case of superconductivity, for example, where different physical principles operate—see Merzbacher (1962)). Instead, the multivalued elementary solutions

$$\exp\{i(l+\alpha)\theta\}J_{|l|}(kr) \quad \text{and} \quad \exp\{i(l+\alpha)\theta\}H_{|l|}^{(1)}(kr) \quad (8)$$

must be replaced by the single-valued solutions

$$\exp(il\theta)J_{|l-\alpha|}(kr) \quad \text{and} \quad \exp(il\theta)H_{|l-\alpha|}^{(1)}(kr). \quad (9)$$

Equation (6) can now be modified to give the correct solution, which we write as

$$\psi(\mathbf{r}) = \psi_{AB}(\mathbf{r}) - \psi_R(\mathbf{r}) \quad (10)$$

where

$$\psi_{AB}(\mathbf{r}) = \sum_{l=-\infty}^{\infty}(-i)^{|l-\alpha|}\exp(il\theta)J_{|l-\alpha|}(kr) \quad (11)$$

and

$$\psi_R(\mathbf{r}) = \sum_{l=-\infty}^{\infty}(-i)^{|l-\alpha|}\exp(il\theta)\frac{J_{|l-\alpha|}(kR)}{H_{|l-\alpha|}^{(1)}(kR)}H_{|l-\alpha|}^{(1)}(kr). \quad (12)$$

Since this solution is a superposition of functions of the form of solution (9), it obviously satisfies conditions (i) and (ii); moreover it also satisfies (iii). As the cylinder radius R tends to zero, ψ_R vanishes, leaving ψ_{AB} as the wave in the presence of a single impenetrable flux line. But ψ_{AB} is the wavefunction derived by Aharonov and Bohm (1959), so this argument justifies their claim to have found the correct limiting form as $R \to 0$. A solution more general than equation (10), in that it allows the cylinder to be penetrable, can be found in the appendix of a paper by Kretschmar (1965b).

Now we must show that the wave ψ satisfies the asymptotic condition (iv). Consider first ψ_R: because of the Bessel functions with argument kR, the series converges rapidly for $|l-\alpha| > kR$, and its behaviour as $r \to \infty$ can be found using standard asymptotic forms for $H^{(1)}_{|l-\alpha|}(kr)$, giving

$$\psi_R(\boldsymbol{r}) \xrightarrow{r\to\infty} \left(\frac{2}{\pi i k r}\right)^{1/2} \exp(ikr) \times \sum_{l=-\infty}^{\infty} \exp\{i(l\theta - \pi|l-\alpha|)\} \frac{J_{|l-\alpha|}(kR)}{H^{(1)}_{|l-\alpha|}(kR)}. \quad (13)$$

This represents a purely outgoing wave, consistent with condition (iv).

The asymptotics of ψ_{AB} are more subtle. In view of the absence of 'outgoing' Bessel functions $H^{(1)}$ from equation (11), it is far from obvious that ψ_{AB} can represent just the wave scattered by the flux line in addition to the incident wave. That this is indeed the case is shown in the Appendix using arguments simpler than those in the original AB paper. The result is

$$\psi_{AB}(\boldsymbol{r}) \xrightarrow[r\to\infty]{|\theta|<\pi - O((kr)^{-1/2})} \exp(-ikr\cos\theta + i\alpha\theta) + \frac{\exp(ikr)\sin(\pi\alpha)}{(2\pi i k r)^{1/2}\cos(\theta/2)} (-1)^{[\alpha]} \exp\{i([\alpha]+\tfrac{1}{2})\theta\}, \quad (14)$$

where $[\alpha]$ denotes the integer part of α.

To see that the first term of equation (14) correctly represents the incident wave, recall that this must correspond to a probability current of particles directed along $-x$. The current is

$$\boldsymbol{j}(\boldsymbol{r}) = \frac{\hbar}{m}\operatorname{Im}(\psi^*\nabla\psi) - \frac{q}{m}\mathbf{A}(\boldsymbol{r})|\psi|^2, \quad (15)$$

which is the expectation value of the velocity density operator

$$\boldsymbol{v}_{op}(\boldsymbol{r}) \equiv \frac{1}{2m}\{(\boldsymbol{p}_{op} - q\mathbf{A}_{op})\delta(\boldsymbol{r}-\boldsymbol{r}_{op}) + \delta(\boldsymbol{r}-\boldsymbol{r}_{op})(\boldsymbol{p}_{op} - q\mathbf{A}_{op})\}. \quad (16)$$

When applied to equation (14), equation (15) yields the correct result

$$\boldsymbol{j}_{AB}(\boldsymbol{r}) \xrightarrow{r\to\infty} -\frac{\hbar k}{m}\hat{\boldsymbol{x}}. \quad (17)$$

The second term of equation (14) describes a purely outgoing wave. Therefore the solution (10) does indeed satisfy all the conditions (i)–(iv).

It appears from equation (14) that the incident and scattered waves are multivalued functions of θ, but this is not the case because a narrow sector near the forward direction $|\theta| = \pi$ is excluded. We show in the Appendix that within this sector ψ_{AB} cannot be separated into incident and scattered parts, and that in the forward direction itself

$$\psi_{AB}(r, \pm\pi) \xrightarrow{r\to\infty} \exp\{i(kr + [\alpha]\pi)\}\cos\{\pi(\alpha - [\alpha])\}. \quad (18)$$

Thus when $\alpha = N + \frac{1}{2}$, where N is an integer, ψ_{AB} vanishes at $|\theta| = \pi$. This is actually a general result, valid for all r and for both ψ_{AB} and ψ_R, i.e.

$$\psi(r, \pm\pi) = 0 \qquad \text{when } \alpha = N + \tfrac{1}{2}. \quad (19)$$

An exact representation of ψ_{AB} when $\alpha = N + \frac{1}{2}$, valid for all r and θ, is given by equation (A9) of the Appendix. The existence of a *nodal line* stretching from 0 to ∞ when $\alpha = N + \frac{1}{2}$, and the stitching together of incident and scattered waves near $|\theta| = \pi$ for all α, will play an important part in §3.

It is easily verified from solutions (10), (11) and (12) that ψ has the following symmetry properties:

$$\psi(r, \theta; \alpha + N) = \exp(iN\theta)\psi(r, \theta; \alpha), \quad (20)$$

$$\psi(r, \theta; -\alpha) = \psi(r, -\theta; \alpha). \quad (21)$$

These imply that all measurements of the intensity $|\psi|^2$ are periodic in α, that what is observed at θ when $\alpha = \frac{1}{2} + \delta$ will be observed at $-\theta$ when $\alpha = \frac{1}{2} - \delta$, and that

$$|\psi(r, \theta)|^2 = |\psi(r, -\theta)|^2 \qquad \text{when } 2\alpha = N. \quad (22)$$

Therefore when studying $|\psi|^2$ it is sufficient to consider $0 \leq \alpha \leq \frac{1}{2}$.

As α varies, the interference pattern of waves scattered by the cylinder changes. When $kR \to 0$ this pattern is simply the AB scattering cross section obtained from equation (14), namely

$$\sigma_{AB}(\theta) = \frac{\sin^2(\pi\alpha)}{2\pi k\cos^2(\theta/2)}. \quad (23)$$

When $kR \gg 1$, however, the interference pattern is a complicated superposition of contributions from direct, reflected and creeping rays (Keller 1958). In the general case, the primitive notion of 'interference fringes' does not apply. But sometimes, if the pattern is locally the resultant of two contributions of approximately equal strength, it is legitimate to speak about fringes, as in elementary discussions of the AB effect. Then the naive superposition of the contributions, with the magnetic phase difference $\exp\{iq\int^{\boldsymbol{r}} \mathbf{A}(\boldsymbol{r}')\cdot d\boldsymbol{r}'/\hbar\}$ as in equation (7), can yield the correct result. For example, consider angular fringes whose contributions have wave number m

and have gone round opposite sides of the cylinder (paths 1 and 2 on figure 1). The wave beyond the cylinder is then

$$\psi \propto \exp\{i(m\theta+\alpha\theta)\} + \exp\{i(-m(-2\pi+\theta)+\alpha(-2\pi+\theta))\} \quad (24)$$

so that

$$|\psi|^2 \propto 4\cos^2\{m(\theta-\pi)+\alpha\pi\}. \quad (25)$$

This angular fringe pattern shifts periodically with α in precisely the manner observed in experiments.

3 Wavefront dislocations

From now on we ignore ψ_R and consider only ψ_{AB}, in particular its wavefronts. These are defined as lines of constant phase χ of the complex wave

$$\psi_{AB}(\mathbf{r}) \equiv |\psi_{AB}(\mathbf{r})| \exp(i\chi(\mathbf{r})) \quad (26)$$

in the plane r, θ. Wave crests are particular wavefronts, defined by

$$\chi(\mathbf{r}) = 2M\pi, \quad (27)$$

where M is an integer. Even though $\psi_{AB}(\mathbf{r})$ is a single-valued function, $\chi(\mathbf{r})$ may be multivalued, in the following sense: during a circuit C in the $\mathbf{r}$ plane, χ may change by an integer multiple S_C of 2π, i.e.

$$S_C = \frac{1}{2\pi}\oint_C d\chi = \frac{1}{2\pi}\oint_C \nabla\chi\cdot d\mathbf{r}. \quad (28)$$

Within C, S_C wave crests must come to an end. Points at which this happens are singularities of $\chi(\mathbf{r})$, called wavefront dislocations by Nye and Berry (1974) by analogy with dislocations of atomic planes in crystals. At a wavefront dislocation, the modulus $|\psi_{AB}|$ must vanish. S_C will be called the dislocation strength within C.

We now study the dislocation structure of ψ_{AB}, beginning with a calculation of S_C for an anticlockwise circuit C consisting of a very large circle surrounding the flux line. When the flux parameter is an integer, that is $\alpha = N$, there is no scattered wave and ψ_{AB} is given exactly for all θ (and also in this case for all r) by the first term of equation (14). The phase is then

$$\chi = -kr\cos\theta + N\theta \qquad (\alpha = N) \quad (29)$$

whose total change round C is $2\pi N$, so that from equation (28) $S_C = N$. When α is not an integer, the first term of equation (14) dominates everywhere except a narrow angular sector centred on $\theta = \pi$ (see the Appendix, equation (A8)), with width $2\Delta\theta$. Between $-\pi+\Delta\theta$ and $\pi-\Delta\theta$ the phase accumulation is $2\pi\alpha$, which is not an integer. Within this sector, however, χ is not given by equation (14), and when $\theta = \pi$ equation (18) shows that

$$\left.\begin{aligned} \chi - kr &= [\alpha]\pi \bmod 2\pi \\ &\quad \text{when } \theta=\pi,\ \alpha-[\alpha]<\tfrac{1}{2}, \\ &= ([\alpha]+1)\pi \bmod 2\pi \\ &\quad \text{when } \theta=\pi,\ \alpha-[\alpha]>\tfrac{1}{2}. \end{aligned}\right\} \quad (30)$$

Figure 2 shows the phase functions implied by this result, defining $\chi + kr$ as zero when $\theta = 0$ and interpreting the 'mod 2π' additions to give continuity with equation (29). It follows that the total phase change round C is $2\pi l_\alpha$, where l_α is the integer closest to α, so that

$$S_C = l_\alpha. \quad (31)$$

Now let C be a very small circle surrounding the flux line. Then $r \to 0$ and the series (11) for ψ_{AB} is dominated by the Bessel function of lowest order, for which l is the integer closest to α. All the angular dependence is in the term $\exp(il\theta)$, so that S_C is again given by equation (31).

The simplest picture consistent with these results is of a wavefront dislocation at the flux line, where S_C wave crests emerge and extend out to the far field. All other wave crests are continuous (from $\theta = -\pi/2$ to $\theta = +\pi/2$). As α varies through $N+\frac{1}{2}$, l_α and hence S_C jump by unity, so that a new wave crest appears or disappears at $r = 0$. The mechanism of this change in S_C when $\alpha = N+\frac{1}{2}$ is a disconnection and reconnection of wavefronts on a nodal line along the negative x axis (equation (19)).

Wave crest patterns are illustrated in figure 3(*a*) for some integer values of α, and in figure 3(*b*) for a sequence of values of α between 1 and 2. The pattern for $\alpha = 1\frac{1}{2}$ was drawn with the aid of the exact representation (A9) in the Appendix.

It is interesting to study the probability current $\mathbf{j}$ close to the flux line. According to equations (15), (26) and (3),

$$\mathbf{j} = \frac{\hbar}{m}|\psi_{AB}|^2\left(\nabla\chi - \frac{\alpha\hat{\boldsymbol{\theta}}}{r}\right), \quad (32)$$

so that the current is not perpendicular to the wavefronts. Calculations based on equation (11) give

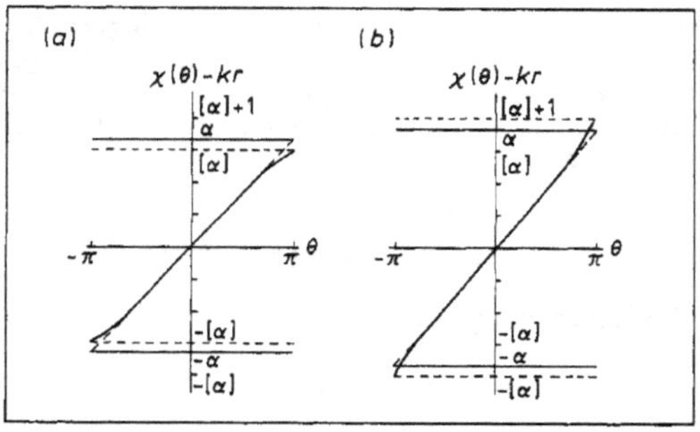

Figure 2 Asymptotic phase $\chi - kr$ for ψ_{AB} as a function of θ. (*a*) $\alpha-[\alpha]<\frac{1}{2}$. (*b*) $\alpha-[\alpha]>\frac{1}{2}$.

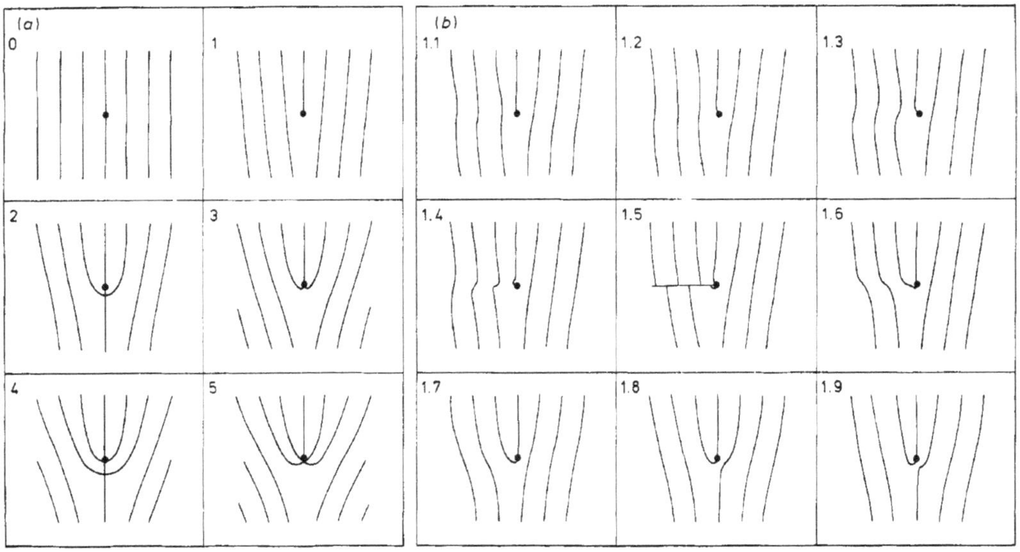

Figure 3 Wave crests of $\exp\{-\mathrm{i}\alpha/2\}\psi_{AB}$. (*a*) Integer values of α from 0 to 5. (*b*) Values of α between 1 and 2, showing unlinking and reconnection as α passes through $1\frac{1}{2}$. In all cases the waves are incident from the right, and the flux parameter corresponds to a magnetic field at the origin and pointing out of the paper. The wave crests were drawn by interpolating between calculated positions for ($r\to\infty$, $\theta=\pm\pi/2$) and on the x axis, taking account of the calculated directions of the wave crests at $r=0$.

$$\left.\begin{aligned} \boldsymbol{j} &\to \frac{K(l_\alpha-\alpha)\hat{\boldsymbol{\theta}}}{r^{1-2|l_\alpha-\alpha|}} \\ &\qquad \text{when } r\to 0,\ \alpha \text{ not near } N+\tfrac{1}{2}; \\ &\to 2K(\alpha-[\alpha]-\tfrac{1}{2})\hat{\boldsymbol{\theta}}\ln(1/kr) \\ &\qquad \text{when } r\to 0,\ \alpha \text{ near } N+\tfrac{1}{2}, \end{aligned}\right\} \tag{33}$$

where K is a constant. Therefore the particle current circulates around the flux line, clockwise when $[\alpha]<\alpha<[\alpha]+\frac{1}{2}$ and anticlockwise when $[\alpha]+\frac{1}{2}<\alpha<[\alpha]+1$; $\boldsymbol{j}$ vanishes for integral or half-integral values of the flux parameter.

The currents just described are invariant under gauge transformations to a different vector potential satisfying equation (1). The wavefront patterns (figure 3) change, however, but their topology is invariant: gauge transformations can cause arbitrary deformations of the wavefronts, but the strength S_C of the dislocation on the flux line does not change. S_C can be regarded as a 'topological quantum number', conserved under any classically permissible gauge transformation. The disappointing fact, however, is that this striking feature of ψ_{AB} is *unobservable*. This conclusion is not a consequence of the dislocation being a phase property, because phase in the position representation can affect amplitude in the momentum representation. Rather, it is a consequence of the following version of an argument given by Wu and Yang (1975).

Experiments can measure only expectation values of operators corresponding to observables. The canonical momentum is not an observable, but the velocity

$$\boldsymbol{v}=\boldsymbol{p}-q\boldsymbol{A} \tag{34}$$

is. In the state $|\psi\rangle$, any function $f(\boldsymbol{v}_{op})$ has expectation value

$$\langle\psi|\,f(\boldsymbol{v}_{op})\,|\psi\rangle=\int \mathrm{d}\boldsymbol{r}\psi^*(\boldsymbol{r})f(-\mathrm{i}\hbar\nabla-q\boldsymbol{A}(\boldsymbol{r}))\psi(\boldsymbol{r}). \tag{35}$$

In the case we have been considering, $\boldsymbol{A}$ is given by equation (3). Now let the flux parameter change from α to $\alpha'=\alpha+1$. According to the solutions (10), (11) and (12) and equation (34), ψ and $\boldsymbol{v}_{op}$ change to

$$\left.\begin{aligned} \psi' &= \exp(\mathrm{i}\theta)\psi \\ \boldsymbol{v}' &= \boldsymbol{p}-q\boldsymbol{A}'=\boldsymbol{p}-q\boldsymbol{A}-\hbar\hat{\boldsymbol{\theta}}/r \end{aligned}\right\} \tag{36}$$

The expectation value of $f(\boldsymbol{v}'_{op})$ is now

$$\langle\psi'|\,f(\boldsymbol{v}'_{op})\,|\psi'\rangle=\int \mathrm{d}\boldsymbol{r}\psi^*(\boldsymbol{r})\exp(-\mathrm{i}\theta) \times f(-\mathrm{i}\hbar\nabla-q\boldsymbol{A}(\boldsymbol{r})-\hbar\hat{\boldsymbol{\theta}}/r)\exp(\mathrm{i}\theta)\psi(\boldsymbol{r}), \tag{37}$$

which is easily shown to equal equation (35).

This means that not only the intensity $|\psi|^2$ but all observable quantities are unaffected by changing α by an integer. All that can be measured is the deviation of the flux parameter from the nearest integer. The integer itself, which is precisely the dislocation strength S_C, cannot be observed: there is no 'dislocation strength operator' whose eigenvalues are S_C.

4 The Aharonov–Bohm effect for water waves

In seeking a system where the dislocation in ψ_{AB} can be observed, we first note an analogy between waves in the presence of a vector potential and waves in a moving medium. Such an analogy (albeit somewhat different from the one we shall present) was originally suggested to us by J H Hannay.

Let waves with frequency Ω and wavevector $\boldsymbol{k}$ propagate in a stationary isotropic medium (which may be inhomogeneous). The dispersion relation (i.e. the Hamiltonian) is

$$\Omega = \omega(k, \boldsymbol{r}), \tag{38}$$

and depends only on the length k of $\boldsymbol{k}$. In a field with vector potential $\mathbf{A}$, the dispersion relation becomes

$$\Omega = \omega(|\boldsymbol{k} - q\mathbf{A}(\boldsymbol{r})/\hbar|, \boldsymbol{r}). \tag{39}$$

If instead the medium is moving, with flow velocity $\boldsymbol{U}(\boldsymbol{r})$, then a plane wave is described locally not by

$$\psi = \exp\{\mathrm{i}(\boldsymbol{k} \cdot \boldsymbol{r} - \omega(k, \boldsymbol{r})t)\} \tag{40}$$

but by

$$\begin{aligned}\psi &= \exp\{\mathrm{i}(\boldsymbol{k} \cdot (\boldsymbol{r} - \boldsymbol{U}(\boldsymbol{r})t) - \omega(k, \boldsymbol{r})t)\} \\ &= \exp\{\mathrm{i}(\boldsymbol{k} \cdot \boldsymbol{r} - (\omega(k, \boldsymbol{r}) + \boldsymbol{k} \cdot \boldsymbol{U}(\boldsymbol{r}))t)\},\end{aligned} \tag{41}$$

so that the dispersion relation in the moving medium is

$$\Omega = \omega(k, \boldsymbol{r}) + \boldsymbol{k} \cdot \boldsymbol{U}(\boldsymbol{r}). \tag{42}$$

To lowest order, this can be written as

$$\Omega \approx \omega\left(\left|\boldsymbol{k} + \frac{k\boldsymbol{U}(\boldsymbol{r})}{v_g(k, \boldsymbol{r})}\right|, \boldsymbol{r}\right) \tag{43}$$

where v_g is the group velocity

$$v_g = \partial\omega(k, \boldsymbol{r})/\partial k. \tag{44}$$

The step from equation (42) to equation (43) is valid provided

$$|\boldsymbol{U}| \ll v_g. \tag{45}$$

Comparing equations (43) and (39), we see that the effect of a slowly moving medium is the same as that of a vector potential, the precise analogy being

$$\frac{q\mathbf{A}(\boldsymbol{r})}{\hbar} \leftrightarrow \frac{-k\boldsymbol{U}(\boldsymbol{r})}{v_g(k, \boldsymbol{r})}. \tag{46}$$

From equations (1) and (2), the analogue of the quantum flux parameter, in the case where the stationary medium is homogeneous, is

$$\alpha \leftrightarrow \frac{-\oint \boldsymbol{U} \cdot \mathrm{d}\boldsymbol{r}}{\lambda v_g} = \frac{-\omega\oint \boldsymbol{U} \cdot \mathrm{d}\boldsymbol{r}}{v_p v_g} \tag{47}$$

where λ is the wavelength and v_p the phase velocity ω/k.

In the AB effect, the magnetic field vanishes outside the central flux line. Therefore the velocity field $\boldsymbol{U}$ analogous to $\mathbf{A}$ must be curl-free, i.e. the waves must propagate on a medium that is flowing irrotationally but with non-zero circulation. This can be achieved with surface waves on swirling water, if the swirling takes the form of an irrotational ('bathtub') vortex. If the density, surface tension and depth of the water are denoted by ρ, γ and d respectively, and if g denotes the acceleration due to gravity, the dispersion law is (Lamb 1945)

$$\omega = \left\{\left(gk + \frac{\gamma k^3}{\rho}\right)\tanh(kd)\right\}^{1/2}. \tag{48}$$

This analogy provides a means of testing the predictions of §3 concerning wavefront dislocations, because for water waves, as opposed to quantum mechanical waves, the crests can be observed. For the analogy to hold good, the condition (45) must be satisfied, but this will always be the case far from the vortex because $|\boldsymbol{U}| \to 0$ there; the dislocation at $r = 0$ can be identified by counting wave crests at large r. According to equation (47), the patterns in figure 3 correspond to water circulating clockwise, so that the wavefronts should be more closely spaced where the waves travel against the current than when they travel with the current, as expected on the basis of the Doppler effect.

5 Experiment†

A rectangular perspex tank was constructed, with dimensions 1.0 m × 0.6 m × 0.15 m. Surface waves were excited on water in the tank by vibrating a straight horizontal dipper 0.15 m long connected to a variable-speed electric motor. The useful frequency range was from 7 to 70 Hz. For the lower frequencies the gravity-wave term gk in equation (48) dominated, whilst for the higher frequencies the surface-tension ripple term $\gamma k^3/\rho$ dominated; the water was sufficiently deep for the term $\tanh(kd)$ to be set equal to unity for all frequencies.

A vortex formed spontaneously on letting the water pour out through a hole 6 mm in diameter in the middle of the bottom of the tank. For the wave-making dipper to operate efficiently it was essential to maintain a constant water level; this

† The results reported in this section were obtained by two of us (MDL and JCW) in an undergraduate research project.

was achieved by pumping the outflowing water back into the tank, taking care to let the pumped water re-enter via submerged perforated tubes, so as to inhibit the development of bulk rotation.

In order to 'freeze' the motion of the wave crests, the water surface was illuminated from below by a stroboscope set to the wave frequency. The light refracted by the wave crests formed bright caustic lines on a screen just above the water. The patterns on the screen were photographed; figure 4 shows a series of such pictures for dislocation strengths ranging from 0 to 3.

According to §3 the dislocation structure of the waves depends on the flux parameter α. For water waves this quantity is given by equation (47), which shows that in order to predict the dislocation strength it is necessary to know not only the angular frequency ω of the waves but also the circulation $\oint \boldsymbol{U} \cdot \mathrm{d}\boldsymbol{r}$ of the vortex. To measure the circulation, small paper discs were allowed to circulate on the water surface near the vortex, and photographed under continuous illumination with an exposure of 0.5 s. The circulating paper discs gave tracks whose lengths were used to estimate the water velocity and hence the circulation. With the dipper running, the streamlines were approximately circular to a distance of about 50 mm from the vortex core. Each photograph contained about ten useful tracks, and measurement gave the same circulation for these tracks to an accuracy of about 20%, which enabled α to be estimated with a standard error of less than 10%. On the other hand, inspection of the photographs of wave crests, and comparison with the results of §3, enabled α to be estimated with an absolute accuracy of about 0.25 (figure 4, cf figure 3). More than fifty comparisons of the values of α obtained by these two different methods were made, using frequencies between 8 Hz and 67 Hz to generate the waves, and a range of water outflow rates to generate the vortex. The values of α ranged from 0 to 2, and in every case the two methods gave complete agreement within the quoted accuracy, thus confirming the theory of §3 and §4.

6 Discussion

From a mathematical point of view, there can be no doubt that ψ_{AB}, given by equation (11), is the correct solution of the AB wave equation (4). Moreover ψ_{AB} correctly models observable features of waves in physical systems: the periodic shifting of interference fringes as the flux varies, confirmed by Chambers (1960) and Möllenstedt and Bayh (1962), and the changing of the wavefront topology as the circulation varies, which we have confirmed and studied (§5). But the physical interpretation of the mathematics and the experiments continues to cause controversy. The question is, are the predicted and observed fringe shifts really effects of potentials in the absence of electromagnetic fields accessible to the particles, or is it possible to explain them entirely in terms of accessible fields?

Weisskopf (1961) considers the flux to be switched on from zero to its final value Φ. He shows that the evolution of the wavefunction to ψ_{AB} can be explained entirely in terms of the accessible electric field induced outside the cylinder, even if this field is made arbitrarily small by changing the flux very slowly. Casati and Guarneri (1979) consider the cylinder to be slightly penetrable, and modelled by a high potential V for $r<R$. They show that for any finite V, however large, the fringe shifts can be explained entirely in terms of the magnetic field within the cylinder, which is now accessible to the particles (their analysis is in terms of the hydrodynamical formulation of quantum mechanics, rather than the wave equation, but their conclusions do not depend on this). Roy (1980) considers the cylinder to be a solenoid of finite length L. He shows that no matter how large L is, a gauge can be found in which the vector potential can be expressed entirely in terms of the weak magnetic field leaking out of the ends of the cylinder into the space accessible to the particles.

In view of these arguments we must agree that if any electromagnetic fields, however small, are or have ever been accessible to the particles, then the state of the particles, in quantum as well as classical mechanics, can be described entirely in terms of these fields. It could be argued that since such fields always exist in practice, there is no AB effect, i.e. no observable consequence of inaccessible fields. But in view of the fact that the wave always tends to ψ_{AB}, no matter how the limiting process is performed, those who argue thus must also believe that infinitesimal causes (fields) can produce finite effects (fringe shifts). On the other hand, if we consider the vector potential (as embodied in the gauge-invariant and quantum-mechanically observable deviation of the flux parameter from an integer) as a primary causative agent, then there is no infinite discordance between the magnitudes of causes and effects in the limit of inaccessibility. We therefore consider the question of the reality of the AB effect to be metaphysical and devoid of observavational implication.

Acknowledgments
We thank G Casati, J H Hannay and F J Wright for helpful discussions and suggestions.

Appendix

The first step in studying the asymptotics of ψ_{AB} as defined by equation (11) is to replace the Bessel functions by the integral representation

$$J_\nu(z) = \frac{1}{2\pi}\int_C \exp\{i(\nu t - 2\sin t)\}\,dt \qquad (A1)$$

(Gradshteyn and Ryzhik 1965), where C is the contour shown in figure A1(a). Then equation (11) becomes

M V Berry et al—Aharonov–Bohm wavefront dislocations (facing page 160)

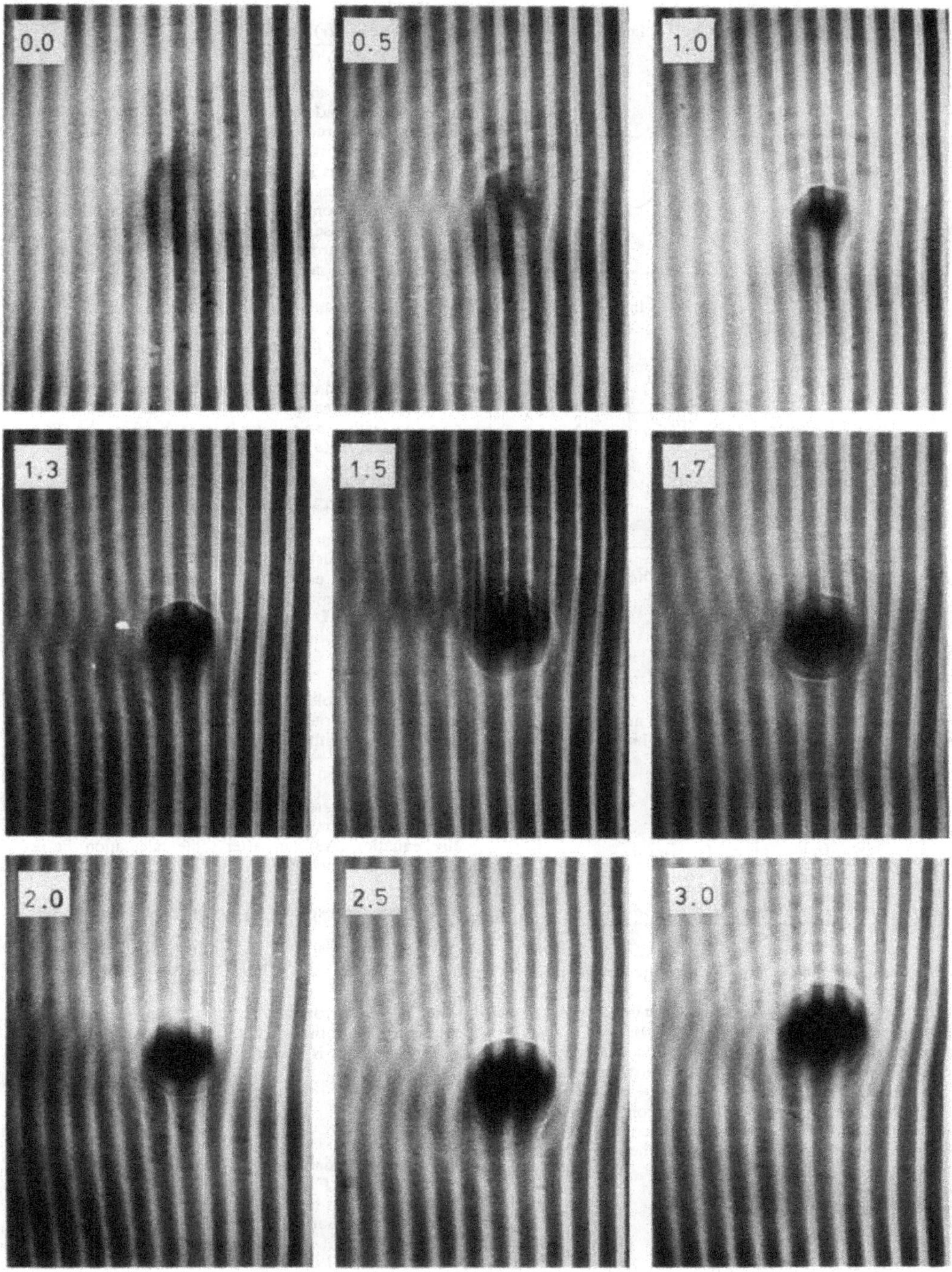

Figure 4 Water wave crests passing an irrotational vortex, giving rise to different strengths of wavefront dislocation. Waves are incident from the right and the water is circulating clockwise. Estimated values of the flux parameter α are indicated.

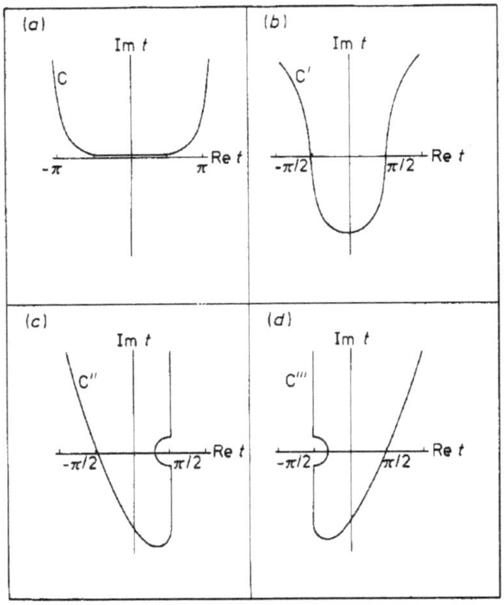

Figure A1 Integration contours in the t plane.

$$\psi_{AB}=\frac{1}{2\pi}\int_C dt\,\exp(-ikr\sin t)\times\sum_{l=-\infty}^{\infty}\exp\left\{i\left(l\theta-|l-\alpha|\frac{\pi}{2}+|l-\alpha|t\right)\right\}. \quad (A2)$$

The sum converges because $\mathrm{Im}\, t>0$, and can be evaluated by splitting the range into $l\leqslant[\alpha]$ and $l\geqslant[\alpha]+1$, to give

$$\psi_{AB}=\frac{1}{2\pi}\int_C dt\,\exp(-ikr\sin t)\times\left(\frac{\exp\{i([\alpha]\theta+(t-\pi/2)(\alpha-[\alpha]))\}}{1-\exp(i(t-\pi/2-\theta))}+\frac{\exp\{i([\alpha]+1)\theta+(t-\pi/2)([\alpha]-\alpha+1))\}}{1-\exp(i(t-\pi/2+\theta))}\right). \quad (A3)$$

For large kr the integral is dominated by its poles and saddle points. The saddle points occur where $(d/dt)(kr\sin t)=0$, i.e. at $t=\pm\pi/2$. The integration contour can be deformed into C′ (figure A1(b)) so as to pass through these. During the deformation, the contour crosses a pole, whose contribution must be included. The first term in (A3) has a pole if $t-\pi/2-\theta=0$, and this contributes if $-\pi<\theta<0$; the second term in (A3) has a pole if $t-\pi/2+\theta=0$, and this contributes if $0<\theta<\pi$. Both contributions are of the same form, namely

$$\psi_{AB}(\text{pole})=\exp\{i(-kr\cos\theta+\alpha\theta)\}\qquad(0<|\theta|<\pi). \quad (A4)$$

This is precisely the incident wave in equation (14).

Next, the contributions to the integral over C′ from the saddle points at $t=\pm\pi/2$ must be evaluated by the method of steepest descents (De Bruijn 1970). A straightforward calculation for $t=-\pi/2$ gives

$$\psi_{AB}(\text{saddlepoint at } t=-\pi/2)=\frac{\exp(ikr)\sin(\pi\alpha)}{(2\pi ikr)^{1/2}\cos(\theta/2)}(-1)^{[\alpha]}\exp\{i([\alpha]+\tfrac{1}{2})\theta\}. \quad (A5)$$

This is precisely the outgoing scattered wave in equation (14). The saddle point at $t=+\pi/2$ would give an incoming 'scattered' wave, but substitution into equation (A3) shows that the coefficient of this wave is zero.

Our arguments leading to the asymptotic form (14) for ψ_{AB} fail when θ is close to 0 or π, because then the pole is close to one of the saddle points and the contributions of these two points cannot be separated.

In the backward direction $\theta=0$, the saddle point at $\pi/2$, which gave no incoming wave, coincides with a pole in each term of (A3); the contour C″ on figure A1(c) gives a pole contribution identical with (A4) (with $\theta=0$), as well as a principal-value integral which vanishes as $kr\to\infty$. Therefore equation (14) remains valid as $\theta\to 0$.

In the forward direction $\theta=\pi$, the saddle point at $t=-\pi/2$, which gave rise to the outgoing wave (A5), coincides with a pole in each term of (A3); the contour C‴ on figure A1(d) gives a pole contribution

$$\psi_{AB}(\text{saddle point and pole at } t=-\pi/2)=\exp\{i(kr+\pi[\alpha])\}\cos\{\pi(\alpha-[\alpha])\},\qquad\theta=\pi, \quad (A6)$$

together with a principal-value integral which vanishes as $r\to\infty$. This justifies equation (18).

Next we must estimate the angular width of the 'diffraction shadow', near the forward direction, within which (A4) and (A5) are not valid. The 'domain' Δt of the stationary point at $t=-\pi/2$ is given by

$$\Delta(kr\sin t)\sim 1\text{ radian},\quad\text{i.e.}\quad\Delta t\sim(2/kr)^{1/2}. \quad (A7)$$

In order for a pole of (A3) to give a separate contribution, it must lie outside the domain Δt. This excludes the sector of width $2\Delta\theta$ defined by

$$\Delta\theta=\pi-|\theta|<O(kr)^{-1/2}, \quad (A8)$$

hence the restriction on $|\theta|$ in equation (14).

For the important transitional case of half-integer flux, equation (A3) can, after some calculation, be reduced to the simple form

$$\psi_{AB}(r,\theta)=\frac{2}{\pi^{1/2}}\exp\{i((N+\tfrac{1}{2})\theta-kr\cos\theta-\pi/4)\}\times\int_0^{(2kr)^{1/2}\cos(\theta/2)}\exp(it^2)\,dt\qquad\text{when } \alpha=N+\tfrac{1}{2}, \quad (A9)$$

which is (apart from a slight correction) the same as equation (23) in the original paper by Aharonov and Bohm (1959). It is evident from (A9) that ψ_{AB} is single-valued.

References

Abramowitz M and Stegun I A (eds) 1964 *Handbook of Mathematical Functions* (New York: Dover)

Aharonov Y and Bohm D 1959 *Phys. Rev.* **115** 485–91

Casati G and Guarneri I 1979 *Phys. Rev. Lett.* **42** 1579–81

Chambers R G 1960 *Phys. Rev. Lett.* **5** 3–5

De Bruijn N G 1970 *Asymptotic Methods in Analysis* (Amsterdam: North-Holland)

Gradshteyn I S and Ryzhik I M 1965 *Tables of Integrals, Series, and Products* (New York: Academic)

Eur. J. Phys. **1** (1980) 162–168. Printed in Northern Ireland

Ingraham R L 1972 *Am. J. Phys.* **40** 1449–52
Keller J B 1958 *Proc. Symp. Appl. Math.* **8** 27–52
Kretschmar M 1965a *Z. Phys.* **185** 73–83
—— 1965b *Z. Phys.* **185** 84–96
Lamb H 1945 *Hydrodynamics* (Cambridge: CUP)
Merzbacher E 1962 *Am. J. Phys.* **30** 237–47
Möllenstedt G and Bayh W 1962 *Naturwiss.* **49** 81–2
Nye J F and Berry M V 1974 *Proc. R. Soc.* A **336** 165–90
Roy S M 1980 *Phys. Rev. Lett.* **44** 111–4
Tassie L J and Peshkin M 1961 *Ann. Phys., NY* **16** 177–84
Weisskopf V F 1961 *Lectures in Theoretical Physics* vol III, ed W E Brittin (New York: Interscience) p67
Wu T T and Yang C N 1975 *Phys. Rev.* D **12** 3845–57

Introductory article

Much ado about nothing: optical dislocation lines (phase singularities, zeros, vortices...)

Michael Berry

H H Wills Physics Laboratory, Tyndall Avenue, Bristol BS8 1TL, United Kingdom

Often, light can be represented, approximately or exactly, by a complex scalar wave ψ, smoothly varying in space and/or time. The field ψ could be a cartesian component of the electromagnetic field or of the vector potential, or one of several scalar potentials appropriate to different circumstances[1]. This meeting has been concerned with the line singularities of the phase of ψ. Here, I wish to make some general remarks about these lines. Except where stated, all the remarks are independent of the particular wave equation that ψ satisfies.

They are lines because the only way that the phase of a smooth function can be singular is for its modulus to vanish; this in turn requires that two conditions are satisfied - $\mathrm{Re}\,\psi = 0$ and $\mathrm{Im}\,\psi = 0$ - and these define a line (think of the simplest local model $\psi = x + iy$, where the singular line of phase is the z axis). Three interpretations of the lines are: as wavefront dislocations, since the patterns of surfaces of constant phase (mod 2π) mirror those of dislocations in the arrangements of atoms in crystals[2,3]; as vortices, since the phase gradient direction (that is, the direction of the current, or Poynting vector) swirls about the singular line like fluid in an irrotational vortex[4-6]; and as zeros, that is threads of darkness.

From a broader perspective, these singularities of faint light are complementary (in the sense introduced by Bohr) to the singularities of bright light. The latter are caustics, that is, envelopes of families of rays in geometrical optics[7,8]. Apart from the obvious complementarity of bright and dark, there is the less obvious fact that measurement of each inhibits measurement of the other. To see a caustic singularity requires seeing the light on a scale large compared with the wavelength; then the dislocations, which are sub-wavelength fine structures, are too small to perceive. On the other hand, under the high magnifications required to see dislocations the caustic singularities are smoothed away by diffraction.

Continuing the comparison, I note that each singularity is a window to a deeper theory. The most striking prediction of geometrical optics is the infinite intensity at a caustic. But this is precisely where ray theory breaks down dramatically, because wave effects soften the divergence, and interference fringes are biggest - for wavelength λ (small compared with other relevant dimensions), they scale as $\lambda^{2/3}$ rather than λ itself (that is why light waves can be seen in the sky, enormously magnified, as supernumerary rainbows[9] close to the main arc which is an angular caustic). So, caustic singularities lead naturally into waves, in which there is an additional field, namely phase, and the singularities of phase are the singularities of wave optics. But these are precisely where wave optics itself breaks down, in the sense that the dark light of a dislocation reveals the photon fluctuations of quantum optics - the next (and to this day the last) level in the hierarchy of physical theories of light[10].

SPIE Vol. 3487 • 0277-786X/98/$10.00

Reprint of 'Much ado about nothing: optical dislocation lines (phase singularities, zeros, vortices...)', Proc. SPIE 3487, International Conference on Singular Optics, 1 (August 3 1998).

Around a dislocation, the phase changes by a multiple of 2π. There are two natural ways to specify the strength of a dislocation in terms of this phase change. I need to discuss them carefully, for a reason that will emerge later. In the first, we let s label points on the dislocation and denote by $\mathbf{n}(s)$ one of the two (opposite) unit vectors along the dislocation. Then the strength S is $1/2\pi$ times the phase increase round the dislocation in a positive (clockwise) circuit defined by $\mathbf{n}$. Thus defined, S is a signed integer, usually ± 1. A convenient formula[11] is

$$S = sign \mathrm{Im} \nabla \psi^* \times \nabla \psi \cdot \mathbf{n} \qquad (1)$$

The choice of $\mathbf{n}$, rather than $-\mathbf{n}$, is arbitrary; often, but not always, it is convenient to make the choice at one point on the dislocation, and define $\mathbf{n}(s)$ elsewhere by continuity.

In the second way of describing the strength of a dislocation, the direction of $\mathbf{n}$ is chosen so that the phase increases in a positive circuit; the strength can then be specified simply by an arrow on the dislocation. Higher-order dislocations can be represented by multiple arrows.

Dislocation lines can move, either in time (e.g. for wave pulses) or as parameters vary. Then they can collide and interact in various ways[11-13]. These interactions are constrained - independently of any wave equation - by a topological conservation law[2] involving the dislocation strength. Consider an arbitrary curve C (stationary or moving) in space. It may be threaded by dislocation lines, in the sense that these may pierce an arbitrary surface Σ spanning C. Let a sense be defined on C, so that the two sides of Σ ('front' and 'back') can be distinguished. Using the second specification of dislocation strength, we define the total strength S_C as the signed number of dislocation arrows piercing Σ, with arrows pointing to the front counting +1 and arrows pointing to the back counting -1; S_C is independent of Σ. The conservation law is that S_C does not change unless a dislocation line crosses C.

Dislocation interactions are also constrained by a second, and more subtle, conservation law[14]. In three dimensions, the wavefronts, on which the phase is constant (mod 2π) are surfaces. Wavefronts intersect each spanning surface Σ in lines - phase contours. Consider the pattern of phase contours on Σ, without regard to the phase labels they carry. The patterns have singularities, of three kinds: not only dislocations but also phase saddles and phase extrema (maxima or minima). For each singularity, a Poincaré index can be defined: in a circuit of the singularity, the index is the number T of complete rotations of the direction of the phase contours in the same direction that the circuit is traversed. For dislocations, from which phase contours on Σ radiate like spokes of a wheel, they rotate in the same sense as the circuit; therefore $T = +1$ for first-order dislocations, irrespective of whether their strength is $S = +1$ or $S = -1$. For phase maxima and minima, the phase contours degenerate to points surrounded by loops, and $T = +1$ for these too. For phase saddles, the contours are hyperbolas that rotate in the opposite direction to the circuit, so $T = -1$. We define the total strength T_C as the sum of the Poincaré indices T for all the singularities on Σ. T_C is independent of Σ. The conservation law is that T_C does not change unless a dislocation line or phase saddle or phase extremum crosses C.

The subtlety arises on considering a continuous family of spanning surfaces Σ, nonintersecting except on C. On each of these surfaces, there can be phase saddles and phase extrema, and it is tempting to connect these points on different Σs, and regard the loci as line singularities of phase saddles and extrema. But this would be wrong, because, unlike dislocations, these lines have no independent existence; they are artefacts of the choice of family of Σs: for a different choice, the lines would be different. In fact, the phase saddles and extrema involved in the T_C conservation law are essentially point singularities in two dimensions (phase saddles and extrema can exist as points in three dimensions, where all three components of phase gradient vanish, but these do not appear to have any significant association with dislocations).

Notwithstanding this appearance of artificiality, the T_C conservation law is useful because it implies that phase saddles must be implicated in certain sorts of dislocation interaction. Consider for example the commonly occurring situation shown in figure 1, of a dislocation curved like a hairpin. As the height z of the surface Σ is increased, the two points where the dislocation intersects Σ collide and annihilate. Since each dislocation point has $T = +1$, the total T for the pair is +2. After the annihilation there are no dislocation points, that is the total T for dislocations is zero. By the conservation law, the collision must involve at least two saddles, whose contribution of -2 cancels that of the dislocations. This argument holds irrespective of the possible presence of phase extrema, because phase saddles are the only source of negative T.

Sometimes, the strength of a dislocation is called its topological charge[15-16]. This terminology is justified for the intersections of dislocation lines with a surface Σ, because the intersections are points analogous to electric charges in two dimensions. However, it is misleading because it can obscure essential features of the dislocation line in three dimensions. To see this, consider again the hairpin of figure 1. Along this curved dislocation, it is natural to choose **n** as illustrated, so that the strength is the same at all points (+1 in this case). But for the plane surface Σ it is natural to use a contrary convention, and define **n** at the two intersections so that both vectors point to the same side of Σ (e.g. to the 'front', with the larger value of z, as indicated by the dotted arrows). Now the two intersections appear to represent dislocations with different strengths (+1 and -1), and their collision as Σ is moved looks like the annihilation of opposite charges. Such a description obscures the fact that the collision - which simply occurs where the curved dislocation line touches one of the family of surfaces Σ - is an artefact of the choice of the family: for surfaces with a different orientation, the collision occurs at a different place.

In a frequently investigated special case, ψ represents a monochromatic wave with a well defined propagation direction ($+z$, say) - for example, that generated by a paraxial superposition of laser beams[17,18]. Then it is natural to consider the surfaces Σ as planes of constant z (the circuits C being regarded as at infinity), and think of increasing z as analogous to increasing 'time'. The stationary dislocation lines appear as 'moving' points on Σ, and this is one of the situations, referred to above, where the designation topological charge is legitimate - although the essential nature of dislocations as lines in three dimensions should never be forgotten.

In addition to the topological indices S and T, and the direction **n**, there are geometrical measures of dislocations. Three of these are the local wavevector (whose orientation with respect to **n** measures the edge-screw character of the dislocation), the local frequency (useful for non-monochromatic waves) and the phase of the dislocation[3].

As a final remark about scalar waves, it is worth noting that although there are no phase extrema in strictly two-dimensional monochromatic waves satisfying a wave equation[14], in paraxial three-dimensional waves these extrema can occur. For example, in a gaussian beam, the concave/convex wavefronts form phase maxima/minima in planes Σ before/after the focus.

In vector electromagnetic fields, several other sorts of line singularities exist, analogous to disclinations in liquid crystals. Their properties have been studied in detail[19-25]. From the general perspective I outlined earlier, these are singularities of the new physical property introduced in the generalization from scalar to vector waves, namely polarization.

ACKNOWLEDGEMENTS

I thank Professor M.S. Soskin for generous hospitality at the conference on Singular Optics, and Professors N.R. Heckenberg, J.F. Nye, M.S. Soskin, and M.V. Vasnetsov for illuminating discussions.

REFERENCES

1. Born, M. and Wolf, E., 1959, *Principles of Optics* (Pergamon, London).
2. Nye, J. F. and Berry, M. V., 1974, Dislocations in wave trains, *Proc. Roy. Soc. Lond.* **A336,** 165-90.
3. Nye, J. F., 1981, The motion and structure of dislocations in wavefronts, *Proc. Roy. Soc. Lond.* **A378,** 219-239.
4. Hirschfelder, J. O., Christoph, A. C. and Palke, W. E., 1974, Quantum mechanical streamlines. 1. Square potential barrier, *J. Chem. Phys.* **61,** 5435-5455.
5. Hirschfelder, J. O. and Tang, K. T., 1976, Quantum mechanical streamlines. III Idealized reactive atom-diatomic molecule collision, *J. Chem. Phys.* **64,** 760-785.
6. Hirschfelder, J. O. and Tang, K. T., 1976, Quantum mechanical streamlines. IV. Collision of two spheres with square potential wells or barriers, *J. Chem. Phys.* **65,** 470-486.
7. Berry, M. V. and Upstill, C., 1980, Catastrophe optics: morphologies of caustics and their diffraction patterns, *Progress in Optics* **18,** 257-346.
8. Berry, M. V., 1981, Singularities in Waves and rays, in *Les Houches Lecture Series Session 35*, eds. Balian, R., Kléman, M. & Poirier, J.-P. (North-Holland: Amsterdam), pp. 453-543.
9. Nussenzveig, H. M., 1992, *Diffraction effects in semiclassical scattering* (University Press, Cambridge).
10. Park, D., 1997, *The fire within the eye: a historical essay on the nature and meaning of light* (University Press, Princeton).
11. Berry, M. V., 1997, Dislocation reactions in nonparaxial gaussian beams, *J. Mod. Opt.*, (in the press).
12. Karman, G. P., Beijersbergen, M. W., van Duijl, A. and Woerdman, J. P., 1997, Creation and annihilation of phase singularities in a focal field, *Optics Letters* **19,** 1503-1505.
13. Nye, J. F., 1997, Unfolding of higher-order wave dislocations, *J. Opt. Soc. Amer. A.*, submitted.
14. Nye, J. F., Hajnal, J. V. and Hannay, J. H., 1988, Phase saddles and dislocations in two-dimensional waves such as the tides, *Proc. Roy. Soc. Lond.* **A417,** 7-20.
15. Soskin, M. S., Gorshkov, V. N., Vasnetsov, M. V., Malos, J. T. and Heckenberg, N. R., 1997, Topological charge and angular momentum of light beams carrying optical vortices, *Phys. Rev. A* **56,** 4064-4075.
16. Beijersbergen, M., 1996, Phase singularities in optical beams, *in Huygens Laboratory* (Leiden), p. 98.
17. Basistiy, I. V., Bazhenov, V. Y., Soskin, M. S. and Vasnetsov, M. V., 1993, Optics of light beams with screw dislocations, *Opt. Comm.* **103,** 422-428.
18. Basistiy, I. V., Soskin, M. S., and Vasnetsov, M. V., 1995, Optical wavefront dislocations and their properties, *Opt. Comm.* **119,** 604-612.
19. Nye, J. F., 1983, Polarization effects in the diffraction of electromagnetic waves: the role of disclinations, *Proc. Roy. Soc. Lond.* **A387,** 105-132.
20. Nye, J. F., 1983, Lines of circular polarization in electromagnetic wave fields, *Proc. Roy. Soc. Lond.* **A389,** 279-290.
21. Nye, J. F. and Hajnal, J. V., 1987, The wave structure of monochromatic electromagnetic radiation, *Proc. Roy. Soc. Lond.* **A409,** 21-36.
22. Nye, J. F., 1997, Singularities in wave fields, *Phil. Trans. Roy. Soc. Lond.* **355,** 2065-2069.
23. Hajnal, J. V., 1987, Singularities of the transverse fields of electromagnetic waves. I. Theory, *Proc. Roy. Soc. Lond.* **A414,** 433-446.
24. Hajnal, J. V., 1987, Singularities in the transverse fields of electromagnetic waves. II. Observations on the electric field., *Proc. Roy. Soc. Lond.* **A414,** 447-468.
25. Hajnal, J. V., 1990, Observation of singularities in the electric and magnetic fields of freely propagating microwaves., *Proc. Roy. Soc. Lond.* **A430,** 413-421.

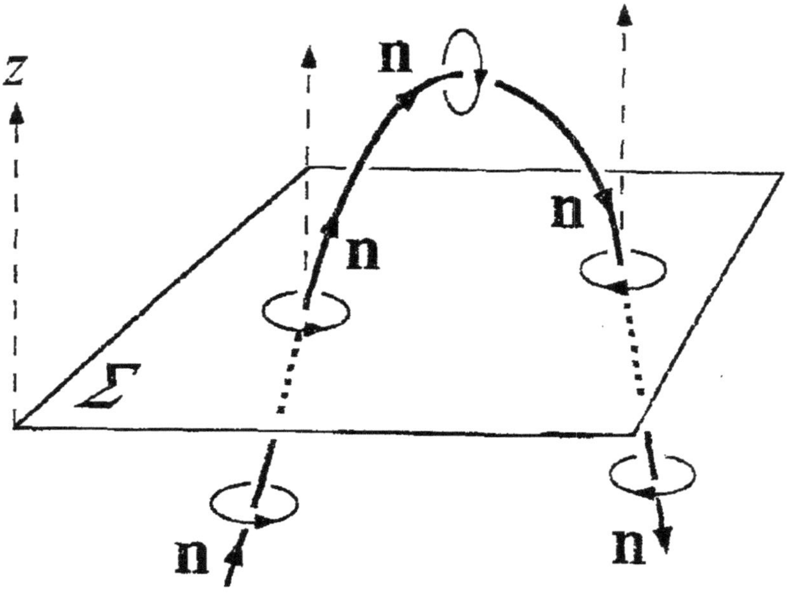

Figure 1. Curved dislocation line piercing a surface Σ. Loops show the sense in which phase increases, and **n** labels the choice of unit vector, continuous along the dislocation.

Geometry of phase and polarization singularities, illustrated by edge diffraction and the tides

Michael Berry

H.H. Wills Physics Laboratory, Tyndall Avenue, Bristol BS8 1TL, United Kingdom

ABSTRACT

In complex scalar fields, singularities of the phase (optical vortices, wavefront dislocations) are lines in space, or points in the plane, where the wave amplitude vanishes. Phase singularities are illustrated by zeros in edge diffraction and amphidromies in the heights of the tides. In complex vector waves, there are two sorts of polarization singularity. The polarization is purely circular on lines in space or points in the plane (C singularities); these singularities have index $\pm 1/2$. The polarization is purely linear on lines in space for general vector fields, and surfaces in space or lines in the plane for transverse fields (L singularities); these singularities have index ± 1. Polarization singularities (C points and L lines) are illustrated in the pattern of tidal currents.

Keywords: amphidromies, edge diffraction, fields, phase, polarization, Sommerfeld solution, singularities, waves.

1. INTRODUCTION

Recently, experimental and theoretical interest in phase singularities has revived and intensified [1, 2]. This is part of what has come to be called (following Soskin) *singular optics*. Interpreted properly in its full generality, singular optics is the study of the geometrical singularities that arise at each level of description in the physics of light (or - still more generally, in any branch of the physics of waves). At the geometrical level, the singularities are caustics, that is, envelopes of families of rays. At the level of scalar optics, the singularities are wavefront dislocations of the phase (otherwise called optical vortices or topological charges). And in vector waves the singularities are in the patterns of polarization. A general review, with many references, has been published by Nye [3]. Here I will concentrate on the phase and polarization singularities, with a twofold aim.

The first purpose is to review these singularities and some associated concepts, and provide a convenient collection of (mostly) known formulas. Section 2 deals with scalar waves (where in the general theory we follow the notation in [4]) and section 5 deals with vector waves (where in the general theory we follow the notation in [25]).

The second purpose is to provide some historical perspective by illustrating how far back in the history of science these ideas can be traced. Section 3 deals with edge diffraction [5]; the story spans more than three centuries, starting with Grimaldi and Newton and culminating in the currently-popular Madelung ('Bohmian') formalism for waves. Edge diffraction also illustrates some important features of phase singularities. Sections 4 and 6 deal with the tides, which are not only important physical phenomena in their own right [6] but also provide transparent illustrations of the physics of complex scalar and vector fields, and the associated singularities.

For simplicity I will concentrate only on monochromatic waves, so there will be a (usually) implicit time factor $\exp(-\mathrm{i}\omega t)$. However, the ideas apply equally to waves that are changing with time, for example signals - indeed phase singularities were discovered in this way [7].

2. PHASE SINGULARITIES IN SCALAR FIELDS

As a function of position $\mathbf{r}$ in two or three dimensions (that is, $\mathbf{r}=\{x,y,z\}$ or $\mathbf{r}=\{x,y\}$, we consider complex scalar waves

$$\psi(\mathbf{r}) = \rho(\mathbf{r})\exp\{\mathrm{i}\chi(\mathbf{r})\} = \xi(\mathbf{r}) + \mathrm{i}\eta(\mathbf{r}), \tag{1}$$

Reprint of 'Geometry of phase and polarization singularities, illustrated by edge diffraction and the tides', Proc. SPIE 4403, Second International Conference on Singular Optics (Optical Vortices): Fundamentals and Applications, 1 (May 30 2001).

with modulus ρ and phase χ, and real and imaginary parts ξ and η (we will usually omit the **r** dependence). Our interest here is in the wavefronts, defined as the surfaces (or in two dimensions the curves) where the phase $\chi(\mathbf{r})$=constant (mod 2π), and in particular the singularities of the wavefronts. In free space, or in a medium without singularities, $\xi(\mathbf{r})$ and $\eta(\mathbf{r})$, and hence $\psi(\mathbf{r})$, are smooth functions of position. Therefore the phase can be singular only if the modulus ρ vanishes, and since $\rho=0$ implies two conditions ($\xi=0$ and $\eta=0$), the phase singularities are lines in space (threads of darkness), and points in the plane.

During a circuit of a phase singularity, the phase changes by a multiple of 2π; generically, that is almost always, the multiple is ±1. This change can be understood in terms of the current vector

$$\mathbf{j} = \mathrm{Im}\,\psi^*\nabla\psi = \rho^2\nabla\chi = \xi\nabla\eta - \eta\nabla\xi. \tag{2}$$

Close to a singularity, lines of **j** are circles centred on the singularity [4], giving rise to the term *optical vortices* to describe the singular lines. The direction **n** of a vortex line is that of the associated vorticity, that is

$$\boldsymbol{\Omega} = \mathbf{n}|\boldsymbol{\Omega}| = \tfrac{1}{2}\nabla\wedge\mathbf{j} = \tfrac{1}{2}\mathrm{Im}\nabla\psi^*\wedge\nabla\psi = \nabla\xi\wedge\nabla\eta. \tag{3}$$

In three dimensions, $\boldsymbol{\Omega}$ defines a natural sense for the singularity, with respect to which the phase increases (that is, the phase change is $+2\pi$). In the plane, it is more natural to define the phase change relative to the fixed normal to the plane, leading to the notion of the strength (*topological charge*) of the singularity, whose value ±1 is given by

$$S = \mathrm{sgn}\left(\partial_x\xi\partial_y\eta - \partial_y\xi\partial_x\eta\right) \tag{4}$$

(for further discussion of this point, see [8, 9]).

Although the **j** lines are circles surrounding the singularity, the phase usually varies nonuniformly. This variation - the local structure of the singularity - is naturally described [4] by the same ellipse that gives the local form of the ρ contours, determined by the quadratic form

$$\rho^2 = (\nabla\xi\cdot\mathbf{R})^2 + (\nabla\eta\cdot\mathbf{R})^2, \tag{5}$$

where **R** describes position in a plane transverse to the singularity, with **R**=0 at the singularity. The phase lines are concentrated near the major axis of this ellipse, and sparse near the minor axis.

In three dimensions, the phase may change along the singular line as well as around it. Then the wavefront surfaces close to the singularity are helicoids, analogous to surfaces defined by atomic planes in a crystal, and this is why lines of phase singularity are also called *wavefront dislocations*. A singularity where χ does not vary along the line is an edge dislocation. If χ does vary along the line, the singularity has screw character. For waves in free space, a local measure of screwness (developing an idea in [10]) is the rate at which χ varies along the direction **n** of the dislocation (we call this the z direction), measured in units of the free-space wavenumber k. Since this variation can depend on the azimuth ϕ round the line, it is convenient to average over ϕ. Thus a possible definition of the screwness σ is

$$\sigma = \frac{1}{k}\langle\chi'\rangle_{\phi\,\mathrm{average}} \equiv \lim_{\mathbf{R}\to0}\frac{\int_0^{2\pi}\mathrm{d}\phi\,\rho^2(\mathbf{R})\chi'(\mathbf{R})}{k\int_0^{2\pi}\mathrm{d}\phi\,\rho^2(\mathbf{R})} = \lim_{\mathbf{R}\to0}\frac{\int_0^{2\pi}\mathrm{d}\phi\,j_z(\mathbf{R})}{k\int_0^{2\pi}\mathrm{d}\phi\,\rho^2(\mathbf{R})} = \lim_{\mathbf{R}\to0}\frac{\int_0^{2\pi}\mathrm{d}\phi\left(\xi(\mathbf{R})\eta'(\mathbf{R}) - \eta(\mathbf{R})\xi'(\mathbf{R})\right)}{k\int_0^{2\pi}\mathrm{d}\phi\left(\xi^2(\mathbf{R}) + \eta^2(\mathbf{R})\right)}, \tag{6}$$

where the primes denote z derivatives. The limit is delicate because **R** must be taken close to the dislocation line but the derivatives must be taken along the direction of the line itself; otherwise, the component j_z will vanish because **j** is perpendicular to the local vorticity $\boldsymbol{\Omega}$(cf. (2) and (3)). Exploiting the fact that ψ varies linearly away from its zero on the dislocation line, we can evaluate the average and take the limit explicitly, and obtain, after a short calculation, the nonsingular expressions

$$\sigma = \frac{\nabla\xi\cdot\nabla\eta' - \nabla\eta\cdot\nabla\xi'}{k\left((\nabla\xi)^2 + (\nabla\eta)^2\right)} = \frac{\mathrm{Im}\nabla\psi^*\cdot\nabla\psi'}{k|\nabla\psi|^2}. \tag{7}$$

For an edge dislocation, σ=0, while for a pure screw dislocation, $\sigma=\pm 1$. It is important to appreciate that the concept of screwness is essentially three-dimensional, and has no meaning for dislocation points in the plane. Thus, the continuously variable quantity σ, describing how χ varies along the dislocation, is unrelated to the (integer) dislocation strength S, which describes how χ varies around the disclocation.

3. NEWTON AND EDGE DIFFRACTION: A NEAR MISS AND AN ASTONISHING ANTICIPATION

In 1717, Isaac Newton [11] speculated on a matter that had disturbed him for half a century, namely the discordance between his corpuscular theory of light and observations that we now interpret as clear evidence for the wave nature of light, in particular Grimaldi's experiments in the 1660s on diffraction by an edge. Newton conjectured that the edge exerts an attractive force on the light, bending its rays into the geometrical shadow. To explain the observation of several fringes near the geometrical shadow, he asked:

Query 3: Are not the rays of Light in passing by the edges and sides of Bodies, bent several times backward and forwards, with a motion like that of an Eel? And do not the three Fringes of colour'd Light above-mentioned arise from three such bendings?

From a modern perspective, Newton's explanation is exactly correct, if his rays are interpreted as the streamlines of ψ, that is, the lines of current $\mathbf{j}$. It has taken a long time to reach this understanding. First, two centuries elapsed before Sommerfeld's exact solution, in 1896 [5], of the diffraction of waves by a half-plane. For incident waves travelling in the positive x direction from $x=-\infty$, with the diffracting half-plane defined by x=0, y <0 (i.e. the edge is at $x=y=0$), Sommerfeld's solution is, in units of 2π/wavelength (i.e. k=1)

$$\begin{gathered}\psi(\mathbf{r}) = \exp\left(-\tfrac{1}{4}\mathrm{i}\pi\right)\left[\exp(\mathrm{i}x)F\{u_-(\mathbf{r})\} + \alpha\exp(-\mathrm{i}x)F\{u_+(\mathbf{r})\}\right] \\ \alpha = +1(\text{Neumann boundary conditions}); \alpha = -1(\text{Dirichlet boundary conditions}); \alpha = 0\ (\text{black screen}),\end{gathered} \tag{8}$$

where F denotes the Fresnel integral

$$F(u) = \int_u^\infty \mathrm{d}t\,\exp\{\mathrm{i}\pi t^2\}, \tag{9}$$

and, in polar coordinates r,ϕ, defined so the edge is at $\phi=-\pi/2$ and $+3\pi/2$,

$$u_-(\mathbf{r}) = -\sqrt{\frac{2r}{\pi}}\sin\tfrac{1}{2}\phi,\quad u_+(\mathbf{r}) = \sqrt{\frac{2r}{\pi}}\cos\tfrac{1}{2}\phi. \tag{10}$$

The first and second terms in (8) represent the incident and reflected waves respectively, together with associated diffracted edge waves (though we note that other interpretations are possible [12]).

A further half-century elasped before detailed computations [13] revealed the rich structure of Sommerfeld's solution, in the Neumann case. Figure 1a shows the wave intensity in the Dirichlet case, clearly indicating the fringes that so intrigued Newton. Figure 1b shows the streamlines, displaying the eel-like undulations precisely as Newton predicted. (These streamlines are conveniently computed as contours of the stream function - whose existence is guaranteed by $\nabla\cdot\mathbf{j}=0$ - namely $s(r,\phi) = -\int_0^r \mathrm{d}r'\mathrm{Im}\psi^*(r',\phi)\partial_\phi\psi(r',\phi)/r'$.)

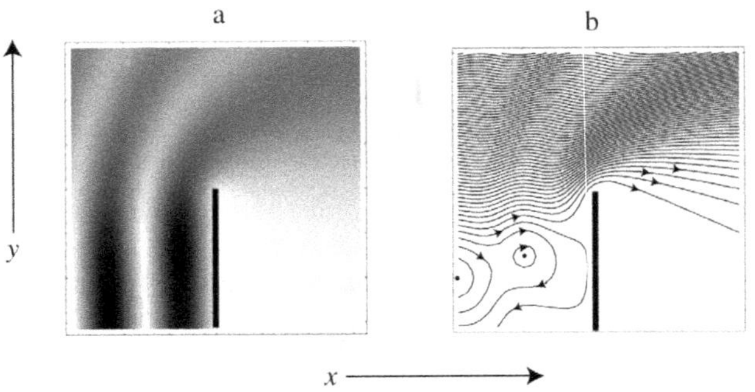

Figure 1. Sommerfeld edge-diffracted wave (8), for Dirichlet boundary conditions. (a) Density plot of wave intensity, proportional to darkness of shading, so the dislocation is visible as a bright spot in front of the screen; (b) streamlines, showing newton's 'eel' undulations and the loop surrounding the dislocation point.

What of the 'force' postulated by Newton to account for the deflection and undulations of the rays? This was discovered only in 1926, in the context of the 'hydrodynamic' formalism for quantum mechanics [14] - though the explanation applies to waves of all sorts. If wave equations are reformulated in terms of $\mathbf{j}$ and ρ, rather than ψ, the streamlines are influenced nonlocally by a 'quantum potential' in addition to any local forces incorporated in the hamiltonian. (Recently, the Madelung formalism has become popular as an interpretation of quantum theory, under the name 'Bohmian mechanics'[15].)

So, Newton was right in every respect. However, he appears not to have imagined that the forces could be strong enough to bend the rays into closed orbits. If he had taken this taken this further step, he would be celebrated as the discoverer of optical vortices. Closed streamlines do not occur in diffraction from a black (nonreflecting) screen, because the interference

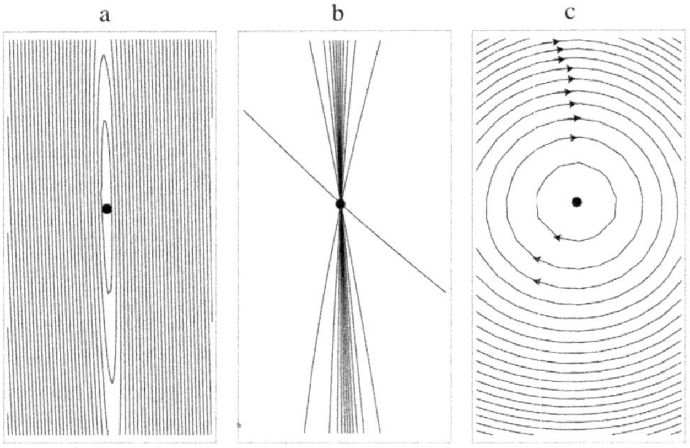

Figure 2. Magnification of dislocation in figure 1. (a) Contour plot of wave intensity, showing highly anisotropic elliptic structure near the zero; (b) wavefronts at intervals of 15°, showing concentration along the ellipse axis; (c) streamlines, showing circular structure near the dislocation.

between the incident and edge-diffracted waves is too weak to generate zeros. However, with reflecting screens (Dirichlet or Neumann), the additional reflected wave produces points of complete destructive interference. One of these is visible on figure 1b. From the magnifications in figure 2, it is clear that although the streamlines are circles the local intensity and phase structures are highly anisotropic. In the present case, this anisotopy has its origin in the fact that in the absence of the edge (that is, for an infinite screen rather than a half-plane), destructive interference between incident and reflected waves would occur on degenerate dark fringes in the form of lines (in two dimensions); the perturbation from the edge-diffracted waves breaks the degeneracy, leaving long narrow low-intensity valleys, reaching zero only at isolated points.

4. WHEWELL AND THE TIDES: THE FIRST PHASE SINGULARITY

In the world's oceans, the tides are waves driven by the gravitation of the moon and the sun. This gives rise to complicated astronomical periodicities, with many fourier components. The fundamental explanation of the tides was given by Newton, and the dynamical equations governing the water, in particular the displacement of its surface, were given by Laplace [6]. By the early nineteenth century, good explanations had been constructed for the tides on an ideal ocean of constant depth, covering an imaginary spherical earth with no land, and for the different periodicities. Here we need consider only the dominant contribution, called the M_2 tide, which is the semidiurnal tide from the moon, with a period of 12.42h.

On the real oceans, the heights of the tides vary in complicated ways from place to place, and in the 1830s this was the subject of extensive studies [16, 17] by William Whewell. The aim was

"...to connect the actual tides of all the different parts of the world - and to account for their varieties and seeming anomalies..."

Whewell realised that much of the complication is due to interference between waves reaching the same part of the ocean by different routes, and perceived that an intelligible picture of the tides would be provided by what we would nowadays call the wavefronts of the tide wave. He called these cotidal lines, and defined them as follows:

"...we may draw a line through all the adjacent parts of the ocean which have high water at the same time; for instance, at 1 o'clock on a given day. We might draw another line through all the places which have high water at 2 o'clock on the same day. Such lines may be called *cotidal lines*; and these will be the principal subject of the present essay."

A particular problem arose in connecting the observations of tides on the British and continental sides of the North Sea (then called the German Ocean) in order to discover the form of the wavefronts over this sea. Here is Whewell's brilliant solution:

"It appears that we may best combine all the facts into a consistent scheme, by dividing this ocean into two *rotatory* systems of tide-waves...[in each space] the cotidal lines may be supposed to rotate round [a point] where there is no tide, for it is clear that at a point where all the cotidal lines meet, it is high water equally at all hours, that is, the tide vanishes."

In other words, the North Sea contains two points of phase singularity! Whewell's map is reproduced in figure 3. These oceanic dislocations were later called *amphidromies*. Their identification and calculation (using Laplace's dynamical equations together with observations of tides round the coasts of the world's oceans) are central in modern studies of the tides [6]. In oceanographic models, the positions of amphidromies are extraordinarily sensitive to changes in the depth of the oceans [18], reflecting the fact (well known in optical applications [19]) that phase singularities are the most delicate features of waves, even though they are generic, that is, structurally stable in the mathematical sense. All the main bodies of water on the earth possess amphidromies; figure 4 shows a perfect specimen close to Crimea, where the singular optics conferences of 1997 and 2000 were held.

The mathematical interpretation of Whewell's argument gives a fine illustration of the formalism developed in section 2, with the water surface $z(\mathbf{r},t)$ written in terms of the complex wave $\psi(\mathbf{r})$ as

$$z(\mathbf{r},t) = \mathrm{Re}\left[\psi(\mathbf{r},t)\exp(-\mathrm{i}\omega t)\right] = \rho(\mathbf{r})\cos(\chi(\mathbf{r}) - \omega t). \tag{11}$$

Why must ψ be complex? Because the high tide must move across the ocean, according to the law $\chi(\mathbf{r})$-ωt=0 (mod 2π). If $\psi(\mathbf{r})$ were a real function, the time-dependence would be just $\cos(\omega t)$, and the phase of the tidal oscillations in the height of the

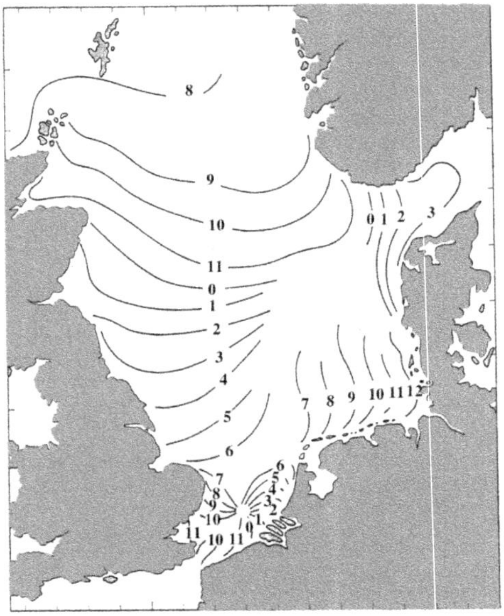

Figure 3. Whewell's map of the cotidal lines in the North Sea, showing two amphidromies (after [17]).

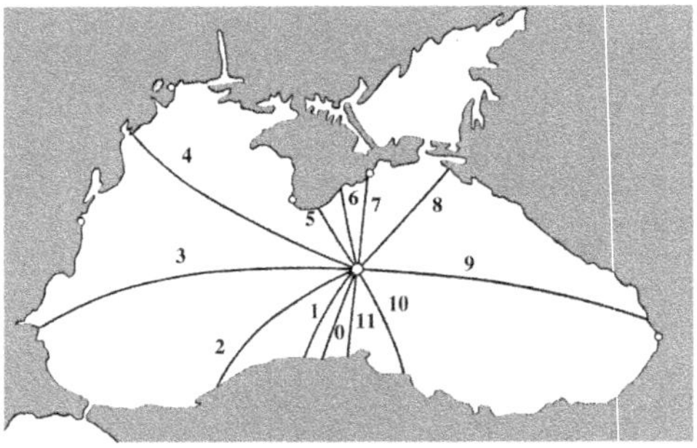

Figure 4. Amphidromy in the Black Sea (after [18]).

water would be the same everywhere: the tide wave would not travel. Precisely such correlated oscillations occur in, for example, the vibrations of a membrane, where the spatial variation of a mode is described by a real function $\psi(\mathbf{r})$; for such a mode in two dimensions, the zeros are nodal lines rather than dislocation points. Both type of wave motion are *steady*, in the sense that the time-dependence is harmonic. In addition, the modal vibration of a membrane is a *standing* wave: its time-dependence is the same everywhere. By contrast, the tide wave is steady but not standing.

Still the fundamental question can be asked: what is the physical reason why the tide waves are merely steady, while the vibrations of a membrane are not just steady but also standing. The answer is an extension of an argument more familiar in quantum mechanics, and involves time-reversal symmetry (T). Complex waves are inevitable whenever T is broken. In membranes, the physics possesses T, and the modes are real. In the tides, however, the rotation of the earth breaks T (the earth rotates from west to east, not vice versa), and so the waves are complex. (A quantum example of stationary states where the modes are represented by complex functions $\psi(\mathbf{r})$ occurs in the presence of magnetic fields, and explicit calculations [20, 21] for such waves in the plane show the expected presence of dislocation points rather than nodal lines.)

Whewell's notion of amphidromes in the pattern of cotidal lines was not immediately accepted. Airy [22] regarded the tideless point singularities as mathematically impossible, and proposed an incoherent alternative (figure 5), involving crossing cotidal lines, that was reproduced uncritically for many decades. The source of Airy's scepticism is not clear, especially in view of his high level of mathematical sophistication. It has been suggested [6] that his misunderstanding was based on a confusion between tide heights and tidal currents (see section 6); another possibility is that Airy had in mind interference as it occurs in membrane vibrations, involving real waves and therefore nodal lines rather than dislocation points. In any case, Whewell's map of the cotidal lines in the North Sea, complete with amphidromies, has been completely vindicated by modern calculations [23] and direct measurements of $z(\mathbf{r},t)$ from satellites.

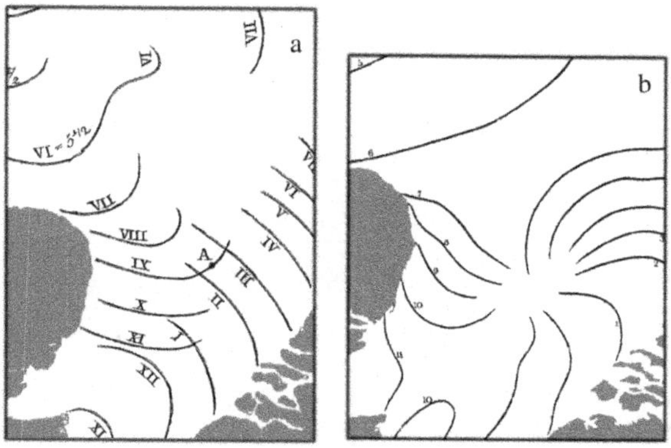

Figure 5. Airy's erroneous alternative [22](a) to Whewell's map of cotidal lines (b) (magnification of figure 3).

5. POLARIZATION SINGULARITIES IN VECTOR FIELDS

Consider now the complex vector field

$$\mathbf{E}(\mathbf{r}) = \mathbf{P}(\mathbf{r}) + \mathrm{i}\mathbf{Q}(\mathbf{r}) \tag{12}$$

constructed from the two real fields **P** and **Q**. **E** could represent the electric or magnetic field of light, or, as we shall see in the next section, the current of the tide. From **E** can be generated the real time-harmonic field

$$\mathrm{Re}\big(\mathbf{E}\exp(-\mathrm{i}\omega t)\big) = \mathbf{P}\cos\omega t + \mathbf{Q}\sin\omega t. \tag{13}$$

At each point **r** in two or three dimensions, this vector describes an ellipse in the plane of $\mathbf{P}(\mathbf{r})$ and $\mathbf{Q}(\mathbf{r})$. This is the *polarization ellipse* of **E**. It is characterized by its normal vector

$$\mathbf{N}_\mathrm{e} \equiv |\mathbf{N}_\mathrm{e}|\mathbf{n}_\mathrm{e} = \tfrac{1}{2}\mathrm{Im}\,\mathbf{E}^*\wedge\mathbf{E} = \mathbf{P}\wedge\mathbf{Q} \tag{14}$$

and by the orthogonal directions of its semiaxes, for which a short calculation gives the formulas

$$\mathbf{E}_+ = \frac{\mathrm{Re}\,\mathbf{E}^*\sqrt{\mathbf{E}\cdot\mathbf{E}}}{\left|\sqrt{\mathbf{E}\cdot\mathbf{E}}\right|} = \frac{\mathrm{Re}(\mathbf{P}-\mathrm{i}\mathbf{Q})\sqrt{P^2-Q^2+2\mathrm{i}\mathbf{P}\cdot\mathbf{Q}}}{\left(\left(P^2-Q^2\right)^2+4(\mathbf{P}\cdot\mathbf{Q})^2\right)^{1/4}},$$
$$\mathbf{E}_- = \frac{\mathrm{Im}\,\mathbf{E}^*\sqrt{\mathbf{E}\cdot\mathbf{E}}}{\left|\sqrt{\mathbf{E}\cdot\mathbf{E}}\right|} = \frac{\mathrm{Im}(\mathbf{P}-\mathrm{i}\mathbf{Q})\sqrt{P^2-Q^2+2\mathrm{i}\mathbf{P}\cdot\mathbf{Q}}}{\left(\left(P^2-Q^2\right)^2+4(\mathbf{P}\cdot\mathbf{Q})^2\right)^{1/4}}. \quad (15)$$

(These expressions are invariant under change of the overall phase of **E**, as they must be in order to represent a fixed ellipse.) Alternatively stated, $\mathbf{E}_\pm$ are the two eigenvectors of the matrix $P_iP_j+Q_iQ_j$ that correspond to nonzero eigenvalues, where i and j denote the components of **P** and **Q** (in three dimensions there is a zero eigenvalue). The three orthogonal vectors $\mathbf{N}_e$, $\mathbf{E}_+$, $\mathbf{E}_-$ form a frame at each point **r**.

Since a complex vector field is represented by a field of ellipses, the vector counterpart of phase singularities are the polarization singularities that can occur in fields of ellipses. In the general three-dimensional case these have been classified by Nye [3, 24, 25]. There are two sorts of singularity.

The first occurs at places **r** where the ellipses degenerate to circles, that is, where the vectors **P** and **Q** have the same length and are perpendicular. These are called C singularities, characterized by

$$\mathbf{E}\cdot\mathbf{E} = P^2+Q^2+2\mathrm{i}\mathbf{P}\cdot\mathbf{Q} = 0 \quad \text{(C singularity)}. \quad (16)$$

For a circle, the axes are undefined, and indeed the formulas (15) become degenerate when $\mathbf{E}\cdot\mathbf{E}=0$. The condition (16) consists of two equations, so C singularities are *lines* in three dimensions and *points* in two dimensions. This is reminiscent of the phase singularities of scalar fields, and indeed the C singularities in **E** are the dislocations of the scalar field $\psi=\mathbf{E}\cdot\mathbf{E}$.

A C singularity is characterized by the pattern of (almost circular) polarization ellipses close to it. The possible patterns are more complicated than the wavefronts that radiate from a dislocation. There are three geometrically distinct generic (that is, stable under perturbation) patterns, called the star, lemon, and monstar [24, 26, 27]. To illustrate them, it is convenient to draw the orthogonal net of lines corresponding to the two direction fields $\mathbf{E}_\pm$ in the plane perpendicular to $\mathbf{N}_e$, and these are depicted in figure 6. Obviously the C singularity is a singularity in this net, because the ellipse axes are undefined there. In transverse (two-dimensional) vector waves, the lemon and star singularities (but not, curiously, the monstar) are familiar in the patterns of isogyres [28] (see also [29] for a simple example). Analogous to the index +1 of a dislocation is the index of a C singularity, defined as the number of rotations of the associated directions in a circuit of the singularity. For the lemon and monstar, the index is +1/2, and for the star the index is -1/2. (In scalar waves, the index -1 corresponds to phase saddles [30]). An additional characterization of C singularities, with no counterpart for dislocations, is the number of lines (in each of the two orthogonal families) passing through the singularity. For the lemon this is 1, and for the star and monstar it is 3. As an example, the following field displays all three C singularities at **r**=0:

$$\mathbf{E} = \{1+\mathrm{i}by, \mathrm{i}(1-x)\} \quad (17)$$

(the singularities are star for $b<0$, lemon for $0<b<2$, and monstar for $b>2$). The C analogue of the strength S of a dislocation is the sense of rotation of the vector (13) round its ellipse, that is, whether the circular polarization is right- or left-handed.

In three dimensions, a C line need not be parallel to the ellipse normal $\mathbf{N}_e$. The direction of the C line (cf. equation 3 for the vorticity defining the direction of a dislocation) is [25]

$$\mathbf{N}_C = \tfrac{1}{2}\mathrm{Im}\nabla(\mathbf{E}^*\cdot\mathbf{E}^*)\wedge\nabla(\mathbf{E}\cdot\mathbf{E}) = \nabla\left(P^2-Q^2\right)\wedge\nabla(\mathbf{P}\cdot\mathbf{Q}). \quad (18)$$

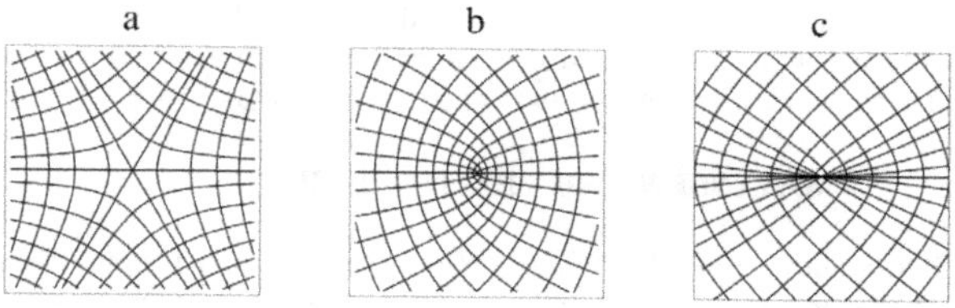

Figure 6. Patterns of axes of ellipses near a C singularity. (a) star (index -1/2); (b) lemon (index +1/2); (c) monstar (index +1/2).

The second type of polarization singularity occurs at places **r** where the ellipses degenerate to lines, that is, where the vectors **P** and **Q** are parallel. These are called L singularities, characterized by

$$\mathrm{Im}\,\mathbf{E}^* \wedge \mathbf{E} = 2\mathbf{P} \wedge \mathbf{Q} = 0 \quad (\text{L singularity}). \tag{19}$$

For a line in space, the normal is undefined, and indeed $\mathbf{N}_e$ in (14) vanishes when (19) holds; in (14), the short-axis vector $\mathbf{E}_-$ vanishes, and the long axis $\mathbf{E}_+$ is parallel to **P** or **Q**. Parallelism of two three-dimensional vectors corresponds to two conditions, so for general complex vector fields in space the L singularities are *lines*. However, parallelism of two two-dimensional vectors corresponds to only one condition, so for general complex vector fields in two dimensions the L singularities are also *lines*, and for transversely polarized fields in space (e.g. paraxial optical fields) the L singularities are *surfaces*.

An L singularity is characterized by the pattern of (very thin) polarization ellipses close to it. In the general three-dimensional case, this can be described [24] in terms of the short axis $\mathbf{E}_-$ of the ellipses in a plane transverse to the L line and perpendicular to **P** or **Q**. The stable patterns are the Poincaré singularities (figure 7) near a zero of a vector field in the plane, with index +1 (lines that are elliptical or radial), or -1 (lines that are hyperbolic) (the patterns can be generated in the $\mathbf{R}=(x, y)$ plane from (14) by choosing $\mathbf{P}=(\mathbf{A}\cdot\mathbf{R},1)$, $\mathbf{Q}=(\mathbf{B}\cdot\mathbf{R},1)$ where **A** and **B** are 2x2 matrices).

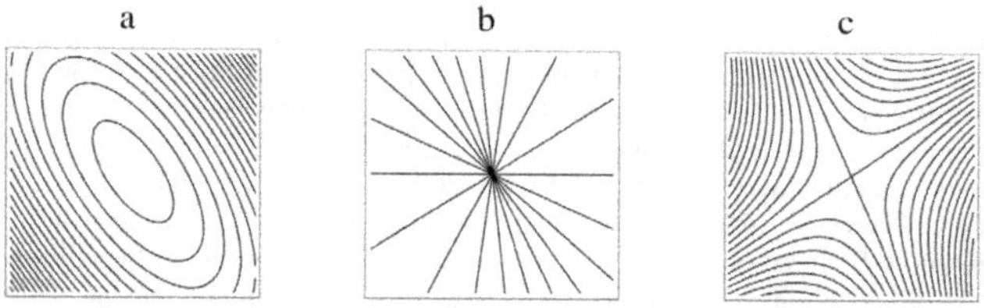

Figure 7. Pattern of short axis of ellipses near an L line in three dimensions. (a) and (b) have index +1, (c) has index -1.

In three dimensions, an L line need not be parallel to the ellipse axis $\mathbf{E}_+$ (i.e. parallel to **P** or **Q**). The direction of the L line is [25]

$$\mathbf{N}_L = \nabla_a \wedge \nabla_b \left(\mathbf{N}_{e,a} \wedge \mathbf{N}_{e,b} \cdot \mathbf{e_P}\right), \tag{20}$$

where $\mathbf{N}_e$ is given by (14), the suffixes a and b indicate the quantities on which the gradient operators act, and $\mathbf{e_P}$ is the common direction of **P** and **Q**.

For both C and L singularities, alternative interpretations are possible, in terms of extreme values of an associated spin, or as matrix degeneracies [25].

The dislocations and polarizations of scalar and vector fields are related in several ways. We have already remarked that the C singularities of any vector field $\mathbf{E}$ can be regarded as dislocations in the associated scalar field $\psi = \mathbf{E}\cdot\mathbf{E}$. Similarly, any scalar field ψ possesses C and L singularities, generated by the associated vector field $\mathbf{E} = \nabla\psi$.

6. HANSEN AND THE TIDAL CURRENTS

Water is incompressible, so in the tides the vertical motion $z(\mathbf{r},t)$ (equation 11) must be accompanied by a horizontal flow; this is the tidal current. By continuity, the velocity $\mathbf{u}(\mathbf{r},t)$ of the tidal current (assumed uniform over the depth $D(\mathbf{r})$ of the ocean, for simplicity) is related to $z(\mathbf{r},t)$:

$$\partial_t z(\mathbf{r},t) = -\nabla \cdot \left[D(\mathbf{r}) \mathbf{u}(\mathbf{r},t) \right]. \tag{21}$$

For harmonic motion, as in the M_2 tide, $\mathbf{u}$ can be generated by a complex vector field $\mathbf{E}$ according to (13), in the same way that z is related to the complex scalar ψ according to (11). Thus (21) can be written

$$\psi(\mathbf{r}) = \frac{\mathrm{i}}{\omega} \nabla \left[D(\mathbf{r}) \mathbf{E}(\mathbf{r}) \right]. \tag{22}$$

Evidently the tidal currents determine the height of the tide, but not conversely; therefore the tidal currents contain additional information, and it is interesting to study them.

The complex vector field $\mathbf{E}(\mathbf{r})$ of the tides determines the ellipse repeatedly described by the velocity of the tidal current at the point $\mathbf{r}$. If $\mathbf{E}$ did not depend on $\mathbf{r}$, and if there were no secular ocean currents, this velocity ellipse would drive each water particle in a similar ellipse, but with the long and short axes reversed. However, $\mathbf{E}$ does depend on $\mathbf{r}$, and the variation would cause water particles to drift nonadiabatically according to the ellipses at different points, in ways that would be quite complicated, even chaotic - if it were not for the secular currents (e.g. the Gulf Stream), that in fact dominate the long-time motion of the water in the real oceans.

The pattern of velocity ellipses for the tidal currents of the North Sea was studied extensively by Hansen [31] (see also [30]). In a tour de force combining observations with Laplace's dynamical equations and (22), he was able to plot crosses representing the magnitudes and directions of the axes of the ellipses at points on a grid covering the sea. To get a clearer picture of the currents, he also plotted contours of a quantity, related to the eccentricity of the ellipses, equivalent to the local expectation value of photon spin in optics [25], namely $S_z = \mathbf{e}_z \cdot \mathbf{N}_e / \mathbf{E}^* \cdot \mathbf{E}$. Although he did not emphasise the polarizaton singularities, he did note the existence of points where $S_z = \pm 1$ - in singularity terminology, these are the C points with left- and right-handed circular polarization - and the lines where $S_z = 0$ - these are the L lines, where the polarization is linear. From this information, we can attempt to draw the orthogonal net of directions of $\mathbf{E}_\pm$ in the ellipse field (figure 8). In the region shown, there are three polarization singularities, two with index -1/2, that is stars, and one with index +1/2; I have drawn this as a lemon although the possibility of its being a monstar cannot be ruled out (on probabilistic grounds [26] we expect monstars to be rare).

Hansen went further, and realized (cf. the remarks at the end of section 5) that there is another field of ellipses, distinct from that associated with the tidal currents, associated with the complex slope vector $\nabla\psi$ of the tide wave. He called this the inclination field of the tide, and plotted its crosses and eccentricity contours.

7. ACKNOWLEDGEMENTS

These recent perspectives on phase and polarization singularities have been reached in close collaboration with Mark Dennis, for which I thank him; I also thank John Nye for helpful conversations. This paper was written during a generously supported visit to the physics department of the Technion, Haifa.

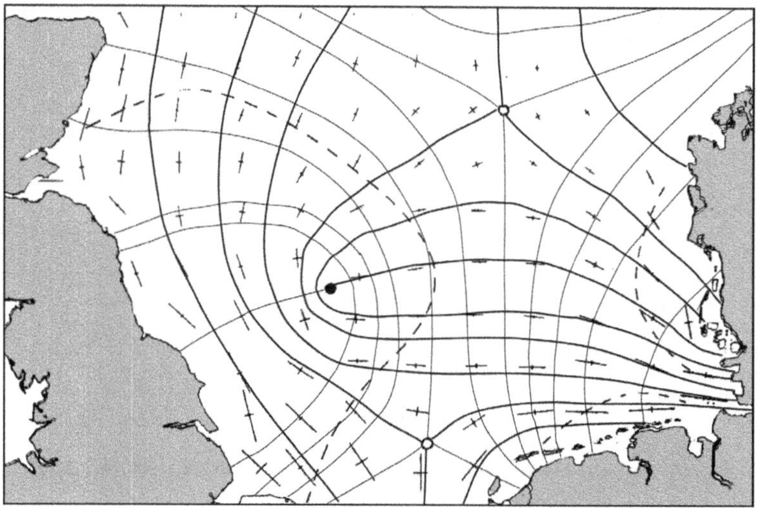

Figure 8. Orthogonal net of directions of the long axes (thick lines) and short axes (thin lines) of ellipse field (crosses) of the tidal currents in the North Sea (adapted from data of Hansen [31]). Circles (open and filled for the two senses of polarization) denote the C singularities (a lemon and two stars), and the dashed lines denote the L singularities.

8. REFERENCES

1. M. S. Soskin (ed), "Singular Optics," in *SPIE*, vol. 3487. Washington: Optical Society of America, 1998.
2. M. Vasnetsov and K. Staliunas, *Optical vortices*. Commack, New York: Nova Science Publications, 1999.
3. J. F. Nye, *Natural focusing and fine structure of light: Caustics and wave dislocations*. Bristol: Institute of Physics Publishing, 1999.
4. M. V. Berry and M. R. Dennis, "Phase singularities in isotropic random waves," *Proc. Roy. Soc. Lond. A456*, pp. 2059-2079, 2000.
5. A. Sommerfeld, *Optics: Lectures on Theoretical Physics, vol. 4*. New York: Academic Press, 1950.
6. D. E. Cartwright, *Tides: a scientific history*. Cambridge: University Press, 1999.
7. J. F. Nye and M. V. Berry, "Dislocations in wave trains," *Proc. Roy. Soc. Lond. A336*, pp. 165-90, 1974.
8. M. V. Berry, "Much ado about nothing: optical dislocation lines (phase singularities, zeros, vortices...)," in ref. 2.
9. I. Freund, "Optical vortex trajectories," *Optics Communs. 181*, pp. 19-33, 2000.
10. J. F. Nye, "The motion and structure of dislocations in wavefronts," *Proc. Roy. Soc. Lond. A378*, pp. 219-239, 1981.
11. I. Newton, *Opticks: or a Treatise of the Reflections, Inflections and Colours of Light*, 2nd ed. Mineola, N.Y.: Dover, 1952.
12. A. I. Khishniak, S. Anokhov, P., R. A. Lymarenko, M. S. Soskin, and M. V. Vasnetsov, "The structure of edge-dislocated wave originated in plane-wave diffraction by a half-plane," *J. Opt. Soc. Amer. A17*, In press, 2000.
13. W. Braunbek and G. Laukien, "Features of refraction by a semi-plane," *Optik. 9*, pp. 174-179, 1952.
14. E. Madelung, "Quantentheorie in hydrodynamische Form," *Z. für Phys. 40*, pp. 322-6, 1926.
15. P. Holland, *The Quantum Theory of Motion. An Account of the De Broglie-Bohm Causal Interpretation of Quantum Mechanics*. Cambridge: University Press, 1993.
16. W. Whewell, "Essay towards a first approximation to a map of cotidal lines," *Phil. Trans. Roy. Soc. Lond.*, pp. 147-236, 1833.
17. W. Whewell, "On the results of an extensive series of tide observations," *Phil. Trans. Roy. Soc. Lond.*, pp. 289-307, 1836.
18. A. Defant, *Physical Oceanography*, vol. 2. Oxford: Pergamon, 1961.
19. M. V. Berry, "Wave dislocations in nonparaxial Gaussian beams," *J. Modern Optics. 45*, pp. 1845-1858, 1998.

20. M. V. Berry and M. Robnik, "Quantum states without time-reversal symmetry: wavefront dislocations in a nonintegrable Aharonov-Bohm billiard," *J. Phys. A*, vol. 19, pp. 1365-1372, 1986.
21. R. J. Mondragon and M. V. Berry, "The quantum phase 2-form near degeneracies: two numerical studies," *Proc. Roy. Soc. Lond. A424*, pp. 263-278, 1989.
22. G. B. Airy, "Tides and Waves," in *Encyclopedia Metropolitana*, vol. 5. London, 1845, pp. 241-396.
23. E. W. Schwiderski, "Global ocean tides, part II: The semidiurnal principal lunar tide (M_2). Atlas of tidal charts and maps," Naval surface weapons center, Silver Spring, Maryland NSWC TR 79-414, 1979.
24. J. F. Nye and J. V. Hajnal, "The wave structure of monochromatic electromagnetic radiation," *Proc. Roy. Soc. Lond. A409*, pp. 21-36, 1987.
25. M. V. Berry and M. R. Dennis, "Polarization singularities in isotropic random waves," *Proc. Roy. Soc. Lond.A457*, pp. 141-155, 2001.
26. M. V. Berry and J. H. Hannay, "Umbilic points on Gaussian random surfaces," *J. Phys. A10*, pp. 1809-21, 1977.
27. I. R. Porteous, *Geometric differentiation: for the intelligence of curves and surfaces*. Cambridge: University Press, 1994.
28. M. Born and E. Wolf, *Principles of Optics*. London: Pergamon, 1959.
29. M. V. Berry, R. Bhandari, and S. Klein, "Black plastic sandwiches demonstrating biaxial optical anisotropy," *Eur. J. Phys. 20*, pp. 1-14, 1999.
30. J. F. Nye, J. V. Hajnal, and J. H. Hannay, "Phase saddles and dislocations in two-dimensional waves such as the tides," *Proc. Roy. Soc. Lond. A417*, pp. 7-20, 1988.
31. W. Hansen, *Gezeiten und Gezeitenströme der halbtägigen Hauptmondtide M_2 in der Nordsee*. Hamburg: Deutsche Hydrographisches Institut, 1952.

10.1098/rspa.2001.0826

THE ROYAL SOCIETY

Knotted and linked phase singularities in monochromatic waves

By M. V. Berry and M. R. Dennis

H. H. Wills Physics Laboratory, Tyndall Avenue, Bristol BS8 1TL, UK

Received 22 November 2000; accepted 29 March 2001

Exact solutions of the Helmholtz equation are constructed, possessing wavefront dislocation lines (phase singularities) in the form of knots or links where the wave function vanishes ('knotted nothings'). The construction proceeds by making a non-generic structure with a strength n dislocation loop threaded by a strength m dislocation line, and then perturbing this. In the resulting unfolded (stable) structure, the dislocation loop becomes an (m, n) torus knot if m and n are coprime, and N linked rings or knots if m and n have a common factor N; the loop or rings are threaded by an m-stranded helix. In our explicit implementation, the wave is a superposition of Bessel beams, accessible to experiment. Paraxially, the construction fails.

Keywords: phase; singularities; dislocations; knots; links; paraxiality

1. Introduction

Wavefront dislocations, that is, phase singularities in complex scalar waves (Nye & Berry 1974; Berry 1981; Nye 1999) (also known as optical vortices), are lines in three dimensions on which the wave intensity vanishes and around which the phase changes by 2π times an integer (the strength of the singularity). For any wave in space, the set of its dislocation lines is a skeleton, supporting the phase structure of the whole field.

It is already known that solutions of the wave equation (monochromatic or time dependent) can be constructed to represent dislocation lines that are straight or curved, or form closed loops. In view of the recent revival of interest in these singular structures (Soskin 1997; Vasnetsov & Staliunas 1999), it is desirable to explore all their possible geometries. Our purpose here is to address the natural question: can wavefront dislocations be knotted or linked?

In reaction–diffusion equations describing chemical waves, wave functions with knotted dislocations have been created as initial conditions, and their structure and evolution have been studied in detail (Winfree 1987; Winfree & Strogatz 1984; Winfree *et al.* 1985; T. Poston & A. T. Winfree 1992, unpublished work). We impose the more demanding condition that the knots remain stationary. Specifically, we seek knots and links in complex scalar wave solutions $\Psi(\boldsymbol{r})$ of the Helmholtz equation

$$\nabla^2\Psi(\boldsymbol{r}) + \Psi(\boldsymbol{r}) = 0. \tag{1.1}$$

(We measure distances in units of wavelength/2π, or, equivalently, choose the wave number $k = 1$.) For $\boldsymbol{r}$, we will use cylindrical polar coordinates $\boldsymbol{r} = (R, \phi, z)$ or Cartesians $\boldsymbol{r} = (x, y, z)$.

Reprinted from Proceedings of the Royal Society A, Volume 457, (2001).

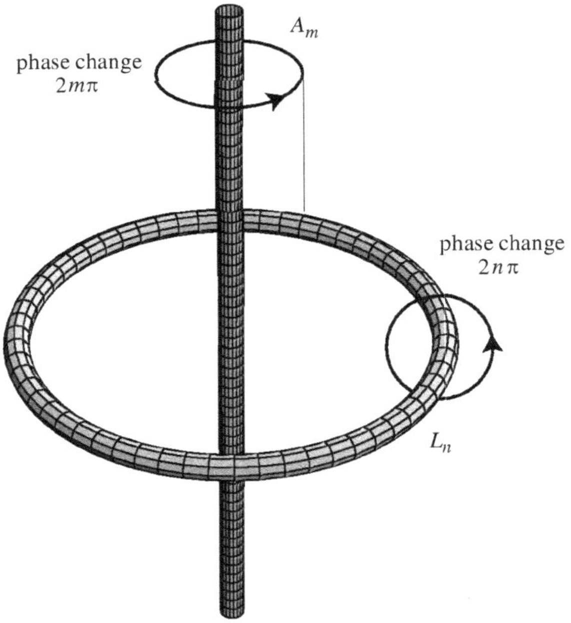

Figure 1. Unstable strength m axial dislocation A_m threading an unstable strength n dislocation loop L_n.

We answer our question in the affirmative, by constructing explicit exact solutions of (1.1) containing dislocation knots and links (zeros of Ψ), using a combination of topology and analysis. In the basic strategy (§ 2), we envisage the knot or link collapsed into a structurally unstable high-order dislocation ring, threaded by an axial high-order dislocation, and then argue that under perturbation the ring will unfold into the desired knot or link. To implement this strategy requires explicit solutions with definite properties, and we choose, from many possible types of solution, a particularly convenient set (§ 3) consisting of superpositions of Bessel beams.

A surprising aspect of the analysis is that although it works for the Helmholtz equation (1.1), it fails for the paraxial approximation to (1.1). This is a consequence of the result (Appendix D), which is of independent interest, that paraxial wave equations cannot possess solutions with dislocation loops whose strength exceeds unity. Two other results arising from this work, which are also of independent interest, are a clarification (Appendix A) of the local structure of higher-order dislocations, and a topological theorem (Appendix B) showing that screw dislocation loops must be threaded by other dislocations.

2. Unstable dislocation structures and their unfoldings

The starting point of our construction will be an unstable (non-generic) structure consisting of a circular dislocation loop L_n of strength $n > 0$, lying in the plane $z = 0$ and with radius R^*, centrally threaded by a strength m dislocation line A_m along the z-axis (figure 1). Then we will perturb this to get the desired knot or link as the stable unfolding of L_n. To create the unstable structure, we first note that near A_m

the local field must be

$$KR^m \exp(\mathrm{i}m\phi), \tag{2.1}$$

where K is a constant. In Appendix A we show that, near L_n, the local field must be

$$(K_+(R-R^*+\mathrm{i}z)^n + K_-(R-R^*-\mathrm{i}z)^n)\exp(\mathrm{i}m\phi), \tag{2.2}$$

where $K_\pm$ are constants with $|K_+|>|K_-|$.

Thus we write the unperturbed wave in the form

$$\psi(\boldsymbol{r}) = \exp(\mathrm{i}m\phi)F(R,z) \tag{2.3}$$

and seek to create the desired structure by imposing conditions on F. For A_m, we require, from (2.1),

$$\left.\begin{aligned} &\partial_R^p\psi(0,z)=0, \quad 0\leqslant p\leqslant m-1,\\ &\partial_R^m\psi(0,z)\neq 0, \end{aligned}\right\} \tag{2.4}$$

where, here and hereafter, we denote derivatives with respect to a variable ξ by the symbol ∂_ξ. For the loop L_n, we require, from (2.2), the $\frac{1}{2}n(n+1)$ conditions

$$\left.\begin{aligned} &\partial_R^q\partial_z^{p-q}\psi(R^*,0)=0, \quad 0\leqslant q\leqslant p, \quad 0\leqslant p\leqslant n-1,\\ &\partial_R^q\partial_z^{n-q}\psi(R^*,0)\neq 0, \quad 0\leqslant q\leqslant n. \end{aligned}\right\} \tag{2.5}$$

In effect, we are envisaging (imagining the unfolding in reverse) that n strength 1 dislocations coalesce into a single strength n dislocation. This requires $\frac{1}{2}n(n+1)$ conditions, rather than just $n-1$; the greater number arises because a strength n dislocation has Poincaré index +1 for any n, whose conservation (in addition to dislocation strength) during the coalescence implies the simultaneous involvement of $n-1$ phase saddles (each with index −1) (Nye *et al.* 1988; Berry 1998).

Around any circuit threaded by L_n, the phase χ changes by $2\pi n$. Alternatively stated, for each value of χ (mod 2π), there are n wavefront surfaces emerging from L_n; any transverse section of L_n cuts these surfaces in n lines, issuing from the intersection point, comprising what we call the *phase star* (parts (*a*)–(*c*) of figure 2). Around any circuit threaded by A_m, the phase changes by $2\pi m$. By taking such a circuit to be a loop just inside L_m, we can interpret this as a constraint on the phase star: along L_n, the phase star must turn m/n times. This implies a helicoidal structure in the wavefronts issuing from L_n, and illustrates a general theorem (Appendix B): the phase change along a dislocation loop equals the total strength of the dislocations threading it.

Now let this A_m, L_n structure be perturbed by an additional weak wave $\psi_p(\boldsymbol{r})$ that does not itself possess any dislocation lines threading L_n (for example, ψ_p could be a plane wave). Thus the total wave is

$$\Psi(\boldsymbol{r}) = \psi(\boldsymbol{r}) + \varepsilon\psi_p(\boldsymbol{r}) = \exp(\mathrm{i}m\phi)F(R,z) + \varepsilon\psi_p(\boldsymbol{r}). \tag{2.6}$$

This splits L_n into n separate dislocation strands with strength 1 (zeros of Ψ), and in each cross-section the phase star splits into n strength 1 stars (parts (*d*)–(*f*) of figure 2) and $n-1$ phase saddles (if the unfolding is incomplete, the saddles can be degenerate, as in figure 2*e*).

If m and n are coprime, these strands cannot take the form of n separate dislocation loops, because each of the n phase stars will be unable to match smoothly with its

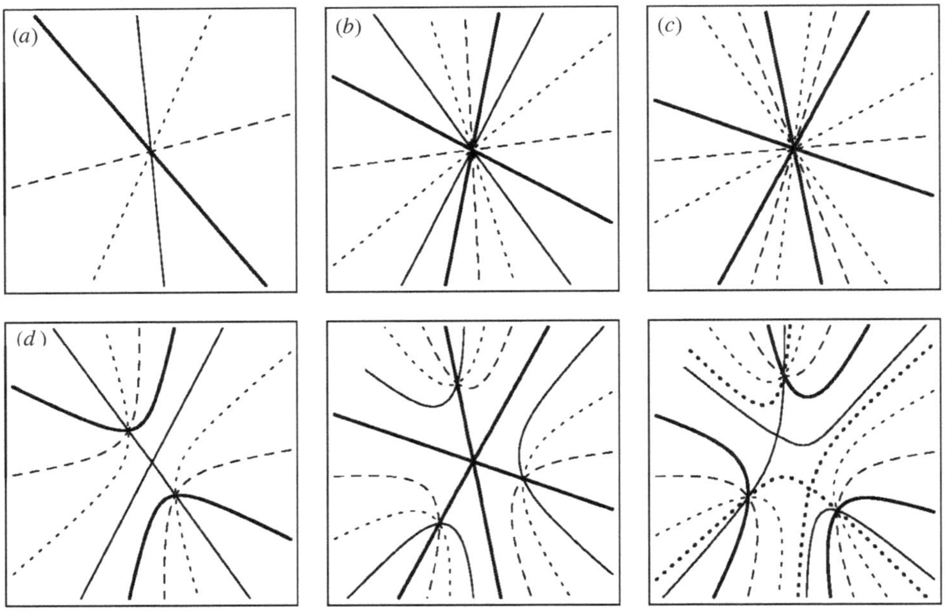

Figure 2. Phase stars. Thick lines: $\chi = 0$ and π (mod 2π) (crests and troughs of $\mathrm{Re}\,\psi$); dashed lines: $\chi = \frac{1}{8}\pi$ and $\frac{5}{8}\pi$ (mod 2π); thin lines: $\chi = \frac{1}{2}\pi$ and $\frac{3}{2}\pi$ (mod 2π); light dotted lines: $\chi = \frac{3}{8}\pi$ and $\frac{7}{8}\pi$ (mod 2π); heavy dotted lines in (*f*): $\chi = 2.671\pi/8$. (*a*) Strength 1 dislocation L_1; (*b*) unstable strength 2 dislocation L_2; (*c*) unstable strength 3 dislocation L_3; (*d*) stable unfolding of (*b*); (*e*) partial unfolding of (*c*); (*f*) stable unfolding of (*c*).

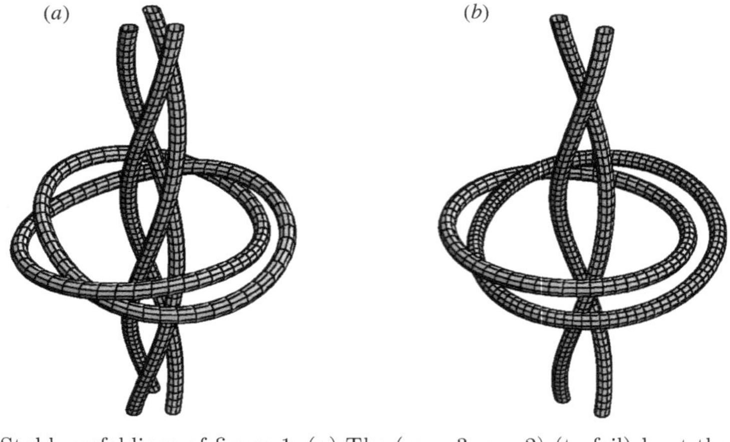

Figure 3. Stable unfoldings of figure 1. (*a*) The ($m = 3, n = 2$) (trefoil) knot threaded by a triple-stranded helix. (*b*) The ($m = 2, n = 2$) link, threaded by a double-stranded helix.

beginning after turning m/n times in a circuit of the axial dislocation. Thus the strands form a single loop that turns n times round the z-axis before closing. Now, we show in Appendix C that the cluster of n unfolded phase stars must be convected by the phase pattern that it carries. It follows that the n strands must also twist m times round the original loop L_n before closing. Therefore, the unfolding of L_n is an

(m, n) torus knot (Adams 1994). Similarly, A_m must unfold into m separate strength 1 dislocation lines, and we will soon see that they form an m-stranded helix. The fully unfolded structure, of an (m, n) torus knot threaded by an m-stranded helix, is illustrated for the trefoil in figure 3*a*.

If m and n are not coprime, but share a common multiple, so that $(m, n) = N(m_0, n_0)$, with m_0, n_0 coprime, the 'knot' into which L_n unfolds consists of N identical linked loops, each of which is an (m_0, n_0) knot, and the whole set of N loops is threaded by an m-stranded helix. Figure 3*b* illustrates this for the simplest link. It is possible for the individual linked dislocations to be knotted; the simplest situation where this occurs is $m = 6$, $n = 4$, corresponding to two linked trefoil knots.

These topological arguments can be confirmed and extended by explicit calculations of the zeros of (2.6) for small ε. It will suffice to consider perturbations ψ_p with circular symmetry, that is, ψ_p depends only on R and z. Consider first the neighbourhood of L_n. Using the local form (2.2) for the unperturbed wave, and defining the constant

$$B_L \equiv \psi_p \quad (\boldsymbol{r} \text{ on } L_n), \tag{2.7}$$

we have, as the equation for the unfolding of L_n,

$$K_+ \rho^n \exp(\mathrm{i}n\gamma) + K_- \rho^n \exp(-\mathrm{i}n\gamma) = -\varepsilon B_L \exp(-\mathrm{i}n\phi), \tag{2.8}$$

where ρ, γ are polar coordinates in azimuthal sections of L_n defined by

$$R - R^* + \mathrm{i}z = \rho \exp(\mathrm{i}\gamma). \tag{2.9}$$

According to (2.8), and remembering that $|K_+| > |K_-|$, $n\gamma$ increases by 2π as $m\phi$ decreases by 2π. It follows that along each strand (ϕ changing by 2π), γ changes by $2\pi m/n$, so along the whole knot (ϕ changing by $2\pi n$), γ changes by $2\pi m$. Thus the knot is indeed an (m, n) torus knot if m and n are coprime, and a link otherwise. On each azimuthal section specified by ϕ, the n solutions γ (of (2.8)), corresponding to the different strands, lie on a circle with radius $\rho = O(\varepsilon^{1/n})$; the union of all these circles is the torus, with coordinates γ and ϕ, on which the dislocation knot is wound.

Consider now the neighbourhood of A_m, and define the function

$$B_A(z) = \psi_p \quad (\boldsymbol{r} \text{ on } A_m). \tag{2.10}$$

Then an analogous argument based on (2.1) shows that the m strength 1 dislocations into which A_m unfolds must lie on the surface

$$R(z) = \varepsilon^{1/m} |B_A(z)/K|^{1/m}. \tag{2.11}$$

This is a tube whose cross-section, with radius of order $\varepsilon^{1/m}$, varies with z. For the m strands, we obtain

$$\phi_j(z) = \frac{\arg(-B_A(z)/K)}{m} + \frac{2\pi j}{m}, \quad 1 \leqslant j \leqslant m. \tag{2.12}$$

Since $B_A(z)$, in fact, must vary with z (if ψ_p satisfies the wave equation and has circular symmetry), the strands rotate and are therefore braided into an m-stranded helix, as claimed. (The sense in which the helix twists with z is unrelated to the sense in which the strands of the knot twist with ϕ.)

3. Bessel knots

At first thought, it would seem simplest—both for theoretical calculation and for experimental realization—to implement the above procedure with the Laguerre–Gauss beams of paraxial optics (Allen *et al.* 1992). However, this attempt fails, because conditions (2.5) for the circular dislocation L_n are impossible to satisfy with any paraxial waves. This follows from the general result, proved in Appendix D, that there can be no higher-order dislocation loops in solutions of the paraxial wave equation.

From the many classes of exact solutions of (1.1) in the form (2.3), we choose the m-Bessel beams (Durnin 1987; Durnin *et al.* 1987)

$$F_{mb}(R,z) = J_m(bR)\exp(\mathrm{i}z\sqrt{1-b^2}). \tag{3.1}$$

Here, b is a constant and $0 \leqslant b \leqslant 1$ for the non-evanescent waves we are interested in. Such beams automatically satisfy (2.4) and so possess the desired axial dislocation A_m. To satisfy (2.5), we choose real constants $b_1, \dots, b_{n(n+1)/2}$ and construct the superposition

$$F(R,z) = \sum_{l=1}^{n(n+1)/2} a_l F_{mb_l}(R,z), \tag{3.2}$$

with real constants a_l. For simplicity, we can choose $b_1 = 1$ and, without loss of generality, we can take $a_1 = 1$. Then, for a fixed choice of the remaining b_l, we can adjust the remaining a_l and R^* to ensure that (2.5) is satisfied, thereby making n zeros of the Bessel superposition coalesce, creating the desired loop L_n.

As examples, we now create a (3, 2) trefoil knot and a (2, 2) link. In both cases, construction of the loop L_2 involves three Bessel functions (of order 3 for the knot, 2 for the link), for which we choose the scaling factors

$$b_1 = 1, \qquad b_2 = \tfrac{1}{3}, \qquad b_3 = \tfrac{2}{3}. \tag{3.3}$$

The coefficients a_2, a_3 (with a_1=1), and the radius R^*, are determined by the equations (2.5), which for the knot become, incorporating (3.1),

$$\left.\begin{aligned} F(R^*,0) &= J_3(R^*) + a_2J_3(\tfrac{1}{3}R^*) + a_3J_3(\tfrac{2}{3}R^*) = 0,\\ \partial_R F(R^*,0) &= J_3'(R^*) + \tfrac{1}{3}a_2J_3'(\tfrac{1}{3}R^*) + \tfrac{2}{3}a_3J_3'(\tfrac{2}{3}R^*) = 0,\\ \partial_z F(R^*,0) &= \tfrac{1}{3}\sqrt{8}a_2J_3(\tfrac{1}{3}R^*) + \tfrac{1}{3}\sqrt{5}a_3J_3(\tfrac{2}{3}R^*) = 0 \end{aligned}\right\} \tag{3.4}$$

(for the link, the equations are the same, but with the Bessel indices 3 replaced by 2). These equations are easy to solve numerically. Indeed, there are many solutions, and we choose those corresponding to coalescence of the two smallest zeros of the Bessel superpositions. The results are

$$\left.\begin{aligned} &a_2 = 10.0302, \quad a_3 = -3.18960, \quad R^* = 5.44992 \quad ((3,2)\text{ trefoil knot}),\\ &a_2 = 4.73341, \quad a_3 = -2.70176, \quad R^* = 4.32636 \quad ((2,2)\text{ link}) \end{aligned}\right\} \tag{3.5}$$

and constitute the data needed to construct the loops L_2.

As the perturbation that converts the unstable L_2 into the stable knot or link, we take the zero-Bessel beam

$$\psi_p(\boldsymbol{r}) = J_0(\tfrac{1}{4}R)\exp(\tfrac{1}{4}\mathrm{i}\sqrt{15}z). \tag{3.6}$$

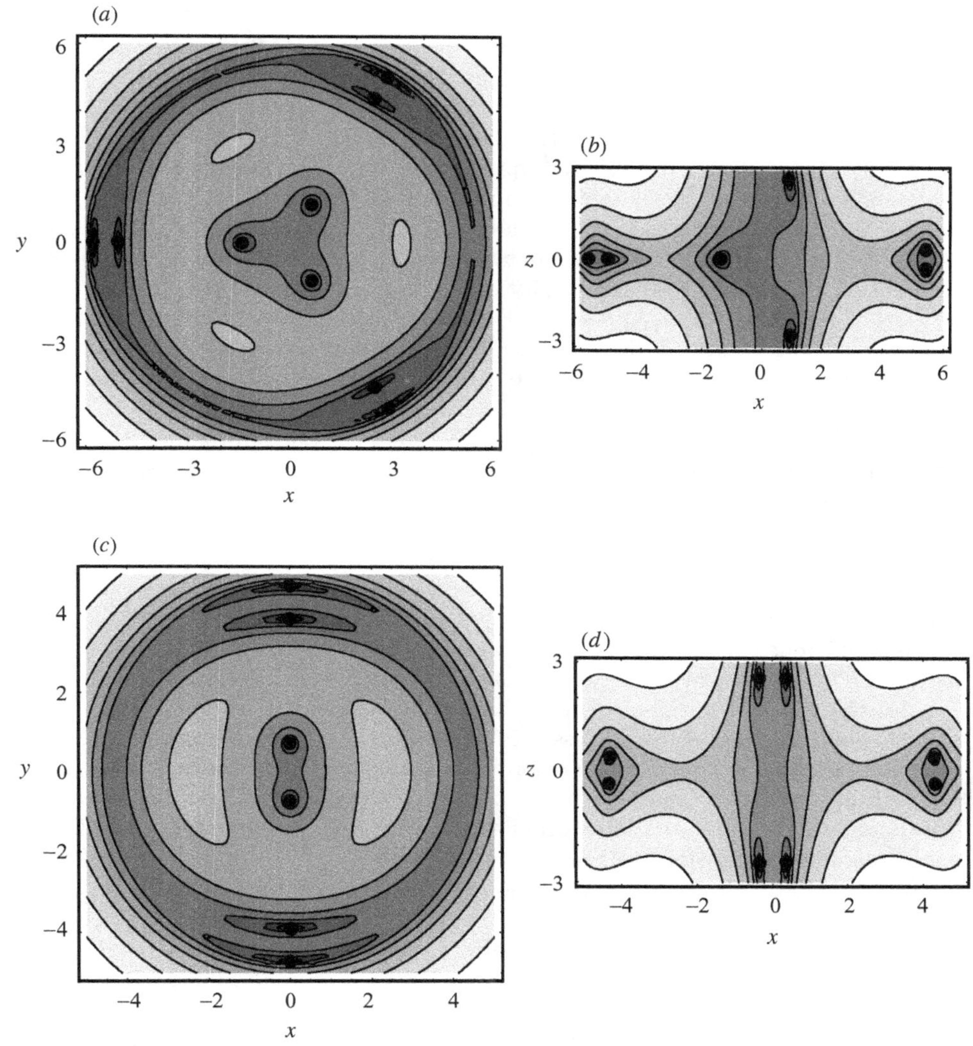

Figure 4. Density plots of wave intensity for the superposition (2.6), (3.2) and (3.5), with ψ_p given by (3.6) and $\varepsilon = 0.02$, in the planes (*a*) (*c*) $z = 0$ and (*b*), (*d*) $x = 0$, for the (*a*), (*b*) ($m = 3, n = 2$) trefoil knot and the (*c*), (*d*) ($m = 2, n = 2$) link. Black dots indicate the points where dislocation lines intersect these planes. The dislocation knot is threaded by a triple-stranded dislocation helix, and the dislocation link by a double-stranded dislocation helix.

Figure 4 shows sections of the total (that is, perturbed) waves for the two cases, one with its dislocation knotted, the other linked.

4. Concluding remarks

By constructing an unstable dislocation structure with high symmetry and then perturbing it, we are able to create waves with a dislocation in the form of any knot or link that can be wound round a torus. For the explicit construction, we chose the superposition (3.2) of Bessel beams (3.1). However, there are many alternative sets of exact solutions of (1.1), of the form (2.3) with definite angular momentum, that have sufficient flexibility to satisfy conditions (2.5). For example, we can take 'polynomial waves' (Nye 1998; Berry 1998), where $F(R, z) = \exp(\mathrm{i}z) \times$ (polynomial in R, z), and the calculations are almost identical to those in § 3.

Optical fields satisfy the Helmholtz equation in the scalar approximation, raising the possibility that the spatially fixed and geometrically stable dislocation knots and links that we have created theoretically could be manufactured experimentally. Bessel beams are good candidates for such realization, since they can be manipulated experimentally (Durnin *et al.* 1987). But the essentially non-paraxial nature of the construction (see below and Appendix D) presents a challenge to experimentalists, especially since the scalar approximation is problematic for non-paraxial light (of course, this particular difficulty would not arise for longitudinal fields such as sound in a fluid).

It was surprising to discover (Appendix D) that paraxial approximations to the wave equation are inadequate to implement our strategy for creating knots. We cannot exclude the possibility that paraxial fields can support knots generated by other means, but there would be a prohibition against collapsing the n strands of such a paraxial knot into a single strength n dislocation loop. The prohibition could originate, for example, in an extraneous dislocation passing between the strands; we see no way to construct such an exotic object.

The non-existence of paraxial loops with strength greater than unity is interesting in its own right, and for at least two reasons. First, it is unusual for a wave equation to restrict the possible morphologies of phase singularities.

Second, non-paraxial beams possess axial dislocation rings of strength unity and all of the same sign (Berry 1994), which can be mimicked paraxially (Karman *et al.* 1997, 1998) by diffraction of a Gaussian beam by an aperture. In both paraxial and non-paraxial cases, these rings can be manipulated by varying a parameter, to undergo a series of singular events (for example, a $+1$ dislocation loop colliding with a loop of saddles and exploding into a strength -1 loop flanked by two $+1$ loops). However, one 'supersingular' event that can occur non-paraxially (Nye 1998; Berry 1998), but that is paraxially impossible, is the (two-parameter) steering of loops of strength unity to coincide pairwise into loops of strength 2. Detailed examination of the algebra describing the supersingular event shows that in the equivalent paraxial situation there is always an additional extraneous loop, of opposite strength to the two that are being steered into coincidence, which arrives simultaneously and prevents the formation of the strength 2 loop.

Phase singularities in three dimensions are vortex lines in the current $\boldsymbol{j} = \mathrm{Im}\,\psi^* \nabla \psi$, so our constructions are vortex knots, as studied in hydrodynamics (Moffatt 1969; Moffatt & Ricca 1992). These waves could describe stationary configurations of electrons in free space (because in this situation the Schrödinger equation reduces to the Helmholtz equation); similar solutions exist in the presence of the Coulomb field from an atomic nucleus (Berry 2001), and these knotted vortices can describe states of electrons in atoms, perhaps as envisaged by Kelvin (1867, 1869).

Although the (m, n) construction includes a wide variety of knots and links, there are many knots that cannot be smoothly wound on a torus, and so the strategy described here does not generate them. The obvious question arises whether every topology of dislocation can be sculpted in a monochromatic wave. For example, is it possible to make the Borromean rings (Cromwell *et al.* 1998), where three loops are entangled but no two are linked?

Dislocation knots can occur, but are they typical? For example, in the sound waves in a room filled with conversation the air contains a forest of dislocation lines—threads of silence—and it is reasonable to suppose that some of the dislocations are knotted. But how many? A precise formulation is as follows. In a Gaussian random wave field with prescribed power spectrum, what fraction of the points where dislocations pierce any two-dimensional section corresponds to dislocation lines that are part of a knot? This is a difficult problem in statistical geometry, but techniques exist (Berry & Dennis 2000) that may be adaptable to deal with topological questions.

M.V.B.'s research is supported by The Royal Society; in addition, he is grateful for the hospitality of the Technion, Haifa, where the first draft of this paper was written. M.R.D. is supported by a University of Bristol postgraduate scholarship. We thank Professor Arthur Winfree for helpful discussions, and for sending us his unpublished paper with Professor Timothy Poston.

Note added in proof

As a result of computations kindly communicated to us by Professor Miles Padgett and Dr Johannes Courtial, we now appreciate that the paraxial prohibition against higher-order dislocation loops does not prevent the formation of knots in paraxial waves. In terms of our strategy, it suffices to apply a 'perturbation' to a wave with n neighbouring parallel dislocation loops of strength 1, rather than a single loop of strength n, provided the 'perturbation' is large enough. Such n-loop complexes can be created easily by replacing the set of radial wavenumbers b_l by very small, even paraxial, values with the same ratios, but using the same parameters a_l that we have specified (e.g. in (3.5)).

Appendix A. Local structure of higher-order Helmholtz dislocations

Let the strength n dislocation pass through the plane $y = 0$ in the positive y-direction, and choose polar coordinates ρ, γ (cf. (2.9)) in the x, z-plane. Therefore, the phase of ψ must increase by $2\pi n$ as γ increases by 2π. It follows that the Fourier expansion of ψ begins with the terms $\exp(\pm in\gamma)$, and smoothness demands that ψ vanishes as ρ^n on the dislocation. This, in turn, implies that $\partial_x^2\psi$ and $\partial_z^2\psi$ vanish as ρ^{n-2}, and so dominate the term ψ in the Helmholtz equation (1.1); they also dominate the term $\partial_y^2\psi$ in (1.1), because the derivative along the dislocation vanishes to higher order (i.e. n). Thus, close to the dislocation, the transverse dependence of ψ must satisfy Laplace's equation

$$\partial_x^2\psi + \partial_z^2\psi = 0. \tag{A 1}$$

The most general solution proportional to ρ^n is

$$\psi = K_+(x + \mathrm{i}z)^n + K_-(x - \mathrm{i}z)^n = \rho^n[K_+ \exp(\mathrm{i}n\gamma) + K_- \exp(-\mathrm{i}n\gamma)]. \tag{A 2}$$

Around the dislocation, that is, as γ increases by 2π, the term with coefficient K_+ winds n times positively in the complex plane of ψ, and the term with K_- winds negatively. For ψ to represent a strength $+n$ dislocation, the term K_+ must dominate, that is, $|K_+| > |K_-|$ (otherwise, ψ represents a strength $-n$ dislocation). Translating to coordinates near the circular dislocation created in § 2, and incorporating the required azimuthal dependence, we obtain (2.2).

The local behaviour (A 2) is a strong restriction on ψ, imposed by the Helmholtz equation. It implies that, around a circle centred on the dislocation, the wave ψ describes an ellipse in the complex ψ-plane, repeated n times, whose eccentricity depends on K_+/K_-. Without the Helmholtz requirement, ψ would be a superposition of monomials of the form $x^j z^{n-j}$, whose variation in its complex plane is more complicated than an ellipse.

Appendix B. Screw dislocation loops must be threaded (see also Winfree (1987) and Winfree *et al.* (1985))

Recall that the strength S of a dislocation is defined as the number of cycles of phase χ in a circuit C around it,

$$2\pi S = \oint_C \mathrm{d}\chi. \tag{B 1}$$

This defines a direction along the singular line, namely that which is right handed with respect to C when $S > 0$.

Imagine now an unknotted dislocation loop L of strength 1. Along L, the phase star (figure 2*a*) usually rotates (that is, it has screw character (Nye & Berry 1974)), and by continuity, the number of rotations must be a signed integer; this is the *screw number* m. For convenience, we define $m > 0$ (< 0) if the screw is left (right) handed with respect to the dislocation direction. If the wavefront surface $\chi = \text{const.}$ ($\mathrm{mod}\, 2\pi$) near L is regarded as a ribbon, then the integer m is the linking number of its edges.

On a closed curve C just inside L (so that there are no dislocations between C and L), the phase cycles m times as the star rotates, that is,

$$2\pi m = \oint_C \mathrm{d}\chi, \tag{B 2}$$

showing that the screw number m must be equal to the dislocation strength threading C and also L.

If the dislocation strength of L is $n > 1$, so that its phase star is multiple (parts (*b*) and (*c*) of figure 2), then the screw number (still defined as the number of rotations) is quantized in units of $1/n$, and the dislocation strength threading the loop is equal to the number of such $1/n$ rotations of the multiple phase star about L. If the screw number is m/n, then m is the linking number of the wavefront ribbon of the torus knot produced by the perturbation described in § 2.

If the threading dislocation A is itself a loop (that is, if A and L are linked), then it has a screw number equal to the strength n of L: each loop's strength equals the other's screw number. In our construction, the Bessel superpositions (3.2) possess an infinite number of dislocation rings in addition to the degenerate structure that unfolds to our knot, so the straight threading dislocation, regarded as a loop closing at infinity, must have an infinite screw number.

Each wavefront $\chi = \text{const.}$ ($\mathrm{mod}\, 2\pi$) is a (possibly infinite) Seifert surface (Adams 1994), smoothly connected at the knotted and helical dislocations with its counterpart $\chi + \pi$.

Appendix C. Perturbed dislocation stars are convected

After perturbation, L_n splits into a dislocation consisting of n strands, each with strength 1. If m and n are coprime, then the strands must form a single dislocation loop, winding n times round the original z-axis. Here we show that each strand must turn m times around the others before closing. Consider a section of the knot, labelled by its azimuth ϕ. The section will contain n phase stars. Far from the strands, the pattern of phase contours is unchanged by the perturbation, and so must turn m/n times during a circuit of A_m, that is, as ϕ increases by 2π. To show that the pattern of stars must also rotate m/n times, we must look more closely.

Each pair of stars will be separated by one or more distinct phase contours, each of which arrives from afar and hits one of the phase saddles between the stars, as in parts (*d*)–(*f*) of figure 2. Consider the phase $\chi(\phi)$ corresponding to one of these saddles, as ϕ increases by 2π. If $\chi(\phi)$ is single valued, that is, if this phase does not change by a multiple of 2π, then the phase contours issuing from this saddle must also turn m/n times, in synchrony with the phase contours far away. Therefore, the stars separated by the phase contours issuing from the saddle must rotate too. An appropriate image is of a 'saddle paddle', convecting the phase stars. The result of this process is that the strands must twist, as claimed.

To show the single valuedness of $\chi(\phi)$ during the circuit, consider the model (cf. (2.2), (2.6), (2.8))

$$\psi(\boldsymbol{r}) = \exp(\mathrm{i}m\phi)[K_+(x+\mathrm{i}y)^n + K_-(x-\mathrm{i}y)^n] + \varepsilon\psi_p(\boldsymbol{r}), \tag{C 1}$$

where x and y are local coordinates centred on the place where L_n pierces the ϕ section. At the saddles, the current $\boldsymbol{j} = \operatorname{Im}\psi^*\nabla\psi = 0$ but $\psi \neq 0$. To lowest order in ε, there is a saddle at $x = y = 0$ (if $n > 2$, this is a degenerate saddle, as in figure 2*e*, corresponding to a partial unfolding, but this does not affect the argument). The phase of this saddle is

$$\chi(\phi) = \arg(\psi_p(x=0, y=0, \phi)), \tag{C 2}$$

and this is single valued as ϕ changes by 2π, from our assumption that ψ_p possesses no dislocations threading L_n.

The stipulation that ψ_p possesses no dislocations threading L_n, so that the phase of ψ_p is single valued on L_n, is essential. Without it, we would, for example, be free to choose ψ_p proportional to $\exp(\mathrm{i}m\phi)$, and then, simply by reversing the tuning of parameters a_l that created the strength n dislocation L_n, we could unfold it into n separate unlinked loops, that is, the convection necessary for the formation of a knot would not occur.

Appendix D. No higher-order paraxial dislocation loops

Let the paraxial direction be z. Any loop, of whatever shape, must have at least two points where its direction is perpendicular to the z-axis. This is obvious geometrically, and also follows from the fact that the tangent vector $\boldsymbol{t}$ to any closed loop varies periodically round the loop, with mean zero, so that any component of $\boldsymbol{t}$ must pass through zero at least twice. Therefore, the absence of paraxial loops with strength $|n| > 1$ follows, if it can be shown that the paraxial wave equation

$$2\mathrm{i}\partial_z\psi + \partial_x^2\psi + \partial_y^2\psi = 0 \tag{D 1}$$

prohibits a strength $|n| > 1$ dislocation perpendicular to the z-axis.

To establish this prohibition, let the dislocation pass through the origin in the y-direction. Then the y variation (along the dislocation) must be slower than the x and z variations, so the term $\partial_y^2\psi$ in (D 1) must be dominated by the other two. Thus, in the x, z-plane, ψ must satisfy

$$2\mathrm{i}\partial_z\psi + \partial_x^2\psi = 0. \tag{D 2}$$

It is immediately clear that there is no solution proportional to $[\sqrt{(x^2+z^2)}]^n$, but it is instructive to examine in more detail the requirements for a dislocation of strength n. Not only must all derivatives $\partial_x^j\partial_z^{p-j}\psi$ of order $p < |n|$ vanish, but all derivatives of $\partial_x^j\partial_z^{n-j}\psi$ of order n must not vanish (cf. (2.5)). However, repeated differentiation of (D 2) shows that this is impossible: if the derivatives less than n vanish, then at least one higher derivative with respect to x must vanish too, spoiling the construction.

The prohibition operates in subtle ways. For example, in (3.1), the innocent replacement of $\sqrt{(1-b^2)}$ by its paraxial equivalent $1-\frac{1}{2}b^2$ results in equations similar to (3.4), which can be solved for the coefficients in the (now paraxial) superposition with $\psi = \partial_R\psi = \partial_z\psi = 0$ at $R = R^*$. However, this solution also has $\partial_R^2\psi = 0$ at $R = R^*$, so what has been created is a degenerate singular ring with strength unity, rather than the desired strength 2 dislocation ring; another singularity, whose strength is opposite to each of the ones seeking to combine, has arrived, producing a cancellation. Such unwanted guests are paraxially unavoidable. Reflecting this, the numerical solution of (3.4) gets progressively more difficult as the transverse wave numbers b in (3.1) get smaller, that is, as the limit of paraxiality is approached.

References

Adams, C. C. 1994 *The knot book*. San Francisco, CA: Freeman.

Allen, L., Beijersbergen, M. W., Spreeuw, R. J. C. & Woerdman, J. P. 1992 Orbital angular momentum of light and the transformation of Laguerre–Gaussian laser modes. *Phys. Rev.* A **45**, 8185–8189.

Berry, M. V. 1981 Singularities in waves and rays. In *Les Houches Lecture Series Session 35* (ed. R. Balian, M. Kléman & J.-P. Poirier), pp. 453–543. Amsterdam: North-Holland.

Berry, M. V. 1994 Evanescent and real waves in quantum billiards and Gaussian beams. *J. Phys.* A **27**, L391–L398.

Berry, M. V. 1998 Wave dislocations in nonparaxial Gaussian beams. *J. Mod. Opt.* **45**, 1845–1858.

Berry, M. V. 2001 Knotted zeros in the quantum states of hydrogen. *Found. Phys.* **31**, 659–667.

Berry, M. V. & Dennis, M. R. 2000 Phase singularities in isotropic random waves. *Proc. R. Soc. Lond.* A **456**, 2059–2079.

Cromwell, P., Beltrami, E. & Rampicini, M. 1998 The Borromean rings. *Math. Intelligencer* **20**, 53–62.

Durnin, J. 1987 Exact solutions for nondiffracting beams. I. The scalar theory. *J. Opt. Soc. Am.* A **4**, 651–654.

Durnin, J., Miceli Jr, J. J. & Eberly, J. H. 1987 Diffraction-free beams. *Phys. Rev. Lett.* **58**, 1499–1501.

Karman, G. P., Beijersbergen, M. W., van Duijl, A. & Woerdman, J. P. 1997 Creation and annihilation of phase singularities in a focal field. *Opt. Lett.* **22**, 1503–1505.

Karman, G. P., Beijersbergen, M. W., van Duijl, A., Bouwmeester, D. & Woerdman, J. P. 1998 Airy pattern reorganization and sub-wavelength structure in a focus. *J. Opt. Soc. Am.* A **15**, 884–899.

Kelvin, Lord 1867 On vortex atoms. *Phil. Mag.* **34**, 15–24.

Kelvin, Lord 1869 On vortex motion. *Trans. R. Soc. Edinb.* **25**, 217–260.

Moffatt, H. K. 1969 The degree of knottedness of tangled vortex lines. *J. Fluid Mech.* **35**, 117–129.

Moffatt, H. K. & Ricca, R. L. 1992 Helicity and the Calugareanu invariant. *Proc. R. Soc. Lond.* A **439**, 411–429.

Nye, J. F. 1998 Unfolding of higher-order wave dislocations. *J. Opt. Soc. Am.* A **15**, 1132–1138.

Nye, J. F. 1999 *Natural focusing and fine structure of light: caustics and wave dislocations.* Bristol: Institute of Physics.

Nye, J. F. & Berry, M. V. 1974 Dislocations in wave trains. *Proc. R. Soc. Lond.* A **336**, 165–90.

Nye, J. F., Hajnal, J. V. & Hannay, J. H. 1988 Phase saddles and dislocations in two-dimensional waves such as the tides. *Proc. R. Soc. Lond.* A **417**, 7–20.

Soskin, M. S. (ed.) 1997 *Singular optics.* Washington, DC: Optical Society of America.

Vasnetsov, M. & Staliunas, K. (eds) 1999 *Optical vortices.* Commack, NY: Nova Science.

Winfree, A. T. 1987 *When time breaks down.* Princeton, NJ: Princeton University Press.

Winfree, A. T. & Strogatz, S. H. 1984 Singular filaments organize chemical waves in three dimensions. *Physica* D **13**, 221–233.

Winfree, A. T., Winfree, E. M. & Seifert, H. 1985 Organizing centers in a cellular excitable medium. *Physica* D **17**, 109–115.

Proc. R. Soc. Lond. A **392**, 45–57 (1984)
Printed in Great Britain

Quantal phase factors accompanying adiabatic changes

BY M. V. BERRY, F.R.S.
H. H. Wills Physics Laboratory, University of Bristol, Tyndall Avenue, Bristol BS8 1TL, U.K.

(*Received* 13 *June* 1983)

A quantal system in an eigenstate, slowly transported round a circuit C by varying parameters $\boldsymbol{R}$ in its Hamiltonian $\hat{H}(\boldsymbol{R})$, will acquire a geometrical phase factor $\exp\{\mathrm{i}\gamma(\mathrm{C})\}$ in addition to the familiar dynamical phase factor. An explicit general formula for $\gamma(\mathrm{C})$ is derived in terms of the spectrum and eigenstates of $\hat{H}(\boldsymbol{R})$ over a surface spanning C. If C lies near a degeneracy of $\hat{H}, \gamma(\mathrm{C})$ takes a simple form which includes as a special case the sign change of eigenfunctions of real symmetric matrices round a degeneracy. As an illustration $\gamma(\mathrm{C})$ is calculated for spinning particles in slowly-changing magnetic fields; although the sign reversal of spinors on rotation is a special case, the effect is predicted to occur for bosons as well as fermions, and a method for observing it is proposed. It is shown that the Aharonov–Bohm effect can be interpreted as a geometrical phase factor.

1. INTRODUCTION

Imagine a quantal system whose Hamiltonian $\hat{H}$ describes the effects of an unchanging environment, and let the system be in a stationary state. If the environment, and hence $\hat{H}$, is slowly altered, it follows from the adiabatic theorem (Messiah 1962) that at any instant the system will be in an eigenstate of the instantaneous $\hat{H}$. In particular, if the Hamiltonian is returned to its original form the system will return to its original state, apart from a phase factor. This phase factor is observable by interference if the cycled system is recombined with another that was separated from it at an earlier time and whose Hamiltonian was kept constant.

My purpose here is to explain how the phase factor contains a circuit-dependent component $\exp(\mathrm{i}\gamma)$ in addition to the familiar dynamical component $\exp(-\mathrm{i}Et/\hbar)$ which accompanies the evolution of any stationary state. A general formula for γ in terms of the eigenstates of $\hat{H}$ will be obtained in § 2. If the circuit is close to a degeneracy in the spectrum of $\hat{H}$, γ takes a particularly simple form which will be derived in § 3; this contains, as a special case, the sign change around a degeneracy of the eigenstates of a system whose Hamiltonian is real as well as Hermitian (Herzberg & Longuet-Higgins 1963; Longuet-Higgins 1975; Mead 1979; Mead & Truhlar 1979; Mead 1980*a*, *b*; Berry & Wilkinson 1984).

A particle of any spin in an eigenstate of a slowly-rotated magnetic field is another case where γ can be calculated explicitly (§ 4), and gives predictions that could be

Reprinted from Proceedings of the Royal Society A, Volume 392, (1984).

tested experimentally. This phase factor exists for bosons as well as fermions. A special case is the sign change of spinors slowly rotated by 2π, predicted by Aharonov & Susskind (1967); this will be shown to be different from the dynamical phase factors measured in experiments on precessing neutrons (reviewed by Silverman 1980).

Finally, it is shown in § 5 that physical effects of magnetic vector potentials in the absence of fields, predicted by Aharonov & Bohm (1959) and observed by Chambers (1960), can be understood as special cases of the geometrical phase factor.

2. General formula for phase factor

Let the Hamiltonian $\hat{H}$ be changed by varying parameters $\boldsymbol{R} = (X, Y, \ldots)$ on which it depends. Then the excursion of the system between times $t = 0$ and $t = T$ can be pictured as transport round a closed path $\boldsymbol{R}(t)$ in parameter space, with Hamiltonian $\hat{H}(\boldsymbol{R}(t))$ and such that $\boldsymbol{R}(T) = \boldsymbol{R}(0)$. The path will henceforth be called a circuit and denoted by C. For the adiabatic approximation to apply, T must be large.

The state $|\psi(t)\rangle$ of the system evolves according to Schrödinger's equation

$$\hat{H}(\boldsymbol{R}(t))\,|\psi(t)\rangle = \mathrm{i}\hbar\,|\dot{\psi}(t)\rangle. \tag{1}$$

At any instant, the natural basis consists of the eigenstates $|n(\boldsymbol{R})\rangle$ (assumed discrete) of $\hat{H}(\boldsymbol{R})$ for $\boldsymbol{R} = \boldsymbol{R}(t)$, that satisfy

$$\hat{H}(\boldsymbol{R})\,|n(\boldsymbol{R})\rangle = E_n(\boldsymbol{R})\,|n(\boldsymbol{R})\rangle, \tag{2}$$

with energies $E_n(\boldsymbol{R})$. This eigenvalue equation implies no relation between the phases of the eigenstates $|n(\boldsymbol{R})\rangle$ at different $\boldsymbol{R}$. For present purposes any (differentiable) choice of phases can be made, provided $|n(\boldsymbol{R})\rangle$ is single-valued in a parameter domain that includes the circuit C.

Adiabatically, a system prepared in one of these states $|n(\boldsymbol{R}(0))\rangle$ will evolve with $\hat{H}$ and so be in the state $|n(\boldsymbol{R}(t))\rangle$ at t.

Thus $|\psi\rangle$ can be written as

$$|\psi(t)\rangle = \exp\left\{\frac{-\mathrm{i}}{\hbar}\int_0^t \mathrm{d}t'\,E_n(\boldsymbol{R}(t'))\right\}\exp\left(\mathrm{i}\gamma_n(t)\right)|n(\boldsymbol{R}(t))\rangle. \tag{3}$$

The first exponential is the familiar dynamical phase factor. In this paper the object of attention is the second exponential. The crucial point will be that its phase $\gamma_n(t)$ is *non-integrable*; γ_n cannot be written as a function of $\boldsymbol{R}$ and in particular is not single-valued under continuation around a circuit, i.e. $\gamma_n(T) \neq \gamma_n(0)$.

The function $\gamma_n(t)$ is determined by the requirement that $|\psi(t)\rangle$ satisfy Schrödinger's equation, and direct substitution of (3) into (1) leads to

$$\dot{\gamma}_n(t) = \mathrm{i}\langle n(\boldsymbol{R}(t))\,|\,\nabla_{\boldsymbol{R}}n(\boldsymbol{R}(t))\rangle\cdot\dot{\boldsymbol{R}}(t). \tag{4}$$

The total phase change of $|\psi\rangle$ round C is given by

$$|\psi(T)\rangle = \exp(\mathrm{i}\gamma_n(\mathrm{C}))\exp\left\{\frac{-\mathrm{i}}{\hbar}\int_0^T \mathrm{d}t\, E_n(\boldsymbol{R}(t))\right\}|\psi(0)\rangle, \tag{5}$$

where the *geometrical phase change* is

$$\gamma_n(\mathrm{C}) = \mathrm{i}\oint_{\mathrm{C}} \langle n(\boldsymbol{R})|\,\nabla_{\boldsymbol{R}}\, n(\boldsymbol{R})\rangle\cdot \mathrm{d}\boldsymbol{R}. \tag{6}$$

Thus $\gamma_n(\mathrm{C})$ is given by a circuit integral in parameter space and is independent of how the circuit is traversed (provided of course that this is slow enough for the adiabatic approximation to hold). The normalization of $|n\rangle$ implies that $\langle n|\nabla_{\boldsymbol{R}} n\rangle$ is imaginary, which guarantees that γ_n is real.

Direct evaluation of $|\nabla_{\boldsymbol{R}} n\rangle$ requires a locally single-valued basis for $|n\rangle$ and can be awkward. Such difficulties are avoided by transforming the circuit integral (6) into a surface integral over any surface in parameter space whose boundary is C. In order to employ familiar vector calculus, parameter space will be considered as three-dimensional, and this will turn out to be the important case in applications; the generalization to higher dimensions will be outlined at the end of this section.

Stokes's theorem applied to (6) gives, in an obvious abbreviated notation.

$$\gamma_n(\mathrm{C}) = -\mathrm{Im}\iint_{\mathrm{C}} \mathrm{d}\boldsymbol{S}\cdot\nabla\times\langle n|\,\nabla n\rangle, \tag{7a}$$

$$= -\mathrm{Im}\iint_{\mathrm{C}} \mathrm{d}\boldsymbol{S}\cdot\langle\nabla n|\times|\nabla n\rangle, \tag{7b}$$

$$= -\mathrm{Im}\iint_{\mathrm{C}} \mathrm{d}\boldsymbol{S}\cdot\sum_{m\neq n}\langle\nabla n|\,m\rangle\times\langle m\,|\nabla n\rangle, \tag{7c}$$

where $\mathrm{d}\boldsymbol{S}$ denotes area element in $\boldsymbol{R}$ space and the exclusion in the summation is justified by $\langle n|\,\nabla n\rangle$ being imaginary. The off-diagonal elements are obtained from (2) as

$$\langle m|\,\nabla n\rangle = \langle m|\,\nabla\hat{H}\,|n\rangle/(E_n - E_m), \quad m\neq n. \tag{8}$$

Thus γ_n can be expressed as

$$\gamma_n(\mathrm{C}) = -\iint_{\mathrm{C}} \mathrm{d}\boldsymbol{S}\cdot \boldsymbol{V}_n(\boldsymbol{R}), \tag{9}$$

where

$$\boldsymbol{V}_n(\boldsymbol{R}) \equiv \mathrm{Im}\sum_{m\neq n}\frac{\langle n(\boldsymbol{R})|\,\nabla_{\boldsymbol{R}}\hat{H}(\boldsymbol{R})|\,m(\boldsymbol{R})\rangle\times\langle m(\boldsymbol{R})|\,\nabla_{\boldsymbol{R}}\hat{H}(\boldsymbol{R})|\,n(\boldsymbol{R})\rangle}{(E_m(\boldsymbol{R})-E_n(\boldsymbol{R}))^2}. \tag{10}$$

Obviously $\gamma_n(\mathrm{C})$ is zero for a circuit which retraces itself and so encloses no area.

Equations (9) and (10) embody the central results of this paper. Because the dependence on $|\nabla n\rangle$ has been eliminated, phase relations between eigenstates with different parameters are now immaterial, and (as is evident from the form of (10)), it is no longer necessary to choose $|m\rangle$ and $|n\rangle$ to be single-valued in $\boldsymbol{R}$: any solutions of (2) may be employed without affecting the value of $\boldsymbol{V}_n$. This is a surprising conclusion, as can be seen by comparing (9) with (7a) which show that $\boldsymbol{V}_n$ is the curl of a vector, $\langle n|\nabla n\rangle$, and $\langle n|\nabla n\rangle$ certainly does depend on the choice of phase of the (single-valued) eigenstate $|n(\boldsymbol{R})\rangle$. The dependence on phase is of the following kind: if $|n\rangle \rightarrow \exp\{\mathrm{i}\mu(\boldsymbol{R})\}|n\rangle$ then $\langle n|\nabla n\rangle \rightarrow \langle n|\nabla n\rangle + \mathrm{i}\nabla\mu$ (in another context the importance of such gauge transformations has been emphasized by Wu & Yang (1975)). Thus the vector is not unique but its curl is. The quantity $\boldsymbol{V}_n$ is analogous to a 'magnetic field' (in parameter space) whose 'vector potential' is Im $\langle n|\nabla n\rangle$. In Appendix A it is shown directly from (10) that $\nabla \cdot \boldsymbol{V}_n$ vanishes, thus confirming that (9) gives a unique value for $\gamma_n(\mathrm{C})$.

Using perturbation theory, Mead & Truhlar (1979) obtained essentially the formulae (9) and (10) for an infinitesimal circuit, in a study of molecular electronic states which (in the Born–Oppenheimer approximation) depend parametrically on nuclear coordinates. Their phase factor was not intended to apply to a $|\psi\rangle$ that evolves slowly under the time-dependent Schrödinger equation, but to the variation of eigenstates $|n\rangle$ under a particular phase-continuation rule in $\boldsymbol{R}$-space which can be shown to give the same result.

In parameter spaces of higher dimension, Stokes's theorem cannot be employed to transform the circuit integral (6). The appropriate generalization, provided by the theory of differential forms (see, for example, Arnold 1978, chap. 7), transforms (6) into the integral of a 2-form over a surface bounded by C. The surprising result (10) can now be expressed as follows: independently of the choice of phases of the eigenstates, there exists in parameter space a *phase 2-form*, which gives $\gamma(\mathrm{C})$ when integrated over any surface spanning C. This 2-form is obtained from (10) by replacing ∇ by the exterior derivative d and $\times$ by the wedge product $\wedge$. The validity of this generalization is consistent with the observation that in the three-dimensional version there are infinitely many choices of interpolating Hamiltonian (and hence of parameter spaces) on the surfaces bounded by C, and the geometrical phase factor is independent of the choice.

Professor Barry Simon (1983), commenting on the original version of this paper, points out that the geometrical phase factor has a mathematical interpretation in terms of holonomy, with the phase two-form emerging naturally (in the form (7*b*)) as the curvature (first Chern class) of a Hermitian line bundle.

3. Degeneracies

The energy denominators in (10) show that if the circuit C lies close to a point $\boldsymbol{R}^*$ in parameter space at which the state n is involved in a degeneracy, then $\boldsymbol{V}_n(\boldsymbol{R})$, and hence $\gamma_n(\mathrm{C})$, is dominated by the terms m corresponding to the other states involved. We shall consider the commonest situation, where the degeneracy involves only two states, to be denoted $+$ and $-$, where $E_+(\boldsymbol{R}) \geqslant E_-(\boldsymbol{R})$. For $\boldsymbol{R}$ near $\boldsymbol{R}^*$, $\hat{H}(\boldsymbol{R})$ can be expanded to first order in $\boldsymbol{R}-\boldsymbol{R}^*$, and

$$\boldsymbol{V}_+(\boldsymbol{R}) = \mathrm{Im}\,\frac{\langle +(\boldsymbol{R})|\,\nabla\hat{H}(\boldsymbol{R}^*)|-(\boldsymbol{R})\rangle \times \langle -(\boldsymbol{R})|\,\nabla\hat{H}(\boldsymbol{R}^*)|+(\boldsymbol{R})\rangle}{(E_+(\boldsymbol{R})-E_-(\boldsymbol{R}))^2}. \qquad (11)$$

Obviously $\boldsymbol{V}_-(\boldsymbol{R}) = -\boldsymbol{V}_+(\boldsymbol{R})$, so that $\gamma_-(\mathrm{C}) = -\gamma_+(\mathrm{C})$.

Without essential loss of generality we can take $E_{\pm}(\boldsymbol{R}^*) = 0$ and $\boldsymbol{R}^* = 0$. $H(\boldsymbol{R})$ can be represented by a 2×2 Hermitian matrix coupling the two states. The most general such matrix satisfying the given conditions depends on three parameters X, Y, Z which will be taken as components of $\boldsymbol{R}$, and by linear transformation in $\boldsymbol{R}$-space can be brought into the following standard form

$$\hat{H}(\boldsymbol{R}) = \frac{1}{2}\begin{bmatrix} Z & X-\mathrm{i}Y \\ X+\mathrm{i}Y & -Z \end{bmatrix}. \tag{12}$$

The eigenvalues are

$$E_+(\boldsymbol{R}) = -E_-(\boldsymbol{R}) = \tfrac{1}{2}(X^2+Y^2+Z^2)^{\frac{1}{2}} = \tfrac{1}{2}R. \tag{13}$$

Thus the degeneracy is an isolated point at which all three parameters vanish. This illustrates an old result of Von Neumann & Wigner (1929): for generic Hamiltonians (Hermitian matrices), it is necessary to vary three parameters in order to make a degeneracy occur accidentally, that is, not on account of symmetry. Alternatively stated, degeneracies have co-dimension three.

The form (12) was chosen to exploit the fact that

$$\nabla\hat{H} = \tfrac{1}{2}\hat{\boldsymbol{\sigma}}, \tag{14}$$

where the components $\hat{\sigma}_X$, $\hat{\sigma}_Y$, $\hat{\sigma}_Z$ of the vector operator $\hat{\boldsymbol{\sigma}}$ are the Pauli spin matrices. When evaluating the matrix elements in (11) it greatly simplifies the calculations to take advantage of the isotropy of spin and temporarily rotate axes so that the Z-axis points along $\boldsymbol{R}$, and to employ the following relations, which come from the commutation laws between the components of $\hat{\boldsymbol{\sigma}}$:

$$\hat{\sigma}_X|\pm\rangle = |\mp\rangle, \quad \hat{\sigma}_Y|\pm\rangle = \pm\mathrm{i}|\mp\rangle, \quad \hat{\sigma}_Z|\pm\rangle = \pm|\pm\rangle. \tag{15}$$

With these rotated axes, (11) gives

$$\left.\begin{aligned} V_{X+} &= (\mathrm{Im}\langle+|\hat{\sigma}_Y|-\rangle\langle-|\hat{\sigma}_Z|+\rangle)/2R^2 = 0, \\ V_{Y+} &= (\mathrm{Im}\langle+|\hat{\sigma}_Z|-\rangle\langle-|\hat{\sigma}_X|+\rangle)/2R^2 = 0, \\ V_{Z+} &= \mathrm{Im}\langle+|\hat{\sigma}_X|-\rangle\langle-|\hat{\sigma}_Y|+\rangle = 1/2R^2. \end{aligned}\right\} \tag{16}$$

Reverting to unrotated axes, we obtain

$$\boldsymbol{V}_+(\boldsymbol{R}) = \boldsymbol{R}/2R^3. \tag{17}$$

Now use of (9) shows that the phase change $\gamma_+(\mathrm{C})$ is the flux through C of the magnetic field of a monopole with strength $-\frac{1}{2}$ located at the degeneracy. Thus we obtain the pleasant result, valid for the natural choice (12) of standard form for $\hat{H}$, that the geometrical phase factor associated with C is

$$\exp\{\mathrm{i}\gamma_{\pm}(\mathrm{C})\} = \exp\{\mp\tfrac{1}{2}\mathrm{i}\Omega(\mathrm{C})\}, \tag{18}$$

where $\Omega(\mathrm{C})$ is the *solid angle* that C subtends at the degeneracy; Ω is, in a sense, a measure of the *view* of the circuit as seen from the degeneracy. The phase factor is independent of the choice of surface spanning C, because Ω can change only in multiples of 4π (when the surface is deformed to pass through the degeneracy).

An important special case of (18) occurs when C consists entirely of *real* Hamiltonians and so is confined to the plane $Y = 0$ (cf. (12)). The energy levels $E_{\pm}$ intersect conically in the space E, X, Z, whose origin, where the degeneracy occurs, is a 'diabolical point' of the type recently studied by Berry & Wilkinson (1984) in the spectra of triangles. This illustrates the result that for real symmetric matrices, degeneracies have co-dimension two: see Appendix 10 of Arnold 1978. If C encloses the degeneracy, $\Omega = \pm 2\pi$; if not, $\Omega = 0$. Thus the phase factor (18) is

$$\exp\{\mathrm{i}\gamma_{\pm}(\mathrm{C})\} = -1, \quad \text{if C encircles the degeneracy,}$$
$$= +1, \quad \text{otherwise,} \tag{19}$$

which expresses the sign changes of real wavefunctions as a degeneracy involving them is encircled, a phenomenon first described by Herzberg & Longuet-Higgins (1963). (Sign changes are not restricted to circuits involving real Hamiltonians: (18) shows that the phase factor is -1 if C lies in *any* plane through the degeneracy and encircles it.)

Confirmation of the correctness of (17) can be obtained without the rotation-of-axes trick, by a lengthy calculation of (11) involving explicit formulae for the eigenvectors $|\pm(\boldsymbol{R})\rangle$ of the matrix (12). Alternatively, direct continuation of the eigenvectors may be attempted. This cannot be accomplished for all circuits by means of (6) because it is not possible to construct eigenvectors that are globally single-valued continuous functions of $\boldsymbol{R}$; multivaluedness can be reduced to singular lines connecting the degeneracy with infinity, and in the analogue $\boldsymbol{V}(\boldsymbol{R})$ these appear as Dirac strings attached to the monopole. Such approaches obscure the simplicity and essential isotropy of the solid-angle result (17).

Using topological arguments not involving explicit formulae for $\gamma_n(\mathrm{C})$, Stone (1976) proved that if C is expanded from one point $\boldsymbol{R}$ and contracted on to another so as to sweep out a surface enclosing a degeneracy, then the geometrical phase factor traverses a circle in its Argand plane. This property (which follows easily from (18)), is the Hermitian generalization of the sign-reversal test for degeneracy.

4. Spins in magnetic fields

A particle with spin s (integer or half-integer) interacts with a magnetic field $\boldsymbol{B}$ via the Hamiltonian

$$\hat{H}(\boldsymbol{B}) = \kappa\hbar\boldsymbol{B}\cdot\hat{\boldsymbol{s}}, \tag{20}$$

where κ is a constant involving the gyromagnetic ratio and $\hat{\boldsymbol{s}}$ is the vector spin operator with $2s+1$ eigenvalues n with integer spacing and that lie between $-s$ and $+s$. The eigenvalues are

$$E_n(\boldsymbol{B}) = \kappa\hbar Bn, \tag{21}$$

and so there is a $(2s+1)$-fold degeneracy when $\boldsymbol{B} = 0$. (The special case $s = \frac{1}{2}$ reproduces the two-fold degeneracy considered in the last section.) We consider the components of $\boldsymbol{B}$ as the parameters $\boldsymbol{R}$ in our previous analysis, and calculate the phase change $\gamma_n(\mathrm{C})$ of an eigenstate $|n, s(\boldsymbol{B})\rangle$ of $\hat{\boldsymbol{s}}$ in the direction along $\boldsymbol{B}$, as $\boldsymbol{B}$ is slowly varied (and hence the spin rotated) round a circuit C.

The vector $\boldsymbol{V}_n(\boldsymbol{B})$ as given by (10) can be expressed by using (20) and (21) as

$$\boldsymbol{V}_n(B) = \frac{\mathrm{Im}}{B^2} \sum_{m \neq n} \frac{\langle n, s(\boldsymbol{B})|\hat{\boldsymbol{s}}|\, m, s(\boldsymbol{B})\rangle \times \langle m, s(\boldsymbol{B})|\hat{\boldsymbol{s}}|\, n, s(\boldsymbol{B})\rangle}{(m-n)^2}. \tag{22}$$

To evaluate the matrix elements we again temporarily rotate axes so that the Z-axis points along $\boldsymbol{B}$, and employ the following generalization of (15):

$$\left.\begin{aligned} (\hat{s}_X + \mathrm{i}\hat{s}_Y)\,|n, s\rangle &= [s(s+1) - n(n+1)]^{\frac{1}{2}}\,|n+1, s\rangle, \\ (\hat{s}_X - \mathrm{i}\hat{s}_Y)\,|n, s\rangle &= [s(s+1) - n(n-1)]^{\frac{1}{2}}\,|n-1, s\rangle, \\ s_Z\,|n, s\rangle &= n\,|n, s\rangle. \end{aligned}\right\} \tag{23}$$

It is clear that only states with $m = n \pm 1$ are coupled with $|n\rangle$ in (22), and that V_x and V_y are zero because they involve off-diagonal elements of $\hat{s}_Z$. To find V_Z, we make use of (23) to obtain

$$\left.\begin{aligned} \langle n \pm 1, s|\, s_X\,|n, s\rangle &= \tfrac{1}{2}[s(s+1) - n(n \pm 1)]^{\frac{1}{2}}, \\ \langle n \pm 1, s|\, s_Y\,|n, s\rangle &= \mp \tfrac{1}{2}\mathrm{i}[s(s+1) - n(n \pm 1)]^{\frac{1}{2}}, \end{aligned}\right\} \tag{24}$$

then (22) gives

$$\begin{aligned} V_{Zn} &= \frac{\mathrm{Im}}{B^2}\{\langle n|\, s_X\,|n+1\rangle\langle n+1\,|s_Y\,|n\rangle - \langle n|\, s_Y\,|n+1\rangle\langle n+1|\, s_X\,|n\rangle \\ &\quad + \langle n|\, s_X\,|n-1\rangle\langle n-1|\, s_Y\,|n\rangle - \langle n|\, s_Y\,|n-1\,|\rangle\langle n-1|\, s_X\,|n\rangle\} \\ &= \frac{n}{B^2}. \end{aligned} \tag{25}$$

Reverting to unrotated axes, we obtain

$$\boldsymbol{V}_n(\boldsymbol{B}) = n\boldsymbol{B}/B^3. \tag{26}$$

Now, use of (9) shows that $\gamma_n(\mathrm{C})$ is the flux through C of the 'magnetic field' of a monopole $-n$ located at the origin of magnetic field space. Thus the geometrical phase factor is

$$\exp\{\mathrm{i}\gamma_n(\mathrm{C})\} = \exp\{-\mathrm{i}n\Omega(\mathrm{C})\}, \tag{27}$$

where $\Omega(\mathrm{C})$ is the solid angle that C subtends at $\boldsymbol{B} = 0$. Note that γ_n depends only on the eigenvalue n of the spin component along $\boldsymbol{B}$ and not on the spin s of the particle, so that γ_n is insensitive to the strength $2s+1$ of the degeneracy at $\boldsymbol{B} = 0$.

It follows from (27) that any phase change can be produced by varying $\boldsymbol{B}$ round a suitable circuit. For fermions (half-integer n), a whole turn of $\boldsymbol{B}$ (rotation through 2π in a plane, giving $\Omega = 2\pi$) produces a phase factor -1. In the special case $n = \frac{1}{2}$ this shows that the sign change of spinors on rotation and the sign change of wavefunctions round a degeneracy have the same mathematical origin. For bosons (integer n), a whole turn of $\boldsymbol{B}$ produces a phase factor $+1$. To produce a sign change, different circuits are required; if $n = 1$, for example, varying $\boldsymbol{B}$ round a cone of semiangle 60° will give $\Omega = \gamma = \pi$ and hence a phase factor -1.

The following experiment could be carried out to test the predictions embodied in (27). A polarized monoenergetic beam of particles in spin state n along a magnetic field $\boldsymbol{B}$ is split into two. Along the path of one beam $\boldsymbol{B}$ is kept constant. Along the path of the other beam, $\boldsymbol{B}$ is kept constant in magnitude but its direction is varied slowly (in comparison with the dynamical precession frequency) round a circuit C subtending a solid angle Ω, the field being generated by an arrangement enabling Ω to be changed. The beams are then combined and the count rate at a detector is measured as a function of Ω. The dynamical phase factor (the second exponential in (5) is the same for both beams because the energy $E_n(\boldsymbol{B})$ (21) is insensitive to the direction of $\boldsymbol{B}$. There will in addition be a propagation phase factor which can be made unity by adjusting the path-length of one of the beams when $\Omega = 0$. The resulting fringes occur as a consequence of the geometrical phase factor. If C is a circuit round a cone of semiangle θ, the predicted intensity contrast is

$$I(\theta) = \cos^2(n\pi(1-\cos\theta)). \tag{28}$$

I wish to emphasize that this proposed experiment is different from those carried out by Rauch *et al.* (1975, 1978) and Werner *et al.* (1975) (see Silverman 1980) with *unpolarized* neutrons in a *constant* magnetic field. Those neutrons were not in an eigenstate, and their phase changed dynamically, rather than geometrically, under the Hamiltonian (20) (with $\boldsymbol{B}$ along Z and $\hat{\boldsymbol{\sigma}}$ replacing $\hat{\boldsymbol{s}}$) according to the evolution operator

$$\exp(-\mathrm{i}\hat{H}t/\hbar) = \exp(-B\kappa t\hat{\sigma}_Z) = \cos\tfrac{1}{2}\kappa Bt\begin{bmatrix}1 & 0\\ 0 & 1\end{bmatrix} + \mathrm{i}\sin\tfrac{1}{2}\kappa Bt\begin{bmatrix}1 & 0\\ 0 & -1\end{bmatrix}. \tag{29}$$

The sign changed whenever $\frac{1}{2}\kappa Bt$ was an odd multiple of π, and this was interpreted on the basis of precession theory as corresponding to odd numbers of complete rotations about $\boldsymbol{B}$.

5. Aharonov–Bohm effect

Consider a magnetic field consisting of a single line with flux Φ. For positions $\boldsymbol{R}$ not on the flux line, the field is zero but there must be a vector potential $\boldsymbol{A}(\boldsymbol{R})$ satisfying

$$\oint_{\mathrm{C}} \boldsymbol{A}(\boldsymbol{R})\cdot\mathrm{d}\boldsymbol{R} = \Phi, \tag{30}$$

for circuits C threaded by the flux line. Aharonov & Bohm (1959) showed that in quantum mechanics such vector potentials have physical significance even though they correspond to zero field. I shall now show how their effect can be interpreted as a geometrical phase change of the type described in § 2.

Let the quantal system consist of particles with charge q confined to a box situated at $\boldsymbol{R}$ and not penetrated by the flux line (figure 1). In the absence of flux

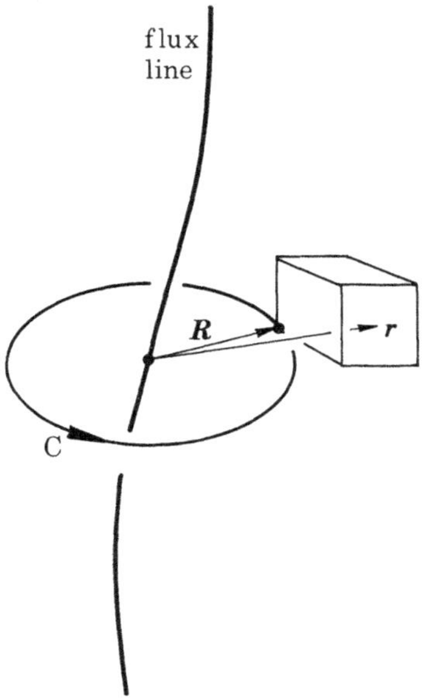

FIGURE 1. Aharonov–Bohm effect in a box transported round a flux line.

($\boldsymbol{A} = 0$), the particle Hamiltonian depends on position $\hat{\boldsymbol{r}}$ and conjugate momentum $\hat{\boldsymbol{p}}$ as follows:

$$\hat{H} = H(\hat{\boldsymbol{p}}, \hat{\boldsymbol{r}} - \boldsymbol{R}), \tag{31}$$

and the wavefunctions have the form $\psi_n(\boldsymbol{r} - \boldsymbol{R})$ with energies E_n independent of $\boldsymbol{R}$. With non-zero flux, the states $|n(\boldsymbol{R})\rangle$ satisfy

$$H(\hat{\boldsymbol{p}} - q\boldsymbol{A}(\hat{\boldsymbol{r}}), \hat{\boldsymbol{r}} - \boldsymbol{R})|\, n(\boldsymbol{R})\rangle = E_n\, |n(\boldsymbol{R})\rangle, \tag{32}$$

an equation whose exact solutions are obtained by multiplying ψ_n by an appropriate Dirac phase factor, giving

$$\langle \boldsymbol{r}|\, n(\boldsymbol{R})\rangle = \exp\left\{\frac{\mathrm{i}q}{\hbar}\int_{\boldsymbol{R}}^{\boldsymbol{r}} \mathrm{d}\boldsymbol{r}' \cdot \boldsymbol{A}(\boldsymbol{r}')\right\} \psi_n(\boldsymbol{r} - \boldsymbol{R}). \tag{33}$$

These solutions are single-valued in $\boldsymbol{r}$ and (locally) in $\boldsymbol{R}$. The energies are unaffected the vector potential.

Now let the box be transported round a circuit C threaded by the flux line; in this particular case it is not necessary that the transport be adiabatic. After completion of the circuit there will be a geometrical phase change that can be calculated from (6) and (33) by using

$$\begin{aligned} \langle n(\boldsymbol{R})|\, \nabla_{\boldsymbol{R}} n(\boldsymbol{R})\rangle &= \iiint \mathrm{d}^3\boldsymbol{r}\, \psi_n^*(\boldsymbol{r} - \boldsymbol{R}) \left\{\frac{-\mathrm{i}q}{\hbar} A(\boldsymbol{R})\, \psi_n(\boldsymbol{r} - \boldsymbol{R}) + \nabla_{\boldsymbol{R}} \psi_n(\boldsymbol{r} - \boldsymbol{R})\right\} \\ &= -\mathrm{i}q\boldsymbol{A}(\boldsymbol{R})/\hbar. \end{aligned} \tag{34}$$

(The vanishing of the second term in braces follows from the normalization of ψ_n.) Evidently in this example the analogy between $\mathrm{Im}\,\langle n|\nabla n\rangle$ and a magnetic vector potential becomes a reality. Thus

$$\gamma_n(\mathrm{C}) = \frac{q}{\hbar}\oint_{\mathrm{C}} \boldsymbol{A}(\boldsymbol{R})\cdot \mathrm{d}\boldsymbol{R} = q\Phi/\hbar, \tag{35}$$

which shows that the phase factor is independent of n, and also of C if this winds once round the flux line. The phase factor could be observed by interference between the particles in the transported box and those in a box which was not transported round the circuit.

In elementary presentations of the Aharonov–Bohm effect (including its anticipation by Ehrenburg & Siday 1949), the Dirac phase factor is often invoked in comparing systems passing opposite sides of a flux line. Such invocations are subject to the objection that the wavefunction thus obtained is not single-valued. One way to avoid this objection is by summation over all contributions (whirling waves) representing different windings round the flux line (Schulman 1981; Berry 1980; Morandi & Menossi 1984). Another way, adopted in the original paper by Aharonov & Bohm, is to solve Schrödinger's equation exactly for scattering in the flux line's vector potential. The argument of the preceding paragraphs, which employs the geometrical phase factor, is a third way of obtaining the Aharonov–Bohm effect by using only single-valued wavefunctions.

Mead (1980*a*, *b*), employs the term 'molecular Aharonov–Bohm effect' in a different context, to describe how degeneracies in electron energy levels affect the spectrum of nuclear vibrations. He explains two options, both leading to the same vibration spectrum. The first option is to continue the electronic states round degeneracies (in the space of nuclear coordinates) in the manner described in this paper, thus causing the electronic wavefunctions to be multi-valued, with a compensating multi-valuedness in the nuclear states, which must be incorporated into their boundary conditions. The alternative is to enforce single-valuedness on the electronic (and hence also the nuclear) states, and this introduces a vector potential into the Schrödinger equation for nuclear motion. In general one may expect such effects whenever an isolated system is considered as being divided into two interacting parts, each slaved to a different aspect of the other (in the molecular case, electron states are slaved to nuclear coordinates, and nuclear states are slaved to the electronic states and wavefunctions). The systems considered in this paper might be regarded as a special case, in which the coupling is with 'the rest of the Universe' (including us as observers). The only role of the rest of the Universe is to provide a Hamiltonian with slowly-varying parameters, thus forcing the system to evolve adiabatically with phase continuation governed by the time-dependent Schrödinger equation.

6. Discussion

A system slowly transported round a circuit will return in its original state; this is the content of the adiabatic theorem. Moreover its internal clocks will register the passage of time; this can be regarded as the meaning of the dynamical phase factor. The remarkable and rather mysterious result of this paper is that in addition the system records its history in a deeply geometrical way, whose natural formulation (9) and (10) involves phase functions hidden in parameter-space regions which the system has not visited.

The total phase of the transported state (5) is dominated by the dynamical part, because $T \to \infty$ in the adiabatic limit, and it might be thought that this must overwhelm the geometrical phase γ_n and make its physical effects difficult to detect. This objection can be met by observing that the strengths of non-adiabatic transitions are exponentially small in T if $\hat{H}$ changes smoothly (Hwang & Pechukas 1977), so that essentially adiabatic evolution can occur even when the dynamical phase is only a few times greater than 2π.

As we saw in § 3, degeneracies in the spectrum of $\hat{H}(\boldsymbol{R})$ are the singularities of the vector $\boldsymbol{V}(\boldsymbol{R})$ (equation (10)) in parameter space, and so have an important effect on the geometrical phase factor. This is reminiscent of the part played by singularities of an analytic function, but the analogy is imperfect: if $\gamma(\mathrm{C})$ were completely singularity-determined, $\boldsymbol{V}(\boldsymbol{R})$ would be the sum of the 'magnetic fields' of 'monopoles' situated at the degeneracies (cf. (17)) and so would have zero curl, which is not the case (zero curl, unlike zero divergence, is not a property which is invariant under deformations of $\boldsymbol{R}$ space, and in the general case the sources of $\boldsymbol{V}$ are not just monopoles but also 'currents' distributed continuously in parameter space). A closer analogy is with wavefront dislocation lines, which are phase singularities of complex wavefunctions in three-dimensional position space (Nye & Berry 1974; Nye 1981; Berry 1981), that dominate the geometry of wavefronts without completely determining them.

In view of the emphasis on degeneracies as organizing centres for phase changes, it is worth remarking that close approach of energy levels is not a necessary condition for the existence of nontrivial geometrical phase factors. Indeed, our examples have shown that $\gamma(\mathrm{C})$ can be non-zero even if C involves isospectral deformations of $\hat{H}(\boldsymbol{R})$ (in the Aharonov–Bohm illustration, the levels E_n do not depend on $\boldsymbol{R}$ at all).

The results obtained here are not restricted to quantum mechanics, but apply more generally, to the phase of eigenvectors of any Hermitian matrices under a natural continuation in parameter space. Therefore they have implications throughout wave physics. For example, the electromagnetic field of a single mode travelling along an optical fibre will change sign if the cross section of the fibre is slowly altered so that its path (in the space of shapes) surrounds a shape for which the spectrum of the Helmholtz equation is degenerate (such as one of the diabolical triangles discovered by Berry & Wilkinson 1984).

I thank Dr J. H. Hannay, Dr E. J. Heller and Dr B. R. Pollard for several suggestions, and Professor Barry Simon for showing me, before publication, his paper which comments on this one. This work was not supported by any military agency.

References

Aharonov, Y. & Bohm, D. 1959 *Phys. Rev.* **115**, 485–491.
Aharonov, Y. & Susskind, L. 1967 *Phys. Rev.* **158**, 1237–1238.
Arnold, V. I. 1978 *Mathematical methods of classical dynamics*. New York: Springer.
Berry, M. V. 1980 *Eur. J. Phys.* **1**, 240–244.
Berry, M. V. 1981 Singularities in waves and rays. In *Les Houches Lecture Notes for session* XXXV (ed. R. Balian, M. Kléman & J-P. Poirier), pp. 453–543. Amsterdam: North-Holland.
Berry, M. V. & Wilkinson, M. 1984 *Proc. R. Soc. Lond.* A **392**, 15–43.
Chambers, R. G. 1960 *Phys. Rev. Lett* **5**, 3–5.
Ehrenburg, W. & Siday, R. E. 1949 *Proc. phys. Soc.* B **62**, 8–21.
Herzberg, G. & Longuet-Higgins, H. C. 1963 *Discuss. Faraday Soc.* **35**, 77–82.
Hwang, J.-T. & Pechukas, P. 1977 *J. chem. Phys.* **67**, 4640–4653.
Longuet-Higgins, H. C. 1975 *Proc. R. Soc. Lond.* A **344**, 147–156.
Mead, C. A. 1979 *J. chem. Phys.* **70**, 2276–2283.
Mead, C. A. 1980*a* *Chem. Phys.* **49**, 23–32.
Mead, C. A. 1980*b* *Chem. Phys.* **49**, 33–38.
Mead, C. A. & Truhlar, D. G. 1979 *J. chem. Phys.* **70**, 2284–2296.
Messiah, A. 1962 *Quantum mechanics*, vol. 2. Amsterdam: North-Holland.
Morandi, G. & Menossi, E. 1984 *Nuovo Cim.* B (Submitted.)
Nye, J. F. 1981 *Proc. R. Soc. Lond.* A **378**, 219–239.
Nye, J. F. & Berry, M. V. 1974 *Proc. R. Soc. Lond.* A **336**, 165–190.
Rauch, H., Wilfing, A., Bauspiess, W. & Bonse, U. 1978 *Z. Phys.* B **29**, 281–284.
Rauch, H., Zeilinger, A., Badurek, G., Wilfing, A., Bauspiess, W. & Bonse, U. 1975 *Physics Lett.* A **54**, 425–427.
Schulman, L. S. 1981 Techniques and Applications of Path Integration. New York: John Wiley.
Silverman, M. P. 1980 *Eur. J. Phys.* **1**, 116–123.
Simon, B. 1983 *Phys. Rev. Lett.* (In the press.)
Stone, A. J. 1976 *Proc. R. Soc. Lond.* A **351**, 141–150.
Von Neumann, J. & Wigner, E. P. 1929 *Phys. Z.* **30**, 467–470.
Werner, S. A., Colella, R., Overhauser, A. W. & Eagen, C. F. 1975 *Phys. Rev. Lett.* **35**, 1053–1055.
Wu, T. T. & Yang, C. N. 1975 *Phys. Rev.* D **12**, 3845–3857.

Appendix A

To show that $\gamma(C)$ is independent of the surface spanning C, it is necessary to prove that $\boldsymbol{V}(\boldsymbol{R})$ (equation (10)) has zero divergence. This can be accomplished by expressing $\boldsymbol{V}$ in terms of the vector Hermitian operator $\hat{\boldsymbol{B}}$ defined by

$$\hat{\boldsymbol{B}} \equiv -\mathrm{i} \sum_n |\nabla n\rangle \langle n|. \tag{A 1}$$

From (8), the off-diagonal elements of $\hat{\boldsymbol{B}}$ are

$$\langle n| \hat{\boldsymbol{B}} |m\rangle = -\mathrm{i} \langle m| \nabla H |n\rangle / (E_n - E_m), \quad m \neq n. \tag{A 2}$$

Thus (10) becomes

$$V = \operatorname{Im}\langle n|\hat{\boldsymbol{B}}\times\hat{\boldsymbol{B}}|\,n\rangle. \tag{A 3}$$

Now we can calculate the divergence:

$$\nabla\cdot V = \operatorname{Im}\{\langle\nabla n|\cdot\hat{\boldsymbol{B}}\times\hat{\boldsymbol{B}}\,|n\rangle+\langle n|\,\boldsymbol{B}\times\boldsymbol{B}\cdot|\nabla n\rangle+\langle n|\,\nabla\cdot(\hat{\boldsymbol{B}}\times\hat{\boldsymbol{B}})\,|n\rangle\}, \tag{A 4}$$

Use of a consequence of (A 1), namely

$$|\nabla n\rangle = \mathrm{i}\hat{\boldsymbol{B}}\,|n\rangle \tag{A 5}$$

gives

$$\nabla\cdot V = n(-\hat{\boldsymbol{B}}\cdot\hat{\boldsymbol{B}}\times\hat{\boldsymbol{B}}+\hat{\boldsymbol{B}}\times\hat{\boldsymbol{B}}\cdot\hat{\boldsymbol{B}})\,|n\rangle+\operatorname{Im}\langle n|\,(\nabla\times\hat{\boldsymbol{B}}\cdot\hat{\boldsymbol{B}}-\hat{\boldsymbol{B}}\cdot\nabla\times\boldsymbol{B}\,|n\rangle. \tag{A 6}$$

For the curl of $\boldsymbol{B}$, (A 1) and (A 5) give

$$\nabla\times\hat{\boldsymbol{B}} = +\mathrm{i}\sum_n |\nabla n\rangle\times\langle\nabla n| = \mathrm{i}\sum_n \hat{\boldsymbol{B}}\,|n\rangle\times\langle n|\hat{\boldsymbol{B}} = \mathrm{i}\hat{\boldsymbol{B}}\times\hat{\boldsymbol{B}}\rangle, \tag{A 7}$$

whence $\nabla\cdot V$ vanishes by the dot-cross rule for triple products.

This result is valid everywhere except at the 'monopole' singularities arising from degeneracies.

Geometric phase memories

Michael Berry

The moment of conception of the geometric phase can be pinpointed precisely, but related ideas had been formulated before, in various guises. Not less varied were the ramifications that became clear once the concept was identified formally.

A little more than 25 years ago, I introduced the geometric phase: when the parameters of a quantum mechanical wavefunction are slowly cycled around a circuit, then the phase of the wavefunction need not return to its original value[1]. In recalling the events surrounding the publication of that paper, I should back up to the early 1970s, to Ian Percival's pioneering insight[2] that classical chaos — then unfamiliar to physicists — must have implications in quantum mechanics, in particular for the spectrum of discrete energy levels of systems whose classical motion would be chaotic, such as atoms in strong magnetic fields or molecules whose atoms interact anharmonically.

My colleague Balazs Gyorffy suggested that such 'irregular spectra' might be described by random-matrix theory, which had been developed to understand the statistics of energy levels in nuclei[3]. A central feature of spectra of random matrices is that energy levels repel. For the universality class of systems with time-reversal symmetry (where spin is unimportant) level repulsion is described quantitatively by the probability distribution of spacings S between nearest-neighbour levels: the probability of finding a given spacing vanishes linearly for near-degeneracies, that is, as S goes to zero. Following insights from von Neumann and Wigner[4], it soon became clear that linear repulsion could be understood as the shadow of true degeneracies (where $S = 0$) in nearby systems in which the system under consideration is imagined to be embedded. These degeneracies require two parameters — one is in general not sufficient to produce a degeneracy — and in terms of these parameters the energy levels are sheets in the form of a double cone. The double cone is also called a diabolo (after a spinning toy of the same shape), so I called the intersections 'diabolical points'.

But how can we know that the two sheets really touch, rather than avoiding each other as energy levels typically do when just one parameter is varied? In 1978 I found the criterion: while encircling a diabolical point in the space of parameters, each of the two wavefunctions, when smoothly continued round its sheet, must change sign[5]. This simple topological result was very satisfying — a property of 2 × 2 matrices that should be in every textbook. But apparently it was in none. Alas, I quickly learned that my 'discovery' was not original: several years earlier, the mathematician Karen Uhlenbeck had written about the sign change[6], and, nearly two decades before that, Christopher Longuet-Higgins[7] and others had found it in a study of energy levels in molecules (the 'others' were physicists in Bristol, but more about this later).

I wanted to see these theoretical constructs — diabolical points, and the associated sign change — in computations for a concrete class of systems. Michael Wilkinson and I chose to explore the spectra of triangular quantum billiards: energy states of particles confined in a triangular domain with hard walls, equivalent to the vibration modes of a triangular drum. Triangles are indeed described by two parameters, namely any two angles. We found several diabolical points[8] and confirmed their existence by calculating the sign change.

The moment of conception

In the spring of 1983, I talked about this work at the Georgia Institute of Technology, emphasizing the importance of time reversal. If this symmetry is broken, the spectra would fall into a different, more general universality class, in which degeneracies would typically require the explorations of three parameters, not two (ref. 4). So, if a weak magnetic field were added to the particle in the triangles, the diabolical points would disappear. At the end of the talk, Ronald Fox (at that time the chairman of the physics department) asked what happens to the sign change when the magnetic field is switched on.

This was the trigger, the moment of conception. My immediate response, "I suppose it's a phase change different from π", was followed by the premature "I'll work it out tonight and tell you tomorrow". In fact it took several weeks to understand the geometric phase properly. The outcome was an unexpected general feature of quantum mechanics: although wavefunctions are single-valued in the space of variables on which they depend, they need not be single-valued under continuation around a circuit of parameters that drive them. And when the driving is slow — 'adiabatic' — the mathematically natural continuation was exactly the one that is enforced by the time-dependent Schrödinger equation. After checking every book on quantum mechanics in our library, and failing to find the geometric phase effect described or even suggested, I decided to write up the work for publication.

John Hannay, who had made useful suggestions while the research was progressing, urged that my proposed title, which included the term 'topological phase factors', was misleading. In the general case, the phase depends continuously on the shape of the circuit in the space of parameters, so the phenomenon I had identified was geometric, not topological. Only in special cases — systems with time-reversal symmetry, or the Aharonov–Bohm effect — is the phase topological.

Anticipations and ramifications

After writing the paper[1], but before submitting it for publication, I enjoyed a long drive through the English countryside with Eric Heller, during which I told him enthusiastically about the phase, and asked if he knew of anything similar. He did. Several years before, Alden Mead and Don Truhlar had identified 'the molecular Aharonov–Bohm effect' in quantum chemistry[9]. Back in Bristol, I anxiously looked at their paper, and found to my dismay that they had indeed anticipated several features of the geometric phase. Fortunately I was able to cite their work in my paper, hoping it retained sufficient novelty and generality to survive the referees' scrutiny.

I sent it to *Proceedings of the Royal Society of London*, my journal of choice for work I was particularly pleased with (at present, I am the editor). The paper was received on

Reprinted from Nature Physics, Volume 6, (2010).

Figure 1 | A polarization bull's-eye singularity photographed in the sky above Bristol University. Between the camera lens and the sky was a 'black-light sandwich', consisting of a sheet of overhead-projector transparency film (which happens to be biaxially anisotropic) sandwiched between two orthogonal polarizing filters[49]. The black stripe is a manifestation of the geometric phase.

13 June 1983. As the journal later informed me, one referee eventually confessed to having lost the manuscript, and it was only in 1984 that it was finally published. Meanwhile, I had met Barry Simon in Australia, and told him about the phase. Overnight, he recognized it as a manifestation of anholonomy as known in fibre bundle theory: the failure of some quantities to return when others, that drive them, are cycled. His paper[10], describing these connections with mathematics and also with earlier work on the quantum Hall effect[11], was published in late 1983. After it appeared in *Physical Review Letters*, and he coined the term 'Berry's phase', the concept, and my subsequent publication, became widely known.

Generalizations and extensions soon followed: from Wilczek and Zee[12] to the non-Abelian case where a collection of degenerate states, rather than a single one, is cycled; from John Hannay[13,14] to an analogous phase in classical mechanics; from Yakir Aharonov and Jeeva Anandan[15] to a formulation that did not require adiabaticity; from John Garrison and Ewan Wright[16] to non-Hermitian evolution.

A particularly far-reaching extension was Hiroshi Kuratsuji and Shini Iida's interpretation[17] of the driving parameters as dynamical variables, whose evolution was influenced by the same geometrical objects as the phase. It became clear[18] that the reaction of the geometric phase on the parameters takes the form of an abstract magnetic field, which I explored with Jonathan Robbins[19]; we called it 'geometric magnetism'. We now know that this is the first of a hierarchy of geometric reaction forces on a slow system coupled to a fast one. Very recently, I have returned to this subject, in a study, with Pragya Shukla, of the infinite series of reaction forces[20]; several features of the separation between fast and slow variables, however, remain mysterious.

I soon learned about other anticipations of the phase than Mead and Truhlar's. While visiting India in 1986, Rajaram Nityananda and Sivaraj Ramaseshan showed me the paper they had written[21] reviving the discovery, by Shivaramakrishnam Pancharatnam[22] 30 years before, of a geometric phase in beams of light whose polarization state is cycled (see Fig. 1). On the long flight home, I made the connection between Pancharatnam's optical picture and my general quantum formalism[23,24], and realized that he had formulated the phase for two-state systems. Pancharatnam was one of the several physicist nephews of Chandrasekhara Venkata Raman; in 1956, when he discovered his polarization phase, he was only 22 years old. Another early version was the discovery by Kenneth Budden and Martin Smith[25] of 'additional phase memory' in radiowave propagation. These anticipations illustrate that in retrospect what we call 'discovery' sometimes looks more like emergence into the air from subterranean intellectual currents[26].

The phase emerged from my earlier interest in semiclassical physics, but influenced my subsequent intellectual trajectory in a very different way. The adiabatic formulation, in which the state is driven slowly, raised the question of corrections to the phase, of higher order in slowness. A calculation, in 1987, of the infinite series of these higher orders, involved the recognition that the series must diverge if there are to be any non-adiabatic transitions between the cycled state and other states[27]. As transitions between states represent the way in which quantum mechanics describes events (in contrast to eigenstates, which describe things), there is a sense in which the divergence of the series is necessary in order for anything to happen. This was the beginning of a series of developments[28,29] continuing into the 1990s and building on earlier seminal insights by Robert Dingle[30] (who had been my doctoral supervisor in the 1960s), in which increasingly sophisticated ways of summing divergent series were devised[31,32], and the inevitability of divergence in series arising throughout theoretical physics[33,34] became clear.

An unavoidable discovery?

It took several years to appreciate that although my understanding of the geometric phase involved a series of accidents — the quantum physics of classical chaos, Ronald Fox's question, the drive with Eric Heller, and others — there was a certain inevitability about it emerging from the physics department of Bristol University. The reason is that physics associated with circuits was embedded in the culture of the department (see also ref. 35). Its inspiration was the deeply geometric personality of Charles Frank, who worked in Bristol from 1946 until his death in 1998. In 1951, he gave the definitive understanding[36] of dislocations in solids in terms of a 'Burgers circuit' of a defect in the real crystal, whose image in a fictitious ideal crystal failed to close. In 1958 he applied similar ideas[37] to classify defects in liquid crystals. In the following year, Longuet-Higgins and a group including Maurice Pryce, then the head of the physics department in Bristol, understood the sign change associated with molecular electronic degeneracies[7]. At the same time, Yakir Aharonov and David Bohm discovered their celebrated eponymous effect[38], interpretable as the phase change commentary

that a quantum electron acquires during a circuit of a line of magnetic flux; soon afterwards, Robert Chambers[39] gave the first experimental demonstration. In 1974, John Nye and I identified phase singularities[40] (also termed wavefront dislocations, nodal lines or wave vortices) as ubiquitous features of waves, classical or quantum. And in the 1980s, Nye gave a similar characterization of polarization singularities[41,42], associated with singularities in vector waves. With this perspective, finding the geometric phase at Bristol appears unsurprising.

After 1983, my interests shifted. I did not follow the many applications of geometric phases in different areas of physics, and so cannot review them. But others have[43–48]. ❒

Michael Berry is in the H. H. Wills Physics Laboratory, Tyndall Avenue, Bristol BS8 1TK, UK.
e-mail: Tracie.Anderson@bristol.ac.uk

References

1. Berry, M. V. *Proc. R. Soc. Lond. A* **392,** 45–57 (1984).
2. Percival, I. C. *J. Phys. B* **6,** L229–L232 (1973).
3. Porter, C. E. *Statistical Theories of Spectra: Fluctuations* (Academic Press, 1965).
4. Von Neumann, J. & Wigner, E. *Phys. Z.* **30,** 467–470 (1929).
5. Berry, M. V. *Ann. Phys.* **131,** 163–216 (1981).
6. Uhlenbeck, K. *Am. J. Math.* **98,** 1059–1078 (1976).
7. Longuet-Higgins, H. C., Öpik, U., Pryce, M. H. L. & Sack, R. A. *Proc. R. Soc. Lond. A* **244,** 1–16 (1959).
8. Berry, M. V. & Wilkinson, M. *Proc. R. Soc. Lond. A* **392,** 15–43 (1984).
9. Mead, C. A. & Truhlar, D. G. *J. Chem. Phys.* **70,** 2284–2296 (1979).
10. Simon, B. *Phys. Rev. Lett.* **51,** 2167–2170 (1983).
11. Thouless, D. J., Kohmoto, M., Nightingale, M. P. & den Nijs, M. *Phys. Rev. Lett.* **49,** 405–409 (1982).
12. Wilczek, F. & Zee, A. *Phys. Rev. Lett.* **52,** 2111–2114 (1984).
13. Hannay, J. H. *J. Phys. A* **18,** 221–230 (1985).
14. Berry, M. V. *J. Phys. A* **18,** 15–27 (1985).
15. Aharonov, Y. & Anandan, J. *Phys. Rev. Lett.* **58,** 1593–1596 (1987).
16. Garrison, J. C. & Wright, E. M. *Phys. Lett. A* **128,** 177–181 (1988).
17. Kuratsuji, H. & Iida, S. *Prog. Theo. Phys.* **74,** 439–445 (1985).
18. Berry, M. V. in *Geometric Phases in Physics* (eds Shapere, A. & Wilczek, F.) 7–28 (World Scientific, 1989).
19. Berry, M. V. & Robbins, J. M. *Proc. R. Soc. Lond. A* **442,** 641–658 (1993).
20. Berry, M. V. & Shukla, P. *J. Phys. A* **43,** 045102 (2010).
21. Ramaseshan, S. & Nityananda, R. *Curr. Sci. India* **55,** 1225–1226 (1986).
22. Pancharatnam, S. *Proc. Ind. Acad. Sci. A* **44,** 247–262 (1956).
23. Berry, M. V. *J. Mod. Opt.* **34,** 1401–1407 (1987).
24. Berry, M. V. *Curr. Sci. India* **67,** 220–223 (1994).
25. Budden, K. G. & Smith, M. S. *Proc. R. Soc. Lond. A* **350,** 27–46 (1976).
26. Berry, M. V. *Phys. Today* **43,** 34–40 (December 1990).
27. Berry, M. V. *Proc. R. Soc. Lond. A* **414,** 31–46 (1987).
28. Berry, M. V. *Proc. R. Soc. Lond. A* **422,** 7–21 (1989).
29. Berry, M. V. *Proc. R. Soc. Lond. A* **429,** 61–72 (1990).
30. Dingle, R. B. *Asymptotic Expansions: Their Derivation and Interpretation* (Academic Press, 1973).
31. Berry, M. V. & Howls, C. J. *Proc. R. Soc. Lond. A* **434,** 657–675 (1991).
32. Berry, M. V. in *Asymptotics Beyond All Orders* (eds Segur, H. & Tanveer, S.) 1–14 (Plenum, 1992).
33. Berry, M. V. & Howls, C. J. *Phys. World* **6,** 35–39 (June 1993).
34. Berry, M. V. in *Proc. 9th Int. Cong. Logic Method. Phil. Sci.* (eds Prawitz, D., Skyrms, B. & Westerståhl, D.) 597–607 (Elsevier, 1994).
35. Berry, M. V. in *Sir Charles Frank OBE FRS, An Eightieth Birthday Tribute* (eds Chambers, R. G., Enderby, J. E., Keller, A., Lang, A. R. & Steeds, J. W.) 207–219 (Adam Hilger, 1991).
36. Frank, F. C. *Phil. Mag.* **42,** 809–819 (1951).
37. Frank, F. C. *Farad. Sci. Disc.* **25,** 19–28 (1958).
38. Aharonov, Y. & Bohm, D. *Phys. Rev.* **115,** 485–491 (1959).
39. Chambers, R. G. *Phys. Rev. Lett.* **5,** 3–5 (1960).
40. Nye, J. F. & Berry, M. V. *Proc. R. Soc. Lond. A* **336,** 165–190 (1974).
41. Nye, J. F. & Hajnal, J. V. *Proc. R. Soc. Lond. A* **409,** 21–36 (1987).
42. Nye, J. F. *Natural Focusing and Fine Structure of Light: Caustics and Wave Dislocations* (Institute of Physics Publishing, 1999).
43. Shapere, A. & Wilczek, F. *Geometric Phases in Physics* (World Scientific, 1989).
44. Zwanziger, J. W., Koenig, M. & Pines, A. *Adv. Chem. Phys.* **41,** 601–646 (1990).
45. Markovski, B. & Vinitsky, S. I. *Topological Phases in Quantum Theory* (World Scientific, 1989).
46. Bohm, A. *The Geometric Phase in Quantum systems: Foundations, Mathematical Concepts, and Applications in Molecular and Condensed-Matter Physics* (Springer, 2003).
47. Li, H.-Z. *Global Properties of Simple Quantum Systems – Berry's Phase and Others* (Shanghai Scientific and Technical Publishers, 1998).
48. Chruscinski, D. & Jamiolkowski, A. *Geometric Phases in Classical and Quantum Mechanics* (Birkhäuser, 2004).
49. Berry, M. V., Bhandari, R. & Klein, S. *Eur. J. Phys.* **20,** 1–14 (1999).

Indistinguishability for quantum particles: spin, statistics and the geometric phase

By M. V. Berry[1] and J. M. Robbins[2]

[1]*H. H. Wills Physics Laboratory, University of Bristol, Tyndall Avenue, Bristol BS8 1TL, UK*

[2]*Basic Research Institute in the Mathematical Sciences, Hewlett-Packard Laboratories Bristol, Filton Road, Stoke Gifford, Bristol BS12 6QZ, UK and School of Mathematics, University of Bristol, University Walk, Bristol BS8 1TW, UK*

The quantum mechanics of two identical particles with spin S in three dimensions is reformulated by employing not the usual fixed spin basis but a transported spin basis that exchanges the spins along with the positions. Such a basis, required to be smooth and parallel-transported, can be generated by an 'exchange rotation' operator resembling angular momentum. This is constructed from the four harmonic oscillators from which the two spins are made according to Schwinger's scheme. It emerges automatically that the phase factor accompanying spin exchange with the transported basis is just the Pauli sign, that is $(-1)^{2S}$. Singlevaluedness of the total wavefunction, involving the transported basis, then implies the correct relation between spin and statistics. The Pauli sign is a geometric phase factor of topological origin, associated with non-contractible circuits in the doubly connected (and non-orientable) configuration space of relative positions with identified antipodes. The theory extends to more than two particles.

1. Introduction

The status of the relation between the spin and the statistics of identical particles in non-relativistic quantum mechanics has been unsatisfactory, for three reasons. First, 'It appears to be one of the few places in physics where there is a rule which can be stated very simply, but for which no one has found a simple and easy explanation.... This probably means that we do not have a complete understanding of the fundamental principle involved' (Feynman *et al.* 1965, pp. 3–4). Second, it represents a departure from the principle in the simplest form of quantum mechanics that indistinguishable situations (e.g. positions with angle coordinates ϕ and $\phi + 2\pi$) are described by the same wavefunction: for identical fermions, the exchanged and unexchanged states differ by a sign. (Of course it is always possible to use multivalued wavefunctions, but at the price of introducing gauge potentials into the Hamiltonian.) Third, the existing proofs incorporate ideas beyond elementary quantum mechanics: relativistic quantum field theory (Pauli 1940; Streater & Wightman 1964), topological solitons (Finkelstein & Rubinstein 1968; Mickelson 1984), or the existence of antiparticles (Tscheuschner 1989; Balachandran *et al.* 1993).

Here we will argue that 'the fundamental principle' that Feynman sought is the

Proc. R. Soc. Lond. A (1997) **453**, 1771–1790
Printed in Great Britain TEX Paper

Reprinted from Proceedings of the Royal Society A, Volume 453, (1997).

correct incorporation of identity into an augmented quantum kinematics in which the space of wavefunctions has, built into it, the indistinguishability of states related by exchange of positions and spins. When this is done, the physics of exchange emerges naturally from the non-relativistic Schrödinger equation with singlevalued wavefunctions (§ 3). The construction of a configuration space in which exchanged configurations are identified (e.g. the points $(\boldsymbol{r}_1, \boldsymbol{r}_2)$ and $(\boldsymbol{r}_2, \boldsymbol{r}_1)$) is not a new idea (Leinaas & Myrheim 1977; Laidlaw & DeWitt 1971), but we complete it by incorporating spin (§§ 2, 4). In order to accomplish this, the spin must be embedded into a larger Hilbert space. The unexpected result is that for particles with spin quantum number S the exchange ('Pauli') sign $(-1)^{2S}$ emerges automatically, as an unfamiliar type of geometric phase (§ 5).

Our essentially three-dimensional argument was inspired by, and can be regarded as a mathematization of, a well-known geometrical trick with a belt (Hartung 1979; Feynman 1987; Gould 1995; Guerra & Marra 1984), suggestive of the Pauli principle. Consider first a single object, tied to one end of a belt, with the other end held fixed. If the object is turned by 4π (about any axis) this introduces a double twist in the belt, which can, however, be eliminated by translating the object with its orientation fixed. Such 'tethered' rotations, where an even number of turns is equivalent to no turns but an odd number is not, are regarded as analogous to fermions, where the wavefunction changes sign after one turn. (Ordinary untethered objects, for which any number of turns is equivalent to no turns, are analogous to bosons.) Now consider two objects, tethered to each other by a belt. Exchanging them introduces a twist into the belt, which can be eliminated by turning one of them once. This suggests that exchange of identical fermions is equivalent to a single turn of one of them, that is to a sign change.

We consider the essence of the connection between spin and statistics to lie in the exchange of two particles, and our argument will be presented in detail for this case. However, the central ideas generalize to permutations of N particles, as will be explained in § 6.

2. Transported spin basis

The wavefunction for two identical particles with spin S depends on their positions $\boldsymbol{r}_1$ and $\boldsymbol{r}_2$. Exchange involves the vector $\boldsymbol{r} \equiv \boldsymbol{r}_2 - \boldsymbol{r}_1$ of relative position—it is unnecessary to consider the centre of mass vector. Under exchange of positions, $\boldsymbol{r}$ becomes $-\boldsymbol{r}$. The spins must be exchanged as well. We will describe the $(2S+1)^2$ spin states of the two particles with the quantum numbers m_1 and m_2 ($|m_1, m_2| \leqslant S$) representing their spin components in the z-direction, and employ the convenient notations

$$M \equiv \{m_1, m_2\}, \quad \bar{M} \equiv \{m_2, m_1\}. \tag{2.1}$$

To incorporate spin exchange, we employ an $\boldsymbol{r}$-dependent ('transported') basis $|M(\boldsymbol{r})\rangle$ rotated (in a sense to be described in § 4) from the commonly employed fixed basis $|M\rangle$. Thus

$$|M(\boldsymbol{r})\rangle = \mathsf{U}(\boldsymbol{r})|M\rangle, \tag{2.2}$$

where $\mathsf{U}(\boldsymbol{r})$ is a unitary operator. The notations $|M(\boldsymbol{r})\rangle$ and $\mathsf{U}(\boldsymbol{r})$ imply that the basis is uniquely determined by $\boldsymbol{r}$; we are therefore excluding the inconvenience of multivalued bases.

We require the basis (2.2) to possess the following properties.

(a) Smoothness: the basis must be a smooth and non-singular function for all $\boldsymbol{r} \neq 0$, e.g. there must be no Dirac strings.

(b) Exchange:

$$|M(-\boldsymbol{r})\rangle = (-1)^K |\bar{M}(\boldsymbol{r})\rangle, \tag{2.3}$$

where K is an integer. It is possible to envisage more general phase factors, but the restriction to a sign is forced by the fact that the basis is uniquely determined by $\boldsymbol{r}$, applied after a double exchange. The sign cannot depend on $\boldsymbol{r}$, because this would imply discontinuities (where the sign switches), and it cannot depend on M because (as an easy argument shows) this would imply a preferred quantization axis: for example, eigenstates of the x-component of spin would not be exchanged.

(c) Parallel transport:

$$\langle M'(\boldsymbol{r})|\nabla M(\boldsymbol{r})\rangle = 0 \tag{2.4}$$

for arbitrary values of the quantum numbers M, M'. This is the simplest rule for the transport of spins. With parallel transport, and the fact that the basis is a function of $\boldsymbol{r}$, there are no local geometric phases (abelian or non-abelian) associated with contractible circuits of $\boldsymbol{r}$.

It is important to emphasize at this point that parallel transport rules out the possibility of constructing operators $\mathsf{U}(r)$ that generate exchange according to (2.2) and (2.3) by using just the usual spin operators $\mathbf{S}_1$ and $\mathbf{S}_2$. The reason is that then the states $|M(\boldsymbol{r})\rangle$ in (2.2) would span the whole Hilbert space of spins, and (2.4) would imply that $\mathsf{U}(r)$ is constant. Therefore we will need to work with a representation of spin that incorporates the usual $(2S+1)^2$-dimensional one but is larger, in that it allows additional operations that can generate exchange while preserving parallel transport ('flat exchange'). In § 4 we shall construct a transported basis that satisfies these requirements. It is far from obvious *a priori* that this can be done—for example, it is impossible to satisfy the analogue of a for the eigenstates $|m(\boldsymbol{r})\rangle$ of $\boldsymbol{r}\cdot\mathbf{S}$ representing the ordinary rotation of a single spin. Having constructed the basis we will derive (in § 4) the centrally important relation

$$K = 2S, \quad \text{i.e. } |\bar{M}(-\boldsymbol{r})\rangle = (-1)^{2S}|M(\boldsymbol{r})\rangle. \tag{2.5}$$

In a sense this encapsulates the belt trick, because it asserts that $\boldsymbol{r} \to -\boldsymbol{r}$ and $M \to \bar{M}$, that is exchange of positions and spins, is equivalent to the sign change $(-1)^{2S}$ from the rotation of one spin.

3. Identifications

With the spins thus attached to the positions, we can represent any spin state $|\Psi(\boldsymbol{r})\rangle$ of the two particles by the $(2S+1)^2$-dimensional vector $\psi_M(\boldsymbol{r})$, where

$$|\Psi(\boldsymbol{r})\rangle = \sum_{m_1,m_2} \psi_M(\boldsymbol{r})|M(\boldsymbol{r})\rangle. \tag{3.1}$$

Now we must identify the points $\boldsymbol{r}$ and $-\boldsymbol{r}$, since these correspond to complete interchange of the particles (positions and spins) and so are indistinguishable. Singlevaluedness of the wavefunction—applied here as elsewhere in quantum mechanics—requires

$$|\Psi(\boldsymbol{r})\rangle = |\Psi(-\boldsymbol{r})\rangle. \tag{3.2}$$

Now, from (2.5),

$$|\Psi(-\boldsymbol{r})\rangle = \sum_M \psi_M(-\boldsymbol{r})(-1)^{2S}|\bar{M}(\boldsymbol{r})\rangle$$
$$= \sum_M \psi_{\bar{M}}(-\boldsymbol{r})(-1)^{2S}|M(\boldsymbol{r})\rangle, \tag{3.3}$$

so that singlevaluedness implies

$$\psi_{\bar{M}}(-\boldsymbol{r}) = (-1)^{2S}\psi_M(\boldsymbol{r}). \tag{3.4}$$

This resembles the usual spin-statistics relation. However, before we can assert that it is the same as the usual relation we must show that the coefficients

$$\psi_M(\boldsymbol{r}) = \langle M(\boldsymbol{r})|\,\Psi(\boldsymbol{r})\rangle, \tag{3.5}$$

in (3.1) satisfy the same Schrödinger equation as the coefficients in the fixed basis, where the wavefunctions are

$$|\,\Psi(\boldsymbol{r})\rangle_{\text{fixed}} = \sum_M \psi_{M,\text{fixed}}(\boldsymbol{r})|M\rangle. \tag{3.6}$$

Thus

$$\psi_{M,\text{fixed}}(\boldsymbol{r}) = \langle M|\,\Psi(\boldsymbol{r})\rangle_{\text{fixed}}. \tag{3.7}$$

In order to show that the quantities defined by (3.5) and (3.7) are the same, we must first define dynamical variables (e.g. momentum and spin) in the transported basis. Of course, these must satisfy the same commutation relations as in the fixed basis. This can be accomplished by generating the transported dynamical variables from their counterparts in the fixed basis by the same unitary transformation $\mathsf{U}(\boldsymbol{r})$ that generates the basis itself. In particular, the momentum operator, which in the fixed basis has the usual form

$$\mathbf{P}_{\text{fixed}} = -\mathrm{i}\hbar\nabla, \tag{3.8}$$

becomes, in the transported basis,

$$\mathbf{P}(\boldsymbol{r}) = \mathsf{U}(\boldsymbol{r})\mathbf{P}_{\text{fixed}}\mathsf{U}^\dagger(\boldsymbol{r}). \tag{3.9}$$

Similarly, the transported spin operators $\mathbf{S}(\boldsymbol{r})$ $(= \{\mathbf{S}_1, \mathbf{S}_2\})$ will be defined in terms of the usual fixed spins $\mathbf{S}_{\text{fixed}}$ by

$$\mathbf{S}(\boldsymbol{r}) = \mathsf{U}(\boldsymbol{r})\mathbf{S}_{\text{fixed}}\mathsf{U}^\dagger(\boldsymbol{r}). \tag{3.10}$$

$\mathbf{P}(\boldsymbol{r})$ and $\mathbf{S}(\boldsymbol{r})$ must be employed instead of $\mathbf{P}_{\text{fixed}}$ and $\mathbf{S}_{\text{fixed}}$ when constructing the Hamiltonian to express the Schrödinger equation in the transported basis. For the momenta, an easy calculation leads to

$$\langle M(\boldsymbol{r})\mid \mathbf{P}(\boldsymbol{r})|\,\Psi(\boldsymbol{r})\rangle = -\mathrm{i}\hbar\nabla\psi_M(\boldsymbol{r}), \tag{3.11}$$

while of course

$$\langle M\mid \mathbf{P}_{\text{fixed}}|\,\Psi(\boldsymbol{r})\rangle_{\text{fixed}} = -\mathrm{i}\hbar\nabla\psi_{M,\text{fixed}}(\boldsymbol{r}) \tag{3.12}$$

and similarly for spins. Therefore the 'transported' and 'fixed' quantities defined by (3.5) and (3.7) do satisfy the same Schrödinger equation (including boundary conditions) and so are the same function. It follows that (3.4) is indeed the usual spin-statistics relation, here derived by requiring the wavefunction to be singlevalued.

In effect, we have shown that although $|\Psi(\boldsymbol{r})\rangle_{\mathsf{fixed}}$ need not be singlevalued under exchange, $|\Psi(\boldsymbol{r})\rangle = \mathsf{U}(\boldsymbol{r})|\Psi(\boldsymbol{r})\rangle_{\mathsf{fixed}}$ must be.

Mathematically, what we are doing is setting up quantum mechanics on a 'two-spin bundle', whose six-dimensional base is the configuration space $\boldsymbol{r}_1$, $\boldsymbol{r}_2$ with exchanged configurations identified and coincidences $\boldsymbol{r}_1 = \boldsymbol{r}_2$ excluded (to make the base a manifold). The fibres are the two-spin Hilbert spaces spanned by the transported basis $|M(\boldsymbol{r})\rangle$. The full Hilbert space consists of global sections of the bundle, i.e. singlevalued wavefunctions. The base manifold has non-trivial topology. It can be regarded as the product of the centre of mass with the space of relative coordinates $\boldsymbol{r}$, with the latter parametrized by the separation distance $r \equiv |\boldsymbol{r}| > 0$ and a point on the projective plane (two spheres with identified antipodes) of relative directions. Exchanges of positions correspond to non-contractible closed loops in this (non-orientable) configuration space.

4. Exchange rotation

As we saw in § 2, we need to construct an enlarged representation of spin that incorporates the usual one but allows additional operations that can generate flat exchange. To the extent that these additional operations are unphysical (e.g. by allowing the spins of the two particles to differ) they must be unobservable: their only role is to accomplish the exchange.

This can be achieved, and the exchange sign calculated, by adapting Schwinger's representation of spin in terms of harmonic oscillators (Schwinger 1965; Sakurai 1994). For a single spin, two independent (that is, commuting) oscillators are required—the a oscillator, with annihilation and creation operators a and $\mathsf{a}^\dagger$, and the b oscillator, with b and $\mathsf{b}^\dagger$. From these can be constructed $\mathbf{S} = (\mathsf{S}_x, \mathsf{S}_y, \mathsf{S}_z)$:

$$\mathbf{S} = \tfrac{1}{2}(\mathsf{a}^\dagger \mathsf{b}^\dagger)\boldsymbol{\sigma}\begin{pmatrix}\mathsf{a}\\ \mathsf{b}\end{pmatrix},$$

$$\text{i.e. } \mathsf{S}_z = \tfrac{1}{2}(\mathsf{a}^\dagger\mathsf{a} - \mathsf{b}^\dagger\mathsf{b}), \tag{4.1}$$

$$\mathsf{S}_+ \equiv \mathsf{S}_x + \mathrm{i}\mathsf{S}_y = \mathsf{a}^\dagger\mathsf{b}, \quad \mathsf{S}_- \equiv \mathsf{S}_x - \mathrm{i}\mathsf{S}_y = \mathsf{b}^\dagger\mathsf{a},$$

where $\boldsymbol{\sigma}$ denotes the vector of Pauli matrices. The components of $\mathbf{S}$ satisfy the commutation rules for angular momentum:

$$\mathbf{S} \times \mathbf{S} = \mathrm{i}\mathbf{S}. \tag{4.2}$$

(Here and hereafter we omit $\hbar$ in expressions involving spins, so that, for example, the eigenvalues of S_z are spin quantum numbers—integer or half-integer—rather than dynamical spins.) In this representation, the eigenstates of $\mathbf{S}^2$ and S_z, with quantum numbers S and m, are number states of the oscillators: if there are n_a quanta in the a oscillator and n_b quanta in the b oscillator, then it follows from (4.1) that

$$S = \tfrac{1}{2}(n_a + n_b), \quad m = \tfrac{1}{2}(n_a - n_b). \tag{4.3}$$

For two spins, we require four oscillators: a_1, b_1, a_2, b_2. The individual spin operators $\mathbf{S}_1$ and $\mathbf{S}_2$ are constructed by analogy with (4.1). To create exchange, we mix the 1 and 2 oscillators rather than the a and b oscillators. The rationale behind this is that since an ordinary spin rotation from z to $-z$, generated by $\mathbf{S}$, changes the sign of the m quantum number and so, by (4.3), interchanges the quanta in the a

and b oscillators, so rotations generated by operators where a and b are replaced by 1 and 2 will interchange the spins. There are two ways of mixing 1 and 2, involving the a and b operators separately, yielding the operator $\mathbf{E}_a$, given by

$$\mathbf{E}_a = \tfrac{1}{2}(\mathsf{a}_1^\dagger \mathsf{a}_2^\dagger)\boldsymbol{\sigma}\begin{pmatrix}\mathsf{a}_1\\ \mathsf{a}_2\end{pmatrix},$$

$$\text{i.e. } \mathsf{E}_{az} = \tfrac{1}{2}(\mathsf{a}_1^\dagger \mathsf{a}_1 - \mathsf{a}_2^\dagger \mathsf{a}_2), \tag{4.4}$$

$$\mathsf{E}_{a+} \equiv \mathsf{E}_{ax} + \mathrm{i}\mathsf{E}_{ay} = \mathsf{a}_1^\dagger \mathsf{a}_2, \quad \mathsf{E}_{a-} \equiv \mathsf{E}_{ax} - \mathrm{i}\mathsf{E}_{ay} = \mathsf{a}_2^\dagger \mathsf{a}_1$$

and similarly the operator $\mathbf{E}_b$. Obviously $[\mathbf{E}_a, \mathbf{E}_b] = 0$. The components of $\mathbf{E}_a$ satisfy angular momentum commutation relations, as do those of $\mathbf{E}_b$. The linear combination

$$\mathbf{E} = \mathbf{E}_a + \mathbf{E}_b \tag{4.5}$$

uniquely shares this property, namely

$$\mathbf{E} \times \mathbf{E} = \mathrm{i}\mathbf{E}. \tag{4.6}$$

Moreover, by elementary calculations, it can be shown that

$$[\mathsf{E}_z, \mathbf{S}_1] = 0, \quad [\mathsf{E}_z, \mathbf{S}_2] = 0, \quad [\mathbf{E}, \mathbf{S}_{\text{tot}}] = 0, \tag{4.7}$$

where $\mathbf{S}_{\text{tot}} = \mathbf{S}_1 + \mathbf{S}_2$. However, $\mathbf{S}_1$ and $\mathbf{S}_2$ do not commute with $\mathsf{E}_\pm = \mathsf{E}_x \pm \mathrm{i}\mathsf{E}_y$, and nor do S_1^2 and S_2^2. In addition, one can show that

$$\mathbf{E}^2 = \mathbf{S}_{\text{tot}}^2. \tag{4.8}$$

We will call $\mathbf{E}$ *exchange angular momentum*, because the group of rotations it generates—*exchange rotations*—can be chosen to satisfy the requirement (2.3). We begin with a simple geometrically motivated construction: as the line joining the particles is turned from $\boldsymbol{e}_z$ to $\boldsymbol{r}$, the two-spin state is turned by the corresponding exchange rotation. A symmetrical choice for the turn is about an axis $\boldsymbol{n}(\boldsymbol{r})$ perpendicular to both $\boldsymbol{e}_z$ and $\boldsymbol{r}$. If (θ, ϕ) are the polar angles of the direction of $\boldsymbol{r}$, then

$$\boldsymbol{n}(\boldsymbol{r}) = -\boldsymbol{e}_x \sin\phi + \boldsymbol{e}_y \cos\phi. \tag{4.9}$$

With this choice, we let the transported basis be generated by

$$\mathsf{U}(\boldsymbol{r}) = \exp\{-\mathrm{i}\theta \boldsymbol{n}(\boldsymbol{r}) \cdot \mathbf{E}\}. \tag{4.10}$$

U acts on an enlarged space of spin states, whose dimension, the number of ways the $4S$ quanta can be distributed among the four oscillators ($\mathbf{E}$ leaves $\mathbf{S}_{\text{tot}}$ invariant), is $\frac{1}{6}(4S+1)(4S+2)(4S+3)$.

Now consider the state of the two spins corresponding to $n_{1a}, n_{1b}, n_{2a}, n_{2b}$, namely

$$|n_{1a}, n_{2a}, n_{1b}, n_{2b}\rangle = C(\mathsf{a}_1^\dagger)^{n_{1a}}(\mathsf{a}_2^\dagger)^{n_{2a}}(\mathsf{b}_1^\dagger)^{n_{1b}}(\mathsf{b}_2^\dagger)^{n_{2b}}|0,0,0,0\rangle, \tag{4.11}$$

where C is a normalization constant. By an obvious extension of (4.3), this can be written in terms of the more familiar quantum numbers for the individual spins:

$$|n_{1a}, n_{2a}, n_{1b}, n_{2b}\rangle = |S_1 + m_1, S_2 + m_2, S_1 - m_1, S_2 - m_2\rangle \equiv |S_1, S_2; M\rangle. \tag{4.12}$$

It then follows from (4.4) and (4.5) that

$$\mathsf{E}_z|S_1, S_2; M\rangle = (S_1 - S_2)|S_1, S_2; M\rangle. \tag{4.13}$$

For identical spins, $S_1 = S_2 = S$, and when S need not be written explicitly we will revert to our previous notation, namely

$$|S_1, S_2; M\rangle \equiv |M\rangle. \tag{4.14}$$

From (4.13), we find, for identical spins, the important result

$$\mathsf{E}_z|M\rangle = 0. \tag{4.15}$$

This ensures gauge invariance of U: any ineffective rotations about the z-axis, applied before U is used to generate the transported basis from the fixed basis, will not introduce phase factors, since

$$\exp(-\mathrm{i}\alpha(\boldsymbol{r})\mathsf{E}_z)|M\rangle = |M\rangle. \tag{4.16}$$

Now we have to show that U does indeed generate spin exchange according to (2.3). To evaluate the effect of the exchange rotation operator (4.10), we first note that

$$\mathsf{U}(\boldsymbol{r}) = \exp\{-\mathrm{i}\theta\boldsymbol{n}(\boldsymbol{r})\cdot\mathbf{E}_a\}\exp\{-\mathrm{i}\theta\boldsymbol{n}(\boldsymbol{r})\cdot\mathbf{E}_b\} \equiv \mathsf{U}_a(\boldsymbol{r})\mathsf{U}_b(\boldsymbol{r}) \tag{4.17}$$

and U_a and U_b commute, so we can consider their actions separately. For this we use the fact that in the Schwinger representation the actions of U_a and U_b on states of arbitrary spin can be evaluated in terms of 2×2 matrices multiplying the vectors of creation operators. Thus $\mathsf{U}_a(\boldsymbol{r})$ induces the transformation

$$\begin{aligned}(\mathsf{a}_1^\dagger\mathsf{a}_2^\dagger) \to (\mathsf{a}_1'^\dagger\mathsf{a}_2'^\dagger) &= \mathsf{U}_a(\boldsymbol{r})(\mathsf{a}_1^\dagger\mathsf{a}_2^\dagger)\mathsf{U}_a^\dagger(\boldsymbol{r})\\ &= (\mathsf{a}_1^\dagger\mathsf{a}_2^\dagger)\exp(-\tfrac{1}{2}\mathrm{i}\theta\boldsymbol{n}(\boldsymbol{r})\cdot\boldsymbol{\sigma})\\ &= (\mathsf{a}_1^\dagger\mathsf{a}_2^\dagger)\begin{pmatrix}\cos\frac{1}{2}\theta & -\exp(-\mathrm{i}\phi)\sin\frac{1}{2}\theta\\ \exp(\mathrm{i}\phi)\sin\frac{1}{2}\theta & \cos\frac{1}{2}\theta\end{pmatrix}\end{aligned} \tag{4.18}$$

and similarly for U_b.

It is instructive first to allow $\mathsf{U}(\boldsymbol{r})$ to act on the general number state (4.11), where the spins need not be the same. From (4.17) and (4.18),

$$\begin{aligned}&(\mathsf{a}_1^\dagger)^{n_{1a}}(\mathsf{a}_2^\dagger)^{n_{2a}}(\mathsf{b}_1^\dagger)^{n_{1b}}(\mathsf{b}_2^\dagger)^{n_{2b}}\\ &\to (\cos\tfrac{1}{2}\theta\,\mathsf{a}_1^\dagger + \exp(\mathrm{i}\phi)\sin\tfrac{1}{2}\theta\,\mathsf{a}_2^\dagger)^{n_{1a}} \times (-\exp(-\mathrm{i}\phi)\sin\tfrac{1}{2}\theta\,\mathsf{a}_1^\dagger + \cos\tfrac{1}{2}\theta\,\mathsf{a}_2^\dagger)^{n_{2a}}\\ &\quad\times(\cos\tfrac{1}{2}\theta\,\mathsf{b}_1^\dagger + \exp(\mathrm{i}\phi)\sin\tfrac{1}{2}\theta\,\mathsf{b}_2^\dagger)^{n_{1b}} \times (-\exp(-\mathrm{i}\phi)\sin\tfrac{1}{2}\theta\,\mathsf{b}_1^\dagger + \cos\tfrac{1}{2}\theta\,\mathsf{b}_2^\dagger)^{n_{2b}}.\end{aligned} \tag{4.19}$$

Similarly, under the action of $\mathsf{U}(-\boldsymbol{r})$, where θ is replaced by $\pi-\theta$ and ϕ by $\phi+\pi$,

$$\begin{aligned}&(\mathsf{a}_1^\dagger)^{n_{1a}}(\mathsf{a}_2^\dagger)^{n_{2a}}(\mathsf{b}_1^\dagger)^{n_{1b}}(\mathsf{b}_2^\dagger)^{n_{2b}}\\ &\to (\sin\tfrac{1}{2}\theta\,\mathsf{a}_1^\dagger - \exp(\mathrm{i}\phi)\cos\tfrac{1}{2}\theta\,\mathsf{a}_2^\dagger)^{n_{1a}} \times (\exp(-\mathrm{i}\phi)\cos\tfrac{1}{2}\theta\,\mathsf{a}_1^\dagger + \sin\tfrac{1}{2}\theta\,\mathsf{a}_2^\dagger)^{n_{2a}}\\ &\quad\times(\sin\tfrac{1}{2}\theta\,\mathsf{b}_1^\dagger - \exp(\mathrm{i}\phi)\cos\tfrac{1}{2}\theta\,\mathsf{b}_2^\dagger)^{n_{1b}} \times (\exp(-\mathrm{i}\phi)\cos\tfrac{1}{2}\theta\,\mathsf{b}_1^\dagger + \sin\tfrac{1}{2}\theta\,\mathsf{b}_2^\dagger)^{n_{2b}}.\end{aligned} \tag{4.20}$$

Common factors in (4.19) and (4.20) can be identified by pulling out phase factors and there follows, in an obvious extension of our previous notation for the transported basis,

$$\begin{aligned}|n_{1a}, n_{2a}, n_{1b}, n_{2b}(\boldsymbol{r})\rangle &\equiv \exp\{-\mathrm{i}\theta\boldsymbol{n}(\boldsymbol{r})\cdot\mathbf{E}\}|n_{1a}, n_{2a}, n_{1b}, n_{2b}\rangle\\ &= (-1)^{n_{2a}+n_{2b}}(\exp(\mathrm{i}\phi))^{(-n_{2a}-n_{2b}+n_{1a}+n_{1b})}|n_{2a}, n_{1a}, n_{2b}, n_{1b}(-\boldsymbol{r})\rangle,\end{aligned} \tag{4.21}$$

showing that the operator $\mathbf{E}$ does indeed generate exchange of spins.

An alternative way of writing (4.21) is (cf. equation (4.12))

$$|S_1, S_2; M(\boldsymbol{r})\rangle = (-1)^{2S_2} \exp\{2\mathrm{i}(S_1 - S_2)\phi\}|S_2, S_1; \bar{M}(-\boldsymbol{r})\rangle, \tag{4.22}$$

from which the desired exchange relation (2.3) follows at once on setting $S_1 = S_2 = S$, as does the Pauli sign (2.5). This sign can be regarded as arising from the rotation of the second spin by 2π (if $\phi = 0$), or equivalently from the rotation of the first spin by 2π (if $\phi = \pi$), or equivalently from the rotation of both spins by π (as in the simplest version of the argument relating exchange to spin rotation (Feynman 1987)).

Similar techniques establish that the transported basis $|M(\boldsymbol{r})\rangle$ is a smooth function of $\boldsymbol{r}$, notwithstanding the Dirac string singularity in $\mathsf{U}(\boldsymbol{r})$ at the south pole, arising from the half-angles in (4.19) (details are given in Appendix A), and that it is parallel-transported according to (2.4) (details are given in Appendix B). Therefore all the requirements laid down in § 2 are satisfied by the transported basis generated by the exchange angular momentum $\mathbf{E}$ defined by (4.4) and (4.5).

In the foregoing analysis, the Pauli sign was derived using the particular choice (4.10) of $\mathsf{U}(\boldsymbol{r})$. In fact, this sign is a consequence of any exchange rotation satisfying the conditions (a) and (b) in § 2 (the parallel-transport condition is automatically satisfied); the argument is given in Appendix C. Thus the implications of the conditions are essentially topological. (An example of a more general exchange rotation is (4.10), multiplied on the left by a smooth exchange rotation even in $\boldsymbol{r}$.)

As we have seen, parallel transport implies that the spin space must be enlarged. If such enlargement is abandoned (and with it flat exchange), a 'counterconstruction' can be devised, satisfying the conditions a and b in § 2 but leading to bizarre relations between spin and statistics, that depend on the value of spin in the four classes $S(\text{mod}\, 2)$. The counterconstruction is described in Appendix D.

Now we elucidate the meaning of the relation (4.15) by writing it in the transported basis. Since

$$\mathbf{E} \cdot \boldsymbol{r}/r = \mathsf{U}(\boldsymbol{r})\mathsf{E}_z\mathsf{U}^\dagger(\boldsymbol{r}), \tag{4.23}$$

(4.15) implies

$$\mathbf{E} \cdot \boldsymbol{r}|M(\boldsymbol{r})\rangle = 0. \tag{4.24}$$

Therefore $\mathbf{E} \cdot \boldsymbol{r}$ annihilates the transported basis states, that is, the transported basis lies in the subspace of eigenvectors of $\mathbf{E} \cdot \boldsymbol{r}$ with eigenvalue zero (indeed, this property could have been used to define the transported basis). Eigenvectors where this eigenvalue is not zero correspond to unphysical spin states with $S_1 \neq S_2$. The definition of transported operators (e.g. equations (3.9) and (3.10)) ensure that these unphysical states never arise during quantum exchanges. The relation (4.24) also ensures that the transported operators are invariant under gauge transformations generated by exchange rotations about the z-axis: if $\mathsf{U}(\boldsymbol{r})$ is preceded by the exchange rotation in (4.16), $\mathbf{P}(\boldsymbol{r})$ transforms as

$$\mathbf{P}(\boldsymbol{r}) \to \mathbf{P}(\boldsymbol{r}) + \hbar\nabla\alpha(\boldsymbol{r})\mathbf{E} \cdot \boldsymbol{r}/r \tag{4.25}$$

and the additional term vanishes for physical states $|M(\boldsymbol{r})\rangle$.

We note that in ordinary spin rotation, generated by $\mathbf{S}$, the states analogous to the transported spin basis are the $m = 0$ eigenstates of the spin along $\boldsymbol{r}$, that is of $\mathbf{S} \cdot \boldsymbol{r}$; in the next section we will have more to say about this analogy.

5. Geometric phases

To see the relation between the way in which we treat quantum identity and the conventional way, consider the coefficents $\psi_{M,\text{fixed}}(\boldsymbol{r})$ (equation (3.7)) of the fixed-basis wavefunction $|\Psi(\boldsymbol{r})\rangle_{\text{fixed}}$ (equation (3.6)). We saw in § 3 that these coefficients are the same as the coefficients $\psi_M(\boldsymbol{r})$ in the transported-basis wavefunction $|\Psi(\boldsymbol{r})\rangle$ (equation (3.1)). According to (3.4) , then, $\psi_{M,\text{fixed}}(\boldsymbol{r})$ acquires the familiar Pauli sign under exchange of positions and spins, that is $\boldsymbol{r} \to -\boldsymbol{r}$, $M \to \bar{M}$; this is the usual formulation of spin and statistics. In the transported basis, however, the sign change of the coefficients is compensated by the sign change (2.5) of the transported basis, and the total wave $|\Psi(\boldsymbol{r})\rangle$ is singlevalued.

With this observation, the Pauli sign $(-1)^{2S}$ is revealed as a geometric phase factor, arising from parallel transport generated by exchange rotation. As with all geometric phases (Berry 1984; Shapere & Wilczek 1989), this sign is the result of dividing a system into two parts; here they are space and spin, and the space wavefunctions $\psi_M(\boldsymbol{r})$ inherit the phase factor from the transported spin kets, to keep the total state singlevalued. An analogous phenomenon is molecular pseudorotation (Longuet-Higgins *et al.* 1959; Delacrétaz *et al.* 1986), where a geometric sign change in electronic wavefunctions, when transported round a cycle of the nuclear coordinates that encloses an electronic degeneracy P, forces the nuclear wavefunctions to change sign when continued around P.

However, the geometric phase in particle exchange does not arise in the familiar way, from the line integral of a vector potential or the flux of a two-form, because these local quantities are zero (equation (2.4)). Rather, the phase is of a different kind: global, and associated with non-contractible circuits in the doubly connected (and non-orientable) configuration space. As we saw at the end of the last section, the transported basis states are analogous to the transported $m = 0$ states in ordinary spin rotation, and these too are known to possess spin-dependent geometric sign changes $(-1)^j$, where j is the total spin (Robbins & Berry 1994) for non-contractible circuits in the projective plane, with observable consequences.

Mathematically, the familiar geometric phase (produced for example by causing a spin to turn in a cycle) is associated with the Chern class, that is with monopole singularities of the two-form whose flux is the phase. On the other hand, the exchange phase is associated with the first Stiefel–Whitney class (Milnor & Stasheff 1974; Nash & Sen 1983; Nakahara 1990) of the two-spin bundle for non-contractible loops in the doubly connected space that incorporates identified states.

To explore the connection with the conventional formulation in more detail, we write the state in the transported basis explicitly for two spin-$\frac{1}{2}$ particles. Denoting quantum numbers $\pm\frac{1}{2}$ by $\pm$, (3.1) is

$$|\Psi(\boldsymbol{r})\rangle = \psi_{++}(\boldsymbol{r})|++(\boldsymbol{r})\rangle + \psi_{+-}(\boldsymbol{r})|+-(\boldsymbol{r})\rangle + \psi_{-+}(\boldsymbol{r})|-+(\boldsymbol{r})\rangle + \psi_{--}(\boldsymbol{r})|--(\boldsymbol{r})\rangle. \tag{5.1}$$

As we have seen, the requirement that this be singlevalued leads to the conditions (3.4), which we can write as follows, using E to denote even functions and O to denote odd functions:

$$\left.\begin{aligned} \psi_{++}(\boldsymbol{r}) &= -\psi_{++}(-\boldsymbol{r}) \equiv O_1(\boldsymbol{r}), \\ \psi_{--}(\boldsymbol{r}) &= -\psi_{--}(-\boldsymbol{r}) \equiv O_{-1}(\boldsymbol{r}), \\ \psi_{+-}(\boldsymbol{r}) &= -\psi_{-+}(-\boldsymbol{r}) \equiv E(\boldsymbol{r}) + O_0(\boldsymbol{r}), \\ \psi_{-+}(\boldsymbol{r}) &\equiv -E(\boldsymbol{r}) + O_0(\boldsymbol{r}). \end{aligned}\right\} \tag{5.2}$$

Thus

$$|\Psi(\boldsymbol{r})\rangle = O_1(\boldsymbol{r})|++(\boldsymbol{r})\rangle + O_{-1}(\boldsymbol{r})|--(\boldsymbol{r})\rangle$$
$$+O_0(\boldsymbol{r})(|+-(\boldsymbol{r})\rangle + |-+(\boldsymbol{r})\rangle) + E(\boldsymbol{r})(|+-(\boldsymbol{r})\rangle - |-+(\boldsymbol{r})\rangle). \quad (5.3)$$

The conventional representation would employ the fixed basis (3.6), where the spin kets lack the $\boldsymbol{r}$ dependence. Since we have already seen that the coefficient functions, $\psi_{++}(\boldsymbol{r})$, etc., are the same in the fixed and transported bases, we can then recognise in (5.3), with transported kets replaced by fixed ones, the familiar decomposition into the triplet of states that are space-odd and spin-even, and the singlet state that is space-even and spin-odd. The traditional form of the exclusion principle follows from the vanishing of $|\Psi(\boldsymbol{r})\rangle$ when the two spin states are the same ($++$ or $--$) and the two positions are the same ($\boldsymbol{r}=0$). It follows from the exchange relation (2.5) that the transformation from the fixed to the transported bases reverses the parity of the basis states under spin exchange, thereby restoring the singlevaluedness of the total wavefunction $|\Psi(\boldsymbol{r})\rangle$.

Because the states $|m_1m_2(\boldsymbol{r})\rangle = |M(\boldsymbol{r})\rangle$ are not carried into themselves under $\boldsymbol{r}\to-\boldsymbol{r}$, but rather into $(-1)^{2S}|m_2m_1(\boldsymbol{r})\rangle$, the Pauli geometric phase factor $(-1)^{2S}$ is non-abelian. The alternative transported basis $|j\mu(\boldsymbol{r})\rangle$ of eigenstates of total spin $(\mathbf{S}_1+\mathbf{S}_2)^2$ and its z component $(\mathbf{S}_1+\mathbf{S}_2)_z$ (e.g. the triplet and singlet states of (5.3)) abelianizes the basis $|m_1m_2(\boldsymbol{r})\rangle$ that we have been using. The states $|j\mu(\boldsymbol{r})\rangle$ are related to $|m_1m_2(\boldsymbol{r})\rangle$ by the Clebsch–Gordan coefficients in the usual way. Unlike m_1 and m_2, j and μ are good quantum numbers under exchange rotations (cf. equation (4.7)), so that $|j\mu(\boldsymbol{r})\rangle$ is carried into itself under $\boldsymbol{r}\to-\boldsymbol{r}$. The geometric phase it acquires is just like the $m=0$ phase for ordinary spin; $|j\mu(\boldsymbol{r})\rangle$ is an eigenstate of $\boldsymbol{r}\cdot\mathbf{E}$, with eigenvalue zero, and is also (unlike $|m_1m_2(\boldsymbol{r})\rangle$) an eigenstate of $\mathbf{E}^2$, with eigenvalue $j(j+1)$ (cf. equation (4.8)), and thus acquires a sign $(-1)^j$ under exchange.

The relation between $(-1)^j$ and the Pauli sign is provided by the Clebsch–Gordan coefficients, which under exchange of spins change by $(-1)^{2S-j}$. This latter sign was conjectured by Leinaas & Myrheim (1977) to be implicated in the spin-statistics relation, which was completed by regarding the additional sign $(-1)^j$ as arising from the parity under position exchange of the spherical harmonics describing the states of total spin.

6. More than two particles

For N identical particles with spin S, it is convenient to represent the positions, and the quantum numbers representing the spin components in the z direction, by (cf. equation (2.1))

$$\boldsymbol{R}\equiv\{\boldsymbol{r}_1,\ldots,\boldsymbol{r}_N\},\quad \boldsymbol{M}\equiv\{m_1,\ldots,m_N\}. \quad (6.1)$$

The elements in these lists (e.g. $\boldsymbol{r}_2$, m_3) denote particle properties, and places in the lists denote particle labels. The spin $\mathbf{S}_i$ of each particle is represented by the pair a_i and b_i of Schwinger oscillators, in terms of whose operators the state $|\boldsymbol{M}\rangle$ is, in an obvious notation (cf. equation (4.11)),

$$|\boldsymbol{M}\rangle = C\prod_{i=1}^{N}(\mathsf{a}_i^\dagger)^{S+m_i}(\mathsf{b}_i^\dagger)^{S-m_i}|\mathbf{0}\rangle, \quad (6.2)$$

where C is a normalization constant.

For $N > 2$, we must consider general permutations of positions $\boldsymbol{R}$ and spins $\boldsymbol{M}$, not only two-particle exchanges. Permutations will be denoted by g, and we adopt the convention that properties of particle i are transferred to particle $g(i)$. Thus, for the three-particle permutation $g(1) = 3$, $g(2) = 1$, $g(3) = 2$,

$$|g\{m_1, m_2, m_3\}\rangle = |m_{g^{-1}(1)}, m_{g^{-1}(2)}, m_{g^{-1}(3)}\rangle = |m_2, m_3, m_1\rangle. \qquad (6.3)$$

The general transformation is conveniently expressed in terms of permutation matrices, that is

$$|g\boldsymbol{M}\rangle = |P_{1j}(g)m_j, P_{2j}(g)m_j, \ldots\rangle, \quad \text{where } P_{ij}(g) \equiv \delta_{i,g(j)}. \qquad (6.4)$$

Every g can be factored into a product of exchanges, and has a parity $\varepsilon(g)$, defined, independently of the chosen sequence, as 0 or 1 if the number of exchanges is even (i.e. $\det P_{ij} = +1$) or odd (i.e. $\det P_{ij} = -1$).

As in § 2 we represent the spin states with a transported basis $|\boldsymbol{M}(\boldsymbol{R})\rangle$ that depends on the positions, obtained (cf. equation (2.2)) from the fixed basis $|\boldsymbol{M}\rangle$ by

$$|\boldsymbol{M}(\boldsymbol{R})\rangle = \mathsf{U}(\boldsymbol{R})|\boldsymbol{M}\rangle. \qquad (6.5)$$

The basis must be a singlevalued and smooth function of $\boldsymbol{R}$, chosen so that the spins are permuted along with the positions, that is (cf. equation (2.3))

$$|g\boldsymbol{M}(g\boldsymbol{R})\rangle = (-1)^{K(g)}|\boldsymbol{M}(\boldsymbol{R})\rangle. \qquad (6.6)$$

It is possible to envisage more general phase factors (we do not consider parastatistics), but the restriction to a sign follows from the argument given after (2.3), and the fact that any g can be decomposed into exchanges. As also explained after (2.3), $K(g)$ is independent of $\boldsymbol{R}$ and $\boldsymbol{M}$. The basis must be parallel-transported, that is (cf. equation (2.4))

$$\langle \boldsymbol{M}'(\boldsymbol{R})|\nabla_{\boldsymbol{R}}\boldsymbol{M}(\boldsymbol{R})\rangle = 0. \qquad (6.7)$$

The unitary operator $\mathsf{U}(\boldsymbol{R})$ will be a 'permutation rotation', a generalization of exchange rotations to be defined later.

As we will show, any permutation rotation satisfying the conditions implies that the sign in (6.6) is

$$K(g) = 2S\varepsilon(g), \quad \text{i.e. } |g\boldsymbol{M}(g\boldsymbol{R})\rangle = (-1)^{2S\varepsilon(g)}|\boldsymbol{M}(\boldsymbol{R})\rangle, \qquad (6.8)$$

from which will follow the spin-statistics relation. There remains the problem of finding an explicit general construction, for all $N > 2$, of a $\mathsf{U}(\boldsymbol{R})$ that generates a smooth transported basis satisfying the above conditions, as we did for $N = 2$ with (4.9) and (4.10). This is a difficult problem (one reason being that permutations do not commute) and we have not solved it; in our view the difficulties are technical rather than fundamental. We do have an explicit construction for $N = 3$ (it is rather elaborate), and we envisage several possibilities for the general case. These will be reported more fully elsewhere; below, we formulate the mathematical problem involved in such constructions.

Given the sign (6.8), we proceed as in § 3, and represent any state $|\Psi(\boldsymbol{R})\rangle$ of the N particles by the $(2S+1)^N$-dimensional vector $\psi_{\boldsymbol{M}}(\boldsymbol{R})$ (cf. equation (3.1)), where

$$|\Psi(\boldsymbol{R})\rangle = \sum_{\boldsymbol{M}} \psi_{\boldsymbol{M}}(\boldsymbol{R})|\boldsymbol{M}(\boldsymbol{R})\rangle. \qquad (6.9)$$

With this representation we must identify the configurations $\boldsymbol{R}$ and $g\boldsymbol{R}$, since these correspond to a permutation of spins as well as positions and so are indistinguishable; therefore the wavefunction must be singlevalued. Imposing this condition, and using (6.8), gives

$$|\Psi(\boldsymbol{R})\rangle = |\Psi(g\boldsymbol{R})\rangle = \sum_{\boldsymbol{M}} \psi_{\boldsymbol{M}}(g\boldsymbol{R})|\boldsymbol{M}(g\boldsymbol{R})\rangle = \sum_{\boldsymbol{M}} \psi_{g\boldsymbol{M}}(g\boldsymbol{R})|g\boldsymbol{M}(g\boldsymbol{R})\rangle$$
$$= (-1)^{2S\varepsilon(g)} \sum_{\boldsymbol{M}} \psi_{g\boldsymbol{M}}(g\boldsymbol{R})|\boldsymbol{M}(\boldsymbol{R})\rangle. \tag{6.10}$$

From (6.9) now follows

$$\psi_{g\boldsymbol{M}}(g\boldsymbol{R}) = (-1)^{2S\varepsilon(g)}\psi_{\boldsymbol{M}}(\boldsymbol{R}). \tag{6.11}$$

By an identical argument to that following (3.4), this can be interpreted as the spin-statistics connection in its familiar form.

To obtain the sign (6.8), we write each permutation g as a product over L exchanges e_l:

$$g = e_L e_{L-1} \cdots e_1. \tag{6.12}$$

Applying these exchanges in sequence gives

$$|g\boldsymbol{M}(g\boldsymbol{R})\rangle = |e_L(e_{L-1}\cdots e_1\boldsymbol{M})(e_L(e_{L-1}\cdots e_1\boldsymbol{R}))\rangle$$
$$= (-1)^{K(e_L)}|e_{L-1}\cdots e_1\boldsymbol{M}(e_{L-1}\cdots e_1\boldsymbol{R})\rangle$$
$$= (-1)^{K(e_L)+K(e_{L-1})+\cdots+K(e_1)}|\boldsymbol{M}(\boldsymbol{R})\rangle, \tag{6.13}$$

where $K(e_l)$ is the sign associated with the lth exchange. From invariance under relabelling, all these signs are the same, so

$$(-1)^{K(g)} = (-1)^{\sum_{l=1}^{L} K(e_l)} = (-1)^{LK(e_1)} = (-1)^{K(e_1)\varepsilon(g)}. \tag{6.14}$$

At the end of Appendix C we show that $K(e_1) = 2S$, which is the sign obtained in § 4 for the exchange of two isolated particles; thence (6.8).

Before discussing the construction of the transported basis, we must define permutation rotations. Let $\mathsf{E}^{(ij)}$ be exchange angular momentum operators for the particle pairs i, j, defined (cf. equations (4.4) and (4.5)) in terms of the Schwinger oscillators a_i, b_i for each of the spins S_i by

$$\mathsf{E}^{(ij)} = \mathsf{E}_a^{(ij)} + \mathsf{E}_b^{(ij)}, \tag{6.15}$$

where

$$\mathsf{E}_a^{(ij)} = \tfrac{1}{2}(\mathsf{a}_i^\dagger \mathsf{a}_j^\dagger)\boldsymbol{\sigma}\begin{pmatrix}\mathsf{a}_i\\ \mathsf{a}_j\end{pmatrix},$$
$$\text{i.e. } \mathsf{E}_{az}^{(ij)} = \tfrac{1}{2}(\mathsf{a}_i^\dagger \mathsf{a}_i - \mathsf{a}_j^\dagger \mathsf{a}_j), \tag{6.16}$$
$$\mathsf{E}_{a+}^{(ij)} \equiv \mathsf{E}_{ax}^{(ij)} + \mathrm{i}\mathsf{E}_{ay}^{(ij)} = \mathsf{a}_i^\dagger \mathsf{a}_j, \quad \mathsf{E}_{a-}^{(ij)} \equiv \mathsf{E}_{ax}^{(ij)} - \mathrm{i}\mathsf{E}_{ay}^{(ij)} = \mathsf{a}_j^\dagger \mathsf{a}_i$$

and similarly for $\mathsf{E}_b^{(ij)}$. Permutation rotations are the unitary operators generated by the exchange angular momenta $\mathsf{E}^{(ij)}$, namely

$$\mathsf{U}(\boldsymbol{R}) = \exp\Big\{-\mathrm{i}\sum_{\mu<\nu=1}^{N} \boldsymbol{c}_{\mu\nu}(\boldsymbol{R})\cdot\mathsf{E}^{(\mu\nu)}\Big\}. \tag{6.17}$$

The parallel-transport requirement is automatically satisfied by the basis generated by this operator, as follows from an easy generalization of Appendix B. As with (4.10) U acts on an enlarged space of spin states; now the dimension is the number of ways the $2NS$ quanta can be distributed among the $2N$ oscillators, namely $[2N(S+1)-1]!/\{(2NS)!(2N-1)!\}$ (the dimension of the space of fixed spin states $|\boldsymbol{M}\rangle$ is $(2S+1)^N$).

Naturally associated with $\mathsf{U}(\boldsymbol{R})$ is the $N \times N$ matrix

$$U_{ij} = \left[\exp\left(-\tfrac{1}{2}\mathrm{i}\sum_{\mu<\nu=1}^{N} \boldsymbol{c}_{\mu\nu}(\boldsymbol{R})\cdot\boldsymbol{\sigma}^{(\mu\nu)}\right)\right]_{ij}. \tag{6.18}$$

Here the $\boldsymbol{\sigma}^{(\mu\nu)}$ are generalized Pauli matrices, defined as three-vectors of $N \times N$ traceless hermitian matrices labelled by $\mu < \nu$, whose only non-zero elements are

$$[\boldsymbol{\sigma}^{(\mu\nu)}]_{\mu\mu} = \boldsymbol{\sigma}_{11}, \quad [\boldsymbol{\sigma}^{(\mu\nu)}]_{\mu\nu} = \boldsymbol{\sigma}_{12}, \quad [\boldsymbol{\sigma}^{(\mu\nu)}]_{\nu\mu} = \boldsymbol{\sigma}_{21}, \quad [\boldsymbol{\sigma}^{(\mu\nu)}]_{\nu\nu} = \boldsymbol{\sigma}_{22}. \tag{6.19}$$

The $\boldsymbol{\sigma}^{(\mu\nu)}$ span the space of $N \times N$ hermitian matrices. Therefore permutation rotations can be regarded as a representation of $SU(N)$ (the group of $N \times N$ unitary matrices with determinant unity), just as exchange rotations can be regarded as a representation of $SU(2)$. In the Schwinger representation, the action of $\mathsf{U}(\boldsymbol{R})$ on the states $|\boldsymbol{M}\rangle$ is effected by the asociated matrix U_{ij} acting to the right on the creation operators, i.e. (cf. equation (4.18))

$$\mathsf{a}_i^\dagger \to (\mathsf{a}_i^\dagger)' = \mathsf{a}_j^\dagger U_{ji}, \tag{6.20}$$

and similarly for the b oscillators.

In Appendix C we show that the permutation condition (6.6) implies the following relation for the associated matrices

$$U_{ij}(g\boldsymbol{R}) = U_{ik}(\boldsymbol{R})Q_{kl}(g^{-1})D_{lj}(g,\boldsymbol{R}). \tag{6.21}$$

Here, Q is a rephased permutation matrix with $\det Q = 1$, that is (cf. equation (6.4))

$$Q_{ij}(g) = \delta_{i,g(j)}\exp\{\mathrm{i}\nu_i\}, \quad \text{where } \sum_{i=1}^{N}\nu_i = \pi\varepsilon(g), \tag{6.22}$$

and D is a diagonal matrix in $SU(N)$. Therefore the construction of the transported basis reduces to finding a $U_{ij}(\boldsymbol{R})$ in $SU(N)$ satisfying (6.21).

One possible construction is to take the columns of $U_{ij}(\boldsymbol{R})$ to be the eigenvectors $|n(\boldsymbol{R})\rangle$, written as column vectors $\langle i|n(\boldsymbol{R})\rangle$, of an $N \times N$ hermitian matrix $H(\boldsymbol{R})$ depending smoothly on $\boldsymbol{R}$. Thus

$$H(\boldsymbol{R}) = \sum_{n=1}^{N}\lambda_n(\boldsymbol{R})|n(\boldsymbol{R})\rangle\langle n(\boldsymbol{R})|, \quad \text{i.e. } H_{ij}(\boldsymbol{R}) = \sum_{n=1}^{N}\lambda_n(\boldsymbol{R})U_{in}(\boldsymbol{R})U_{nj}^\dagger(\boldsymbol{R}), \tag{6.23}$$

where the eigenvalues $\lambda_n(\boldsymbol{R})$ are chosen to be symmetric under permutations g. Orthogonality and completeness of the $|n(\boldsymbol{R})\rangle$ guarantee that $U_{ij}(\boldsymbol{R})$ is unitary. Our hope that $H(\boldsymbol{R})$ will be simpler than $U_{ij}(\boldsymbol{R})$ springs from the fact that for $N = 2$, where $U_{ij}(\boldsymbol{R})$ is the matrix in (4.18) (cf. equations (4.9) and (4.10)), the expression (6.23) gives the simple formula

$$H(\boldsymbol{R}) = \boldsymbol{r}\cdot\boldsymbol{\sigma}, \tag{6.24}$$

with $\lambda_\pm(\boldsymbol{r}) = \pm r$.

Calculations based on (6.21) and (6.22) lead to the requirements

$$[H(\boldsymbol{R}), H(g\boldsymbol{R})] = 0, \tag{6.25}$$

showing that the matrices related by permutation of positions commute, and

$$|n(g\boldsymbol{R})\rangle\langle n(g\boldsymbol{R})| = |g^{-1}(n)(\boldsymbol{R})\rangle\langle g^{-1}(n)(\boldsymbol{R})|, \tag{6.26}$$

showing that a permutation of positions leads the same permutation of the eigenstates $|n(\boldsymbol{R})\rangle$ (and possibly a change of phase). Another way to write this last result is

$$H(g\boldsymbol{R}) = \sum_{n=1}^{N} \lambda_{g(n)}(\boldsymbol{R})|n(\boldsymbol{R})\rangle\langle n(\boldsymbol{R})|; \tag{6.27}$$

that is, permutation of positions changes the eigenvalue associated with each state $|n(\boldsymbol{R})\rangle$ from $\lambda_n(\boldsymbol{R})$ to its permuted counterpart $\lambda_{g(n)}(\boldsymbol{R})$. These general conclusions are illustrated by (6.24) where g is the exchange $\boldsymbol{r} \to -\boldsymbol{r}$, whose effect on H is to interchange either the two states $|\pm(\boldsymbol{r})\rangle$ or the two eigenvalues $\pm r$.

A consequence of (6.26), together with the orthogonality of the states $|n(\boldsymbol{R})\rangle$, is that for positions $\boldsymbol{R} = g\boldsymbol{R}$, where two or more particles coincide, the states $|n(\boldsymbol{R})\rangle$, and hence the matrix $U_{ij}(\boldsymbol{R})$, are singular, so $H(\boldsymbol{R})$ is degenerate (as in (6.24) where $\boldsymbol{r} = 0$). The number of degenerating eigenvalues must equal the number of coinciding particles. So, finding a non-singular transported basis reduces to the mathematical problem of finding an $N \times N$ hermitian matrix $H_{ij}(\boldsymbol{R})$ with degeneracies only where $\boldsymbol{R} = g\boldsymbol{R}$; degeneracies at other points would lead to singularities in the basis constructed from $H_{ij}(\boldsymbol{R})$. Here the codimension of degeneracies gives cause for optimism. To make two particles coincide, it is necessary to vary three coordinates (to make the components of the interparticle vector vanish), and this is precisely the codimension of degeneracies (that is, the number of parameters in a generic complex hermitian matrix that must be varied in order to produce a degeneracy of two eigenvalues). To make N particles coincide, it is necessary to vary $3N-3$ coordinates. For $N > 2$ this is less than the codimension $N^2 - N$ of a degeneracy of N eigenvalues, showing that the matrices we seek are special rather than generic, and there are insuffient parameters to lead us to expect unwanted degeneracies on higher-dimensional manifolds containing the points where $\boldsymbol{R} = g\boldsymbol{R}$.

7. Discussion

It is important to be precise about the sense in which we claim to have derived the connection between spin and statistics, embodied in the Pauli sign $(-1)^{2S}$. The form of quantum mechanics that we have used to describe identical particles, involving the transported spin basis and wavefunctions invariant under $\boldsymbol{r} \to -\boldsymbol{r}$, was not itself derived, but postulated as being closest in spirit to the more familiar quantum mechanics without exchange. This is physics, not mathematics, and so it can be tested by experiment. But since this quantum mechanics leads to the same physics (e.g. the exclusion principle) as more conventional formulations, the experiments have already been carried out, and support the theoretical predictions.

However, given this form of quantum mechanics, and our particular implementation in terms of exchange angular momentum, the Pauli principle is inevitable and we have derived it. The obvious question now is whether the result is unique. We are not claiming that the implementation is unique: there could be ways to define exchange angular momentum that do not involve harmonic oscillators, and, more generally, ways to augment spin so as to incorporate exchange without introducing exchange angular momentum. Nevertheless, we conjecture that the Pauli sign will emerge from any transported basis that satisfies the specified conditions. Certainly this is true for any transported basis generated by exchange rotations. More generally, we suggest that the relation between spin and statistics could be determined by the first Stiefel–Whitney class of the two-spin bundle, which in turn could be determined just by the conditions (a)–(c) in § 2.

The main role of the Schwinger representation has been to provide a way of embedding the spin space in a larger Hilbert space (eigenstates of harmonic oscillators), to enable it to be parallel-transported. It is natural to ask whether the exchange rotation could be accomplished without this larger space, using the individual spin operators $\mathbf{S}_1$ and $\mathbf{S}_2$ alone. It could not, at least while maintaining the requirement of parallel transport (the abandonment of which leads to the bizarre consequences explored in Appendix D). An analogous enlargement of spin space has been noted before (Berry 1987) in the interpretation of experiments on polarized light in a coiled optical fibre. Photons are spin-1 particles, which because of transversality are confined to the two-state subspace of spinors with spin quantum numbers $m = \pm 1$ along the propagation direction. However, in order to accommodate within a fixed basis the changing propagation direction in a coiled fibre, three states are necessary. It is curious that in fibres the local $m = 0$ state is excluded, whereas in spin exchange it is the $m \neq 0$ states—unphysical because they correspond to particles with different spins—that are excluded. However, as J. H. Hannay has pointed out to us, linearly polarized light can be regarded as being in the $m = 0$ state of the component of spin along the polarization, rather than propagation, direction.

We thank Professor J. P. Keating for helpful comments on the paper. J.M.R. was supported by the SERC (UK) and the NSF (USA) (grant number Int-9203313) during some of this work.

Appendix A. The transported basis is smooth

Using the Schwinger representation, we need to show that the transported function on the right-hand side of (4.19) is a smooth function of $\boldsymbol{r}$ for $r > 0$. Obviously it is necessary to examine only the neighbourhood of the poles. Near the north pole, we can set up local cartesian coordinates

$$\xi = \theta \cos\phi, \quad \eta = \theta \sin\phi, \quad (\theta \ll 1). \tag{A 1}$$

Then the local approximation to (4.19) is

$$(\mathsf{a}_1^\dagger)^{n_{1a}}(\mathsf{a}_2^\dagger)^{n_{2a}}(\mathsf{b}_1^\dagger)^{n_{1b}}(\mathsf{b}_2^\dagger)^{n_{2b}} \to (\mathsf{a}_1^\dagger + \tfrac{1}{2}(\xi + \mathrm{i}\eta)\mathsf{a}_2^\dagger)^{n_{1a}}(\mathsf{a}_2^\dagger - \tfrac{1}{2}(\xi - \mathrm{i}\eta)\mathsf{a}_1^\dagger)^{n_{2a}} \\ \times(\mathsf{b}_1^\dagger + \tfrac{1}{2}(\xi + \mathrm{i}\eta)\mathsf{b}_2^\dagger)^{n_{1b}}(\mathsf{b}_2^\dagger - \tfrac{1}{2}(\xi - \mathrm{i}\eta)\mathsf{b}_1^\dagger)^{n_{2b}}, \tag{A 2}$$

which is obviously a smooth function of ξ and η.

Near the south pole, we can set up local cartesian coordinates

$$\xi = (\pi - \theta)\cos\phi, \quad \eta = (\pi - \theta)\sin\phi, \quad ((\pi - \theta) \ll 1). \tag{A 3}$$

Now the local approximation to (4.19) is

$$(\mathsf{a}_1^\dagger)^{n_{1a}}(\mathsf{a}_2^\dagger)^{n_{2a}}(\mathsf{b}_1^\dagger)^{n_{1b}}(\mathsf{b}_2^\dagger)^{n_{2b}} \to (\exp\{\mathrm{i}\phi\})^{n_{1a}+n_{1b}}(-\exp\{-\mathrm{i}\phi\})^{n_{2a}+n_{2b}} \\ \times(\mathsf{a}_2^\dagger + \tfrac{1}{2}(\xi - \mathrm{i}\eta)\mathsf{a}_1^\dagger)^{n_{1a}}(\mathsf{a}_1^\dagger - \tfrac{1}{2}(\xi + \mathrm{i}\eta)\mathsf{a}_2^\dagger)^{n_{2a}} \\ \times(\mathsf{b}_2^\dagger + \tfrac{1}{2}(\xi - \mathrm{i}\eta)\mathsf{b}_1^\dagger)^{n_{1b}}(\mathsf{b}_1^\dagger - \tfrac{1}{2}(\xi + \mathrm{i}\eta)\mathsf{b}_2^\dagger)^{n_{2b}}. \tag{A 4}$$

Smoothness is threatened by the phase factors, which can be written as

$$(\exp\{i\phi\})^{n_{1a}+n_{1b}}(-\exp\{-i\phi\})^{n_{2a}+n_{2b}} = (-1)^{2S_2}\exp\{2i\phi(S_1 - S_2)\}. \tag{A 5}$$

However, the ϕ dependence disappears for identical spins, so the basis depends smoothly on ξ and η near the south pole. Note the contrast with the more familiar transport of spin-$\frac{1}{2}$ states by ordinary spin rotation, where the dependence on $\frac{1}{2}\theta$ leads to a singularity at the south pole. Exchange rotation avoids this because the singularities cancel from the commuting parts $\mathbf{E}_a$ and $\mathbf{E}_b$ of $\mathbf{E}$ (equation (4.4))—but only for the physical states $S_1 = S_2$.

Smoothness can also be demonstrated using the gauge invariance (4.16) under exchange rotations about the z axis, as follows. We have that

$$|M(\boldsymbol{r})\rangle = \mathsf{U}(\boldsymbol{r})|M\rangle = \mathsf{U}'(\boldsymbol{r})|M\rangle, \tag{A 6}$$

where U' is any operator that differs from U by an arbitrary gauge rotation, that is

$$\mathsf{U}'(\boldsymbol{r}) = \mathsf{U}(\boldsymbol{r})\exp(i\alpha(\boldsymbol{r})\mathsf{E}_z). \tag{A 7}$$

Now, we can choose U as in (4.10), whose smoothness near the north pole guarantees that $|M(\boldsymbol{r})\rangle$ is smooth there, and we can choose U' to be the alternative exchange rotation—from $\boldsymbol{e}_z$ to $-\boldsymbol{e}_z$ and then from $-\boldsymbol{e}_z$ to $\boldsymbol{r}$—

$$\mathsf{U}'(\boldsymbol{r}) = \exp\{-i(\pi - \theta)\boldsymbol{n}(\boldsymbol{r})\cdot\mathbf{E}\}\exp\{-i\pi\mathsf{E}_y\}, \tag{A 8}$$

the obvious smoothness of which near the south pole guarantees that $|M(\boldsymbol{r})\rangle$ is smooth there also.

Appendix B. The transported basis is parallel-transported

To prove the vanishing of the connection (2.4), we begin by writing

$$\langle M'(\boldsymbol{r})|\nabla M(\boldsymbol{r})\rangle = \langle M|\mathsf{U}^\dagger(\boldsymbol{r})\nabla\mathsf{U}(\boldsymbol{r})|M\rangle. \tag{B 1}$$

Since $\mathsf{U}(\boldsymbol{r})$ and $\mathsf{U}(\boldsymbol{r} + d\boldsymbol{r})$ are infinitesimally different exchange rotations,

$$\mathsf{U}^\dagger(\boldsymbol{r})\nabla\mathsf{U}(\boldsymbol{r}) = \mathbf{A}(\boldsymbol{r})\mathsf{E}_x + \mathbf{B}(\boldsymbol{r})\mathsf{E}_y + \mathbf{C}(\boldsymbol{r})\mathsf{E}_z. \tag{B 2}$$

Thus

$$\langle M'(\boldsymbol{r})|\nabla M(\boldsymbol{r})\rangle = \mathbf{A}(\boldsymbol{r})\langle M'|\mathsf{E}_x|M\rangle + \mathbf{B}(\boldsymbol{r})\langle M'|\mathsf{E}_y|M\rangle + \mathbf{C}(\boldsymbol{r})\langle M'|\mathsf{E}_z|M\rangle. \tag{B 3}$$

The matrix element involving E_z vanishes by (4.15). Those involving E_x and E_y can be written in terms of E_{a+}, E_{a-}, E_{b+}, E_{b-} defined in (4.4). These operators shift quanta from the (a_1, b_1) oscillator pair to the (a_2, b_2) oscillator pair and therefore change (S_1, S_2) to $(S_1 \pm \frac{1}{2}, S_2 \mp \frac{1}{2})$, so that all matrix elements such as those in (B 3), with $S_1 = S_2$, vanish, thereby proving (2.4).

(To demonstrate (B 2) explicitly, we can use the result, valid for any operator $\mathbf{C}(\boldsymbol{r})$, that

$$\exp[-\mathbf{C}(\boldsymbol{r})]\nabla\exp[\mathbf{C}(\boldsymbol{r})] = \nabla\mathbf{C} + (1/2!)[\nabla\mathbf{C}, \mathbf{C}] + (1/3!)[[\nabla\mathbf{C}, \mathbf{C}], \mathbf{C}] + \cdots \tag{B 4}$$

For the particular operator (4.10), a calculation gives

$$\mathsf{U}^\dagger(\boldsymbol{r})\nabla\mathsf{U}(\boldsymbol{r}) = -i\frac{\boldsymbol{e}_\theta\boldsymbol{n}\cdot\mathbf{E}}{r} + i\frac{\boldsymbol{e}_\phi}{r\sin\theta}\left(\frac{\boldsymbol{r}\cdot\mathbf{E}}{r} + (1 - 2\cos\theta)\mathbf{E}_z\right), \tag{B 5}$$

which has the form (B 2).)

Parallel transport can also be proved by a lengthy calculation based on the representation of Bargmann (1962), in which creation operators are replaced by complex variables and use is made of

$$\langle M'(\boldsymbol{r})|\nabla M(\boldsymbol{r})\rangle \propto \iint \mathrm{d}a_1\mathrm{d}a_2 \iint \mathrm{d}b_1\mathrm{d}b_2 \exp\{-(|a_1|^2+|a_2|^2+|b_1|^2+|b_2|^2)\}$$
$$\times (a_1'^*)^{S+m_1'}(a_2'^*)^{S+m_2'}(b_1'^*)^{S-m_1'}(b_2'^*)^{S-m_2'}$$
$$\times \nabla (a_1'^*)^{S+m_1}(a_2'^*)^{S+m_2}(b_1'^*)^{S-m_1}(b_2'^*)^{S-m_2}, \qquad \text{(B 6)}$$

where the primed variables are defined in terms of the unprimed variables by (4.18), and the integrations are over the complex planes of the variables, e.g.

$$\mathrm{d}a_1 \equiv \mathrm{d}\,\mathrm{Re}\,a_1\,\mathrm{d}\,\mathrm{Im}\,a_1. \qquad \text{(B 7)}$$

We do not give the details.

Appendix C. Pauli sign from general exchange rotation

We assume that $\mathsf{U}(\boldsymbol{r})$ is an exchange rotation satisfying (2.2) and (2.3). It follows that

$$\mathsf{U}^\dagger(\boldsymbol{r})\mathsf{U}(-\boldsymbol{r})|M\rangle = (-1)^K|\bar{M}\rangle. \qquad \text{(C 1)}$$

$|\bar{M}\rangle$ can be expressed in terms of $|M\rangle$ through a fixed exchange rotation, e.g. an exchange rotation $\mathsf{R}_y(\pi)$ by π about y. Indeed, employing the Schwinger representation in the form (4.18), with $\theta=\pi$, $\phi=0$, we obtain

$$(\mathsf{a}_1^\dagger\mathsf{a}_2^\dagger) \to (\mathsf{a}_1^\dagger\mathsf{a}_2^\dagger)(-\mathrm{i}\sigma_y) = (\mathsf{a}_1^\dagger\mathsf{a}_2^\dagger)\begin{pmatrix} 0 & -1 \\ 1 & 0 \end{pmatrix} = (\mathsf{a}_2^\dagger \;\; -\mathsf{a}_1^\dagger). \qquad \text{(C 2)}$$

From (4.11) and (4.19), we obtain

$$|\bar{M}\rangle = (-1)^{2S}\mathsf{R}_y(\pi)|M\rangle. \qquad \text{(C 3)}$$

Thus (C 1) can be written as

$$\mathsf{D}(\boldsymbol{r})|M\rangle = (-1)^{K-2S}|M\rangle, \qquad \text{(C 4)}$$

where

$$\mathsf{D}(\boldsymbol{r}) = \mathsf{R}_y^\dagger(\pi)\mathsf{U}^\dagger(\boldsymbol{r})\mathsf{U}(-\boldsymbol{r}). \qquad \text{(C 5)}$$

As $\mathsf{D}(\boldsymbol{r})$ is a product of exchange rotations, it is itself an exchange rotation, and thus may be expressed in the form $\exp\{-\mathrm{i}\boldsymbol{c}(\boldsymbol{r}).\mathbf{E}\}$. The eigenvector equation (C 4) then implies (cf. equations (4.11 and (4.12)) that

$$(\mathsf{a}_1'^\dagger)^{S+m_1}(\mathsf{a}_2'^\dagger)^{S+m_2}(\mathsf{b}_1'^\dagger)^{S-m_1}(\mathsf{b}_2'^\dagger)^{S-m_2}$$
$$= (\text{phase factor}) \times (\mathsf{a}_1^\dagger)^{S+m_1}(\mathsf{a}_2^\dagger)^{S+m_2}(\mathsf{b}_1^\dagger)^{S-m_1}(\mathsf{b}_2^\dagger)^{S-m_2}, \qquad \text{(C 6)}$$

where, as in (4.18), the primed and unprimed oscillator pairs are related by

$$(\mathsf{a}_1'^\dagger\mathsf{a}_2'^\dagger) = (\mathsf{a}_1^\dagger\mathsf{a}_2^\dagger)\exp\{-\mathrm{i}\boldsymbol{c}(\boldsymbol{r})\cdot\boldsymbol{\sigma}\} \qquad \text{(C 7)}$$

and similarly for b and b'. Since (C 6) must hold for all M, we deduce that

$$\mathsf{a}_j' = (\text{phase factor}) \times \mathsf{a}_j, \quad \mathsf{b}_j' = (\text{phase factor}) \times \mathsf{b}_j. \qquad \text{(C 8)}$$

Thus $\exp\{-\mathrm{i}\boldsymbol{c}(\boldsymbol{r}).\boldsymbol{\sigma}\}$ is diagonal, (that is $\boldsymbol{c}(\boldsymbol{r})$ is along the z direction), so

$$\mathsf{D}(\boldsymbol{r}) = \exp\{-\mathrm{i}\alpha(\boldsymbol{r})\mathsf{E}_z\}. \qquad \text{(C 9)}$$

Then (4.16) implies that the eigenvalues of the states $|M\rangle$ in (C 4) are in fact unity, and (2.5) follows. (It is not the case that the operator $\mathsf{D}(\boldsymbol{r})$ is equal to unity: its action on unphysical states, where the particles would have different spins, would introduce phases.)

This derivation of the Pauli sign generalizes to $N > 2$ particles. As argued in § 6, it suffices to consider the exchange e of a single pair, say particles 1 and 2. The permutation condition (6.6) implies

$$\mathsf{U}^\dagger(\boldsymbol{R})\mathsf{U}(e\boldsymbol{R})|\boldsymbol{M}\rangle = (-1)^{K(e)}|e\boldsymbol{M}\rangle. \tag{C 10}$$

$|e\boldsymbol{M}\rangle$ is related to $|\boldsymbol{M}\rangle$ by, for example, the exchange rotation $\mathsf{R}_y^{(12)}(\pi)$, the associated matrix (6.18) of which is

$$\exp\{-\tfrac{1}{2}\mathrm{i}\pi\sigma_y^{(12)}\} = \begin{pmatrix} -\mathrm{i}\sigma_y & 0 \\ 0 & I \end{pmatrix}, \tag{C 11}$$

where I is the $N-2$-dimensional identity matrix. Its effect in the Schwinger representation (cf. equation (6.20)) is to replace $(\mathsf{a}_1^\dagger \mathsf{a}_2^\dagger)$ by $(\mathsf{a}_2^\dagger \; -\mathsf{a}_1^\dagger)$ while leaving the other a_j unchanged, and similarly for b_j. It follows that

$$|e\boldsymbol{M}\rangle = (-1)^{2S}\mathsf{R}_y^{(12)}(\pi)|\boldsymbol{M}\rangle. \tag{C 12}$$

Thus (C 10) can be written as

$$\mathsf{D}(e,\boldsymbol{R})|\boldsymbol{M}\rangle = (-1)^{K(e)-2S}|\boldsymbol{M}\rangle, \tag{C 13}$$

where

$$\mathsf{D}(e,\boldsymbol{R}) = \mathsf{R}_y^{(12)\dagger}(\pi)\mathsf{U}^\dagger(\boldsymbol{R})\mathsf{U}(e\boldsymbol{R}). \tag{C 14}$$

By an argument identical to that for $N = 2$ (equations (C 6)–(C 9)), it follows that the matrix associated with D is diagonal, and thence, using the generalization

$$\mathsf{E}_z^{(\mu\nu)}|\boldsymbol{M}\rangle = 0, \tag{C 15}$$

of (4.15), that $K(e) = 2S$, which is what we wanted to show. Further, the relation (6.21) for a general permutation now follows by repeated application of (C 14), in the form

$$\mathsf{U}(e\boldsymbol{R}) = \mathsf{U}(\boldsymbol{R})\mathsf{Q}(e)\mathsf{D}(e,\boldsymbol{R}), \tag{C 16}$$

with $\mathsf{Q}(e) = \mathsf{R}_y^{(12)}(\pi)$.

Appendix D. The counterconstruction: spin statistics without parallel transport

Without the enlargement of the spin space forced by the parallel-transport requirement (2.4), the operator $\mathsf{U}(\boldsymbol{r})$ in (2.2) can be represented as a $(2S+1)^2 \times (2S+1)^2$ matrix. Then the exchange condition (2.3) can be written as

$$\mathsf{U}(-\boldsymbol{r}) = \mathsf{U}(\boldsymbol{r})(-1)^K\mathbf{P}, \tag{D 1}$$

where $\mathbf{P}$ is the permutation matrix, satisfying

$$\mathbf{P}|M\rangle = |\bar{M}\rangle, \tag{D 2}$$

that exchanges the labels of the $(2S+1)^2$ pairs of spin quantum numbers m_1 and m_2. It is not hard to show that

$$\det \mathbf{P} = (-1)^{S(2S+1)} \tag{D 3}$$

(one way is to order the pairs so that those with the same spin quantum numbers are listed together, and those with different quantum numbers are adjacent to their exchanged partner pairs; then the matrix P consists of a $(2S+1)$-dimensional unit diagonal block, followed by $S(2S+1)$ unit off-diagonal 2×2 blocks, whose determinant is easy to calculate). Thus

$$\det \mathsf{U}(-\boldsymbol{r}) = \det \mathsf{U}(\boldsymbol{r})(-1)^{(K+S)(2S+1)}. \tag{D 4}$$

Now we incorporate the smoothness condition (a) of § 2. It implies that $\mathsf{U}(\boldsymbol{r})$ can be defined on a closed loop and remain smooth as the loop is continuously contracted to a point only if $\det \mathsf{U}(\boldsymbol{r})$ has zero winding number round the loop. Considering the loop to be any great circle in a sphere with constant $r = |\boldsymbol{r}|$, we see that this is impossible (i.e. winding is unavoidable) if the sign in (D 4) is negative. Therefore

$$(K+S)(2S+1) \text{ is even.} \tag{D 5}$$

When this condition holds, $\mathsf{U}(\boldsymbol{r})$ can be constructed explicitly by exploiting the implication of (D 1) that $(-1)^K\mathsf{P}$ is unitary with unit determinant, and so can be written as

$$(-1)^K\mathsf{P} = \exp\{\mathrm{i}\mathsf{H}\}, \quad [(-1)^K\mathsf{P}]^2 = \exp\{2\mathrm{i}\mathsf{H}\} = 1, \tag{D 6}$$

where H is hermitian and traceless. Then we define $\mathsf{U}(\boldsymbol{r})$ on the equator $\theta = \frac{1}{2}\pi$ (using polar coordinates for $\boldsymbol{r}$) as

$$\mathsf{U}(r, \tfrac{1}{2}\pi, \phi) = \exp\{\mathrm{i}(\phi/\pi)\mathsf{H}\}. \tag{D 7}$$

This is a continuous family of unitary matrices with unit determinant, satisfying (D 1) and beginning (at $\phi = 0$) and ending (at $\phi = 2\pi$) at the identity. Since $SU(2S+1)$ is simply connected, the closed loop $\mathsf{U}(r, \frac{1}{2}\pi, \phi)$ can be continuously contracted to the identity, and we can define $\mathsf{U}(\boldsymbol{r})$ in the northern hemisphere so that on circles of latitude it interpolates between the loop (D 7) on the equator and the identity at the north pole. $\mathsf{U}(\boldsymbol{r})$ is then defined in the southern hemisphere by continuation using (D 1).

For each S, we can examine the implications of the condition (D 5) for K even (boson statistics) and K odd (fermion statistics). There are four cases:

(i) if $S = 0, 2, 4, \ldots$, (D 4) requires K even, that is bose statistics;

(ii) if $S = 1, 3, 5, \ldots$, (D 4) requires K odd, that is fermi statistics;

(iii) if $S = \frac{1}{2}, \frac{5}{2}, \frac{9}{2}, \ldots$, (D 4) cannot be satisfied for any integer K, so the counterconstruction is incompatible with any statistics; and

(iv) if $S = \frac{3}{2}, \frac{7}{2}, \frac{11}{2}, \ldots$, (D 4) can be satisfied for any integer K, so the counterconstruction is compatible with both bose and fermi statistics.

The results of this counterconstruction are not only bizarre but also incoherent, in the sense that the construction cannot be carried out for case (iii) and does not lead to a unique spin-statistics relation in case (iv) (as well as giving the wrong relation for case (ii)).

References

Balachandran, A. P., Daughton, A., Gu, Z.-C., Sorkin, R. D., Marmo, G. & Srivastava, A. M. 1993 Spin-statistics theorems without relativity or field theory. *Int. J. Mod. Phys.* A **8**, 2993–3044.

Bargmann, V. 1962 On the representations of the rotation group. *Rev. Mod. Phys.* **24**, 829–845.

Berry, M. V. 1984 Quantal phase factors accompanying adiabatic changes. *Proc. R. Soc. Lond.* A **392**, 45–57.

Berry, M. V. 1987 The adiabatic phase and Panacharatnam's phase for polarized light. *J. Mod. Opt.* **34**, 1401–1407.

Delacrétaz, G., Grant, E. R., Whetten, R. L., Wöste, L. & Zwanziger, J. W. 1986 Fractional quantization of molecular pseudorotation in Na3. *Phys. Rev. Lett.* **56**, 2598–2601.

Feynman, R. P. 1987 In *The 1986 Dirac memorial lectures* (ed. R. P. Feynman & S. Weinberg). Cambridge University Press.

Feynman, R. P., Leighton, R. B. & Sands, M. 1965 *The Feynman lectures on physics.* Reading, MA: Addison-Wesley.

Finkelstein, D. & Rubinstein, J. 1968 Connection between spin, statistics and kinks. *J. Math. Phys.* **9**, 1762–1779.

Gould, R. R. 1995 Answer to question #7 (Neuenschwander, D. E. 1994 The spin-statistics theorem *Am. J. Phys.* **62**, 972.) *Am. J. Phys.* **63**, 109.

Guerra, F. & Marra, R. 1984 A remark on a possible form of spin-statistics theorem in non-relativistic quantum mechanics. *Phys. Lett.* B **141**, 93–94.

Hartung, R. W. 1979 Pauli principle in Euclidean geometry. *Am. J. Phys.* **47**, 900–910.

Laidlaw, M. G. G. & DeWitt, C. M. 1971 Feynman functional integrals for systems of indistinguishable particles. *Phys. Rev.* D **3**, 1375–1378.

Leinaas, J. M. & Myrheim, J. 1977 On the theory of identical particles. *Nuovo Cim.* B **37**, 1–23.

Longuet-Higgins, H. C., Öpik, U., Pryce, M. H. L. & Sack, R. A. 1959 Studies of the Jahn–Teller effect. II. The dynamical problem. *Proc. R. Soc. Lond.* A **244**, 1–16.

Mickelson, J. 1984 Geometry of spin and statistics in classical and quantum mechanics. *Phys. Rev.* D **30**, 1843–1845.

Milnor, J. W. & Stasheff, J. D. 1974 *Characteristic classes.* Princeton University Press.

Nakahara, M. 1990 *Geometry, topology and physics.* Bristol: Adam Hilger.

Nash, C. & Sen, S. 1983 *Topology and geometry for physicists.* London: Academic.

Pauli, W. 1940 Connection between spin and statistics. *Phys. Rev.* **58**, 716–722.

Robbins, J. M. & Berry, M. V. 1994 A geometric phase for $m = 0$ spins. *J. Phys.* A **27**, L435–L438.

Sakurai, J. J. 1994 *Modern quantum mechanics.* Reading, MA: Addison-Wesley.

Schwinger, J. 1965 In *Quantum theory of angular momentum* (ed. L. C. Biedenharn & H. Van Dam), pp. 229-279. New York: Academic.

Shapere, A. & Wilczek, F. (eds) 1989 *Geometric phases in physics.* Singapore: World Scientific.

Streater, F. W. & Wightman, A. S. 1964 *PCT, spin and statistics, and all that.* New York: Benjamin.

Tscheuschner, R. D. 1989 Topological spin-statistics relation in quantum field theory. *Int. J. Theor. Phys.* **28**, 1269–1310.

Received 21 May 1997; accepted 23 June 1997

Quantum Indistinguishability: Spin-statistics without Relativity or Field Theory?

Michael Berry* and Jonathan Robbins†

*H H Wills Physics Laboratory, Tyndall Avenue,
Bristol BS8 1TL, United Kingdom
† Basic Research Institute in the Mathematical Sciences,
Hewlett-Packard Laboratories Bristol,
Filton Road, Stoke Gifford, Bristol BS12 6QZ, United Kingdom, and
School of Mathematics, University Walk, Bristol BS8 1TW, United Kingdom

Abstract. We review the formulation of quantum mechanics for identical spinning particles with wavefunctions that are singlevalued when permuted configurations are identified. The identification requires the spins to be smoothly permuted along with position variables, so spin is represented in a position-dependent 'transported basis', rather than the usual fixed basis. The simplest transported basis, constructed in terms of spins represented as pairs of commuting harmonic oscillators, gives the correct connection between spin and statistics. More complicated constructions can give the wrong exchange sign. The theory is generalized to incorporate additional properties such as isospin, colour and strangeness. Some remarks about the relation between this approach and those based on relatitivity and/or field theory are given.

1. INTRODUCTION

In the nonrelativistic quantum mechanics of a fixed number of identical particles, the relation between spin and statistics (SS) sits awkwardly on top of the theory, as a separate postulate: it is simply asserted that the wavefunctions of particles with integer-plus-half spin change sign when the variables describing their spin and position are exchanged, whereas wavefunctions for particles whose spin is an integer do not change

CP545, *Spin-Statistics Connection and Commutation Relations,* edited by R. C. Hilborn and G. M. Tino sign. Here we will investigate the possibility that SS is already contained in the theory, as a hidden consequence of imposing geometrical requirements that follow naturally from indistinguishability. We will give brief nontechnical summaries of, and comment on, the ideas in two recent papers [1] [2].

If SS is an awkward addition to nonrelativistic quantum theory, there is awkwardness too in the attempt to be described here. The explanation of any phenomenon on the basis of a well-established physical theory must begin with assumptions about how the theory is to be applied in the particular case under consideration, followed by deductions that lead unambiguously and precisely to the phenomenon. An explanation is more convincing if it is fruitful, in the sense of successfully predicting phenomena that have not yet been observed.

Here the 'phenomenon' is already known: it is SS, the principle determining how wavefunctions behave under exchange. Of course, this principle has many consequences (e.g. the Pauli exclusion principle) that are fundamental to our description of the world, but it is hard to see how an explanation of SS itself can have any new experimental consequences (though we live in hope that such pessimism will prove wrong). Thus, the degree to which any purported explanation of SS is accepted - once it has been agreed that the deductive part is technically correct - must hinge on the naturalness of the assumptions, and so involves elements of subjectivity and aesthetics. In [1, 2], 'naturalness' was interpreted as: constructing the quantum theory of identical particles using the same principles that are accepted and routinely applied in the quantum physics of nonidentical particles (or individual identical particles).

A further awkwardness is that SS involves not some experimental number whose prediction with increasing accuracy might increase confidence in the theory (like the fine-structure constant), but is simply a sign.

Nevertheless, SS cries out for understanding. There are many derivations based on relativity, beginning with [3], that have been comprehensively reviewed [4], and we will comment a little on these in section 6. And there have been many attempts at derivations that do not involve relativity, based on a variety of different assumptions [5]. We think that previous derivations have lacked a crucial geometrical ingredient, described in section 2, that follows from indistinguishability.

For simplicity, we will concentrate on the understanding of SS for two identical particles with spin S. However, the extension to N particles involved interesting technical challenges that have resulted in beautiful mathematical constructions by Atiyah [6], and we will mention one of these briefly in section 4.

2. SINGLEVALUEDNESS UNDER EXCHANGE

In the position representation, the wavefunction $|\Psi\rangle$ describing the state of two identical particles with spin S depends on the position vectors $\mathbf{r}_1$ and $\mathbf{r}_2$. Only the relative position $\mathbf{r}=\mathbf{r}_2-\mathbf{r}_1$ is relevant here, so we write $|\Psi(\mathbf{r})\rangle$. Exchange of positions corresponds to $\mathbf{r}\to-\mathbf{r}$.

The central assumption in [2] was that the wavefunction for identical particles must be singlevalued under exchange. Thus

$$|\Psi(\mathbf{r})\rangle=|\Psi(-\mathbf{r})\rangle. \tag{1}$$

At first this seems absurd, but our $|\Psi\rangle$ differs from the more familiar wavefunction in a crucial respect, namely that it has, built into it, the property that exchange of positions $\mathbf{r}\to-\mathbf{r}$ is automatically accompanied by exchange of spin states, so that $\mathbf{r}\to-\mathbf{r}$ corresponds to complete exchange. Only then can indistinguishability be demanded, and the condition (1) imposed. To invoke singlevaluedness for identical particles is not a new idea [7-9]; but the characteristic feature of our approach is the systematic incorporation of spin exchange along with exchange of positions.

To incoporate spin exchange, it is necessary to represent the spin part of the state in a way that is unusual but unavoidable. Following [1], we write the complete state as

$$|\Psi(\mathbf{r})\rangle=\sum_M \psi_M(\mathbf{r})|M(\mathbf{r})\rangle. \tag{2}$$

Here $M\equiv\{m_1,m_2\}$ labels the spin state of the particles, with m denoting the z component of spin; exchange of spins corresponds to $M\to\overline{M}\equiv\{m_2,m_1\}$. $|M(\mathbf{r})\rangle$ is the *transported spin basis*, that is, a basis for representing spins in a way that depends on the relative position of the particles, with the exchange requirement

$$|M(-\mathbf{r})\rangle=(-1)^K|\overline{M}(\mathbf{r})\rangle, \tag{3}$$

where K is an integer (see [1, 2]), implying that exchange generates a sign rather than some more general phase factor. The coefficients $\psi_M(\mathbf{r})$ describe the spatial dependence of the state.

The representation (2) is to be contrasted with the familiar expansion in terms of a fixed spin basis $|M\rangle$, namely

$$|\Psi(\mathbf{r})\rangle_{\text{fixed}} = \sum_M \psi_M(\mathbf{r})|M\rangle. \tag{4}$$

As was shown in [1], the coefficients $\psi_M(\mathbf{r})$ – which are, after all, the physically measurable quantities (up to a single overall phase) - are the same as those in (2) (see also section 3 later). With $|\Psi(\mathbf{r})\rangle_{\text{fixed}}$, there is no spin exchange accompanying position exchange, so indistinguishability is not incorporated and there is no justification for imposing the singlevaluedness requirement (1). With (2), however, the application of (1) implies that any sign change (3) of the transported basis is compensated by a sign change in the coefficients:

$$\psi_{\overline{M}}(-\mathbf{r}) = (-1)^K \psi_M(\mathbf{r}). \tag{5}$$

This is the usual form in which SS is assumed. Of course, to reproduce *the* SS, rather than a generic form of SS, it is necessary to show that

$$K = 2S. \tag{6}$$

This requires consideration of the transported basis $|M(\mathbf{r})\rangle$, as will be described in section 3.

It is important to emphasize that with the representation (2), SS, in the form (5), emerges as a quantization condition implied by singlevaluedness. This brings SS into the same framework as other derivations within elementary quantum mechanics. For example the quantization of a component $m\hbar$ of orbital angular momentum using wavefunctions requires singlevaluedness of exp($im\phi$) under $\phi \to \phi + 2\pi$, reflecting the fact that these two angles represent the same point. And in the Aharonov-Bohm effect [10, 11] a similar application of singlevaluedness is required to get a definite (and experimentally confirmed) prediction for quantum scattering by inaccessible magnetic flux. In fact, in every situation that we know in elementary quantum mechanics, wavefunctions representing the same configuration are the same (up to choice of gauge).

In effect, the incorporation of the transported basis as in (2) enables $\mathbf{r}$ and $-\mathbf{r}$ to be regarded as the same point in the configuration space of the two particles. This identification changes the topology of the configuration space, making it nonorientable and non-simply-connected. The space is the direct product of the centre of mass, the separation distance $r=|\mathbf{r}|$, and the projective plane (2-sphere with antipodal points identified) that represents directions $\mathbf{r}/r$. An intrinsic procedure would be to construct quantum mechanics by erecting two-spin bundles on this base space, and this has been systematically carried out [12]. However, we will here continue to use the more elementary approach of regarding **r** as a euclidean vector and then imposing singlevaluedness under **r**→-**r**. (This is analogous to the common and convenient procedure of regarding azimuth angles ϕas variables with values on the real line $-\infty < \phi < \infty$, and then insisting that functions are periodic, rather than considering functions whose domain is the circle.)

3. TRANSPORTED BASIS

The basis $|M(\mathbf{r})\rangle$ is a set of $(2S+1)^2$ spinors, because the z quantum numbers m_1 and m_2 of the two spins can range from -S to +S. Each spinor is required to be a singlevalued and smooth function of **r**. In addition, we impose the parallel-transport requirement

$$\mathbf{A}_{M,M'} \equiv \mathrm{i}\langle M'(\mathbf{r})|\nabla M(\mathbf{r})\rangle = 0, \tag{7}$$

to guarantee the vanishing of the curvature of the connection between neighbouring positions **r** ('flat exchange'). (For further discussion, see [12].)

Parallel transport implies that $|M(\mathbf{r})\rangle$ inhabits an ambient space, within which it is smoothly transported, that is larger than the $(2S+1)^2$-dimensional space of the fixed basis $|M\rangle$. Without enlargement, (7) would imply that $|M(\mathbf{r})\rangle$ is independent of **r** and so unable to satisfy the fundamental exchange requirement (3). We find it necessary to emphasize that this enlargement is in no way undesirable or unphysical. Nor is it unfamiliar: in [1] we give the analogy of light in a coiled optical fibre, where transversality implies that in a frame whose z axis is along the the local propagation direction the electric field vector can be described with only two components (x and y) whereas a fixed basis requires all three components. An even simpler analogy is that in a space $\mathbf{r}=\{x, y, z\}$, each vector in a field $\mathbf{v}(\mathbf{r})$ possesses only one component when described in a local frame $\{x_1, x_2, x_3\}$ whose x_3 axis is directed along **v**, but requires three components in the ambient space $\{x, y, z\}$.

The transformation between the transported and fixed bases is described by a unitary operator $\mathbf{U}(\mathbf{r})$, such that

$$|\Psi(\mathbf{r})\rangle = \mathbf{U}(\mathbf{r})|\Psi(\mathbf{r})\rangle_{\text{fixed}}, \quad |M(\mathbf{r})\rangle = \mathbf{U}(\mathbf{r})|M\rangle. \tag{8}$$

An immediate consequence is the equality

$$\psi_M(\mathbf{r}) = \langle M(\mathbf{r})|\Psi(\mathbf{r})\rangle = \langle M|\Psi(\mathbf{r})\rangle_{\text{fixed}} \tag{9}$$

asserted after (4). As described in [1, 2], **U**(**r**) also generates dynamical variables (e.g. momentum and spin) in the transported basis from their more familiar fixed counterparts, thereby guaranteeing that all local physics (for example Schrödinger equations derived from hamiltonians) is the same as in the fixed basis.

The main technical content of [1] was a construction of **U**(**r**). In this, each of the two spins was represented by the formalism of Schwinger [13], by two harmonic oscillators: a_1 and b_1 for one particle, and a_2 and b_2 for the other. Each of these four oscillators had its own creation and annihilation operator, with operators corresponding to different oscillators commuting. The dimension of the ambient space, on which **U** acts and within which the $(2S+1)^2$ states $|M(\mathbf{r})\rangle$ are smoothly transported, is the number of ways that $4S$ quanta can be distributed among four oscillators, namely $(4S+1)(4S+2)(4S+3)/6$; for S=1/2, there are 10 such states, in contrast to the four fixed-basis and transported-basis states.

Exchange was incorporated using the following insight. For a single spin, interchanging the number of quanta in its a and b oscillators corresponds to replacement of m by $-m$, equivalent to a rotation of the axis of quantization from z to - z. Therefore, interchanging the quanta in the 1 and 2 oscillators corresponds to exchanging the spin states of the two particles and can be used to define an 'exchange angular momentum', analogous to spin, which generates 'exchange rotations' **U**(**r**) from z to **r**, whose effect is precisely to generate a transported basis with the desired exchange property (3). With this construction, in which each spin is decomposed into its 'atomic spin bosons', it was possible to make an explicit calculation of the exchange sign in (3), and the result was the correct SS sign (6). The calculation was extended in I from two to N identical particles.

In [1] we suggested that any construction of the transported basis that was smooth, singlevalued and parallel-transported would lead to the correct exchange sign. This was wrong. In [2] we exhibited two 'perverse'constructions that satisfy these requirements but which, unlike those based on the Schwinger formalism, lead to the wrong sign. The first perverse construction applies to spin S=0, and reflects a question frequently asked by people sceptical of the arguments in [1]: can the exchange of two spinless particles be accompanied by a fermionic sign? In this perverse construction, they can. The single transported state is represented as the unit vector

$$|M(\mathbf{r})\rangle = |\{0,0\}(\mathbf{r})\rangle = \mathbf{r}/|\mathbf{r}|. \tag{10}$$

This is singlevalued, smooth, and parallel-transported, and involves the extended spin space spanned by the three basis states $\mathbf{e}_x$, $\mathbf{e}_y$ and $\mathbf{e}_z$, of which only one (e.g. $\mathbf{e}_z$) corresponds to the fixed-spin state $|\{0,0\}\rangle$. The operator $\mathbf{U}(\mathbf{r})$ is then rotation from $\mathbf{e}_z$ to $\mathbf{r}$. Under $\mathbf{r}\to-\mathbf{r}$, $|M(\mathbf{r})\rangle$ changes sign fermionically, rather than being bosonically invariant.

The second perverse construction in [2] applies to spin S=1/2, and consists in replacing all commutators in the Schwinger formalism by anticommutators. For this 'anti-Schwinger' construction, the exchange sign is +1, rather than the fermionic -1.

We have not found a general principle to exclude these perverse constructions, and others that generate the wrong exchange sign. However, all the perverse constructions we have found are defective in one or more ways, described in [2]. For example, the spin-zero fermion construction fails the test of simplicity, because (10) is decomposable into a constant (unity) that satisfies the requirements - and is what Schwinger gives - and a superfluous factor with no intrinsic connection to spin. And the anti-Schwinger construction is special in that it applies only to S=1/2 and so fails to describe the statistics of composite objects that can have any spin (there are generalizations of anti-Schwinger for higher spins, but they are cumbersome).

By contrast, the Schwinger construction applies for all spins (and also for all N - see section 4), generates the transported basis without superfluous factors, and also is intrinsically related to spin. In the reformulation of our approach by constructing N-spin bundles on the identified configuration space [12], the Schwinger construction emerges as the simplest implementation of the geometrical requirements.

4. ATIYAH'S CONSTRUCTION for *N* PARTICLES

In [1] we extended the Schwinger construction to the general case of$N$ particles (with permutations instead of exchanges). The N spins are built from 2N oscillators, from whose creation and annihilation operators it is possible to construct 'permutation angular momenta', generating 'permutation rotations' and thence the transported basis states. We showed that any such construction must yield the correct SS sign. But we were unable to exhibit an explicit construction, analogous to the exchange rotation from z to $\mathbf{r}$ for two particles. We reduced the problem to that of finding a unitary NxN matrix $U_{ij}(\mathbf{R})$, smoothly dependent on the positions $\mathbf{R}=\{\mathbf{r}_1,...\mathbf{r}_N\}$, with the property that any permutation of the $\mathbf{r}_i$ results in the corresponding permutation of its columns, up to an overall phase.

Recently, Atiyah [6] has produced several such constructions. Here we will describe the simplest. The matrix $U_{ij}(\mathbf{R})$ is obtained from the polar decomposition of an NxN matrix $V_{ij}(\mathbf{R})$, each of whose columns $\mathbf{v}_j(\mathbf{R})$ is associated with the jth particle, so that the required permutation property is assured. It suffices to explain $\mathbf{v}_1(\mathbf{R})$.

Let the directions $(\mathbf{r}_2-\mathbf{r}_1)/|\mathbf{r}_2-\mathbf{r}_1|,...(\mathbf{r}_N-\mathbf{r}_1)/|\mathbf{r}_N-\mathbf{r}_1|$ of the other particles, as seen from 1, be described by their complex stereographic coordinates $\zeta_2,... \zeta_N$ (that is, the real and imaginary parts of ζ_j are the cartesian coordinates of the intersection with the equatorial plane of the line joining the south pole of the unit sphere centred on $\mathbf{r}_1$ to the point where the vector connecting $\mathbf{r}_1$ to $\mathbf{r}_j$ intersects the sphere). Then the components $v_{m,1}$ of $\mathbf{v}_1$ are the coefficients in the expansion

$$P_1(z)=\prod_{n=2}^{N}\left(z-\zeta_n\right)=\sum_{m=1}^{N}\frac{z^{m-1}}{\sqrt{(m-1)!(N-m)!}}v_{m,1}. \tag{11}$$

The orthogonalization leading to $U_{ij}(\mathbf{R})$ requires the columns $\mathbf{v}_j(\mathbf{R})$ to be independent, that is $\det V_{ij}(\mathbf{R})\neq 0$ for any configuration $\mathbf{R}$ where no two particles coincide. At present this is a plausible conjecture, proved for N=3 and some special cases (e.g. N particles in a line), but not generally; this problem remains open. Independence of the columns has however been shown for a more elaborate version of this construction [6].

5. EXTENDED SPIN-STATISTICS RELATIONS FOR PARTICLES WITH ADDITIONAL PROPERTIES

Returning now to two particles, we incorporate into the theory the fact that particles can be characterised not only by position and spin but by one or more further quantum properties, that we denote by P. Examples of P are isospin, strangeness and colour. We denote the values of P by p (assumed discrete), and the pair of values for two particles - and the associated exchanged pair - by

$$P\equiv\left\{p_1,p_2\right\},\quad \overline{P}\equiv\left\{p_2,p_1\right\}. \tag{12}$$

If we regard P as describing different states of identical particles, the argument we employed in [1] to derive the spin-statistics relation can be extended by requiring the state to be singlevalued under full exchange, including $P\rightarrow\overline{P}$ as well as $\mathbf{r}\rightarrow-\mathbf{r}$.

To implement this idea, we write the state of the two particles as

$$|\Psi_P(\mathbf{r})\rangle = \sum_M \psi_{M,P}(\mathbf{r})|M(\mathbf{r})\rangle, \tag{13}$$

in which $|M(\mathbf{r})\rangle$ is the same transported spin basis as before, with the exchange sign (3) and (6). Singlevaluedness, that is

$$|\Psi_P(\mathbf{r})\rangle = |\Psi_{\overline{P}}(-\mathbf{r})\rangle, \tag{14}$$

leads to the extended spin-statistics relation

$$\psi_{\overline{M},\overline{P}}(-\mathbf{r}) = (-1)^{2S}\psi_{M,P}(\mathbf{r}). \tag{15}$$

This is consistent with the requirement that the original spin-statistics relation must hold when the P state of both particles is the same, that is $P = \overline{P}$.

In the above argument, P has been treated differently from spin, notwithstanding the fact that the operators representing P (e.g. isospin) can have the same mathematical structure as angular momenta. The reason is that such mathematical resemblance conceals a physical difference: it is spin, and not any other property P, that is uniquely related to spatial rotations, because of its connection (section 6) with galilean or Lorentz invariance.

An argument similar to that leading to (15) has been given [14] in the context of Kaluza-Klein theory.

The decision to regard the particles as identical, embodied in (14), needs further discussion. An alternative possibility would be to regard the different values p_1, p_2 as distinguishing the particles. It seems absurd to consider macroscopic objects such as apples and pears as identical particles in different states of quantum fruitiness (P). Nevertheless, it is possible to choose to do this - but the choice is inconsequential, because as is well known it leaves unconstrained the symmetry of the space-spin part of the state - the symmetry of the P part of the state can always be adjusted to satisfy (15). The extended spin-statistics relation has consequences only when superpositions of states with different p are meaningful, and the interactions are such that transitions can occur between them (so that the P physics is coherently entangled with the space-spin physics).

6. SPIN AND RELATIVITY

There is a complicated history of derivations of SS[4] using arguments that rely on relativity, involving successively more refined postulates (causality, absence of negative-energy states, hermitean fields...). This raises the question of the relation between relativistic approaches and our nonrelativistic formulation. On this subject we can make only scattered remarks.

First we should point out that our derivation was nonrelativistic in the sense that it made no use of relativity, and not in the sense of being a low-velocity approximation. Since time never entered our considerations, the exchanges we considered (involving the variables $\mathbf{r}$ and M) can be regarded as taking place at fixed time. But fixed time is not relativistically invariant. Regarded relativistically, our exchanges were spacelike. This makes our arguments appear complementary to the some of the quantum field theoretic ones [15, 16], which involve the creation of pairs of antiparticles, and therefore are based on timelike exchanges.

Second, although we considered only the relation between spin and statistics, and not the origin of spin itself, the widespread belief that spin is unavoidably relativistic has led to doubts about our arguments involving exchange. But the existence of spin is equally a consequence of galilean relativity as of einsteinian relativity. This point has been well made before [17, 18], and it is not necessary to repeat the general arguments. However, we think it worth outlining the galilean spin-1/2 case in the simplest and least technical way, in an argument attributed to Feynman [19]. This is done in the Appendix.

Third, there is the intriguing possibility that the field-theoretic arguments could be made to operate in reverse, in the following sense. Suppose that the nonrelativistic programme outlined here is eventually completed, so that it would become clear that SS is embedded in quantum mechanics in a fundamental way, more primitive than field theory. Then instead of deriving SS by demanding that field theory satisfy certain requirements, such as causality and energy positivity, the knowledge that SS must be true might be invoked to show that field theory already possesses these desirable properties. This would be much more satisfactory, after all, than having to impose them.

Fourth, we are not aware that there exists any relativistic field theory, for particles with spin, that involves the configuration space we use here, where indistinguishability is incorporated geometrically by the identification of permuted configurations.

Fifth, we note that Anandan [14] has presented a relativistic generalization of our construction, in what may be a first step in establishing a bridge to the field-theoretic arguments.

APPENDIX. GALILEAN TRANSFORMATIONS AND SPIN 1/2

For a free particle without spin, with hamiltonian $H=p^2/2m$, Schrödinger's equation

$$i\hbar\frac{\partial}{\partial t}\psi(\mathbf{r},t)=-\frac{\hbar^2}{2m}\nabla_{\mathbf{r}}^2\psi(\mathbf{r},t) \tag{A1}$$

is Galilean-invariant in the following sense. Under the transformation to

$$\begin{gathered} t\to t_1\equiv t-T,\quad \mathbf{r}\to\mathbf{r}_1\equiv\mathbf{R}\mathbf{r}-\mathbf{v}t-\mathbf{a}, \\ \psi(\mathbf{r},t)\to\psi_1(\mathbf{r}_1,t_1)\equiv\psi(\mathbf{r},t)\exp\left\{-i\frac{m}{\hbar}\left(\mathbf{v}\cdot\mathbf{r}_1+\tfrac{1}{2}v^2t_1\right)\right\}, \end{gathered} \tag{A2}$$

where T is a constant scalar, $\mathbf{a}$ and $\mathbf{v}$ are constant vectors, and $\mathbf{R}$ is a constant rotation matrix, the equation preserves its form:

$$i\hbar\frac{\partial}{\partial t_1}\psi_1(\mathbf{r}_1,t_1)=-\frac{\hbar^2}{2m}\nabla_{\mathbf{r}_1}^2\psi_1(\mathbf{r}_1,t_1). \tag{A3}$$

For a particle with spin 1/2, this invariance is obviously shared by the two-spinor Schrödinger equation generated by the free 2x2 matrix Hamiltonian

$$\mathbf{H}=\frac{1}{2m}(\mathbf{S}\cdot\mathbf{p})^2=\frac{1}{2m}p^2\mathbf{1}, \tag{A4}$$

where $\mathbf{S}$ is the vector of Pauli matrices.

In both cases, external fields with potentials $\mathbf{A}(\mathbf{r},t)$, $V(\mathbf{r},t)$ can then be introduced by minimal coupling to the particle's charge q through

$$\mathbf{p}\to\mathbf{p}-q\mathbf{A}(\mathbf{r},t). \tag{A5}$$

and addition of $qV(\mathbf{r},t)$ to $\mathbf{H}$. In the spin 1/2 case, coupling to the first equation in (A4) leads to

$$\begin{aligned} \mathsf{H} &= \frac{1}{2m}\left(\mathsf{S}\cdot(\mathbf{p}-q\mathbf{A})\right)^2 + qV(\mathbf{r},t) \\ &= \frac{1}{2m}\left[\mathbf{p}-q\mathbf{A}(\mathbf{r},t)\right]^2\mathbf{1} - \frac{q\hbar}{2m}\mathsf{S}\cdot\mathbf{B}(\mathbf{r},t) + qV(\mathbf{r},t). \end{aligned} \tag{A6}$$

where $\mathbf{B}=\nabla\times\mathbf{A}$. This is the Pauli equation, with the spin operator $\mathsf{s}=h\mathsf{S}/4\pi$ coupled to the magnetic field with a magnetic moment **m**, that is

$$\frac{q\hbar}{2m}\mathsf{S}\cdot\mathbf{B} = \mathbf{m}\cdot\mathbf{B}, \quad \text{where} \quad \mathbf{m} = \frac{q\hbar}{2m}\mathsf{S} = \frac{q\mathsf{s}}{m}. \tag{A7}$$

This **m**, originating in a free equation that is invariant under galilean transformations, is the same – that is, it has the same gyromagnetic ratio - as that in the corresponding Dirac equation, which is Lorentz-invariant.

REFERENCES

1. Berry, M. V., and Robbins, J. M., Indistinguishability for quantum particles: spin, statistics and the geometric phase. *Proc. Roy. Soc. Lond.* **A453**, 1771-1790 (1997).
2. Berry, M. V., and Robbins, J. M., Quantum indistinguishability: alternative constructions of the transported basis. *J. Phys. A (Letters)* **33**, L207-L214.
3. Pauli, W., Connection between spin and statistics. *Phys. Rev.* **58**, 716-722 (1940).
4. Duck, I., and Sudarshan, E. C. G., *Pauli and the spin-statistics theorem* , World Scientific, Singapore, 1997.
5. Duck, I., and Sudarshan, E. C. G., Toward an understanding of the spin-statistics theorem. *Am. J. Phys.* **66**, 284-303 (1998).
6. Atiyah, M., Geometry of classical particles. *Asian Journal of Mathematics, in press* (2000).
7. Laidlaw, M. G. G., and DeWitt, C. M., Feynman functional integrals for systems of indistinguishable particles. *Phys. Rev. D.* **3**, 1375-1378 (1971).
8. Leinaas, J. M., and Myrheim, J., On the theory of identical particles. *Nuovo Cim.* **37B**, 1-23 (1977).
9. Sorkin, R., Particle statistics in three dimensions. *Phys. Rev. D.* **27**, 1787-1792 (1983).
10. Aharonov, Y., and Bohm, D., Significance of electromagnetic potentials in the quantum theory. *Phys. Rev.* **115**, 485-491 (1959).
11. Olariu, S., Popescu, I., and I.I., I., The quantum effects of electromagnetic fluxes. *Revs. Mod. Phys.* **57**, 339-436 (1985).
12. Robbins, J. M., *in preparation* (2000).
13. Schwinger, J., "On angular momentum", in *Quantum theory of angular momentum* (L. C. Biedenharn, and H. Van Dam, Eds.), Academic Press, New York, 1965, pp. 229-279.
14. Anandan, J., Spin-statistics connection and relativistic Kaluza-Klein space-time. Physics Letters A **248**, 124-130 (1998).
15. Balachandran, A. P., Daughton, A., Gu, Z.-C., Sorkin, R. D., Marmo, G., and Srivastava, A. M., Spin-statistics theorems without relativity or field theory. *Int. J. Mod. Phys.* **A8**, 2993-3044 (1993).

16. Feynman, R. P., "The reason for antiparticles", in *The 1986 Dirac memorial lectures* (R. P. Feynman, and S. Weinberg, Eds.), Cambridge University Press, New York, 1987.

17. Levy-Leblond, J.-M., Nonrelativistic particles and wave equations. *Commun. Math. Phys* **6**, 286-311 (1967).

18. Levy-Leblond, J.-M., The pedagogical role and epistemological significance of group theory in quantum mechanics. *Riv. Nuovo. Cim* **4**, 99-143 (1974).

19. Mackintosh, A. R., The Stern-Gerlach experiment, electron spin and intermediate quantum mechanics. *Eur. J. Phys.* **4**, 97-106 (1983).

Chapter 2

Caustics and Related Optics

My first teaching assignment, while still a Ph.D. student at St Andrews, was to give the graduate course in general relativity. In this daunting task, I was greatly assisted by the eloquent presentations in the books by J. L. Synge, so I seized the chance to hear his lectures in London the following year. But his topic was not relativity; instead, he spoke about the Hamiltonian theory of rays and waves. This subject, and Synge's presentation of it, enchanted me, and I immediately conceived a postgraduate research project: to marry the largely geometric descriptive approach of Synge with the refined asymptotics of my St. Andrews supervisor Bob Dingle (see Chapter 4 and [7.8]). The resulting 'physical asymptotics' has animated much of my research.

The earliest fruits of this inspiration from Synge were papers from the late 1960s [B6, B8, B11] on semiclassical quantum scattering. In this work, I recognised the central importance of caustics as the envelopes (focal sets) of families of classical trajectories, and the need for more sophisticated 'uniform asymptotics' [B10] to accommodate them.

But it was only in the mid-1970s, after John Ziman showed me the book on catastrophe theory by the mystical and visionary mathematician René Thom, and Christopher Zeeman helped deconstruct it for me, that I recognised the central importance for natural caustics of the property of genericity (equivalent terms are structural stability and universality). This emphasis on what is typical, rather than exactly solvable special cases, was emerging elsewhere in physics at the same time: in statistical mechanics, as the 'universality classes' of critical behaviour associated with phase transitions, and in chaotic dynamics (see Chapter 3 and [B76]). Thom's mathematics, and its extension by Vladimir Arnold [7.10], provided a classification of stable caustics. A bonus was that this mathematics, initially rooted in the geometry of classical trajectories or the light rays of geometrical optics, also provided a framework for describing 'diffraction catastrophes': the delicate and characteristic patterns of wave interference that decorate the geometrical caustics. The mathematical representation of diffraction catastrophes by oscillatory integrals led to a new class of special functions [B326, B421], explored in detail theoretically and experimentally (see e.g. [B79]).

The first application of this insight [2.1] was to the scattering of beams of molecules from the periodically undulating surfaces of crystals. The usefulness of what seemed at that time abstruse mathematics was confirmed by its prediction of an unexpected far-field caustic when the surface profile was non-separable. A more general exposition [2.2] followed.

For some time, I had realised that many features of quantum waves have their more accessible counterparts in optics. Applied to caustics, this flowered into the extensive development of 'catastrophe optics' in Bristol in the late 1970s, in collaborations with Colin Upstill, Francis Wright and John Nye. This activity culminated in several reviews ([B89], [B105]). The richest application [B58] of this circle of ideas, which turned out to require the full hierarchy of diffraction catastrophes, was to the statistics of fluctuations in random short waves (e.g. strongly twinkling starlight). The predictions of this theory still await detailed experimental investigation (notwithstanding our early attempt [B114]).

In 1979, the late Nandor Balazs and I found the 'Airy packet' [B78] solution of the free-space time-dependent Schrödinger equation, representing a wave that evolves without spreading and while accelerating. Both properties seem to contradict insights about waves and classical trajectories on which no forces act; but there is no paradox because what is accelerating are not the individual (straight) rays but their caustic. Optical scientists then realised that when implemented via the paraxial wave equation, the same solution represents a wave that bends as it propagates diffractionlessly in a uniform medium: an 'Airy beam', now understood as the simplest member of a large class of curved ('accelerating') optical beams (see also [B496]).

My fascination with optics, and especially its history ([2.3], [2.4]), widened. I came to realise [2.5] the importance of the 1830s. In that decade, the three main singularities of light (and waves more generally) began to be understood: singularities of ray families (i.e. caustics), of phase, and of polarization.

A new look ([2.6], [2.7]) at the 'conoscopic figures' — interference patterns in thin sheets of anisotropic material viewed through crossed polarizers — with Rajendra Bhandari and Susanne Klein, led to a reformulation of classical crystal optics [B355], with Mark Dennis, in which we unified birefringence, chirality and anisotropic absorption in terms of singularities in direction space. In turn, this led, in collaboration with Mike Jeffrey, to a comprehensive understanding of Hamilton's conical refraction [B360, B386, B387, B393, B423], and a review [2.8] of this astonishing phenomenon.

Extreme interference, associated with the superposition of many waves with arithmetically-related phases that give rise to striking coherence phenomena, was explored in a mathematical study with Joel Goldberg ([B171], published as the first paper in the new journal *Nonlinearity* (whose name I had coined). Several years later, I realised that essentially the same mathematics — generalising the Gauss sums of number theory — provides a comprehensive understanding of the optical Talbot effect (in [B274] with Susanne Klein) and its formal counterpart in quantum physics, namely revivals of a periodic initial state [B275]. An unexpected consequence was the prediction, with Eberhard Bodenschatz [B304], that coherent interference of waves can generate patterns identical to diffraction catastrophes, even where there are no geometrical rays. This body of work culminated in a review [2.9] with Irene Marzoli and Wolfgang Schleich.

I was, and remain, delighted by unusual or striking optical phenomena, and finding, or commenting on, simple explanations of them: extreme non-paraxiality in reflections from complete spheres [2.10], bright-edged 'shadows' (actually caustics) below rippling water ([B111], with Jo Hajnal), oxymoronic black-and-white fringes from white light ([B256], with Anna Wilson), 'Laplacian' images in oriental magic mirrors [2.11] and their transmission counterpart 'magic windows' (see [B497]), Anderson localization controlling reflections from stacks of plastic overhead-projector sheets ('Transparent mirrors' [B281], with Susanne Klein), polarization singularities in the blue sky ([B373], with Mark Dennis and Raymond Lee), and Raman's curious error in explaining mirages ([B465], [7.1]). A collection of claritons concerning the optics of nature [2.12] was my celebration of 2015 as the International Year of Light.

Another dramatic wave phenomenon, in surface waves on water rather than light, is the 'bore', on the River Severn near Bristol, in which the incoming tide arrives suddenly and the downstream flow reverses direction. I was privileged to witness the world's biggest bore: the 'Silver Dragon' on the Qiantang river near Hangzhou [2.13].

J. Phys. A: Math. Gen., Vol. 8, No. 4, 1975. Printed in Great Britain. © 1975

Cusped rainbows and incoherence effects in the rippling-mirror model for particle scattering from surfaces

M V Berry
H H Wills Physics Laboratory, Bristol University, Tyndall Avenue, Bristol BS8 1TL, UK

Received 9 September 1974

Abstract. We consider scattering from a corrugated hard surface Σ with random moving perturbations (a 'rippling mirror'). Kirchhoff's approximation enables the classical limit, diffraction effects and incoherence to be treated within the same framework. The classical rainbow is a curve $\mathscr{C}$ in the two-dimensional space of deflections $\boldsymbol{G}$; we study the topology of $\mathscr{C}$ and show that it has cusps whose positions are sensitive to the form of Σ. Classically the scattering is singular on Σ but diffraction softens the singularities; we give the diffraction functions to be used near and on smooth parts and cusps of $\mathscr{C}$, and derive criteria for the observability of rainbow structure (taking account of surface periodicity which quantizes $\boldsymbol{G}$). Random thermal perturbations of Σ blur the diffracted beams; we introduce a simple approximation for the blurring function, and this suggests a simple method for inverting experimental data to obtain the 'surface phonon spectrum', even in cases where 'multiphonon processes' dominate.

1. Introduction

The scattering of beams of particles (atoms, molecules or ions) from solid surfaces can give useful information about particle–solid interactions (Toennies 1974). To extract this information from the experimental results, however, one requires a workable theory, that is a theory that is neither so simple that it fails to describe a wide range of phenomena nor so complicated that detailed predictions cannot easily be made. From this point of view the most useful theory so far assumes that the interaction potential between particle and solid is zero outside a surface Σ, and rises suddenly to infinity as the particle approaches the solid through Σ. This totally reflecting surface has predominantly the two-dimensional periodicity of the surface of a perfect solid, but is perturbed in a random manner by thermal effects. To describe approximately the scattering from this 'rippling mirror' the theory employs Kirchhoff's diffraction integral, rather than giving an exact treatment based on the wave equation. We use the expression 'rippling mirror' instead of the more common 'corrugated hard surface' because we wish explicitly to consider the inelastic effects of the time dependence of the random perturbations of Σ.

The exact quantum scattering from the rippling mirror Σ has been discussed in detail by Beeby (1972, 1973), while the additional Kirchhoff approximation has been used by Garibaldi *et al* (1974), who call it the eikonal approximation. Using a special form for Σ, these latter authors give a quantum-mechanical analysis of the classical surface rainbow discovered by McClure (1970) and Smith *et al* (1969). They also give a formalism for the inelastic and diffuse incoherent scattering that arises from the space-and-time-dependent parts of the random perturbation of Σ, but they do not discuss the nature of this incoherent scattering.

In this paper we use exactly the same scattering model as Garibaldi *et al* (1974), but we go further, and obtain some rather simple general results. The first concerns rainbows: by examining more realistic surfaces Σ, which are less symmetrical than that of Garibaldi *et al*, we establish the topology of the classical rainbow in the image domain, that is, in the two-dimensional space of directions of the scattered particles. In this space the rainbow consists of two closed curves ('caustics'), one inside the other; the inner curve has several cusps (the number depends on the symmetry of the lattice). Our second result concerns the diffraction functions that must be used to describe the quantum-mechanical softening of the classical rainbow singularities near smooth and cusped caustics. We discuss the conditions under which these phenomena could be observed. Our final result concerns incoherent scattering: we derive a simple approximate formula, of apparently rather general applicability, giving the diffuse and inelastic incoherent scattering *explicitly* in terms of the spectrum of the random perturbations of Σ; the formula gives a very simple method, in principle, of determining this spectrum from scattering data, and may solve a puzzle concerning 'multiphonon processes' (Beeby 1973).

Before beginning our main argument we discuss briefly the limitations of the model used. By approximating the actual smooth particle–solid potential by the (repulsive) hard mirror Σ, we are neglecting the effects of the attractive potential well beyond the hard core. The principal such effects are a modification of the incident-beam wavevector near the mirror (Beeby 1971, Garibaldi *et al* 1974), and the existence of surface bound states into which the incident particles may fall (Lennard-Jones and Devonshire 1937, Cabrera *et al* 1970, Wolken 1973). By employing Kirchhoff diffraction theory, we are neglecting multiple reflections between different parts of the surface (Beeby 1972, Beckmann and Spizzichino 1963). Such reflections will be insignificant if the total variation $\Delta\theta$ of surface slopes is small and if the incident beam does not graze the surface; in fact $\Delta\theta$ does not exceed about 10° (Nahr, private communication), so that the use of Kirchhoff's integral should not involve any serious approximation. In addition, our scattering model implicitly neglects recoil effects, that is, any influence of the projectiles on the solid; this will be justified if the projectiles are light and the solid atoms heavy and tightly bound.

2. Kirchhoff diffraction theory

We employ Cartesian coordinates $\boldsymbol{r} = (x, y, z)$ to locate points in space. In the 'horizontal' plane $z = 0$ we locate points by $\boldsymbol{R} = (x, y)$. The form of the rippling mirror Σ at time t is defined by its height $h(\boldsymbol{R}, t)$ above the plane $\boldsymbol{R}$. The function h is the sum of a periodic stationary part $h_p(\boldsymbol{R})$ and a time-dependent random perturbation $h_r(\boldsymbol{R}, t)$, ie

$$h(\boldsymbol{R}, t) = h_p(\boldsymbol{R}) + h_r(\boldsymbol{R}, t), \tag{2.1}$$

where

$$h_p(\boldsymbol{R}+m\boldsymbol{a}+n\boldsymbol{b}) = h_p(\boldsymbol{R}), \tag{2.2}$$

$\boldsymbol{a}$ and $\boldsymbol{b}$ being unit vectors in the surface lattice and m and n being integers.

The incident beam is represented quantum-mechanically by a single plane wave $\psi_{inc}(\boldsymbol{r}, t)$ with frequency ω_0 (corresponding to an energy $E_0 = \hbar\omega_0$) and wavevector $\boldsymbol{k}_0$.

The wavelength is $\lambda_0 = 2\pi/k_0 = h/(2mE_0)^{1/2}$, where m is the mass of the particles. The horizontal component of $\boldsymbol{k}_0$ is $\boldsymbol{K}_0$ and the vertical component k_{0z} is negative because the wave is approaching Σ from above. Obviously $K_0 < k_0$ for physically interesting waves. The scattered particles emerge in different directions $\boldsymbol{k}$ and with different frequencies ω (because of the Doppler shifts caused by the time dependence of h_r); therefore we represent them by a wavefunction $\psi_{sc}(\boldsymbol{r}, t)$ which is a sheaf of plane waves receding from Σ, so that the vertical component k_z of $\boldsymbol{k}$ is positive if the magnitude K of the horizontal component $\boldsymbol{K}$ is less than k, and positive imaginary (corresponding to evanescent waves) if $K > k$. Thus we can write ψ_{inc} and ψ_{sc} in the form

$$\psi_{inc}(\boldsymbol{r}, t) = \exp[\mathrm{i}(\boldsymbol{K}\,.\,\boldsymbol{R} - |k_{0z}|z - \omega_0 t)]$$
$$\psi_{sc}(\boldsymbol{r}, t) = \iint \mathrm{d}\boldsymbol{K} \int_{-\infty}^{\infty} \mathrm{d}\omega a(\boldsymbol{K}_0, \omega_0; \boldsymbol{K}, \omega) \exp[\mathrm{i}(\boldsymbol{K}\,.\,\boldsymbol{R} + k_z z - \omega t)], \tag{2.3}$$

where a is the amplitude of the wave specified by $\boldsymbol{K}$ and ω (we could specify each wave by its three-dimensional wavevector $\boldsymbol{k}$, but the form written is more useful).

We wish to calculate the amplitudes a. This we do by using the boundary condition that the total wave $\psi_{inc} + \psi_{sc}$ vanishes on Σ. Strictly speaking the form (2.3) for ψ_{sc} is exact only if z exceeds the largest value of $h(\boldsymbol{R}, t)$; for smaller z (but still above Σ) there will be small-amplitude waves that have been scattered from the hills down into the valleys of Σ. This effect will cause multiple scattering, and we neglect it. Thus the boundary condition on Σ gives the following integral equation for a:

$$\exp[\mathrm{i}(\boldsymbol{K}_0\,.\,\boldsymbol{R} - |k_{0z}|h(\boldsymbol{R}, t) - \omega_0 t)]$$
$$= -\iint \mathrm{d}\boldsymbol{K} \int_{-\infty}^{\infty} \mathrm{d}\omega a(\boldsymbol{K}_0, \omega_0; \boldsymbol{K}, \omega) \exp[\mathrm{i}(\boldsymbol{K}\,.\,\boldsymbol{R} + k_z h(\boldsymbol{R}, t) - \omega t)]. \tag{2.4}$$

This holds for all $\boldsymbol{R}$ and t. Unfortunately this equation cannot be solved exactly for a in closed form. But we notice that if h is zero or varies linearly with $\boldsymbol{R}$ and t we can use Fourier's theorem, to obtain

$$a(\boldsymbol{K}_0, \omega_0; \boldsymbol{K}, \omega)$$
$$= \frac{-1}{(2\pi)^3} \iint \mathrm{d}\boldsymbol{R} \int \mathrm{d}t \exp\{\mathrm{i}[(\boldsymbol{K}_0 - \boldsymbol{K})\,.\,\boldsymbol{R} - (|k_{0z}| + k_z)h(\boldsymbol{R}, t) - (\omega_0 - \omega)t]\}. \tag{2.5}$$

This is exact if

$$h(\boldsymbol{R}, t) = \boldsymbol{P}\,.\,\boldsymbol{R} + vt \tag{2.6}$$

which corresponds to a flat surface inclined at an angle $\tan^{-1}|\boldsymbol{P}|$ (to the horizontal), whose contours are perpendicular to $\boldsymbol{P}$ and which moves upwards at the constant speed v. Then (2.5) gives

$$a(\boldsymbol{K}_0, \omega_0; \boldsymbol{K}, \omega) = -\delta(\boldsymbol{K}_0 - \boldsymbol{K} + (|k_{0z}| + k_z)\boldsymbol{P})\delta(\omega - \omega_0 - (|k_{0z}| + k_z)v). \tag{2.7}$$

It can be verified that this correctly describes specular reflection from $\boldsymbol{K}_0$ into a direction $\boldsymbol{K}$ determined by the slope $\boldsymbol{P}$ of Σ, with the frequency Doppler shifted from ω_0 to ω by the motion of Σ.

Kirchhoff diffraction theory consists in using (2.5) for general surfaces Σ, which do not have the simple form (2.6). This will obviously be a better approximation, the smaller are the curvatures and accelerations of Σ. The principal advantage of (2.5) is the variety of phenomena it can describe. It is the simplest expression describing scattering from surfaces whose deviations from flatness may range from small perturbations to hills and dales many wavelengths in extent, for which classical mechanics gives a good description for the reflection (see § 3). This is particularly useful in particle–surface scattering, because it is often the case that reflection of h_p (cf (2.1)) may be treated classically or semiclassically while h_r is so small that its effects must be calculated using diffraction theory. Another property of (2.5) is that it gives at least a first approximation to the evanescent waves, for which $K > k$.

Experimentally, what is measured is the current of particles (far from the surface) travelling in a small range of directions about a chosen direction $\boldsymbol{K}$ with energies lying within a small range about a chosen energy $\hbar\omega$. If the surface is infinite in extent, or the experiment infinite in duration, then this current will be infinite; therefore we calculate the current I per unit area of Σ per unit time. If Σ has a (large) illuminated area $\mathscr{A}$ and the experiment lasts for a (long) time $\mathscr{T}$, then the current is proportional to

$$I(\boldsymbol{K}_0, \omega_0; \boldsymbol{K}, \omega) = \frac{|a(\boldsymbol{K}_0, \omega_0; \boldsymbol{K}, \omega)|^2(2\pi)^3}{\mathscr{A}\mathscr{T}}, \tag{2.8}$$

apart from purely kinematic factors (for a discussion of these, see Garibaldi *et al* 1974). We shall find that this expression always gives sensible results. For example, in the case of the flat surface (2.6), $|a|^2$ would involve squares of delta functions (see (2.7)). These simplify as follows: we write, symbolically,

$$\begin{aligned}\delta^2(\boldsymbol{K})\delta^2(\omega) &= \delta(\boldsymbol{K})\delta(\omega) \lim_{\mathscr{A}\mathscr{T}\to\infty} \left(\frac{1}{(2\pi)^2}\iint_{\mathscr{A}} \mathrm{d}\boldsymbol{R}\, \mathrm{e}^{\mathrm{i}\boldsymbol{K}.\boldsymbol{R}} \times \frac{1}{2\pi}\int_{\mathscr{T}} \mathrm{d}t\, \mathrm{e}^{\mathrm{i}\omega t}\right) \\ &= \delta(\boldsymbol{K})\delta(\omega)\mathscr{A}\mathscr{T}/(2\pi)^3. \end{aligned} \tag{2.9}$$

Thus (2.8) gives, for this case,

$$I = \delta(\boldsymbol{K} - \boldsymbol{K}_0 + (|k_{0z}| + k_z)\boldsymbol{P})\delta(\omega_0 - \omega - (|k_{0z}| + k_z)v), \tag{2.10}$$

an expression corresponding to a finite total intensity.

3. Perfect periodicity: topology of classical rainbows

If Σ is periodic, that is, if we are ignoring the effects of thermal disturbances, we can set h_r in (2.1) equal to zero, and expand in a Fourier series the function of $h_p(\boldsymbol{R})$ that appears in the diffraction integral (2.5). Then, using (2.8), the intensity I becomes

$$I(\boldsymbol{K}_0, \omega_0; \boldsymbol{K}, \omega) = \delta(\omega - \omega_0) \sum_{\boldsymbol{G}} |S_{\boldsymbol{G}}|^2 \delta(\boldsymbol{K} - \boldsymbol{K}_0 - \boldsymbol{G}). \tag{3.1}$$

We have reproduced here the well known result that in this case the scattering is elastic ($\omega = \omega_0$), and the emergent particles appear as a series of *diffracted beams* in directions $\boldsymbol{K} = \boldsymbol{K}_0 + \boldsymbol{G}$, where $\boldsymbol{G}$ are the two-dimensional vectors of the reciprocal surface lattice. The strength of the $\boldsymbol{G}$th diffracted beam is $|S_{\boldsymbol{G}}|^2$, where the diffraction amplitude $S_{\boldsymbol{G}}$ is given by

$$S_{\boldsymbol{G}} = \frac{1}{A}\iint_{\text{unit cell}} \mathrm{d}\boldsymbol{R} \exp\{-\mathrm{i}[\boldsymbol{G}.\boldsymbol{R} + (|k_{0z}| + k_z)h_p(\boldsymbol{R})]\}. \tag{3.2}$$

(A is the area of the unit cell.)

The hemisphere of directions of scattered particles corresponds to the circle $|\boldsymbol{K}| < k_0$ in the 'image' plane $\boldsymbol{K}$. This circle has area πk_0^2. Each diffracted beam 'occupies' a unit reciprocal lattice cell in $\boldsymbol{K}$ space. Each such cell has area $4\pi^2/A$. Thus the number $\mathcal{N}$ of observed diffracted beams is

$$\mathcal{N} = \frac{Ak_0^2}{4\pi} = \frac{\pi A}{\lambda_0^2} = \frac{mAE_0}{2\pi\hbar^2}. \tag{3.3}$$

Now, under nearly classical conditions we may regard λ_0 and $\hbar$ as small, or m and k_0 as large. Thus $\mathcal{N}$ is large, the diffracted beams are densely distributed in direction, and the 'deflection' $\boldsymbol{G} = \boldsymbol{K} - \boldsymbol{K}_0$ in (3.2) may be regarded as a quasi-continuous variable. We wish to discuss the form of $S_{\boldsymbol{G}}$ under these *semiclassical* conditions.

The important point is that when $k_0(= k)$ is large the integrand of (3.2) oscillates rapidly as $\boldsymbol{R}$ traverses the unit cell, and most of the area of integration gives no contribution, because of destructive interference. However at isolated points $\boldsymbol{R}_i(\boldsymbol{K}_0, \boldsymbol{G})$ where the *phase of the integrand is stationary* this cancellation does not occur, and we obtain contributions to $S_{\boldsymbol{G}}$. The stationary points $\boldsymbol{R}_i$ are given by

$$\boldsymbol{G} = -(|k_{0z}| + k_z)\nabla h_p(\boldsymbol{R}_i). \tag{3.4}$$

But this is exactly the condition for a surface point $\boldsymbol{R}$ to reflect a classical particle specularly from $\boldsymbol{K}_0$ to $\boldsymbol{K}_0 + \boldsymbol{G}$, so that we have found that only the classical paths contribute to the quantum diffraction integral (3.2) when k_0 is large. In interpreting all our subsequent classical and semiclassical formulae, it should not be forgotten that (3.4) is really an implicit equation for $\boldsymbol{G}$, since this vector appears also in

$$k_z(= [k^2 - (\boldsymbol{K}_0 + \boldsymbol{G})^2]^{1/2});$$

for gently-varying surfaces Σ this dependence of k_z on $\boldsymbol{G}$ is weak and in the general case it can be checked (in some cases laboriously) that the dependence does not invalidate any of our conclusions (eg figure 6 for the form of the 'rainbow line').

If the points $\boldsymbol{R}_i$ are sufficiently well separated for the integrand in (3.2) to oscillate many times between them, we may use the method of stationary phase to approximate $S_{\boldsymbol{G}}$. This involves expanding the phase in (3.2) about each point $\boldsymbol{R}_i$ up to terms in $(\boldsymbol{R} - \boldsymbol{R}_i)^2$, and evaluating the resulting Gaussian integrals. We arrive at the following 'simple semiclassical' formula for the diffraction amplitudes:

$$S_{\boldsymbol{G}} \simeq \frac{2\pi}{A(|k_{0z}| + k_z)} \sum_i \gamma_i \frac{\exp\{-\mathrm{i}[\boldsymbol{G} \cdot \boldsymbol{R}_i + (|k_{0z}| + k_z)h_p(\boldsymbol{R}_i)]\}}{(|\mathscr{H}(\boldsymbol{R}_i)|)^{1/2}}, \tag{3.5}$$

where the summation is over all points $\boldsymbol{R}_i$ reflecting particles with deviation $\boldsymbol{G}$, and

$$\gamma_i \equiv \begin{cases} +i & \\ -i & \text{if the phase } [\ldots] \text{ in (3.2) has at } \boldsymbol{R}_i \text{ a} \\ 1 & \end{cases} \begin{matrix} \text{minimum} \\ \text{maximum} \\ \text{saddle point,} \end{matrix} \tag{3.6}$$

and where $\mathscr{H}(\boldsymbol{R})$ is the *Hessian* of h_p at $\boldsymbol{R}$, defined by

$$\mathscr{H}(\boldsymbol{R}) \equiv \frac{\partial^2 h_p}{\partial x^2}\frac{\partial^2 h_p}{\partial y^2} - \left(\frac{\partial^2 h_p}{\partial x \partial y}\right)^2. \tag{3.7}$$

For the gently-varying surfaces Σ with which we are concerned, it is helpful to visualize $\mathscr{H}(\boldsymbol{R})$ as the *Gaussian curvature* of Σ at $\boldsymbol{R}$, since this quantity differs from $\mathscr{H}$ only by a factor $[1 + (\nabla h_p)^2]^{-3/2}$, which in turn differs from unity by only a few per cent. The Gaussian curvature is the product of the two principal curvatures at $\boldsymbol{R}$, and is positive where Σ is concave or convex, and negative where Σ is saddle-shaped.

Experimentally what is observed is $|S_{\boldsymbol{G}}|^2$ (equation (3.1)) and the simple semiclassical result (3.5) shows that this quantity consists of a set of 'steady' terms from each classical path, plus cross terms describing interferences between different classical paths. In the extreme classical limit these interference oscillations (of I as a function of $\boldsymbol{G}$ or $\boldsymbol{K}_0$) are too rapid to be detected, and any experiment would detect only their average, which is zero; in this case (3.5) gives

$$|S_{\boldsymbol{G}}|^2_{\text{classical}} = \frac{4\pi^2}{A^2(|k_{0z}|+k_z)^2}\sum_i \frac{1}{|\mathscr{K}(\boldsymbol{R}_i)|}. \tag{3.8}$$

Thus the Gaussian curvature of Σ at $\boldsymbol{R}_i$ is an inverse measure of the *strength of the contribution* from the path i. This can be seen in a purely classical way as follows: from (3.7) and the specular condition (3.4), it follows that $\mathscr{K}(\boldsymbol{R})$ is proportional to the *Jacobian of the mapping* from 'surface space' $\boldsymbol{R}$ onto 'deflection space' $\boldsymbol{G}$, $\boldsymbol{K}_0$ being kept constant; since in the incident beam particles are uniformly distributed over $\boldsymbol{R}$, this Jacobian $|\mathrm{d}\boldsymbol{G}/\mathrm{d}\boldsymbol{R}|$ is an inverse measure of the density of particles scattered into direction $\boldsymbol{G}$.

For a given $\boldsymbol{K}_0$, the intensity scattered with deflection $\boldsymbol{G}$ will be *infinite* whenever any contributing surface point $\boldsymbol{R}_i(\boldsymbol{G}, \boldsymbol{K}_0)$ lies at a place where $\boldsymbol{K}(\boldsymbol{R})$ vanishes. Now the equation

$$\mathscr{K}(\boldsymbol{R}) = 0 \tag{3.9}$$

defines a *line* $\mathscr{L}$ on the surface; each point $\boldsymbol{R}$ of $\mathscr{L}$ defines an 'image' point $\boldsymbol{G}$ according to (3.4), so that the image of $\mathscr{L}$ is also a line, $\mathscr{C}$, in the deflection space $\boldsymbol{G}$. We call $\mathscr{C}$ the *rainbow line*; any of the (densely distributed) diffraction spots lying near $\mathscr{C}$ will be very intense, so that $\mathscr{C}$ should show up clearly on intensity plots across the $\boldsymbol{G}$ plane. In optical terminology $\mathscr{C}$ is a 'caustic' of the reflected 'rays'—it is the locus of rays for which *angular focusing* occurs. Mathematically, $\mathscr{C}$ is a *singularity* in the mapping from $\boldsymbol{G}$ back to $\boldsymbol{R}$. As we cross $\mathscr{C}$ by varying $\boldsymbol{G}$, then two (or sometimes three) surface points $\boldsymbol{R}_i$ coalesce; thus the simple method of stationary phase leading to (3.5) is not applicable, and $S_{\boldsymbol{G}}$ is actually not infinite but merely large, as we shall discuss in more detail in § 4.

Thus according to classical mechanics the observed scattering should be dominated by the line $\mathscr{C}$ on $\boldsymbol{G}$. What is the form of this rainbow line? To answer this we need first to find the form of the line $\mathscr{L}$ on $\boldsymbol{R}$, defined by (3.9), that generates $\mathscr{C}$. Let the centres of the atoms in the top layer of the solid define the lattice points in $\boldsymbol{R}$. Then the surface Σ, considered as a landscape above the $\boldsymbol{R}$ plane, will have *summits* (full circles on figure 1) at the lattice points, because of the strong repulsive forces that the surface atoms exert on the incoming particles. At the corners of each Wigner–Seitz lattice cell—that is at the point farthest from atoms—Σ will have minima, or *immits* (a terminology introduced by Cayley 1859; see also Maxwell 1870; immits are marked open circles on figure 1). On each side of a lattice cell (broken line in figure 1) there must be a *saddle point* (crosses on figure 1). Let us assume that we are dealing with the simplest case where these are the *only* extrema of Σ; it is always possible to introduce more summits, immits or saddles by introducing more atoms into each unit cell of the surface layer. Figure 1 has been drawn for a rectangular lattice; for non-rectangular lattices the Wigner–Seitz cell would be hexagonal, and each summit would be surrounded by six immits and six saddles.

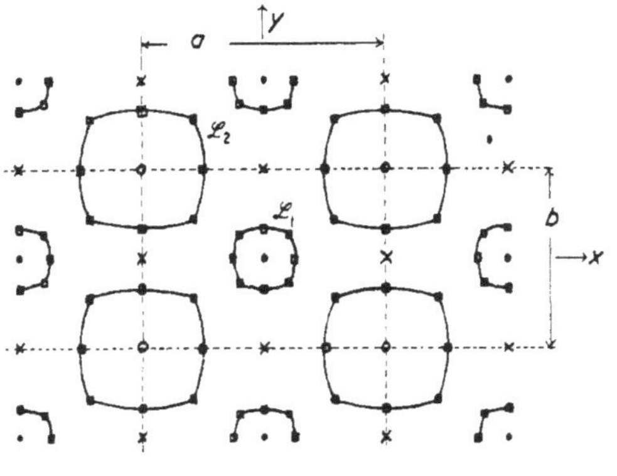

Figure 1. Features of the periodic surface Σ: ------ lattice lines; ● summits; ○ immits; × saddles; —— locus $\mathscr{L}$ of points of zero Gaussian curvature; □ points that must lie on $\mathscr{L}$.

Now the required line $\mathscr{L}$ is the locus of points of zero Gaussian curvature, that is the locus of points where at least one of the two principal curvatures of Σ vanishes. At a summit, both curvatures are (say) negative, at an immit both are positive, at a saddle one curvature is positive and one is negative. Therefore between a saddle and a summit, or a saddle and an immit, we must cross $\mathscr{L}$ $2n-1$ times, where n is an integer; between a summit and an immit both curvatures must change an odd number of times so we must cross $\mathscr{L}$ $2n$ times. The simplest case is $n = 1$, and points satisfying these conditions are marked by open squares on figure 1, and they are joined by a possible line $\mathscr{L}$ (marked full curve). Thus $\mathscr{L}$ consists of closed curves surrounding the regions of positive $\mathscr{K}$ containing summits and immits, while the open region of negative curvature, containing saddles, extends through the lattice. The curves $\mathscr{L}_1$, around summits (figure 1), will generally not have the same shape as the curves $\mathscr{L}_2$, around immits, because summits and immits are not symmetrical features of Σ—they correspond respectively to repulsive regions near atoms and relatively attractive 'interstitial' regions, so that Σ varies more gently near immits than near summits. (Note that it is not possible to satisfy the above conditions on the curvatures if $\mathscr{L}$ surrounds the saddles.)

The rainbow line $\mathscr{C}$ in $\boldsymbol{G}$ is generated from $\mathscr{L}$ by (3.4). The two closed curves $\mathscr{L}_1$ and $\mathscr{L}_2$ (figure 1) will generate two closed curves $\mathscr{C}_1$ and $\mathscr{C}_2$ in $\boldsymbol{G}$. We expect $\mathscr{C}_2$ to lie within $\mathscr{C}_1$, because the slopes $\nabla h_p(\boldsymbol{R})$ of Σ are smaller near immits (ie near $\mathscr{L}_2$) than near summits (ie near $\mathscr{L}_1$). To find the form of $\mathscr{C}_1$ and $\mathscr{C}_2$ we first introduce the special surface

$$h_p^0(\boldsymbol{R}) = h_0\left(\cos\left(\frac{2\pi x}{a}\right) + \cos\left(\frac{2\pi y}{b}\right)\right), \tag{3.10}$$

in which summits and immits are symmetrical. For this form of Σ it is easy to solve (3.10) for $\mathscr{L}$, and we find (as do Garibaldi *et al* 1974) that $\mathscr{L}_1$ and $\mathscr{L}_2$ touch in this special case, and form the following set of intersecting lines across $\boldsymbol{R}$:

$$\begin{aligned} x &= (m+\tfrac{1}{2})a/2 \\ y &= (n+\tfrac{1}{2})b/2 \end{aligned} \tag{3.11}$$

where m and n are integers. Now we use (3.4) and find that the image $\mathscr{C}$ is a single rectangle given by

$$G_x = \pm\frac{2\pi h_0}{a}(|k_{0z}|+k_z) \qquad \left(|G_y| < \frac{2\pi h_0}{b}(|k_{0z}|+k_z)\right)$$
$$G_y = \pm\frac{2\pi h_0}{b}(|k_{0z}|+k_z) \qquad \left(|G_x| < \frac{2\pi h_0}{a}(|k_{0z}|+k_z)\right). \tag{3.12}$$

In the general case, $h_p(\boldsymbol{R})$ is given not by (3.10) but by

$$h_p(\boldsymbol{R}) = h_p^{(0)}(\boldsymbol{R}) + \epsilon h_p^{(1)}(\boldsymbol{R}), \tag{3.13}$$

where ϵ is a perturbation parameter and $h_p^{(1)}(\boldsymbol{R})$ is any periodic surface not symmetrical in summits and immits. Then our general topological arguments tell us that the rectangular rainbow line (3.12) must split into two as soon as ϵ departs from zero. How does this splitting occur? To answer this we use 'catastrophe theory'; this is a branch of differential topology based on a theorem by Thom (1969, 1972) concerning singularities of mappings defined by gradients (in our case the mapping is $\boldsymbol{G} \to \boldsymbol{R}$, defined by the inverse of (3.4)). The theorem states that the singularities can only be of certain restricted types. In our two-dimensional case the singularities in the plane $\boldsymbol{G}$ are *lines* $\mathscr{C}$, as we know, which are smooth except at isolated points where they may have *cusps*. A cusp is a point where a curve reverses direction as it is traversed; the simplest example of a cusp is the point $x = y = 0$ on the curve $y^2 = x^3$. Thus we expect the rainbow line $\mathscr{C}$ to have cusps. To see how these cusps arise we consider the *rainbow surface* $\mathscr{C}$ generated by adding the third dimension ϵ to the space $\boldsymbol{G}$; this corresponds to looking at all the rainbows from the family of Σ's defined by (3.13). The rainbow surface must have two sheets $\mathscr{C}_1$ and $\mathscr{C}_2$ which touch at $\epsilon = 0$ where it has a rectangular section. Each corner corresponds to a singularity in three dimensions, and the only permitted singularity in which two sheets touch at a corner is, by Thom's theorem, the *hyperbolic umbilic*, illustrated in figure 2. Away from the singularity, that is for nonzero values of ϵ where

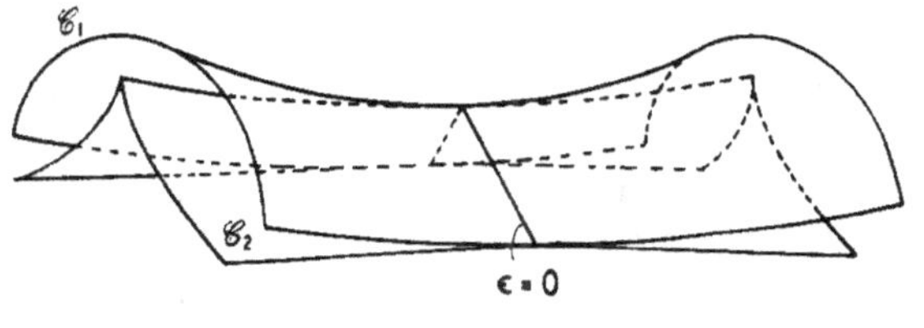

Figure 2. Hyperbolic umbilic catastrophe.

the symmetry of Σ between summits and immits is broken, $\mathscr{C}$ has split into two curves, as expected, and the inner one has a cusp where $h_p^{(0)}(\boldsymbol{R})$ gave a corner. Thus we expect the rainbow lines to take the form shown on figure 3, which is the main result of this section. (For non-rectangular lattices the Wigner–Seitz cell is hexagonal and there would be six cusps.)

These conclusions are supported by a detailed analysis of the special case for which the symmetry-breaking surface modification $h_p^{(1)}(\boldsymbol{R})$ of equation (3.13) is given by

$$h_p^{(1)}(\boldsymbol{R}) = h_0 \cos\left(\frac{2\pi x}{a}\right)\cos\left(\frac{2\pi y}{b}\right); \tag{3.14}$$

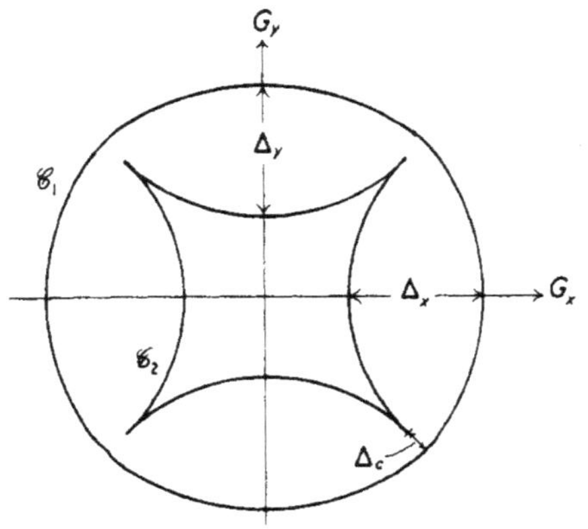

Figure 3. General form of rainbow lines: $\mathscr{C}_1$ and $\mathscr{C}_2$ originate near summits and immits of Σ respectively.

the rainbow lines $\mathscr{C}$ have precisely the form shown on figure 3. In addition we find that the separation Δ_c in the $\boldsymbol{G}$ plane between $\mathscr{C}_1$ and $\mathscr{C}_2$ at a corner (figure 3) is, for small ϵ

$$\Delta_c = 4\pi\left(\frac{1}{a^2}+\frac{1}{b^2}\right)^{1/2}\epsilon^2 h_0(|k_{0z}|+k_z) \sim \frac{16\pi^2\epsilon^2 h_0\sqrt{2}}{a\lambda_0}. \tag{3.15}$$

It is also easy to calculate the separations Δ_x and Δ_y between $\mathscr{C}_1$ and $\mathscr{C}_2$ (figure 3) along the G_x and G_y axes, and we find

$$\begin{aligned}\Delta_x &= \frac{4\pi}{a}\epsilon h_0(|k_{0z}|+k_z) \sim \frac{16\pi^2\epsilon h_0}{a\lambda_0}\\ \Delta_y &= \frac{4\pi}{b}\epsilon h_0(|k_{0z}|+k_z) \sim \frac{16\pi^2\epsilon h_0}{b\lambda_0}.\end{aligned} \tag{3.16}$$

The computations of McClure (1970, 1971) which gave the first theoretical evidence for rainbows in the classical scattering of particles from surfaces, were insufficiently detailed to show the cusp structure clearly. McClure presents intensity maps of the scattering as a function of polar angles θ and ϕ (in our notation, $G_x = k\sin\theta\cos\phi - K_{0x}$, $G_y = k\sin\theta\sin\phi - K_{0y}$). He uses a Monte Carlo procedure which averages over paths emerging in angular 'bins' whose widths are $\Delta\theta = 1^0$ and 3^0, $\Delta\phi = 5^0$. This is a little too coarse, and obscures much of the detail in figure 3; nevertheless, his intensity maps do show 'ridge' and 'tentpole' features, probably corresponding to smooth parts and cusps of the rainbow line. Strictly speaking the ridges and tentpoles should be infinitely high, because the classical rainbow strength is infinite (although integrable!); however the Monte Carlo procedure does not use the formula (3.8) for the contribution from each path but instead calculates directly the Jacobian $|\mathrm{d}(\theta, \phi)/\mathrm{d}\boldsymbol{R}|$, and thus averages over the angular bins. (In McClure's calculations (3.8) would not apply, because he uses a realistic smooth atom–surface potential rather than the mirror Σ.)

The shape of the rainbow line (figure 3) could in principle give detailed information about the surface Σ; this is clear from (3.15) and (3.16), which depend differently on ϵ. In practice the details of $\mathscr{C}$ will be blurred, to a greater or lesser degree by diffraction and by thermal motion of Σ; we consider these effects in more detail in the next two sections.

4. Perfect periodicity: sewing the quantum flesh on the classical bones

There are two kinds of quantum or diffraction effect which conspire to obscure the details of $\mathscr{C}$. The first arises because $\boldsymbol{G}$ is not a continuous variable but consists of reciprocal lattice points; thus patterns in the $\boldsymbol{G}$ plane (eg figure 3) are sampled at discrete points, and these will usually miss the rainbow line. The second is that the function $|S_G|^2$ (equation (3.1)) which is sampled is not given precisely by the classical or semiclassical formulae (3.8) or (3.5), which diverge on $\mathscr{C}$, but by the diffraction integral (3.2), which is large but finite on $\mathscr{C}$. Expressing these effects in another way, we can say that interference between waves emerging from different points in the same lattice cell in $\boldsymbol{R}$ blurs out the rainbow line into a diffraction pattern in $\boldsymbol{G}$, while interference between waves emerging from equivalent points in different cells quantizes $\boldsymbol{G}$ so that this diffraction pattern is sampled at discrete points.

First we discuss the blurring of $\mathscr{C}$ by diffraction. Near to a smooth portion of $\mathscr{C}$ the semiclassical formula (3.5) breaks down because two contributing surface points (say $\boldsymbol{R}_1$ and $\boldsymbol{R}_2$) coalesce, thus violating the condition for the applicability of the method of stationary phase. In these circumstances we require a *uniform approximation* to S_G, that is, a formula for (3.2) which is valid on and near $\mathscr{C}$ and which reduces to (3.5) far from $\mathscr{C}$. Uniform approximations to integrals were invented by Chester *et al* (1957), introduced into scattering theory by Berry (1966) (see also Berry and Mount 1972), and shown to be numerically extremely accurate by Mount (1973).

In the present problem the result is that the contribution $S_G^{(\text{rainbow})}$ from $\boldsymbol{R}_1$ and $\boldsymbol{R}_2$ in (3.5) must be replaced by the following formula, involving *Airy functions* Ai (Abramowitz and Stegun 1964) and their derivatives Ai′:

$$S_G^{(\text{rainbow})} = \frac{2\pi\sqrt{\pi}\exp[\frac{1}{2}\mathrm{i}(\Phi_1+\Phi_2-\frac{3}{2}\pi)]}{A(|k_{0z}|+k_z)}\Bigg[\left(\frac{1}{\mathscr{K}_1^{1/2}}+\frac{1}{(-\mathscr{K}_2)^{1/2}}\right)\left(\frac{3(\Phi_2-\Phi_1)}{4}\right)^{1/6}$$
$$\times \mathrm{Ai}[-(\tfrac{3}{4}(\Phi_2-\Phi_1)^{2/3}]-\mathrm{i}\left(\frac{1}{\mathscr{K}_1^{1/2}}-\frac{1}{(-\mathscr{K}_2)^{1/2}}\right)\left(\frac{4}{3(\Phi_2-\Phi_1)}\right)^{1/6}$$
$$\times \mathrm{Ai}'[-(\tfrac{3}{4}(\Phi_2-\Phi_1))^{2/3}]\Bigg]. \tag{4.1}$$

Here Φ denotes the phase

$$\Phi \equiv -[\boldsymbol{G}.\boldsymbol{R}+(|k_{0z}|+k_z)h_p(\boldsymbol{R})] \tag{4.2}$$

in (3.5), and the subscripts 1 and 2 refer to the two contributing points $\boldsymbol{R}_1$ and $\boldsymbol{R}_2$. *On the illuminated side* of the rainbow $\mathscr{C}$ the two Gaussian curvatures $\mathscr{K}_1$ and $\mathscr{K}_2$ will have opposite signs, and we have chosen $\boldsymbol{R}_1$ so that $\mathscr{K}_1$ is positive and assumed that Φ_1 is a minimum. The real positive root of $(\Phi_2-\Phi_1)^{2/3}$ is taken, so that the Airy functions have a negative argument and are thus oscillatory functions (figure 10.6 of Abramowitz and Stegun 1964); equation (4.1) then describes the 'supernumerary rainbows' (Airy 1838, Ford and Wheeler 1959). For deflections $\boldsymbol{G}$ *on the dark side of* $\mathscr{C}$ there are no real paths $\boldsymbol{R}_1$ and $\boldsymbol{R}_2$, and we require complex solutions of (3.4); we can then take a real negative root of $(\Phi_2-\Phi_1)^{2/3}$, and the Airy functions have a positive argument, and decay exponentially 'into the shadow'. On $\mathscr{C}$ the expression (4.1) is finite, and a little analysis shows that $|S_G|^2$ rises to a value of order $(h_0/\lambda_0)^{1/3}$ larger than in the 'classical' region away from $\mathscr{C}$ (h_0 is a measure of the maximum excursion of Σ from the $\boldsymbol{R}$ plane). By analysing the special surface (3.10) Garibaldi *et al* (1974) also discuss the diffractive softening of the rainbow singularity; they obtain an expression for $S_G^{(\text{rainbow})}$ in terms of Bessel functions of large order; these can, however, be uniformly approximated by Airy functions (Abramowitz and Stegun 1964) so that the formalism of these authors is a special case of ours.

Near a *cusp* of $\mathscr{C}$, even (4.1) breaks down, because not two but three points $\boldsymbol{R}_i$ coalesce. Instead of Airy functions we must use the following function to describe the diffraction (Pearcey 1946, Berry and Nye 1975):

$$C(X, Y) \equiv \int_{-\infty}^{\infty} \mathrm{d}t \exp\left[\mathrm{i}\left(\frac{t^4}{8} - \frac{t^2 X}{2} + tY\right)\right]. \tag{4.3}$$

The variables X and Y are smooth distortions of G_x and G_y, the manner in which the distortions must be carried out (to obtain a uniform approximation) being explained by Connor (1973). It turns out that on the cusp itself $|S_G|^2$ rises to a value of order $(h_0/\lambda_0)^{1/2}$ larger than in the 'classical' region away from $\mathscr{C}$, so that the cusps are the most strongly diffracting parts of the rainbow line. A contour map of $|C(X, Y)|^2$ constitutes figure 4; it is seen how the Airy oscillations appear as we cross $\mathscr{C}$ far from the cusp at $X = Y = 0$.

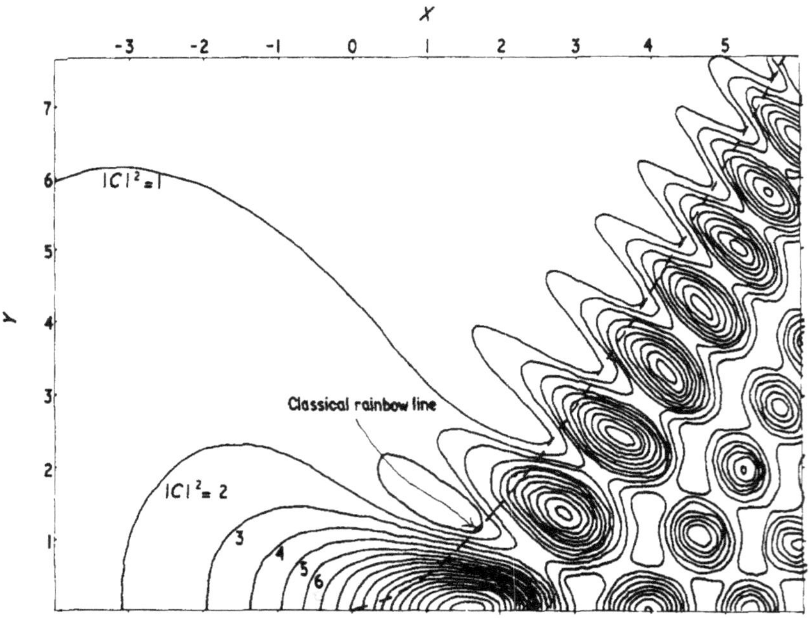

Figure 4. Contours of the cusp diffraction function $|C|^2$ (equation (4.3)).

These results concerning the uniform approximation of the diffraction integral (3.2) are summarized in figure 5.

Now we consider how the visibility of rainbows is affected by the quantization of $\boldsymbol{G}$. The principle is that to observe clearly any feature in the $\boldsymbol{G}$ plane it must be densely covered with diffraction spots. Consider for example, the separation Δ_x between rainbow lines $\mathscr{C}_1$ and $\mathscr{C}_2$ along the G_x axis (figure 3); Δ_x is given by equation (3.16). The G_x spacing

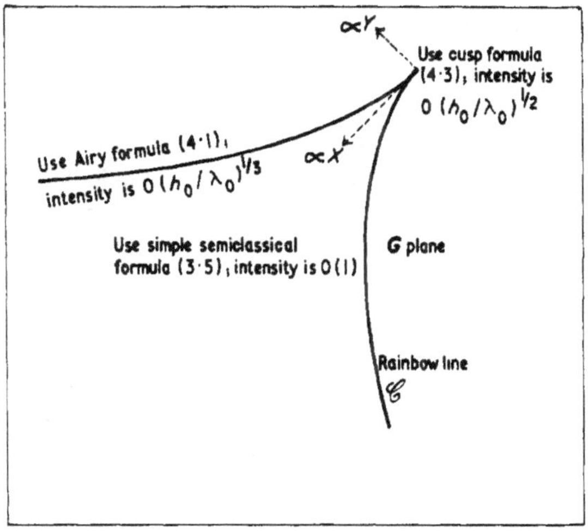

Figure 5. Summary of semiclassical diffraction formulae.

of diffraction spots is $2\pi/a$. Therefore the number $\mathcal{M}_x$ of diffraction spots between $\mathscr{C}_1$ and $\mathscr{C}_2$ is

$$\mathcal{M}_x = 2\epsilon h_0(|k_{0z}|+k_z) \sim \frac{4\pi\epsilon h_0}{\lambda_0}(\cos\theta_0 + \cos\theta), \tag{4.4}$$

where θ_0 and θ are the angles made with the z direction by the incident and scattered particles. For $\mathscr{C}_1$ and $\mathscr{C}_2$ to be distinguishable, $\mathcal{M}_x$ must be large, ie

$$\frac{\lambda_0}{h_0} \ll 4\pi\epsilon(\cos\theta_0 + \cos\theta). \tag{4.5}$$

Another 'feature' is the separation Δ_c between a cusp of $\mathscr{C}_2$ and the nearest 'corner' of $\mathscr{C}_1$ (figure 3); from (3.16), the number $\mathcal{M}_c$ of diffraction spots in Δ_c is

$$\mathcal{M}_c = 2\epsilon^2 h_0(|k_{0z}|+k_z) \sim \frac{4\pi\epsilon^2 h_0}{\lambda_0}(\cos\theta_0 + \cos\theta). \tag{4.6}$$

This must be large, so that we require

$$\frac{\lambda_0}{h_0} \ll 4\pi\epsilon^2(\cos\theta_0 + \cos\theta) \tag{4.7}$$

which is more restrictive than (4.5) if $\epsilon < 1$.

Conditions (4.5) and (4.7) for the distinguishability of two rainbow lines are necessary but not sufficient, because we must also demand that the diffraction spots are sufficiently densely packed for the *individual* rainbow lines to be resolved. A reasonable condition for this is that the number of spots $\mathcal{M}_r$ in the largest Airy-function maximum of equation (4.1) is large. This maximum spans the argument range from about 0 to -2 (Abramowitz and Stegun 1964), so that, from (4.1), we require that the deviation Δ_r in $\boldsymbol{G}$ space from the rainbow line is such that the difference in phase between the two contributing paths is

$$\Phi_2 - \Phi_1 = \frac{4\sqrt{8}}{3}. \tag{4.8}$$

Since the problem of the width of a single rainbow maximum is essentially one-dimensional, we can confine our attention to the G_x axis, and calculate the width Δ_r by treating h_p as varying only with x, so that the phase Φ in (4.2) depends only on G_x and x. Now x is related to G_x via the path equation (3.4), which now reads

$$\frac{\partial \Phi(G_x, x)}{\partial x} = 0. \tag{4.9}$$

Thus we can write the phase of a path as $\Phi(G_x, x(G_x))$; we wish to expand this function about the rainbow direction G_{xr} for which in addition to (4.9) we must have (3.9), which in one dimension is

$$\frac{\partial^2 \Phi(G_x, x)}{\partial x} = 0. \tag{4.10}$$

The expansion is tricky; for $G_x < G_{xr}$ there are two contributing paths. Their phase difference turns out to be

$$\Phi_2 - \Phi_1 = \frac{4\sqrt{2}}{3}\left(\frac{(\partial^2 \Phi(G_{xr}, x_r)/\partial x \partial G_x)^3}{\partial^3 \Phi(G_{xr}, x_r)/\partial x^3}(G_{xr} - G_x)^3\right)^{1/2}, \tag{4.11}$$

where x_r is $x(G_r)$, that is the coordinate for which $h_p(x)$ has a point of inflexion, giving a maximum deviation. Putting in the explicit form (4.2) for Φ and using (4.8) we obtain for the width $\Delta_r = |G_{xr} - G_x|$ the expression

$$\Delta_r = \frac{[4|d^3 h_p(x_r)/dx^3| k(\cos\theta_0 + \cos\theta)]^{1/3}}{1 - (dh_p(x_r)/dx)\tan\theta}. \tag{4.12}$$

It is not hard to show that the denominator never vanishes. For the cosine surface (3.10) Δ_r becomes

$$\Delta_r = \frac{(2\pi/a)(8\pi(\cos\theta_0 + \cos\theta)h_0/\lambda)^{1/3}}{1 + (2\pi h_0 \tan\theta/a)} \simeq \frac{4\pi}{a}\left(\frac{2\pi h_0}{\lambda_0}\right)^{1/3}. \tag{4.13}$$

For the number $\mathcal{M}_r$ of diffraction spots this gives

$$\mathcal{M}_r = \frac{(8\pi(\cos\theta_0 + \cos\theta)h_0/\lambda)^{1/3}}{1 + (2\pi h_0 \tan\theta/a)} \simeq 2\left(\frac{2\pi h_0}{\lambda_0}\right)^{1/3}. \tag{4.14}$$

In the most favourable case $\theta_0 = \theta = 0$ (ie $\boldsymbol{K}_0 = \boldsymbol{G} = 0$) the criterion for clearly resolving the rainbow structure is

$$\frac{h_0}{\lambda_0} \ll 16\pi. \tag{4.15}$$

In practical cases the value of h_0 might be about 0·5 Å. Then if the incoming particles have $\lambda_0 \sim 0{\cdot}1$ Å, (4.14) gives $\mathcal{M}_r \sim 6$ so that the rainbow structure should be clearly resolved. If in addition the 'asymmetry parameter' ϵ is 0·1, (4.4) gives $\mathcal{M}_x \sim 12$, so that $\mathscr{C}_1$ and $\mathscr{C}_2$ could just be distinguished along the G_x axis of figure 3. However (4.6) gives $\mathcal{M}_c \sim 1$, so that the cusp structure would be confused in this case.

5. Inelastic and diffuse incoherence effects from random perturbations

When a random perturbation $h_r(\boldsymbol{R}, t)$ is added to $h_p(\boldsymbol{R})$, the scattered intensity I is no longer given by the series (3.1) of elastic diffracted beams with quantized directions $\boldsymbol{K} = \boldsymbol{K}_0 + \boldsymbol{G}$. Instead we use the Kirchhoff formalism of § 2; equations (2.5), (2.8) and (3.2) give

$$I(\boldsymbol{K}_0, \omega_0; \boldsymbol{K}, \omega) = \frac{\Sigma_{G_1} \Sigma_{G_2} S_{G_1} S^*_{G_2}}{(2\pi)^3 \mathscr{A}\mathscr{T}} \iint \mathrm{d}\boldsymbol{R}_1 \iint \mathrm{d}\boldsymbol{R}_2 \int \mathrm{d}t_1 \int \mathrm{d}t_2 J$$

where

$$J \equiv \exp\{\mathrm{i}[(\boldsymbol{K}_0 - \boldsymbol{K}) \cdot (\boldsymbol{R}_1 - \boldsymbol{R}_2) - (\omega_0 - \omega)(t_1 - t_2) + \boldsymbol{G}_1 \cdot \boldsymbol{R}_1 - \boldsymbol{G}_2 \cdot \boldsymbol{R}_2 - (|k_{0z}| + k_z)(h_r(\boldsymbol{R}_1, t_1) - h_r(\boldsymbol{R}_2, t_2))]\}. \tag{5.1}$$

Now we must *average* I over the ensemble of random functions $h_r(\boldsymbol{R}, t)$. Because the perturbation of Σ is the summation of a multitude of small independent effects ('surface phonons'), h_r is *Gaussian random* (Rice 1944, 1945, Longuet-Higgins 1956). Denoting ensemble averages by $\langle \cdots \rangle$, we can now use standard noise theory to give, for the average of the function of h_r appearing in (5.1),

$$\langle \exp[-\mathrm{i}(|k_{0z}| + k_z)(h_r(\boldsymbol{R}_1, t_1) - h_r(\boldsymbol{R}_2, t_2))] \rangle = \exp[-H_2(|k_{0z}| + k_z)^2(1 - C(\boldsymbol{R}_1 - \boldsymbol{R}_2, t_1 - t_2))], \tag{5.2}$$

where H is the RMS value of h_r, namely

$$H = (\langle h_r^2(\boldsymbol{R}, t) \rangle)^{1/2}, \tag{5.3}$$

and C is the autocorrelation function of h_r, namely

$$C(\boldsymbol{R}_1 - \boldsymbol{R}_2, t_1 - t_2) \equiv \frac{\langle h_r(\boldsymbol{R}_1, t_1) h_r(\boldsymbol{R}_2, t_2) \rangle}{\langle h_r^2(\boldsymbol{R}, t) \rangle}. \tag{5.4}$$

The mean value $\langle h_r \rangle$ is zero by definition. Why have we chosen random noise theory for ensemble averaging, rather than a rigorous thermal method based on the density matrix (eg that of Glauber 1955)? For two reasons: first, it is not clear how the coordinates and potentials of the surface atoms define the rippling mirror surface Σ; and, second, the crystal structure, and vibratory departures from that structure, are both different at the surface from the bulk (indeed it is the aim of atomic scattering experiments to discover these differences).

Several of the integrations in (5.1) can now be performed, and we get

$$I = \sum_{\boldsymbol{G}} \frac{|S_G|^2}{(2\pi)^3} \mathrm{e}^{-H^2 q^2} \iint \mathrm{d}\boldsymbol{R} \int \mathrm{d}t \exp\{\mathrm{i}[(\boldsymbol{K}_0 - \boldsymbol{K} + \boldsymbol{G}) \cdot \boldsymbol{R} - (\omega_0 - \omega)t]\} \, \mathrm{e}^{H^2 q^2 C(\boldsymbol{R}, t)}, \tag{5.5}$$

where we introduce the notation

$$q \equiv |k_{0z}| + k_z = (k_0^2 - K_0^2)^{1/2} + (k^2 - K^2)^{1/2} = k_0 \cos\theta_0 + k \cos\theta. \tag{5.6}$$

For large values of $\boldsymbol{R}$ or t, C vanishes (there can be no correlation between widely separated events); therefore we separate the 'tail' of the integrals in (5.5) by writing

$$\mathrm{e}^{H^2 q^2 C} = 1 + (\mathrm{e}^{H^2 q^2 C} - 1). \tag{5.7}$$

This gives

$$I(\boldsymbol{K}_0, \omega_0; \boldsymbol{K}, \omega) = \delta(\omega-\omega_0)\sum_{\boldsymbol{G}}|S_{\boldsymbol{G}}|^2 \, \mathrm{e}^{-H^2q^2}\delta(\boldsymbol{K}-\boldsymbol{K}_0+\boldsymbol{G}) + \sum_{\boldsymbol{G}}|S_{\boldsymbol{G}}|^2 I^{\mathrm{r}}(\boldsymbol{K}_0-\boldsymbol{K}+\boldsymbol{G}, \omega-\omega_0), \tag{5.8}$$

where

$$I^{\mathrm{r}}(\boldsymbol{Q}, \Omega) \equiv \frac{\mathrm{e}^{-H^2q^2}}{(2\pi)^3}\iint \mathrm{d}\boldsymbol{R}\int \mathrm{d}t \exp[\mathrm{i}(\boldsymbol{Q}\cdot\boldsymbol{R}-\Omega t)](\mathrm{e}^{H^2q^2C(\boldsymbol{R},t)}-1). \tag{5.9}$$

These results are exact (on our model). They show that the scattered particles emerge in the form of: (*a*) *coherent* elastically diffracted beams whose strengths $|S_{\boldsymbol{G}}|^2$ (cf (3.14)) are reduced by the familiar Debye–Waller factor; and (*b*) *incoherent* fans of radiation whose shape is determined by the correlations of h_{r} by the function I^{r} defined by (5.9). If h_{r} is time-independent (a 'frozen random surface'), then C depends only on $\boldsymbol{R}$ and I^{r} contains the factor $\delta(\Omega)$, so that the diffraction is all elastic and the only effect of incoherence is *diffuse* scattering between the beams at $\boldsymbol{K}_0+\boldsymbol{G}$. If on the other hand h_{r} is space-independent (so that the periodic surface 'shivers' up and down as a rigid whole), then C depends only on t and I^{r} contains the factor $\delta(\boldsymbol{Q})$, so that all diffraction appears in the beams $\boldsymbol{K}_0+\boldsymbol{G}$ and the only effect of incoherence is *inelastic* scattering into frequencies ω different from ω_0.

If $Hq \sim 4\pi H/\lambda$ is small the exponential involving C in (5.9) can be expanded, and I^{r} is easy to evaluate in terms of the phonon spectrum; this is the 'one-phonon case' (Beeby 1972). However, H is of the same order of magnitude as the amplitude of thermal vibrations in the solid, that is, about 0·1 Å; thus, even if λ is as large as 1 Å, Hq exceeds unity and we have to consider 'multiphonon processes'. The novel contribution of this section is an approximation to the integrand of (5.9) that enables I^{r} to be evaluated in simple closed form for any value of Hq. We introduce the 'exponential substitution'

$$\mathrm{e}^{H^2q^2C(\boldsymbol{R},t)}-1 \simeq (\mathrm{e}^{H^2q^2}-1)C\left(\frac{Hq\boldsymbol{R}}{(1-\mathrm{e}^{-H^2q^2})^{1/2}}, \frac{Hqt}{(1-\mathrm{e}^{-H^2q^2})^{1/2}}\right). \tag{5.10}$$

This has remarkable properties (Berry 1973); it is exact when R or t is infinite, and when $\boldsymbol{R}$ and t are zero; it correctly describes the quadratic departure of the left-hand side from $\mathrm{e}^{H^2q^2}-1$ when $\boldsymbol{R}$ or t is small; it is correct for all $\boldsymbol{R}$ and t if Hq is small ('one-phonon case'). Thus for an (unphysical) 'step' correlation function, falling abruptly from unity to zero for one value of $|\boldsymbol{R}|$ and t, (5.10) is exact. For the Gaussian function C, whose 'isotropic' form is

$$C(\boldsymbol{R}, t) = \mathrm{e}^{-R^2/2R_0^2}\,\mathrm{e}^{-t^2/2t_0^2}, \tag{5.11}$$

(5.10) is quite accurate, as figure 6 shows. Where the exponential substitution is poor is when Hq is large and C is negative, but this kind of 'anticorrelation' would only occur if h_{r} were quasi-periodic; however, the dominant periodicity of Σ is already incorporated in $h_{\mathrm{p}}(\boldsymbol{R})$, and negative values for C are unlikely.

Now we can evaluate the incoherence function I^{r}, in terms of the *spectrum* $\bar{C}(\boldsymbol{Q}, \Omega)$ of the correlations of h_{r}; this function is defined as

$$\bar{C}(\boldsymbol{Q}, \Omega) \equiv \frac{1}{(2\pi)^3}\iint \mathrm{d}\boldsymbol{R}\int \mathrm{d}t \exp[\mathrm{i}(\boldsymbol{Q}\cdot\boldsymbol{R}-\Omega t)]C(\boldsymbol{R}, t). \tag{5.12}$$

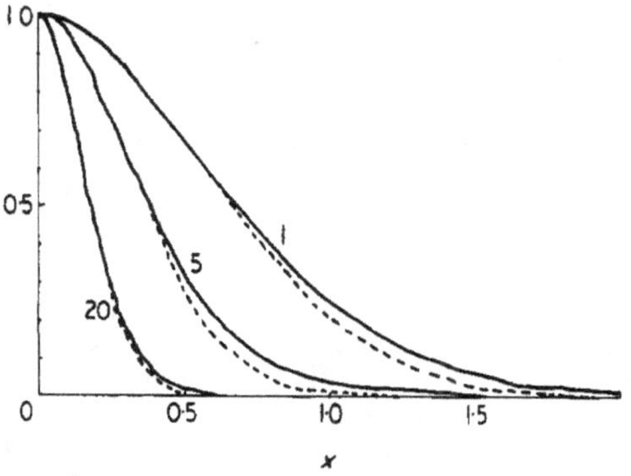

Figure 6. Test of exponential substitution (5.10) for Gaussian autocorrelation function: we plot $(e^{A\exp(-x^2)}-1)/(e^A-1)$ (full lines) and $e^{-Ax^2/(1-e^{-A})}$ (dotted lines) against x for $A = 1$, 5 and 20.

$\bar{C}$ measures the strength of the $\boldsymbol{Q}, \Omega$ Fourier component of $h_r(\boldsymbol{R}, t)$; it can be called the 'surface phonon spectrum'. Using (5.10), (5.9) now becomes

$$I^r(\boldsymbol{Q}, \Omega) \simeq \frac{(1-e^{-H^2q^2})^{5/2}}{H^3q^3}\bar{C}\left(\frac{\boldsymbol{Q}(1-e^{-H^2q^2})^{1/2}}{Hq}, \frac{\Omega(1-e^{-H^2q^2})^{1/2}}{Hq}\right). \tag{5.13}$$

This is the main result of this section.

Two limiting cases are of interest. If Hq is small, we can expand the exponentials, and we get

$$I^r(\boldsymbol{Q}, \Omega) \xrightarrow{Hq\ \text{small}} H^2q^2\bar{C}(\boldsymbol{Q}, \Omega). \tag{5.14}$$

This corresponds to h_r being a weak perturbation of Σ. Almost all the scattering is into the elastic diffracted beams; incoherence effects are small, and the scattering deviating by $\boldsymbol{Q}$ and Ω from an elastic beam comes from the 'surface phonon' with wavevector $\boldsymbol{Q}$ and frequency Ω.

The other limit is large Hq; then the exponentials are negligible and

$$I^r(\boldsymbol{Q}, \Omega) \xrightarrow{Hq\ \text{large}} \frac{1}{H^3q^3}\bar{C}\left(\frac{\boldsymbol{Q}}{Hq}, \frac{\Omega}{Hq}\right). \tag{5.15}$$

In this case h_r is so large (in comparison with λ) that even the incoherent scattering is classical. The elastic beams now have negligible intensity, because of the Debye–Waller factor. In this case the scattering into $\boldsymbol{Q}$ and Ω comes not from the 'phonon' with $\boldsymbol{Q}$ and Ω, but from that with wavevector $\boldsymbol{Q}/Hq$ and frequency Ω/Hq. This is understandable classically in terms of the atoms bouncing specularly off the moving phonons: if $\boldsymbol{Q}$ corresponds to a deviation $\Delta\theta = |\boldsymbol{Q}|/k$ from the diffracted beam direction, then

$$\frac{|\boldsymbol{Q}|}{Hq} \sim \frac{k\Delta\theta}{2Hk} = \frac{\Delta\theta}{2H}; \tag{5.16}$$

a 'phonon' with this wavenumber and amplitude H has a maximum slope $\Delta\theta/2$, and so can reflect specularly by $\Delta\theta$. If Ω corresponds to an energy change $\Delta E = \hbar\Omega$ from E_0, then

$$\frac{\Omega}{Hq} \sim \frac{\Delta E\hbar}{2H\hbar \times \text{momentum}} = \frac{\Delta v}{2H}, \tag{5.17}$$

where Δv is the change in speed of the atoms; a 'phonon' with this frequency and amplitude H has a maximum (vertical) speed $\Delta v/2$, and so can change the speed of a reflected particle by Δv.

The general result (5.13) shows that the incoherence function $I^{\rm r}$ for scattering from a given surface always has the same shape; only the scale changes with Hq. This might explain a puzzle described by Beeby (1973): the incoherent scattering seems to have the 'one-phonon' form, even when 'multiphonon processes' are known to dominate.

Perhaps the most useful property of (5.13) is that it suggests that the blurring in direction and energy of the diffracted beams gives directly the 'surface phonon spectrum' $\tilde{C}(\boldsymbol{Q}, \Omega)$ (apart from the scaling factor $Hq(1-e^{-H^2q^2})^{1/2}$ which could be obtained by measuring the diminution $e^{-H^2q^2}$ of the strengths of the diffracted beams). For this to be possible the diffracted beams should not be so broadened that they overlap significantly. We can examine this with the aid of the model autocorrelation function (5.11), for which the 'phonon spectrum' (5.12) is given by

$$\tilde{C}(\boldsymbol{Q}, \Omega) = \frac{t_0 R_0^2}{(2\pi)^{3/2}} \exp\left(-\frac{Q^2 R_0^2}{2} - \frac{\Omega^2 t_0^2}{2} \right). \tag{5.18}$$

From (5.13), we have, for the 'breadth' $Q_{\rm b}$ of the diffuse scattering near each diffracted beam,

$$Q_{\rm b} = \frac{Hq\sqrt{2}}{R_0(1-e^{-H^2q^2})^{1/2}} \to \begin{cases} \sqrt{2}/R_0 & Hq \text{ small} \\ \dfrac{Hq\sqrt{2}}{R_0} \sim \dfrac{4\pi H\sqrt{2}}{\lambda_0 R_0} & Hq \text{ large.} \end{cases} \tag{5.19}$$

($Q_{\rm b}$ is the '1/e halfwidth' of the $\boldsymbol{Q}$ dependence of $I^{\rm r}$.) In order to be able to invert experimental data to find $\tilde{C}(\boldsymbol{Q}, \Omega)$, $Q_{\rm b}$ must be small in comparison with the separation $2\pi/a$ of the diffracted beams.

Finally we recognize that diffuse scattering will blur the rainbow structure arising under semiclassical conditions. To resolve an individual Airy maximum, $Q_{\rm b}$ must be small compared with $\Delta_{\rm r}$ (equation (4.13)), ie

$$\frac{Ha\sqrt{2}}{R_0 \lambda_0^{2/3}(2\pi h_0)^{1/3}} \ll 1. \tag{5.20}$$

For sufficiently small λ_0 this condition is always violated. To resolve the separation between rainbow lines $\mathscr{C}_1$ and $\mathscr{C}_2$ (figure 3) along G_x, $Q_{\rm b}$ must be small compared with Δ_x or Δ_y (equation (3.16)), ie

$$\frac{aH\sqrt{2}}{4\pi\epsilon h_0 R_0} \ll 1. \tag{5.21}$$

To resolve the separation between a cusp of $\mathscr{C}_2$ and the nearest part of $\mathscr{C}_1$, $Q_{\rm b}$ must be small compared with $\Delta_{\rm c}$ (equation (3.15)), ie

$$\frac{H\sqrt{2}}{4\pi\epsilon^2[(1/a^2)+(1/b^2)]^{1/2} h_0 R_0} \ll 1. \tag{5.22}$$

These three conditions assume that the scattering from $h_{\rm r}$ is nearly classical, so that the 'large Hq' limiting form in (5.19) can be used; if, however, the scattering from $h_{\rm r}$ is weak, we must use the 'small Hq' form for $Q_{\rm b}$.

This page is left blank intentionally.

J. Phys. A: Math. Gen., Vol. 8. © 1975—*M V Berry*

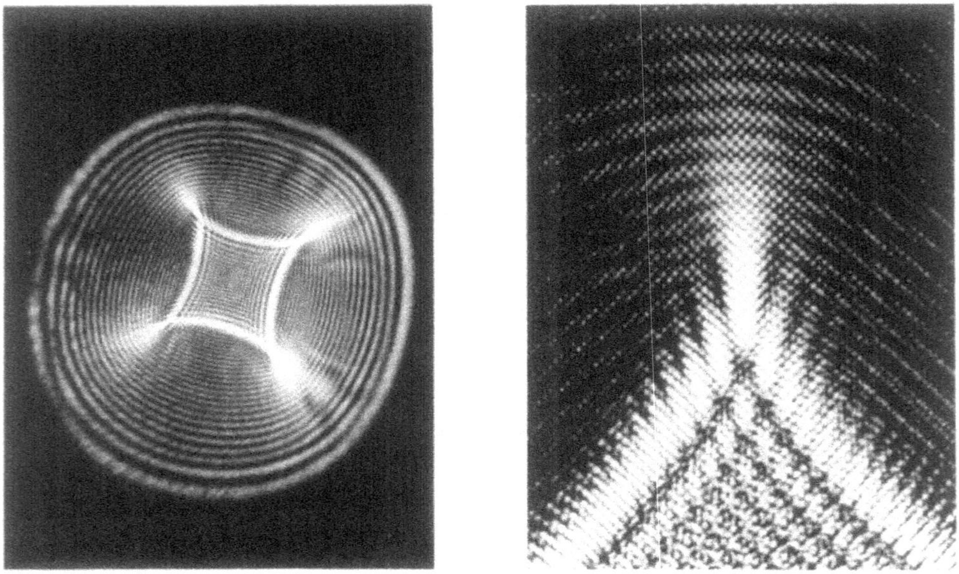

(a) (b)

Figure 7. (Plate) (*a*) Caustics of refraction from periodically frosted glass surface ($a = \frac{1}{2}$ mm), using laser light ($\lambda = 6328$ Å). Compare with figures 3 and 4. (*b*) detail near cusp showing quantization into diffraction spots *G*.

[*facing page* 583]

6. Discussion

The properties of rainbows predicted in §§ 3 and 4 are strikingly confirmed by an optical analogue experiment based on refraction. Laser light transmitted through periodically frosted glass (such as is often used for bathroom windows) is observed on a distant screen. Figure 7(*a*) (plate) is a photograph of the resulting pattern. The two rainbow lines $\mathscr{C}_1$ and $\mathscr{C}_2$ are clearly seen to have the form of figure 3, 'Airy' fringes (equation (4.1)) are clearly seen near smooth parts of $\mathscr{C}$, and diffraction near each cusp of $\mathscr{C}_2$ is clearly of the same form as figure 4; moreover, on a very fine scale (figure 7(*b*) (plate)) it is possible to see the beginning of the quantization of $\boldsymbol{G}$ into diffraction spots, which occurs because the laser beam illuminates several unit cells of the glass surface. Cusped rainbows can also be seen if a distant point source of light is viewed with the eye placed close behind an irregular droplet 'lens' on a glass surface (for example a spectacle lens or car windscreen on a rainy night, or a half-empty wineglass), as explained by Berry and Nye (1975).

In particle scattering from surfaces it would probably not be easy to resolve all the details of the rainbow structure. Thermal diffuse scattering should be minimized by operating at a low temperature, and to see diffraction effects the beam should be as nearly mono-energetic as possible (unless energy-selective detectors are used, so that the 'colours of the rainbow' can be discerned). Finally, the de Broglie wavelength should be as small as possible; this should be achieved with fast light particles rather than heavy particles, to minimize the effects of their impact on the surface. Despite these difficulties, the fine details of surface rainbows ought to be studied experimentally, because they are very sensitive to the details of Σ, as we showed in § 3 (this is generally true of classical effects, the principle being that smaller wavelengths probe finer details).

Now we discuss the method based on equation (5.13) which we suggest might be used to measure the 'phonon spectrum'; of course the theory is not exact: we have used the 'exponential substitution' (5.10), and the rippling-mirror model is itself an approximation, as is the use of Kirchhoff's diffraction integral (2.5). Nevertheless (5.13) should be qualitatively correct for a wide variety of conditions, and at worst would lead to an 'effective surface phonon spectrum' being reconstructed from the experimental measurements.

Acknowledgments

I am happy to thank Professor J P Toennies for the hospitality of the Max-Planck Institüt in Göttingen, Professor H Nahr for introducing me to surface scattering and for several helpful discussions, Professor J A Barker for suggesting the frosted-glass optical analogue and Mr T Osman for making the glass surface used to obtain figure 7 (plate).

References

Abramowitz H and Stegun A 1964 *Handbook of Mathematical Functions* (Washington: US National Bureau of Standards)

Airy G B 1838 *Proc. Camb. Phil. Soc.* **6** 379–402

Beckmann P and Spizzichino A 1963 *The Scattering of Electromagnetic Waves from Rough Surfaces* (Oxford and New York: Pergamon)

Beeby J L 1971 *J. Phys.* C: *Solid St. Phys.* **4** L359–62
—— 1972 *J. Phys.* C: *Solid St. Phys.* **5** 3438–61
—— 1973 *J. Phys.* C: *Solid St. Phys.* **6** 1229–41
Berry M V 1966 *Proc. Phys. Soc.* **89** 479–90
—— 1973 *Phil. Trans. R. Soc.* A **273** 611–58
Berry M V and Mount K E 1972 *Rep. Prog. Phys.* **35** 315–97
Berry M V and Nye J F 1975 to be published
Cabrera N, Celli V, Goodman F O and Manson R 1970 *Surf. Sci.* **19** 67–92
Cayley 1859 *Phil. Mag.* S4 **18** 264–8
Chester C, Friedman B and Ursell F 1957 *Proc. Camb. Phil. Soc.* **53** 599–611
Connor J N L 1973 *Molec. Phys.* **26** 1217–31
Ford K W and Wheeler J A 1959 *Ann. Phys., NY* **7** 259–322
Garibaldi U, Levi A G, Spadacini R and Tommei G E 1974 *Surf. Sci.* to be published
Glauber R J 1955 *Phys. Rev.* **98** 1692–8
Lennard-Jones J E and Devonshire A F 1937 *Proc. R. Soc.* A **158** 253–68
Longuet-Higgins M S 1956 *Phil. Trans. R. Soc.* A **249** 321–87
Maxwell J C 1870 *Phil. Mag.* S4 **40** 421–7 (also in *Collected Works* vol 2)
McClure J D 1970 *J. Chem. Phys.* **52** 2712–8
—— 1971 *J. Chem. Phys.* **57** 2810–22
Mount K E 1973 *J. Phys.* B: *Atom. Molec. Phys.* **6** 1397–411
Pearcey T 1946 *Phil. Mag.* **37** 311–7
Rice S O 1944 *Bell Syst. Tech. J.* **23** 282–332
—— 1945 *Bell Syst. Tech. J.* **24** 46–156
Smith J N Jr, O'Keefe D R, Saltsburg H and Palmer R L 1969 *J. Chem. Phys.* **50** 4667–71
Thom R 1969 *Topology* **8** 313–35
—— 1972 *Stabilité Structurelle et Morphogenèse* (New York: Benjamin)
Toennies J P 1974 *Appl. Phys.* **3** 91–114
Wolken G Jr 1973 *J. Chem. Phys.* **58** 3047–64

ADVANCES IN PHYSICS, 1976, VOL. 25, NO. 1, 1–26

Waves and Thom's theorem

By M. V. BERRY
H. H. Wills Physics Laboratory, Bristol BS8 1TL, U.K.

[Received 30 December 1975]

ABSTRACT

Short-wave fields can be well approximated by families of trajectories. These families are dominated by their singularities, i.e. by caustics, where the density of trajectories is infinite. Thom's theorem on singularities of mappings can be rigorously applied and shows that structurally stable caustics—that is those whose topology is unaltered by 'generic' perturbation—can be classified as 'elementary catastrophes'. Accurate asymptotic approximations to wave functions can be built up using the catastrophes as skeletons: to each catastrophe there corresponds a canonical diffraction function. Structurally unstable caustics can be produced by special symmetries, and the detailed form of the caustic that results from symmetry-breaking can often be determined by identifying the structurally unstable caustic as the special section of a higher-dimensional catastrophe. Sometimes it is clear that the unstable caustic is a special section of a catastrophe of infinite co-dimension; these fall outside the scope of Thom's theorem and suggest new directions for mathematical investigation. The discussion is illustrated with numerous examples from optics and quantum mechanics.

CONTENTS

§ 1. INTRODUCTION

Underlying wave theories are their short-wave limits in the form of Hamiltonian descriptions involving families of trajectories. In the case of quantum mechanics these trajectories are the paths of classical particles, while for electromagnetism they are the rays of geometrical optics. A wave field corresponds to a *family* of trajectories rather than a single trajectory. The family is defined by an action function S from which its individual trajectories can be derived; in isotropic media (where the Hamiltonian function depends on the canonical momentum $\mathbf{p}$ only through its length $|\mathbf{p}|$) the trajectories are normal to the contour surfaces of S, i.e. to the wavefronts. General reviews of Hamiltonian theories have been given by Synge (1954, 1960) and Goldstein (1951).

A family can, and usually does, exhibit a property that does not reside in any of its individual trajectories, namely *focusing*. This occurs on the contact of neighbouring trajectories, i.e. on their *envelope*. The general term for a focal region is a *caustic*, and it is caustics that form the subject of this paper.

Caustics are doubly important in wave theory. In the first place, the intensity, that is the density of trajectories, is infinite on caustics. Therefore under short-wave conditions, where the trajectory picture is a good approximation, wave fields will be dominated by caustics, which will form the significant structure of images formed on screens or photographic plates placed in the field. This is well expressed by Stavroudis (1972) :

" The caustic is one of the few things in geometrical optics that has any physical reality. Wavefronts and rays are not realizable ; they are just convenient symbols on which we can hang our ideas. The caustic on the other hand is real and becomes visible by blowing a cloud of smoke in the region of the focus of a lens."

In the second place, it is precisely on caustics that the elementary trajectory picture must break down because solutions of wave equations always have finite intensity provided the wavelength is not exactly zero. Therefore caustics are ' windows ' through which wave effects can still be discerned when the wavelength has become so short that interference fringes, etc. are too fine to be seen elsewhere in the wave field. It has been known for some time (Berry 1966, 1969 a, b, Berry and Mount 1972) that refined trajectory pictures can be devised for various types of caustic giving short-wave asymptotic approximations to wave fields, with the property that they are valid *uniformly* through the caustic regions.

Until recently, however, it was not at all clear what forms caustics could take. Now, however, the celebrated theorem of Thom (Wassermann 1974, Bröcker 1975) enables certain caustics—those which are ' structurally stable ' —to be classified according to their topology. Thom's theorem is a piece of pure mathematics ; it also forms the basis of a more speculative system of ideas called ' catastrophe theory ' (Thom 1975). The applicability of the theorem to certain caustics of families of trajectories has been proved by Jänich (1974), and the derivation of uniform asymptotic approximations to wave fields in the presence of these caustics has been put on a rigorous basis by Maslov (1972) (see also Kravtsov (1968) and Duistermaat (1974)). There is no need to repeat these derivations here.

The purpose of this article is twofold. First, we explain how Thom's theorem can be used descriptively and predictively to solve problems in wave physics. Secondly, we point out some important classes of ' non-generic ' caustic which lie outside the scope of Thom's theorem. A fully general treatment of all these subjects does not yet exist. Therefore throughout most of this paper we employ for illustrative purposes a model system which is very simple but which nevertheless comprehends a variety of non-trivial wave phenomena ; this model is introduced in § 2. In § 3 we state Thom's theorem. In § 4 we give a series of examples of caustics that are ' elementary catastrophes ', and also show how the theorem can predict the form of caustics resulting from certain kinds of symmetry-breaking. Section 5 is devoted to wave functions : we show how uniform asymptotic approximations can be obtained in terms of canonical 'diffraction functions' which can be derived in simple generic cases from Thom's catastrophes. Finally, in § 6 we investigate some caustics that are catastrophes of infinite co-dimension, resulting from high symmetries whose mathematical implications are not yet understood.

§ 2. Model system: the far field from a perturbed plane wavefront

Consider a monochromatic plane wave travelling along the direction $+z$ in a homogeneous isotropic medium. Let the wavefront at $z=0$ be deformed into a surface W by lifting the part at $\mathbf{R}\equiv(x, y)$ from $z=0$ to $z=f(\mathbf{R})$ (fig. 1). We assume that all radii of curvature of W are large in comparison with the wavelength λ, and that all slopes are small (i.e. $|\nabla f| \ll 1$); this does not mean that $f(\mathbf{R})$ is small, and indeed we are particularly interested in cases where f/λ can be large so that a trajectory description is valid. The restriction on slopes is not necessary but greatly simplifies the algebra.

Fig. 1

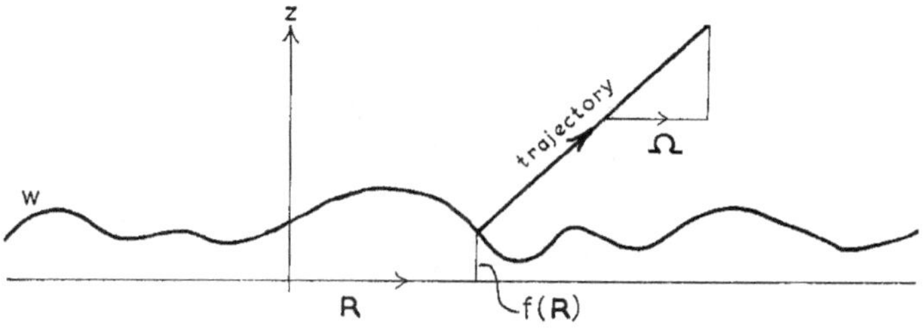

Notation for initial wavefront W and trajectories.

A variety of smooth phase objects will produce such wavefronts, and we give a few examples: (i) Undulating refracting surfaces such as the panes of glass used for bathroom windows. If n is the refractive index, then to produce W by transmission the profile of the undulating surface must be $f(\mathbf{R})n/n-1$. (ii) Irregular water droplet 'lenses' on inhomogeneously dirty flat glass surfaces. Again the profile is $f(\mathbf{R})n/n-1$, where n is now the refractive index of water. For small droplets the internal pressure p is constant, and $f(\mathbf{R})$ is restricted by the surface tension γ to obey the equation

$$\nabla^2 f(\mathbf{R}) = -\frac{(n-1)p}{n\gamma}. \tag{1}$$

This has the general solution

$$f(\mathbf{R}) = \mathrm{Re}\; g(x+iy) - \frac{(n-1)p}{4n\gamma} R^2, \tag{2}$$

where g is any analytic function and $R \equiv |\mathbf{R}|$. The optics of such droplets have been briefly described by Minnaert (1954) and Larmor (1891) (see also the night scene in Fellini 1963). The caustics can be clearly seen by viewing a distant street lamp with the eye close behind small raindrops on a window pane. (iii) Undulating reflecting surfaces ($n=-1$), such as crystal surfaces illuminated by beams of atoms (Berry 1975). (iv) Scattering of a beam of particles of mass m and (high) energy E by a spherically symmetric field of force with potential $V(r)$, where the wavefront beyond the scatterer is given by

$$f(R)=\int_R^\infty dr\,\frac{r}{\sqrt{(r^2-R^2)}}\,\frac{V(r)}{E} \tag{3}$$

which, being cylindrically symmetric, involves only R. This type of scattering is discussed by Glauber (1958).

We discuss only the *far field* of the waves from W, to which only the *directions* of the trajectories and waves contribute. We specify directions by projections $\mathbf{\Omega}$ on the xy plane (fig. 1), i.e. by $\Omega_x=\sin\theta\cos\phi$, $\Omega_y=\sin\theta\sin\phi$, where θ and ϕ are the usual polar angles with z as axis. The intensity $I(\mathbf{\Omega})$ is defined by the flux through $d\mathbf{\Omega}$ far from W when there is unit flux through unit area of W itself. (Flux is energy flow in optics, particle flow in mechanics.)

Each point $\mathbf{R}$ on W gives rise to a trajectory normal to W, whose direction is

$$\mathbf{\Omega}(\mathbf{R})=-\nabla f(\mathbf{R}). \tag{4}$$

For given $\mathbf{\Omega}$ there may be several points $\mathbf{R}$ satisfying this equation ; we label them $\mathbf{R}_i(\mathbf{\Omega})$. Then the intensity I on the trajectory picture is

$$I(\mathbf{\Omega})=\sum_i\left|\frac{d\mathbf{\Omega}}{d\mathbf{R}_i(\mathbf{\Omega})}\right|^{-1}, \tag{5}$$

where $|d\mathbf{\Omega}/d\mathbf{R}|$ is the Jacobian of the *mapping* from $\mathbf{R}$ to $\mathbf{\Omega}$ that each trajectory defines in accordance with eqn. (4). Explicitly,

$$\left|\frac{d\mathbf{\Omega}}{d\mathbf{R}}\right|=\left|\frac{\partial^2 f}{\partial x^2}\frac{\partial^2 f}{\partial y^2}-\left(\frac{\partial^2 f}{\partial x\partial y}\right)^2\right|\equiv|K(\mathbf{R})|, \tag{6}$$

where $K(\mathbf{R})$ is the *Gaussian curvature* of W at $\mathbf{R}$.

Fig. 2

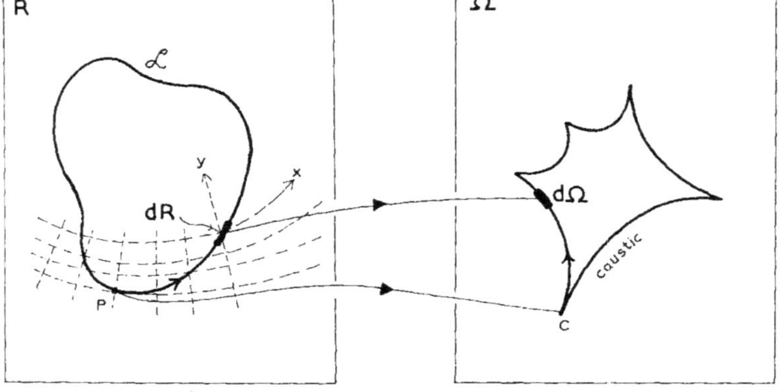

Caustic as image of line $\mathscr{L}$ where Gaussian curvature of W vanishes.

The caustics of the family of trajectories travelling to infinity from W are the *singularities* of the mapping $\mathbf{R}\rightarrow\mathbf{\Omega}$; they occur where $|d\mathbf{\Omega}/d\mathbf{R}|$ vanishes and I is infinite. Therefore caustics are the images in $\mathbf{\Omega}$ of the lines $\mathscr{L}$ on W where K vanishes (fig. 2). These directional caustics at infinity are of course the asymptotes of the more complicated spatial caustics beyond W. Thom's theorem, to be discussed in § 3, tells us about the forms the caustics may take.

Crucial to the application of Thom's theorem is the fact that the trajectories (4) can be derived from a *generating function* $\Phi(\mathbf{R}, \mathbf{\Omega})$, of the form

$$\Phi(\mathbf{R}, \mathbf{\Omega})=\mathbf{\Omega}\,.\,\mathbf{R}+f(\mathbf{R}), \tag{7}$$

by the gradient conditions

$$\nabla_{\mathbf{R}}\phi=0. \tag{8}$$

The generating function Φ is a single-valued non-singular function of its arguments $\mathbf{R}$ and $\mathbf{\Omega}$, whereas the *action function* of the family of trajectories, namely

$$S_i(\mathbf{\Omega})=\Phi(\mathbf{R}_i(\mathbf{\Omega}), \mathbf{\Omega}), \tag{9}$$

is a many-valued function of $\mathbf{\Omega}$ because there may be many points $\mathbf{R}_i$ sending trajectories to $\mathbf{\Omega}$. The different ' branches ' S_i join on the caustic. That (9) is indeed the ' directional action ' can easily be verified by calculating the optical distance along a trajectory from W to some finite point and then letting the point recede to infinity in direction $\mathbf{\Omega}$.

The existence of a generating function is not confined to our simple example ; Jänich (1974) and Maslov (1972) (see also Duistermaat (1974)) show that all Hamiltonian families of trajectories can be derived from generating functions. Introducing the language of catastrophe theory, we say that $\mathbf{\Omega}$ are the *control parameters*, on which the observable intensity I depends. In the general case $\mathbf{\Omega}$ need not represent direction ; it could include position in space, time, parameters affecting the shape of scattering potentials or diffracting objects, the refractive index of a medium, etc. The auxiliary quantities $\mathbf{R}$, on which I does not depend, are called the *state variables*. The vectors $\mathbf{R}$ and $\mathbf{\Omega}$ may have any number of components. It is usually not difficult to find the generating function for a particular situation and there are general techniques (described by the authors cited) involving the introduction of variables canonically conjugate to those in which the caustics appear, the generating function being the solution of an appropriate Hamilton–Jacobi equation in mixed coordinate-momentum representation (see also Kravtsov 1968).

On the *wave theory* the intensity for the simple model system can be found to a close approximation by methods described by Berry (1975). The result (which is almost obvious) is the Fraunhofer diffraction integral

$$I(\mathbf{\Omega})=|\psi(\mathbf{\Omega})|^2=\left|\frac{k}{2\pi}\iint d\mathbf{R}\exp\left[ik\Phi(\mathbf{R}, \mathbf{\Omega})\right]\right|^2, \tag{10}$$

where ψ is the wave amplitude, Φ is given by eqn. (7), and $k\equiv 2\pi/\lambda$.

In § 5 we shall show how to evaluate diffraction integrals of this type. At this point we note that the exponent in the integrand is proportional to precisely the function Φ just introduced to generate the trajectories. Maslov (1972) shows that this is a general feature of integrals that are asymptotic approximations (valid as $k \to \infty$) to wave functions. (For an example of such an integral representation in scattering theory see § 6 of Berry and Mount (1972).)

§ 3. Thom's theorem

Caustics, then, are singularities of gradient maps of the form (8) derived from generating functions $\Phi(\mathbf{R}, \mathbf{\Omega})$. Thom's theorem concerns caustics with the property of *structural stability*. This means that a perturbation (e.g. of the Hamiltonian or of the initial wavefront W) leaves the local structure of the caustic unchanged in the sense that the perturbed and unperturbed caustics are related by a smooth reversible transformation (a 'diffeomorphism'). A situation in which the caustics are structurally stable is 'generic'—there is 'nothing special' about it. A structurally unstable caustic, on the other hand, whose topological type is changed by perturbation, results from a 'non-generic' situation where some special symmetry exists.

The theorem states (Wassermann 1974, Bröcker 1975) that there exists only a finite number of types of structurally stable caustic for each value of the dimensionality n of the control parameter space $\mathbf{\Omega}$, and it gives explicit standard forms for the generating functions when n is less than 7. n is called the *co-dimension* of the caustic. Thom calls the standard caustics the 'elementary catastrophes'. It is remarkable that for fixed n the structurally stable caustics are not affected by adding extra state variables $\mathbf{R}$; depending on the caustic type, only one component x, or two components x, y, of $\mathbf{R}$, determine the caustic, and any other components can always be transformed so as to enter $\Phi(\mathbf{R}, \mathbf{\Omega})$ quadratically, and then (cf. eqn. (8)) they cannot produce singularities.

Structurally stable caustics—'elementary catastrophes'—with co-dimensions 1, 2 and 3. x and y are components of the state vector $\mathbf{R}$ and Ω_1, Ω_2 and Ω_3 are control parameters Ω.

Co-dimension	Name	Generating function $\Phi(\mathbf{R}, \mathbf{\Omega})$	Singularity index σ
1	Fold	$x^3/3 + \Omega_1 x$	$\frac{1}{6}$
2	Cusp	$x^4/4 + \Omega_1 x^2/2 + \Omega_2 x$	$\frac{1}{4}$
3	Swallow-tail	$x^5/5 + \Omega_3 x^3/3 + \Omega_2 x^2/2 + \Omega_1 x$	$\frac{3}{10}$
3	Elliptic umbilic	$x^3 - 3xy^2 - \Omega_3(x^2 + y^2) - \Omega_1 x - \Omega_2 y$	$\frac{1}{3}$
3	Hyperbolic umbilic	$x^3 + y^3 + \Omega_3 xy - \Omega_1 x - \Omega_2 y$	$\frac{1}{3}$

The table gives the names and generating functions of the structurally stable caustics for $n \leqslant 3$, and on fig. 3 are sketched the forms of these caustics in $\mathbf{\Omega}$ space. What is peculiar to each catastrophe is the local nature of its highest singularity. Consider, for example, the cusp (fig. 3 *b*); a generic section such as the randomly chosen line $\mathscr{L}$ will have at most the 'fold' point catastrophes characteristic of one control dimension, and this will be structurally stable. However, a non-generic section, passing through the

Fig. 3

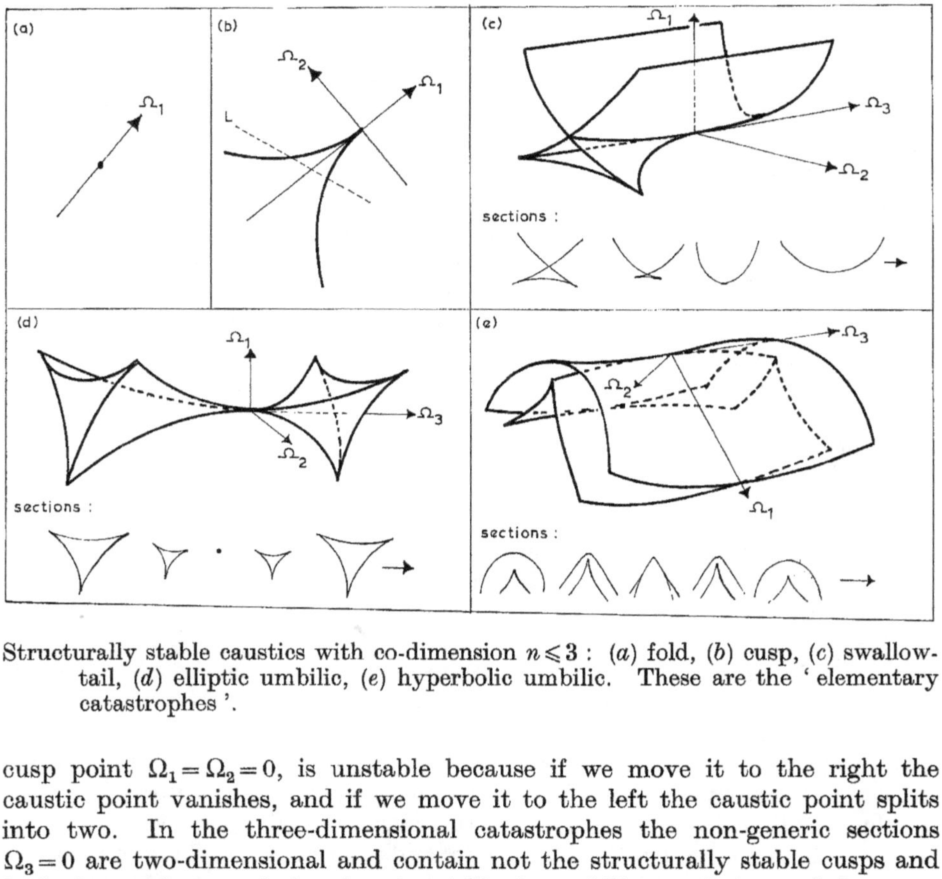

Structurally stable caustics with co-dimension $n \leqslant 3$: (*a*) fold, (*b*) cusp, (*c*) swallow-tail, (*d*) elliptic umbilic, (*e*) hyperbolic umbilic. These are the 'elementary catastrophes'.

cusp point $\Omega_1 = \Omega_2 = 0$, is unstable because if we move it to the right the caustic point vanishes, and if we move it to the left the caustic point splits into two. In the three-dimensional catastrophes the non-generic sections $\Omega_3 = 0$ are two-dimensional and contain not the structurally stable cusps and folds but either an isolated point (elliptic umbilic), a finite-angled corner (hyperbolic umbilic), or a point where the caustic curvature diverges smoothly (swallow-tail), all three of these points being structurally unstable. The pattern of development from instability to stability as the section moves from non-generic to generic is called the *smooth unfolding of the singularity*. As n increases, so does the number of catastrophes, a fact appreciated as early as 1873 by Maxwell (Campbell and Garnett 1882).

We emphasize that the standard forms of the table describe only the *local* caustic structure, and leave the *global* topology completely unspecified. This is illustrated in fig. 4, which shows a hyperbolic umbilic (singularity at P) globally different from fig. 3 (*e*).

Fig. 4

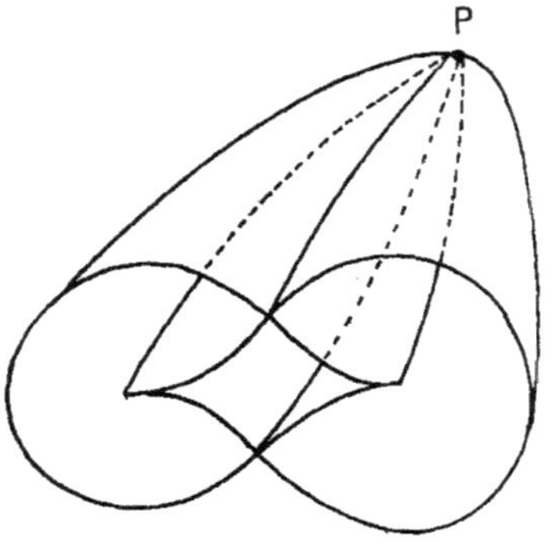

Hyperbolic umbilic with global topology different from fig. 3 (*e*) (after Zeeman).

Armed with Thom's theorem, one can quite easily recognize non-generic caustics. Their appearance always indicates the presence of some symmetry. If the symmetry is broken generically by parameters $\boldsymbol{\epsilon}$, these may be added to the control parameters $\boldsymbol{\Omega}$. If $(\boldsymbol{\Omega}, \boldsymbol{\epsilon})$ is of sufficiently small dimensionality, the now structurally stable unfolding of the caustic is an elementary catastrophe. This ability to predict the details of symmetry-breaking is a very powerful feature of Thom's theorem ; we shall see an application of it in § 4. Often, however, $\boldsymbol{\epsilon}$ is of infinite dimension—there are infinitely many topologically different ways of breaking the symmetry—and Thom's theorem is inapplicable ; we shall give several examples of this in § 6.

§ 4. Caustics of finite co-dimension

For the simple model system of § 2, the caustics appear in the two-dimensional space of directions $\boldsymbol{\Omega}$, as lines on a distant screen whose Cartesian coordinates are proportional to the control parameters $\Omega_1 = \Omega_x$ and $\Omega_2 = \Omega_y$. Reference to the table shows that the only structurally stable caustics are smooth fold lines which may be interrupted only at cusp points where the caustic curve reverses direction and crosses its tangent (i.e. these caustics may not come to an end or have finite-angled corners or be isolated points). To verify this experimentally, simply shine a laser beam through a generic object (e.g. examples (i) or (ii) of § 2—the irregular glass called ' Atlantic ' made by Pilkington's is excellent for this purpose as fig. 5 shows) and let the image fall on a light-coloured wall in a darkened room ; the size of the irregularities of the objects should be about the same as the width of the beam. It is difficult to lay down conditions guaranteeing genericity ; usually it is sufficient for no special circumstances to have attended the manufacture of the refracting objects. We shall amplify this point later.

How do folds and cusps arise in this model system ? Recall that caustics in $\boldsymbol{\Omega}$ are images under the mapping (4) of the line $\mathscr{L}$ on W where the Gaussian curvature K vanishes (fig. 2). $K = 0$ implies that at least one of the two principal curvatures of W vanishes at each point on $\mathscr{L}$. Set up local Cartesian coordinates $\mathbf{R} = \mathbf{i}x + \mathbf{j}y$ whose origin is at 0 on $\mathscr{L}$, and let $\mathbf{i}$ be locally parallel to the principal direction corresponding to that curvature which vanishes. Then $\partial^2 f/\partial x^2$ and $\partial^2 f/\partial x \partial y$ both vanish at 0, and the line element $d\boldsymbol{\Omega}$ of caustic

Fig. 5

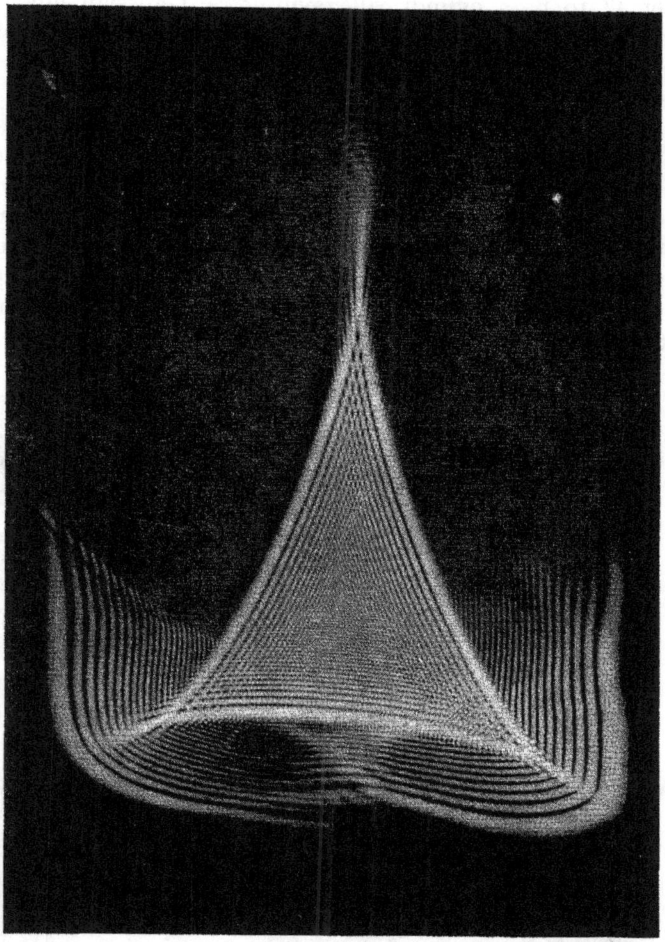

Far-field caustic from irregular window pane of Pilkington's 'Atlantic' glass, showing partially unfolded elliptic and hyperbolic umbilics.

resulting from the line element $d\mathbf{R} = \mathbf{i}\,dx + \mathbf{j}\,dy$ on $\mathscr{L}$ near 0 is easily calculated from (4) to be

$$d\mathbf{\Omega} = -\mathbf{i}\,\frac{\partial^2 f}{\partial y^2}\,dy. \tag{11}$$

Thus the caustic is locally perpendicular to the direction on W of the principal curvature whose vanishing generates it; this is a special case of a result proved in a text by Eisenhart (1960).

As 0 moves round $\mathscr{L}$, the caustic is generated by adding all the $d\mathbf{\Omega}$ given by (11). The caustic will be smooth (i.e. a fold catastrophe) at almost all points. The exceptions are the cusps (C on fig. 2) which arise from points P (fig. 2) where $\mathscr{L}$ touches the direction $\mathbf{i}$ of vanishing curvature, since there $dy = 0$ and (11) shows that $d\mathbf{\Omega}$ reverses direction.

It is quite easy to produce caustics with no cusps. For example, choose the shape of W to be a cylindrically symmetric Gaussian hill. Then $\mathscr{L}$ is the circle corresponding to the radius R at which the profile inflects. The direction of vanishing curvature is radial, and therefore never touches $\mathscr{L}$, so that there are no cusps. The caustic is a circle in $\mathbf{\Omega}$. For case (iv) of § 2 this example corresponds precisely to potential scattering of 'rainbow' type (Ford and Wheeler 1959). The optical rainbow is produced by reflections as well as refractions (Descartes 1637) but it too is a circular directional caustic, and as such is the simplest natural example of an elementary catastrophe. (An end-point on a smooth curve is not a two-dimensional catastrophe, so that if the crock of gold exists it must be structurally unstable.)

For irregular droplet lenses (case (ii) of § 2), on the other hand, we can show that the caustics (fig. 6) *must* have cusps. First, recall that the profiles

Fig. 6

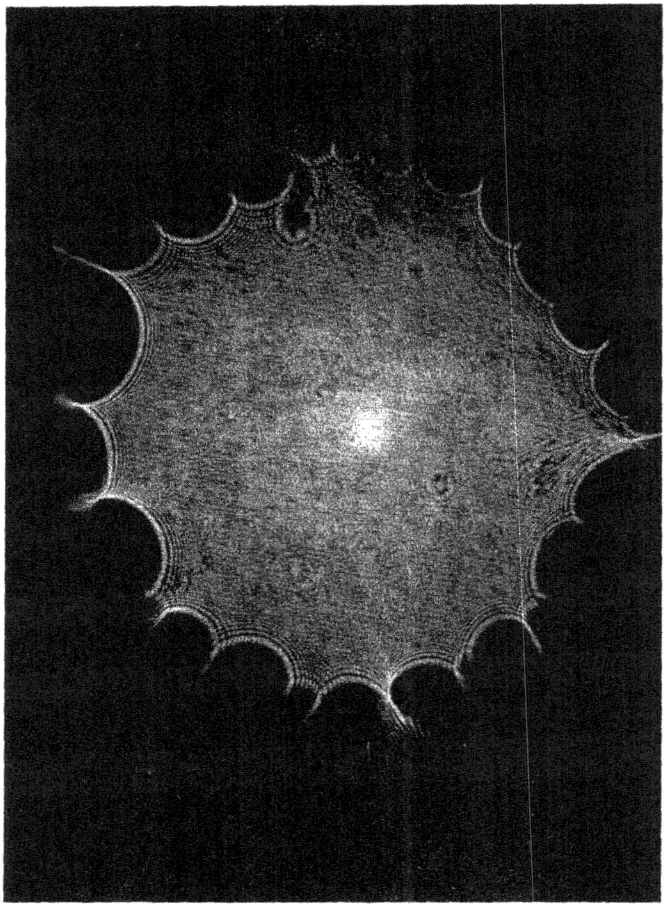

(*a*)

Far-field caustics from irregular water droplet on flat glass plate. (*c*) and (*e*) are tracings of (*b*) and (*d*). Swallow-tails are marked with arrows.

Fig. 6 (*continued*)

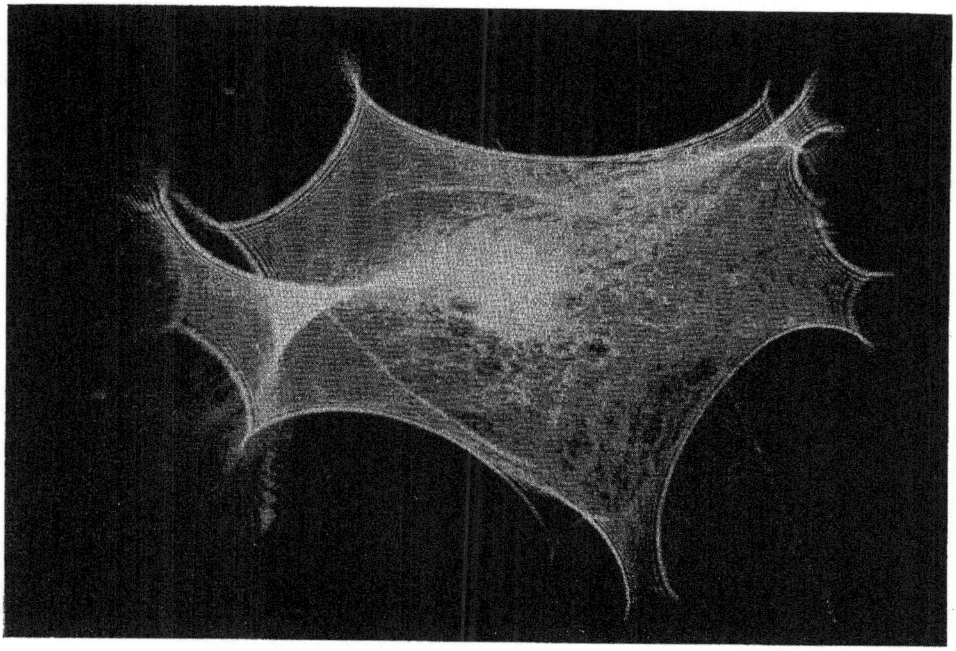

(*b*)

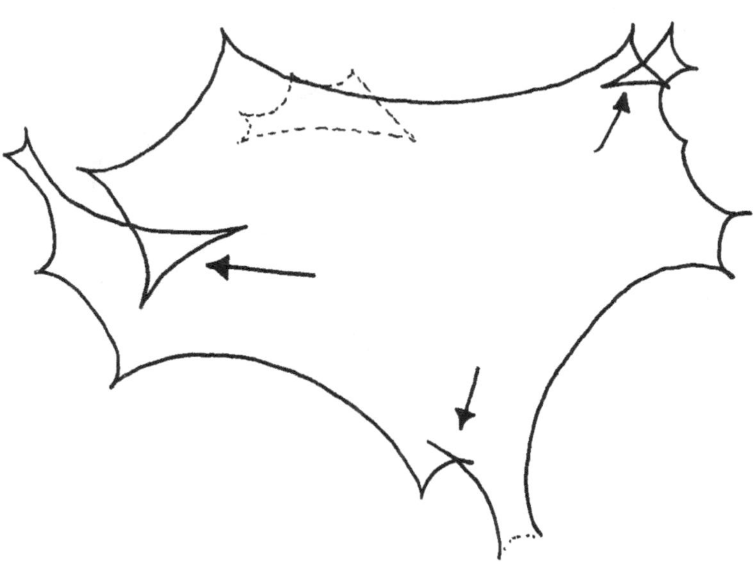

(*c*)

Fig. 6 (*continued*)

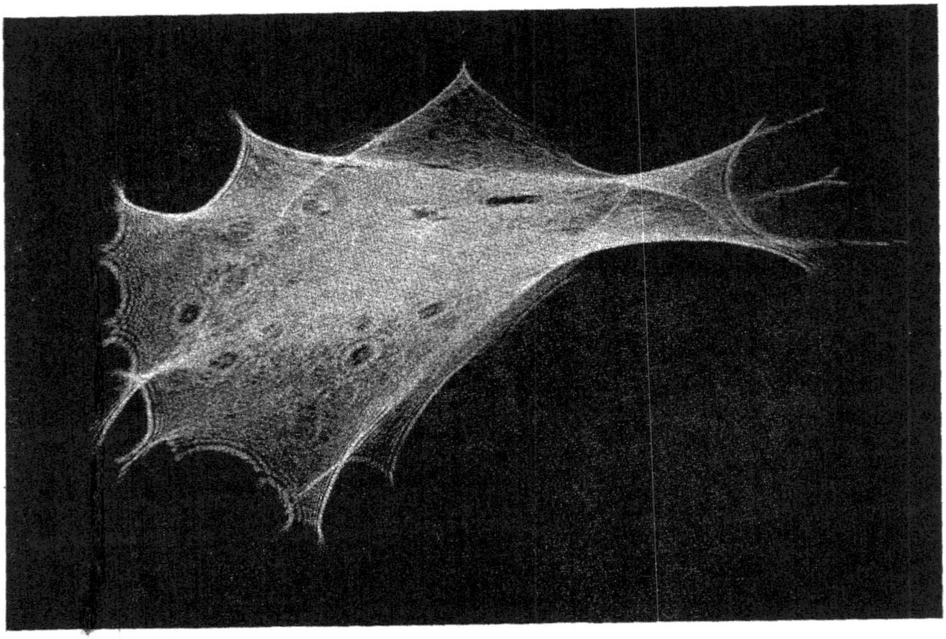

(*d*)

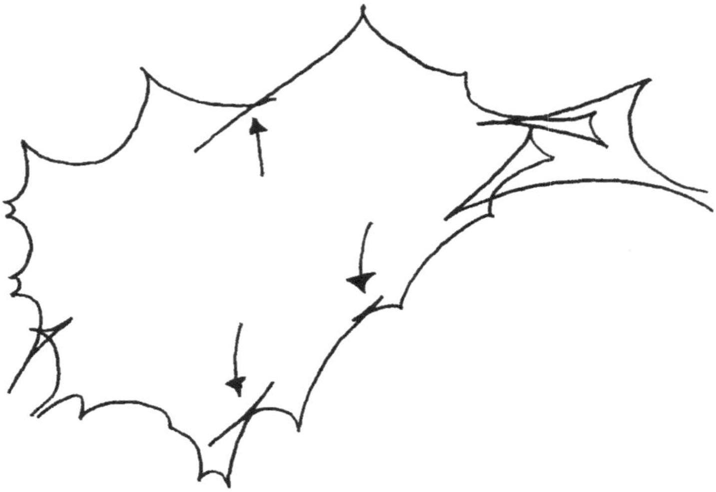

(*e*)

$f(\mathbf{R})$ are limited by surface tension to have the form of eqn. (2). Combined with (6) this gives, as the equation of the line $\mathscr{L}$,

$$|g''(\zeta)| = \frac{(n-1)p}{2n\gamma}, \tag{12}$$

where $\zeta = x + iy$, x and y being now global and not local Cartesians. Therefore we may write g'' on $\mathscr{L}$ as

$$g''(\zeta) = \frac{(n-1)p}{2n\gamma} \exp\,[i\chi(\zeta)]. \tag{13}$$

During one circuit of $\mathscr{L}$ the phase χ increases monotonically (Titchmarsh 1949), by an amount

$$\Delta\chi \equiv \oint_{\mathscr{L}} d\chi = \oint_{\mathscr{L}} d(\operatorname{Im} \ln g'') = \oint_{\mathscr{L}} \frac{d\zeta g'''}{g''} = 2m\pi, \tag{14}$$

where m is the number of zeros of g'' inside $\mathscr{L}$, non-simple zeros being counted according to their multiplicity.

Next we require the angle ψ made with the x-axis by the direction of the vanishing curvature. From (2), the curvature C_ψ of the line in W with direction ψ is easily found to be

$$C_\psi = |g''| \cos\,(2\psi + \chi) - \frac{(n-1)p}{2n\gamma}. \tag{15}$$

The two principal directions are the values of ψ for which this expression is extremal, and the direction of vanishing curvature (for which (12) holds) is

$$\psi = -\frac{\chi}{2}. \tag{16}$$

The difference between this angle and the angle ϕ made by $\mathscr{L}$ with the x-axis is

$$\phi - \psi = \phi + \frac{\chi}{2} \tag{17}$$

and cusps occur whenever $\phi - \psi$ passes through a multiple of π. Now, in one circuit of $\mathscr{L}$, ϕ changes by 2π and $\chi/2$ by $m\pi$. Therefore if $\phi + \chi/2$ is monotonic round $\mathscr{L}$ there are $m+2$ cusps. If $\phi + \chi/2$ were not monotonic, a rotation of coordinates in $\mathbf{R}$ would change ϕ but not χ and hence the number of times $\phi + \chi/2$ passed through a multiple of π would be altered. But the number of cusps is invariant under rotation of coordinates, and so, by contradiction, $\phi + \chi/2$ must be monotonic.

Finally, we observe that m must exceed zero for non-trivial cases, since any analytic function $g''(\zeta)$ with no zeros inside $\mathscr{L}$ must be a constant and this corresponds to droplets which are caps of ellipsoids (cf. eqn. (2)) which give no far-field caustics. Therefore we have shown that the caustic of the droplet defined by $g(\zeta)$ has two more cusps than $g''(\zeta)$ has zeros in $\mathscr{L}$. Thus the smallest possible number of cusps is three. It is possible to prove that the caustics from these surface tension dominated droplets must always be concave outwards (fig. 6); occasionally apparent exceptions to this are observed in the form of caustics with points of inflexion, but the corresponding drops are always larger and visibly affected by gravity so that the theory based on eqn. (1) does not apply.

A family of model droplets illustrating these properties is that generated by

$$g(\zeta)=\zeta^n \tag{18}$$

where $n=3, 4, 5 \ldots$, $\mathscr{L}$ is a circle, and the caustic is a hypocycloid of n cusps (Lawrence 1972). By comparing an observed caustic with n cusps with one of these 'standard caustics', it might be possible to estimate the variation of the contact angle around the droplet perimeter, and hence obtain information about inhomogeneity of wetting (Dr. E. Zichy of I.C.I. is investigating this). The droplets $f(\mathbf{R})$ corresponding to (18) have fluted maxima. Their caustics should form the basis of a new class of 'hyperelliptic' catastrophes (the elliptic umbilic corresponds to $n=3$).

An interesting possibility not covered by eqn. (18) occurs when the m zeros of g'' are isolated instead of coincident. Then, if g'' is a polynomial function of ζ, $\mathscr{L}$ will enclose all the zeros for sufficiently large p (eqn. (12)) and there will be $m+2$ cusps on the caustic. For very small p, on the other hand, $\mathscr{L}$ consists of m disconnected small closed curves, one round each zero, and there will be m closed caustics each with three cusps. Therefore there must in general be $m-1$ intermediate values of p at which $\mathscr{L}$ separates and two new cusps are born. The cusps appear in the manner shown in fig. 7, which is drawn for the droplet generated by

$$g''(\zeta)=1-\zeta^2; \tag{19}$$

this can have a maximum at $\mathbf{R}=0$ that splits into two as p diminishes through $2n\gamma/n-1$. We can regard p as a control parameter additional to Ω_x and Ω_y. However, the birth of two cusps in this fashion is not a higher catastrophe, since, as fig. 8 shows, $p=2n\gamma/n-1$ is simply a particular section touching a cusp edge that forms a smooth curve.

Fig. 7

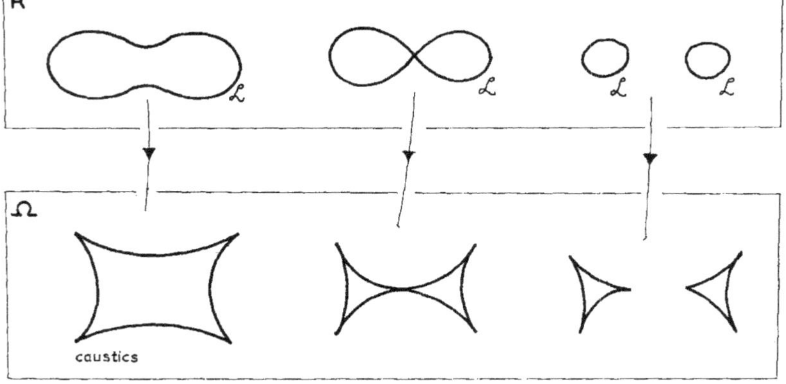

Birth of two cusps for droplet generated by (19) as p passes through $2n\gamma/n-1$.

Fig. 8

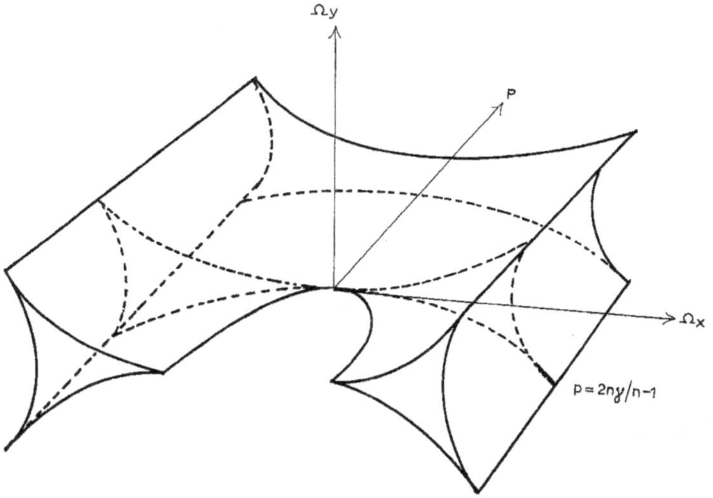

Caustic in $\mathbf{\Omega}$, p control space from droplet generated by (19).

For a given $f(\mathbf{R})$ defining a wavefront W, the control parameter space $\mathbf{\Omega}$ is two-dimensional. However, as the last example showed, we can think of $f(\mathbf{R})$ as embedded in a family of functions generated by one or more parameters $\boldsymbol{\epsilon}$ (e.g. pressure in a droplet). In this higher-dimensional control space the catastrophes are not restricted to folds and cusps, and changing $\boldsymbol{\epsilon}$ may produce caustics in $\mathbf{\Omega}$ which are singular two-dimensional sections of umbilics, swallow-tails and higher catastrophes.

Often there is no possibility of varying $\boldsymbol{\epsilon}$; then, however, it may be the case that in ' function space ' $f(\mathbf{R})$ is ' close to ' a higher catastrophe, whose slightly unfolded section will be recognizable in $\mathbf{\Omega}$. For hyperbolic and elliptic umbilics, the table and eqn. (7) show that it is natural to take $\Omega_1 = \Omega_x$ and $\Omega_2 = \Omega_y$, so that Ω_3 is the fixed parameter in $f(\mathbf{R})$. Such sections with Ω_3 fixed (figs. 3 (*d*) and 3 (*e*)) are often seen in caustics from irregular glass; fig. 5 is an example showing both umbilics. For water droplet caustics the elliptic umbilic is generated by eqn. (18) with $n = 3$; hyperbolic umbilics cannot occur because the outer branch would have to be convex outwards thus violating the theorem that the droplet caustics must always be concave outwards.

Swallow-tails are less common than umbilics in caustics generated by our simple model. The state space is one-dimensional and there is no natural identification of Ω_1, Ω_2 or Ω_3 with Ω_x and Ω_y; in addition a partially unfolded swallow-tail would correspond to two close tangencies of $\mathscr{L}$ with the direction of vanishing curvature, a circumstance unlikely often to occur. The author has scanned several hundred undulations of irregular glass with a laser beam without seeing a single unmistakeable swallow-tail. Thom (1975, fig. 4) gives a rather unconvincing example. However, it is possible to produce swallow-tails in the far field of droplet lenses whose perimeters are highly convoluted. Figures 6 (*b*) and 6 (*d*) show examples of this; the swallow-tails are marked on the tracings of the caustics on figs. 6 (*c*) and 6 (*e*). Figure 6 (*e*) also shows three 'specimens' of sections through the four-dimensional 'butterfly' catastrophe (Thom 1975). (On fig. 6 (*b*) there appears the faint image of a second caustic superimposed on the principal one; it arises from a second droplet dimly illuminated by the periphery of the laser beam.) This sort of search for catastrophes resembles not so much physics as specimen-collecting in botany. Individual specimens of a species of plant can vary enormously in size but share topological features (in animal species, by contrast, individuals are more nearly congruent); similarly, individual water droplet lenses vary enormously, and until the droplet hits the dusty glass surface one never knows exactly what its caustic will look like, only that some kind of catastrophe (species) will be involved. Liesegang (1953) shows beautiful swallow-tails formed in the beam of an electron microscope.

By deliberately considering wavefronts W whose form $f(\mathbf{R})$ is non-generic it is possible to produce a structurally unstable caustic. If this can be identified as the singular section of a higher catastrophe then the topology of the caustic that would be produced by a generic perturbation of W can be deduced simply by inspecting the unfolding of the catastrophe. Thus can Thom's theorem be used predictively.

For example, consider the non-generic case where $f(\mathbf{R})$ is *separable*, that is

$$f(\mathbf{R}) = g(\mathbf{a} \,.\, \mathbf{R}) + h(\mathbf{b} \,.\, \mathbf{R}), \tag{20}$$

where $\mathbf{a}$ and $\mathbf{b}$ are arbitrary non-parallel unit vectors on $\mathbf{R}$. It is not hard to show from the ray condition (4) and the caustic condition $K = 0$ that the caustics in $\boldsymbol{\Omega}$ are straight lines parallel to $\mathbf{a}$ and $\mathbf{b}$, meeting in corners but not crossing. In particular, if g and h are periodic functions whose topology is that of a sinusoid (i.e. two inflexions per period), then the caustic will be a parallelogram. The author tested this by sending a laser beam through two panes of glass on each of which a one-dimensional pattern of ridges had been impressed when the glass was softened by heating. The two panes were placed with the ridged faces together and the two sets of ridges not parallel. Figure 9 shows the result. Each non-generic corner is identified by inspection of fig. 3 as the section $\Omega_3 = 0$ of a hyperbolic umbilic, and this is confirmed by expanding $f(\mathbf{R})$ near the points on W that generate the corners, whereupon the functional form given in the table is obtained.

If now we imagine $f(\mathbf{R})$ (given by (20) with periodic f and g) perturbed by adding a non-separable term $\epsilon f_1(\mathbf{R})$, then each corner will unfold according to fig. 3 (*e*) so that the caustic will now be a distorted four-cusped astroid surrounded by a smooth convex closed curve. This prediction was first made during a study of the reflection of atomic beams from crystal surfaces; for a full description of this problem, and a photograph of an optical analogue confirming the prediction, see Berry (1975). Another application of catastrophe theory explains a curious feature of the near-field caustic lines painted by the sun on the sand beneath the sea (Minnaert 1954, p. 33): they frequently meet in triple junctions. These are non-generic and moreover do not correspond to sections of any three-dimensional catastrophe (fig. 3). This suggests that the triple junctions are really much more complicated generic structures imperfectly resolved or blurred by the $\frac{1}{2}°$ divergence caused by the finite

Fig. 9

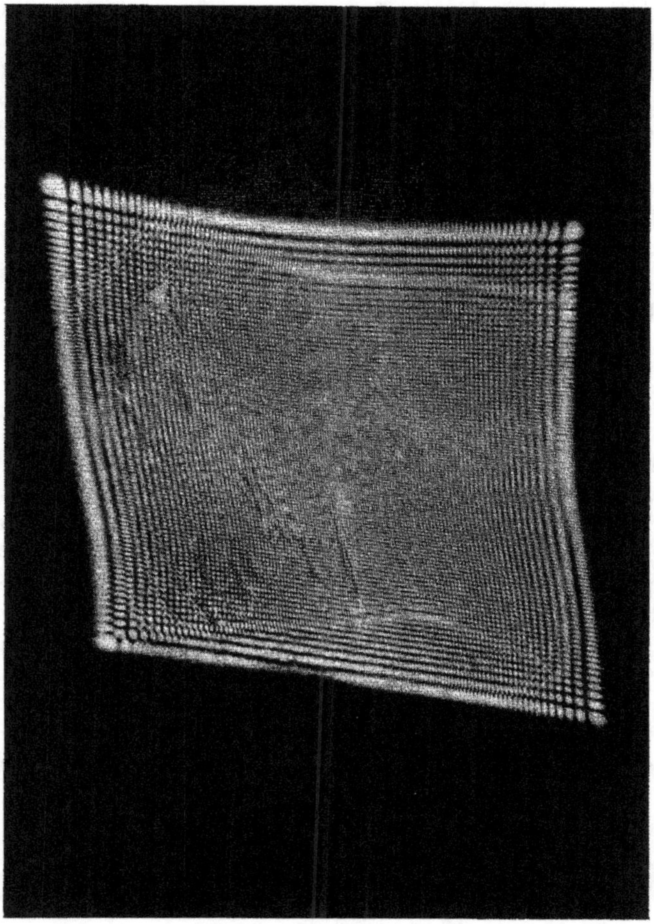

Far-field caustic from two superposed panes of ridged glass, showing singular sections of four hyperbolic umbilics.

disc of the sun. And, indeed, experiment confirms this (see Berry and Nye 1976).

§ 5. Caustics and wave functions

Near caustics the wave intensity shows dramatic diffraction effects. These are obvious on figs. 5, 6 and 9. In this section we give an analytical description of these phenomena, based on the generalization of the diffraction integral (10), namely

$$\psi(\mathbf{\Omega}) \equiv \left(\frac{k}{2\pi}\right)^{m/2} \iint d^m\mathbf{R} \exp\left[ik\Phi(\mathbf{R}, \mathbf{\Omega})\right]a(\mathbf{R}, \mathbf{\Omega}), \tag{21}$$

where a is a slowly varying function of its arguments (see Duistermaat 1974), and m is the dimensionality of the state space $\mathbf{R}$.

For short wavelengths k is large so that the exponential oscillates rapidly with $\mathbf{R}$ and the principal contributions come from those points where the phase $k\Phi$ is stationary. The stationary phase condition is eqn. (8), and this simply tells us that the stationary points $\mathbf{R}_i(\mathbf{\Omega})$ contributing to $\psi(\mathbf{\Omega})$ are those sending *trajectories* to $\mathbf{\Omega}$. For a general choice of control parameters $\mathbf{\Omega}$ it will usually be the case that the different $\mathbf{R}_i(\mathbf{\Omega})$ are well separated, so that the method of stationary phase can be used to evaluate (21) asymptotically for large k. This involves the $m \times m$ matrix

$$M_\Phi(\mathbf{R}) \equiv \left\{ \frac{\partial^2 \Phi}{\partial R_i \partial R_j} \right\}; \tag{22}$$

the method yields ψ as

$$\psi(\mathbf{\Omega}) \approx \sum_i \frac{\exp\,(i[kS_i(\mathbf{\Omega}) + \alpha_i \pi/4])}{|\det M_\Phi(\mathbf{R}_i(\mathbf{\Omega}))|^{1/2}}\, a(\mathbf{R}_i, \mathbf{\Omega}), \tag{23}$$

where α_i is the signature of M at $\mathbf{R}_i$ and S_i is the action (9).

The determinant in (23) is simply the Jacobian of the mapping from $\mathbf{R}$ to $\mathbf{\Omega}$. Taking the absolute square of (23), we obtain the intensity $I(\mathbf{\Omega})$ given by the pure trajectory picture (eqn. (5)), plus interference terms which for large k oscillate rapidly in $\mathbf{\Omega}$ (see § 6.2 of Berry and Mount (1972)).

Variation of $\mathbf{\Omega}$ will usually result in caustics being encountered ; then the mapping $\mathbf{R} \to \mathbf{\Omega}$ is singular, $\det M$ vanishes, two or more stationary points $\mathbf{R}_i$ coalesce, the formula (23) diverges and the simple stationary-phase method fails to give correctly the contribution of the trajectories to the asymptotic form of ψ. The nature of the divergence depends on the manner in which the stationary points coalesce, that is, on the topological type of the caustic.

In these circumstances a uniform approximation to (21), valid whether or not $\mathbf{\Omega}$ is near a caustic, can be found by comparison with a simpler integral that we may write as

$$j(\boldsymbol{\omega}) \equiv \left(\frac{k}{2\pi}\right)^{m/2} \iint d^m\mathbf{r} \exp\,[ik\phi(\mathbf{r}, \boldsymbol{\omega})]. \tag{24}$$

The function ϕ is chosen to be simpler in form than Φ but identical so far as the topology of coalescence of its stationary points is concerned. If the caustic produced by Φ is generic, then it is obvious that a sensible choice for ϕ is the standard form (table) given by Thom for the corresponding elementary catastrophe, with $\mathbf{r}$ as state variables and $\boldsymbol{\omega}$ as control variables.

Therefore Thom's theorem does far more than classify caustics : it also classifies the diffraction functions $j(\boldsymbol{\omega})$ that clothe the skeleton of structurally stable caustics when k is large but not infinite. The simplest of these functions, clothing the fold catastrophe, was introduced by Airy (1838) (see also Abramowitz and Stegun (1964)), while that clothing the cusp was introduced by Pearcey (1946) (see also Berry (1975)). Airy and Pearcey diffraction patterns are clearly visible on figs. 5 and 6. The pattern across the singular section of the hyperbolic umbilic (caustic corner) can be seen on fig. 9 ; analytically the wave function for this case is simply the product of two Airy functions. Mr. F. J. Wright is currently computing plane sections of the diffraction patterns for the swallow-tail and the elliptic and hyperbolic umbilics.

Before showing how to express (21) in terms of the comparison integrals (24), we introduce the *singularity index* σ, which gives a gross indication of the strength of the caustic. The greatest amplitude ψ occurs for control parameters $\mathbf{\Omega}_0$ at which the highest-order coalescence of stationary points occurs, and σ is defined by

$$|\psi(\mathbf{\Omega}_0)| = 0(k^{\sigma}). \tag{25}$$

For the elementary catastrophes (table) the singularity occurs at $\mathbf{\Omega} = 0$ and it is not hard to find σ; the values are given in the table. It is clear that σ increases with the co-dimension n. This is not surprising since the more control parameters there are the greater is the number of stationary points that can be made to coalesce. Arnol'd (1973) has calculated σ for some non-elementary castastrophes with finite co-dimension; the general problem is extremely difficult. For caustics of infinite co-dimension (§ 6) only a few results exist: glory scattering has $\sigma = \frac{1}{2}$ and the forward diffraction peak formed by scattering from a potential whose long range tail decays as (distance)$^{-\mu}$ has $\sigma = (\mu + 1)/(\mu - 1)$ (Berry 1969 a). Finally, we note that σ does not always exist; scattering from an exponentially decaying potential, for example, gives a forward amplitude proportional to $k \ln^2 k$ (Keller and Levy 1963).

Now we return to the problem of evaluating (21). Change the variables from $\mathbf{R}$ to $\mathbf{r}$ by the mapping

$$\Phi(\mathbf{R}, \mathbf{\Omega}) = C(\mathbf{\Omega}) + \phi(\mathbf{r}, \boldsymbol{\omega}(\mathbf{\Omega})), \tag{26}$$

to make (21) as similar as possible to (24). Choose $C(\mathbf{\Omega})$ and the comparison parameters $\boldsymbol{\omega}(\mathbf{\Omega})$ in such a way that the mapping is one-to-one. This can be achieved by making the stationary points $\mathbf{R}_i(\mathbf{\Omega})$ of Φ map onto the stationary points $\mathbf{r}_i(\boldsymbol{\omega})$ of ϕ, as can be seen from

$$\nabla_{\mathbf{R}}\Phi \,.\, d\mathbf{R} = \nabla_{\mathbf{r}}\phi \,.\, d\mathbf{r}. \tag{27}$$

For co-dimension n there can be $n+1$ stationary points involved in the coalescence (because each new parameter Ω enables another stationary point to be moved into coincidence with any pre-existing cluster). Therefore the $n+1$ equations

$$\Phi(\mathbf{R}_i(\mathbf{\Omega}), \mathbf{\Omega}) \equiv S_i(\mathbf{\Omega}) = \phi(\mathbf{r}_i(\boldsymbol{\omega}(\mathbf{\Omega})), \boldsymbol{\omega}(\mathbf{\Omega})) + C(\mathbf{\Omega}) \tag{28}$$

are just sufficient to determine the $n+1$ quantities C, $\boldsymbol{\omega}$. For some values of $\mathbf{\Omega}$ (on the 'dark side' of caustics) there will be fewer than $n+1$ real points $\mathbf{R}_i$, but by analytic continuation it is always possible to find *complex trajectories* $\mathbf{R}_i$, so that we always have $n+1$ equations (see, e.g., Berry 1966).

In terms of the new variables $\mathbf{r}$, (21) becomes, exactly,

$$\psi(\mathbf{\Omega}) = \left(\frac{k}{2\pi}\right)^m \exp[ikC(\mathbf{\Omega})] \iint d^m\mathbf{r}\, g(\mathbf{r}, \mathbf{\Omega}) \exp[ik\phi(\mathbf{r}, \boldsymbol{\omega}(\mathbf{\Omega}))], \tag{29}$$

where

$$g(\mathbf{r}, \mathbf{\Omega}) \equiv \left|\frac{d\mathbf{R}}{d\mathbf{r}}\right| a(\mathbf{R}(\mathbf{r}), \mathbf{\Omega}). \tag{30}$$

Next we approximate by replacing g by the first term of an expansion about the stationary points $\mathbf{r}_i$, on the grounds that it is the stationary points that dominate the behaviour of (29). The expansion can be generated by writing

$$g(\mathbf{r}, \mathbf{\Omega}) = g_0(\boldsymbol{\omega}) + \mathbf{g}_1(\omega) \cdot \nabla_\omega \phi(\mathbf{r}, \boldsymbol{\omega}) + \mathbf{h}(\mathbf{r}, \boldsymbol{\omega}) \cdot \nabla_r \phi(\mathbf{r}, \boldsymbol{\omega}), \tag{31}$$

where $\boldsymbol{\omega} = \boldsymbol{\omega}(\mathbf{\Omega})$, $\mathbf{g}_1$ is an n-component vector and $\mathbf{h}$ an m-component vector. At the stationary points, $\mathbf{h} \cdot \nabla_r \phi$ vanishes and so we neglect this term altogether. (If we included it, and expanded each component of $\mathbf{h}$ in a similar fashion to (31), the result would be an asymptotic series for (29) involving inverse powers of k. The derivation of the complete asymptotic series, and their interpretation as exact representations of wave functions ψ (Dingle 1973) is a challenging task for the future.) Neglecting $\mathbf{h}$, we can write (29) in terms of the $n+1$ quantities g_0, $\mathbf{g}_1$, and the comparison integrals $j(\boldsymbol{\omega})$ and their derivatives :

$$\psi(\mathbf{\Omega}) = \exp\,[ikC(\mathbf{\Omega})] \left[g_0(\boldsymbol{\omega}(\mathbf{\Omega}))j(\boldsymbol{\omega}(\mathbf{\Omega})) + \frac{\mathbf{g}_1(\boldsymbol{\omega}(\mathbf{\Omega}))}{ik} \cdot \nabla_\omega j(\boldsymbol{\omega}(\mathbf{\Omega})) \right]. \tag{32}$$

To find g_0 and $\mathbf{g}_1$ we use the $n+1$ equations obtained by evaluating (31) at the stationary points. The left-hand side requires a knowledge of $|d\mathbf{R}/d\mathbf{r}|$ at each stationary point (cf. (30)), and differentiation of (27) leads to

$$\left|\frac{d\mathbf{R}}{d\mathbf{r}}\right|_i = \left|\frac{\det M_\phi(\mathbf{r}_i(\boldsymbol{\omega}(\mathbf{\Omega})))}{\det M_\Phi(\mathbf{R}_i(\mathbf{\Omega}))}\right|^{1/2}, \tag{33}$$

where M is defined by (22). Therefore g_0 and $\mathbf{g}_1$ are determined by solving

$$\left|\frac{\det M_\phi(\mathbf{r}_i(\boldsymbol{\omega}(\mathbf{\Omega})))}{\det M_\Phi(\mathbf{R}_i(\mathbf{\Omega}))}\right|^{1/2} a(\mathbf{R}_i(\mathbf{\Omega}), \mathbf{\Omega}) = g_0 + \mathbf{g}_1 \cdot \nabla_\omega \phi(\mathbf{r}_i(\boldsymbol{\omega}(\mathbf{\Omega})), \boldsymbol{\omega}(\mathbf{\Omega})). \tag{34}$$

The wave function is now given by (32) in terms of $j(\boldsymbol{\omega})$ defined by (24). This technique gives a *uniform approximation* : for $\mathbf{\Omega}$ far from caustics the comparison integrals themselves may be expanded by the stationary phase method and we quickly recover (23). When $\mathbf{\Omega}$ is very close to the highest singularity $\mathbf{\Omega}_0$, on the other hand, we can neglect the second term in (32) and set $\boldsymbol{\omega}$ proportional to $\mathbf{\Omega}$ in the first term, thus obtaining a *transitional approximation* valid very close to $\mathbf{\Omega}_0$ (for examples see Airy (1838) and Ford and Wheeler (1959)). The uniform approximation for the fold catastrophe was first obtained by Chester, Friedman and Ursell (1957), and introduced into wave theory by Berry (1966), and Ludwig (1966, 1967). This latter author also devised uniform approximations for some other elementary catastrophes, as did Connor (1973, 1974, 1976). The whole subject has been put on a rigorous basis, so far as structurally stable caustics are concerned, by Duistermaat (1974). However, we emphasize that the result (32) is not restricted to these cases ; in fact Berry (1969 a) obtained uniform approximations for two structurally unstable caustics of infinite co-dimension (see § 6). Numerically, uniform approximations are very accurate (Mount 1973, Berry and Mount 1972), and they have been employed in the inversion of scattering data to determine interaction potentials (Buck 1971, 1974, Mullen and Thomas 1973).

§ 6. Caustics of infinite co-dimension

It sometimes happens that a wave system is strongly constrained to have a high degree of symmetry, so that the caustic is structurally unstable and extra parameters $\boldsymbol{\epsilon}$ are necessary to restore genericity, that is, to produce all the different structurally stable caustics 'near' to the original one. We are concerned here with cases where the number of components of $\boldsymbol{\epsilon}$ is infinite, so that the caustic is of infinite co-dimension. There is as yet no mathematical scheme classifying caustics of this type; Thom's theorem is inapplicable because it is restricted to co-dimension less than 7. We discuss three examples.

The first occurs when the wavefront W (§ 2) has cylindrical symmetry and its profile $f(\mathbf{R})$ has a maximum not at the origin. This maximum forms a circle ($R=R_0$ say) in $\mathbf{R}$, near which W is locally part of a torus. This situation corresponds to *glory scattering* in quantum mechanics (Berry 1969 a) or meteorology (Van de Hulst 1957, Nussenzveig 1969) where a trajectory with non-zero impact parameter interacts with a spherical object and emerges in the forward or backward direction. It can also be produced by an annular droplet on a glass surface, $f(R)$ being given by (2) with $g=\ln(x+iy)$.

The Gaussian curvature K vanishes on $R=R_0$, so that (§ 2) this circle generates the caustic. When $R=R_0$ the slope ∇f vanishes, and so according to eqn. (4) the corresponding trajectories all have $\boldsymbol{\Omega}=0$: the far-field caustic is a single point in the forward direction. A point is structurally unstable in $\boldsymbol{\Omega}$, and the possibility suggests itself that we might be seeing the singularity $\Omega_3=0$ of the elliptic umbilic (fig. 3 (*d*)).

That this is not the case can be seen in two ways. First, the generating function $\Phi(\mathbf{R},\boldsymbol{\Omega})$ (eqn. (7)), when written in simple standard form obtained by expansion about R_0, is

$$\Phi(\mathbf{R},\boldsymbol{\Omega})=-\frac{(\mathbf{R}-\mathbf{R}_0)^2}{2a}+\Omega R\cos\theta \tag{35}$$

where θ is the polar angle in $\mathbf{R}$, measured with respect to the direction of $\boldsymbol{\Omega}$. This function has different symmetry from the elliptic umbilic (table), and cannot be transformed into it. Second, the comparison integral (24) for this case is, for large k and small Ω,

$$j(\boldsymbol{\Omega})=R_0(2\pi a)^{1/2}k^{1/2}\exp(-i\pi/4)J_0(kR_0\Omega), \tag{36}$$

where J_0 is the zero-order Bessel function of the first kind. Therefore the singularity index, defined by eqn. (25) and found by setting $\boldsymbol{\Omega}=0$, is $\sigma=\frac{1}{2}$, so that the glory is a stronger singularity than the elliptic umbilic or indeed any other elementary catastrophe.

The focusing into $\boldsymbol{\Omega}=0$ arises from the circular symmetry of W. If this is broken by perturbations $f_1(\mathbf{R},\boldsymbol{\epsilon})$ with parameters $\boldsymbol{\epsilon}$ of the profile $f(\mathbf{R})$, then the point caustic should break up into closed cusped figures, the number of cusps depending on the topology of f_1. But there is an infinity of topologically different ways in which the (continuous group) symmetry of a circle can be broken, so that to find all the structurally stable caustics 'near' the glory requires infinitely many parameters $\epsilon_1, \epsilon_2 \ldots$. We conjecture that such a generic unfolding of the function (35) is obtained by a series of circular harmonic perturbations, namely

$$f_1(\mathbf{R}, \boldsymbol{\epsilon}) = \sum_{n=2}^{\infty} \epsilon_{n-1} R^n \cos(n\theta + \alpha_n), \tag{37}$$

involving a countable infinity of parameters.

Our second caustic of infinite co-dimension is the *forward diffraction peak* from localized objects such as a wavefront deformation $f(\mathbf{R})$ vanishing smoothly as $\mathbf{R} \to \infty$. This caustic is also the isolated point $\mathbf{\Omega} = 0$, produced by the flat 'region at infinity' for which $\nabla f = 0$. K vanishes over an area rather than on a line, so the forward peak is an even stronger structurally unstable caustic than the glory.

To evaluate the singularity index σ for this case it is first necessary to make the diffraction integral (10) converge by subtracting the forward delta function that would occur if W were flat ($f = 0$). The resulting differential scattering cross section is obtained from (10) by replacing $\exp(ik\Phi)$ by $\exp(ik\Phi) - 1$. A detailed analysis (Berry 1969 a) then shows that σ depends on the manner in which $f(\mathbf{R})$ goes to zero at infinity. For example, in the case of potential scattering (eqn. (3)) when $V(r)$ has the Van der Waals' tail, $-r^{-6}$, we found $\sigma = 1{\cdot}4$. This is powerful focusing indeed, and the forward peak is the strongest directional caustic yet discovered.

As we might expect, the unfolding of the forward peak is more complicated than that of the glory. The 'symmetry' to be broken is the asymptotic vanishing of $f(\mathbf{R})$, so that a typical symmetry-breaking perturbation $\epsilon f_1(\mathbf{R})$ must have maxima, minima and saddles right out to $\mathbf{R} = \infty$. Each undulation gives a caustic (perhaps resembling fig. 5 or 6) and in $\mathbf{\Omega}$ space these will all be superposed. As $\epsilon \to 0$ this mass of caustics shrinks down to $\mathbf{\Omega} = 0$ giving the forward peak. To examine this process in detail, let the forward peak be produced by a cylindrically symmetrical function $f_0(R)$ of Gaussian form, and let the perturbation $\epsilon f_1(\mathbf{R})$ be a periodic undulation with square unit cells containing a minimum at the centre and maxima at each corner, and consider

$$f(\mathbf{R}) = (1 - \epsilon) f_0(R) + \epsilon f_1(\mathbf{R}) \tag{38}$$

as ϵ diminishes from 1 to 0. The progression is shown on fig. 10. When $\epsilon = 1$ only f_1 acts, and the caustic (fig. 10 (*a*)) is an astroid with surrounding smooth curve (Berry 1975). Slightly diminishing ϵ introduces $f_0(R)$ which

Fig. 10

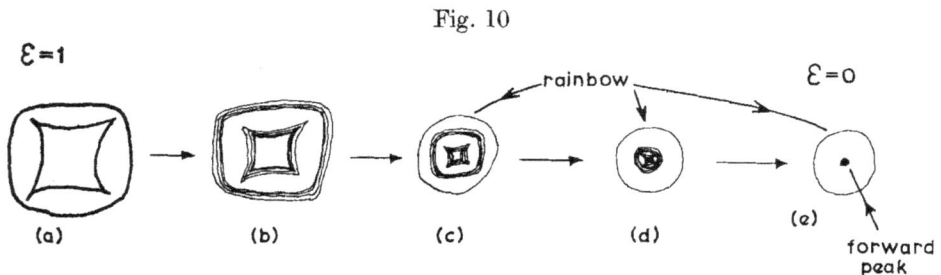

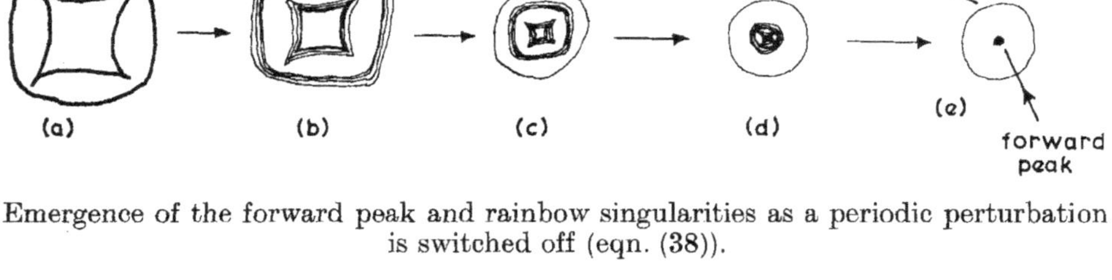

Emergence of the forward peak and rainbow singularities as a periodic perturbation is switched off (eqn. (38)).

slightly perturbs the unit cells near $\mathbf{R}=0$ so that the caustic splits into an infinite concentric sequence (fig. 10 (*b*)) with ' limit lines ' from the unit cells at infinity. When ϵ is further decreased the diminishing slopes of $f_1(\mathbf{R})$ cause this caustic cluster (fig. 10 (*c*)) to shrink in $\mathbf{\Omega}$, while the growth of $f_0(R)$ exposes its inflexion line as a growing ' rainbow ' smooth caustic surrounding all the others. Finally, as $\epsilon \to 0$ the caustic cluster from $f_1(\mathbf{R})$ collapses (fig. 10 (*d*)) onto the forward peak (fig. 10 (*e*)) at $\mathbf{\Omega}=0$, leaving the rainbow circle at some finite $|\mathbf{\Omega}|$. Of course there are infinitely many ways apart from this in which ' superposition and collapse ' can occur, so that the infinite co-dimensionality of the forward peak is probably uncountable. (Perhaps this is one of the ' lump ' catastrophes discussed by Thom (1975, pp. 103–4).)

The third class of caustics with infinite co-dimension are those occurring at point *sources*. Of course the radiation fields are singular at source points, but we are primarily interested in the pattern of emerging trajectories. Obviously a vibrating source at rest relative to the propagation medium (or to the observer in the case of light) is an isolated point caustic (a ' focus '). No perturbation of the medium will break this even though it is structurally unstable. Motion of the source can produce fold caustics emanating from it (e.g. Cerenkov radiation, or Kelvin's ship wave pattern—see Lamb 1945), but this situation is still non-generic because the folds meet at a finite angle and not, say, at a cusp. To understand the unfolding it is necessary to imagine the time-reversed situation where there is no source and the caustic point arises from the collapse of a wavefront with a high degree of symmetry (e.g. a circle in a homogeneous medium with source at rest). Then any of an infinite-parameter class of perturbations of this wavefront will break the caustic into a structurally stable form (e.g. a cusped line in two dimensions). In the real-time problem, however, the presence of the source is a strong constraint suppressing all such perturbations, and the unfolding cannot be realized.

§ 7. Concluding remarks

Thom's theorem has enabled us to make powerful general statements about the caustics that dominate wave fields when the wavelength is short. Firstly, it lists the canonical forms that caustics in generic systems may take. Secondly, it enables canonical diffraction functions to be derived that give accurate representations of the wave fields that clothe these caustics. And, thirdly, departures from genericity, caused by special conditions corresponding to particular values of a small number of parameters describing the system, give rise to caustics that are structurally unstable ; these are easily identified because they are not elementary catastrophes with co-dimension equal to the dimension of the space in which the caustics are observed, but special sections of higher catastrophes, whose unfoldings lead to predictions of the forms of the structurally stable caustics that would result if genericity were restored by removing the special conditions.

We emphasize that structural stability of a caustic does not imply physical stability of the system generating the caustic, and vice versa. For example, a spherical raindrop is rendered physically stable by the force of surface tension, and yet it gives rise to the glory which is so structurally unstable as to require infinitely many unfolding parameters. Attempts are being made to develop extensions of Thom's theorem to deal with such cases, by studying singularities which are structurally stable in a restricted sense, namely under perturbations of special types (e.g. those preserving spherical symmetry). Interesting as such studies undoubtedly are, they should not deflect too much effort from the more difficult task of discovering unfoldings that are structurally stable under all perturbations ; in the case of the glory, for example, generic perturbations are certainly possible, in the form of oscillations of raindrops away from sphericity.

Next, we mention two aspects of short-wave theory to which Thom's theory has not so far been applied. The first is the determination of the spectrum of normal modes of bound systems (e.g. energy levels in quantum mechanics) ; this problem is dominated by the topologies of closed trajectories (Balian and Bloch 1974, Berry and Tabor 1976), and caustics play a relatively minor role, although they do affect the spectrum.

The second arises when the pattern of trajectories is so complicated that the caustics are separated by less than a wavelength, on the average. This case is realized, for example, when waves are reflected by a surface that is rough on all scales (Berry 1972) or when trajectories are ergodic (Percival 1973). We conjecture that in such situations no short-wave limit exists. This is an important problem that should richly repay detailed study.

It is possible that catastrophes penetrate even deeper into wave theory than this article suggests. On the finest scale, the most delicate features of waves are the singularities of their phase maps, christened ' wavefront dislocations ' by Nye and Berry (1974). Now, the phase map for the cusp diffraction function, plotted by Pearcey (1946), shows a striking pattern of edge dislocation points. The three-dimensional catastrophe diffraction functions must have a pattern of dislocation lines with highly interesting topology. I conjecture that each generic caustic has an associated dislocation structure that is structurally stable. If this is correct, Thom's theorem would also classify wavefront dislocations (the classification would be incomplete because there are simple ways of creating dislocations that are not associated with caustics). This subject is now being studied in collaboration with Mr. F. J. Wright.

To conclude, let us compare the application of Thom's theorem in wave theory, as considered in this article, with the application of the theorem in statistical mechanics, to phase transitions of infinite systems. In both cases catastrophe theory gives an overall description : geometrical optics in the wave case, ' mean field theory ' (e.g. Van der Waals' equation) in the statistical case. Very close to the singularities, however, more refined methods must be employed, because of diffraction in the wave case (requiring the integration techniques of § 5) and fluctuations in the statistical case (requiring special techniques for calculating ' critical exponents '). This point is well made by Schulman (1974).

Acknowledgments

I particularly wish to thank Professor John Nye for many long discussions about irregular droplet lenses. In addition, I thank Dr. Manfred Faubel for an experimental suggestion, Mr. George Keene for photography, Mr. Tony

Osman for making the undulating glass plates, and Dr. Tim Poston for some helpful comments.

References

Abramowitz, H., and Stegun, I. A., 1964, *Handbook of Mathematical Functions* (Washington : U.S. National Bureau of Standards).
Airy, G. B., 1838, *Proc. Camb. phil. Soc. math. phys. Sci.*, **6,** 379.
Arnol'd, V. I., 1973, *Russ. math. Survs*, **28,** No. 5, 19 (English translation).
Balian, R., and Bloch, C., 1974, *Ann. Phys.*, **85,** 514.
Berry, M. V., 1966, *Proc. phys. Soc.*, **89,** 479 ; 1969 a, *J. Phys.* B, **2,** 381 ; 1969 b, *Sci. Prog., Oxf.*, **57,** 43 ; 1972, *J. Phys.* A, **5,** 272 ; 1975, *Ibid.*, **8,** 566.
Berry, M. V., and Mount, K. E., 1972, *Rep. Prog. Phys.*, **35,** 315.
Berry, M. V., and Nye, J. F., 1976 (to be published).
Berry, M. V., and Tabor, M., 1976 *Proc. R. Soc.* (in the press).
Bröcker, Th., 1975, *Differential Germs and Catastrophes* (Cambridge University Press).
Buck, U., 1971, *J. chem. Phys.*, **54,** 1923 ; 1974, *Rev. mod. Phys.*, **46,** 369.
Campbell, L., and Garnett, W., 1882, *The Life of James Clerk Maxwell* (London : Macmillan), pp. 434–44.
Chester, C., Friedman, B., and Ursell, F., 1957, *Proc. Camb. phil. Soc. math. phys. Sci.*, **53,** 599.
Connor, J. N. L., 1973, *Molec. Phys.*, **26,** 1217 ; 1974, *Ibid.*, **27,** 853 ; 1976, *Ibid.*, **31,** 33.
Descartes, R., 1637, *Discours de La Méthode* (Appendix).
Dingle, R. B., 1973, *Asymptotic Expansions : Their Derivation and Interpretation* (London : Academic Press).
Duistermaat, J. J., 1974, *Communs pure appl. Maths.*, **27,** 207.
Eisenhart, L. P., 1960, *A Treastise on the Differential Geometry of Curves and Surfaces* (Dover), p. 180.
Fellini, F., 1963, 8½ (Film distributed by Cineriz, Italy).
Ford, K. W., and Wheeler, J. A., 1959, *Ann. Phys.*, **7,** 259, 287.
Glauber, R. J., 1958, *Lectures in Theoretical Physics*, Vol. 1, edited by W. E. Brittin and L. G. Dunham (New York : Interscience), p. 315.
Goldstein, H., 1951, *Classical Mechanics* (Addison–Wesley).
Jänich, K., 1974, *Math. Annln*, **209,** 161.
Kravtsov, Yu. A., 1968, *Soviet Phys. Acoust.*, **14,** 1 (English translation).
Keller, J. B., and Levy, B. R., 1963, *Proc. Conf. Electromagn. Scattering*, 1962, edited by M. Kerker (Oxford : Pergamon), p. 3.
Lamb, H., 1945, *Hydrodynamics* (Cambridge University Press), p. 438.
Larmor, J., 1891, *Proc. Camb. phil. Soc. math. phys. Sci.*, **7,** 131.
Lawrence, J. Dennis, 1972, *A Catalog of Special Plane Curves* (New York : Dover).
Liesegang, S., 1953, *Optik*, **10,** 5.
Ludwig, D., 1966, *Communs pure appl. Math.*, **19,** 215 ; 1967, *Ibid.*, **20,** 103.
Maslov, V. P., 1972, *Theorié des Perturbations et Méthodes Asymptotiques* (Paris : Dunod). (Original Russian publication, 1965.)
Minnaert, M., 1954, *The Nature of Light and Colour in the Open Air* (Dover).
Mount, K. E., 1973, *J. Phys.* B, **6,** 1397.
Mullen, J. M., and Thomas, B. S., 1973, *J. chem. Phys.*, **58,** 5216.
Nussenzveig, H. M., 1969, *J. Math. Phys.*, **10,** 82.
Nye, J. F., and Berry, M. V., 1974, *Proc. R. Soc.* A, **336,** 365.
Pearcey, T., 1946, *Phil. Mag.*, **37,** 311.
Percival, I. C., 1973, *J. Phys.* B, **6,** L229.
Schulman, L., 1975, *Functional Integration and its Applications*, edited by A. M. Arthurs (Oxford : Clarendon Press), p. 144.
Stavroudis, O. N., 1972, *The Optics of Rays, Wavefronts and Caustics* (New York : Academic Press), p. 79.

SYNGE, J. L., 1954, *Geometrical Mechanics and de Broglie Waves* ; 1960, *Encyclopedia of Physics*, edited by S. Flügge (Berlin : Springer-Verlag), **311,** 1.
THOM, R., 1975, *Structural Stability and Morphogenesis* (Reading, Mass. : Benjamin).
TITCHMARSH, E. C., 1949, *The Theory of Functions* (Oxford University Press), p. 122.
VAN DE HULST, H. C., 1957, *Light Scattering by Small Particles* (New York : John Wiley).
WASSERMANN, G., 1974, *Stability of Unfoldings*, Vol. 393 (Springer Mathematical Notes).

Michael Berry

Slippery as an eel

The Fire Within the Eye: A Historical Essay on the Nature and Meaning of Light
David Park
1997 Princeton University Press 378pp £25.00hb

With passion and poetry, David Park sets out to get behind the optical science we are familiar with as physicists. He tells us how attempts to understand light have been at the heart of people's efforts to make sense of the world ever since they began to reflect on it (note how the natural choice of metaphor reflects this). It is a fascinating story, beginning with the "immense fact [that] we can see", and ending...well, it has not ended yet, as we will "see".

Early theories are discussed in detail. Empedocles' "visual ray" is "like a long finger projecting from the eye, and sight is a kind of touch". It was believed that the ray can occasionally be seen, for example in the gleam of an animal's eye from the darkness near a campfire, and that it can have powerful effects, as in the "evil eye". Then there was the "eidolon", conceived by Leucippus: "...under the influence of light the surface of any visible object continually produces thin veils of matter, perhaps only one atom thick, which peel off and retain their shape as they fly with immense speed in every direction". These notions persisted for centuries, and much ingenuity was expended in overcoming the problems they raised, such as "How does the eidolon get into the little hole in my eye?".

We learn how Euclid was the first to try "to catch Nature in the web of mathematically exact reasoning" with his studies of the optics of mirrors based on the laws of specular reflection. Much later, Islamic scientists pursued the mathematical study of light, in constructing the beginnings of optics as we recognize it today. Apparently, Ibn Sahl in 984 knew the law of refraction, soon forgotten and rediscovered half a millennium later by Harriot, Snel and Descartes. At about the same time, the Egyptian scientist Alhazen began to understand how the eye works, and (following Ptolemy) appreciated that proposed explanations can be tested by experiment.

Inevitably, many pages are devoted to Newton, and here I read two sentences that knocked me flat. As everyone knows, Newton imagined light as a stream of particles travelling along rays, but pondered intensely on effects that seemed discordant with this picture and were convincingly explained much later in terms of light waves. One of these was Grimaldi's observation in 1665 of the fringes in light diffracted from an edge. In one of the famous "queries" in his *Opticks*, Newton asked: "Are not the rays of Light in passing by the edges and sides of Bodies, bent several times backwards and forwards, with a motion like that of an Eel? And do not the three Fringes of colour'd Light above-mention'd arise from three such bendings?" Park thinks that "science still awaits a mathematical theory of the eel". He is wrong, and in an interesting way.

One way of writing wave equations, discovered in the context of quantum mechanics by Madelung and emphasized by Bohm and his followers, is in terms of the local current vector rather than the function describing the wavefield. The lines of current can be regarded as analogous to the rays of geometrical optics, but survive into wave optics. Where propagating waves interfere, these rays indeed wriggle like an eel, as the result of non-Newtonian forces acting from edges etc. Although (perhaps for reasons of historical contingency) this is not the interpretation of wave physics that most of us use, all wave phenomena can be regarded as the effect of these generalized rays. So, Newton was right after all!

Any study of light must include colour, and Park gives an excellent account of the familiar story leading from Newton's prism through Fraunhofer's spectroscopy to the quantum mechanics of today. I was, however, astonished by Euler's "clairvoyant insight" that coloured light originates in internal vibrations of atoms in the emitting body. *Seeing* colours is very different, of course. Eyes are not spectroscopes, but respond to the excitation of three colour receptors – that is, colour space is three- (rather than infinite-) dimensional. Young's pioneering understanding of this fact is well described, as is Goethe's impassioned dissidence from this developing consensus. Although Goethe's unscientific modes of expression made his view unpopular, modern studies – particularly by Edwin Land – have rehabilitated his emphasis on the importance of the surroundings. This can be restated in a way that resonates with our contemporary thinking: while Newton, Young and Maxwell pioneered the local theory of colour, Goethe pioneered the non-local theory.

Relativity is introduced in terms of the rise and demise of the ether, as the medium in which light was supposed to wave. Here is Park on the Michelson–Morley experiment. "It is a rare thing when the Lord bends down and speaks to his children, but on this occasion he did. Clothing his word in the language of Nature, he told those two men that they had blown the ether away; but they didn't hear him."

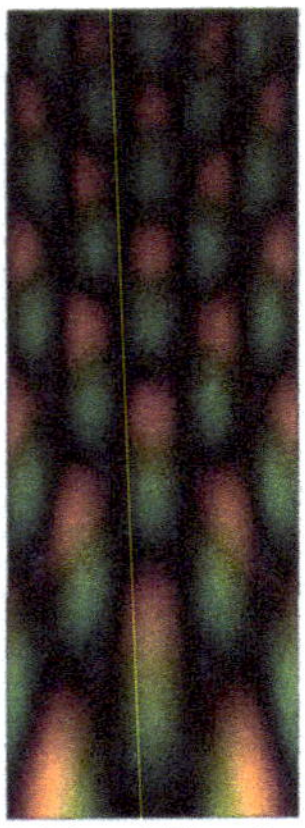

Rainbow skittles – the interference of three waves of white light

I was disappointed by two omissions. There is no explicit mention of the development of caustics, that is generalized focal curves and surfaces. Interference near caustics produces the largest and brightest wave effects; their most dramatic manifestations are supernumerary rainbows. There is no evidence that Newton noticed these; if he had, the development of physics might have been very different. But Young knew about them, and pointed out that here was a natural phenomenon that Newton's theory (without the eels) was unable to explain, but his could. And I would have liked to have read something about Hamilton's theory of conical refraction, which, as well as getting right to the heart of light's transversality, also laid the groundwork for the mathematical unification of particle and wave motion that was so important in the construction of quantum mechanics.

That is where we are now. Light is waves. Light is particles. Interactions are nonlocal. Is this the end? Park thinks not. As he says: "Some scientific questions are interesting and some uninteresting, where the words are defined as follows. After an uninteresting question has been settled, it is settled. After an interesting question has been settled, it keeps popping up again. The question of non-locality, in the opinion of many physicists, is interesting." I could not agree more.

Michael Berry is a professor of physics in the Department of Physics, Bristol University, UK

REVIEWS

Michael Berry

Deconstructing rainbows

The Rainbow Bridge: Rainbows in Art, Myth and Science
Raymond L Lee and Alastair B Fraser
2001 The Pennsylvania State University Press and SPIE Press 393pp $65.00hb

A single ray of light has a pathetic repertoire, limited to bending and bouncing (into water, glass or air, and from mirrors). But when rays are put together into a family – sunlight, for example – the possibilities get dramatically richer. This is because a family of rays has the holistic property, not inherent in any individual ray, that it can be focused so as to concentrate on caustic lines and surfaces. Caustics are the brightest places in an optical field. They are the singularities of geometrical optics. The most familiar caustic is the rainbow, a grossly distorted image of the Sun in the form of a giant arc in the skyspace of directions, formed by the angular focusing of sunlight that has been twice refracted and once reflected in raindrops.

This explanation of the rainbow, given by Descartes in 1637, was both a pioneering exercise of computational theoretical physics and the culmination of several millennia of what Raymond Lee and Alastair Fraser call "the slow and convoluted evolution of physical reasoning about the rainbow".

We learn, for example, how Aristotle – in about 350 BC – understood the rainbow as redirected sunlight, but misunderstood the redirection as mirror reflection from clouds. We learn how Qutb al-Din in Persia and Theodoric in Freiberg, in about 1300, understood the importance of reflection and refraction within individual raindrops but were unable to explain why the light is concentrated near the rainbow angle.

Equally "slow and convoluted" was our understanding of a rainbow's colours, which culminated (at least at the level of ray optics) with Newton's incorporation of the spectral dispersion of water into Descartes' theory. An original feature of the book is the detailed quantification of the colours of several natural rainbows, depicted as curves in the diagram of chromaticity co-ordinates representing hue and saturation. This exercise has the unexpected and startling result that the range of rainbow colours is tiny – less than 5% of the gamut of colours on a TV screen.

Yet this miserable smudge of wan and unsaturated colour "sticks in the popular imagination as a paragon of color variety". Moreover, although "popular imagination" also envisages the rainbow's colours as a pure prismatic sequence, Newton's theory predicts a more complicated pattern, incorporating light far from the caustic and smoothed over the finite angular size of the Sun's disc.

R BISHOP
Seeing the light – the supernumerary rainbow, as seen here above Newton's birthplace just inside the main arc, helped usher in the wave theory of light.

The Descartes–Newton picture explains much, but fails at a fundamental level by ignoring the wave nature of light. Often the interference fringes in rainbows are obscured by decoherence, but they can sometimes be seen as one or more "supernumerary bows" – a "faintly reproving name, one that has persisted long after we know that they are an integral part of the rainbow, not a vexing corruption of it".

"Deeper theories may not give superior explanations."

Lee and Fraser explain in detail how, through Thomas Young's 1803 analysis in terms of interference, "the supernumerary rainbows proved to be the midwife that delivered the wave theory of light to its place of dominance in the nineteenth century". Sadly, the authors do not include the celebrated photograph by Roy Bishop that shows a rainbow over the house in Woolsthorpe where Isaac Newton was born (see above). The magnificent irony is that this rainbow shows a clear supernumerary, magnifying the inadequacy of Newton's theory for all to see, and showing how its ray optics must be replaced by the completely different concepts of wave physics.

Young's interference theory was itself an approximation. The full diffraction theory of light near a rainbow caustic was developed more than 30 years later by George Airy. I would have liked to see more discussion of this, in particular how the "Airy integral" has proved seminal throughout wave physics – for example in nuclear, atomic and molecular rainbows in quantum scattering involving spherically symmetric interaction potential fields of massive particles. The simple Airy theory gives the canonical description of waves very close to a caustic but incorrectly predicts the positions for the supernumeraries far from the rainbow angle; however, in a development not described here, mathematics from the 1950s enables the Airy integral to be adapted to apply everywhere.

Even Airy's theory is not the last word, because it neglects polarization effects, which require a full vector-wave analysis based on Maxwell's electromagnetic equations. This was provided by Gustav Mie's exact calculation in 1908 of the scattering of light by a homogeneous refracting sphere. Extensive developments by Moyses Nussenzveig in the 1960s uncovered the Airy integral concealed in the intricacies of Mie's mathematics.

However, as the authors wisely point out, the deeper theories do not necessarily give superior explanations of the natural rainbow. The reason is that their distinctive predictions (for example of the delicate interferences "marbling" the bow) are obscured in practice by the width of the Sun's disc and by the range of different sizes of raindrops in a shower.

As the book's title implies, the authors' ambitions extend well beyond the physics of rainbows. They are aware of at least 150 myths and contradictory religious invocations in which the rainbow is sometimes benign, sometimes ominous. They catalogue the struggles of artists over the centuries to depict the rainbow realistically, often painting the arch in oblique perspective because it was misconceived as an object rather than an image, and – in what seem simple failures of observation – with the elusive colours of this "chameleon of the air" in reverse order.

Lee and Fraser are masters of prose, and their book is sumptuously produced and abundantly illustrated. They are to be congratulated for producing not only a definitive rainbow scholarship but also a gorgeous work of cultural synthesis.

Michael Berry is Royal Society research professor at the H H Wills Physics Laboratory, Bristol University, UK millennium essay

Making waves in physics

Three wave singularities from the miraculous 1830s.

Michael Berry

Singularities are places where mathematical quantities become infinite, or change abruptly. In waves — of light, for example — there can be singularities in the intensity, in the phase, or in the polarization. This is a modern view, sharply different from the traditional approach where waves were simply the solutions of wave equations, and singularities — if considered at all — were regarded as awkward places where the usual treatments fail. And yet the foundations of wave singularities were laid in three astonishing papers published as early as the 1830s, a decade whose intellectual significance we are only now beginning to appreciate.

Reversing historical order, we start in 1838, with a paper by George Biddell Airy. The immediate stimulus was the theory of the rainbow. Two centuries earlier, Descartes had understood the bright bow as Sun rays directionally focused by raindrops. This geometrical theory gives a good first approximation but fails to account for the delicate supernumerary bows sometimes seen just inside the main arc. In 1801 Thomas Young realized that by regarding light as waves it is possible to understand supernumaries as interference fringes, but could not give a precise mathematical description. Airy's contributions were: to appreciate that the rainbow is a particular example of a caustic, that is, a line where light rays are focused (the bright lines on the bottom of swimming pools are also caustics); to see that caustics are singularities, where ray optics predicts infinite brightness; to realize that wave physics will soften the singularities; and to discover the precise mathematical description of this softening, in the form of his rainbow integral (pictured above).

Airy's paper was doubly influential. First, because refined techniques soon devised by George Gabriel Stokes to study the rainbow integral established divergent infinite series as an important tool in bridging gaps between physical theories (in this case ray and wave optics) and uncovered a mathematical phenomenon whose ramifications are still being explored. Second, because the rainbow integral, describing wave interference decorating the simplest kind of caustic, was found to be the first in a hierarchy of 'diffraction catastrophes'. These more elaborate wave singularities are now classified using catastrophe theory — mathematics whose application greatly advances the physics of caustics.

Next are two papers from 1833 and 1836 by the polymath William Whewell (to whom

Airy's rainbow integral.

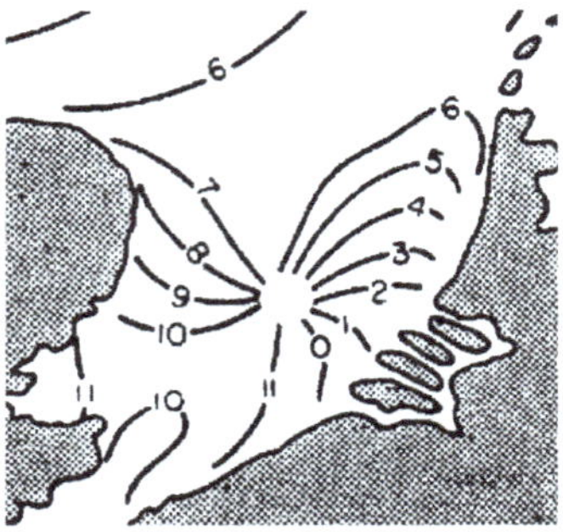

Whewell's amphidromy between England and Holland.

we owe the word 'physicist', with its "four sibilant consonants that fizz like a squib"). He was studying the tides in the oceans, and seeking to "connect the actual tides of all the different parts of the world — and to account for their varieties and seeming anomalies". From Young he learned to concentrate on the cotidal lines connecting places where the tide is high at a given time. Cotidal lines are wavefronts of the tide, regarded as a wave of twelve-hour period; they are contours of the phase of this wave. Whewell appreciated that a map of cotidal lines would render intelligible the pattern of tides around the coasts, and coordinated hundreds of new observations, in an early international scientific collaboration.

Extrapolating away from the coasts into the 'German Ocean' he reached the extraordinary conclusion that there must be "rotatory systems of tide-waves [where] the cotidal lines ... revolve around [a point] where there is no tide, for it is clear that at a point where all the cotidal lines meet, it is high-water equally at all times". One of Whewell's 'amphidromies' can be seen above. What Whewell discovered were the phase singularities of the tide waves.

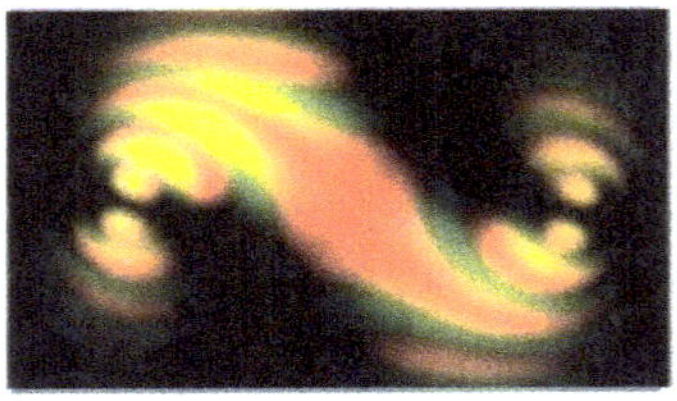

Hamilton's diabolical points (bullseyes) in several square centimetres of overhead-projector transparency foil viewed obliquely through crossed polarizers; in each bullseye, the interference rings are contours of difference of wave speeds, centred on an optic axis, and the black stripes reflect geometric phases.

Phase singularities are now recognized as important features of all waves; in three dimensions they are lines rather than points. At a phase singularity, the wave intensity is zero — in contrast to caustics where the intensity is (geometrically) infinite. In acoustics, the singularities are threads of silence; in light, they are optical vortices; in superfluids, quantized vortices; and in superconductors, quantized lines of magnetic flux.

In addition to intensity and phase, waves of light exhibit polarization: they are vector waves. Polarization has its singularities too, discovered by our third author, William Rowan Hamilton, in 1832, as an unexpected consequence of Augustin Fresnel's theory of the optics of crystals. In a general anisotropic material, two waves propagate in each direction, with different speeds and polarizations. There are, however, two singular directions, optic axes, where the speeds are the same. As functions of direction, the two speeds can be represented by surfaces forming a diabolo (double cone) at each optic axis. Hamilton deduced that at such 'diabolical points' the wave direction corresponds to a cone of rays. This unprecedented 'conical refraction' was soon observed, confirming that light is a transverse wave. Diabolical points can easily be seen directly (pictured above).

Hamilton's diabolical point was the first physical example of degeneracy between eigenvalues of a real symmetric matrix. Its descendants thrive in many areas of science today, for example, as conical intersections between energy levels in quantum chemistry, as Bloch wave degeneracies in the quantum Hall effect, and as the simplest geometric phases. ■

Michael Berry is in the Department of Physics, University of Bristol, Bristol BS8 1TL, UK.

Reprinted from Nature, Volume 403, (2000).

Eur. J. Phys. **20** (1999) 1–14. Printed in the UK PII: S0143-0807(99)95676-3

Black plastic sandwiches demonstrating biaxial optical anisotropy

Michael Berry†, Rajendra Bhandari‡ and Susanne Klein†

† H H Wills Physics Laboratory, Univeristy of Bristol, Tyndall Avenue, Bristol BS8 1TL, UK
‡ Raman Research Institute, Bangalore 560 080, India

Received 7 July 1998

Abstract. Transparent overhead-projector foil is an anisotropic material with three different principal refractive indices. Its properties can be demonstrated very simply by sandwiching the foil between crossed polarizers and looking through it at any diffusely lit surface (e.g., the sky). Coloured interference fringes are seen, organized by a pattern of rings centred on two 'bullseyes' in the directions of the two optic axes. The fringes are difference contours of the two refractive indices corresponding to propagation in each direction, and the bullseyes are degeneracies where the refractive-index surfaces intersect conically. Each bullseye is crossed by a black 'fermion brush' reflecting the sign change (geometric phase) of each polarization in a circuit of the optic axis. Simple observations lead to the determination of the three refractive indices, up to an ordering ambiguity.

1. Introduction

Hamilton's discovery of conical refraction in 1830 (Born and Wolf 1959) was a milestone in the development of the classical picture of light as a transverse wave. The phenomenon was an unexpected consequence of Fresnel's theoretical analysis of the propagation of polarized light in transparent anisotropic materials (in those days, the only such materials were crystals). According to Fresnel's theory, in any given direction two waves can propagate through the material without change. The waves have different refractive indices and orthogonal linear polarizations (we will not consider chirality here). However, there are four special directions, specified by two optic axes (along which light can propagate either way), where the two refractive indices are the same and the polarizations arbitrary, so that the material behaves as if it is isotropic. Such general materials are called biaxial; in the special case where there is one symmetry direction, the two axes coincide, and the material is uniaxial. If the refractive indices are represented as two sheets in a polar plot (the wave surface), the optic axes are singular directions where the sheets are connected. Locally the connection is like that of a double cone (a diabolo). Hamilton pointed out that this singularity gives rise to several physical phenomena, which were soon observed and studied in detail, and later incorporated into the electromagnetic description of light.

Although well understood for more than a century (Pockels 1906, Hartshorne and Stuart 1960, Bloss 1961), the phenomena are not widely known to physicists, and are regarded as obscure. This is a pity, because the connections of the cones are singularities of polarization

0143-0807/99/010001+14$19.50

optics, and so are, in a sense, at the heart of the subject. One reason for the lack of emphasis on the conical effects might be that they are rarely seen; this in turn could be because their observation was thought to require plates of biaxial crystal. One of our purposes here is to describe an extremely simple demonstration (section 2), based on an observation made by one of us (RB), by means of which some of the physics of optic axes and the associated cones can be demonstrated and explored easily.

Notwithstanding the venerable physics, there are several reasons why such a demonstration is timely, and the explanation of these is our other purpose. First, with light regarded as a stream of photons, polarization phenomena provide fine illustrations of the fundamental ideas of quantum state preparation, orthogonality, measurement, completeness and evolution; even at the classical level, it is helpful to employ a notation (section 4) reminiscent of quantum mechanics. Second, conical intersections are now being understood as organizing a wide range of quantum phenomena in solid-state physics (Simon 1983) and chemistry (Herzberg and Longuet-Higgins 1963, Teller 1937, Mead and Truhlar 1979), and it is good to recall the optical context in which they first appeared. Third, the demonstration gives immediate reality to some mathematical phenomena associated with matrices depending on parameters: degeneracies of eigenvalues, singularities in the pattern of eigenfunctions, and eigenfunction anholonomy (geometric phases). In our explanations, we will emphasize these phenomena, and give only the outline of the standard optical theory (Born and Wolf 1959, Landau *et al* 1984) of biaxial materials.

2. Experiment

The apparatus is a 'sandwich' (figure 1). Its 'bread' is two squares of polaroid sheet with their transmission directions perpendicular. Its 'filling' is a square of overhead-projector plastic transparency foil. For individual viewing, the squares can be 3 cm $\times$ 3 cm. For lecture demonstrations, the squares should be about 30 cm$\times$30 cm, and the whole sandwich kept flat and rigid by a 'wrapping' of clear acrylic or glass panes.

Without the filling, the sandwich is simply a pair of crossed polaroids, and transmits no light. But the inserted plastic foil is optically anisotropic—and biaxial because it has been stretched two ways during manufacture (Keller 1998). Therefore the filled sandwich is not black. The experiment consists in viewing it by transmitted diffuse light over a range of directions. An individual viewer can do this by holding the sandwich close to one eye and looking through it at a light-coloured wall, or the sky (figure 2(*a*)).

Brilliantly coloured patterns are seen. They are called conoscopic figures (Hartshorne and Stuart 1960). Looking off to the side, it becomes clear that the colours are interference

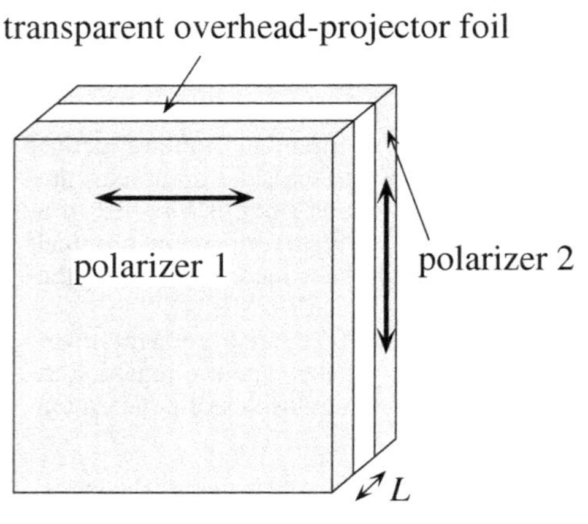

Figure 1. Structure of the black plastic sandwich.

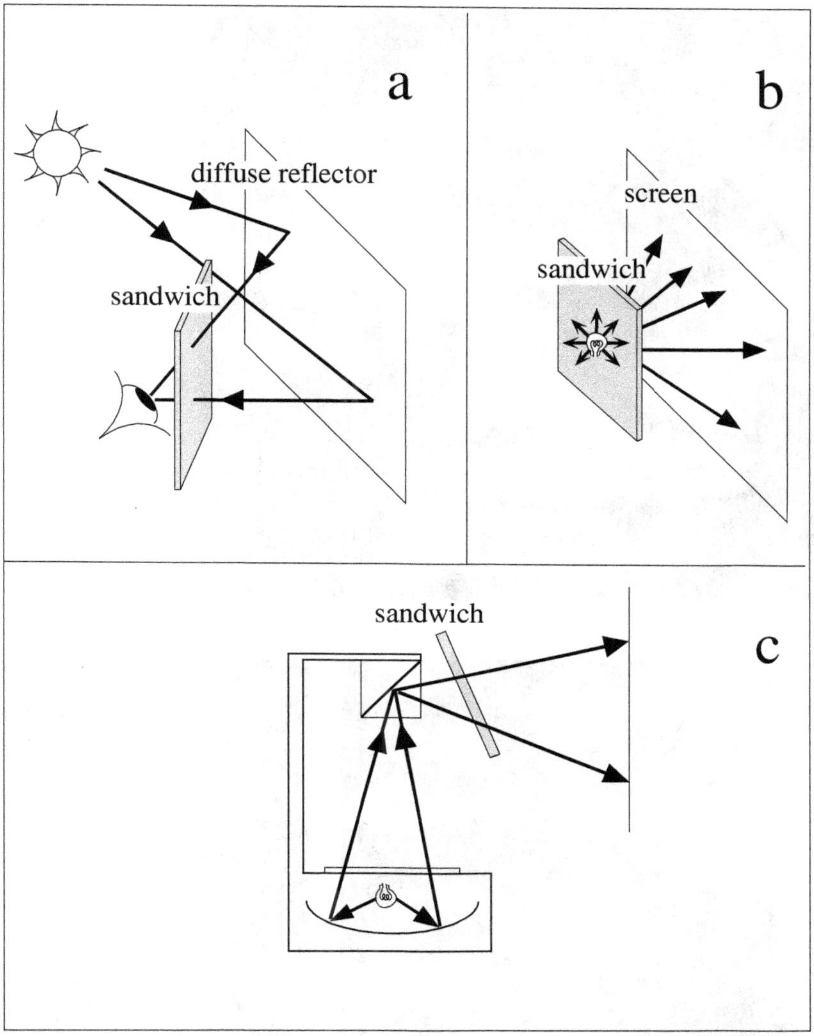

Figure 2. Observing the interference fringes. (*a*) Looking directly through the sandwich (the diffuse reflector can be the sky); (*b*) by projection from a small bright source; (*c*) to an audience, using an overhead projector.

fringes, organized about two sets of circles ('bullseyes') symmetrically disposed about the normal to the foil, separated by several tens of degrees (the precise angle depends on the type of foil). The bullseyes can be moved to the forward direction, where they can be seen more easily, by tilting the sandwich so that its normal points away from the forward direction (figure 2(*a*)). The complete pattern can be viewed by holding a small bright source of light close to the sandwich, and projecting the image onto white paper on the other side (figure 2(*b*)); the bulb from a Mini Maglite™ torch, with its lens removed, is an ideal source. (Alternatively, the complete pattern can be seen in the traditional way, with a polarizing microscope.) The individual bullseyes can be demonstrated to a large audience using the (slightly) divergent light of an overhead projector, by placing the (large) sandwich obliquely in the beam (figure 2(*c*)).

The patterns can be recorded in several ways. For their global structure, it is convenient

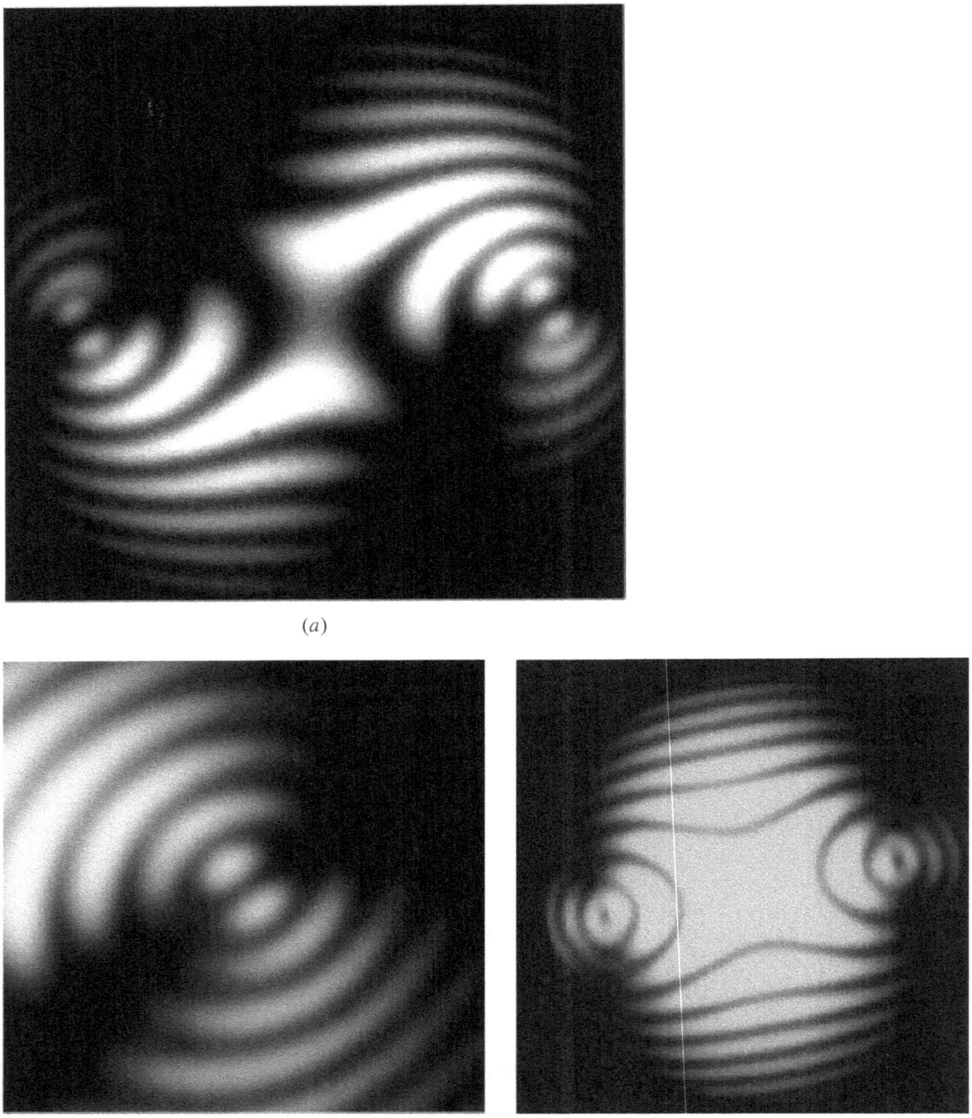

(*a*) (*b*) (*c*)

Figure 3. Conoscopic interference figures in monochromatic light. (*a*) Global structure, with the fringes centred on two 'bullseyes' crossed by black 'fermion brushes'; (*b*) a single bullseye; (*c*) disconnection of brushes near bullseyes for polarizer orientation near 45°.

(A colour image of a single bullseye is included in the electronic version of this article as an additional figure 9 after the list of references on page 14; see `http://www.iop.org`)

to use an arrangement like that of figure 2(*b*), with the screen replaced by a camera attached to a microscope with a wide-aperture objective lens; the details of the pattern—for example, the individual bullseyes—can be seen by magnification. Alternatively, the individual bullseyes can be photographed simply by replacing the eye in figure 2(*a*) by a camera aimed at the sky. Figure 3(*a*) shows the pattern produced by monochromatic light, and figure 3(*b*) shows a magnification of one of the bullseyes. The main features, in addition to the interference fringes, are broad black stripes passing through the bullseyes. For a reason to be explained later, we call these the *fermion brushes*. The inclination of the brushes, relative to the line joining the bullseyes, can be altered by rotating the foil relative to the crossed polaroids.

For some foils, and certain ranges of orientation of the polarizers, the fermion brushes close to the bullseyes were disconnected (figure 3(*c*)). Superficially this resembles the effects of optical activity (Gibbs 1882, Goldhammer 1892, Pocklington 1901), in addition to the birefringence, which would break the degeneracy of the refractive indices at the optic axes. However, the details seem incompatible with this hypothesis, both in the sensitivity to polarizer orientation and the fact that the centre of the bullseye is always dark. J F Nye has suggested that the disconnection might be caused by variation of the orientation of the optic axes along the propagation path through the foil. This puzzling phenomenon (which we have not seen described in the literature) would repay investigation; we do not consider it further here.

With plastic foils we did not succeed in seeing the original phenomenon of conical refraction, namely the transformation of a narrow initial beam into a hollow cone (Born and Wolf 1959). This is because the foil is too thin: a simple calculation shows that the broadening is much less than the width of the laser beam we used, and attempts to use a stack of many foils failed because of the gaps between them and the difficulty of aligning the optic axes of successive foils.

3. Geometry and notation

Let the normal to the sandwich-filling foil define the z direction, and consider light travelling within the foil in a direction specified by the unit vector $\boldsymbol{s} = \{s_x, s_y, s_z\}$ (parallel to the wavevector), with polar angles θ, ϕ. The direction outside is obtained from $\boldsymbol{s}$ by a simple refraction correction, to be given later. It will be convenient also to denote directions by points on the plane $\boldsymbol{R} = \{X, Y\}$, with polar coordinates R, ϕ, obtained from $\boldsymbol{s}$ by stereographic projection from the south pole of the $\boldsymbol{s}$ sphere (figure 4). The equations are

$$\boldsymbol{s} = \frac{1}{1+R^2}\{2X, 2Y, 1-R^2\} \qquad R = \tan\tfrac{1}{2}\theta. \tag{1}$$

The two waves, + and −, with frequency ω and free-space wavenumber $k = \omega/c = 2\pi/\lambda$, which travel in the $\boldsymbol{s}$ direction, will be written as functions of position $\boldsymbol{r}$ and time t in terms of their electric vectors $\boldsymbol{D}$ (transverse to $\boldsymbol{s}$, unlike the electric field vector $\boldsymbol{E}$) as follows:

$$\boldsymbol{D}_\pm(\boldsymbol{s}, \boldsymbol{r}, t) = \boldsymbol{d}_\pm(\boldsymbol{s}) \exp\{ikn_\pm(\boldsymbol{s})\boldsymbol{s}\cdot\boldsymbol{r} - \omega t\} \tag{2}$$

where $n_\pm(\boldsymbol{s})$ and $\boldsymbol{d}_\pm(\boldsymbol{s})$ are, respectively, the refractive indices and (orthogonal) polarizations of the waves. When we use the stereographic representation of directions, the same symbols

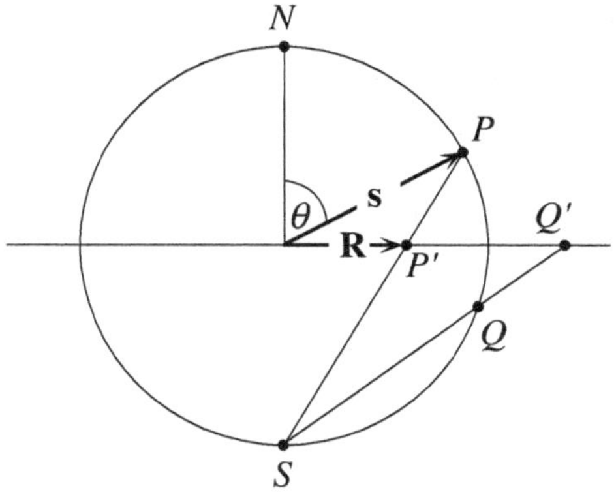

Figure 4. Stereographic projection of the wave direction from the sphere of unit vectors $\boldsymbol{s}$ to the plane $\boldsymbol{R}$, illustrated for two points P and Q (images P' and Q').

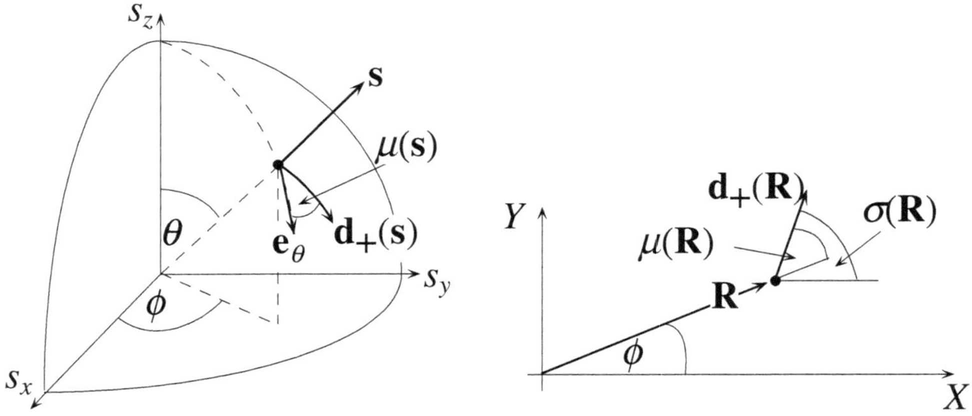

Figure 5. Directions of eigenpolarization d_+. (a) On the s sphere, making an angle $\mu(s)$ with the unit vector e_θ. (b) In the R plane, making an angle $\sigma(R)$ with the X axis.

will be used, that is $n_\pm(R)$ and $d_\pm(R)$, with the understanding that $d_\pm(R)$ means the stereographically projected (and still orthogonal) polarizations. Where no confusion arises, we will drop the labels + and −. The polarization vector $d_+(s)$ makes an angle $\mu(s)$ with the unit vector e_θ on the s sphere (figure 5(a)), and its projection onto R makes an angle $\sigma(R) = \mu(R) + \phi$ with the X axis (figure 5(b)), where $\mu(R) = \mu(s)$ (because the projection preserves angles) (for d_-, $\pi/2$ must be added to these angles). Thus

$$d_+(s) = e_\theta \cos\mu + e_\phi \sin\mu \qquad d_+(R) = e_X \cos\sigma + e_Y \sin\sigma. \tag{3}$$

Since d and $-d$ represent the same polarization, all physical quantities derived from d must be invariant under changes of μ and ϕ by π.

The foil is an anisotropic transparent dielectric of thickness L, specified by its constitutive relation, which we write in the form

$$D = \varepsilon_0 \mathsf{n}^2 E \tag{4}$$

where n^2 is the dimensionless dielectric tensor, written in terms of the three principal refractive indices as

$$\mathsf{n}^2 = \begin{pmatrix} n_x^2 & 0 & 0 \\ 0 & n_y^2 & 0 \\ 0 & 0 & n_z^2 \end{pmatrix}. \tag{5}$$

We choose the y axis such that n_y is the intermediate-valued principal index; this will make the bullseyes lie on the X axis. Moreover, we make the asumption $n_x < n_z$, and will discuss this at the end of section 6. Thus

$$n_x < n_y < n_z. \tag{6}$$

It will be convenient to define

$$\alpha \equiv \frac{1}{n_x^2} - \frac{1}{n_y^2} \qquad \beta \equiv \frac{1}{n_y^2} - \frac{1}{n_z^2}. \tag{7}$$

Although we shall work with exact formulae, we note that in the foil $\alpha \ll 1$ and $\beta \ll 1$.

Finally, the polarizers will be specified by their orientations γ and $\gamma + \pi/2$ with respect to the X axis.

4. Theory: intensity variation

The observed pattern is the intensity $I(\boldsymbol{R},\gamma)$ of the light when it emerges from the sandwich after travelling in the direction R, with the orientation of the polarizers specified by γ. When calculating $I(\boldsymbol{R},\gamma)$, it is convenient to use vectors $\boldsymbol{V}$ to represent the $\boldsymbol{D}$ vector of the light as it traverses the foil; we think of $\boldsymbol{V}$ as the column vector representing $\boldsymbol{D}$, and denote by $\boldsymbol{V}^{\mathrm{T}}$ the corresponding row vector (transpose). $\boldsymbol{V}$ and $\boldsymbol{V}^{\mathrm{T}}$ are analogous to the ket and bra in quantum mechanics; thus the dot product $\boldsymbol{U}\cdot\boldsymbol{V}$ can alternatively be written $\boldsymbol{U}^{\mathrm{T}}\boldsymbol{V}$.

The first polarizer projects $\boldsymbol{V}$ onto a state of linear polarization, represented by a vector $\boldsymbol{P}$. In quantum mechanics, this step is the preparation of the state. As the light enters the anisotropic foil, we resolve its state $\boldsymbol{P}$ into components along the two eigenpolarizations $\boldsymbol{d}_{\pm}(\boldsymbol{R})$. These are complete and orthonormal, i.e.

$$\begin{aligned}&\boldsymbol{d}_{+}(\boldsymbol{R})\boldsymbol{d}_{+}^{\mathrm{T}}(\boldsymbol{R})+\boldsymbol{d}_{-}(\boldsymbol{R})\boldsymbol{d}_{-}^{\mathrm{T}}(\boldsymbol{R})=\mathsf{I}\\&\boldsymbol{d}_{\pm}(\boldsymbol{R})\cdot\boldsymbol{d}_{\pm}(\boldsymbol{R})=1\qquad \boldsymbol{d}_{\pm}(\boldsymbol{R})\cdot\boldsymbol{d}_{\mp}(\boldsymbol{R})=0.\end{aligned}\tag{8}$$

The resolution is

$$\boldsymbol{P}=\left(\boldsymbol{d}_{+}(\boldsymbol{R})\boldsymbol{d}_{+}^{\mathrm{T}}(\boldsymbol{R})+\boldsymbol{d}_{-}(\boldsymbol{R})\boldsymbol{d}_{-}^{\mathrm{T}}(\boldsymbol{R})\right)\boldsymbol{P}=\boldsymbol{d}_{+}(\boldsymbol{R})\cdot\boldsymbol{P}\,\boldsymbol{d}_{+}(\boldsymbol{R})+\boldsymbol{d}_{-}(\boldsymbol{R})\cdot\boldsymbol{P}\,\boldsymbol{d}_{-}(\boldsymbol{R}).\tag{9}$$

The polarizations + and − propagate independently through the foil, and acquire phases determined by their refractive indices $n_{\pm}(\boldsymbol{R})$. The graphs of these two functions, either as polar plots (that is, as radial distances for direction $\boldsymbol{s}$), or perpendicular to the $\boldsymbol{R}$ plane, are the sheets of the refractive-index (wave) surface. Since the distance travelled in the foil is

$$L(\boldsymbol{s})=\frac{L}{\cos\theta}\equiv L(\boldsymbol{R})=L\left(\frac{1+R^2}{1-R^2}\right)\tag{10}$$

the polarizations become

$$\boldsymbol{d}_{\pm}(\boldsymbol{R})\to\boldsymbol{d}_{\pm}(\boldsymbol{R})\exp\{\mathrm{i}kn_{\pm}(\boldsymbol{R})L(\boldsymbol{R})\}\equiv\boldsymbol{d}_{\pm}(\boldsymbol{R})\exp\{\mathrm{i}\chi_{\pm}(\boldsymbol{R})\}.\tag{11}$$

In quantum mechanics, this step is the evolution of the state, governed by the Schrödinger equations (here Maxwell's equations). Thus the state after propagation through the foil, and before entering the second polarizer, is

$$\boldsymbol{V}(\boldsymbol{R})=\boldsymbol{d}_{+}(\boldsymbol{R})\cdot\boldsymbol{P}\,\boldsymbol{d}_{+}(\boldsymbol{R})\exp\{\mathrm{i}\chi_{+}(\boldsymbol{R})\}+\boldsymbol{d}_{-}(\boldsymbol{R})\cdot\boldsymbol{P}\,\boldsymbol{d}_{-}(\boldsymbol{R})\exp\{\mathrm{i}\chi_{-}(\boldsymbol{R})\}.\tag{12}$$

The second polarizer projects $\boldsymbol{V}$ onto a state represented by a vector $\overline{\boldsymbol{P}}$, perpendicular to the first polarizer $\boldsymbol{P}$, i.e.

$$\overline{\boldsymbol{P}}\cdot\boldsymbol{P}=0.\tag{13}$$

In quantum mechanics, this step is the measurement of the state. The desired intensity is

$$I(\boldsymbol{R},\gamma)=\left|\overline{\boldsymbol{P}}\cdot\boldsymbol{V}(\boldsymbol{R})\right|^2\tag{14}$$

and a short calculation using (8) and (13) leads to

$$I(\boldsymbol{R},\gamma)=4\left[\boldsymbol{d}_{+}(\boldsymbol{R})\cdot\overline{\boldsymbol{P}}\right]^2\left[\boldsymbol{d}_{+}(\boldsymbol{R})\cdot\boldsymbol{P}\right]^2\sin^2\left\{\tfrac{1}{2}\Delta\chi(\boldsymbol{R})\right\}\tag{15}$$

where

$$\Delta\chi(\boldsymbol{R})\equiv\chi_{+}(\boldsymbol{R})-\chi_{-}(\boldsymbol{R})=kL(\boldsymbol{R})\left[n_{+}(\boldsymbol{R})-n_{-}(\boldsymbol{R})\right].\tag{16}$$

To calculate this intensity explicitly, it is convenient to imagine the foil as fixed, and the orientation of the polarizers as variable. Let the first polarizer have orientation γ: this means that it transmits light whose electric $\boldsymbol{D}$ vector is oriented at an angle γ to the X axis in the $\boldsymbol{R}$ plane. The second polarizer transmits light in the perpendicular direction, i.e. $\gamma+\pi/2$. Thus in equation (15)

$$P=(\cos\gamma,\ \sin\gamma)\qquad \overline{\boldsymbol{P}}=(\sin\gamma,\ -\cos\gamma).\tag{17}$$

With the polarization $\boldsymbol{d}_+(\boldsymbol{R})$ as defined in (3), equation (15) gives

$$I(\boldsymbol{R},\gamma)=\sin^2\left\{2\left(\gamma-\sigma(\boldsymbol{R})\right)\right\}\sin^2\left\{\tfrac{1}{2}\Delta\chi(\boldsymbol{R})\right\}. \tag{18}$$

In equation (18), the second factor describes the fringes resulting from the interference of the two polarizations that emerge from the foil, when projected onto the common state $\overline{\boldsymbol{P}}$. From equation (16), these fringes are approximately contours of difference of the two refractive indices (the $\boldsymbol{R}$ dependence of L gives only a small correction). The thicker the foil, or the shorter the wavelength λ, the closer the fringes. Because of the λ dependence, the fringes are coloured (the λ dependence of the refractive indices, i.e. dispersion of the optic axes, is a much smaller effect). Close to the conical singularities of the two refractive-index sheets where $n_+=n_-$, the contours are closed loops (actually circles); these are the bullseyes that reveal the singularities and dominate the images.

The first factor in (18) vanishes along the lines where $\sigma(\boldsymbol{R})=\gamma\ (\mathrm{mod}\ \pi/2)$. These are the black brushes, called isogyres, which therefore reveal the polarization directions at $\boldsymbol{R}$: as the crossed polarizers are rotated relative to the foil, the brushes sweep through all these directions. We shall see that in a circuit of the singularity (in $\boldsymbol{R}$ space) σ changes by π; this implies that for each γ the brush crosses the bullseye in a single smooth line. Note that equation (18) is unaltered if γ or σ is changed by $\pi/2$, reflecting invariance of the physics under exchange of the two polarizers $\boldsymbol{P}$ and $\overline{\boldsymbol{P}}$ or the two polarizations $\boldsymbol{d}_+$ and $\boldsymbol{d}_-$.

5. Theory: fringes, isogyres and the fermion brush

With the electric vector $\boldsymbol{D}$ (2), and the constitutive equation (4), Maxwell's equations reduce to

$$\boldsymbol{d}(\boldsymbol{s})+n^2(\boldsymbol{s})\boldsymbol{s}\times\boldsymbol{s}\times\mathbf{n}^{-2}\boldsymbol{d}(\boldsymbol{s})=0 \tag{19}$$

where $\boldsymbol{d}$ denotes $\boldsymbol{d}_+$ or $\boldsymbol{d}_-$. Although this represents three linear equations, transversality implies that the consistency condition (vanishing determinant of the operator in (19)) leads to two, not three refractive indices $n(\boldsymbol{s})$ for each $\boldsymbol{s}$ (the third would be infinite, corresponding to electrostatic fields rather than waves). These two indices $n_\pm(\boldsymbol{s})$ are given by the eigenvalues of the part of the matrix $\mathbf{n}^{-2}$ transverse to $\boldsymbol{s}$; the corresponding eigenvectors are the polarizations $\boldsymbol{d}_\pm(\boldsymbol{s})$.

A long calculation, using the first equation in (3) to represent $\boldsymbol{d}$, and the definitions (5) and (7), leads to a formula in which the refractive index surfaces are conveniently expressed in terms of the positions of the four optic axes. In the $\boldsymbol{R}$ plane, these are at

$$Y=0\qquad X=\pm X_{\mathrm{c}},\ \pm\frac{1}{X_{\mathrm{c}}}\qquad X_{\mathrm{c}}\equiv\sqrt{1+\frac{\beta}{\alpha}}-\sqrt{\frac{\beta}{\alpha}}. \tag{20}$$

On the $\boldsymbol{s}$ sphere, the axes are

$$\phi=0,\ \pi\qquad \theta=\theta_{\mathrm{c}},\ \pi-\theta_{\mathrm{c}}\qquad \tan\theta_{\mathrm{c}}=\sqrt{\frac{\alpha}{\beta}}. \tag{21}$$

The indices are

$$\begin{aligned}\frac{1}{n_\pm^2(\boldsymbol{s})}&=\frac{1}{n_y^2}-\frac{\alpha}{2(1+R^2)^2}\left\{H\pm\sqrt{G}\right\}\\ H&\equiv(X^2-X_{\mathrm{c}}^2)\left(X^2-\frac{1}{X_{\mathrm{c}}^2}\right)-4Y^2\frac{\beta}{a}\\ G&\equiv64X^2Y^2\frac{\beta}{a}\left(1+\frac{\beta}{a}\right)+\left[H+Y^2\left(2(1+X^2)+8\frac{\beta}{\alpha}+Y^2\right)\right]^2.\end{aligned} \tag{22}$$

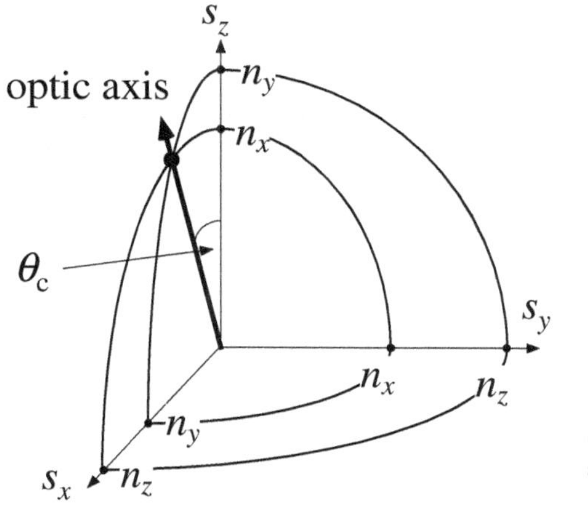

Figure 6. Refractive index surfaces in direction space, showing the optic axis (degeneracy) where the surfaces intersect conically.

This formula incorporates the following special cases, illustrated in figure 6 for an octant of the $\boldsymbol{s}$ sphere:

$$\begin{aligned}
&X = X_c,\ Y = 0 \ \text{(waves along optic axes)}: && n_+ = n_- = n_y \\
&X = Y = 0 \ \text{(waves along } \boldsymbol{e}_z\text{)}: && n_+ = n_y,\ n_- = n_x \\
&X = 1,\ Y = 0 \ \text{(waves along } \boldsymbol{e}_x\text{)}: && n_+ = n_z,\ n_- = n_y \\
&X = 0,\ Y = 1 \ \text{(waves along } \boldsymbol{e}_y\text{)}: && n_+ = n_z,\ n_- = n_x.
\end{aligned} \tag{23}$$

Close to the optic axes, equation (22) can be simplified to display the form of the conical intersections explicitly. We find

$$\begin{aligned}
n_\pm(\boldsymbol{R}) &\approx n_y\left[1 + C\left\{\pm\sqrt{(X - X_c)^2 + Y^2} - (X - X_c)\right\}\right] \\
C &\equiv \frac{\sqrt{(n_y^2 - n_x^2)(n_z^2 - n_y^2)}}{\sqrt{(n_z^2 - n_x^2)}\left[\sqrt{(n_z^2 - n_x^2)} - \sqrt{(n_z^2 - n_y^2)}\right]}.
\end{aligned} \tag{24}$$

The fact that the coefficients of $(X - X_c)^2$ and Y^2 in the square root are the same shows that the cones are circular near their intersection (in both the $\boldsymbol{R}$ and $\boldsymbol{s}$ spaces).

The polarizations (isogyres) $\boldsymbol{d}$ are the eigenvectors defined by (19). Geometrically, these are the principal axes of the ellipse defined by the intersection of the plane normal to s with the indicatrix (Nye 1995). On the $\boldsymbol{s}$ sphere (figure 5(a)) these are specified by the angle $\mu(\boldsymbol{s})$. A calculation from (19) gives

$$\tan 2\mu(\boldsymbol{s}) = \frac{\sin 2\phi \cos\theta}{(\beta/\alpha)\sin^2\theta - \cos 2\phi \cos^2\theta}. \tag{25}$$

This gives both polarizations (with μ values differing by 2π). At the optic axes (21), the numerator and denominator vanish, so that the pattern of isogyres is singular.

In the projective plane, where $\boldsymbol{d}$ is specified by the angle $\sigma(\boldsymbol{R})$, equation (25) can be cast in an interesting form by introducing the complex direction variable

$$\zeta \equiv X + \mathrm{i}Y \tag{26}$$

and the following complex function with zeros at each of the optic axes:

$$A(\zeta) \equiv (1 - \zeta^2)^2 - 4\zeta^2\frac{\beta}{\alpha} = (\zeta^2 - X_c^2)\left(\zeta^2 - \frac{1}{X_c^2}\right). \tag{27}$$

Then the isogyres (now denoted $\sigma(\zeta)$) satisfy

$$\mathrm{Im}\, A(\zeta) \exp\{-2\mathrm{i}\sigma(\zeta)\} = 0. \tag{28}$$

From this we can find a function $B(\zeta)$ whose real and imaginary parts have the isogyres as their contours. For if we choose $\exp -\mathrm{i}\sigma = B'(\zeta) = 1/\sqrt{A}$, then $\nabla \mathrm{Re}\, B = \{\cos\sigma, \sin\sigma\}$ and $\nabla \mathrm{Im}\, B = \{-\sin\sigma, \cos\sigma\}$. The function is

$$B(\zeta) = \int_0^\zeta \frac{\mathrm{d}\zeta}{\sqrt{A(\zeta)}} = X_\mathrm{c} F\left(\arcsin \frac{\zeta}{X_\mathrm{c}},\ X_\mathrm{c}^4\right) \tag{29}$$

where F denotes the elliptic integral (we use the convention in *Mathematica*™ (Wolfram 1991)).

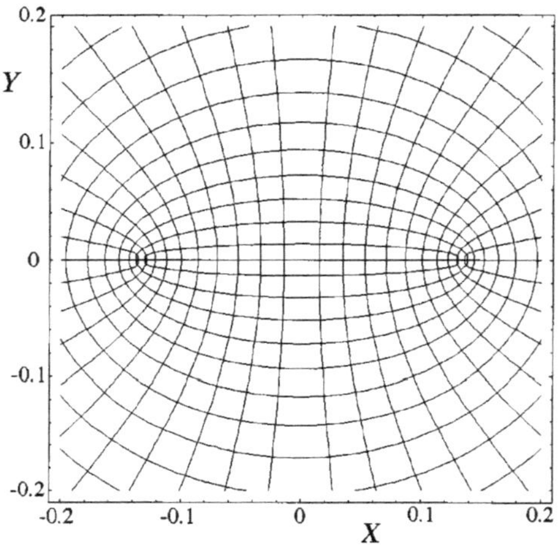

Figure 7. Isogyres in the projective plane, calculated for $n_x = 1.563$, $n_y = 1.58$, $n_z = 1.883$; the optic axes are at $X_\mathrm{c} = 0.133$, corresponding to an angle $\theta_\mathrm{c} = 13.4°$.

Figure 7 shows the orthogonal net of polarization directions, computed from (29), for a range that includes the two optic axes in the northern hemisphere of the s plane. The pattern is dominated by the singularities at $\pm X_\mathrm{c}$. Around a circuit of each singularity, each of the two polarizations rotates by π in the same sense as the circuit. This means that each singularity has index $+1/2$; it cannot be otherwise, because the total index of singularities of a line field on a sphere must, by the 'hairy sphere' theorem (Spivak 1975, Iyanaga and Kawada 1968), be $+2$ (the Euler–Poincaré characteristic of a sphere), and so each of the four identical optic-axis singularities must have index $+1/2$. In one classification of the singularities of line fields (Berry and Hannay 1977, Berry and Upstill 1980), these are 'lemons'; the terminology arose in connection with umbilic points on surfaces (Darboux 1896, see also the historical remarks in Berry 1989).

In the image intensity (18), the first factor can be written explicitly by means of (28) as

$$\sin^2\{2(\gamma - \sigma(\boldsymbol{R}))\} = \frac{1}{2}\left\{1 + \frac{\left[\mathrm{Im}^2 A(\zeta) - \mathrm{Re}^2 A(\zeta)\right]\cos 4\gamma - 2\mathrm{Re}\, A(\zeta)\mathrm{Im}\, A(\zeta)\sin 4\gamma}{|A(\zeta)|^2}\right\}. \tag{30}$$

Now the full intensity can be calculated from (18), (16), (10) and (22), and rendered as density plots to simulate the observations (e.g., those in figure 3). Three such simulations, for different values of the polarizer orientation γ, are shown in figure 8(a–c); they should be compared with the experimental figure 3(a), whose conditions they have been chosen to match, as will be explained in section 6.

The most striking features are the two bullseyes, centred on the optic axes. Figure 8(d) shows a magnification of one of these. We have already mentioned the broad black brush

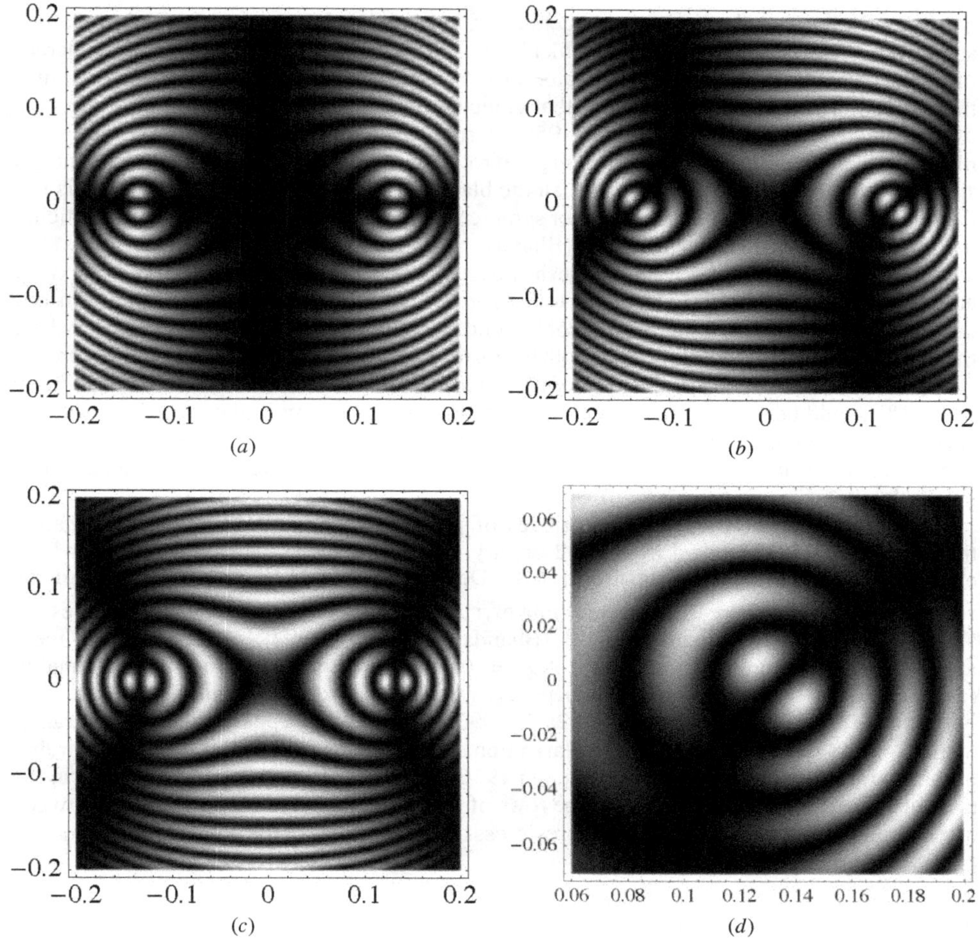

Figure 8. Simulations of the conoscopic figures, calculated from equation (18) as density plots of the intensity $I(\boldsymbol{R}, \gamma)$ for the same conditions as in figure 7, with $\lambda = 0.546\ \mu$m, $L = 100\ \mu$m. (*a*) $\gamma = 0$; (*b*) $\gamma = \pi/8$; (*c*) $\gamma = \pi/4$. A magnification of one of the bullseyes in (*b*) is shown in (*d*).

through the centre. This has its origin in the first factor in (18). During a circuit of the bullseye, the black brush will be encountered whenever $\sigma(R) = \gamma$ (modulo $\pi/2$). Because the singularity of the pattern of polarizations has index 1/2, $\sigma(\boldsymbol{R})$ changes by π around such a circuit, so the brush condition will be satisfied twice. Now, close to the bullseye the change in $\sigma(\boldsymbol{R})$ is uniform for a circular circuit; this follows from the square root singularity in (29) with (27). Therefore these two encounters will be on opposite sides of the bullseye, resulting in in a single black line passing through it smoothly. In local polar coordinates R_{local}, ϕ_{local}, the intensity of light emerging near the bullseye is

$$I(\boldsymbol{R}, \gamma) \approx \sin^2(2\gamma - \phi_{\text{local}}) \sin^2(D R_{\text{local}}) \qquad (\boldsymbol{R} \approx \{X_c, 0\}) \tag{31}$$

where in the second factor D is a constant and the circularity of the cones (equation (24)) has been invoked.

So, the black brushes are consequences of the fact that the eigenvectors $\boldsymbol{d}$ of Maxwell's equation (19) are not single-valued, but reverse round a circuit of each bullseye. The reversal corresponds to a sign change in each eigenvector, i.e. a phase change of π. Such sign changes are universally present (Arnold 1978, Uhlenbeck 1976) during a circuit of a simple degeneracy of a real symmetric matrix (the operator in (19) can be expressed this way by a simple transformation), with two eigenvalues that coincide linearly. They are the simplest example of a geometric phase (Pancharatnam 1956, Berry 1984, Shapere and Wilczek 1989), and mathematically identical to the sign change on rotation of quantum particles with half-integer spin (Silverman 1980); that is why we call the black lines fermion brushes. In optics this fact that the polarization states have 4π spinor symmetry, even though photons are spin-1 particles, has been demonstrated experimentally (Bhandari 1993a, 1997).

We emphasize that the fermion brushes occur here because the foil we use is biaxially anisotropic. If the material were uniaxial, as in plastic threads made by stretching in one direction, rather than two, this would correspond to confluence of the two optic axes, where the singularities would each have strength +1 instead of +1/2 (the total index on the s sphere would still be +2). Then the eigenvectors s would be single-valued, and the brush condition from (18) would be satisfied four times, rather than twice, in a circuit of each axis. Therefore each bullseye would be traversed by a black cross rather than a single black line, as is well known for uniaxial materials (Born and Wolf 1959) (in (31) this situation can be described locally by replacing $\phi_{\rm local}$ by $2\phi_{\rm local}$).

Bullseyes are not the only singularities of polarization optics. Consider propagation through the foil as described by a 2×2 unitary (Jones) matrix $\mathbf{U}$, depending on $\boldsymbol{R}$, whose eigenangles are $\pm\frac{1}{2}\Delta\chi(\boldsymbol{R})$ (equation (16)). Degeneracies of $\mathbf{U}$ are the dark rings of the conoscopic figures, where $\frac{1}{2}\Delta\chi$ is a multiple of π; if the multiple is even $\mathbf{U}$ has eigenvalues +1, and if it is odd $\mathbf{U}$ has eigenvalues -1 (Bhandari and Love 1994). Special among these are the bullseyes: points in $\boldsymbol{R}$ where $\Delta\chi = 0$ (eigenvalues +1) because of the equality (degeneracy) of the two refractive indices $n_\pm$. A different sort of singularity, observed in interference experiments (Bhandari 1992a, b, 1993b) and yielding a phase change of 2π rather than π, occurs when two interfering polarization states become orthogonal, so that according to Pancharatnam's definition (Pancharatnam 1956) a relative phase cannot be defined. In the present context these singularities are the rows of bright points of conoscopic figures between crossed polarizers (where the 'anti-isogyres' cross the bright rings) or the points of zero intensity between parallel polarizers.

6. Characterizing the foil

We seek the three refractive indices n_x, n_y, n_z. Many precise methods are available for this, described in detail in standard texts (Pockels 1906, Hartshorne and Stuart 1960, Bloss 1961). Our aim here is to show how a rough estimate of n_x, n_y and n_z can be obtained with simple naked-eye observations. The three indices differ by small amounts, i.e. in equation (7) $\alpha \ll 1$ and $\beta \ll 1$. Therefore the average index, which determines optical effects not involving polarization, can be chosen as n_y. We measured this by the longitudinal shift of an image viewed through a stack of foils, with the films wetted with oil to reduce reflections from the interfaces; the result was $n_y = 1.57 \pm 0.01$, identical to the tabulated value for polycarbonate (Kroschwitz 1987).

The remaining two refractive indices can be determined as follows. First, by measuring the angle $2\theta_{\rm external}$ between the bullseyes, the ratio α/β can be obtained from (21), using Snell's law for the refraction correction relating $\theta_{\rm external}$ to the direction $\theta_{\rm c}$ of the optic axes in the foil:

$$\sin\theta_{\rm external} = n_y \sin\theta_{\rm c}. \tag{32}$$

For the foil that generated the bullseyes in figure 3(a), we measured $\theta_{\rm external} = (24 \pm 1)^\circ$, giving $\alpha/\beta = 0.07$. Second, the phase difference between the two waves travelling normal to the foil can be determined from the number N of fringes between the bullseye ($X = X_{\rm c}$) and the centre of the pattern (at $X = 0$) where the two waves have zero phase difference. N need not be an integer, and is easiest to estimate by counting dark fringes. From equations (15) and

(16), this gives

$$n_y - n_x = \frac{N\lambda}{L}. \tag{33}$$

From figure 3(*a*), $N \approx 3.1$, and since the foil has thickness 0.1 mm we find, for $\lambda = 0.546\ \mu$m, $n_y - n_x = 0.017$.

Collecting these results, for the foil we used (Niceday Write-on OHP film) we find

$$n_x = 1.553 \qquad n_y = 1.57 \qquad n_z = 1.88. \tag{34}$$

Although two of the values are close together, the rather large angle between the optic axes indicates that the foil is far from uniaxial. The large index corresponds to waves whose $\boldsymbol{D}$ vector is normal to the foil.

At this point we encounter an annoying ambiguity, related to the ordering of the refractive indices n_x, n_y, n_z, and known in the literature as the problem of the optic sign (Bloss 1961). The observation of bullseyes rules out the possibility that the intermediate index corresponds to the direction normal to the foil (if it did, the optic axes would lie in the plane of the foil); this motivated our conventional choice of n_y as the intermediate index. For the remaining indices, we made the assumption $n_x < n_z$ (equation (6)), that is, the smallest refractive index corresponds to waves polarized in the plane of the foil. But what if the opposite is true, i.e. $n_x > n_z$? Then α and β would both be negative; but this would leave unaffected the ratio determining the axis θ_c according to (21). Incorporating the measurement of the fringe number N, we find that if a and b are small but otherwise arbitrary, the following two situations cannot be distinguished by the pattern of bullseyes they produce:

$$\begin{aligned} &\{(n_y - a)(= n_x),\ n_y,\ (n_y + b)(= n_z)\} \\ &\{(n_y - b)(= n_z),\ n_y,\ (n_y + a)(= n_x)\}. \end{aligned} \tag{35}$$

If a and b are different, these represent physically different materials. We think that this ambiguity could be resolved only by a much more accurate experiment than those envisaged here, equivalent to directly measuring the indices n_x and n_y by the lateral shifts of images seen through a stack of foils, using x- and y-polarized light, and seeing which is the larger. In the present case, the alternative determination gives physically unacceptable indices, namely 1.60, 1.58 and 1.28 (the lowest value—less than that of water—is too small for a plastic).

Acknowledgment

We thank Professor J F Nye for a careful reading of the manuscript.

References

Arnold V I 1978 *Mathematical Methods of Classical Mechanics* (Berlin: Springer)
Berry M V 1984 Quantal phase factors accompanying adiabatic changes *Proc. R. Soc.* A **392** 45–57.
——1989 *Geometric Phases in Physics* ed A Shapere and F Wilczek (Singapore: World Scientific) pp 7–28
Berry M V and Hannay J H 1977 Umbilic points on Gaussian random surfaces *J. Phys. A: Math. Gen.* **10** 1809–21
Berry M V and Upstill C 1980 Catastrophe optics: morphologies of caustics and their diffraction patterns *Progr. Opt.* **18** 257–346
Bhandari R 1992a Observation of Dirac singularities with light polarization. I *Phys. Lett.* A **171** 262–6
——1992b Observation of Dirac singularities with light polarization. II *Phys. Lett.* A **171** 267–70
——1993a 4p spinor symmetry—some new observations *Phys. Lett.* A **180** 15–20
——1993b Interferometry without beam splitters—a sensitive technique for spinor phases *Phys. Lett.* A **180** 21–4
——1997 Polarization of light and topological phases *Phys. Rep.* **281** 1–64
Bhandari R and Love J 1994 Polarization eigenmodes of a QHQ retarder—some new features *Opt. Commun.* **10** 479–84
Bloss F D 1961 *An Introduction to the Methods of Optical Crystallography* (New York: Holt, Rinehart and Winston)
Born M and Wolf E 1959 *Principles of Optics* (Oxford: Pergamon)
Darboux G 1896 *Lecons sur la theorie des surfaces* (Paris: Gauthier-Villars)

Gibbs J W 1882 On double refraction in perfectly transparent media which exhibit the phenomenon of circular polarization *Am. J. Sci. (ser. 3)* **23** 460–76
Goldhammer M D 1892 Theorie electromagnetique de la polarisation roratiore naturelle des corps transparents *J. Physique Theor. Appl. (ser. 3)* **1** 205–9
Hartshorne N H and Stuart A 1960 *Crystals and the Polarising Microscope* (London: Arnold)
Herzberg G and Longuet-Higgins H C 1963 Intersection of potential-energy surfaces in polyatomic molecules *Disc. Faraday Soc.* **35** 77–82
Iyanaga S and Kawada Y (ed) 1968 *Encyclopedic Dictionary of Mathematics* (Cambridge, MA: MIT Press)
Keller A 1998 Private communication
Kroschwitz J I (ed) 1987 *Encyclopedia of Polymer Science and Engineering* (New York: Wiley)
Landau L D, Lifshitz E M and Pitaevskii L P 1984 *Electrodynamics of Continuous Media* (Oxford: Pergamon)
Mead C A and Truhlar D G 1979 On the determination of Born–Oppenheimer nuclear motion wave functions including complications due to conical intersections and identical nuclei *J. Chem. Phys.* **70** 2284–96
Nye J F 1995 *Physical Properties of Crystals* (Oxford: Clarendon) ch 13 and appendix H
Pancharatnam S 1956 Generalized theory of interference, and its applications. Part I. Coherent pencils *Proc. Ind. Acad. Sci.* A **44** 247–62
Pockels F 1906 *Lehrbuch der Kristalloptik* (Leipzig: Teubner)
Pocklington H C 1901 On rotatory polarization in biaxial crystals *Lond., Edin. Dublin Phil. Mag. J. Sci. (ser. 2)* **2** 361–70
Shapere A and Wilczek F (ed) 1989 *Geometric Phases in Physics* (Singapore: World Scientific)
Silverman M 1980 The curious problem of spinor rotation *Eur. J. Phys.* **1** 116–23
Simon B 1983 Holonomy, the quantum adiabatic theorem, and Berry's phase *Phys. Rev. Lett.* **51** 2167–70
Spivak M 1975 *A Comprehensive Introduction to Differential Geometry* (San Francisco, CA: Publish or Perish)
Teller E 1937 The crossing of potential surfaces *J. Phys. Chem.* **41** 109–16
Uhlenbeck K 1976 Generic properties of eigenfunctions *Am. J. Math.* **98** 1059–78
Wolfram S 1991 *Mathematica* (Reading, MA: Addison-Wesley)

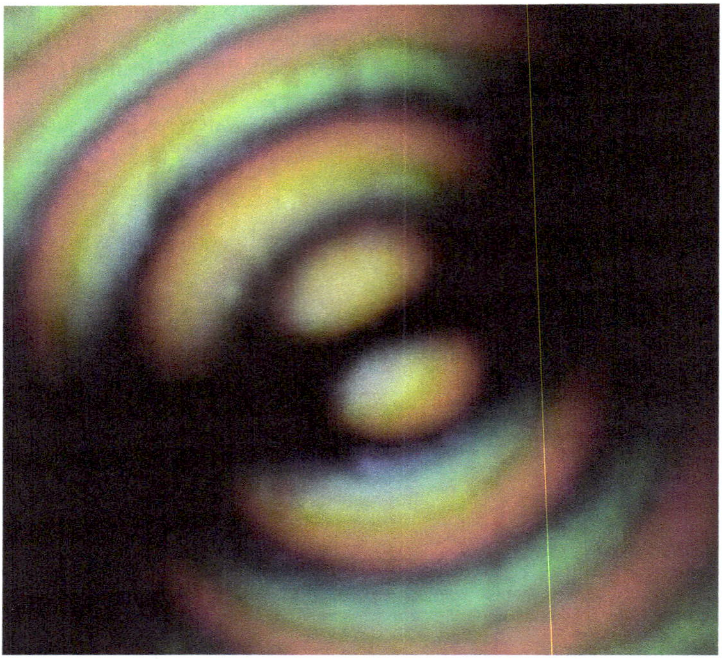

Figure 9. Colour image of conoscopic interference in white light for a single bullseye.

INSTITUTE OF PHYSICS PUBLISHING JOURNAL OF OPTICS A: PURE AND APPLIED OPTICS

J. Opt. A: Pure Appl. Opt. **6** (2004) S24–S25 PII: S1464-4258(04)66577-9

Black polarization sandwiches are square roots of zero

M V Berry and M R Dennis

H H Wills Physics Laboratory, University of Bristol, Tyndall Avenue, Bristol BS8 1TL, UK

Received 29 July 2003, accepted for publication 28 August 2003
Published 24 February 2004
Online at stacks.iop.org/JOptA/6/S24 (DOI: 10.1088/1464-4258/6/3/004)

Abstract
In the 2 × 2 matrices representing retarders and ideal polarizers, the eigenvectors are orthogonal. An example of the opposite case, where eigenvectors collapse onto one, is matrices **M** representing crystal plates sandwiched between a crossed polarizer and analyser. For these familiar combinations, $\mathbf{M}^2 = 0$, so black sandwiches can be regarded as square roots of zero. Black sandwiches illustrate physics associated with degeneracies of non-Hermitian matrices.

Keywords: polarization, degeneracy, nilpotence, matrix optics

Tudor [1, 2] has recently remarked that the common classification of polarizing devices into retarders, represented by unitary operators, and ideal polarizers, represented by projection operators, fails to include some commonly used combinations of optical elements. Both retarders and ideal polarizers are represented by 2 × 2 Jones matrices [3–5] whose eigenvectors are orthogonal; the difference lies in the eigenvalues: for retarders, two complex eigenvalues lie on the unit circle, and for ideal polarizers, one eigenvalue is zero. Tudor gives the example of a linear polarizer followed by a quarter-wave plate, where the eigenvectors are not orthogonal. He also gives an example of the extreme case of nonorthogonal eigenvectors, namely *parallel* eigenvectors: a linear polarizer sandwiched between two identical quarter-wave plates oriented at 45° to the axes of the polarizer; this example falls into a class described by de Lang [6].

Here we extend Tudor's remark in a very simple way, by pointing out that this extreme case of parallel eigenvectors is realized optically by a much wider class of devices including some of the most familiar combinations employed in polarization optics, namely any specimen (e.g. a crystal plate) between a crossed polarizer and analyser—which we call a *black sandwich*, for obvious reasons. The matrix for a black sandwich is

$$\mathbf{M} = \mathbf{P}_+\mathbf{A}\mathbf{P}_-, \tag{1}$$

where

$$\mathbf{P}_\pm = |\pm\rangle\langle\pm| \tag{2}$$

are the projection matrices corresponding to the ideal polarizer (−) and analyser (+) that select orthogonal states represented by the column (e.g. Jones) vectors $|\pm\rangle$, and **A**, representing the specimen, can be any 2 × 2 matrix.

Because of its unsymmetrical form (polarizer and analyser different), **M** is non-Hermitian, and explicit calculation shows that both of its eigenvalues are zero. Therefore the two eigenvectors, that are different for the general case, have here collapsed onto one, namely $|+\rangle$. A polarization $|\psi\rangle$ entering the black sandwich emerges in the state

$$\mathbf{M}|\psi\rangle = (\langle+|\mathbf{A}|-\rangle\langle-|\psi\rangle)|+\rangle. \tag{3}$$

In contrast to the ideal polarizer $\mathbf{P}_+$, from which the emerging light is also in the state $|+\rangle$ but which extinguishes incident light in the orthogonal state $|-\rangle$, the black sandwich extinguishes its own eigenstate. This is obvious when **M** is written in the form

$$M = (\langle+|A|-\rangle)|+\rangle\langle-|. \tag{4}$$

It follows immediately that

$$\mathbf{M}^2 = 0, \tag{5}$$

i.e. **M** is nilpotent, reflecting the obvious fact that the combination of two black sandwiches extinguishes all light. Therefore black sandwiches (1), incorporating general matrices **A**, can be regarded as nontrivial square roots of zero. (The trivial square root $\mathbf{M} = 0$ corresponds to $\mathbf{A} = 1$—simply a crossed polarizer and analyser, i.e. a sandwich with no filling.)

A familiar black sandwich consists of a transparent crystal plate between a linear polarizer and analyser. If the polarizer and analyser are

$$|-\rangle = \begin{pmatrix}0\\1\end{pmatrix} \quad \text{(polarizer)}, \qquad |+\rangle = \begin{pmatrix}1\\0\end{pmatrix} \quad \text{(analyser)}, \tag{6}$$

and the orthogonal eigenpolarizations and eigenvalues of the crystal are

$$|\phi_1\rangle = \begin{pmatrix} u \\ v \end{pmatrix}, \qquad \text{eigenvalue } \lambda_0 + \lambda,$$
$$|\phi_2\rangle = \begin{pmatrix} v^* \\ -u^* \end{pmatrix}, \qquad \text{eigenvalue } \lambda_0 - \lambda, \tag{7}$$

with $|u|^2 + |v|^2 = 1$, then $\mathbf{A}$ is the unitary matrix

$$\mathbf{A} = \exp(\mathrm{i}\lambda_0)[|\phi_1\rangle\langle\phi_1|\exp(\mathrm{i}\lambda) + |\phi_2\rangle\langle\phi_2|\exp(-\mathrm{i}\lambda)]$$
$$= \exp(\mathrm{i}\lambda_0)\left[\cos\lambda\mathbf{I} + \mathrm{i}\sin\lambda\begin{pmatrix} |u|^2 - |v|^2 & 2uv^* \\ 2u^*v & |v|^2 - |u|^2 \end{pmatrix}\right], \tag{8}$$

and the black sandwich matrix is

$$\mathbf{M} = \langle +|\mathbf{A}|-\rangle|+\rangle\langle -| = 2\mathrm{i}uv^*\exp(\mathrm{i}\lambda_0)\sin\lambda\begin{pmatrix} 0 & 1 \\ 0 & 0 \end{pmatrix}. \tag{9}$$

For an anisotropic material, the polarization components u and v depend on direction, and any diffuse light (e.g. the sky, or white paper) viewed through the sandwich will exhibit the familiar conoscopic figures [7, 8]. An easy way to see the conoscopic figures displaying the polarization singularity at the optic axis of a biaxial material is with a 'crystal' consisting of a sheet of overhead-projector transparency film [9], so this 'black plastic sandwich' is a square root of zero. For conoscopic figures corresponding to more general polarization singularities, with $\mathbf{A}$ representing crystals that are gyrotropic and dichroic as well as birefringent, see [10].

The de Lang class [6] of nilpotent devices, mentioned at the end of the first paragraph, is more restricted, being of the form $\mathbf{M} = \mathbf{UPU}$, where $\mathbf{U}$ is a quarter-wave retarder whose eigenpolarizations are directions on the Poincaré sphere perpendicular to that selected by $\mathbf{P}$. This class forms a four-parameter family of devices (including overall phase), whereas the black sandwiches form a ten-parameter family (or, if $\mathbf{A}$ is restricted to the class of retarders, a six-parameter family).

The polarization optics of black sandwiches joins a growing class of physics associated with the collapse of two eigenvectors onto one at a degeneracy of eigenvalues of non-Hermitian matrices. Other examples occur in the diffraction of atoms by 'crystals of light' [11–13], in nuclear physics [14–17], and in the linewidths of unstable lasers [18]. Such degeneracies also occur in the optics of absorbing crystals [8, 10, 19–21], for light travelling along a 'singular axis' (to avoid confusion, we emphasize that in (1) these degeneracies occur in the crystal matrix $\mathbf{A}$, not the black sandwich matrix $\mathbf{M}$).

Acknowledgments

MVB is supported by the Royal Society, and MRD is supported by the Leverhulme Trust.

References

[1] Tudor T 2003 Generalized observables in polarization optics *J. Phys. A: Math. Gen.* **36** 9577–90

[2] Tudor T 2003 *ICO Topical Meeting on Polarization Optics (Physics Department Selected Papers* 8) ed A A Friesem and J Turunen (Polvijärvi: University of Joensuu) pp 46–47

[3] Azzam R M A and Bashara N M 1977 *Ellipsometry and Polarized Light* (Amsterdam: North-Holland)

[4] Berry M V and Klein S 1996 Geometric phases from stacks of crystal plates *J. Mod. Opt.* **43** 165–80

[5] Brosseau C 1998 *Fundamentals of Polarised Light: a Statistical Optics Approach* (New York: Wiley)

[6] de Lang H 1967 Polarization properties of optical resonators passive and active *Philips Res. Rep.* **37** Suppl. 1–67

[7] Born M and Wolf E 1959 *Principles of Optics* (London: Pergamon)

[8] Ramachandran G N and Ramaseshan S 1961 Crystal optics *Handbuch der Physik* vol XXV/I, ed H Flügge (Berlin: Springer)

[9] Berry M V, Bhandari R and Klein S 1999 Black plastic sandwiches demonstrating biaxial optical anisotropy *Eur. J. Phys.* **20** 1–14

[10] Berry M V and Dennis M R 2003 The optical singularities of birefringent dichroic chiral crystals *Proc. R. Soc.* A **459** 1261–92

[11] Oberthaler M K, Abfalterer R, Bernet S, Schmiedmayer J and Zeilinger A 1996 Atom waves in crystals of light *Phys. Rev. Lett.* **77** 4980–3

[12] Berry M V and O'Dell D H J 1998 Diffraction by volume gratings with imaginary potentials *J. Phys. A: Math. Gen.* **31** 2093–101

[13] Berry M V 1998 Lop-sided diffraction by absorbing crystals *J. Phys. A: Math. Gen.* **31** 3493–502

[14] Heiss W D 2000 Repulsion of resonance states and exceptional points *Phys. Rev.* E **61** 929–32

[15] Heiss W D and Harney H L 2001 The chirality of exceptional points *Eur. Phys. J.* D **17** 149–51

[16] Rotter I 2001 Correlations in quantum systems and branch points in the complex plane *Phys. Rev.* C **64** 034301

[17] Rotter I 2002 Branch points in the complex plane and geometric phases *Phys. Rev.* E **65** 026217

[18] Berry M V 2003 Mode degeneracies and the Petermann excess-noise factor for unstable lasers *J. Mod. Opt.* **50** 63–81

[19] Pancharatnam S 1955 The propagation of light in absorbing biaxial crystals—I. Theoretical *Proc. Indian Acad. Sci.* **42** 86–109

[20] Pancharatnam S 1955 The propagation of light in absorbing biaxial crystals—II. Experimental *Proc. Indian Acad. Sci.* **42** 235–48

[21] Berry M V 1994 Pancharatnam, virtuoso of the Poincaré sphere: an appreciation *Curr. Sci.* **67** 220–23

E. Wolf, Progress in Optics 50

Chapter 2

Conical diffraction: Hamilton's diabolical point at the heart of crystal optics

by

M.V. Berry, M.R. Jeffrey

H.H. Wills Physics Laboratory, Tyndall Avenue, Bristol BS8 1TL, UK

ISSN: 0079-6638 DOI: 10.1016/S0079-6638(07)50002-8

Reprinted from Progress in Optics, Volume 50, (2007).

Contents

§ 1. Introduction.

The bicentenary of the birth in 1805 of William Rowan Hamilton (fig. 1) comes at a time of renewed interest in the phenomenon of conical refraction (Born and Wolf [1999], Landau, Lifshitz and Pitaevskii [1984]), in which a narrow beam of light, incident along the optic axis of a biaxial crystal, spreads into a hollow cone within the crystal, and emerges as a hollow cylinder. The prediction of the effect in 1832 (Hamilton [1837]), and its observation by Lloyd (fig. 1) soon afterwards (Lloyd [1837]), caused a sensation. For a non-technical account, see Lunney and Weaire [2006].

Our purpose here is to describe how the current understanding of conical refraction has been reached after nearly two centuries of theoretical and experimental study. Although our treatment will be roughly historical, we do not adhere to the practice, common in historical research, of describing each episode using only sources and arguments from the period being studied. For the early history of conical refraction, this has already been done, by Graves [1882] and O'Hara [1982]. Rather, we will weave each aspect of the theory into the historical development in ways more concordant with the current style of theoretical physics, hoping thus to bring out connections with other phenomena in mathematics and physics.

There are many reasons why conical refraction is worth revisiting:

(i) It was an early (perhaps the first) example of a qualitatively new phenomenon predicted by mathematical reasoning. By the early 1800s, it was widely appreciated that mathematics is essential to understanding the natural world. However, the phenomena to which mathematics had been applied were already familiar (e.g., tides, eclipses, and planetary orbits). Prediction of qualitatively new effects by mathematics may be commonplace today, but in the 1830s it was startling.

(ii) With its intimate interplay of position and direction, conical refraction was the first non-trivial application of phase space, and of what we now call dynamics governed by a Hamiltonian.

(iii) Its observation provided powerful evidence confirming that light is a transverse wave.

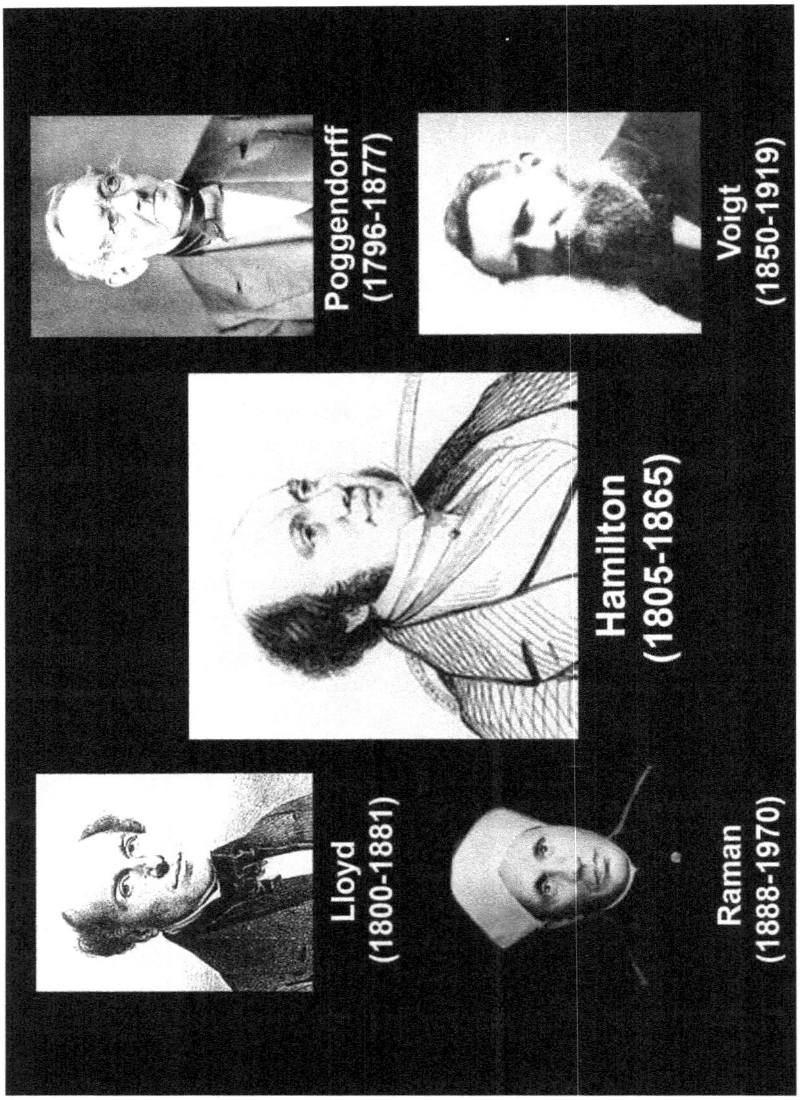

Fig. 1. Dramatis personae.

(iv) Recently, it has become popular to study light through its singularities (Berry [2001], Nye [1999], Soskin and Vasnetsov [2001]). In retrospect, we see conical refraction as one of the first phenomena in singular polarization optics; another is the pattern of polarization in the blue sky (Berry, Dennis and Lee [2004]).

(v) It was the first physical example of a conical intersection (diabolical point) (Berry [1983], Uhlenbeck [1976], Berry and Wilkinson [1984]) involving a degeneracy. Nowadays, conical intersections are popular in theoretical chemistry, as spectral features indicating the breakdown of the Born–Oppenheimer separation between fast electronic and slow nuclear freedoms (Cederbaum, Friedman, Ryaboy and Moiseyev [2003], Herzberg and Longuet-Higgins [1963], Mead and Truhlar [1979]). By analogy, conical refraction can be reinterpreted as an exactly solvable model for quantum physics in the presence of a degeneracy.

(vi) The effect displays a subtle interplay of ray and wave physics. Although its original prediction was geometrical (Sections 3 and 4), there are several levels of geometrical optics (Sections 5 and 7), of which all except the first require concepts from wave physics, and waves are essential to a detailed understanding (Section 6). That is why we use the term conical diffraction, and why the effect has taken so long to understand.

(vii) Analysis of the theory (Berry [2004b]) led to identification of an unexpected universal phenomenon in mathematical asymptotics: when exponential contributions to a function compete, the smaller exponential can dominate (Berry [2004a]).

(viii) There are extensions (Section 8) of the case studied by Hamilton, and their theoretical understanding still presents challenges. Effects of chirality (optical activity) have only recently been fully understood (Belsky and Stepanov [2002], Berry and Jeffrey [2006a]), and further extensions incorporate absorption (Berry and Jeffrey [2006b], Jeffrey [2007]) and nonlinearity (Indik and Newell [2006]).

(ix) Conical diffraction is a continuing stimulus for experiments. Although the fine details of Hamilton's original phenomenon have now been observed (Berry, Jeffrey and Lunney [2006]), predictions of new structures that appear in the presence of chirality, absorption and nonlinearity remain untested.

(x) The story of conical diffraction, unfolding over 175 years, provides an edifying contrast to the current emphasis on short-term science.

Although all results of the theory have been published before, some of our ways of presenting them are original. In particular, after the exact treatment in

Sections 2–4 we make systematic use of the simplifying approximation of paraxiality. This is justified by the small angles involved in conical diffraction (Appendix 1), and leads to what we hope are the simplest quantitative explanations of the various phenomena.

§ 2. Preliminaries: electromagnetism and the wave surface

Hamilton's prediction was based on a singular property of the wave surface describing propagation in an anisotropic medium. Originally, this was formulated in terms of Fresnel's elastic-solid theory. Today it is natural to use Maxwell's electromagnetic theory.

For the physical fields in a homogeneous medium, we write plane waves with wavevector $\mathbf{k}$ and frequency ω as

$$\mathrm{Re}\big[\{\mathbf{D_k}, \mathbf{E_k}, \mathbf{B_k}, \mathbf{H_k}\}\exp\{\mathrm{i}(\mathbf{k}\cdot\mathbf{r}-\omega t)\}\big], \tag{2.1}$$

in which the vectors $\mathbf{D_k}$, etc., are usually complex. From Maxwell's curl equations,

$$\omega\mathbf{D_k} = -\mathbf{k}\times\mathbf{H_k}, \qquad \omega\mathbf{B_k} = \mathbf{k}\times\mathbf{E_k}. \tag{2.2}$$

A complete specification of the fields requires constitutive equations. For a transparent nonmagnetic nonchiral biaxial dielectric, these can be written as

$$\mathbf{E_k} = \boldsymbol{\varepsilon}^{-1}\mathbf{D_k}, \qquad \mathbf{B_k} = \mu_0\mathbf{H_k}. \tag{2.3}$$

$\boldsymbol{\varepsilon}^{-1}$ is the inverse dielectric tensor, conveniently expressed in principal axes as

$$\boldsymbol{\varepsilon}^{-1} = \frac{1}{\varepsilon_0}\begin{pmatrix} 1/n_1^2 & 0 & 0 \\ 0 & 1/n_2^2 & 0 \\ 0 & 0 & 1/n_3^2 \end{pmatrix}, \tag{2.4}$$

where n_i are the principal refractive indices, all different, with the conventional ordering

$$n_1 < n_2 < n_3. \tag{2.5}$$

For biaxiality (all n_i different), the microscopic structure of the material must have sufficiently low symmetry; in the case of crystals, this is a restriction to the orthorhombic, monoclinic or triclinic classes (Born and Wolf [1999]). Conical refraction depends on the differences between the indices, so we define

$$\alpha \equiv \frac{1}{n_1^2} - \frac{1}{n_2^2}, \qquad \beta \equiv \frac{1}{n_2^2} - \frac{1}{n_3^2}. \tag{2.6}$$

It will be convenient to express the wavevector **k** in terms of the refractive index $n_{\mathbf{k}}$, through the definitions

$$\mathbf{k} \equiv k\mathbf{e_k} \equiv k_0 n_{\mathbf{k}} \mathbf{e_k}, \quad k_0 \equiv \frac{\omega}{c}, \tag{2.7}$$

incorporating

$$c = \frac{1}{\sqrt{\varepsilon_0 \mu_0}}. \tag{2.8}$$

It is simpler to work with the electric vector **D** rather than **E**, because **D** is always transverse to the propagation direction $\mathbf{e_k}$. Then eqs. (2.2) and (2.3) lead to the following eigenequation determining the possible plane waves:

$$\frac{1}{n_{\mathbf{k}}^2}\mathbf{D_k} = -\mathbf{e_k} \times \mathbf{e_k} \times \left(\varepsilon_0 \boldsymbol{\varepsilon}^{-1} \cdot \mathbf{D_k}\right). \tag{2.9}$$

This involves a 3×3 matrix whose determinant vanishes, and the two eigenvalues $\lambda_\pm$ of the 2×2 inverse dielectric tensor transverse to **k** give the refractive indices

$$n_{\mathbf{k}\pm} = \frac{1}{\sqrt{\lambda_\pm(\mathbf{e_k})}}. \tag{2.10}$$

Of several different graphical representations (Born and Wolf [1999], Landau, Lifshitz and Pitaevskii [1984]) of the propagation governed by $n_{\mathbf{k}\pm}$, we choose the two-sheeted polar plot in direction space; this is commonly called the wave surface, though there is no universally established terminology. The wave surface has the same shape as the constant ω surface in **k** space, that is [cf. eq. (2.7)], the contour surface of the dispersion relation

$$\omega(\mathbf{k}) \equiv \frac{ck}{n_{\mathbf{k}}}; \tag{2.11}$$

$\omega(\mathbf{k})$ is the Hamiltonian generating rays in the crystal, with **k** as canonical momentum.

An immediate application of the wave surface, known to Hamilton and central to his discovery, is that for each of the two waves with wavevector **k**, the ray direction, that is, the direction of energy transport, is perpendicular to the corresponding sheet of the surface. This can be seen from the first Hamilton equation, according to which the group (ray) velocity is

$$\mathbf{v}_{\mathrm{g}} = \nabla_{\mathbf{k}} \omega(\mathbf{k}) \tag{2.12}$$

(for this case of a homogeneous medium, the second Hamilton equation simply asserts that **k** is constant along a ray). Alternatively (Landau, Lifshitz and

Pitaevskii [1984]), the ray direction can be regarded as that of the Poynting vector,

$$\mathbf{S} = \mathrm{Re}\,\mathbf{E}^* \times \mathbf{H}, \tag{2.13}$$

because for any displacement $\mathbf{dk}$ with ω constant – that is, any displacement in the wave surface –

$$\mathbf{S} \cdot \mathrm{d}\mathbf{k} = \frac{1}{2}\omega\,\mathrm{Re}\big[\mathbf{E}^* \cdot \mathrm{d}\mathbf{D} - \mathrm{d}\mathbf{E}^* \cdot \mathbf{D} + \mathbf{H} \cdot \mathrm{d}\mathbf{B}^* - \mathrm{d}\mathbf{H} \cdot \mathbf{B}^*\big] = 0, \tag{2.14}$$

where the first equality is a consequence of Maxwell's equations and the second follows from the linearity and Hermiticity of the constitutive relations (2.3). (This argument is slightly more general than that of Landau, Lifshitz and Pitaevskii [1984], because it includes transparent media that are chiral as well as biaxial.)

§ 3. The diabolical singularity: Hamilton's ray cone

Figure 2 is a cutaway representation of the wave surface, showing four points of degeneracy where $n_{\mathbf{k}+} = n_{\mathbf{k}-}$, located on two optic axes in the k_1, k_3 plane in $\mathbf{k}$ space. Each intersection of the surfaces takes the form of a double cone, that is, a diabolo, and since these are the organizing centres of degenerate behaviour we call them diabolical points, a term adopted in quantum (Ferretti, Lami and

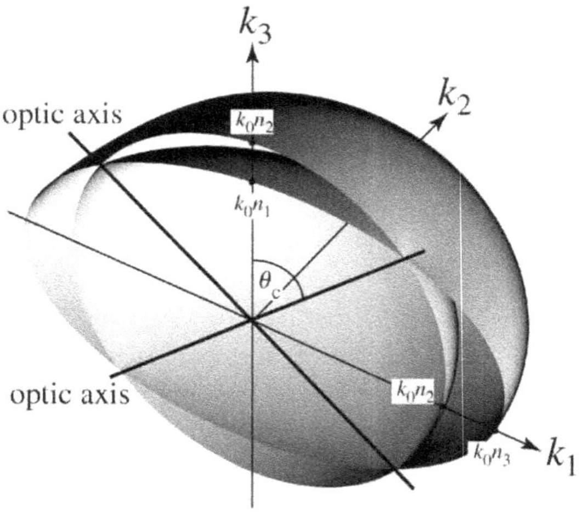

Fig. 2. Wave surface for $n_1 = 1.1$, $n_2 = 1.4$, $n_3 = 1.8$, showing the four diabolical points on the two optic axes.

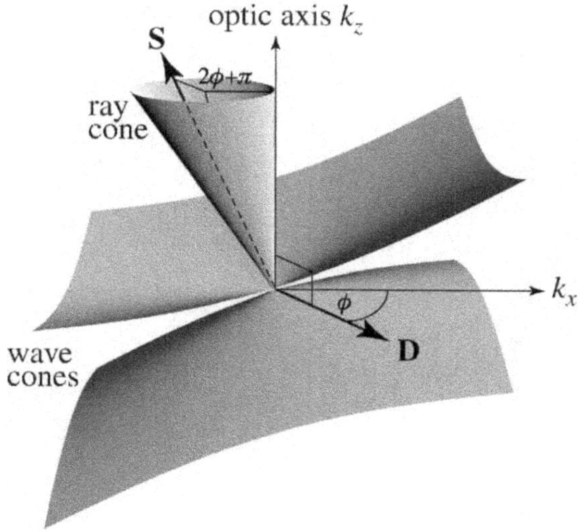

Fig. 3. The ray cone is a slant cone normal to the wave cone.

Villani [1999]) and nuclear physics (Chu, Rasmussen, Stoyer, Canto, Donangelo and Ring [1995]) as well as optics. In fact, such degeneracies are to be expected, because of the theorem of Von Neumann and Wigner [1929] that degeneracies of real symmetric matrices [such as that in (2.9)] have codimension two, and indeed we have two parameters, representing the direction of **k**.

Hamilton's insight was that at a diabolical point the normals (rays) to the surfaces are not defined, so there are infinitely many normals (rays), not two as for all other **k**. These normals to the wave cone define another cone. This is the ray cone (fig. 3), about whose structure – noncircular and skewed – we can learn by extending an argument of Born and Wolf [1999].

We choose **k** along an optic axis, and track the Poynting vector as **D** rotates in the plane transverse to **k**. From (2.13) and Maxwell's equations, **S** can be expressed as

$$\mathbf{S} = \frac{c}{n_{\mathbf{k}}} \mathrm{Re}\big[\mathbf{E}^* \cdot \mathbf{D}\mathbf{e}_{\mathbf{k}} - \mathbf{E}^* \cdot \mathbf{e}_{\mathbf{k}}\mathbf{D}\big]. \tag{3.1}$$

Using coordinates k_x, k_y, k_z, with $\mathbf{e}_z$ along **k** and $\mathbf{e}_x$ in the k_1, k_3 plane (i.e., $\mathbf{e}_y = \mathbf{e}_2$),

$$\mathbf{D} = \begin{pmatrix} D_x \\ D_y \\ 0 \end{pmatrix}, \qquad \mathbf{E} = \begin{pmatrix} E_x \\ E_y \\ E_z \end{pmatrix} = \frac{1}{\varepsilon_0}\big(\boldsymbol{\varepsilon}'\big)^{-1}\mathbf{D}. \tag{3.2}$$

Here $\boldsymbol{\varepsilon}'$ is the rotated dielectric matrix, determined by four conditions: rotation about the y axis does not change components involving the 2 axis, the choice of $\mathbf{k}$ along an optic axis (degeneracy) implies that the xx and yy elements are the same, and the trace and determinant of the matrices $\boldsymbol{\varepsilon}$ and $\boldsymbol{\varepsilon}'$ are the same. Thus

$$\varepsilon_0(\boldsymbol{\varepsilon}')^{-1} = \frac{1}{n_2^2}\begin{pmatrix}1 & 0 & 0\\ 0 & 1 & 0\\ 0 & 0 & 1\end{pmatrix} + \begin{pmatrix}0 & 0 & \sqrt{\alpha\beta}\\ 0 & 0 & 0\\ \sqrt{\alpha\beta} & 0 & \alpha-\beta\end{pmatrix}. \tag{3.3}$$

It follows that, in eq. (3.1),

$$\mathbf{E}^* \cdot \mathbf{D} = \frac{1}{\varepsilon_0 n_2^2}\left(|D_x|^2 + |D_y|^2\right),$$
$$E_z = \mathbf{E}^* \cdot \mathbf{e_k} = \frac{1}{\varepsilon_0}\sqrt{\alpha\beta}\,D_x. \tag{3.4}$$

Denoting the direction of $\mathbf{D}$ in the k_x, k_y plane by ϕ, that is

$$D_x = D\cos\phi, \qquad D_y = D\sin\phi, \tag{3.5}$$

we obtain a ϕ-parametrized representation of the surface swept out by $\mathbf{S}$:

$$\mathbf{S} = \frac{cD^2}{\varepsilon_0 n_2^3}\left(\mathbf{e}_z - \frac{1}{2}\tan 2A(\mathbf{e}_x + \mathbf{e}_x\cos 2\phi + \mathbf{e}_y \sin 2\phi)\right). \tag{3.6}$$

This is the ray surface, in the form of a skewed noncircular cone (fig. 3), with half-angle A given by

$$\tan 2A = n_2^2\sqrt{\alpha\beta}, \tag{3.7}$$

From eq. (3.6) it is clear that in a circuit of the ray cone ($2\phi = 2\pi$), the polarization direction ϕ rotates by half a turn, illustrating the familiar 'fermionic' property of degeneracies (Berry [1984], Silverman [1980]).

The direction of the optic axis, that is, the polar angle θ_c in the xz plane (fig. 2) can be determined from the rotation required to transform $\boldsymbol{\varepsilon}$ into $\boldsymbol{\varepsilon}'$. Thus

$$\boldsymbol{\varepsilon}' = \mathbf{R}\boldsymbol{\varepsilon}\mathbf{R}^{-1}, \tag{3.8}$$

where

$$\mathbf{R} = \begin{pmatrix}\sin\theta_c & 0 & -\cos\theta_c\\ 0 & 1 & 0\\ \cos\theta_c & 0 & \sin\theta_c\end{pmatrix}, \tag{3.9}$$

whence identification with eq. (3.3) fixes θ_c as

$$\tan\theta_c = \sqrt{\frac{\alpha}{\beta}}. \tag{3.10}$$

§ 4. The bright ring of internal conical refraction

In a leap of insight that we now recognise as squarely in the spirit of singular optics, Hamilton [1837] realised that the ray cone would appear inside a crystal slab, on which is incident a narrow beam directed along the optic axis. The hollow cone would refract into a hollow cylinder outside the slab (fig. 4). This is a singular situation because a beam incident in any other direction would emerge, doubly refracted, into just two beams, not a cylinder of infinitely many rays. This is internal conical refraction, so-called because the cone is inside the crystal.

Hamilton also envisaged external conical refraction, in which a different optical arrangement results in a cone outside the crystal. This is associated with a circle of contact between the wave surface and a tangent plane, "somewhat as a plum can be laid down on a table so as to touch and rest on the table in a whole circle of contact" (Graves [1882]). Since the theory is similar for the two effects,

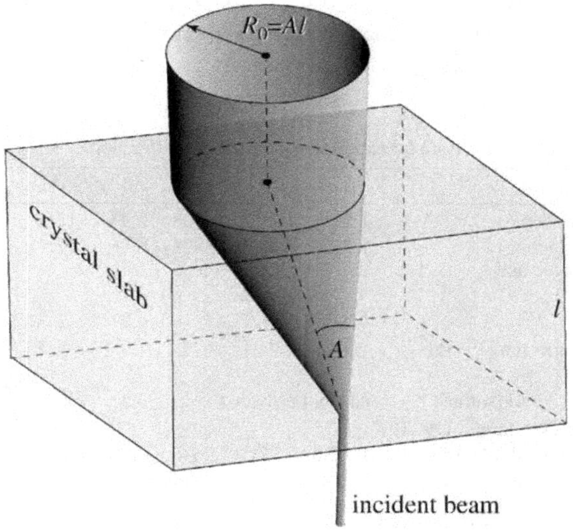

Fig. 4. Schematic of Hamilton's prediction of internal conical refraction.

our emphasis henceforth will be on internal, rather than external, conical refraction.

Hamilton's research attracted wide attention. According to Graves [1882], Airy called conical refraction "perhaps the most remarkable prediction that has ever been made"; and Herschel, in a prescient anticipation of singular optics, wrote "of theory actually remanding back experiment to read her lesson anew; informing her of facts so strange, as to appear to her impossible, and showing her all the singularities she would observe in critical cases she never dreamed of trying". In particular, the predictions fascinated Lloyd [1837], who called them "in the highest degree novel and remarkable", and regarded them as "singular and unexpected consequences of the undulatory theory, not only unsupported by any facts hitherto observed, but even opposed to all the analogies derived from experience. If confirmed by experiment, they would furnish new and almost convincing proofs of the truth of that theory."

Using a crystal of aragonite, Lloyd succeeded in observing conical refraction. The experiment was difficult because the cone is narrow: the semi-angle A is small. If the slab has thickness l, the emerging cylinder, with radius

$$R_0 = Al, \tag{4.1}$$

is thin unless l is large. To see the cylinder clearly, R_0 must be larger than the width w of the incident beam. Lloyd used a beam narrowed by passage through small pinholes with radii $w \leqslant 200$ µm. As we will see, this interplay between R_0 and w is important. However, large l brings the additional difficulty of finding a

Table 1
Data for experiments on conical diffraction

Experiment	n_1, n_2, n_3	A (°)	l (mm)	w (µm)	ρ_0
Lloyd [1837] (aragonite)	1.5326, 1.6863, 1.6908	0.96	12	$\leqslant 200$	$\geqslant 1.0$
Potter [1841] (aragonite)	1.5326, 1.6863, 1.6908	0.96	12.7	12.7	16.7
Raman, Rajagopalan and Nedungadi [1941] (naphthalene)	1.525, 1.722, 1.945	6.9	2	0.5	500
Schell and Bloembergen [1978a] (aragonite)	1.530, 1.680, 1.695	1.0	9.5	21.8	7.8
Mikhailychenko [2005] (sulfur)	data not provided	3.5	30	17	56
Fève, Boulanger and Marnier [1994] (sphere of KTP = $KTiOPO_4$)	1.7636, 1.7733, 1.8636	0.92	2.56	53.0	1210
Berry, Jeffrey and Lunney [2006] (MDT = $KGd(WO_4)_2$	2.02, 2.06, 2.11	1.0	25	7.1	60

For the experiments of Lloyd and Raman, w is the pinhole radius; for the other experiments, w is the $1/e$ intensity half-width of the laser beam.

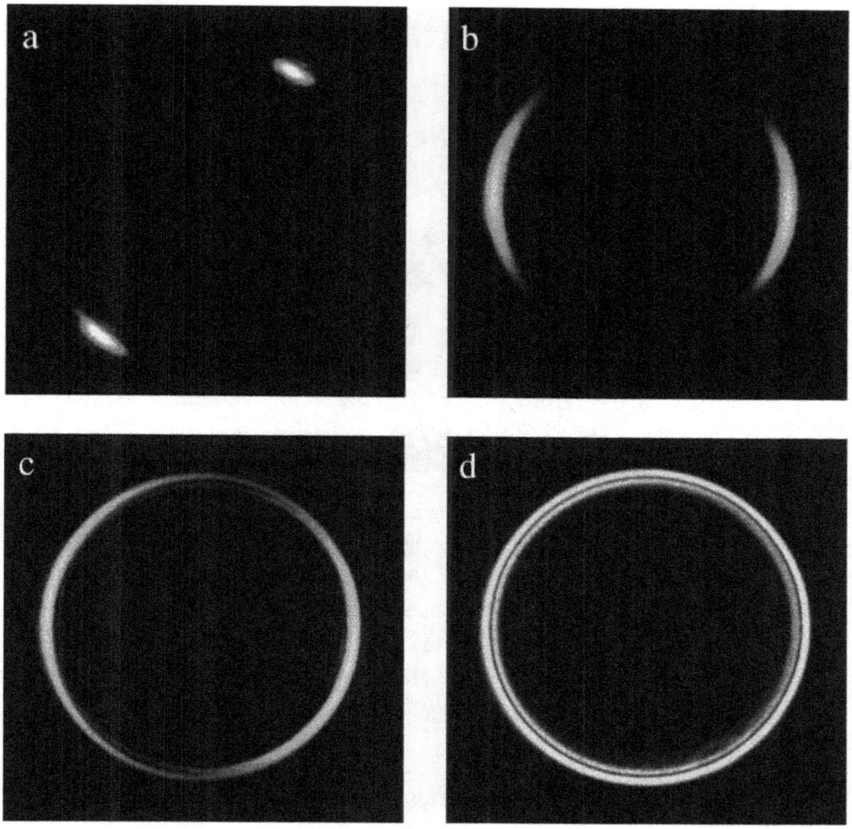

Fig. 5. The transition (a–d) from double to conical refraction as the incident beam direction approaches the optic axis. In (d) two rings are visible, separated by the Poggendorff dark ring studied in Section 5. This figure is taken from Berry, Jeffrey and Lunney [2006] (see also Table 1).

clean enough length of crystal; Lloyd describes how he explored several regions of his crystal before being able to detect the effect.

Nowadays it is simpler to use a laser beam with waist width w rather than a pinhole with radius w. Nevertheless, observation of the effect remains challenging. Table 1 summarises the conditions of the several experiments known to us; we will describe some of them later.

Figure 5 illustrates how double refraction transforms into conical refraction as the direction of the incident light beam approaches an optic axis. Of this transformation, Lloyd [1837] wrote: "This phenomenon was exceedingly striking. It looked like a small ring of gold viewed upon a dark ground; and the sudden and

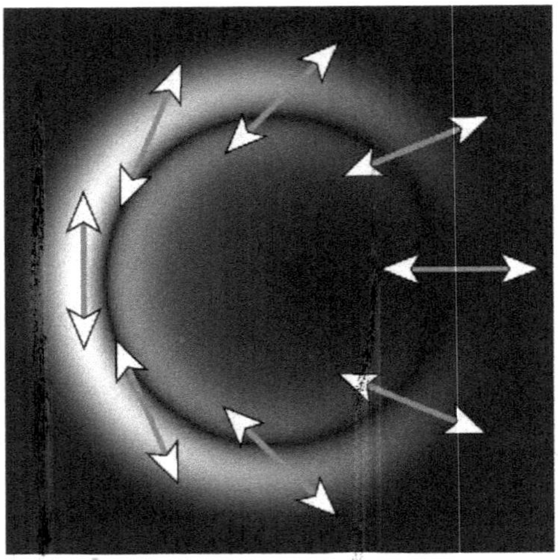

Fig. 6. Simulation of rings for incident beam linearly polarized horizontally, with polarization directions superimposed, showing the 'fermionic' half-turn of the polarization.

almost magical change of the appearance, from two luminous points to a perfect luminous ring, contributed not a little to enhance the interest."

Lloyd also observed a feature that Hamilton had predicted: the half-turn of polarization round the ring (fig. 6).

§ 5. Poggendorff's dark ring, Raman's bright spot

Hamilton was aware that his geometrical theory did not give a complete description of conical refraction: "I suspect the *exact* laws of it depend on things yet unknown" (Graves [1882]). And indeed, as fig. 5(d) shows, closer observation reveals internal structure in Hamilton's ring, that he did not predict and Lloyd did not detect: there is not one bright ring but two, separated by a dark ring. This was first reported in a brief but important paper by Poggendorff [1839] (fig. 1), who noted "...a bright ring that encompasses a coal-black sliver" ("einem hellen Ringe vereinigen, der ein kohlschwarzes Scheibchen einschliefst"). The existence of two bright rings, rather than one, was independently discovered soon afterwards by Potter [1841] (Table 1).

After more than 65 years, the origin of Poggendorff's dark ring was identified by Voigt [1905a] (fig. 1) (see also Born and Wolf [1999]), who pointed out that Hamilton's prediction involved the idealization of a perfectly collimated infinitely narrow beam. This is incompatible with the wave nature of light: a beam of spatial width w must contain transverse wavevector components (i.e., k_x, k_y) extending over at least a range $1/w$ (more, if the light is incoherent); this is just the optical analogue of the uncertainty principle. Therefore the incident beam will explore not just the diabolical point itself but a neighbourhood of the optic axis in **k** space. In fact Hamilton knew about the off-axis waves: "it was in fact from considering them and passing to the limit that I first deduced my expectation of conical refraction". Lloyd too was aware of "the angle of divergence produced by diffraction in the minutest apertures".

Voigt's observation was that the strength of the light generated by the off-axis waves inside and outside the cylinder is proportional to the radius $\sqrt{k_x^2 + k_y^2}$ of the circumference of contributing rings on the wave cone. At the diabolical point this vanishes, so the intensity is zero on the geometrical cylinder itself. Thus, the Poggendorff dark ring is a manifestation of the area element in plane polar coordinates in **k** space.

The elementary quantitative theory of the Poggendorff ring is based on geometrical optics, incorporating the finite **k** width of the beam – which of course is a consequence of wave physics. Rigorous geometrical-optics treatments were given by Ludwig [1961] and Uhlmann [1982]; here, our aim is to obtain the simplest explicit formulae. To prepare for later analysis of experiments, we formulate the theory so as to describe the light beyond the crystal. Appropriately enough, we will use Hamilton's principle.

We begin with the optical path lengths from a point $\mathbf{r}_i$ on the entrance face $z = 0$ of the crystal to a point $\mathbf{r}$ beyond the crystal at a distance z from the entrance face, for the two waves with transverse wavevector components k_x, k_y:

$$\begin{aligned}\text{path length} &= k_x(x - x_\text{i}) + k_y(y - y_\text{i}) + l\sqrt{k_0^2 n_{\mathbf{k}\pm}^2 - k_x^2 - k_y^2} \\ &\quad + (z - l)\sqrt{k_0^2 - k_x^2 - k_y^2}.\end{aligned} \tag{5.1}$$

The exit face of the crystal is $z = l$, but all our formulae are also valid for $z < l$, corresponding to observations, made with lenses outside the crystal, of the virtual field inside the crystal, as was appreciated long ago by Potter [1841], and later by Raman [1941] and Raman, Rajagopalan and Nedungadi [1941].

For the geometrical theory and all subsequent analysis, we will use the following dimensionless variables (fig. 7):

$$\begin{aligned}\boldsymbol{\rho} &= \{\xi, \eta\} = \rho\{\cos\phi, \sin\phi\} \equiv \frac{1}{w}\{x + R_0, y\}, \\ \boldsymbol{\kappa} &= \{\kappa_x, \kappa_y\} = \kappa\{\cos\phi_\kappa, \sin\phi_\kappa\} \equiv w\{k_x, k_y\}, \\ \rho_0 &\equiv \frac{R_0}{w}.\end{aligned} \tag{5.2}$$

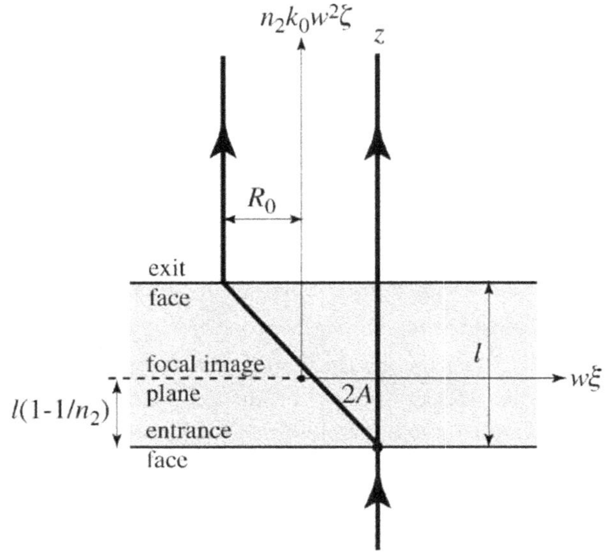

Fig. 7. Dimensionless coordinates for conical diffraction theory.

Here, transverse position $\boldsymbol{\rho}$ and transverse wavevectors $\boldsymbol{\kappa}$ are measured in terms of the beam width w, with $\boldsymbol{\rho}$ measured from the axis of the cylinder. Especially important is the parameter ρ_0, giving the radius of the cylinder in units of w; this single quantity characterises the field of rays – and also of waves, as we shall see – replacing the five quantities n_1, n_2, n_3, l, w. Well-developed rings correspond to $\rho_0 \gg 1$.

Near the diabolical point, κ is small, so we can write the sheets of the wave surface as

$$n_{\mathbf{k}\pm} = n_2\left(1 + \frac{A(-\kappa_x \pm \kappa)}{k_0 n_2 w}\right). \tag{5.3}$$

Now comes an important simplification, to be used in all subsequent analysis: because all angles are small (Table 1), we use the paraxial approximation (Appendix 1) to expand the square roots in eq. (5.1). This leads to

$$\text{path length} = k_0(n_2 l + z - l) + \Phi_\pm(\boldsymbol{\kappa}, \boldsymbol{\rho}, \boldsymbol{\rho}_i), \tag{5.4}$$

where

$$\Phi_\pm(\boldsymbol{\kappa}, \boldsymbol{\rho}, \boldsymbol{\rho}_\mathrm{i}) \equiv \boldsymbol{\kappa} \cdot (\boldsymbol{\rho} - \boldsymbol{\rho}_\mathrm{i}) \pm \kappa\rho_0 - \frac{1}{2}\kappa^2\zeta \tag{5.5}$$

and

$$\zeta \equiv \frac{l + n_2(z-l)}{n_2 k_0 w^2}. \tag{5.6}$$

Here ζ is a dimensionless propagation parameter, measuring distance from the 'focal image plane' $z = l(1 - 1/n_2)$ (fig. 7) where the sharpest image of the incident beam (pinhole or laser waist) would be formed if the crystal were isotropic (i.e., if A were zero). The importance of the focal image plane $\zeta = 0$ was first noted in observations by Potter [1841].

By Hamilton's principle, rays from $\{\boldsymbol{\rho}_\mathrm{i}, 0\}$ to $\{\boldsymbol{\rho}, z\}$ correspond to waves with wavenumber $\boldsymbol{\kappa}$ for which the optical distance is stationary. Thus

$$\nabla_{\kappa}\Phi = \boldsymbol{\rho} - \boldsymbol{\rho}_\mathrm{i} \pm \rho_0 \mathbf{e}_\kappa - \boldsymbol{\kappa}\zeta = 0, \tag{5.7}$$

that is

$$\boldsymbol{\rho} - \boldsymbol{\rho}_\mathrm{i} = (\kappa\zeta \mp \rho_0)\mathbf{e}_\kappa. \tag{5.8}$$

Consider for the moment rays from $\boldsymbol{\rho}_\mathrm{i} = 0$. Squaring eq. (5.8) and using $\rho > 0$, $\kappa > 0$ leads to

$$\begin{aligned} +\text{: two solutions: (a): } \boldsymbol{\kappa} &= \frac{(\rho+\rho_0)}{\zeta}\mathbf{e}_\rho, \\ \text{(b): } \boldsymbol{\kappa} &= \frac{(\rho_0-\rho)}{\zeta}\mathbf{e}_\rho \quad (\rho \leqslant \rho_0), \\ -\text{: one solution: (c): } \boldsymbol{\kappa} &= \frac{(\rho-\rho_0)}{\zeta}\mathbf{e}_\rho \quad (\rho \geqslant \rho_0). \end{aligned} \tag{5.9}$$

For each ρ, there are two solutions: (a); and either (b) (if $\rho \leqslant \rho_0$) or (c) (if $\rho \geqslant \rho_0$). As fig. 8 illustrates, these correspond to minus the slopes of the surface

$$\frac{1}{2}\kappa^2\zeta \mp \kappa\rho_0. \tag{5.10}$$

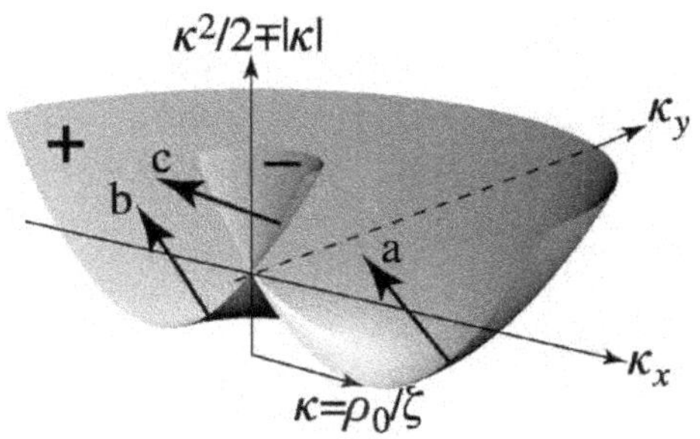

Fig. 8. The 'Hamiltonian' surface (5.10) whose normals generate the paraxial rays.

A ray bundle $\mathrm{d}\boldsymbol{\kappa}$ will reach $\mathrm{d}\boldsymbol{\rho}$ at ζ with intensity proportional to $|J|^{-1}$ where J is the Jacobian

$$J = \det \frac{\mathrm{d}\boldsymbol{\rho}}{\mathrm{d}\boldsymbol{\kappa}} = \det \frac{\partial(\xi, \eta)}{\partial(\kappa_x, \kappa_y)} = \zeta\left(\zeta \mp \frac{\rho_0}{\kappa}\right). \tag{5.11}$$

The ray equation (5.8) gives

$$|J|^{-1} = \frac{\kappa}{\zeta\rho} = \frac{|\rho \pm \rho_0|}{\zeta^2 \rho}, \tag{5.12}$$

which for the minus sign vanishes linearly on the cylinder $\rho = \rho_0$, where, from (5.9), the contributing wavevector is the diabolical point $\kappa = 0$. Thus the Poggendoff dark ring is an 'anticaustic'.

This geometrical theory also explains an important observation made by Raman, Rajagopalan and Nedungadi [1941] (fig. 1), who emphasized that the ring pattern changes dramatically beyond the crystal; in our notation, the pattern depends strongly on the distance ζ from the focal image plane. Raman saw that as ζ increases, a bright spot develops at $\rho = 0$. This is associated with the factor $1/\rho$ in the inverse Jacobian (5.12), corresponding to a singularity on the cylinder axis $\rho = 0$ – a line caustic, resulting from the ring of normals where the surface (5.10) turns over at $\kappa = \rho_0/\zeta$ (fig. 8). At the turnover, the surface is locally toroidal, so the Raman spot is analogous to the optical glory (Nussenzveig [1992], van de Hulst [1981]) – another consequence of an axial caustic resulting from circular symmetry. Raman, Rajagopalan and Nedungadi [1941] pointed out that the turnover is related to the circle of contact in external conical refraction, so that the two effects discovered by Hamilton cannot be completely separated. The central spot had been observed earlier by Potter [1841], who however failed to understand its origin.

Since the factor $1/\rho$ in (5.12) applies to all z, why does the Raman spot appear only as ζ increases? The reason is that in the full geometrical-optics intensity I_{geom}, the inverse Jacobian must be modulated by the angular distribution of the incident beam. Let the incident transverse beam amplitude (assumed circular) have Fourier transform $a(\kappa)$. Important cases are a pinhole with radius w, and a Gaussian beam with intensity $1/e$ half-width w, for which [in the scaled variables (5.2)]

$$\begin{aligned} &\text{circular pinhole: } a_{\mathrm{P}}(\kappa) = \frac{J_1(\kappa)}{\kappa}, \\ &\text{Gaussian beam: } a_{\mathrm{G}}(\kappa) = \exp\left(-\frac{1}{2}\kappa^2\right). \end{aligned} \tag{5.13}$$

Incorporating the Jacobian and the ray equations gives

$$I_{\text{geom}}(\rho, \zeta) = \frac{1}{2\zeta^2 \rho}\Big[|\rho - \rho_0|\big|a\big(|\rho - \rho_0|/\zeta\big)\big|^2 + (\rho + \rho_0)\big|a\big((\rho + \rho_0)/\zeta\big)\big|^2\Big]. \tag{5.14}$$

The first term represents the bright rings, separated by the Poggendorff dark ring; in this approximation, the rings are symmetric. The second term is weak except near $\rho = 0$, where it combines with the first term to give the Raman central spot, with strength proportional to

$$\frac{1}{\rho\zeta^2}\big|a(\rho_0/\zeta)\big|^2, \tag{5.15}$$

which (for a Gaussian beam, for example) is exponentially small unless ζ approaches ρ_0, and decays slowly with ζ thereafter.

Although I_{geom} captures some essential features of conical refraction, it fails to describe others. Like all applications of geometrical optics, it neglects interference and polarization, and fails where there are geometrical singularities. Here the singularities are of two kinds: a zero at the anticaustic cylinder $\rho = \rho_0$, and focal divergences on the axial caustic $\rho = 0$ and in the focal image plane $\zeta = 0$.

Interference and polarization can be incorporated by adding the geometrical amplitudes (square roots of Jacobians) rather than intensities, with phases given by the values of $\Phi_\pm$ (from eq. (5.5)) at the contributing $\boldsymbol{\kappa}$ values. We will return to improvements of geometrical-optics theory in Section 7, and compare a more sophisticated version with exact wave theory (see fig. 12 later).

The singularity at $\zeta = 0$ is associated with our choice of $\boldsymbol{\rho}_{\text{i}} = 0$. Thus, although I_{geom} incorporates the effect of w on the angular spectrum of the incident beam, it neglects the more elementary lateral smoothing of the cylinder of conical refraction. One possible remedy is to average I_{geom} over points $\boldsymbol{\rho}_{\text{i}}$, that is, across the incident beam. In the focal image plane, where the unsmoothed I is singular, such averaging would obscure the Poggendorff dark ring. However, averaging over $\boldsymbol{\rho}_{\text{i}}$ is correct only for incoherent illumination, which does not correspond to Lloyd's and later experiments, where the light is spatially coherent. The correct treatment of the focal image plane, and of other features of the observed rings, requires a full wave theory.

§ 6. Belsky and Khapalyuk's exact paraxial theory of conical diffraction

Nearly 40 years after Raman, the need for a full wave treatment, based on an angular superposition of plane waves, was finally appreciated. Following (and correcting) an early attempt by Lalor [1972], Schell and Bloembergen [1978a] supplied such a theory, but this was restricted to the exit face of the crystal (that is, it did not incorporate the ζ dependence of the ring pattern); moreover, it was unnecessarily complicated because it did not exploit the simplifying feature of paraxiality. The breakthrough was provided by Belsky and Khapalyuk [1978], who did make use of paraxiality and gave definitive general formulae, obtained after "quite lengthy calculations", which they omitted. We now give an elementary derivation of the same formulae.

The field $\mathbf{D} = \{D_x, D_y\}$ outside the crystal is a superposition of plane waves $\boldsymbol{\kappa}$, each of which is the result of a unitary 2×2 matrix operator $\mathsf{U}(\boldsymbol{\kappa})$ acting on the initial vector wave amplitude $\mathbf{a}(\boldsymbol{\kappa})$. Thus

$$\mathbf{D} = \frac{1}{2\pi} \iint \mathrm{d}\boldsymbol{\kappa} \exp\{\mathrm{i}\boldsymbol{\kappa} \cdot \boldsymbol{\rho}\} \mathsf{U}(\boldsymbol{\kappa}) \mathbf{a}(\boldsymbol{\kappa}). \tag{6.1}$$

$\mathsf{U}(\boldsymbol{\kappa})$ is determined from two requirements: its eigenphases must be the $\boldsymbol{\rho}$-independent part of the phases $\Phi_\pm$ (5.5), involving the refractive indices $n_\pm(\boldsymbol{\kappa})$, and because the diabolical point at $\boldsymbol{\kappa} = 0$ is a degeneracy, its eigenvectors (the eigenpolarizations) must change sign as ϕ_κ changes by 2π. These are satisfied by

$$\mathsf{U}(\boldsymbol{\kappa}) = \exp\{-\mathrm{i}\mathsf{F}(\boldsymbol{\kappa})\}, \tag{6.2}$$

where

$$\begin{aligned} \mathsf{F}(\boldsymbol{\kappa}) &= \frac{1}{2}\kappa^2\zeta \begin{pmatrix} 1 & 0 \\ 0 & 1 \end{pmatrix} + \rho_0\kappa \begin{pmatrix} \cos\phi_\kappa & \sin\phi_\kappa \\ \sin\phi_\kappa & -\cos\phi_\kappa \end{pmatrix} \\ &= \frac{1}{2}\kappa^2\zeta\mathbf{1} + \rho_0\boldsymbol{\kappa} \cdot \mathsf{S}, \end{aligned} \tag{6.3}$$

in which the compact form involves two of the Pauli spin matrices

$$\boldsymbol{\kappa} \cdot \mathsf{S} = \boldsymbol{\sigma}_3\kappa_x + \boldsymbol{\sigma}_1\kappa_y. \tag{6.4}$$

Evaluating the matrix exponential (6.2) gives the explicit form

$$\mathsf{U}(\boldsymbol{\kappa}) = \exp\left\{-\frac{1}{2}\mathrm{i}\kappa^2\zeta\right\}\left[\cos\rho_0\kappa\mathbf{1} - \mathrm{i}\frac{\sin\rho_0\kappa}{\kappa}\boldsymbol{\kappa} \cdot \mathsf{S}\right]. \tag{6.5}$$

We learned from A. Newell of a connection between the evolution associated with $\mathsf{U}(\boldsymbol{\kappa})$ and analytic functions. Although we do not make use of this connection, it is interesting, and we describe it in Appendix 2.

It is not hard to show that $\mathsf{U}(\boldsymbol{\kappa})$ possesses the required eigenstructure, namely

$$\mathsf{U}(\boldsymbol{\kappa})\mathbf{d}_\pm(\boldsymbol{\kappa}) = \lambda_\pm(\boldsymbol{\kappa})\mathbf{d}_\pm(\boldsymbol{\kappa}), \tag{6.6}$$

where

$$\lambda_\pm(\boldsymbol{\kappa}) = \exp\left\{\mathrm{i}\left(-\frac{1}{2}\kappa^2\zeta \pm \rho_0\kappa\right)\right\},$$
$$\mathbf{d}_+(\boldsymbol{\kappa}) = \begin{pmatrix} \cos\frac{1}{2}\phi_\kappa \\ \sin\frac{1}{2}\phi_\kappa \end{pmatrix}, \qquad \mathbf{d}_-(\boldsymbol{\kappa}) = \begin{pmatrix} \sin\frac{1}{2}\phi_\kappa \\ -\cos\frac{1}{2}\phi_\kappa \end{pmatrix}. \tag{6.7}$$

Without significant loss of generality, we can regard the incident beam as uniformly polarized and circularly symmetric, that is

$$\mathbf{a}(\boldsymbol{\kappa}) = a(\kappa)\begin{pmatrix} d_{0x} \\ d_{0y} \end{pmatrix} = a(\kappa)\mathbf{d}_0, \tag{6.8}$$

where $a(\kappa)$ is the Fourier amplitude introduced in the previous section [cf. eq. (5.13)], related to the transverse incident beam profile $D_0(\rho)$ by

$$\mathbf{D}_0 = \mathbf{d}_0 D_0(\rho) = \mathbf{d}_0 \int_0^\infty \mathrm{d}\kappa\,\kappa J_0(\kappa\rho)a(\kappa). \tag{6.9}$$

Combining eqs. (6.1), (6.5) and (6.9), we obtain, after elementary integrations, and recalling $\boldsymbol{\rho} = \rho\{\cos\phi, \sin\phi\}$,

$$\mathbf{D} = \begin{pmatrix} B_0 + B_1\cos\phi & B_1\sin\phi \\ B_1\sin\phi & B_0 - B_1\cos\phi \end{pmatrix}\mathbf{d}_0, \tag{6.10}$$

where

$$B_0(\rho,\zeta;\rho_0) = \int_0^\infty \mathrm{d}\kappa\,\kappa a(\kappa)\exp\left\{-\frac{1}{2}\mathrm{i}\zeta\kappa^2\right\}J_0(\kappa\rho)\cos(\kappa\rho_0),$$
$$B_1(\rho,\zeta;\rho_0) = \int_0^\infty \mathrm{d}\kappa\,\kappa a(\kappa)\exp\left\{-\frac{1}{2}\mathrm{i}\zeta\kappa^2\right\}J_1(\kappa\rho)\sin(\kappa\rho_0). \tag{6.11}$$

These are the fundamental integrals of the Belsky–Khapalyuk theory.

For unpolarized or circularly polarized incident light, eq. (6.10) gives the intensity as

$$I = \mathbf{D}^* \cdot \mathbf{D} = |B_0|^2 + |B_1|^2. \tag{6.12}$$

A useful alternative form for $\mathbf{D}$, in terms of the eigenvectors $\mathbf{d}_\pm$ evaluated at the direction ϕ of $\boldsymbol{\rho}$, and involving

$$A_+ \equiv B_0 + B_1, \qquad A_- \equiv B_0 - B_1, \tag{6.13}$$

is

$$\begin{aligned}\mathbf{D} = {} & A_+\left(d_{0x}\cos\frac{1}{2}\phi + d_{0y}\sin\frac{1}{2}\phi\right)\mathbf{d}_+(\rho) \\ & + A_-\left(d_{0x}\sin\frac{1}{2}\phi - d_{0y}\cos\frac{1}{2}\phi\right)\mathbf{d}_-(\rho).\end{aligned} \tag{6.14}$$

$\mathbf{D}$ is a single-valued function of $\boldsymbol{\rho}$, although $\mathbf{d}_\pm(\boldsymbol{\rho})$ change sign around the origin. The corresponding intensity,

$$\begin{aligned}I = {} & |A_+|^2\left|d_{0x}\cos\frac{1}{2}\phi + d_{0y}\sin\frac{1}{2}\phi\right|^2 \\ & + |A_-|^2\left|d_{0x}\sin\frac{1}{2}\phi - d_{0y}\cos\frac{1}{2}\phi\right|^2,\end{aligned} \tag{6.15}$$

contains no oscillations resulting from interference between A_+ and A_-, because $\mathbf{d}_+$ and $\mathbf{d}_-$ are orthogonally polarized. But, as we will see in the next section, the exact rings do possess oscillations, from A_+ and A_- individually.

Associated with the polarization structure of the beam (6.14) (see also fig. 6) is an interesting property of the angular momentum (Berry, Jeffrey and Mansuripur [2005]): for well-developed rings (large ρ_0), the initial angular momentum, which is pure spin, is transformed by the crystal into pure orbital, and reduced in magnitude, the difference being imparted to the crystal.

§ 7. Consequences of conical diffraction theory

We begin our explanation of the rich implications of the paraxial wave theory by obtaining an explicit expression for the improved geometrical-optics theory anticipated in Section 5. The derivation proceeds by replacing the Bessel functions in eq. (6.11) by their asymptotic forms, and then evaluating the integrals by their stationary-phase approximations. The stationary values of κ specify the rays (5.9), and the result, for the quantities $A_\pm$, is:

$$\begin{aligned}A_{+\mathrm{geom}} &= \frac{\sqrt{|\rho_0-\rho|}}{\zeta\sqrt{\rho}}a\left(\frac{|\rho_0-\rho|}{\zeta}\right)\exp\left\{\mathrm{i}\frac{(\rho_0-\rho)^2}{2\zeta}\right\}\times\begin{pmatrix}1 \text{ if } \rho<\rho_0 \\ -\mathrm{i} \text{ if } \rho>\rho_0\end{pmatrix}, \\ A_{-\mathrm{geom}} &= -\mathrm{i}\frac{\sqrt{\rho_0+\rho}}{\zeta\sqrt{\rho}}a\left(\frac{\rho_0+\rho}{\zeta}\right)\exp\left\{\mathrm{i}\frac{(\rho_0+\rho)^2}{2\zeta}\right\}.\end{aligned} \tag{7.1}$$

For an incident beam linearly polarized in direction γ, that is

$$\mathbf{d}_0 = \begin{pmatrix}\cos\gamma \\ \sin\gamma\end{pmatrix}, \tag{7.2}$$

the resulting electric field (6.14) is

$$\mathbf{D} = A_{+\mathrm{geom}} \cos\left(\frac{1}{2}\phi - \gamma\right)\mathbf{d}_+(\phi) + A_{-\mathrm{geom}} \sin\left(\frac{1}{2}\phi - \gamma\right)\mathbf{d}_-(\phi), \quad (7.3)$$

and the intensity is

$$\begin{aligned} I_{\mathrm{geom1}} &= \mathbf{D}^* \cdot \mathbf{D} \\ &= \left[|A_{+\mathrm{geom}}|^2 \cos^2\left(\frac{1}{2}\phi - \gamma\right) + |A_{-\mathrm{geom}}|^2 \sin^2\left(\frac{1}{2}\phi - \gamma\right)\right]. \quad (7.4) \end{aligned}$$

The elementary geometrical-optics intensity I_{geom} (5.14) is recovered for unpolarized light by averaging over γ, or by superposing intensities for any two orthogonal incident polarizations. Without such averaging, the first term in I_{geom1} gives the lune-shaped ring structure observed by Lloyd (fig. 6) with linearly polarized light, and both terms combine to give the unpolarized geometrical central spot.

Next, we examine the detailed structure of the Poggendorff rings, starting with the focal image plane $\zeta = 0$ where the rings are most sharply focused. Figure 9

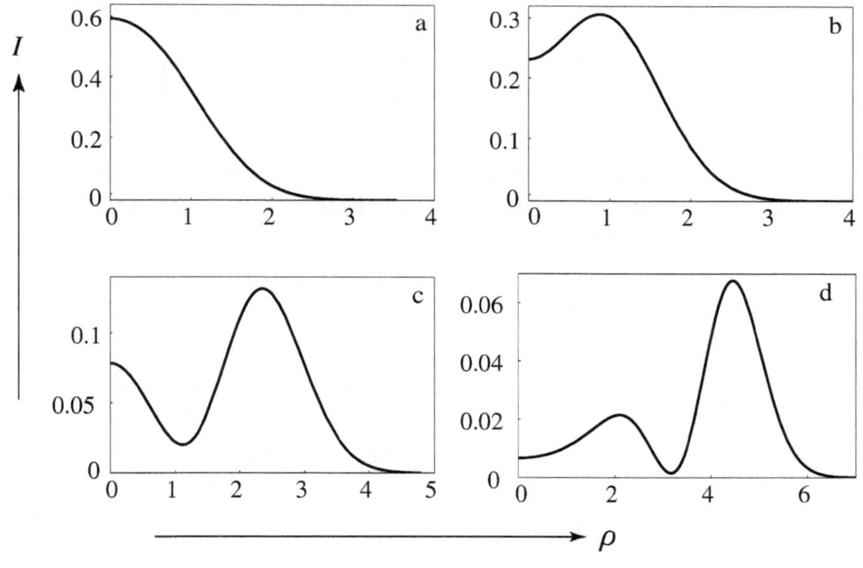

Fig. 9. Emergence of double ring structure in the focal image plane $\zeta = 0$ as ρ_0 increases, computed from eq. (6.10) for an unpolarized Gaussian incident beam. (a) $\rho_0 = 0.5$; (b) $\rho_0 = 0.8$; (c) $\rho_0 = 2$; (d) $\rho_0 = 4$.

shows how the double ring emerges as ρ_0 increases. Significant events, corresponding to sign changes in the curvature of I at $\rho = 0$, are: the birth of the first ring when $\rho_0 = 0.627$; the birth of a central maximum when $\rho_0 = 1.513$; and the birth of the second bright ring when $\rho_0 = 2.669$.

As ρ_0 increases further, the rings become localized near $\rho = \rho_0$, but their shape is independent of ρ_0 and depends only on the form of the incident beam. To obtain a formula for this invariant shape, we cannot use geometrical optics for the large-ρ_0 asymptotics. The reason is that although eq. (7.1) is a good approximation to A_-, it fails for the function A_+ that determines the rings, because the ray contribution comes from the neighbourhood of the end-point $\kappa = 0$ of the integrals. So, although the Bessel functions can still be replaced by their asymptotic forms, the stationary-phase approximation is invalid. Near the rings, with

$$\Delta\rho \equiv \rho - \rho_0, \tag{7.5}$$

Bessel asymptotics gives A_+ as

$$A_{+\text{rings}} = \frac{1}{\sqrt{\rho}} f(\Delta\rho, \zeta), \tag{7.6}$$

where

$$f(\Delta\rho, \zeta) = \sqrt{\frac{2}{\pi}} \int_0^\infty \mathrm{d}\kappa\, \sqrt{\kappa} a(\kappa) \cos\left\{\kappa\Delta\rho - \frac{1}{4}\pi\right\} \exp\left\{-\frac{1}{2}\mathrm{i}\kappa^2\zeta\right\}. \tag{7.7}$$

In the focal image plane $\zeta = 0$, f can be evaluated analytically for a pinhole incident beam, in terms of elliptic integrals E and K. With the conventions in Mathematica (Wolfram [1996]),

$$\begin{aligned} f_{0\mathrm{p}}(\Delta\rho) &\equiv f(\Delta\rho, 0) \\ &= \begin{cases} \dfrac{2\sqrt{2}}{\pi}\left[\sqrt{1-\Delta\rho}\,E\left(\dfrac{2}{1-\Delta\rho}\right) + \dfrac{\Delta\rho}{\sqrt{1-\Delta\rho}} K\left(\dfrac{2}{1-\Delta\rho}\right)\right] \\ \quad (\Delta\rho < -1), \\ \dfrac{2}{\pi}\left[-K\left(\dfrac{1}{2}(1-\Delta\rho)\right) + 2E\left(\dfrac{1}{2}(1-\Delta\rho)\right)\right] \\ \quad (|\Delta\rho| < 1), \\ 0 \quad (\Delta\rho > 1). \end{cases} \end{aligned} \tag{7.8}$$

Figure 10(a) shows the corresponding intensity. To our knowledge, this unusual focused image has not been seen in any experiment.

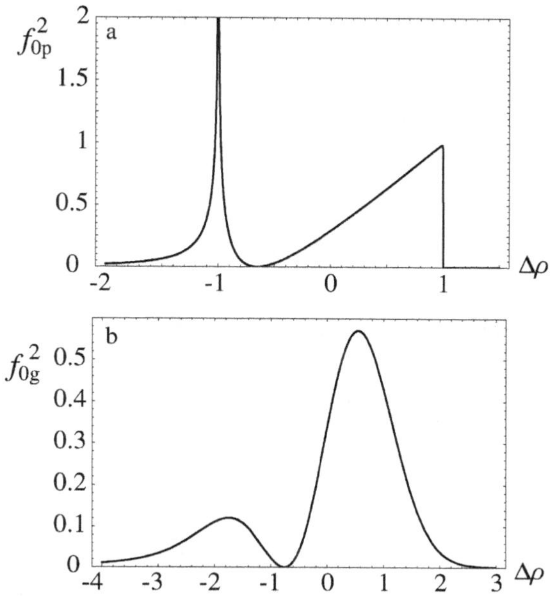

Fig. 10. Intensity functions $f_0^2(\Delta\rho)$ of rings in the focal image plane, for (a) a circular pinhole (7.8), and (b) a Gaussian beam (7.9).

For a Gaussian beam, the focal image can be expressed in terms of Bessel functions:

$$\begin{aligned} f_{0g}(\Delta\rho) &\equiv f(\Delta\rho, 0) \\ &= \frac{|\Delta\rho|^{3/2}\exp(-\frac{1}{4}\Delta\rho^2)}{2\sqrt{2\pi}}\left[K_{\frac{3}{4}}\left(\frac{1}{4}\Delta\rho^2\right) + \operatorname{sgn}\Delta\rho K_{\frac{1}{4}}\left(\frac{1}{4}\Delta\rho^2\right)\right. \\ &\quad \left. + \pi\sqrt{2}\Theta(-\Delta\rho)\left(I_{\frac{3}{4}}\left(\frac{1}{4}\Delta\rho^2\right) - I_{\frac{1}{4}}\left(\frac{1}{4}\Delta\rho^2\right)\right)\right]. \end{aligned} \tag{7.9}$$

Figure 10(b) shows the corresponding intensity, and fig. 11 shows how the approximation f_{0g} gets better as ρ_0 increases. The focal image functions f_{0p} and f_{0g} have been discussed in detail by Belsky and Stepanov [1999], Berry [2004b] and Warnick and Arnold [1997]; of several equivalent representations, eqs. (7.8) and (7.9) are the most convenient.

Away from the focal image plane, that is as ζ increases from zero, secondary rings develop, in the form of oscillations within the inner bright ring [the solid curves in fig. 12(b–e)]; these were discovered in numerical computations by Warnick and Arnold [1997]. Using asymptotics based on eq. (7.7), the secondary

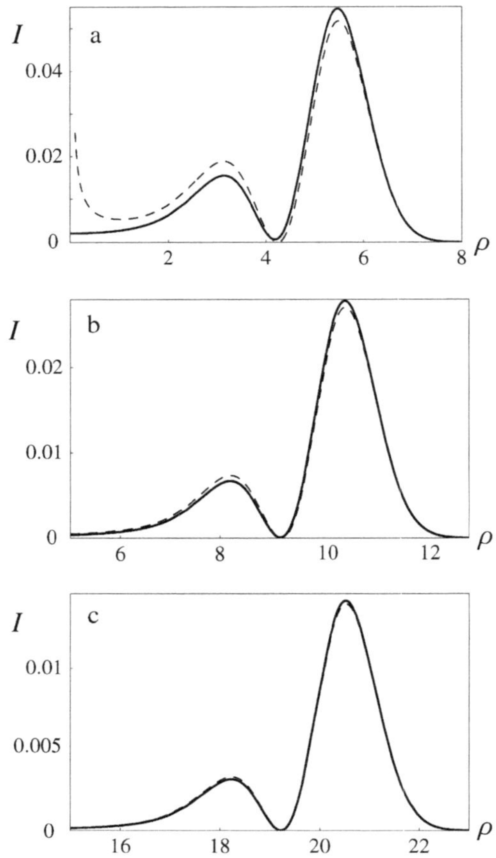

Fig. 11. Ring intensity in the focal image plane for a Gaussian incident beam, calculated exactly (solid curves), and in the local approximation (7.6) and (7.9) (dashed curves), for (a) $\rho_0 = 5$; (b) $\rho_0 = 10$; (c) $\rho_0 = 20$. [The divergences at $\rho = 0$ for the approximate rings come from the factor $1/\sqrt{\rho}$ in eq. (7.6).]

rings can be interpreted as interference between a geometrical ray and a wave scattered from the diabolical point in **k** space (Berry [2004b]); the associated mathematics led to a surprising general observation in the asymptotics of competing exponentials (Berry [2004a]).

All the features so far discussed are displayed in the simulated image and cutaway in fig. 13, which can be regarded as a summary of the main results of conical diffraction theory. The parameters ($\rho_0 = 20$, $\zeta = 8$) are chosen to display the two bright rings, the Poggendorff dark ring, the nascent Raman spot (whose intensity will increase for larger ζ), and the secondary rings.

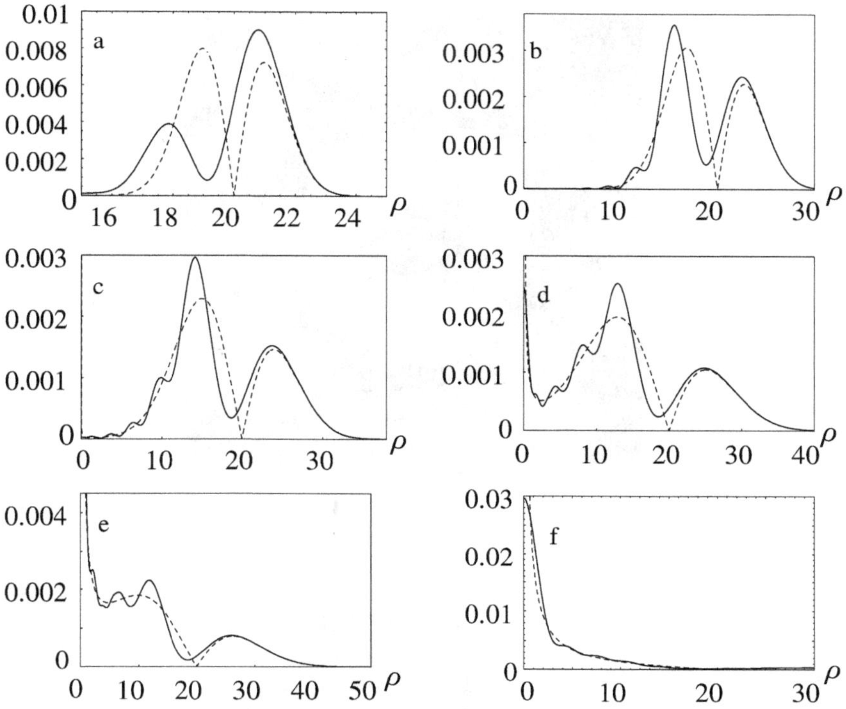

Fig. 12. Intensity for a Gaussian incident beam for $\rho_0 = 20$ and (a) $\zeta = 1$; (b) $\zeta = 4$; (c) $\zeta = 6$; (d) $\zeta = 8$; (e) $\zeta = 10$; (f) $\zeta = 20$. Solid curves: exact theory; dashed curves: refined geometrical-optics theory (7.13).

As ζ increases further, the inner rings approach the Raman spot [solid curves in fig. 12(e,f)], and are then described by Bessel functions (Berry [2004b]):

$$B_0(\rho,\zeta;\rho_0) \approx \rho_0\sqrt{\frac{\pi}{2\zeta^3}}\exp\left\{\mathrm{i}\left(\frac{\rho_0^2}{2\zeta}-\frac{1}{4}\pi\right)a\left(\frac{\rho_0}{\zeta}\right)J_0\left(\frac{\rho\rho_0}{\zeta}\right)\right\},$$

$$B_1(\rho,\zeta;\rho_0) \approx \rho_0\sqrt{\frac{\pi}{2\zeta^3}}\exp\left\{\mathrm{i}\left(\frac{\rho_0^2}{2\zeta}-\frac{3}{4}\pi\right)a\left(\frac{\rho_0}{\zeta}\right)J_1\left(\frac{\rho\rho_0}{\zeta}\right)\right\},$$

$$(\rho \ll \rho_0,\ \zeta \gg 1). \tag{7.10}$$

Figure 14 illustrates how well the approximation $J_0^2 + J_1^2$ approximates the true intensity. The weak oscillations (shoulders at zeros of $J_1(\rho\rho_0/\zeta)$) are the result of interference involving the next large-ρ_0 correction term to the geometrical-optics approximation (7.1).

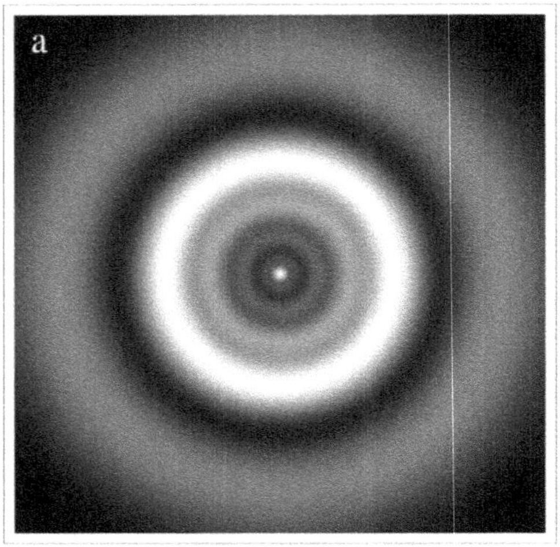

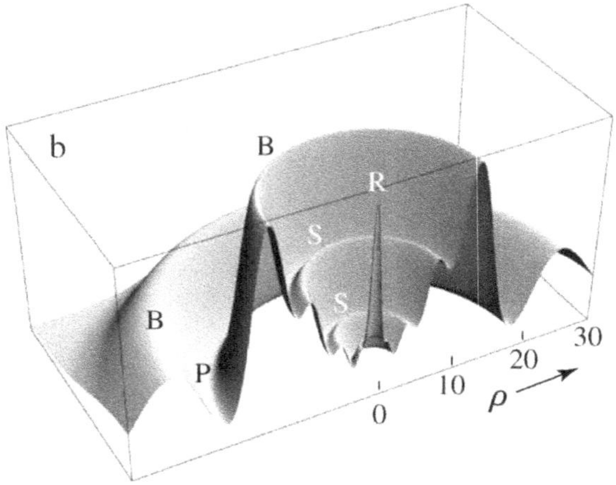

Fig. 13. (a) Density plot and (b) cutaway 3D plot, of conical diffraction intensity for $\rho_0 = 20$, $\zeta = 8$, showing bright rings B, Poggenforff dark ring P, Raman spot R and secondary rings S.

In the important special case of a Gaussian incident beam, the conical diffraction integrals (6.11) for general ζ can be obtained by complex continuation from the focal image plane $\zeta = 0$. This useful simplification is a variant of the complex-source trick of Deschamps [1971]. It is based on the observation

$$\exp\left\{-\frac{1}{2}\kappa^2\right\}\exp\left\{-\frac{1}{2}\mathrm{i}\kappa^2\zeta\right\} = \exp\left\{-\frac{1}{2}\kappa^2(1+\mathrm{i}\zeta)\right\}, \tag{7.11}$$

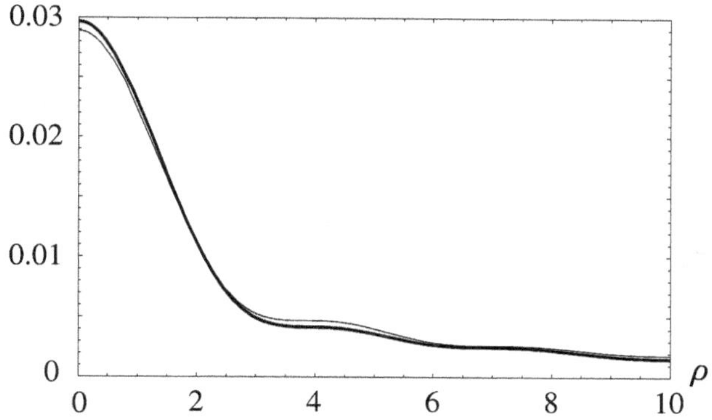

Fig. 14. Weak oscillations decorating the Raman spot for a Gaussian incident beam; thick curve: exact intensity; thin curve: Bessel approximation (7.10), for $\rho_0 = 20$, $\zeta = 20$.

and leads to

$$B_{0,1}(\rho, \zeta; \rho_0) = \frac{1}{1 + \mathrm{i}\zeta} B_{0,1}\left(\frac{\rho}{\sqrt{1 + \mathrm{i}\zeta}}, 0, \frac{\rho_0}{\sqrt{1 + \mathrm{i}\zeta}}\right). \tag{7.12}$$

In geometrical optics, the same trick leads to a further refinement, incorporating $a(\kappa)$ into the stationary-phase approximation (so that the rays are complex):

$$\begin{aligned} A_{+\mathrm{geom1}} &= \frac{\sqrt{|\rho_0 - \rho|}}{(\zeta - \mathrm{i})\sqrt{\rho}} \exp\left\{\mathrm{i}\zeta \frac{(\rho_0 - \rho)^2}{2(\zeta^2 + 1)}\right\} \\ &\quad \times \exp\left\{-\frac{(\rho_0 - \rho)^2}{2(\zeta^2 + 1)}\right\} \times \begin{pmatrix} 1 \text{ if } \rho < \rho_0 \\ -\mathrm{i} \text{ if } \rho > \rho_0 \end{pmatrix}, \\ A_{-\mathrm{geom1}} &= -\mathrm{i}\frac{\sqrt{\rho_0 + \rho}}{(\zeta - \mathrm{i})\sqrt{\rho}} \exp\left\{\mathrm{i}\zeta \frac{(\rho_0 + \rho)^2}{2(\zeta^2 + 1)}\right\} \exp\left\{-\frac{(\rho_0 + \rho)^2}{2(\zeta^2 + 1)}\right\}. \end{aligned} \tag{7.13}$$

Figure 12 shows how accurately this reproduces the oscillation-averaged rings and the Raman spot when ζ is not small. Near the focal plane, however, the geometrical approximation is only rough, even though the refinement eliminates the singularity at $\zeta = 0$.

§ 8. Experiments

Experimental studies of conical refraction are few, probably because of the difficulty of finding, or growing, crystals of sufficient quality and thickness. Table 1 lists the investigations known to us. We have already mentioned the pioneering observations of Lloyd, Poggendorff, Potter and Raman.

Lloyd [1837] stated that he used several pinholes, but gave only the size of the largest (which he used as a way of measuring A), so we do not know the values of w, and hence ρ_0, corresponding to his rings. Poggendorff [1839] gave no details of his experiment, except that it was performed with aragonite. Potter [1841] gave a detailed description of his experiments with aragonite, but misinterpreted his observation of the double ring as evidence against the diabolical connection of the sheets of the wave surface, leading to polemical criticisms of Hamilton and Lloyd. Raman, Rajagopalan and Nedungadi [1941] chose naphthalene, whose crystals have a large cone angle A and which they could grow in sufficient thickness. They emphasize that they did not see the Poggendorff dark ring in their most sharply focused images; but in the focal image plane the two bright rings are very narrow (comparable with $w \approx 0.5$ μm), so they might not have resolved them.

Schell and Bloembergen [1978a] compared measured ring profiles with numerically integrated wave theory and with the stationary-phase (geometrical optics) approximation, in a plane corresponding to the exit face of their aragonite crystal. From the data in Table 1, it follows that this corresponds to $\zeta \approx 1.3$, a regime in which there is no central spot and no secondary rings, and, as they report, geometrical optics is a reasonable approximation.

Perkal'skis and Mikhailychenko [1979] and Mikhailychenko [2005] report large-scale demonstrations of internal and external conical refraction with rhombic sulfur.

In an ingenious investigation, Fève, Boulanger and Marnier [1994] used KTP in the form of a ball rather than a slab. The theory of Section 6 applies, provided the parameters are interpreted as follows:

$$\rho_0 \to \rho_{\text{ball}} = \rho_0\left(1 - 2(n_2 - 1)\frac{d}{l}\right),$$
$$\zeta \to \zeta_{\text{ball}} = \frac{l + d(2 - n_2)}{n_2 k w^2}, \tag{8.1}$$

where l is the diameter of the ball and d is the distance between the exit face of the ball and the observation plane. In this case, a cone emerges from the ball, giving rings whose radius $w\rho_{\text{ball}}$ depends on d; the radius vanishes at a point, close to the ball, where the generators of the cone (that is, the rays) cross. They obtain good agreement between the measured ring profile and the geometrical-optics intensity, which is a good approximation in their regime of enormous effective ρ_0 and relatively modest ζ (Table 1).

The many wave-optical and geometrical-optical phenomena predicted by the detailed theory of conical diffraction have recently been observed by Berry, Jef-

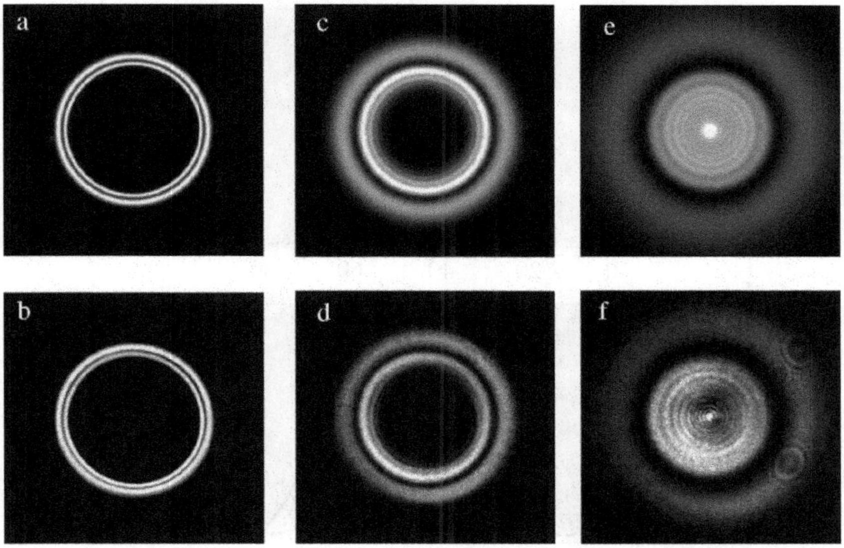

Fig. 15. Theoretical and experimental images for a monoclinic double tungstate crystal illuminated by a Gaussian beam (Berry, Jeffrey and Lunney [2006]), for $\rho_0 = 60$. (a,b) $\zeta = 3$; (c,d) $\zeta = 12$; (e,f) $\zeta = 30$; (a,c,e) theory; (b,d,f) experiment.

frey and Lunney [2006] in a crystal of the monoclinic double tungstate material $KGd(WO_4)_2$, obtained from Vision Crystal Technology (VCT [2006]). The agreement, illustrated in figs. 15 and 16, is quantitative as well as qualitative. Their measurements confirm predictions for the ζ-dependence of the radii and separation of the main rings, and of the sizes of the interference fringes: the secondary rings decorating the inner bright ring, and the rings decorating the Raman spot.

§ 9. Concluding remarks

Although conical diffraction exemplifies a fundamental feature of crystal optics, namely the diabolical point, it can also be regarded as a curiosity, because the effect seems to occur nowhere in the natural universe, and no practical application seems to have been found. To forestall confusion, we should immediately qualify these assertions.

It is Hamilton's idealized geometry (collimated beam, parallel-sided crystal slab, etc.) that does not occur in nature. In generic situations, for example an anisotropic medium that is also inhomogeneous, it is likely for a ray to encounter a

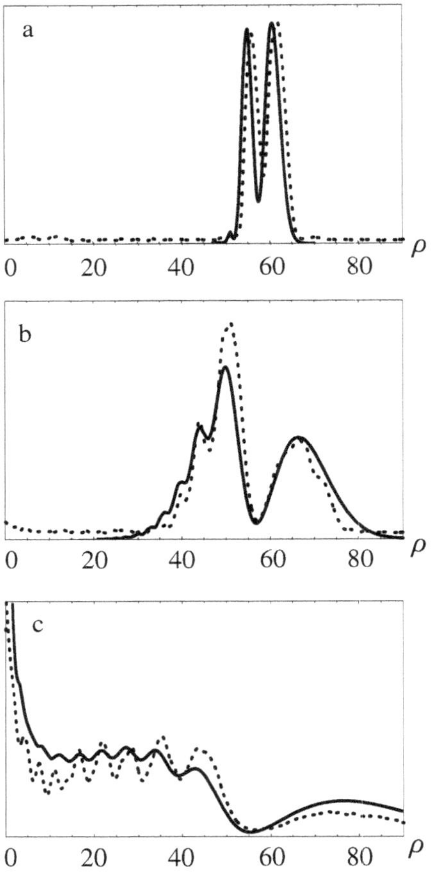

Fig. 16. As fig. 15, with (a) $\zeta = 3$; (b) $\zeta = 12$; (c) $\zeta = 30$. Solid curves: theory; dashed curves: angular averages of experimental images. (After Berry, Jeffrey and Lunney [2006].)

point where its direction is locally diabolical, corresponding to a local refractive-index degeneracy (Naida [1979]); or, in quantum mechanics, a coupled system with fast and slow components (e.g., a molecule) can encounter a degeneracy of the adiabatically evolving fast sub-system. In understanding such generic situations, the analysis of the idealized case will surely play a major part.

On the practical side, it is possible that the bright cylinder of conical refraction can be applied to trap and manipulate small particles; this is being explored, but the outcome is not yet clear. And a related phenomenon associated with the diabolical point, namely the conoscopic interference figures seen under polarized illumination of very thin crystal plates, is a well-established identification technique in mineralogy (Liebisch [1896]).

As we have tried to explain, our understanding of the effect predicted by Hamilton is essentially complete. We end by describing several generalizations that are less well understood.

Although the theory of Section 6 applies to any paraxial incident beam, detailed explorations have been restricted to pinhole and Gaussian beams. A start has been made by King, Hogervorst, Kazak, Khilo and Ryzhevich [2001] and Stepanov [2002] in the exploration of other types of beam, for example Laguerre–Gauss and Bessel beams.

A further extension is to materials that are chiral (optically active) as well as biaxially birefringent. This alters the mathematical framework, because chirality destroys the diabolical point by separating the two sheets of the wave surface, reflecting the change of the dielectric matrix from real symmetric to complex Hermitian. There have been several studies incorporating chirality (Belsky and Stepanov [2002], Schell and Bloembergen [1978b], Voigt [1905b]), leading to the recent identification of the central new feature: the bright cylinder of conical refraction is replaced by a 'spun cusp' caustic (Berry and Jeffrey [2006a]). So far this has not been seen in any experiment.

The introduction of anisotropic absorption (dichroism) brings a more radical change: each diabolical point splits into two branch-points, reflecting the fact that the dielectric matrix is now non-Hermitian (Berry [2004c], Berry and Dennis [2003]). The dramatic effects of absorption on the pattern of emerging light have been recently described by Berry and Jeffrey [2006b]. The combined effects of dichroism and chirality have been described by Jeffrey [2007].

The final generalization incorporates nonlinearity. Early results were reported by Schell and Bloembergen [1977] and Shih and Bloembergen [1969], and the subject has been revisited by Indik and Newell [2006].

Acknowledgements

Our research is supported by the Royal Society. We thank Professor Alan Newell for permission to reproduce the argument in Appendix 2.

Appendix 1: Paraxiality

The paraxial approximation requires small angles, equivalent to replacing $\cos\theta$ by $1 - \theta^2/2$, that is, to assuming $\theta^4/24 \ll 1$ for all wave deflection angles θ. In conical refraction, deflections are determined by the half-angle of the ray cone, which from eq. (3.7) is

$$A = \frac{1}{2}\arctan\left(n_2^2\sqrt{\alpha\beta}\right). \tag{A.1}$$

This is indeed small in practice, because of the near-equality of the three refractive indices. For the Lloyd [1837] experiment on aragonite, the Berry, Jeffrey and Lunney [2006] experiment on MDT, and the Raman, Rajagopalan and Nedungadi [1941] experiment on naphthalene (whose cone angle is the largest yet reported), the data in Table 1 give

$$\begin{aligned} \frac{1}{24}A_{\text{Lloyd}}^4 &= 3.3\times10^{-9}, \\ \frac{1}{24}A_{\text{Berry}}^4 &= 9.1\times10^{-9}, \\ \frac{1}{24}A_{\text{Raman}}^4 &= 8.5\times10^{-6}. \end{aligned} \tag{A.2}$$

As is often emphasized, internal and external conical refraction are associated with different aspects of the geometry of the wave surface. But in the paraxial regime the difference between the cone angles (A and A_{ext} respectively) disappears. To explore this, we first note that (Born and Wolf [1999])

$$A_{\text{ext}} = \frac{1}{2}\arctan\left(n_1 n_3\sqrt{\alpha\beta}\right). \tag{A.3}$$

In terms of the refractive-index differences

$$\mu_1 \equiv \frac{n_2-n_1}{n_2}, \qquad \mu_3 \equiv \frac{n_3-n_2}{n_2}, \tag{A.4}$$

we have, to lowest order,

$$A \approx A_{\text{ext}} \approx \sqrt{\mu_1\mu_3}, \tag{A.5}$$

which is proportional to the refractive-index differences. The difference between the angles is

$$A - A_{\text{ext}} \approx \sqrt{\mu_1\mu_3}\left[\mu_1 - \mu_3 + \frac{1}{4}\left(3\mu_1^2 - 2\mu_1\mu_3 + 3\mu_3^2\right)\right], \tag{A.6}$$

which is proportional to the square of the index differences, except when the two differences are equal, when it is proportional to the cube. For the aragonite, MDT and naphthalene experiments,

$$\begin{aligned} A_{\text{Lloyd}} - A_{\text{ext,Lloyd}} &= 8.9\times10^{-2}A_{\text{Lloyd}}, \\ A_{\text{Berry}} - A_{\text{ext,Berry}} &= -4.4\times10^{-3}A_{\text{Berry}}, \\ A_{\text{Raman}} - A_{\text{ext,Raman}} &= -2.7\times10^{-4}A_{\text{Raman}}. \end{aligned} \tag{A.7}$$

Appendix 2: Conical refraction and analyticity

We seek the paraxial differential equation for the evolution of the wave inside the crystal. Within the framework of Section 6, the wave can be described by setting $z = l$ and regarding $\zeta = l/k_0 w^2$ as a variable, enabling the evolution operator (6.2) and (6.3) to be written as

$$\mathbf{U}(\kappa) = \exp\left\{-\mathrm{i}\zeta\left(\frac{1}{2}\kappa^2\mathbf{1} + \Gamma\kappa\cdot\mathbf{S}\right)\right\}, \tag{B.1}$$

where

$$\Gamma \equiv Ak_0 w. \tag{B.2}$$

Writing $\boldsymbol{\kappa}$ as the differential operator

$$\boldsymbol{\kappa} = -\mathrm{i}\nabla = -\mathrm{i}\{\partial_\xi, \partial_\eta\}, \tag{B.3}$$

and differentiating eq. (B.1) with respect to ζ leads to

$$\mathrm{i}\partial_\zeta\begin{pmatrix} D_\xi \\ D_\eta \end{pmatrix} = \left[-\frac{1}{2}\nabla^2\begin{pmatrix} 1 & 0 \\ 0 & 1 \end{pmatrix} - \mathrm{i}\Gamma\begin{pmatrix} \partial_\xi & \partial_\eta \\ \partial_\eta & -\partial_\xi \end{pmatrix}\right]\begin{pmatrix} D_\xi \\ D_\eta \end{pmatrix}. \tag{B.4}$$

The differential operator connects the components of **D** in the same way as the Cauchy–Riemann conditions for analytic functions. The relation is clearer when **D** is expressed in a basis of circular polarizations and ξ and η are replaced by complex variables, as follows

$$D_+ \equiv \frac{1}{\sqrt{2}}(D_\xi - \mathrm{i}D_\eta), \qquad D_- \equiv \frac{1}{\sqrt{2}}(D_\xi + \mathrm{i}D_\eta),$$
$$w_+ \equiv \xi + \mathrm{i}\eta, \qquad w_- \equiv \xi - \mathrm{i}\eta. \tag{B.5}$$

Then eq. (B.4) becomes

$$\mathrm{i}\partial_\zeta\begin{pmatrix} D_+ \\ D_- \end{pmatrix} = -2\left[\partial_{w_+}\partial_{w_-}\begin{pmatrix} 1 & 0 \\ 0 & 1 \end{pmatrix} + \mathrm{i}\Gamma\begin{pmatrix} 0 & \partial_{w_+} \\ \partial_{w_-} & 0 \end{pmatrix}\right]\begin{pmatrix} D_+ \\ D_- \end{pmatrix}. \tag{B.6}$$

Thus D_+ and D_- propagate unchanged if D_+ is a function of w_+ alone and D_- is a function of w_- alone, that is if the field components are analytic or anti-analytic functions. For such fields, there is no conical refraction, and (for example) a pattern of zeros in the incident field propagates not conically but as a set of straight optical vortex lines parallel to the ζ direction. However, these analytic functions do not represent realistic optical beams, which must decay in all directions ϕ as $\rho \to \infty$.

References

Belsky, A.M., Khapalyuk, A.P., 1978, Internal conical refraction of bounded light beams in biaxial crystals, Opt. Spectrosc. (USSR) **44**, 436–439.

Belsky, A.M., Stepanov, M.A., 1999, Internal conical refraction of coherent light beams, Opt. Commun. **167**, 1–5.

Belsky, A.M., Stepanov, M.A., 2002, Internal conical refraction of light beams in biaxial gyrotropic crystals, Opt. Commun. **204**, 1–6.

Berry, M.V., 1983, Semiclassical Mechanics of regular and irregular motion, in: Iooss, G., Helleman, R.H.G., Stora, R. (Eds.), Les Houches Lecture Series, vol. **36**, North-Holland, Amsterdam, pp. 171–271.

Berry, M.V., 1984, Quantal phase factors accompanying adiabatic changes, Proc. Roy. Soc. Lond. A **392**, 45–57.

Berry, M.V., 2001, Geometry of phase and polarization singularities, illustrated by edge diffraction and the tides, in: Soskin, M. (Ed.), Singular Optics 2000, vol. **4403**, SPIE, Alushta, Crimea, pp. 1–12.

Berry, M.V., 2004a, Asymptotic dominance by subdominant exponentials, Proc. Roy. Soc. A **460**, 2629–2636.

Berry, M.V., 2004b, Conical diffraction asymptotics: fine structure of Poggendorff rings and axial spike, J. Optics A **6**, 289–300.

Berry, M.V., 2004c, Physics of nonhermitian degeneracies, Czech. J. Phys. **54**, 1040–1047.

Berry, M.V., Dennis, M.R., 2003, The optical singularities of birefringent dichroic chiral crystals, Proc. Roy. Soc. A **459**, 1261–1292.

Berry, M.V., Dennis, M.R., Lee, R.L.J., 2004, Polarization singularities in the clear sky, New J. of Phys. **6**, 162.

Berry, M.V., Jeffrey, M.R., 2006a, Chiral conical diffraction, J. Opt. A **8**, 363–372.

Berry, M.V., Jeffrey, M.R., 2006b, Conical diffraction complexified: dichroism and the transition to double refraction, J. Opt. A **8**, 1043–1051.

Berry, M.V., Jeffrey, M.R., Lunney, J.G., 2006, Conical diffraction: observations and theory, Proc. R. Soc. A **462**, 1629–1642.

Berry, M.V., Jeffrey, M.R., Mansuripur, M., 2005, Orbital and spin angular momentum in conical diffraction, J. Optics A **7**, 685–690.

Berry, M.V., Wilkinson, M., 1984, Diabolical points in the spectra of triangles, Proc. Roy. Soc. Lond. A **392**, 15–43.

Born, M., Wolf, E., 1999, Principles of Optics, seventh ed., Pergamon, London.

Cederbaum, L.S., Friedman, R.S., Ryaboy, V.M., Moiseyev, N., 2003, Conical intersections and bound molecular states embedded in the continuum, Phys. Rev. Lett. **90**, 013001-1-4.

Chu, S.Y., Rasmussen, J.O., Stoyer, M.A., Canto, L.F., Donangelo, R., Ring, P., 1995, Form factors for two-neutron transfer in the diabolical region of rotating nuclei, Phys. Rev. C **52**, 685–696.

Deschamps, G.A., 1971, Gaussian beam as a bundle of complex rays, Electronics Lett. **7**, 684–685.

Ferretti, A., Lami, A., Villani, G., 1999, Transition probability due to a conical intersection: on the role of the initial conditions of the geometric setup of the crossing surfaces, J. Chem. Phys. **111**, 916–922.

Fève, J.P., Boulanger, B., Marnier, G., 1994, Experimental study of internal and external conical refraction in KTP, Opt. Commun. **105**, 243–252.

Graves, R.P., 1882, Life of Sir William Rowan Hamilton, vol. **1**, Hodges Figgis, Dublin.

Hamilton, W.R., 1837, Third Supplement to an Essay on the Theory of Systems of Rays, Trans. Royal Irish Acad. **17**, 1–144.

Herzberg, G., Longuet-Higgins, H.C., 1963, Intersection of potential-energy surfaces in polyatomic molecules, Disc. Far. Soc. **35**, 77–82.

Indik, R.A., Newell, A.C., 2006, Conical refraction and nonlinearity, Optics Express **14**, 10614–10620.

Jeffrey, M.R., 2007, The spun cusp complexified: complex ray focusing in chiral conical diffraction, J. Opt. A. **9**, 634–641.

King, T.A., Hogervorst, W., Kazak, N.S., Khilo, N.A., Ryzhevich, A.A., 2001, Formation of higher-order Bessel light beams in biaxial crystals, Opt. Commun. **187**, 407–414.

Lalor, E., 1972, An analytical approach to the theory of internal conical refraction, J. Math. Phys. **13**, 449–454.

Landau, L.D., Lifshitz, E.M., Pitaevskii, L.P., 1984, Electrodynamics of Continuous Media, Pergamon, Oxford.

Liebisch, T., 1896, Grundriss der Physikalischen Krystallographie, Von Veit & Comp, Leipzig.

Lloyd, H., 1837, On the phenomena presented by light in its passage along the axes of biaxial crystals, Trans. R. Ir. Acad. **17**, 145–158.

Ludwig, D., 1961, Conical refraction in crystal optics and hydromagnetics, Comm. Pure. App. Math. **14**, 113–124.

Lunney, J.G., Weaire, D., 2006, The ins and outs of conical refraction, Europhysics News **37**, 26–29.

Mead, C.A., Truhlar, D.G., 1979, On the determination of Born–Oppenheimer nuclear motion wave functions including complications due to conical intersections and identical nuclei, J. Chem. Phys. **70**, 2284–2296.

Mikhailychenko, Y.P., 2005, Large scale demonstrations on conical refraction, http://www.demophys.tsu.ru/Original/Hamilton/Hamilton.html.

Naida, O.N., 1979, "Tangential" conical refraction in a three-dimensional inhomogeneous weakly anisotropic medium, Sov. Phys. JETP **50**, 239–245.

Newell, A., 2005. Private communication.

Nussenzveig, H.M., 1992, Diffraction Effects in Semiclassical Scattering, University Press, Cambridge.

Nye, J.F., 1999, Natural Focusing and Fine Structure of Light: Caustics and Wave Dislocations, Institute of Physics Publishing, Bristol.

O'Hara, J.G., 1982, The prediction and discovery of conical refraction by William Rowan Hamilton and Humphrey Lloyd (1832–1833), Proc. R. Ir. Acad. **82A**, 231–257.

Perkal'skis, B.S., Mikhailychenko, Y.P., 1979, Demonstration of conical refraction, Izv. Vyss. Uch. Zav. Fiz. **8**, 103–105.

Poggendorff, J.C., 1839, Ueber die konische Refraction, Pogg. Ann. **48**, 461–462.

Potter, R., 1841, An examination of the phaenomena of conical refraction in biaxial crystals, Phil. Mag. **8**, 343–353.

Raman, C.V., 1941, Conical refraction in naphthalene crystals, Nature **147**, 268.

Raman, C.V., Rajagopalan, V.S., Nedungadi, T.M.K., 1941, Conical refraction in naphthalene crystals, Proc. Ind. Acad. Sci. A **14**, 221–227.

Schell, A.J., Bloembergen, N., 1977, Second harmonic conical refraction, Opt. Commun. **21**, 150–153.

Schell, A.J., Bloembergen, N., 1978a, Laser studies of internal conical diffraction. I. Quantitative comparison of experimental and theoretical conical intensity dirstribution in aragonite, J. Opt. Soc. Amer. **68**, 1093–1098.

Schell, A.J., Bloembergen, N., 1978b, Laser studies of internal conical diffraction. II. Intensity patterns in an optically active crystal, α-iodic acid, J. Opt. Soc. Amer. **68**, 1098–1106.

Shih, H., Bloembergen, N., 1969, Conical refraction in second harmonic generation, Phys. Rev. **184**, 895–904.

Silverman, M., 1980, The curious problem of spinor rotation, Eur. J. Phys. **1**, 116–123.

Soskin, M.S., Vasnetsov, M.V., 2001, Singular optics, in: Progress in Optics, vol. **42**, North-Holland, Amsterdam, pp. 219–276.

Stepanov, M.A., 2002, Transformation of Bessel beams under internal conical refraction, Opt. Commun. **212**, 11–16.

Uhlenbeck, K., 1976, Generic properties of eigenfunctions, Am. J. Math. **98**, 1059–1078.
Uhlmann, A., 1982, Light intensity distribution in conical refraction, Commn. Pure. App. Math. **35**, 69–80.
van de Hulst, H.C., 1981, Light Scattering by Small Particles, Dover, New York.
VCT, 2006, Home page of Vision Crystal Technology, http://www.vct-ag.com/.
Voigt, W., 1905a, Bemerkung zur Theorie der konischen Refraktion, Phys. Z. **6**, 672–673.
Voigt, W., 1905b, Theoretisches unt Experimentelles zur Aufklärung des optisches Verhaltens aktiver Kristalle, Ann. Phys. **18**, 645–694.
Von Neumann, J., Wigner, E., 1929, On the behavior of eigenvalues in adiabatic processes, Phys. Z. **30**, 467–470.
Warnick, K.F., Arnold, D.V., 1997, Secondary dark rings of internal conical refraction, Phys. Rev. E **55**, 6092–6096.
Wolfram, S., 1996, The Mathematica Book, University Press, Cambridge.

It is 200 years since Thomas Young performed his famous double-slit experiment but the interference of waves that weave rich tapestries in space and space–time continues to provide deep insights into geometrical optics and semi-classical limits

Quantum carpets, carpets of light

Michael Berry, Irene Marzoli and Wolfgang Schleich

IN 1836 Henry Fox Talbot, an inventor of photography, published the results of some experiments in optics that he had previously demonstrated at a British Association meeting in Bristol (figure 1*a*). "It was very curious to observe that though the grating was greatly out of the focus of the lens…the appearance of the bands was perfectly distinct and well defined…the experiments are communicated in the hope that they may prove interesting to the cultivators of optical science."

1 Talbot and his diffraction effect

(*a*) Henry Fox Talbot (1800–1877) (Picture courtesy of The National Trust and the Fox Talbot Museum and kindly supplied by Michael Gray). (*b*) The Talbot effect from a Ronchi grating – a grating that has equally spaced transparent and opaque slits. The intensity pattern at the grating ($z = 0$) is reconstructed at the Talbot distance $z = z_T$ but is shifted by half a period. The images that can be viewed with the aid of a lens (not shown) occur at intermediate distances given by rational fractions p/q of z_T. These "fractional Talbot images" consist of q overlapping copies of the grating; several values of p/q are illustrated here. Neighbouring images are shifted by a/q, and coherently superposed with phases given by the Gauss sums of number theory (see box).

Talbot made his remarkable observation as he inspected a coarsely ruled diffraction grating, illuminated with white light, through a magnifying lens. With the lens held close to the grating, the rulings appeared in sharp focus, as expected. Moving the lens away, so that the grating was no longer in focus, the images should have become blurred but instead remained sharp. Moreover, the images consisted of alternating bands, the complementary colours (e.g. red and green) of which depended on the distance of the lens from the grating.

As the lens receded further, the sequence of colours repeated several times over a distance of the order of metres. With monochromatic light, the rulings of the grating first appeared blurred, as expected, but then mysteriously reappeared in sharp focus at multiples of a particular distance, z_T (figure 1*b*).

Forgotten for decades

This "Talbot effect" – the repeated self-imaging of a diffraction grating – was forgotten for nearly half a century until it was rediscovered by Lord Rayleigh in 1881. Rayleigh showed that the so-called Talbot distance, z_T, is given by a^2/λ, where a is the slit spacing of the grating and λ is the wavelength of the light. For red light with a wavelength of 632.8 nm and a grating with 50 slits per inch ($a = 0.508$ mm) the Talbot distance is 407.8 mm. Rayleigh pointed out that the Talbot effect could be employed practically to reproduce diffraction gratings of different sizes by exposing photographic film to a Talbot-reconstructed image of an original grating. Then the Talbot effect was forgotten again.

2 Quantum revivals

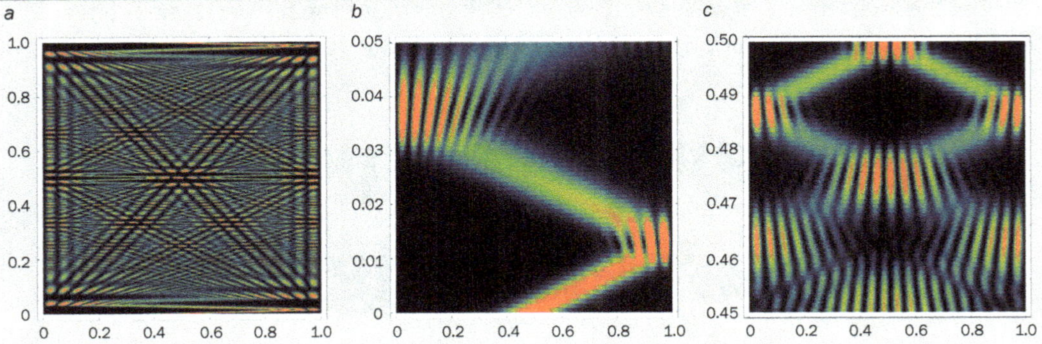

Quantum carpets, i.e. plots of probability density, for the propagation of a Gaussian wavepacket in a 1-D box with length $x = 1$. The horizontal axes are the coordinate x and the vertical axes are t/T_r where t is time and T_r is the revival time. (*a*) One period of the quantum carpet; (*b*) magnification of (*a*) for short times, showing the initial packet bouncing in the box as it spreads; (c) magnification near $T_r/2$, showing fractional revival of two copies of the initial packet. The intensity increases with brightness and saturation, and is colour-coded with hue (maxima represented by red). These carpets are calculated using the equation in the box, for $w = 1/10$, $k = 20$ (thus giving a classical period $T_c = T_r/20$).

Recent investigations have revealed that the Talbot effect is far more than a mere optical curiosity. First, the effect is one of a class of phenomena involving the extreme coherent interference of waves. Second, these phenomena have deep and unexpected roots in classical number theory. And third, they illustrate something we are starting to appreciate more and more, namely the rich and intricate structure of limits in physics.

And who was Talbot? He is best known for his invention of photography, independently of French physicist Louis-Jacques-Mandé Daguerre, rather than the physics of diffraction gratings. And it is Talbot's process, involving a negative from which many positive copies can be made, that we use in all non-digital photography today. But Talbot did much more: as an English landowner and parliamentarian, he ran the family estate at Lacock; he was one of the inventors of the polarizing microscope that has been indispensable to mineralogists; he proved theorems about elliptic integrals; and he contributed to the transcription of Syrian and Chaldean inscriptions.

Quantum revivals

There is a phenomenon in quantum physics that is closely related to the Talbot effect. A quantum wave packet – representing an electron in an atom, for example – can be constructed from a superposition of highly excited stationary states so that it is localized near a point on the electron's classical orbit. If the packet is released, it starts to propagate around the orbit. This propagation is guaranteed by the correspondence principle: for highly excited states, quantum and classical physics must agree. The packet then spreads along the orbit, and eventually fills it. (It also spreads transversely – that is away from the orbit – but that is not important in this context.)

Over very long periods of time, however, something extraordinary happens: the wave packet contracts and after a time, T_r, returns from the dead and reconstructs its initial form. This is a quantum revival. As time goes on, the revivals repeat. In a wide class of circumstances, the reconstructions are almost perfect.

In the Talbot effect, the pattern of intensity is doubly periodic, both across the grating and as a function of distance from the grating. Similarly, for quantum revivals the probability density is both periodic around the nucleus and as a function of the time evolution of the wave packet. These periodicities are the result of coherent interference. Rayleigh understood this for gratings (where the interference is between the wavelets from the different slits), and there is a similar result for revivals. For a packet consisting of hydrogen-like states close to the nth energy level, E_n, the revival time is of the order of nT_c, where $T_c = hn^3/2E_1$ is the orbital period of a classical electron and h is Planck's constant. Because of these periodicities, the intensities can be depicted as repeating patterns in the plane that resemble exotic carpets or alien landscapes (figures 2 and 3).

Arithmetic

Now comes the number theory. In addition to the reconstructed Talbot images at $z = z_T$ and quantum revivals at $t = T_r$, there are more complicated reconstructions at fractional multiples of the Talbot distance and the revival time. The Talbot image at distance $(p/q)z_T$ is a superposition of q copies of the initial grating separated by a/q (figure 1*b*). Thus if $p/q = 3/5$, there are five superposed images. This is the *fractional Talbot effect*. And the quantum wave at time $(p/q)T_r$ is a superposition of q copies of the initial wave packet, separated by an angular distance $2\pi/q$. These are the *fractional revivals*.

At the heart of the rather complicated mathematical explanation of this remarkable phenomenon is the simple identity $(-1)^{n^2} = (-1)^n$, where n is an integer that labels the contributing wavelets. The application of this identity greatly simplifies the sums over the contributing wavelets. Each image in a reconstruction has the same amplitude, $1/\sqrt{q}$, and its phase is given by beautiful sums discovered by Gauss at the end of the 18th century (see box). In words, the phase of the nth image is the direction of the resultant vector when q unit vectors are added in the complex plane. The angle between the first pair of vectors is $2\pi n/q$ (plus a constant that does not depend on n) and decreases by $2\pi p/q$ for each successive pair. The rich patterns of the carpets originate in the complicated patterns of

3 Talbot carpets and mountains

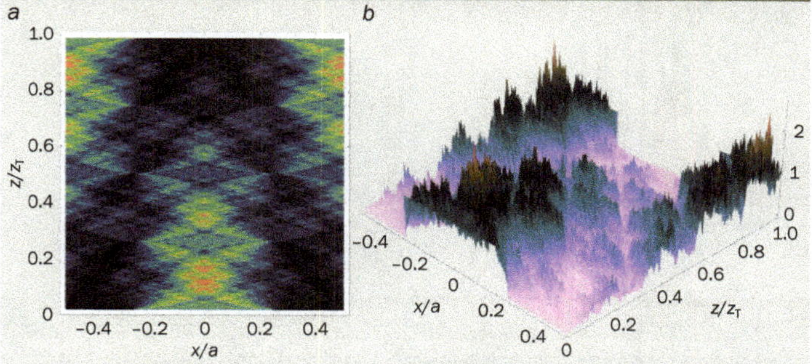

(*a*) A carpet of light from a Ronchi grating. A density plot of Talbot intensity in a single period transversely (across the grating) and longitudinally (along the direction of the incident light). The colour coding is the same as in figure 2. Note that the grating and associated interference patterns have reappeared at $z = z_T$ but shifted by a half-period. (*b*) The mountains of Talbot. A visualization of the intensity in (*a*) as a surface.

4 Gauss sums

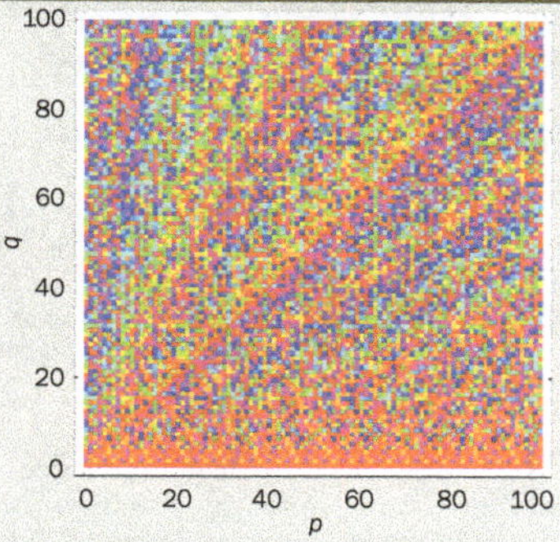

Phases of Gauss sums for the $n = 2$ Talbot image (see box), colour-coded by hue. Red represents a phase of zero, while light blue represents phases of $\pm\pi$.

the phases of the Gauss sums (figure 4).

On closer inspection, diagonal "canals" can be seen criss-crossing the carpets. In Talbot carpets, the canals are minima of intensity that are correlated between neighbouring image planes. In quantum carpets, they are space–time structures corresponding to features of the wave that move at fixed velocities (unrelated to the speed of the classical packet). The canals can be understood in several equivalent ways: in terms of destructive interference between positive- and negative-order diffracted beams (that is Fourier components of the patterns); by the transverse structures that remain after averaging along the anticipated directions of the canals; and by arguments involving position and momentum simultaneously (through the evolution of Wigner functions in phase space).

Fractal carpets

Both Talbot and quantum carpets can have intricate fine structure. Consider the case where the transparency of the grating is discontinuous, as it is in the Ronchi grating where the rulings consist of sharp opaque and transparent bars. Alternatively imagine that the quantum wave is discontinuous, as for a particle in a box where the initial wave amplitude plummets to zero at the walls, even though it varies smoothly inside the box.

In Talbot terminology the reconstructed images in the plane $z = (p/q)z_T$ consist of q superposed copies of the grating, complete with discontinuities. Although there is an infinite number of images at fractional distances, they still represent an infinitesimal subset of all possible images.

What happens in planes that are irrational fractions of the Talbot distance, for example $z_T/\sqrt{2}$? The light intensity is then a *fractal function* of distance x across the image. To be precise, this means that the graph of intensity as a function of x, regarded as a curve in the plane, is continuous but not differentiable. In other words, although the intensity has a definite value at every point, the curve has no definite slope and so is infinitely jagged. Such curves are described as having a "fractal dimension", D, between one (corresponding to a smooth curve) and two (corresponding to a curve so irregular that it occupies a finite area). Carpets generated by Ronchi gratings, for example, have a fractal dimension of $3/2$ (figure 5).

Carpets are functions of the longitudinal coordinate z (along the propagation direction) as well as the transverse coordinate x, so the wave intensities can be plotted as surfaces on the x–z plane (see figure 3*b*, for example).

What is the fractal dimension of these Talbot or quantum landscapes? As these landscapes are surfaces, they have a dimension that takes a value between two and three. For a surface where all directions are equivalent, the dimension of the surface is one greater than the dimension, D, of a curve that cuts through it. For a Talbot landscape, where the fractal image curves have $D = 3/2$, we might expect the dimension to be $1 + 3/2 = 5/2$.

But Talbot landscapes are not isotropic. For fixed x, the intensity also varies as a function of distance z from the grating, with a fractal dimension of $7/4$ (figure 5). Therefore the longitudinal fractals are more irregular than the transverse ones. Finally, the intensity is more regular along the diagonal canals because of the cancellation of large Fourier components and has fractal dimension of $5/4$ (figure 5). The landscape is dominated by the largest of these fractal dimensions, and so is a surface with dimension $1 + 7/4 = 11/4$.

This fractal structure also applies to the evolving wave from a particle in a box with uniform initial wave amplitude. There are space fractals (varying x with constant t), where the graph of the probability density has $D = 3/2$; time fractals (varying t with constant x), where the graph of the probability density has $D = 7/4$; and space–time fractals (along the diagonal canals), where $D = 5/4$. Some of these results about fractals can be generalized to waves evolving in enclosures of any dimensions, even with shapes that do not give rise to revivals, provided the initial state is discontinuous, even if only at the boundary.

Asymptotic physics

There is, however, a sense in which the carpets – at least as we have described them – are fictions. Like all calculations in physics, they rely on approximations that are valid only in certain limiting circumstances. For a start, the perfect Talbot reconstructions, and the ideal fractals in the case of gratings with sharp-edged slits, depend on the illuminating wave being perfectly plane and infinite in extent.

5 Fractal Talbot intensities

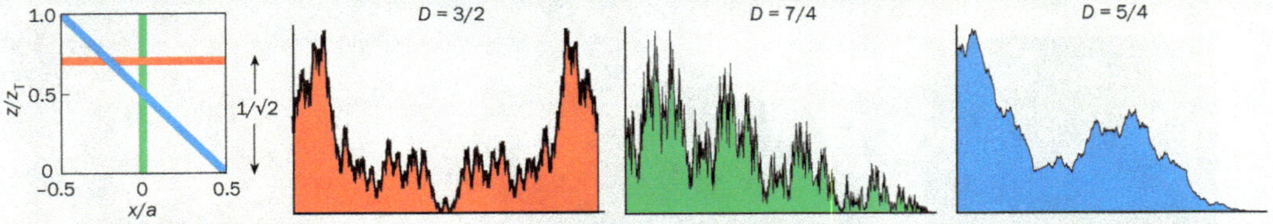

Talbot fractals from a Ronchi grating. The graphs of intensity have fractional dimensions D. Transverse fractals have $D = 3/2$ (in planes of constant z that are separated from the grating by distances that are irrational fractions of z_T). Longitudinal fractals have $D = 7/4$ (in sections where x is constant), while diagonal fractals have $D = 5/4$.

Formulas used to calculate quantum revivals and carpets

- Quantum revivals (figure 2): for an initial packet with width w starting from $x = 1/2$ and with momentum k, the wave is (up to a constant factor)

$$\psi = \sum_{l=-\infty}^{\infty} (-\mathrm{i})^l \sin(\pi l x) \exp[-\tfrac{1}{2}\pi^2 w^2 (l-k)^2 - \mathrm{i}\pi l^2 t/T_r]$$

- Talbot carpet (figures 1*b*, 3 and 5): the wave from a Ronchi grating is

$$\psi = \tfrac{1}{2} + (2/\pi) \sum_{k=0}^{\infty} (-1)^k \cos[2\pi(2k+1)x/a] \exp[-\mathrm{i}\pi(2k+1)^2 z/z_T]/(2k+1)$$

- Talbot carpet (figure 6): the wave from a phase grating producing a sinusoidal wavefront with amplitude H is

$$\psi = \sum_{n=-\infty}^{\infty} \mathrm{i}^n J_n(2H/\lambda) \exp[\mathrm{i}\pi(2nx/a - n^2 z/z_T)]$$

where the J_n are Bessel functions.

- Gauss sums (figure 4): For the nth superposed Talbot image at distance $z = (p/q)z_T$, the phase is $\chi(n;p,q)$, defined by

$$\exp[\mathrm{i}\chi(n;p,q)] = \frac{1}{\sqrt{q}} \sum_{s=1}^{q} \exp\left(\mathrm{i}\pi\{-s^2 p/q + s[2n/q + e(p)]\}\right)$$

where $e(p) = 0$ if p is even and $e(p) = 1$ if p is odd.

Obviously this is not true in practice, and corrections can be calculated. For $\mathcal{N}$ illuminated slits, the finest details of the reconstructed images and of the fractals are blurred transversely by $\Delta x \sim a/\mathcal{N}$ and longitudinally (i.e. in the depth of focus of Talbot images) by $\Delta z \sim z_T/\mathcal{N}^2$.

More fundamental is the fact that the finite wavelength of the light, λ, must limit the fine detail in the optical field, so that even with infinitely many slits the images would still be blurred.

Perfect reconstruction is an artefact of the *paraxial approximation*, in which the deflections θ of all relevant diffracted beams are assumed to be small, so that $\cos\theta \sim 1 - \theta^2/2$. Deviations from this approximation get smaller as the ratio λ/a decreases, that is in the short-wave limit, or for coarse gratings. The blurring in both the transverse and longitudinal directions is of the order of $\sqrt{(a\lambda)}$. However, the blurring in z is much smaller relative to the periodicity in the pattern because the detail is stretched over the Talbot distance, which greatly exceeds the slit spacing.

These blurrings should not be confused with the more familiar Fresnel diffraction of light from a single edge of a slit, which would be of order a in the transverse direction at $z = z_T$. An alternative way to state the Talbot effect is that the Fresnel blurring from the edges of all the slits is cancelled by the coherent interference of the waves.

In the quantum case, the perfection of the revivals is degraded by an effect that seems different but is mathematically very similar. Analogous to the paraxial approximation is the assumption that the energy can be approximated by terms that are both linear and quadratic functions of the quantum number n, over the range of energies included in the initial state. (If the variation were merely linear, then the wave packet would not spread at all.)

In the case of a non-relativistic particle in a box, the revivals are perfect because the variation in energy is quadratic. However, quadratic variation only approximately describes an electron in an atom: the approximation is *semi-classical*, that is it gets better as the quantum number, n, corresponding to the central energies in the packet increases. Non-quadratic corrections to the spectrum degrade the revivals, in ways that are now well understood.

Paradoxical limits

Quantum and optical carpets provide a dramatic illustration of how limits in physics that seem familiar can in fact be complicated and subtle. It is no exaggeration to say that perfect Talbot images, and infinite detail in the Talbot fractals, are emergent phenomena: they emerge in the paraxial limit as λ/a approaches zero.

At first this seems paradoxical, because the short-wave approximation is usually regarded as one in which interference can be neglected, whereas the Talbot effect depends entirely on interference. The paradox is dissolved by noting that the Talbot distance increases as the wavelength approaches zero, so here we are dealing with the combined limit of short wavelength and long propagation distance: in the short-wavelength limit, the Talbot reconstructed images recede to infinity.

In a similar way, perfect quantum revivals emerge from the semi-classical limit of highly excited states. Again it might seem paradoxical that an interference effect persists in a semi-classical limit (and indeed emerges perfectly in that limit). The dissolution of the paradox is analogous to that in the Talbot case: the semi-classical limit here is combined with the long time limit (recall that $T_r \sim nT_c$), so the revivals occur in

6 Talbot reconstructed caustics

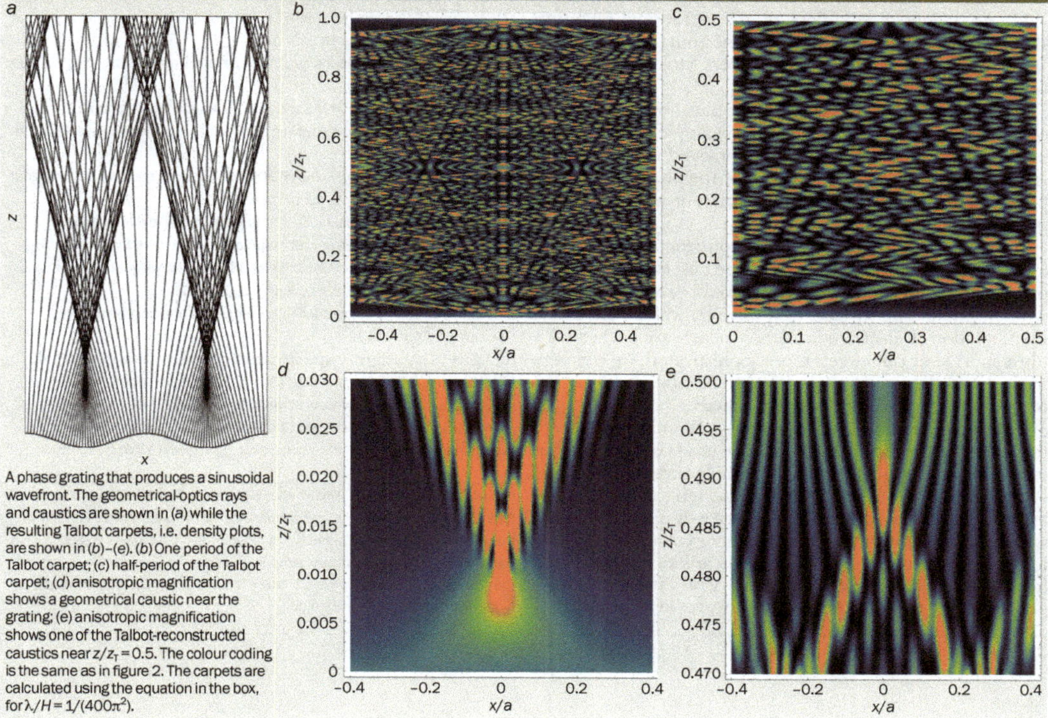

A phase grating that produces a sinusoidal wavefront. The geometrical-optics rays and caustics are shown in (*a*) while the resulting Talbot carpets, i.e. density plots, are shown in (*b*)–(*e*). (*b*) One period of the Talbot carpet; (*c*) half-period of the Talbot carpet; (*d*) anisotropic magnification shows a geometrical caustic near the grating; (*e*) anisotropic magnification shows one of the Talbot-reconstructed caustics near $z/z_T = 0.5$. The colour coding is the same as in figure 2. The carpets are calculated using the equation in the box, for $\lambda/H = 1/(400\pi^2)$.

the infinite future.

Even within the paraxial approximation, a clash of limits can generate very rich phenomena. Consider a transparent phase grating, that is a diffraction grating that transmits light with uniform amplitude but introduces a sinusoidal phase variation with period a.

In the geometrical-optics approximation, light rays passing through the grating are deflected so that they are normal to the sinusoidal wavefront that is generated immediately beyond the grating. These rays generate the pattern shown in figure 6*a*, which is dominated by caustics, i.e. curves onto which the light is focused and in which the intensity is concentrated.

Pairs of caustics join at cusps located near the centres of curvature of the wavefronts. The resulting pattern of rays is not periodic in z, and is thus discordant with the notion that the pattern of waves must repeat with period z_T for the Talbot effect to occur. Again the paradox is dissolved by the observation that the Talbot distance is of the order of $1/\lambda$, and so recedes to infinity in the geometrical-optics limit where λ tends to zero.

However, when λ/a is small but non-zero, both the caustics and the Talbot repetitions must coexist. This means that the geometrical caustics, including the cusps for distances much less than the Talbot distance, must be regenerated at both integer multiples and fractional multiples of z_T. Figures 6*b* and *c* show the Talbot carpet for this sinusoidal grating, with apparently no trace of the geometrical-optics caustics and their cusps.

However, appropriate magnification shows that the original and regenerated caustics (and associated diffraction fringes associated with a single sinusoid) are indeed hidden in the detail of the carpet (figures 6*d* and *e*). It is worth emphasizing what a remarkable phenomenon this regeneration is: geometrical-optics caustics have been constructed entirely by interference, in regions where there are no focusing rays. They are the ghosts of caustics. There is a curious analogy between these "caustics without rays" and Moiré patterns, which are "fringes without waves".

Experiments and outlook

These predictions – Gauss sums, waves reconstructing themselves, fractal carpets and canals – are not just arcana from the imaginations of theorists. Many of the results have been confirmed by experiments.

Fractal aspects of Talbot interference were seen in optical experiments by one of us (MB) and Susanne Klein of Bristol University in 1996. Fractional Talbot images were seen in matter waves (a beam of helium atoms) diffracted by a microfabricated grating, in experiments carried out at the University of Konstanz in 1997 by Stephan Nowak, Christian Kurtsiefer, Christian David and Tilman Pfau.

And fractional quantum revivals have been seen in electrons in potassium atoms, which were knocked into appropriate initial states by laser pulses using a "pump-probe" technique. These experiments were performed by John Yeazell and Carlos Stroud at the University of Rochester in 1991. Analogous observations of the vibrational quantum states of bromine molecules were carried out in 1996 by Max Vrakking, David Villeneuve and Albert Stolow at the University of Ottawa in Canada.

There are also the beginnings of applications for technology. Following Rayleigh's pioneering example, the fine detail in Talbot wavefunctions is being employed in lithography. And the number-theory properties of the phases have been proposed as the basis of a means to carry out arithmetic computations based on interference.

However, the predicted Talbot reconstruction of caustics for smooth phase gratings has yet to be observed. An appropriate grating for such experiments could be a sheet of glass with an undulating surface, or a wave on water. And as the manipulation of electronic and molecular wave packets with ultra-short laser pulses gets more sophisticated, we envisage more detailed experimental exploration of the fine structure of the carpets, such as the diagonal canals.

The optical field behind a coherently illuminated diffraction grating, and the quantum particle in a box with uniform initial state, are two of the most venerable and widely studied systems in physics. It was quite unexpected to discover such richness hidden in them, and also to see it displayed by such a wide range of physical systems. Further exploration of these intricate manifestations of the coherent addition of waves is a fitting way to celebrate the bicentenary of Thomas Young's great discovery of the interference of waves of light.

Further reading

M V Berry 1996 Quantum fractals in boxes *J. Phys.* **A 26** 6617–6629

M V Berry and E Bodenschatz 1999 Caustics, multiply-reconstructed by Talbot interference *J. Mod. Optics* **46** 349–365

M V Berry and S Klein 1996 Integer, fractional and fractal Talbot effects *J. Mod. Optics* **43** 2139–2164

J H Eberly, N B Narozhny and J J Sanchez-Mondragon 1980 Periodic spontaneous collapse and revival in a simple quantum system *Phys. Rev. Lett.* **44** 1323–1326

O Friesch, I Marzoli and W P Schleich 2000 Quantum carpets woven by Wigner functions *New J. Phys.* **2** 4.1–4.11

A E Kaplan *et al.* 2000 Multi-mode interference: highly regular pattern formation in quantum wave packet evolution *Phys. Rev.* **A 61** 032101-1-6

C Leichtle, I S Averbukh and W P Schleich 1996 Multilevel quantum beats: an analytical approach *Phys. Rev.* **A 54** 5299–5312

S Nowak *et al.* 1997 Higher-order Talbot fringes for atomic matter waves *Opt. Lett.* **22** 1430–1432

H F Talbot 1836 Facts relating to optical science no. IV *Phil. Mag.* **9** 401–407

M J J Vrakking, D M Villeneuve and A Stolow 1996 Observation of fractional revivals of a molecular wavepacket *Phys. Rev.* **A 54** R37–40

J A Yeazell and C R Stroud Jr 1991 Observation of fractional revivals in the evolution of a Rydberg atomic wave packet *Phys. Rev.* **A 43** 5153–5156

Michael Berry is in the H H Wills Physics Laboratory, Tyndall Avenue, Bristol BS8 1TL, UK. **Irene Marzoli** is in the Dipartimento di Matematica e Fisica and Unità INFM, Università di Camerino, Via Madonna delle Carceri, 62032 Camerino, Italy, e-mail marzoli@campus.unicam.it. **Wolfgang Schleich** is in the Abteilung für Quantenphysik, Universität Ulm, D-89069 Ulm, Germany, e-mail schleich@physik.uni-ulm.de

Reflections on a Christmas-tree bauble

M V Berry
H H Wills Physics Laboratory
Bristol

1 Introduction

Totally reflecting spheres are most commonly encountered as decorations hanging from Christmas trees, and as steel ball-bearings. In earlier times they were hung in the windows of country cottages to ward off evil spirits, and known as witch balls; these are now highly prized as antiques. On the Continent large silvered glass globes are sometimes employed to decorate gardens, because of the attractive images of cloudscapes that they reflect. The imaging properties of these spheres must surely have been analysed many times during the past few hundred years (if not before), but a fairly extensive search through the literature has failed to uncover any published work on the subject, except for the book by Minnaert (1940) which contains a brief description of the properties of garden globes, together with a number of applications to the study of atmospheric optical phenomena. Every book on optics deals with spherical mirrors, of course, but the treatments seem universally restricted to the paraxial rays from mirrors of small aperture; even in these works, it is often the properties of *concave* reflectors that are emphasized (because of their usefulness as magnifiers), convex mirrors being relegated to brief mention as a special case to which the formulae remain applicable if the sign of the radius of curvature is reversed.

As far as the teaching of optics at an elementary level is concerned, this neglect of the properties of complete spheres is a pity. Certainly the restriction to paraxial rays is necessary in simple optical instruments, but we shall show in §3 that it is unnecessary when dealing with observations with the naked eye alone. The main purpose of this article is to draw the attention of science teachers to a number of visually appealing phenomena, which are analysed in the first part of §2 in a way that should be understandable by most pupils at O level or even earlier; there is no need to introduce mathematical approximations, such as 'sin θ can be replaced by θ', which may confuse the application of the law of specular reflection—perhaps the most intuitively immediate of all the topics in elementary physics. The deeper analysis of the structure of the images, in §3, might help A level pupils to understand more precisely why the paraxial approximation is necessary for telescopes, etc. (the contents of that section could perhaps form the basis of a 'research project').

2 The appearance of the images

Consider a sphere of radius r (figure 1) whose centre C lies at a distance d from the eye E of an observer. We shall ignore diffraction and stereoscopic effects throughout, and defer until §3 the implications of the finite size of the pupil of the eye. The rays reaching E from the surface must lie within a cone of semiangle $\phi_{\max}$, given by

$$\sin \phi_{\max} = r/d \,. \tag{1}$$

Rays making an angle ϕ at the eye, measured from the 'forward direction' *EC*, have been reflected at a point P after coming from a direction making an angle θ with *EC*. As ϕ *decreases* from its maximum value $\phi_{\max}$, θ *increases* from a minimum, equal to $\phi_{\max}$ (the case of grazing incidence), to a maximum value of 180° (when ϕ is zero). This means that rays from *all incident directions* θ are imaged in the sphere, except those lying in the cone surrounding the forward direction which is intercepted by the sphere itself.

It is this *wide-angle property* of the reflection from a complete sphere which is responsible for the fascination of the images that may be seen in it (see figure 2, and two paintings by Escher (1967)). By holding the sphere against the window of a room, for instance, an observer looking outwards can see the interior of the room (which is behind him) reflected in the central regions, surrounded by the (weirdly curved) image of the window frame, which in turn is surrounded by most of the scene outside the room, reflected at near-grazing incidence around the periphery of the sphere (which may therefore be thought of as a sort of 'topological synthesiser', enabling the inside and outside of a room to be seen together). Out-of-doors, the whole sky can be seen reflected in a portion of the globe, and the resulting increase of contrast gradients often enables cloud detail to be discerned which is not immediately apparent to the unaided eye (this is an interesting example, inasmuch as it shows the occasional advantage of a minifier over a magnifier—the detail concerned would not be visible in a telescope). The spheres can also be used to take wide-angle photographs over a range of nearly 360° (actually $360° - 2\phi_{max}$), whereas conventional wide-angle lenses (also used in spyholes in front doors to inspect potentially unwelcome callers) have a range that rarely exceeds 180°; the only disadvantage of the reflectors is the unavoidable appearance of the camera lens in the centre of the picture. The motion of objects is dramatically distorted by reflection in the spheres—a car approaching from the front appears on the edge, moves inwards, turns around, and disappears by receding into the centre of the sphere, while overhanging foliage seems to 'pour' into the centre as the observer passes underneath.

Of course an incomplete sphere will still show the wide-angle property to a certain degree, which explains the common use of convex mirrors for rear viewing on cars and bicycles, as an anti-theft precaution in supermarkets, and simply for room decoration (for an early example, see the painting 'The Arnolfini Wedding', by Jan van Eyck 1434, in the National Gallery, London).

To proceed further, it is necessary to know the precise relation between the angle of incidence θ and the angle ϕ at the eye (figure 1). An elementary trigonometrical calculation, based on the law of specular reflection at P, leads to the result

$$d \sin \phi = r \cos\left(\frac{\theta - \phi}{2}\right). \quad (2)$$

As d increases, and the eye recedes from the sphere, the images (which are localized within the sphere, see §3) appear to be distributed across a disc—the projection of the sphere on a plane perpendicular to EC. In this limit, when ϕ_{max} is small, virtually the whole sphere of incident directions θ is imaged onto this disc. We specify positions on the disc by their distance a from the centre; from figure 1, we have

$$a = EP \times \sin\phi \simeq d \sin \phi \ (d \gg r)$$

and substitution into (2), together with neglect of ϕ compared with θ on the right-hand side of (2) leads to

$$a = r \cos\frac{\theta}{2}. \quad (3)$$

This relation, which may also be derived directly, shows how the image moves from the edge of the disc to the centre as the object moves from the forward to the backward directions. The limit we are discussing, which is valid in the commonly-occuring situation where the spheres subtends a small angle at the eye, is not the paraxial case, because although the rays

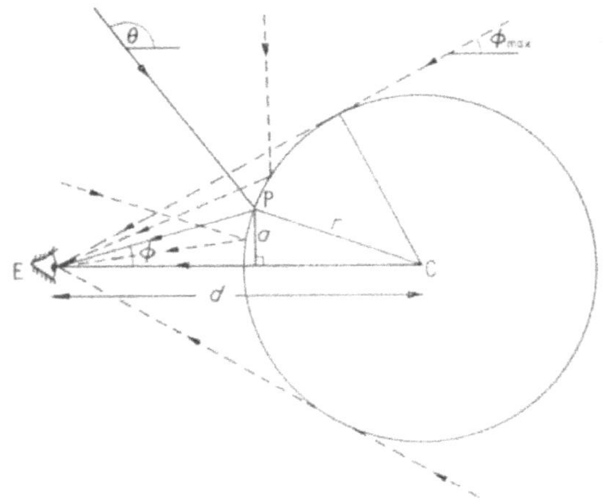

Figure 1 *Geometry of basic reflections.*

Figure 2

entering the eye are nearly parallel, the rays incident on the sphere come from all directions. An interesting special case of (3) occurs when looking down on a globe out-of-doors (figure 3). The horizon, which is the locus of points making an angle θ equal to 90° with the vertical EC, is imaged as a circle whose radius is

$$a = r\cos 45° = \frac{r}{\sqrt{2}}$$

within this circle is the reflection of the sky, while outside it can be seen the ground.

To include the three-dimensional aspects of the problem, we rotate figure 1 about the axis EC, specifying the rotation by an azimuth angle ψ which runs from 0° to 360°. We can then regard the reflection as *mapping* a *solid angle* $d\Omega$ (figure 4) of incident rays, given by

$$d\Omega = \sin\theta\, d\theta\, d\psi$$

onto an *area* dA of the image disc, given by

$$dA = a\, da\, d\psi.$$

The *mapping ratio* from solid angle to area is found, by making use of (3), to be

$$\left|\frac{dA}{d\Omega}\right| = \frac{a}{\sin\theta}\left|\frac{da}{d\theta}\right| = \frac{r\cos(\theta/2)\times r\sin(\theta/2)}{2\sin\theta} = \frac{r^2}{4} \qquad (4)$$

a relation which can be checked by noting that the total area of the disc is πr^2, while the total solid angle of the sphere of directions is 4π. The important point is that the mapping is *uniform*—equation (4) is independent of θ and ψ, so that a given $d\Omega$ is mapped onto the same dA, no matter where the incident ray bundle is situated in relation to the forward direction.

This uniformity of mapping has interesting implications: for instance, we know that any variations of sky brightness that are observed in a globe are real, and are not artefacts of the reflection process. Further, the strikingly distorted images seen in the globe nevertheless subtend the same relative areas as do the original objects, only angles and the straightness of lines being changed. This suggests that an *area-preserving* map of the whole world would be seen on the image disc if a small silvered globe were situated at the centre of the earth (assumed transparent), those countries being mapped with least distortion which lie on OE produced (ie behind the observer). Figure 5 shows the parallels and meridians for polar and equatorial viewing positions. This mapping is known to cartographers as 'Lambert's azimuthal equal-area projection', introduced in 1772 (see Raisz 1948); it is widely employed to show world air routes.

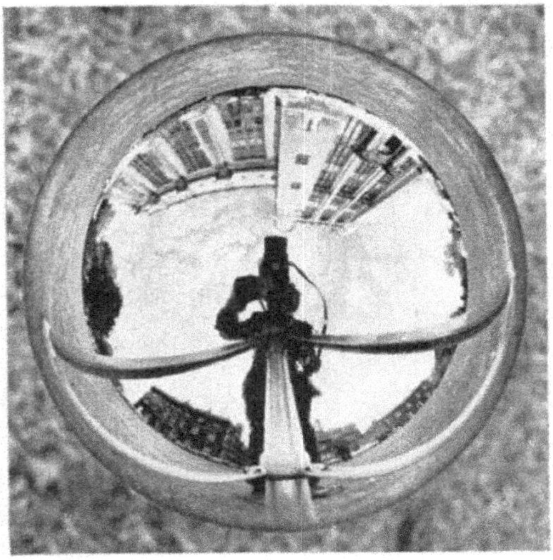

Figure 3

3 The structure of the images

We now ask: where are the images localized, that are formed by reflection in a silvered globe? This question played no part in our treatment of §2, because there we considered the eye pupil as a point, which in general admits only a single ray from a point object, whose image is therefore not localized at all (this is why a hole, pierced with a fine needle in a piece of metal foil held close to the eye, enables the long-sighted to read in comfort at close range, and the short-sighted to see distant scenes clearly). Consider all those rays hitting the sphere which have emerged from a point object O (see figure 6, which is the same as figure 1 with the arrows on the rays reversed). They will be reflected into all directions, and if the emergent rays are produced backwards into the sphere, they will intersect, not at a single image point, but on a *caustic surface*, so named because on the corresponding surface for a concave reflector or convex lens the sun's heat may be concentrated.

The eye will see the 'image' of O at different points

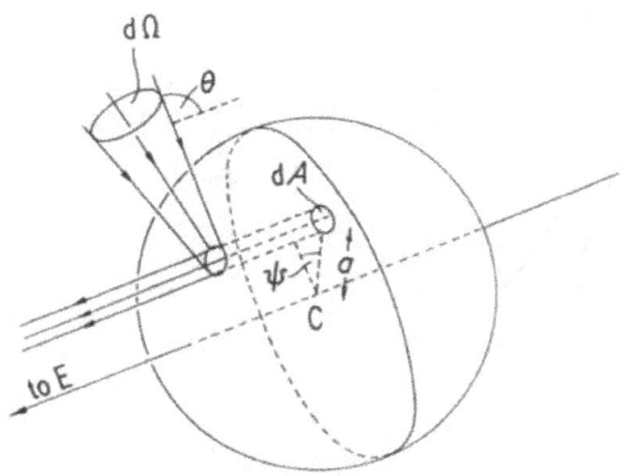

Figure 4 Mapping of incident rays onto image disc.

on this surface (or curve, if we restrict ourselves to a plane), depending (figure 6) on the angle of observation θ, which has the same meaning as in §2. It is easily seen that the caustic always lies within the sphere (touching it only for the grazing rays), so that the images are all virtual. The point nearest the centre is the cusp K on the line CO corresponding to the directly-reflected ray ($\theta = 180°$). The closest possible approach of this cusp to the centre occurs when the object lies at infinity, in which case KC is $r/2$—the well-known value for the focal length of a convex mirror for paraxial rays. Thus, whatever the positions of object and eye, we know that the image will always lie between two spheres, centred on C, whose radii are r and $r/2$.

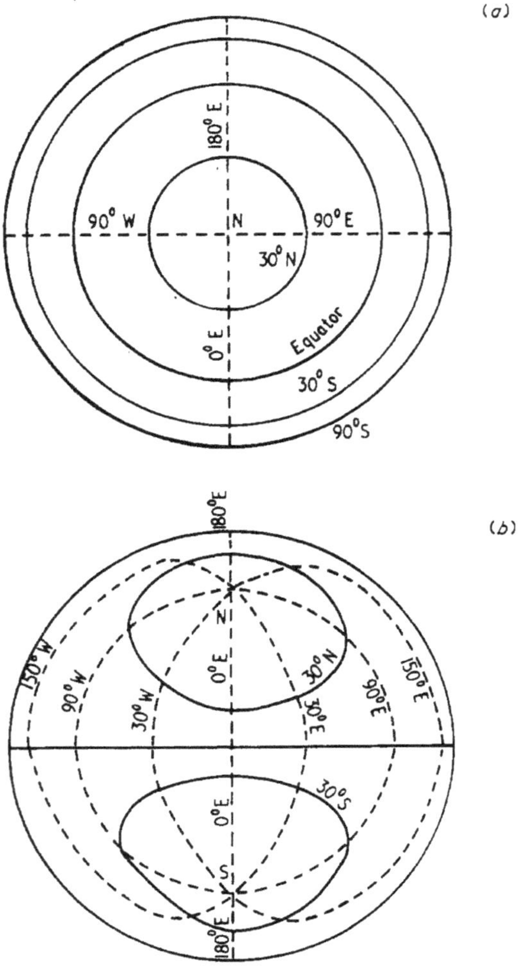

Figure 5 Sketch of parallels (———) and meridians (— — —) on an equal-area map of the world for (a) polar projection (b) equatorial projection.

This blurring-out into caustics of the images of point objects in wide-aperture mirrors is the reason why most treatments are restricted to the paraxial case. Yet the images seen in spheres—and indeed in the 'crazy mirrors' of varying curvature found at fairgrounds—are frequently crystal clear, so that we must investigate why it is that caustics are not often observed directly with the unaided eye. (The caustics seen in teacups are no exception to this rule, because they are observed not by looking along the rays whose envelope they are, but by scattered light.) The basic reason is that the small aperture e of the eye pupil admits rays from only a short arc-length dS of the caustic, while the least distance of distinct vision g sets a limit to the closeness with which this length can be examined.

To estimate the angle α subtended at the eye by the 'image' of height h which this portion of the caustic essentially constitutes, we consider figure 7, which shows (greatly exaggerated) those rays with angles of emergence ranging from θ to $\theta + d\theta$ which enter the eye. The size h is clearly given by

$$h = \frac{dS d\theta}{2} = \frac{dS}{d\theta}\frac{d\theta^2}{2} = R\frac{d\theta^2}{2}$$

where R is the radius of curvature of the caustic at the point considered. The angle α subtended at the eye is

$$\alpha = \frac{h}{g} = \frac{Rd\theta^2}{2a} \qquad (5)$$

while $d\theta$ is given by (figure 7)

$$e = h + gd\theta \approx gd\theta \qquad (6)$$

(the term h has been neglected because it only becomes

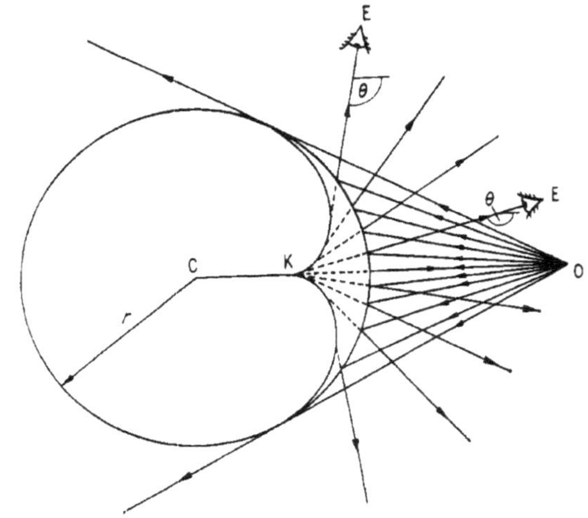

Figure 6 Formation of virtual caustic of reflection.

comparable with e for huge spheres exceeding several metres in diameter—in other words, the 'image' is generally much smaller than the eye pupil, whose diameter it can never exceed under any circumstances). Substituting for $d\theta$ in (5) gives

$$\alpha = \frac{Re^2}{2g^3}.$$

Caustics are observable if α exceeds the smallest angle β that the eye can resolve, which is given by the Rayleigh criterion as

$$\beta = 1{\cdot}22\lambda/e$$

λ being the wavelength of the light used. The critical ratio α/β is therefore

$$\frac{\alpha}{\beta} = \frac{R}{2{\cdot}44\lambda}\left(\frac{e}{g}\right)^3. \qquad (7)$$

This formula applies to observation in any curved surface. We can estimate R for a sphere by inspecting figure 6, from which it is clear that R is of order r; an exact calculation gives

$$R = \frac{3r}{4}\cos\frac{\theta}{2}$$

for the case where the object point is very far away. The ratio (7) now becomes

$$\frac{\alpha}{\beta} = \frac{0{\cdot}31r}{\lambda}\left(\frac{e}{g}\right)^3\cos\frac{\theta}{2} \qquad (8)$$

from which we can immediately see that the cusp K is unobservable (this corresponds to back reflection of the paraxial rays from the front of the sphere, with $\theta = 180°$).

Under normal daytime conditions the aperture e of the pupil is about 3×10^{-3}m, while the least distance g of comfortably distinct vision is about 0·2m; taking the wavelength of yellow light as 5×10^{-7}m, we get

$$\frac{\alpha}{\beta} \simeq 2r\cos\frac{\theta}{2}$$

where r, the sphere radius, is measured in metres. This ratio only exceeds unity for spheres more than about a metre across, when the additional complication arises that the caustic is buried deep inside the sphere (except at near-grazing incidence), and cannot be approached to within the distance g. Thus we do not expect to see caustics easily in daylight.

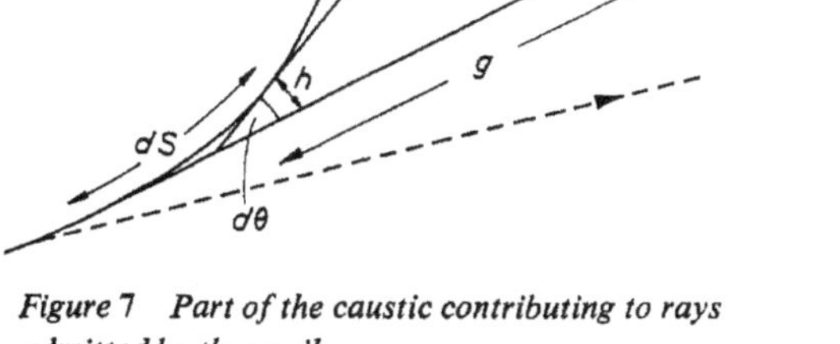

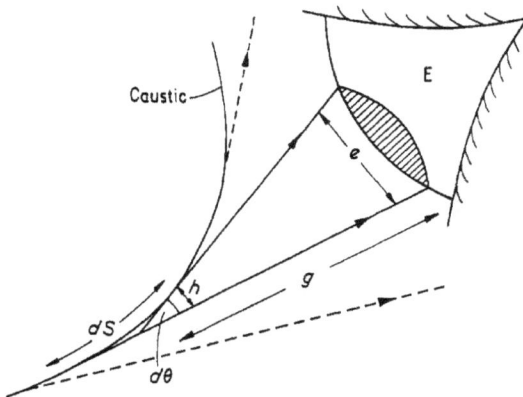

Figure 7 Part of the caustic contributing to rays admitted by the pupil.

A dramatic increase in the magnitude of α/β can be obtained by using the facts that a dark-adapted pupil can expand to at least double its normal size, while most people can with a certain discomfort focus on objects as close as 0·12m. The ratio α/β now becomes

$$\frac{\alpha}{\beta} \simeq 75\,r\cos\frac{\theta}{2}$$

and we predict that in these new circumstances the distortion of the images of point objects should be observable in spheres down to a few centimetres across.

These conclusions are borne out by experiment. In a darkened room the caustic is clearly seen in a sphere of 0·1 m radius, but it vanishes rapidly—ie the image contracts almost to a point—when a light is switched on, causing the pupil to contact. Observation is best carried out at about $\theta = 90°$, when $\cos\theta/2 \simeq 0{\cdot}7$, the slight improvement in resolution at more glancing angles (smaller θ) being outweighed by the greater effects of surface imperfections, etc. An adequate point source is a pinhole in a piece of kitchen foil, wrapped over the shade of a desk lamp.

In optical instruments, such as telescopes, where the image is greatly magnified by an eyepiece, the caustic would be clearly visible, and would constitute a serious aberration of the system, if observation were not restricted to the narrow bundle of paraxial rays for which $\cos\theta/2$ is negligible (of equation (8)).

4 Conclusions

It seems that silvered globes may be a useful aid in the teaching at various levels of a wide range of concepts, including the laws of reflection, plane trigonometry, solid angle, mapping of a sphere, image formation, and caustics. An easy way of producing the globes is to silver the inside surface of a round-bottomed flask, whose neck provides a useful handle (care should be taken to select a flask whose glass is relatively free from blemishes, and it is advisable to stop up the end of the neck with a rubber bung to exclude dust.)

Acknowledgments

I am happy to thank Dr Derek Greenwood and Dr Michael Hart for their helpful suggestions, Mr Tony Osman for silvering the first flask, Mr George Keene for taking the photographs, and Professor Charles Frank for correcting a mistake in one of the figures.

References

Escher M C 1967 *The graphic work of M C Escher* (London: Oldbourne)
Minnaert M, 1940 *Light and colour in the open air* (London: G Bell and Sons)
Raisz E, 1948 *General cartography* (New York: McGraw-Hill)

INSTITUTE OF PHYSICS PUBLISHING EUROPEAN JOURNAL OF PHYSICS

Eur. J. Phys. **27** (2006) 109–118 doi:10.1088/0143-0807/27/1/012

Oriental magic mirrors and the Laplacian image

M V Berry

H H Wills Physics Laboratory, Tyndall Avenue, Bristol BS8 1TL, UK

Received 26 September 2005, in final form 20 October 2005
Published 24 November 2005
Online at stacks.iop.org/EJP/27/109

Abstract

The pattern embossed on the back of an oriental magic mirror appears in the patch of light projected onto a screen from its apparently featureless reflecting surface. In reality, the embossed pattern is reproduced in low relief on the front, and analysis shows that the projected image results from pre-focal ray deviation. In this interesting regime of geometrical optics, the image intensity is given simply by the Laplacian of the height function of the relief. For patterns consisting of steps, this predicts a characteristic effect, confirmed by observation: the image of each step exhibits a bright line on the low side and a dark line on the high side. Laplacian-image analysis of a magic-mirror image indicates that steps on the reflecting surface are about 400 nm high and laterally smoothed by about 0.5 mm.

1. Introduction

Cast and polished bronze mirrors, made in China and Japan for several thousand years, exhibit a curious property [1–4], long regarded as magical. A pattern embossed on the back (figure 1(b)) is visible in the patch of light projected onto a screen from the reflecting face (figure 2), when this is illuminated by a small source, even though no trace of the pattern can be discerned by direct visual inspection of the reflecting face (figure 1(a)). The pattern on the screen is not the result of the focusing responsible for conventional image formation, because its sharpness is independent of distance, and also because the magic mirrors are slightly convex. It was established long ago [2] that the effect results from the deviation of rays by weak undulations on the reflecting surface, introduced during the manufacturing process and too weak to see directly, that reproduce the much stronger relief embossed on the back. Such 'Makyoh imaging' (from the Japanese for 'wonder mirror') has been applied to detect small asperities on nominally flat semiconductor surfaces [5–8].

My aim here is to draw attention (section 2) to a simple and beautiful fact, central to the optics of magic mirrors, that has not been emphasized—either in the qualitative accounts [9–11] or in an extensive geometrical-optics analysis [12]: in the optical regime relevant to magic mirrors, the image intensity is given, in terms of the height function $h(\mathbf{r})$ of the relief

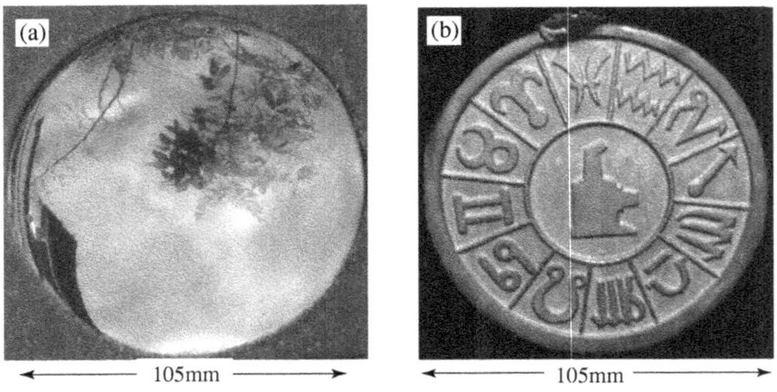

Figure 1. (a) Convex reflecting face of magic mirror, (b) pattern embossed on back face of magic mirror. These and subsequent images were photographed with a Fuji 610F digital camera, and saved as jpeg files, the only manipulation being conversion from RGB to greyscale.

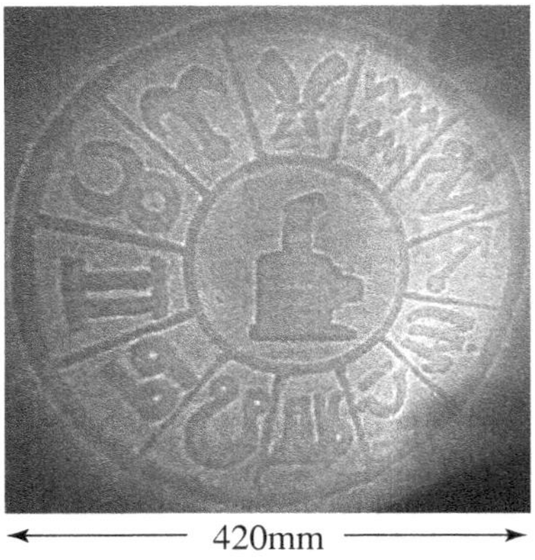

Figure 2. Magnified magic-mirror image reflected onto a screen by the illuminated front face.

on the reflecting surface, by the Laplacian $\nabla^2 h(\mathbf{r})$ (here $\mathbf{r}$ denotes position in the mirror plane: $\mathbf{r} = \{x, y\}$). The Laplacian image predicts striking effects for patterns, such as those on magic mirrors, that consist of steps (section 3); these predictions are supported by experiment (section 4). The detailed study of reflection from steps throws up an unresolved problem (section 5) concerning the relation between the pattern embossed on the back and the relief on the reflecting surface.

The Laplacian image is an approximation to geometrical optics, which is itself an approximation to physical optics. The appendix contains a discussion of the Laplacian image starting from the wave integral representing Fresnel diffraction from the mirror surface.

2. Geometrical optics and the Laplacian image

If we measure the height $h(\mathbf{r}')$ from the convex surface of the mirror (figure 3), assumed to have radius of curvature R_0, then the deviation of the surface undulations from a reference plane (figure 3) is

$$\eta(\mathbf{r}') = -\frac{r'^2}{2R_0} + h(\mathbf{r}'). \tag{1}$$

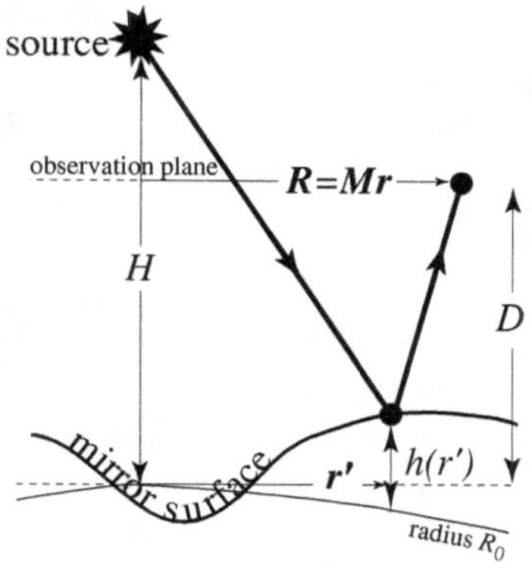

Figure 3. Geometry and coordinates for formation of magic-mirror image. For clarity, the surface elevation $h(\mathbf{r}')$ (measured from the convex surface with radius of curvature R_0) is exaggerated; in reality, the surface radii of curvature can be comparable with or smaller than R_0, so the mirror's undulating surface can be entirely convex.

The specularly reflected rays of geometrical optics are determined by the stationary value(s) of the optical path length L from the source (distance H from the reference plane) to the position $\mathbf{R}$ on the screen (distance D from the reference plane) via the point $\mathbf{r}'$ on the mirror. This is

$$\begin{aligned} L &= \sqrt{(H-\eta(\mathbf{r}'))^2 + r'^2} + \sqrt{(D-\eta(\mathbf{r}'))^2 + (\mathbf{R}-\mathbf{r}')^2} \\ &\approx H + D + \Lambda(\mathbf{r}', \mathbf{R}), \end{aligned} \tag{2}$$

where in the second line we have employed the paraxial approximation (all ray angles small), with

$$\Lambda(\mathbf{r}', \mathbf{R}) = \frac{r'^2}{2H} + \frac{(\mathbf{R}-\mathbf{r}')^2}{2D} + \frac{r'^2}{R_0} - 2h(\mathbf{r}'). \tag{3}$$

In applying the stationarity condition

$$\nabla_{\mathbf{r}'}\Lambda(\mathbf{r}', \mathbf{R}) = 0, \tag{4}$$

it is convenient to define the magnification M, the reduced distance Z, and the demagnified observation position $\mathbf{r}$ referred to the mirror surface:

$$M \equiv 1 + \frac{D}{H} + \frac{2D}{R_0}, \qquad Z \equiv \frac{2D}{M}, \qquad \mathbf{r} \equiv \frac{\mathbf{R}}{M}. \tag{5}$$

We note an effect of the convexity that will be important later: as the source and screen distance increase, Z approaches the finite asymptotic value R_0.

With these variables, the position $\mathbf{r}'(\mathbf{r},Z)$, on the mirror, of rays reaching the screen position $\mathbf{r}$, is the solution of

$$\mathbf{r} = \mathbf{r}' - Z\nabla h(\mathbf{r}'). \tag{6}$$

The focusing and defocusing responsible for the varying light intensity at $\mathbf{r}$ involves the Jacobian determinant of the transformation from $\mathbf{r}'$ to $\mathbf{r}$, giving, after a short calculation,

$$\begin{aligned} I_{\text{geom}}(\mathbf{r}, Z) &= \text{constant} \times \left(\frac{\partial x}{\partial x'}\frac{\partial y}{\partial y'} - \frac{\partial x}{\partial y'}\frac{\partial y}{\partial x'}\right)^{-1}_{\mathbf{r}'\to\mathbf{r}'(\mathbf{r},Z)} \\ &= \left(1 - Z\nabla^2 h(\mathbf{r}') + Z^2\left(\frac{\partial h(\mathbf{r}')}{\partial x'^2}\frac{\partial h(\mathbf{r}')}{\partial y'^2} - \left(\frac{\partial h(\mathbf{r}')}{\partial x'\partial y'}\right)^2\right)\right)^{-1}_{\mathbf{r}'\to\mathbf{r}'(\mathbf{r},Z)}, \end{aligned} \tag{7}$$

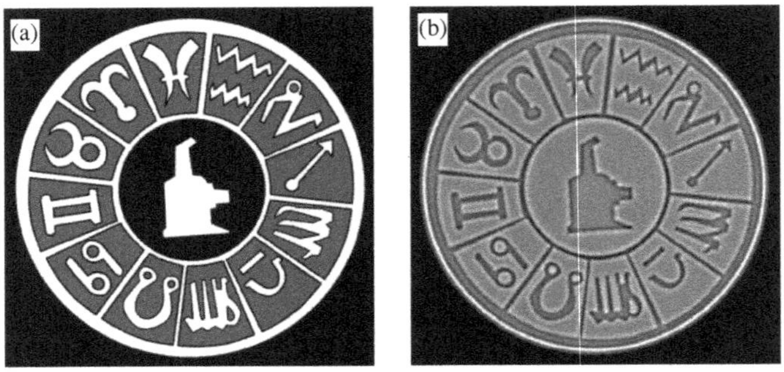

Figure 4. (a) Tracing of relief on the back of the mirror, with step heights shaded according to elevation (lowest black, highest white); (b) Laplacian image of (a), smoothed by $l = 0.5$ mm according to equations (10), (12) and (13).

where the result has been normalized to $I_{\text{geom}} = 1$ for the convex mirror without surface relief (i.e. $h(\mathbf{r}) = 0$).

So far, this is standard geometrical optics [12]. In general, more than one ray can reach $\mathbf{r}$—that is, (6) can have several solutions $\mathbf{r}'$—and the boundaries of regions reached by different numbers of rays are caustics [13, 14]. In magic mirrors, however, we are concerned with a limiting regime satisfying

$$\frac{Z}{R_{\min}} \ll 1, \tag{8}$$

where $R_{\min}$ is the smallest radius of curvature of the surface irregularities. Then there is only one ray, (6) simplifies to

$$\mathbf{r}' \approx \mathbf{r}, \tag{9}$$

and the intensity simplifies to

$$I_{\text{Laplacian}}(\mathbf{r}, Z) = 1 + Z\nabla^2 h(\mathbf{r}). \tag{10}$$

This is the Laplacian image. Changing Z affects only the contrast of the image and not its form, so (10) explains why the sharpness of the image is independent of screen position, provided (8) holds. The intensity is a linear function of the surface irregularities h, which is not the case in general geometrical optics (i.e. when (8) is violated), where, as has been emphasized [12] the relation (7) is nonlinear. And, as already noted, for a distant source and screen Z approaches the value R_0, implying that (8) holds for any distance of the screen if $R_0 \ll R_{\min}$, that is, provided the irregularities are sufficiently gentle or the mirror is sufficiently convex. Alternatively stated, the convexity of the mirror can compensate any concavity of the irregularity h, in which case there are no caustics for any screen position.

3. Laplacian images of steps

The relief h_{back} on the back of magic mirrors commonly consists of a pattern of steps (h_{back}, like h, is measured outwards from the mid-plane of the mirror, so increasing step heights on both the front and back correspond to increasing h and h_{back}). Figure 4(a) shows a tracing of the pattern of figure 1(b), with step heights shaded according to elevation. It seems that during the manufacturing process this is reproduced on the reflecting surface, with the steps greatly diminished—by a factor a, say—and slightly smoothed—by a distance l, say—so that, modelling the smoothing as Gaussian,

$$h(\mathbf{r}', l) = \frac{a}{2\pi l^2} \iint \mathrm{d}^2\mathbf{r}''\, h_{\mathrm{back}}(\mathbf{r}'') \exp\left\{-\frac{(\mathbf{r}''-\mathbf{r}')^2}{2l^2}\right\}. \tag{11}$$

Then the Laplacian image can be implemented with the transformation

$$\nabla^2 h(\mathbf{r}', l) = \iint \mathrm{d}^2\mathbf{r}''\, h_{\mathrm{back}}(\mathbf{r}'') K(\mathbf{r}''-\mathbf{r}'), \tag{12}$$

where the kernel is

$$K(\mathbf{r}''-\mathbf{r}') = \frac{a}{2\pi l^6}[(\mathbf{r}''-\mathbf{r}')^2 - 2l^2] \exp\left\{-\frac{(\mathbf{r}''-\mathbf{r}')^2}{2l^2}\right\}. \tag{13}$$

(In image processing, this transformation is commonly employed for edge detection [15–17].)

It is easy to implement the Laplacian image (10) using the transformation (12) and (13). In Mathematica^{TM}, for example [18], this involves essentially only three lines of code: one to import the image as a list, one to define the kernel K, and one to define the convolution. Figure 4(b) shows a magic-mirror image simulated in this way; it should be compared with the observation in figure 2. The essential features of the image, correctly reproduced by the theory, are associated with the steps: each step on the back appears in the image as a bright line on the low side, where the concavity of h leads to a concentration of rays, and a dark line on the high side, where the convexity of h leads to a depletion of rays.

To examine the image in more detail, we model the l-smoothed step, with height h_0, by

$$h(x) = \frac{h_0}{2}\operatorname{erf}\left(\frac{x}{l}\right) = \frac{h_0}{\sqrt{\pi}} \int_0^{x/l} \mathrm{d}t \exp(-t^2) \tag{14}$$

and introduce the dimensionless position and distance variables

$$\xi \equiv \frac{x}{Ml}, \qquad \xi' \equiv \frac{x'}{l}, \qquad \zeta \equiv Z\frac{h_0}{l^2}. \tag{15}$$

Then the exact ray equation (6) becomes

$$\xi = \xi' - \frac{\zeta}{\sqrt{\pi}} \exp(-\xi'^2) \Rightarrow \xi'(\xi, \zeta), \tag{16}$$

and the geometrical intensity is

$$I_{\mathrm{geom}}(\xi, \zeta) = \left[1 + \frac{2\zeta\xi'}{\sqrt{\pi}} \exp\{-\xi'(\xi,\zeta)^2\}\right]^{-1}. \tag{17}$$

The Laplacian image (10) is simply

$$I_{\mathrm{Laplacian}}(\xi, \zeta) = 1 - \frac{2\zeta\xi}{\sqrt{\pi}} \exp\{-\xi^2\}. \tag{18}$$

4. Experiment

Equation (18) is the prediction of the Laplacian theory for the image of a smoothed step. To compare it with observation, we first extract a part of the image (figure 2), corresponding to a prominent step; this is shown in figure 5(a). Next, we reduce the noise by smoothing along the step (figure 5(b)). The intensity profile of the image is the full curve in figure 6.

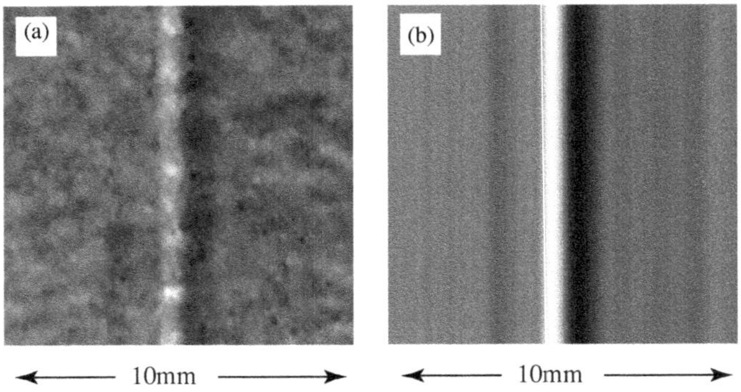

Figure 5. (a) Magnification of part of left-hand vertical step near the centre of magic-mirror image in figure 2; (b) as (a), after averaging along the step. The length scales denote distances on the mirror surface, not the image.

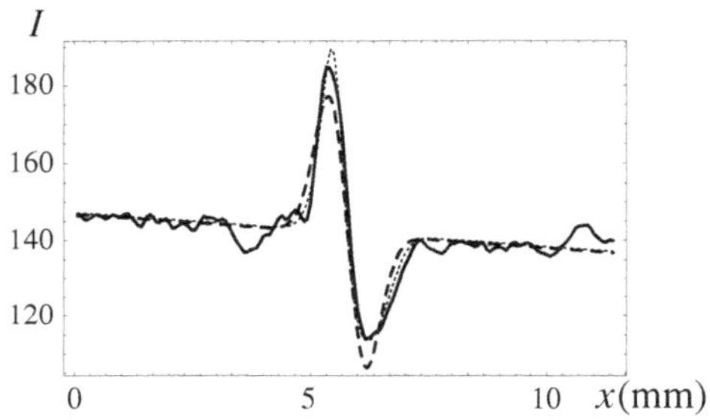

Figure 6. Full curve: intensity across the step in figure 5(b) (arbitrary units); dashed curve: Laplacian-image fit from equation (18), with a straight line of finite slope replacing the constant term; dotted curve: full geometrical-optics fit, incorporating the ray displacement (16) and intensity (17).

Measurements on the curve give the intensity contrast as

$$C_{\exp} \equiv \frac{2(I_{\max} - I_{\min})}{(I_{\max} + I_{\min})} = 0.467. \tag{19}$$

Comparison with the theoretical contrast from the extrema of (18) (at $\xi = \pm 1/\surd 2$), namely

$$C_{\text{theory}} = 2\zeta\sqrt{\frac{2}{\pi\,\mathrm{e}}}, \tag{20}$$

leads to the identification $\zeta = 0.482$.

In the experiment, the source (a halogen lamp) and screen were at the same distance from the mirror, also chosen to coincide with R_0: $D = H = R_0 = 800$ mm. Thus, from (5), and also as observed (cf the scale in figure 2), the magnification is $M = 4$, leading to $Z = H/2 = 400$ mm. Fitting the observed step profile to (18) (dashed curve in figure 6), gives the step width $l = 0.560$ mm. The relation (15) now gives the step height $h_0 = \zeta l^2/Z = 378$ nm. This value is substantially less than the wavelengths of visible light, so it is not surprising that the steps cannot be seen directly.

The Laplacian-image fit in figure 6 is good, but fails to incorporate a slight asymmetry between the two sides of the step: on the bright side, the intensity rises higher above the mean than it falls below the mean on the dark side. To understand this, we investigate the degree to which the condition (8) is satisfied. In the dimensionless variables (15), (8) corresponds to

$$\zeta \ll \zeta^* \equiv \sqrt{\frac{\pi\,\mathrm{e}}{2}} = 2.067\ldots \tag{21}$$

The value $\zeta = 0.482$ derived from observation is substantially smaller than ζ^*—well within the no-caustic regime that we have identified as corresponding to magic mirror imaging. However, fitting with the full geometrical-optics theory (16) and (17) (dotted curve in figure 6) gives a small correction that reduces the discrepancy by introducing an asymmetry in the correct sense. (The asymmetry between the bright and dark sides of the step increases with ζ until $\zeta = \zeta^*$ where the caustic is born.)

5. Concluding remarks

The theory based on the Laplacian image accords well with observation, at least for the mirror studied here. The key insight is that the image of a step is neither a dark line nor a bright line, as sometimes reported [11], but is bright on one side and dark on the other. It is possible that there are different types of magic mirror, where for example the relief is etched directly onto the reflecting surface and protected by a transparent film [11], but these do not seem to be common. Sometimes, the pattern reflected onto a screen is different from that on the back, but this is probably a trick, achieved by attaching a second layer of bronze, differently embossed, to the back of the mirror.

Pre-focal ray concentrations leading to Laplacian images are familiar in other contexts, though they are not always recognized as such. An example based on refraction occurs in old windows, where a combination of age and poor manufacture has distorted the glass. The distortion is not evident in views seen through the window when standing close to it. However, when woken by the low morning sun shining through a gap in the curtains onto an opposite wall, one often sees the distortions magnified as a pattern of irregular bright and dark lines. If the equivalent of (8) is satisfied, that is if the distortions and propagation distance are not too large, the intensity is the Laplacian image of the window surface. (When the condition is not satisfied, the distortions can generate caustics.)

Only the optics of the mirror has been studied here. The manner in which the pattern embossed on the back gets reproduced on the front has not been considered. Referring to (11), this involves the sign of the coefficient a in the relation between h_{back} and h. There have been several speculations about the formation of the relief. One is that the relief is generated while the mirror is cooling, by unequal contraction of the thick and thin parts of the pattern [10]; it is not clear what sign of a this leads to. Another [4] is that cooling generates stresses, and that during vigorous grinding and polishing the thin parts yield more than the thick parts, leading to the thick parts being worn down more; this leads to $a < 0$. However, this seems to contradict the observations, which point firmly to $a > 0$: bright (dark) lines on the image, indicating low (high) sides of the steps on the reflecting face, are associated with the low (high) sides of the steps on the back (figure 7(a)), not the reverse (figure 7(b)). This suggests two avenues for further research. First, the sign of a should be determined by direct measurement of the profile of the reflecting surface; I predict $a > 0$. Second, whatever the result, the mechanism should be investigated by which the process of manufacture reproduces onto the reflecting surface the pattern on the back.

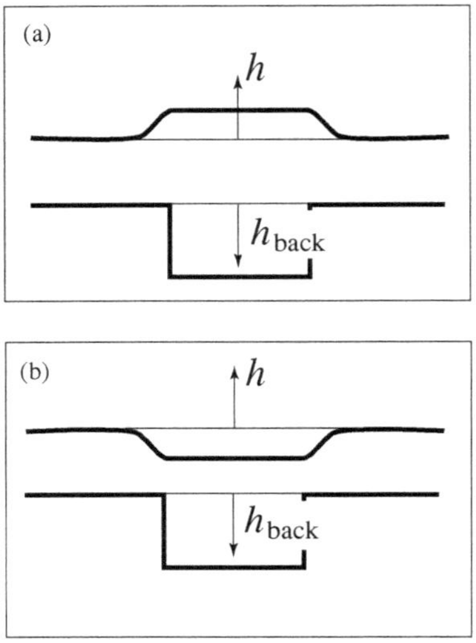

Figure 7. (a) Relief h of the reflecting surface with the same sign as the relief h_{back} embossed on the back, i.e. $a > 0$ in (11); this is the sign supported by observation. (b) As (a), but with $a < 0$, suggested by the stress-release theory but not by observation.

Acknowledgment

My research is supported by the Royal Society of London.

Appendix. Diffraction

For magic mirrors, the Laplacian-image intensity (10) is a good approximation to the full geometrical-optics theory (7). How accurate is geometrical optics as an approximation to physical (wave) optics? To investigate this, we represent the reflected light wave ψ, with wavenumber $k = 2\pi/\lambda$ for light of wavelength λ, as a Fresnel (paraxial) diffraction integral. From the optical path length (3), and in terms of the variables (5), the integral, normalized to unity when $h(\mathbf{r}) = 0$, is

$$\psi(\mathbf{r}, Z) = -\mathrm{i}\frac{k}{\pi Z}\iint_{\text{mirror}} \mathrm{d}\mathbf{r}' \exp\{\mathrm{i}k[(\mathbf{r}' - \mathbf{r})^2/Z - 2h(\mathbf{r}')]\}. \tag{A.1}$$

Geometrical optics emerges in the familiar way, as the large k asymptotic approximation obtained by the stationary-phase method [19], which selects the rays (6) corresponding to the values of $\mathbf{r}'$ that contribute coherently to the integral.

To investigate the quality of the approximation, we integrate (A.1) numerically, with the profile (14) corresponding to a single step. With the dimensionless variables (15), and

$$\kappa \equiv kh_0, \tag{A.2}$$

(A.1) becomes, in terms of the variable $\tau = \xi' - \xi$,

$$\psi(\xi, \zeta, \kappa) = \frac{\exp\left(-\frac{1}{4}\mathrm{i}\pi\right)}{\sqrt{\pi}}\int_{-\infty}^{\infty} \mathrm{d}\tau \exp\{\mathrm{i}[\tau^2 - \kappa\,\mathrm{erf}(\xi + \tau\sqrt{\zeta/\kappa})]\}. \tag{A.3}$$

The integral converges fast enough for convenient numerical evaluation if the contour is deformed into a complex path with $\tau = \sigma \exp(\mathrm{i}\pi/8)$ $(-\infty < \sigma < +\infty)$. Choosing $\zeta = 0.482$

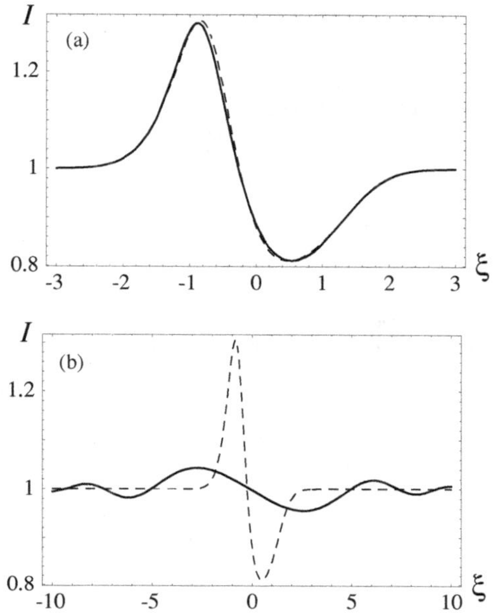

Figure 8. (a) Comparison of wave intensity $I = |\psi|^2$, computed from (A.3) (full curve) with the geometrical-optics intensity (17) (dashed curve), for the empirical values $\zeta = 0.482$, $\kappa = 3.65$. (b) As (a), with $\kappa = 0.05$, corresponding to a step height $h_0 = 5.2$ nm, and with the wave intensity computed from (A.4) (indistinguishable from (A.3) for this case).

(section 4), and representing visible light by wavelength $\lambda = 650$ nm, so that (A.2) and the height $h_0 = 378$ nm give $\kappa = 3.65$, we obtain the image shown in figure 8(a). Evidently geometrical optics is an excellent approximation.

The fact that $h_0 = 378$ nm is smaller than the wavelengths in visible light does not imply that the Laplacian image is the small-κ limit of (A.3), namely the perturbation limit corresponding to infinitely weak relief. Indeed it is not: the perturbation limit, obtained by expanding the exponential in (A.3) and evaluating the integral over τ, with a renormalized denominator to incorporate the known limit $I = 1$ for $\xi = \pm\infty$, is

$$\psi_{\text{pert}}(\xi, \zeta, \kappa) = \frac{1 - \mathrm{i}\kappa \,\mathrm{erf}(\xi/\sqrt{1 + \mathrm{i}\zeta/\kappa})}{\sqrt{1 + \kappa^2}}. \tag{A.4}$$

For the gentlest steps, this predicts low-contrast oscillatory images, very different from the Laplacian images of geometrical optics; this is illustrated in figure 8(b), calculated for $k = 0.05$, corresponding to $h_0 = 5.2$ nm.

References

[1] Auckland G 2001 Magic Mirrors, or Through the Looking Glass http://www.grand-illusions.com/magicmirror/magmir1.htm
[2] Ayrton W E and Perry J 1878–9 The magic mirror of Japan *Proc. R. Soc.* **28** 127–48
[3] Thompson S P 1897 *Light, Visible and Invisible* (London: Macmillan)
[4] Bragg W L 1933 *The Universe of Light* (London: G Bell)
[5] Kugimiya K 1990 'Makyoh': the 2000 year old technology still alive *J. Cryst. Growth* **103** 420–2
[6] Hahn S, Kugimiya K and Yamashita M 1990 Characterization of mirror-like wafer surfaces using the magic-mirror method *J. Cryst. Growth* **103** 423–32

[7] Tokura S, Fujino N, Ninomiya M and Masuda K 1990 Characterization of mirror-polished silicon wafers by Makyoh method *J. Cryst. Growth* **103** 437–42
[8] Shiue C-C, Lie K-H and Blaustein P R 1992 Characterization of deformations and texture defects on polished wafers of III-V compound crystals by the magic mirror method *Semicond. Sci. Technol.* **7** A95–7
[9] Swinson D B 1992 Chinese 'magic' mirrors *Phys. Teach.* **30** 295–9
[10] Yan Y-L 1992 Three demonstrations from ancient Chinese bronzeware *Phys. Teach.* **30** 341–3
[11] Mak S-Y and Yip D-Y 2001 Secrets of the Chinese magic mirror replica *Phys. Educ.* **36** 102–7
[12] Riesz F 2000 Geometrical optical model of the image formation in Makyoh (magic-mirror) topography *J. Phys. D: Appl. Phys.* **33** 3033–40
[13] Berry M V and Upstill C 1980 Catastrophe optics: morphologies of caustics and their diffraction patterns *Prog. Opt.* **18** 257–346
[14] Nye J F 1999 *Natural Focusing and Fine Structure of Light: Caustics and Wave Dislocations* (Bristol: Institute of Physics Publishing)
[15] Marr D and Hildreth E 1980 Theory of edge detection *Proc. R. Soc. B* **207** 187–217
[16] Chen J S and Medioni G 1989 Detection, localization and estimation of edges *IEEE Trans. Pattern Anal. Mach. Intell.* **11** 191–8
[17] Baraniak R G 1995 Laplacian Edge Detection http://www.owlnet.rice.edu/~elec539/Projects97/morphjrks/laplacian.html
[18] Wolfram S 1996 *The Mathematica Book* (Cambridge: Cambridge University Press)
[19] Born M and Wolf E 1959 *Principles of Optics* (London: Pergamon)

Contemporary Physics, 2015
Vol. 56, No. 1, 2–16, http://dx.doi.org/10.1080/00107514.2015.971625

Taylor & Francis
Taylor & Francis Group

Nature's optics and our understanding of light

M.V. Berry*

H H Wills Physics Laboratory, University of Bristol, Bristol BS8 1TL, UK

(Received 8 September 2014; accepted 23 September 2014)

Optical phenomena visible to everyone have been central to the development of, and abundantly illustrate, important concepts in science and mathematics. The phenomena considered from this viewpoint are rainbows, sparkling reflections on water, mirages, green flashes, earthlight on the moon, glories, daylight, crystals and the squint moon. And the concepts involved include refraction, caustics (focal singularities of ray optics), wave interference, numerical experiments, mathematical asymptotics, dispersion, complex angular momentum (Regge poles), polarisation singularities, Hamilton's conical intersections of eigenvalues ('Dirac points'), geometric phases and visual illusions.

Keywords: refraction; reflection; interference; caustics; focusing; polarisation

1. Introduction

Natural optical phenomena have been the subject of many studies over many centuries, and have been described many times in the technical [1,2] and popular [3–5] literature. Yet another general presentation would be superfluous. Instead, as my way of celebrating the International Year of Light, this article will have a particular intellectual emphasis: to bring out connections between what can be seen with the unaided, or almost unaided, eye and general explanatory concepts in optics and more widely in physics and mathematics – to uncover the arcane in the mundane. In addition, I will make some previously unpublished observations concerning several curious optical effects.

Each section will describe a particular phenomenon, or class of phenomena, which I try to present in the simplest way compatible with my theme of underlying concepts. The sections are almost independent and can be read separately.

A disclaimer: The historical elements in what follows should not be interpreted as scientific history as practised by professionals, where it is usual to study the contributions of scientists in the light of the times in which they lived. My approach is different: to consider the past in the light of what we know today – for the simple reason that the scientific contributions we remember are those that have turned out to be fruitful in later years or even (as we will see) later centuries. The significance of the past changes over time.

2. Rainbows: the power of numerical experiments

Figure 1 is Roy Bishop's photograph of a primary rainbow accompanied by a faint secondary bow. It is iconic because the house in the picture is Isaac Newton's birthplace and it was Newton who gave the first explanation of the rainbow's colours [6]. We will see later why the picture is ironic as well as iconic.

To a physicist, the colours are a secondary feature, associated mainly with the dependence of refractive index of water on wavelength (optical dispersion). More fundamental is the very existence of a bright arc in the sky. This had been explained by Descartes in 1638 [7]. To calculate the paths of light rays (Figure 2(a)) refracted into and out of a raindrop, with one reflection inside, he used the law of refraction that he probably discovered independently, though we associate it with Snel, who knew it already (and it was known to Harriot several decades before, and to Ibn Sahl half a millennium earlier [8]).

Nowadays, we would use elementary trigonometry to find the deviation D of a ray incident on the drop with impact parameter x, if the refractive index is n:

$$D(x) = \pi - \left(4\sin^{-1}\frac{x}{n} - 2\sin^{-1}x\right). \quad (1)$$

The calculation reveals a minimum deviation (Figure 2(b)) given by

$$\frac{\mathrm{d}D}{\mathrm{d}x} = \frac{2\left(\sqrt{1-x^2} - 2\sqrt{n^2-x^2}\right)}{\sqrt{(1-x^2)(n^2-x^2)}} = 0 \quad \text{for } x = x_{\min} = \sqrt{\frac{4-n^2}{3}}, \quad (2)$$

corresponding to the deviation of the rainbow ray, namely

*Email: asymptotico@bristol.ac.uk

Figure 1. Rainbow over Isaac Newton's birthplace, showing the primary bow decorated by a supernumerary bow, and a faint secondary bow. Reproduced by kind permission of Professor Roy Bishop.

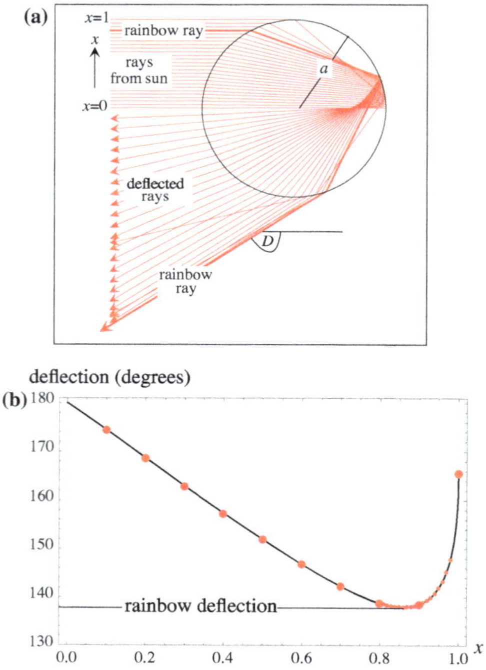

Figure 2. (a) Rays in a raindrop; showing the minimum deviation (rainbow) ray. (b) Ray deflection function, with dots indicating the points calculated by Descartes.

$$D_{\min} = D(x_{\min}) = 2\cos^{-1}\left(\frac{(4-n^2)^{3/2}}{3^{3/2}n^2}\right)$$
$$= 180° - 42.03° \text{ for } n = \frac{4}{3}. \qquad (3)$$

Descartes proceeded differently. Our facility with trigonometric calculations was not available in his time; instead, he used a geometric version of Snel's law to compute the rays laboriously, one by one. As we would say now, he performed a *numerical experiment*. The dots in Figure 2(b) correspond to the rays he calculated [7].

But why should the rainbow ray, emerging at $D_{\min}$, be bright? What about the other, more deviated, rays, illuminating the sky inside the bow? Descartes understood that although the drop is lit uniformly in x, the rays emerge non-uniformly in D. In particular, rays incident in an interval dx near $x_{\min}$ emerge concentrated into a range dD = 0. This is *angular focusing*: a lot goes into a little. The rays emerge as a *directional caustic*. The rainbow caustic is a bright cone emerging from each droplet; and we, looking up at the rain, see, brightly lit, in the form of an arc, all the drops on whose cones our eyes lie. We will encounter the concept of a caustic repeatedly in later sections of this paper; it denotes the envelope of a family of rays, that is the focal line or surface touched by each member of the family. A caustic is a holistic property of a ray family, not inherent in any individual ray. Caustics are the singularities of geometrical optics [9].

The intensity I corresponding to deviation D, given by equating the light entering in an annulus around x and emerging in a solid angle around D, is

$$|2\pi \sin D \mathrm{d}D| I \propto |2\pi x| \mathrm{d}x, \text{ i.e. } I \propto \left|\frac{x}{\sin D}\left(\frac{\mathrm{d}D}{\mathrm{d}x}\right)^{-1}\right|. \qquad (4)$$

This diverges at the rainbow angle (3), predicting, on this geometrical-optics picture, infinite intensity where dD/dx = 0. The singularity would be softened by the 1/2° width of the sun's disc and the colour dispersion.

Now look more closely at Figure 1, and notice the bright line just inside the main arc. This *supernumerary bow* did not fit into the Newton–Descartes scheme, and there seems no evidence that Newton noticed it (the term 'supernumerary means 'surplus to requirements', i.e. 'unwanted'). The explanation had to wait for nearly a century, when Young [10–12] pointed out, as an example of his wave theory of light, that supernumerary bows are *interference fringes*, resulting from the superposition of the waves associated with the two rays that emerge in each direction D away from the minimum. It is remarkable that just by looking up in the sky at this fine detail (visible in about half of natural rainbows), one sees directly the replacement of the theory of light in terms of rays – geometrical optics – by the deeper and more fundamental wave theory. This is the irony of Figure 1: interference fringes, that Newton could not explain, hovering over his house, as if in mockery.

Young understood that his two-wave picture could not be a precise description of the light near a rainbow because the intensity would still diverge at the rainbow angle, and he had the insight that – using modern terminology – a wave should be described by a smooth wave function. It took nearly 40 years for Airy [13] to provide the definitive formula for the smooth wave function near a caustic. He calculated that the intensity (Figure 3(a)) is the square of an integral, now named after him:

$$\mathrm{Ai}(x) = \frac{1}{\pi}\int_0^\infty \mathrm{d}t \cos\left(\frac{1}{3}t^3 + xt\right). \tag{5}$$

For light wavelength λ and a drop with radius a, the 'rainbow-crossing variable' x is [14]

$$x = (D_{\min} - D)\left(\frac{4\pi a}{3\lambda}\right)^{2/3} \frac{\sqrt{n^2-1}}{(4-n^2)^{1/6}}. \tag{6}$$

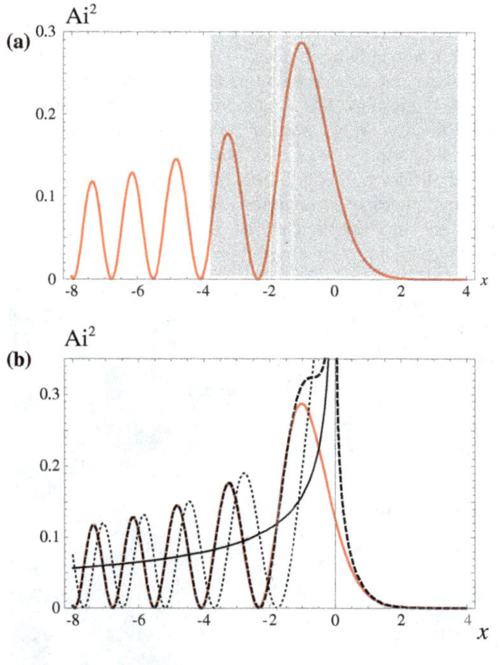

Figure 3. (a) Airy function intensity, with shading indicating the range that Airy computed by numerical integration. (b) Red curve: Airy intensity; black curve: geometrical optics approximation; dotted curve: Young's interfering ray approximation; Dashed curve: lowest-order Stokes asymptotics.

Airy realised that Ai(x) describes the wave close to any caustic, not just that associated with a rainbow.

Notwithstanding repeated attempts to understand (5), Airy did 'not succeed in reducing it to any known integral'. Therefore, he resorted to what Descartes had done two centuries before and what we continue to do today when faced with a mathematically intractable theory. He performed a *numerical experiment*: evaluating the integral by approximate summation of the integrand in increments dt – a far from trivial task given the oscillatory nature and slow convergence. The result was that he could calculate Ai(x) over the range $|x| < 3.748$ shown shaded on Figure 3(a) [13], including just two peaks of $\mathrm{Ai}^2(x)$.

This restriction to barely two intensity maxima was frustrating, because 30 supernumerary fringes had been observed in laboratory experiments with transparent spheres. What was lacking was the *asymptotics* of Ai(x): a precise description of the oscillations for $x \ll -1$ and the decay into the geometrically forbidden region $x \gg +1$. This was supplied 10 years later by Stokes [15], who showed that (Figure 3(b))

$$\frac{\cos\left(\frac{2}{3}(-x)^{3/2} + \frac{1}{4}\pi\right)}{\sqrt{\pi}(-x)^{1/4}} \underset{x\ll -1}{\leftarrow} \mathrm{Ai}(x) \underset{x\gg 1}{\leftarrow} \frac{\exp\left(-\frac{2}{3}x^{3/2}\right)}{2\sqrt{\pi}x^{1/4}}. \tag{7}$$

Stokes's paper was technically remarkable; using the differential equation satisfied by Ai(x), he 'pre-invented' what later came to be known as the WKB method and, to identify certain constants, he anticipated the method of stationary phase for oscillatory integrals. But his insight was far deeper, leading him to identify a difficulty and contribute to its solution, in a way that has proved central to contemporary mathematics.

The asymptotics (7) shows that for $x \gg +1$, Ai(x) is described by one exponential function, while for $x \ll -1$, there are two exponentials (because $\cos\theta = (\exp(\mathrm{i}\theta) + \exp(-\mathrm{i}\theta))/2$). One of these is the analytic continuation, through complex $z = x + \mathrm{i}y$, of the exponential for $x \gg +1$. But where did the other come from? The key was identified by Stokes a further decade later [16,17]: the exponentials in (7) are the first terms in formally exact expressions as *divergent infinite series*. For Re $(z) \gg +1$, i.e. on the dark side, the series is

$$\mathrm{Ai}(z) = \frac{\exp\left(-\frac{2}{3}z^{3/2}\right)}{2\sqrt{\pi}z^{1/4}} \sum_{n=0}^{\infty} (-1)^n \frac{\left(n-\frac{1}{6}\right)!\left(n-\frac{5}{6}\right)!}{2\pi n!\left(\frac{4}{3}z^{3/2}\right)^n}. \tag{8}$$

The divergence arises from the factorials: two in the numerator dominating one in the denominator. Nevertheless, for large z, the terms start by getting smaller, making the series practical for accurate numerical evaluation.

Stokes thought that the greatest accuracy is obtained by truncating the series at the least term (after which, the

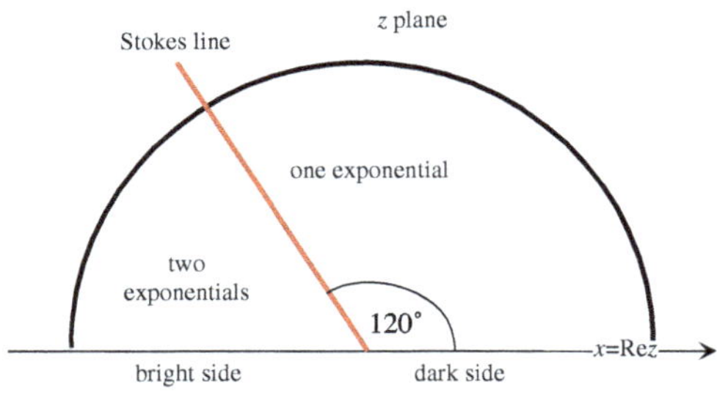

Figure 4. Complex plane of Ai(z), with the Stokes line across which the second exponential is born.

series starts to diverge), leaving a small remainder representing an irreducible vagueness in the representation of Ai(z) by the series. In a path in the upper z half-plane from z positive real to negative real (Figure 4), the small exponential on the right of (7) becomes exponentially large when $\arg z = 120°$. Near this 'Stokes line', the second exponential on the left of (7) is exponentially small – smaller, indeed, than the remainder of the truncated series, allowing it to enter Ai(z) unnoticed and then to grow into the previously problematic oscillatory second exponential for z negative real, giving the cosine interference fringes on the bright side of the rainbow.

Although Stokes was wrong in thinking that the accuracy of factorially divergent series is limited by the smallest term, his Stokes lines appear in a wide variety of functions and are seminal to our modern understanding of divergent series. We now know [18–21] that the second exponential is born from the resummed divergent tail of the series multiplying the first exponential, and in a manner that is the same for all factorially divergent series. This universality, first emphasised by Dingle [22], enables repeated resummation and computation of functions with unprecedented accuracy [23,24]. The associated technicalities are now being applied in quantum field theory and string theory [25].

We see that understanding the rainbow has been an intellectual thread linking numerical experimentation, evidence for wave optics superseding ray optics and the mathematics of divergent series.

This far from exhausts the physics associated with the rainbow. The λ dependence of the rainbow Airy argument (6) contributes diffraction colours [26] in addition to the dispersion colours explained by Newton [7]. The electromagnetic vector nature of light explains subtle polarisation detail [14]. The function Ai(x), that in its original form, e.g. with the variable (6), describes only the close neighbourhood of a caustic, can be stretched to provide a *uniform approximation* [27–29] extending the accuracy to regions far from the caustic. Analogues of rainbow scattering occurs in quantum [28,30,31] and condensed matter [32] physics. Finally, Ai(x) is now understood as the simplest member of a hierarchy of *diffraction catastrophes* [9,33,34], describing waves near caustics of increasing geometrical complexity.

3. Sparkling seas: twinkling and lifelong fidelity

Figure 5 shows images of the sun reflected by wavy water. The images we see correspond to places where the water surface is sloped to reflect light into our eye. As the waves move and the shape of the surface changes, the images move too. They appear and disappear in pairs. Such events, called 'twinkles', correspond to caustic surfaces (Figure 6(a)) in the air above the water, passing through the eye. The succession of twinkles, often too rapid for us to follow in detail, gives rise to the sparkling appearance of the water. The intricate topology of the reflections and the coalescence events, and the associated statistics for waves represented by Gaussian random functions were studied in pioneering papers by Longuet-Higgins [35–37]. The techniques he developed were applied to the statistics of caustics [9] and extended to describe phase [38] and polarisation singularities [39] in random waves.

Here, I draw attention to a statistical problem hinted at by Longuet-Higgins [35,37] but still unsolved. Each specular point is born accompanied by a partner, prompting this question: For a random moving water surface, what fraction R of specular points dies by annihilation with the partner it was born with? This property can be called 'lifelong fidelity'. The fraction R is different for different types of randomness. It seems a hard problem even for the simple case of isotropic monochromatic

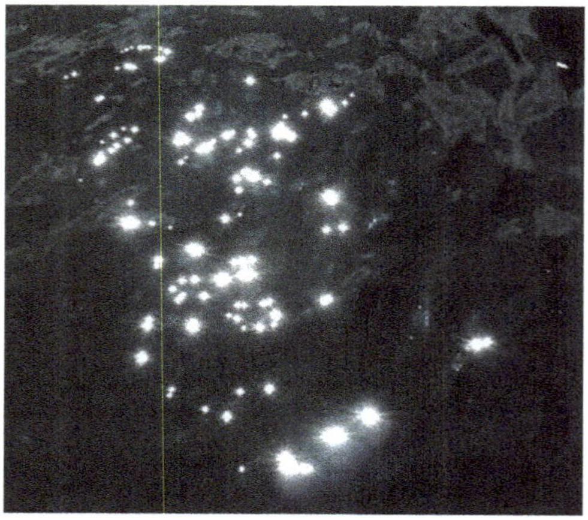

Figure 5. Images of the sun (specular points) sparkling on the water in Bristol docks.

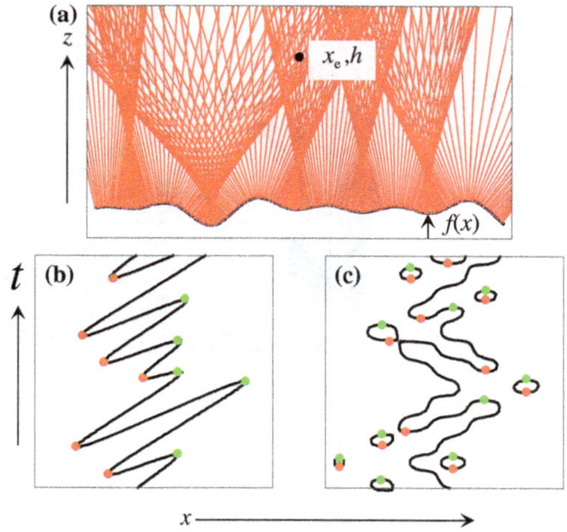

Figure 6. (a) Reflection caustics above a water surface (13). (b) Space–time plot of evolving images, for the rigidly moving surface (14), with births of image pairs indicated by red dots and deaths by green dots; (c) As (b) for the interfering waves (15), with the loops indicating images enjoying lifelong fidelity.

Gaussian randomness, because it requires statistics non-local in both space and time.

To get a little insight, I illustrate the problem by considering the simpler case of a corrugated water surface generating a reflected wavefront with height $f(x)$ depending on a single variable x (the height of the water surface is proportional to $f(x)$). For the eye located at (x_e, h), the optical path distance from a point x on the water surface is (for surfaces with gentle slopes),

$$\Phi(x; x_e, h, t) = \sqrt{(h - f(x,t))^2 + (x - x_e)^2} \approx h - f(x,t) + \frac{(x - x_e)^2}{2h}. \tag{9}$$

The specular points x_n, i.e. the rays, are the paths for which this function is stationary:

rays:
$$\partial_x \Phi = 0 \Rightarrow x - hf'(x,t) = x_e \Rightarrow x = \{x_n(x_e, h, t)\}. \tag{10}$$

Therefore, the moving images are represented by the locus

$$x - hf'(x,t) = x_e \tag{11}$$

in the (x, t) plane. The caustics at time t are the curves

$$\partial_x^2 \Phi = 0 \Rightarrow z = \frac{1}{f''(x,t)} \Rightarrow z(x_e, h, t). \tag{12}$$

To illustrate lifelong fidelity, or lack of it, I start with a stationary surface given by a random superposition of N sinusoids:

$$f_0(x) = \sum_{n=0}^{N} \cos(k_n x + \phi_n). \tag{13}$$

Figure 6(a) illustrates the caustics for a sample function of this type, with $N = 5$. If the water moves rigidly, represented by

$$f_1(x,t) = f_0(x - t), \tag{14}$$

then the caustics translate rigidly sideways and it is obvious that there is no lifelong fidelity (Figure 6(b)). Non-rigid motion can be represented by two interfering oppositely moving waves:

$$f_2(x,t) = \frac{1}{2}(f_0(x - t) + f_0(x + t)). \tag{15}$$

Now, some of the caustics move back and forth or up and down, and most images enjoy lifelong fidelity (closed loops in Figure 6(b)); for a single sinusoid ($N = 1$), *all* caustics in $f_2(x, t)$ move up and down and lifelong fidelity is universal.

For the two-dimensional case, Longuet-Higgins presents a time exposure showing paths of specular points on the water surface [35]. Some of the paths are closed loops, but in contrast to the space–time plots in the corrugated case, these do not always correspond to lifelong fidelity. This is a rich subject for further analytical study and numerical experiment.

4. Mirages, green flashes: light bent and dispersed by air

Air bends light. One consequence, again involving caustics in an essential way, is the mirage, most commonly seen on a hot day when distant cars appear reflected from the surface of a long, straight road. As is well known [3,4,40], this is really refraction masquerading as reflection: bending by the gradient of refractive index, increasing from the hot road surface to the cooler air above (Figure 7).

On the simplest theory, the index depends weakly on height z:

$$n(z) = 1 + \Delta n(z), \quad \Delta n(z) \ll 1. \tag{16}$$

From Snel's law, the direction and curvature of a ray are given by

$$n(z)\cos(\theta(z)) = n(z_{\min}) \Rightarrow z''(x) \approx \Delta n'(z), \tag{17}$$

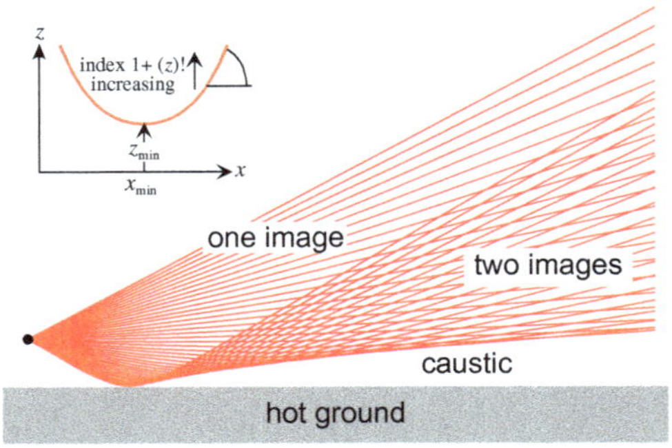

Figure 7. Mirage rays (20) for the monotonic index profile (20).

where z_{min} is the height where the ray is horizontal. For constant index gradient, i.e. $n(z)$ locally linear, this leads to parabolic ray paths:

$$\Delta n(z) = Az \Rightarrow z = z_{\text{min}} + \frac{1}{2}A(x - x_{\text{min}})^2. \tag{18}$$

(A curious sidelight on history: this simple explanation has been challenged several times over more than two centuries, on the grounds, based on a misunderstanding of Snel's law, that once a ray becomes horizontal, it could never curve upwards again [41].)

Figure 7 shows the family of ray paths from a point source, with a more realistic index function, increasing from $1 + \Delta_0$ at the ground to $1 + 2\Delta_0$ above:

$$\Delta n(z) = \Delta_0\left(2 - \exp\left(-\frac{z}{L}\right)\right). \tag{19}$$

The rays can be determined analytically:

$$x - x_{\text{min}} = \frac{1}{\sqrt{2\Delta_0}} \exp\left(-\frac{z_{\text{min}}}{L}\right) \times \left(z - z_{\text{min}} + 2L \log\left(1 + \sqrt{1 - \exp\left(-\frac{z - z_{\text{min}}}{L}\right)}\right)\right) \tag{20}$$

Each ray from the source corresponds to a different choice of x_{min} and z_{min} determined by its initial slope. The caustic of the family is clearly visible. It separates eye positions below the caustic, from which the source is invisible, from positions above, where two images can be seen. (Some rays from the source hit the ground, giving rise to an additional boundary in Figure 7: a shadow edge separating the two-image region above the caustic from a one-image region higher still.)

The caustic has a further significance. Each point of a distant object emits its own family of rays, so the complete object emits a family of families of rays. With the eye in a fixed position, the parts of the object that can be seen, bounded below by the 'vanishing line', are those whose caustics lie below the eye.

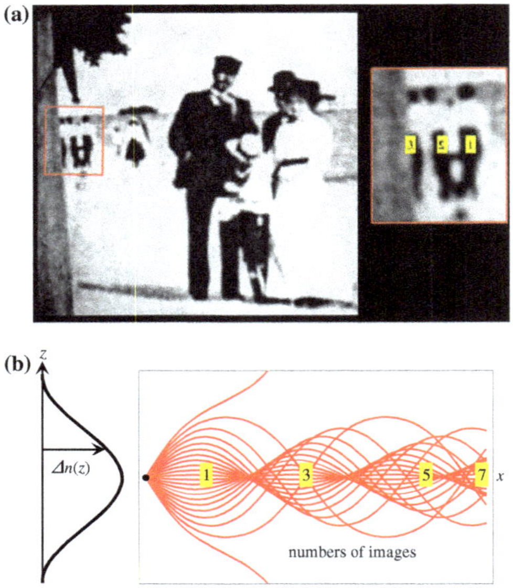

Figure 8. (a) Mirage with three images, formed by refraction near a hot wall (adapted from W. Hillers, *Phys. Z.* 14 (1913), 719–723, reproduced in [3]). (b) Mirage rays (22) in a duct, for the index profile (21) possessing a maximum.

Sometimes, more than two mirage images are visible; Figure 8(a) shows a case where there are three [3]. Such multiple images can arise when the index is not a monotonic function of height, or from undulations of the surface; refractive indices with a maximum generate a duct, where in principle any number of images can be seen. A solvable example is

$$\begin{gathered}\Delta n(z) = A\cos\left(\tfrac{z}{a}\right) \Rightarrow z''(x) = -\tfrac{A}{a}\sin\left(\tfrac{z(x)}{a}\right) \\ \{|z| < a\pi,\ z(0) = 0,\ z'(0) \equiv \theta_0\},\end{gathered} \tag{21}$$

for which, the rays, illustrated in Figure 8(b), involve the Jacobian elliptic functions sn and am [42]:

$$z(x) = 2a\sin^{-1}\operatorname{sn}\left(\frac{x\theta_0}{2a}\middle|\frac{2\sqrt{A}}{\theta_0}\right) = 2a\,\operatorname{am}\left(\frac{x\theta_0}{2a}\middle|\frac{2\sqrt{A}}{\theta_0}\right). \tag{22}$$

For more on the topology of images, see [43,44].

Mirages involve local bending of light, with the layer of heated air a few centimetres high and horizontal distances of a few hundred metres. On a much larger scale, the entire atmosphere of the spherical earth can be regarded as a lens, with index decaying to unity over a height of a few kilometres. The resulting bending elevates the image of the sun by slightly more than its own diameter (about 1/2°), so we can see it immediately before sunrise and after sunset, when it is just below the horizon.

The earth-lens suffers from chromatic aberration because the bending is weakly dispersive, with the blue being refracted more than the red. This causes the green image of the setting sun to be above the red image by about 10 arcsec, giving rise to the occasionally visible striking phenomenon of the green flash [45,46] (the blue sunlight has been scattered to make the blue sky). Important modifications of this basic explanation arise from the fact that the dispersion need not be a monotonic function of height (e.g. when there is an inversion layer) [46].

More effects of the earth-lens will be examined in the next section.

5. Earthlight on the moon: astronomical coincidences

There are two circumstances in which we see the moon lit from the earth. Near new moon (Figure 9(a)), the part of the moon's disc that is hidden from the sun can be seen in the pale ghostly light reflected by the earth – also called 'earthshine' or 'the old moon in the new moon's arms'. And in the opposite situation, during a lunar eclipse, the moon can still be seen, even though it is within the earth's shadow, in light refracted onto it by

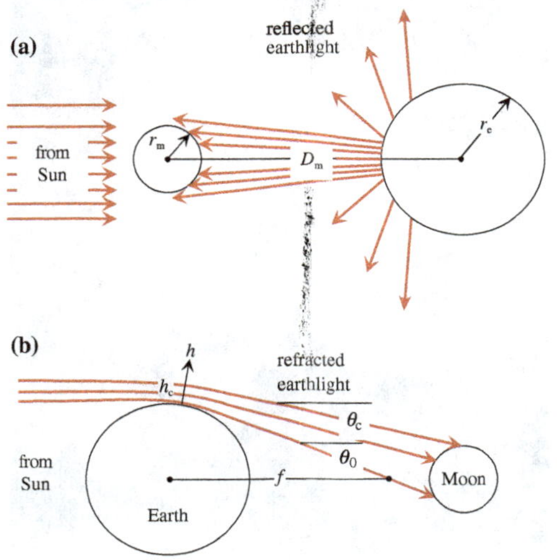

Figure 9. (a) The old moon in the new moon's arms: earthlight reflected onto the moon at new moon. (b) earthlight refracted onto the moon during a lunar eclipse.

the earth's atmosphere (Figure 9(b)). This eclipse light corresponds to all the simultaneous sunrises and sunsets on the earth, blood-reddened by double passage through the atmosphere.

These two earthlights are of very different origins. Nevertheless, casual viewing suggests that the brightness of the earth-lit moon is roughly similar in the two cases. In this section, I outline calculations supporting this observation.

At new moon, the sun illuminates the disc of the earth that faces the moon. The earth's albedo A denotes the fraction of this light that is diffusely reflected into (approximately) a hemisphere of the sky. Most of it disappears into space, but a small fraction hits the moon. Neglecting obliquity effects, the fraction of sunlight incident on the earth that lights up the new moon is (Figure 9(a))

$$F_{\text{new}} = A\frac{\pi r_{\text{m}}^2}{2\pi D_{\text{m}}^2} = \frac{A r_{\text{m}}^2}{2D_{\text{m}}^2}. \tag{23}$$

During a lunar eclipse, the sunlight reaching the moon corresponds to a thin annulus of height h_c, below which rays passing close to the earth's surface are bent onto the moon (Figure 9(b)). During its passage through the atmosphere, this light is attenuated by a factor α. Thus, the fraction of incident sunlight that lights up the eclipsed moon is

$$F_{\text{eclipse}} = \alpha\frac{2\pi h_c r_e}{\pi r_e^2} = \frac{2\alpha h_c}{r_e}. \tag{24}$$

We need to calculate the ratio of these two earthlights, namely

$$R \equiv \frac{F_{\text{new}}}{F_{\text{eclipse}}} = \frac{A r_{\text{m}}^2 r_e}{4\alpha h D_{\text{m}}^2}. \tag{25}$$

Some of the required numbers are

$$r_e = 6371\,\text{km},\ r_{\text{m}} = 1737\,\text{km}, \quad D_{\text{m}} = 384{,}000\,\text{km},\ A \sim 0.3. \tag{26}$$

In addition, we need h_c and α.

To calculate h_c, we need the angular deflection of sunrays passing the earth at height h above the ground. This can be calculated from Snel's law, but it is simpler to use the observed deviation θ_0 of a ground-grazing ray (twice the elevation of the setting sun) and assume an exponential atmosphere with scale height H. Thus

$$\theta(h) = \theta_0 \exp\left(-\frac{h}{L}\right),\ L = 8\,\text{km},\ \theta_0 = 1.18^\circ = 0.0206. \tag{27}$$

An interesting related fact (apparently first noticed by John Herschel) is that the focal length of the earth-lens, determined by the distance from the earth that a grazing ray crosses the sun–earth axis, is relatively close to the moon:

$$f = \frac{r_e}{\theta_0} = 309{,}349\,\text{km} = D_m - 74{,}650\,\text{km}. \qquad (28)$$

It is easy to show that this grazing ray hits the moon after crossing the axis at the focus. Non-grazing rays do not cross at the focus (the earth-lens has powerful spherical aberration), and h_c is determined by the deflection which hits the moon's edge (Figure 9(b)):

$$\theta_c = \frac{r_e - r_m}{D_m} \Rightarrow h_c = L \log\left(\frac{\theta_0 D_m}{r_e - r_m}\right) = 4.28\,\text{km}. \qquad (29)$$

The final number we need is the attenuation α. This is the square of the attenuation of the setting sun, namely the brightness of the setting sun relative to the unattenuated (roughly noonday) sun. We know it is a small number because we can gaze at the setting sun but not the noonday sun. α is small, but it is not zero, indicating (if proof were needed) that the earth is not flat: if it were, the sun setting over the faraway edge of the world would be invisible.

In reality, α is enormously variable and involves both scattering and absorption. But we can estimate it from first principles using the following formula for the ground-level attenuation per unit distance, resulting from scattering in clear air [47], in terms of λ, refractive index deviation $\Delta_0 = n - 1$ and the molecular particle density N:

$$\gamma_0 = \frac{2}{3\pi}\left(\frac{2\pi}{\lambda}\right)^4 \frac{\Delta_0^2}{N}. \qquad (30)$$

Reasonable numbers give

$$\lambda = 5 \times 10^{-7}\,\text{m},\ \Delta_0 = 0.000292,\ N = 2.5 \times 10^{25}\,\text{m}^{-3}$$
$$\Rightarrow \gamma_0 = 1.8 \times 10^{-2}\,\text{km}^{-1}. \qquad (31)$$

Using this to calculate the exponential attenuation of a ray passing at height h and averaging from $h = 0$ to $h = h_c$, gives

$$\alpha = \frac{1}{h_c}\int_0^{h_c} dh \exp\left(-\gamma_0\sqrt{2\pi L r_e}\exp\left(-\frac{h}{L}\right)\right)$$
$$= 6.8 \times 10^{-4} = (0.0261)^2. \qquad (32)$$

(The square root 0.0261 is an estimate – apparently reasonable – of the attenuation of the setting sun: about 16 dB.)

Thus, finally we get, from (25), the ratio of earthlight intensities:

$$R = \frac{A r_m^2 r_e}{4\alpha D_m^2 L \log\left(\frac{\theta_0 D_m}{r_e - r_m}\right)} \sim 3.4. \qquad (33)$$

This number should not be taken seriously as a precise estimate. But it does indicate that the two very different illuminations of the moon by the earth – at new moon from the brightly lit 'full earth' when most of the light is squandered into space, and during a lunar eclipse when the thin annulus of deflected light is efficiently focused onto the moon – are of comparable strengths. For an interesting related discussion of the visibility of the horizon at different heights, see [48].

6. The glory: focusing that vanishes geometrically

The optical glory is a halo around the shadow of an illuminated observer's head cast by the sun on a cloud or mist-bank (Figure 10). Nowadays, it is most commonly seen while flying in sunshine, looking down at the airplane's shadow on a cloud below. It took several centuries for this beautiful phenomenon to be fully understood [14,49], following its first recorded observation [4],

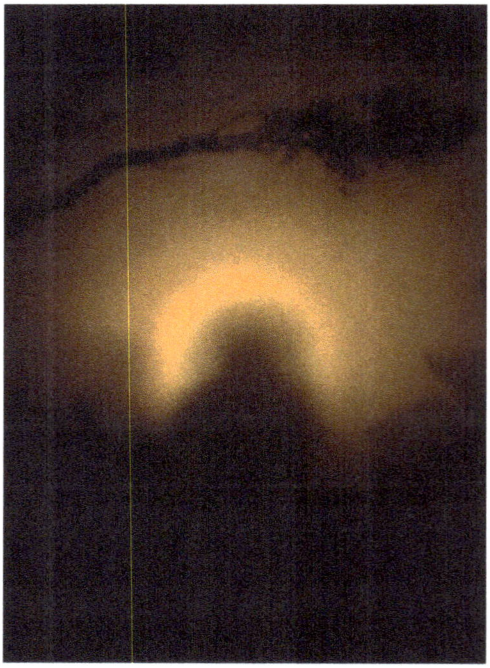

Figure 10. Glory around the shadow of the author's head on a cloud below Erice, Sicily, from light illuminating a temple in the village above.

because its explanation involves subtle connections between concepts usually regarded as separate.

First, note that since the glory appears at the edge of the observer's shadow on a cloud, it must be a back-scattering phenomenon associated with water droplets. It involves reflection from individual drops and so should not be confused with the much weaker Anderson localisation, which is back-scattering enhancement arising from the coherent interference of light multiply scattered by many droplets [50,51].

Imagine (counterfactually as it will turn out) that there is a light ray, incident non-axially (i.e. with finite impact parameter x) that enters the droplet, gets reflected once inside, and then emerges precisely backwards, i.e. with deflection $D(x) = \pi$ (Figure 11(a)). Rays near this x will cross the symmetry axis, and associated with this rotational symmetry would be an entire ring of such rays, giving rise to focusing in the backward direction [47]: an *axial caustic*, represented by the singularity from $1/\sin D$ in (4) when $D = \pi$. It follows from (1) that such a ray exists with impact parameter x if the index n is

$$D(x) = \pi \Rightarrow n(x) = \frac{x\sqrt{2}}{\sqrt{1-\sqrt{1-x^2}}}. \tag{34}$$

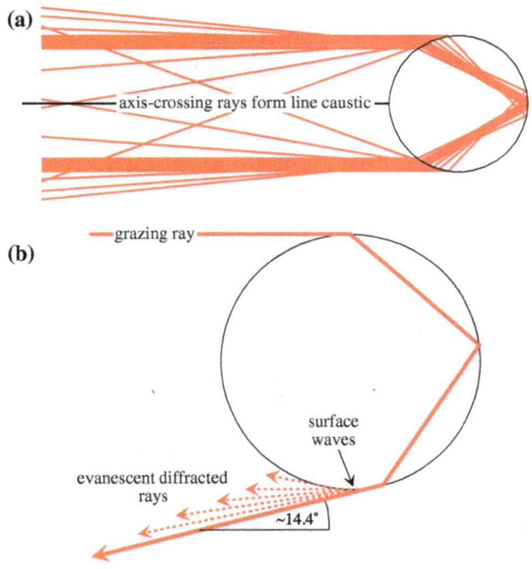

Figure 11. (a) Rays emerging backwards from a transparent sphere with index $n = 1.7$, and crossing the symmetry axis to form a focal line. (b) Diffracted rays creeping round the surface of a water droplet, generating the dominant contribution to the glory when they cross the symmetry axis.

As x increases from 0 to 1, $n(x)$ decreases from 2 to $\sqrt{2}$. But the refractive index of water, approximately 4/3, lies outside this range, so there are no axis-crossing rays: the glory enhancement cannot be explained as a focusing effect within geometrical optics. (Backward ray focusing from spheres of glass or plastic with indices in the range $\sqrt{2} < n < 2$, finds application in retro-reflecting paint.)

To understand the true mechanism of nature's glory, we note that although the rays do not reach the backward direction for water, they get close. For the grazing rays, with $x = 1$, the deflection is

$$D(x=1) = \pi - \left(4\sin^{-1}\left(\frac{1}{n}\right) - \pi\right)$$
$$\Rightarrow 180° - 14.36° \text{ if } n = \frac{4}{3}. \tag{35}$$

To accommodate the 14° shortfall, some of the emerging light gets trapped into a *surface wave* that creeps around the surface of the droplet, radiating tangentially while doing so (Figure 11(b)) and reaching $D = 180°$ and beyond. These axis-crossing evanescent surface waves form the backward caustic and contribute to the glory. According to this mechanism, the back-scattered intensity I_b, for light wavelength λ and drop radius a, is [14,49]

$$I_b \sim \left(\frac{a}{\lambda}\right)^{8/3} \exp\left(-\text{constant}\left(\frac{a}{\lambda}\right)^{1/3}\right), \tag{36}$$

in which the first factor describes axial focusing, the exponential represents the decay as the creeping wave skips the 14° while radiating, and the powers 1/3 come from the fact that the surface is a caustic of the creeping waves.

A full analysis of electromagnetic waves scattered from transparent spheres [49,52] reveals great complexity beyond the creeping-and-focusing picture. This includes: waves skipping many times inside the drop before emerging, causing high-order backward rainbows; very fine angle- and size-dependent interference oscillations; and delicate polarisation effects. But the exponential angular decay is a dominant feature, described analytically in terms of *complex angular momentum*: an unexpected application to this natural phenomenon of the *Regge poles* (complex angular-momentum singularities) devised to explain the quantum scattering of elementary particles.

From (36), we see that the intensity vanishes for large droplets because the focusing enhancement is cancelled by the evanescence (decay during the 14° skip) and also for very small droplets, where the focusing is softened by diffraction. That is why the glory is observed in the small droplets occurring in clouds or mist, but not in rain where the droplets are bigger.

Summing up: *the glory is a focusing effect that vanishes in the geometrical-optics limit.*

7. Hidden daylight: polarisation singularities in the sky

A fundamental aspect of light waves is that they are electromagnetic. Therefore, they can be polarised. But we (unlike some other animals) possess only the most rudimentary perception of polarisation, blinding us to a beautiful polarisation pattern decorating the daylight sky above us [53]. Although sunlight arrives unpolarised, the Rayleigh (dipole) scattering from air molecules that is responsible for the blue sky induces polarisation [3]. Scattered sunlight is strongly polarised perpendicular to the sun (as can immediately be verified with a polarising sheet), whereas forward and back-scattered sunlight remains unpolarised.

If sunlight were scattered only once, daylight would be unpolarised in the direction of the sun and in the opposite direction, namely the anti-sun (visible before sunrise and after sunset). But in reality, the number of unpolarised points (directions) in the sky is not two, but four: several degrees above and below the sun and the anti-sun. Three were observed in the nineteenth century and the fourth, below the anti-sun, was seen only recently, from a balloon [54]. The reason for four is that each single-scattering unpolarised direction is split into two by multiple scattering.

The sky is decorated by a pattern of polarisation directions (e.g. of the electric vector of daylight) organised by these four points, only two of which are visible at any time. Figure 12(a) shows the pattern at a time when the sun is in the indicated position. The key to understanding it, discovered only relatively recently [55] by emphasising something lacking in a tradition of elaborate multiple-scattering theory [56–58], is geometric: the realisation that the unpolarised points are *polarisation singularities*: places where the direction of polarisation is undetermined.

Near each singularity, the geometry is that of a 'fingerprint', around which the polarisation direction turns by 180°. The turn is 180°, rather than 360°, because a half-turn leaves polarisation unchanged: polarisation is not a vector, but a direction without a sense (though, of course, it is a consequence of the vector nature of light). And the turn, in the same sense as each unpolarised point is encircled, means that the singularity index is +1/2, rather than −1/2. Since there are four unpolarised points, the total index on the sphere of sky directions is +2, consistent with the Poincaré–Hopf theorem [59]: any smooth direction field on a sphere must have index +2.

The quantitative description of the pattern is provided by representing the polarisation by a complex function of sky direction, with zeros at the four unpolarised points. The simplest such function is a quartic polynomial and leads to the following theory [55]. The sky direction corresponding to elevation θ and azimuth ϕ is also represented, in stereographic projection, by a complex number:

$$\zeta = x + \mathrm{i}y = \frac{\left(1 - \tan\frac{1}{2}\theta\right)}{\left(1 + \tan\frac{1}{2}\theta\right)} \exp(\mathrm{i}\phi). \tag{37}$$

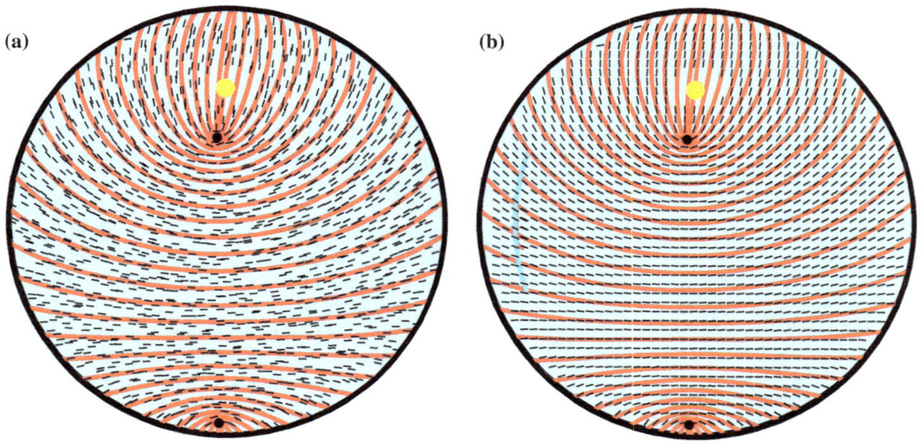

Figure 12. Line segments: observed directions of polarised daylight in the day sky; full curves: theory (39). (a) Line segments randomised within each cell of the measurement grid. (b) Line segments before randomisation, as originally published [55].

In this projection of the sky onto a plane, the visible hemisphere is represented by the unit disc whose centre is the zenith and boundary is the horizon. For sun elevation α and splitting δ of each singularity pair, we define

$$y_s = \frac{(1-\tan\frac{1}{2}\alpha)}{(1+\tan\frac{1}{2}\alpha)}, \quad A = \tan\frac{1}{4}\delta, \quad \zeta_t = \frac{\zeta + iy_s}{1+i\zeta y_s}. \quad (38)$$

Then, there is a function $f(x, y)$ whose contours are parallel to the polarisation direction at the sky point (x, y), namely

$$f(x,y) = \mathrm{Im}\int^{\zeta_t} \frac{du}{\sqrt{-(u^2+A^2)(u^2+A^{-2})}} = A\,\mathrm{Im}F\left(\sin^{-1}\left(\frac{i\zeta_t}{A}\right), A^2\right), \quad (39)$$

in which F is the Legendre elliptic integral of the first kind [44]. As Figure 12(a) illustrates, this theory, based on *the elliptic integral in the sky*, gives a very accurate description of the observed polarisation pattern.

Unfortunately, the picture published in the paper [55] reporting the theory and experiment was not Figure 12(a) but Figure 12(b), which looks rather different. In fact, both pictures represent exactly the same data. In Figure 12(b), the experimental line segments representing the polarisation directions are centred on the point of the rectangular grid on which they were measured; the eye is drawn to the grid, rather than the directions of the line segments. A simple way to eliminate this misleading perception (understood only after Figure 12(b) was published) is to randomise the positions of the line segments within each unit cell of the grid. The result is Figure 12(a): the grid is no longer visible, and the agreement between theory and experiment is much clearer.

An interesting and controversial speculation [60,61] is that the Vikings, in their tenth-century voyages between Norway, Iceland and what is now Canada, might have used the sky polarisation pattern in conjunction with natural birefringent crystals (e.g. Iceland spar), as an aid to navigation.

8. Light in crystals: Hamilton's cone

Many natural transparent crystals are optically anisotropic and so are obvious materials for light to exhibit its polarisation properties. In each direction in such a crystal, two plane waves can travel, with orthogonal polarisations and different refractive indices [62]. A polar plot of the refractive indices in direction space generates the two-sheeted 'Fresnel wave surface', or, as we would call it now, the momentum-space contour surfaces of the Hamiltonians governing each of the two waves. In the most general crystals, all three principal dielectric constants are different, and, as Hamilton discovered [63,64], they intersect at four points on two 'optic axes'. The intersections take the form of double cones (diabolos).

There is beautiful physics associated with Hamilton's cones. In 1830, Hamilton himself made the first physical prediction based on his concept of phase space: that light incident on a thick slab of crystal along an optic axis would spread into a cone and emerge as a hollow cylinder. The immediate experimental observation of 'conical refraction' by his colleague Lloyd [65–67] created a sensation and brought instant fame to the young Hamilton. This story is described elsewhere [68], together with its modern developments.

A different (and more easily reproduced) demonstration of the cone geometry is illustrated in Figure 13. This shows the diffuse light of the sky viewed through a 'black light sandwich' [69]. The 'bread' consists of two crossed polarising sheets, and would allow no light to pass if there were nothing between them. But between the sheets is the 'filling' of the sandwich, consisting of a sheet of overhead projector transparency film: a material which, though not crystalline, is biaxially anisotropic. The simplest theory explaining the 'conoscopic' Figure 13 is the following.

Light passing through the transparency 'crystal' in directions, close to the optic axis, specified by coordinates $(x, y) = r(\cos\phi, \sin\phi)$ and propagating along z, can be described by a two-component vector $|\psi\rangle$ representing the linear polarisation amplitudes along directions perpendicular to z. The evolution of the polarisation as the light passes through the transparency, whose thickness is l, is determined by a Schrödinger-lookalike equation, written for wavenumber k,

$$\frac{i}{k}\partial_z|\psi\rangle = \hat{H}|\psi\rangle,$$
$$\hat{H} = \begin{pmatrix} y & x \\ x & -y \end{pmatrix} = r\begin{pmatrix} \cos\phi & \sin\phi \\ \sin\phi & -\cos\phi \end{pmatrix} \quad 0 \leqslant z \leqslant l. \quad (40)$$

The Hamiltonian $\hat{H}$ describes the light in direction (x, y), whose eigenvalues $\pm r$ describe the refractive index diabolo, and the eigenvectors are the orthogonal polarizations

$$\hat{H}|\pm\rangle = \pm r|\pm\rangle, \quad |+\rangle = \begin{pmatrix} \cos\frac{1}{2}\phi \\ \sin\frac{1}{2}\phi \end{pmatrix}, \quad |-\rangle = \begin{pmatrix} \sin\frac{1}{2}\phi \\ -\cos\frac{1}{2}\phi \end{pmatrix}. \quad (41)$$

Passing through the sandwich in the sequence 'polarizer-propagation-analyzer' is formally analogous to the sequence 'preparation-evolution-measurement' for quantum states, and the resulting direction-dependent intensity, representing the view through the sandwich, is

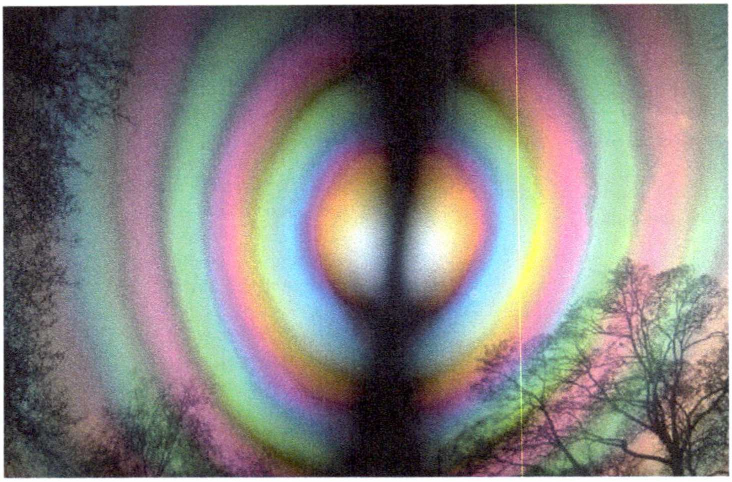

Figure 13. Bull's eye in the sky above Bristol, photographed through a black light sandwich: biaxially anisotropic overhead-transparency foil between crossed polarizers.

$$\begin{aligned} I(x,y) &= \left|\langle \text{analyzer}|\text{propagation matrix } \exp(-\mathrm{i}kl\hat{H})|\text{polarizer}\rangle\right|^2 \\ &= \left|\begin{pmatrix} 0 & 1 \end{pmatrix}\begin{pmatrix} \cos lr - \mathrm{i}y\frac{\sin klr}{r} & -\mathrm{i}x\frac{\sin klr}{r} \\ -\mathrm{i}x\frac{\sin klr}{r} & \cos lr + \mathrm{i}y\frac{\sin klr}{r} \end{pmatrix}\begin{pmatrix} 1 \\ 0 \end{pmatrix}\right|^2 \\ &= \frac{x^2\sin^2 klr}{r^2}. \end{aligned} \tag{42}$$

This clearly explains Figure 13. The $\sin^2$ factor describes circular interference fringes as the loci of constant separation of the sheets of the diabolo, centred on the 'bull's eye' where the two indices are degenerate. The $(x/r)^2$ factor describes the vertical black 'brush', which is ultimately a consequence of the half-angles in (41), according to which the eigenpolarizations change sign (π phase change) in a circuit of the optic axis $r = 0$.

I regard this π-phase change, observed in Lloyd's 1831 conical refraction experiment, as the first *geometric phase*, anticipating all of those being studied today [70,71]. And Hamilton's cone is the prototype of all the *conical intersections* now being studied in theoretical chemistry [72] and condensed matter physics, and popularly referred to as 'Dirac cones' [73]. In my opinion, this is a misnomer: Dirac never drew or even mentioned cones in this context, and the linear dependence of eigenvalues on momentum, leading to the Dirac equation, is exactly the geometry that Hamilton emphasised in the1830s.

Conoscopic figures, like Figure 13, are familiar to mineralogists. Several new intensity patterns have been predicted [74], but not yet observed, for crystals that are chiral and anisotropically absorbing as well as biaxially birefringent, and with different combinations of polarizer and analyzer.

9. The squint moon: a projection illusion

It is possible to see the sun and moon in the sky simultaneously at different periods of the day during each month, provided the sky is clear. The part of the moon's disc that we see, waxing from crescent to half to gibbous as the moon's phase changes from new to full, corresponds to our view of the hemisphere that is lit by the sun (Figure 14(a)). Therefore, we might expect the normal to the lit face (i.e. the normal to the line joining the horns of the moon) to point towards the sun. But it does not: the lit face points above the sun, to an extent that increases between new moon and full moon. This is the unexpected and striking 'squint moon' phenomenon [75–77]. Its explanation is a combination of geometry and perception.

It is obvious that in three-dimensional space, the normal to the lit hemisphere points directly to the sun. Therefore, the squint must be an illusion. It has alternatively been called the *New* Moon Illusion [78], to distinguish it from the unrelated more familiar moon illusion in which the moon appears large near the horizon. The squint illusion can be immediately dispelled [3] by stretching a string taut close to the eye and orienting it from the moon to the sun.

To understand the illusion, we first note that the sun and moon are too distant for our stereoscopic vision to

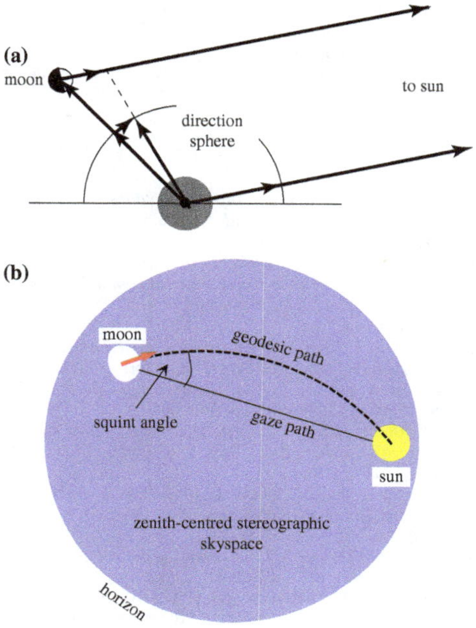

Figure 14. (a) Moon–sun geometry in space and on the direction sphere. (b) Simulation of the squint moon as seen in 'skyspace', modelled by zenith-centred stereographic projection from the direction sphere.

operate, so we cannot perceive their arrangement in three-dimensional space. We cannot see distances; all we can perceive are directions. Each direction can be regarded as a point on an earth- (or eye-) centred unit sphere. On this sphere, the straight line in space joining the sun and moon projects onto a great circle: a geodesic curve. But we do not perceive this sphere directly. Instead, we see objects in the sky, and relations between them, as though projected onto an imaginary screen, roughly flat. Call this 'skyspace'. The squint moon illusion is related to the projection from the direction sphere onto skyspace.

I do not know which projection our visual system chooses, or even if it is the same at different times and for different people. But the illusion is remarkably stable against the change from one projection to another. To illustrate this, Figure 14(b) shows the stereoscopic projection of skyspace according to (37), mimicking how we might see the sky when lying on our back looking up. The squint is very clear: the geodesic connecting the moon and sun on the direction sphere projects onto a circular arc in skyspace, whose tangent at the position of the moon points in a different direction to the straight 'gaze path' from the moon to the sun in skyspace.

Zenith-centred stereographic projection is particularly interesting because on this skyspace the horizon is curved; yet the squint is strong, apparently contradicting claims [75] that the illusion depends on seeing the horizon as straight. It would be interesting to see if the squint persists in space far above the earth, where there is no horizon.

In most skyspaces, geodesics on the direction sphere appear projected as curves; stereographic projection is just one example. But Professor Zeev Vager (personal communication) points out an important class of projections for which great circles on the sphere appear as straight lines, namely the perspective ('pinhole') projections. For these skyspaces, the squint arises because perspective does not preserve angles: it is not conformal. In particular, the angle in skyspace, between the horns of the moon and the (now straight) line from the sun to the moon, is not 90°.

10. Concluding remarks: the sky as an optics laboratory

It should be clear from my eclectic series of examples that natural optical phenomena illustrate, and have been implicated in the development of, a surprisingly large number of scientific concepts:

- *Caustics* – the singularities of geometrical optics – are central to understanding rainbows, mirages, and the sparkling of light on water.
- *Numerical experiments* played a central part in understanding both the ray and wave aspects of rainbows. Replacing numerical experiments by analysis led to a seminal insight into
- *Mathematical asymptotics*, still being developed now, with new classes of applications.
- *Dispersion* explains the green flash, as well as the rainbow colours in geometrical optics.
- *Complex angular momentum*, in the form of
- *Regge poles*, supplied the key to understanding the glory. Several numerical
- *Coincidences* underlie the comparable brightnesses of the two earthlights on the moon (less familiar than the coincidentally similar angular sizes of the sun and moon, responsible for the spectacle of the solar eclipse).
- The blue day sky and natural crystals exhibit the *Polarisation singularities* being intensively explored as one of the pillars of singular optics.
- *Geometric phases* and
- *Conical intersections* first appeared in crystal optics. And the squint moon exemplifies the
- *Geometry of visual illusions* that mislead our perception.

Acknowledgements

I thank Professor Roy Bishop for permission to reproduce Figure 1, Professor Zeev Vager and Mr. Omer Abramson for introducing me to the squint moon illusion and a helpful correspondence, Professor Andrew Young for advice on mirages, and Professors Mark Dennis and Pragya Shukla for their careful readings of the first draft and helpful suggestions.

Notes on contributor

Sir M.V. Berry is a theoretical physicist at the University of Bristol, where he has been for nearly twice as long as he has not. His research centres on the relations between physical theories at different levels of description (classical and quantum physics, and ray optics and wave optics). He delights in finding familiar phenomena illustrating deep concepts: the arcane in the mundane.

References

[1] J.M. Pernter and F.M. Exner, *Meterologische Optik*, W. Braunmüller, Wien, 1922.
[2] R.A.R. Tricker, *Introduction to Meteorological Optics*, Elsevier, New York, 1970.
[3] M. Minnaert, *The Nature of Light and Colour in the Open Air*, Dover, New York, 1954.
[4] R. Greenler, *Rainbows, Halos and Glories*, University Press, Cambridge, 1980.
[5] D.K. Lynch and W. Livingston, *Color and Light in Nature*, University Press, Cambridge, 1995.
[6] R. Lee and A. Fraser, *The Rainbow Bridge: Rainbows in Art, Myth and Science*, Pennsylvania State University and SPIE press, Bellingham, WA, 2001.
[7] C.B. Boyer, *The Rainbow from Myth to Mathematics*, Princeton University Press, Princeton, NJ, 1987.
[8] D. Park, *The Fire within the Eye: A Historical Essay on the Nature and Meaning of Light*, University Press, Princeton, NJ, 1997.
[9] M.V. Berry and C. Upstill, *Catastrophe optics: morphologies of caustics and their diffraction patterns*, Prog. Opt. 18 (1980), pp. 257–346.
[10] T. Young, *The Bakerian lecture: on the theory of light and colours*, Phil. Trans. Roy. Soc. 92 (1802), pp. 12–48.
[11] T. Young, *The Bakerian lecture: experiments and calculations relative to physical optics*, Phil. Trans. Roy. Soc. Lond. 94 (1804), pp. 1–16.
[12] M.V. Berry, *Exuberant interference: rainbows, tides, edges, (de)coherence*, Phil. Trans. Roy. Soc. Lond. A 360 (2002), pp. 1023–1037.
[13] G.B. Airy, *On the intensity of light in the neighbourhood of a caustic*, Trans. Camb. Phil. Soc. 6 (1838), pp. 379–403.
[14] H.M. Nussenzveig, *Diffraction Effects in Semiclassical Scattering*, University Press, Cambridge, 1992.
[15] G.G. Stokes, *On the numerical calculation of a class of definite integrals and infinite series*, Trans. Camb. Phil. Soc. 9 (1847), pp. 379–407.
[16] G.G. Stokes, *On the discontinuity of arbitrary constants which appear in divergent developments*, Trans. Camb. Phil. Soc. 10 (1864), pp. 106–128.
[17] G.G. Stokes, *On the discontinuity of arbitrary constants that appear as multipliers of semi-convergent series*, Acta Math. 26 (1902), pp. 393–397.
[18] M.V. Berry, *Uniform asymptotic smoothing of Stokes's discontinuities*, Proc. Roy. Soc. Lond. A422 (1989), pp. 7–21.
[19] M.V. Berry, *Stokes' phenomenon; smoothing a Victorian discontinuity*, Publ. Math. Institut des Hautes Études scientifique 68 (1989), pp. 211–221.
[20] M.V. Berry, *Asymptotics, superasymptotics, hyperasymptotics*, in *Asymptotics Beyond All Orders*, H. Segur and S. Tanveer, eds., Plenum, New York, 1992, pp. 1–14.
[21] M.V. Berry and C.J. Howls, *Divergent series: taming the tails*, in *The Princeton Companion to Applied Mathematics*, N. Higham, ed., University Press, Princeton, NJ, 2015, in press.
[22] R.B. Dingle, *Asymptotic Expansions: their Derivation and Interpretation*, Academic Press, New York, 1973.
[23] M.V. Berry and C.J. Howls, *Hyperasymptotics*, Proc. Roy. Soc. Lond. A430 (1990), pp. 653–668.
[24] M.V. Berry and C.J. Howls, *Hyperasymptotics for integrals with saddles*, Proc. Roy. Soc. Lond. A434 (1991), pp. 657–675.
[25] G.V. Dunne and M. Ünsal, Uniform WKB, multi-instantons, and resurgent trans-series, Phys. Rev. D 89 (2014), p. 105009.
[26] R.L. Lee, *What are 'all the colors of the rainbow'?* Appl. Opt. 30 (1991), pp. 3401–3407.
[27] C. Chester, B. Friedman, and F. Ursell, *An extension of the method of steepest descents*, Proc. Camb. Phil. Soc. 53 (1957), pp. 599–611.
[28] M.V. Berry, *Uniform approximation for potential scattering involving a rainbow*, Proc. Phys. Soc. 89 (1966), pp. 479–490.
[29] R. Wong, *Asymptotic Approximations to Integrals*, Academic Press, New York, 1989.
[30] K.W. Ford and J.A. Wheeler, *Semiclassical description of scattering*, Ann. Phys. (NY) 7 (1959), pp. 259–286.
[31] K.W. Ford and J.A. Wheeler, *Application of semiclassical scattering analysis*, Ann. Phys. (NY) 7 (1959), pp. 287–322.
[32] M.V. Berry, *Cusped rainbows and incoherence effects in the rippling-mirror model for particle scattering from surfaces*, J. Phys. A 8 (1975), pp. 566–584.
[33] M.V. Berry and C.J. Howls, *Integrals with coalescing saddles*. Chapter 36, in *NIST Digital Library of Mathematical Functions*, F.W.J. Olver, D.W. Lozier R.F. Boisvert and C.W. Clark, eds., University Press, Cambridge, 2010, pp. 775–793.
[34] J.F. Nye, *Natural Focusing and Fine Structure of Light: Caustics and Wave Dislocations*, Institute of Physics Publishing, Bristol, 1999.
[35] M.S. Longuet-Higgins, *Reflection and refraction at a random moving surface. I. Pattern and paths of specular points*, J. Opt. Soc. Amer. 50 (1960), pp. 838–844.
[36] M.S. Longuet-Higgins, *Reflection and refraction at a random moving surface. II. Number of specular points in a Gaussian surface*, J. Opt. Soc. Amer. 50 (1960), pp. 845–850.
[37] M.S. Longuet-Higgins, *Reflection and refraction at a random moving surface. III. Frequency of twinkling in a Gaussian surface*, J. Opt. Soc. Amer. 50 (1960), pp. 851–856.
[38] M.V. Berry and M.R. Dennis, *Phase singularities in isotropic random waves*, Proc. Roy. Soc. A 456 (2000), pp. 2059–2079, corrigenda in A456 p3048.

[39] M.V. Berry and M.R. Dennis, *Polarization singularities in isotropic random vector waves*, Proc. Roy. Soc. Lond. A457 (2001), pp. 141–155.
[40] A.T. Young, *An introduction to mirages*, 2012. Available at: http://mintaka.sdsu.edu/GF/mirages/mirintro.html.
[41] M.V. Berry, *Raman and the mirage revisited: confusions and a rediscovery*, Eur. J. Phys. 34 (2013), pp. 1423–1437.
[42] DLMF, *NIST Handbook of Mathematical Functions,* University Press, Cambridge, 2010. Available at: http://dlmf.nist.gov.
[43] W. Tape, *The topology of mirages*, Sci Am. 252 (1985), pp. 120–129.
[44] M.V. Berry, *Disruption of images: the caustic-touching theorem*, J. Opt. Soc. Amer. A4 (1987), pp. 561–569.
[45] D.J.K. O'Connell, *The Green Flash and Other Low Sun Phenomena*, Vatican Observatory, Rome, 1958.
[46] A.T. Young, *Annotated bibliography of atmospheric refraction, mirages, green flashes, atmospheric refraction, etc.*, 2012. Available at: http://mintaka.sdsu.edu/GF/bibliog/bibliog.html.
[47] H.C. van de Hulst, *Light Scattering by Small Particles*, Dover, New York, 1981.
[48] C.F. Bohren and A.B. Fraser, *At what altitude does the horizon cease to be visible?* Am. J. Phys. 54 (1986), pp. 222–227.
[49] H.M. Nussenzveig, *High-frequency scattering by a transparent sphere. II. Theory of the rainbow and the Glory*, J. Math. Phys. 10 (1969), pp. 125–176.
[50] R. Lenke, U. Mack, and G. Maret, *Comparison of the 'glory' with coherent backscattering of light in turbid media*, J. Opt. A 4 (2002), pp. 309–314.
[51] C.M. Aegerter and G. Maret, *Coherent backscattering and Anderson localization of light*, Prog. Opt. 52 (2009), pp. 1–62.
[52] H.M. Nussenzveig, *High-frequency scattering by a transparent sphere. I. Direct reflection and transmission*, J. Math. Phys. 10 (1969), pp. 82–124.
[53] G. Horvath, J. Gal, and I. Pomozi, *Polarization portrait of the Arago point: video-polarimetric imaging of the neutral points of skylight polarization*, Naturwiss 85 (1998), pp. 333–339.
[54] G. Horváth, B. Bernáth, B. Suhai, and A. Barta, *First observation of the fourth neutral polarization point in the atmosphere*, J. Opt. Soc. Amer. A. 19 (2002), pp. 2085–2099.
[55] M.V. Berry, M.R. Dennis, and R.L.J. Lee, *Polarization singularities in the clear sky*, New J. Phys. 6 (2004), p. 162.
[56] S. Chandrasekhar and D. Elbert, *Polarization of the sunlit sky*, Nature 167 (1951), pp. 51–55.
[57] S. Chandrasekhar and D. Elbert, *The illumination and polarization of the sunlit sky on Rayleigh scattering*, Trans. Amer. Philos. Soc. 44 (1954), pp. 643–728.
[58] J.H. Hannay, *Polarization of sky light from a canopy atmosphere*, New. J. Phys 6 (2004), p. 197.
[59] V. Guillemin and A. Pollack, *Differential Topology*, Prentice-Hall, Englewood Cliffs, NJ, 1974.
[60] R. Hegedüs, S. Åkesson, R. Wehner, and G. Horváth, *Could Vikings have navigated under foggy and cloudy conditions by skylight polarization? On the atmospheric optical prerequisites of polarimetric Viking navigation under foggy and cloudy skies*, Proc. Roy. Soc. A 463 (2007), pp. 1081–1095.
[61] L.K. Karlsen, *Secrets of the Viking Navigators*, One Earth Press, Seattle, WA, 2003.
[62] M. Born and E. Wolf, *Principles of Optics*, Pergamon, London, 2005.
[63] W.R. Hamilton, *Third supplement to an essay on the theory of systems of rays*, Trans. Roy. Irish. Acad. 17 (1837), pp. 1–144.
[64] J.G. Lunney and D. Weaire, *The ins and outs of conical refraction*, Europhysics News 37 (2006), pp. 26–29.
[65] H. Lloyd, *On the phaenomena presented by light in its passage along the axes of biaxial crystals*, Phil. Mag. 2 (1833), pp. 112–120 and 207–210.
[66] H. Lloyd, *Further experiments on the phaenomena presented by light in its passage along the axes of biaxial crystals*, Phil. Mag. 2 (1833), pp. 207–210.
[67] H. Lloyd, *On the phenomena presented by light in its passage along the axes of biaxial crystals*, Trans. Roy. Irish. Acad. 17 (1837), pp. 145–158.
[68] M.V. Berry and M.R. Jeffrey, *Conical diffraction: Hamilton's diabolical point at the heart of crystal optics*, Prog. Opt. 50 (2007), pp. 13–50.
[69] M.V. Berry, R. Bhandari, and S. Klein, *Black plastic sandwiches demonstrating biaxial optical anisotropy*, Eur. J. Phys. 20 (1999), pp. 1–14.
[70] M.V. Berry, *Quantal phase factors accompanying adiabatic changes*, Proc. Roy. Soc. Lond. A392 (1984), pp. 45–57.
[71] A. Shapere and F. Wilczek, *Geometric Phases in Physics*, World Scientific, Singapore, 1989.
[72] L.S. Cederbaum, R.S. Friedman, V.M. Ryaboy, and N. Moiseyev, *Conical intersections and bound molecular states embedded in the continuum*, Phys. Rev. Lett 90 (2003), pp. 013001-1-4.
[73] E. Kalesaki, C. Delerue, C. Morais Smith, W. Beugeling, G. Allan, and D. Vanmaekelbergh, *Dirac cones, topological edge states, and nontrivial flat bands in two-dimensional semiconductors with a honeycomb nanogeometry*, Phys. Rev. X 4 (2014), p. 011010.
[74] M.V. Berry and M.R. Dennis, *The optical singularities of birefringent dichroic chiral crystals*, Proc. Roy. Soc. A. 459 (2003), pp. 1261–1292.
[75] B. Schölfopf, *The moon tilt illusion*, Perception 27 (1998), pp. 1229–1232.
[76] H.J. Schlichting, *Schielt der Mond?* Spektrum der Wissenschaft (2012), pp. 56–58.
[77] G. Glaeser and K. Schott, *Geometric considerations about seemingly wrong tilt of crescent moon*, KoG (Croatian Soc. Geom. Graph.) 13 (2009), pp. 19–26.
[78] B.J. Rogers and S.M. Anstis, *The new moon illusion*, in *The Oxford Compendium of Visual illusions*, A.G. Shapiro and D. Todorovic, eds., Oxford University Press, New York and Oxford, 2014, in press.

Chasing the Silver Dragon

Michael Berry

Occurring on only a few dozen rivers around the world, tidal bores are as rare as they are intriguing. **Michael Berry** outlines the science behind this natural phenomenon and describes his sighting in China of one of the most spectacular bores of them all

One of the great sights of the Bristol area in the UK is the tidal bore on the River Severn – a wave that steepens and grows as the tide advances inland towards the city of Gloucester. Marking the beginning of the incoming tide, this giant wave is pulled up-river by the Moon's gravity. I like to see the bore at night. As it approaches, the effect of the distant but then growing roar is magical – and there is plenty of light from Gloucester (and from the Moon when it is visible) to watch the bore as it passes by.

We can enjoy the spectacle of the bore, but understanding it raises some questions. Why are there two tides each day – not just one, in the water on the side of the Earth that faces the Moon? Of the more than 700 tides each year, why does a bore on the Severn occur during at most 50 of them, rather than all? And why does the Moon play a role anyway, given that the Sun's gravity is nearly 200 times stronger?

The fundamental fact is that the force that raises the tide is not simply the gravity from the Moon and the Sun. It would be, if the Earth were held fixed in space by an imaginary cosmic vice. But the solid Earth is not fixed; it moves in an orbit round the Sun, perturbed by a slight wobble caused by the Moon. The tide-raising force is the difference between the force on the water and the force on the solid Earth. And the force attracting the water on the side of the Earth facing the Moon is greater than that on the solid Earth, while the force on the water on the side of the Earth away from the Moon is less than that on the solid Earth (figure 1*a*).

The tide-raising force therefore points outwards, leading to two tidal bulges on opposite sides of the Earth (figure 1*b*). As the Earth turns, a given geographical location experiences a bulge twice each day, leading to the two tides. Quantitatively, the tide-raising force is the gradient of the forces from the Sun and Moon: it is therefore an inverse-cube force

Michael Berry is professor emeritus at the H H Wills Physics Laboratory, University of Bristol, UK, michaelberryphysics.wordpress.com

1 What makes a tidal bore?

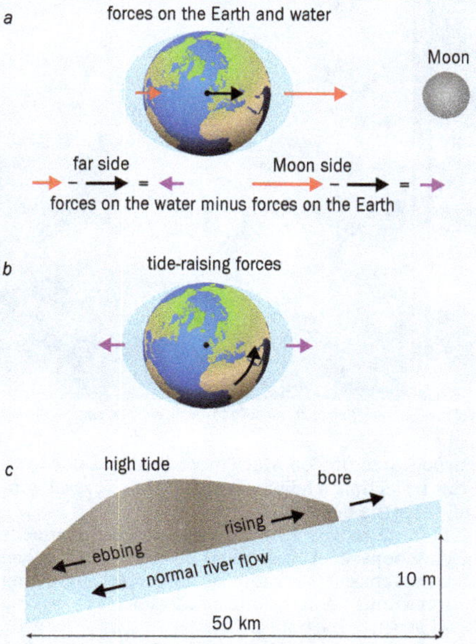

(*a*) Water on our planet is pulled more strongly towards the Moon on the side of the Earth facing the Moon than on the side facing away because the latter is further away (red arrows). The lunar pull on the solid Earth lies between these extremes (black arrow).
(*b*) Subtracting the pull on the Earth from the pull on the water leads to net outward forces (purple), which cause twice-daily tidal bulges. As the Earth turns (black arrow), different places experience the tides at different times. (*c*) The profile of a typical bore flowing along a river at high tide (brown) shown not to scale. At the places where the arrows are, the tide is either rising (flowing in) or ebbing (flowing out).

from each, rather than the familiar inverse-square. And this tide-raising force from the Moon is about twice that from the Sun.

The half-strength tidal force from the Sun is far from negligible and makes the height of the tides sensitive to astronomical alignments. Tides are stronger when the Sun and Moon are almost in line: at full and new Moon, and when the inclined orbit of the Moon intersects that of the Earth along a line pointing to the Sun. The effect is enhanced when the Moon and Earth are closest to their parent bodies along their elliptical orbits. Unlike eclipses and other astronomical phenomena, the arrival times and heights of bores cannot be predicted precisely, because these are sensitive to the amount of water already flowing downriver. (Don't therefore be fooled by seemingly precise arrival-time predictions you can find online.) The most impressive bores occur at particularly high tides after a period when there has been no rain for several days.

The familiar tides at the seashore rise and fall gradually, over several hours. (The waves we see breaking on the shore are caused by winds, not tides.) What is different about rivers that host tidal bores – making bores sensitive to geography as well as astronomy – is that they are open to a large ocean and get narrower and shallower over a long distance upstream. The speed of waves on water is limited by the depth: $\sqrt{(gh)}$, where g is the acceleration due to gravity and h the depth. The crest of the tide as it travels upriver gets deeper than the trough, so it travels faster, generating a wave that gets steeper, eventually with a fairly sharp front (figure 1*c*).

This difference of speeds between the crest and trough is a simple example of nonlinearity – the mathematical phenomenon that underlies all fluid-mechanical analyses of the formation and shape of the bore. But no mathematical treatment has captured the variety of shapes of the bore at different places along the river. The wave can break, tumbling as it crashes against the banks. Elsewhere, the bore advances upstream in a stately procession of smooth waves behind the front; under moonlight, its mirror surface gleams like liquid mercury. Mathematical models emphasize different features, for example as a moving shock wave, smoothed with a few undulations behind.

One question people often ask is whether a bore is a soliton. The answer is no, because a soliton is a wave whose height is the same on both sides of its peak, whereas the arrival of a bore is the beginning of the tide: the water is deeper behind the bore than in front of it. A striking aspect of a bore is that after it passes, as the tide continues to rise for about an hour, the river flows backwards – upstream but downwards. The smooth downstream flow before the bore passes contrasts with the wildly turbulent tide rushing upstream afterwards.

The arrival times and heights of bores cannot be predicted precisely, as these are sensitive to the amount of water already flowing downriver

Destination China

The tidal bore on the River Severn is caused by the extraordinary tidal range in the Bristol Channel into which it flows. It is the world's second highest, and can be as much as 17 m. The associated bore starts about 20 km upstream from the road bridges linking England and Wales at the mouth of the Severn and continues for another 20 km, and its height rarely exceeds 1 m. In the UK there are several other bores – on the rivers Parrett near Bridgwater and Dee near Chester, for example. In France, there used to be a bore on the Seine near Caudebec-en-Caux, *le mascaret*, but this largely disappeared after extensive dredging in the 1960s. In Brazil, the *pororoca* is a dramatic bore on the Amazon and some of its tributaries, holding the world record for long-distance

Dramatic sight Huge crowds watch the Silver Dragon bore from the banks of the Qiantang (left) but you have to get up close to see details such as reflection (right).

surfing: about 12 km.

In September last year, however, I was privileged to see the world's largest tidal bore: the "Silver Dragon" near Hangzhou in south-east China, on the Qiantang river, the mouth of which opens out towards Shanghai. My host was Huan-Qiang Zhou, a physicist from Chongqing, a thousand miles to the west, whom I knew through the *Journal of Physics A*, which is published by the Institute of Physics (along with *Physics World*). Hangzhou is his home city and Zhou's generous arrangements involved finding a local "fixer" who booked our hotels and meals, selected prime locations to view the bore, and decided on the best times to do so.

Tidal bores occur roughly twice a day, separated by a gap of 12 hours and 25 minutes. We saw the Silver Dragon six times: by day and by night over three days. It is a major tourist attraction; local media estimated that on the day of the biggest bore, more than 100 000 people lined the banks to watch it, and at prime locations our host had to buy tickets to reach the river bank. I saw no non-Chinese people other than the Australian physicist who accompanied us.

As with the Severn bore, you hear the low roar of the wave before you see it. The Qiantang river is almost 3 km wide, in contrast with the mere 50 m width of the Severn near Gloucester, so the wave is louder, and the roar is audible a full 20 minutes before the bore is glimpsed as a thin white line in the distance. There is great anticipation as the angry wave approaches, before a shout goes up as it rushes by. At night, there were fewer people watching the Qiantang bore and, because of the way sound refracts differently when the ground is colder, its roar could be heard even earlier.

Several long walls jut out perpendicular to the river bank. Standing on one of them, we saw the wave approach head-on, crash into the wall and then reflect. This reflected wave was an awesome sight, hugely amplified as it receded and interfered coherently with the still-advancing tide. The Severn bore can be reflected too, after it hits the weir upstream from Maisemore, which is usually the limit of its journey. Standing on Maisemore Bridge, I once saw a tiny reflection, a few centimetres high, several minutes after the bore had passed.

The bore moves up the Qiantang river at speeds varying between 10 km/h and 20 km/h. On the last night, we chased it for tens of kilometres, following it by car until 3 a.m. The fixer chose a final viewing location where the bore crashed against another wall. As we parked, seconds before the bore arrived, police rushed alongside on motorcycles, screaming at us to shift our car, and ourselves, several tens of metres away. With my mistrust of authority, I thought they were being unnecessarily officious. But they were right, because the reflected wave smashed through the protective chain-link fence, and would have drenched us and the car, probably knocking us over.

With the interval of 12 hours and 25 minutes between bores, and long drives between viewings in several different places, our sleep patterns were disrupted, leaving us permanently exhausted. Our discomfort was somewhat alleviated by the luxurious hotels chosen by the fixer. One deserves mention as an attraction in itself: the Ningbo Hiatian Yizhou Hotel, which is spectacularly located on the Qiantang river as it opens out into its estuary. The building lies in the middle of a bridge spanning the river, claimed to be the world's longest over clear water – nearly 40 km.

Experiencing unification

A primary aim of physics is to unify the different fundamental forces. A giant step in this direction was Maxwell's unification of electricity, magnetism and light, in his electromagnetic theory that now underpins our communications technology. The effort continues, with attempts to unify gravity with the strong and electroweak interactions that act on microscopic scales. But we should not forget the first unification, which was Isaac Newton's discovery that the force that holds us to the ground, the force that keeps the Moon in its orbit, and the force that drives the tides, is in fact one force: gravity.

Witnessing a tidal bore, we experience this unification directly. ■

Chapter 3
Quantum Chaology

The earliest phase of quantum asymptotics, described in the previous chapter, culminated in a wide-ranging review of semiclassical mechanics [B23], written with Kate Mount in 1972. I now regard this as 'prehistoric semiclassical mechanics', because it was written before two important developments. The first, already described in Chapter 2, was the catastrophe classification of stable caustics. The second, brought to my attention by Ian Percival, was the then new area of deterministic chaos in classical dynamics. Percival insisted that new insights, beyond our existing semiclassical formulations, were required to understand quantum phenomena where the corresponding classical motion is chaotic (i.e. irregular or non-integrable).

Sidestepping the familiar WKB formalism, but using elementary semiclassical ideas, I conjectured [3.1] that quantum wavefunctions in classically chaotic systems can be modelled as Gaussian random functions. Boundary conditions can be incorporated ([B340], and [B344] with H. Ishio). The Gaussian random wave hypothesis has been supported and developed in many papers by others; these include the important observation that some chaotic states are decorated ('scarred') by classical periodic orbits.

Understanding the quantum eigenvalues (energy levels) in classically chaotic systems took somewhat longer. In 1975, when my late colleague Balazs Gyorffy made me aware of random-matrix theory, it was immediately clear that this would be a good model for the statistics of the eigenvalues. A prediction, soon confirmed in numerical calculations, was that neigbouring levels would repel each other. The conjecture that quantum energy-level statistics in the classically chaotic case would display random-matrix universality was later formulated precisely, and more generally, by Oriol Bohigas, Marie-Joys Giannoni and Charles Schmit. In the contrary case of classically integrable (i.e. non-chaotic] systems, Michael Tabor and I argued [B61] that the statistics would be very different: Poisson, with no level repulsion.

Understanding statistics in the chaotic case required two new ingredients. The first was the formulation of semiclassical mechanics, in terms of periodic classical orbits, by Martin Gutzwiller [7.11]. Our 1972 review [B23] concluded that this approach, then newly-published, was likely to be important, but at that time we could not understand that it would be central to quantum chaology. Once we did appreciate the importance of periodic orbits, an early application, with John Hannay [B95], was the analytic quantization of the classically chaotic 'Arnold cat' linear map on the torus. But this was not generic, because of arithmetic relations between the actions of the periodic orbits, leading to too much coherent interference for random-matrix statistics to apply.

The second ingredient was the discovery by John Hannay and Alfredo Ozorio de Almeida that in generic cases (in particular when there are no arithmetic relations between actions) the contributions of the long periodic orbits obey a universal sum rule.

Combining the two [3.2] explained why random-matrix universality applies in quantum chaology. It is always gratifying when a theory yields insights beyond those that inspired it. In this case, periodic-orbit theory delivered more than the reason for random-matrix universality. It showed that although the short-range correlations between energy levels are described by random-matrix theory, this model must fail for the longer-range level correlations, because these are controlled by the short periodic orbits, which are not universal.

Extensions soon followed: to systems without time-reversal symmetry ('Aharonov-Bohm billiards' [B145], [B148], with Marko Robnik), and to relativity ('Neutrino billiards' [B161], with Raul Mondragon). My 1987 Bakerian lecture [3.3] reviewed these and other developments (and also explained the term 'chaology' — see also [B191]). Since then, many other people have improved, generalised and applied the periodic-orbit theory of quantum chaology in many directions.

In January 1985, on a journey from Bielefeld to Heidelberg in weather so cold that the windows of the normally well-heated German train froze on the inside, the clariton struck me that quantum chaology should also describe the heights of the zeros of the Riemann zeta function (if the Riemann hypothesis is true). A celebrated conjecture a decade earlier had proposed that some statistics of the zeros would be described by random-matrix theory, but the connection with quantum chaology predicted much more: the zeros are energy levels of a quantum system whose classical counterpart has well-defined properties, and the random-matrix distribution of the zeros must break down in a precisely specified way.

I described the zeta/classical analogy in [3.2] and [3.3], and in more detail in [3.4]; the breakdown of random-matrix theory for distant Riemann zeros was confirmed in [3.5]. Jon Keating and I reviewed this subject [3.6], and he and others have independently extended and deepened the analogy in many directions. But it depends on the existence of a chaotic classical dynamical system underlying the zeros, and we do not know what this is. We know some of its properties ([3.6] and [B401]) — for example, that it has periodic orbits whose lengths are multiples of logarithms of primes — and a hint [B306] that the system might be related to an unstable oscillator. But the connection remains enigmatic and tantalising, exemplifying Piet Hein's aphorism: "Problems worthy of attack prove their worth by hitting back".

It would be possible to see the Riemann zeros in an optical or microwave experiment, because these separate the side lobes of radiation patterns of suitably-designed antennas [B451, B481]. Complementing this 'eyemath' is 'earmath': hearing 'the music of the primes' and the Riemann zeros [B454] and the different random-matrix ensembles ([B456], with Pragya Shukla).

J. Phys. A: Math. Gen., Vol. 10, No. 12, 1977. Printed in Great Britain. © 1977

Regular and irregular semiclassical wavefunctions

M V Berry
H H Wills Physics Laboratory, Bristol University, Tyndall Avenue, Bristol BS8 1TL, UK

Received 19 July 1977, in final form 11 August 1977

Abstract. The form of the wavefunction ψ for a semiclassical regular quantum state (associated with classical motion on an N-dimensional torus in the $2N$-dimensional phase space) is very different from the form of ψ for an irregular state (associated with stochastic classical motion on all or part of the $(2N-1)$-dimensional energy surface in phase space). For regular states the local average probability density Π rises to large values on caustics at the boundaries of the classically allowed region in coordinate space, and ψ exhibits strong anisotropic interference oscillations. For irregular states Π falls to zero (or in two dimensions stays constant) on 'anticaustics' at the boundary of the classically allowed region, and ψ appears to be a Gaussian random function exhibiting more moderate interference oscillations which for ergodic classical motion are statistically isotropic with the autocorrelation of ψ given by a Bessel function.

1. Introduction

In generic classical Hamiltonian bound systems with $N(\geqslant 2)$ degrees of freedom some orbits wind smoothly round N-dimensional tori in the $2N$-dimensional phase space, and some orbits explore $(2N-1)$-dimensional regions of the energy 'surface' in a stochastic manner (Arnol'd and Avez 1968, Ford 1975, Whiteman 1977, Berry 1978). Percival (1973) took up an old idea of Einstein (1917) and suggested that in the semiclassical limit (i.e. as $\hbar \to 0$) there would be 'regular' and 'irregular' quantum states corresponding to these two sorts of classical motion. In the limiting case of a completely integrable system the whole phase space is filled with tori and all states are regular, and in the opposite limit of a completely ergodic system almost all orbits wander stochastically over the whole energy surface and all states are irregular. According to Percival regular and irregular states could be distinguished by their behaviour under perturbation; this distinction obviously involves the matrix elements between *different* states.

Here I make conjectures about *individual* energy eigenstates. It appears that the nature of the wavefunction $\psi(\boldsymbol{q})$ is very different as $\hbar \to 0$ for regular and irregular states. The main differences are in the behaviour near boundaries of classically allowed regions (§ 2) and in the nature of the oscillations of $\psi(\boldsymbol{q})$ (§ 3). These differences suggest simple ways in which a quantum state numerically computed or graphically displayed can be shown to be regular or irregular.

The quantities to be calculated are *local averages* over coordinates $\boldsymbol{q}(\equiv q_1 \ldots q_N)$ of functions $f(\boldsymbol{q})$ that depend on $\psi(\boldsymbol{q})$, denoted by $\overline{f(\boldsymbol{q})}$ and defined by

$$\bar{f}(q_1 \ldots q_N) \equiv \frac{1}{\Delta^N} \int_{q_1-\frac{1}{2}\Delta}^{q_1+\frac{1}{2}\Delta} \mathrm{d}Q_1 \ldots \int_{q_N-\frac{1}{2}\Delta}^{q_N+\frac{1}{2}\Delta} \mathrm{d}Q_N f(Q_1 \ldots Q_N), \tag{1}$$

where

$$\lim_{\hbar\to 0} \Delta = 0 \qquad \text{but} \qquad \lim_{\hbar\to 0} (\hbar/\Delta) = 0. \tag{2}$$

The conditions on Δ ensure that the average $\bar{f}$ is taken over many oscillations of the wavefunction, since the scale of these oscillations is of order $\hbar$. I shall study the local average probability density $\Pi(\boldsymbol{q})$, namely

$$\Pi(\boldsymbol{q}) \equiv \overline{|\psi(\boldsymbol{q})|^2}, \tag{3}$$

and the autocorrelation function $C(\boldsymbol{X};\boldsymbol{q})$ of $\psi(\boldsymbol{q})$, namely

$$C(\boldsymbol{X};\boldsymbol{q}) \equiv \overline{\psi(\boldsymbol{q}+\tfrac{1}{2}\boldsymbol{X})\psi^*(\boldsymbol{q}-\tfrac{1}{2}\boldsymbol{X})}/\Pi(\boldsymbol{q}). \tag{4}$$

The principal tool for this study is Wigner's function $\Psi(\boldsymbol{q},\boldsymbol{p})$ corresponding to the state $\psi(\boldsymbol{q})$. This is defined as

$$\Psi(\boldsymbol{q},\boldsymbol{p}) \equiv \frac{1}{h^N}\int \mathrm{d}\boldsymbol{X}\, \mathrm{e}^{-\mathrm{i}\boldsymbol{p}\cdot\boldsymbol{X}/\hbar}\, \psi(\boldsymbol{q}-\tfrac{1}{2}\boldsymbol{X})\psi^*(\boldsymbol{q}+\tfrac{1}{2}\boldsymbol{X}). \tag{5}$$

Detailed studies of the semiclassical behaviour of Ψ have been made by Berry (1977a) and Voros (1976, 1977). In terms of Ψ, the local average probability density is

$$\Pi(\boldsymbol{q}) = \int \mathrm{d}\boldsymbol{p}\, \bar{\Psi}(\boldsymbol{q},\boldsymbol{p}) \tag{6}$$

and the autocorrelation function is

$$C(\boldsymbol{X};\boldsymbol{q}) = \int \mathrm{d}\boldsymbol{p}\, \mathrm{e}^{\mathrm{i}\boldsymbol{p}\cdot\boldsymbol{X}/\hbar}\, \bar{\Psi}(\boldsymbol{q},\boldsymbol{p})/\Pi(\boldsymbol{q}). \tag{7}$$

For the present purposes it is sufficient to take for the averaged Wigner function $\bar{\Psi}$ the crudest classical approximation, namely the density in the classical phase space $\boldsymbol{q}, \boldsymbol{p}$ over the manifold explored by the classical orbit corresponding to the quantum state ψ being considered. For an *integrable* system this is a torus, specified by the actions $\boldsymbol{I}_\psi = (I_1 \ldots I_N)$ round the N irreducible cycles. Quantum conditions (reviewed by Percival 1977) select the $\boldsymbol{I}_\psi$ that can correspond to quantum states. Any point $(\boldsymbol{q},\boldsymbol{p})$ in phase space can be specified by the actions $\boldsymbol{I}(\boldsymbol{q},\boldsymbol{p})$ of the torus through $(\boldsymbol{q},\boldsymbol{p})$ and the conjugate angle variables $\boldsymbol{\theta}(\boldsymbol{q},\boldsymbol{p})$ locating the position of $(\boldsymbol{q},\boldsymbol{p})$ on this torus. Then the averaged Wigner function is (Berry 1977a):

$$\bar{\Psi}(\boldsymbol{q},\boldsymbol{p}) = \frac{\delta(\boldsymbol{I}(\boldsymbol{q},\boldsymbol{p})-\boldsymbol{I}_\psi)}{(2\pi)^N}. \tag{8}$$

(Note that this involves an N-dimensional delta function.)

For an *ergodic* system I assume that the relevant classical orbits are the typical ones, that pass close to all points on the energy surface corresponding to the energy E of the state ψ. Then if $H(\boldsymbol{q},\boldsymbol{p})$ denotes the classical Hamiltonian it is shown under reasonable assumptions by Voros (1976, 1977) that the averaged Wigner function is

$$\bar{\Psi}(\boldsymbol{q},\boldsymbol{p}) = \frac{\delta(E-H(\boldsymbol{q},\boldsymbol{p}))}{\int \mathrm{d}\boldsymbol{q} \int \mathrm{d}\boldsymbol{p}\, \delta(E-H(\boldsymbol{q},\boldsymbol{p}))}. \tag{9}$$

(Note that this involves only a one-dimensional delta function.) The 'microcanonical' assumption (9) differs radically from that made by Gutzwiller (1971), who considers that the relevant orbits in this case are the individual unstable periodic trajectories. Although dense, these are of total measure zero and this makes it unlikely that they could support quantum states (any attempt to use orbits in the neighbourhood of the periodic trajectories will be frustrated by their instability).

For a *quasi-integrable* system the existence of some tori is guaranteed by the 'KAM' theorem of Kolmogoroff (1954), Arnol'd (1963) and Moser (1962). Although these tori are distributed pathologically in phase space there are infinitely many of them near one with actions $\boldsymbol{I}_\psi$ and this, taken together with a smoothing on scales small in comparison with $\hbar$, is probably sufficient to give meaning to the function $\boldsymbol{I}(\boldsymbol{q}, \boldsymbol{p})$ and hence to (8). There will be gaps in the system of tori; in these gaps motion is stochastic and fills a $(2N-1)$-dimensional region smaller than the whole energy surface. Usually the orbits do not fill such stochastic regions uniformly but for simplicity I shall assume that they do. Then the averaged Wigner function will be given by (9) restricted to the region explored by the motion.

The Hamiltonian will be taken as

$$H(\boldsymbol{q}, \boldsymbol{p}) = \frac{p^2}{2m} + V(\boldsymbol{q}) \tag{10}$$

where m is the mass of the system and $V(\boldsymbol{q})$ the potential in which it moves. This ensures that $\psi(\boldsymbol{q})$ can be considered real. The presence of a magnetic field or of anisotropy in the momentum terms introduces complications but no essential differences in the results.

2. Caustics and anticaustics

According to (6) the averaged probability density is the projection of the averaged Wigner function 'down' the $\boldsymbol{p}$ directions. The singularities of this projection are interesting because they show how $\Pi(\boldsymbol{q})$ behaves near the boundary of the classically allowed region in $\boldsymbol{q}$ space.

For integrable systems the projections of tori are singular on the well known *caustics* where $\Pi(\boldsymbol{q})$ becomes infinite. Explicitly, (6) and (8) give

$$\Pi(\boldsymbol{q}) = \frac{1}{(2\pi)^N} \sum_i |\mathrm{d}\boldsymbol{I}(\boldsymbol{q}, \boldsymbol{p}_i(\boldsymbol{q}))/\mathrm{d}\,\boldsymbol{p}|^{-1} = \frac{1}{(2\pi)^N} \sum_i |\mathrm{d}\,\boldsymbol{\theta}(\boldsymbol{q}, \boldsymbol{p}_i(\boldsymbol{q}))/\mathrm{d}\,\boldsymbol{q}|, \tag{11}$$

where the derivatives denote Jacobian determinants and $\boldsymbol{p}_i(\boldsymbol{q})$ is the ith intersection of a fibre through $\boldsymbol{q}$ with the torus $\boldsymbol{I}_\psi$ (there is always a finite number of these intersections). As $\boldsymbol{q}$ moves onto a caustic $|\mathrm{d}\boldsymbol{\theta}/\mathrm{d}\boldsymbol{q}|$ diverges as two or more intersections i coincide (figure 1).

It is the presence of caustics that gives regular wavefunctions their striking and distinctive properties. The forms of the caustics in generic cases are governed by the catastrophe theory of Thom (1972) and Arnol'd (1975). This is because for the part of the torus for which $\boldsymbol{q}$ lies near the caustic it is possible to define a local generating function $G(\boldsymbol{p}; \boldsymbol{q})$ by

$$G(\boldsymbol{p}; \boldsymbol{q}) \equiv \int_{\boldsymbol{p}_0}^{\boldsymbol{p}} \boldsymbol{q}(\boldsymbol{p}') \cdot \mathrm{d}\boldsymbol{p}' - \boldsymbol{p} \cdot \boldsymbol{q}, \tag{12}$$

where $\boldsymbol{q} = \boldsymbol{q}(\boldsymbol{p})$ is the equation of the torus near its 'edge' (figure 1). In terms of G the

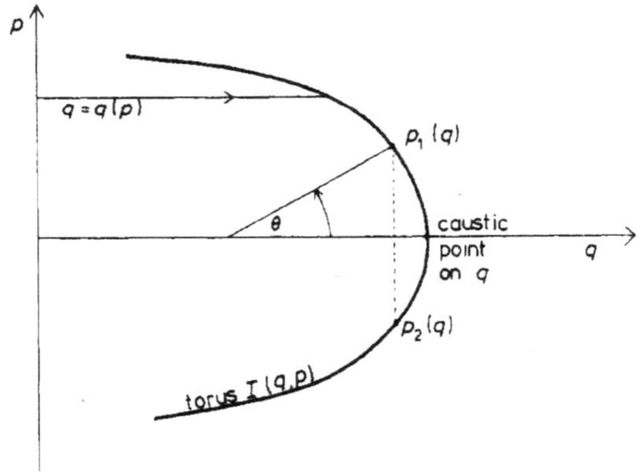

Figure 1. Coordinates and momenta near 'edge' of a torus in phase space.

gradient map

$$\nabla_{p} G = \boldsymbol{q}(\boldsymbol{p}) - \boldsymbol{q} = 0 \tag{13}$$

defines the torus locally and the singularities of the map, where

$$|\det \partial^2 G/\partial p_i \partial p_j| = |\mathrm{d}\boldsymbol{q}/\mathrm{d}\boldsymbol{p}| = 0, \tag{14}$$

define the caustics in $\boldsymbol{q}$ space. This follows from the second term of (11) on realising that $\boldsymbol{p}$ varies smoothly with $\boldsymbol{\theta}$ when $\boldsymbol{q}$ is near a caustic (figure 1). (It is not possible to define a *global* generating function of the form (12), because (12) fails near caustics in momentum space where $\boldsymbol{q}(\boldsymbol{p})$ is not defined.)

When $N = 2$ the possible catastrophes are fold lines and cusp points. If $V(\boldsymbol{q})$ is a simple potential well with circular symmetry, for example, there are no cusps and the caustic is two circular fold lines with radii determined by the libration points of the orbit. To see how such caustics are produced by projection it is simplest to visualise the tori in the three-dimensional energy surface $\mathscr{E}$ with coordinates q_1, q_2 and the angle ϕ made by $\boldsymbol{p}$ with the p_1 axis; because of the periodicity in ϕ the energy surface has the topology of a solid torus. Figure 2 shows $\mathscr{E}$ and also a two-dimensional torus corresponding to motion with constant angular momentum.

Other forms of potential well can give rise to caustics with cusps, as illustrated by figure 3(e) which shows the caustic of an orbit computed by Marcus (private communication). One way this might arise can be introduced by first considering an isolated stable triangle orbit (figure 3(a)) on a non-circular billiard table (i.e. a potential zero for q_1, q_2 inside a boundary and infinite outside). This will be surrounded by tori in q_1, q_2, ϕ (Lazutkin 1973, Dvorin and Lazutkin 1973), one of which is shown in figure 3(b). The projection (caustic) is shown in figure 3(c); there are no cusps, and the non-generic triple junctions result from the discontinuity in $V(\boldsymbol{q})$ at the boundary. The 'generification' of the torus (figure 3(d)) that results from softening $V(\boldsymbol{q})$ at the boundary corresponds to pumping air into a flat tyre and on projection gives the cusped caustic of figure 3(e).

For $N > 2$ the caustics can typically form higher catastrophes. In each case $\Pi(\boldsymbol{q})$ rises to infinity at the caustic. For the fold, if x is normal distance from the caustic into

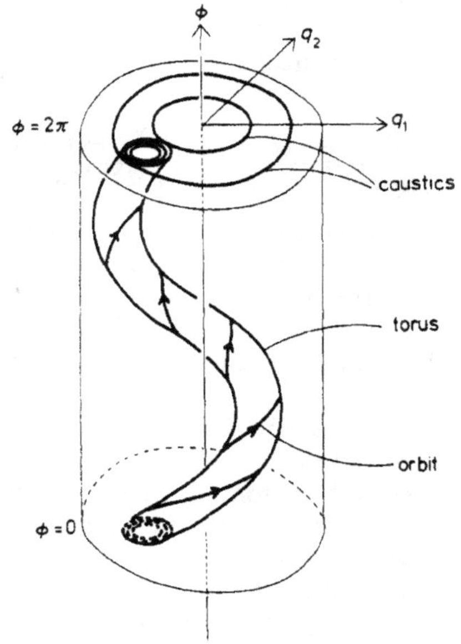

Figure 2. Energy 'surface' $\mathscr{E}$ with coordinates $(q_1, q_2, \phi)(\phi = 0$ and $\phi = 2\pi$ are identified). The torus corresponding to an orbit with constant angular momentum is shown, and the caustics which envelop its projection along ϕ.

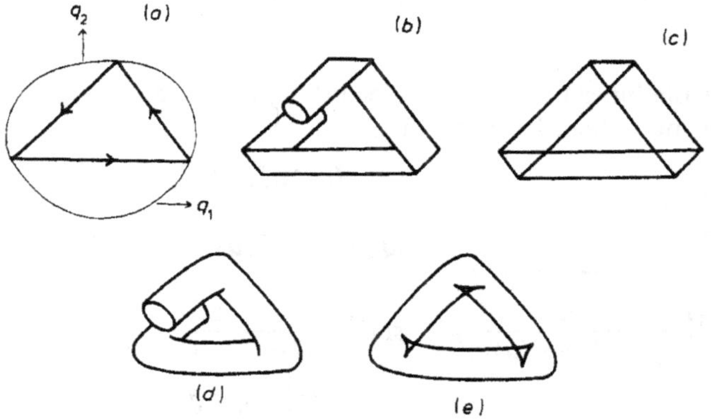

Figure 3. (a) Stable isolated closed orbit on billiard table; (b) torus in q_1, q_2, ϕ space (ϕ perpendicular to paper) inhabited by nearby quasi-periodic orbit; (c) caustic resulting from projection of (b); (d) generification of torus in (b) resulting from softening potential at the boundary; (e) caustic resulting from projection of (d).

the classically allowed region, the divergence has the form

$$\Pi \propto \operatorname{Re} x^{-1/2}. \tag{15}$$

At higher catastrophes the divergence is stronger.

For non-zero $\hbar$ the divergences are softened by quantum effects and ψ rises to a value of order $\hbar^{-\beta}$ where β is the 'singularity index' defined by Arnol'd (1975). The caustics are clothed by striking diffraction patterns (Berry 1976) characteristic of the particular catastrophe involved.

This behaviour is very different from what happens in an ergodic system. There, equations (8), (9) and (10) give

$$\Pi(\boldsymbol{q})=\frac{\int \mathrm{d}\boldsymbol{p}\,\delta(E-H(\boldsymbol{q},\boldsymbol{p}))}{\int \mathrm{d}\boldsymbol{p}\int \mathrm{d}\boldsymbol{q}\,\delta(E-H(\boldsymbol{q},\boldsymbol{p}))}=\frac{(E-V(\boldsymbol{q}))^{\frac{1}{2}N-1}\Theta(E-V(\boldsymbol{q}))}{\int \mathrm{d}\boldsymbol{q}(E-V(\boldsymbol{q}))^{\frac{1}{2}N-1}\Theta(E-V(\boldsymbol{q}))}, \tag{16}$$

where Θ denotes the unit step function. When $N>2$, $\Pi(\boldsymbol{q})$ vanishes at the boundary $E=V(\boldsymbol{q})$ of the classically allowed region. For $N=2$, $\Pi(\boldsymbol{q})$ is constant over the allowed region. In no case does Π diverge on the boundary (except in the trivial situation $N=1$ when the system is integrable *and* ergodic and the boundary points are caustics of fold type). Therefore I shall call the boundaries of these irregular wavefunctions *anticaustics*.

In a quasi-integrable system the boundaries of the projections of stochastic regions corresponding to irregular states will also have anticaustics. As an example of this let $N=2$ and consider a potential well perturbed from circularity. Surrounding each 'unperturbed' torus that supported closed orbits there will be a gap between the tori of the perturbed system. Let one such gap span the angular momenta L_1 to $L_2(>L_1)$ and assume that the stochastic trajectories fill this gap uniformly. Then $\Pi(\boldsymbol{q})$ for a corresponding irregular quantum state will vary with radial coordinate q as

$$\begin{aligned}\Pi(\boldsymbol{q})&\propto\int_0^{2\pi}\mathrm{d}\phi\int_0^{\infty}\mathrm{d}p\,p\delta\Big(E-V(q)-\frac{p^2}{2m}\Big)\Theta(L_2-pq\sin\phi)\Theta(pq\sin\phi-L_1)\\&\propto\int_0^{2\pi}\mathrm{d}\phi\;\Theta(L_2-q[2m(E-V(q))]^{1/2}\sin\phi)\Theta(q[2m(E-V(q))]^{1/2}\sin\phi-L_1).\end{aligned} \tag{17}$$

If q_1^-, q_2^- are the inner libration radii for orbits with L_1 and L_2, and q_1^+, q_2^+ are the corresponding outer libration radii, then

$$\Pi(\boldsymbol{q})\propto\begin{cases}0 & (q<q_1^-,\,q>q_1^+)\\ \cos^{-1}\Big(\dfrac{L_1}{q[2m(E-V(q))]^{1/2}}\Big) & (q_1^-<q<q_2^-,\,q_2^+<q<q_1^+)\\ \sin^{-1}\Big(\dfrac{L_2}{q[2m(E-V(q))]^{1/2}}\Big)-\sin^{-1}\Big(\dfrac{L_1}{q[2m(E-V(q))]^{1/2}}\Big) & \\ & (q_2^-<q<q_2^+)\end{cases} \tag{18}$$

The form of this expression is sketched in figure 4; it can be seen that there are indeed anticaustics at the outer boundaries of the classically allowed region, where $\Pi(q)$ rises as $(q-q_1^-)^{1/2}$ and falls as $(q_1^+-q)^{1/2}$.

3. Autocorrelation of the wavefunction

The function $C(\boldsymbol{X};\boldsymbol{q})$ defined by (4) gives the scale and directionality of the pattern of oscillations of ψ near $\boldsymbol{q}$. For an integrable system, (7) and (8) give (cf (11))

$$C(\boldsymbol{X};\boldsymbol{q})=\frac{\sum_i|\mathrm{d}\boldsymbol{\theta}(\boldsymbol{q},\boldsymbol{p}_i(\boldsymbol{q}))/\mathrm{d}\boldsymbol{q}|\,\mathrm{e}^{\mathrm{i}\boldsymbol{p}_i(\boldsymbol{q})\cdot\boldsymbol{X}/\hbar}}{\sum_i|\mathrm{d}\boldsymbol{\theta}(\boldsymbol{q},\boldsymbol{p}_i(\boldsymbol{q}))/\mathrm{d}\boldsymbol{q}|}. \tag{19}$$

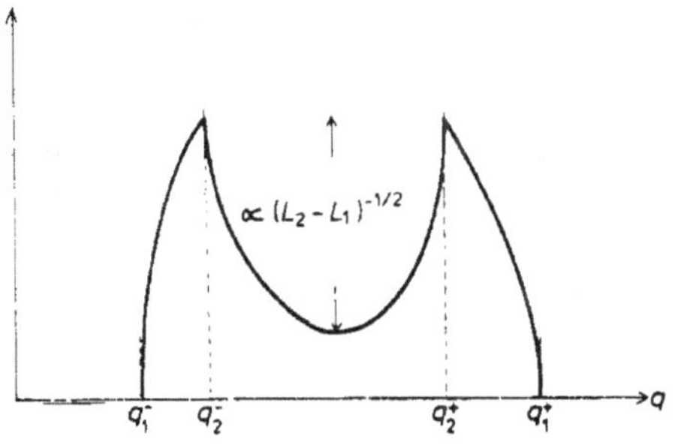

Figure 4. Local average probability density arising from projection of phase space region with constant energy between two tori corresponding to motions with angular momenta L_1 and L_2.

The wavevectors $\boldsymbol{p}_i/\hbar$ in the 'spectrum' of ψ all have the same length $[2m(E-V(\boldsymbol{q}))]^{1/2}/\hbar$ but different directions. There is a finite number of such wavevectors, so $C(\boldsymbol{X})$ is anisotropic. This anisotropy is most marked near a typical point on the caustic, where only two vectors (equal in magnitude and opposite in direction) contribute significantly to (19), giving rise to 'Airy' fringes parallel to the caustic.

Again this behaviour is very different from what happens in the ergodic case, where there are infinitely may contributing $\boldsymbol{p}$ vectors and the spectrum of ψ is continuous. When H is given by (10), equations (7) and (9) give

$$C(\boldsymbol{X};\boldsymbol{q})=\frac{\int \mathrm{d}\boldsymbol{\Omega}\,\exp\{\mathrm{i}\boldsymbol{\Omega}\,.\,\boldsymbol{X}[2m(E-V(\boldsymbol{q}))]^{1/2}/\hbar\}}{\int \mathrm{d}\boldsymbol{\Omega}}, \tag{20}$$

where $\boldsymbol{\Omega}$ is the unit vector along $\boldsymbol{p}$. This integral can be evaluated in terms of standard Bessel functions to give

$$C(\boldsymbol{X};\boldsymbol{q})=\Gamma(\tfrac{1}{2}N)\frac{J_{\frac{1}{2}N-1}(X[2m(E-V(\boldsymbol{q}))]^{1/2}/\hbar)}{\{X[2m(E-V(\boldsymbol{q}))]^{1/2}/2\hbar\}^{\frac{1}{2}N-1}}. \tag{21}$$

(For $N=2$ and $N=3$ this expression is simply $J_0(\xi)$ and $\sin\xi/\xi$ respectively, where ξ is the argument of the Bessel function.) Just as in the integrable case all oscillations of ψ have the de Broglie wavelength $\hbar/[2m(E-V(\boldsymbol{q}))]^{1/2}$. Now, however, the oscillations near $\boldsymbol{q}$ are statistically isotropic, even close to the anticaustics.

The autocorrelation function is not by itself sufficient to determine all statistical properties of ψ. However it is likely that for stochastic classical motion the phases of the different contributions $\boldsymbol{p}$ to ψ are uncorrelated, because the orbit would accumulate many action units $\hbar$ in its 'unpredictable' wanderings between passages through the neighbourhood of $\boldsymbol{q}$. This would imply that ψ is a *Gaussian random function* of $\boldsymbol{q}$ (Rice 1944, 1945, Longuet-Higgins 1956), whose spectrum at $\boldsymbol{q}$ is simply the local average of the Wigner function $\Psi(\boldsymbol{q},\boldsymbol{p})$. For ergodic motion Ψ is given by (9) leading to the isotropic statistics described by (21), while for non-ergodic stochastic motion in a gap between KAM tori the contributing $\boldsymbol{p}$ are continuously distributed over a limited range of directions resulting in ψ being a random wave whose statistics have some anisotropy. All statistical porperties of a Gaussian random function (probability distribution of ψ and its derivatives, correlations between ψ at two or more points, etc) are determined by $\Pi(\boldsymbol{q})$ and $C(\boldsymbol{X};\boldsymbol{q})$.

Of course ψ for an integrable system cannot be a Gaussian random function because its spectrum of wavevectors is discrete. However there exist in quasi-integrable systems stable closed orbits of arbitrarily complicated topology, which will be surrounded by tori whose projections onto $\boldsymbol{q}$ will be crossed by many caustics not confined to the boundary of the classically allowed region. It is meaningful to think of the wave ψ in the presence of such caustics as having statistical properties but these will be highly non-Gaussian as I have explained elsewhere (Berry 1977b).

4. Conclusions

I have suggested that in the semiclassical limit quantum energy eigenstates separate into two universality classes distinguished by the morphology of their wavefunctions. States in Percival's regular spectrum, associated with tori in classical phase space, have vivid patterns of regular interference fringes and violent fluctuations in intensity associated with caustics of the classical motion. In sharp contrast, states in Percival's irregular spectrum, associated with stochastic motion in phase space, have random patterns of interference maxima and minima (statistically isotropic in the ergodic case) with more temperate intensity fluctuations of Gaussian random type and 'anticaustics' at boundaries of the classical motion.

Closely analogous behaviour of wavefunctions is currently being studied in optics, in connection with Gaussian and non-Gaussian laser speckle patterns (Jakeman and Pusey 1975) and Gaussian and non-Gaussian twinkling of starlight (Jakeman *et al* 1976, Berry 1977b). Gaussian wavefunctions arise when waves traverse a medium producing an irregular wavefront whose topography varies rapidly on a wavelength scale, and non-Gaussian wavefunctions (with caustics, etc) arise when the wavefront varies smoothly on a wavelength scale. In both the optical and the quantum cases the different behaviour of ψ arises from the same cause: regular waves have underlying trajectories in phase space that are smoothly distributed on the scale of wavelength or $\hbar$, while in irregular waves the trajectories show structure down to scales smaller than wavelength or $\hbar$. Therefore $\hbar$ gives quantum oscillatory detail to regular wavefunctions but plays the completely different role of a quantum smoothing parameter in irregular wavefunctions.

The regular and irregular behaviour described here should be obvious on computer-generated contour maps of eigenfunctions, provided these are 'semiclassical' enough. For a system with two degrees of freedom this would probably require eigenfunctions with about a hundred extrema (ten nodes in each direction), whose computation using suitable basis functions would involve diagonalising 200×200 matrices and is feasible with current technology.

For a quasi-integrable system the semiclassical limit will be more complicated than I have described here, because there will be some states not clearly identifiable as regular or irregular, associated with stochastic regions of small measure resulting from the destruction of tori whose frequency ratios are high-order rational numbers. A description of the different regimes expected as $\hbar$ gets smaller is given by Berry (1977a, 1978).

Acknowledgments

The arguments presented here were put into their final form as a result of discussions with Dr A Voros, and I would like to express my gratitide to him. I also thank Dr M Tabor for many conversations on this subject.

References

Arnol'd V I 1963 *Usp. Mat. Nauk* **18** No. 5 13–39, No. 6 91–196 (1963 *Russ. Math. Surv.* **18** No. 5 9–36, No. 6 85–191)
—— 1975 *Usp. Mat. Nauk* **30** No. 5 3–65 (1975 *Russ. Math. Surv.* **30** No. 5 1–75)
Arnol'd V I and Avez A 1968 *Ergodic Problems of Classical Mechanics* (New York: Benjamin)
Berry M V 1976 *Adv. Phys.* **25** 1–26
—— 1977a *Phil. Trans. R. Soc.* A **287** 237–71
—— 1977b *J. Phys. A: Math. Gen.* **10** 2061–81
—— 1978 *Integrable and Stochastic Systems: a Primer* ed. R Helleman to be published
Dvorin M M and Lazutkin V F 1973 *Funkt. Anal. Ego Pril.* **7** No. 2 20–7 (1974 *Funct. Anal. Applic.* 103–9)
Einstein A 1917 *Verh. Dt. Phys. Ges.* **19** 82–92
Ford J 1975 *Fundamental Problems in Statistical Mechanics* ed. E G D Cohen, vol. 3 (Amsterdam: North-Holland) pp 215–55
Gutzwiller M C 1971 *J. Math. Phys.* **12** 343–58
Jakeman E and Pusey P N 1975 *J. Phys. A: Math. Gen.* **8** 369–91
Jakeman E, Pike E R and Pusey P N 1976 *Nature* **263** 315–7
Kolmogoroff A N 1954 *Dokl. Akad. Nauk* **98** 527–30
Lazutkin V F 1973 *Izv. Akad. Nauk, Mat. Ser.* **37** 186–216, 437–64 (1973 *Math. USSR Izv.* **7** 185–214, 439–66)
Longuet-Higgins M S 1956 *Phil. Trans. R. Soc.* A **249** 321–87
Moser J 1962 *Nachr. Akad. Wiss. Göttingen* **1** 1–20
Percival I C 1973 *J. Phys. B: Atom. Molec. Phys.* **6** L229–32
—— 1977 *Adv. Chem. Phys.* **36** 1–61
Rice S O 1944 *Bell Syst. Tech. J.* **23** 282–332
—— 1945 *Bell Syst. Tech. J.* **24** 46–156
Thom R 1972 *Stabilité Structurelle et Morphogenèse* (Reading, Mass.: Benjamin) (English translation 1975)
Voros A 1976 *Ann. Inst. Henri Poincaré* A **24** 31–90
—— 1977 *Ann. Inst. Henri Poincaré* A **26** 343–403
Whiteman K J 1977 *Rep. Prog. Phys.* **40** 1033–69

Proc. R. Soc. Lond. A **400**, 229–251 (1985)
Printed in Great Britain

Semiclassical theory of spectral rigidity

By M. V. Berry, F.R.S.
H. H. Wills Physics Laboratory, Tyndall Avenue, Bristol BS8 1TL, U.K.

(*Received* 20 *February* 1985)

The spectral rigidity $\Delta(L)$ of a set of quantal energy levels is the mean square deviation of the spectral staircase from the straight line that best fits it over a range of L mean level spacings. In the semiclassical limit $(\hbar \rightarrow 0)$, formulae are obtained giving $\Delta(L)$ as a sum over classical periodic orbits. When $L \ll L_{max}$, where $L_{max} \sim \hbar^{-(N-1)}$ for a system of N freedoms, $\Delta(L)$ is shown to display the following universal behaviour as a result of properties of very long classical orbits: if the system is classically integrable (all periodic orbits filling tori), $\Delta(L) = \frac{1}{15}L$ (as in an uncorrelated (Poisson) eigenvalue sequence); if the system is classically chaotic (all periodic orbits isolated and unstable) and has no symmetry, $\Delta(L) = \ln L/2\pi^2 + D$ if $1 \ll L \ll L_{max}$ (as in the gaussian unitary ensemble of random-matrix theory); if the system is chaotic and has time-reversal symmetry, $\Delta(L) = \ln L/\pi^2 + E$ if $1 \ll L \ll L_{max}$ (as in the gaussian orthogonal ensemble). When $L \gg L_{max}$, $\Delta(L)$ saturates non-universally at a value, determined by short classical orbits, of order $\hbar^{-(N-1)}$ for integrable systems and $\ln(\hbar^{-1})$ for chaotic systems. These results are obtained by using the periodic-orbit expansion for the spectral density, together with classical sum rules for the intensities of long orbits and a semiclassical sum rule restricting the manner in which their contributions interfere. For two examples $\Delta(L)$ is studied in detail: the rectangular billiard (integrable), and the Riemann zeta function (assuming its zeros to be the eigenvalues of an unknown quantum system whose unknown classical limit is chaotic).

1. Introduction

Several statistical measures of the regularity of sequences of eigenvalues were introduced to describe the energy levels of many-particle systems such as nuclei (Porter 1965). Recently these spectral measures have been employed for bound systems with few freedoms, to explore the ways in which the distribution of quantal energy levels reflects integrability or chaos in the underlying classical trajectories. It was expected, and found, that classically integrable systems have levels that are locally uncorrelated and well described by a Poisson distribution; in contrast, classically chaotic systems have levels with strong local repulsion, well described by the eigenvalues of matrices drawn randomly from appropriate ensembles (for reviews see Berry 1983, 1984; Bohigas & Giannoni 1984).

My purpose in this paper is to explain the semiclassical origin and the limits of validity of these two types of spectral universality, by deriving theoretical

Reprinted from Proceedings of the Royal Society A, Volume 400, (1985).

expressions for one of the spectral statistics, namely the rigidity. This will be defined in (5) in terms of the spectral staircase

$$\mathcal{N}(E) \equiv \sum_n \Theta(E-E_n), \tag{1}$$

where $E_n = E_1, E_2 \dots$ is the eigenvalue sequence and Θ denotes the unit step function. We also require the spectral density

$$d(E) \equiv \mathrm{d}\mathcal{N}(E)/\mathrm{d}E = \sum_n \delta(E-E_n). \tag{2}$$

For a system with N freedoms the local averages of these functions are

$$\langle \mathcal{N}(E) \rangle = \Omega(E)/h^N; \quad \langle d(E) \rangle = (d\Omega(E)/dE)/h^N, \tag{3}$$

where $\Omega(E)$ is the classical phase-space volume enclosed by the surface with energy E, given in terms of the Hamiltonian $H(q,p)$ by

$$\Omega(E) = \int \mathrm{d}^N q \int \mathrm{d}^N p \, \Theta(E - H(q,p)). \tag{4}$$

In the cases of interest here, $N \geqslant 2$.

The averages denoted by $\langle \quad \rangle$ in (3) refer to all levels in an energy range that is classically small, i.e. small in comparison with E, but semiclassically large, i.e. large in comparison with the mean level spacing $\langle d \rangle^{-1} \sim \hbar^N$. This mean level spacing will play an important role in what follows, and will be called the inner energy scale.

The rigidity $\Delta(L)$ is now defined as the local average of the mean square deviation of the staircase from the best fitting straight line over an energy range corresponding to L mean level spacings, namely

$$\Delta(L) \equiv \left\langle \min_{(A,B)} \frac{\langle d(E) \rangle}{L} \int_{-L/2\langle d \rangle}^{L/2\langle d \rangle} \mathrm{d}\epsilon \, [\mathcal{N}(E+\epsilon) - A - B\epsilon]^2 \right\rangle. \tag{5}$$

This function was introduced by Dyson & Mehta (1963) (they called it Δ_3 to distinguish it from two less useful statistics). Minimizing over A and B leads to

$$\Delta(L) = \left\langle \left\{ \frac{\langle d \rangle}{L} \int_{-L/2\langle d \rangle}^{L/2\langle d \rangle} \mathrm{d}\epsilon \, \mathcal{N}^2(E+\epsilon) - \left[\frac{\langle d \rangle}{L} \int_{-L/2\langle d \rangle}^{L/2\langle d \rangle} \mathrm{d}\epsilon \, \mathcal{N}(E+\epsilon) \right]^2 - 12 \left[\frac{\langle d \rangle^2}{L^2} \int_{-L/2\langle d \rangle}^{L/2\langle d \rangle} \mathrm{d}\epsilon \, \epsilon \, \mathcal{N}(E+\epsilon) \right]^2 \right\} \right\rangle. \tag{6}$$

When $L \ll 1$, the fact that $\mathcal{N}(E)$ is a staircase leads to the limit $\Delta \to \frac{1}{15}L$ whatever distribution the levels have (provided this is non-singular). Therefore the spectral rigidity gives no information about the very finest scales corresponding to the spacings between neighbouring levels. Its usefulness lies in the way it describes correlations over level sequences longer than the inner energy scale (which corresponds to $L = 1$).

We shall demonstrate the existence of two universality classes of rigidity, extending from $L \sim 1$, corresponding to the inner energy scale, to a value $L_{\max}$

corresponding to an outer energy scale h/T_{min}, where T_{min} is the period of the shortest classical closed orbit. Thus

$$L_{\text{max}} \equiv h\langle d\rangle / T_{\text{min}} \sim \hbar^{-(N-1)} \tag{7}$$

greatly exceeds unity, in spite of the fact that the outer energy scale is of order $\hbar$ and hence classically small. Now we can be more precise about the local averages denoted by $\langle\ \ \rangle$: these correspond to energy ranges much larger than the outer scale but still classically small, for example energy ranges of order $\hbar^{\frac{1}{2}}$.

The first universality class occurs for classically integrable systems. In these, as will be shown in §3, the Poisson form $\frac{1}{15}L$ extends from $L = 0$ to L_{max}. For $L > L_{\text{max}}$, $\Delta(L)$ reaches a saturation value (not universal), Δ_{∞}, which can be regarded as a measure of the totality of the spectral fluctuations on all scales. This extends earlier work (Berry & Tabor 1977*a*) in which Poisson statistics were shown to describe local fluctuations, and, as will be shown explicitly in §4 it explains recent numerical results of Casati *et al.* (1985) on the rectangular billiard.

The second universality class occurs for classically chaotic systems. In these, as will be shown in §6, $\Delta(L)$ increases only logarithmically in the range $1 \lesssim L < L_{\text{max}}$, which indicates long-range rigidity in the level distribution. The fact that for systems with time-reversal symmetry the coefficient of the logarithm is twice what it is when there is no such symmetry is given a simple semiclassical explanation in §8. For $L > L_{\text{max}}$, $\Delta(L)$ reaches a saturation value (not universal), Δ_{∞}, much smaller than in the integrable case. When applied to the 'level sequence' consisting of the imaginary parts of the zeros of the Riemann zeta function (§7), our results are consistent with what little is known and conjectured about the asymptotics of this sequence.

In deriving these results from (6), semiclassical methods are essential. The reason is that for local spectral statistics to attain well defined limiting values, classically small energy ranges must contain many levels, and this happens only as $\hbar \to 0$ (for scaling systems such as billiards, this is equivalent to $E \to \infty$). The semiclassical technique employed here (in §2) is the representation of the spectral density $d(E)$ as a sum over all the periodic orbits of the classical system, introduced and developed by Gutzwiller (1967, 1969, 1970, 1971, 1978) and Balian & Bloch (1972, 1974).

When applied to the calculation of spectral rigidity, the periodic-orbit technique requires two further ingredients. The first is a classical sum rule for the orbit intensities, recently discovered by Hannay & Ozorio de Almeida (1984), and the second is a new semiclassical sum rule derived in §5.

The arguments and conclusions of this paper complement those of Pechukas (1983). He obtained all the spectral statistics, but made use of a statistical assumption about the wavefunction. I make no such statistical assumption, but discuss only $\Delta(L)$.

2. Rigidity in terms of periodic orbits

Periodic-orbit theory gives the semiclassical spectral density as

$$d(E) = \langle d(E)\rangle + d_{osc}(E), \tag{8}$$

where $d_{osc}(E)$ is a sum over classical periodic orbits, each of which contributes an oscillatory function and whose combined effect is to produce, by constructive interference, a sequence of singularities (such as the δ functions in (2) or approximations to them) minus $\langle d\rangle$. We shall write d_{osc} in a form appropriate to systems that are completely integrable or completely chaotic. In an integrable system, closed orbits are not isolated but form $(N-1)$–parameter families filling N-dimensional phase-space tori. A chaotic system we define as one that is ergodic and for which, in addition, all closed orbits are isolated and therefore unstable. The oscillatory contribution in (8) is

$$d_{osc}(E) = \frac{1}{\hbar^{\mu+1}} \sum_j A_j(E) \exp\{iS_j(E)/\hbar\}. \tag{9}$$

Detailed descriptions and discussions of this formula were given by Berry (1983, 1984); here we need the following facts. In the sum, j labels all distinct periodic orbits including all multiple traversals, positive and negative (but not zero). It will be of crucial importance that negative traversals correspond to retracings, where the orbit is followed backwards in time, and not to time-reversed orbits (the latter exist only for systems with time-reversal symmetry). The exponent μ is $\frac{1}{2}(N-1)$ for integrable systems and zero for chaotic ones. This difference is important and for integrable systems the periodic orbits on a torus combine coherently to produce much stronger spectral oscillations than an isolated orbit of a chaotic system, as well as giving the first hint that the level statistics will be different too. The (real) amplitudes A_j will be discussed later. In the exponent, the phase contains the action $S_j(E)$, defined for m traversals as

$$S_j(E) \equiv m\left\{\oint \boldsymbol{p}\cdot d\boldsymbol{q} + \alpha\hbar\right\}, \tag{10}$$

where the integral is over a single traversal and $\alpha\hbar$, which will play no part in what follows, gives focusing corrections such as Maslov indices. Because of the negative traversals, d_{osc} is real.

In (9) the energy dependence of the oscillations is determined by the orbit periods $T_j(E)$ because

$$T_j(E) = dS_j(E)/dE. \tag{11}$$

The longest oscillation, giving the largest scale of spectral fluctuations, comes from the shortest orbit and has 'wavelength' given by the outer scale h/T_{min}, already defined. Thus the constructive interference that gives δ functions, whose mean spacing is the inner scale $\langle d\rangle^{-1} \approx \hbar^N$, is determined by very long orbits, with periods $T \sim \hbar^{-(N-1)}$.

To incorporate the oscillations (9) into the rigidity formula (6) we use the fact that the energy range $L/\langle d\rangle$ is classically small (although it may be semiclassically large) to write

$$S_j(E+\epsilon) \approx S_j(E) + \epsilon T_j(E) \tag{12}$$

and ignore the ϵ dependences of A_j and $\langle d\rangle$. Thus the spectral staircase is

$$\mathcal{N}(E) = \langle \mathcal{N}(E)\rangle + \mathcal{N}_{\rm osc}(E),$$

where

$$\mathcal{N}_{\rm osc}(E+\epsilon) = \frac{-i}{\hbar^{\mu}} \sum_j \frac{A_j}{T_j} \exp\{i(S_j+\epsilon T_j)/\hbar\}. \tag{13}$$

The integrals in (6) are now elementary and give

$$\Delta(L) = \left\langle \frac{1}{\hbar^{2\mu}} \sum_i \sum_j \frac{A_i A_j}{T_i T_j} \exp\{i(S_i - S_j)/\hbar\} \times [F(y_i - y_j) - F(y_i)\,F(y_j) - 3F'(y_i)\,F'(y_j)] \right\rangle, \tag{14}$$

where

$$y_j \equiv LT_j/2\hbar\langle d\rangle, \tag{15}$$

$$F(y) \equiv \sin y/y, \tag{16}$$

and primes denote differentiation.

To arrive at (14) we made use of the result

$$\langle \exp\{iS_j/\hbar\}\rangle \to 0 \quad \text{as} \quad \hbar \to 0, \tag{17}$$

which holds because the local averaging is over an energy range much greater than the outer scale. It is tempting to think that the same principle of destructive interference will eliminate the non-diagonal terms $i \neq j$ in (14). This is the case for integrable systems (as will be shown in §5), but for chaotic systems the proliferation of pairs of very long orbits with action differences $(S_i - S_j) < \hbar$ is important and we shall see that local averaging does not diagonalize the sum.

However, the restriction to pairs with $(S_i - S_j)/\hbar < 1$ does have the effect that in the functions F in (14) we can set $y_i = y_j$. The reason is that, for long orbits,

$$S_j \to T_j\, N\Omega/(\mathrm{d}\Omega/\mathrm{d}E) \tag{18}$$

(Hannay & Ozorio de Almeida 1984), and together with (3) this implies

$$|y_i - y_j| \to \frac{|S_i - S_j|}{\hbar}\,\frac{L}{2N\mathcal{N}}, \tag{19}$$

which vanishes because $L \ll \mathcal{N}$. Thus the rigidity becomes

$$\Delta(L) = \frac{2}{\hbar^{2\mu}} \int_0^{\infty} \frac{\mathrm{d}T}{T^2}\, \phi(T)\, G(LT/2\langle d\rangle\hbar), \tag{20}$$

where

$$G(y) \equiv 1 - F^2(y) - 3(F'(y))^2 \tag{21}$$

and

$$\phi(T) \equiv \left\langle \sum_i \sum_j{}^{+} A_i A_j \cos\{(S_i - S_j)/\hbar\}\, \delta\{T - \tfrac{1}{2}(T_i + T_j)\} \right\rangle \tag{22}$$

in which the + on the summation denotes restriction to positive traversals (i.e. $T_j > 0$).

Figure 1 shows the important function $G(y)$. Because G is small if $y \leqslant 1$, it selects from the sum (22) only those pairs of orbits whose average period exceeds $2\langle d\rangle\hbar/L$; therefore G will be called the orbit selection function. Such selection is physically reasonable because $\Delta(L)$ is defined by (5) in terms of deviations from a linear approximation to the staircase over an energy range $L/\langle d\rangle$: the periodic-orbit sum (13) shows that this linear approximation is determined by orbits with $T_j < \langle d\rangle\hbar/L$, and deviations from it by orbits with $T_j > \langle d \rangle \hbar / L$. The effects of this orbit selection depend strongly on the value of L in comparison with unity and $L_{\max}$ (equation 7). For the shortest closed orbit, (7) shows that when $L = L_{\max}$ the argument of G is $y = \pi$.

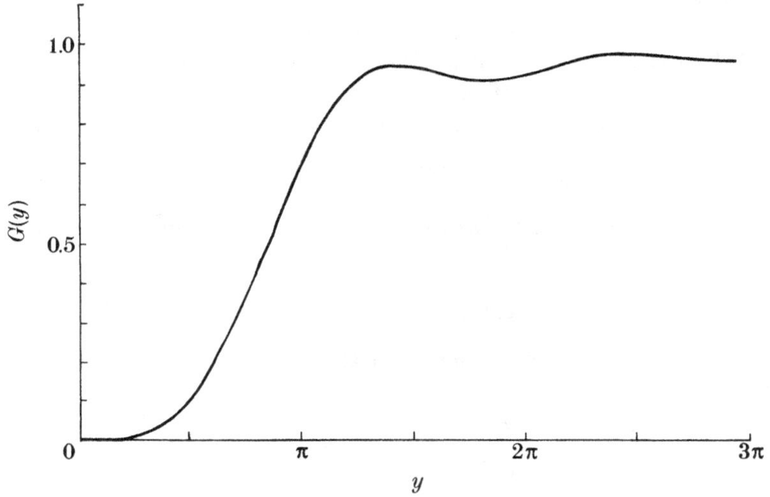

FIGURE 1. Orbit selection function defined by (21).

It will be important to know the large-T limiting form of the diagonal sum in (22), namely

$$\phi_D(T) \equiv \langle \Sigma^+ A_j^2 \delta(T - T_j) \rangle . \tag{23}$$

This is the number density of orbits with periods near T, weighted with intensities A_j^2. As $T \to \infty$, the density of periodic orbits increases and the intensities decrease: as a power-law for integrable systems and exponentially for chaotic ones. The results of the competition between these two tendencies was calculated by Hannay & Ozorio de Almeida (1984) as a consequence of their extension of the important idea that very long periodic orbits are uniformly distributed in phase space (see, for example, Parry and Pollicot 1983; Parry 1984). They found that

$$\phi_D(T) \to (\mathrm{d}\Omega/\mathrm{d}E)/(2\pi)^{N+1} \quad \text{(integrable)} \tag{24}$$

and

$$\phi_D(T) \to T/4\pi^2 \quad \text{(chaotic)}. \tag{25}$$

(The integrable result (24) had previously been found by Berry & Tabor (1977*a*), in a different way.)

These formulae correspond to the amplitudes A_j combining incoherently as in (23), which is the correct procedure if no symmetry enforces strict degeneracy among the orbit actions S_j. When such degeneracy does exist, the appropriate amplitudes must be combined coherently, and we shall see in §8 that this has important repercussions for time-reversal symmetry. In writing (25) we have, for simplicity, ignored multiple traversals ($M > 1$ in 10), because these give contributions to ϕ_D that vanish as $T \to \infty$ (this will be illustrated in §7).

3. Integrable systems

For integrable systems, global action-angle variables exist, with each set of actions $\boldsymbol{I} = \{I_1 \ldots I_N\}$ denoting a phase-space torus. The Hamiltonian can be written $H(\boldsymbol{I})$ and the frequencies $\boldsymbol{\omega} = \{\omega_1 \ldots \omega_N\}$ on the torus $\boldsymbol{I}$ are given by $\boldsymbol{\omega} = \nabla_I H(\boldsymbol{I})$. The periodic orbits at energy E are knots on the torus, with winding numbers $\boldsymbol{M} = \{M_1 \ldots M_N\}$ for the N irreducible cycles. These winding numbers constitute the label j in the periodic-orbit sums of the previous section. Each $\boldsymbol{M}$ defines a resonant torus $\boldsymbol{I_M}$, which is one whose frequencies are commensurable, and hence a period $T_{\boldsymbol{M}}$, by

$$\boldsymbol{\omega}(\boldsymbol{I_M}) = 2\pi \boldsymbol{M}/T_{\boldsymbol{M}}; \quad H(\boldsymbol{I_M}) = E. \tag{26}$$

A convenient form for the orbit amplitudes $A_{\boldsymbol{M}}$ is that given by Berry & Tabor (1977*b*):

$$A_{\boldsymbol{M}}^2 = \frac{(2\pi)^{N-1}}{T_{\boldsymbol{M}}^N |\boldsymbol{\omega}\cdot\partial \boldsymbol{I_M}/\partial T_{\boldsymbol{M}} \det\{\partial\omega_i/\partial I_j\}_{\boldsymbol{M}}|}. \tag{27}$$

For the rigidity, (20)–(22), together with the diagonal average to be justified later, give the topological sum

$$\Delta(L) = \frac{2}{\hbar^{N-1}} \sum_{\boldsymbol{M}}^{+} \frac{A_{\boldsymbol{M}}^2}{T_{\boldsymbol{M}}^2} G(LT_{\boldsymbol{M}}/2\langle d\rangle \hbar). \tag{28}$$

Thus $\Delta(L)$ is a sum of weighted scaled orbit selection functions (figure 1), one for each resonant torus.

When $L \ll L_{\max}$, G selects only long orbits and the topological sum can be evaluated by using the continuum limit (24). This gives

$$\begin{aligned}\Delta(L) &= \frac{\hbar}{(2\pi\hbar)^N \pi} \frac{\mathrm{d}\Omega}{\mathrm{d}E} \int_0^\infty \frac{\mathrm{d}T}{T^2} G(LT/2\langle d\rangle\hbar) \\ &= \frac{L}{2\pi}\int_0^\infty \frac{\mathrm{d}y}{y^2} G(y).\end{aligned} \tag{29}$$

The integral, from (21) and (16), equals $\frac{2}{15}\pi$, so that

$$\Delta(L) = \tfrac{1}{15} L \quad (L \ll L_{\max}). \tag{30}$$

This supports the claim that the local spectra of classically integrable systems belong to the universality class of uncorrelated level sequences; of course 'local' means $L \ll L_{\max}$.

When $L \gg L_{\max}$, the orbit selection function in (28) is unity for all orbits $\boldsymbol{M}$, and Δ attains a saturation value given by the convergent sum

$$\Delta_\infty = \frac{2}{\hbar^{N-1}} \sum_{\boldsymbol{M}}^{+} \frac{A_{\boldsymbol{M}}^2}{T_{\boldsymbol{M}}^2}. \tag{31}$$

Although this is semiclassically large, the r.m.s. fluctuations $\Delta_\infty^{\frac{1}{2}}$ in the staircase are still much smaller than the mean height $\langle\mathcal{N}\rangle$ of the staircase itself (equation (3)), by a factor $\hbar^{\frac{1}{2}(N+1)}$. Saturation of the rigidity has been observed in numerical calculations by Seligman *et al.* (1985) (for an integrable polynomial Hamiltonian) and Casati *et al.* (1985) (for an integrable billiard (see the next section)).

In the crossover region $L \sim L_{max}$, (28) predicts a few weak oscillations as L increases to reveal the contribution of the shortest orbit with period T_{min}. The slowest oscillations in $\Delta(L)$ have an L-wavelength of L_{max}.

4. An example: billards in a rectangle

For a particle of mass m moving freely within a rectangle with sides a, b and impenetrable walls, the quantal energy levels are

$$E_{l,n} = \hbar^2 \pi^2 (l^2 \alpha^{-\frac{1}{2}} + n^2 \alpha^{\frac{1}{2}})/2mab, \tag{32}$$

where

$$\alpha \equiv a^2/b^2. \tag{33}$$

We assume without loss of generality that $a \geqslant b$, i.e. $\alpha \geqslant 1$. Classically, the action Hamiltonian for this two-dimensional integrable system is

$$H(I_1, I_2) = \pi^2 (I_1^2/a^2 + I_2^2/b^2)/2m \tag{34}$$

with frequencies

$$\boldsymbol{\omega} = \pi^2 (I_1/a^2, I_2/b^2)/m. \tag{35}$$

It follows from (26) that the period of the closed orbit with topology $\boldsymbol{M} = (M_1, M_2)$ is

$$T_{\boldsymbol{M}} = [2m(M_1^2 a^2 + M_2^2 b^2)/E]^{\frac{1}{2}}. \tag{36}$$

Three of these orbits are illustrated in figure 2. The resonant tori whose orbits have these periods have actions

$$\boldsymbol{I}_{\boldsymbol{M}} = 2m\,(M_1 a^2, M_2 b^2)/\pi T_{\boldsymbol{M}}. \tag{37}$$

In terms of these quantities it is easy to calculate the torus amplitudes (27)

$$A_{\boldsymbol{M}}^2 = m^2 a^2 b^2/\pi^3 E T_{\boldsymbol{M}}. \tag{38}$$

It might be thought that time-reversal symmetry of the periodic orbits might cause them to contribute twice to the sum (28) for the rigidity. But this is not the case, because although each orbit on the torus $\boldsymbol{I}_{\boldsymbol{M}}$ is distinct from its time-reverse if neither M_1 nor M_2 is zero, both these orbits are included on the same torus, which is a four-sheeted geometric object with inversion symmetry about $\boldsymbol{p} = 0$ in momentum space, and hence time-reversal symmetry. If either M_1 or M_2 is zero (for example the orbit (0, 1) in figure 2) the orbit is self-retracing and hence is its own time-reverse. The corresponding torus has only two sheets rather than four and so counts $\frac{1}{2}$ in amplitude and $\frac{1}{4}$ in intensity (for a detailed discussion of this phenomenon see Appendix C of Richens & Berry (1981)).

It is natural to express $\Delta(L)$ in terms of the scaled energy

$$\mathscr{E} \equiv E\langle d\rangle \;(= \langle \mathscr{N}(E)\rangle), \tag{39}$$

corresponding to a mean level spacing of unity, which by (28) gives

$$\Delta(L) = \frac{\mathscr{E}^{\frac{1}{2}}}{\pi^{\frac{5}{2}}} \sum_{M_1=0}^{\infty} \sum_{M_2=0}^{\infty} \frac{\delta_{\boldsymbol{M}}\, G(y_{\boldsymbol{M}})}{(M_1^2 \alpha^{\frac{1}{2}} + M_2^2 \alpha^{-\frac{1}{2}})^{\frac{3}{2}}}, \tag{40}$$

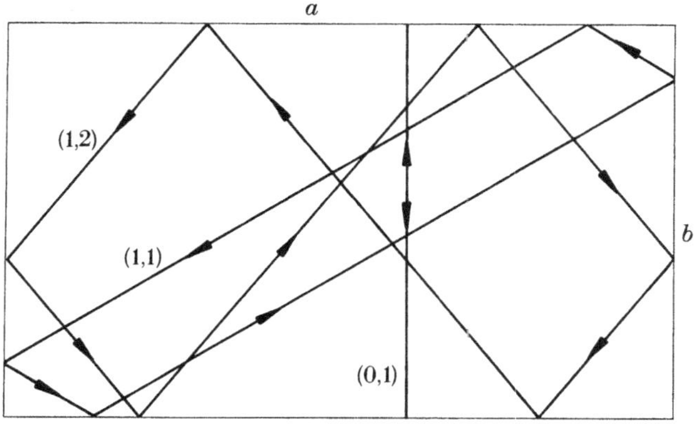

FIGURE 2. Rectangular billiard showing closed orbits with three different pairs of winding numbers (M_1, M_2). The orbit $(0, 1)$ is self-retracing; the other two are not.

where
$$\delta_{\boldsymbol{M}} \equiv \begin{cases} 0 & \text{if } M_1 = M_2 = 0, \\ \frac{1}{4} & \text{if one of } M_1 \text{ and } M_2 \text{ is zero,} \\ 1 & \text{otherwise} \end{cases} \tag{41}$$

and
$$y_{\boldsymbol{M}} = L\{\pi(M_1^2\alpha^{\frac{1}{2}} + M_2^2\alpha^{-\frac{1}{2}})/\mathscr{E}\}^{\frac{1}{2}}. \tag{42}$$

The shortest periodic orbit has winding numbers $(0, 1)$ and (7) leads to

$$L_{\max} = (\pi\mathscr{E})^{\frac{1}{2}}\alpha^{\frac{1}{4}}. \tag{43}$$

When $L \gg L_{\max}$, $G \approx 1$ for all $\boldsymbol{M}$ and so the saturation rigidity is

$$\Delta_\infty = \frac{\mathscr{E}^{\frac{1}{2}}}{\pi^{\frac{5}{2}}} \sum_{M_1=0}^{\infty} \sum_{M_2=0}^{\infty} \frac{\delta_{\boldsymbol{M}}}{(M_1^2\alpha^{\frac{1}{2}} + M_2^2\alpha^{-\frac{1}{2}})^{\frac{3}{2}}}. \tag{44}$$

This convergent series is easy to sum numerically. For α close to unity it varies slowly with α and so can be approximated by its value when $\alpha = 1$, which is (Zucker 1974)

$$\Delta_\infty \overset{(\alpha=1)}{=} \mathscr{E}^{\frac{1}{2}}\pi^{-\frac{5}{2}}[\zeta(\tfrac{3}{2})\beta(\tfrac{3}{2}) - \tfrac{1}{2}\zeta(3)] = 0.0947\,\mathscr{E}^{\frac{1}{2}}, \tag{45}$$

where ζ and β denote the number-theoretic series

$$\zeta(s) \equiv \sum_{n=1}^{\infty} \frac{1}{n^s}; \quad \beta(s) \equiv \sum_{n=1}^{\infty} \frac{(-1)^{n+1}}{(2n-1)^s}. \tag{46}$$

These results will now be compared with quantal calculations of $\Delta(L)$ by Casati *et al.* (1985). They used the levels $\alpha l^2 + n^2$ and so our $\mathscr{E}$ is their energy multiplied by $\pi/4\alpha^2$. Figure 3 shows the comparison for two energies, between $\Delta(L)$ computed from (40) for $\alpha = 1$ and their quantal calculations for an ensemble of α-values between 0.9 and 1.2 (a range in which the different theoretical Δ-curves are almost indistinguishable). The agreement is good in the universal crossover and saturation ranges of L, although their oscillations are slightly stronger (this might be the result of their different averaging procedure).

Casati *et al.* also show (non-local) averages of $\Delta(L)$ over the whole energy range from zero to $\mathscr{E}$, for $\alpha = \frac{1}{3}\pi$. From (45), their curves ought to saturate at $\frac{2}{3} \times 0.0947\, \mathscr{E}^{\frac{1}{2}}$, and comparison shows that they do.

Finally, Casati *et al.* show a graph of $\Delta(L)$ in which L consists of the range from the ground state to the Lth level. In our notation this corresponds to choosing $\mathscr{E} = \frac{1}{2}L$ so that L is always in the saturation range and (45) predicts

$$\Delta = 0.0947\, L^{\frac{1}{2}}/\sqrt{2} = 0.067L^{\frac{1}{2}}. \tag{47}$$

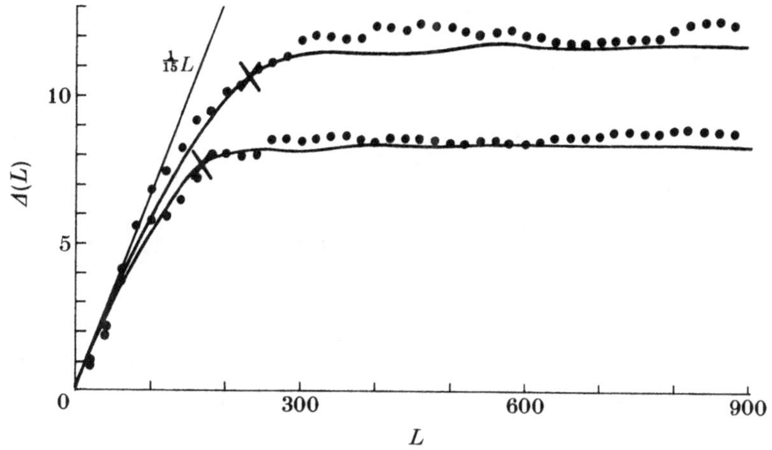

FIGURE 3. The full curves show $\Delta(L)$ computed from (40) for $\alpha = 1$ and $\mathscr{E} = 10\,500 \times \frac{1}{4}\pi$ (lower curve) and $\mathscr{E} = 20\,500 \times \frac{1}{4}\pi$ (upper curve) by using 1250 closed orbits. The circular points show data from Casati *et al.* (1985). The theoretical crossover values $L_{\max}$ are 161 and 225 and are indicated by crosses. The straight line shows the local universal Poisson rigidity $\frac{1}{15}L$.

Their curve is fitted by $0.063L^{\frac{1}{2}}$, which is a close agreement when one considers that the theoretical formulae have here been applied to values of L that are certainly not classically small.

5. A semiclassical sum rule

The periodic-orbit sum (9) can be at best conditionally convergent because it represents the spectral density, with δ singularities at the energy levels. The delicate conspiracy of amplitudes A_j and phases $S_j/\hbar$ by which this is achieved is not fully understood, but there is some theoretical evidence that the semiclassical approximation (9) is in fact capable of reproducing singularities when infinitely long orbits are included. For integrable systems, we refer to Norcliffe & Percival (1968), Balian & Bloch (1972), and Berry & Tabor (1976). For chaotic systems, we refer to the Selberg identity (reviewed by Hejhal (1976) and McKean (1972)), which shows that for manifolds of constant negative curvature (9) is exact, and also to a study by Gutzwiller (1980) and an example to be given in §7 here. The purpose of the present section is to derive an identity that must be satisfied by the function $\phi(T)$ (equation (22)) that appears in the rigidity formula (20), to ensure that the periodic-orbit sum (9) has the correct density of singularities.

By analytically continuing the energy to $E \to E + i\eta$ and using the representation of the spectral density (2) as the imaginary part of the causal Green function, $d(E)$ can be expressed as the limit $\eta \to 0$ of the Lorentzians

$$d_\eta(E) = -\frac{1}{\pi} \operatorname{Im} \sum_n \frac{1}{E - E_n + i\eta}. \tag{48}$$

If

$$\eta \langle d \rangle \ll 1 \tag{49}$$

the Lorentzians do not overlap, and so

$$d_\eta^2(E) = \frac{\eta^2}{\pi^2} \sum_n \frac{1}{[(E-E_n)^2 + \eta^2]^2}. \tag{50}$$

It now follows from

$$\frac{2\eta^3}{\pi} \int_{-\infty}^{\infty} \frac{\mathrm{d}x}{(x^2+\eta^2)^2} = 1 \tag{51}$$

that

$$d(E) = \lim_{\eta \to 0} 2\pi\eta \, d_\eta^2(E). \tag{52}$$

By taking local averages and using the representation (8),

$$\langle d(E) \rangle = \lim_{\eta \to 0} 2\pi\eta \langle d^2_{\mathrm{osc},\,\eta}(E) \rangle. \tag{53}$$

The semiclassical formula for $d_{\mathrm{osc},\,\eta}(E)$, analogous to (9) for $d_{\mathrm{osc}}(E)$, involves only positive traversals of periodic orbits, and gives

$$\langle d(E) \rangle = \lim_{\eta \to 0} \frac{4\pi\eta}{\hbar^{2\mu+2}} \langle \sum_i \sum_j{}^{+} A_i A_j \cos\{(S_i - S_j)/\hbar\} \exp\{-\eta(T_i + T_j)/\hbar\} \rangle \tag{54}$$

$$= \lim_{\eta \to 0} \frac{4\pi\eta}{\hbar^{2\mu+2}} \int_0^{\infty} \mathrm{d}T \, \phi(T) \exp\{-2\eta T/\hbar\}, \tag{55}$$

where (12) has been used with $\epsilon = i\eta$ and where ϕ is defined by (22).

Asymptotic inversion of the Laplace transform and use of (49) now gives

$$\phi(T) \to \langle d \rangle \hbar^{2\mu+1}/2\pi \quad \text{if} \quad T \gg \hbar\langle d \rangle. \tag{56}$$

This is the semiclassical sum rule. It guarantees that the amplitudes and phases of very long orbits generate the mean level density, and hence shows how late terms in the representation (8) and (9) determine the first term: an 'analytic bootstrap' reminiscent of that introduced in one dimension by Voros (1983).

If we define a new variable to measure time in relation to the inner energy scale, i.e.

$$\tau \equiv T/h\langle d \rangle, \tag{57}$$

and write

$$\phi(T) \equiv \langle d \rangle \hbar^{2\mu+1} K(\tau)/2\pi. \tag{58}$$

then the sum rule (56) gives

$$K(\tau) \to 1 \quad \text{when} \quad \tau \gg 1. \tag{59}$$

The physical interpretation of $K(\tau)$ is that this function is the spectral form factor, defined as the Fourier transform of the correlation function of the spectral density:

$$K(\tau) = \langle d\rangle^{-2} \int_{-\infty}^{\infty} \mathrm{d}L \, \langle d(E-L/2\langle d\rangle)\, d(E+L/2\langle d\rangle)\rangle \exp\{2\pi i L\tau\}. \tag{60}$$

(apart from a δ function at $\tau = 0$). In terms of K, the pair correlation function of the levels is

$$g(L) = 1 - \frac{1}{\pi L}\int_0^{\infty} \mathrm{d}\tau \sin\{2\pi L\tau\}\, K'(\tau). \tag{61}$$

For *integrable* systems, (56) can be written

$$\phi(T) \to (\mathrm{d}\Omega/\mathrm{d}E)/(2\pi)^{N+1} \quad \text{if} \quad T \gg \hbar\langle d\rangle, \tag{62}$$

which is identical to the asymptotic value (24) of the diagonal sum. Thus neglect of off-diagonal terms in ϕ is indeed justified for these systems, as asserted previously. Moreover, the form factor $K(\tau)$ is unity not only when $\tau \gg 1$ (as in (59)), but also down to the much smaller value

$$\tau_{\min} = T_{\min}/h\langle d\rangle \quad (\ll 1). \tag{63}$$

It then follows from (61) that the pair correlation is unity, which implies lack of level correlation, for sequences of length $L \ll L_{\max}$, consistent with the behaviour already found for $\Delta(L)$ in §3.

6. Chaotic systems without time-reversal symmetry

For chaotic systems the diagonal approximation to $\phi(T)$, which must be valid for any given $T \gg T_{\min}$ if $\hbar$ is small enough, is (25). Together with (59) and (63) this implies that the form factor $K(\tau)$ defined by (57) and (58) has the behaviour

$$K(\tau) \to \begin{cases} \tau & (\tau_{\min} \ll \tau \ll 1), \\ 1 & (\tau \gg 1). \end{cases} \tag{64}$$

The rigidity is given by (20) as

$$\Delta(L) = \frac{1}{2\pi^2}\int_0^{\infty} \frac{\mathrm{d}y}{y} \frac{K(y/\pi L)}{y/\pi L}\, G(y). \tag{65}$$

When $L \ll 1$, K can be replaced by unity (because of (64)) to give for $\Delta(L)$ the correct limiting form $\frac{1}{15}L$ (cf. (29) and (30) and the discussion below (6)). This result could not have been obtained without the semiclassical sum rule.

When $1 \ll L \ll L_{\max}$, it is possible to divide the integration range of (65) into two parts by choosing a value of Y that satisfies $1 \ll Y \ll L$, so that, from (64) and figure 1,

$$\left.\begin{aligned} \frac{K(y/\pi L)}{y/\pi L} &\approx 1 \quad \text{if} \quad y < Y \\ \text{and} \qquad G(y) &\approx 1 \quad \text{if} \quad y > Y. \end{aligned}\right\} \tag{66}$$

Thus
$$\Delta(L) = \frac{1}{2\pi^2}\left[\int_0^Y \frac{dy}{y}\,G(y) + \int_Y^\infty \frac{dy}{y}\,\frac{K(y/\pi L)}{y/\pi L}\right]. \tag{67}$$

The definitions (16) and (21) lead to
$$\int_0^Y dy\, G(y)/y = \ln Y + \gamma + \ln 2 - \tfrac{9}{4}, \tag{68}$$

where γ is the Euler constant 0.577.... Integration by parts gives
$$\int_Y^\infty \frac{dy}{y}\,\frac{K(y/\pi L)}{y/\pi L} = -\ln(Y/\pi L) - \int_0^\infty d\tau\, \ln\tau\, \frac{d}{d\tau}\left(\frac{K(\tau)}{\tau}\right). \tag{69}$$

Thus the rigidity is
$$\Delta(L) = (\ln L)/2\pi^2 + D \quad (1 \ll L \ll L_{max}), \tag{70}$$

where
$$D = \frac{1}{2\pi^2}\left[\ln 2\pi + \gamma - \tfrac{9}{4} - \int_0^\infty d\tau\, \ln\tau\, \frac{d}{d\tau}\left(\frac{K(\tau)}{\tau}\right)\right]. \tag{71}$$

Equation (70) is precisely the asymptotic rigidity of the gaussian unitary ensemble (g.u.e.) of random-matrix theory (Mehta 1967), i.e. $\Delta(L)$ averaged over the spectra of large Hermitian matrices whose elements are gaussian random variables with statistics invariant under unitary transformations. The logarithmic dependence and correct prefactor are a direct consequence of the diagonal sum rule (25) given by Hannay & Ozorio de Almeida (1984), which applies when all closed orbits are isolated and unstable, with no action degeneracies. Without the semiclassical sum rule, however, the additive constant D would be infinite (because (69) would then be illegitimate).

Without specifying $K(\tau)$ (or what is equivalent, $\phi(T)$) more closely than (64), the value of the constant D cannot be determined, and I know no direct semiclassical arguments based on the definition (22) by which this can be achieved. However, with the simplest interpolation, namely
$$K_0(\tau) = \begin{Bmatrix} \tau & (\tau \leqslant 1), \\ 1 & (\tau \geqslant 1), \end{Bmatrix} \tag{72}$$

the integral in (71) is -1 and
$$D = (\ln 2\pi + \gamma - \tfrac{5}{4})/2\pi^2 = 0.0590, \tag{73}$$

which is exactly the correct constant given by random-matrix theory for the g.u.e.! The reason is that (72) is the exact form factor of the g.u.e. (Mehta 1967). However, this must be regarded as a remarkable coincidence, because where there is time-reversal symmetry we shall see (§8) that the simplest interpolation fails to give the exact result. In any case, D is small and not very sensitive to $K(\tau)$. To illustrate this, the interpolations
$$K_1(\tau) = \tau/(1+\tau), \quad K_2(\tau) = \tau/(1+\tau^2)^{\frac{1}{2}}, \quad K_3(\tau) = 2\tau\pi^{-1}\arctan(\pi/2\tau) \tag{74}$$

give
$$D_1 = 0.0083, \quad D_2 = 0.0434, \quad D_3 = 0.0312. \tag{75}$$

The discontinuity in slope of the 'correct' form factor (71) is very surprising. It implies that as $\hbar \to 0$ the double sum (22) for $\phi(T)$ has an abrupt transition at $T = h\langle d\rangle$, between the diagonal (25) and that given by the semiclassical sum rule (56). The origin of this 'semiclassical phase phase-transition' is at present obscure. In the next section we shall present a curious example of it. (It should also be remarked that the g.u.e. repulsion between *neighbouring* levels, which causes $g(L)$ (equation (61)) to vanish as L^2 when $L \to 0$, cannot be obtained from the semiclassical arguments leading to (64), but is of course implicit in (72).)

The preceding results explain local universality of the rigidity when $L \ll L_{\max}$, and the logarithmic behaviour (70) has recently been observed in computations for a chaotic system without time-reversal symmetry by Seligman *et al.* (1985).

When $L \gg L_{\max}$, short orbits, and hence $\tau \sim \tau_{\min}$ (equation (63)) give important contributions to the saturation value, which is given by (20) as

$$\Delta_\infty = 2\int_0^\infty \mathrm{d}T\,\phi(T)/T^2. \tag{76}$$

This can be expressed in terms of the short (non-universal) orbits and the (universal) density of the long ones by introducing an 'intermediate' period T_{I} such that

$$T_{\min} \ll T_{\mathrm{I}} \ll h\langle d\rangle. \tag{77}$$

Then

$$\begin{aligned}\Delta_\infty &= 2\sum_{T_j<T_{\mathrm{I}}}\frac{A_j^2}{T_j^2}+\frac{1}{2\pi^2}\int_{T_{\mathrm{I}}/h\langle d\rangle}^\infty \frac{\mathrm{d}\tau}{\tau^2}K(\tau)\\ &= 2\sum_{T_j<T_{\mathrm{I}}}\frac{A_j^2}{T_j^2}+\frac{1}{2\pi^2}\ln\left\{\frac{h\langle d\rangle}{T_{\mathrm{I}}}\right\}-\frac{1}{2\pi^2}\int_0^\infty \mathrm{d}\tau\,\ln\tau\frac{\mathrm{d}}{\mathrm{d}\tau}\left(\frac{K(\tau)}{\tau}\right).\end{aligned} \tag{78}$$

A useful approximation to Δ_∞, valid up to an additive constant, can be obtained by replacing T_{I} by $T_{\min}$, thereby extrapolating the continuous approximation (25) down to the shortest orbits. This gives (by also using (7), (72) and (3))

$$\Delta_\infty \approx \frac{1}{2\pi^2}\ln\{\mathrm{e}\,L_{\max}\} = \frac{(N-1)}{2\pi^2}\ln\left\{\frac{1}{\hbar}\left(\frac{e}{T_{\min}}\frac{\mathrm{d}\Omega}{\mathrm{d}E}\right)^{1/(N-1)}\right\}, \tag{79}$$

and shows that for chaotic systems the semiclassical spectral fluctuations increase only logarithmically with $\hbar^{-1}$ and so are much weaker than for integrable systems (cf. equation (31)).

7. Example: Riemann's zeta function

According to the Riemann hypothesis (Edwards 1974), the non-trivial zeros of the function $\zeta(s)$, defined in (46), all lie on the line $\mathrm{Re}\,s = \frac{1}{2}$. It is a natural conjecture, apparently first made by Hilbert and Polya, that the imaginary parts of these zeros are the eigenvalues of a linear operator (for a discussion, see Hejhal 1976). In this section I shall present evidence supporting the view that if this operator is regarded as the Hamiltonian of some (unknown) bound quantum-mechanical system, then in the classical limit the corresponding (unknown) dynamical system has trajectories that are chaotic and without time-reversal symmetry. By applying the semiclassical theory developed in this paper it will then be possible: to explain the local g.u.e. statistics that the Riemann zeros display (as was originally conjectured by Montgomery (1973) and as observed in computations reported by Bohigas & Giannoni (1984) and attributed by them to Odlyzko); to obtain an interesting formula involving prime numbers; and to predict the mean square fluctuations of the Riemann staircase.

Pavlov & Fadeev (1975) (see also Lax & Phillips (1976) and Gutzwiller (1983)) have discovered a scattering system (a leaky surface of constant negative curvature) whose phase shifts are given by the zeta function with Res = 1, and whose classical limit is chaotic. It is not clear what relation, if any, exists between their work and that described here.

The starting point is the formula for $\zeta(s)$ as a product over primes p:

$$\zeta(s) = \prod_p (1-p^{-s})^{-1}. \tag{80}$$

Defining
$$s \equiv \tfrac{1}{2}+iE, \quad \zeta(\tfrac{1}{2}-iE) \equiv D(E), \tag{81}$$

we have
$$\ln D(E) = -\sum_p \ln(1-\exp\{iE\ln p\}/p^{\frac{1}{2}}). \tag{82}$$

Assuming the Riemann hypothesis, and noting that $\ln D$ vanishes as $\mathrm{Im}\,E\to+\infty$, we see that this function jumps by $-i\pi$ as each Riemann zero $E = E_n$ is traversed from above, so that the oscillatory part of the spectral staircase is

$$\begin{aligned} \mathcal{N}_{\mathrm{osc}}(E) &= -\pi^{-1} \lim_{(\eta\to 0)} \mathrm{Im}\ln D(E+i\eta) \\ &= -\frac{1}{\pi}\,\mathrm{Im}\sum_p\sum_{m=1}^{\infty}\frac{\exp\{im\ln p\}}{mp^{\frac{1}{2}m}}. \end{aligned} \tag{83}$$

The oscillatory part of the spectral density is therefore

$$d_{\mathrm{osc}}(E) = -\frac{1}{\pi}\sum_p \frac{\ln p[\cos(E\ln p)-p^{-\frac{1}{2}}]}{1-2\cos(E\ln p)/p^{\frac{1}{2}}-p^{-1}} \tag{84}$$

or
$$d_{\mathrm{osc}}(E) = -\frac{1}{2\pi}\sum_p\sum_{m=-\infty}^{\infty}{}' \ln p\exp\{-|m|\ln p/2\}\exp\{iEm\ln p\}, \tag{85}$$

where the prime on the second summation means that $m = 0$ is omitted.

In the form (85) the analogy with the periodic-orbit formula (9) is clear: each primitive orbit corresponds to a prime p and is traversed m times; the action is

$$S_{m,p} = Em\ln p. \tag{86}$$

If E is regarded as the energy, the period is

$$T_{m,p} = \mathrm{d}S/\mathrm{d}E = m\ln p \tag{87}$$

and the amplitude is

$$A_{m,p} = -(\ln p\exp\{-|m|\ln p/2\})/2\pi. \tag{88}$$

Equation (86) shows that in this unknown dynamical system $\hbar^{-1}$ scales with E, so that we can set $\hbar = 1$ and regard $E\to\infty$ as the semiclassical limit. The amplitudes decay exponentially with period, as they must for a chaotic system (in contrast to an integrable one). Moreover the diagonal part $\phi_D(T)$ of the sum (22) is, by the prime number theorem,

$$\begin{aligned}\phi_D(T) &= \sum_{m=1}^{\infty} \sum_p A^2_{m,p}\, \delta(T - T_{m,p}) \\ &= \frac{1}{4\pi^2} \sum_{m=1}^{\infty} \sum_p \frac{\ln^2 p}{p^m}\, \delta(T - m \ln p) \\ &\xrightarrow{(T\to\infty)} \frac{T}{4\pi^2} \sum_{m=1}^{[T/\ln 2]} \exp\{-T(1-m^{-1})\}/m^2 \approx \frac{T}{4\pi^2},\end{aligned} \tag{89}$$

which agrees exactly with the diagonal sum rule (25) for periodic orbits in a chaotic non-degenerate system.

The local average of the Riemann staircase is (Montgomery 1976)

$$\langle \mathcal{N}(E) \rangle = \frac{E}{2\pi}\left(\ln\left\{\frac{E}{2\pi}\right\} - 1\right) - \frac{7}{8}, \tag{90}$$

corresponding to the average density

$$\langle d(E) \rangle = \frac{1}{2\pi} \ln\left\{\frac{E}{2\pi}\right\}. \tag{91}$$

The inner energy scale is thus $2\pi/\ln\{E/2\pi\}$, which vanishes in the semiclassical limit $E \to \infty$, in contrast with the outer energy scale $2\pi/T_{\min} = 2\pi/\ln 2$.

Figure 4*a*, *b* illustrates how well the periodic-orbit sum, with (equation (84)) and without (equation (85)) with $|m| = 1$) repetitions, is capable of reproducing δ functions at the Riemann zeros and $-\langle d(E) \rangle$ between them. (I am not claiming that (84) or (85) is an efficient way to calculate the positions of the zeros, because it is not: to roughly locate a zero near E requires about $(E/2\pi)/\ln\{E/2\pi\}$ primes, compared to $(E/2\pi)^{\frac{1}{2}}$ terms of the Riemann–Siegel formula (Edwards 1974), which would locate the same zero much more accurately.)

If on the basis of the identifications (86)–(88) and the diagonal sum (89) the existence of a chaotic non-degenerate dynamical system underlying the Riemann zeros is accepted, the arguments of preceding sections can be applied, and show that the zeros have the spectral rigidity of the g.u.e. In particular, for the spectral form factor (58), the semiclassical analysis gives the limits in (64). These limits agree with what was proved (for $\tau < 1$) and conjectured (for $\tau > 1$) by Montgomery (1973), namely that the Riemann zeros have a $K(\tau)$ that tends to the g.u.e. form (72) as $E \to \infty$.

When expressed in terms of closed orbits (primes and powers of primes) by using (58) and (22) this leads to a formula involving prime numbers that can be tested numerically. The test is most easily made by using not $K(\tau)$ but its integral, which from (72) is expected to be

$$S(\tau) \equiv 2\int_0^{\tau} d\tau'\, K_0(\tau') = \begin{cases} \tau^2 & (\tau \leqslant 1), \\ 2\tau - 1 & (\tau \geqslant 1). \end{cases} \tag{92}$$

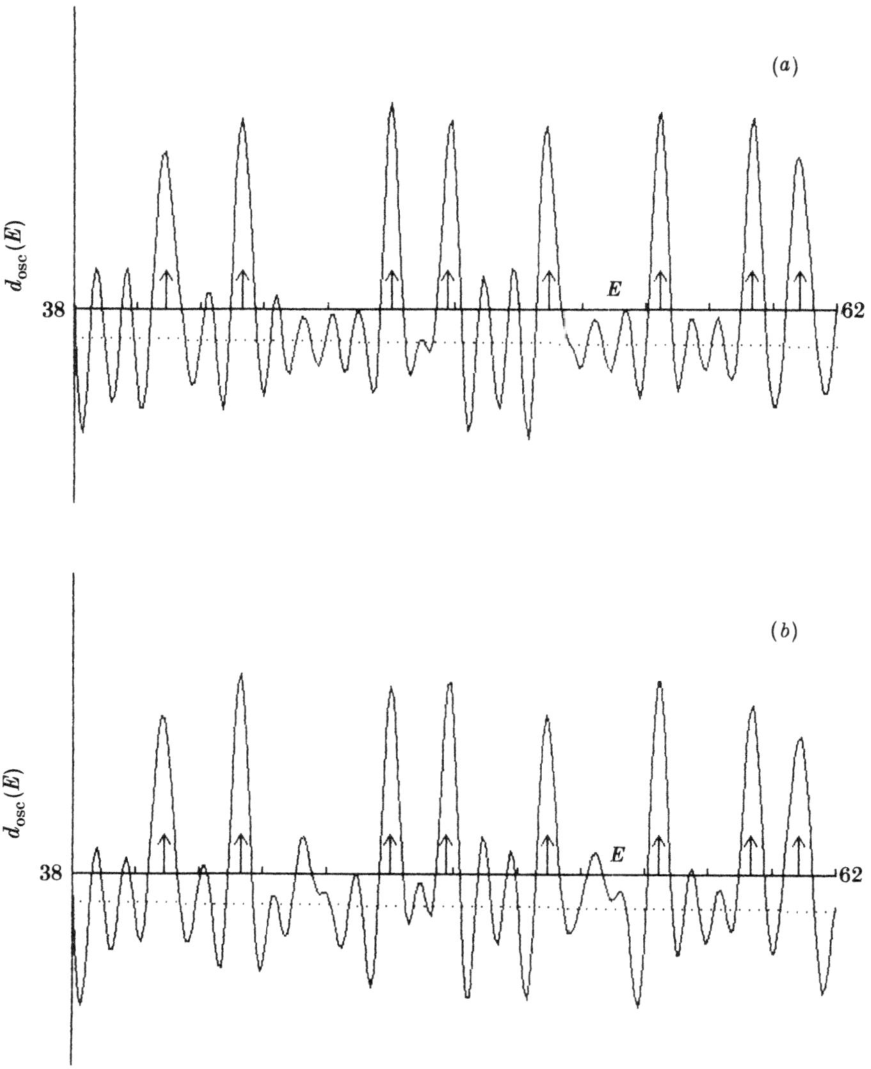

FIGURE 4. Spectral density of the Riemann zeros calculated with 150 primes (primitive closed orbits) from (*a*) equation (84) (i.e. including all repetitions, (*b*) equation (85) with $|m| = 1$ (i.e. without repetitions). In both figures the arrows denote the exact positions of the zeros and the dotted curve below the E-axis is $-\langle d(E) \rangle$.

The local spectral average $\langle\ \rangle$ in (22) is conveniently implemented by gaussian smoothing over an energy range η that is classically small but semiclassically large, i.e.

$$2\pi | \ln 2 \ll \eta \ll E. \tag{93}$$

The closed-orbit formula can now be written down as

$$S(\tau) = \lim_{(E\to\infty)} \frac{2}{(\ln\{E/2\pi\})^2} \sum_{p_1,\, m_1,\, p_2,\, m_2}^{p_1^{m_1} p_2^{m_2} < (E/2\pi)^{2\tau}} \sum \frac{\ln p_1 \ln p_2}{p_1^{\frac{1}{2}m_1} p_2^{\frac{1}{2}m_2}} \cos(E \ln\{p_1^{m_1}/p_2^{m_2}\})$$
$$\times \exp(-\eta^2 \ln^2\{p_1^{m_1}/p_2^{m_2}\}). \tag{94}$$

Of particular interest is the 'phase phase-transition' that (92) predicts at $\tau = 1$, where incoherence causes the double sum to depart from its diagonal form τ^2 and adopt the linear form $2\tau - 1$ dictated by the semiclassical sum rule. Figure 5 shows a preliminary numerical test of (94) without repetitions, supporting the view that this phase transition does exist.

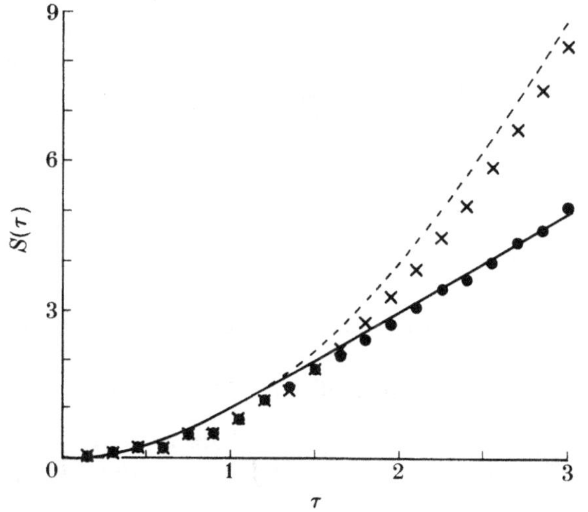

FIGURE 5. Integrated spectral form factor $S(\tau)$ for the Riemann zeros. The circular points are computed from (94) without repetitions, i.e. $m_1 = m_2 = 1$, for $E = 90$, and by using primes $p < 3517$ and a smoothing of $\eta = 4\pi/\ln 2 = 18.3$; the full line is the prediction (92). The crosses show the diagonal sum in (94) and the broken line is the prediction τ^2.

The rigidity $\Delta(L)$ should rise logarithmically according to (70) and (73) until $L \sim L_{\max}$ where, from (7), (87) and (91)

$$L_{\max} = \ln(E/2\pi)/\ln 2. \tag{95}$$

When $L \gg L_{\max}$, the semiclassical analogy predicts that $\Delta(L)$ saturates at a value given by (78) and (88) as

$$\begin{aligned}\Delta_\infty &= \Big(\sum_{p<p_{\mathrm{I}}} p^{-1} + \ln\ln\{E/2\pi\} - \ln\ln p_{\mathrm{I}} + 1\Big)/2\pi^2 \\ &= (\ln\ln\{E/2\pi\} + 1.2615)/2\pi^2.\end{aligned} \tag{96}$$

This shows that the r.m.s. staircase fluctuations grow as $(\ln\ln E)^{\frac{1}{2}}$ (it has been conjectured (Montgomery 1976, 1977) that the largest fluctuations grow as $(\ln E/\ln\ln E)^{\frac{1}{2}}$).

The saturation fluctuations predicted by (96) grow very slowly: for $E = 10^3$, $\Delta_\infty = 0.146$ (and $L_{\max} = 7.3$); for $E = 10^6$, $\Delta_\infty = 0.190$ (and $L_{\max} = 17.3$). To reach $\Delta_\infty = 0.5$ requires $E = 10^{2379}$, and to reach $\Delta_\infty = 1$ – an r.m.s. staircase deviation from $\langle \mathcal{N}(E) \rangle$ of only one level – requires $E = 10^x$ where $x \approx 5 \times 10^7$. This slow growth, indicating very slow approach to the semiclassical limit, would seem to rule out any direct test of (96).

8. Chaotic systems with time-reversal symmetry

When there is time-reversal symmetry, every orbit that is not self-retracing must be combined coherently with its time-reverse in the sum (22), because both orbits have the same action and period. The counterpart of the chaotic classical diagonal sum rule (25) must incorporate this coherence. Thus instead of $A_j^2+A_j^2=2A_j^2$ we have $(A_j+A_j)^2=4A_j^2$, so the right-hand side of (25) must be multiplied by two. This means that the form factor $K(\tau)$ defined by (58) has the limiting form 2τ, rather than τ, when $\tau_{\text{min}} \ll \tau \ll 1$. The semiclassical sum rule is, however, unaffected by the symmetry, so that instead of (6) the conditions on $K(\tau)$ are

$$K(\tau) \to \begin{cases} 2\tau & (\tau_{\text{min}} \ll \tau \ll 1), \\ 1 & (\tau \gg 1). \end{cases} \tag{97}$$

The rigidity $\Delta(L)$ is given by (65) with this new form factor, and the argument parallels that in §6.

When $L \ll 1$, the limit $\frac{1}{15}L$ is regained, and is as before a consequence of the semiclassical sum rule.

When $1 \ll L \ll L_{\text{max}}$, the analogues of (67), (70) and (71) are

$$\begin{aligned} \Delta(L) &= \frac{1}{\pi^2}\left[\int_0^Y \frac{\mathrm{d}y}{y}\,G(y)+\frac{1}{2}\int_Y^\infty \frac{\mathrm{d}y}{y}\,\frac{K(y/\pi L)}{y/\pi L}\right] \\ &= \frac{1}{\pi^2}\ln L+E \end{aligned} \tag{98}$$

with

$$E=\frac{1}{\pi^2}\left[\ln 2\pi+\gamma-\frac{9}{4}-\frac{1}{2}\int_0^\infty \mathrm{d}\tau\,\ln\tau\,\frac{\mathrm{d}}{\mathrm{d}\tau}\left(\frac{K(\tau)}{\tau}\right)\right]. \tag{99}$$

Equation (98) is precisely the asymptotic rigidity of the gaussian orthogonal ensemble (g.o.e.) of random-matrix theory (Dyson & Mehta 1963), that is $\Delta(L)$ averaged over the spectra of large real symmetric matrices whose elements are gaussian random variables with statistics invariant under orthogonal transformations. The logarithmic dependence and correct prefactor again follow from the Hannay & Ozorio de Almeida (1984) sum rule when this is modified to include the orbital degeneracy.

Without specifying $K(\tau)$ more closely than (97), the additive constant E cannot be determined and, as in §6, there is no obvious semiclassical way of doing this. Moreover, the simplest interpolation, analogous to (72), namely

$$K_1(\tau)=\begin{cases} 2\tau & (\tau \leqslant \frac{1}{2}), \\ 1 & (\tau \geqslant \frac{1}{2}), \end{cases} \tag{100}$$

gives

$$E=0.067, \tag{101}$$

which is not the correct g.o.e. constant. To obtain this, it is necessary to use the correct form factor, which is (Mehta 1967)

$$K_0(\tau)=\begin{cases} 2\tau-\tau\ln\{1+2\tau\} & (\tau \leqslant 1), \\ 2-\tau\ln\left\{\dfrac{1+2\tau}{2\tau-1}\right\} & (\tau \geqslant 1), \end{cases} \tag{102}$$

which gives
$$E = (\ln 2\pi + \gamma - \tfrac{5}{4} - \tfrac{1}{8}\pi^2)/\pi^2 = -0.00695. \tag{103}$$

For comparison, the smooth interpolation
$$K_2(\tau) = 2\tau/(1+2\tau) \tag{104}$$
gives
$$E = -0.054. \tag{105}$$

The 'correct' form factor (102) is also discontinuous, but in contrast to (72) the discontinuity is only in the third derivative. Therefore, the sum (22) for $\phi(t)$ also has a 'semiclassical phase phase-transition' when there is time-reversal symmetry, with a higher order than the phase transition when there is no such symmetry.

The result (98) explains the local universality observed in many numerical experiments on chaotic systems with time-reversal symmetry (see the review by Bohigas & Giannoni 1984), which holds for $L \ll L_{\text{max}}$. When $L \gg L_{\text{max}}$, the rigidity saturates non-universally, at a value approximately given by the analogue of (79), namely
$$\begin{aligned}\varDelta_\infty &= \pi^{-2}\ln\{\mathrm{e}L_{\text{max}}\} - \tfrac{1}{8}\\ &= \frac{(N-1)}{\pi^2}\ln\left\{\frac{1}{\hbar}\left(\frac{\mathrm{e}}{T_{\text{min}}}\frac{\mathrm{d}\Omega}{\mathrm{d}E}\right)^{1/(N-1)}\right\} - \frac{1}{8}.\end{aligned} \tag{106}$$

9. Discussion

The study reported here suggests a number of questions and directions for further investigation.

We have considered only second-order statistics ($\varDelta(L)$ and $K(\tau)$). It is natural to enquire about higher statistics such as many-level distributions and the distribution of spacings between neighbouring levels. Direct generalization of the methods used here would involve multiple sums analogous to (22), and diagonal and partially diagonal analogues of the sum rules (24), (25) and (56). At present it is not clear how this could be accomplished.

Even the rigidity has been studied only in the integrable and chaotic extremes. In view of the recent interest in spectral statistics of systems that show a transition to chaos as a parameter is varied (Robnik 1984, Seligman *et al.* 1985; Meyer *et al.* 1984; Berry & Robnik 1984), it is desirable to extend the theory to cover these more general cases. In the range $L \ll L_{\text{max}}$, it follows as a rough approximation from the periodic orbit theory that $\varDelta(L)$ is the sum of two contributions, one from the isolated unstable orbits in the chaotic region of phase space and one from the closed orbits in the region where there are tori (because of the finite resolution imposed by $\hbar$, orbits in this region can be considered as filling resonant 'tori' even though in reality they are isolated with near-marginal stability). The conjecture that $\varDelta(L)$ will be approximately additive has also been made by Seligman *et al.* (1985). As L increases, the contribution of the orbits in the integrable component should increase in relative importance, and ought to completely dominate fluctuations in the saturation régime $L \gg L_{\text{max}}$.

Another interesting case is that of non-integrable systems with non-isolated periodic orbits. One important example is the stadium billiard of Bunimovich

(1974) (see also Berry 1981 *a*), a chaotic system in which all periodic orbits are isolated with the exception of those bouncing perpendicularly between the straight sides, which form a one-parameter family. This single orbit (and its repetitions) will not spoil the logarithmic universality of $\Delta(L)$ for $L \ll L_{max}$ because the orbit selection function $G(y)$ (figure 1) will eliminate it. But its contribution will increase rapidly with L and it will dominate the saturation régime $L \gg L_{max}$; arguments from §4 then give, for a stadium whose parallel sides have length a and are separated by b,

$$\Delta_\infty = (2\pi E)^{\frac{1}{2}} b^2 \zeta(3)/8\hbar\pi^3 a \tag{107}$$

(the factor $\zeta(3)$ incorporates repetitions). Analogous behaviour is predicted for the Sinai billiard (Berry 1981 *b*), for which there are non-isolated orbits whose finite number depends on the radius of the central disc. The emergence and dominance of particular non-isolated orbits as L increases through L_{max} is a good example of transition from universal to non-universal spectral behaviour.

Another example, whose theoretical treatment is less clear, is billiards in irrational-angled polygons. These systems have been conjectured to be ergodic (Hobson 1976), but they have zero Kolmogorov entropy (Sinai 1976) and so are not chaotic. This behaviour stems from their closed orbits, which are almost all non-isolated and moreover marginally stable. The arguments of this paper strongly suggest that the spectral fluctuations should be much stronger than for the chaotic systems considered in §§6, 8, i.e. $\Delta(L)$ should rise faster than logarithmically, in spite of a numerical study of the level spacings distribution of triangles (Berry & Wilkinson 1984) suggesting g.o.e. behaviour.

It is at first sight surprising that for the chaotic systems considered in §§6, 8 the Kolmogorov entropy S did not appear in the rigidity formulae. S is a non-universal quantity and so might be expected to contribute to $\Delta(L)$ when $L \sim L_{max}$. However, S^{-1}, being the time for the separation between two nearby orbits to grow by a factor e, is of the same order of magnitude as T_{min} (for billiards the ratio is a geometrical factor), which does appear in the formulae (cf. (79) and (106)). Of course there remains the possibility that S might contribute directly to a higher-order statistic.

Another problem is the derivation of the semiclassical sum rule (56) directly from the definition (22). The derivation in §5 was based on (53), which is the condition for the spectral density to contain the correct density of singularities. A direct derivation from (22) would require detailed knowledge, at present lacking, of correlations between the actions of long orbits, which conspire in the off-diagonal terms of the sum to cancel the growth in the diagonal terms and cause $\phi(T)$ to saturate at the value (56). Such knowledge might also explain the 'phase phase-transitions' in the form factors (72) and (102).

The semiclassical sum rule is only the first in an infinite hierarchy of similar relations, obtained as the result of generalizing (53) by expressing a δ function as the limit of its analytic continuation raised to a power, namely

$$\langle d(E)\rangle = \lim_{\eta\to 0} \frac{(4\pi\eta)^{l-1}\,\Gamma^2(l)}{\Gamma(2l-1)} \langle d^l_{\mathrm{osc},\,\eta}(E)\rangle. \tag{108}$$

These higher-order sum rules might help in understanding higher-order spectral statistics. Their existence is related to a more fundamental and very remarkable 'bootstrap' property of the periodic-orbit sum (9): as a consequence of the spectrum being determined by the singularities of d_{osc}, which in turn are generated by very long classical periodic orbits, any finite number of terms may be deleted from (9) without destroying the spectral information it contains.

I thank Dr J. Hannay and Dr M. Robnik for helpful discussions.

Note added in proof (4 *June* 1985). Dr A. Voros has pointed out to me that (80), on which the closed-orbit sum (85) for the density of Riemann zeros is based, does not converge if Re$s < 1$. Therefore the ability of (85) to discriminate individual zeros, as illustrated in figure 4, might deteriorate as E increases, and could fail altogether when $E > 2\pi \exp(4\pi) \sim 2 \times 10^6$.

References

Balian, R. & Bloch, C. 1972 *Ann. Phys.* **69**, 76–160.
Balian, R. & Bloch, C. 1974 *Ann. Phys.* **85**, 514–545.
Berry, M. V. 1981*a* *Eur. J. Phys.* **2**, 91–102.
Berry, M. V. 1981*b* *Ann. Phys.* **131**, 163–216.
Berry, M. V. 1983 Semiclassical mechanics of regular and irregular motion. In *Chaotic behavior of deterministic systems* (Les Houches Lectures, vol. XXXVI, ed. G. Iooss, R. H. G. Helleman & R. Stora, pp. 171–271. Amsterdam: North-Holland.
Berry, M. V. 1984 Structures in semiclassical spectra: a question of scale. In *The wave–particle dualism* (ed. S. Diner, D. Fargue, G. Lochak & F. Selleri), pp. 231–252. Dordrecht: D. Reidel.
Berry, M. V. & Tabor, M. 1976 *Proc. R. Soc. Lond.* A **349**, 101–123.
Berry, M. V. & Tabor, M. 1977*a* *Proc. R. Soc. Lond.* A **356**, 375–394.
Berry, M. G. & Tabor, M. 1977*b* *J. Phys.* A **10**, 371–379.
Berry, M. V. & Robnik, M. 1984 *J. Phys.* A **17**, 2413–2421.
Berry, M. V. & Wilkinson, M. 1984 *Proc. R. Soc. Lond.* A **392**, 15–43.
Bohigas, O. & Giannoni, M. J. 1984 Chaotic motion and random-matrix theories. In *Mathematical and computational methods in nuclear physics* (ed. J. S. Dehesa, J. M. G. Gomez & A. Polls). *Lecture Notes in Physics* vol. 209, pp. 1–99. New York: Springer-Verlag.
Bunimovich, L. A. 1974 *Funct. Anal. Appl.* **8**, 254–255.
Casati, G., Chirikov, B. V. & Guarneri, I. 1985 *Phys. Rev. Lett.* **54**, 1350–1353.
Dyson, F. J. & Mehta, M. L. 1963 *J. Math. Phys.* **4**, 701–712.
Edwards, H. M. 1974 *Riemann's Zeta Function.* New York and London: Academic Press.
Gutzwiller, M. C. 1967 *J. Math. Phys.* **8**, 1979–2000.
Gutzwiller, M. C. 1969 *J. Math. Phys.* **10**, 1004–1020.
Gutzwiller, M. C. 1970 *J. Math. Phys.* **11**, 1791–1806.
Gutzwiller, M. C. 1971 *J. Math. Phys.* **12**, 343–358.
Gutzwiller, M. C. 1978 In *Path integrals and their applications in quantum, statistical and solid-state physics* (ed. G. J. Papadopoulos & J. T. Devreese), pp. 163–200. New York: Plenum.
Gutzwiller, M. C. 1980 *Phys. Rev. Lett.* **45**, 150–153.
Gutzwiller, M. C. 1983 *Physica* **7**D, 341–355.
Hannay, J. H. & Ozorio de Almeida, A. M. 1984 *J. Phys. A* **17**, 3429–3440.
Hejhal, D. A. 1976 *Duke math. J.* **43**, 441–482.
Hobson, A. 1976 *J. Math. Phys.* **16**, 2210–2214.
Lax, P. D. & Phillips, R. S. 1976 *Scattering theory for automorphic functions.* Princeton University Press.
McKean, H. P. 1972 *Communs pure. appl. Math.* **25**, 225–246.

Mehta, M. L. 1967 *Random matrices and the statistical theory of energy levels*. New York and London: Academic Press.
Meyer, H.-D., Haller, E., Köppel, H. & Cederbaum, L. S. *J. Phys.* A **17**, L831–836.
Montgomery, H. L. 1973 *Proc. Symp. pure Math.* **24**, 181–193.
Montgomery, H. L. 1976 *Proc. Symp. pure Math.* **38**, 307–310.
Montgomery, H. L. 1977 *Communs Math. Helv.* **52**, 511–523.
Norcliffe, A. & Percival, I. C. 1968 *J. Phys.* B **1**, 774–83.
Parry, W. 1984 *Ergod. Th. Dynam. Syst.* **4**, 117–134.
Parry, W. & Pollicott, M. 1983 *Ann. of Math* **118**, 573–591.
Pavlov, B. S. & Fadeev, L. D. 1975 *Soviet Math.* **3**, 522–548.
Pechukas, P. 1983 *Phys. Rev. Lett.* **51**, 943–946.
Porter, C. E. 1965 *Statistical theories of spectra: fluctuations*. New York: Academic Press.
Richens, P. J. & Berry, M. V. 1981 *Physica* **1***D*, 495–512.
Robnik, M. 1984 *J. Phys.* A **17**, 1049–1074.
Seligman, T. H. & Verbaarschot, J. J. M. 1985 *Phys. Lett.* A **108**, 183–187.
Seligman, T. H., Verbaarschot, J. J. M. & Zirnbauer, M. R. 1985 *J. Phys.* A (In the press.)
Sinai, Ya. G. 1976 *Introduction to ergodic theory*. Princeton University Press.
Voros, A. 1983 *Ann. Inst. H. Poincaré* **39**, 211–338.
Zucker, I. J. 1974 *J. Phys.* A **7**, 1568–1575.

Proc. R. Soc. Lond. A **413**, 183–198 (1987)
Printed in Great Britain

The Bakerian Lecture, 1987

Quantum chaology

By M. V. Berry, F.R.S.
H. H. Wills Physics Laboratory, Tyndall Avenue, Bristol BS8 1TL, U.K.

(*Lecture delivered* 5 *February* 1987 – *Typescript received* 2 *March* 1987)

Bounded or driven classical systems often exhibit chaos (exponential instability that persists), but their quantum counterparts do not. Nevertheless, there are new régimes of quantum behaviour that emerge in the semiclassical limit and depend on whether the classical orbits are regular or chaotic, and this motivates the following definition.

Definition. Quantum chaology is the study of semiclassical, but non-classical, behaviour characteristic of systems whose classical motion exhibits chaos.

This is illustrated by the statistics of energy levels. On scales comparable with the mean level spacing (of order h^N for N freedoms), these fall into universality classes: for classically chaotic systems, the statistics are those of random matrices (real symmetric or complex hermitian, depending on the presence or absence of time-reversal symmetry); for classically regular ones, the statistics are Poisson. On larger scales (of order h, i.e. classically small but semiclassically large), universality breaks down. These phenomena are being explained by representing spectra in terms of classical closed orbits: universal spectral behaviour has its origin in very long orbits; non-universal behaviour depends only on short ones.

In Henry Baker's day, 'chaology' meant 'The history or description of *the* chaos' (O.E.D. 1893). *The* chaos was the state of the world before creation ('without form, and void') so that chaology was a theological term. That area of theology has not been very active for the past two centuries (unless we extend its scope to include some recent speculations in cosmology) and so we are justified in reviving the term chaology, which will now refer to the study of unpredictable motion in systems with causal dynamics, as exemplified by the contributions at the meeting on 'dynamical chaos' of which this lecture is a part.

But what is 'quantum chaology'? One obstacle to a definition is the growing understanding that quantum systems are not chaotic in the way that classical systems are. (I am speaking of unpredictability in the evolution of the expectation values of observable quantities, and not of the quite different randomness unavoidably encoded in the wavefunction.)

As an example, consider ionizing a hydrogen atom by shining microwaves on it. This is well modelled by the quantum mechanics of an electron in two electric fields: Coulomb, from the nucleus, and oscillatory, from the radiation. If the atom

Reprinted from Proceedings of the Royal Society A, Volume 413, (1987).

is highly excited to begin with, we might be justified, on the basis of the correspondence principle, in thinking of the electron as moving classically. If in addition the illuminating microwaves are intense, the classical progress towards ionization is not a smooth outward spiralling but an erratic diffusion: the fields make the electron orbits chaotic (Leopold & Percival 1979; Jensen 1985). Exactly this behaviour (or rather the ionization probabilities that follow from it) has been observed in experiments (Bayfield *et al.* 1977). (We are here very far from the perturbation régime of one-photon ionization, the photoelectric effect, that was so important at the birth of quantum mechanics.) Surely these experiments illustrate 'quantum chaos'? They do not, because chaos is unpredictability that *persists* (strictly for infinite times) and in these experiments the atoms traverse only a short stretch of microwave field and so diffuse for only a short time.

The surprise comes in quantum calculations for longer times. These show that although initially the highly excited quantum electron absorbs energy in the classical way (that is, diffusively), after a long time there is a transition to a new régime in which the quantum electron absorbs energy more slowly. The first calculations (Casati *et al.* 1979) were for a model system, in which a particle on a ring (a rotator) is kicked periodically with an impulse that depends on where it is. For strong kicks the classical rotator momentum diffuses (energy grows linearly). But the quantum energy almost always eventually stops growing (usually it oscillates quasiperiodically). The analogous régime for the ionization problem (Casati *et al.* 1984; Casati *et al.* 1986; Blümel & Smilansky 1987) has not yet been probed experimentally, although I understand that it soon will be.

These calculations are important because they illustrate a general phenomenon: the quantum suppression of classical chaos (Chirikov *et al.* 1981; Fishman *et al.* 1982; Grempel *et al.* 1984). To see easily that this suppression must occur, observe first that classical chaos can be regarded as the emergence of complexity on infinitely fine scales in classical phase space: smooth curves representing families of orbits develop elaborate convolutions, like cream spreading on coffee. But quantum mechanics involves Planck's constant h, which is an area in phase space (momentum times distance) below which structure is smoothed away (for an illustration see Korsch & Berry 1981).

Although we do not have chaotic quantum evolution, we do have here a *new quantum phenomenon* that emerges in the semiclassical limit in systems that classically *are* chaotic, and this motivates the following definition.

Definition. Quantum chaology is the study of semiclassical, but non-classical, behaviour characteristic of systems whose classical motion exhibits chaos.

'Semiclassical' here means 'as $h \to 0$'. (Of course Planck's constant is not dimensionless and so can take any value, depending on the choice of units; what is meant is that the ratio of h to some classical quantity with the same dimensions – action – tends to zero.)

Here I will concentrate not on time evolution but on the quantum chaology of *spectra*, that is eigenvalues of the energy operator for systems whose classical counterparts are chaotic. This is important because these eigenvalues are the energies of stationary states, which are the quantum mechanical way of describing *things*, that is persisting objects like atoms and molecules, whose properties do not depend on when we measure them. We will be concerned not with the ground state but with the description of many highly excited states; this is the semiclassical limit.

My main aim is to bring to your attention a remarkable quantum chaotic property of spectra, and describe the first step towards explaining it. Before doing so, I must point out that these semiclassical quantum problems are but one example of the asymptotics of eigenvalues. Essentially the same mathematics describes the modes of vibration of elastic membranes, or sound waves (in a lecture hall my voice excites modes near the 20000th, which is surely 'asymptotic'), and much else. The 'classical limits' of these non-quantum problems involve the 'rays' of elasticity or sound; geometrically the rays are geodesics: straight line trajectories reflected specularly, like billiard balls, at the boundaries of the domain. The two-dimensional billiard domain of figure 1*a* has chaotic geodesics: it is the stadium of Bunimovich. The domain of figure 1*b*, the circle, does not. In mechanical terminology, the stadium orbits are *ergodic* (they possess no constants of motion other than the energy) while the circle orbits are *integrable* (because of symmetry, their angular momentum is conserved as well). For a quantum particle of mass m in a billiard domain D, eigenvalues E are determined by

$$\left.\begin{aligned}\nabla^2\psi + (2mE/\hbar^2)\psi &= 0 \text{ in D},\\ \psi &= 0 \text{ on the boundary of D.}\end{aligned}\right\} \tag{1}$$

The remarkable quantum chaotic property is that the distribution of the eigenvalues displays *universality*. This is the slightly pretentious way in which physicists denote identical behaviour in different systems. The most familiar example is thermodynamics near critical points (of, say, fluids and magnets).

To see the universality we need to magnify the spectrum so that the mean spacing of the levels is unity. The required magnification is the mean level density $\langle d \rangle$. What is $\langle d \rangle$? The answer comes from the roughest eigenvalue asymptotics, initiated by Pockels in 1891, developed by Rayleigh and Jeans who needed to count cavity modes for the theory of black-body radiation, and given a firm mathematical foundation by Herman Weyl in 1913 (for a review see Baltes & Hilf 1976). Their result was that if the classical system has N freedoms (e.g. $N = 2$ for billiards) then

$$\langle \mathrm{d}(E) \rangle \rightarrow \frac{\mathrm{d}\Omega(E)/\mathrm{d}E}{h^N} \quad \text{as} \quad h \rightarrow 0, \tag{2}$$

where $\Omega(E)$ is the volume of that part of classical phase space whose points have energies less than E. (These ideas have been refined and extended in several directions: see for example Kac 1966; Simon 1983*a*,*b*; Berry 1987.)

The level spacing is thus of order h^N, so that we need a microscope with power h^{-N}. What do we see with it? Of course we see the individual scaled levels, call them x_j, instead of the original levels E_j. Ideally we would have an asymptotic theory to predict these levels with an error that gets semiclassically small in comparison with the mean spacing h^N. For integrable (non-chaotic) classical motion we do have such a theory, in the form of the W.K.B. method and its

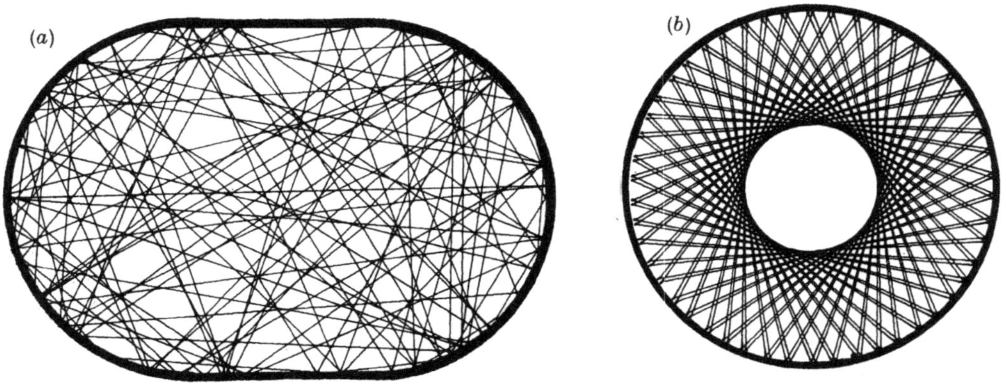

FIGURE 1. Classical orbits (bouncing geodesics) in billiards: (*a*) stadium of Bunimovich (chaotic), (*b*) circle (regular). For more details see, for example, Berry 1981*a*.

refinements and descendants (see, for example, Berry & Mount 1972; Percival 1977; Berry 1983). And these methods can be extended far into the chaotic régime if there is some residual order in phase space ('vague tori') and under not too semiclassical conditions (Reinhardt & Dana, this symposium). But for fully chaotic systems no fully asymptotic eigenvalue theory exists: we must make do with *statistics* of levels, and it is these that exhibit universality.

One such statistic, a short-range one, is the *level spacings probability distribution* $P(S)$, that is, the distribution of $S_j = x_{j+1} - x_j$. Figure 2 shows $P(S)$ computed from several hundred levels of the stadium, superimposed on $P(S)$ for another

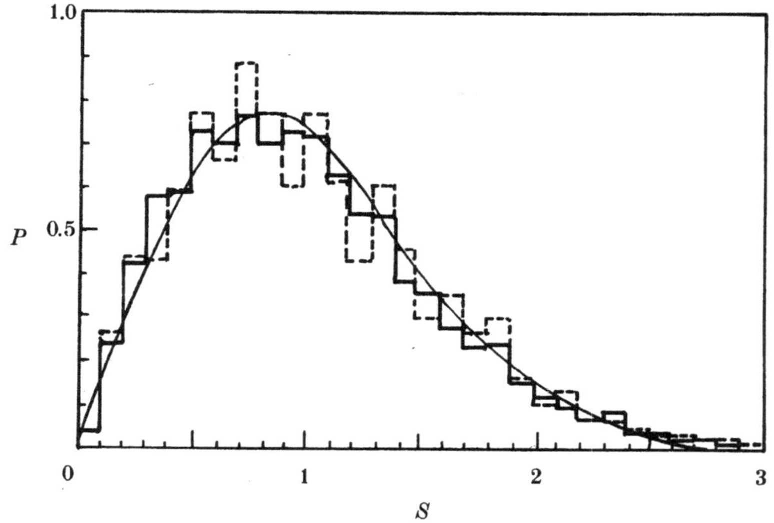

FIGURE 2. Level spacing histograms $P(S)$ for eigenvalues of the stadium billiard (full lines, after Bohigas 1984*a*) and the Sinai billiard (dashed lines, after Bohigas *et al.* 1984*b*), and the level spacings distribution for random real symmetric matrices (smooth curve), closely approximated by $P(S) = (\frac{1}{2}\pi \exp\{-\frac{1}{4}\pi S^2\}$.

classically chaotic billiard system: the billiard of Sinai, which is a square with a circular obstacle at its centre. These are two different systems, but the distributions are evidently the same; this is universality.

Another statistic – a long-range one – is the *spectral rigidity* $\Delta(L)$. This measures the fluctuations of the spectral staircase $\mathcal{N}(x)$, whose treads are at the eigenvalues x_j and whose risers have unit height ($\mathcal{N}(x)$ counts the number of levels below x). The rigidity (Dyson & Mehta 1963) is the mean-square deviation of the staircase from the straight line that best fits it over a range L, that is

$$\Delta(L) = \left\langle \min_{A,B} \frac{1}{L} \int_{-\frac{1}{2}L}^{\frac{1}{2}L} L \,\mathrm{d}x \,[\mathcal{N}(x) - Ax - B]^2 \right\rangle. \tag{3}$$

Figure 3 shows the rigidities for the same two chaotic billiards; again they are the same, illustrating universality.

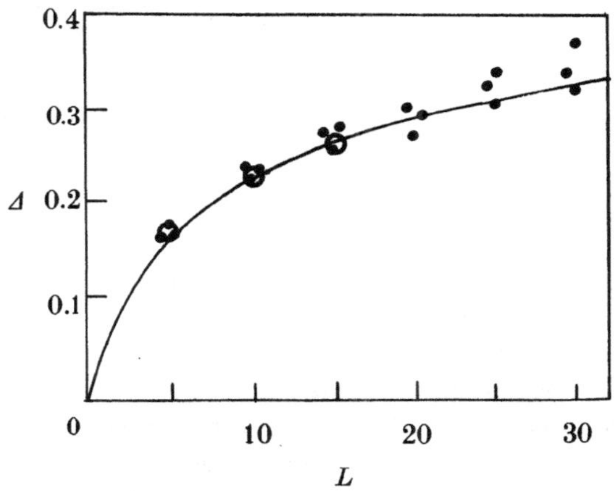

FIGURE 3. Spectral rigidity $\Delta(L)$ for eigenvalues of the stadium billiard (filled circles, after Bohigas *et al.* 1984*a*) and the Sinai billiard (open circles, after Bohigas *et al.* 1984*b*), and the rigidity for random real symmetric matrices (smooth curve), whose asymptote is $\Delta(L) \rightarrow (1/\pi^2) \ln L + \text{const.}$ as $L \rightarrow \infty$.

Now, it is clear from figures 2 and 3 that these data are accurately fitted by smooth curves representing the eigenvalue statistics of infinite real symmetric *matrices whose elements are random numbers*. Random-matrix theory (Porter 1965) was developed in the 1960s to model the complicated many-body energy operators for atomic nuclei (whose observed spectra they describe very well (Haq *et al.* 1982)). Ten years ago we (Berry & Tabor 1977) began to suspect it might also describe systems which although simple (like billiards) have chaotic classical orbits, and this has turned out to be so (Bohigas & Giannoni 1984).

Contrast this universality class with the spectral statistics of systems whose classical motion is not chaotic. Figure 4*a* shows the spacings distribution, and figure 4*b* the rigidity, for that most humble of regular systems, the particle in a two-dimensional rectangular box. It was surprising (ten years ago) to predict (Berry & Tabor 1977) and then find the statistics to be those of a set of *random*

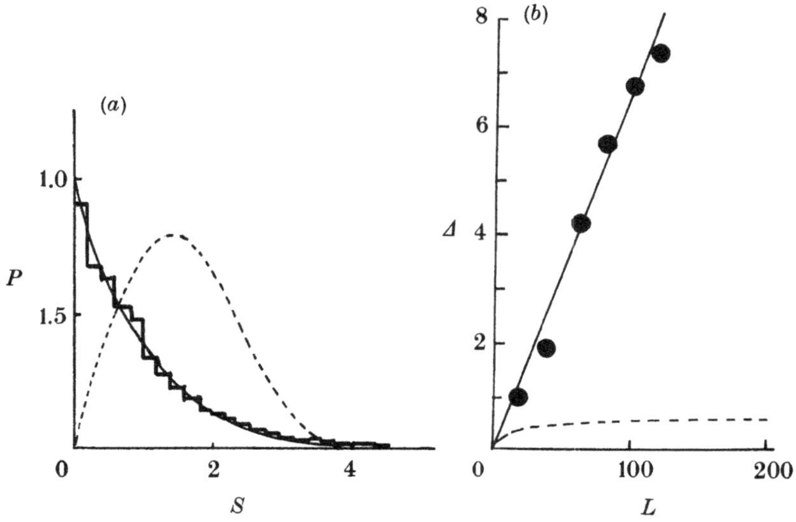

FIGURE 4. Level spacings distribution $P(S)$ (histogram in (a) and spectral rigidity $\Delta(L)$ (circles in (b)) for eigenvalues of the rectangular billiard. The full curves are the statistics for Poisson-distributed eigenvalues ($P(S) = \exp(-S)$ and $\Delta(L) = \frac{1}{15}L$) and the dashed curves are the statistics for random real symmetric matrices.

numbers (that is, poissonian). These behave very differently from the eigenvalues of random matrices, which are more well ordered in that they repel each other: for example, $P(S)$ vanishes linearly as $S \to 0$ instead of tending to a constant, and the asymptote of $\Delta(L)$ rise only logarithmically rather than linearly.

So far we have two universality classes, one for classically chaotic systems and one for classically regular systems, with spectra generated by random real symmetric matrices and Poisson processes respectively. Now, the matrices of quantum mechanics need not be real symmetric. The most general case is achieved for systems which, unlike billiards (or, more generally, particles in scalar potentials), *do not possess time-reversal symmetry* (T). For these, the energy operators are represented by complex hermitian, rather than real symmetric, matrices. The spectra of such random matrices, and also of the corresponding quantized chaotic systems, fall into a *third universality class*.

To illustrate it we break T by applying an external magnetic field to a charged particle moving chaotically. It is very instructive to concentrate the field into a single line of magnetic flux Φ. This is the chaotic equivalent of the effect discovered nearly thirty years ago in Bristol by Aharonov & Bohm (1959): the flux line does not alter the classical trajectories but does affect the quantum mechanics, in this case by changing the eigenvalues (Berry & Robnik 1986*a*). These are determined not by (1) but by

$$\left.\begin{aligned}(\nabla - \mathrm{i}q\boldsymbol{A}(\boldsymbol{r})/\hbar)^2\psi + (2mE/\hbar^2)\psi = 0 \text{ in D},\\ \psi = 0 \quad \text{on the boundary of D},\end{aligned}\right\} \tag{4}$$

where $\boldsymbol{A}(\boldsymbol{r})$ is any vector potential satisfying $\nabla \times \boldsymbol{A} = \Phi\delta(\boldsymbol{r})$.

Figure 5 shows the spectral statistics of an Aharonov–Bohm billiard ('Africa') with chaotic trajectories (Africa is a cubic conformal image of the unit disc, illustrated in figure 6). Evidently $P(S)$ now vanishes quadratically as $S \to 0$, rather than linearly. The rigidity is different too: its logarithmic asymptote is only half that for chaotic systems with T. Thus T-breaking induces a *spectral phase transition*, to the third universality class. (Additional symmetries can mimic the effect of T, as explained by Robnik & Berry 1986). The Aharonov–Bohm chaotic billiard might appear contrived, but might be capable of realization with a tiny solenoid and the essentially two-dimensional electrons in certain semiconductor interfaces (M. Pepper, personal communication). Exact sum rules for Aharonov–Bohm eigenvalues are given by Berry (1986*a*).

It is instructive to digress and look at the *wavefunctions* of these systems without T, and particularly at their zeros (Berry & Robnik 1986*b*). *With T*, wavefunctions

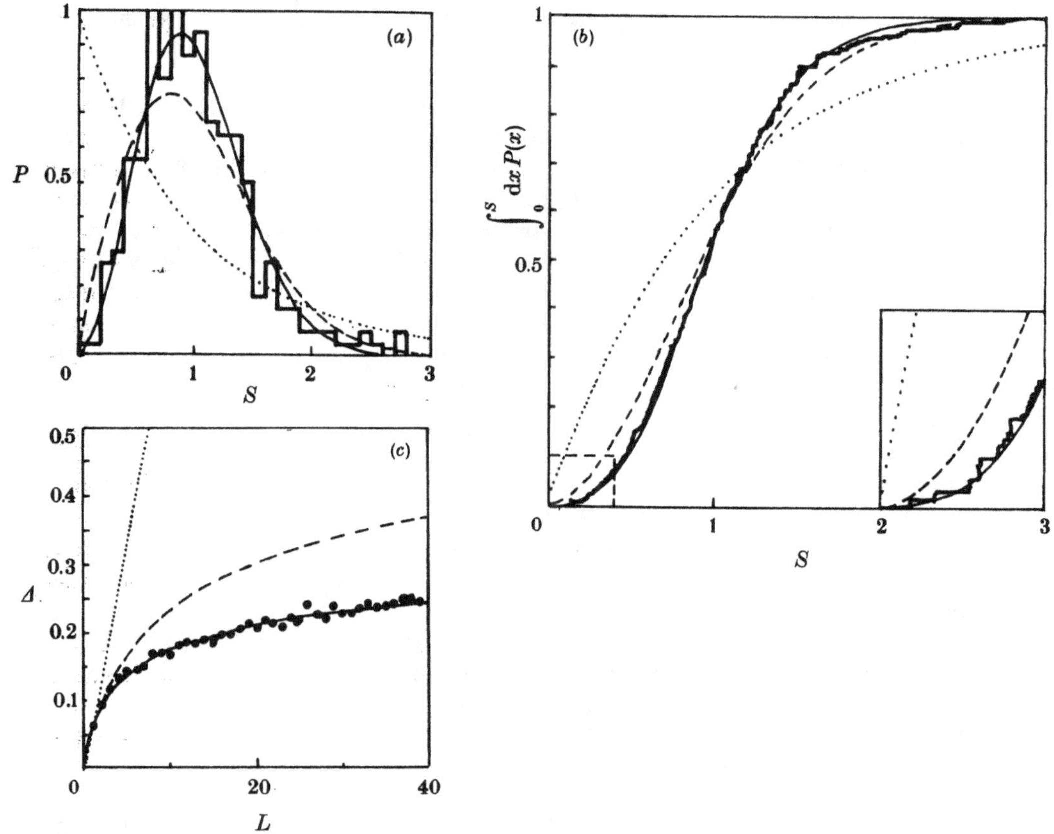

FIGURE 5. Level spacings distribution $P(S)$ (histogram in (*a*)), cumulative level spacings distribution $\int_0^S dx P(x)$ (histogram in (*b*)), and spectral rigidity $\Delta(L)$ (circles in (*c*)) for eigenvalues of the Aharonov–Bohm 'Africa' billiard with flux $q\Phi/h = \frac{1}{2}(\surd 5 - 1)$. The full curves are the statistics for random complex hermitian matrices, for which $P(S) \approx (32/\pi^2)\exp(-4S^2/\pi)$ and $\Delta(L) \to (1/2\pi^2)\ln L + \text{const.}$ as $L \to \infty$. The dashed curves are the statistics for random real symmetric matrices and the dotted curves are Poisson statistics.

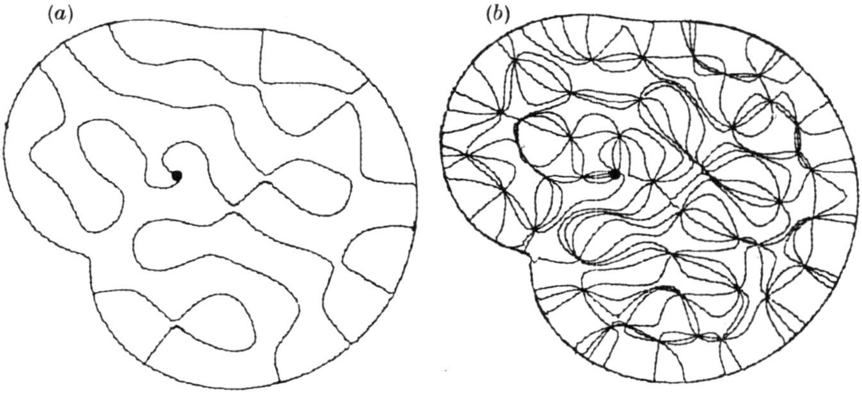

FIGURE 6. 50th eigenstate ψ of 'Africa' Aharonov–Bohm billiard with flux $q\Phi/h = \frac{1}{2}(\sqrt{5}-1)$. (a) Nodal lines of Re ψ, which are very similar to the nodal lines of ψ with zero flux; (b) wavefronts (contours of phase of ψ) at intervals of $\frac{1}{4}\pi$.

are real and so in two dimensions their zeros are the familiar *nodal lines* ($\psi(x,y) = 0$), which in quantum chaotic systems wander irregularly (figure 6*a*) with average spacing equal to the de Broglie wavelength $\lambda = \hbar/\sqrt{2mE}$ (McDonald & Kaufman 1979; Berry 1983; see also Heller 1984, 1986). Without T, wavefunctions are inescapably complex and so their zeros are points ($\mathrm{Re}\,\psi(x,y) = 0$, $\mathrm{Im}\,\psi(x,y) = 0$). Each of these points is a singularity of the wavefronts (contours of the phase of ψ) (figure 6*b*), which radiate from it like spokes from an axle (Nye & Berry 1974; Berry 1981*b*). Classical waves, like those on the surface of the sea, are of course real, but share the properties of 'inescapably complex' ones if their patterns are stationary but not standing, that is $\mathrm{Re}\,\psi$, where

$$\psi(\boldsymbol{r},\, t) = F(\boldsymbol{r})\, \mathrm{e}^{-\mathrm{i}\omega t} \tag{5}$$

with $F(\boldsymbol{r})$ complex; thus the nodal lines of $\mathrm{Re}\,\psi$ move. The *tide waves* are like this, because of the symmetry-breaking caused by the Earth's rotation (relative to the Moon), and the phase singularities are the *amphidromic points* where the cotidal lines (wavefronts) meet (figure 7), as described by Whewell (1833, 1836) (see also Defant 1961).

Back now to eigenvalues. There is a set of numbers of great mathematical importance whose statistics precisely mimic the energy levels of a quantum chaotic system without T, namely the imaginary parts of the *zeros of Riemann's zeta function*. This function is defined (Edwards 1974) by analytically continuing to the whole complex z-plane Euler's product over primes p:

$$\zeta(z) = \prod_p \frac{1}{1-p^{-z}}. \tag{6}$$

Riemann showed that the zeros of $\zeta(z)$ determine the fluctuations in the density of primes (that is their importance) and conjectured that they all have real part $\frac{1}{2}$; thus

$$\zeta(\tfrac{1}{2}+\mathrm{i}E_j) = 0, \tag{7}$$

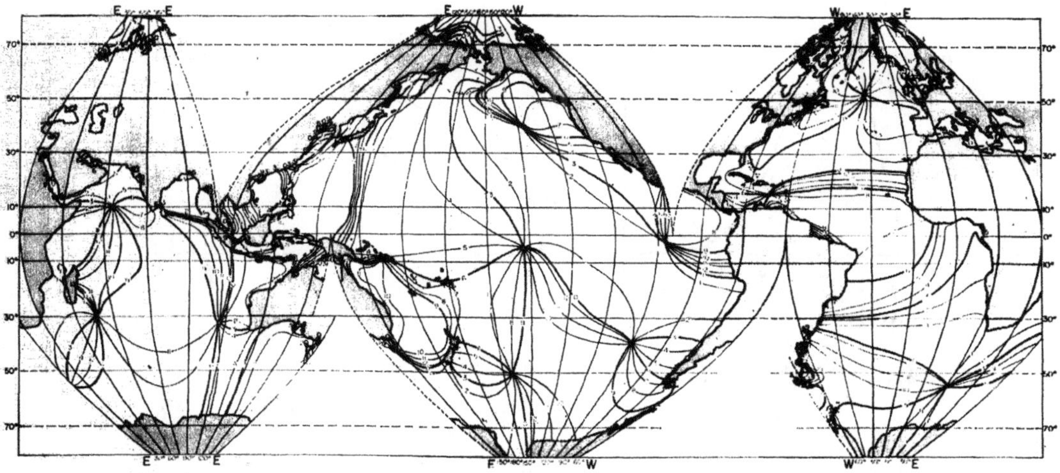

FIGURE 7. Cotidal lines in the oceans. These are wavefronts of the 12h tide wave, a forced vibration of the water of the whole earth; each line connects points where the tide is high at a given time. The singularities are amphidromic points, where there is no tide. (From Defant 1961.)

where $\{E_j\}$ are real. This conjecture has been verified by computation for the first 1.5×10^9 zeros (Van de Lune *et al.* 1986). It is an old idea (going back at least to Hilbert & Polya) that the Riemann conjecture would be confirmed if it could be shown that $\{E_j\}$ are the eigenvalues of some hermitian operator, but this has not been found.

Recently Odlyzko (1987) has computed some statistics for spectacularly high E_j. Figure 8*a* shows the spacings distribution for 10^5 zeros near the 10^{12}th; agreeing very closely with $P(S)$ for random complex hermitian matrices and so with that of some unknown quantum system without T whose unknown classical limit is chaotic. He also computed the number variance (figure 8*b*), a quantity closely related to the rigidity, and discovered that the three- and four-zero correlations (figure 8*c*, *d*) agrees perfectly with the corresponding complex random-matrix statistics. Riemann's conjecture thus acquires, in addition to its number-theoretic importance, a further significance (Berry 1986*b*): when (if) the operator with eigenvalues E_j is found, it will surely be simple, and will provide a paradigm for quantum chaology comparable with the harmonic oscillator for quantum non-chaology.

Here is a way of breaking T without magnetic fields, in *relativistic quantum chaology*. Take a massless particle ('neutrino') moving in the plane and described by the equation of Dirac (who gave this lecture in the year of my birth), but with a four-scalar potential $V(x, y)$ rather than the usual electric potential. For such a particle, the wave is a two-component spinor satisfying (Berry & Mondragon 1987)

$$\begin{pmatrix} V(x,y) & -i\hbar c(\partial_x - i\partial_y) \\ -i\hbar c(\partial_x + i\partial_y) & -V(x,y) \end{pmatrix} \begin{pmatrix} \psi_1 \\ \psi_2 \end{pmatrix} = E \begin{pmatrix} \psi_1 \\ \psi_2 \end{pmatrix}. \tag{8}$$

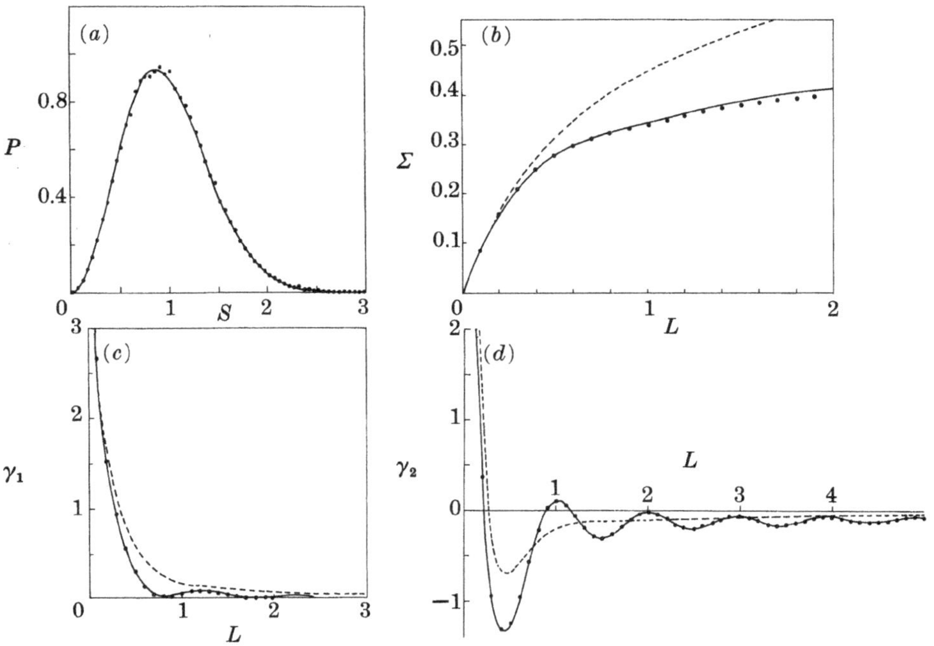

FIGURE 8. Statistics of imaginary parts of Riemann zeros, computed by A. M. Odlyzko. (*a*) $P(S)$ (from Odlyzko 1987); (*b*) number variance $\Sigma(L) \equiv \langle (n-L)^2 \rangle$, where n is the actual number of zeros in an interval where the average number is L (the interval is $2\pi L/\ln(E/2\pi e)$) (from data kindly supplied by A. M. Odlyzko); (*c*) Skewness $\gamma_1(L) \equiv \langle (n-L)^3 \rangle / \langle (n-L)^2 \rangle^{\frac{3}{2}}$ (kindly supplied by O. Bohigas); (*d*) excess $\gamma_2(L) \equiv \langle (n-L)^4 \rangle / \langle (n-L)^2 \rangle^2 - 3$ (kindly supplied by O. Bohigas). Full curves, random complex hermitian matrices; dashed curves, random real symmetric matrices.

This equation does not possess time-reversal invariance. Figure 9 shows the spectral statistics when $V(x,y)$ represents a hard wall (neutrino billiards), showing once again the statistics of complex hermitian random matrices if the billiard is classically chaotic, and Poisson statistics if it is regular.

Originally I hoped, following a suggestion of Professor Atiyah, that this kind of relativistic quantum chaology might help in the search for the elusive Riemann operator, but this has not yet proved to be so. However, Volkov & Pankratov (1985) and Pankratov *et al.* (1987) have recently discovered that an equation very similar to (7) appears to describe peculiar electron states localized in the interface between certain pairs of semiconductors (e.g. PbTe and SnTe, and HgTe and CdTe).

There is a *fourth* universality class, associated with chaotic systems that have time-reversal symmetry and also half-integer total spin (Porter 1965), but I will not speak about it.

So far we have seen that on fine scales the statistics of spectra fall into universality classes that depend on whether the classical motion is regular or chaotic, and on the symmetry of the energy operator. Now I have to explain how this universality is compromised in two ways.

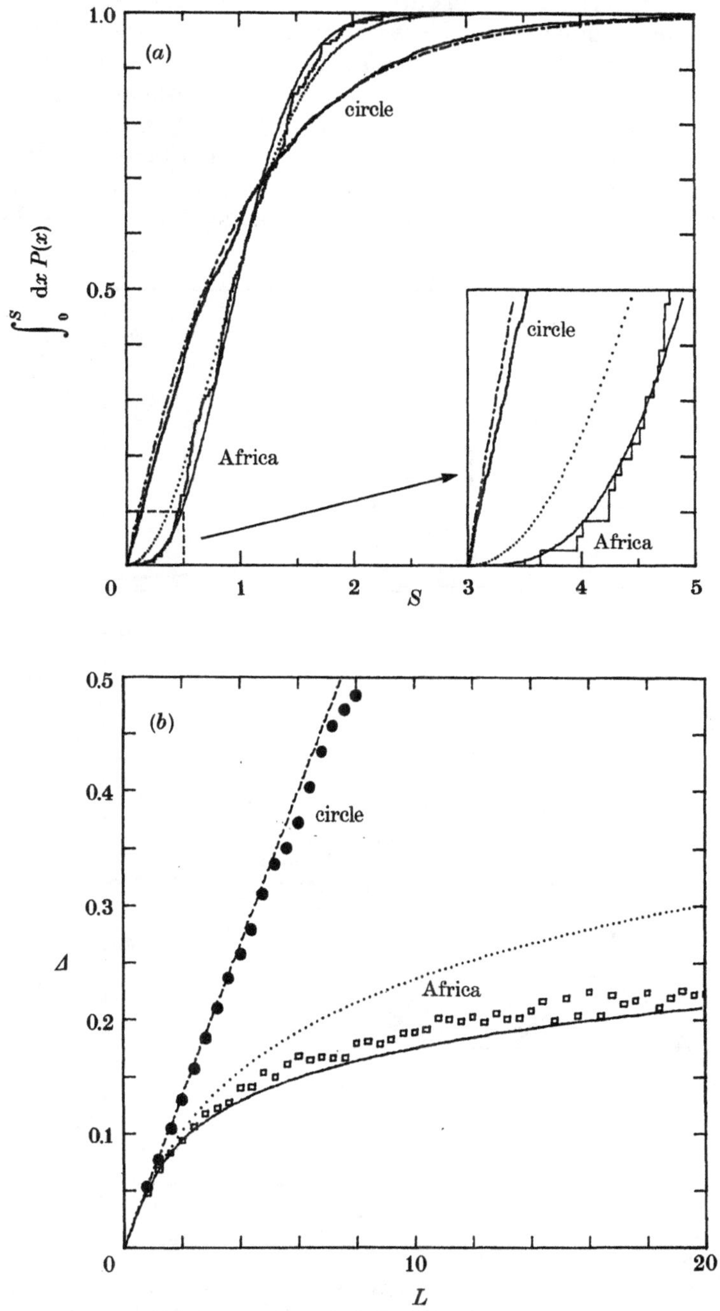

FIGURE 9. (a) Cumulative level spacings distribution $\int_0^S \mathrm{d}x P(x)$ and (b) spectral rigidity $\Delta(L)$, for neutrino 'Africa' and neutrino circle billiards. Full curves, random complex hermitian matrices; dotted curves, random real symmetric matrices; dashed and chain curves, Poisson statistics.

First, some very important systems are partly regular and partly chaotic in their classical motion; vibrating molecules, for example. Their spectral statistics can be understood as those of a *superposition* of spectra from different universality classes, each spectrum being associated with a different chaotic or regular region in classical phase space (Berry & Robnik 1984). Figure 10 shows some recent calculations by Wunner *et al.* (1986), of the spacings distribution of the zero-angular-momentum, even-parity electron levels of a hydrogen atom in a very strong magnetic field ($6T$), in three different energy ranges. The point is that the

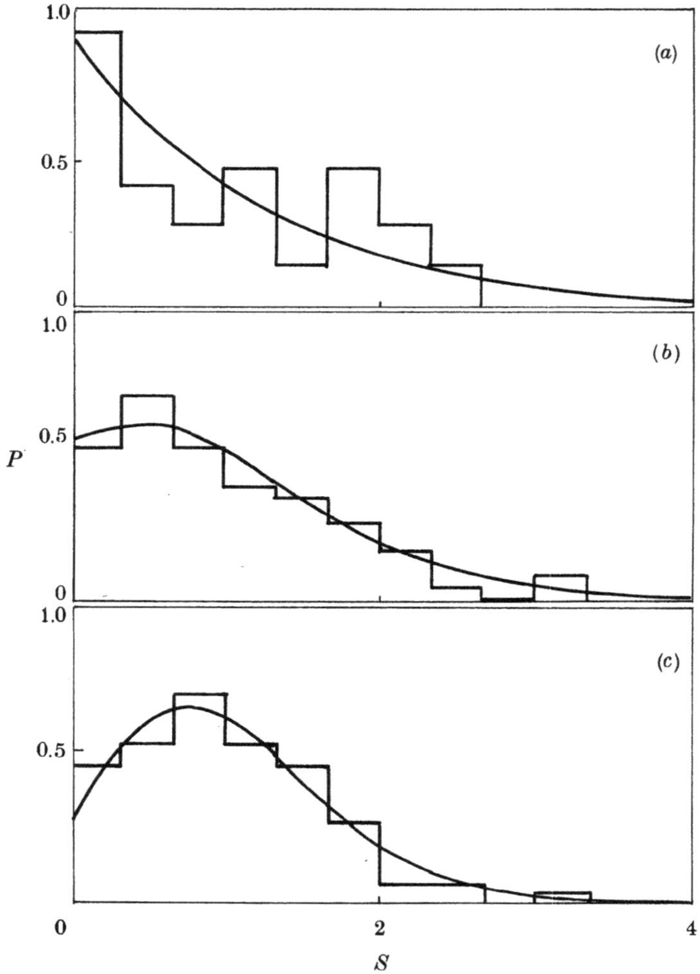

FIGURE 10. Level spacings distributions $P(S)$ for even-parity, zero-angular momentum energy levels of a hydrogen atom in a 6T uniform magnetic field, for three different energy ranges with different phase-space fractions q of regular orbits: (*a*) $-130\ \text{cm}^{-1} < E < -100\ \text{cm}^{-1}$ ($q = 0.71$, 47 levels); (*b*) $-100\ \text{cm}^{-1} < E < -70\ \text{cm}^{-1}$ ($q = 0.32$, 71 levels); (*c*) $-70\ \text{cm}^{-1} < E < -40\ \text{cm}^{-1}$ ($q = 0.16$, 116 levels); the smooth curves are $P(S)$ for superpositions of Poisson and random real symmetric matrix spectra. (From Wunner *et al.* 1986.)

corresponding classical motion gets more chaotic as the energy increases. These régimes are now within the reach of experiment (and are of course far removed from the familiar low-field 'perturbation' domain of the Zeeman effect).

The second compromise, of deep theoretical significance, is that universality is only local: for correlations involving very many levels, it breaks down. Recall the h^{-N} microscope that magnified the energies E_j to the numbers x_j with mean spacing unity, and note that N, the number of classical freedoms, is at least two for non-trivial cases. Now reduce the microscope's power to h^{-1}. (These gedankenmagnifications are strongly reminiscent of the 'non-standard analysis' used nowadays to describe infinitesimals (Harnik 1986).) We will see energy ranges that are still *classically* small (of order h) but *semiclassically large* in that they include many levels (a number of order $h^{-(N-1)}$). At these magnifications, energy-level statistics are not universal: they depend on classical details.

To illustrate the breakdown of universality at long range, figure 11*a* shows the rigidity for the (classically regular) particle in a rectangular box, computed by Casati *et al.* (1985). When L is not too large we see the straight line of the 'universal' Poisson statistics (this was figure 4*b*), but when L approaches the square root of the number of the highest level included in the calculation (which for this case corresponds to an energy range of order h), $\Delta(L)$ oscillates and then saturates at a value that depends on this number and also on the aspect ratio of the rectangle; that is, non-universally. Figure 11*b* shows the number variance for the Riemann zeros (underlying which there appears to be a chaotic classical system). When L is not too large we see the logarithmic curve of the 'universal' statistic for random complex matrices (this was figure 8*b*), but for larger L the variance oscillates about a value which depends on the number of zeros, that is, non-universally.

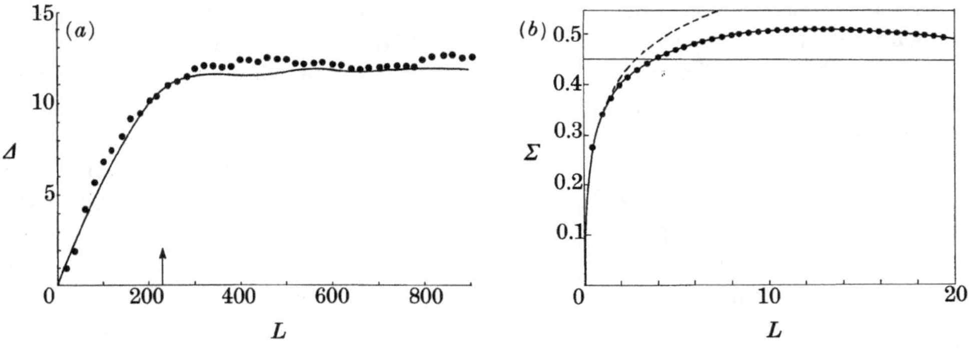

FIGURE 11. (*a*) Spectral rigidity $\Delta(L)$ for rectangular billiard, continuing figure 4*b* to larger L. The circles were computed from the eigenvalues near the 20000th Casati *et al.* 1985); the smooth curve was obtained from the sum over closed orbits by Berry (1985); the arrow is the L corresponding to an energy range $h/T_{\min}$, where $T_{\min}$ is the period of the shortest closed orbit. (*b*) Number variance $\Sigma(L)$ for 10^5 Riemann zeros near the 10^{12}th, continuing figure 8*b* to larger L; the circles are plotted from data kindly supplied by A. M. Odlyzko; the smooth curve is the 'semiclassical' theory (adapted from Berry 1985), which predicts oscillation about the horizontal line $\Sigma(\infty) = 0.4518$ $(= [\ln\ln(E/2\pi)+1.2615]/\pi^2$, where $(E/2\pi)\ln(E/2\pi e) = 10^{12})$; the dashed curve is for random complex matrices.

Until now I have spoken of spectral universality as an unexplained observation based on numerical experiments inspired by guesses. And so it was until recently, but now we have the beginnings of a theory (Berry 1985). Because of the h-magnifications involved, the theory has to be semiclassical: we must 'sew the quantum flesh on the classical bones'.

What are these bones? According to a beautiful picture developed by Gutzwiller (1971, 1978) and by Balian & Bloch (1972), they are the *classical closed orbits*, in terms of which an asymptotic formula can be given for the density of quantum eigenvalues (for a review, see Berry 1983; for the application to integrable systems, see Berry & Tabor 1976). These ideas can be traced back to de Broglie who in 1923 conceived of quantization as the constructive self-inteference of waves accompanying orbiting particles (think of Ouroboros, the mythical self-swallowing snake). For some mathematical systems (the Laplace–Beltrami operator on surfaces of constant negative curvature), the relation between spectra and closed geodesics is exact rather than asymptotic, and is called the Selberg trace formula (McKean 1972; Hejhal 1976; Balazs & Voros 1986; Series, this symposium.)

With the exception of some simple cases, the quantum levels are not in one-to-one correspondence with closed orbits (for an illustration, see Keating & Berry 1988; if they were, we would have a general formula for semiclassical quantization. Instead, each classical orbit describes an *oscillatory clustering* of the levels on a scale ΔE determined by its period T_n: this scale is just what would be expected from the uncertainty principle:

$$\Delta E = h/T_n. \tag{9}$$

Thus longer orbits give spectral information on finer scales, and it is this observation that gives the key to understanding the universality of the statistics (Berry 1985). With the h^{-N} microscope, we are concerned with the finest scales of spectral structure, of the order of the mean level spacing, so $\Delta E \sim h^N$. These scales depend on classical orbits with periods $T_n \sim h/\Delta E \sim 1/h^{(N-1)}$, that is, on *extremely long orbits*. Now, the distribution of these long orbits in phase space is very different for integrable and chaotic systems. For integrable systems, the orbits form continuous families whose number grows with period as T^N. For chaotic systems, the orbits are isolated and unstable and their number proliferates exponentially (as $\exp(HT)/HT$ where H is the Kolmogorov entropy – instability exponent of the orbit). In an important paper, Hannay & Ozorio de Almeida (1984) have shown that the way these long orbits contribute to one form of the asymptotic spectral formula is *universal*: it depends only on whether the orbits are chaotic or not, and on no other feature of the classical motion. It is this *classical* universality that begets the *quantum* universality, for it is possible to employ it as one ingredient in a derivation (Berry 1985) of the spectral rigidity $\Delta(L)$ (but not, so far, the spacings distribution), yielding precisely the Poisson and random-matrix formulae that so accurately fit the numerical computations.

The same arguments explain why universality breaks down at the larger energy scales $\Delta E \sim h$: the spectral fluctuations in this range are determined by orbits with period $T \sim h/\Delta E \sim h^0$, which are not long and so differ from system to system. The quantitative theory of this breakdown of universality (Berry 1985) works rather well, as figure 11 shows.

In summary, the vigorous development of quantum chaology during the last decade has been stimulated by the interplay of two factors: the realization that chaotic motion is ubiquitous in classical mechanics, and the discovery of associated new régimes of quantum behaviour. But the mathematical difficulties in understanding these régimes are severe, fundamentally because the semiclassical limit $\hbar \to 0$ is highly singular. At the risk of sounding slightly paradoxical, I would say that we are discovering the connections between classical mechanics and quantum mechanics to be richer and more subtle than either mechanics is when considered on its own.

References

Aharonov, Y. & Bohm, D. 1959 *Phys. Rev.* **115**, 485–491.
Balazs, N. L. & Voros, A. 1986 *Physics Rep.* **143**, 109–240.
Balian, R. & Bloch, C. 1972 *Ann. Phys.* **69**, 76–160.
Baltes, H. P. & Hilf, E. R. 1976 *Spectra of finite systems*. Mannheim: B. I. Wissenschaftsverlag.
Bayfield, J. E., Gardner, L. D. & Koch, P. M. 1977 *Phys. Rev. Lett.* **39**, 76–79.
Berry, M. V. 1981*a* *Eur. J. Phys.* **2**, 91–102.
Berry, M. V. 1981*b* Singularities in waves and rays. In *Physics of defects* (*Les Houches Lectures* 34) (ed. R. Balian, M. Kleman & J.-P. Poirier), pp. 453–543. Amsterdam: North Holland.
Berry, M. V. 1983 Semiclassical mechanics of regular and irregular motion. In *Chaotic behaviour of deterministic systems* (*Les Houches Lectures* 36) (ed. G. Iooss, R. H. G. Helleman & R. Stora), pp. 171–271. Amsterdam: North Holland.
Berry, M. V. 1985 *Proc. R. Soc. Lond.* A **400**, 229–251.
Berry, M. V. 1986*a* *J. Phys.* A **19**, 2281–2296.
Berry, M. V. 1986*b* In *Quantum chaos and statistical nuclear physics* (ed. T. H. Seligman & H. Nishioka) (Springer lecture notes in physics no. 263), pp. 1–17.
Berry, M. V. 1987 *J. Phys. A* **20**, 2389–2403.
Berry, M. V. & Mondragon, R. J. 1987 *Proc. R. Soc. Lond.* A **412**, 53–74.
Berry, M. V. & Mount, K. E. 1972 *Rep. Prog. Phys.* **35**, 315–397.
Berry, M. V. & Robnik, M. 1984 *J. Phys. A* **17**, 2413–2421.
Berry, M. V. & Robnik, M. 1986*a* *J. Phys. A* **19**, 649–668.
Berry, M. V. & Robnik, M. 1986*b* *J. Phys. A* **19**, 1365–1372.
Berry, M. V. & Tabor, M. 1976 *Proc. R. Soc. Lond.* A **349**, 101–123.
Berry, M. V. & Tabor, M. 1977 *Proc. R. Soc. Lond.* A **356**, 375–394.
Blümel, R. & Smilansky, U. 1987 *Z. Phys.* (In the press.)
Bohigas, O. & Giannoni, M. J. 1984 Chaotic motion and random-matrix theories. In *Mathematical and computational methods in nuclear physics* (ed. J. S. Dehesa, J. M. G. Gomez & A. Polls) (Lecture Notes in Physics 209), pp. 1–99. New York: Springer Verlag.
Bohigas, O., Giannoni, M. J. & Schmit, C. 1984*a* *J. Phys. Lett.* **45**, L1015–L1022.
Bohigas, O., Giannoni, M. J. & Schmit, C. 1984*b* *Phys. Rev. Lett* **52**, 1–4.
Casati, G., Chirikov, B. V., Ford, J. & Izraelev, F. M. 1979 In *Stochastic behaviour in classical and quantum hamiltonian systems* (ed. G. Casati & J. Ford) (Springer lecture notes in physics 93), pp. 334–352.
Casati, G., Chirikov, B. V. & Guarneri, I. 1985 *Phys. Rev. Lett.* **54**, 1350–1353.
Casati, G., Chirikov, B. V. & Shepelyansky, D. L. 1984 *Phys. Rev. Lett.* **53**, 2525–2528.
Casati, G., Chirikov, B. V., Shepelyansky, D. L. & Guarneri, I. 1986 *Phys. Rev. Lett.* **57**, 823–826.
Chirikov, B. V., Izraelev, F. M. & Shepelyansky, D. L. 1981 *Soviet Sci. Rev.* C **2**, 209–267.
Defant, A. 1961 *Physical oceanography*, vol. 2. London: Pergamon.
Dyson, F. J. & Mehta, M. L. 1963 *J. math. Phys.* **4**, 701–712.
Edwards, H. M. 1974 *Riemann's zeta function*, New York and London: Academic Press.
Fishman, Shmuel, Grempel, D. R. & Prange, R. E. 1982 *Phys. Rev. Lett.* **49**, 509–512.
Grempel, D. R., Fishman, Shmuel & Prange, R. E. 1984 *Phys. Rev.* A **29**, 1639–1647.
Gutzwiller, M. C. 1971 *J. math. Phys.* **12**, 343–358.

Gutzwiller, M. C. 1978 In *Path integrals and their applications in quantum statistical and solid-state physics* (ed. G. J. Papadopoulos & J. T. Devreese), pp. 163–200. New York: Plenum.
Hannay, J. H. & Ozorio de Almeida, A. M. 1984 *J. Phys.* A **17**, 3429–3440.
Haq, R. U., Pandey, A. & Bohigas, O. 1982 *Phys. Rev. Lett.* **48**, 1086–1089.
Harnik, V. 1986 *Math. Intell.* **8**, 41–47, 63.
Hejhal, D. A. 1976 *Duke math. J* **43**, 441–482.
Heller, E. J. 1984 *Phys. Rev. Lett.* **53**, 1515–1518.
Heller, E. J. 1986 In *Quantum chaos and statistical nuclear physics* (ed. T. H. Seligman & H. Nishioka) (Springer lecture notes in physics no. 263), pp. 162–181.
Jensen, R. V. 1985 In *Chaotic behaviour in quantum systems* (ed. G. Casati), pp. 171–186. New York: Plenum.
Kac, M. 1966 *Am. Math. Mon.* **73**, no. 4, part II, 1–23.
Keating, J. P. & Berry, M. V. 1988 (In preparation.)
Korsch, H. J. & Berry, M. V. 1981 *Physical* **3D**, 627–636.
Leopold, J. G. & Percival, I. C. 1979 *J. Phys.* B **12**, 709–721.
McDonald, S. W. & Kaufman, A. N. 1979 *Phys. Rev. Lett.* **42**, 1189–1191.
McKean, H. P. 1972 *Communs pure appl. Math.* **25**, 225–246.
Nye, J. F. & Berry, M. V. 1974 *Proc. R. Soc. Lond.* A **336**, 165–90.
Odlyzko, A. M. 1987 *Maths Comput.* **48**, no. 4, 273–308.
O.E.D. 1893 *A new English dictionary on historical principles*, vol. 2 (ed. J. A. H. Murray). Oxford: Clarendon Press.
Pankratov, O. A., Pakhomov, S. V. & Volkov, B. A. 1987 *Solid St. Commns* **61**, 93–96.
Percival, I. C. 1977 *Adv. chem. Phys.* **36**, 1–61.
Porter, C. E. 1965 *Statistical theories of spectra: fluctuations*, New York: Academic Press.
Robnik, M. & Berry, M. V. 1986 *J. Phys.* A **19**, 669–682.
Simon, B. 1983*a* *J. funct. Anal.* **53**, 84–98.
Simon, B. 1983*b* *Ann. Phys.* **146**, 209–220.
Van de Lune, J., te Riele, H. J. J. & Winter, D. T. 1986 *Maths Comput.* **46**, no. 74, 667–681.
Volkov, B. A. & Pankratov, O. A. 1985 *JETP Lett* **42**, 178–181.
Whewell, W. 1833 *Phil. Trans. R. Soc. Lond.* **123**, 147–236.
Whewell, W. 1836 *Phil. Trans. Roy. Soc. Lond.* **126**, 289–307.
Wunner, G., Woelk, U., Zech, I., Zeller, G., Ertl, T., Geyer, F., Schweizer, W. & Ruder, H. 1986 *Phys. Rev. Lett.* **57**, 3261–3264.

RIEMANN'S ZETA FUNCTION: A MODEL FOR QUANTUM CHAOS?

M.V.Berry
H.H.Wills Physics Laboratory, Tyndall Avenue,
Britol BS8 1TL, U.K.

1 INTRODUCTION

The celebrated hypothesis of Riemann[1] is that all the complex zeros of his function $\zeta(z)$ have real part 1/2, so that the quantities $\{E_j\}$ defined by

$$\zeta(\tfrac{1}{2} - iE_j) = 0 \tag{1}$$

are all real. There is evidence supporting the hypothesis: the first few million E_j have been computed and are all real, and it has been proved that uncountably many E_j are real. My purpose in this speculative paper is to extend the old suggestion that the E_j are real because they are eigenvalues of some Hermitian operator $\hat{H}$. The extensions are that if $\hat{H}$ is regarded as the Hamiltonian of a quantum-mechanical system then

(i) $\hat{H}$ has a classical limit

(ii) the classical orbits are all chaotic (unstable)

(iii)the classical orbits do not possess time-reversal symmetry.

To make these assertions plausible I will combine two sorts of evidence. The first (section 2) concerns largely numerical results connecting $\{E_j\}$ with the spectra of infinite random complex Hermitian matrices. The second (section 3) concerns analogies between a (divergent) representation of the number of Riemann zeros with $0<E_j<E$ and an asymptotic formula expressing the number of quantum energy levels in any given interval as a sum over classical closed orbits. Finally (section 4), I will give a semiclassical interpretation of the Riemann-Siegel formula (the basis of a powerful method for computing the $\{E_j\}$[1]) leading to a conjectured generalization that would be a quantization formula for classically chaotic systems.

The most useful formulae for $\zeta(z)$ will be the familiar ones: as a product over primes p or a sum of inverse powers of integers n, namely

$$\zeta(z) = \prod_p \left(1 - p^{-z}\right)^{-1} \qquad (\mathrm{Re}\, z > 1) \tag{2a}$$

$$= \sum_{n=1}^{\infty} n^{-z} \qquad (\mathrm{Re}\, z > 1) \tag{2b}$$

Reprinted from 'Quantum chaos and statistical nuclear physics', Springer Lecture Notes in Physics No. 263, (1986).

even though neither representation converges on the line Rez=1/2 where the zeros are. Of course there are many representations which are valid for Rez=1/2, such as this resummation of (2b):

$$\zeta(z) = \frac{1}{1-2^{1-z}} \sum_{n=1}^{\infty} (-1)^{n+1} n^{-z} \qquad (\mathrm{Re}\, z > 0) \tag{3}$$

The symmetry [1] relating $\zeta(z)$ and $\zeta(1-z)$ implies that each of the zeros (1) with real E_j has a counterpart with $-E_j$, and with this understanding we henceforth regard $\{E_j\}$ as a set of positive numbers with $E_1<E_2<E_3\ldots$.

2 RIEMANN ZEROS AND RANDOM MATRICES

A useful separation of the $\{E_j\}$ into an average part and a fluctuating part can be achieved by means of the <u>Riemann staircase.</u> This is defined as

$$N_R(E) \equiv \sum_{j=1}^{\infty} \Theta(E-E_j) \tag{4}$$

where Θ denotes the unit step function. $N_R(E)$ is simply the number of zeros with $E_j<E$. The average $\langle N_R(E)\rangle$ is a smooth approximation to the staircase, whose form is known [1] to be

$$\langle N_R(E)\rangle = \frac{E}{2\pi}\left(\ln\left\{\frac{E}{2\pi}\right\} - 1\right) + \frac{7}{8} . \tag{5}$$

Fig.1a shows just how close an approximation this is, even for small E. Figs 1b,c show that it is necessary to include the term 7/8.

Deviations from $\langle N_R(E)\rangle$ constitute the fluctuations in $\{E_j\}$. The statistics of these fluctuations can be studied numerically, and it is found that with high accuracy they coincide with the statistics of the eigenvalues of a typical member of the 'Gaussian unitary ensemble' (GUE)[2] of complex Hermitian matrices whose elements are Gauss -distributed in a way that is invariant under unitary transformations.

One such statistic is the probability distribution of the normalized spacings $\{S_j\}$ between adjacent zeros; these are defined by

$$S_j \equiv (E_{j+1}-E_j)\Big/\langle d_R((E_j+E_{j+1})/2)\rangle \tag{6}$$

where $\langle d_R(E)\rangle$ is the average density of zeros, given by (5) as

3

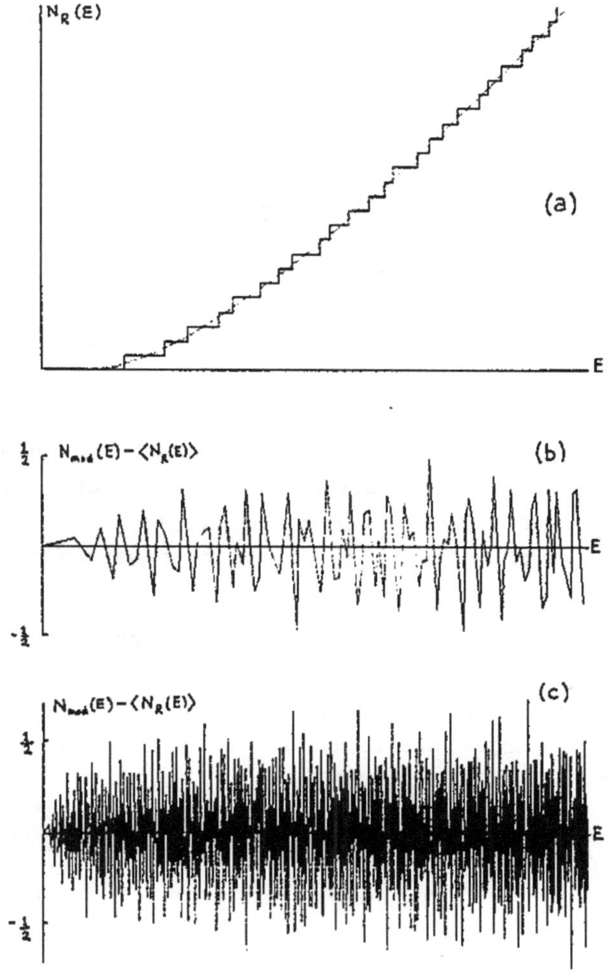

Fig.1 a) Riemann staircase $N_R(E)$, and (dotted) its average $\langle N_R(E)\rangle$, for the lowest 25 zeros; b) deviation $N_{mod}(E)-\langle N_R(E)\rangle$ for the lowest 100 zeros; c) $N_{mod}(E)-\langle N_R(E)\rangle$ for the lowest 1000 zeros ($N_{mod}(E)$ is $N_R(E)$ made continuous by replacing the N'th step by the straight line joining E_N, $N-1/2$ and E_{N+1}, $N+1/2$).

$$\langle d_R(E)\rangle = \frac{d}{dE}\langle N_R(E)\rangle = \frac{1}{2\pi}\ln\left\{\frac{E}{2\pi}\right\} \tag{7}$$

The spacings distribution P(S) is shown in fig.2, together with $P_{GUE}(S)$, for which random-matrix theory [2] gives the close approximation

$$P_{GUE}(S) = \frac{32}{\pi^2} S^2 \exp\left\{-4S^2/\pi\right\}. \tag{8}$$

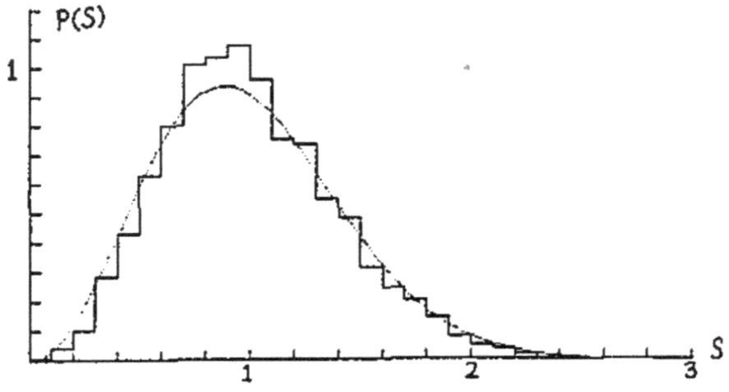

Fig.2 Histogram of distribution P(S) for the first 5000 Riemann zeros, together with $P_{GUE}(S)$ dotted.

Far more extensive computations have been performed by Odlyzko; these are unpublished, but preliminary reports have been given by Dyson [3] and Bohigas and Giannoni [4]. Odlyzko calculated sequences of up to 10^5 Riemann zeros, reaching to the 10^{11}th, and studied not only P(S) but also correlation functions betwen pairs, triplets and quartets of zeros. Apart from one apparent exception, to which we will return in section 3, his statistics are in excellent agreement with those of the GUE.

A theorem of Montgomery [5] supports the conclusion suggested by Odlyzko's results, that the statistics of Riemann zeros are precisely those of the GUE. The theorem concerns the form factor $K(\tau)$ of the zeros; this is the Fourier transform of the pair correlation function, and is defined as

$$K(\tau) = \lim_{M\to\infty} \left\{ \frac{1}{M} \sum_{j=1}^{M} \sum_{k=1}^{M} \exp\{2\pi i \tau (x_j - x_k)\} - \frac{\sin M\pi\tau}{\pi\tau} \right\} \quad (9)$$

where $\{x_j\}$ are the Riemann zeros, scaled so as to have unit mean spacing, that is

$$x_j \equiv \langle N_R(E_j) \rangle. \quad (10)$$

Montgomery proves that for $|\tau|<1$, $K(\tau)$ coincides with the GUE form factor [6]

$$\left.\begin{aligned} K_{GUE}(\tau) &= |\tau| \qquad (|\tau|<1) \\ &= 1 \qquad (|\tau|>1) \end{aligned}\right\}, \quad (11)$$

and conjectures that this agreement continues to hold when $|\tau|>1$.

Now, GUE statistics are distinctive [4]: they differ sharply, for example, from the Poisson statistics of numbers generated sequentially by a random process, and from the Gaussian orthogonal ensemble (GOE) statistics of eigenvalues of random real symmetric (as opposed to complex Hermitian) matrices. It would seem difficult to simulate GUE statistics by a process not involving eigenvalues of complex Hermitian matrices, so from now on I will regard the experimental observation that the Riemann zeros obey GUE statistics as evidence that there is indeed a nontrivial complex infinite Hermitian matrix $\hat{H}$ with eigenvalues E_j. ('Nontrivial' means that $\hat{H}$ must not contain the $\{E_j\}$ explicitly - as in the diagonal matrix $E_i \delta_{ij}$ or any simple unitary transform thereof.)

Among the class of operators $\hat{H}$ which are quantum Hamiltonians with classical limits, those with discrete energy spectra obeying GUE statistics have classical orbits which are chaotic and without time-reversal symmetry. This has been demonstrated numerically by Seligman and Verbaarschot [7], Berry and Robnik [8], and Robnik and Berry [9], and explained theoretically by Berry [10]. Both of the above conditions are necessary for GUE statistics: for systems whose orbits are not chaotic but integrable (with or without time-reversal symmetry), Berry and Tabor [11] showed the energies to be Poisson-distributed; and for systems which are chaotic but which do have time-reversal symmetry, Bohigas, Giannoni and Schmit [12] and others [13,4,10] have shown the energies to be GOE-distributed.

Of course the observation that GUE statistics are shared by both the Riemann zeros E_j and the eigenvalues of classically chaotic systems without time-reversal symmetry does not imply that the $\{E_j\}$ come from a classically chaotic $\hat{H}$, but we now turn to analytical evidence which strongly suggests that they do.

3 OUROBOROLOGY

Ouroboros [14] was the mythical snake that swallowed its tail, and serves to symbolize the constructive interference of quantum waves associated with classical orbits (fig.3). Such interference forms the basis of a technique for generating the energy spectrum, developed by Gutzwiller [15-17] and Balian and Bloch [18]; for an elementary review, see [19]. The technique has been used to explain why classically integrable systems have Poisson-distributed energy levels

6

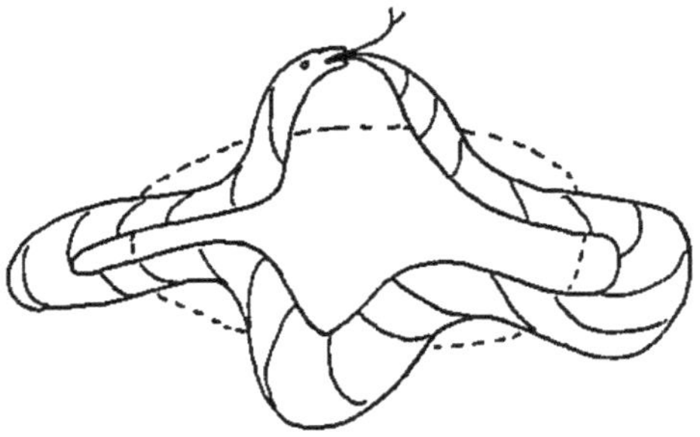

Fig.3 A quantum wave (Ouroboros) interfering constructively round a classical closed orbit.

[11] and why chaotic systems have GOE- or GUE- distributed levels [10].

Ouroborology is based on representing the spectral density as the imaginary part of the trace of the Fourier transform of the propagator (Green function of the time-dependent Schrödinger equation), which is expressed semiclassically (i.e. for small Planck's constant $\hbar$) as a sum over classical paths. The spectral staircase N(E), defined by (4) with $\{E_j\}$ now being the energy levels, is just the integral of the spectral density, so N(E) can similarly be expressed in terms of closed classical paths with energy E, the relation being that the fluctuating part of the staircase is

$$N_{osc}(E) \equiv N(E) - \langle N(E) \rangle = \mathrm{Im} \sum_p \sum_{m=1}^{\infty} B_{pm} \exp\{i S_{pm}(E)/\hbar + \phi_{pm}\}. \tag{12}$$

In this formula, $\langle N(E) \rangle$ is the average staircase (cf (5)) and will be discussed later. The double sum is over all closed orbits, each being an m-fold traversal of a primitive orbit labelled p. S_{pm} is the action of the orbit m, given in terms of the canonical phase-space variables q_μ, p_μ by

$$S_{pm}(E) = \oint p_\mu dq_\mu \quad (= m S_{p1}(E)). \tag{13}$$

The phases ϕ_{pm} and amplitudes B_{pm} depend on the focusing and stability of a bundle of (non-closed) orbits centred on the closed one. We require only the formulae for systems with two freedoms whose closed orbits are all isolated and unstable (making the dynamics chaotic) and without focal points. Then $\phi_{pm}=0$, and [15,16]

$$B_{pm} = \left[2\pi m \sinh\{m\lambda_p(E)/2\}\right]^{-1}, \tag{14}$$

where $\lambda_p(E)(>0)$ is the instability exponent of the primitive orbit p at energy E (i.e. $\exp\{\pm\lambda_p\}$ are the eigenvalues of the 2x2 matrix M_p of linearized phase-space deviations transverse to p, and

$$2\sinh\{m\lambda_p/2\} = \left[-\det\{M_p^m - 1\}\right]^{1/2}.$$

Thus for such systems the semiclassical asymptotic formula for the spectral fluctuations is

$$N_{osc}(E) = \frac{1}{2\pi}\sum_p\sum_{m=1}^{\infty}\frac{\sin\{m S_{p1}(E)/\hbar\}}{m\sinh\{m\lambda_p(E)/2\}}. \tag{15}$$

Next, we note that $N_{osc}(E)$, as defined by (4) and the first member of (12), has unit discontinuities at each eigenvalue. Therefore the closed-orbit sum (15) can at best be conditionally convergent, with the discontinuities determined by the very long orbits. For these, $m\lambda_p/2$ is large, and so we can replace the series by

$$N_{osc}(E) \approx \frac{1}{\pi}\sum_p\sum_{m=1}^{\infty}\frac{1}{m}\exp\{-m\lambda_p(E)/2\}\sin\{m S_{p1}(E)/\hbar\}. \tag{16}$$

This replacement will be reconsidered later.

Let us now turn to $\zeta(z)$ with $z=\frac{1}{2}-iE$. Just above the real E axis, the phase of decreases by as ReE passes each Riemann E_j. Moreover $\zeta(z)\to 1$ as $\mathrm{Im}(E)\to+\infty$ (i.e. as $\mathrm{Re}z\to+\infty$). Therefore the fluctuating part of the Riemann staircase is

$$N_{R,osc}(E) \equiv N_R(E) - \langle N_R(E)\rangle = -\frac{1}{\pi}\lim_{\eta\to 0}\mathrm{Im}\ln\zeta\left(\tfrac{1}{2}-i(E+i\eta)\right). \tag{17}$$

Now pretend that the product formula (2a) can be used when Rez=1/2 (more about this later), substitute into (17) and expand the logarithms. This gives

$$N_{R,osc}(E) \approx -\frac{1}{\pi}\sum_p\sum_{m=1}^{\infty}\frac{\sin\{mE\ln p\}}{m\,p^{m/2}} \tag{18}$$

which apart from a sign, to be discussed later, has the same form as the semiclassical expression (16) if the following identifications are made: the label p for primitive closed orbits denotes prime numbers; the actions of the closed orbits are

$$S_{pm} = mE\ln p\ ; \tag{19}$$

Planck's constant $\hbar$ is unity, so that the semiclassical limit is $E\to\infty$; and the instability exponents (independent of E) are

$$\lambda_p = \ln p\,. \tag{20}$$

8

It is worth remarking that it follows from (19) that the periods of the closed orbits would be

$$T_{pm} = \frac{dS_{pm}}{dE} = m \ln p = \ln p^m. \tag{21}$$

There is thus a formal analogy between fluctuations of the Riemann staircase and fluctuations of the spectral staircase of a classically chaotic system. Because of the distinctive GUE statistics (and for other reasons [10]) the classical orbits must lack time-reversal symmetry. Mathematicians have noticed essentially the same analogy as that between (15) or (16) and (18), in the context of a special case for which a certain transform of (15) is exact rather than asymptotic. This is the Selberg trace formula, which equates a sum over eigenvalues of the Laplace-Beltrami operator (playing the role of $\hat{H}$)on a manifold of constant negative curvature to a sum over closed geodesics on this manifold (all unstable, i.e. chaotic). But despite extensive study (see McKean [20] and Hejhal [21]) this analogy has not led to the identification of the mysterious 'Riemann' classical system with the properties (19-21). The 'closest approach' has been the discovery by Pavlov and Faddeev [22] and Gutzwiller [23] of a scattering (rather than bound) system whose phaseshifts (rather than energy levels) are given by $\zeta(z)$ with Rez=1 (rather than 1/2).

An apparently anomalous outcome [3] of Odlyzko's computation of the form factor $K(\tau)$(eq.9) of the Riemann zeros gives further support to the semiclassical analogy. Although he finds good overall agreement with the GUE formula (11), close examination of the difference $K-K_{GUE}$ reveals a series of spikes for small τ . Such spikes are predicted by the semiclassical theory, because as I have shown elsewhere [10] the universality of the GUE statistics ceases to hold for large energy scales, that is short time scales. For $K(\tau)$ the semiclassicl formula for $|\tau|<1$, expressed in terms of the closed-orbit amplitudes and periods, is [10]

$$K(\tau) = \pi^2 \tau^2 \sum_p \sum_{m=1}^{\infty} B_{pm}^2 \, \delta(\tau - T_{pm}/2\pi\hbar\langle d\rangle). \tag{22}$$

Now as $\hbar \to 0$, $\hbar\langle d\rangle \sim \hbar^{-(D-1)}$ for a system with D freedoms (D>1), so that the spike associated with a given orbit slides towards $\tau=0$ as $\hbar \to 0$. Near any finite τ, then, spikes are semiclassically thickly clustered and it is their average [10] which gives $K \approx |\tau|$ as in (11). But an accurate evaluation of $K(\tau)$ should reveal at least the first few spikes. In the Riemann case, (21) shows that the spikes should occur at τ-values proportional to logarithms of powers of primes. If the first of Odlyzko's spikes occurs at $\tau = K\ln 2$, the others occur at positions which on his picture are indistinguishable from $K\ln 3$, $K\ln 4$, $K\ln 5$ and $K\ln 7$, but as expected there is no spike at $K\ln 6$ because 6 is not a power of a prime.

Four objections may be raised against the chaos analogy for the Riemann zeros.

Objection 1: the Riemann closed-orbit formula (18) depends on the product (2a) which does not converge when Rez=1/2. But the analogous semiclassical formulae (15) and (16) almost certainly do not converge either. The physical reason for this is that the number of closed orbits of a chaotic system proliferates exponentially as their period (or action) increases, overwhelming the effect of the instability exponents λ_r in making the amplitudes decay exponentially. The mathematical reason was explained to me by Dr.A.Voros in the context of the Selberg trace formula: the fundamental quantum object for which semiclassical techniques give an expression in terms of closed orbits is not $N_{osc}(E)$ (or its derivative which is the fluctuating part of the spectral density) but the trace of the resolvent

$$g(E) \equiv \sum_j \frac{1}{E - E_j} \tag{23}$$

for which the closed-orbit formula makes sense only when $\eta \equiv \mathrm{Im}E$ exceeds some finite value. This means that in the formula

$$N(E) = -\frac{1}{\pi} \lim_{\eta \to 0} \mathrm{Im} \int_0^{E+i\eta} dE' g(E') \tag{24}$$

the limit $\eta \to 0$ cannot be taken when using ourolorology for g(E). Retaining finite η has the effect of introducing further factors $\exp[-i\eta T_{pm}(E)/\hbar]$ into (15) and (16), making the sums converge but at the price of giving a staircase whose steps are smoothed by η, thereby frustrating attempts to discriminate individual eigenvalues. Objection 1 therefore disappears because both the Riemann and semiclassical closed-orbit formulae share the disadvantage of not converging for real E. One could argue that this shared disadvantage strengthens the analogy.

It is interesting to look numerically at the divergence of the product (2a), especially in view of earlier computations [10] of the spectral density (derivative of (18) which with small numbers of primes showed pronounced peaks at the lowest few zeros, nicely simulating the delta-functions that the exact spectral density must possess. It is simplest to calculate the truncated product

$$|\zeta_M(\tfrac{1}{2}-iE)| \equiv \prod_{p<M}\left|1-\frac{e^{iE\ln p}}{p^{1/2}}\right|^{-1}. \tag{25}$$

As fig.4a shows, very few factors suffice to discriminate the lowest zeros. As M increases, however, $|\zeta_M|$ oscillates increasingly fast between the zeros, in contrast to the exact $|\zeta|$ which has only one maximum between each pair of zeros (cf fig.6 later); and when M=10000 (fig.4b) the oscillations are threatening to obscure the first zero. That such

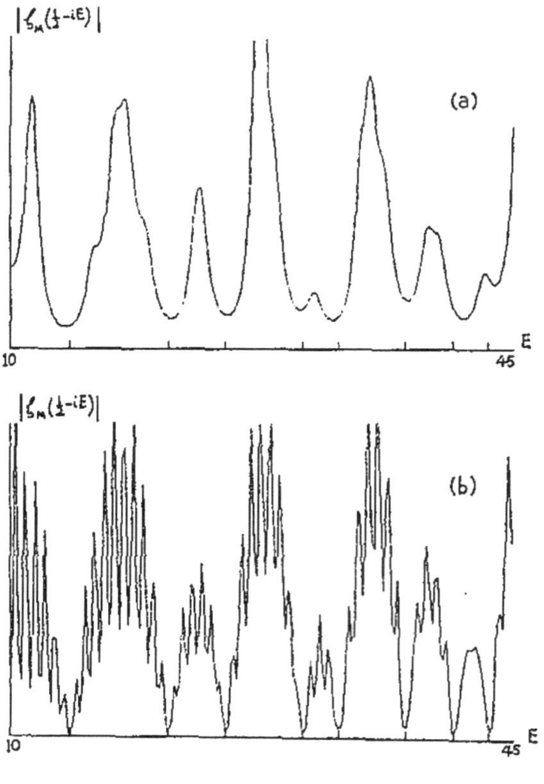

Fig.4a Truncated Riemann product $\zeta_M(\frac{1}{2}-iE)$ as a function of E for M=5 (three factors in (25)), with ticks marking the exact Riemann zeros E_j; b) as a) but with M=10000

obscuration will eventually occur is illustrated in fig.5, which shows $|\zeta_M|$ as a function of M evaluated at the exact positon of the first Riemann zero E_1=14.135... : at first $|\zeta_M|$ decreases, apparently indicating convergence onto E_1, but, when M exceeds about 2000, $|\zeta_M|$ begins to oscilalte with increasing amplitude. Rough asymptotics shows

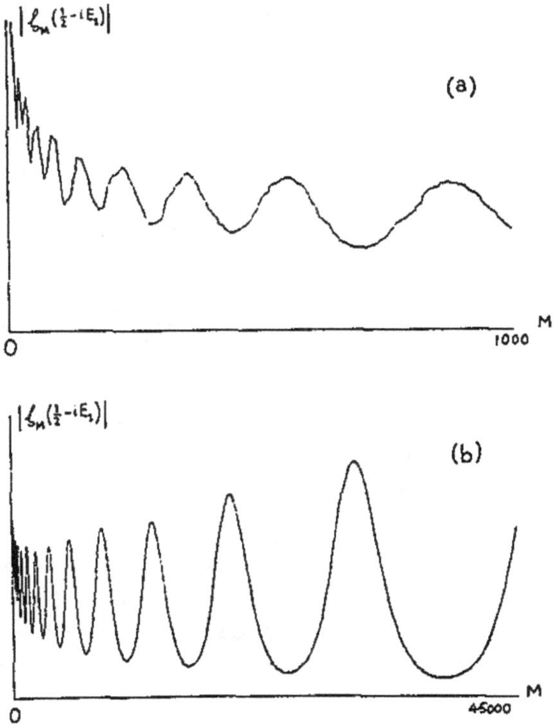

Fig.5 Truncated Riemann product $|\zeta_M(\frac{1}{2}-E_1)|$ as a function of M, evaluated at the lowest Riemann zero E_1, for a) M<1000; b) M<45000.

that $|\zeta_M|$ eventually diverges as

$$|\zeta_M| \sim \exp\{M^{1/2}\sin\{E\ln M\}/E\ln M\}. \qquad (26)$$

The fact that (18) relies on evaluating the Riemann product on the line Rez=1/2, which is displaced by 1/2 from the nearest line on which it converges (z=1), suggests that E=1/2 is the smallest interval over which the Riemann zeros can be discriminated in this way, and hence that ouroborology might fail altogether when the mean separation of zeros is about 1/2. From (7), this occurs when $E_j \sim 2\pi\exp(4\pi) \approx 2\times10^6$, i.e. $j \sim 3\times10^6$, and preliminary numerical exploration in this region indeed suggests the beginning of a failure of (18) to discriminate individual zeros.

Objection 2: the passage from (15) to (16), in which $\sinh(m\lambda/2)$ was replaced by $\exp(m\lambda/2)/2$, was a swindle, implying that for a proper analogy the Riemann formula (18) ought to involve not $p^{m/2}=\exp\{m\ln p/2\}$

but $2\sinh\{m\ln p/2\}$. One answer lies in considering not ζ itself but the product

$$P(E) \equiv \prod_{k=0}^{\infty} \zeta\left(\tfrac{1}{2}+k-iE\right). \tag{27}$$

For real E this converges and has the same zeros as $\zeta(\frac{1}{2}-iE)$, and so can be employed instead of $\zeta(\frac{1}{2}-iE)$ to approximate the fluctuations of the Riemann staircase. An easy calculation (cf.(17-18)) gives

$$N_{P,osc}(E) = -\frac{1}{\pi}\lim_{\eta\to 0} \operatorname{Im}\ln P(E+i\eta) \approx -\frac{1}{2\pi}\sum_p \sum_{m=1}^{\infty} \frac{\sin\{mE\ln p\}}{m\sinh\{m\ln p/2\}} \tag{28}$$

which obviously is analogous to (15).

Objection 3: the average density of Riemann zeros (7) has a logarithmic form which is not easy to interpret as the average density $\langle d\rangle$ of eigenvalues of an operator $\hat{H}$ with a classical limit $H(q_\mu, p_\mu)$. For example, in a finite D-dimensional 'billiard' enclosure, $\langle d\rangle \sim E^{(D/2-1)}$; replacing the enclosure by a binding potential similarly fails to give a logarithm. However, Simon [24,25] draws attention to a class of Hamiltonians which are classically unbound but have discrete quantal energy spectra. Of these, a planar billiard with a channel reaching to infinity whilst narrowing hyperbolically does have a logarithmic average level density (and moreover displays intermittent chaos). This can be seen in an elementary way using the Weyl formula

$$\langle d(E)\rangle \sim \frac{\text{area}}{4\pi} \tag{29}$$

(defining energy as $E \equiv k^2$ where k is the de Broglie wavenumber), together with the idea that waves do not penetrate (except with exponential evanescence) where the channel is narrower than a wavelength, that is narrower than about k^{-1}. If the channel boundary has equation $y=B/x$, the area for waves with energy E is thus

$$\text{area} \sim B\int^{y\sim 1/k} y\,dx = B\int^{Bk} dx/x \sim B\ln B\sqrt{E} \sim \ln E \tag{30}$$

giving $\langle d\rangle \sim \ln E$ as claimed. This disposes of objection 3, although there is of course no suggestion that the Riemann $\hat{H}$ really is a billiard of this type (even with magnetic field, to break time-reversal symmetry).

Objection 4: the semiclassical and Riemann formulae (16) and (18) have opposite signs. I have no clear answer to this. It is possible to get negative signs for some of the B_{pm} in (14), when the unstable orbits have focal points [15,16]: for a primitive orbit with n_p focal points, the sign is $(-1)^{[mn_p+1)/2]}$, where [ ] denotes integer part. But no choice of n_p gives negative signs for all the B_{pm}. However, this determination of the sign implicitly assumes an $\hat{H}$ of the form 'Kinetic energy + potential energy', acting on scalar states; without these restrictions it might be possible to get all-negative amplitudes.

4 THE RIEMANN-SIEGEL FORMULA: A RULE FOR QUANTIZING CHAOS?

It follows from the functional equation for $\zeta(z)$ [1] that the following function Z(E) is real and even for real E:

$$Z(E) \equiv \exp\{-i\theta(E)\}\,\zeta(\tfrac{1}{2}-iE) \tag{31}$$

where

$$\theta(E) = \arg\Gamma(\tfrac{1}{4}+\tfrac{E}{2}) - \tfrac{E}{2}\ln\pi \approx \tfrac{E}{2}\ln\tfrac{E}{2\pi} - \tfrac{E}{2} - \tfrac{\pi}{8}$$
$$= \pi\left(\langle N_R(E)\rangle + 1\right) \tag{32}$$

(cf.5). The Riemann-Siegel formula is an asymptotic representation of Z(E) for large E:

$$Z(E) = -2\sum_{n=1}^{Q(E)} \frac{\cos\{\pi\langle N_R(E)\rangle - E\ln n\}}{n^{1/2}} + R(E) \tag{33}$$

where

$$Q(E) \equiv \left[\sqrt{(E/2\pi)}\right] \tag{34}$$

and R(E) is a series of remainder terms [1] whose main effect is to cancel the discontinuities of the main sum arising from the E-dependence of the limit Q. Fig.6 shows how accurate the formula is, even for small E.

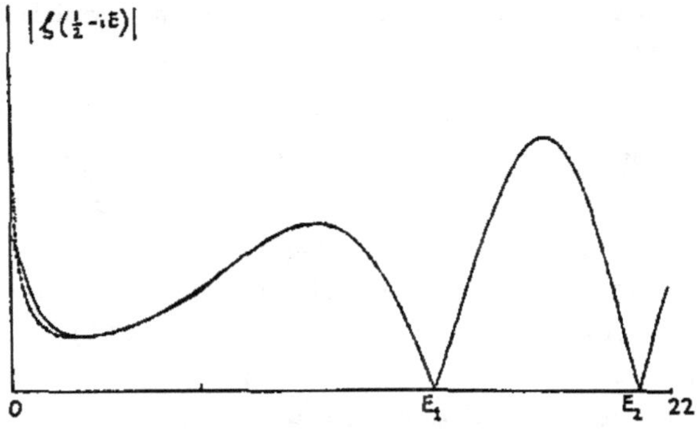

Fig.6. Comparison of $|\zeta(\frac{1}{2}-iE)|$ (full line) with Riemann-Siegel formula plus one correction term (dotted line). (The number Q(E) of terms in the sum (33) changes from zero to one at $E=2\pi$, indicated by a tick)

14

Now I will outline how the Riemann-siegel formula can be obtained from the series (2b) by an argument with a semiclassical interpretation suggesting a generalization. First note that in obtaining (2b) by expanding the products in (2a), use is made of the factorization theorem that for any integer n we can write

$$\ln n = \sum_p m_p \ln p \tag{35}$$

with a unique choice of the set of integers $\{m_p = 0,1,2\ldots\}$. On the semiclassical analogy, the sum index n can thus be interpreted as runing over <u>all possible combinations of orbit periods</u> (cf.21). Each such combination $\ln n$ will be called a <u>pseudoperiod</u> (pseudoperiods where all m_p except one are zero are periods of actual orbits).

Next, split the sum (2b) into two, with pseudoperiods with n less than and greater than some initially arbitrary value Q, and apply Poisson's summation formula to the second sum:

$$\begin{aligned}\zeta(\tfrac{1}{2}-iE) &= \sum_{n=1}^{Q} \frac{\exp\{iE\ln n\}}{n^{1/2}} + \sum_{n=Q+1}^{\infty} \frac{\exp\{iE\ln n\}}{n^{1/2}} \\ &= \sum_{n=1}^{Q} \frac{\exp\{iE\ln n\}}{n^{1/2}} + \sum_{m=-\infty}^{\infty} \int_{Q+\delta}^{\infty} dn \, \frac{\exp\{i(E\ln n - 2\pi mn)\}}{n^{1/2}}\end{aligned} \tag{36}$$

where $0<\delta<1$. For large E the integrals may be approximated by the method of stationary phase. The stationary point of the m'th integral is at $n=E/2\pi m$, which lies in the integration range only if $1\leqslant m \leqslant [E/2\pi(Q+\delta)]$. The choice $Q=Q(E)$ (equation 34) and $\delta \to 0$ gives the same limits for the sums over n and m, and then stationary phase leads to

$$\zeta(\tfrac{1}{2}-iE) \approx \sum_{n=1}^{Q(E)} \left(\frac{\exp\{iE\ln n\}}{n^{1/2}} + \frac{\exp\{i(-\frac{\pi}{4} + E\ln\{E/2\pi n\} - E)\}}{n^{1/2}} \right) \tag{37}$$

which gives the Riemann-Siegel sum (33) when combined with the definitions (31) and (32).

What the Poisson technique has achieved in (37) is a <u>resummation of the orbits with long pseudoperiods</u>, to give a series of the same form as the sum over the short pseudoperiods, with precisely the correct phase relation to make $\zeta(\frac{1}{2}-iE)$ have the required analytic structure (31). To see what immense advantage this resummation has produced as compared with the naive ouroborology result (18), consider the oscillations of the terms in (33). The first term, with n=1 (corresponding to the zero pseudoperiod with all m_p equal to zero in 35), oscillates fastest. This term alone gives zeros

$$\langle N_R(E_j)\rangle = j - \tfrac{1}{2} \quad (j = 1,2,3\ldots), \tag{38}$$

with the correct density but which are (asymptotically) uniformly distributed instead of GUE-distributed. The terms with n>1 (nonzero pseudoperiods) oscillate more slowly, and the highest term n=Q(E) is almost constant:

$$\frac{d}{dE}\left(\pi\langle N_R(E)\rangle - E\ln Q\right) = \pi\langle d_R(E)\rangle - \ln Q(E)$$
$$= \ln\{\surd(E/2\pi)\} - \ln\{[\surd(E/2\pi)]\} \approx 0 \qquad (39)$$

By contrast, the terms given by unresummed ouroborology oscillate ever faster to give the noise and divergence visible in figs. 4b and 5b.

It is natural to speculate that a similar resummation might be possible for semiclassical quantum chaos. For the simplest nontrivial case, embodied in equations (14-16), this speculation generates, by an argument to be outlined in a moment, a real function W(E) whose zeros would be the eigenvalues. The pseudoorbits are simpler than in the Riemann case, in that each primitive orbit contributes at most one traversal. Thus the actions, periods and instability exponents may be written

$$\left.\begin{array}{l} S^{(k)}(E) = \sum_p \ell_p S_{p1} \;;\; T^{(k)}(E) = \sum_p \ell_p dS_{p1}/dE \;;\; \lambda^{(k)}(E) = \sum_p \ell_p \lambda_p \\ \text{with } \{\ell_p = 0 \text{ or } 1\} \text{ labelled } k \text{ in order of increasing } T^{(k)} \end{array}\right\} \qquad (40)$$

Then the analogue of the Riemann-Siegel formula (33) would be

$$W(E) = -2\sum_{k=1}^{k_{max}} (-1)^{g_k} \exp\{-\lambda^{(k)}(E)/2\} \cos\{\pi\langle N(E)\rangle - S^{(k)}(E)/\hbar\} \qquad (41)$$

where g_k is the number of terms (nonzero ℓ_p's) in the sums (40), and k_{max} is given by the condition that the highest term is non-oscillatory:

$$T^{(k_{max})}(E) = \pi\hbar\,\langle d(E)\rangle \qquad (42)$$

Because $\langle d\rangle \sim \hbar^{-D}$ for a system with D freedoms, the longest pseudoperiod is of order $\hbar^{-(D-1)}$, so the number of terms in (41) grows exponentially as $\hbar$ decreases or E increases; nevertheless (41) should at least give sensible results for individual eigenvalues in the semiclassical limit, unlike the ouroborology formula (16).

The conjectured Riemann-Siegel analogue (41) arises from considering the function

$$w(E) = \exp\left\{\int_0^E dE'\left(g(E') - \langle g(E')\rangle\right)\right\}, \qquad (43)$$

where g(E) is the resolvent (23) and <g(E)> its average. For real E approaching the real axis from above, use of the average of (24) gives

$$\lim_{\eta\to 0} \omega(E+i\eta) = \exp\left\{-\int_{-\infty}^{\infty} dE' \ln\left|1-\frac{E}{E'}\right| \langle d(E')\rangle\right\} \exp\{i\pi\langle N(E)\rangle\} \prod_j \left(1-\frac{E}{E_j}\right) \quad (44)$$

Therefore $\omega(E)$ has zeros at the eigenvalues E_j. On the other hand, for sufficiently large Im E the fluctuations g-<g> can be approximated semiclassically by the analogue of (16) (which replaces the analogue of (15) by an argument similar to that centred on (27) and (28)). Thus

$$\omega(E) \approx \exp\left\{-\sum_p \sum_{m=1}^{\infty} \frac{\exp\{-m\lambda_p/2\}}{m} \exp\{imS_{p1}/\hbar\}\right\}$$

$$= \prod_p \left(1 - \exp\{-\lambda_p/2\} \exp\{iS_{p1}(E)/\hbar\}\right), \quad (45)$$

giving $\omega(E)$ as a product over primitive orbits, analogous to the product (2a) for $\zeta(z)$ with the difference that the factors do not appear as reciprocals (this originates in the mysterious sign difference described in objection 4 at the end of section 3).

Expanding the product in (45) we obtain an analogue of the sum (2a), involving the pseudoorbits (40). Then the conjectured Riemann-Siegel analogue (41), with W(E) identified from (44) as proportional to the product $\prod_j(1 - E/E_j)$, would follow, if only (!) one could resum the high pseudoorbits in a way similar to that which gave the true Riemann-Siegel formula (33).

Exactly the dynamical zeta function $\omega(E)$, in the form (45), has been written down by Gutzwiller [17]. Ruelle [26,27] introduced analogous zeta functions to study chaotic dissipative dynamical systems and these have been further investigated by Parry [28,29] (who introduced what I here call pseudoorbits) and Parry and Pollicott [30]; in these dissipative zeta functions, orbit actions and instability exponents are (by implication) considered as proportional to periods and entropy. None of these authors appear to have considered the possibility that a Riemann-Siegel formula might exist for dynamical zeta functions.

ACKNOWLEDGMENT No military agency supported this research.

REFERENCES

[1] H.M.Edwards, Riemann's Zeta Function (Academic Press: New York and London) (1974)

[2] C.E.Porter, Statistical theories of Spectra: Fluctuations (Academic Press: New York (1965)
[3] F.J.Dyson, in 'Symmetries in particle physics' (eds: I.Bars, A.Chodos and C-H Tze; Plenum: New York and London) pp 279-281 (1984)
[4] O.Bohigas and M.J.Giannoni, Chaotic motion and Random-matrix Theories in Mathematical and Computational Methods in Nuclear Physics eds.J.S.Dehesa, J.M.G.Gomez and A.Polls, Lecture Notes in Physics 209 (Springer-Verlag, N.Y.) pp.1-99 (1984)
[5] H.L.Montgomery, Proc.Symp.Pure Math. 24 181-193 (1973)
[6] M.L.Mehta, Random Matrices and the Statistical Theory of Energy Levels (Academic Press: New York and London) (1967)
[7] T.H.Seligman, and J.J.M.Verbaarschot, Phys.Lett. 108A 183-187 (1985)
[8] M.V.Berry and M.Robnik, J.Phys.A. (1986) In press
[9] M.Robnik and M.V.Berry, J.Phys.A. (1986) In press
[10] M.V.Berry, Proc.Roy.Soc.Lond. A400 229-251 (1985)
[11] M.V.Berry and M.Tabor, Proc.Roy.Soc.Lond. A356 375-394 (1977)
[12] O.Bohigas, M.J.Giannoni and C.Schmit, Phys.Rev.Lett. 52 1-4 (1984)
[13] M.V.Berry, Ann.Phys. (N.Y) 131 163-216 (1981)
[14] Ouroboros: see Encyclopaedia Brittanica (15th ed.) vol.9 p13 (1985)
[15] M.C.Gutzwiller, J.Math.Phys. 12 343-358 (1971)
[16] M.C.Gutzwiller, in 'Path Integrals and their Applications in Quantum, Statistical and Solid-State Physics' (Eds. G.J.Papadopoulos and J.T.Devreese) Plenum, N.Y. 163-200 (1978)
[17] M.C.Gutzwiller, in 'Stochastic Behaviour in Classical and Quantum Hamiltonian Systems' (eds. G.Casati and J.Ford) Lecture Notes in Physics 93 (Springer: Berlin)(1979)
[18] R.Balian and C Bloch, Ann.Phys.(N.Y) 69 76-160 (1972)
[19] M.V.Berry, Semiclassical Mechanics of Regular and Irregular Motion in Chaotic Behavior of Deterministic Systems (Les Houches Lectures XXXVI, eds.G.Iooss, R.H.G.Helleman and R.Stora (North-Holland: Amsterdam) pp171-271 (1983)
[20] H.P.McKean, Comm.Pure App. Math. 25 225-246 (1972)
[21] D.A.Hejhal, Duke Math.J. 43 441-482 (1976)
[22] B.S.Pavlov and L.D.Faddeev, J.Soviet Math 3 522-548 (1975)
[23] M.C.Gutzwiller, Physica 7D 341-355 (1983)
[24] B.Simon, Ann.Phys. 146 209-220 (1983)
[25] B.Simon, J.Funct.Anal. 53 84-98 (1983)
[26] D.Ruelle, Inventiones Math 34 231-242 (1976)
[27] D.Ruelle, Ergod.Th.and Dynam.Syst. 2 99-107 (1982)
[28] W.Parry Israel J.Math. 45 41-52 (1983)
[29] W.Parry Ergod Th.and Dynam.Syst. 4 117-134 (1984)
[30] W.Parry and M.Pollicott, Annals of Math. 118 573-591 (1983)

Nonlinearity 1 (1988) 399–407. Printed in the UK

Semiclassical formula for the number variance of the Riemann zeros

M V Berry
H H Wills Physics Laboratory, Tyndall Avenue, Bristol BS8 1TL, UK

Received 9 December 1987, in final form 20 March 1988
Accepted by J D Gibbon

Abstract. By pretending that the imaginary parts E_m of the Riemann zeros are eigenvalues of a quantum Hamiltonian whose corresponding classical trajectories are chaotic and without time-reversal symmetry, it is possible to obtain by asymptotic arguments a formula for the mean square difference $V(L;x)$ between the actual and average number of zeros near the xth zero in an interval where the expected number is L. This predicts that when $L \ll L_{\max} = \ln(E/2\pi)/2\pi \ln 2$ (where $x = (E/2\pi)(\ln(E/2\pi) - 1) + \frac{7}{8}$), V is the variance of the Gaussian unitary ensemble (GUE) of random matrices, while when $L \gg L_{\max}$, V will have quasirandom oscillations about the mean value $\pi^{-2}(\ln \ln(E/2\pi) + 1.4009)$. Comparisons with $V(L;x)$ computed by Odlyzko from 10^5 zeros E_m near $x = 10^{12}$ confirm all details of the semiclassical predictions to within the limits of graphical precision.

1. Introduction

It was realised long ago [1] that the truth of the Riemann hypothesis would be established if it could be shown that the imaginary parts E_m of the non-trivial zeros of $\zeta(z)$ are eigenvalues of a self-adjoint operator. Montgomery [2] suggested that the statistics of the E_m are those of the eigenvalues of an infinite complex Hermitian matrix drawn randomly from the Gaussian unitary ensemble (GUE) [3]. In a recent numerical study, Odlyzko [4] showed that while short-range statistics (such as the distribution of the spacings $E_{m+1} - E_m$ between neighbouring zeros) accurately conform to GUE predictions, long-range statistics (such as the correlations between distant spacings) do not, and are better described in terms of primes.

I have argued elsewhere [5, 6] that exactly this behaviour would be expected if the E_m were eigenvalues not of a random matrix but of the Hamiltonian operator obtained by quantising some still-unknown dynamical system without time-reversal symmetry, whose phase-space trajectories are chaotic. The theory is based on asymptotics of the semiclassical limit, in which Planck's constant $\hbar \to 0$. Its central result [7] is that statistics of eigenvalues separated by less than $O(\hbar)$ are universal (that is independent of the details of the Hamiltonian) and given (in the absence of time-reversal symmetry) by the GUE, while the statistics of eigenvalues with larger

0951-7715/88/010399 + 09\$02.50

separations are non-universal and depend on the details of the closed trajectories of the corresponding classical system. The connection with $\zeta(z)$ comes from an analogy between the Von Mangoldt formula [1] for $\ln(\zeta(z))$ and an expression for the semiclassical eigenvalue density fluctuations in terms of closed orbits.

My purpose here is to illustrate how the semiclassical theory can give a uniformly accurate description of both the long-range (non-universal) and short-range (universal) statistics of the E_m, by studying the number variance of the zeros. This is the mean square difference between the actual number of E_m and the expected number, in an interval where the expected number is L. Although the semiclassical analogy has not led to an identification of the elusive Riemann operator (if indeed there is one), this illustration does suggest that certain tantalising hints [5] about the underlying dynamical system deserve to be taken seriously.

2. Number variance

The mean number of zeros with height less than E is [1]

$$\mathcal{N}(E) = (E/2\pi)\{\ln(E/2\pi) - 1\} + \tfrac{7}{8} \tag{1}$$

so that the numbers

$$x_m \equiv \mathcal{N}(E_m) \tag{2}$$

form a sequence with mean spacing unity. Such a sequence can be regarded as a singular density

$$\mathrm{d}(x) = \sum_{m=1}^{\infty} \delta(x - x_m) \tag{3}$$

concentrated at the points x_m on the x axis. We will employ the notion of asymptotic averaging, that is averaging over a range Δx satisfying $1 \ll \Delta x \ll x$, and denote the operation by $\langle\ \rangle$. Obviously $\langle d\rangle = 1$. The fluctuating part of the level density, defined as

$$\tilde{\mathrm{d}}(x) \equiv \mathrm{d}(x) - 1 \tag{4}$$

provides the entry point for the semiclassical theory to be described in §3.

In the range of $x - L/2$ to $x + L/2$ the number of zeros is

$$n(L;x) = \int_{x-L/2}^{x+L/2} \mathrm{d}x_1\, \mathrm{d}(x_1). \tag{5}$$

Obviously $\langle n(L;x)\rangle = L$. The number variance $V(L;x)$ is thus

$$V(L;x) = \langle [n(L;x) - L]^2\rangle = \left\langle \int_{x-L/2}^{x+L/2} \mathrm{d}x_1 \int_{x-L/2}^{x+L/2} \mathrm{d}x_2\, \tilde{\mathrm{d}}(x_1)\, \tilde{\mathrm{d}}(x_2)\right\rangle. \tag{6}$$

This is conveniently expressed in terms of the form factor (Fourier transform of the pair correlation of the density fluctuations)

$$K(\tau;x) \equiv \int_{-\infty}^{\infty} \mathrm{d}\xi\, \exp(2\pi\mathrm{i}\xi\tau)\langle \tilde{\mathrm{d}}(x-\xi/2)\, \tilde{\mathrm{d}}(x+\xi/2)\rangle \tag{7}$$

because

$$\langle \tilde{d}(x_1)\,\tilde{d}(x_2)\rangle = \int_{-\infty}^{\infty} d\tau K(\tau; (x_1+x_2)/2)\exp[2\pi i\tau(x_1-x_2)]. \tag{8}$$

Substituting into (6) and setting $(x_1+x_2)/2$ equal to x for $x \gg 1$ and $L \ll x$ gives

$$V(L;x) = \frac{2}{\pi^2}\int_0^{\infty} d\tau \frac{K(\tau;x)}{\tau^2}\sin^2(\pi L\tau). \tag{9}$$

3. Long and short orbits

Now pretend that E_m are eigenvalues of a quantum Hamiltonian with a classical limit that is chaotic in the sense that all closed orbits are isolated and unstable. From any such set of E_m we can construct the scaled energies x_m from (2) using the appropriate counting function $\mathcal{N}(E)$. Thence we obtain the level density fluctuations $\tilde{d}(x)$ from (3) and (4). Gutzwiller [8, 9] and Balian and Bloch [10] have shown that in the semiclassical limit $\tilde{d}(x)$ can be expressed as a sum over all closed orbits at that energy E which corresponds to x according to (2). The sum is over all primitive orbits (labelled p) and their repetitions (labelled r where $1 \leqslant r \leqslant \infty$). Each orbit gives an oscillatory contribution to $\tilde{d}(x)$, with a phase $rS_p/\hbar$ where S_p is the classical action of the primitive orbit, and an amplitude A_{rp} that depends on the instability exponents of the orbit [8].

An important role is played by the periods of the orbits, given by

$$T_{rp} = r\,\partial S_p/\partial E. \tag{10}$$

These enter the form factor $K(\tau;x)$, which can be obtained [5] from (7) as

$$K(\tau;x) \simeq \frac{2\pi}{\hbar\rho}\left\langle \sum_{r_1p_1}\sum_{r_2p_2} A_{r_1p_1}A_{r_2p_2}\cos\{(r_1S_{p_1}-r_2S_{p_2})/\hbar\}\delta\{T-\tfrac{1}{2}(T_{r_1p_1}+T_{r_2p_2})\}\right\rangle \tag{11}$$

In this formula, ρ is the mean level density $d\mathcal{N}(E)/dE$, and T is a time variable related to τ by the scaling

$$T = \tau\hbar\rho. \tag{12}$$

For understanding the double sum (11) it is important to note that $\hbar\rho$ increases as $\hbar$ decreases for systems capable of displaying chaos (for example, in a classical billiard with N freedoms, $\hbar\rho \sim \hbar^{-(N-1)}$). This means that the period $T_{\min}$ of the shortest closed orbit corresponds to $\tau \ll 1$, while $\tau = 1$ corresponds to an extremely long orbit.

Choose an intermediate value τ^* satisfying

$$T_{\min}/2\pi\hbar\rho \ll \tau^* \ll 1. \tag{13}$$

For $\tau < \tau^*$, asymptotic averaging will remove the non-diagonal terms $r_1 \neq r_2$, $p_1 \neq p_2$ because of incoherence in the trigonometric factors (S_p depends on x), leaving

$$K(\tau;x) \simeq \frac{2\pi}{\hbar\rho}\sum_p\sum_{r=1}^{\infty} A_{rp}^2\delta(T-T_{rp}) \qquad (\tau < \tau^*). \tag{14}$$

Obviously the positions and strengths of the δ spikes in the sum depend on the details of the classical dynamics of the particular system being studied: because of this, $K(\tau;x)$ is described as non-universal when $\tau<\tau^*$.

On the other hand, for $\tau>\tau^*$ only very long orbits are involved; these proliferate exponentially with increasing T for chaotic systems and so contribute as though distributed continuously. Their contribution can be calculated [5] using a remarkable sum rule of Hannay and Ozorio de Almeida [11] relating the orbit amplitudes A_{rp} to their increasing number, and some heuristic semiclassical arguments. The result is that $K(\tau;x)$ becomes universal, that is independent of the detailed dynamics (provided these are chaotic and without time-reversed symmetry). Moreover, the expression obtained is precisely the form factor of the GUE, namely [12]

$$K(\tau;x)\simeq K_{\mathrm{GUE}}(\tau)=\tau\theta(1-\tau)+\theta(\tau-1)\qquad(\tau>\tau^*)\tag{15}$$

where θ denotes the unit step.

Substituting the non-universal and universal formulae (14) and (15) for $K(\tau;x)$ into (9) gives the number variance as

$$V(L;x)=8\sum_p\sum_{r=1}^{T_{rp}<2\pi\hbar\rho\tau^*}\frac{A_{rp}^2}{T_{rp}^2}\sin^2(LT_{rp}/2\hbar\rho)+\frac{2}{\pi^2}\int_{\tau^*}^{\infty}\mathrm{d}\tau\frac{K_{\mathrm{GUE}}(\tau)}{\tau^2}\sin^2(\pi L\tau).\tag{16}$$

This applies to any classically chaotic system, and arguments similar to those given elsewhere [5] for a related statistic (the spectral rigidity) show that $V(L;x)$ grows logarithmically according to the universal GUE formula until $L\sim L_{\max}\equiv\hbar\rho/T_{\min}$ and then oscillates non-universally whilst remaining bounded.

To apply (16) to the Riemann zeros we should of course know the underlying classical system. Without this knowledge we can only proceed by analogy, identifying T_{rp} and A_{rp} from the Von Mangoldt formula as explained in [5], ignoring difficulties [6] with the analogy. The results associate primitive closed orbits with primes p:

$$T_{rp}=r\ln p\qquad A_{rp}=-\ln p\exp(-r\ln p/2)/2\pi.\tag{17}$$

For this system $\hbar$ does not appear explicitly but can be regarded as concealed in E (as with quantum billiards); the semiclassical limit is $E\to\infty$. We also require the mean density of zeros $\rho(E)$, which from (1) is

$$\rho(E)=\frac{1}{2\pi}\ln(E/2\pi).\tag{18}$$

Now we can substitute into (16) and evaluate the integral, to get the number variance of the zeros:

$$\begin{aligned}V(L;x)=&\frac{1}{\pi^2}\{\ln(2\pi L)-\mathrm{Ci}(2\pi L)-2\pi L\mathrm{Si}(2\pi L)+\pi^2L-\cos(2\pi L)+1+\gamma\}\\&+\frac{1}{\pi^2}\left\{2\sum_p\sum_{r=1}^{p^r<(E/2\pi)^{\tau^*}}\frac{\sin^2(\pi Lr\ln p/\ln(E/2\pi))}{r^2p^r}+\mathrm{Ci}(2\pi L\tau^*)-\ln(2\pi L\tau^*)-\gamma\right\}.\end{aligned}\tag{19}$$

Here Si and Ci are the sine and cosine integrals [13], γ is Euler's constant

0.577 215 . . . and x is related to E by (2). This formula is our main result. As $x \to \infty$, the dependence on τ^* disappears provided τ^* satisfies (13), i.e. $\tau^* \gg \ln 2/\ln(E/2\pi)$.

The terms in the first set of braces give the GUE number variance, with limiting behaviour

$$V_{\mathrm{GUE}}(L) \simeq \begin{cases} L & \text{if } L \ll 1 \\ \dfrac{1}{\pi^2}[\ln(2\pi L) + 1 + \gamma] & \text{if } L \gg 1. \end{cases} \tag{20}$$

The second set of braces involves the sum over prime powers. Because of the upper limit, the largest value of the argument of the $\sin^2$ function is $\pi L \tau^*$. If $L \ll 1/\tau^*$ the sum is negligible and so are the remaining terms in these braces. Thus the first set of braces dominates, and $V \simeq V_{\mathrm{GUE}}$, if $L \ll 1/\tau^*$; this is the universal regime.

If $L \gg 1/\tau^*$, asymptotics of Si and Ci give

$$V(L;x) \simeq \frac{1}{\pi^2}\left(2 \sum_p \sum_{r=1}^{p^r < (E/2\pi)^{\tau^*}} \frac{\sin^2(\pi L r \ln p/\ln(E/2\pi))}{r^2 p^r} - \ln \tau^* + 1\right) \qquad \text{if } L \gg 1/\tau^*. \tag{21}$$

This describes asymptotic oscillations with amplitudes $(2\pi r^2 p^r)^{-1}$ and wavelengths

$$\Delta L = \ln(E/2\pi)/r \ln p \tag{22}$$

Because these wavelengths are incommensurable we expect the oscillations to have a quasirandom character. Obviously the oscillations depend on the detailed 'dynamics' (prime-period closed orbits), so that $L \gg 1/\tau^*$ is the non-universal regime.

The mean of the asymptotic oscillations is

$$\begin{aligned} \bar{V} &= \frac{1}{\pi^2}\left(\sum_{p=2}^{(E/2\pi)^{\tau^*}} p^{-1} + \sum_{r=2}^{\infty}\sum_{p=2}^{\infty} (r^2 p^r)^{-1} - \ln \tau^* + 1\right) \\ &= \frac{1}{\pi^2}[\ln \ln(E/2\pi) + 1.4009]. \end{aligned} \tag{23}$$

To leading order this is $\ln \ln E/\pi^2$, in exact agreement with the estimate by Selberg [14, 15] that the mean square part of the fluctuating part of the counting function is

$$\left\langle \left[\int_0^{x(E)} dx\, \tilde{d}(x)\right]^2 \right\rangle \sim \ln \ln E/2\pi^2. \tag{24}$$

This follows from (6) when written as

$$\begin{aligned} V(L;x) &= \left\langle \left[\int_0^{x+L/2} dx_1\, \tilde{d}(x_1) - \int_0^{x-L/2} dx_2\, \tilde{d}(x_2)\right]^2 \right\rangle \\ &\simeq 2\left\langle \left[\int_0^{x(E)} dx\, \tilde{d}(x)\right]^2 \right\rangle - 2\left\langle \int_0^{x+L/2} dx_1\, \tilde{d}(x_1) \int_0^{x-L/2} dx_2\, \tilde{d}(x_2) \right\rangle \\ &\to 2\left\langle \left[\int_0^{x(E)} dx\, \tilde{d}(x)\right]^2 \right\rangle \qquad \text{as } L \to \infty,\ L/x \ll 1 \end{aligned} \tag{25}$$

if it is assumed that the counting fluctuations at $x \pm L/2$ become uncorrelated as $L \to \infty$, $L/x \ll 1$.

4. Comparison with computation

Using techniques explained elsewhere [4], Odlyzko computed 10^5 consecutive zeros E_m, starting with $m = 10^{12} + 1$. Thus $x \simeq 10^{12}$ and, from (1) and (2), $E = 2.677 \times 10^{11}$. From these zeros he computed the number variance $V(L; x)$ for $0 \leqslant L \leqslant 1000$ in steps of 0.1.

Odlyzko's data are to be compared with the semiclassical values of $V(L; x)$ computed using (19). This involves τ^* which from (13), (17) and (18) must satisfy

$$\ln 2/\ln(E/2\pi) \ll \tau^* \ll 1 \qquad \text{i.e.} \qquad 0.028 \ll \tau^* \ll 1. \tag{26}$$

I chose $\tau^* = \frac{1}{4}$, so that the sum in (19) included prime powers $p^r \leqslant 449$, but checked that the curves of $V(L; x)$ against L were unaffected by reducing τ^* to 0.15 (i.e. $p^r \leqslant 37$). GUE universality should break down near $L_{\max} = \hbar\rho/T_{\min} = \ln(E/2\pi)/2\pi \ln 2 = 5.62$, and be replaced by the asymptotic oscillations, whose longest-wavelength component has $\Delta L = \ln(E/2\pi)/\ln 2 = 35.31$ (equation 22) and amplitude $1/2\pi^2 = 0.051$. The asymptotic mean (23) is $\bar{V} = 0.4659$; this differs slightly from the value $\bar{V} = 0.4663$ obtained by substituting $\sin^2 = \frac{1}{2}$ in (21), indicating residual dependence on τ^* which, however, does not affect the resolution of the graphs presented here.

Figures 1–4 show the results. In figure 1 the range is $0 \leqslant L \leqslant 2$. The exact and semiclassical V are indistinguishable, and GUE is a good approximation except near $L = 2$. Note how poor a fit to the data is the variance of the Gaussian *orthogonal* ensemble of real symmetric matrices (which would correspond to a dynamical system with time-reversal symmetry).

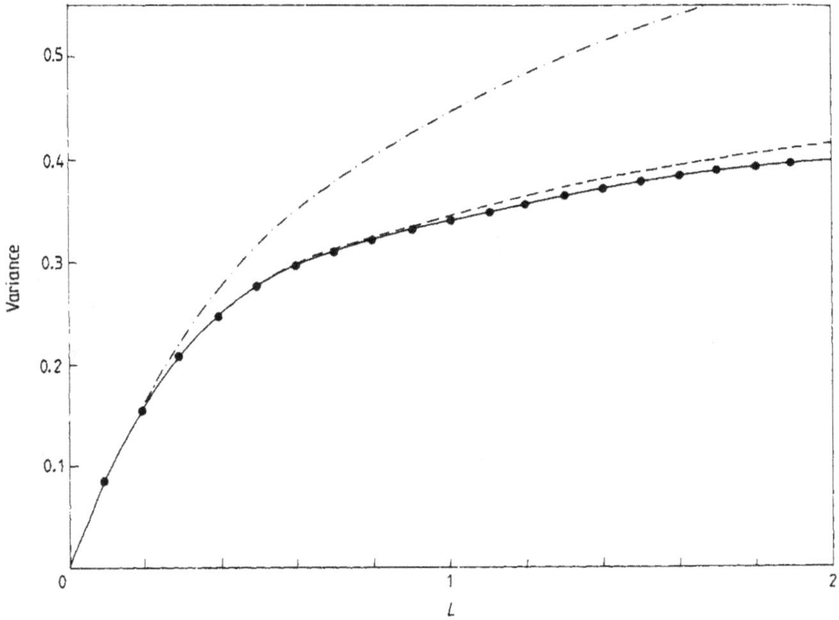

Figure 1. Number variance $V(L; x)$ of the Riemann zeros, for $0 \leqslant L \leqslant 2$ and $x = 10^{12}$. Dots: computed from the zeros by Odlyzko; full curve: semiclassical formula (19) with $\tau^* = \frac{1}{4}$; broken curve: number variance of the GUE; chain curve: number variance of the Gaussian orthogonal ensemble (GOE) of real symmetric random matrices.

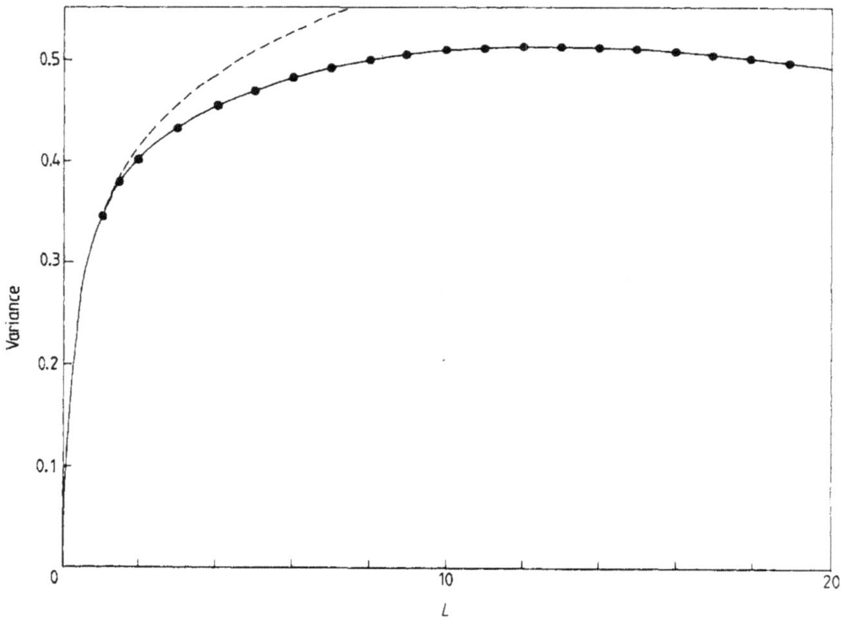

Figure 2. As figure 1 but for $0 \leqslant L \leqslant 20$ (GOE not shown).

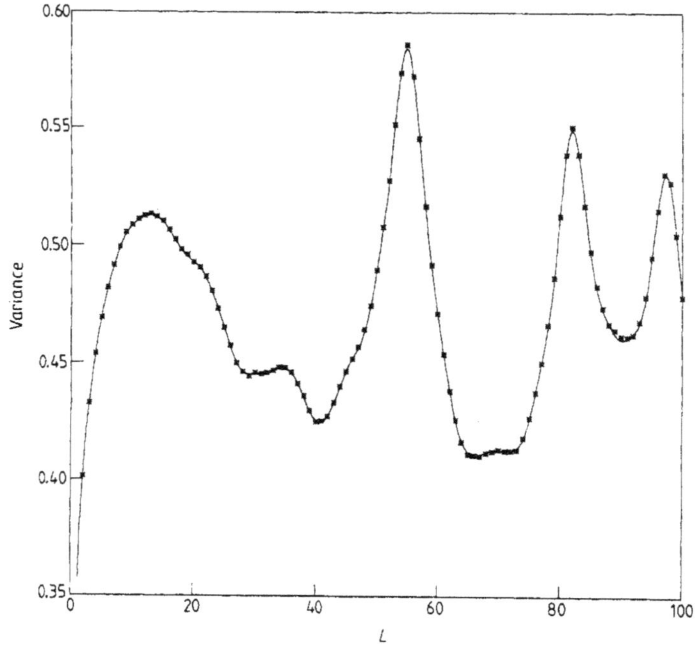

Figure 3. As figure 2 but for $0 \leqslant L \leqslant 100$ and with stars rather than dots for the variance computed from the zeros (GUE not shown).

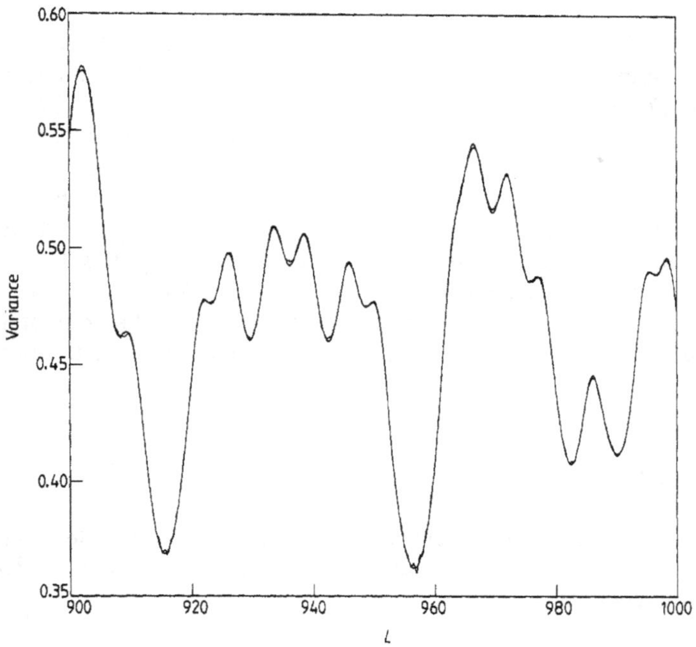

Figure 4. As figure 2 but for $900 \leqslant L \leqslant 1000$, with the 'exact' and semiclassical variances drawn as smooth curves (the 'exact' curve is more jagged) (GUE not shown).

In figure 2 the range is $0 \leqslant L \leqslant 20$. Again 'theory' and 'experiment' agree. The GUE variance ceases to be even a rough approximation when $L \sim 5$, as expected, and the maximum near $L = 13$ signals the onset of asymptotic oscillation.

In figure 3 the range is $0 \leqslant L \leqslant 100$. The asymptotic oscillations are now obvious, and quasirandom as expected with the predicted amplitudes, wavelengths and mean values. (The oscillations come from the double sum in (19), and not from the Ci and Si functions.) Slight discrepancies are visible: the 'theoretical' peaks are noticeably lower than the 'experimental' ones.

In figure 4 the range is $900 \leqslant L \leqslant 1000$. Again the quasirandom oscillations agree very well with semiclassical theory. There are, however, some rapid oscillations with $\Delta L \simeq 1$ which the semiclassical formula cannot reproduce (from (22) this would require $p^r \sim E/2\pi \sim 4 \times 10^{10}$ which is excluded from (19) because it would imply $\tau^* = 1$ in violation of (13)). The rapid oscillations are probably an artefact of averaging, because when $L = 1000$ there are only 100 independent samples.

Over the whole range $0 \leqslant L \leqslant 1000$ the largest difference between the 'exact' and semiclassical variances is 0.003 (over the range $0 \leqslant L \leqslant 100$ it is 0.002). It therefore appears that semiclassical theory gives a uniformly accurate description of the correlations in the heights of the zeros in both the universal and non-universal ranges—at least for this statistic. It is worth remarking that the transition from universal to non-universal spectral statistics has not yet been seen in any honestly quantum Hamiltonian with a chaotic classical limit, because not enough eigenvalues have been computed (the transition has been seen for an integrable system [5]).

To forestall premature optimism it must be explained why zeros near $r = 10^{12}$ might not be in the fully asymptotic regime. The whole analysis was based on pretending that $\zeta(\frac{1}{2} + iE)$ is described for real E by the Euler product over primes, whereas of course this does not converge unless $\mathrm{Im}\, E < -\frac{1}{2}$. This could mean that pairs of zeros with $E_{m+1} - E_m < \frac{1}{2}$ will not be separated by the Euler formula. At height E the number $C(E)$ of zeros thus confused would be $C(E) \approx (\frac{1}{2})\rho(E) = \ln(E/2\pi)/4\pi$. For these computations, $C \sim 2$ which is not large. To reach the fully asymptotic regime of large C it is necessary to go to much greater heights: thus $C = 10$ when $E = 2 \times 10^{55}$ (level number 5×10^{56}) and $C = 100$ when $E = 4 \times 10^{546}$ (level number 8×10^{548}). Note however that C increases in the same way as the limit $L_{\max}$ of GUE universality, so the asymptotic oscillations in $V(L; x)$ should survive as $x \to \infty$. The best hope is that in spite of being based on the Euler product the semiclassical formula (19) nevertheless gives the variance correctly, by being the analytic continuation to real x of some complex-x generalisation of the definition (6). (Titchmarsh's theorem 14.21 in [15] might provide the starting point for such a justification.)

Acknowledgment

It is a pleasure to thank Dr Andrew Odlyzko of the AT and T Bell Laboratories for his kindness in making available his unpublished computations of the Riemann zeros and their number variance.

References

[1] Edwards H M 1974 *Riemann's zeta function* (New York: Academic)
[2] Montgomery H L 1973 The pair correlation of the zeros of the Riemann zeta function *Proc. Symp. Pure Math.* **24** (Providence, RI: AMS) 181–93
[3] Porter C E (ed) 1965 *Statistical theories of spectra: fluctuations* (New York; Academic)
[4] Odlyzko A M 1987 On the distribution of spacings between zeros of the zeta function *Math. Comp.* **48** 273–308
[5] Berry M V 1985 Semiclassical theory of spectral rigidity *Proc. R. Soc.* A **400** 229–51
[6] Berry M V 1986 Riemann's zeta function: a model for quantum chaos? *Quantum chaos and statistical nuclear physics* (*Springer Lecture Notes in Physics No.* 263) eds T H Seligman and H Nishioka (Berlin: Springer) pp 1–17
[7] Berry M V 1987 Quantum Chaology *Proc. R. Soc.* A **413** 183–98
[8] Gutzwiller M C 1971 Periodic orbits and classical quantization conditions *J. Math. Phys.* **12** 343–58
[9] Gutzwiller M C 1978 Path integrals and the relation between classical and quantum mechanics *Path integrals and their applications in quantum, statistical and solid-state physics* ed G J Papadopoulos and J T Devreese (New York: Plenum) pp 163–200
[10] R Balian and C Bloch 1972 Distribution of eigenfrequencies for the wave equation in a finite domain: III. Eigenfrequency density oscillations *Ann. Phys., NY* **69** 76–160
[11] Hannay J H and Ozorio de Almeida A M 1984 Periodic orbits and a correlation function for the semiclassical density of states *J. Phys. A: Math. Gen.* **17** 3429–40
[12] Mehta M L 1967 *Random matrices and the statistical theory of energy levels* (New York: Academic)
[13] Abramowitz M and Stegun I A 1964 *Handbook of mathematical functions* (Washington, DC: US Govt Printing Office)
[14] Selberg A 1946 Contributions to the theory of the Riemann zeta-function *Arch. Math. Naturvid.* **B 48** 89–155
[15] Titchmarsh E C 1951 *The theory of the Riemann zeta-function* (Oxford: Oxford University Press)

SIAM Review
Vol. 41, No. 2, pp. 236–266

The Riemann Zeros and Eigenvalue Asymptotics*

M. V. Berry†
J. P. Keating‡

Abstract. Comparison between formulae for the counting functions of the heights t_n of the Riemann zeros and of semiclassical quantum eigenvalues E_n suggests that the t_n are eigenvalues of an (unknown) hermitean operator H, obtained by quantizing a classical dynamical system with hamiltonian H_{cl}. Many features of H_{cl} are provided by the analogy; for example, the "Riemann dynamics" should be chaotic and have periodic orbits whose periods are multiples of logarithms of prime numbers. Statistics of the t_n have a similar structure to those of the semiclassical E_n; in particular, they display random-matrix universality at short range, and nonuniversal behaviour over longer ranges. Very refined features of the statistics of the t_n can be computed accurately from formulae with quantum analogues. The Riemann-Siegel formula for the zeta function is described in detail. Its interpretation as a relation between long and short periodic orbits gives further insights into the quantum spectral fluctuations. We speculate that the Riemann dynamics is related to the trajectories generated by the classical hamiltonian $H_{\mathrm{cl}} = XP$.

Key words. spectral asymptotics, number theory

AMS subject classifications. 11M26, 11M06, 35P20, 35Q40, 41A60, 81Q10, 81Q50

PII. S0036144598347497

1. Introduction. Our purpose is to report on the development of an analogy, in which three areas of mathematics and physics, usually regarded as separate, are intimately connected. The analogy is tentative and tantalizing, but nevertheless fruitful. The three areas are eigenvalue asymptotics in wave (and particularly quantum) physics, dynamical chaos, and prime number theory. At the heart of the analogy is a speculation concerning the zeros of the Riemann zeta function (an infinite sequence of numbers encoding the primes): the Riemann zeros are related to the eigenvalues (vibration frequencies, or quantum energies) of some wave system, underlying which is a dynamical system whose rays or trajectories are chaotic.

Identification of this dynamical system would lead directly to a proof of the celebrated Riemann hypothesis. We do not know what the system is, but we do know many of its properties, and this knowledge has brought insights in both directions: from mathematics to physics, by stimulating the development of new spectral asymptotics, and from physics to mathematics, by indicating previously unsuspected correlations between the Riemann zeros. We have reviewed some of this material before

*Received by the editors October 16, 1998; accepted for publication (in revised form) November 21, 1998; published electronically April 23, 1999.
http://www.siam.org/journals/sirev/41-2/34749.html

†H. H. Wills Physics Laboratory, Tyndall Avenue, Bristol BS8 1TL, United Kingdom.
‡School of Mathematics, University of Bristol, University Walk, Bristol BS8 1TW, United Kingdom and Basic Research Institute in the Mathematical Sciences, Hewlett-Packard, Laboratories Bristol, Filton Road, Stoke Gifford, Bristol BS12 6QZ, United Kingdom (J.P.Keating@bristol.ac.uk).

[1, 2, 3, 4, 5, 6], but these accounts do not include several recent developments to be described here, especially those in the last part of section 4 and all of sections 5 and 6.

To motivate the approach from physics, we begin with the counting function for the primes, $\pi(x)$, defined as the number of primes less than x (thus $\pi(3.5) = 2$); this is a staircase function, with unit steps at the primes p. The density of primes is the distribution

$$\pi'(x) \equiv \sum_p \delta(x - p). \tag{1.1}$$

At the roughest level of description, and with the distribution appropriately smoothed,

$$\pi'(x) \sim \frac{1}{\log x} \tag{1.2}$$

(as implied by the prime number theorem: $\pi(x) \sim x/\log x$).

One of Riemann's great achievements [7, 8] was to give an exact formula for $\pi'(x)$, constructed as follows. First, $\pi'(x)$ is expressed in terms of a function $J(x)$ [7, Chap. 1] that has jumps at prime powers:

$$\pi'(x) = \frac{1}{x}\sum_{k=1}^{\infty} \frac{\mu_k x^{1/k}}{k^2} J'\left(x^{1/k}\right). \tag{1.3}$$

In this formula, μ_k are the Möbius numbers $(1, -1, -1, 0, -1, 1, \ldots)$ [7]. Each of the partial densities J' is the sum of a smooth part (dominated by (1.2)) and an infinite series of oscillations:

$$J'(x) = \frac{1}{\log x}\left(1 - \frac{1}{x\left(x^2 - 1\right)}\right) - \frac{2}{\sqrt{x}\log x}\sum_{\mathrm{Re}t_n > 0} \frac{\cos\{\mathrm{Re}(t_n)\log x\}}{x^{\mathrm{Im}t_n}} \tag{1.4}$$

(see section 1.18 of [7]). Here the numbers t_n in the oscillatory contributions are related to the complex Riemann zeros, defined as follows.

Riemann's zeta function, depending on the complex variable s, is defined as

$$\zeta(s) \equiv \prod_p \left(1 - p^{-s}\right)^{-1} = \sum_{n=1}^{\infty} n^{-s} \quad (\mathrm{Re}s > 1) \tag{1.5}$$

and by analytic continuation elsewhere in the s plane. It is known that the complex zeros (i.e., those with nonzero imaginary part) of $\zeta(s)$ lie in the "critical strip" $0 < \mathrm{Re}s < 1$, and the Riemann hypothesis states that in fact all these zeros lie on the "critical line" $\mathrm{Re}s = 1/2$ (see Figure 1). The numbers t_n in (1.4) are defined by

$$\zeta\left(\tfrac{1}{2} + it_n\right) = 0 \quad (\mathrm{Re}t_n \neq 0). \tag{1.6}$$

If the Riemann hypothesis is true, all the (infinitely many) t_n are real, and are the heights of the zeros above the real s axis. It is known by computation that the first 1,500,000,001 complex zeros lie on the line [9], as do more than one-third of all of them [10].

Each term in the sum in (1.4) describes an oscillatory contribution to the fluctuations of the density of primes, with larger $\mathrm{Re}t_n$ corresponding to higher frequencies.

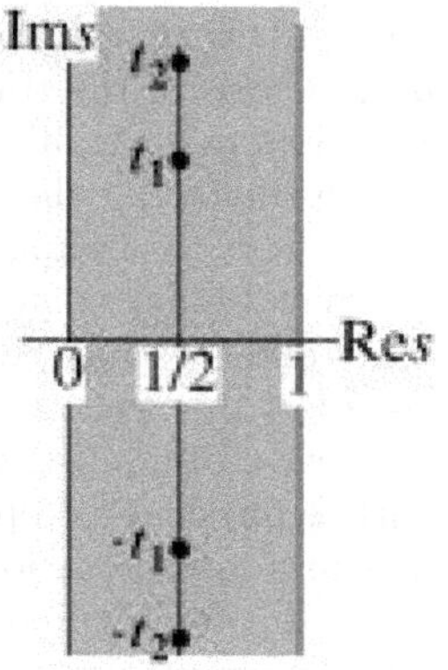

Fig. 1 *Complex s plane, showing the critical strip (shaded) and the complex Riemann zeros (there are trivial zeros at $s = -2, -4, \ldots$).*

Because of the logarithmic dependence, each oscillation gets slower as x increases. This slowing-down can be eliminated by the change of variable $u = \log x$; thus

$$\begin{aligned} f(u) &\equiv \tfrac{1}{2}\exp\left(\tfrac{1}{2}u\right)\left[u\pi'(\exp u) - 1\right] + \tfrac{1}{4} \\ &= -\sum_{\mathrm{Re}t_n > 0} \cos\{\mathrm{Re}(t_n)\,u\}\exp\{-\mathrm{Im}(t_n)\,u\} + O\left(\exp\left(-\tfrac{1}{6}u\right)\right). \end{aligned} \tag{1.7}$$

If the Riemann hypothesis is true, $\mathrm{Im}t_n = 0$ for all n, and the function $f(u)$, constructed from the primes, has a discrete spectrum; that is, the support of its Fourier transform is discrete. If the Riemann hypothesis is false, this is not the case. The frequencies t_n are reminiscent of the decomposition of a musical sound into its constituent harmonics. Therefore there is a sense in which we can give a one-line nontechnical statement of the Riemann hypothesis: "The primes have music in them."

However, readers are cautioned against thinking that it would be easy to hear this prime music by constructing $f(u)$ as defined in (1.7) and then converting it into an audio signal. In order for the human ear to hear the lowest Riemann zero, with $t_1 = 14.13\ldots$, it would be necessary to play $N \approx 100$ periods of $\cos(t_1 u)$, requiring primes in the range $0 < x < \exp(2\pi N/t_1) \approx \exp(45) \approx 10^{19}$.

On this acoustic analogy, the heights t_n (hereinafter referred to simply as "the zeros") are frequencies. This raises the compelling question: frequencies of what? A natural answer would be: frequencies of some vibrating system. Mathematically, such frequencies—real numbers—are discrete eigenvalues of a self-adjoint (hermitean) operator. That the search for such an operator might be a fruitful route to proving the Riemann hypothesis is an old idea, going back at least to Hilbert and Polya [7]; what is new is the physical interpretation of this operator and the detailed information now available about it.

The mathematics of almost all eigenvalue problems encountered in wave physics is essentially the same, but the richest source of such problems is quantum mechanics, where the eigenvalues are the energies of stationary states ("levels"), rather than frequencies as in acoustics or optics, and the operator is the hamiltonian. Reflecting this catholicity of context, we will refer to the t_n interchangeably as energies or frequencies, and the operator as H (Hilbert, Hermite, Hamilton...).

To help readers navigate through this review, here is a brief description of the sections. In section 2 we describe the basis of the Riemann-quantum analogy, which is an identification of the periodic orbits in the conjectured dynamics underlying the Riemann zeros, made by comparing formulae for the counting functions of the t_n and of asymptotic quantum eigenvalues. Section 3 explains the significance of the long periodic orbits in giving rise to universal (that is, system-independent) behaviour in classical and semiclassical mechanics and, by analogy, the Riemann zeros. The application of these ideas to the statistics of the zeros and quantum eigenvalues is taken up in section 4. Section 5 is a description of a powerful method for calculating the t_n (the Riemann-Siegel formula), with a physical interpretation in terms of resurgence of long periodic orbits that implies new interpretations of the periodic-orbit sum for quantum spectra. The properties of the conjectured dynamical system are listed in section 6, where it is speculated that the zeros are eigenvalues of some quantization of the dynamics generated by the hamiltonian $H_{\text{cl}} = XP$.

2. The Analogy. The basis of the analogy is a formal similarity between representations for the fluctuations of the counting functions for the Riemann zeros t_n and for vibration frequencies associated with a system whose rays are chaotic. For the t_n (assumed real), the counting function is defined for $t > 0$ as

$$\mathcal{N}(t) \equiv \sum_{n=1}^{\infty} \Theta(t - t_n), \tag{2.1}$$

where Θ denotes the unit step. Central to our arguments is the fact that $\mathcal{N}(t)$ can be decomposed as follows [11]:

$$\mathcal{N}(t) = \langle \mathcal{N}(t) \rangle + \mathcal{N}_{\text{fl}}(t), \tag{2.2}$$

where

$$\begin{aligned} \langle \mathcal{N}(t) \rangle &\equiv \frac{\theta(t)}{\pi} + 1 = \frac{1}{\pi}\left[\arg\Gamma\left(\frac{1}{4} + \frac{1}{2}it\right) - \frac{1}{2}t\log\pi\right] + 1 \\ &= \frac{t}{2\pi}\log\left(\frac{t}{2\pi e}\right) + \frac{7}{8} + O\left(\frac{1}{t}\right) \end{aligned} \tag{2.3}$$

and

$$\mathcal{N}_{\text{fl}}(t) = \frac{1}{\pi}\lim_{\varepsilon\to 0}\operatorname{Im}\log\zeta\left(\frac{1}{2} + it + \varepsilon\right). \tag{2.4}$$

(The branch of the logarithm is chosen to be continuous, with $\mathcal{N}_{\text{fl}}(0) = 0$.)

These two components can be interpreted as the smooth and fluctuating parts of the counting function. Here and hereinafter the notation $\langle \cdots \rangle$ denotes a local average of a fluctuating quantity, over a range large compared with the length scales of the fluctuations but small compared with any secular variation. Implicit in such averaging is an asymptotic parameter; in the present case this is t, and the averaging range is large compared with the mean spacing of the zeros but small compared with t itself.

The formula for $\langle \mathcal{N} \rangle$ can be obtained from the functional equation for $\zeta(s)$ [7]. It follows by differentiating the last member of (2.3) that the asymptotic density of the zeros is

$$\langle d(t) \rangle = \frac{1}{2\pi}\log\left(\frac{t}{2\pi}\right) + O\left(\frac{1}{t^2}\right) \tag{2.5}$$

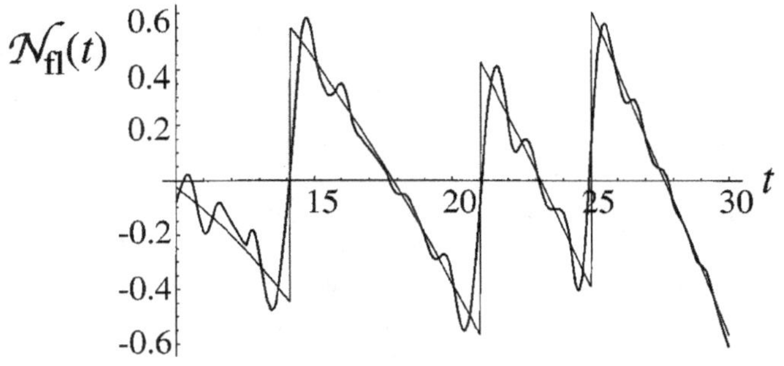

Fig. 2 *Thick line: Divergent series* (2.6) *for the counting function fluctuations* $\mathcal{N}_{\text{fl}}$ *of the Riemann zeros, including all values of* m *and the first* 50 *primes* p. *Thin line: Exact calculation of* $\mathcal{N}_{\text{fl}}$ *from* (2.4).

and therefore that the mean spacing between the zeros decreases logarithmically with increasing t. Underlying the formula for $\mathcal{N}_{\text{fl}}$ are the observations that the phase of a function jumps by π on passing close to a zero, and that $\zeta(s) \to 1$ as $\text{Re}s \to \infty$, so that between the jumps in $\mathcal{N}_{\text{fl}}$ this function varies smoothly, implying that its average value is zero.

Now we substitute into (2.4) the Euler product (1.5), disregarding the fact that this does not converge in the critical strip, and obtain the divergent but formally exact expression

$$\begin{aligned}\mathcal{N}_{\text{fl}}(t) &= -\frac{1}{\pi}\text{Im}\sum_p \log\left\{1 - \frac{\exp(-it\log p)}{\sqrt{p}}\right\} \\ &= -\frac{1}{\pi}\sum_p \sum_{m=1}^{\infty} \frac{\exp\left(-\frac{1}{2}m\log p\right)}{m} \sin\{tm\log p\}.\end{aligned} \tag{2.6}$$

This formula gives the fluctuations as a series of oscillatory contributions, each labelled by a prime p and an integer m, corresponding to the prime power p^m. Terms with $m > 1$ are exponentially smaller than those with $m = 1$. The oscillation corresponding to p has a "wavelength" (that is, t-period)

$$\tau_p = \frac{2\pi}{\log p}. \tag{2.7}$$

In order to discriminate individual zeros, sufficiently many terms must be included in the sum for this wavelength to be less than the mean spacing; from (2.5), this gives $p < t/2\pi$. When truncated in this way, the sum (2.6) can reproduce the jumps quite accurately for low-lying zeros, as Figure 2 shows, even though the complete sum diverges.

Consider now a classical dynamical system [12] in a configuration space with D freedoms, coordinates $\mathbf{q} = \{q_1, \ldots, q_D\}$, and momenta $\mathbf{p} = \{p_1, \ldots, p_D\}$. Trajectories are generated by a hamiltonian function $H(\mathbf{q}, \mathbf{p})$ on the two-dimensional phase space $\{\mathbf{q}, \mathbf{p}\}$, whose conserved value is the energy E. In quantum physics, $\mathbf{q}$ and $\mathbf{p}$ are operators, with commutation relation $[\mathbf{q}, \mathbf{p}] = i\hbar$, where $\hbar \equiv h/2\pi$ is Planck's constant. Then $H(\mathbf{q}, \mathbf{p})$, augmented by boundary conditions, becomes a hermitean wave operator, whose eigenvalues, discrete if the system is bound, are the quantum energy levels E_n. More generally, this formalism applies to any wave system (e.g., water waves [13]) with coordinates $\mathbf{q}$ and wavenumber $\mathbf{k}$, defined by a dispersion relation $\omega(\mathbf{q}, \mathbf{k})$, the connection between the quantum and wave formalisms being

$$\mathbf{p} = \hbar\mathbf{k}, \quad H(\mathbf{q}, \mathbf{p}) = \hbar\omega(\mathbf{q}, \mathbf{p}/\hbar). \tag{2.8}$$

Familiar wave equations appear when the commutation relations are implemented with $\mathbf{k} = -i\nabla$, and Hamilton's equations are the corresponding ray equations (in optics these are the rays generated by Snell's law or Fermat's principle). For example, a locally uniform medium (H independent of $\mathbf{q}$) with impenetrable walls corresponds to "quantum billiards," where waves are governed by the Helmholtz equation with Dirichlet boundary conditions, and the (straight) rays are reflected specularly at the walls [14]. Of special interest to us is the asymptotics of the eigenvalues E_n in the semiclassical limit $\hbar \to 0$, which from (2.8) is equivalent to the short-wavelength or high-frequency limit.

Waves, in particular the eigenfunctions of H, usually depend not on individual trajectories but on families of trajectories, whose global structure is an important determinant of the energy-level asymptotics. Of interest here is the case where the trajectories are chaotic [15, 16, 17], that is, where E is the only globally conserved quantity and neighbouring trajectories diverge exponentially. Then on a given energy shell (that is, for given E), the usual structure—and the one we will consider here—is that all initial conditions generate trajectories that explore the $(2D-1)$-dimensional energy surface ergodically, except for a set, dense but of zero measure, of (one-dimensional) isolated unstable periodic orbits.

An important result of modern mathematical physics, central to the Riemann-quantum analogy, is that these isolated periodic trajectories determine the fluctuations in the counting function $\mathcal{N}(E)$ of the energy levels [18, 19, 20, 21]. Using the notation (2.2), with E replacing t, we can separate $\mathcal{N}(E)$ into its smooth and fluctuating parts $\langle\mathcal{N}(E)\rangle$ and $\mathcal{N}_{\mathrm{fl}}(E)$. The averaging is over an energy interval large compared with the mean level spacing but classically small, that is, vanishing with $\hbar$. We state the formula for $\mathcal{N}_{\mathrm{fl}}(E)$ and then explain it:

$$N_{\mathrm{fl}}(E) \sim \frac{1}{\pi}\sum_p \sum_{m=1}^{\infty} \frac{\sin\left\{mS_p(E)/\hbar - \frac{1}{2}\pi m\mu_p\right\}}{m\sqrt{\left|\det\left(\mathsf{M}_p^m - I\right)\right|}}. \tag{2.9}$$

The symbol $\sim$ indicates that the formula applies asymptotically, that is, for small $\hbar$. (In the special case of the Selberg trace formula [21], corresponding to waves on a compact surface of constant negative curvature, the formula is exact.) The index p labels primitive periodic orbits, that is, orbits traversed once. The index m labels their repetitions. Therefore, the two sums together include all periodic orbits. $S_p(E)$ is the action of the primitive orbit p, that is,

$$S_p(E) = \oint_p \mathbf{p}\cdot d\mathbf{q}. \tag{2.10}$$

In terms of S_p, the period of the orbit is

$$T_p = \frac{\partial S_p}{\partial E}. \tag{2.11}$$

The hyperbolic symplectic matrix M_p (the monodromy matrix) describes the exponential growth of deviations from p of nearby (linearized) trajectories, between successive crossings of a Poincaré surface of section transverse to p. μ_p is the Maslov phase, determined [22] by the winding round p of the stable and unstable manifolds containing the orbit.

Physically, the appearance of periodic orbits is not surprising. The levels E_n, counted by $\mathcal{N}$, are associated with stationary states, that is, states or modes that are time-independent. By the correspondence principle, their asymptotics should depend on phase space structures unchanged by evolution along rays, that is, the invariant manifolds with energy E. In the type of chaotic dynamics we are considering, there are two types of invariant manifold: the whole energy surface, which determines $\langle \mathcal{N}(E) \rangle$ as we will see, and, decorating this, the tracery of periodic orbits, which determines the finer details of the spectrum as embodied in the fluctuations $\mathcal{N}_{\text{fl}}(E)$.

For long orbits, the determinant is dominated by its expanding eigenvalues, and, for large T_p,

$$\det\left(\mathsf{M}_p^m - 1\right) \sim \exp\left(m\lambda_p T_p\right), \tag{2.12}$$

where λ_p is the Liapunov (instability) exponent of the orbit p. Thus, approximately,

$$\mathcal{N}_{\text{fl}}(E) \sim \frac{1}{\pi} \sum_p \sum_{m=1}^{\infty} \frac{\exp\left(-\frac{1}{2} m\lambda_p T_p\right)}{m} \sin\left\{ m S_p(E)/\hbar - \frac{1}{2}\pi m \mu_p \right\}. \tag{2.13}$$

Now we can make the formal analogy with the corresponding formula (2.6) for the counting function fluctuations of the Riemann zeros:

(2.14)

	Quantum	Riemann
Dimensionless actions	$\frac{mS_p}{\hbar}$	$mt \log p$
Periods	mT_p	$m \log p$
Stabilities	$\frac{1}{2}\lambda_p T_p$	$\frac{1}{2}\log p \Rightarrow \lambda_p = 1$
Asymptotics	$\hbar \to 0$	$t \to \infty$

.

The nonappearance of $\hbar$ on the "Riemann" side indicates that the dynamical system underlying the zeros is scaling, in the sense that the trajectories are the same for all "energies" t, as in the most familiar scaling system, namely, quantum billiards, where, for a particle of mass m, energy scales according to the combination $k = \sqrt{(2mE)}/\hbar$, and, for an orbit of length L_p, $S_p/\hbar = kL_p$. With the analogy, primes acquire a new significance, as primitive periodic orbits, whose periods are log$p$. The index m in (2.6) then labels their repetitions.

The fact that all orbits have the same instability exponent (unity) indicates that the Riemann dynamics is homogeneously unstable, that is, uniformly chaotic. Moreover, the dynamics does not possess time-reversal symmetry. If it did, degeneracy of actions between each orbit and its time-reversed partner would lead to their contributing coherently to $\mathcal{N}(t)$, so that for most orbits (those that are not self-retracing) the prefactor in (2.6) would be $2/\pi$ rather than $1/\pi$.

An alternative form of the periodic-orbit sum (2.9), which will be useful later, is in terms of the level density

$$d(E) = \frac{d\,\mathcal{N}(E)}{dE}. \tag{2.15}$$

Denoting primitive and repeated periodic orbits by the common index j $(= \{p, m\})$, we can write

$$d_{\mathrm{fl}}(E) = \frac{1}{\pi\hbar}\sum_j A_j \cos\{S_j(E)/\hbar\}, \tag{2.16}$$

where for convenience we have absorbed the Maslov indices into the actions, and the amplitude A_j is

$$A_j \sim \frac{T_j}{m\sqrt{\det|(\mathsf{M}_j - \mathsf{I})|}} \tag{2.17}$$

as $\hbar \to 0$. For the Riemann zeros, the corresponding formula, from (2.6) and (2.14), has $\mu_j = 0$,

$$A_j = -\frac{\log p}{p^{m/2}} = -\frac{T_j}{m}\exp\left\{-\tfrac{1}{2}T_j\right\}, \tag{2.18}$$

and is an identity rather than an asymptotic approximation.

There are two discordant features of the analogy [1], to which we will return. First, the exponential decay of long orbits in the quantum formula (2.13) is an approximation to the determinant in (2.9), whereas for the Riemann zeros the exponential in (2.6) is exact. Second, the negative sign in (2.6) indicates that when the Maslov phases $\pi m\mu_p/2$ are reinstated in (2.13) their value should be π for all orbits, but this is hard to understand because if the index is π for a given orbit it should be 2π for the same orbit traversed twice.

The smooth part $\langle\mathcal{N}(E)\rangle$ of the counting function is, to leading order in $\hbar$, the number of phase space quantum cells (volume h^D) in the volume $\Omega(E)$ of the energy surface $H = E$; thus $\langle\mathcal{N}(E)\rangle \approx \Omega(E)/h^D$. For billiards, Ω is proportional to the spatial volume confining the system (this is Weyl's asymptotics [23]). The mean level density is thus

$$\langle d(E)\rangle \sim \frac{\Omega'(E)}{h^D}. \tag{2.19}$$

In the quantum formula (2.13), each orbit contributes an oscillation to $\mathcal{N}_{\mathrm{fl}}(E)$, with energy "wavelength" (cf. (2.7))

$$\varepsilon_p = \frac{h}{T_p(E)}. \tag{2.20}$$

This should be compared with the mean spacing of the eigenvalues, which is the reciprocal of the mean level density and so (from (2.19)) of order $\hbar^D$. An important implication is that the oscillation contributed by a given orbit has, asymptotically, a wavelength much larger than the mean level spacing. Thus in order to have a chance of resolving individual levels it is necessary to include at least all those orbits with periods up to

$$T_H(E) = 2\pi\hbar\langle d\rangle = O\left(\frac{1}{\hbar^{D-1}}\right). \tag{2.21}$$

This evokes the time-energy uncertainty relation, so T_H is called the Heisenberg time. Asymptotically, T_H corresponds to very long orbits, or, in the Riemann case, large primes $p_H(t) = t/2\pi$ (cf. the discussion following (2.7)). In what follows, this emphasis on long orbits will play a key role.

3. Long Orbits and Universality. In a classically chaotic system, the periodic orbits proliferate exponentially as their period increases [24], with density

$$
\begin{aligned}
\rho(T) &\equiv \frac{\text{number of orbits with periods between } T \text{ and } T+dT}{dT} \\
&\sim \frac{\exp(\lambda T)}{T} \quad \text{as } T \to \infty.
\end{aligned}
\tag{3.1}
$$

Here, λ is the topological entropy of the system. In the cases we are interested in, λ can be identified with a suitable average of the instability exponents of long periodic orbits (cf. (2.12)). In the Riemann case, where according to (2.14) the periodic orbits correspond to primes, (3.1) nicely reproduces the prime number theorem (1.2) and thereby reinforces the analogy (the repetitions, labelled by m, give exponentially smaller corrections).

From (2.18), the proliferation in (3.1) cancels the decay of the intensities A_j^2 for long orbits. One way to write this is

$$
\lim_{T\to\infty} \frac{1}{T} \sum_j A_j^2 \delta(T - T_j) = 1. \tag{3.2}
$$

This is the sum rule of Hannay and Ozorio de Almeida [25]. Its importance is threefold: first, it does not contain $\hbar$ and so is a *classical* sum rule. Second, the amplitudes A_j nevertheless have significance in *quantum* (i.e., wave) asymptotics, because they give the strengths of the contributions to spectral density fluctuations. Third, the rule is universal: (3.2) contains no specific feature of the dynamics—it holds for all systems that are ergodic. One way to appreciate the naturalness of this universality is to imagine that a long orbit with energy E, inscribed on the constant-energy surface $H = E$, forms an intricate tracery that, with the slightest smoothing, could cover the surface uniformly with respect to the microcanonical (Liouville) measure. This "phase-space democracy" is the basis of Hannay and Ozorio de Almeida's derivation.

Expressed mathematically, this ergodicity-related sum rule corresponds to an eigenvalue $\nu_0 = 1$ (associated with the invariant measure) of the Perron–Frobenius operator that generates the classical flow in phase space. Equivalently [26, 27], it corresponds to a simple pole at $s = 0$ of the dynamical zeta function $\zeta_D(s)$, defined (for two-dimensional systems, for example) by

$$
\frac{1}{\zeta_D(s)} \equiv \prod_p \prod_{m=0}^{\infty} \left(1 - \frac{\exp(sT_p)}{|\Lambda_p|\Lambda_p^m}\right)^{m+1}, \tag{3.3}
$$

where Λ_p is the larger eigenvalue ($|\Lambda_p| > 1$) of the monodromy matrix M. The rest of the spectrum of the Perron–Frobenius operator, or equivalently the analytic structure of $\zeta_D(s)$ away from $s = 0$, determines the rate of approach to ergodicity—that is, it is related to the system-specific short-time dynamics.

Now recall that according to (2.21) the long orbits determine spectral fluctuations on the scale of the mean level separation. The universality of the classical sum rule suggests that the spectral fluctuations should also show universality on this scale. And by the Riemann-quantum analogy, we expect this spectral universality to extend to the Riemann zeros t_n.

It is in the *statistics* of the levels and Riemann zeros that the universality appears. This is to be expected, since ergodicity is a statistical property of long orbits.

It is important to note that we are here considering individual systems and not ensembles, so statistics cannot be defined in the usual way, as ensemble averages. Instead, we rely on the presence of an asymptotic parameter (see the remarks after (2.4), and before (2.9)): high in the spectrum (or for large t in the Riemann case), there are many levels (or zeros) in a range where there is no secular variation, and it is this large number that enables averages to be performed. Universality then emerges in the limit $\hbar \to 0$ (or $t \to \infty$) for correlations between fixed numbers of levels or zeros.

A mathematical theory of universal spectral fluctuations already exists in the more conventional context where statistics are defined by averaging over an ensemble. This is *random-matrix theory* [28, 29, 30, 31, 32], where the correlations between matrix eigenvalues are calculated by averaging over ensembles of matrices whose elements are randomly distributed, in the limit where the dimension of the matrices tends to infinity. Here the relevant ensemble is that of complex hermitean matrices: the "Gaussian unitary ensemble" (GUE). As will be discussed in the next section, it is precisely these statistics that apply to high eigenvalues of individual chaotic systems without time-reversal symmetry, and also to high Riemann zeros, in the sense that the spectral or Riemann-zero averages described in the previous paragraph coincide with GUE averages.

First, however, we give a very simple argument [33] showing that the approach to universality must be nonuniform. The classical sum rule (3.2) applies to long orbits but not to short ones, because these will reflect the specific dynamics of the system whose spectrum is being considered. Therefore, spectral features that depend on short orbits can be expected to be nonuniversal. From (2.20), these are fluctuations on the energy scale $\varepsilon_0 = h/T_0$, where T_0 is the period of the shortest orbit. This scale is asymptotically small but still large compared with the separation of order h^D between neighbouring eigenvalues. On this basis, we expect universality to be a good approximation for correlations between eigenvalues separated by up to $O(1/h^{(D-1)})$ mean spacings, but not for larger separations. For the Riemann zeros, $T_0 = \log 2$ (equation (2.14)), whereas the mean separation between zeros is $2\pi/\log(t/2\pi)$. Therefore universality for zeros near t should break down beyond about $\log(t/2\pi)/\log 2$ mean spacings. We regard the observation of the breakdown of random-matrix universality for the Riemann zeros [34], in accordance with this prediction, as giving powerful support to the analogy with quantum or wave eigenvalues.

4. Periodic-Orbit Theory for Spectral Statistics. In discussing statistics, it will be simplest to measure intervals between eigenvalues or Riemann zeros in units of the local mean spacing. We denote such intervals by x, and the corresponding levels or zeros, referred to a local origin, by x_n; in these units, $\langle d(x)\rangle = 1$. We will mainly be concerned with statistics that are bilinear in the level density, the simplest being the *pair correlation* of the density fluctuations, defined in [31], in the sense of a distribution, as

$$\begin{aligned} R(x;y) &\equiv \text{probability density of separations } x \text{ of levels or zeros} \\ &\quad\ \text{close to a scaled position } y \\ &= \frac{1}{N}\sum_{m\neq n} \delta\left(x_m - x_n - x\right) \\ &= \left\langle d\left(y-\tfrac{1}{2}x\right) d\left(y+\tfrac{1}{2}x\right)\right\rangle - \delta\left(x\right) \\ &= 1 + \left\langle d_{\mathrm{fl}}\left(y-\tfrac{1}{2}x\right) d_{\mathrm{fl}}\left(y+\tfrac{1}{2}x\right)\right\rangle - \delta\left(x\right). \end{aligned} \tag{4.1}$$

(In the second member, the sum is over a stretch of N levels near y, with $N >> 1$.) R gives the correlation between levels near E, or, correspondingly, Riemann zeros near

t; for simplicity of notation, we will henceforth not indicate these base levels (denoted y in (4.1)).

Closely related to R is the *form factor* $K(\tau)$ (the name comes from crystallography), defined as

$$\begin{aligned} K(\tau) &= 1 + \int_{-\infty}^{\infty} dx \exp\{2\pi i x \tau\}(R(x) - 1) \\ &= -\delta(\tau) + \frac{1}{N}\sum_m \sum_n \exp\{2\pi i \tau (x_m - x_n)\}, \end{aligned} \tag{4.2}$$

where the sum is as in (4.1). Here the variable τ (conjugate to x) is the scaled time

$$\tau = \frac{T}{T_H}, \tag{4.3}$$

where T_H is the Heisenberg time (equation (2.21)). With the definitions given, both R and K tend to 1 at long range; the term $\delta(x)$ in R ensures that this requirement is compatible with (4.2).

Other statistics that are bilinear in d can be expressed in terms of K or R. A useful one is the *number variance*:

$$\begin{aligned} &\Sigma^2(x) \equiv \text{variance of number of levels or zeros in} \\ &\text{an interval where the mean number is } x \\ &= \left\langle \left[\mathcal{N}\left(y + \tfrac{1}{2}x\right) - \mathcal{N}\left(y - \tfrac{1}{2}x\right) - x\right]^2 \right\rangle \\ &= \frac{2}{\pi^2}\int_0^{\infty} d\tau \frac{K(\tau)}{\tau^2}\sin^2(\pi x \tau) \\ &= x + 2\int_0^x dy\,(x - y)\,[R(y) - 1]. \end{aligned} \tag{4.4}$$

The correlation function (4.1) is determined by the spectral density fluctuations, for which there is the semiclassical formula (2.16). Our aim in this section is to explain how to employ this observation to calculate these bilinear statistics, obtaining not only the universal random-matrix limit but also the corrections to this corresponding to large eigenvalue or zero separations, or short times. The argument is subtle and has several levels of refinement, of which we start with the simplest [3, 5, 33].

We will calculate $K(\tau)$. The first step is to substitute (2.16) into (4.1), thereby obtaining a double sum over periodic orbits. Since all the actions are positive, we can simplify the averages (over a small interval of eigenvalues or along the critical line) using

$$\langle \cos\{S_j/\hbar\}\cos\{S_k/\hbar\}\rangle = \tfrac{1}{2}\langle \cos\{(S_j - S_k)/\hbar\}\rangle. \tag{4.5}$$

The dimensionless intervals x that we will be considering may be large but must correspond to classically small energy ranges, so we can approximate the actions using

$$S_j\left(E_0 \pm \frac{x}{2\langle d\rangle}\right) = S_j \pm \frac{xT_j}{2\langle d\rangle} + O\left(\frac{x^2}{\langle d\rangle^2}\right), \tag{4.6}$$

where S_j, T_j, and d are evaluated at E_0. Elementary manipulations, and evaluating the integral in (4.2), give the asymptotic (that is, small-$\hbar$) form factor as the double sum

$$(4.7)\ K(\tau) = \frac{1}{4(\pi\langle d\rangle\hbar)^2}\left\langle\sum_j\sum_k A_jA_k\cos\{(S_j-S_k)/\hbar\}\,\delta\left(|\tau|-\frac{T_j+T_k}{4\pi\langle d\rangle\hbar}\right)\right\rangle.$$

It is convenient now to consider separately the diagonal part K_{diag} of the sum (terms with $j=k$) and the off-diagonal part K_{off} (terms with $j\neq k$). For K_{diag}, we have

$$K_{\text{diag}}(\tau) = \frac{1}{4(\pi\langle d\rangle\hbar)^2}\sum_j A_j^2\delta\left(|\tau|-\frac{T_j}{2\pi\langle d\rangle\hbar}\right). \tag{4.8}$$

In the limit $\hbar\to 0$, τ fixed, the sum over orbits can be evaluated using the Hannay-Ozorio sum rule (3.2), giving

$$\lim_{\hbar\to 0} K_{\text{diag}}(\tau) = |\tau|. \tag{4.9}$$

This is universal: all details of the specific dynamics have disappeared. Because of the Riemann-quantum analogy, the same behaviour should hold for the pair correlation of the Riemann zeros. Here we make contact with the seminal work of Montgomery [35], who indeed proved (4.9) in that case.

Now we observe that in random-matrix theory the exact form factor of the GUE is

$$K_{\text{GUE}}(\tau) = |\tau|\,\Theta(1-|\tau|)+\Theta(|\tau|-1). \tag{4.10}$$

(Θ is the unit step.) For later reference, the GUE pair distribution function, obtained from (4.2), is

$$R_{\text{GUE}}(x) = 1-\left(\frac{\sin(\pi x)}{\pi x}\right)^2. \tag{4.11}$$

Evidently the approximation (4.9), based on periodic orbits, captures exactly the random-matrix behaviour for $|\tau|<1$, without invoking any random matrices. This led Montgomery [35] to conjecture (following a suggestion of Dyson and independently of any semiclassical argument) that for the Riemann zeros $K(\tau)=K_{\text{GUE}}(\tau)$ in the limit $t\to\infty$.

Clearly, (4.9) does not give the random-matrix result when $|\tau|>1$. Indeed it fails drastically by not satisfying the requirement, necessary for any form factor representing a discrete set of points (eigenvalues or zeros), that $K(\tau)\to 1$ as $\tau\to\infty$. This failure reflects the importance of K_{off}, and implies that for large τ (long orbits) the off-diagonal terms in the double sum (4.7) cannot vanish through incoherence, as might naively be thought, but must conspire by destructive coherent interference to cancel the term τ from K_{diag} and replace it by 1. This is consistent with the Montgomery conjecture, which implies

$$K_{\text{off}}(\tau) = \Theta(|\tau|-1)(1-|\tau|) \tag{4.12}$$

in the limit $t\to\infty$.

One reason why K_{diag} alone is inadequate is the proliferation of orbits: for sufficiently long times, there will be many pairs of orbits whose actions differ by less than $\hbar$, so that they cannot be regarded as incoherent in (4.7). This phenomenon, that in some appropriate sense the large-τ limit of the double sum must be 1, is the *semiclassical sum rule*. Originally [33] the rule was obtained by a different argument, and was mysterious. Now there is a better understanding of the mechanism by which the cancellation occurs [36, 37]; we will discuss it later.

Indeed, for the Riemann zeros, (4.12) can be derived [4] using a conjecture of Hardy and Littlewood [38] concerning the pair distribution of the prime numbers. These correlations are important because if the logarithms of the primes (primitive orbit periods) were pairwise uncorrelated, K_{off}, being the average of a sum of random phases, would be zero. The Hardy-Littlewood conjecture is that $\pi_2(k;X)$, defined as the number of primes $p \leq X$ such that $p+k$ is also a prime, has the following asymptotic form for large X:

$$\pi_2(x) \sim \frac{X}{\log^2 X} C(k) \tag{4.13}$$

with

$$C(k) = \begin{cases} 0 & \text{if } k \text{ is odd} \\ 2\prod\limits_{q>2}\left(1-\dfrac{1}{(q-1)^2}\right)\prod\limits_{\substack{p>2\\ p|k}}\left(\dfrac{p-1}{p-2}\right) & \text{if k is even,} \end{cases} \tag{4.14}$$

where the q-product includes all odd primes, and the p-product includes all odd prime divisors of k. Pairwise randomness would correspond to $C(k)=1$. It can be demonstrated [4] that as $K\to\infty$

$$\sum_{k=1}^{K} C(k) \sim K - \tfrac{1}{2}\log K \tag{4.15}$$

and so on average

$$C(k) \sim 1 - \frac{1}{2|k|} \tag{4.16}$$

for large k. This in turn was shown to imply (4.12) in the limit $t\to\infty$ [4].

We have seen that K_{diag} is universal in the limit $\hbar\to 0$; that is, it is independent of the specific features of the dynamics. These reappear—in a dramatically nonuniform way—in the approach to the limit. To see this, note first that it is only for short orbits, that is, when $\tau << 1$, that universality breaks down. Next, choose a τ^* corresponding to a time much longer than the shortest period T_0 and shorter than the Heisenberg time T_H, that is,

$$\frac{T_0}{2\pi\langle d\rangle\hbar} << \tau^* < 1. \tag{4.17}$$

We continue to use the Hannay-Ozorio sum rule for $\tau > \tau^*$, the limit (4.12) for K_{off} ensuring the correct GUE formula (4.10) for $\tau > 1$, but take the contributions from orbits with period $T_j < 2\pi\langle d\rangle\hbar\tau^*$ directly from (4.8). Thus

$$\begin{aligned} K(\tau) &\approx K_{\text{GUE}}(\tau) \\ &+ \frac{1}{4(\pi\langle d\rangle\hbar)^2}\sum_{T_j<2\pi\langle d\rangle\hbar\tau^*} A_j^2\delta\left(|\tau| - \frac{T_j}{2\pi\langle d\rangle\hbar}\right) - |\tau|\,\Theta(\tau^* - |\tau|) \end{aligned} \tag{4.18}$$

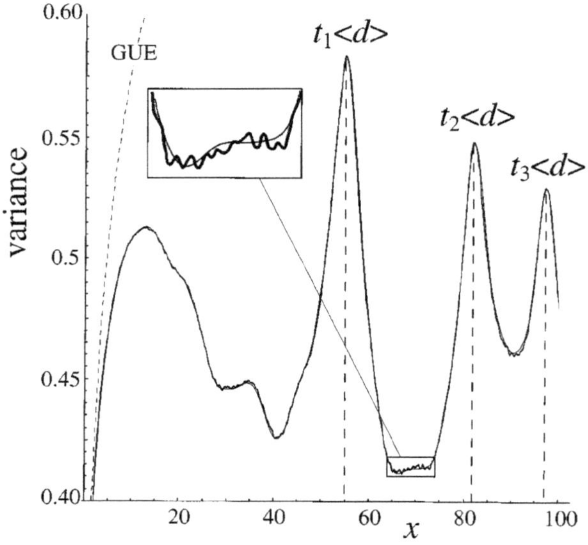

Fig. 3 *Number variance $\sum^2(x)$ (4.4) of the Riemann zeros t_n near $n = 10^{12}$, calculated from (4.18) (with $\tau^* = 1/4$), (2.14), and (2.18) (thin line), compared with $\sum^2(x)$ computed from numerically calculated zeros by Odlyzko [39, 40] (thick line); all the zeros are close to $t = 2.677 \times 10^{11}$, and their smoothed density is $\langle d \rangle = 3.895\ldots$. Note the resurgence resonances (cf. (4.23)) associated with the lowest zeros t_1, t_2, and t_3, and that the theory fails to capture small, fast oscillations in the data.*

is a candidate for a semiclassical formula for the form factor. Later we will see that this is not quite correct: the proper incorporation of the off-diagonal terms in the double sum introduces a small but important modification near $\tau = 1$. For the moment, we continue to discuss (4.18).

This formula for $K(\tau)$, applied to the Riemann zeros, is extremely accurate. When employed in conjunction with (4.4) to calculate the number variance of the zeros [34], it reproduces almost perfectly this statistic as computed from numerical values of high zeros [39, 40]. Figure 3 shows that the agreement extends from the random-matrix regime (small x) to the far nonuniversal regime. Note however the tiny oscillatory deviations; we will return to these later.

For the pair correlation, we have

$$R(x) = R_{\text{GUE}}(x) + R_c(x). \tag{4.19}$$

Remarkably, it is possible to calculate the correction R_c explicitly and in closed form at this level of approximation. The formula was obtained for both the Riemann zeros and for general systems in [41], and independently in [42] for the Riemann zeros. From (4.18), (4.2), and (2.14), we get

$$\begin{aligned} R_c(x) \approx R_c^1(x) = \frac{1}{2(\pi\langle d\rangle)^2} \sum_{\substack{m,p \\ p^m < \exp(2\pi\langle d\rangle\tau^*)}} \frac{\log^2 p}{p^m} \cos\left\{\frac{xm\log p}{\langle d\rangle}\right\} \\ -2\int_0^{\tau^*} d\tau \cos\{2\pi x\tau\}, \end{aligned} \tag{4.20}$$

where $\langle d\rangle$ is given by (2.15). The sum is insensitive to the value of τ^* provided this is not too small, so we set $\tau^* = \infty$. Next, we write

$$\log^2 p = \frac{\log^2(p^m)}{m} + (1-m)\log^2 p. \tag{4.21}$$

This corresponds to separating R_c^1 into contributions from primitive orbits (first term) and repetitions (second term). In the repetitions, the sum over m can be evaluated explicitly. For the first term, we use [7]

$$J'(w) = \sum_{p,m}\frac{\delta(w-p^m)}{m} = \frac{1}{2\pi i}\int_{a-i\infty}^{a+i\infty} ds w^{s-1}\log\zeta(s) \quad (a>1). \tag{4.22}$$

Some tricky but elementary manipulations now give

$$\begin{aligned} R_c^1(x) = \frac{1}{2(\pi\langle d\rangle)^2}\Bigg[\frac{1}{\xi^2} - \partial_\xi^2 \mathrm{Re}\log\zeta(1-i\xi) \\ -\mathrm{Re}\sum_{p=2}^{\infty}\frac{\log^2 p}{(p\exp\{i\xi\log p\}-1)^2}\Bigg], \end{aligned} \tag{4.23}$$

where

$$\xi \equiv \frac{x}{\langle d\rangle}. \tag{4.24}$$

This formula has a very interesting structure, worth discussing in detail. First, $\xi \to 0$ in the limit $t\to\infty$ for any fixed x, and so the pole in the zeta function cancels the singularity $1/\xi^2$. Second, the prefactor $1/\langle d\rangle^2$ ensures that the correction R_c^1 is asymptotically small in comparison with R_{GUE} (equation (4.11)). Third, the dependence on ξ shows that R_c^1 involves the separation between zeros in the original variable $t = \mathrm{Im}s$ (heights of zeros along the critical line), rather than the scaled separation x; this means that structural features of R_c^1 appear asymptotically at larger x than the oscillations in R_{GUE}, as expected for nonuniversal features of correlations. Fourth, the contributions from repetitions (the sum over p in (4.23)) are less significant than those from primitive orbits (first two terms), as Figure 4 shows. Fifth, and most important, the appearance of $\zeta(1-i\xi)$ indicates an astonishing resurgence property of the zeros: in the pair correlation of high Riemann zeros, the low Riemann zeros appear as resonances. This is illustrated in Figure 5. The resonances also appear as peaks in the nonuniversal part of the number variance (Figure 3).

For generic dynamical systems without time-reversal symmetry, it can be verified directly that the analogue of (4.23) is [41, 43]

$$\begin{aligned} R_c^1(x) &= \frac{1}{2(\pi\hbar\langle d\rangle)^2} \\ &\times\left[\frac{1}{\xi^2} - \partial_\xi^2\mathrm{Re}\log\zeta_D(i\xi) - \mathrm{Re}\sum_p\sum_{m=0}^{\infty}\frac{(m+1)T_p^2}{\left(|\Lambda_p|\Lambda_p^m\exp\{-i\xi T_p\}-1\right)^2}\right], \end{aligned} \tag{4.25}$$

where ζ_{D} is the dynamical zeta function defined in (3.3), and now $\xi = x/\hbar\langle d\rangle$. Again, the pole in the zeta function (now at $s=0$) cancels the singularity $1/\xi^2$. In this case, the resonances discussed above are caused by singularities of $\log\zeta_{\mathrm{D}}(s)$ away from $s=0$, that is, by subdominant eigenvalues of the Perron–Frobenius operator.

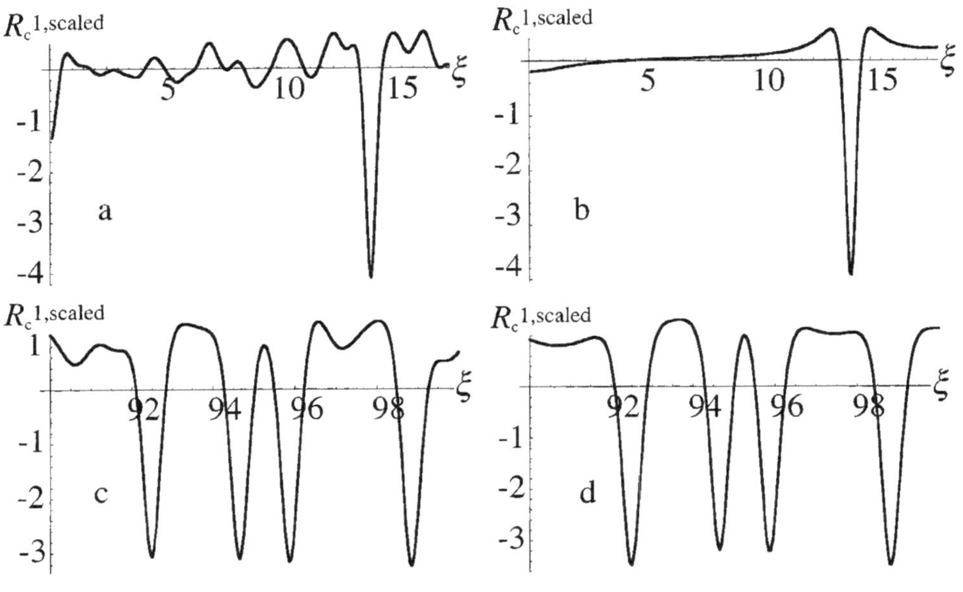

Fig. 4 *Nonuniversal correction to the pair correlation of the Riemann zeros, calculated from* (4.23) *as* $R_c^{1\ \text{scaled}}(\xi) \equiv 2(\pi\langle d\rangle)^2 R_c(x)$. *Parts* (a) *and* (c) *include repetitions;* (b) *and* (d) *omit repetitions.*

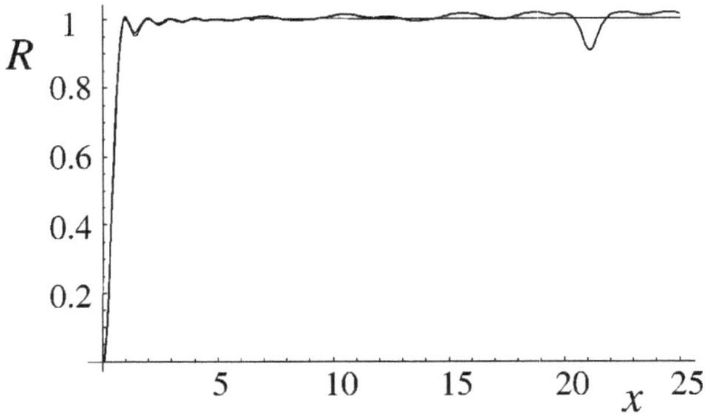

Fig. 5 *Pair correlation* $R(x)$ *of the Riemann zeros, calculated "semiclassically" (thick line) from* (4.19) *and* (4.23), *for zeros near* $n = 10^5$, *and random-matrix behaviour* $R_{GUE}(x)$ *(thin line); note the first nonuniversal resurgence resonance near* $x = 21$.

Now we return to the tiny oscillatory deviations noticeable in Figure 3, reflecting small errors in (4.19) and (4.23). These are again associated with the approach to the $t \to \infty$ limit of the form factor, rather than the limit itself: whereas (4.23) captures the appropriate large-t asymptotics of K_{diag}, the GUE-motivated replacement (4.12) incorporates only the $t \to \infty$ limit of K_{off}.

For the Riemann zeta function, this can be corrected as follows. We have already noted above that the formula (4.12) for K_{off} can be derived using the smoothed expression (4.16) for the Hardy-Littlewood conjecture. The large-t asymptotics we

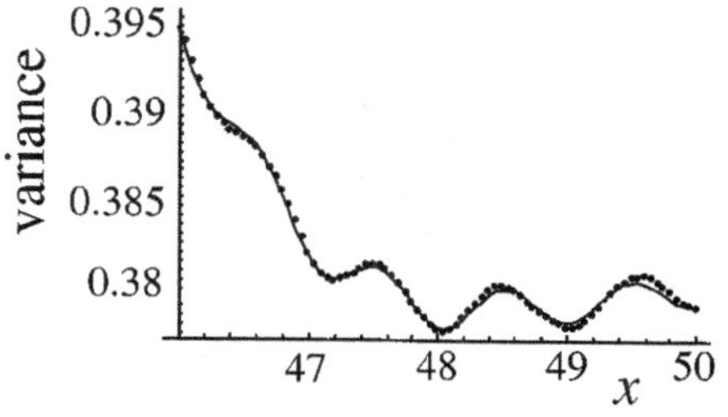

Fig. 6 *Number variance $\sum^2(x)$ (4.4) of the Riemann zeros t_n near $n = 10^9$, calculated from (4.19) and (4.26), including the off-diagonal correction (4.27)–(4.28) [70] (full line), compared with $\sum^2(x)$ computed from numerically calculated zeros by Odlyzko [39, 40] (dots); all the zeros are close to $t = 3.719 \times 10^8$, and their smoothed density is $\langle d \rangle = 2.848\ldots$.*

seek comes from using the original unsmoothed form (4.14) [5, 41]. The result is that

$$R_c(x) = R_c^1(x) + R_c^2(x), \tag{4.26}$$

in which R_c^1 is given by (4.23), and

$$R_c^2(x) \approx \frac{1}{2(\pi\langle d\rangle)^2}\left[-\frac{\cos(2\pi x)}{\xi^2} + |\zeta(1+i\xi)|^2 \operatorname{Re}\{\exp(2\pi i x)\, b(\xi)\}\right], \tag{4.27}$$

where

$$b(\xi) = \prod_p \left(1 - \frac{(p^{i\xi}-1)^2}{(p-1)^2}\right) \tag{4.28}$$

is a convergent product over the primes. As with the diagonal term (cf. the discussion after (4.24)), convergence as $\xi \to 0$ is ensured by the pole of the zeta function.

This second correction, although small, does incorporate the small oscillations, through the trigonometric functions with argument $2\pi x$. Asymptotically (that is, as $t \to \infty$), these oscillations are fast (cf. (4.23)–(4.24)) in comparison with the variations from the resonances of the zeta function. When employed in conjunction with (4.4), the correction accurately reproduces the oscillatory deviation (Figure 3) in the number variance of the zeros; this is illustrated in Figure 6.

Unfortunately, this derivation of (4.27) for the Riemann zeros cannot be imitated for general chaotic dynamical systems because we have no a priori knowledge of the correlations between the actions of different periodic orbits, analogous to the Hardy-Littlewood conjecture for the primes. It is possible to get some information by working backwards and, assuming that the GUE expression (4.10) or (4.11) describes the pair correlation of eigenvalues in generic chaotic systems without time-reversal symmetry, deriving the universal limiting form of the implied action correlations [37]. (This procedure essentially follows an analogous derivation for the primes themselves, assuming the Montgomery conjecture [44]). An interesting feature of this approach is that it leads to predictions about *classical* trajectories based on the distribution of *quantum* energy levels. However, it gives no information about the deviations from random-matrix universality that are the focus of our concern here.

Recently a theory has been developed that overcomes these difficulties [5, 41]. It is based on two observations.

First, as already noted above, quantum eigenvalues (or Riemann zeros) are resolved by the trace formula if the sum (2.9) over periodic orbits is truncated near the Heisenberg time T_H (this will be made more precise in the next section). Hence, if the trace formula thus truncated generates the approximation

$$\tilde{\mathcal{N}}(E) = \langle \mathcal{N}(E) \rangle + \tilde{\mathcal{N}}_{\mathrm{fl}}(E) \tag{4.29}$$

to the counting function, the quantities $\tilde{E}_n$ defined by

$$\tilde{\mathcal{N}}\left(\tilde{E}_n\right) = n + \tfrac{1}{2} \tag{4.30}$$

should be good semiclassical approximations to the exact eigenvalues. The theory is based on calculating the correlations in this approximate spectrum.

Second, the diagonal terms $K_{\mathrm{diag}}(\tau)$ are asymptotically dominant in the form factor for $\tau < 1$, corresponding to times less than T_H. This implies that orbits with periods less than T_H make contributions that are effectively uncorrelated; treating them in this way allows the correlations in the $\tilde{E}_n$ spectrum to be computed exactly.

For chaotic systems without time-reversal symmetry, the result is that when $x >> 1$ the deviations from the GUE formula can also be represented in the form (4.26), (4.25), and (4.27), where $\zeta(1 + i\xi)$ is replaced by $\zeta_{\mathrm{D}}(i\xi)$ (defined by (3.3)) and, in (4.28), $b(\xi)$ is replaced by

$$\begin{aligned} b(\xi) = \frac{1}{\gamma^2} \prod_p {}_2\phi_1 \{&\exp(-iT_p\xi), \exp(-iT_p\xi); \\ &\Lambda_p^{-1}; \Lambda_p^{-1}, \left|\Lambda_p^{-1}\right| \exp(iT_p\xi)\} \frac{\left|\zeta_D^{(p)}(0)\right|^2}{\left|\zeta_D^{(p)}(i\xi)\right|^2} \end{aligned} \tag{4.31}$$

(again $\xi = x/\hbar\langle d \rangle$). Here γ is the residue of the pole at $s = 0$ of $\zeta_D(s)$, ${}_2\phi_1$ is the q-hypergeometric function [45], and ζ_D^p is the pth element of the product over primitive orbits in (3.3).

The formal similarity between the results for the Riemann zeros and for the semiclassical eigenvalues is striking, and reinforced by the fact that the derivation of (4.31) just outlined leads precisely to (4.28) when applied to the zeros. Indeed, by Fourier-transforming (4.26) with respect to t, this can be regarded as a heuristic derivation of the Hardy-Littlewood conjecture. In the same way, Fourier-transforming the corresponding result for dynamical systems with respect to $1/\hbar$ leads to a classical periodic orbit correlation function corresponding directly to the Hardy-Littlewood conjecture and reducing to the universal form conjectured in [37] in the long-time limit. It is a challenge to derive these correlations within classical mechanics.

We finish this section on connections between statistics of the Riemann zeros and quantum eigenvalues by remarking that the results for pair correlations extend to correlations of higher order. Thus Montgomery's conjecture for the two-point correlation of the Riemann zeros generalizes to all n-point correlations. Specifically, the irreducible n-point correlation function

$$\tilde{R}_n(x_1, x_2, \ldots, x_n) \equiv \frac{1}{\langle d \rangle^n} \left\langle \prod_{i=1}^n d_{fl}\left(t + \frac{x_i}{\langle d \rangle}\right) \right\rangle \tag{4.32}$$

tends asymptotically to the corresponding GUE expression:

$$\lim_{t\to\infty} \tilde{R}_n\left(x_1, x_2 \ldots, x_n\right) = \det \mathbf{S}, \tag{4.33}$$

where the elements S_{ij} of the $n \times n$ matrix $\mathbf{S}$ are given by

$$S_{ij} = s\left(x_i, x_j\right) = \frac{\sin\left\{\pi\left(x_i - x_j\right)\right\}}{\pi\left(x_i - x_j\right)}\left(1 - \delta_{ij}\right). \tag{4.34}$$

The analogue of Montgomery's theorem for the diagonal contributions to $\tilde{R}_n$ was proved for $n = 3$ [46] and then for all $n \geq 2$ [47]. The off-diagonal contributions were calculated using a generalization of the Hardy-Littlewood conjecture for $n = 3$ and $n = 4$ [48] and then for all $n \geq 2$ [49]. In all cases the results confirm the conjecture (4.33) and (4.34). The nonuniversal deviations from the GUE formulae (4.33)–(4.34) were calculated for $n = 3$ and $n = 4$ [41] using the method outlined above, and take a form (related to the structure of $\zeta(s)$ as $s \to 1$) directly analogous to that already discussed. As expected, this extends to the higher order correlations of quantum eigenvalues.

5. Riemann-Siegel Formulae. A powerful stimulus to the development of analogies between quantum eigenvalues and the Riemann zeros has been the Riemann-Siegel formula for $\zeta(s)$. As explained in [7], this very effective way of computing the zeros (especially high ones)—employed in most numerical computations nowadays—was discovered by Siegel in the 1920s among papers left by Riemann after his death 60 years earlier. We present the formula in an elementary way, chosen to facilitate our subsequent exploration of its intricate interplay with quantum mechanics. Riemann's derivation [11, 50] was different, and a remarkable achievement, because although it was one of the first applications of his method of steepest descent for integrals it was more sophisticated than most applications today, in that the saddle about which the integrand is expanded is accompanied by an infinite string of poles.

It is a consequence of the functional equation satisfied by $\zeta(s)$ [11] that the following function $Z(t)$ is even, and real for real t:

$$Z\left(t\right) \equiv \exp\left\{i\theta\left(t\right)\right\}\zeta\left(\tfrac{1}{2} + it\right). \tag{5.1}$$

Here $\theta(t)$ is the function appearing in the smoothed counting function (2.3) for the zeros. Naive substitution of the Dirichlet series (1.5) gives the formal expression

$$Z\left(t\right) = \exp\left\{i\theta\left(t\right)\right\}\sum_{n=1}^{\infty}\frac{\exp\left\{-it\log n\right\}}{n^{1/2}}. \tag{5.2}$$

This is doubly unsatisfactory. First, it does not converge—a defect shared with its relative (2.6) for $\mathcal{N}_{\text{fl}}(t)$ (cf. (2.4)) and similarly originating in the inadmissibility of (1.5) in the critical strip. Second, it is not manifestly real as $Z(t)$ must be.

Both defects can be eliminated by truncating the series (5.2) at a finite $n = n^*(t)$ and resumming the tail. The truncation $n^*(t)$ is chosen to be the term whose phase $\theta(t) - t\log n$ is stationary with respect to t; the asymptotic formula for θ (last member of (2.3)) gives

$$n^*\left(t\right) = \text{Int}\left(\sqrt{\frac{t}{2\pi}}\right). \tag{5.3}$$

A crude resummation [1] using the Poisson summation formula leads to a result equivalent to the "approximate functional equation" [11]:

$$Z(t) = 2\sum_{n=1}^{n^*(t)} \frac{\cos\{\theta(t) - t\log n\}}{n^{1/2}} + \cdots. \tag{5.4}$$

This is a remarkable example of resurgence: the resummed terms in the tail $n > n^*(t)$ are the complex conjugates of the early terms $1 \le n \le n^*(t)$, so that the series in (5.4)—called the "main sum" of the Riemann-Siegel expansion—is real, like the exact $Z(t)$. The zeros generated by the first term alone ($n = 1$), that is, $\cos\theta(t) = 0$, have the correct mean density (cf. (2.3)). Higher terms shift the zeros closer to their true positions, and introduce the random-matrix fluctuations. It is worth mentioning that the zeros obtained by including successive terms in (5.4) cannot be regarded as the eigenvalues of hermitean operators that approximate the still-unknown Riemann operator, because these partial sums of the main sum each have zeros for complex t [6].

Unfortunately, the truncation (5.3) introduces another defect: the sum is a discontinuous function of t, unlike $Z(t)$, which is analytic. The discontinuities can be eliminated by formally expanding the difference between (5.2) and the sum in (5.4) about the truncation limit $N(t)$, to obtain the correction terms in (5.4). This will depend on the fractional part of $\sqrt{(t/2\pi)}$ as well as its integer part n^*, so it is convenient to define

$$a(t) \equiv \sqrt{\frac{t}{2\pi}} \equiv n^*(t) + \frac{1}{2}(1 - z(t)). \tag{5.5}$$

The expansion is in powers of $1/a$ (henceforth we do not write the t-dependences explicitly), and gives

$$Z(t) = 2\sum_{n=1}^{n^*} \frac{\cos\{\theta(t) - t\log n\}}{n^{1/2}} + \frac{(-1)^{n^*+1}}{a^{1/2}} \sum_{r=0}^{\infty} \frac{C_r(z)}{a^r}. \tag{5.6}$$

This procedure was devised in [4], where it was used to calculate the first correction term $C_0(z)$, and elaborated in [51] in a study of the higher corrections.

The sum over r is the Riemann-Siegel expansion. Its terms $C_r(z)$ are constructed from derivatives (up to the $3r$th) of

$$C_0(z) = \frac{\cos\left\{\frac{1}{2}\pi\left(z^2 + \frac{3}{4}\right)\right\}}{\cos\{\pi z\}} \tag{5.7}$$

with coefficients determined by an explicit recurrence relation involving the coefficients (Bernoulli numbers) in the Stirling expansion of $\theta(t)$ for large t. The next few coefficients are

$$\begin{aligned}
C_1(z) &= \frac{C_0^{(3)}(z)}{12\pi^2},\\
C_2(z) &= \frac{C_0^{(2)}(z)}{16\pi^2} + \frac{C_0^{(6)}(z)}{288\pi^4},\\
C_3(z) &= \frac{C_0^{(1)}(z)}{32\pi^2} + \frac{C_0^{(5)}(z)}{120\pi^4} + \frac{C_0^{(9)}(z)}{10368\pi^6},\\
C_4(z) &= \frac{C_0(z)}{128\pi^2} + \frac{19C_0^{(4)}(z)}{1536\pi^4} + \frac{11C_0^{(8)}(z)}{23040\pi^6} + \frac{C_0^{(12)}(z)}{497664\pi^8}
\end{aligned} \tag{5.8}$$

(superscripts in brackets denote derivatives). Gabcke [50] calculated $C_r(z)$ for $r \le 12$. Later terms get very complicated; for example,

$$
\begin{aligned}
C_{20}(z) = {} & \frac{332727711 C_0(z)}{274877906944\,\pi^{10}} + \frac{117753804989 C_0^{(4)}(z)}{3298534883328\,\pi^{12}} \\
& + \frac{13899745416281 C_0^{(8)}(z)}{692692325498880\pi^{14}} + \frac{311274631265011 C_0^{(12)}(z)}{16458369653853388\,\pi^{16}} \\
& + \frac{24311037030485304 17 C_0^{(16)}(z)}{449313491550197514 24000\pi^{18}} \\
& + \frac{232544268738862214941 C_0^{(20)}(z)}{37318694855326404968448 0000\pi^{20}} \\
& + \frac{361888761444289010497 C_0^{(24)}(z)}{1064899933783461120599654 40000\pi^{22}} \\
& + \frac{6654063104532271592 3177 C_0^{(28)}(z)}{6843046974492521160973379174400000\pi^{24}} \\
& + \frac{391261681973226653 C_0^{(32)}(z)}{25057539453517190072893440000000\pi^{26}} \\
& + \frac{1259995823308801 C_0^{(36)}(z)}{85717193786692288213155840000000\pi^{28}} \\
& + \frac{713214794639 C_0^{(40)}(z)}{85717193786692288213155840000000\pi^{30}} \\
& + \frac{50407933481 C_0^{(44)}(z)}{17650884544555675988853050572800000\pi^{32}} \\
& + \frac{1039499 C_0^{(48)}(z)}{17683632011243163328349995008000000\pi^{34}} \\
& + \frac{22411 C_0^{(52)}(z)}{32139197378979392841852100018176 0000\pi^{36}} \\
& + \frac{59 C_0^{(56)}(z)}{13636202316509828105757248150568960000\pi^{38}} \\
& + \frac{C_0^{(60)}(z)}{93271623844927224243379577349891686400 00\pi^{40}} .
\end{aligned} \tag{5.9}
$$

An elaborate asymptotic analysis [51] shows that the high orders ("asymptotics of the asymptotics") can be represented compactly as a "decorated factorial series" whose terms are

$$
C_r(z) = \frac{\Gamma\left(\frac{1}{2}r\right)}{\left(\pi\sqrt{2}\right)^{r+1}} f(r, z), \tag{5.10}
$$

where for large r

$$
\begin{aligned}
f(r, z) \sim {} & \sum_{m=0}^{\infty} (-1)^{m(m-1)/2} \exp\left\{-\left(m + \frac{1}{2}\right)^2\right\} \\
& \times \left\{ \begin{array}{ll} \sin\left\{(2m+1)\sqrt{r}\right\} \cos\left\{\left(m+\frac{1}{2}\right)\pi z\right\} & (r \text{ even}) \\ \cos\left\{(2m+1)\sqrt{r}\right\} \sin\left\{\left(m+\frac{1}{2}\right)\pi z\right\} & (r \text{ odd}) \end{array} \right\}.
\end{aligned} \tag{5.11}
$$

Comparison with numerically computed $C_r(z)$ (up to $r = 50$, using special techniques to evaluate the derivatives of $C_0(z)$) shows that these formulae capture the fine details of the Riemann-Siegel coefficients, even for small r.

The factorial in (5.10) means that the sum over r in (5.6) is divergent in the manner familiar in asymptotics: the terms get smaller and then diverge. Asymptotics folkore suggests, and Borel summation (implemented analytically and checked numerically) confirms, that optimal accuracy obtainable from the Riemann-Siegel formula (without further resummation) corresponds to truncating the sum at the least term. This has

$$
r^* = \mathrm{Int}\,(2\pi t) \tag{5.12}
$$

and the resulting error is of order

$$Z(t) - 2\sum_{n=1}^{n^*} \frac{\cos\{\theta(t) - t\log n\}}{n^{1/2}} - \frac{(-1)^{n^*+1}}{a^{1/2}} \sum_{r=0}^{r^*} \frac{C_r(z)}{a^r} = O(\exp\{-\pi t\}). \tag{5.13}$$

The accuracy is very high: even for the lowest Riemann zero, $r^*(t_1) = 89$ and $\exp\{-\pi t_1\} \sim 10^{-20}$. Nevertheless, it is possible to do better, as we shall see later.

Now we turn to the quantum analogues of the Riemann-Siegel formula for classically chaotic systems with $D > 1$, as envisaged in [1], explored in detail in [52], and derived in [53]. These studies are motivated by the hope that such an effective method of computing Riemann zeros might lead to a useful way to calculate quantum eigenvalues.

First, the counterpart of $Z(t)$ in (5.1) is a function with zeros at the quantum energy levels E_n; this is the quantum spectral determinant

$$\begin{aligned} \Delta(E) &= \prod_n A(E, E_n)(E - E_n) = \det\{A(E,H)(E-H)\} \\ &= \det A \exp\{\operatorname{trlog}(E-H)\}, \end{aligned} \tag{5.14}$$

where H is the hermitean wave operator (section 2) and the real factor A is introduced to make the product converge. Hermiticity implies that Δ is real for real E; this "quantum functional equation" is analogous to the functional equation for $\zeta(s)$, which implies that $Z(t)$ is real for real t.

To find the counterpart of the Dirichlet series (5.2), we note that the quantum eigenvalue counting function can be written (cf. (2.4)) as

$$\mathcal{N}(E) = -\frac{1}{\pi} \lim_{\varepsilon \to 0} \operatorname{Im}\operatorname{Trlog}\{1 - (E + i\varepsilon)/H\}. \tag{5.15}$$

Now the decomposition into smooth and fluctuating parts, together with the periodic-orbit formula (2.9), leads to

$$\Delta(E) \sim B(E)\exp\{-i\pi\langle\mathcal{N}(E)\rangle\} \prod_p \exp\left\{-\sum_{m=1}^{\infty} \frac{\exp\{imS_p(E)/\hbar\}}{m\sqrt{\left|\det\left(\mathsf{M}_p^m - \mathsf{I}\right)\right|}}\right\}, \tag{5.16}$$

where $B(E)$ is real and nonzero for real E and where we have absorbed the Maslov indices into S.

Expanding the product over primitive orbits p and the exponential of the sum over repetitions m, we obtain a series of terms that can be labelled by

$$n = \{0, 1, 2 \ldots\} \Leftrightarrow \{m_p\} = \{m_1, m_2 \ldots\}. \tag{5.17}$$

Here m_p represents the number of repetitions of the orbit p. Each term corresponds to a sum over actions:

$$S_n(E) = \sum_p m_p S_p(E). \tag{5.18}$$

The expansions lead to

$$\Delta(E) \sim B(E)\exp\{-i\pi\langle\mathcal{N}(E)\rangle\} \sum_{n=0}^{\infty} D_n(E)\exp\{iS_n(E)/\hbar\} \tag{5.19}$$

with an explicit form for the coefficients D_n that we do not give here [52]. As (5.18) indicates, the terms n correspond to composite orbits, or pseudo-orbits, consisting of combinations of repetitions of different periodic orbits. We label the composite orbits so that increasing n corresponds to increasing period

$$\mathcal{T}_n(E) = \frac{\partial \mathcal{S}_n(E)}{\partial E} \tag{5.20}$$

with $n = 0$ representing no orbit at all, that is, $m_p = 0$ (for which the coefficient $D_0 = 1$).

The sum (5.19) is the counterpart of the Dirichlet series (5.2) for $Z(t)$, with composite orbits n related to primitive orbits p in the same way that the integers n are related to the primes p (cf. (1.5)). Moreover (5.19) diverges, like the sum (2.9) from which it was obtained, and it is not manifestly real as the exact $\Delta(E)$ must be. Our interpretation of the Riemann-Siegel formula suggests a similar resummation of the tail of the series (5.19) after truncation at the term whose phase is stationary with respect to E. This term—the counterpart of $n^*(t)$ in (5.3)—represents the composite orbit defined by

$$\frac{d}{dE}\left[\mathcal{S}_n(E)/\hbar - \pi\langle\mathcal{N}(E)\rangle\right] = 0. \tag{5.21}$$

The corresponding period $\mathcal{T}^*(E)$ is

$$\mathcal{T}^*(E) = \pi\hbar\langle d(E)\rangle = \tfrac{1}{2}T_H(E), \tag{5.22}$$

where $T_H(E)$ is the Heisenberg time (2.20).

Comparison with the Riemann-Siegel main sum in (5.4) suggests that the sum of the composite orbits with $\mathcal{T}_n > \mathcal{T}^*$ is, approximately, the complex conjugate of the sum of the orbits with $\mathcal{T}_n < \mathcal{T}^*$. In fact, this relation can be derived using arguments based on analytic continuation with respect to E [53]. These arguments also indicate a more detailed correspondence: between the sums of groups of terms with periods $\mathcal{T}^* + X$ and $\mathcal{T}^* - X$. The resulting "Riemann-Siegel lookalike" formula is

$$\Delta(E) \sim 2B(E) \sum_{\mathcal{T}_n < \mathcal{T}^*(E)} D_n(E)\cos\left\{\mathcal{S}_n(E)/\hbar - \pi\langle\mathcal{N}(E)\rangle\right\} + \cdots. \tag{5.23}$$

(For a different derivation, see [54].)

With (5.23) it is possible to reproduce some low-lying quantum eigenvalues, and of course the fact that the sum is finite is a major advantage over the infinite divergent series (2.9) and (5.19). However, for a chaotic system with $D > 1$ the number of terms with $\mathcal{T}_n < \mathcal{T}^*$ is exponentially large in $1/\hbar$, so the Riemann-Siegel lookalike is not as useful for calculating high quantum eigenvalues as (5.4) is for calculating Riemann zeros. The origin of the difference is the exponential proliferation of periodic orbits (and composite orbits), together with the fact that $\langle d\rangle$ increases as $1/\hbar^D$, whereas for the Riemann zeros, whose classical counterpart appears to be quasi-one-dimensional, $\langle d\rangle$ increases as $\log t$. Moreover, (5.23) is discontinuous at the energies of composite orbits with period $\mathcal{T}^*$.

No way has yet been found to implement the obvious suggestion of cancelling the discontinuities in the quantum formula (5.23) by a series of corrections analogous to the terms involving $C_r(z)$ in the Riemann-Siegel expansion (5.6). However, a different completion of the Riemann-Siegel main sum was discovered ([55], generalizing an idea in [4]), that does have a quantum analogue.

In this alternative approach to the resummed Dirichlet series, the abrupt truncation is replaced by a smoothed cutoff involving the complementary error function and an optimization parameter K. An argument involving analytic continuation in t leads to

$$
\begin{aligned}
Z(t) = 2\mathrm{Re}\sum_{n=1}^{\infty}\Bigg[&\frac{\exp\{i[\theta(t)-t\log n]\}}{n^{1/2}}\\
&\times\frac{1}{2}\mathrm{Erfc}\left\{(\log n-\theta'(t))\sqrt{\frac{t}{2(K^2-i\theta''(t))}}\right\}\Bigg]+\cdots
\end{aligned}
\tag{5.24}
$$

with an explicit expression for the correction terms. With K chosen appropriately, this smoothed sum can reproduce $Z(t)$ to an accuracy equivalent to that of the Riemann-Siegel main sum together with several correction terms. The corrections in (5.24) form an explicit asymptotic series enabling $Z(t)$ to be calculated with an accuracy of order $\exp(-t^2)$; this improvement over the Riemann-Siegel $\exp(-\pi t)$ is possible because (5.24) involves the higher transcendental function Erfc, whereas the Riemann-Siegel expansion involves only elementary functions. Several related representations of $Z(t)$ are now known [56, 57, 58].

The improved representation (5.24), together with the explicit correction terms, can readily be adapted to the quantum spectral determinant. The smoothed version of the Riemann-Siegel lookalike (5.23) is obtained by an argument involving analytic continuation with respect to $1/\hbar$, leading to

$$
\begin{aligned}
\Delta(E) = 2B(E)\,\mathrm{Re}\sum_{n=0}^{\infty}\Bigg[&D_n(E)\exp\{i[\pi\langle\mathcal{N}(E)\rangle-S_n(E)/\hbar]\}\\
&\times\frac{1}{2}\mathrm{Erfc}\left\{\frac{S_n(E)-\pi\langle\mathcal{N}_1(E)\rangle}{2(K^2\hbar-i\pi\langle\mathcal{N}_2(E)\rangle)}\right\}\Bigg]+\cdots,
\end{aligned}
\tag{5.25}
$$

where $\mathcal{N}_i$ denotes the ith derivative of $\mathcal{N}$ with respect to $1/\hbar$. A numerical test of this formula for the hyperbola billiard (a classically chaotic system with $D=2$) shows that it can reproduce quantum eigenvalues with high accuracy, even resolving near-degenerate pairs of levels [59].

Finally, we note an important clue to the Riemann dynamics, hidden in the asymptotics (5.10), (5.11) of the Riemann-Siegel expansion (5.6). It concerns the implied small exponential $\exp\{-\pi t\}$ (cf. the error in (5.13)). The same exponential appears in the asymptotics of the gamma functions in $\theta(t)$ (equation (2.3)). Quantum mechanics suggests this is the "phase factor" corresponding to a periodic orbit with imaginary action (an "instanton" in physics jargon). If we write

$$\exp\{iS\}=\exp\{-\pi t\} \tag{5.26}$$

(remembering $\hbar=1$ for the Riemann zeros), the implied period is

$$T=\frac{\partial S}{\partial'\mathrm{energy}'}=\frac{\partial S}{\partial t}=i\pi. \tag{5.27}$$

So, it seems that as well as the real periodic orbits in (2.14), with periods $m\log p$, there are complex periodic orbits, with periods that are multiples of $i\pi$.

6. Spectral Speculations. Although we do not know the conjectured Riemann operator H whose eigenvalues (all real) are the heights t_n of the Riemann zeros, the analogies presented so far suggest a great deal about it. To summarize:

a. H has a classical counterpart (the "Riemann dynamics"), corresponding to a hamiltonian flow, or a symplectic transformation, in a phase space.

b. The Riemann dynamics is chaotic, that is, unstable and bounded.

c. The Riemann dynamics does not have time-reversal symmetry. In addition, we note the recent discovery [60, 61] of modified statistics of the low zeros for the ensemble of Dirichlet L-functions, associated with a symplectic structure.

d. The Riemann dynamics is homogeneously unstable.

e. The classical periodic orbits of the Riemann dynamics have periods that are independent of "energy" t, and given by multiples of logarithms of prime numbers. In terms of symbolic dynamics, the Riemann dynamics is peculiar, and resembles Chinese: each primitive orbit is labelled by its own symbol (the prime p) in contrast to the usual situation where periodic orbits can be represented as words made of letters in a finite alphabet.

f. The Maslov phases associated with the orbits are also peculiar: they are all π. The result appears paradoxical in view of the relation between these phases and the winding numbers of the stable and unstable manifolds associated with periodic orbits [22], but finds an explanation in a scheme of Connes [62].

g. The Riemann dynamics possesses complex periodic orbits (instantons) whose periods are multiples of $i\pi$.

h. For the Riemann operator, leading-order semiclassical mechanics is exact: as in the case of the Selberg trace formula [21], $\zeta(1/2 + it)$ is a product over classical periodic orbits, without corrections.

i. The Riemann dynamics is quasi-one-dimensional. There are two indications of this. First, the number of zeros less than t increases as $t \log t$; for a D-dimensional scaling system, with energy parameter $\alpha(E)$ proportional to $1/\hbar$, the number of energy levels increases as $\alpha(E)^D$. Second, the presence of the factor $p^{-m/2}$ in the counting function fluctuation formula (2.6), rather than the determinant in the more general Gutzwiller formula (2.9), suggests that there is a single expanding direction and no contracting direction.

j. The functional equation for $\zeta(s)$ resembles the corresponding relation—a consequence of hermiticity—for the quantum spectral determinant.

We have speculated [6] that the conjectured Riemann operator H might be some quantization of the following extraordinarily simple classical hamiltonian function $H_{\text{cl}}(X, P)$ of a single coordinate X and its conjugate momentum P:

$$H_{\text{cl}}(X, P) = XP. \tag{6.1}$$

Now we outline the reasons for this tentative association of XP with $\zeta(s)$.

At the *classical* level, (6.1) has a hyperbolic point at the origin in the infinite-phase (X, P) plane, and generates the following equations of motion and trajectories:

$$\dot{X} = X, \quad \text{i.e.,} \quad X(t) = X(0)\exp(t); \quad \dot{P} = -P, \quad \text{i.e.,} \quad P(t) = P(0)\exp(-t). \tag{6.2}$$

Thus classical evolution is uniformly unstable, with stretching in X and contraction in P. Furthermore, the motion has the desired lack of time-reversal symmetry: velocity cannot be reversed ($\dot{X}$ is tied to X in (6.2)) and so the orbit cannot be retraced.

At the *semiclassical* level, we can try to estimate the smoothed counting function $\langle \mathcal{N}(E) \rangle$ of energy levels E_n generated by the quantum version of (6.1). For this it is necessary to specify a value of Planck's constant $\hbar$. We choose $\hbar = 1$; other choices simply rescale the energies. $\langle \mathcal{N}(E) \rangle$ is the area $\mathcal{A}$ under the constant-energy hyperbola $E = XP$, measured in units of the "Planck cell" area $2\pi\hbar = 2\pi$, with a

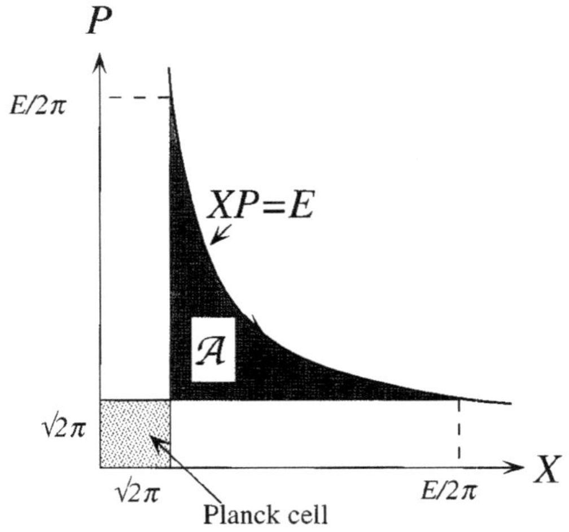

Fig. 7 *Phase space for $H_{cl} = XP$, with cutoffs for semiclassical regularization.*

Maslov index correction given by $\alpha/4\pi$, where α is the angle turned through along the orbit in phase space (this correction gives the "1/2" in the quantization of the harmonic oscillator). We encounter the immediate difficulty that $\mathcal{A}$ is infinite: motion generated by $H = XP$ is unbounded, and so does not give discrete quantum energies. As will be clear later, closing the phase space to make the motion bounded is a central unsolved problem. In the interim, a simple (perhaps the simplest) expedient is to regularize by truncating in X and P as indicated in Figure 7. The result (unaltered by representing the Planck cell by a rectangle instead of a square) is that $\langle \mathcal{N}(E) \rangle$ is precisely the asymptotics of the smoothed counting function for the Riemann zeros (last member of (2.3)), including the term 7/8, with t replaced by the energy E.

At the *quantum* level, the simplest formally hermitean operator corresponding to (6.1) is

$$H = \frac{1}{2}(XP + PX) = -i\left(X\frac{d}{dX} + \frac{1}{2}\right). \tag{6.3}$$

The formal eigenfunctions, satisfying

$$H\psi_E(X) = E\psi_E(X) \tag{6.4}$$

are

$$\psi_E(X) = \frac{A}{X^{1/2-iE}}. \tag{6.5}$$

We note the appearance of the power X^{-s} appearing in the Dirichlet series for $\zeta(s)$ (as integer^{-s}) and the Euler product (as prime^{-s}), with the symmetrization (6.3) placing s on the critical line.

It is evident that XP is simply a canonically rotated version of the inverted harmonic oscillator $P^2 - X^2$, which in turn is a complexified version of the usual harmonic oscillator $P^2 + X^2$. Some of these connections have been noted before [63, 64, 65, 66, 67]. The first-order operator XP is the simplest representative of this class, with the monomials (6.5) avoiding the complications of the parabolic cylinder eigenfunctions of $P^2 - X^2$.

To evaluate the corresponding momentum eigenfunction $\phi_E(P)$ (Fourier transform of (6.5)), it is necessary to specify a continuation across $X = 0$. The simplest choice, for a reason to be given later, is to make the wavefunction even in X, that is, to replace X by $|X|$. Then

$$
\begin{aligned}
\phi_E(P) &\equiv \frac{1}{\sqrt{2\pi}} \int_{-\infty}^{\infty} dX \psi_E(X) \exp(-iPX) \\
&= \frac{A}{|P|^{1/2+iE}} 2^{iE} \frac{\Gamma\left(\frac{1}{4}+\frac{1}{2}iE\right)}{\Gamma\left(\frac{1}{4}-\frac{1}{2}iE\right)} \\
&= \frac{A}{\sqrt{2\pi}\,|P/2\pi|^{1/2+iE}} \exp\{2i\theta(E)\}.
\end{aligned}
\tag{6.6}
$$

It follows that, up to factors that can easily be made symmetrical, the position and momentum eigenfunctions are each other's time-reverses. Thus we find a physical interpretation of the function $\theta(t)$ (defined in (2.3)) at the heart of the functional equation (cf. (5.1)) for $\zeta(s)$.

The major problem remaining is to find boundary conditions that would convert XP into a well-defined hermitean operator with discrete eigenvalues. This is equivalent to specifying the way in which parts of the (X, P) plane are connected so as to compactify the (quantum and classical) motion. Some hints in this direction follow.

Our observations about the complex periodic orbits of the Riemann dynamics (see the last paragraph of section 5) suggest that X and $-X$ should be identified. The reason is that the complex orbits of X, obtained by replacing t by $i\tau$ in (6.2), have period $2\pi i$, which becomes the desired $i\pi$ (equation (5.27)) on identifying $\pm X$.

To proceed further, we consider the symmetries of XP, in the hope (so far unrealized) of superposing solutions of (6.4) acted on by operations in the symmetry group, with each solution multiplied by the appropriate group character. An obvious symmetry is dilation: XP is invariant under

$$
X \to KX, \quad P \to P/K. \tag{6.7}
$$

From (6.2), K corresponds to evolution after time $\log K$. This implies that the operator (6.3) generates dilations, in the same way that the momentum operator generates translations, and the following series of transformations makes this obvious:

$$
\begin{aligned}
f(KX) &= f(\exp\{\log K + \log X\}) = \exp\left\{(\log K)\frac{d}{d\log X}\right\} f(X) \\
&= \exp\left\{(\log K) X \frac{d}{dX}\right\} f(X) = K^{X\frac{d}{dX}} f(X) = \frac{1}{K^{1/2-iH}} f(X).
\end{aligned}
\tag{6.8}
$$

One possibility is to choose the integer dilations $K = m$, and the characters unity. Then the superposition of solutions (6.5) does contain $\zeta(1/2-iE)$ as a factor, but there seems no reason to impose the condition that this must vanish. Moreover, the set of integer dilations does not form a group (the inverse multiplications $1/m$ are missing).

Another possibility, closely related to the ideas of [62], is to use not all integers but the group of integers under multiplication (mod k) [68]. This would have two advantages. First, it involves only integer dilations. Second, including the characters $\chi(n)$ of this group (sets of k complex numbers with unit modulus) opens the possibility of widening the interpretation as eigenvalues of XP, to include the zeros of Dirichlet

L-functions. These are defined by the series

$$L_{\chi}(s) \equiv \sum_{n=1}^{\infty} \frac{\chi(n)}{n^{s}}. \tag{6.9}$$

(The special case $\chi = 1$ corresponds to $\zeta(s)$.) It is conjectured that for all these L-functions the complex zeros lie on the line $\text{Re}s = 1/2$. On this interpretation, each L-function corresponds to a different self-adjoint extension of XP under identification of positions X that are related by dilations in the group of integers under multiplication (mod k). An analogy is with the quantum mechanics of a particle in a periodic potential (e.g., an electron in a crystal): from the Bloch-Floquet theorem, solutions of the underlying differential equation are all periodic up to a phase factor $\exp(i\alpha)$; each choice of α is a different self-adjoint extension, and generates a discrete spectrum. The analogy is imperfect, because α is continuous, whereas the L-functions cannot be continuously parameterized. A closer analogy is with quantization on a torus phase space [69], where for topological reasons the permited phases are discrete.

The dynamics (6.2) suggests that the system might be closed by connecting the asymptotic positions with the asymptotic momenta. Then particles flowing out at $X = \pm\infty$ would be reinjected at $P = \pm\infty$. Related to this is a class of dilations where K is H-dependent (of course these are still symmetries of H). Specifically, the choice $K = 2\pi/(XP)$ yields the canonical transformation

$$X \to X_1 = \frac{2\pi}{P}, \quad P \to P_1 = \frac{XP^2}{2\pi}, \tag{6.10}$$

corresponding to exchange of X and P (the more familiar $X \to P$, $P \to -X$ does not leave XP invariant). A short calculation gives the transformed quantum wavefunction $\psi_1(X_1)$ in terms of the untransformed momentum wavefunction ϕ as

$$\psi_1(X_1) = \frac{(2\pi)^{1/4}}{|X_1|}\phi\left(\frac{\sqrt{2\pi}}{|X_1|}\right). \tag{6.11}$$

We do not know how to convert this "quantum exchange" into an effective boundary condition, but note its connection with the following intriguing identity, obtained from the momentum wavefunction formula (6.6) and the functional equation for $\zeta(s)$:

$$\begin{aligned} &X^{1/2}\zeta\left(\tfrac{1}{2} - iE\right)\psi_E(X) - P^{1/2}\zeta\left(\tfrac{1}{2} + iE\right)\phi_E(P) = 0, \\ &\text{where } PX = 2\pi \ (= h). \end{aligned} \tag{6.12}$$

If (only) the minus were a plus, this would be a condition generating the Riemann zeros.

We can sum up these scattered remarks about XP by returning to the properties listed at the beginning of this section. XP is consistent with point a, part of b (XP dynamics is unstable but not bounded), and c, d, g, h, i, and j. Concerning point e, the appearance of times that are logarithms of integers begins to be plausible in view of the association between dilation and evolution, but primes do not appear in any obvious way. We have no explanation of property f.

REFERENCES

[1] M. V. Berry, *Riemann's zeta function: A model for quantum chaos?*, in Quantum Chaos and Statistical Nuclear Physics, T. H. Seligman and H. Nishioka, eds., Lecture Notes in Phys. 263, Springer-Verlag, New York, 1986, pp. 1–17.

[2] M. V. Berry, *Quantum chaology (the Bakerian lecture)*, Proc. Roy. Soc. Lond. Ser. A, 413 (1987), pp. 183–198.
[3] M. V. Berry, *Some quantum-to-classical asymptotics*, in Chaos and Quantum Physics, Les Houches Lecture Series 52, M.-J. Giannoni, A. Voros, and J. Zinn-Justin, eds., North-Holland, Amsterdam, 1991, pp. 251–304.
[4] J. P. Keating, *The Riemann zeta-function and quantum chaology*, in Quantum Chaos, G. Casati, I. Guarneri, and V. Smilansky, eds., North-Holland, Amsterdam, 1993, pp. 145–185.
[5] J. P. Keating, *Periodic orbits, spectral statistics, and the Riemann zeros*, in Supersymmetry and Trace Formulae: Chaos and Disorder, J. P. Keating, D. E. Khmelnitskii, and I. V. Lerner, eds., Plenum, New York, 1998, pp. 1–15.
[6] M. V. Berry and J. P. Keating, *$H = xp$ and the Riemann zeros*, in Supersymmetry and Trace Formulae: Chaos and Disorder, J. P. Keating, D. E. Khmelnitskii, and I. V. Lerner, eds., Plenum, New York, 1998, pp. 355–367.
[7] H. M. Edwards, *Riemann's Zeta Function*, Academic Press, New York, London, 1974.
[8] D. Zagier, *The first 50 million prime numbers*, Math. Intelligencer, 0 (1977), pp. 7–19.
[9] J. van de Lune, H. J. J. te Riele, and D. T. Winter, *On the zeros of the Riemann zeta function in the critical strip*. IV, Math. Comp., 46 (1986), pp. 667–681.
[10] N. Levinson, *More than one third of the zeros of Riemann's zeta-function are on $\sigma = 1/2$*, Adv. Math., 13 (1974), pp. 383–436.
[11] E. C. Titchmarsh, *The theory of the Riemann zeta-function*, Clarendon Press, Oxford, UK, 1986.
[12] V. I. Arnold, 1978, *Mathematical Methods of Classical Mechanics*, Springer-Verlag, New York, 1978.
[13] J. L. Synge, *The Hamiltonian method and its application to water waves*, Proc. Roy. Irish Acad. A, 63 (1963), pp. 1–34.
[14] M. V. Berry, *Regularity and chaos in classical mechanics, illustrated by three deformations of a circular billiard*, European J. Phys., 2 (1981), pp. 91–102.
[15] M. V. Berry, *Regular and irregular motion*, in Topics in Nonlinear Mechanics, S. Jorna, ed., AIP Conf. Proc. 46, 1978, pp. 16–120.
[16] H. G. Schuster, *Deterministic Chaos. An Introduction*, VCH Verlagsgesellschaft, Weinheim, Germany, 1988.
[17] A. M. Ozorio de Almeida, 1988, *Hamiltonian Systems: Chaos and Quantization*, Cambridge, University Press, Cambridge, UK, 1988.
[18] M. C. Gutzwiller, *Periodic orbits and classical quantization conditions*, J. Math. Phys., 12 (1971), pp. 343–358.
[19] M. C. Gutzwiller, *Chaos in classical and quantum mechanics*, Springer-Verlag, New York, 1990.
[20] R. Balian and C. Bloch, *Distribution of eigenfrequencies for the wave equation in a finite domain:* III. *Eigenfrequency density oscillations*, Ann. Physics, 69 (1972), pp. 76–160.
[21] N. L. Balazs and A. Voros, *Chaos on the pseudosphere*, Phys. Rep., 143 (1986), pp. 109–240.
[22] J. M. Robbins, *Maslov indices in the Gutzwiller trace formula*, Nonlinearity, 4 (1991), pp. 343–363.
[23] H. P. Baltes and E. R. Hilf, *Spectra of Finite Systems*, B-I Wissenschaftsverlag, Mannheim, Germany, 1976.
[24] Y. G. Sinai, *Introduction to Ergodic Theory*, Princeton University Press, Princeton, NJ, 1976.
[25] J. H. Hannay and A. M. Ozorio de Almeida, *Periodic orbits and a correlation function for the semiclassical density of states*, J. Phys. A., 17 (1984), pp. 3429–3440.
[26] W. Parry and M. Pollicott, *Zeta-functions and the periodic orbit structure of hyperbolic dynamics*, Asterisque, 187/188 (1990), pp. 9 et seq.
[27] P. Cvitanovic and B. Eckhardt, J. Phys. A., 24 (1991), pp. L237–L241.
[28] M. L. Mehta, *Random Matrices and the Statistical Theory of Energy Levels*, Academic Press, New York, London, 1967.
[29] F. J. Dyson, *A Brownian-motion model for the eigenvalues of a random matrix*, J. Math. Phys., 3 (1962), pp. 1191–1198.
[30] F. J. Dyson and M. L. Mehta, *Statistical theory of the energy levels of complex systems* IV, J. Math. Phys., 4 (1963), pp. 701–712.
[31] C. E. Porter, *Statistical Theories of Spectra Fluctuations*, Academic Press, New York, 1965.
[32] O. Bohigas and M. J. Giannoni, *Chaotic Motion and Random-Matrix Theories*, in Mathematical and Computational Methods in Nuclear Physics, J. S. Dehesa, J. M. G. Gomez, and A. Polls, eds., Lecture Notes in Phys. 209, Springer-Verlag, 1984, pp. 1–99.
[33] M. V. Berry, *Semiclassical theory of spectral rigidity*, Proc. Roy. Soc. Lond. Ser. A, New York, 400 (1985), pp. 229–251.

[34] M. V. BERRY, *Semiclassical formula for the number variance of the Riemann zeros*, Nonlinearity, 1 (1988), pp. 399–407.

[35] H. L. MONTGOMERY, *The pair correlation of zeros of the zeta function*, Proc. Sympos. Pure Math., 24 (1973), pp. 181–193.

[36] J. P. KEATING, *The semiclassical sum rule and Riemann's zeta function*, in Adriatico Research Conference on Quantum Chaos, H. Cerdeira, R. Ramaswamy, M. Gutzwiller, and G. Casati, eds., World Scientific, River Edge, NJ, 1991, pp. 280–294.

[37] N. ARGAMAN, F. M. DITTES, E. DORON, J. P. KEATING, A. KITAEV, M. SIEBER, AND U. SMILANSKY, *Correlations in the actions of periodic orbits derived from quantum chaos*, Phys. Rev. Lett., 71 (1992), pp. 4326–4329.

[38] G. H. HARDY AND J. E. LITTLEWOOD, *Some problems in "Partitio Numerorum"* III: *On the expression of a number as a sum of primes*, Acta Math., 44 (1923), pp. 1–70.

[39] A. M. ODLYZKO, *On the distribution of spacings between zeros of the zeta function*, Math. Comp., 48 (1987), pp. 273–308.

[40] A. M. ODLYZKO, *The* 10^{20}*th zero of the Riemann zeta function and* 175 *million of its neighbours*, AT&T Bell Laboratory preprint, 1990.

[41] E. B. BOGOMOLNY AND J. P. KEATING, *Gutzwiller's trace formula and spectral statistics: Beyond the diagonal approximation*, Phys. Rev. Lett., 77 (1996), pp. 1472–1475.

[42] A. FUJII, *On the Berry Conjecture*, preprint, 1997.

[43] A. V. ANDREEV, O. AGAM, B. D. SIMONS, AND B. L. ALTSHULER, *Quantum chaos, irreversible classical dynamics, and random matrix theory*, Phys. Rev. Lett., 76 (1996), pp. 3947–3950.

[44] D. GOLDSTON AND H. L. MONTGOMERY, *Pair correlation of zeros and primes in short intervals*, in Analytic Number Theory and Diophantine Problems, Proc. 1984 Stillwater Conference at Oklahoma State University, pp. 183–203.

[45] G. GASPER AND M. RAHMAN, *Basic Hypergeometric Series*, Cambridge University Press, Cambridge, UK, 1990.

[46] D. A. HEJHAL, *On the Triple Correlation of Zeros of the Zeta Function*, IMRN, 1994, pp. 293–302.

[47] Z. RUDNICK AND P. SARNAK, *Zeros of principal L-functions and random-matrix theory*, Duke Math. J., 81 (1996), pp. 269–322.

[48] E. B. BOGOMOLNY AND J. P. KEATING, *Random matrix theory and the Riemann zeros* I: *Three- and four-point correlations*, Nonlinearity, 8 (1995), pp. 1115–1131.

[49] E. B. BOGOMOLNY AND J. P. KEATING, *Random-matrix theory and the Riemann zeros* II: *n-point correlations*, Nonlinearity, 9 (1996), pp. 911–935.

[50] W. GABCKE, *Neue Herleitung und Explizite Restabschätzung der Riemann-Siegel-Formel*, Ph.D. Thesis, University of Göttingen, Germany, 1979.

[51] M. V. BERRY, *The Riemann-Siegel formula for the zeta function: High orders and remainders*, Proc. Roy. Soc. Lond. Ser. A, 450, 1995, pp. 439–462.

[52] M. V. BERRY AND J. P. KEATING, *A rule for quantizing chaos?*, J. Phys. A, 23 (1990), pp. 4839–4849.

[53] J. P. KEATING, *Periodic-orbit resummation and the quantization of chaos*, Proc. Roy. Soc. Lond. Ser. A, 436 (1992), pp. 99–108.

[54] E. B. BOGOMOLNY, *Semiclassical quantization of multidimensional systems*, Nonlinearity, 5 (1992), pp. 805–866.

[55] M. V. BERRY AND J. P. KEATING, *A new approximation for* $\zeta(1/2+ it)$ *and quantum spectral determinants*, Proc. Roy. Soc. Lond. Ser. A, 437 (1992), pp. 151–173.

[56] R. B. PARIS, *An asymptotic representation for the Riemann zeta function on the critical line*, Proc. Roy. Soc. Lond. Ser. A, 446 (1994), pp. 565–587.

[57] R. B. PARIS AND S. CANG, *An exponentially-improved Gram-type formula for the Riemann zeta function*, Methods Appl. Anal., 4 (1997), pp. 326–338.

[58] R. B. PARIS AND S. CANG, *An asymptotic representation for* $\zeta(1/2+ it)$, Methods. Appl. Anal., 4 (1997), pp. 449–470.

[59] J. P. KEATING AND M. SIEBER, *Calculation of spectral determinants*, Proc. Roy. Soc. Lond. Ser. A, 447 (1994), pp. 413–437.

[60] N. KATZ AND P. SARNAK, *Zeros of Zeta Functions, Their Spacings and Their Spectral Nature*, preprint, 1997.

[61] P. SARNAK, *Quantum chaos, symmetry and zeta functions*, Curr. Dev. Math. (1997), pp. 84–115.

[62] A. CONNES, *Formule de trace en géométrie non-commutative et hypothèse de Riemann*, C.R. Acad. Sci. Paris, 323 (1996), pp. 1231–1236.

[63] S. NONNEMACHER AND A. VOROS, *Eigenstate structures around a hyperbolic point*, J. Phys. A., 30 (1997), pp. 295–315.

[64] R. K. Bhaduri, A. Khare, and J. Law, *Phase of the Riemann zeta function and the inverted harmonic oscillator*, Phys. Rev. E, 52 (1995), pp. 486–491.
[65] A. Khare, *The phase of the Riemann zeta function*, Pramana, 48 (1997), pp. 537–553.
[66] J. V. Armitage, *The Riemann hypothesis and the Hamiltonian of a quantum mechanical system*, in Number Theory and Dynamical Systems, M. M. Dodson and J. A. G. Vickers, eds., London Math. Soc. Lecture Note Ser. 134, Cambridge University Press, Cambridge, UK, 1989, pp. 153–172.
[67] S. Okubo, *Lorentz-invariant hamiltonian and Riemann hypothesis*, J. Phys. A., 31 (1998), pp. 1049–1057.
[68] T. M. Apostol, *Introduction to Analytic Number Theory*, Springer-Verlag, New York, 1976.
[69] J. P. Keating, F. Mezzadri, and J. M. Robbins, *Quantum boundary conditions for torus maps*, Nonlinearity, 1999, in press.
[70] E. Bogomolny and J. P. Keating, 1999, *Asymptotics of the Pair Correlation of Riemann Zeros*, in preparation.

Chapter 4
Asymptotics

While in St Andrews in the early 1960s, I learned from Bob Dingle about his theory of the high orders of divergent series. This inspired an enthusiasm for asymptotics, but I did not then realise how seminal his achievements were. Indeed, for two decades afterwards, I declared myself as a 'first terms asymptotist', arguing that high orders are unnecessary: getting the leading order right, as in the uniform approximations near caustics (Chapter 2) is accurate enough for most applications.

My first step beyond this restricted view of asymptotics came in 1987, when I showed [B164], in a theory of corrections to the geometric phase in powers of slowness, that the high-order terms have the 'factorial divided by power' structure, which I had learned from Dingle was a common ('universal') feature of asymptotic expansions: the terms initially get smaller and then diverge, with the highest accuracy achieved by truncating the series at its smallest term.

Central to Dingle's research was the phenomenon discovered by Stokes in 1847, in which a given function, expanded as factorially-divergent series in terms of a large parameter, is described by different asymptotic series in different parameter regimes (for example, different sectors in the complex plane of the parameter). The conventional understanding was that the transition from one such series to another is discontinuous, within the error of the approximation. In 1989, in my first development of Dingle's ideas, I found ([4.1], see also [B190]) that the transition can be described more precisely by formally resumming the terms of the divergent tail of one of the series. The result is a universal scaling function, in which the transition between different series, each truncated at their least term, is smooth, and described by an error function. This theory, applied to the time-dependent Schrödinger equation, describes the history of quantum transitions driven by slowly-changing forces ([B201], and [B221] with Richard Lim). It also inspired a poem [4.2].

Dingle's deepest discovery is 'resurgence' (the name was coined by Jean Écalle who rediscovered and generalised the concept). Resurgence is the appearance, in the high orders of any of the component series representing a function, of the low orders of all the other series, together with their correction terms. Dingle envisaged an approximation scheme based on resurgence, in the form of a sequence of asymptotic series, and he implemented the leading resurgence series in this sequence. Christopher Howls and I ([4.3, 4.4, 4.5], also [B234]) developed his idea into a systematic 'hyperasymptotic' iteration procedure, in which a sequence of series represents the function with exponentially increasing accuracy. Howls and others have extended our original procedure in several directions and with more mathematical rigour. One of Dingle's unpublished results, that I discovered among his papers after he died, is 'self-resurgence', relating the high and low orders of the same divergent series [B498].

Calculations of the heights of the Riemann zeros (see Chapter 3) are usually based on the Riemann-Siegel formula, consisting of a 'main sum' [B483] plus a series of corrections. In 1986, I had understood the main sum by resumming (section 4 of [3.4]) the long periodic orbits in the quantum chaology interpretation of zeta (Chapter 3); only later did I understand this as an example of resurgence. A few Riemann-Siegel correction terms give sufficient accuracy for most purposes; the high-order corrections have an intricate structure [4.6], in which the universal factorial-divided-by-power is decorated with theta functions.

A variant of the Riemann-Siegel formula, developed with Jon Keating [4.7] and explicitly based on resurgence, gives a main sum with considerably increased accuracy, and also, in a reversal of the quantum/Riemann analogy, provides a consistent resummation of the Gutzwiller series of periodic orbits in quantum chaology.

Proc. R. Soc. Lond. A **422**, 7–21 (1989)
Printed in Great Britain

Uniform asymptotic smoothing of Stokes's discontinuities

BY M. V. BERRY, F.R.S.
H. H. Wills Physics Laboratory, Tyndall Avenue, Bristol BS8 1TL, U.K.

(*Received* 14 *June* 1988)

Across a Stokes line, where one exponential in an asymptotic expansion maximally dominates another, the multiplier of the small exponential changes rapidly. If the expansion is truncated near its least term the change is not discontinuous but smooth and moreover universal in form. In terms of the singulant F – the difference between the larger and smaller exponents, and real on the Stokes line – the change in the multiplier is the error function

$$\pi^{-\frac{1}{2}}\int_{-\infty}^{\sigma} \mathrm{d}t \exp(-t^2) \quad \text{where} \quad \sigma = \operatorname{Im} F/(2\operatorname{Re} F)^{\frac{1}{2}}.$$

The derivation requires control of exponentially small terms in the dominant series; this is achieved with Dingle's method of Borel summation of late terms, starting with the least term. In numerical illustrations the multiplier is extracted from Dawson's integral (erfi) and the Airy function of the second kind (Bi): the small exponential emerges in the predicted universal manner from the dominant one, which can be 10^{10} times larger.

1. INTRODUCTION

Stokes's phenomenon concerns the behaviour of small exponentials while hidden by large ones (Stokes 1864, 1871, 1889, 1902). Such exponentials occur commonly in the asymptotic approximation of functions $y(X;k)$ defined by integrals or differential equations and dependent on a large parameter k and variables $X = (X_1, X_2, \ldots)$. In the simplest case there are just two exponentials, and the lowest-order approximation incorporating both can be written

$$y(X;k) \approx M_+(X;k)\exp\{k\phi_+(X)\} + \mathrm{i}S(X;k)M_-(X;k)\exp\{k\phi_-(X)\}$$
$$(\operatorname{Re}\phi_+(X) > \operatorname{Re}\phi_-(X)), \quad (1)$$

where the dominant and subdominant contributions are labelled $+$ and $-$, and the factor i is inserted for later convenience. The prefactors $M_\pm$ vary slowly with X and their k-dependences are simple powers. Attention will here focus on the *Stokes multiplier function* $S(X;k)$ weighting the subdominant exponential; this varies rapidly when X is near the *Stokes line* of y, where

$$\operatorname{Im}[\phi_+(X) - \phi_-(X)] = 0. \quad (2)$$

On the Stokes line (a set of codimension 1) it can be said that the dominance of $+$ over $-$ is maximal.

Reprinted from Proceedings of the Royal Society A, Volume 422, (1989).

The need to retain the subdominant term, even though it is numerically insignificant in (1) as $k \to \infty$, and the need for Stokes's multiplier, both spring from a common cause; maintaining the validity of (1) when X crosses *anti*-Stokes lines (far from Stokes's lines) on which $\operatorname{Re}[\phi_+ - \phi_-] = 0$ and the exponential previously called – becomes the dominant one. (Readers are warned that some authors employ the term Stokes line to denote what we here call an anti-Stokes line, and vice versa.)

A simple illustrative example, to which we shall later return, due to Stokes (1864) and well described by Dingle (1973, hereinafter called I) is the complex error function

$$\left.\begin{aligned} y(X;k) &= \int_{-\mathrm{i}\infty}^{Z} \mathrm{d}t \exp(kt^2), \\ (Z &= X_1 + \mathrm{i}X_2). \end{aligned}\right\} \tag{3}$$

Near the positive real Z axis the dominant contribution to y as $k \to \infty$ comes from the end-point of integration $t = Z$;

$$y \sim (2kZ)^{-1} \exp(kZ^2) \quad (Z \text{ positive real}). \tag{4}$$

Thus $\phi_+ = Z^2$ and $M_+ = (2kZ)^{-1}$. Near the positive imaginary axis this would predict that y is exponentially small, which is false because the integral is then dominated by the stationary point at $t = 0$, giving

$$y \sim \mathrm{i}(\pi/k)^{\frac{1}{2}} \quad (Z \text{ positive imaginary}). \tag{5}$$

This would suggest the asymptotics

$$y \sim (2kZ)^{-1} \exp(kz^2) + \mathrm{i}(\pi/k)^{\frac{1}{2}} \tag{6}$$

involving $\phi_- = 0$ and $M_- = (\pi/k)^{\frac{1}{2}}$. Thus the Stokes line (2) is the real axis $X_2 = 0$, and the anti-Stokes lines, where dominance is exchanged, are the diagonals $X_1 = \pm X_2$. But (6) fails near the *negative* imaginary axis because it would predict $y \sim \mathrm{i}(\pi/k)^{\frac{1}{2}}$ whereas it is obvious from (3) that y is exponentially small and given by the continuation of (4). To encompass the three regions discussed, we must write

$$y \sim (2kZ)^{-1} \exp(kZ^2) + \mathrm{i}S(Z;k)(\pi/k)^{\frac{1}{2}}, \tag{7}$$

incorporating Stokes's multiplier S which must change from 0 to 1 between the anti-Stokes lines $X_2 = -X_1$ and $X_2 = +X_1$. This change in S is Stokes's phenomenon.

The conventional view (Stokes 1864) is powerfully (and unconventionally) argued by Dingle in I. It asserts that the change in S is discontinuous and localized at the Stokes line: on one side, S takes a value, S_- say; on the other, $S = S_- + 1$; on the line itself, $S = S_- + \frac{1}{2}$. For the example (3) the intuition behind this view is illustrated by figure 1, which shows how the steepest-descent contours of the integral ($\operatorname{Im} t^2 = \operatorname{Im} Z^2$) change discontinuously across the Stokes line, suddenly bringing in the subdominant contribution from the stationary point at $t = 0$

Smoothing Stokes's discontinuities **9**

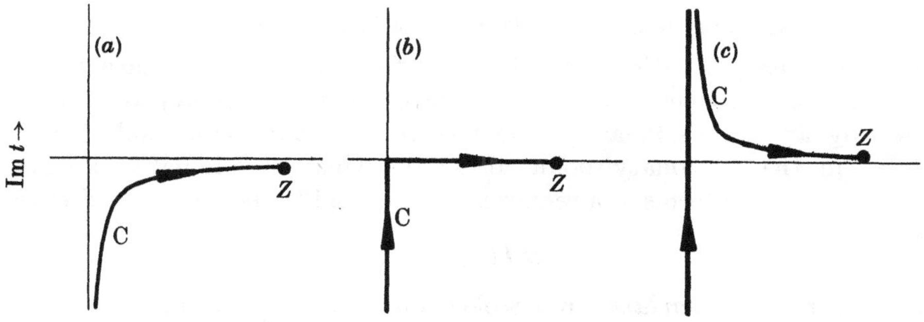

FIGURE 1. Steepest-descent contours of (3) from $t = -\mathrm{i}\infty$ to $t = Z = X_1 + \mathrm{i}X_2$, for three values of Z near the Stokes line $X_2 = 0$.

($S_- = 0$ in this case). It is worth repeating Stokes's description of the asymptotic emergence of his discontinuity. As a Stokes line is crossed,

> ...the inferior term enters as it were into a mist, is hidden for a little from view, and comes out with its coefficient changed. The range during which the inferior term remains in a mist decreases indefinitely as [the asymptotic parameter] increases indefinitely. (Stokes 1902)

From the context it is clear that Stokes is referring to asymptotic series interpreted by truncation near their least term. My aim here is to dispel Stokes's mist and show that his discontinuity is an artefact of poor resolution: with the appropriate magnification, S changes smoothly. Moreover, with the appropriate variable to describe the crossing of Stokes's line the change in the function $S(X;k)$ is *universal*, that is, the same for all problems in a wide class.

Obtaining this result requires control of the magnitude of the dominant exponential contribution with error small compared to the size of the subdominant exponential. Such control will be achieved by analysing the dominant asymptotic series (in descending powers of k) of which only the first term is included in (1). The series is

$$\left.\begin{aligned} y(X;k) &= M_+ \exp(k\phi_+) \sum_{r=1}^{\infty} a_r, \\ (a_0 &= 1; \quad a_r \propto k^{-r}). \end{aligned}\right\} \tag{8}$$

This diverges and so is numerically meaningless, but Dingle (I) explains how it can nevertheless be regarded as a coded representation of y, which can be reconstructed exactly (in principle and sometimes in practice) by proper interpretation of the late terms $r \gg 1$. The interpretation reveals how the subdominant exponential (together with S) originates from the divergence of the late terms. My derivation of the leading-order functional form of S across a Stokes line is a simple development within Dingle's interpretative scheme.

2. Derivation of the Stokes multiplier function

The obstruction preventing the series in (8) from converging is the existence of the subdominant exponential. (In the example (3) the series appended to (4) is obtained by an expansion about the integration limit $t = Z$, and divergence originates in the stationary point at $t = 0$.) This subdominant exponential engenders in the late terms a_r a remarkable universality, best expressed in terms of the complex quantity

$$F \equiv k(\phi_+ - \phi_-). \tag{9}$$

Dingle calls this the *singulant*; on a Stokes line, it is real and positive. He shows that

$$a_r \rightarrow \frac{M_-(r-\beta)!}{2\pi M_+ F^{r-\beta+1}} \quad \text{as} \quad r \rightarrow \infty. \tag{10}$$

For example, if y is defined by an integral for which (as in the example (3)) the dominant exponential comes from a limit of integration and the subdominant one from a stationary point, then $\beta = \frac{1}{2}$ (I, pp. 111, 145). If y is defined by an integral for which both exponentials are associated with stationary points, then $\beta = 1$ (I, pp. 135, 145). If y is a solution of a second-order differential equation, say

$$\partial_z^2 y = k^2 Q(Z)\, y \tag{11}$$

with $\operatorname{Re} Q > 0$ in the region of $Z = X_1 + \mathrm{i}X_2$ being studied, for which the exponentials come from the two primitive phase-integral (JWKB) approximations (giving

$$M_\pm = Q^{-\frac{1}{4}} \quad \text{and} \quad \phi_\pm = \pm \int_a^Z Q^{\frac{1}{2}}\, \mathrm{d}z'$$

where a is a simple zero of $Q^2(Z)$), then again $\beta = 1$ (I, p. 299) (for an nth-order zero, (10) is multiplied by $2\cos(\pi/(n+2))$).

To interpret (8) Dingle employs Borel summation, not for the whole series as is customary, but for the nth term and beyond, where n is close to the value $r \sim |F|$ for which a_r is least. Assuming this formal procedure is valid, we obtain, by using (10),

$$y \approx M_+ \exp(k\phi_+) \sum_{r=0}^{n-1} a_r + \mathrm{i} M_- S_n(F) \exp(k\phi_-), \tag{12}$$

where

$$S_n(F) \equiv \frac{-\mathrm{i}}{2\pi} \exp(F) \sum_{r=n}^{\infty} \frac{(r-\beta)!}{F^{r-\beta+1}}. \tag{13}$$

The interpretation is obtained by writing the factorial as an integral and performing the summation

$$\begin{aligned} S_n(F) &= \frac{-\mathrm{i}\exp(F)}{2\pi F^{1-\beta}} \int_0^\infty \mathrm{d}S \exp(-S)\, S^{-\beta} \sum_{r=n}^{\infty} \left(\frac{S}{F}\right)^r \\ &= \frac{-\mathrm{i}}{2\pi} \int_0^\infty \mathrm{d}t \frac{t^{n-\beta} \exp\{F(1-t)\}}{1-t}. \end{aligned} \tag{14}$$

To complete the interpretation it is necessary to specify the t-contour relative to the pole at $t = 1$. This corresponds to specifying the contour of integration (as will be illustrated later), or the desired solution of a differential equation, when defining y. Different choices differ by real constants, corresponding to the value of the quantity S_- in section 1. We specify that the contour passes above $t = 1$, so that

$$S_n(F) = \tfrac{1}{2} - \frac{\mathrm{i}}{2\pi} ⨍_0^\infty \mathrm{d}t \frac{t^{n-\beta} \exp\{F(1-t)\}}{1-t}, \tag{15}$$

where now the principal value of the integral is taken. This choice corresponds to the situation in example (3), where the Stokes multiplier switches on from zero ($S_- = 0$) as $\mathrm{Im}\, F$ increases through zero.

Now we identify S_n as the Stokes multiplier and determine its dominant asymptotics when F is large and nearly real. A crucial simplification (corresponding to the evaluative interpretation of asymptotic series adopted by Stokes) occurs if we truncate the r-sum near its least term, that is at

$$n - 1 = \mathrm{Int}\,(|F| + \alpha), \tag{16}$$

where α is of order unity (as is β). With this truncation, the stationary point of (15) almost coincides with its pole $t = 1$, whose neighbourhood therefore dominates the integral. Let

$$F \equiv A + \mathrm{i}B \tag{17}$$

(where A and B are real with $A \gg 1$ and $B \ll A$) and

$$n - \beta \equiv A + \mu, \quad \text{i.e.} \quad \mu = \mathrm{Int}\,(|F| + \alpha) - \beta - A + 1, \tag{18}$$

(so μ is of order unity) and change variables to $x \equiv t - 1$. Then expanding the integrand in (15) to third order about $x = 0$ gives

$$\begin{aligned} S_n(F) &= \tfrac{1}{2} + \frac{\mathrm{i}}{2\pi} ⨍_{-1}^\infty \frac{\mathrm{d}x}{x} \exp\{(A+\mu)\ln(1+x) - Ax - \mathrm{i}Bx\} \\ &\approx \tfrac{1}{2} + \frac{1}{2\pi} \int_{-\infty}^\infty \mathrm{d}x (1 + \mu x + \tfrac{1}{3}Ax^3) \exp(-\tfrac{1}{2}Ax^2)(\sin Bx + \mathrm{i}\cos Bx)/x \\ &= \frac{1}{2}\left(1 + \frac{1}{\pi}\int_{-\infty}^\infty \frac{\mathrm{d}x}{x} \exp(-\tfrac{1}{2}Ax^2)\sin Bx\right) \\ &\quad + \frac{\mathrm{i}}{2\pi}\int_{-\infty}^\infty \mathrm{d}x \exp(-\tfrac{1}{2}Ax^2)(\mu + \tfrac{1}{3}Ax^2)\cos Bx \\ &= \frac{1}{\sqrt{\pi}} \int_{-\infty}^{B/(2A)^{\frac{1}{2}}} \mathrm{d}t \exp(-t^2) - \mathrm{i}(2\pi A)^{-\frac{1}{2}}(\mathrm{Fract}\{|F| + \alpha\} \\ &\quad + \beta - \alpha - \tfrac{4}{3} - B^2/6A) \times \exp(-B^2/2A). \end{aligned} \tag{19}$$

The real part dominates, and comparison with (1) and (12) yields the change in the Stokes multiplier as

$$S(\sigma) = \frac{1}{\sqrt{\pi}} \int_{-\infty}^\sigma \mathrm{d}t \exp(-t^2) \tag{20}$$

involving the *Stokes variable*

$$\sigma(X;k) = B/(2A)^{\frac{1}{2}} = \operatorname{Im} F/(2\operatorname{Re} F)^{\frac{1}{2}}$$
$$= k^{\frac{1}{2}} \operatorname{Im}(\phi_+ - \phi_-)/\{2\operatorname{Re}(\phi_+ - \phi_-)\}^{\frac{1}{2}}. \quad (21)$$

Equations (20) and (21) carry the following implication: under a magnification of order $k^{\frac{1}{2}}$, the multiplier varies smoothly from S_- to $S_- + 1$ across the Stokes line, the functional dependence on the natural variable σ being that of the error function.

According to (19), the imaginary part of $S_n(F)$ is smaller than the subdominant exponential in (12) by a factor $A^{-\frac{1}{2}} \approx |F|^{-\frac{1}{2}} \sim k^{-\frac{1}{2}}$. This gives the assurance that the dominant series has been controlled to better-than-exponential accuracy. Thus to detect the multiplier numerically, as we will do in the next section, it is sufficient to subtract from the exact function y the dominant series taken up to its nearly least term, i.e.

$$\left[y\exp(-k\phi_-) - M_+ \exp(F) \sum_{r=0}^{\operatorname{Int}(|F|+\alpha)} a_r\right] \Big/ M_- \to \mathrm{i}\,(S(\sigma) + S_-) \quad \text{as} \quad |F| \to \infty \quad (22)$$

independently of α provided α is of order unity. This equation embodies our main result.

Any alteration in α changes the number of terms included in the sum (cf. 16), and is compensated by a change in the imaginary part of (19) which is small compared with $S(\sigma)$. Two natural choices for the 'best' α are (i) that which minimizes the imaginary part of (19) on the average, which gives (because the average of Fract$\{x\}$ is $\frac{1}{2}$) $\alpha = \beta - \frac{5}{6}$; (ii) that for which the smallest a_r has $r = n$ or $r = n+1$, which is the case if $\alpha = \beta - \frac{1}{2}$.

The imaginary terms in (19) constitute the lowest-order approximation in the technique of 'terminants' or 'converging factors', which has been employed (see I and Olver (1974)) to correct the dominant series representing y *on* the Stokes line. Here, of course, we are studying the variation of y *across* a Stokes line.

3. Numerical illustrations

According to (22), the Stokes multiplier function, which is of order unity, is the difference of two large quantities (of order $\exp(F)$). To detect S numerically, two conditions must be satisfied: y must be calculable to better-than-exponential precision, and the coefficients a_r in the asymptotic expansion must be known (or calculable) for large r. There follow two examples for which these conditions are met.

The first is *Dawson's integral* (Abramowitz & Stegun 1964)

$$y \equiv \int_0^Z \mathrm{d}t \exp(kt^2) = k^{-\frac{1}{2}} \operatorname{erfi}(k^{\frac{1}{2}} Z) \quad (23)$$

which differs by $\mathrm{i}\,(\pi/k)^{\frac{1}{2}}$ from the example (3). We are interested in $Z = X_1 + \mathrm{i} X_2$ near the positive real axis. Obviously y is real on the Stokes line, so that in (7) S must vanish when $X_2 = 0$, implying in turn that in the general expression (14) the principal value must be taken and that $S_- = -\frac{1}{2}$ in (22). Figure 2 shows the

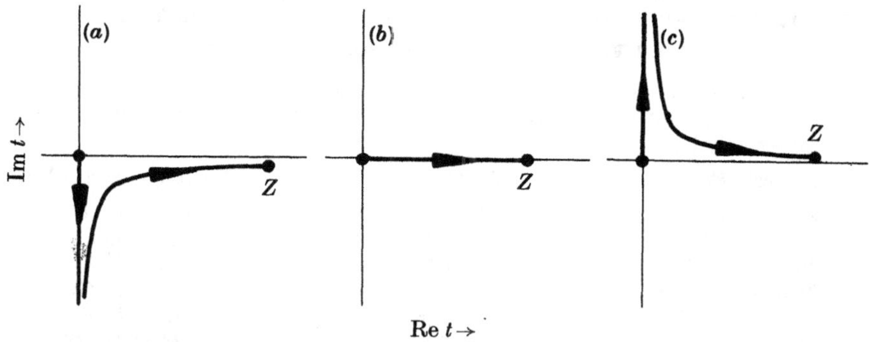

FIGURE 2. Steepest-descent contours of Dawson's integral (23) from $t = 0$ to $t = Z = X_1 + iX_2$, for three values of Z near the Stokes line $X_2 = 0$.

steepest-descent contours, for comparison with those in figure 1. The singulant is given by (9) and (7) as

$$F = kZ^2 \tag{24}$$

so that the Stokes variable is

$$\sigma = (2k)^{\frac{1}{2}} X_1 X_2/(X_1^2 - X_2^2)^{\frac{1}{2}} \approx (2k)^{\frac{1}{2}} X_2. \tag{25}$$

In the expansion (8) the coefficients a_r can be found by elementary asymptotics, for example (I, p. 5) changing the integration variable in (23) by $kt^2 = kZ^2 + u$ and expanding the jacobian about $u = 0$, with the result

$$a_r = (r - \tfrac{1}{2})!\, F^{-r}/\sqrt{\pi}. \tag{26}$$

Thus the limiting form (10) is here exact for all r, and $\beta = \frac{1}{2}$.

Incorporating (24)–(26) into the general result (22) with $S_- = -\frac{1}{2}$ we obtain, as the formula for the asymptotic emergence of the multiplier,

$$\operatorname{Im}\left\{\operatorname{erfi}(F^{\frac{1}{2}})/\sqrt{\pi} - (2\pi)^{-1} \exp(F) \sum_{r=0}^{\operatorname{Int}(|F|+\alpha)} (r - \tfrac{1}{2})!/F^{r+\frac{1}{2}}\right\}$$

$$\to \frac{1}{\sqrt{\pi}} \int_0^{\sigma} \mathrm{d}t \exp(-t^2) \quad \text{as} \quad |F| \to \infty. \tag{27}$$

We shall here regard σ and $|F|$ as given, and obtain X_1 and X_2 by inversion of (24) and (25):

$$\left.\begin{aligned} k^{\frac{1}{2}} X_1 = |F|^{\frac{1}{2}} \cos\theta; \quad k^{\frac{1}{2}} X_2 = |F|^{\frac{1}{2}} \sin\theta \\ \text{where} \qquad \theta = \tfrac{1}{2} \arccos\{[1 + (\sigma^2/|F|)^2]^{\frac{1}{2}} - \sigma^2/|F|\}. \end{aligned}\right\} \tag{28}$$

Tables 1–4 show the results of a numerical test of (27) over a range of values of σ, for singulants $|F| = 5$ and $|F| = 25$ and truncation variables $\alpha = -\frac{1}{3}$ and $\alpha = 0$ (these are the two 'best' choices of α described at the end of §2). The integral for erfi was evaluated along a straight-line contour from X_1 to $X_1 + iX_2$ (the integral from 0 to X_1 being real) by the extended Simpson's rule; sufficient accuracy was achieved with 200 steps for $|F| = 5$ and 2000 steps for $|F| = 25$, and checked by evaluating the convergent series for erfi.

Table 1. Comparison of theory and 'experiment' for the Stokes multiplier function for Dawson's integral (23)

The singulant modulus is $|F| = 5$ and the truncation variable is $\alpha = -\frac{1}{3}$. σ is the Stokes variable (21) and $F = k(X_1 + iX_2)^2$ with X_1 and X_2 given by (28). (LHS and RHS are left-hand and right-hand sides respectively.)

σ	$2\,\mathrm{Im\,erfi}\,(F^{\frac{1}{2}})/\sqrt{\pi}$	$2\times$ LHS of (27)	$2\times$ (LHS $-$ RHS) of (27)
0.2	21.532 335 635 1	0.220 432 726 808	−0.002 277 090 460 5
0.4	32.406 989 074	0.421 905 998 99	−0.006 499 177 409 1
0.6	30.369 785 607 3	0.591 079 840 706	−0.012 791 995 848
0.8	21.029 192 239 2	0.722 780 152 631	−0.019 336 679 001 1
1.0	11.224 112 560 1	0.818 898 722 383	−0.023 815 907 806 3
1.2	4.455 199 927 81	0.885 393 233 85	−0.024 931 438 183 8
1.4	0.939 754 805 283	0.929 422 218 948	−0.022 870 317 935 5
1.6	−0.434 973 606 01	0.957 562 519 129	−0.018 790 516 202
1.8	−743 819 764 526	0.975 062 006 02	−0.014 031 146 904
2.0	−0.634 984 356 598	0.985 737 130 522	−0.009 586 512 135 48
2.2	−0.411 962 286 068	0.992 177 544 661	−0.005 960 263 230 16
2.4	−0.190 731 569 089	0.996 048 680 562	−0.003 263 089 936 87

Table 2. As table 1 with $\alpha = 0$

σ	$2\,\mathrm{Im\,erfi}\,(F^{\frac{1}{2}})/\sqrt{\pi}$	$2\times$ LHS of (27)	$2\times$ (LHS $-$ RHS) of (27)
0.2	21.532 335 635 1	0.242 465 868 574	0.019 756 051 305 7
0.4	32.406 989 074	0.463 805 332 974	0.035 400 156 575
0.6	30.369 785 607 3	0.648 329 058 716	0.044 457 222 162 2
0.8	21.029 192 239 2	0.788 860 592 493	0.046 743 760 860 5
1.0	11.224 112 560 1	0.886 578 386 719	0.043 863 756 53
1.2	4.455 199 927 81	0.948 451 996 877	0.038 127 324 843 3
1.4	0.939 754 805 283	0.983 881 887 394	0.031 589 350 511 1
1.6	−0.434 973 606 01	1.001 915 070 31	0.025 562 034 975 8
1.8	−0.743 819 764 526	1.009 685 586 97	0.020 592 434 041 3
2.0	−0.634 984 356 598	1.012 031 282 78	0.016 707 640 125
2.2	−0.411 962 286 068	1.011 844 149 78	0.013 706 341 889 7
2.4	−0.190 731 569 089	1.010 670 520 05	0.011 358 749 547 4

Table 3. As table 1 with $|F| = 25$

σ	$2\,\mathrm{Im\,erfi}\,(F^{\frac{1}{2}})/\sqrt{\pi}$	$2\times$ LHS of (27)	$2\times$ (LHS $-$ RHS) of (27)
0.2	7 833 454 975.39	0.222 258 567 81	−0.000 451 249 458 156
0.4	2 631 901 621.35	0.427 119 255 066	−0.001 285 921 333 19
0.6	−4 826 024 257.27	0.600 670 814 514	−0.003 201 022 040 09
0.8	−3 211 113 836.27	0.737 760 543 823	−0.004 356 287 808 85
1.0	1 495 138 180.67	0.840 577 602 386	−0.002 137 027 802 68
1.2	1 999 214 434.28	0.908 263 206 482	−0.002 061 465 551 5
1.4	137 285 256.727	0.946 264 266 968	−0.006 028 269 915 31
1.6	−647 807 326.949	0.970 328 092 575	−0.006 024 942 755 88
1.8	−311 385 089.924	0.987 966 120 243	−0.001 127 032 680 75
2.0	42 695 984.0144	0.997 133 061 29	0.001 809 418 632 56
2.2	98 642 118.0045	0.999 141 067 266	0.001 003 259 375 05
2.4	39 463 275.4458	0.998 938 336 968	−0.000 373 433 530 283

Table 4. As table 1 with $|F| = 25$ and $\alpha = 0$

σ	$2\,\mathrm{Im\,erfi}\,(F^{\frac{1}{2}})/\sqrt{\pi}$	$2 \times$ LHS of (27)	$2 \times$ (LHS − RHS) of (27)
0.2	7833454975.39	0.226700782776	0.00399096550766
0.4	2631901621.35	0.435605049133	0.00719987273419
0.6	−4826024257.27	0.612331390381	0.00845955382661
0.8	−3211113836.27	0.751268863678	0.00915203204589
1.0	1495138180.67	0.854352474213	0.0116378440235
1.2	1999214434.28	0.920836210251	0.0105115382174
1.4	137285256.727	0.956623911858	0.00433137497452
1.6	−647807326.949	0.978073358536	0.00172032320481
1.8	−311385089.924	0.993239462376	0.00414630945182
2.0	42695984.0144	1.00040954351	0.00508590085702
2.2	98642118.0045	1.00099922717	0.00286141927476
2.4	39463275.4458	0.999897457659	0.00058568716054

Successive columns clearly show the decreasing orders of magnitude of the first term on the left-hand side of (27) (i.e. the integral), whose order is $\exp(|F|)/|F|^{\frac{1}{2}}$ (at least for small σ); the whole left-hand side of (27) (i.e. the 'experimental' Stokes multiplier), whose order is unity; and the difference between the two sides of (27), whose order can be shown (by easy extension of the argument leading to (19)) to be $|F|^{-1}$. The two sides of (27) are shown plotted against σ for $|F| = 5$ and 25 in figures 3*a*, *b*, for the two values of α. Evidently the agreement between 'experimental' and 'theoretical' multipliers improves with increasing $|F|$, as it should. The two curves for the 'best' α bracket the theoretical curve.

For the second example we take the *Airy function of the second kind* (Abramowitz & Stegun 1964),

$$y = \int_C \mathrm{d}t \exp\{k(-\tfrac{1}{3}t^3 + tZ)\} = 2\pi k^{-\frac{1}{3}} \mathrm{Bi}\,(Zk^{\frac{2}{3}}) \tag{29}$$

where C is the contour shown in figure 4*a*. We are interested in $Z = X_1 + \mathrm{i}X_2$ near the positive real axis. The integrand has stationary points at $t_\pm = \pm Z^{\frac{1}{2}}$, so that the dominant and subdominant exponents in (1), and the singulant, are

$$\phi_\pm = \pm\tfrac{2}{3}Z^{\frac{3}{2}}, \quad F = \tfrac{4}{3}kZ^{\frac{3}{2}}, \tag{30}$$

where the roots are positive real when Z is positive real, that is on the Stokes line. Thus the Stokes variable (21) is

$$\sigma = (\tfrac{2}{3}k)^{\frac{1}{2}}\,\mathrm{Im}\,(X_1 + \mathrm{i}X_2)^{\frac{3}{2}}/[\mathrm{Re}\,(X_1 + \mathrm{i}X_2)^{\frac{3}{2}}]^{\frac{1}{2}} \approx (\tfrac{3}{2}k)^{\frac{1}{2}}X_2/X_1^{\frac{1}{4}}. \tag{31}$$

This shows that for fixed k the 'width' of the Stokes zone increases slowly (as $X^{\frac{1}{4}}$) away from the origin.

Figures 4*b–d* show the steepest-descent contours for Z nearly positive real. Evidently the contribution of the subdominant exponential reverses across the Stokes line and vanishes on it, as for Dawson's integral, so again $S_- = -\frac{1}{2}$. After taking into account the double contour through the dominant stationary point, the prefactors $M_\pm$ can be obtained from the simplest steepest-descent argument, giving

$$M_+ = 2M_- = 2(\pi^2/k^2Z)^{\frac{1}{4}}. \tag{32}$$

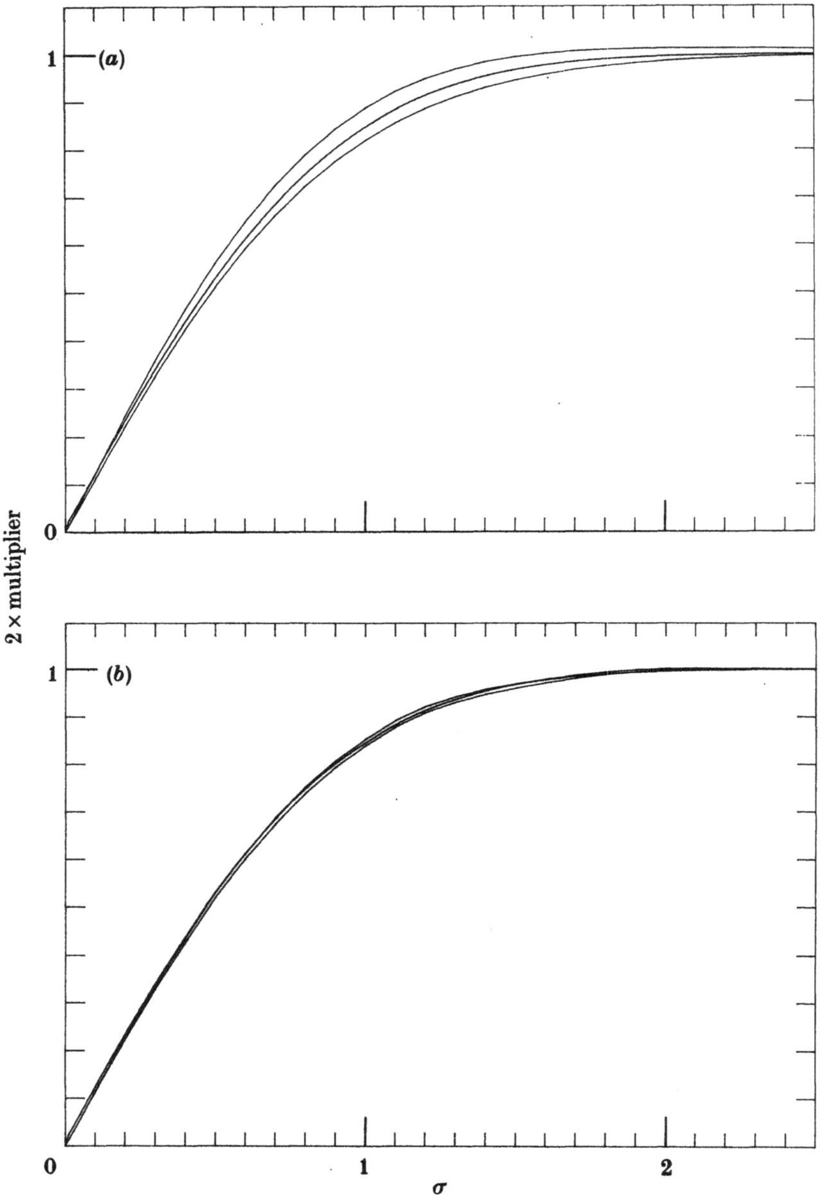

FIGURE 3. Stokes multiplier function for Dawson's integral erfi (equation (23)) for (a) singulant $|F| = 5$; (b) singulant $|F| = 25$. The middle curve is the 'theoretical' multiplier (RHS of (27)) and the upper and lower curves are the 'experimental' multipliers (LHS of (27)) for $\alpha = 0$ and $\alpha = -\frac{1}{2}$ respectively.

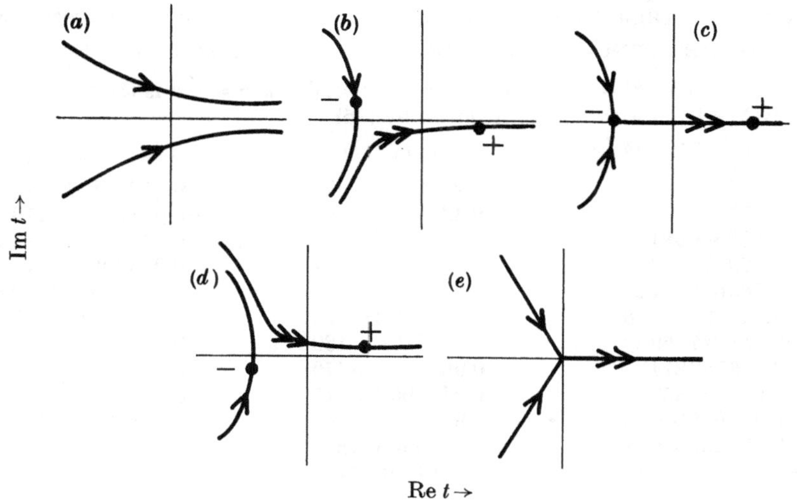

FIGURE 4. Integration contours for the Airy function Bi (equation (29)). (*a*) generic contour; (*b*) steepest-descent contour for $\operatorname{Im} Z < 0$; (*c*) steepest-descent contour for $\operatorname{Im} Z = 0$; (*d*) steepest-descent contour for $\operatorname{Im} Z > 0$; (*e*) contour for numerical integration of Bi. In *b–d* the subdominant and dominant stationary points are labelled + and −. Double arrows mean contours counted twice.

Further steepest-descent analysis (or phase-integral solution of Airy's equation $\partial_Z^2 y = k^2 Z y$) gives the expansion coefficients as

$$a_r = (r-\tfrac{1}{6})!\,(r-\tfrac{5}{6})!/(2\pi r!\,F^r). \tag{33}$$

This satisfies the initial condition $a_0 = 1$ (because $(-\tfrac{1}{6})!\,(-\tfrac{5}{6})! = 2\pi$), and conforms to the limit (10) with $\beta = 1$ (because $(r-\mu)!\,(r-\nu)!/r! \to (r-\mu-\nu)!$ as $r \to \infty$).

Incorporating (29)–(33) into the general result (22) with $S_- = -\tfrac{1}{2}$, including the contour doubling, we obtain, as the formula for the asymptotic emergence of the multiplier,

$$2\operatorname{Im}\left\{\pi^{\frac{1}{2}}(\tfrac{3}{4}F)^{\frac{1}{6}}\exp(\tfrac{1}{2}F)\operatorname{Bi}[(\tfrac{3}{4}F)^{\frac{2}{3}}] - \exp(F)\sum_{r=0}^{\operatorname{Int}[|F|+\alpha]}(r-\tfrac{1}{6})!\,(r-\tfrac{5}{6})!/(2\pi r!\,F^r)\right\}$$

$$\to \frac{2}{\sqrt{\pi}}\int_0^\sigma \mathrm{d}t\exp(-t^2) \quad \text{as} \quad |F| \to \infty. \tag{34}$$

Again we shall regard σ and $|F|$ as given, obtaining F by solving (30) and (31) for X_1 and X_2;

$$\left.\begin{aligned} k^{\frac{2}{3}}X_1 &= (\tfrac{3}{4}|F|)^{\frac{2}{3}}\cos\phi; \quad k^{\frac{2}{3}}X_2 = (\tfrac{3}{4}|F|)^{\frac{2}{3}}\sin\phi,\\ \text{where}\quad \phi &= \tfrac{2}{3}\arccos\{[1+(\sigma^2/|F|)^2]^{\frac{1}{2}} - \sigma^2/|F|\}.\end{aligned}\right\} \tag{35}$$

Tables 5 and 6 show the results of a numerical test of (34) over a range of values of σ, for singulants $|F| = 5$ and $|F| = 25$. For these integer $|F|$ the two 'best' truncation variables, $\alpha = \tfrac{1}{6}$, and $\alpha = \tfrac{1}{2}$, give identical sums in (34). The function Bi was computed from the extended Simpson's rule, after deforming the contour C

TABLE 5. COMPARISON OF THEORY AND 'EXPERIMENT' FOR THE STOKES MULTIPLIER FUNCTION FOR THE MODIFIED AIRY FUNCTION (29)

The singulant modulus is $|F| = 5$ and the truncation variable is $\alpha = \frac{1}{6}$. σ is the Stokes variable (21) and $F = \frac{4}{3}k(X_1 + \mathrm{i}X_2)^{\frac{3}{2}}$ with X_1 and X_2 given by (35).

σ	$\mathrm{Im}\{\pi^{\frac{1}{2}}(\frac{3}{4}F)^{\frac{1}{6}}\mathrm{e}^{\frac{1}{2}F}\,\mathrm{Bi}((\frac{3}{4}F)^{\frac{2}{3}})\}$	LHS of (34)	(LHS − RHS) of (34)
0.2	86.4604996724	0.221345661477	−0.0121781510106
0.4	123.97607385	0.424404297191	−0.013568051778
0.6	105.332800481	0.595782204695	−0.0159029679234
0.8	59.8291753225	0.729510930471	−0.0184963021774
1.0	18.7836794412	0.826574068164	−0.0202398523799
1.2	−5.02003487888	0.892523281855	−0.0204348576503
1.4	−13.9548273995	0.934750647189	−0.0191036079774
1.6	−14.786787214	0.960427535479	−0.0167804344415
1.8	−12.6179983574	0.975396073945	−0.0141291135887
2.0	−9.96309293078	0.98386023376	−0.0116649481508
2.2	−7.69826258958	0.98856662358	−0.00965797161679
2.4	−5.97784412023	0.991179082126	−0.00816718750941

TABLE 6. AS TABLE 5 WITH $|F| = 25$

σ	$\mathrm{Im}\{\pi^{\frac{1}{2}}(\frac{3}{4}F)^{\frac{1}{6}}\mathrm{e}^{\frac{1}{2}F}\,\mathrm{Bi}((\frac{3}{4}F)^{\frac{2}{3}})\}$	LHS of (34)	(LHS − RHS) of (34)
0.2	68717981224.8	0.218231201172	−0.0152926113155
0.4	19591748977.8	0.422904968262	−0.0150673807073
0.6	−44424281459.3	0.599472045898	−0.01221312672
0.8	−24788677708.8	0.739608764648	−0.00839846799997
1.0	16363698359.3	0.840934753418	−0.00587916712617
1.2	16545841811.1	0.908313751221	−0.00464438828493
1.4	−1014224788.75	0.949258327484	−0.00459592768219
1.6	−6181168808.05	0.972358703613	−0.00484926630728
1.8	−2073873258.79	0.984602928162	−0.00492225937189
2.0	856749591.829	0.990872144699	−0.00465303721144
2.2	909857385.36	0.993932247162	−0.00429234803532
2.4	237699764.863	0.995340585709	−0.00400568392631

(figure 4*a*) of the integral (29) to that in figure 4*e*, and checked by evaluating the convergent series for Bi.

As for Dawson's integral, successive columns clearly show the expected decreasing orders of magnitude. In the second column, containing the values of the first term on the left-hand side of (34), the order is now $\exp(|F|)$, which is larger by $|F|^{\frac{1}{2}}$ than the corresponding order for Dawson's integral. The left and right sides of (34) are plotted against σ for $|F| = 5$ and $|F| = 25$ in figures 5*a*, *b*.

It is again evident that the agreement between 'experimental' and 'theoretical' multipliers is excellent and improves with increasing $|F|$. A feature of this example, not present for Dawson's integral, is that it illustrates the general case in which the limit (10), forming the basis of our theory of the multiplier, is an approximation valid for the late terms a_r, rather than being exact for all r.

Stokes (1964) himself illustrated the change in the multiplier, by computing an Airy function at two complex arguments with phases $\pm 30°$ from a Stokes line, and a common modulus for which $|F| = \sqrt{128}$. Then (30), (21) and (20) give $\sigma = -2$ and $S(\sigma) = 0.005$, for which the change has barely begun, and $\sigma = +2$ and $S(\sigma) = 0.995$, for which it is virtually complete.

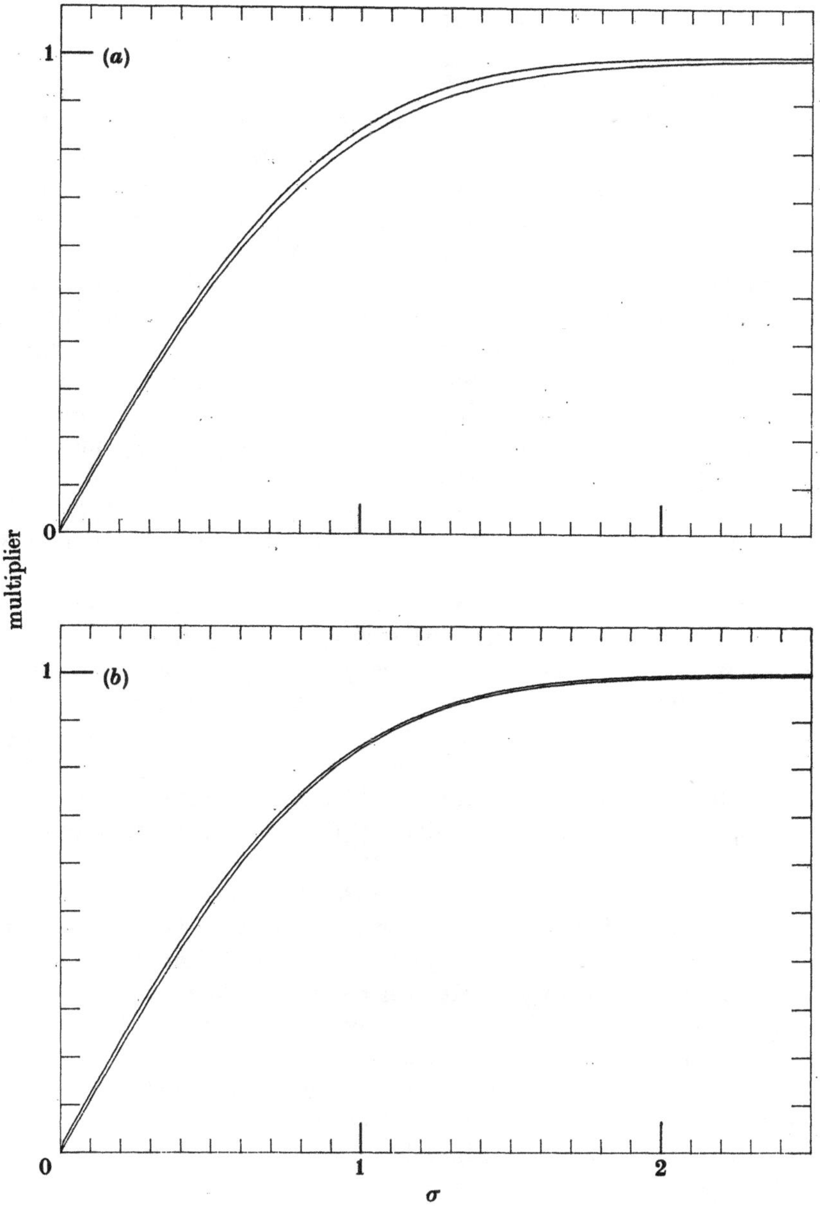

FIGURE 5. Stokes multiplier function for Airy function of the second kind (Bi) (equation (29)) for (*a*) singulant $|F| = 5$; (*b*) singulant $|F| = 25$. The upper curves are the 'theoretical' multipliers (RHS of (34)) and the lower curves are the 'experimental' multipliers (LHS of (34)) for $\alpha = \frac{1}{6}$.

4. Concluding remarks

We have found that across the Stokes line, on which the large exponential is maximally dominant, the multiplier of the small exponential varies rapidly but smoothly in a universal manner that is almost the simplest imaginable; the error function of a natural variable (equation (21)) depending on the singulant (difference between exponents). In numerical computations the Stokes multiplier (of order unity) emerged stably, and in agreement with theory, as the difference between two large quantities (of order up to 10^{10}, cf. table 6).

The universality class appears to be large. Our derivation depended on Dingle's (I) factorial formula (10) for the late terms a_r of the dominant asymptotic series. This type of asymptotics applies (at least) to integrals of functions involving $\exp(k\phi)$ as $k \rightarrow \infty$, whether dominated by an end-point or a stationary point; and to the solutions of ordinary differential equations with first-order turning points (for a turning point of order n the multiplier (20) is simply magnified by $2\cos(\pi/(n+2))$). But the results remain valid for some kinds of superfactorial divergence, for example $a_r \rightarrow (r!)^m/k^r$ as $r \rightarrow \infty$ (as can be shown by multiple Borel summation or, more easily, with multiplication formulae for factorials). It would be interesting to establish the limits of universality.

A trivial way to lose both the universality and simplicity of the main result is truncate the dominant series far from the least term, so that in (16) $|\alpha| \gg 1$. In these circumstances a general theory is still possible provided the limit (10) remains valid for all the Borel-summed terms, that is for all $r > n$, but is unsatisfactory in two respects. First, the multiplier S acquires an imaginary part comparable to its real part. Second, the multiplier now depends on $D \equiv (|F|-n)/(2\,\mathrm{Re}\,F)^{\frac{1}{2}}$ as well as σ, and when $|D|$ is large S rises to values of order $\exp(D^2)$ and oscillates with period of order D^{-1} over a total σ-range of order $|D|$ (rather than unity).

Our treatment is limited by being restricted to the situation where there is just one dominant and one subdominant exponential (equation (1)). What happens when there are more? The following conjecture is natural. Let the exponents be $k\phi_1, k\phi_2, \ldots$, ordered so that $\mathrm{Re}\,\phi_i > \mathrm{Re}\,\phi_j$ if $i < j$, and define for each pair the singulant $F_{ij} \equiv k(\phi_i - \phi_j)$ with $i < j$. Then across the line $\mathrm{Im}\,F_{ij} = 0$ the jth exponential is maximally dominated by the ith and its switching-on is described by the universal Stokes multiplier function that we have obtained, the appropriate singulant being $F = F_{ij}$. A proof (or disproof) is desirable.

It is worth pointing out that although in our examples (erfi and Bi) the parameters X were components of a complex variable $X_1 + \mathrm{i}X_2$, the parameter space need not possess a complex structure. To illustrate this, consider the oscillatory integral describing the cusp diffraction catastrophe (Pearcey's integral)

$$y(X_1, X_2; k) = \int_{-\infty}^{\infty} \mathrm{d}t \exp\{\mathrm{i}k(\tfrac{1}{4}t^4 + \tfrac{1}{2}X_1 t^2 + X_2 t)\} \tag{36}$$

as studied by Wright (1980). The asymptotics are dominated by real stationary points giving oscillatory contributions (waves). In the (real) parameter space X_1, X_2 there are three real stationary points inside the cusp $27X_2^2 + 4X_1^3 = 0$, and one real and two complex stationary points outside. In this outside region the real stationary point always contributes to (36), and one of the complex ones never does. Wright discovered that the (exponentially small) contribution of the second complex stationary point switches on across the *Stokes set*, which is the different cusp $27X_2^2-(5+3\surd 3)X_1^3=0$.

An immediate application of the main result is to the *birth of reflected waves* as described by $y''(x)+k^2n^2(x)\,y(x)=0$, where $n(x)$ is a real non-zero refractive index profile. The reflected wave is exponentially small and (if the incident wave is defined by the dominant W.K.B. expansion truncated at its least term) switches on where the Stokes line from the nearest complex turning point crosses the x-axis. The switch occurs over $|F|^{\frac{1}{2}}$ wavelengths, where the singulant $|F|$ is the exponent in the reflection amplitude. (This interpretation originated in conversation with Professor R. G. Littlejohn.)

Finally, there ought to be connections between this work (and, more generally, Dingle's interpretative theory of asymptotic series (I)) and Écalle's recent doctrine of *résurgence* (Écalle 1981, 1984; see also Voros 1983; Pham *et al.* 1989).

I thank Professor F. Pham for the hospitality of the mathematics department of the University of Nice, where this research was carried out.

References

Abramowitz, M. & Stegun, I. A. 1964 *Handbook of Mathematical Functions*. Washington: National Bureau of Standards.

Dingle, R. B. 1973 *Asymptotic expansions: their derivation and interpretation*. New York & London: Academic Press. (Referred to as I in the text.)

Écalle, J. 1984 'Cinq applications des fonctions résurgentes'. (Preprint 84T62 Orsay.)

Écalle, J. 1981 'Les fonctions résurgentes'. (In 3 volumes.) *Publ. Math. Université de Paris-Sud.*

Olver, F. W. J. 1974 *Asymptotics and special functions*, ch. 14. New York & London: Academic Press.

Pham, F., Nosmas, C. & Candelpergher, B. 1989 'Résurgence, quantized canonical transformations and multi-instanton expansions'. In *Prospect in Algebraic Analysis*, dedicated to Professor M. Sato on his sixtieth birthday. (In the press.)

Stokes, G. G. 1864 *Trans. Camb. phil. Soc.* **10**, 106–128. Reprinted in *Mathematical and Physical papers by the late Sir George Gabriel Stokes*. Cambridge University Press 1904, vol. IV, 77–109.

Stokes, G. G. 1871 *Trans. Camb. phil. Soc.* **11**, 412–425. Reprinted in *Mathematical and Physical papers by the late Sir George Gabriel Stokes*. Cambridge University Press 1904, vol. IV, 283–298.

Stokes, G. G. 1889 *Proc. Camb. phil. Soc.* **6** (6). Reprinted in *Mathematical and Physical papers by the late Sir George Gabriel Stokes*. Cambridge University Press 1905, vol. V, 221–225.

Stokes, G. G. 1902 *Acta math., Stockh.* **26**, 393–397. Reprinted in *Mathematical and Physical papers by the late Sir George Gabriel Stokes*. Cambridge University Press 1905, vol. V, 283–287.

Voros, A. 1983 *Ann. Inst. H. Poincaré* **39**, 211–338.

Wright, F. J. 1980 *J. Phys.* A **13**, 2913–2928.

Emotional asymptotics

Michael Berry ~2000

(inspired by the Stokes phemoneon)

Passions rise.
We cross the bifurcation;
something intense and complex is born.
But it is evanescent from the start, and soon decays.
Much later, we meet again.
Now there is another dominant exponential in her life.
As we cross our Stokes line, all passion vanishes.
The whole thing was an error (function).

Hyperasymptotics for integrals with saddles

By M. V. Berry and C. J. Howls
H. H. Wills Physics Laboratory, Tyndall Avenue, Bristol BS8 1TL, U.K.

Integrals involving $\exp\{-kf(z)\}$, where $|k|$ is a large parameter and the contour passes through a saddle of $f(z)$, are approximated by refining the method of steepest descent to include exponentially small contributions from the other saddles, through which the contour does not pass. These contributions are responsible for the divergence of the asymptotic expansion generated by the method of steepest descent. The refinement is achieved by means of an exact 'resurgence relation', expressing the original integral as its truncated saddle-point asymptotic expansion plus a remainder involving the integrals through certain 'adjacent' saddles, determined by a topological rule. Iteration of the resurgence relation, and choice of truncation near the least term of the original series, leads to a representation of the integral as a sum of contributions associated with 'multiple scattering paths' among the saddles. No resummation of divergent series is involved. Each path gives a 'hyperseries', depending on the terms in the asymptotic expansions for each saddle (these depend on the particular integral being studied and so are non-universal), and certain 'hyperterminant' functions defined by integrals (these are always the same and hence universal). Successive hyperseries get shorter, so the scheme naturally halts. For two saddles, the ultimate error is approximately $\epsilon^{2.386}$, where ϵ (proportional to $\exp(-A|k|)$ where A is a positive constant), is the error in optimal truncation of the original series. As a numerical example, an integral with three saddles is computed hyperasymptotically.

1. Introduction

We intend to discuss very accurate asymptotics for a class of integrals commonly occurring in pure and applied mathematics and physics, namely

$$I^{(n)}(k) = \int_{C_n(\theta_k)} \mathrm{d}z\, g(z) \exp\{-kf(z)\}. \tag{1}$$

Here $|k|$ is the large asymptotic parameter, and it will be convenient to regard $k \equiv |k| \exp(\mathrm{i}\theta_k)$ as complex. The functions f and g are analytic in a region which we will specify later. There are several saddles (stationary points) of f, which we assume to be simple zeros of the derivative $f'(z)$. The infinite oriented contour $C_n(\theta_k)$ is the path of steepest descent through the nth saddle, at $z = z_n$, along the two valleys, issuing from z_n, of the real part of the exponential. It is common to encounter integrals where the contour – for example, the real axis – is not a path of steepest descent, but in such cases the contour can be deformed to give the function as a sum of integrals of the type (1).

Thus (1) defines a function of k for each saddle n; if f is an Mth order polynomial, there are $M-1$ saddles. As is well known, the method of steepest descent (De Bruijn

Proc. R. Soc. Lond. A (1991) **434**, 657–675
Printed in Great Britain

Reprinted from Proceedings of the Royal Society A, Volume 434, (1991).

1958), based on expanding the integrand about z_n, generates for each integral an asymptotic series in powers of k^{-1}, multiplying a leading exponential $\exp\{-kf_n\}$, where $f_n \equiv f(z_n)$. As well as yielding extremely useful approximations, the method is fundamental in physics (providing for example connections between wave and ray optics, quantum and classical mechanics, and statistical mechanics and thermodynamics). Nevertheless, the series diverges, and the common view is that it can represent the exact integral $I^{(n)}$ only within an accuracy comparable with the size of the least term.

Here we show how much greater accuracy, and deeper understanding, can be obtained with the aid of the *principle of resurgence*, inspired by the works of Dingle (1973) and Écalle (1981, 1984), which we express in the following way. The reason for the divergence of the asymptotic series for $I^{(n)}$ is the existence of *other saddles* $z_{m \neq n}$, through which C_n does not pass. These contribute smaller exponentials, 'beyond all orders' of k^{-1}. Therefore the divergent part of the series must contain information about these other saddles. This is also true for the other asymptotic series, based on the other saddles z_m. Thus all the asymptotic series are related by a requirement of mutual consistency: each must contain, in its late terms, all the terms of the asymptotic series from all the other saddles. This information is present in coded form, because the series all diverge.

Its systematic decoding, to obtain successive approximations more accurate than the least term of the original series, is what in our recent paper (Berry & Howls 1990*a*, hereinafter called I) we called 'hyperasymptotics'. This terminology was introduced to distinguish the novel procedure from the following two familiar schemes. First, ordinary (Poincaré) asymptotics, namely stopping at a fixed order N independent of $|k|$; as is well known, the error here is of order $|k|^{-(N+1)}$. Second, 'superasymptotics', namely stopping at the least term (a procedure originally used by Stokes (1847)); the order of this term is proportional to $|k|$, and the error is now of order $\exp(-A|k|)$ where A is a positive constant. With hyperasymptotics we achieved an error $\exp(-2.386A|k|)$.

The functions we studied in I were solutions of second-order differential equations, of Schrödinger type, dominated by a single transition point. Such equations have two solutions, each represented in lowest-order WKB theory by an exponential. Hyperasymptotics was based on repeated Borel summation, based on a formal resurgence relation, discovered by Dingle (1973), relating the late terms of the series multiplying one exponential and the early terms of the series multiplying the other. That example was special because it involved only two exponentials and the terms in the two asymptotic series were the same, up to signs.

Hyperasymptotics based on the integral (1) is much more general because many exponentials are involved and the terms in all the associated asymptotic series are different. The treatment differs from that in I, in that it does not involve the resummation of divergent series. Instead, we shall employ the iteration of an exact and finite (i.e. not formal) resurgence formula, giving the remainder of the truncated asymptotic expansion for $I^{(n)}$ in terms of the integrals $I^{(m)}$ through certain other saddles, selected by a rule depending on the topology of $f(z)$. This integral resurgence formula generalizes the Stieltjes-transform relation postulated in recent work by Boyd (1990).

2. Steepest-descent expansion

The steepest path through z_n, that is $C_n(\theta_k)$, is defined by $k(f(z)-f_n)$ real and increasing away from z_n. As the phase θ_k is altered, the steepest path near z_n rotates half as fast, so that when k returns to its original value, after a phase change of 2π, the orientation of $C_n(\theta_k)$ has reversed, and $I^{(n)}(k)$ has changed sign. Therefore the integrals $I^{(n)}(k)$ are double-valued, and the functions $T^{(n)}(k)$, defined by

$$I^{(n)}(k) \equiv k^{-\frac{1}{2}} \exp[-kf_n]\, T^{(n)}(k),$$

i.e.

$$T^{(n)}(k) = k^{\frac{1}{2}} \int_{C_n(\theta_k)} \mathrm{d}z\, g(z) \exp\{-k[f(z)-f_n]\} \tag{2}$$

are single valued.

These functions are not, however, continuous, because the steepest paths jump whenever they pass through one of the other saddles $m \neq n$. One way to see that jumps must occur is to consider the case where $f(z)$ is an Mth-order polynomial, so that there are $M-1$ saddles. Then $f(z)$ grows at infinity like z^M, so that, when θ_k changes, $C_n(\theta_k)$ rotates at infinity $\frac{1}{2}M$ times as fast as near z_n; therefore when $M > 2$ (that is, when there is more than one saddle) the steepest paths must jump if they are merely to reverse orientation during a circuit of k in its plane. Each jump is an example of the Stokes phenomenon, which as we will explain later is in this formalism a discontinuity, rather than a smooth transition (Berry 1989*a*), because we have *chosen* the path to be the one of steepest descent.

We require the coefficients $T_r^{(n)}$ in the formal asymptotic expansion

$$T^{(n)}(k) = \sum_{r=0}^{\infty} \frac{T_r^{(n)}}{k^r}. \tag{3}$$

The expansion is formal because, as is well known and as we shall study in detail later in this section, it diverges.

To obtain the coefficients, and also to develop our resurgence formula, we shall use a representation of $T^{(n)}(k)$ as a double integral. This is based on the transformation to the new integration variable

$$u(z) \equiv k[f(z)-f_n]. \tag{4}$$

For each value of z on $C_n(\theta_k)$, u is real and non-negative. For each value of u (except $u=0$), there are two values of z (figure 1): $z_+(u)$, on the half of the steepest descent path emerging from z_n, and $z_-(u)$, on the half leading into z_n. The transformation gives

$$T^{(n)}(k) = \int_0^{\infty} \mathrm{d}u \frac{\exp(-u)}{k^{\frac{1}{2}}} \left\{ \frac{g(z_+(u))}{f'(z_+(u))} - \frac{g(z_-(u))}{f'(z_-(u))} \right\}. \tag{5}$$

The quantity in curly brackets can be written as the contour integral

$$\left\{ \frac{g(z_+(u))}{f'(z_+(u))} - \frac{g(z_-(u))}{f'(z_-(u))} \right\} = \frac{1}{2\pi i u^{\frac{1}{2}}} \oint_{\Gamma_n(\theta_k)} \mathrm{d}z \frac{g(z)[k\{f(z)-f_n\}]^{\frac{1}{2}}}{f(z)-f_n-u/k}, \tag{6}$$

where $\Gamma_n(\theta_k)$ is the positive (anticlockwise) loop surrounding $C_n(\theta_k)$ (figure 2). The square root, defined as having phase zero on the path $C_n(0)$ emerging from the saddle, and π on the path leading into it, is single-valued on $C_n(\theta_k)$ because of the double-

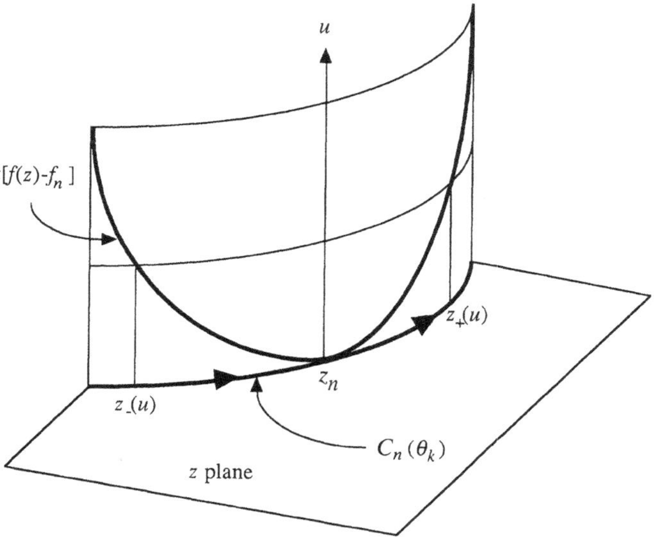

Figure 1. Double-valued mapping (equation (4)) from z to u.

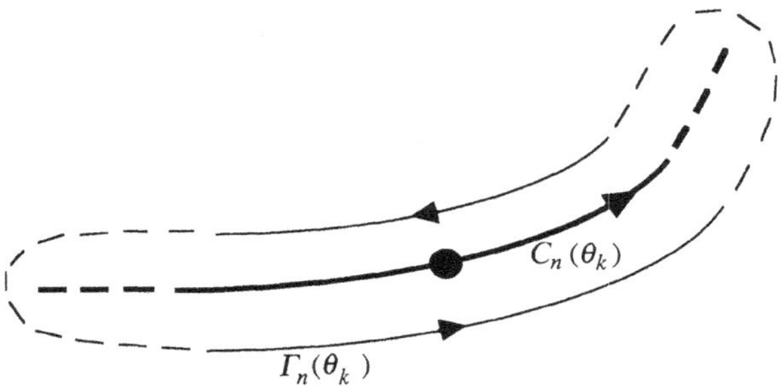

Figure 2. Steepest path $C_n(\theta_k)$ through saddle n, and loop $\Gamma_n(\theta_k)$ enclosing it.

valuedness of (4). Equations (6) and (5) now give the desired representation, which will form the basis of all that follows:

$$T^{(n)}(k) = \frac{1}{2\pi i}\int_0^\infty du \frac{\exp(-u)}{u^{\frac{1}{2}}} \oint_{\Gamma_n(\theta_k)} dz \frac{g(z)[f(z)-f_n]^{\frac{1}{2}}}{f(z)-f_n-u/k}. \tag{7}$$

Expanding the denominator in powers of k^{-1} gives the coefficients in (3) as (cf. Dingle 1973, p. 119)

$$T_r^{(n)} = \frac{(r-\frac{1}{2})!}{2\pi i}\oint_n dz \frac{g(z)}{[f(z)-f_n]^{r+\frac{1}{2}}}, \tag{8}$$

where the subscript n indicates that now the contour $\Gamma_n(\theta_k)$ has been shrunk to a small positive loop around z_n. These integrals can be evaluated exactly in terms of the coefficients in the expansions of f and g about z_n, to yield the explicit (and complicated) expressions (Dingle 1973, p. 119ff) for the terms in the saddle-point expansion. For example, the leading term $r = 0$ is

$$T_0^{(n)} = (2\pi/f_n'')^{\frac{1}{2}} g_n \tag{9}$$

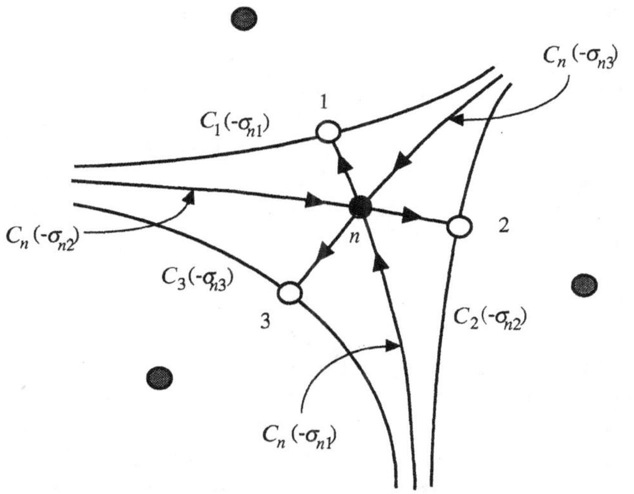

Figure 3. Special phase contours of $f(z)-f_n$ (steepest paths) through saddle n (●), encountering adjacent saddles $m = 1, 2, 3$ (○), with their steepest paths; non-adjacent saddles (shaded circles) are also shown.

(here, as before, primes denote derivatives and subscripts n denote quantities evaluated at z_n).

3. Resurgence relation

We expand the denominator in (7) to isolate the first N terms, using

$$\frac{1}{1-x} = \sum_{r=0}^{N-1} x^r + \frac{x^N}{1-x}. \tag{10}$$

Thus

$$T^{(n)}(k) = \sum_{r=0}^{N-1} \frac{T_r^{(n)}}{k^r} + R^{(n)}(k,N), \tag{11}$$

where $T_r^{(n)}$ are the coefficients (8) and the remainder $R^{(n)}$ is

$$R^{(n)}(k,N) = \frac{1}{2\pi i k^N}\int_0^\infty du \exp(-u)u^{N-\frac{1}{2}} \times \oint_{\Gamma_n(\theta_k)} dz \frac{g(z)}{[f(z)-f_n]^{N+\frac{1}{2}}\{1-u/k(f(z)-f_n)\}}. \tag{12}$$

Next we deform the contour $\Gamma_n(\theta_k)$ in a particular way, which as will now be explained depends on the topology of $f(z)$. Consider all the steepest paths through the saddle n, for different θ_k. As illustrated in figure 3, some of them are special in that they encounter other saddles m. We call these the saddles *adjacent* to n. From now on we will make extensive use of this concept of adjacency.

To specify the paths through the adjacent saddles we define the 'singulants' (a term introduced by Dingle (1973))

$$F_{nm} \equiv |F_{nm}| \exp(i\sigma_{nm}) \equiv f_m - f_n. \tag{13}$$

The special steepest paths are those corresponding to kF_{nm} positive real, that is

$$k = |k| \exp(-\mathrm{i}\sigma_{nm}), \quad \text{i.e. } \theta_k = -\sigma_{nm}. \tag{14}$$

The steepest path $C_n(-\sigma_{nm})$ turns sharply through a right angle at z_m, to continue descending into a valley of $\exp\{-k[f(z)-f_n]\}$ (whether the path turns right or left depends on whether θ_k approaches $-\sigma_{nm}$ from below or above; cf. figure 4 later).

We deform $\Gamma_n(\theta_k)$ by expanding it onto the union of arcs at infinity and arcs through the adjacent saddles m. These arcs are the steepest paths through m with k given by (14), that is $C_m(-\sigma_{nm})$. We obtain, symbolically, for (12),

$$\oint_{\Gamma_n(\theta_k)} \mathrm{d}z \ldots = \sum_m (-1)^{\gamma_{nm}} \int_{C_m(-\sigma_{nm})} \mathrm{d}z \ldots, \tag{15}$$

where the sum is over the adjacent saddles and γ_{nm} is an 'orientation anomaly', 0 if the arc of the expanded $\Gamma_n(\theta_k)$ has the same orientation as $C_m(-\sigma_{nm})$, and 1 otherwise. For the relation (15) to hold, three conditions must be satisfied. First, $|g|/|f|^{N+\frac{1}{2}}$ must decay at infinity faster than $1/|z|$, in order for the infinite arcs to give zero contribution. Second, there must be no zeros of the quantity in braces in the denominator of (12), for any u, in the region of the z plane swept by this contour expansion; in Appendix A we study these 'dangerous zeros' (which for $u = 0$ are also branch points of the integrand), and show that they do lie outside this region. Third, the functions f and g must contain no singularities in this same region; this is the analyticity condition mentioned in §1.

Next, along the arc of the expanded path which passes through m we transform the u integration variable in (12) to v, according to the relation

$$u \equiv v\,[f(z)-f_n]/F_{nm} = v + v\,[f(z)-f_m]/F_{nm}. \tag{16}$$

Like u, v is real and positive. The remainder (12) becomes

$$R^{(n)}(k,N) = \frac{1}{2\pi \mathrm{i} k^N} \sum_m \frac{(-1)^{\gamma_{nm}}}{F_{nm}^{N+\frac{1}{2}}} \int_0^\infty \mathrm{d}v \frac{\exp(-v)}{1-v/(kF_{nm})} v^{N-\frac{1}{2}}$$
$$\times \int_{C_m(-\sigma_{nm})} \mathrm{d}z\, g(z) \exp\{-v[f(z)-f_m]/F_{nm}\}. \tag{17}$$

The z integral is of the form (1), and so, using (2), we see that the remainder has been expressed as a sum over integrals through the adjacent saddles:

$$R^{(n)}(k,N) = \frac{1}{2\pi \mathrm{i}} \sum_m \frac{(-1)^{\gamma_{nm}}}{(kF_{nm})^N} \int_0^\infty \mathrm{d}v \frac{v^{N-1}\exp(-v)}{1-v/(kF_{nm})} T^{(m)}\left(\frac{v}{F_{nm}}\right). \tag{18}$$

Combining this with (11), we obtain the following exact resurgence formula, which will be the basis for hyperasymptotics:

$$T^{(n)}(k) = \sum_{r=0}^{N-1} \frac{T_r^{(n)}}{k^r} + \frac{1}{2\pi \mathrm{i}} \sum_m \frac{(-1)^{\gamma_{nm}}}{(kF_{nm})^N} \int_0^\infty \mathrm{d}v \frac{v^{N-1}\exp(-v)}{1-v/(kF_{nm})} T^{(m)}\left(\frac{v}{F_{nm}}\right). \tag{19}$$

This formula provides an explicit and exact form for the remainder in the method of steepest descent. When we use it for hyperasymptotics we shall choose N to be the order of the least term of the asymptotic series (3). But it holds for any N, and by choosing N fixed and letting $|k| \to \infty$ we can use it to establish at once that (3) is an asymptotic series in the sense of Poincaré, that is the remainder is of order $|k|^{-N}$ (for a development of this idea, see Boyd (1991)).

An immediate application of (19) is to the derivation of formal resurgence relations for the coefficients $T_r^{(n)}$ in the expansion about the nth saddle. Choosing $N = 0$ (so that the sum over r is empty), substituting the series (3) for the saddles n and m into both sides of (19), and identifying coefficients of powers of k^{-1}, we obtain

$$\begin{aligned} T_r^{(n)} &= \frac{1}{2\pi i}\sum_m (-1)^{\gamma_{nm}} \sum_{t=0}^{\infty} \frac{(r-t-1)!}{F_{nm}^{r-t}} T_t^{(m)} \\ &= \frac{1}{2\pi i}\sum_m (-1)^{\gamma_{nm}} \frac{(r-1)!}{F_{nm}^r}\left[T_0^{(m)} + \frac{F_{nm}}{r-1} T_1^{(m)} \ldots\right]. \end{aligned} \tag{20}$$

This gives the late terms ($r \gg 1$) of the series for a given saddle as a sum over the early terms of the series for the adjacent saddles. The leading contribution, from the adjacent saddle m^* with the smallest singulant $|F_{nm}|$, is

$$T_r^{(n)} \approx T_0^{(m^*)} \frac{(-1)^{\gamma_{nm^*}}(r-1)!}{F_{nm^*}^r} \tag{21}$$

and has the familiar 'factorial/power' form. (In the interesting special case where two or more adjacent saddles have the same |singulant|, their contributions add.) The complete resurgence relation (20) is formal because the factorials are infinite when $t > r-1$, but it does give an asymptotic expansion for the late terms. By Borel summation, the formal relation can be converted into an exact one, but this is just (19) (indeed we originally obtained (19) in this way, after conjecturing (20)).

In the special case of two saddles (for which there is just one adjacent saddle m) the resurgence relation (20) was discovered by Dingle (1973) not only for integrals (where it was rediscovered by Balian *et al.* 1979) but also for the analogous case of second-order differential equations (where it was rediscovered by Rakovic & Solov'ev (1989)).

Now we show how the resurgence formula (19) incorporates the Stokes phenomenon (Stokes 1864). This is the appearance of a subdominant exponential (Berry 1989*b*) as the contour $C_n(\theta_k)$ sweeps through one of the adjacent saddles m. As we have seen, one way to make this happen is to vary θ_k through $-\sigma_{nm}$, defined by (13). Let

$$k = |k| \exp\{i(-\sigma_{nm} + \delta)\} \tag{22}$$

and let δ sweep through zero. The behaviour of the integration contour $C_n(\theta_k)$ during this process is illustrated in figure 4, and shows that $T^{(n)}$ has a discontinuity, whose magnitude must be

$$\begin{aligned} &T^{(n)}(|k| \exp\{i(-\sigma_{nm} + 0_+)\}) - T^{(n)}(|k| \exp\{i(-\sigma_{nm} + 0_-)\}) \\ &\qquad = (-1)^{\gamma_{nm}+1} \exp(-kF_{nm})\, T^{(m)}(|k| \exp\{-i\sigma_{nm}\}) \end{aligned} \tag{23}$$

(the extra minus sign can be confirmed by diagrams corresponding to the two possible orientation anomalies). This jump is exponentially small because kF_{nm} is positive real. Exactly the same jump is given by (19); its origin is the pole at $v = kF_{nm}$, which sweeps up through the integration contour (positive real axis) as δ increases through zero (the extra minus comes from the sign of the denominator in (19)). Note that the discontinuity is independent of the truncation order N, a known aspect of the Stokes phenomenon (Dingle 1973).

At first sight, the discontinuity might appear inconsistent with the universal smooth behaviour across Stokes lines, established by Berry (1989*a*). However, the discordance arises from a difference of definition. In this paper it has been convenient

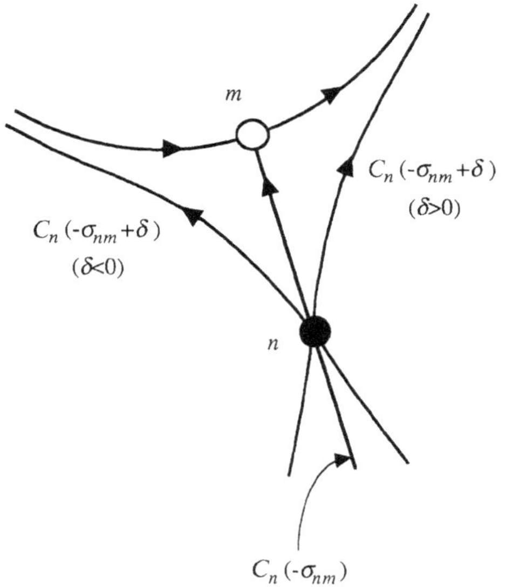

Figure 4. Jump in the steepest path through n as adjacent saddle m is passed, corresponding to the appearance of a subdominant exponential (Stokes's phenomenon).

to define the integration contour in (1) as the single infinite arc through z_n, connecting valleys of the integrand. This contour is k-dependent, and jumps when θ_k passes $-\sigma_{nm}$. On the other hand, it is more usual to define integrals with contours between infinite limits independent of k, which are therefore analytic functions of k near $\theta_k = -\sigma_{nm}$. The conventional definition can easily be incorporated in the present framework by allowing the pole in (19) to drag the integration contour with it when it passes the positive real axis, as we have explained in detail in §5 of I.

4. Hyperasymptotic multiple scattering

The iteration of the resurgence formula (19) is more complicated than in I, because now there can be several adjacent saddles (as embodied in the sum over m in (19)), rather than just two. After iterating (19) s times, we obtain

$$T^{(0)} = \sum_{r=0}^{N_0-1} T_r^{(0)} K_r^{(0)} + \sum_1 \sum_{r=0}^{N_1-1} T_r^{(1)} K_r^{(01)} + \sum_1 \sum_2 \sum_{r=0}^{N_2-1} T_r^{(2)} K_r^{(012)} + \ldots \sum_1 \ldots \sum_s \left(\sum_{r=0}^{N_s-1} T_r^{(s)} K_r^{(01\ldots s)} + R^{(01\ldots s)} \right). \quad (24)$$

To avoid long formulae we have here used an abbreviated notation: 0 stands for the starting saddle n_0, 1 for the saddles n_1, adjacent to n_0, reached in the first iteration, $\ldots s$ for the saddles n_s reached in the sth iteration. For reasons to be explained later, we allow the arbitrary truncation orders N to change at each iteration. As well as the coefficients $T_r^{(s)}$ in all the primitive asymptotic series (3), this formula involves 'hyperterminant' integrals $K_r^{(01\ldots s)}$ generalizing those in I, and the remainders $R^{(01\ldots s)}$.

The hyperterminants and the remainders depend on

$$y_{p-1,p,p+1} \equiv F_{p-1,p}/F_{p,p+1}. \quad (25)$$

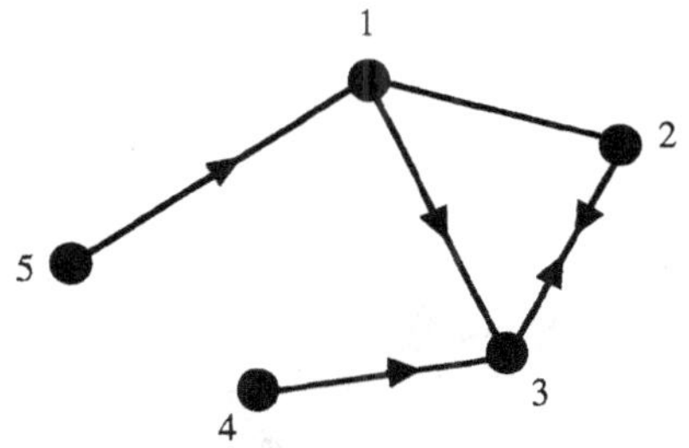

Figure 5. A possible adjacency diagram for five saddles, with arrows pointing to the nearest; thus adjacent to 1 are 2, 3 and 5, with 3 nearest.

The hyperterminants are

$$K_r^{(0)} = 1/k^r$$

$$K_r^{(01\ldots s)} = \left(\prod_{p=0}^{s-1} \int_0^\infty \mathrm{d}v_p\right) J_r^{(01\ldots s)}, \tag{26}$$

where

$$J_r^{(01)} = \frac{(-1)^{\gamma_{01}}}{2\pi \mathrm{i} k^{N_0} F_{01}^{N_0-r}} \frac{v_0^{N_0-r-1} \exp(-v_0)}{(1-v_0/kF_{01})},$$

$$J_r^{(01\ldots s)} = \frac{(-1)^{\gamma_{s-1,s}}}{2\pi \mathrm{i} F_{s-1,s}^{N_{s-1}-r}} \frac{v_{s-1}^{N_{s-1}-r-1} \exp(-v_{s-1})}{(1-(v_{s-1}/v_{s-2})\, y_{s-2,s-1,s})} J_{N_{s-1}}^{(01\ldots s-1)} \quad (s \geqslant 2). \tag{27}$$

The remainder after s iterations is

$$R^{(0\ldots s)} = \left(\prod_{p=0}^{s} \int_0^\infty \mathrm{d}v_p\right) \sum_{s+1} J_0^{(0\ldots s+1)}\, T^{(s+1)}\left(\frac{v_s}{F_{s,s+1}}\right). \tag{28}$$

The scheme (24)–(28) gives an exact representation of the integral (1). It looks complicated, and might appear to give no advantage over the original representation as a single integral, because the hyperterminants are multiple integrals whose order increases with order of iteration. However, the hyperterminants are *universal functions* of the singulants and truncations, which need be evaluated only once; the non-universal aspects of the scheme, distinguishing the particular integral being evaluated, are embodied in the coefficients $T_r^{(s)}$ in (24).

It is convenient to interpret (24) by regarding iteration as multiple scattering between successive groups of adjacent saddles. The first term (involving $K^{(0)}$) corresponds to no scatterings and involves the single saddle 0, the second term (involving $K^{(01)}$) corresponds to single scatterings from 0 to one of the adjacent saddles 1, etc. A diagrammatic representation is appropriate. The first step is to construct an 'adjacency diagram' (figure 5) in which the saddles are depicted as points in a plane (e.g. the original z plane), and each is connected to those adjacent to it (the significance of the arrows will be explained later). A multiple scattering path is now defined as a sequence of saddles, each adjacent to the next. The set of multiple scattering paths can be constructed as illustrated in figure 6. The number of paths proliferates rapidly: in figure 6, which corresponds to five saddles, connected as in figure 5, the first three iterations generate 15 paths. When there are just two saddles (the case considered in I) there is only one scattering path, bouncing back and forth between them. For such backscattering, the definitions (13) and (25) give $y = -1$, so that the denominators in the integrals (26) and (27) do not vanish. In general, a denominator will vanish if the corresponding y is positive real, but we

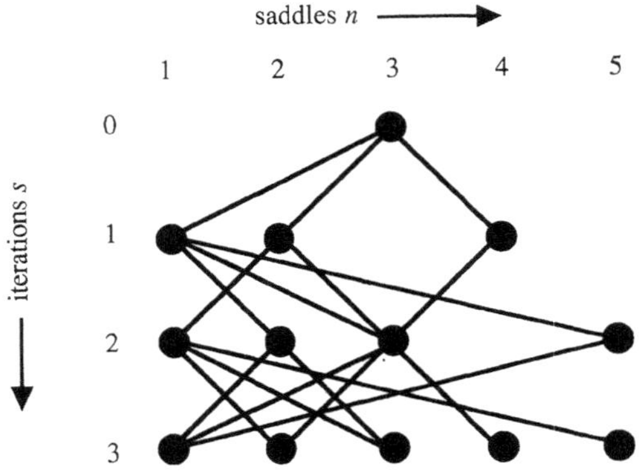

Figure 6. The fifteen multiple scattering paths from saddle 3, generated by the adjacency diagram of figure 5 by hyperasymptotic iteration.

ignore this exceptional case (which is the signature of a 'double Stokes phenomenon', in which a steepest path from one saddle hits a second, turns through a right angle and then encounters a third).

In (24) each scattering path contributes a 'hyperseries' over the orders r of the original asymptotic expansion, whose limits are $r = 0$ and $r = N_s - 1$ (for a path with s scatterings). Unless the truncations N_s are specified, the scheme is not unique. In any truncation scheme, each hyperseries must be shorter than its predecessor, because from (27) the hyperterminant integrals (26) do not converge unless $N_{p+1} < N_p$. This has two consequences. First, the sequence of hyperseries along each scattering path must eventually halt – when it generates a hyperseries containing a single term. Second, each sequence contains only a finite amount of information, because it involves only a finite number of primitive coefficients $T_r^{(n)}$ (the number being the truncation N when the saddle n first appears).

Because of this finite information, hyperasymptotics will halt with a finite remainder. It might seem natural to make the remainder as small as possible by choosing a large starting truncation, and indeed arbitrary accuracy can be obtained in this way, as we showed in I. However, such schemes are unnatural because they lead to series whose terms get very large before decreasing, thereby violating the 'live now, pay later' philosophy underlying asymptotics, where ultimate accuracy is sacrificed for improvement at every stage. (Without this philosophy, one could as well use the *convergent* series in k which can be found in many cases, but which are computationally inefficient when $|k|$ is large.)

To preserve the spirit of asymptotics, we shall consider each scattering path separately and demand that the successive series it generates are optimally truncated; that is, truncated near their least terms, so that the remainders $R^{(0\cdots s)}$ for each s along the path are smallest. In determining the optimal Ns, we shall make use of the concept of the *nearest* saddle to a given saddle n, defined as the adjacent saddle m with the smallest singulant modulus $|F_{nm}|$. As in (21), we shall denote nearest saddles by asterisks, so that, for example, the path 0123* is the path starting at saddle 0, going through given saddles 1 and 2, and ending at the saddle 3* nearest to 2. In an adjacency diagram such as figure 5, nearest saddles can be indicated by arrows.

To determine the optimal truncations we must estimate the remainders $R^{(0\cdots s)}$.

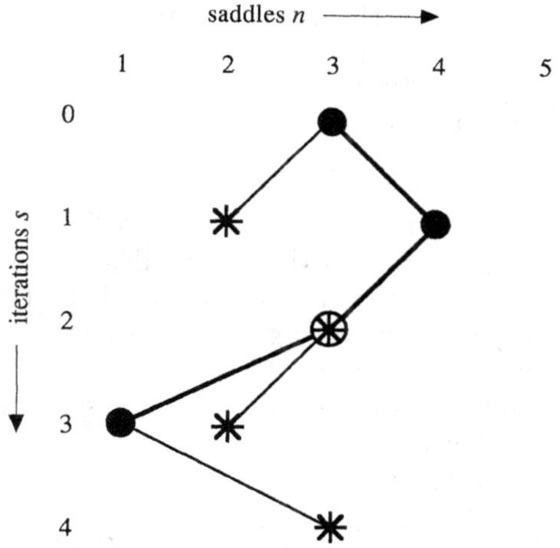

Figure 7. One of the multiple scattering paths in figure 6 (thick line) together with nearest saddles (*) contributing at each level of truncation (shown linked to their progenitors by thin lines). For the second iteration the nearest saddle (3) is also the next saddle on the actual path.

This we do by approximating $T^{(s+1)}$ to lowest order with (21) (thereby replacing the sum over the $(s+1)$th order saddles, adjacent to the sth, by the saddle $(s+1)^*$ nearest to the sth). Moreover, the denominators in (27) are of order unity (this follows from the procedure, used in I, of replacing the vs in the denominators by the maxima of the rest of the integrands), and so we approximate them by unity. The vs are now uncoupled, and the remainders can be evaluated as products of factorials. These can be simplified with the aid of Stirling's approximation, and we obtain

$$\left.\begin{aligned} |R^{(0)}| &\approx \frac{N_0^{N_0-\frac{1}{2}}\exp(-N_0)}{\sqrt{(2\pi)}\,|kF_{01^*}|^{N_0}}|T_0^{(1^*)}|, \\ |R^{(0\ldots s)}| &\approx \frac{N_s^{N_s-\frac{1}{2}}\exp(-N_0)}{(2\pi)^{(s+1)/2}\,|k|^{N_0}\,|F_{s,(s+1)^*}|^{N_s}}\prod_{p=0}^{s-1}\frac{(N_p-N_{p+1})^{(N_p-N_{p+1}-\frac{1}{2})}}{|F_{p,p+1}|^{N_p-N_{p+1}}}\,|T_0^{((s+1)^*)}| \end{aligned}\right\} \tag{29}$$

Successive minimization along the scattering path now gives

$$\left.\begin{aligned} N_0 &= \operatorname{Int}|kF_{01^*}|, \\ N_s &= \operatorname{Int}\frac{N_{s-1}}{1+|y_{s-1,s,(s+1)^*}|} = \operatorname{Int}\frac{N_{s-1}}{1+|F_{s-1,s}/F_{s,(s+1)^*}|}. \end{aligned}\right\} \tag{30}$$

Henceforth we shall omit the 'Int' symbols. The nearest saddle to s, that is $(s+1)^*$, cannot be more distant than the preceding saddle $s-1$, so that

$$|y_{s-1,s,(s+1)^*}| \geqslant 1, \quad \text{and} \quad N_s \leqslant \tfrac{1}{2}N_{s-1}. \tag{31}$$

Thus each hyperseries is at most half the length of its predecessor. Figure 7 shows one of the scattering paths of figure 6, together with the nearest saddles which contribute to the successive truncations.

With these truncations, the optimal hyperasymptotic remainders can be estimated from (29). The zero-stage (superasymptotic) remainder is

$$|R^{(0)}| \approx \frac{\exp(-|kF_{01^*}|)}{\sqrt{(2\pi|kF_{01^*}|)}}|T_0^{(1^*)}|. \tag{32}$$

The improvement from the $(s-1)$th to the sth stage is

$$\left|\frac{R^{(0\ldots s)}}{R^{(0\ldots s-1)}}\right| \approx \frac{1+|y_{s-1,s,(s+1)^*}|}{\sqrt{(2\pi N_{s-1}|y_{s-1,s,(s+1)^*}|)}} \times \exp\left\{-N_{s-1}\ln\left[\frac{|F_{s,(s+1)^*}|+|F_{s-1,s}|}{|F_{s-1,s^*}|}\right]\right\}\left|\frac{T_0^{((s+1)^*)}}{T_0^{(s^*)}}\right|. \tag{33}$$

This does represent an improvement, rather than a degradation, because (31) and the definition of nearness guarantee that the argument of the logarithm exceeds unity, so that the exponent is positive. On average the improvements diminish, because the truncation limits decrease (equation 30). The improvement is greater if s is a saddle far from $s-1$ (rather than the nearest saddle s^*), and less if the saddle $(s+1)^*$ nearest to s is closer to s than $s-1$ was.

In the special case of two saddles, equivalent to that considered in I, the formulae simplify considerably. There is only one adjacent saddle, and it is of course the nearest, so that (cf. (25)) $y_{s-1,s,s+1}=-1$ and (cf. (30)) the sth truncation limit is $N_s=|kF_{01}|/2^s$. The hyperasymptotic improvement (33) is

$$\left|\frac{R^{(0\ldots s)}}{R^{(0\ldots s-1)}}\right| \approx \frac{2^{s/2}}{\sqrt{(\pi|kF_{01}|)}}\exp\left\{-\frac{|kF_{01}|}{2^{s-1}}\ln 2\right\}. \tag{34}$$

It is the accumulation of these improvements which, together with (32), gives the final remainder $\exp\{-|kF_{01}|(1+2\ln 2)\}=\exp\{-2.386|kF_{01}|\}$ obtained in I.

The optimal truncations (30), and the estimate (33) for the improvements in the remainder, hold for a given multiple scattering path. Each path can be followed to its end (when hyperasymptotics halts) and the final remainder estimated. In general the remainders will all be different. Obviously the overall accuracy of the scheme is controlled by the path with the largest final remainder. Therefore it is pointless to evaluate hyperseries along the other paths more precisely than this, so that hyperasymptotics for these other paths should be prematurely terminated at this level.

5. Example: Pearcey's integral

We shall calculate

$$\mathrm{P}(x,y)\equiv\int_C \mathrm{d}z\,\exp\{\mathrm{i}(\tfrac{1}{4}z^4+\tfrac{1}{2}xz^2+yz)\}, \tag{35}$$

where C descends into the valleys at $\infty\exp(\mathrm{i}\pi/8)$ and $\infty\exp(5\mathrm{i}\pi/8)$. This integral was first studied (for real x and y) by Pearcey (1946). Provided the contour passes through just one of the three saddles, this has the form (1), with $k=1$ (any other k can be reduced to 1 by scaling x and y), $g=1$ and

$$f(z;x,y)=-\mathrm{i}(\tfrac{1}{4}z^4+\tfrac{1}{2}xz^2+yz). \tag{36}$$

We choose the complex values

$$x=7,\quad y=1+\mathrm{i} \tag{37}$$

thereby ensuring that the magnitudes of the singulants are all different.

The positions of the saddles (where $z^3+xz+y=0$) are

$$\left.\begin{aligned} z_1&=+0.077621247095613-\mathrm{i}2.574820095698692,\\ z_2&=-0.143675227409104-\mathrm{i}0.142009934077984,\\ z_3&=+0.066053980313492+\mathrm{i}2.716830029776676,\end{aligned}\right\} \tag{38}$$

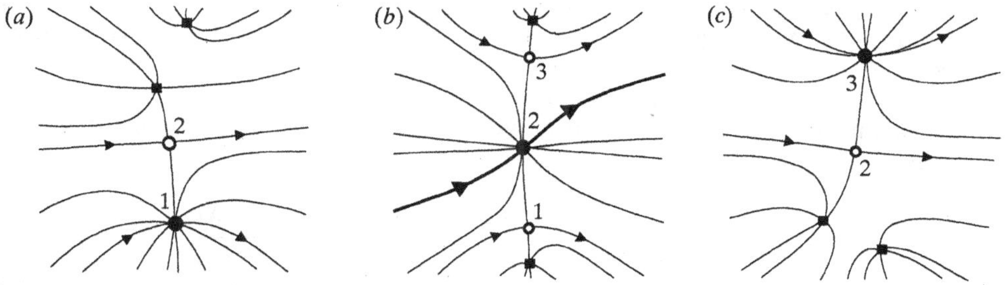

Figure 8. Some phase contours of $f(z)-f_n$, for the Pearcey integral (35)–(37) with (*a*) $n=1$, (*b*) $n=2$, (*c*) $n=3$. These saddles are denoted by •, and adjacent saddles by ○; the bold line contour in (*b*) corresponds to the steepest path deformation of the contour in (35); dangerous zeros (Appendix A) are denoted (for $u=0$) by ■.

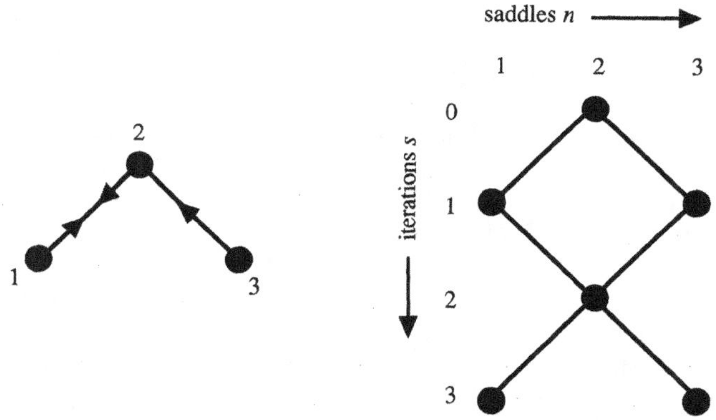

Figure 9. Adjacency diagram, and multiple scattering paths from saddle 2, corresponding to figure 8.

and the contour along the real axis can be deformed into the steepest path through saddle 2. Figure 8 defines the orientations of the contours through the three saddles, and shows the constant-phase lines (cf. figure 3) for the three functions $f(z)-f_n$. Thus 2 is adjacent to 1 and 3, and 1 and 3 are not adjacent. The adjacency diagram, and the multiple scattering paths from 2, are rather simple, as shown in figure 9. The orientation anomalies (cf. equation (15)) are

$$\gamma_{12}=1, \quad \gamma_{21}=0, \quad \gamma_{32}=0, \quad \gamma_{23}=1. \tag{39}$$

The three singulants are

$$\left.\begin{aligned} F_{12}(=-F_{21}) &= +2.429\,559\,462\,904\,937-\mathrm{i}\,9.601\,681\,152\,318\,827, \\ |F_{12}| &= 9.904\,294\,025\,047\,193, \\ F_{32} &= -2.858\,116\,325\,734\,320-\mathrm{i}\,14.897\,069\,623\,055\,830, \\ |F_{32}| &= 15.168\,767\,658\,765\,224, \\ F_{13} &= +5.287\,675\,788\,639\,256+\mathrm{i}\,5.295\,388\,470\,737\,003, \\ |F_{13}| &= 7.483\,358\,490\,796\,506. \end{aligned}\right\} \tag{40}$$

Therefore 1 is the nearer of the saddles adjacent to 2. We shall soon give the reason for including F_{13} even though this singulant cannot contribute to hyperasymptotics because 1 and 3 are not adjacent.

The raw data of the hyperasymptotic scheme (24) are the asymptotic coefficients $T_r^{(n)}$ appearing in (3). From the contour integral (8) we obtain (by expanding the denominator binomially)

$$T_r^{(n)} = \frac{-2^{3r+\frac{1}{2}}\mathrm{i}^{r+\frac{1}{2}}}{z_n^{4r+1}(3+x/z_n^2)^{3r+\frac{1}{2}}}\sum_{t=0}^{r}\frac{(-\frac{1}{8})^t(3+x/z_n^2)^t(3r-t-\frac{1}{2})!}{(2r-2t)!\,t!}. \tag{41}$$

These coefficients can be expressed in closed form in terms of Legendre or Gegenbauer polynomials (Abramowitz & Stegun 1972), as follows:

$$\begin{aligned} T_r^{(n)} &= \frac{\sqrt{(2\pi)}\,\mathrm{i}^{r+\frac{1}{2}}}{2^r z_n^{4r+1}(3+x/z_n^2)^{(5r+1)/2}}\left[\frac{\mathrm{d}^r}{\mathrm{d}\zeta^r}\mathrm{P}_{3r}(\zeta)\right]_{\zeta=\{(3+x/z_n^2)/2\}^{-\frac{1}{2}}} \\ &= \frac{-\sqrt{2}\,\mathrm{i}^{r+\frac{1}{2}}(r-\frac{1}{2})!}{z_n^{4r+1}(3+x/z_n^2)^{2r+\frac{1}{2}}}\mathrm{C}_{2r}^{r+\frac{1}{2}}(\{(3+x/z_n^2)/2\}^{-\frac{1}{2}}). \end{aligned} \tag{42}$$

On the basis of the leading-order late terms formula (21) and the singulants (40) we expect the least terms of the sequences $T_r^{(1)}$, $T_r^{(2)}$, $T_r^{(3)}$ to be near $r = 10$, 10 and 15 respectively. Figure 10 shows that this is the case. If saddles 1 and 3 had been adjacent, the smallest $T_r^{(1)}$ and $T_r^{(3)}$ would have been near $r = 7$; the fact that they are not confirms that the nearest saddle, which dominates the asymptotics, is determined not simply by proximity (smallest $|F|$): adjacency is necessary too.

We also require the hyperterminant integrals, defined by (26) and (27), for the multiple scattering paths of figure 9. In the present case, where $k = 1$ and both paths starting from saddle 2 scatter back to 2, the first and second hyperterminants are

$$\begin{aligned} K_r^{(01)} &= \frac{(-1)^{\gamma_{01}}}{2\pi\mathrm{i}F_{01}^{N_0-r}}\int_0^\infty \mathrm{d}v\frac{v^{N_0-r-1}\exp(-v)}{1-v/F_{01}}, \\ K_r^{(012)} &= \frac{(-1)^{\gamma_{01}+\gamma_{12}}(-1)^{N_1-r-1}}{4\pi^2F_{01}^{N_0-r}} \\ &\quad\times\int_0^\infty \mathrm{d}v_0\frac{v_0^{N_0-N_1-1}\exp(-v_0)}{1-v_0/F_{01}}\int_0^\infty \mathrm{d}v_1\frac{v_1^{N_1-r-1}\exp(-v_1)}{1+v_1/v_0}. \end{aligned} \tag{43}$$

(Note that here the subscripts 0, 1, 2 refer to orders of scattering, as in (24)–(28), and are not labels of saddles.) These integrals were computed as explained in Appendix B of I. We also needed the hyperterminants $K_r^{(0123)}$, which were approximated by replacing the vs in all the denominators by the stationary values of the other factors of the integrands, thereby uncoupling the integrals, which become proportional to factorials.

To the nearest integer, the rules (30) give the optimal truncation limits for the different paths of figure 9 (labelled by their saddle sequences) as

$$\left.\begin{aligned} N(2) = 10,\quad N(21) = N(23) = 5,\quad N(212) = N(232) = 2, \\ N(2121) = N(2123) = N(2321) = N(2323) = 1. \end{aligned}\right\} \tag{44}$$

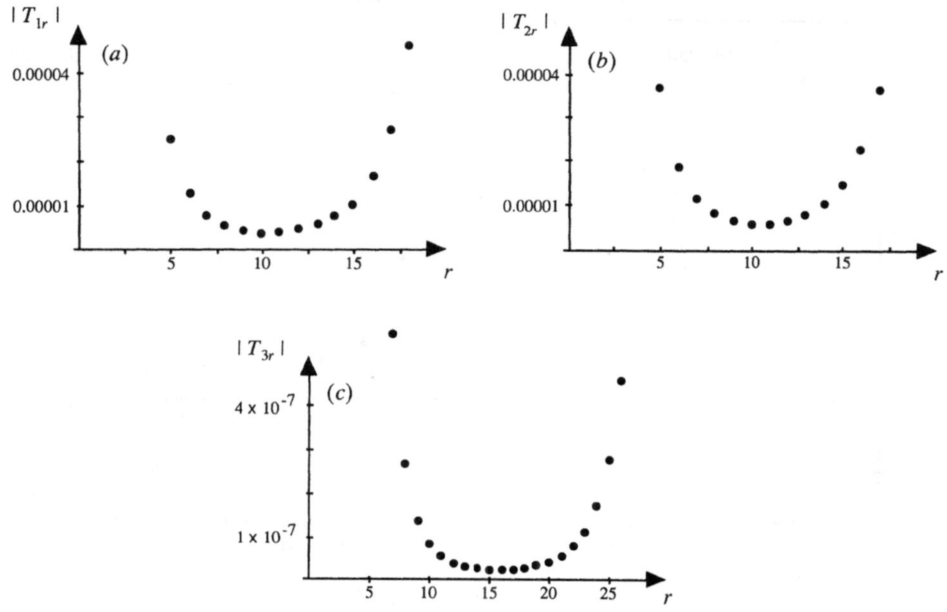

Figure 10. Magnitudes of asymptotic coefficients (41) for the three saddles in Pearcey's integral (35)–(37). Note the different scales for saddle 3.

Figure 11 shows the magnitudes of all the terms on all the scattering paths. This shows that the path with the largest remainder when hyperasymptotics terminates is 2121. As explained at the end of §4, this path controls the accuracy, and we should discard smaller terms from other scattering paths; these are the four terms below the line in figure 11.

Now we can evaluate the contributions to the terms in the hyperseries (24) and compare the partial sums with the 'exact' value of the Pearcey integral; this was computed from the convergent double series in ascending powers of x and y, taking about 3000 terms to ensure sufficient accuracy. Obviously it is sensible to add the terms, in the various hyperseries, in order of decreasing magnitude. Figure 12 shows the relative errors |approximate/exact -1| thus achieved. As expected, the errors decrease uniformly until the term 2121 at the end of the accuracy-controlling path is included. After that, there is no improvement, even though the last four terms decrease by nearly two orders of magnitude – indeed the very next term (2123) increases the error.

Table 1 shows some numerical values and relative errors of computations of $\mathrm{P}(7, 1+\mathrm{i})$. In addition to the 'exact' value, we include the lowest-order saddle-point approximation (that is, the first term of (3)), superasymptotics (that is, (3) truncated at its least term), and the best hyperasymptotic approximation, which includes the path 2121 and all larger terms. The ultimate error agrees well with the theoretical estimates (32) and (33), which give 1.103×10^{-12}.

Although our main aim was analytical understanding rather than the development of computational algorithms, we remark that hyperasymptotics seems to be numerically efficient: computation of the optimal approximation was about three

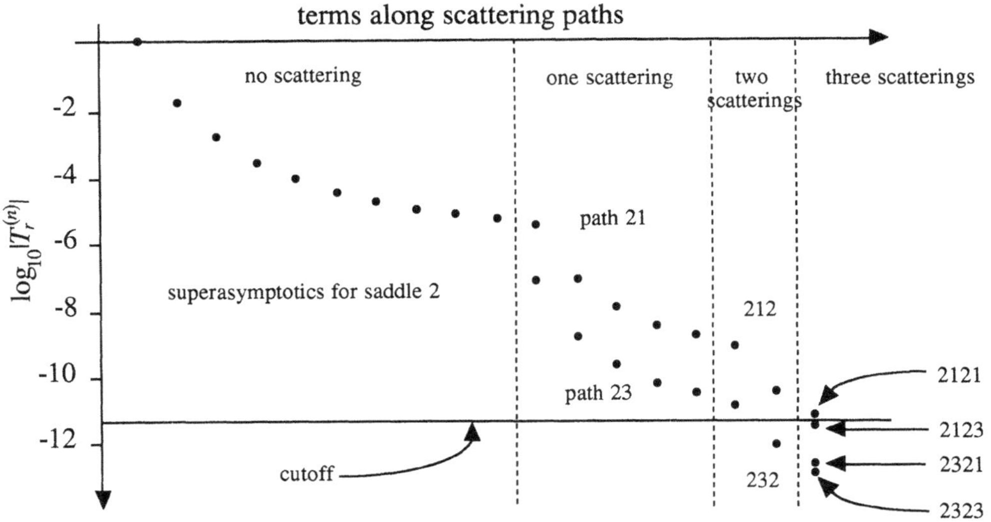

Figure 11. Magnitudes of the terms in the multiple scattering paths of figure 9 in the hyperasymptotics of Pearcey's integral.

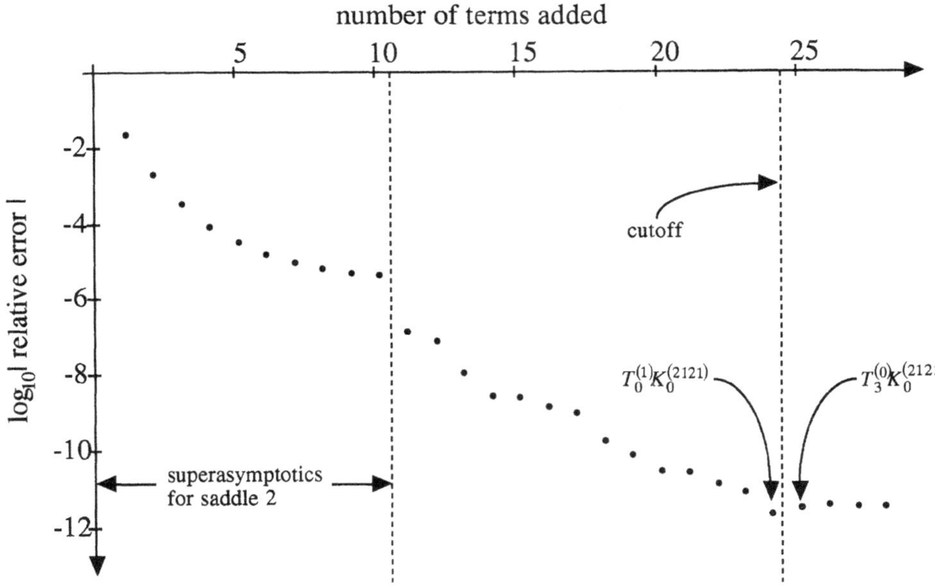

Figure 12. Magnitudes of the relative errors in hyperasymptotics of Pearcey's integral, incorporating successively smaller terms.

Table 1. *Approximations to the Pearcey integral (35)*

level	approximation to P (7, 1+i)	|approx./exact−1|
lowest	0.779703507027512+i0.765551648542315	1.496×10^{-2}
super.	0.788920520763900+i0.752101783262683	2.916×10^{-6}
ultimate hyper.	0.788922837595360+i0.752103959759701	1.535×10^{-12}
exact	0.788922837596969+i0.752103959759243	0

times faster than the 'exact' value. (The absolute time taken to calculate the 'exact' value, using the package Mathematica on an Apple Macintosh IIfx computer, was 400s.)

6. Concluding remarks

We have greatly extended the accuracy of the method of steepest descent, while retaining its numerical effectiveness as an integration technique, by basing it on the exact resurgence relation (19). The resulting iteration scheme (24)–(28) incorporates the effects of successively smaller exponentials, which arise from ever more distant groups of adjacent saddles and repeated scatterings from nearer ones. An attractive feature is the separation of non-universal properties characterizing the particular integral (1) being studied, and embodied in the raw asymptotic coefficients $T_r^{(n)}$, from the universal properties common to all integrals with saddles and embodied in the universal hyperterminant integrals (26)–(27). In a sense the history of asymptotics is repeating itself, albeit at a more refined stage, because as with Stokes's (1847) procedure of optimizing the raw asymptotic series by truncation at its least term, our iteration scheme too comes to a natural halt, representing an optimum level of approximation. We envisage several lines of further enquiry.

First, we hope mathematicians will be able to establish rigorous error bounds (rather than our estimates (32)–(33)). The explicit formula (28) for the hyperasymptotic remainder, which eliminates manipulations of divergent series, provides a natural starting-point for such investigations. It would be necessary to combine the errors for all the multiple scattering paths, which we have considered in isolation.

Second, there could be extensions to integrals (1) involving functions $f(z)$ and $g(z)$ with singularities. Clearly, each class of singularity would require special treatment.

Third, there is the question of 'beating the halting barrier', that is increasing the accuracy of the optimized scheme. Recall that for two saddles the error is roughly $\exp(-2.386|kF|)$. Further systematic improvement of the approximation ('ultra-asymptotics'?) would require evaluation of the roughly $\log_2|kF_{01*}|$-fold multiple-integral remainder (28). One possibility is to substitute the *convergent* series, in rising powers of the argument, for the function $T^{(s+1)}$. The motivation for this procedure is that the shortening of successive optimally truncated hyperseries can be regarded as a 'renormalization' of the original large parameter, which becomes effectively unity when the procedure halts. Preliminary indications are that fully halted hyperasymptotics can be terminated in this way, but that convergence is very slow.

Fourth, the k-dependence of hyperasymptotics should be unravelled. The first hyperseries (superasymptotics) is an exponential multiplied by a finite series in descending powers of k. The higher hyperseries have approximately this form, with smaller exponentials, but establishing the full k-dependence would involve approximating the hyperterminant integrals (26)–(27) to hyperasymptotic accuracy. The result could be that hyperasymptotics is a multiple-scale expansion whose large parameters are the different singulants.

Fifth, there is the extension to integrals with more than one variable. As well as being necessary to beat the halting barrier and approximate the hyperterminants, as just discussed, hyperasymptotics of multiple integrals would be useful in physics. Double integrals occur naturally in diffraction theory, where large k corresponds to the geometrical-optics limit and small exponentials represent complex rays. Infinite-dimensional integrals occur in statistical mechanics and quantum field theory, where small exponentials represent instantons. It is tempting to regard hyperasymptotics as a model for the feeble interaction of particles, represented by saddles. Each lowest saddle-point contribution gives the 'bare mass' of the corresponding particle; higher-order terms give the 'local renormalization' of this mass; the renormalization diverges, and hyperasymptotics describes the ultimate breakdown of locality, when the particles can no longer be considered in isolation.

Sixth, resurgence should be systematically applied to WKB theory for differential equations. In one dimension this would provide the extension of I to equations with effectively more than one transition point. The generalization of adjacency to transition points might justify the WKB topological rules of Knoll & Schaefer (1976), and complement the exact WKB analysis of Voros (1983). In more dimensions, one goal of resurgence would be to understand the semiclassical (small $\hbar$) expansions of quantum mechanics. For example, the expansion of the resolvent operator in powers of $\hbar$ (see Baltes & Hilf (1976) and Stewartson & Waechter (1971) for the case of quantum billiards, i.e. vibrating membranes) usually diverges, and we expect the divergence to contain information about the contributions from the periodic orbits of the corresponding classical system. These contributions are of order $\exp(\mathrm{i}S/\hbar)$, where S is the action of the orbit. Each periodic-orbit contribution is itself the first term of a semiclassical expansion; it is likely that all these expansions diverge and that the divergences are interlocked by resurgence. For classically chaotic systems, the sum over all the periodic orbits itself diverges (for a discussion see Berry (1991)), and this divergence, like a similar one for the Riemann zeta function (Berry & Keating 1990), can also be interpreted by resurgence, albeit of a type different from that derived here.

We are grateful to Dr W. G. C. Boyd for carefully reading the manuscript and making many helpful suggestions. C.J.H. received financial support from SERC.

Appendix A. The dangerous zeros

This concerns places where the quantity in braces in the denominator of (12) vanishes for real non-negative u, that is places where

$$\phi_n(z) \equiv k(f(z)-f_n) \tag{A 1}$$

is real and non-negative. We must show that there are no such 'dangerous zeros' in the region $\mathscr{R}$ defined as that swept by the deformation (by expansion) of the contour $\Gamma_n(\theta_k)$ (figure 2) described in §3. This is the region enclosed by the contours $C_m(-\sigma_{nm})$ through the saddles m adjacent to n (figure 3), together with the arcs at infinity joining their ends, and excluding the contour $C_n(\theta_k)$ itself (where $\phi_n(z)$ is real and non-negative by definition).

Consider the lines of constant phase of $\phi_n(z)$ issuing from the double zero of $\phi_n(z)$ at the saddle z_n; z_n is a singularity of these lines, around which the phase changes by 4π (i.e. a phase dislocation of strength 2 (see Nye & Berry 1974)). Only on two of these lines, those where the phase of $\phi_n(z)$ is 0 and 2π, is $\phi_n(z)$ real and non-negative, and these lines are the two halves of $C_n(\theta_k)$. The remaining phase lines fill a region bounded by the special phase lines (defined in the paragraph after (12)) through the adjacent saddles. This is precisely the region $\mathscr{R}$ which therefore cannot contain dangerous zeros.

The function $\phi_n(z)$ has zeros other than the double one at z_n. These are situated outside $\mathscr{R}$ and can be reached successively by starting from z and passing over the saddles $m \neq n$ along lines of constant phase of $\phi_n(z)$. Typically these other zeros are simple (i.e. phase dislocations of strength 1), with a dangerous zero (for given u) on one line of constant phase of $\phi_n(z)$ issuing from each of them. If $f(z)$ is an Mth order polynomial, there are $M-2$ such simple zeros of $\phi_n(z)$, and hence $M-2$ dangerous zeros outside $\mathscr{R}$. Figure 8 shows an example with $M = 4$, where there are three saddles each with its associated function $\phi_n(z)$ which has two simple zeros.

References

Abramowitz, M. & Stegun, I. A. 1972 *Handbook of mathematical functions.* Washington, D.C.: National Bureau of Standards.

Balian, R., Parisi, G. & Voros, A. 1979 In *Feynman path integrals* (ed. S. Albeverio *et al.*), pp. 337–360. *Lecture notes in physics*, vol. 106. Springer.

Baltes, H. P. & Hilf, E. R. 1976 *Spectra of finite systems.* Mannheim: B.-I. Wissenschaftsverlag.

Berry, M. V. 1989*a* *Proc. R. Soc. Lond.* A **422**, 7–21.

Berry, M. V. 1989*b* *Publ. Math. IHÉS* **68**, 211–221.

Berry, M. V. 1991 Some quantum-to-classical asymptotics. In *Chaos and quantum physics.* Les Houches Lecture Series 52 (ed. M. J. Giannoni & A. Voros). Amsterdam: North-Holland. (In the press.)

Berry, M. V. & Howls, C. J. 1990 *Proc. R. Soc. Lond.* A **430**, 653–668. (I of the text.)

Berry, M. V. & Keating, J. P. 1990 *J. Phys.* A **23**, 4839–4849.

Boyd, W. G. C. 1990 *Proc. R. Soc. Lond.* A **429**, 227–246.

Boyd, W. G. C. 1991 (In preparation.)

De Bruijn, N. J. 1958 *Asymptotic methods in analysis.* Amsterdam: North-Holland.

Dingle, R. B. 1973 *Asymptotic expansions: their derivation and interpretation.* New York and London: Academic Press.

Écalle, J. 1981 Les fonctions résurgentes (3 vols). Publ. Math. Université de Paris-Sud.

Écalle, J. 1984 Cinq applications des fonctions résurgentes. Preprint 84T62, Orsay.

Knoll, J. & Schaefer, R. 1976 *Ann. Phys., New York* **97**, 307–366.

Nye, J. F. & Berry, M. V. 1974 *Proc. R. Soc. Lond.* A **336**, 165–190.

Pearcey, T. 1946 *Phil. Mag.* **37**, 311–317.

Rakovic, M. J. & Solov'ev, E. A. 1989 *Phys. Rev.* A **40**, 6692–6694.

Stewartson, K. & Waechter, R. T. 1971 *Proc. Camb. phil. Soc.* **69**, 353–363.

Stokes, G. G. 1847 *Trans. Camb. phil. Soc.* **9**, 379–407. (Reprinted in *Mathematical and physical papers by the late Sir George Gabriel Stokes* (Cambridge University Press 1904), vol. II, pp. 329–357.)

Stokes, G. G. 1864 *Trans. Camb. phil. Soc.* **10**, 106–128. (Reprinted in *Mathematical and physical papers by the late Sir George Gabriel Stokes* (Cambridge University Press 1904), vol. IV, pp. 77–109.)

Voros, A. 1983 *Ann. Inst. H. Poincaré* **39**, 211–338.

Received 1 March 1991; accepted 17 April 1991

Physics World June 1993 **35**

Research into limits of divergent series has revitalised many areas of physics – as well as studies of asymptotic expansions themselves

Infinity interpreted

MICHAEL BERRY AND CHRISTOPHER HOWLS

Divergent series are the invention of the devil, and it is shameful to base on them any demonstration whatsoever – **Abel, 1828**

DESPITE the denunciations of the mathematician Abel, if the devil did invent divergent series it was because his creator counterpart chose to build our physical universe so that they are among the more useful ways to describe its finite properties.

It is common to consider physical situations as perturbations of an idealised problem that can be solved exactly, and get the solution of the real problem as a series in powers of a small parameter δ describing the strength of the perturbation:

$$S = \sum_{n=0}^{\infty} a_n \delta^n.$$

Sometimes these series converge (i.e. beyond a certain order n the magnitude of successive terms approaches zero and the sum approaches a finite value or asymptote); there is a prejudice in favour of such solutions. But the limit $\delta \to 0$ often turns out to be singular and then difficulties can arise: finite physical quantities in well founded theories can be represented by series that diverge, with successive terms beyond a certain order increasing in magnitude so that the sum tends to infinity.

Physicists have been uneasy about these series, but have tolerated them because truncations after the first few terms give very good approximations to observed quantities. Only recently has it been appreciated that the tails of divergent series have universal properties. These have allowed us to tame the tails and reveal a rich structure of exponentially small terms responsible for a variety of physical phenomena, and moreover answer questions in mathematical asymptotics – the study of limits – that have endured since the nineteenth century.

Reduction of theories

Limits arise in physics at a fundamental level because $\delta \to 0$ can describe the reduction of a general theory to a more restricted (usually earlier) theory. A case where the limit is unproblematic is special relativity, which reduces smoothly to Newtonian mechanics for objects with speed v small compared with the speed of light c. To see this we set $\delta = v/c$. Wherever a Lorentz factor $\gamma = 1/\sqrt{(1-\delta^2)}$ appears in an equation of special relativity it may be expanded as a convergent Taylor series in δ to give the corresponding low-speed formulae, reducing to Newtonian mechanics when $\delta = 0$.

Contrast this with the geometrical (ray) limit of physical optics, for example in the rainbow (figure 1 and front cover); in this limit, for each colour the wavelength is small compared with the radius a of a raindrop (we could choose $\delta = \lambda/a$), so that diffraction can be ignored. Imagine a bow at a single wavelength. The region under the bow is bright because two rays emerge in each direction. No rays reach the region above the bow – the dark side. The boundary is a caustic, where rays coalesce and the bow is most intense.

The illumination on the bright side was calculated by Young in 1804, who moved beyond the geometric optics limit by using an approximate theory in which wave energy travels along the rays. He was able to explain the observation that, beneath the bow, there is a series of "supernumerary" bows that arise from interference oscillations. (These effects are often washed out by waves of different colours and from droplets of different sizes, but supernumerary bows can be discerned beneath the main bow in figure 1.)

The oscillations can be represented as the usual two-wave

1 Supernumerary rainbows (faint fringes below the main arc) above Newton's birthplace (Woolsthorpe Manor, Lincolnshire). The irony is that Newton could not have understood these fringes with his corpuscular theory of light, because they are the result of wave interference. (Photograph courtesy of Professor Roy Bishop, Acadia University, Nova Scotia, Canada)

Airy's rainbow integral

The Airy integral is defined by

$$\mathrm{Ai}(z) \equiv \frac{1}{2\pi}\int_{-\infty}^{\infty} \mathrm{d}t \exp\left\{\mathrm{i}\left(\frac{1}{3}t^3 + zt\right)\right\} \qquad (1)$$

when z is real, and by a similar integral with a modified contour when z is complex. Stokes found that when z is large and positive a formally exact representation is the series

$$\mathrm{Ai}(z) = \frac{\exp\{-\frac{2}{3}z^{3/2}\}}{2z^{1/4}\sqrt{\pi}}\sum_{r=0}^{\infty}(-1)^r T_r, \qquad (2)$$

with coefficients

$$T_r = \frac{1}{(36z^{3/2})^r}\frac{(3r-\frac{1}{2})!}{r!(r-\frac{1}{2})!} \qquad (3)$$

The high orders have the form

$$T_r \approx \frac{(r-1)!}{2\pi(\frac{4}{3}z^{3/2})^r}, \quad (r \gg 1). \qquad (4)$$

When z is large the terms decrease at first and then increase, so that the series (2) diverges. The least term is near $r = r^* = 4z^{3/2}/3$. Truncating at this order and summing the divergent tail (using Borel summation) generates a finite representation valid throughout the upper z half-plane and incorporating the second exponential and the Stokes phenomenon:

$$\mathrm{Ai}(z) = \frac{\exp\{-\frac{2}{3}z^{3/2}\}}{2z^{1/4}\sqrt{\pi}}\sum_{r=0}^{r^*-1}(-1)^r T_r + \mathrm{i}S(z)\frac{\exp\{+\frac{2}{3}z^{3/2}\}}{2z^{1/4}\sqrt{\pi}}\sum_{r=0}^{r^*-1} T_r. \qquad (5)$$

Here $S(z)$ switches rapidly but smoothly from 0 to 1 across the Stokes line $z = |z|\exp\{2\mathrm{i}\pi/3\}$ (figure 2), thereby generating the second exponential, which is subdominant there, and, for negative real z, the supernumerary rainbow oscillations. A good approximation to $S(z)$ involves the error function. □

fringe pattern with intensity $I = \sin^2(2\pi z/\lambda)$ where z is a rainbow-crossing coordinate. This is mathematically singular as $\lambda \to 0$, because the sine oscillates infinitely fast, and a smooth description, reproducing geometrical optics, can only be achieved by averaging. On the dark side, there is a different singularity, typical where waves reach places that rays do not, namely exponential decay: $I = \exp(-Az/\lambda)$ where $A > 0$. Across the caustic $z = 0$, something peculiar happens: one real exponential (on the dark side) has transformed into two complex ones (comprising the sine on the bright side). How has the second exponential appeared?

Sudden appearances of exponential terms are common in physical asymptotics. For example in radioactive decay the above formulae give semiclassical approximations to the

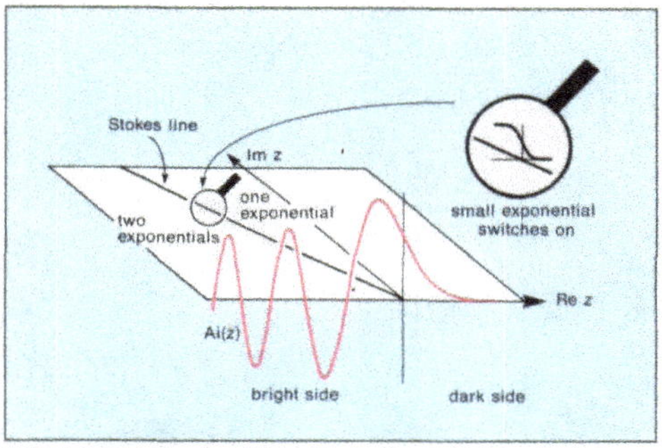

2 Airy's rainbow function Ai(z), with small exponential switching on across Stokes line in the complex plane

quantum tunnelling of a particle from a potential well (inside the nucleus) into a barrier. The character of the wave changes from oscillatory to decaying at the classical turning point (where the particle momentum is zero).

The above approximations based on exponentials are supposed to be valid in the asymptotic regime where δ can be regarded as small (i.e. the wavelength or, in the quantum case, Planck's constant, are small compared with characteristic parameters of the problem). But they fail at caustics and turning points, obscuring the transitions to geometrical optics (i.e. from waves to rays) and classical mechanics.

Among crossovers between theories in physics, smooth transitions – as with special relativity – are exceptional; usually the limits are singular. Other examples are the transitions from statistical mechanics to thermodynamics, where the singularity describes critical phenomena, and from viscous to inviscid flow, where it involves turbulence.

This is not to say that the theories are necessarily incompatible – each is self-consistent and agrees well with observation in its own domain. Singularities mean that on the borderline between the theories there is a wall impenetrable to elementary mathematical interpretation. Only now are nonelementary interpretations being devised, making some of these walls transparent.

Stokes and the rainbow

In 1838, Airy devised a diffraction integral Ai(z) (see box) representing, for each colour, the variation of the wave across a rainbow (the intensity is $\mathrm{Ai}^2(z)$). He was, however, unable to compute Ai(z) far on the bright side (z large and negative) and so compare his theory with measurements of the angles of supernumerary bows (and make the connection with geometrical optics). This was because, although he had found a convergent series, the number of terms needed to get a satisfactory approximation increased with $|z|$, and the integral itself converged very slowly. This situation is similar to that in quantum chromodynamics today, where technical difficulties inhibit the calculation of observed quantities.

In 1847, Stokes devised approximations for Ai(z) in the desired asymptotic regions $|z| \gg 1$ (that is, far from the caustic at $z = 0$), and showed that the function behaves as a damped exponential on the dark side (positive z) and sinusoidally on the bright side. The accuracy increased with $|z|$, and was sufficient to enable comparison with measured supernumerary bows. Stokes attacked the problem posed by the different numbers of exponentials

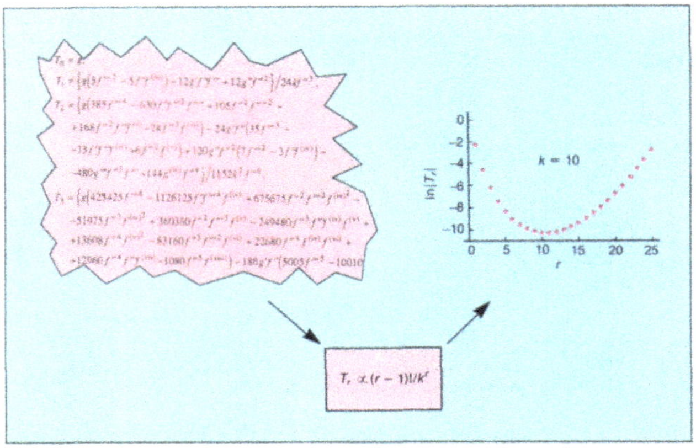

3 Dingle's universal "factorial divided by power" emerges from complicated formulae for late terms of an asymptotic series (in this case for an integral with a saddle)

on the two sides. He realised that the key lies in regarding z as a complex variable (figure 2). The second exponential appears across certain lines (now called Stokes lines) on a path in the z plane from the dark to bright sides (positive to negative real axes) avoiding the caustic singularity at $z=0$. On Stokes lines, the nascent exponential is smallest (subdominant) relative to the other. The birth of the subdominant exponential is called the Stokes phenomenon. (On the bright side, the second exponential becomes the second ray in the interfering pair.)

A major feature of Stokes' analysis was his appreciation that the pure exponentials in his approximation should each be multiplied by a series of corrections (see box) that are powers of $1/z$. He obtained these by solving the differential equation that Ai(z) satisfies, anticipating the WKB method of wave mechanics. The resulting asymptotic series was expected to give better approximations for larger $|z|$. As it turns out, successive terms do get smaller initially, and the accuracy is indeed improved by including them, but then the terms begin to increase, and the series diverges. For numerical purposes the optimal procedure is to truncate the series at its least term.

However, something has gone wrong, because since the series diverges its numerical sum would be infinity rather than Ai(z). The mathematical description of the physics is formally exact – there has been no approximation in the calculation of the correction terms – but we need to interpret the infinite result to rescue something sensible. Stokes appreciated that the divergence of series representing physics in asymptotic regimes is not peculiar to Airy's rainbow integral but is a general phenomenon.

It turns out that, far from being a hindrance, the divergence of the asymptotic series, suitably interpreted, actually explains the Stokes phenomenon. In his prescient but crude analysis Stokes thought that the birth of the second exponential is a sudden event; in fact it occurs smoothly, but it was more than a century before the details were discovered.

Universal divergences

Mathematicians responded to Stokes' work by ignoring it. The theory of asymptotic series developed along different lines, initiated by Poincaré in the 1880s. For each function possessing an asymptotic series, only a fixed number of terms is retained, and the divergent tail discarded and replaced by an exact remainder. The accuracy of the truncation can then be gauged by an estimate of the remainder, which has to be calculated separately for each function. In this work, divergent and convergent series are treated on the same basis, so the main issue of divergence is side-stepped.

In particular, Poincaré asymptotics fails to achieve the accuracy of truncation at the least term. For this reason, we call Stokes' technique of optimal truncation "superasymptotics" (reflecting Kelvin's description of it as "mathematical supersubtlety"). But even superasymptotics cannot provide a detailed description of exponential effects such as the Stokes phenomenon. For this, we need what Martin Kruskal of Rutgers University calls "asymptotics beyond all orders".

In the 1950s, R B Dingle, a physicist at St Andrews, returned to the original ideas of Stokes and confronted the divergences head-on, introducing ideas which have now led to a great simplification in their interpretation. He regarded the divergent tail of a formal expansion not as indicating lack of precision but rather as a source of information to be decoded to reveal the remainder. Dingle traced the divergences of asymptotic series to singularities (for example the caustic at $z=0$ in the rainbow). This led him to the result that whole classes of singularities arising in physics determine the form of the high-order (late) terms in a universal and simple way. This is astonishing because the early terms are not universal and are usually very complicated (figure 3). If k (proportional to $1/\delta$) is the large asymptotic parameter in such a series then Dingle's common form for the late terms T_r is

$$T_r \propto (r-1)!/k^r, \qquad r \gg 1.$$

An example is given in the box (there, $k=4z^{3/2}/3$).

For large k the terms rapidly diminish at first, but the factorial in the numerator eventually dominates the power in the denominator, and the terms increase unboundedly. The least term has $r \approx |k|$ and size $\exp(-|k|)$, i.e. exponentially small. This is a common pattern of divergence, not restricted to the Airy and other mathematical functions but also describing – as was later discovered – the divergences of series with much more complicated structure, such as those in quantum field theory (where, for example, δ could be the fine-structure constant).

Dingle used his late term approximations in conjunction with a procedure called Borel summation, in which the factorial is replaced by a well known integral representation. Thus he interpreted the infinity and obtained a *finite* representation of the divergent tail directly, in the form of readily calculable integrals called terminants. These extend beyond superasymptotics, and constitute a decoding of the information contained in the divergent tail. The universality of the late terms is mirrored in the universality of the terminants, which depend only on k and the order of the term where the series is truncated.

Terminants can also be adapted to give a refined description of the Stokes phenomenon. In 1988 we showed that if an asymptotic series is optimally truncated, i.e. near its least term $r \approx |k|$, the terminant reproduces the subdominant exponential switching on across the

Stokes line. The form of the switching is universal, and consists of an error function (probability integral) multiplying the exponential, and rising smoothly from 0 to 1 over a range proportional to $1/\sqrt{|k|}$ (figure 2). Resorting to paths in imaginary space to explain the births of exponentials might appear to be a mathematical nicety of no relevance to the physical world, but the complex approach enables the asymptotics to be followed smoothly around a seemingly impenetrable singularity.

Moreover, the Stokes phenomenon can occur for real variables. For example, a quantum system driven by slowly changing forces can jump between states, in a transition whose probability diminishes exponentially with slowness. The history of the transition, that is the growth of the occupancy of the new state, is the switching-on of a small exponential. Using an appropriate measurement (the physical counterpart of optimal truncation) the switching would conform to the universal error function. Therefore the Stokes smoothing could be studied experimentally.

The Stokes phenomenon also occurs in inflationary models of cosmology. E Calzetta (Buenos Aires) showed that in a "mixmaster" model of the early Universe the smooth creation of exponentially small numbers of particles is represented by universal error functions. A multitude of such smooth creations stokes the inflation of the Universe.

Resurgence and hyperasymptotics

The simple "factorial divided by power" formula for T_r is only a first approximation to the late terms. Dingle also found an asymptotic series for the corrections. A remarkable and beautiful feature is that Dingle's series connects the terms in the different asymptotic series representing a given function. For example, the late terms of the asymptotic series multiplying a dominant exponential can be expressed systematically by the early terms of a subdominant series, and vice versa. For functions with many exponentials (such as integrals in diffraction theory or statistical mechanics, whose saddles represent rays or equilibrium phases), all associated asymptotic series are similarly related. These formulae are called resurgence relations, after a related idea developed in the 1980s by the mathematician Jean Ecalle of Orsay.

By the sequential use of resurgence relations and Borel summation, it is now possible to get very accurate reconstructions of functions from their divergent series. This is a systematic extension of superasymptotics, developed since 1990; we call it "hyperasymptotics". The zeroth stage of hyperasymptotics is Stokes' superasymptotics. Successive applications of the procedure generate exponentially smaller contributions, in the form of further optimally truncated asymptotic series. Unlike other summation procedures, the successive "hyperseries" automatically incorporate all Stokes phenomena associated with the successive small exponentials.

As an example, consider the integral

$$N(k) = \sqrt{k}\int_{-\infty}^{+\infty} dz \exp\left(-k(z^2-1)^2\right), \quad k > 0$$

where k is the (large) asymptotic parameter. This arises in a field theoretic model of a particle (called an instanton) tunnelling between two wells at $z = \pm 1$ (a slightly modified version counts Feynman diagrams in certain statistical and field theories). When k is large, the asymptotic series derived from $N(k)$ is dominated by two exponentials, related to the wells. There is also a third contribution, a factor $\exp(-k)$ smaller than the other two, arising from the peak at $z = 0$ of the potential barrier separating the wells. It is this small exponential that can be physically interpreted as the tunnelling of an instanton through the barrier.

Hyperasymptotics can discriminate between the three contributions (figure 4) and generate accurate approximations even when the "large parameter" k is smaller than unity, where Poincaré asymptotics and superasymptotics certainly fail. Moreover this integral lies precisely on a Stokes line, where the instanton is created, which is a regime conventionally regarded as not amenable to Borel summation.

Recent applications

Small exponentials are common in nonlinear physics, and it is perhaps here that the new theories could be most fruitful. For example, Kruskal and others have recently used exponential asymptotics to elucidate the influence of surface tension on the selection and stability of viscous fingers when one fluid is forced into another, and to analyse a geometric model for dendritic growth of crystals. And in chaology Vincent Hakim of the Ecole Normale Supérieure in Paris and others have summed divergent perturbation series in accurate calculations of the exponentially small widths of the chaotic regions that open up when a non-chaotic system is perturbed.

A notable success of the new approach has emerged from quantum chaology, where a long-standing problem is to calculate the energies of high-lying states of microscopic

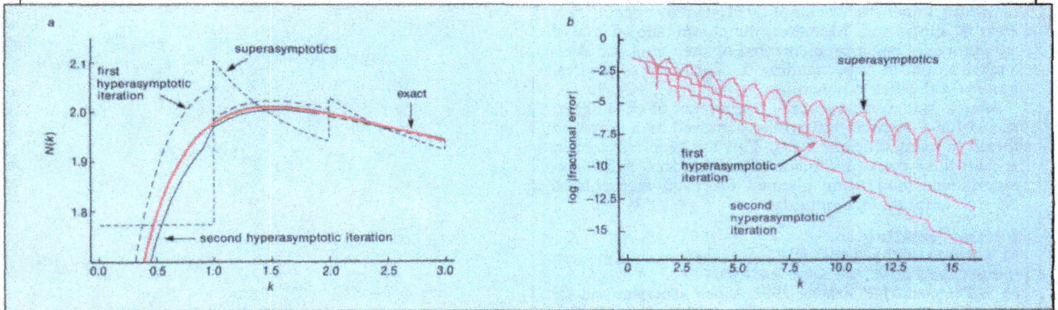

4 Superasymptotics and several levels of hyperasymptotics for the instanton integral

systems whose classical limit is chaotic (e.g. atomic electrons in strong magnetic fields). This requires semiclassical approximations, where the small parameter δ is proportional to Planck's constant $\hbar$. In 1971 Martin Gutzwiller of IBM Yorktown Heights obtained an asymptotic representation of the quantum spectrum as a sum over classical periodic orbits, but this diverges. Now it has been summed exactly, using resurgence to generate exponentials required to make the series behave sensibly. One outcome has been to provide greatly improved energies for quantum mechanics on odd-shaped billiard tables (a natural testing-ground for theories in quantum chaology).

The main achievement, however, has been the application of semiclassical resurgence to the Riemann zeta function (related to deep problems of mathematics through its connection with prime numbers). The zeros of this function are analogous to energy levels in quantum chaology (see "Physics and the queen of mathematics" by Jonathan Keating *Physics World* April 1990 pp46–50, and "Semiclassicists saved from infinity" by Tania Monteiro *Physics World* October 1991 pp21–22). Until recently, the zeta function was computed by a technique – the Riemann–Siegel formula – which, although very accurate, involves discontinuities detrimental to the computation of zeros. However, the Riemann zeta analogue of the new quantum formula is readily calculable, involves no discontinuities, and gives, term for term, approximations orders of magnitude better than its Riemann–Siegel cousin.

Prospects

Asymptotics was conceived in Victorian times, when it provided the fastest and most accurate means of approximation. Now that workstations with powerful numerical routines (e.g. for integration) are standard desktop furniture, the need for some of the purely numerical applications of asymptotics has diminished. On the other hand, computer algebra now makes it practicable to evaluate many terms of asymptotic series which were previously impenetrably complicated, thereby opening new areas of application. Moreover, the computational role of asymptotics is being supplemented by their vital use as an analytical tool in understanding the behaviour of physical systems with small exponentials.

After a sudden emergence – reminiscent of chaology in the 1970s – asymptotics is now perceived as a common thread in many different fields. This has been reflected in at least three recent international conferences, and a six-month workshop being planned for 1995 at the Isaac Newton Institute in Cambridge.

Further reading

M V Berry and C J Howls 1991 Hyperasymptotics for integrals with saddles *Proc. R. Soc. Lond.* **A434** 657–75

M V Berry and J P Keating 1992 A new approximation for $\zeta(1/2+it)$ and quantum spectral determinant *Proc. R. Soc. Lond.* **A437** 151–73

R B Dingle 1973 *Asymptotic Expansions: Their Derivation and Interpretation* (Academic Press, New York and London)

R Lim and M V Berry 1991 Superadiabatic tracking for quantum evolution *J. Phys. A: Math. Gen.* **24** 3255–64

H Segur and S Tanveer (eds) 1991 *Asymptotics Beyond All Orders* (Plenum, New York)

Michael Berry and **Christopher Howls** are in the Department of Physics, H H Wills Physics Laboratory, University of Bristol, Tyndall Avenue, Bristol BS8 1TL, UK

Divergent series: taming the tails

M. V. Berry[1] & C. J. Howls[2]

[1] *H. H. Wills Physics Laboratory, Tyndall Avenue, Bristol BS8 1TL, UK*
[2] *Mathematical Sciences, University of Southampton, Southampton, SO17 1BJ, UK.*

1 Introduction

By the 17th century, in what became the theory of convergent series, it was beginning to be understood how a sum of infinitely many terms could be finite; this is now a fully developed and largely standard element of every mathematician's education. Contrasting with it is the theory of series that do not converge, especially those in which the terms first get smaller but then increase factorially: this is the class of 'asymptotic series', encountered frequently in applications, with which this article is mainly concerned. Although now a vibrant area of research, the development of the theory of divergent series has been tortuous and often accompanied by controversy. As a pedagogical device to explain the subtle concepts involved, we will focus on the contributions of individuals and describe how the ideas developed during several (overlapping) historical epochs, often driven by applications ranging from wave physics to number theory. This article complements the accompanying Companion article by P D Miller on 'Perturbation Theory (including Asymptotic Expansions)'.

2 The Classical Period

In 1747, the Reverend Thomas Bayes (better known for his theorem in probability theory) sent a letter to Mr John Canton, F.R.S; it was published posthumously in 1763. Bayes demonstrated that the series now known as Stirling's expansion for $\log(z!)$, "asserted by some eminent mathematicians," does not converge. Arguing from the recurrence relation relating successive terms of the series, he showed that the coefficients "increase at a greater rate than what can be compensated by an increase of the powers of z, though z represent a number ever so large." As we would say now, this expansion of the factorial function is a factorially divergent asymptotic series. The explicit form of the series, written formally as an equality, is

$$\log(z!) = (z+1/2)\log z + \log\sqrt{2\pi} - z + \frac{1}{2\pi^2 z}\sum_{r=0}^{\infty}(-1)^r \frac{a_r}{(2\pi z)^{2r}},$$

where

$$a_r = (2r)!\sum_{n=1}^{\infty}\frac{1}{n^{2r+2}}.$$

Bayes claimed that Stirling's series "can never properly express any quantity at all" and the methods used to obtain it "are not to be depended upon."

Leonhard Euler, in extensive investigations of a wide variety of divergent series beginning several years after Bayes, took the opposite view. He argued that such series have a precise meaning, to be decoded by suitable resummation techniques (several of which he invented): "if we employ [the] definition ... that ... the sum of a series is that quantity which generates the series, all doubts with respect to divergent series vanish and no further controversy remains."

With the development of rigorous analysis in the 19th century, Euler's view, which as we will see is the modern one, was sidelined and even derided. As Niels Henrik Abel wrote in 1828: "Divergent series are the invention of the devil, and it is shameful to base on them any demonstration whatsoever." Nevertheless, divergent series, especially factorially divergent ones, repeatedly arose in application. Towards the end of the century they were embraced by Oliver Heaviside, who used them in pioneering studies of radio wave propagation. He obtained reliable results using undisciplined semi-empirical arguments that were criticised by mathematicians, much to his disappointment: "It is not easy to get up any enthusiasm after it has been artificially cooled by the wet blanket of rigorists."

3 The Neoclassical Period

In 1886, Henri Poincaré published a definition of asymptotic power series, involving a large parameter z, that was both a culmination of previous

2

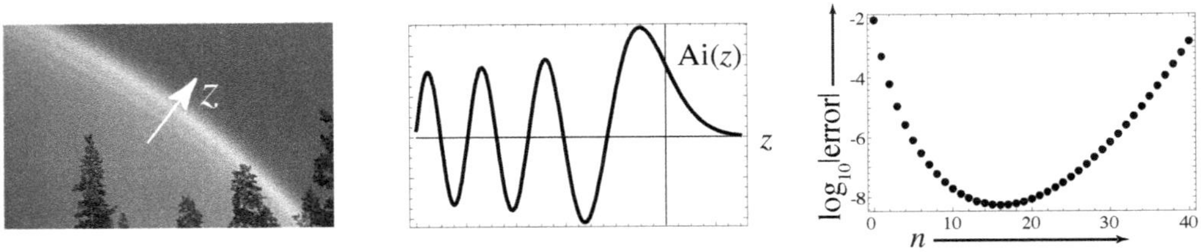

Figure 1: Left: rainbow-crossing variable, along any line transverse to the rainbow curve. Middle: Airy function. Right: error from truncating asymptotic series for $\mathrm{Ai}(z)$ at the term $n-1$, for $z=5.24148$; optimal truncation occurs at the nearest integer to $F=4z^{3/2}/3$, i.e. $n=16$.

work by analysts and the foundation of much of the rigorous mathematics that followed. A series of the form $\sum_{n=0}^{\infty} a_n/z^n$ is defined as asymptotic by Poincaré if the error resulting from truncation at the term $n=N$ vanishes as $K/z^{N+1}(K>0)$ as $|z|\to\infty$ in a certain sector of the complex z plane. In retrospect, Poincaré's definition seems a retrograde step, because although it encompasses convergent as well as divergent series in one theory, it fails to address the distinctive features of divergent series that ultimately lead to the correct interpretation that can also cure their divergence.

It was George Stokes, in research inspired by physics nearly four decades before Poincaré, who laid the foundations of modern asymptotics. He tackled the problem of approximating an integral devised by George Airy to describe waves near caustics, the most familiar example being the rainbow. This is what we now call the Airy function $\mathrm{Ai}(z)$, defined by the oscillatory integral

$$\mathrm{Ai}(z)\equiv\frac{1}{2\pi}\int_{-\infty}^{\infty}\exp\left(\frac{\mathrm{i}}{3}t^3+\mathrm{i}zt\right)\mathrm{d}t,$$

the rainbow-crossing variable being $\mathrm{Re}z$ (Figure 1) and the light intensity being $\mathrm{Ai}^2(z)$. Stokes derived the asymptotic expansion representing the Airy function for $z>0$, and showed that it is factorially divergent. His innovation was to truncate this series not at a fixed order N but at its smallest term (*optimal truncation*), corresponding to an order $N(z)$ that increases with z. By studying the remainder left after optimal truncation, he showed that it is possible to achieve exponential accuracy (Figure 1) far beyond the power-law accuracy envisaged in Poincaré's definition. We will call such optimal truncation *superasymptotics*.

Superasymptotics enabled Stokes to understand a much deeper phenomenon, one that is fundamental to the understanding of divergent series. In $\mathrm{Ai}(z)$, $z>0$ corresponds to the dark side of the rainbow, where the function decays exponentially: physically, this represents an evanescent wave. On the bright side $z<0$, the function oscillates trigonometrically, that is, as the sum of two complex exponential contributions, each representing a wave; the interference of these waves generates the 'supernumerary rainbows' (whose observation was one of the phenomena earlier adduced by Thomas Young in support of his view that light is a wave phenomenon). One of these complex exponentials is the continuation across $z=0$ of the evanescent wave on the dark side. But where does the other originate?

Stokes's great discovery was that this second exponential appears during continuation of $\mathrm{Ai}(z)$ in the complex plane from positive to negative z, across what is now called a 'Stokes line', where the dark-side exponential reaches its maximum size. Alternatively stated, the small (subdominant) exponential appears when maximally hidden behind the large (dominant) one. For $\mathrm{Ai}(z)$ the Stokes line is $\arg z=120°$ (Figure 2).

Stokes thought that the least term in the asymptotic series representing the large exponential constitutes an irreducible vagueness in the description of $\mathrm{Ai}(z)$ in his superasymptotic scheme. By quantitative analysis of the size of this least term, Stokes concluded that only at maximal dominance could this obscure the small expo-

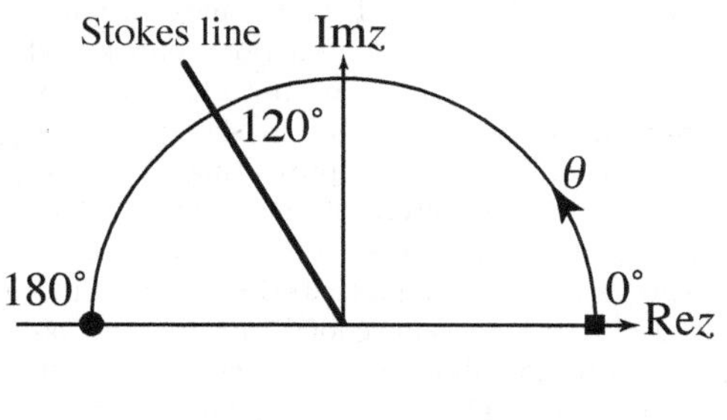

Figure 2: Complex plane of argument z (whose real part is the z of Figure 1) of $\mathrm{Ai}(z)$, showing Stokes line at $\arg z = 120°$.

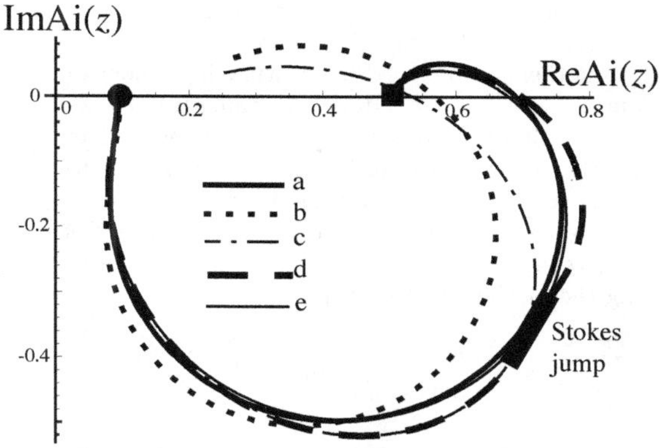

Figure 3: Approximations to $\mathrm{Ai}(z\exp(\mathrm{i}\theta))$ in the Argand plane, $(\mathrm{Re}(\mathrm{Ai}), \mathrm{Im}(\mathrm{Ai}))$ for $z = 1.31\ldots$, i.e. $F = 4z^{3/2}/3 = 2$, plotted parametrically from $\theta = 0°$ (•) to $\theta = 180°$ (■). The curves are: (a) exact Ai; (b) lowest-order asymptotics (no correction terms); (c) optimal truncation without Stokes jump; (d) optimal truncation including Stokes jump; (e) optimal truncation including smoothed Stokes jump. For this value of F, the optimally truncated sum contains only two terms; on the scale shown, the Stokes jump would be invisible for larger F. Note that without the Stokes jump (curves b and c) the asymptotics must deviate from the exact function beyond the Stokes line at $\theta = 120°$.

nential, which could then appear without inconsistency. As we will explain later, Stokes was wrong to claim that superasymptotics—optimal truncation—represents the best approximation that can be achieved within asymptotics. But his identification of the Stokes line with the birth of the small exponential (Figure 3) was correct. Moreover, he also appreciated that the concept was not restricted to $\mathrm{Ai}(z)$ but applies to a wide variety of functions arising from integrals, solutions of differential equations and recurrence equations, etc., for which the associated asymptotic series are factorially divergent.

This *Stokes phenomenon*, connecting different exponentials representing the same function, is central to our current understanding of such divergent series, and is the feature that distinguishes them most sharply from convergent ones. In view of this seminal contribution, it is ironic that George ('G H') Hardy makes no mention of the Stokes phenomenon in his textbook 'Divergent series'. Nor does he exempt Stokes from his devastating assessment of 19th century English mathematics: "there [has been] no first-rate subject, except music, in which England has occupied so consistently humiliating a position. And what have been the peculiar characteristics of such English mathematics ...? ...for the most part, amateurism, ignorance, incompetence, and triviality."

4 The Modern Period

Late in the 19th century, Jean-Gaston Darboux showed that for a wide class of functions the high derivatives diverge factorially. This would become an important ingredient in later research, for the following reason. Asymptotic expansions (particularly those encountered in physics and applied mathematics) are often based on local approximations: the steepest-descent method for approximating integrals is based on local expansion about a saddle-point, the phase-integral method for solving differential equations (e.g. the Wentzel-Kramers-Brillouin (WKB) approximation to Schrödinger's equation in quantum mechanics) is based on local expansions of the coefficients, etc. Therefore successive orders of approximation involve successive derivatives, and

4

the high orders, responsible for the divergence of the series, involve high derivatives.

Another major late 19th century ingredient of our modern understanding was Émile Borel's development of a powerful summation method in which the factorials causing the high orders to diverge are tamed by replacing them by their integral representation. Often this enables the series to be summed 'under the integral sign'. Underlying the method is the formal equality

$$\sum_{r=0}^{\infty}\frac{a_r}{z^r}=\sum_{r=0}^{\infty}\frac{a_r r!}{z^r r!}=\int_0^{\infty}\mathrm{d}t\; e^{-t}\sum_{r=0}^{\infty}\frac{a_r}{r!}\left(\frac{t}{z}\right)^r.$$

Reading this from right to left is instructive. Interchanging summation and integration shows why the series on the left diverges if the a_r increase factorally (as in the cases we are considering): the integral is over a semi-infinite range, yet the sum in the integrand converges only for $|t/z|<1$. Borel summation effectively repairs an analytical transgression that may have caused the divergence of the series. The power of Borel summation is that, as was fully appreciated only later, it can be analytically continued across Stokes lines, where some other summation techniques fail (for example Padé approximants).

Now we come to the central development in modern asymptotics. In a seminal and visionary advance, motivated initially by mathematical difficulties in evaluating some integrals occurring in solid-state physics and developed in a series of papers culminating in a book published in 1973, Robert Dingle synthesized earlier ideas into a comprehensive theory of factorially divergent asymptotic series.

Dingle's starting point was Euler's insight that divergent series are obtained by a sequence of precisely specified mathematical operations on the integral or differential equation defining the function being approximated, so the resulting series must represent the function exactly, albeit in coded form, which it is the task of asymptotics to decode. Next was the realization that Darboux's expression of high derivatives in terms of factorials implies that the high orders of a wide class of asymptotic series diverge similarly. In turn this means that the terms beyond Stokes's optimal truncation—representing the tails of such series beyond superasymptotics—can all be Borel-summed in the same way.

The next insight was Dingle's most original contribution. Consider a function represented by several different formal asymptotic series (for example those corresponding to the two exponentials in $\mathrm{Ai}(z)$), each representing the function differently in sectors of the complex plane separated by Stokes lines. Since each series is a formally complete representation of the function, each must contain, coded into its high orders, information about all the other series. Thus Darboux's factorials are simply the first terms of asymptotic expansions of each of the late terms of the original series. Dingle appreciated that the natural variables implied by Darboux's theory are the differences between the various exponents; usually these are proportional to the large asymptotic parameter. In the simplest case, where there are only two exponentials, there is one such variable, which Dingle called the *singulant*, denoted F. For the Airy function $\mathrm{Ai}(z)$, $F=4z^{3/2}/3$.

We exhibit Dingle's expression for the high orders for an integral with two saddle-points a and b, corresponding to exponentials $\exp(-F_a)$ and $\exp(-F_b)$ with $F_{ab}=F_b-F_a$ and series with terms $T_r^{(a)}$ and $T_r^{(b)}$: for $r\gg 1$, the terms of the a series are related to those of the b series by

$$\begin{aligned}T_r^{(a)} &= K\frac{(r-1)!}{F_{ab}^r}\left(T_0^{(b)}+\frac{F_{ab}}{(r-1)}T_1^{(b)}\right.\\ &\left.\quad+\frac{F_{ab}^2}{(r-1)(r-2)}T_2^{(b)}+\ldots\right),\end{aligned}$$

in which K is a constant. This shows that although the early terms $T_0^{(a)},T_1^{(a)},T_2^{(a)},\ldots$ of an asymptotic series can rapidly get extremely complicated, the high orders display a miraculous functional simplicity.

With Borel's as the chosen summation method, Dingle's late terms formula enabled the divergent tails of series to be summed in terms of certain *terminant* integrals, and then re-expanded to generate new asymptotic series, exponentially small compared with the starting series. He envisaged that "these terminant expansions can themselves be closed with new terminants; and so on, stage after stage." Such resummations, beyond superasymptotics, were later called *hyperasymptotics*.

Thus Dingle envisaged a universal technique

for repeated resummation of factorially divergent series, to obtain successively more accurate exponential improvements far beyond that achievable by Stokes's optimal truncation of the original series. The meaning of universality is that although the early terms—the ones that get successively smaller—can be very different for different functions, the summation method for the tails is always the same, involving terminant integrals that are the same for a wide variety of functions. The method automatically incorporates the Stokes phenomenon. Although Dingle clearly envisaged the hyperasymptotic resummation scheme as described above, he applied it only to the first stage; this was sufficient to illustrate the high improvement in numerical accuracy as compared with optimal truncation.

Like Stokes before him, Dingle presented his new ideas not in the 'lemma, theorem, proof' style familiar to mathematicians, but in the discursive manner of a theoretical physicist. Perhaps this is why it has taken several decades for the originality of his approach to be widely appreciated and accepted. Meanwhile his explicit relation, connecting the early and late terms of different asymptotic series representing the same function, was rediscovered independently by several people. In particular, Jean Écalle coined the term *resurgence* for the phenomenon, and in a sophisticated and comprehensive framework applied it to a very wide class of functions.

5 The Postmodern Period

One of the first steps beyond superasymptotics into hyperasymptotics was an application of Dingle's ideas to give a detailed description of the Stokes phenomenon. In 1988, one of us (Michael Berry) resummed the divergent tail of the dominant series of the expansion, near a Stokes line, of a wide class of functions including $\mathrm{Ai}(z)$, to give birth to the change in the subdominant exponential, occurring not suddenly as in previous accounts of the phenomenon, but smoothly and in a universal manner. In terms of Dingle's singulant F, now defined as the difference between the exponents of the dominant and subdominant exponentials, the Stokes line corresponds to the positive real axis in the complex F plane, asymptotics corresponds to $\mathrm{Re}F \gg 1$, and the Stokes phenomenon corresponds to crossing the Stokes line, that is $\mathrm{Im}F$ passing through zero.

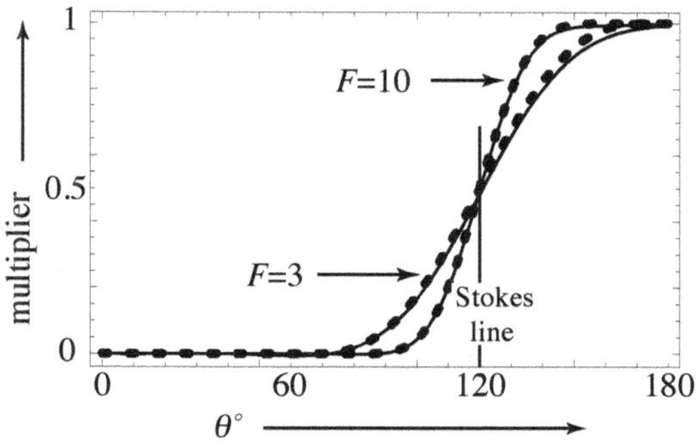

Figure 4: Stokes multiplier (full curve) and error-function smoothing (dashed curve), for $\mathrm{Ai}(z\exp(\mathrm{i}\theta))$, for $z = 1.717\ldots$ (i.e. $F = 4z^{3/2}/3 = 3$) and $z = 3.831\ldots$ (i.e. $F = 4z^{3/2}/3 = 10$).

The result of the resummation is that the change in the coefficient of the small exponential—the Stokes multiplier—is universal for all factorially divergent series, and proportional to

$$\frac{1}{2}\left(1+\mathrm{Erf}\left(\frac{\mathrm{Im}F}{\sqrt{2\mathrm{Re}F}}\right)\right).$$

In the limit $\mathrm{Re}F \to \infty$ this becomes the unit step. For $\mathrm{Re}F$ large but finite, the formula describes the smooth change in the multiplier (Figure 4), and makes precise the description given by Stokes in 1902 after thinking about divergent series for more than half a century: "...the inferior term enters as it were into a mist, is hidden for a little from view, and comes out with its coefficient changed. The range during which the inferior term remains in a mist decreases indefinitely as the [large parameter] increases indefinitely." The smoothing shows that the 'range' referred to by Stokes, that is the effective thickness of the Stokes line, is of order $\sqrt{\mathrm{Re}F}$.

The full hyperasymptotic repeated resummation scheme envisaged by Dingle has been implemented in several ways. We (the present authors) investigated one-dimensional integrals with several saddle-points, each associated with an exponential and its corresponding asymptotic se-

6

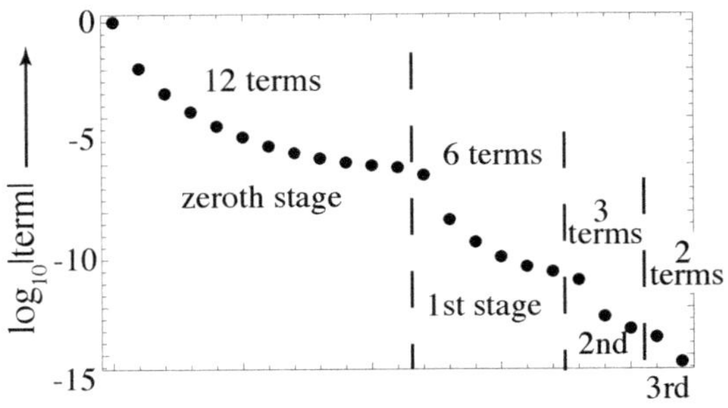

Figure 5: Terms in the first four stages in the hyperasymptotic approximation to $\mathrm{Ai}(4.326\ldots) = 4.496\ldots \times 10^{-4}$, i.e. $F = 4z^{3/2}/3 = 12$, normalized so that the lowest approximation is unity. For the lowest approximation, i.e. no correction terms, the fractional error is $\varepsilon \approx 0.01$; after stage 0 of hyperasymptotics, i.e. optimal trunction of the series (superasymptotics), $\varepsilon \approx 3.6 \times 10^{-7}$; after stage 1, $\varepsilon \approx 1.3 \times 10^{-11}$; after stage 2, $\varepsilon \approx 4.4 \times 10^{-14}$; after stage 3, $\varepsilon \approx 6.1 \times 10^{-15}$. At each stage, the error is of the same order as the first neglected term.

ries. With each 'hyperseries' truncated at its least term, this incorporated all subdominant exponentials and all associated Stokes phenomena; and the accuracy obtained far exceeded superasymptotics (Figure 5) but was nevertheless limited.

It was clear from the start that in many cases unlimited accuracy could, in principle, be achieved with hyperasymptotics, by truncating the hyperseries not at the smallest term but beyond it (although this introduces numerical stability issues associated with the cancellation of larger terms). This version of the hypersymptotic programme was carried out by Adri Olde Daalhuis, who reworked the whole theory, introducing mathematical rigor and effective algorithms for computing Dingle's terminant integrals and their multidimensional generalizations, and applied the theory to differential equations with arbitrary finite numbers of transition points.

There has been an explosion of further developments. Écalle's rigorous formal theory of resurgence has been developed in several ways, based on the Borel (effectively inverse-Laplace) transform. This converts the factorially-divergent series into a convergent one, with radius of convergence determined by singularities on a Riemann sheet. These singularities are responsible for the divergence of the original series, and for integrals discussed above, correspond to the adjacent saddle points. In the Borel plane, complex and microlocal analysis allows the resurgence linkages between asymptotic contributions to be uncovered and exact remainder terms to be established. Notable results include exponentially accurate representations of quantum eigenvalues (R. Balian & C. Bloch, A. Voros, F. Pham, E. Delabaere); this inspired the work of the current authors on quantum eigenvalue counting functions, linking the divergence of the series expansion of smoothed spectral functions to oscillatory corrections involving the classical periodic orbits.

T. Kawai and Y. Takei in Kyoto have extended 'formally exact', exponentially accurate, WKB analysis to several areas, most notably to Painlevé equations. They have also developed a theory of 'virtual turning points' and 'new Stokes curves'. In the familiar WKB situation, with only two wave-like asymptotic contributions, Stokes lines emerge from classical turning points, and never cross. With three or more asymptotic contributions, Stokes lines can cross in the complex plane at points where the WKB solutions are not singular. Local analysis shows that an extra, active, 'new Stokes line' sprouts from one side only of this regular point; this can be shown to emerge from a distant virtual turning point, where, unexpectedly, the WKB solutions are not singular. This discovery has been explained by C.J. Howls, P. Langman and A.B. Olde Daalhuis in terms of the Riemann sheet structure of the Borel plane and linked hyperasymptotic expansions, and independently by S.J. Chapman and D.B. Mortimer in terms of matched asymptotics.

Groups led by S.J. Chapman and J.R. King have developed and applied the work of M. Kruskal and H. Segur to a variety of nonlinear and PDE problems. This involves a local matched-asymptotic analysis near the distant Borel singularities that generate the factorially-divergent terms in the expansion, to identify the form of late terms, thereby allowing for an optimal truncation and exponentially accurate approach. Applications include: selection problems in viscous fluids, gravity-capilliary solitary waves, oscillating shock solutions in Kuramoto-Shivashinsky equations, elastic buckling, nonlinear instabilities in pattern formation, ship wave modelling and the seeking of reflectionless hull profiles. Using a similar approach, O. Costin and S. Tanveer have identified and quantified the effect of 'daughter singularities' not present in initial data of PDE problems, but which are generated at infinitesimally short times. In so doing, they have also found a formally exact Borel representation for small-time solutions of 3D Navier-Stokes equations, offering a promising tool to explore the global existence problem.

Other applications include quantum transitions, quantum spectra, the Riemann zeta function high on the critical line, and even the philosophy of representing physical theories describing phenomena at different scales by singular relations.

Further Reading

1. Batterman, R. W., 2002, *The Devil in the Details: Asymptotic Reasoning in Explanation, Reduction and Emergence* (University Press, Oxford).
2. Berry, M. V., 1989, Uniform asymptotic smoothing of Stokes's discontinuities *Proc. Roy. Soc. Lond.* **A422**, 7-21.
3. Berry, M. V. & Howls, C. J., 1991, Hyperasymptotics for integrals with saddles *Proc. Roy. Soc. Lond.* **A434**, 657-675.
4. Delabaere, E. & Pham, F., 1999, Resurgent methods in semi-classical mechanics *Ann. Int. H. Poincaré* **71**, 1-94.
5. Digital Library of Mathematical Functions, 2010, *NIST Handbook of Mathematical Functions* (University Press, Cambridge) `http://dlmf.nist.gov`, chapters 9 and 36.
6. Dingle, R. B., 1973, *Asymptotic Expansions: their Derivation and Interpretation* (Academic Press, New York and London).
7. Howls, C. J., 1997, Hyperasymptotics for multidimensional integrals, exact remainder terms and the global connection problem *Proc. R. Soc. Lond.* **A453**, 2271–2294.
8. Kruskal, M. D. & Segur, H., 1991, Asymptotics beyond all orders in a model of crystal growth *Stud. App. Math.* **85**, 129-181.
9. Olde Daalhuis, A. B., 1998, Hyperasymptotic solutions of higher order linear differential equations with a singularity of rank one *Proc. R. Soc. Lond.* **A454**, 1-29.
10. Stokes, G. G., 1864, On the discontinuity of arbitrary constants which appear in divergent developments *Trans. Camb. Phil. Soc.* **10**, 106-128.

The Riemann–Siegel expansion for the zeta function: high orders and remainders

By M. V. Berry

H. H. Wills Physics Laboratory, Tyndall Avenue, Bristol BS8 1TL, UK

On the critical line $s = \frac{1}{2} + it$ (t real), Riemann's zeta function can be calculated with high accuracy by the Riemann–Siegel expansion. This is derived here by elementary formal manipulations of the Dirichlet series. It is shown that the expansion is divergent, with the high orders r having the familiar 'factorial divided by power' dependence, decorated with an unfamiliar slowly varying multiplier function which is calculated explicitly. Terms of the series decrease until $r = r^* \approx 2\pi t$ and then increase. The form of the remainder when the expansion is truncated near r^* is determined; it is of order $\exp(-\pi t)$, indicating that the critical line is a Stokes line for the Riemann–Siegel expansion. These conclusions are supported by computations of the first 50 coefficients in the expansion, and of the remainders as a function of truncation for several values of t.

1. Introduction

The Riemann–Siegel series, deciphered by Siegel in the 1920s from Riemann's manuscripts of the 1850s, is a very accurate and widely used method of calculating Riemann's function $\zeta(s)$ on the critical line (Edwards 1974). My aim here is to understand the structure of the series. By 'understanding' I mean three things; first, devising a transparent formalism for obtaining the terms in the expansion, enabling high orders to be calculated; second, establishing the dominant behaviour of the high orders; and third, estimating the dependence of the truncation error on the order of truncation when this is large, and thence the ultimate accuracy that can be obtained with the method. The latter is particularly interesting in view of the recent development of alternative methods for calculating ζ to high accuracy (Berry & Keating 1992; Paris 1994).

On the critical line $s = \frac{1}{2} + it$ (t real), $\zeta(s)$ is complex. However, it follows from the functional equation for $\zeta(s)$ that the function

$$Z(t) = \exp(\mathrm{i}\theta(t))\zeta(\tfrac{1}{2} + \mathrm{i}t), \tag{1}$$

where

$$\theta(t) = \arg\Gamma(\tfrac{1}{4} + \tfrac{1}{2}\mathrm{i}t) - \tfrac{1}{2}t\log\pi \tag{2}$$

is real for real t (and also even). The Riemann–Siegel series is an expansion of $Z(t)$ for large t, whose starting-point is the separation of this function into a 'main sum' plus a remainder. It is convenient first to define

$$a \equiv \surd(t/2\pi), \quad N \equiv \mathrm{Int}(a), \quad a - N \equiv \tfrac{1}{2}(1 - z). \tag{3}$$

Proc. R. Soc. Lond. A (1995) **450**, 439–462
Printed in Great Britain

TEX Paper

Reprinted from Proceedings of the Royal Society A, Volume 450, (1995).

Then the separation is

$$Z(t) = 2\sum_{n=1}^{N} \frac{\cos(\theta(t) - t\log n)}{n^{1/2}} + R(t). \tag{4}$$

The remainder $R(t)$ can be written as a formal power series in $1/a$, namely

$$R(t) = \frac{(-1)^{N+1}}{a^{1/2}} \sum_{r=0}^{\infty} \frac{C_r(z)}{a^r}. \tag{5}$$

This is the Riemann–Siegel expansion, whose coefficients $C_r(z)$ will be our principal concern. In correcting the main sum, the expansion removes the discontinuities where the upper limit jumps, that is at $t = 2\pi N^2$ (N integer), and interpolates between these points: as t increases from $2\pi N^2$ to $2\pi(N+1)^2$, z decreases from 1 to -1, with $z = 0$ corresponding to points $2\pi(N + \frac{1}{2})^2$.

Riemann's technique for obtaining the coefficients $C_r(z)$ (explained for example by Edwards (1974) and Titchmarsh (1986)) is an intricate application of the saddle-point method to an integral representation of $Z(t)$, with the subtlety that the saddle lies on a line containing a string of poles. By contrast, the formalism I use in § 2 is wholly elementary and based on the Dirichlet series $\zeta(s) = \sum_{n=1}^{\infty} n^{-s}$ ($\mathrm{Re}\, s > 1$). Since this series does not converge on the critical line, the method is formal and so cannot be regarded as a substitute for the customary derivation which gives the same results. However, the method has the advantages of exposing the essential algebra of the expansion, in a way that enables new results to be found later, and of generating a series for $Z(t)$ that is automatically real in spite of not explicitly using the functional equation. The coefficient $C_r(z)$ involves derivatives (up to the $3r$th) of the function

$$F(z) = \frac{\cos(\frac{1}{2}\pi(z^2 + \frac{3}{4}))}{\cos(\pi z)}. \tag{6}$$

In § 3 the elementary formalism is used to calculate high orders of the expansion, that is $C_r(z)$ for large r. The main result is

$$C_r(z) = \frac{\Gamma(\frac{1}{2}r)}{(\pi\sqrt{2})^{r+1}} f(r, z), \tag{7}$$

where f is bounded and given by the rapidly convergent series

$$f(r, z) \approx \sum_{m=0}^{\infty} (-1)^{m(m-1)/2} \exp\{-(m + \tfrac{1}{2})^2\} \times \begin{Bmatrix} \sin\{(2m+1)\sqrt{r}\}\cos\{(m+\frac{1}{2})\pi z\} & (r \text{ even}) \\ \cos\{(2m+1)\sqrt{r}\}\sin\{(m+\frac{1}{2})\pi z\} & (r \text{ odd}) \end{Bmatrix} \quad \text{when } r \gg 1. \tag{8}$$

These formulae show that the Riemann–Siegel expansion (5) diverges, with the divergence dominated by the 'factorial divided by a power' typical of asymptotic series. For large t the terms in (5) get smaller before they increase, with the minimum near $r = r^* = \mathrm{Int}(2\pi t)$. This familiar divergence is multiplied by the factor $f(r, z)$, whose oscillations are slow in comparison with the growth of $\Gamma(\frac{1}{2}z)$. Therefore the expansion falls in the class of 'decorated factorial series' now beginning to be encountered in asymptotics; another example is saddle-point expansions whose divergence is dominated by coalescing distant saddles (Berry & Howls 1993). In the Riemann–Siegel case, however, the decoration f has an unfamiliar form.

These results are used in § 4 to estimate the remainder when the Riemann–Siegel series is truncated at some large order R. The remainder is $S_R(t)$, where

$$R(t) = \frac{(-1)^{N+1}}{a^{1/2}} \sum_{r=0}^{R} \frac{C_r(z)}{a^r} + S_R(t). \tag{9}$$

I give two heuristic arguments leading to the expectation that $S_R(t)$ is of order $\exp(-\pi t)$. Then a direct calculation, using a variant of Borel summation, gives

$$S_R(t) \approx \frac{(-1)^N e^{-\pi t}}{a^{1/2}\sqrt{2}} \left\{ \left[-\mathrm{i}\,\mathrm{erf}(\mathrm{i}\sigma) + \frac{1}{3\sqrt{(\pi R)}}(4\sigma^2 + \tfrac{1}{2})e^{\sigma^2} \right] [f(R+2, z) + f(R+1, z)] \right.$$
$$\left. + \frac{1}{2\sqrt{(\pi R)}} e^{\sigma^2} [f(R+2, z) - f(R+1, z)] \right\}, \tag{10}$$

where
$$\sigma = (\tfrac{1}{2}R - \pi t)/\sqrt{R}. \tag{11}$$

Optimal truncation, that is truncation of the Riemann–Siegel series near $R = r^*$ (where $\sigma \approx 0$), thus yields an error of the expected order $\exp(-\pi t)$. This shows that the Riemann–Siegel expansion is less accurate than that of Berry & Keating (1992), whose error is bounded by $\exp(-t^{4/3})$ (now we conjecture that this can be reduced to $\exp(-t^2)$).

The arguments leading to the asymptotic formulae (7) and (10) are formal and non-rigorous, so it is desirable to test them by comparison with direct calculations of the $C_r(z)$ and the remainders $S_R(t)$. This requires high derivatives of $F(z)$ (equation (6)), which are difficult to evaluate. In § 5 methods are given for evaluating these derivatives exactly for the special cases $z = 0$, $z = \pm\frac{1}{2}$ and $z = \pm 1$ (§ 5 *a*), and asymptotically for any z (§ 5 *b*).

In § 6 *a* numerical comparisons are given between the first 50 'experimental' Riemann–Siegel coefficients $C_r(z)$ and the 'theoretical' prediction (7) and (8) for a range of z values. The theory works very well, with the decoration $f(r, z)$ reproducing fine details of the coefficients, even for r as small as 5. In § 6.2 comparisons are given between the 'experimental' and 'theoretical' remainders $S_R(t)$ as functions of truncation R, for several values of t. The theoretical formula (10) passes this highly discriminating test very well. It shows that the Riemann–Siegel formula is capable of astonishing accuracy, even for $t < 2\pi$, when there are no terms in the main sum (i.e. $N = 0$ in (4)). For example, when $t = \pi$ the Riemann–Siegel series (5) starts to diverge when $r \approx 20$ and can generate $Z(t)$ to one part in 10^5. All computations were carried out using MATHEMATICA (Wolfram 1991).

In previous numerical applications (e.g. computations of the zeros by Brent (1979), van de Lune *et al.* (1986), Odlyzko (1987, 1990), Odlyzko & Schönhage (1988)), only a few terms of the Riemann–Siegel series were needed, and the questions addressed here, of the asymptotics for large r, did not arise. The most extensive theoretical study of the Riemann–Siegel formula was by Gabcke (1979). He gave explicit formulae for the $C_r(z)$ for $r \leqslant 12$, and derived strict (and realistic) error bounds for the $S_R(t)$ for $R \leqslant 10$ (e.g. $|S_{10}(t)| < 25\,966t^{-23/4}$). He speculated that the series diverges, and proved this for the special case of the coefficients $C_{2m}(1)$, using a direct method (Appendix B) that sidesteps the complications of the general formalism. We explore this further in Appendix D by determining the explicit form of these particular coefficients for $2m$ large (using a result about asymptotic series established in Appendix C). As it turns out, this special case is misleading, in the sense that the divergence of the $C_{2m}(1)$ is weaker than in the general case and is not captured by the leading-order theory (indeed (8) predicts zero for $C_{2m}(1)$).

2. Derivation of the series

In the function defined by (1) we can formally substitute the Dirichlet series for $\zeta(s)$, and obtain

$$Z(t) = \exp(\mathrm{i}\theta(t)) \sum_{n=1}^{\infty} \frac{\exp(-\mathrm{i}t \log n)}{n^{1/2}}. \tag{12}$$

The sum does not converge and its terms are not real. However, we note that the main sum in (4) comprises the first N terms of the Dirichlet series, together with their complex conjugates. (These complex conjugate terms can be obtained (Berry 1986; Titchmarsh 1986) from the resummation of the divergent tail of (12).) Thus the remainder in (4) can be written as

$$R(t) = \sum_{n=N+1}^{\infty} \frac{\exp\{\mathrm{i}[\theta(t) - t \log n]\}}{n^{1/2}} - \sum_{n=1}^{N} \frac{\exp\{-\mathrm{i}[\theta(t) - t \log n]\}}{n^{1/2}}. \tag{13}$$

The Riemann–Siegel expansion (5) will be obtained by expanding the terms in the two sums about the limits $N+1$ and N. This ingenious procedure, devised by Keating (1993) to calculate the lowest coefficient $C_0(z)$, will be the basis of all that follows.

To obtain the expansion, it is necessary to make use of the asymptotic expansion of $\theta(t)$. Gabcke (1979) shows from (2) that

$$\theta(t) = \tfrac{1}{2}t(\log(t/\pi) - 1) - \tfrac{1}{8}\pi + \chi(t), \tag{14}$$

where $\chi(t)$ has the formal expansion

$$\chi(t) = \sum_{m=1}^{\infty} \frac{b_m}{t^{2m-1}}, \quad \text{where} \quad b_m = \frac{(2^{2m-1} - 1)|B_{2m}|}{2^{2m+1}m(2m-1)}, \tag{15}$$

in which B_{2m} are the Bernoulli numbers. For the first sum in (13), we define the new variable K by

$$n = N + 1 + K = a(1 + Q(K,z)/a), \quad 0 \leqslant K \leqslant \infty, \tag{16}$$

where a and z are defined in (3) and

$$Q(K,z) \equiv \tfrac{1}{2}(1+z) + K. \tag{17}$$

The analogous definition for the second sum is

$$n = N - K = a(1 - Q(K,-z)/a), \quad 0 \leqslant K \leqslant N-1. \tag{18}$$

Now the phase in the first sum in (13) is expanded using (14) and (16). A short calculation gives

$$\begin{aligned}[\theta(t) - t\log n]_{(\mathrm{mod}\, 2\pi)} = {} & (N+1+K)\pi + \tfrac{3}{8}\pi + \chi(2\pi a^2) - \tfrac{1}{2}\pi z^2 \\ & + 2\pi z Q(K,z) + 2\pi a^2 \sum_{m=3}^{\infty} \frac{1}{m}\left(-\frac{Q(K,z)}{a}\right)^m. \end{aligned} \tag{19}$$

With these substitutions, the first sum in (13) becomes

$$\frac{(-1)^{N+1}}{a^{1/2}}T(a,z), \tag{20}$$

where

$$T(a,z) = \exp\{\mathrm{i}\pi(\tfrac{3}{8}-\tfrac{1}{2}z^2)+\mathrm{i}\chi(2\pi a^2)\} \times \sum_{K=0}^{\infty}(-1)^K \frac{\exp\{\mathrm{i}\pi Q(K,z)(2z+2a-Q(K,z))\}}{(1+Q(K,z)a)^{1/2+2\pi \mathrm{i}a^2}}. \tag{21}$$

The second sum in (13) is given by a similar expression with the upper limit replaced by $K = N-1$. This upper limit plays no part in the Riemann–Siegel expansion, and will henceforth be replaced by $K = \infty$ (this amounts to including terms with negative n in the second sum in (13), a point to which we will return). Thus we find

$$R(t) = \frac{(-1)^{N+1}}{a^{1/2}}[T(a,z)+T^*(-a,-z)]. \tag{22}$$

To generate the Riemann–Siegel expansion (5), it is necessary to expand T in powers of $1/a$, i.e.

$$T(a,z) = \sum_{r=0}^{\infty}\frac{T_r(z)}{a^r}. \tag{23}$$

Then the Riemann–Siegel coefficients are

$$C_r(z) = T_r(z) + (-1)^r T_r^*(-z). \tag{24}$$

We shall find that $T_r(-z) = (-1)^r T_r(z)$; thus the $C_r(z)$ are real, as they must be, and satisfy the symmetry relation

$$C_r(-z) = (-1)^r C_r(z). \tag{25}$$

The lowest coefficient is obtained directly from (21) (Keating 1993) as

$$\begin{aligned} C_0(z) &= 2\,\mathrm{Re}\left\{\exp\{\mathrm{i}\pi(\tfrac{3}{8}-\tfrac{1}{2}z^2)\}\sum_{K=0}^{\infty}(-1)^K\exp\{2\mathrm{i}\pi Q(K,z)z\}\right\} \\ &= \cos\{\tfrac{1}{2}\pi(z^2+\tfrac{3}{4})\}/\cos(\pi z) \equiv F(z). \end{aligned} \tag{26}$$

To get the general term in the expansion, it is convenient to use the operator notation

$$D \equiv \frac{\partial}{\partial z} \tag{27}$$

to bring (21) to a symbolic form where the K sum is the same as that in (26), namely

$$T(a,z) = \exp\{\mathrm{i}\pi(\tfrac{3}{8}-\tfrac{1}{2}z^2)+\mathrm{i}\chi(2\pi a^2)\} \times \frac{\exp\{aD+\mathrm{i}D^2/4\pi\}}{(1+D/2\pi\mathrm{i}a)^{1/2+2\pi\mathrm{i}a^2}}\sum_{K=0}^{\infty}(-1)^K\exp\{2\mathrm{i}\pi Qz\}. \tag{28}$$

Thus

$$T(a,z) = \exp\{\mathrm{i}\pi(\tfrac{3}{8}+\tfrac{1}{2}z^2)\} \times \exp(-\mathrm{i}\pi z^2)\left[\left\{\frac{\exp\{\mathrm{i}\chi(2\pi a^2)+aD+\mathrm{i}D^2/4\pi\}}{(1+D/2\pi\mathrm{i}a)^{1/2+2\pi\mathrm{i}a^2}}\right\}\frac{\exp(\mathrm{i}\pi z\zeta)}{2\cos(\pi z)}\right]_{\zeta\to z}. \tag{29}$$

The operator in braces has an expansion in $1/a$ and D. It is easier to find this if the function is real, which we achieve with new variables x and y defined by

$$a \equiv (1/2y)\surd(\mathrm{i}/2\pi), \quad D \equiv \tfrac{1}{2}x\surd(2\pi/\mathrm{i}). \tag{30}$$

Then

$$\left\{\frac{\exp\{\mathrm{i}\chi(2\pi a^2) + aD + \mathrm{i}D^2/4\pi\}}{(1 + D/2\pi\mathrm{i}a)^{1/2+2\pi\mathrm{i}a^2}}\right\} \equiv g(x, y)$$

$$= \frac{1}{(1 - xy)^{1/2-1/4y^2}} \exp\left\{\mathrm{i}\chi\left(\frac{\mathrm{i}}{4y^2}\right) + \frac{x}{4y} + \frac{x^2}{8}\right\}$$

$$= \exp\left\{-\sum_{m=1}^{\infty}(-1)^m b_m (4y^2)^{2m-1} + \sum_{m=1}^{\infty}(xy)^m\left(\frac{1}{2m} - \frac{x^2}{4(m+2)}\right)\right\}. \tag{31}$$

This has an expansion in powers of x and y, with the term in y^r multiplied by x^{3r}, x^{3r-2}, etc.:

$$g(x, y) = \sum_{r=0}^{\infty}\sum_{m=0}^{\mathrm{Int}(3r/2)} g_{rm} x^{3r-2m} y^r. \tag{32}$$

Obviously, $g_{00} = 1$. In Appendix A it is shown that the coefficients satisfy the recurrence relation,

$$g_{r+1,m} = \frac{1}{2(r+1)}\sum_{k=0}^{k_1(r,m)} g_{r-k,m-1-k} - \frac{1}{4(r+1)}\sum_{k=0}^{k_2(r,m)} \frac{(k+1)}{(k+3)} g_{r-k,m-k}$$
$$+ \frac{1}{r+1}\sum_{p=0}^{p_1(m)} (-1)^p 2^{4p+2}(4p+2) b_{p+1}\, g_{r-4p-1,m-3(2p+1)}, \tag{33}$$

where the limits of the sums are

$$k_1(r,m) = \begin{cases} m-1 & (m \leqslant r+1), \\ 3r - 2m + 2 & (m \geqslant r+1), \end{cases}$$
$$k_2(r,m) = \begin{cases} m & (m \leqslant r), \\ 3r - 2m & (m \geqslant r), \end{cases} \tag{34}$$
$$p_1(m) = \mathrm{Int}(\tfrac{1}{6}m - \tfrac{1}{2}).$$

With the expansion (32), after replacing x and y by D and a from (30), we find, for the coefficient of $1/a^r$ in (29),

$$T_r(z) = \exp\{\mathrm{i}\pi(\tfrac{3}{8} - \tfrac{1}{2}z^2)\}(-1)^r \frac{1}{\pi^{2r}} \sum_{m=0}^{\mathrm{Int}(3r/2)} g_{rm}\mathrm{i}^{-m}(\tfrac{1}{2}\pi)^m$$
$$\times \left[D^{3r-2m}\frac{\exp(\mathrm{i}\pi z\zeta)}{2\cos(\pi z)}\right]_{\zeta\to z}. \tag{35}$$

To evaluate the derivatives, we define

$$\Phi(z) \equiv \exp\{\tfrac{1}{2}\mathrm{i}\pi(z^2 + \tfrac{3}{4})\}/2\cos(\pi z) \tag{36}$$

and use Leibniz's rule for differentiating a product. Thus

$$\exp\{\tfrac{1}{2}\mathrm{i}\pi(\tfrac{3}{4}-z^2)\}\left[D^n \frac{\exp(\mathrm{i}\pi z\zeta)}{2\cos(\pi z)}\right]_{\zeta\to z}$$

$$= \exp(-\tfrac{1}{2}\mathrm{i}\pi z^2)[D^n \exp\{\mathrm{i}\pi(z\zeta - \tfrac{1}{2}z^2)\}\,\Phi(z)]_{\zeta\to z}$$

$$= \sum_{s=0}^{n} \frac{n!}{s!(n-s)!}\Phi^{(n-s)}(z)[D^s \exp\{-\tfrac{1}{2}\mathrm{i}\pi(z-\zeta)^2\}]_{\zeta\to z}, \tag{37}$$

where $\Phi^{(m)}(z)$ denotes the mth derivative. The derivatives D^s are zero unless s is even. A little calculation now leads to

$$T_r(z) = \frac{(-1)^r}{\pi^{2r}} \sum_{q=0}^{\mathrm{Int}(3r/2)} (\tfrac{1}{2}\pi)^q \frac{(-\mathrm{i})^q}{(3r-2q)!}\Phi^{(3r-2q)}(z)\sum_{m=0}^{q}\frac{(3r-2m)!}{(q-m)!}g_{rm}. \tag{38}$$

To get the Riemann–Siegel coefficients $C_r(z)$, this must be combined with $T_r^*(-z)$ according to (24). The fact that even and odd derivatives are, respectively, even and odd functions of z (cf. (36)) guarantees that the C_r are real. In (38) the terms with even q combine to give derivatives of

$$2\,\mathrm{Re}\,\Phi(z) = F(z), \tag{39}$$

where Φ is defined by (6), and the terms with odd q combine to give derivatives of

$$2\,\mathrm{Im}\,\Phi(z) = \sin\{\tfrac{1}{2}\pi(z^2+\tfrac{3}{4})\}/\cos(\pi z). \tag{40}$$

Now, Im $\Phi(z)$ has poles at $z = \pm\frac{1}{2}$, whereas Re $\Phi(z)$ is finite at these points (zeros of the numerator and denominator cancel). But $Z(t)$ is a smooth function, so that the poles cannot contribute to the Riemann–Siegel coefficients. Therefore the sum over m in (38) must vanish if q is odd. Extensive computations confirm that this is so, but I have not succeeded in finding a general proof.

Incorporating this observation, we obtain the Riemann–Siegel coefficients in their final form:

$$C_r(z) = \sum_{p=0}^{\mathrm{Int}(3r/4)} \frac{F^{(3r-4p)}(z)}{\pi^{2(r-p)}}d_{rp}, \tag{41}$$

where

$$d_{rp} = \frac{(-1)^{r+p}}{2^{2p}(3r-4p)!}\sum_{m=0}^{2p}\frac{(3r-2m)!}{(2p-m)!}g_{rm}. \tag{42}$$

The multipliers d_{rp} are rational numbers, defined in terms of the g_{rm} which are calculated from the recurrence relation (33).

Table 1 shows some of the d_{rp}, extending the list of Gabcke (1979) which showed these multipliers for $r \leqslant 12$. For small m and p it is possible to calculate g_{rm} and

d_{rp} for all r; the results are

$$\left.\begin{aligned} g_{r0} &= \frac{(-1)^r}{12^r r!}, \quad g_{r1} = \frac{(-1)^{r+1}3r(3r-1)}{12^r r!}, \\ g_{r2} &= \frac{27}{16}\frac{(-1)^r(15r^3-38r^2+29r-6)}{12^r(r-1)!}, \\ d_{r0} &= \frac{1}{12^r r!}, \quad d_{r1} = \frac{9}{5}\frac{(3r-1)}{12^r(r-2)!}. \end{aligned}\right\} \tag{43}$$

3. High orders

The coefficients $T_r(z)$ in (23) can be written as

$$T_r(z) = \frac{1}{2\pi\mathrm{i}}\oint \frac{\mathrm{d}a}{a} a^r T(a,z), \tag{44}$$

in which the contour is a loop at infinity enclosing the origin. For T we substitute the series (21), treating each term separately. Thus we obtain

$$T_r(z) \equiv \sum_{K=0}^{\infty} U_{Kr}(z), \tag{45}$$

where

$$\begin{aligned} U_{Kr}(z) &= \exp\{\mathrm{i}\pi(\tfrac{3}{8} - \tfrac{1}{2}z^2 + 2zQ - Q^2)\} \\ &\quad \times \frac{(-1)^K}{2\pi\mathrm{i}}\oint \frac{\mathrm{d}a}{a} a^r \frac{\exp\{\mathrm{i}(2\pi a Q + \chi(2\pi a^2))\}}{(1+Q/a)^{1/2+2\pi\mathrm{i}a^2}}. \end{aligned} \tag{46}$$

The aim is to find the asymptotic form of this integral for large r, and then perform the summation over K.

Three preliminaries will enable the integral to be cast in a form suitable for asymptotic evaluation. First, the exponent $\mathrm{i}\chi$ will be neglected; this seems drastic, but is justified by a calculation (Appendix D) of the asymptotic series (in $1/a$) for the corresponding exponential. This reveals that although the high orders diverge factorially, the divergence of the terms from the rest of (46) is exponentially larger and so dominates. It now follows from a result on the combination of asymptotic series (Appendix C) that the contribution of χ to the high-order Riemann–Siegel coefficients is negligible.

Second, we rearrange the exponent in (46) using

$$(-\pi Q^2 + 2\pi z Q - \tfrac{1}{2}\pi z^2)_{\mathrm{mod}\ 2\pi} = +\pi Q^2 - \tfrac{1}{2}\pi. \tag{47}$$

This is easily proved from the definition (17) of Q.

Third, we note that without χ the only singularities of the integrand in (46) are branch points at $a=0$ and $a=-Q$ These can be connected with a cut, onto which the contour can be shrunk (figure 1). On the upper (lower) lip, the phase of $(1+Q/a)$ is $-\pi$ $(+\pi)$. With the natural change of variable $a=-Qu$, (46) now becomes

$$\begin{aligned} U_{Kr}(z) &= \exp\{\mathrm{i}\pi(-\tfrac{1}{8} + Q^2)\}\frac{(-1)^{K+r}}{\pi}Q^r \\ &\quad \times \int_0^1 \frac{\mathrm{d}u\, u^r}{\sqrt{u(1-u)}} \exp\{-2\pi\mathrm{i}Q^2u(1+u\log(u^{-1}-1))\}\cosh(2\pi^2Q^2u^2). \end{aligned} \tag{48}$$

Table 1. *Lists of multipliers* d_{rp} *for* $0 \leqslant r \leqslant 14$

(Calculated from equations (42) and (33); (*a*) $p \leqslant 3$, (*b*) $p \geqslant 4$; the first columns list the values of r.)

(*a*)

0	1			
1	$\frac{1}{12}$			
2	$\frac{1}{288}$	$\frac{1}{16}$		
3	$\frac{1}{10\,368}$	$\frac{1}{120}$	$\frac{1}{32}$	
4	$\frac{1}{497\,664}$	$\frac{11}{23\,040}$	$\frac{19}{1\,536}$	$\frac{1}{128}$
5	$\frac{1}{29\,859\,840}$	$\frac{7}{414\,720}$	$\frac{901}{645\,120}$	$\frac{5}{384}$
6	$\frac{1}{2\,149\,908\,480}$	$\frac{17}{39\,813\,120}$	$\frac{18\,889}{232\,243\,200}$	$\frac{367}{122\,880}$
7	$\frac{1}{180\,592\,312\,320}$	$\frac{1}{119\,439\,360}$	$\frac{2\,131}{696\,729\,600}$	$\frac{6\,649}{23\,224\,320}$
8	$\frac{1}{17\,336\,861\,982\,720}$	$\frac{23}{171\,992\,678\,400}$	$\frac{11\,153}{133\,772\,083\,200}$	$\frac{88\,651}{5\,573\,836\,800}$
9	$\frac{1}{1\,872\,381\,094\,133\,760}$	$\frac{13}{7\,223\,692\,492\,800}$	$\frac{8\,503}{4\,815\,794\,995\,200}$	$\frac{66\,727}{111\,476\,736\,000}$
10	$\frac{1}{224\,685\,731\,296\,051\,200}$	$\frac{29}{1\,386\,948\,958\,617\,600}$	$\frac{703}{23\,115\,815\,976\,960}$	$\frac{229\,787}{13\,759\,414\,272\,000}$
11	$\frac{1}{29\,658\,516\,531\,078\,758\,400}$	$\frac{1}{4\,680\,952\,735\,334\,400}$	$\frac{6\,103}{13\,869\,489\,586\,176\,000}$	$\frac{52\,889}{144\,473\,849\,856\,000}$
12	$\frac{1}{4\,270\,826\,380\,475\,341\,209\,600}$	$\frac{1}{513\,567\,385\,819\,545\,600}$	$\frac{229\,741}{41\,941\,336\,508\,596\,224\,000}$	$\frac{40\,459}{6\,164\,217\,593\,856\,000}$
13	$\frac{1}{666\,248\,915\,354\,153\,228\,697\,600}$	$\frac{19}{1\,186\,340\,661\,243\,150\,336\,000}$	$\frac{30\,673}{503\,296\,038\,103\,154\,688\,000}$	$\frac{37\,117}{374\,476\,218\,826\,752\,000}$
14	$\frac{1}{111\,929\,817\,779\,497\,742\,421\,196\,800}$	$\frac{41}{341\,666\,110\,438\,027\,296\,768\,000}$	$\frac{76\,997}{132\,870\,154\,059\,232\,837\,632\,000}$	$\frac{12\,975\,523}{10\,065\,920\,762\,063\,093\,760\,000}$

(*b*)

7	$\frac{407}{81\,920}$	$\frac{5}{1\,024}$					
8	$\frac{26\,405}{33\,030\,144}$	$\frac{427}{65\,536}$	$\frac{41}{32\,768}$				
9	$\frac{21\,543\,701}{326\,998\,425\,600}$	$\frac{3\,781}{2\,064\,384}$	$\frac{2\,603}{393\,216}$				
10	$\frac{2\,320\,279}{678\,218\,956\,800}$	$\frac{26\,988\,341}{118\,908\,518\,400}$	$\frac{32\,851}{9\,437\,184}$	$\frac{2\,603}{524\,288}$			
11	$\frac{413\,542\,091}{3\,296\,144\,130\,048\,000}$	$\frac{13\,891\,121}{850\,195\,906\,560}$	$\frac{31\,587\,481}{47\,563\,407\,360}$	$\frac{5\,377}{983\,040}$	$\frac{2\,603}{1\,048\,576}$		
12	$\frac{8\,283\,026\,179}{2\,373\,223\,773\,634\,560\,000}$	$\frac{135\,149\,833\,261}{171\,399\,494\,762\,496\,000}$	$\frac{1\,681\,009\,907}{25\,113\,479\,086\,080}$	$\frac{7\,040\,165}{4\,227\,858\,432}$	$\frac{351\,937}{50\,331\,648}$	$\frac{2\,643}{4\,194\,304}$	
13	$\frac{733\,972\,733}{9\,492\,895\,094\,538\,240\,000}$	$\frac{7\,804\,440\,109}{280\,471\,900\,520\,448\,000}$	$\frac{2\,927\,256\,839\,861}{685\,597\,979\,049\,984\,000}$	$\frac{71\,350\,619}{298\,969\,989\,120}$	$\frac{15\,136\,141}{4\,227\,858\,432}$	$\frac{88\,645}{12\,582\,912}$	
14	$\frac{965\,416\,409}{683\,488\,446\,806\,753\,280\,000}$	$\frac{4\,636\,378\,387}{6\,094\,204\,258\,222\,080\,000}$	$\frac{1\,890\,353\,754\,197}{9\,872\,610\,898\,319\,769\,600}$	$\frac{3\,706\,405\,599\,341}{182\,826\,127\,746\,662\,400}$	$\frac{227\,026\,787}{304\,405\,807\,104}$	$\frac{1\,760\,589}{268\,435\,456}$	$\frac{88\,645}{16\,777\,216}$

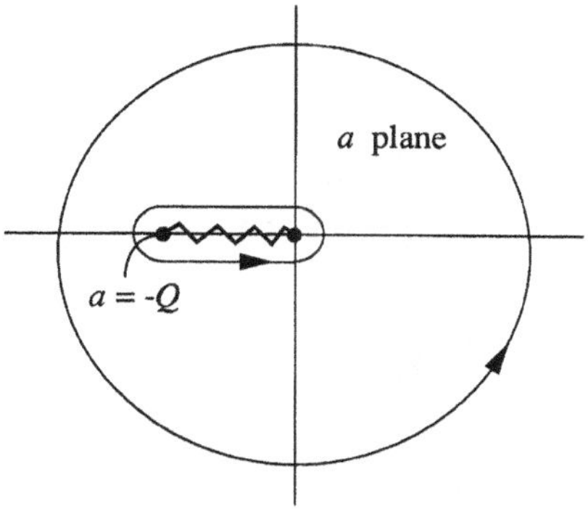

Figure 1. Contours in the a plane for the integral (46).

This integral will be evaluated by the method of steepest descent. The main contribution comes from the part of the cosh function with the negative exponent. The corresponding exponent of the integrand has a stationary point between $u = 0$ and $u = 1$. The value of the exponent at this point depends on Q which is proportional to K. Since it will be necessary to sum over K, the most effective procedure is to determine the stationary point of the whole exponent in (48) with respect to both u and Q. This exponent is

$$E(u, Q; r) = r\log(Qu) - 2\pi Q^2[\pi u^2 + \mathrm{i}(-\tfrac{1}{2} + u + u^2\log(u^{-1} - 1))]. \tag{49}$$

The stationary point (denoted by a superscript 's') is at

$$u^{\mathrm{s}} = \tfrac{1}{2}, \quad Q^{\mathrm{s}} = \sqrt{r}/\pi. \tag{50}$$

(Note that at this point $t = 2\pi a^2 = r/2\pi$, so that $\chi(2\pi a^2) \sim 1/r$, further justifying neglect of this quantity in (46).) The stationary values of E and its second derivatives (denoted by subscripts) are

$$\left.\begin{aligned} E^{\mathrm{s}} &= \tfrac{1}{2}r\log(\tfrac{1}{2}r) - \tfrac{1}{2}r - r\log(\pi\sqrt{2}), \\ E^{\mathrm{s}}_{uu} &= -8r(1 - 2\mathrm{i}\pi), \quad E^{\mathrm{s}}_{uQ} = -4\pi\sqrt{r}, \quad E^{\mathrm{s}}_{QQ} = -2\pi^2. \end{aligned}\right\} \tag{51}$$

Thus we can write

$$E(u, Q; r) = E^{\mathrm{s}} + \tfrac{1}{2}E^{\mathrm{s}}_{uu}(u - \tfrac{1}{2})^2 + E^{\mathrm{s}}_{uQ}(u - \tfrac{1}{2})(Q - \sqrt{r}/\pi) + \tfrac{1}{2}E^{\mathrm{s}}_{QQ}(Q - \sqrt{r}/\pi)^2 + \dots. \tag{52}$$

Evaluating the u integral, and noting that

$$\exp(E^{\mathrm{s}}) = \frac{1}{(\pi\sqrt{2})^r}(\tfrac{1}{2}r)^{r/2}\exp(-\tfrac{1}{2}r) \approx \frac{\sqrt{r}}{(\pi\sqrt{2})^r 2\sqrt{\pi}}\Gamma(\tfrac{1}{2}r) \tag{53}$$

leads, on reinstating K, to

$$U_{Kr}(z) \approx \frac{\Gamma(\frac{1}{2}r)}{(\pi\sqrt{2})^r}\exp(\tfrac{1}{8}\mathrm{i}\pi)\frac{(-1)^{K+r}}{4\pi\sqrt{1 - 2\mathrm{i}\pi^{-1}}}\exp\left\{-\frac{2\pi}{2\pi^{-1} + \mathrm{i}}\left(K + \tfrac{1}{2}(1+z) - \frac{\sqrt{r}}{\pi}\right)^2\right\}. \tag{54}$$

This has a maximum near $K = \sqrt{r}/\pi$, so for large r the lower limit $K = 0$ of the sum in (45) can be replaced by $-\infty$ with an error that is exponentially small. Then the sum $T_r(z)$ is a theta-function series, which by the Poisson summation formula can be transformed into another theta-function series that converges rapidly, namely

$$T_r(z) \approx \frac{(-1)^r\Gamma(\frac{1}{2}r)}{4(\pi\sqrt{2})^{r+1}}\mathrm{i}\sum_{m=-\infty}^{\infty}(-1)^{m(m-1)/2}\exp\{-(m+\tfrac{1}{2})^2\}$$
$$\times\exp\{\mathrm{i}[(m+\tfrac{1}{2})\pi z-(2m+1)\sqrt{r}]\}. \quad (55)$$

Because of the symmetry about $m = -\frac{1}{2}$, the terms with negative and positive m can be combined, thus giving T_r as the real expression:

$$T_r(z) \approx \frac{(-1)^r\Gamma(\frac{1}{2}r)}{2(\pi\sqrt{2})^{r+1}}\sum_{m=0}^{\infty}(-1)^{m(m-1)/2}\exp\{-(m+\tfrac{1}{2})^2\}$$
$$\times\sin\{(2m+1)\sqrt{r}-(m+\tfrac{1}{2})\pi z\}. \quad (56)$$

Finally, adding $T_r(-z)$ according to (24) gives high-order Riemann–Siegel coefficients as

$$C_r(z) = \frac{\Gamma(\frac{1}{2}r)}{(\pi\sqrt{2})^{r+1}}f(r,z), \quad (57)$$

where

$$f(r,z) \approx \sum_{m=0}^{\infty}(-1)^{m(m-1)/2}\exp\{-(m+\tfrac{1}{2})^2\}$$
$$\times\left\{\begin{matrix}\sin\{(2m+1)\sqrt{r}\}\cos\{(m+\frac{1}{2})\pi z\} & (r\text{ even}), \\ \cos\{(2m+1)\sqrt{r}\}\sin\{(m+\frac{1}{2})\pi z\} & (r\text{ odd}),\end{matrix}\right\}\quad\text{when } r\gg 1 \quad (58)$$

as claimed in § 1.

Reinstating a, we see that the terms C_r/a^r in the Riemann–Siegel series (5) behave like

$$\frac{\Gamma(\frac{1}{2}r)}{(a\pi\sqrt{2})^r} = \frac{\Gamma(\frac{1}{2}r)}{(\pi t)^{r/2}}. \quad (59)$$

For fixed large t, the terms decrease rapidly and then increase, in the manner familiar in an asymptotic series (Dingle 1973). The least term is near $r^* = 2\pi t$.

The particular coefficients $C_{2m}(1)$ can be determined solely from the requirement that the discontinuities in the main sum in (4) are removed by the Riemann–Siegel expansion (5), leaving $Z(t)$ continuous as it must be. This argument, by Gabcke (1979), is elaborated in Appendix B, and extended in Appendix D to determine the high-order behaviour of these coefficients. The result is

$$C_{2m}(1) \approx \frac{\Gamma(m)}{(2\pi)^{2m}}\times\left\{\begin{matrix}-\cos(\frac{1}{8}\pi)/48m & (m\text{ even}) \\ \sin(\frac{1}{8}\pi)2\pi & (m\text{ odd})\end{matrix}\right\}\quad\text{for } m\gg 1. \quad (60)$$

The powers in the denominator contain the factor 2, rather than $\sqrt{2}$ as in the general case (57). Therefore these special coefficients are smaller by a factor 2^{-m} than the even coefficients for $z \neq \pm 1$, consistent with (58) predicting zero for the high-order behaviour in this case.

4. Remainders

We seek the form of the remainders $S_R(t)$ of the Riemann–Siegel expansion, truncated at terms $r = R$ close to optimal, that is close to the least term $r \approx r^*$. The remainders are defined by (9). Before proceeding to a direct calculation, I give two heuristic arguments suggesting that they will be of order $\exp(-\pi t)$. The arguments are based on the observation that the divergence of an asymptotic series often originates in terms omitted in its derivation, so the size of these terms limits the accuracy that can be obtained with the expansion.

First, recall that in deriving the Riemann–Siegel expansion we included the terms with negative n in the second sum in (13) (see the remarks following equation (21)). These have the form

$$\exp\{it\log(-n)\} = \exp(-\pi t)\exp(it\log n) \tag{61}$$

and indeed involve the claimed exponential.

Second, the formal expansion (15) for the quantity $\chi(t)$ defined in (14) fails to capture the Stokes phenomenon for the gamma function (Berry 1991) that appears in the definition (2) of $\theta(t)$. Using the reflection and duplication formulae for the gamma function, Gabcke (1979) showed that $\theta(t)$ can be written in terms of $\Gamma(it)$ and $\Gamma(2it)$ through the identity

$$\theta(t) = \tfrac{1}{2}t(\log(t/2\pi) - 1) - \tfrac{1}{8}\pi + \tfrac{1}{2}\operatorname{Im}[\mu(2it) - \mu(it)] + \tfrac{1}{2}\arctan\{\exp(-\pi t)\}, \tag{62}$$

where $\mu(z)$ is defined by

$$\Gamma(z) \equiv \sqrt{2\pi}z^{z-\frac{1}{2}}\exp\{-z + \mu(z)\}. \tag{63}$$

The formal expansion (15), which contributes to the Riemann–Siegel expansion, is obtained by applying Stirling's series to $\mu(it)$ and $\mu(2it)$ in (62). The term involving $\exp(-\pi t)$ in (62) is beyond all orders of the expansion and so does not contribute to the calculation of its terms, although it is of course part of the function $Z(t)$ being approximated. $\exp(-\pi t)$ is the first in a string of small exponentials in the asymptotics of $\zeta(\frac{1}{2} + it)$; elsewhere I will pursue the idea that these correspond to complex periodic orbits ('instantons') in the conjectured associated dynamics.

With this expectation that the remainders will be of order $\exp(-\pi t)$, we now proceed to a direct calculation. $S_R(t)$ can be written formally as the divergent tail of the series (5). If R is large, the asymptotic formulae (57) and (58) can be substituted for the terms, giving (cf. (59))

$$S_R(t) \approx \frac{(-1)^{N+1}}{\pi\sqrt{2a}}\sum_{R+1}^{\infty}\frac{\Gamma(\frac{1}{2}r)}{(\pi t)^{r/2}}f(r,z). \tag{64}$$

Because this is divergent, it must be interpreted. This will be achieved using a variant of Borel summation (Dingle 1973). If the terms with r even and r odd are separated, the functions $f(r,z)$ are slowly varying and can be replaced by their values at the lower limit of the sums. Thus

$$S_R(t) \approx \frac{(-1)^{N+1}}{\pi\sqrt{2a}}\left\{\frac{f(R+1,z)}{(\pi t)^{(R+1)/2}}\sum_{m=0}^{\infty}\frac{\Gamma(\frac{1}{2}(R+1)+m)}{(\pi t)^m}\right.$$
$$\left. + \frac{f(R+2,z)}{(\pi t)^{R/2+1}}\sum_{m=0}^{\infty}\frac{\Gamma(\frac{1}{2}R+1+m)}{(\pi t)^m}\right\}. \tag{65}$$

Replacing the gamma functions by their integral representations, and then evaluating the sums, gives, after an elementary change of variable, the convergent representation

$$S_R(t) \approx \frac{(-1)^{N+1}}{\pi\sqrt{2a}}\{f(R+1,z)\,I(R-1,\pi t)+f(R+2,z)\,I(R,\pi t)\}, \tag{66}$$

where

$$I(R,w) = \fint_0^\infty \mathrm{d}u \frac{\exp(-wu)u^{R/2}}{1-u}. \tag{67}$$

The principal value is chosen to ensure that $S_R(t)$ is real for real t. We seek to evaluate this when R is large and close to the least term $2\pi t = 2w$. Then the integrand has a saddle close to the pole at $u = 1$, and it is natural to expand about $u = 1$. To third order, we obtain

$$\begin{aligned} I(R,w) &= -\mathrm{e}^{-w}\fint_{-\infty}^{\infty} \frac{\mathrm{d}s}{s}\exp\{-\tfrac{1}{4}Rs^2+(\tfrac{1}{2}R-w)s\}(1+\tfrac{1}{6}Rs^3+\ldots) \\ &= -\mathrm{e}^{-w}\fint_{-\infty}^{\infty} \frac{\mathrm{d}v}{v}\exp\{-v^2+2\sigma v\}(1+\frac{4}{3\sqrt{R}}v^3+\ldots), \end{aligned} \tag{68}$$

where

$$\sigma(R,w) \equiv (\tfrac{1}{2}R-w)/\sqrt{R}. \tag{69}$$

The first integral in (68) can be transformed into an error function

$$\fint_{-\infty}^{\infty} \frac{\mathrm{d}v}{v}\mathrm{e}^{-v^2}\sinh(2\sigma v) = 2\sqrt{\pi}\int_0^\sigma \mathrm{d}\sigma\,\mathrm{e}^{\sigma^2} = -\mathrm{i}\pi\,\mathrm{erf}(\mathrm{i}\sigma) \tag{70}$$

(which of course is real) and the second integral is elementary.

Incorporating the difference between $I(R,w)$ and $I(R-1,w)$ in (66) into the term of order $1/\sqrt{R}$, we finally obtain

$$\begin{aligned} S_R(t) \approx \frac{(-1)^N\exp(-\pi t)}{\sqrt{2a}}\Big\{&(-\mathrm{i}\ \mathrm{erf}(\mathrm{i}\sigma)+\frac{\exp(\sigma^2)}{3\sqrt{R\pi}}(4\sigma^2+\tfrac{1}{2})) \\ &\times(f(R+1,z)+f(R+2,z))+\frac{\exp(\sigma^2)}{2\sqrt{R\pi}}(f(R+2,z)-f(R+1,z))\Big\}. \end{aligned} \tag{71}$$

This is the result (10) claimed in § 1. It indeed has the expected leading-order dependence $\exp(-\pi t)$. Moreover it is easily confirmed from (57) that $S_R(t)$ is of the same order as the first term omitted in the truncated Riemann–Siegel series, consistent with Gabcke's rigorous bounds for the first few terms, and asymptotics folklore.

Finally, we consider briefly the case where t is complex, that is when the zeta function is being calculated off the critical line. For optimal truncation, that is $R = \mathrm{Int}(2\pi\,\mathrm{Re}\,t)$, the remainder (71) consists, to leading order, of $\exp(-\pi t)$ times a mutiplier involving

$$-\mathrm{i}\ \mathrm{erf}\left(\frac{\mathrm{Im}\,\pi t}{\sqrt{R}}\right) \approx -\mathrm{i}\ \mathrm{erf}\left(\frac{\mathrm{Im}\,\pi t}{\sqrt{2\,\mathrm{Re}\,\pi t}}\right). \tag{72}$$

This is precisely the universal multiplier describing the smooth switching of the sign of a subdominant exponential across a Stokes line (Berry 1989). Here the subdominant exponential is $\exp(-\pi t)$ (the dominant exponential being unity), and the Stokes line is the critical line t real. An interesting fact is that the 'half-width' of the Stokes line, that is the range over which the argument of the error function increases by unity, is $\mathrm{Im}\, t \sim \sqrt{(2t/\pi)}$, which is larger that the half-width $\mathrm{Im}\, t = \frac{1}{2}$ of the critical strip.

5. Calculation of derivatives contributing to the coefficients

According to (41), the coefficient $C_r(z)$ involves the nth derivatives $F^{(n)}(z)$ of the function $F(z)$ (defined by (6)), for $n \leqslant 3r$. These derivatives are nonsingular functions of z, because the zeros of the numerator and denominator cancel. Direct evaluation is very inefficient, because it gives each high $F^{(n)}(z)$ in terms of powerful singularities that must cancel when summed. In this section I describe two effective methods for evaluating the derivatives, as alternatives, applicable for large r, to the Taylor expansions commonly used for small r (Edwards 1974).

(a) *Special values of z*

For $z = 0$, Leibniz's formula for the derivative of a product gives, for the even derivatives (the odd ones being zero)

$$F^{(2m)}(0) = (2m)!\, \mathrm{Re}\, \exp(\tfrac{3}{8}\mathrm{i}\pi) \times \sum_{s=0}^{m} \frac{1}{(2s)!(2m-2s)!} (\sec \pi z)^{(2s)}_{z=0} (\exp(\tfrac{1}{2}\mathrm{i}\pi z^2))^{(2m-2s)}_{z=0}. \tag{73}$$

Use of

$$\sec \pi z = \sum_{n=0}^{\infty} (-1)^n E_{2n} \frac{(\pi z)^{2n}}{(2n)!}, \tag{74}$$

where E_{2n} are the Euler numbers, leads to

$$F^{(2m)}(0) = (2m)!(\tfrac{1}{2}\pi)^m \sum_{s=0}^{m} (-2\pi)^s E_{2s} \frac{\cos\{\frac{1}{2}\pi(m-s+\frac{3}{4})\}}{(2s)!(m-s)!}. \tag{75}$$

For $z = 1$, we note that

$$F(1+\zeta) = \sin\{\tfrac{1}{2}\pi(\zeta^2 + \tfrac{3}{4})\} + \cos\{\tfrac{1}{2}\pi(\zeta^2 + \tfrac{3}{4})\} \tan \pi\zeta. \tag{76}$$

The first (second) term is even (odd) in ζ, and so gives the even (odd) derivatives. Use of

$$\tan \pi\zeta = \sum_{n=1}^{\infty} (-1)^{n-1} B_{2n} (\pi\zeta)^{2n-1} \frac{2^{2n}(2^{2n}-1)}{(2n)!}, \tag{77}$$

where B_{2n} are the Bernoulli numbers, leads to

$$\left.\begin{aligned} F^{(2m)}(1) &= \pi^{-1/2}(m-\tfrac{1}{2})!(2\pi)^m \sin\{\tfrac{1}{2}\pi(m+\tfrac{3}{4})\}, \\ F^{(2m+1)}(1) &= 4\pi(2m+1)!(\tfrac{1}{2}\pi)^m \\ &\quad \times \sum_{s=0}^{m} (-8\pi)^s B_{2s+2} \frac{(2^{2s+2}-1)\cos\{\frac{1}{2}\pi(m-s+\frac{3}{4})\}}{(2s+2)!(m-s)!}. \end{aligned}\right\} \tag{78}$$

For $z = \frac{1}{2}$, we note that

$$F(\tfrac{1}{2}+\zeta) = \tfrac{1}{2}\cos(\tfrac{1}{2}\pi\zeta^2)\sec(\tfrac{1}{2}\pi\zeta) + \tfrac{1}{2}\sin(\tfrac{1}{2}\pi\zeta^2)\,\mathrm{cosec}(\tfrac{1}{2}\pi\zeta). \tag{79}$$

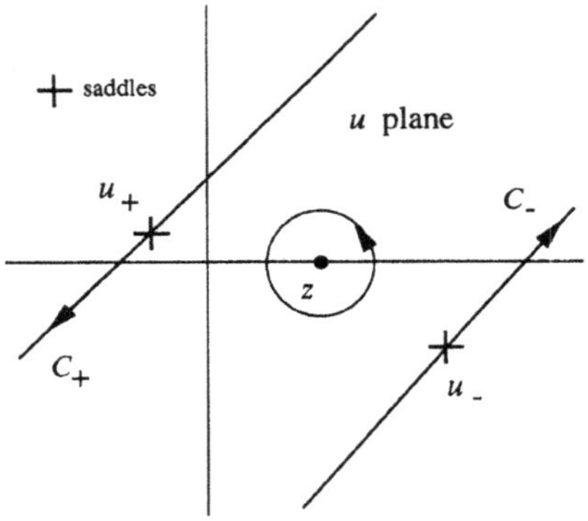

Figure 2. Contours in the u plane for the integral (82).

Again, the first (second) term is even (odd) in ζ, and so gives the even (odd) derivatives. Use of (74) and

$$\operatorname{cosec}(\tfrac{1}{2}\pi\zeta) = 2\sum_{n=0}^{\infty}(-1)^{n+1}B_{2n}(\tfrac{1}{2}\pi\zeta)^{2n-1}\frac{(2^{2n-1}-1)}{(2n)!} \tag{80}$$

leads to

$$\left.\begin{aligned}
F^{(4m)}(\tfrac{1}{2}) &= \tfrac{1}{2}(-1)^m(4m)!(\tfrac{1}{2}\pi)^{2m}\sum_{s=0}^{m}\frac{E_{4s}(-1)^s(\tfrac{1}{2}\pi)^{2s}}{(4s)!(2m-2s)!},\\
F^{(4m+1)}(\tfrac{1}{2}) &= (-1)^{m+1}(4m+1)!(\tfrac{1}{2}\pi)^{2m}\sum_{s=0}^{m}\frac{B_{4s}(-1)^s(\tfrac{1}{2}\pi)^{2s}(2^{4s-1}-1)}{(4s)!(2m-2s+1)!},\\
F^{(4m+2)}(\tfrac{1}{2}) &= \tfrac{1}{2}(-1)^{m+1}(4m+2)!(\tfrac{1}{2}\pi)^{2m+2}\sum_{s=0}^{m}\frac{E_{4s+2}(-1)^s(\tfrac{1}{2}\pi)^{2s}}{(4s+2)!(2m-2s)!},\\
F^{(4m+3)}(\tfrac{1}{2}) &= (-1)^{m}(4m+3)!(\tfrac{1}{2}\pi)^{2m+2}\sum_{s=0}^{m}\frac{B_{4s+2}(-1)^s(\tfrac{1}{2}\pi)^{2s}(2^{4s+1}-1)}{(4s+2)!(2m-2s+1)!}.
\end{aligned}\right\} \tag{81}$$

(b) Asymptotics of high derivatives

A familiar dogma of asymptotics is the invocation of Darboux's theorem (Dingle 1973) to infer that high derivatives of a function at a regular point are determined by the nearest singularity. This fails for $F(z)$, which has no finite singularities. Therefore it is necessary to use a different method. From (6), Cauchy's theorem gives

$$F^{(n)}(z) = \frac{n!}{2\pi}\operatorname{Re}\exp(-\tfrac{1}{8}\mathrm{i}\pi)\oint\frac{\mathrm{d}u}{(u-z)^{n+1}}\frac{\exp(\tfrac{1}{2}\mathrm{i}\pi u^2)}{\cos(\pi u)}, \tag{82}$$

where the contour is a small loop surrounding $u = z$.

The poles at the zeros of the cosine do not contribute: they are cancelled when the real part is taken. Instead, the integral is dominated by saddles. Their contributions can be extracted by expanding the contour into lines C_+ and C_- connecting infinity in the first and third quadrants (figure 2), where the exponential converges. Saddles lie in the second (fourth) quadrants on C_+ (C_-), where the negative (positive) exponential in the cosine dominates. Therefore we can write

$$\frac{1}{\cos(\pi u)} = 2\sum_{p=0}^{\infty}(-1)^p \exp\{\pm i\pi u(2p+1)\} \tag{83}$$

(upper sign: on C_+, in upper half-plane; lower sign: on C_-, in lower half-plane) and express the derivatives as

$$F^{(n)}(z) = \frac{n!}{\pi} \operatorname{Re} \exp(-\tfrac{1}{8}i\pi) \sum_{p=0}^{\infty} (-1)^p [I^+_{np}(z) + I^-_{np}(z)]. \tag{84}$$

Here

$$I^{\pm}_{np}(z) \equiv \int_{C_\pm} du \exp\{-\phi^{\pm}_{np}(u,z)\}, \tag{85}$$

where

$$\phi^{\pm}_{np}(u,z) = (n+1)\log(u-z) - \tfrac{1}{2}i\pi u^2 \mp i\pi u(2p+1). \tag{86}$$

Each integral is dominated by a single saddle, at

$$u = u^{\pm}_{np}(z)\tfrac{1}{2}\{\mp e^{-i\pi/4}\surd[(4/\pi)(n+1) + i(z \pm (2p+1))^2] + z \mp (2p+1)\}. \tag{87}$$

Then the saddle-point method, including the first correction term, gives the approximations

$$I^{\pm}_{np}(z) \approx \sqrt{\frac{2\pi}{\partial^2_u \phi^{\pm}_{np}}} \exp(-\phi^{\pm}_{np}) \left\{1 + \frac{1}{8}\left[\frac{5(\partial^3_u \phi^{\pm}_{np})^2}{3(\partial^2_u \phi^{\pm}_{np})^3} - \frac{\partial^4_u \phi^{\pm}_{np}}{(\partial^2_u \phi^{\pm}_{np})^2}\right]\right\} \tag{88}$$

in which all quantities are evaluated at $u^{\pm}_{np}$. Substitution into (84) gives the derivatives. For large n, the contributions from higher p in (84) and from the saddle-point correction (in square brackets in (88)) diminish rapidly, making this an effective method for computing the high derivatives.

6. Numerical tests of the formulae

(a) *Riemann–Siegel coefficients*

Here we compare the 'experimental' $C_r(z)$ computed from the exact formulae (41) and (42), using coefficients g_{rm} obtained from the recurrence relation (33), with the 'theory' represented by the asymptotic formulae (57) and (58). The coefficients were computed for $0 \leqslant r \leqslant 50$.

For the 'experiment' it is necessary to evaluate the derivatives $F^{(n)}(z)$ up to $n = 150$. This was done using the saddle-point approximations of § 5 *b* with $0 \leqslant p \leqslant 3$. It is convenient to present the data by factoring out the dominant dependence in (57). Therefore the 'experimental' graphs show $f(r,z)$ defined by (57), that is

$$f(r,z) \equiv \frac{(\pi\surd 2)^{r+1}}{\Gamma(\frac{1}{2}r)} C_r(z) \tag{89}$$

and the 'theoretical' curves show the large-r approximation (58).

Figures 3 and 4 are synoptic comparisons for a range of z values between 0 and 1, for r even and r odd. Evidently the theory works well, even for r as small as 5. Figure 5 shows comparisons for some individual z values.

It is necessary to check the unlikely possibility that the good agreement in figures

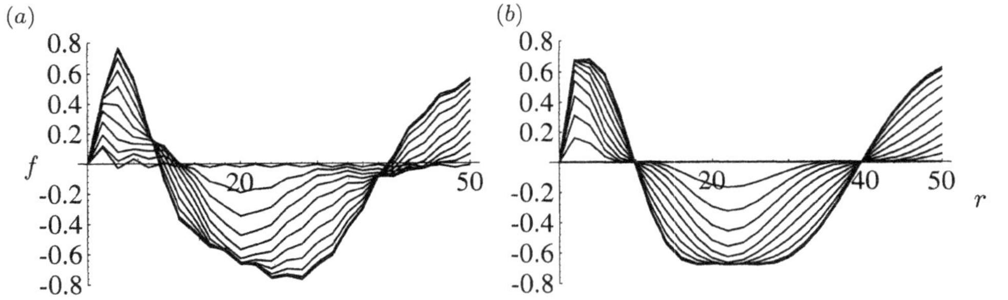

Figure 3. Decoration function $f(r, z)$ (defined by (90)) multiplying factorials in the Riemann–Siegel coefficients, for even r. The curves show $z = 0$ [0.1] 1. (a) 'Experimental' coefficients (41) and (42) with derivatives approximated by the saddle-point method of § 5 b; (b) 'theory' (58).

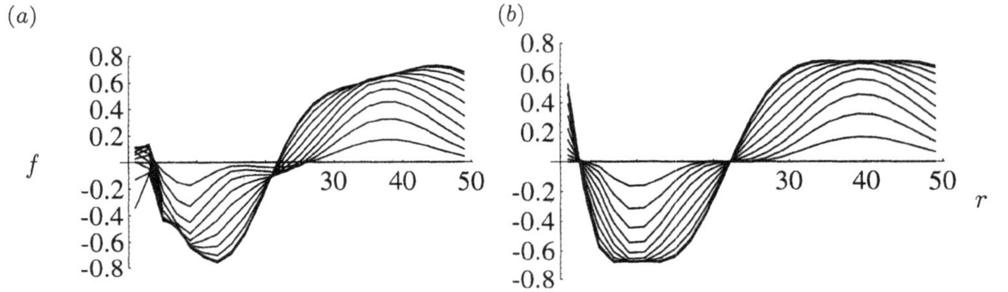

Figure 4. As figure 3, with r odd.

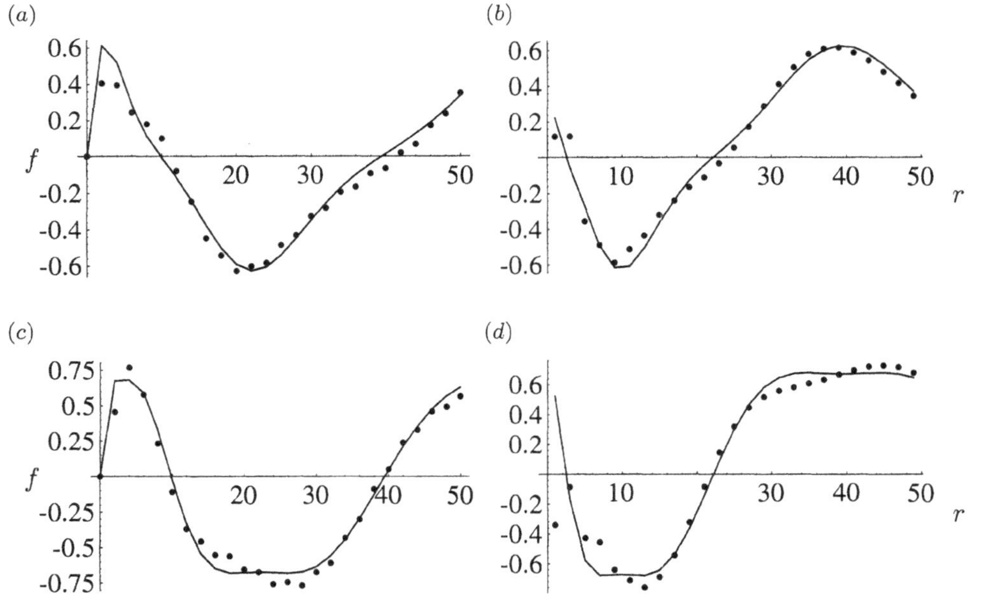

Figure 5. As figure 3, for (a) $z = 0.5$, even r; (b) $z = 0.5$, odd r; (c) $z = 0$, even r; (d) $z = 1$, odd r. Dots, 'experiment'; full lines 'theory'.

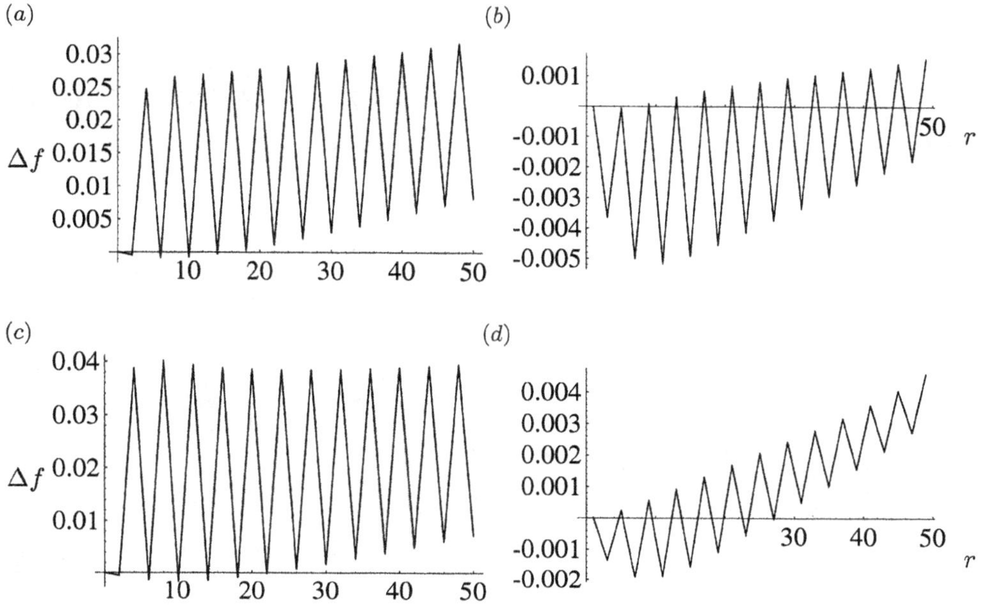

Figure 6. Test of saddle-point approximation of § 5 *b* with $0 \leqslant p \leqslant 3$, for derivatives of $F(z)$, with $\Delta f(r, z) = f_{\text{exact}}(r, z) - f_{\text{saddle}}(r, z)$, for (*a*) $z = 0.5$, even r; (*b*) $z = 0.5$, odd r; (*c*) $z = 0$, even r; (*d*) $z = 1$, odd r.

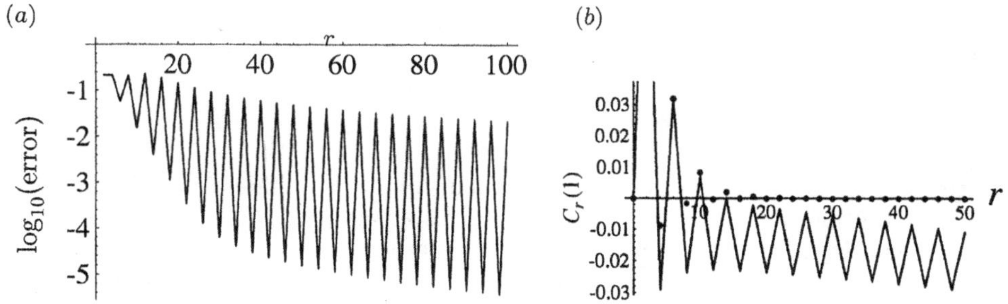

Figure 7. Coefficients $C_r(1)$ against r for r even. (*a*) $\log_{10}$(error) in the high-order approximation (60), where error = (approximate C/exact C) − 1; (*b*) coefficients calculated exactly (dots), and via saddle-point approximation for the derivatives (full line).

3–5 is a fortuitous consequence of the saddle-point approximation for the F derivatives. That this is not the case is clear from figure 6, which shows the difference between the coefficients computed with the exact and approximate derivatives, for the particular cases (§ 5 *a*) where the derivatives can be computed exactly. The errors never exceed 4% and would barely be perceptible in figures 3–5. It is interesting to note that the errors are smaller when r has the form $4m$ than for $r = 4m + 2$, and much smaller still when r is odd.

Figure 7 concerns the particular coefficients $C_{2m}(1)$, which can easily be computed exactly (Appendix B). Figure 7*a* confirms that the formula (60) for the high orders is better for larger r, and also shows that the relative error is much smaller for $r = 4m + 2$ than for $r = 4m$. Figure 7*b* shows that for these coefficients (which are

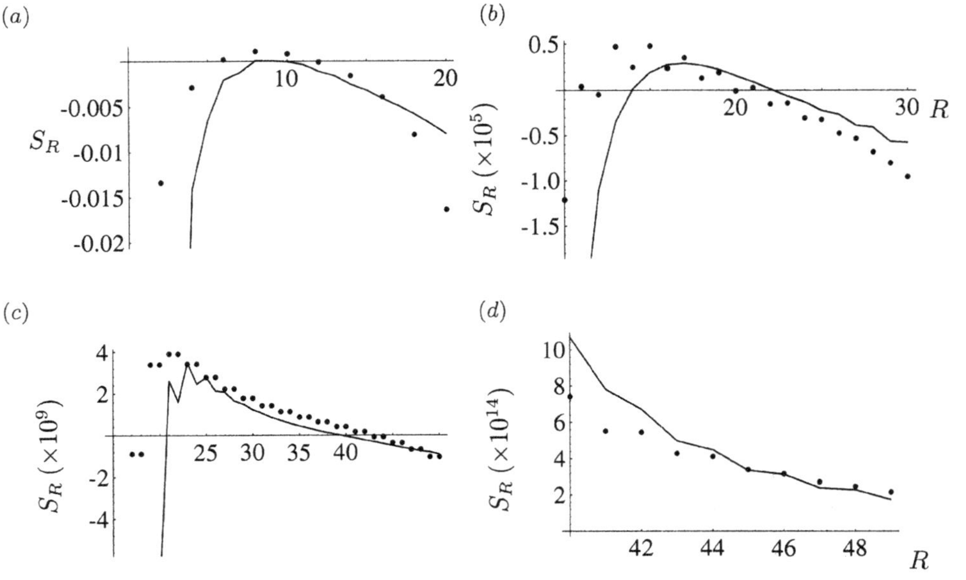

Figure 8. Riemann–Siegel remainders $S_R(t)$, computed exactly from (9) (dots), and from the theory (71) (full lines), for the following values of t (table 2): (*a*) $t = \frac{1}{2}\pi$ (least term near $R = 10$); (*b*) $t = \frac{9}{8}\pi$ (least term near $R = 22$); (*c*) $t = 2\pi$ (least term near $R = 40$); (*d*) $t = \frac{25}{4}\pi$ (least term near $R = 62$).

exponentially smaller than in the general case) large errors result from using the saddle-point approximation to calculate the derivatives.

(*b*) *Exact and approximate remainders*

To test the theory (71) it is necessary to calculate the remainders $S_R(t)$ defined by (9), by subtracting from $Z(t)$ the first $R+1$ terms of the Riemann–Siegel expansion. As explained by Edwards (1974), $Z(t)$ can be computed with arbitrary accuracy, albeit not very efficiently, by a method based on the Euler–Maclaurin sum formula. When calculating the remainders, the Riemann–Siegel coefficients must be evaluated with an accuracy $\exp(-\pi t)$. This cannot be achieved by calculating the F derivatives with the saddle-point approximation of § 5 *b*. Therefore I restricted the tests to $z = 0$, $z = \frac{1}{2}$ and $z = 1$, for which the derivatives can be calculated exactly as explained in § 5 *a*.

Even so, the comparison was necessarily restricted to low values of t; otherwise, the fact that the optimal truncation is near $R = 2\pi t$ would have required the evaluation of very high derivatives. For these, the methods of § 5 *a*, although exact, are inefficient and require very high numerical accuracy. For example, $R = 50$ requires the Euler and Bernoulli numbers up to order 150, and E_{150} is an integer of order 10^{233} and B_{150} is a ratio of integers with numerator of order 10^{149}. Table 2 shows the values of t for which the theory of the remainders was tested.

Figure 8 shows the 'experimental' and 'theoretical' remainders as functions of truncation R. Evidently the theory (71) works well for a large range of near-optimal truncations (including, it seems, figure 8*d*, where $R = r^* = 62$ is out of range of the computations).

Table 2. *Special values of t*

t	N	z	$r^* \approx 2\pi t$	$\Gamma(\frac{1}{2}r^*)/(\frac{1}{2}r^*)^{(r^*+1)/2}$
$\frac{1}{2}\pi$	0	0	10	3.7×10^{-3}
$\frac{9}{8}\pi$	0	$-\frac{1}{2}$	22	3.4×10^{-6}
2π	0	-1	39	3.4×10^{-10}
$\frac{25}{8}\pi$	1	$+\frac{1}{2}$	62	3.3×10^{-15}

It should be noted that although these t values are large in the sense that the optimum order r^* of Riemann–Siegel truncation is large, they are all smaller than the lowest Riemann zero (near $t = 14.1$), and for all except one ($t = 25\pi/8 \approx 9.8$) the main sum is empty. For the lowest Riemann zero, $r^* = 89$, which is out of reach of our computations.

The extraordinary accuracy of the Riemann–Siegel expansion is achieved at a high computational cost. To attain the optimal accuracy $\exp(-\pi t)$, the number of terms that must be included is $2\pi t$; this is $4\pi^2 N^2$ and so increases faster than the number of terms in the main sum in (4). Nevertheless, it is interesting to speculate that it might be possible to attain still higher accuracy, by extending 'hyperasymptotics' (Berry & Howls 1991) to the Riemann–Siegel expansion.

I thank Dr Jonathan Keating and Professor Michael Morgan for very helpful conversations, and useful comments based on their careful reading of the manuscript.

Appendix A. Recurrence relation for g_{rm}

This is the derivation of (33). The first step is to expand $g(x, y)$, defined by (31), in terms of y. Thus we write

$$g(x,y) = \exp\left(\sum_{r=1}^{\infty} \gamma_r(x)y^r\right) = \sum_{r=0}^{\infty} g_r(x)y^r \tag{A 1}$$

with $g_0(x) = 1$. Differentiating the second equation with respect to y and identifying terms in y gives

$$g_{r+1}(x) = \frac{1}{r+1}\sum_{m=0}^{r} (m+1)\gamma_{m+1}(x)g_{r-m}(x). \tag{A 2}$$

Now we note that, from (31),

$$\gamma_r(x) = \frac{x^r}{2r} - \frac{x^{r+2}}{4(r+2)} - [(-1)^{(r+2)/4}2^r b_{(r+2)/4}]_{\text{when } (r+2)/4 \text{ is integer}}. \tag{A 3}$$

The next step is to expand $g_r(x)$ in powers of x (cf. (32)):

$$g_r(x) = \sum_{s=0}^{\mathrm{Int}(3r/2)} g_{rs}x^{3r-2s}. \tag{A 4}$$

Substitution into (A 2) and identifying terms now gives, ignoring for the moment the upper limits of the sums,

$$(r+1)g_{r+1,m} = \frac{1}{2}\sum_{k=0}^{k_1(r,m)} g_{r-k,m-1-k} - \frac{1}{4}\sum_{k=0}^{k_2(r,m)} \frac{(k+1)}{(k+3)} g_{r-k,m-k}$$

$$+\left[\sum_k 2^{k+1}(-1)^{(k-1)/4}(k+1)b_{(k+3)/4}g_{r-k,m-3(k+1)/2}\right]_{(k+3)/4 \text{ integer}}. \qquad \text{(A 5)}$$

In the last sum we change the variable to $k = 4p+1$ and thereby obtain (33).

The upper limits (34) of the sums are determined by the requirement that in each g_{rm} the indices must satisfy $0 \leqslant m \leqslant \mathrm{Int}(3r/2)$. Thus in the first sum, $k_1(r,m)$ follows from

$$0 \leqslant m-1-k \leqslant \mathrm{Int}(\tfrac{3}{2}(r-k)) \qquad \text{(A 6)}$$

and similarly for k_2. For the third sum, the same principle gives

$$0 \leqslant m - 3(2p+1) \leqslant \mathrm{Int}(\tfrac{3}{2}(r-4p-1)) \qquad \text{(A 7)}$$

and $p(m)$ follows from the first inequality.

Appendix B. $C_{2m}(1)$ from discontinuities in the main sum (Gabcke 1979)

Continuity of $Z(t)$ at $t = 2\pi N^2$, in spite of the discontinuity of the main sum in (4), implies that the Riemann–Siegel coefficients in (5) must satisfy

$$0 = \lim_{\varepsilon\to 0}[Z(2\pi N^2+\varepsilon) - Z(2\pi N^2-\varepsilon)]$$

$$= 2N^{-1/2}\cos\{\theta(2\pi N^2) - 2\pi N^2\log N\} + \frac{(-1)^{N+1}}{N^{1/2}}\sum_{r=0}^{\infty}\frac{C_r(+1)+C_r(-1)}{N^r}. \qquad \text{(B 1)}$$

The symmetry relation (25), together with (14), gives

$$\sum_{m=0}^{\infty}\frac{C_{2m}(+1)}{N^{2m}} = (-1)^N\cos\{\theta(2\pi N^2) - 2\pi N^2\log N\}$$

$$= \cos(\tfrac{1}{8}\pi)\cos\{\chi(2\pi N^2)\} + \sin(\tfrac{1}{8}\pi)\sin\{\chi(2\pi N^2)\}. \qquad \text{(B 2)}$$

Noting the expansion (15) for χ, and defining

$$x \equiv \mathrm{i}/N^2, \quad d_m \equiv (-1)^m b_m/(2\pi)^{2m-1} \qquad \text{(B 3)}$$

we see that the desired $C_{2m}(1)$ depend on the real coefficients e_l (with $e_0 = 1$) in the expansion

$$\exp\{\mathrm{i}\chi(2\pi N^2)\} = \exp\left(-\sum_{m=1}^{\infty} d_m x^{2m-1}\right) \equiv \sum_{l=0}^{\infty} e_l x^l. \qquad \text{(B 4)}$$

From (B 2), the coefficients are

$$C_{4l}(1) = (-1)^l\cos(\tfrac{1}{8}\pi)e_{2l}, \quad C_{4l+2}(1) = (-1)^l\sin(\tfrac{1}{8}\pi)e_{2l+1}. \qquad \text{(B 5)}$$

The e_l can be found from the following recurrence relation, which follows from differentiating (B 4):

$$e_{l+1} = -\frac{1}{l+1} \sum_{s=0}^{\mathrm{Int}(l/2)} (2s+1) d_{s+1} e_{l-2s}. \tag{B 6}$$

Gabcke's argument can be extended by requiring that derivatives of $Z(t)$ are continuous at $t = 2\pi N^2$. This leads to (rather cumbersome) equations determining the nth derivatives of the Riemann–Siegel coefficients C_r at $z = 1$, where $n + r$ is even.

Appendix C. Two results on late terms in asymptotic series

These results (which are probably 'well known to those who know well', and are given here for completeness) will be used in Appendix D. The first concerns the exponential of an asymptotic series. If

$$s(x) = \sum_{n=1}^{\infty} s_n x^n \tag{C 1}$$

is a factorially divergent formal asymptotic series, we seek the high-order coefficients e_n defined by

$$\exp\left(\sum_{n=1}^{\infty} s_n x^n\right) \equiv \sum_{n=0}^{\infty} e_n x^n. \tag{C 2}$$

Clearly, $e_0 = 1$.

Differentiating (C 2) gives

$$\begin{aligned} e_{n+1} &= \frac{1}{n+1} \sum_{m=0}^{n} (1+m) s_{m+1} e_{n-m} \\ &= s_{n+1} + \frac{n}{n+1} s_n e_1 + \frac{n-1}{n+1} s_{n-1} e_2 + \dots \end{aligned} \tag{C 3}$$

For $n \gg 1$ the assumed factorial divergence of the s_n implies that the terms in this series diminish as powers of $1/n$. We include the first two terms, to allow for the circumstance (which will occur in Appendix D) that the first might vanish because of symmetry. Thus

$$e_{n+1} \approx s_{n+1} + s_n s_1 \quad \text{for } n \gg 1. \tag{C 4}$$

The second result concerns the product of two asymptotic series. If

$$s(x) = \sum_{n=0}^{\infty} s_n x^n \quad \text{and} \quad t(x) = \sum_{n=0}^{\infty} t_n x^n \tag{C 5}$$

are factorially divergent formal asymptotic series with $s_0 = 1$ and $t_0 = 1$, we seek the high-order coefficients p_n defined by

$$s(x)t(x) = \sum_{n=0}^{\infty} s_n x^n \sum_{n=0}^{\infty} t_n x^n \equiv \sum_{n=0}^{\infty} p_n x^n. \tag{C 6}$$

Clearly, $p_0 = 1$.

Collecting terms with the same power of x gives

$$p_n = \sum_{k=0}^{n} s_k t_{n-k} = t_n + s_1 t_{n-1} + \ldots + s_n + t_1 s_{n-1} + \ldots \tag{C 7}$$

For $n \gg 1$ the assumed factorial divergence of the s_n and t_n implies that the terms in the two series following t_n and s_n in (C 7) diminish as powers of $1/n$. Therefore

$$p_n \approx \max(s_n, t_n) \quad \text{for } n \gg 1. \tag{C 8}$$

In words, the high orders in the product series are simply the high orders of the dominant of the two component asymptotic series.

Appendix D. Two results for the gamma series exp(iχ)

The first result is the derivation of the formula (60) for the high coefficients $C_{2m}(1)$. This is an application of the 'exponential' formula (C 4) to the high orders of the series (B 4) generating $C_{2m}(1)$. Identification of the middle member of (B 4) with the first member of (C 2) gives

$$s_{2m} = 0, \quad s_{2m+1} = -d_{m+1}. \tag{D 1}$$

For the high d_m, we use the definitions (B 3) and (15), together with asymptotics of the Bernoulli numbers, to obtain

$$d_m = \frac{(-1)^m (2^{2m-1} - 1)|B_{2m}|}{(2\pi)^{2m-1} 2^{2m} 2m(2m-1)} \approx \frac{(-1)^m (2m-2)!}{(2\pi)^{4m-1}} \quad \text{for } m \gg 1. \tag{D 2}$$

Equation (C 4) now gives

$$\left.\begin{aligned} e_{2m} &\approx s_{2m-1} s_1 = -\frac{|B_2|}{16\pi} d_m \approx \frac{(-1)^{m+1}(2m-2)!}{96\pi (2\pi)^{4m-1}} \\ e_{2m+1} &\approx s_{2m+1} = -d_{m+1} \approx \frac{(-1)^m (2m)!}{(2\pi)^{4m+3}} \end{aligned}\right\} \quad \text{for } m \gg 1. \tag{D 3}$$

Substitution into (B 5) gives the claimed formulae (60).

The second result is the justification, promised in the second paragraph of § 3, for ignoring terms generated by the factor exp(iχ) when calculating the high orders of the Riemann–Siegel series. From (B 3) and (B 4) we find

$$\exp\{i\chi(2\pi a^2)\} = \sum_{l=0}^{\infty} \frac{e_l i^l}{a^{2l}} \tag{D 4}$$

From (D 4) we have the high-order behaviour

$$\frac{e_l}{a^{2l}} \sim \frac{\Gamma(l)}{(2\pi)^{2l} a^{2l}} \tag{D 5}$$

so the coefficient of $1/a^r$ in the expansion of exp(iχ) is proportional to

$$\frac{\Gamma(\frac{1}{2}r)}{(2\pi)^r}. \tag{D 6}$$

This is smaller by $2^{-r/2}$ than the leading dependence found for the C_r in (57) and so, by the result (C 8), the factor $\exp(\mathrm{i}\chi)$ does not contribute to the asymptotics of the Riemann–Siegel coefficients. Another way to see this is to reinstate t and note that the terms in the expansion of $\exp(\mathrm{i}\chi)$ diverge as

$$\frac{\Gamma(\frac{1}{2}r)}{(2\pi t)^{r/2}}, \tag{D 7}$$

which is smaller than (59) and by Borel summation generates a remainder of order $\exp(-2\pi t)$, rather than the $\exp(-\pi t)$ found in § 4.

References

Berry, M. V. 1986 In *Quantum chaos and statistical nuclear physics* (ed. T. H. Seligman & H. Nishioka), pp. 1–17. Springer Lecture Notes in Physics No. 263.

Berry. M. V. 1989 *Proc. R. Soc. Lond.* A **422**, 7–21.

Berry, M. V. 1991 *Proc. R. Soc. Lond.* A **434**, 465–472.

Berry, M. V. & Howls, C. J. 1991 *Proc. R. Soc. Lond.* A **434**, 657–675.

Berry, M. V. & Howls, C. J. 1993 *Proc. R. Soc. Lond.* A **443**, 107–126.

Berry, M. V. & Keating, J. P. 1992 *Proc. R. Soc. Lond.* A **437**, 151–173.

Brent, R. P. 1979 *Math. Comp.* **33**, 1361–1372.

Dingle, R. B. 1973 *Asymptotic expansions: their derivation and interpretation.* New York and London: Academic Press.

Edwards, H. M. 1974 *Riemann's zeta function.* New York and London: Academic Press.

Gabcke, W. 1979 Neue Herleitung und Explizite Restabschätzung der Riemann–Siegel-Formel. Ph.D. thesis, Göttingen.

Keating, J. P. 1993 In *Quantum chaos* (ed. G. Casati, I. Guarneri & U. Smilansky), pp. 145–185. Amsterdam: North-Holland.

Odlyzko, A. M. 1987 *Math. Comp.* **48**, 273–308.

Odlyzko, A. M. 1990 The 10^{20}th zero of the Riemann zeta function and 175 million of its neighbours. AT&T Bell Laboratory preprint.

Odlyzko, A. M. & Schönhage, A. 1988 *Trans. Am. math. Soc.* **309**, 797–809.

Paris, R. B. 1994 *Proc. R. Soc. Lond.* A **446**, 565–587.

Titchmarsh, E. C. 1986 *The theory of the Riemann zeta-function*, 2nd edn. Oxford: Clarendon Press.

van de Lune, J., te Riele, H. J. J. & Winter, D. T. 1986 *Math. Comp.* **46**, 667–681.

Wolfram, S. 1991 *Mathematica.* Addison-Wesley.

Received 16 November 1994; accepted 4 January 1995

A new asymptotic representation for $\zeta(\frac{1}{2}+\mathrm{i}t)$ and quantum spectral determinants

BY M. V. BERRY AND J. P. KEATING†
H. H. Wills Physics Laboratory, Tyndall Avenue, Bristol BS8 1TL, U.K.

By analytic continuation of the Dirichlet series for the Riemann zeta function $\zeta(s)$ to the critical line $s=\frac{1}{2}+\mathrm{i}t$ (t real), a family of exact representations, parametrized by a real variable K, is found for the real function $Z(t)=\zeta(\frac{1}{2}+\mathrm{i}t)\exp\{\mathrm{i}\theta(t)\}$, where θ is real. The dominant contribution $Z_0(t,K)$ is a convergent sum over the integers n of the Dirichlet series, resembling the finite 'main sum' of the Riemann–Siegel formula (RS) but with the sharp cut-off smoothed by an error function. The corrections $Z_3(t,K)$, $Z_4(t,K)$ ... are also convergent sums, whose principal terms involve integers close to the RS cut-off. For large K, Z_0 contains not only the main sum of RS but also its first correction. An estimate of high orders $m \gg 1$ when $K < t^{\frac{1}{6}}$ shows that the corrections Z_k have the 'factorial/power' form familiar in divergent asymptotic expansions, the least term being of order $\exp\{-\frac{1}{2}K^2t\}$.

Graphical and numerical exploration of the new representation shows that Z_0 is always better than the main sum of RS, providing an approximation that in our numerical illustrations is up to seven orders of magnitude more accurate with little more computational effort. The corrections Z_3 and Z_4 give further improvements, roughly comparable to adding RS corrections (but starting from the more accurate Z_0). The accuracy increases with K, as do the numbers of terms in the sums for each of the Z_m.

By regarding Planck's constant $\hbar$ as a complex variable, the method for $Z(t)$ can be applied directly to semiclassical approximations for spectral determinants $\Delta(E,\hbar)$ whose zeros $E=E_j(\hbar)$ are the energies of stationary states in quantum mechanics. The result is an exact analytic continuation of the exponential of the semiclassical sum over periodic orbits given by the divergent Gutzwiller trace formula. A consequence is that our result yields an exact asymptotic representation of the Selberg zeta function on its critical line.

1. Introduction

Riemann's celebrated function $\zeta(s)$ arises not only in connection with prime numbers (Edwards 1974) but also as a model for spectral determinants in quantum chaology (Berry 1986, 1991; Berry & Keating 1990; Keating 1992*a*, *b*). It is often necessary to be able to represent $\zeta(\frac{1}{2}+\mathrm{i}t)$ in a simple way high in the critical strip ($|\mathrm{Im}\,t|<\frac{1}{2}$, $|\mathrm{Re}\,t| \gg 1$) and calculate it there with great accuracy. Particularly important is the critical line t real, where, according to the Riemann hypothesis, the non-trivial zeros lie. Our purpose here is to describe a way of doing this which has some advantages over the method currently used, and which can be applied directly to the analogous functions in quantum mechanics.

† Present address: Department of Mathematics, The University, Manchester M13 9PL, U.K.

Proc. R. Soc. Lond. A (1992) **437**, 151–173
Printed in Great Britain

Reprinted from Proceedings of the Royal Society A, Volume 437, (1992).

The method is an extension of the formal approximating scheme introduced in Keating (1992*a*) for semiclassical formulae, and applied to $\zeta(s)$ in Keating (1992*b*). Here we go further; first, by eliminating the formal aspect of this approach; second, by pursuing it to its natural conclusion to obtain a complete asymptotic expansion; and third, by fine-tuning a parameter to improve convergence.

It follows from the functional equation for $\zeta(s)$ that the function defined by

$$Z(t) \equiv \exp\{\mathrm{i}\theta(t)\}\,\zeta(\tfrac{1}{2}+\mathrm{i}t) \tag{1}$$

with

$$\exp\{\mathrm{i}\theta(t)\} = \left[\frac{\Gamma(\frac{1}{4}+\frac{1}{2}\mathrm{i}t)}{\Gamma(\frac{1}{4}-\frac{1}{2}\mathrm{i}t)}\right]^{\frac{1}{2}} \exp\{-\tfrac{1}{2}\mathrm{i}t\ln\pi\} \tag{2}$$

is an even function of t (θ is odd). Moreover, it is real when t is real, as is $\theta(t)$. The simplest representation for $\zeta(s)$, namely the Dirichlet series

$$\zeta(s) = \sum_{n=1}^{\infty} \frac{1}{n^s} \tag{3}$$

converges only when $\operatorname{Re} s > 1$, and so fails when used in (1) for real t, generating a $Z(t)$ that is neither obviously real nor obviously even.

Computation of $Z(t)$ for real t using (3) requires analytic continuation. A powerful method, now universally used when t is large (Haselgrove 1963; Brent 1979; Odlyzko 1987), is the Riemann–Siegel formula (Edwards 1974; Titchmarsh 1986), hereinafter called RS:

$$Z(t) = 2\sum_{n=1}^{N(t)} \frac{\cos\{\theta(t)-t\ln n\}}{\sqrt{n}} - (-1)^{N(t)}\left(\frac{2\pi}{t}\right)^{\frac{1}{4}} \sum_{j=0}^{k}\left(\frac{2\pi}{t}\right)^{j/2} \Phi^{(j)}\{p(t)\} + R^{(k)}(t). \tag{4}$$

Here $N(t)$ and $p(t)$ are defined by

$$N(t) \equiv \operatorname{Int}\sqrt{\frac{t}{2\pi}}, \quad p(t) \equiv \sqrt{\frac{t}{2\pi}} - \operatorname{Int}\sqrt{\frac{t}{2\pi}} \tag{5}$$

and the functions $\Phi^{(j)}(p)$ are combinations of derivatives of

$$\Phi^{(0)}(p) = \cos\{2\pi(p^2-p-\tfrac{1}{16})\}/\cos\{2\pi p\}. \tag{6}$$

To understand what follows, it is helpful to consider the following interpretation of RS. The sum over n (the 'main sum'), which usually dominates $Z(t)$, follows from substituting (3) into (1), truncating the resulting divergent series at the term $N(t)$ whose phase is stationary, and adding the complex conjugate of this truncated series as an approximate resummation of the divergence (Titchmarsh 1986, ch. IV; Berry 1986; Keating 1992*a*, *b*). As an approximation to $Z(t)$, the main sum suffers from the defect of being a discontinuous function of t because of the discontinuous upper limit $N(t)$. It is the role of the correction terms, in the sum over j, to remedy this by removing, one by one, the discontinuities in successive derivatives at the truncation point.

Probably, the sum of all these corrections is an asymptotic expansion of $Z(t)$, but we know of no proof. And there appear to have been no studies of the high orders of the expansion, such as would be necessary to estimate the accuracy with which $Z(t)$ could be computed by choosing k in (4) to be the least term. Bounds do exist, however, for some of the remainders $R^{(k)}(t)$, a particularly useful one being (Gabcke 1979)

$$|R^{(4)}(t)| < 0.017/t^{\frac{11}{4}} \quad (t > 200). \tag{7}$$

The representation we derive here (§§2 and 3) superficially resembles RS in being dominated by a sum over n, similar to the main sum in (4), and possessing a series of correction terms. It has, however, several advantages over RS. First, all its terms are analytic functions of t: there are no discontinuities. Second, it is formally exact, unlike RS whose remainders $R^{(k)}(t)$ contain an unspecified exponentially small integral (Edwards 1974). Third, the size of the late terms can be estimated explicitly (§§4 and 5), showing that they have the 'factorial divided by power' form familiar in asymptotic expansions (Dingle 1973). Fourth, numerical studies (§6) suggest that term by term the new series is more accurate than RS. And fifth, the derivation generalizes (§7) directly to the series encountered in the determination of eigenvalues of wave operators associated with chaotic dynamical systems, an example of which is the Selberg zeta function (Balazs & Voros 1986). We emphasize this, because such series also suffer from fundamental convergence problems, and no analogue of RS is known for them.

Although the new formula looks like RS with its discontinuities smoothed away term by term, this appearance is misleading. Our formula involves an additional parameter. Moreover it contains RS in a complicated way; indeed we show (§4) that there is a limiting régime in which our dominant series contains not only the main sum of RS but at least its first correction term as well.

Before embarking on the analysis, we make two remarks. First, according to (2) the function $Z(t)$ has square-root branch points at the zeros of the gamma functions in $\exp\{i\theta(t)\}$. These points complicate the analysis slightly, and it might be thought preferable to work with a different combination of ζ and Γ, namely $\Xi(t)$ (eq. (2.1.16) in Titchmarsh 1986), which as well as being real for real t, and even, is also an entire function. In fact much of the argument to follow can be applied equally to $\Xi(t)$, but the final formulae are numerically far less effective than those for $Z(t)$. The reason is probably that the asymptotics must also be flexible enough to accommodate the rapidly decreasing amplitude factor by which the modulus $|\Xi(t)|$ differs from $|\zeta(t)|$ when t is real, a factor which is uninteresting in studies of the zeros, and which we never encounter because $|Z(t)| = |\zeta(t)|$ when t is real.

The second remark is that for simplicity of exposition we shall restrict ourselves to writing formulae valid for real t, that is on the critical line. However, continuation to complex t in the analyticity strip of $Z(t)$ is not difficult.

2. Analytic continuation

By Cauchy's theorem we have

$$Z(t) = \frac{1}{2\pi i}\int_{C_+ + C_-} \frac{dz}{z}\gamma(z,t)\, Z(z+t), \tag{8}$$

where $C_\pm$ are the contours shown in figure 1, and γ is any function, analytic inside the integration strip, for which the integral converges and $\gamma(0,t) = 1$. Choosing $\gamma(z,t)$ even in z, and using the fact that $Z(t)$ is even, we obtain

$$Z(t) = \frac{1}{2\pi i}\int_{C_-} \frac{dz}{z}\gamma(z,t)\,[Z(z+t)+Z(z-t)]. \tag{9}$$

Except near the branch point $z = -t-\frac{1}{2}i$, the Dirichlet series (3) converges

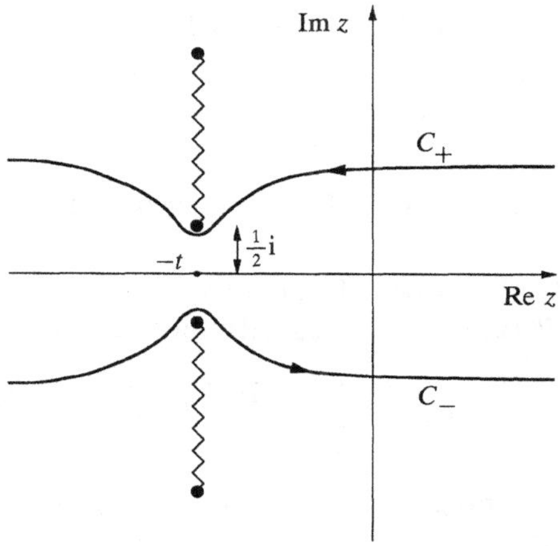

Figure 1. Integration contours C_+ and C_- in the z plane, with cuts (zigzags) connecting the square-root branch points of $Z(t)$.

everywhere on C_-, because $\operatorname{Im} z < -\frac{1}{2}$ (corresponding to $\operatorname{Re} s > 1$). However, the branch point gives a vanishing contribution. Therefore we can substitute (3), and obtain

$$Z(t) = \sum_{n=1}^{\infty} [T_n(t) + T_n(-t)], \tag{10}$$

where

$$T_n(t) = \frac{\exp\{\mathrm{i}(\theta(t) - t\ln n)\}}{\sqrt{n}} \frac{1}{2\pi\mathrm{i}} \int_{C_-} \frac{\mathrm{d}z}{z} \gamma(z,t) \exp\{\mathrm{i}[\theta(z+t) - \theta(t) - z\ln n]\}. \tag{11}$$

This analytic continuation gives $Z(t)$ on the critical line as a manifestly even function. The fact that $\theta(t)$ is odd leads (by an argument involving the deformation of C_- to the real axis plus an infinitesimal semicircle) to the relation

$$T_n(-t) = T_n^*(t) \tag{12}$$

and thence, via (10), to the (necessarily) real even function

$$Z(t) = 2\operatorname{Re} \sum_{n=1}^{\infty} T_n(t). \tag{13}$$

Of course these formulae make sense only if the sums and integrals converge. We achieve this by making the choice

$$\gamma(z,t) = \exp(-z^2K^2/2|t|), \tag{14}$$

where K is a constant whose significance will become clear later. Convergence of the integrals (11) is obvious. In Appendix A we show that the sum over T_n converges too: after an initially rapid decrease as $\exp[-\ln^2(n/N)]$, the ultimate decrease of the terms is very slow, namely as $n^{-1}\ln^{-\frac{3}{2}} n$. This slow convergence of the n-sum will now be hastened by expanding each T_n as a series whose terms can be evaluated explicitly. The form (14) is chosen *ad hoc*. We have not explored the possibility of choosing γ to optimize convergence.

3. Series expansion

Henceforth we consider $t > 0$; this leads to no loss of generality because $z(t)$ is even. In (11) we expand $\exp\{i\theta(z+t)\}$ as

$$\exp\{i[\theta(z+t)-\theta(t)]\} = \exp\{i[z\theta'(t)+\tfrac{1}{2}z^2\theta''(t)]\}\left[1+\sum_{m=3}^{\infty} z^m b_m(t)\right], \tag{15}$$

in which the functions $b_m(t)$ can be calculated recursively from $\theta(t)$ in terms of the polygamma functions $\psi^{(n)}(x)$ ($(n-1)$st logarithmic derivative of $\Gamma(x)$),

$$\sum_{m=3}^{\infty} z^m b_m(t) = \exp\left\{i\sum_{s=3}^{\infty}(\tfrac{1}{2}z)^s \frac{\operatorname{Im} i^s \psi^{(s-1)}(\frac{1}{4}+\frac{1}{2}it)}{s!}\right\}-1. \tag{16}$$

In the exponent of the integrand in (11) we now group the terms linear and quadratic in z by defining

$$\xi(n,t) \equiv \ln n - \theta'(t), \quad Q^2(K,t) \equiv K^2 - it\theta''(t). \tag{17}$$

For each term labelled by m in (15), we can now evaluate the integral in (11) and substitute into (13). In this way we obtain our main result:

$$Z(t) = Z_0(t,K)+Z_3(t,K)+Z_4(t,K)+\dots. \tag{18}$$

The integral for Z_0 has a quadratic exponential and a first-order pole, and can be evaluated in terms of the complementary error function (eq. 7.1.4 of Abramowitz & Stegun 1964):

$$Z_0(t,K) = 2\operatorname{Re}\sum_{n=1}^{\infty}\frac{\exp\{i[\theta(t)-t\ln n]\}}{\surd n}\times\tfrac{1}{2}\operatorname{Erfc}\left\{\frac{\xi(n,t)}{Q(K,t)}\surd(\tfrac{1}{2}t)\right\}. \tag{19}$$

The integrals Z_m ($m \geqslant 3$) have quadratic exponentials multiplied by positive powers, and can be evaluated in terms of Hermite polynomials (eq. 8.951 of Gradshteyn & Ryzhik 1980):

$$Z_m(t,K) = \frac{2}{\surd\pi}(\tfrac{1}{2}t)^{m/2}\operatorname{Re}\frac{(-i)^m b_m(t)}{Q^m(K,t)}\sum_{n=1}^{\infty}\frac{\exp\{i[\theta(t)-t\ln n]\}}{\surd n}$$
$$\times\exp\left\{\frac{-\xi^2(n,t)\,t}{2Q^2(K,t)}\right\}H_{m-1}\left\{\frac{\xi(n,t)}{Q(K,t)}\surd(\tfrac{1}{2}t)\right\} \quad (m \geqslant 3). \tag{20}$$

(An earlier version of this theory failed to incorporate the quadratic term in the exponent of (15), so that the counterpart of the multiplying series began with b_2 rather than b_3. This led to a slightly different representation for $Z(t)$, involving K rather than Q (whose first term – the counterpart of Z_0 – was obtained by Keating (1992*b*)). The low-order approximations to this representation were numerically much less accurate and their subsequent analysis proved more complicated than that of (18)–(20).)

The representation (18)–(20) gives Z, which is independent of K, as a series of contributions Z_m; we call this the m-series. Z_0 is the main term, and $Z_3, Z_4 \dots$ are corrections. Each Z_m is a K-dependent sum of terms labelled n; we call these the n-sums.

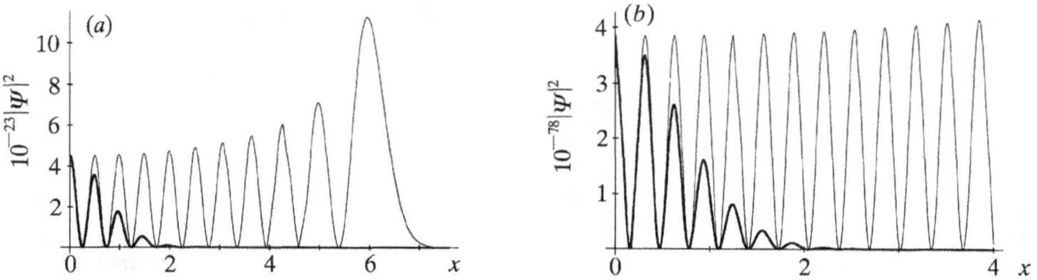

Figure 2. Comparison between squares of the functions $\psi_1 = \exp(-x^2)H_m(x)$ appearing in (20) (thick lines) and the squares of the Hermite functions $\psi_2 = \exp(-\frac{1}{2}x^2)H_m(x)$ (thin lines), for (a) $m = 20$, (b) $m = 50$.

The convergence of the n-sums is crucial to the usefulness of our representation, and depends on the smallness of

$$\exp\{-\xi^2(n,t)\,t/2Q^2(K,t)\} \tag{21}$$

for the following reasons. In (19) the function Erfc is approximately proportional to this factor when ξ is large. And the Gauss–Hermite product in the second line of (20) is also dominated by the gaussian factor. This is not obvious. It follows from the fact that the 'width' of the product can be defined as

$$W_m \equiv \sqrt{\left[\int_0^\infty \mathrm{d}x\, x^2[\exp(-x^2)H_m(x)]^2 \Big/ \int_0^\infty \mathrm{d}x\,[\exp(-x^2)H_m(x)]^2\right]} = \frac{1}{2}\sqrt{\left(\frac{4m-1}{2m-1}\right)} \tag{22}$$

and is, asymptotically, independent of m and only $\surd 2$ greater than W_0 (equation (22) can be derived from eq. 7.375.1 of Gradshteyn & Ryzhik (1980)). This behaviour is quite different from that of the widths of the harmonic oscillator functions, where (cf. eq. 7.375.2 of Gradshteyn & Ryzhik (1980))

$$W_m^{\mathrm{osc}} \equiv \sqrt{\left[\int_0^\infty \mathrm{d}x\, x^2[\exp(-\tfrac{1}{2}x^2)H_m(x)]^2 \Big/ \int_0^\infty \mathrm{d}x\,[\exp(-\tfrac{1}{2}x^2)H_m(x)]^2\right]} = \surd(m+\tfrac{1}{2}). \tag{23}$$

Here the Gaussian factor is weaker, and the widths increase with m. Figure 2 illustrates this striking difference between the functions appearing in (20) and the harmonic oscillator functions. The very useful consequence is that the convergence of the n-sums is the same for all the Z_m.

We now note that for the large t of interest here we may approximate $\theta(t)$ (and the derived quantities ξ, Q and the b_m) with Stirling's asymptotic expansion for the gamma function:

$$\theta(t) = \frac{t}{2}\left(\ln\left\{\frac{t}{2\pi}\right\} - 1\right) - \frac{\pi}{8} + \frac{1}{48t} - \frac{7}{5760t^3} + \ldots. \tag{24}$$

In subsequent analysis we shall make extensive use of this formula.

In particular, we find, to lowest order,

$$\xi(n,t) \approx \ln\{n/\surd(t/2\pi)\} \approx \ln\{n/N(t)\}, \tag{25}$$

where $N(t)$ is the RS cut-off (5). Substituting this into (21) now shows that the n convergence depends on

$$\left|\exp\left\{\frac{-[\ln(n/N)]^2 t}{2Q^2}\right\}\right| = \exp\left\{\frac{-[\ln(n/N)]^2 tK^2}{2(K^4+\frac{1}{4})}\right\}, \tag{26}$$

and so is faster than any power of n but slower than exponential. Note that this rapid convergence now holds for all n; there is no drastic slowing-down for very large n, as with (13). The reason is that there the slowing-down was caused by a branch point of the integrand in (11) (see Appendix A), but there is no analogous contribution associated with the terms in the expansion (15). As we shall see in §5, the price to be paid for this is that the m-series is a divergent asymptotic expansion, but the least term is so small that the divergence has no effect on practical computations.

A consequence of (26) is that the upper limit guaranteeing that neglected terms are smaller than $\exp(-A)$ is

$$\begin{aligned} n = n^* &= \sqrt{\left(\frac{t}{2\pi}\right)}\exp\left\{K\sqrt{\frac{2A(1+1/4K^4)}{t}}\right\} \\ &\approx N\exp\{(K/N)\sqrt{[(A/\pi)(1+1/4K^4)]}\}. \end{aligned} \tag{27}$$

In what follows we shall make frequent use of this estimate. Note that n^* has a minimum value of $N\exp\{\sqrt{(A/\pi)}/N\}$, when $K = 1/\sqrt{2}$.

For our representation to be a serious rival to RS when t is large, we require that the number of n-terms in each Z_m is not much larger than the number N in the main sum of RS. Therefore we impose upon K the restriction

$$K \ll N(t) \sim t^{\frac{1}{2}}. \tag{28}$$

Note that this is compatible with $K \gg 1$, a fact we exploit later. Then

$$n^* \approx N + K\sqrt{[(A/\pi)(1+1/4K^4)]} \tag{29}$$

revealing the meaning of K as proportional to the number of terms by which RS truncation has been smoothed, when t is large. Roughly, the leading sum Z_0 involves terms $n \leqslant N+K$, and the corrections $Z_{m\geqslant 3}$ involve terms $N-K \leqslant n \leqslant N+K$.

4. Quadratic phase approximation

In computations (§6), we shall use the series (18)–(20) with no approximations (except the replacement of $\theta(t)$ and its derivatives using the early terms of (24)), and it is these formulae which we shall generalize in §7. For $\zeta(s)$ we can, however, go much further, by studying the behaviour of the terms near the RS cut-off. By (25), this corresponds to $\xi = 0$. Therefore we write

$$n \equiv N(t) + k \tag{30}$$

and make use of

$$t = 2\pi[N(t)+p(t)]^2 \tag{31}$$

(cf. (5)) to expand $\theta(t) - t\ln n$ in the phases of the summands to second order in k. Thus for large t we have

$$[\theta(t) - t\ln n]_{\mathrm{mod}\,2\pi} = \pi N + \pi k + 2\pi p^2 - \tfrac{1}{8}\pi - 4\pi pk + O\{(p-k)^3/N\}. \tag{32}$$

Similarly,
$$\xi(n,t) = (k-p)/N + O\{((k-p)/N)^2\}. \tag{33}$$

We use these approximations differently for Z_0 and $Z_{m\geqslant 3}$. In Z_0 the aim is to understand the connection with RS. First we write, in (19),

$$\tfrac{1}{2}\mathrm{Erfc}\,(x) = \Theta(-x) + \tfrac{1}{2}\mathrm{Erfc}\,(|x|)\,\mathrm{sgn}\,(x), \tag{34}$$

where Θ denotes the unit step. To obtain the leading-order behaviour of the second term it is necessary in (19) to replace the phase by (32), and make the approximation $\xi = 0$. Then the dependence on K disappears, and we obtain (Keating 1992*b*)

$$\begin{aligned} Z_0(t) &\approx 2\sum_{n=1}^{N} \frac{\cos\{\theta - t\ln n\}}{\sqrt{n}} + (-1)^N \left(\frac{2\pi}{t}\right)^{\frac{1}{4}} \\ &\quad \times \mathrm{Re}\exp\{2\mathrm{i}\pi(p^2 - \tfrac{1}{16})\}\left(1 + \sum_{k=-\infty}^{\infty} \mathrm{sgn}\,(k)(-1)^k \exp\{-4\pi\mathrm{i}kp\}\right) \\ &= 2\sum_{n=1}^{N} \frac{\cos\{\theta - t\ln n\}}{\sqrt{n}} - (-1)^N \left(\frac{2\pi}{t}\right)^{\frac{1}{4}} \Phi^{(0)}(p), \end{aligned} \tag{35}$$

which are precisely the main sum and first correction in RS (equation (4)). It is possible that more – perhaps all – terms of RS are contained in Z_0 and can be extracted by extending the expansion about $n = N$, but we have not pursued this.

In $Z_{m\geqslant 3}$ the ultimate aim will be to construct (§5) a theory of the high orders $m \gg 1$ of the m-series. First we replace the phase in (20) by (32) (this is valid under condition (38) below), but now it is necessary to replace ξ by (33) (rather than $\xi = 0$, which would give $Z_m = 0$). Thus after some reduction we find

$$\begin{aligned} Z_m(t,K) &\approx \frac{2(-1)^N}{\sqrt{\pi}} \left(\frac{t}{2}\right)^{m/2} \left(\frac{2\pi}{t}\right)^{\frac{1}{4}} \mathrm{Re}\,\frac{(-\mathrm{i})^m b_m \exp\{2\pi\mathrm{i}(p^2 - \frac{1}{16})\}}{Q^m} \\ &\quad \times \sum_{k=-\infty}^{\infty} (-1)^k \exp\{-4\pi\mathrm{i}pk\} \exp\left\{\frac{-\pi(k-p)^2}{Q^2}\right\} H_{m-1}\left\{\frac{(k-p)}{Q}\sqrt{\pi}\right\} \quad (m \geqslant 3). \end{aligned} \tag{36}$$

The sum over k can be transformed by the Poisson summation formula, giving a series of integrals which can be evaluated exactly (using eq. 7.374.6 of Gradshteyn & Ryzhik (1980), and $Q^2 \approx K^2 - \frac{1}{2}\mathrm{i}$, which follows from (17) and the leading term of (24)). All phases cancel, and we obtain

$$\begin{aligned} Z_m(t,K) &\approx 2(-1)^N [\mathrm{Im}\, b_m] \left(\frac{t}{2\pi}\right)^{m/2 - \frac{1}{4}} (2\pi)^{m-1} \\ &\quad \times \sum_{l=-\infty}^{\infty} (-1)^{l(l+1)/2} (l + \tfrac{1}{2} - 2p)^{m-1} \exp\{-\pi K^2 (2p - \tfrac{1}{2} - l)^2\} \quad (m \geqslant 3). \end{aligned} \tag{37}$$

This approximation will be valid if the neglected cubic terms in the expansion (32) of $\theta - t\ln n$ are small compared with π, that is if (cf. (29))

$$K\sqrt{[(A/\pi)(1 + 1/4K^4)]} < (\tfrac{3}{2}N)^{\frac{1}{3}} \sim t^{\frac{1}{6}}. \tag{38}$$

This condition supersedes (28) and therefore guarantees that the smoothing range of the n-sums is small compared with the size of the main sum of RS. Like (28), (38) is also compatible with $K \gg 1$ when t is large. A numerical test of this approximation will be presented in §6.

5. Late terms of the *m*-series

We can make use of (37) to estimate high orders ($m \gg 1$) of the expansion (18) if K is chosen to satisfy

$$1 \ll K \ll N^{\frac{1}{3}}. \tag{39}$$

This choice is a sensible compromise between K small (so that the n-sums are not too unwieldy) and K large, when as we shall see the corrections are small.

When K and m are large, the sum over l in (37) is dominated by its biggest term, namely

$$l = \text{nearest integer to } 2p - \tfrac{1}{2} \pm \surd[(m-1)/2\pi K^2]. \tag{40}$$

As m increases, the summand will equal

$$\pm x^{m-1} \exp\{-\pi K^2 x^2\} \tag{41}$$

with $x(=l-2p+\frac{1}{2})$ varying irregularly over unit ranges including $x^* \equiv \pm\surd[(m-1)/2\pi K^2]$. Therefore we can estimate the sum by the average of its biggest term, namely

$$\sum_{l=-\infty}^{\infty} (-1)^{l(l+1)/2} (l+\tfrac{1}{2}-2p)^{m-1} \exp\{-\pi K^2 (2p-\tfrac{1}{2}-l)^2\}$$

$$\approx \pm \int_{x^*-\frac{1}{2}}^{x^*+\frac{1}{2}} \mathrm{d}x\, x^{m-1} \exp\{-\pi K^2 x^2\} \approx \pm \frac{(\frac{1}{2}m)!}{mK^m \pi^{m/2}}. \tag{42}$$

We also require the form of the high expansion coefficients $b_m(t)$ defined by (16). In Appendix B we show that

$$b_m(t) \approx \left(\frac{2}{\pi^3 e m^2 t}\right)^{\frac{1}{4}} \frac{(-1)^m}{(t+\frac{1}{2}\mathrm{i})^m} \exp\{\tfrac{1}{8}\mathrm{i}(2t-\pi)\} \quad (m \gg t). \tag{43}$$

Substituting (42) and (43) into (37), we obtain the estimate that on average

$$Z_m(t,K) \underset{m\to\infty}{\longrightarrow} \pm(-1)^N \left(\frac{2}{tm^3\pi^3\surd e}\right)^{\frac{1}{4}} \sin\{\tfrac{1}{4}t - \tfrac{1}{8}\pi\} \frac{(\frac{1}{2}m)!}{F^m}, \tag{44}$$

where

$$F = -K\surd(\tfrac{1}{2}t). \tag{45}$$

For large t, these terms decrease and then increase in typical 'factorial/power' fashion. It therefore appears that the m-series is a divergent asymptotic expansion, and we can expect (Dingle 1973) that the error is of the same order as the first omitted term. This is smallest when

$$\tfrac{1}{2}m = |F|^2, \quad \text{i.e. } m = m^* \approx K^2 t. \tag{46}$$

Thus the accuracy with which $Z(t)$ could be approximated by our representation (18)–(20), considered as a bare asymptotic series (that is without resummation), is

$$|Z_{m^*}(t,K)| \sim (tm^{*3})^{-\frac{1}{4}} \exp(-\tfrac{1}{2}m^*) \sim (1/t^2K^3) \exp\{-\tfrac{1}{2}K^2 t\}. \tag{47}$$

We have encountered this 'small exponential' before. As shown in Appendix A, it is approximately the size of the term T_n in the series (13) for which the rapid decrease $\exp(-\ln^2 n)$ yields to the slow decrease $n^{-1}\ln^{-\frac{3}{2}} n$. This slow decrease, and the form of the divergence responsible for the least term (47), both originate in the branch

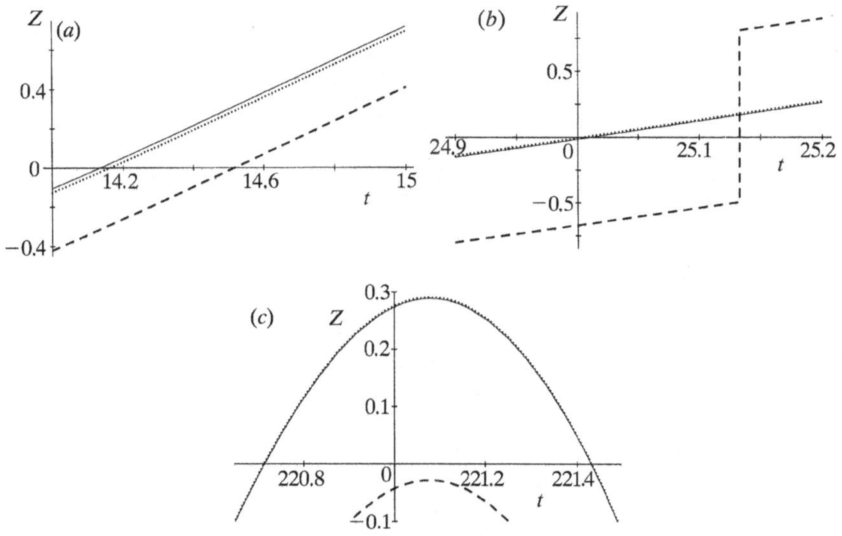

Figure 3. Comparison of lowest term Z_0 (equation (19)) for $K = 1/\sqrt{2}$ (dotted lines), the main sum of RS (dashed lines), and the exact Z (full lines), for different ranges of t. The number of terms used to compute Z_0 were (*a*) $n^* = 4$, (*b*) $n^* = 4$, (*c*) $n^* = 8$.

point of the integrand in (11). Now we can see that the influence of this branch point has been transformed: from the slow convergence of (13), via the expansion (16) whose n-sums (19) and (20) converge much faster, into a divergence of the m-series.

The least term (47) is smallest if K is as large as it can be, consistent with the restriction (38) implied by the quadratic approximation, so

$$|Z_{m^*}(t,K)| \gtrsim t^{-\frac{5}{2}} \exp(-t^{\frac{4}{3}}). \tag{48}$$

We do not expect to be able to achieve this enormous accuracy in practice, because the number m^* of terms in the m-series that would be required is, from (46),

$$m^* \sim N^{\frac{2}{3}} t \sim t^{\frac{4}{3}} \sim N^{\frac{8}{3}}, \tag{49}$$

i.e. much larger than the number N of terms in the n-sum in Z_0 (or the main sum of RS).

We would like to compare (48) with the best accuracy that could be achieved with RS by summing the j-series in (4) to its least term, but are frustrated by lack of knowledge of the late terms $\Phi^{(j)}(p)$.

6. Numerical illustrations

First we present (figure 3) pictorial comparisons between $Z(t)$ and the two lowest-order approximations: $Z_0(t)$ from our representation (19), and the main sum of RS. To make the n-sums as short as possible, we chose $K = 1/\sqrt{2}$ (cf. (27)), and for the convergence exponent we chose $A = 10$ (i.e. accuracy $\exp(-10) = 5 \times 10^{-5}$, which was adequate for pictures). The differences between Z_0 and the RS main sum were obvious under low magnification, but considerable magnification of small t-ranges was necessary to visually separate Z_0 from the exact Z.

Table 1. *Computations of $Z(t)$ for t_1*

t	$Z(t)$	$N(t)$	$p(t)$
18	2.3367997	1	0.693

	Z_{approx}	$Z-Z_{\text{approx}}$
RS main sum	1.9934571	3.4×10^{-1}
$+\Phi^{(0)}$	2.3396565	-2.8×10^{-3}
$+\Phi^{(1)}$	2.3354758	1.3×10^{-3}
$+\Phi^{(2)}$	2.3368160	-1.6×10^{-5}
$+\Phi^{(3)}$	2.3367659	3.4×10^{-5}
$+\Phi^{(4)}$	2.3367962	-3.5×10^{-6}
$K=0.5,\ n^*=14$	Z_{approx}	$Z-Z_{\text{approx}}$
Z_0	2.3031845	3.4×10^{-2}
Z_0+Z_3	2.2885406	4.8×10^{-2}
$Z_0+Z_3+Z_4$	2.3114179	2.5×10^{-2}
$K=1.0,\ n^*=14$	Z_{approx}	$Z-Z_{\text{approx}}$
Z_0	2.3297183	7.1×10^{-3}
Z_0+Z_3	2.3338470	3.0×10^{-3}
$Z_0+Z_3+Z_4$	2.3361610	6.4×10^{-4}
$K=1.5,\ n^*=32$	Z_{approx}	$Z-Z_{\text{approx}}$
Z_0	2.3355051	1.3×10^{-3}
Z_0+Z_3	2.3366315	1.7×10^{-4}
$Z_0+Z_3+Z_4$	2.3367028	9.7×10^{-5}

We chose three ranges which included Riemann zeros. Even as low as the first zero (figure 3*a*), Z_0 gives an excellent approximation. The superiority of Z_0 is particularly striking near the second zero (figure 3*b*), where the main sum of RS has a discontinuity. The RS main sum misses two zeros near the 90th (figure 3*c*), but these are captured by Z_0, which is barely distinguishable from $Z(t)$.

For a more detailed exploration of the representation (18)–(20), we chose three values of t:

$$t_1 = 18, \quad t_2 = 7005.08186, \quad t_3 = 2\pi(200.15)^2 = 251\,704.54477728. \tag{50}$$

For each, we computed Z_0 and the first two corrections Z_3 and Z_4, for several values of K. We used the asymptotic approximation (24) in $\theta(t)$ and in the quantities $Q(K,t)$, $\xi(n,t)$, $b_3(t)$ and $b_4(t)$ dependent upon it (cf. (B 4)). For convenience we show the explicit formulae for Z_3 and Z_4:

$$Z_3(t,K) = -\frac{1}{12\pi}\left(\frac{2\pi}{t}\right)^{\frac{1}{2}} \operatorname{Re}\frac{1}{Q^3(K,t)}\sum_{n=1}^{n^*}\frac{\exp\{i[\theta(t)-t\ln n]\}}{\sqrt{n}}\left[1-\frac{\xi^2(n,t)\,t}{Q^2(K,t)}\right]\exp\left\{\frac{-\xi(n,t)^2\,t}{2Q^2(K,t)}\right\},$$

$$Z_4(t,K) = \frac{1}{8\pi}\left(\frac{2\pi}{t}\right)^{\frac{1}{2}} \operatorname{Re}\frac{1}{Q^5(K,t)}\sum_{n=1}^{n^*}\frac{\exp\{i[\theta(t)-t\ln n]\}}{\sqrt{n}}$$
$$\times\,\xi(n,t)\left[1-\frac{\xi^2(n,t)\,t}{3Q^2(K,t)}\right]\exp\left\{\frac{-\xi(n,t)^2\,t}{2Q^2(K,t)}\right\}. \tag{51}$$

Table 2. *Computations of $Z(t)$ for t_2*

t	$Z(t)$	$N(t)$	$p(t)$
7005.08186	0.00396735727731	33	0.390

	Z_{approx}	$Z-Z_{\text{approx}}$
RS main sum	−0.06600396768502	7.0×10^{-2}
$+\Phi^{(0)}$	0.00393683499809	3.1×10^{-5}
$+\Phi^{(1)}$	0.00396655239702	8.0×10^{-7}
$+\Phi^{(2)}$	0.00396735603114	1.2×10^{-9}
$+\Phi^{(3)}$	0.00396735721927	5.8×10^{-11}
$+\Phi^{(4)}$	0.00396735727731	$<4.4\times10^{-13}$

$K=1$, $n^*=37$	Z_{approx}	$Z-Z_{\text{approx}}$
Z_0	0.00399124165286	-2.4×10^{-5}
Z_0+Z_3	0.00396599955999	1.4×10^{-6}
$Z_0+Z_3+Z_4$	0.00396736308864	-5.8×10^{-9}

$K=3$, $n^*=44$	Z_{approx}	$Z-Z_{\text{approx}}$
Z_0	0.00396145445096	5.9×10^{-6}
Z_0+Z_3	0.00396733343461	2.4×10^{-8}
$Z_0+Z_3+Z_4$	0.00396735718159	9.6×10^{-11}

$K=10$, $n^*=88$	Z_{approx}	$Z-Z_{\text{approx}}$
Z_0	0.00396747696881	-1.2×10^{-7}
Z_0+Z_3	0.00396735693288	3.4×10^{-10}
$Z_0+Z_3+Z_4$	0.00396735728143	-4.1×10^{-12}

The upper limits n^* were chosen according to (27) with $A=33$, thereby ensuring convergence of the sums to $\exp(-33)=5\times10^{-15}$. For t_1, we also evaluated Z_0, Z_3 and Z_4 using the exact formula (2) for $\theta(t)$, and confirmed that even for this small value the errors introduced by (24) are small compared with the deviations from the exact $Z(t)$ of Z_0, Z_0+Z_3 and $Z_0+Z_3+Z_4$.

For comparison, we computed $Z(t)$ by RS, including the main sum and five corrections (that is, $\Phi^{(0)}$ through $\Phi^{(4)}$ in (4)). The bound (7) ensured that in all cases the errors of RS were small compared with those of Z_0, Z_3 and Z_4, and we confirmed this with a more accurate evaluation of $Z(t_1)$.

Tables 1–3 show the results of these computations, which we now discuss. The first value t_1 is very low; it lies between the first two Riemann zeros. As table 1 shows, the RS improves quite slowly as more terms are included. The same is true of our series, especially for $K=0.5$; this is the only case where we approach the least term of the m-series as predicted by (46), which gives $m^*\sim4.5$. For the higher values, $K=1$ and $K=1.5$, Z_3 and Z_4 do improve significantly on Z_0, which is itself better than the main sum of RS in all three cases. There is a price for this improvement: because t_1 is so small, the number n^* of terms in the n-sums of Z_m is always much bigger than the number of terms in the main sum of RS (here $N=1$), i.e. the condition (28) is violated.

The value of t_2 is chosen between two close Riemann zeros, where Z is very small

Table 3. *Computations of $Z(t)$ for t_3*

t	$Z(t)$	$N(t)$	$p(t)$
251 704.544 777 283 6	−1.463 773 120 222 623	200	0.15

	Z_{approx}	$Z-Z_{\text{approx}}$
RS main sum	−1.419 501 811 072 478	-4.4×10^{-2}
$+\Phi^{(0)}$	−1.463 770 667 363 168	-2.5×10^{-6}
$+\Phi^{(1)}$	−1.463 773 114 608 635	-5.6×10^{-9}
$+\Phi^{(2)}$	−1.463 773 120 222 490	-1.3×10^{-13}
$+\Phi^{(3)}$	−1.463 773 120 222 619	-4.6×10^{-15}
$+\Phi^{(4)}$	−1.463 773 120 222 623	$< 2.4\times10^{-17}$

$K=1$, $n^*=203$	Z_{approx}	$Z-Z_{\text{approx}}$
Z_0	−1.463 766 937 951 2	-6.1×10^{-6}
Z_0+Z_3	−1.463 773 108 489 4	-1.2×10^{-8}
$Z_0+Z_3+Z_4$	−1.463 773 120 211 0	-1.2×10^{-11}

$K=3$, $n^*=210$	Z_{approx}	$Z-Z_{\text{approx}}$
Z_0	−1.463 772 335 497 91	-7.8×10^{-7}
Z_0+Z_3	−1.463 773 120 583 50	3.6×10^{-10}
$Z_0+Z_3+Z_4$	−1.463 773 120 221 83	-9.9×10^{-13}

$K=10$, $n^*=235$	Z_{approx}	$Z-Z_{\text{approx}}$
Z_0	−1.463 773 124 118 41	3.9×10^{-9}
Z_0+Z_3	−1.463 773 120 220 92	-1.7×10^{-12}
$Z_0+Z_3+Z_4$	−1.463 773 120 222 64	1.7×10^{-14}

(this is a near counterexample to the Riemann hypothesis). The main sum of RS misses both zeros by predicting (table 2) the wrong sign for Z (cf. figure 3*c*, which shows a similar occurrence for a smaller t); the first correction $\Phi^{(0)}$ gives a much better approximation, and with higher corrections the errors decrease rapidly. For all three values of K, Z_0 is much more accurate than the main sum of RS, and Z_3 and Z_4 give much better approximations still. Note that for $K=1$ and $K=3$ the number of terms n^* is not much larger than the value $N(t_2)=33$ for RS; both satisfy (28), and $K=1$ also satisfies the condition (38) for the validity of the quadratic approximation.

To represent what we expect to happen in the truly asymptotic region, t_3 is chosen so that the main sum of RS has $N(t_2)=200$ terms. The main sum gives an error of a few percent. For all K in table 3, n^* is not much larger than N, yet Z_0 is better than the main sum of RS by between four and seven orders of magnitude. The improvements with Z_3 and Z_4 are each between two and three orders of magnitude, which is comparable to the improvement with extra terms of RS. All three K values satisfy (28), and $K=1$ and $K=3$ also satisfy the condition (38) for the validity of the quadratic approximation.

In all cases, except when $t=t_1$ and $K=0.5$, the already accurate approximation provided by Z_0 was greatly improved by including Z_3 and Z_4. Consistent with this rapid convergence, we noticed that each error was very close to the next omitted term.

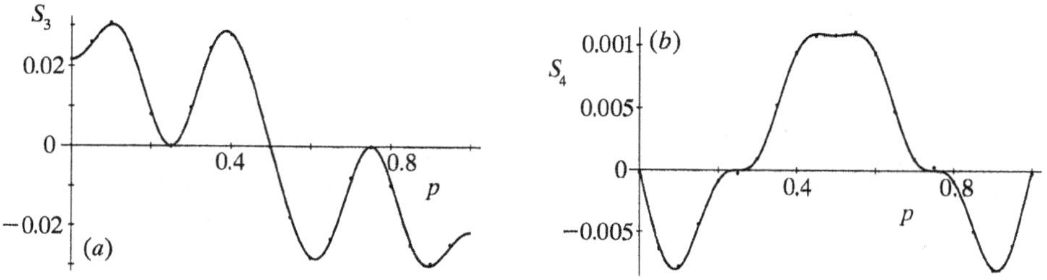

Figure 4. Comparison of S_m (points), defined by the second member of (52), with the quadratic approximation S_m^{quad} (full lines), defined by the third member of (52), with $N = 200$ and $K = 2$, for (*a*) $m = 3$, (*b*) $= 4$.

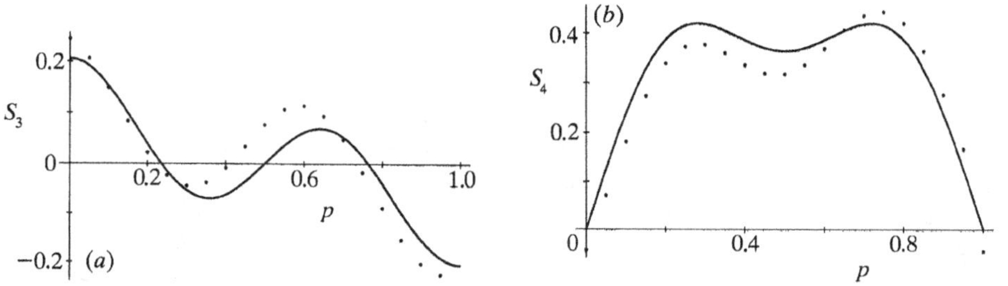

Figure 5. As figure 4, with $N = 2$ and $K = 1/\sqrt{2}$.

The estimates in §5 of late terms in the series Z_m depended on the quadratic phase approximation of §4. In figures 4 and 5 we illustrate this approximation for Z_3 and Z_4 as $p(t)$ ranges from 0 to 1, that is (31) as t ranges from $2\pi N^2$ to $2\pi(N+1)^2$. From (37) and (B 4), the quadratic approximation can be written

$$\begin{aligned} S_m(N,p,K) &\equiv 6(-1)^{N+1}(m-3)(N+p)^{m-\frac{3}{2}} Z_m(2\pi(N+p)^2, K) \\ &\approx \sum_{l=-\infty}^{\infty} (-1)^{l(l+1)/2}(2p-l-\tfrac{1}{2})^{m-1} \exp\{-\pi K^2(2p-\tfrac{1}{2}-l)^2\} \\ &\equiv S_m^{\text{quad}}(p,K) \quad (m = 3,4). \end{aligned} \tag{52}$$

Figure 4 shows S and S^{quad} for $N = 200$ ($t \approx 250000$), and $K = 2$, which lies comfortably within the expected range of validity (38) of the quadratic approximation (for $A = 33$, (37) requires $K < 2.6$). As expected, the approximation is excellent over the whole range of p.

More surprising is figure 5, which shows S and S^{quad} for $N = 2$ ($25.1 < t < 56.5$). For these low values of t there there is no K satisfying (38); nevertheless for $K = 1/\sqrt{2}$, where the left-hand side of (38) is smallest, the quadratic approximation still gives a good qualitative fit for all p.

7. Generalization to quantum spectral determinants

In the study of the spectra of quantum systems whose classical counterparts have chaotic trajectories (Berry 1987; Eckhardt 1988; Gutzwiller 1990), divergent or conditionally convergent series occur which are closely analogous to the Dirichlet series (3) for $\zeta(s)$. Our purpose in this section is to show that, remarkably, precisely the resummation method we have used for $Z(t)$ can be applied directly, to yield an asymptotic series of convergent contributions for an analogous function in quantum mechanics.

This is the *quantum spectral determinant* $\Delta(E,\hbar)$, constructed as follows. In quantum mechanics the energy levels $E_j(\hbar)$ are the (necessarily real) eigenvalues of a hermitian operator (the quantum hamiltonian), obtained by quantizing, with Planck's constant $\hbar$, a classical hamiltonian function $H(\boldsymbol{q},\boldsymbol{p})$ of phase space variables

$$\boldsymbol{q} = \{q_1,\ldots,q_D\}, \quad \boldsymbol{p} = \{p_1,\ldots,p_D\}. \tag{53}$$

Here D is the number of classical freedoms. We confine ourselves to chaotic systems, where the orbits generated by $H(\boldsymbol{q},\boldsymbol{p})$ are all unstable. The spectral determinant of the hamiltonian operator is

$$\Delta(E,\hbar) \equiv \prod_j \{A\{E,E_j(\hbar)\}[E-E_j(\hbar)]\}, \tag{54}$$

where A, whose role is to make the product converge (Voros 1987), is a zero-free function which is real for real E and $\hbar$. Thus $\Delta(E,\hbar)$ is real for real E and $\hbar$, and its zeros are the quantum energies $E_j(\hbar)$. In this context, quantum chaology (Berry 1987) is the study of the small-$\hbar$, or semiclassical, asymptotics of the E_j, and is obviously related to the small-$\hbar$ asymptotics of the real quantum spectral determinant $\Delta(E,\hbar)$.

For wide classes of system, E and $\hbar$ are related by scaling, so that Δ depends only on one variable. Examples are the quantum mechanics of billiards, motion on compact curved surfaces, and particles in potentials which are homogeneous functions of $\boldsymbol{q}$. In such cases, the small-$\hbar$ asymptotics and the large-E asymptotics are the same. In general, however, there is no scaling relation between E and $\hbar$, and the semiclassical and high-energy limits are not the same. Then we shall regard $\Delta(E,\hbar)$ as a function of complex $\hbar$ with E fixed and real. In the analogy with $\zeta(\frac{1}{2}+\mathrm{i}t)$, the variable corresponding to t is $1/\hbar$. This choice of variable might seem odd but is in fact natural, because then we are quantizing a system with fixed classical mechanics, an important simplification for non-scaling systems, where the dynamics depends non-trivially on E (the same procedure has been used by Balian & Bloch (1974) and Berry & Tabor (1977)).

By semiclassical techniques based on the trace formula of Gutzwiller (1971) for $\ln\Delta$, Berry & Keating (1990) obtained the following semiclassical expression for the quantum spectral determinant,

$$\Delta(\hbar,E) \approx \Delta^{\mathrm{sc}}(\hbar,E) \equiv B(E,\hbar)\exp\{\mathrm{i}\pi\bar{\mathcal{N}}(E,\hbar)\}\sum_{n=0}^{\infty} C_n(E)\exp\{-\mathrm{i}\mathcal{S}_n(E)/\hbar\} \tag{55}$$

in which the quantities have the following meanings. The sum is over pseudo-orbits, that is linear combinations of (primitive and repeated) periodic classical orbits (all unstable) with energy E; n labels pseudo-orbits in increasing (pseudo) period $\mathcal{T}$. $\mathcal{S}_n$ is the action of the nth pseudo-orbit, that is the sum of the actions of the periodic orbits of which it is composed. The coefficients C_n involve the stability exponents of the periodic orbits (which do not depend on $\hbar$), and the Maslov phases (Maslov & Fedoriuk 1981; Robbins 1991). The exponent $\bar{\mathcal{N}}(E,\hbar)$ is the smoothed spectral staircase, counting the mean number of levels with $E_j < E$. Semiclassical approximation (see Berry 1983) provides an asymptotic series (the analogue of (24)) for this quantity, whose leading term is

$$\bar{\mathcal{N}}(E,\hbar) \approx \Omega(E)/(2\pi\hbar)^D, \tag{56}$$

where Ω is the classical phase-space volume with energy less than E, namely

$$\Omega(E) = \iint \mathrm{d}\boldsymbol{q}\,\mathrm{d}\boldsymbol{p}\,\Theta\{E - H(\boldsymbol{q},\boldsymbol{p})\}. \tag{57}$$

Finally, $B(E,\hbar)$ is real and non-zero when E and $\hbar$ are real. The quantum analogue of $Z(t)$ is $-\Delta(E,\hbar)/B(E,\hbar)$.

For real $\hbar$, the representation (55) is divergent (or, at best, only conditionally convergent (Sieber & Steiner 1991)). A simplified argument (Eckhardt & Aurell 1988; Keating 1992*b*) revealing the nature of the divergence is that the exponential decrease of the C_n for long orbits, namely

$$C_n(E) \sim \exp\{-\tfrac{1}{2}\lambda(E)\,\mathscr{T}_n(E)\} \quad \text{as } \mathscr{T}_n \to \infty, \tag{58}$$

where λ denotes the metric entropy, is dominated by the exponential proliferation of periodic (and pseudo) orbits, whose number $\nu(E)\,\mathrm{d}\mathscr{T}$ between $\mathscr{T}$ and $\mathscr{T}+\mathrm{d}\mathscr{T}$ is

$$\nu(E) \sim \exp\{+\lambda(E)\,\mathscr{T}(E)\} \quad \text{as } \mathscr{T} \to \infty \tag{59}$$

(we are here assuming for simplicity that the metric and topological entropies are equal). Thus the sum diverges like the integral over $\mathscr{T}$ of

$$\exp\{+\tfrac{1}{2}\lambda(E)\,\mathscr{T}(E)\}. \tag{60}$$

By making $\hbar$ complex, (55) can be made absolutely convergent, the condition being that for long orbits

$$\mathrm{Im}\, 1/\hbar < -\lambda\mathscr{T}/2\mathscr{S}. \tag{61}$$

Now, for the long-period orbits and pseudo-orbits of an ergodic system, $\mathscr{S}$ and $\mathscr{T}$ are proportional (Hannay & Ozorio de Almeida 1984), the precise relationship being

$$\mathscr{S} \approx \mathscr{T} D\Omega/\Omega'. \tag{62}$$

(This follows from

$$S = \int \boldsymbol{p}\cdot\mathrm{d}\boldsymbol{q} = \int \mathrm{d}t\,\boldsymbol{p}\cdot\mathrm{d}\boldsymbol{q}/\mathrm{d}t \to T\langle \boldsymbol{p}\cdot\mathrm{d}\boldsymbol{q}/\mathrm{d}t\rangle,$$

where $\langle\ldots\rangle$ denotes ergodic averaging over the energy surface.) Thus the convergence condition (61) becomes

$$\mathrm{Im}\, 1/\hbar < -\lambda(E)\,\Omega'(E)/2D\Omega(E). \tag{63}$$

This is the familiar 'entropy barrier', here expressed in terms of complex $1/\hbar$ rather than the more usual complex E.

For real $\hbar$, (55) fails, not only by diverging but by being not obviously real as the exact Δ must be. The analogy between this situation and that for the function $Z(t)$ defined by (1) has already been employed by Berry (1986) and Berry & Keating (1990) to conjecture for Δ^{sc} a manifestly real and finite approximate resummation of (55), analogous to the main sum of RS. Keating (1992*a*) has given a formal argument supporting this conjecture, as has Bogomolny (1992), and there is some computational support for the relation (Sieber & Steiner 1991). Now we can go much further, and give an *exact* resummation of (55), analogous to the representation of $Z(t)$ obtained in §3.

To demonstrate this, we use analytic continuation as in §2, in the variable $1/\hbar$. This requires an analogue of the functional equation $Z(t) = Z(-t)$. The analogue is the exact relation

$$\Delta(E,\hbar) = \Delta(E,-\hbar), \tag{64}$$

which holds because the eigenvalues E_j are independent of the sign of $\hbar$. This is true even for a system without time-reversal symmetry: $\hbar$-reversal changes the hermitian operator to its conjugate, leaving the eigenvalues unchanged. Cauchy's theorem can now be used as in (8) and (9), to yield the analogues of (10) and (11). The contour C_- can be taken to lie outside the entropy barrier (63), so (55) (the analogue of (3)) may be used as a valid semiclassical approximation to obtain an analytic continuation to real $1/\hbar$.

To obtain the analogue of the manifestly real expression (13) we need the analogue of (12). This must be

$$C_n(E)\exp\{\mathrm{i}[\pi\mathcal{N}(E,\hbar)-\mathcal{S}_n(E)/\hbar]\} \rightarrow [C_n(E)\exp\{\mathrm{i}[\pi\mathcal{N}(E,\hbar)-\mathcal{S}_n(E)/\hbar]\}]^* \quad \text{if } \hbar\rightarrow-\hbar. \tag{65}$$

Mere substitution of $-\hbar$ for $\hbar$ is inadequate to demonstrate the truth of this formula. That it is the correct continuation of terms in the time-independent semiclassical approximation (including $\mathcal{N}(E,\hbar)$ and the Maslov phases incorporated in C_n) follows from the behaviour under $\hbar$-reversal of the time-dependent Schrödinger equation from which semiclassical approximations can be derived (see Berry 1991). (For the Selberg zeta function (Balazs & Voros 1986), (65) follows from standard formulae.)

Thus we find the exact semiclassical resummation for real $\hbar$:

$$\Delta^{\mathrm{sc}}(E,\hbar) = 2B(E,\hbar)\,\mathrm{Re}\sum_{n=0}^{\infty} U_n(E,\hbar), \tag{66}$$

where (cf. (11))

$$U_n(E,\hbar) = C_n(E)\exp\{\mathrm{i}(\pi\mathcal{N}(E,\hbar)-\mathcal{S}_n(E)/\hbar)\}\frac{1}{2\pi\mathrm{i}}\int_{C_-}\frac{\mathrm{d}z}{z}\gamma(z,\hbar) \times\exp\{\mathrm{i}[\pi\mathcal{N}(E,(\hbar^{-1}+z)^{-1})-\pi\mathcal{N}(E,\hbar)-z\mathcal{S}_n(E)]\}. \tag{67}$$

To convert this into a usable formula we again begin by choosing (*ad hoc*) the function $\gamma(z,\hbar)$ (cf. (14)):

$$\gamma(z,\hbar) = \exp(-\tfrac{1}{2}z^2K^2|\hbar|). \tag{68}$$

Next we follow the procedure of §3 by considering $1/\hbar > 0$ and expanding the exponential in powers of z (cf. (15)), as follows:

$$\exp\{\mathrm{i}\pi[\mathcal{N}(E,(\hbar^{-1}+z)^{-1})-\mathcal{N}(E,\hbar)]\} = \exp\{\mathrm{i}\pi[z\mathcal{N}_1(E,\hbar)+\tfrac{1}{2}z^2\mathcal{N}_2(E,\hbar)]\}\left[1+\sum_{m=3}^{\infty} z^m\beta_m(E,\hbar)\right], \tag{69}$$

where the subscripts on $\mathcal{N}$ denote derivatives with respect to $1/\hbar$. Now we collect together the terms in z and z^2, defining (cf. (17))

$$\left.\begin{aligned}\xi(n,\hbar,E) &\equiv \mathcal{S}_n(E)-\pi\mathcal{N}_1(E,\hbar)\ \left(\sim\mathcal{S}_n(E)-\frac{D\Omega(E)}{2(2\pi\hbar)^{D-1}}\right),\\ Q^2(K,\hbar,E) &\equiv K^2-\frac{\mathrm{i}\pi}{\hbar}\mathcal{N}_2(E,h)\ \left(\sim K^2-\mathrm{i}\frac{D(D-1)\,\Omega(E)}{2(2\pi\hbar)^{D-1}}\right).\end{aligned}\right\} \tag{70}$$

Finally, the integrals over z in (67) can be evaluated exactly, giving (cf. (18)–(20)) the m-series

$$\Delta^{\mathrm{sc}}(E,\hbar) = \Delta_0(E,\hbar,K)+\Delta_3(E,\hbar,K)+\Delta_4(E,\hbar,K)+\dots, \tag{71}$$

where

$$\Delta_0(E,\hbar,K) = 2B(E,\hbar)\,\mathrm{Re}\sum_{n=0}^{\infty} C_n(E) \times \exp\{\mathrm{i}[\pi\mathcal{N}(E,\hbar)-\mathcal{S}_n(E)/\hbar]\}\,\tfrac{1}{2}\,\mathrm{Erfc}\left\{\frac{\xi(n,\hbar,E)}{Q(K,\hbar,E)}\Big/\sqrt{\left(\frac{1}{2\hbar}\right)}\right\} \tag{72}$$

and

$$\Delta_m(E,\hbar,K) = (2B(E,\hbar)/\sqrt{\pi})\,(1/2\hbar)^{m/2} \times \mathrm{Re}\frac{(-\mathrm{i})^m\beta_m(E,\hbar)}{Q^m(K,\hbar,E)}\sum_{n=0}^{\infty} C_n(E)\exp\{\mathrm{i}[\pi\mathcal{N}(E,\hbar)-\mathcal{S}_n(E)/\hbar]\} \times \exp\left\{\frac{-\xi^2(n,\hbar,E)}{2\hbar Q^2(K,\hbar,E)}\right\}H_{m-1}\left\{\frac{\xi(n,\hbar,E)}{Q(K,\hbar,E)}\Big/\sqrt{\left(\frac{1}{2\hbar}\right)}\right\} \quad (m \geqslant 3). \tag{73}$$

Previously, the arguments of Keating (1992*a*) were used in Keating (1992*b*) and Aurich & Steiner (1992) to obtain an approximation to the first term Δ_0, corresponding to replacing Q by K.

We will not comment in detail on these formulae, because their structure is so similar to (18)–(20) for $Z(t)$, which we have already explored, but we do wish to make two remarks. The first concerns Δ_0. This is analogous to the main sum of RS with the sharp cut-off smoothed away. The smoothing is centred on the pseudo-orbit for which ξ, given in (70), is zero. Because of the factor $\hbar^{-(D-1)}$ in the second term, this is a long pseudo-orbit, so $\mathcal{S}$ can be replaced by its approximation (62). Thus when $\mathcal{T}$ is large

$$\begin{aligned}\xi(n,\hbar,E) &\approx \frac{\mathcal{T}_n(E)\,D\Omega(E)}{\Omega'(E)} - \frac{D\Omega(E)}{2(2\pi\hbar)^{D-1}} \\ &= (D\Omega(E)/\Omega'(E))\,[\mathcal{T}_n(E)-\pi\hbar\bar{d}(E,\hbar)],\end{aligned} \tag{74}$$

where

$$\bar{d}(E,\hbar) \approx \Omega'(E)/(2\pi\hbar)^D \tag{75}$$

is the semiclassical smoothed level density. The centre of the smoothing is therefore the pseudo-orbit n^* whose period is

$$\mathcal{T}_{n^*}(E) = \pi\hbar\bar{d}(E,\hbar). \tag{76}$$

It is satisfying that this result, here derived by analytic continuation of $1/\hbar$, is exactly that previously guessed (Berry 1986; Berry & Keating 1990), or obtained by analytic continuation of E (Keating 1992*a*). Note that the above derivation involves the non-trivial result (62), which requires the classical motion to be ergodic.

The second remark is that it would be interesting to study the convergence of the m-series (71) by investigating its late terms (i.e. Δ_m for large m).

We emphasize that (71)–(73) is an exact analytic continuation of the semiclassical formula (55), which completely eliminates the difficulties caused by the lack of convergence of Gutzwiller's trace formula. But in general (71)–(73) does not provide an exact representation for the quantum spectral determinant $\Delta(E,\hbar)$, because (55) is an approximation, valid to lowest order in $\hbar$. In particular, the zeros of (71)–(73) will be semiclassical approximations to the exact eigenvalues $E_j(\hbar)$. To improve the approximation, it would be necessary to incorporate higher $\hbar$-corrections into (55). If this were done, analytic continuation could be carried out as we have done here, for each successive order of semiclassical approximation.

However, there is at least one case where the Gutzwiller trace formula, and the semiclassical quantum spectral determinant (55) derived from it, are not semiclassical approximations but are exact (although of course not absolutely convergent within the entropy barrier). This is the spectral determinant for the Laplace–Beltrami operator on compact surfaces of constant negative curvature, for which the logarithm of (55) is the celebrated Selberg trace formula (Balazs & Voros 1986). For this system, our formulae (71)–(73) should provide an exact analytic continuation, enabling arbitrarily high eigenvalues to be determined from the periodic geodesics with no approximation (except the exponentially small error resulting from truncating the series of Δ_m (cf. (47) and (48)), which can be made arbitrarily small by increasing K). This procedure should converge much more rapidly than gaussian smoothing of the spectral density (logarithmic derivative of $\Delta(E,\hbar)$) as used by Aurich *et al.* (1988) (analogous to Delsarte's (1966) regularization of $\ln\zeta(\frac{1}{2}+it)$), which requires approximately N^2, rather than N, terms.

We do not wish to imply that the exact regularization of the semiclassical spectral determinant solves all problems in the quantum chaology of spectra. Several difficulties remain. First, there is the question of higher-order $\hbar$-corrections; we expect this to be crucial in resolving groups of close-lying eigenvalues. Second, there is the question of whether the regularization guarantees that the approximate eigenvalues will be real, like the exact ones; we believe it does not. Third, there is the difficulty caused by the fact that even with the (approximate) cut-off (76) the sums in the contributions Δ_m involve exponentially many pseudo-orbits and so are cumbersome; here the curvature expansions (Cvitanovic & Eckhardt 1989) and related ideas for pruning the pseudo-orbits (Bogomolny 1992) could prove important. Finally, there is the question of the statistics of the zeros of the semiclassical spectral determinant, and their relation to the universal statistics of random-matrix theory (Bohigas & Giannoni 1984; Berry 1985, 1988; Keating 1992*b*).

We thank P. Boasman, J. H. Hannay and J. M. Robbins for their careful reading of the typescript, and several helpful comments. J. P. K. is grateful to the Royal Society for support during the period of this research.

Appendix A. Convergence of the analytically continued Dirichlet series

Here our aim is to estimate high orders in the n-sum (13), that is T_n for n much larger than the cutoff N in RS (equation (5)), and thereby investigate the convergence of (13). Thus we must study the integral

$$T_n(t) = \frac{\exp\{-it\ln n\}}{\sqrt{n}}\frac{1}{2\pi i}\int_{C_-}\frac{dz}{z}\exp\left\{-iz\ln n-\frac{K^2z^2}{2t}\right\}\exp\{i\theta(z+t)\}. \quad \text{(A 1)}$$

There are two contributions, one of which dominates when $N \ll n \ll N\exp(K^2)$ and the other when $n \gg N\exp(K^2)$; they originate respectively from a saddle and a branch point in the z plane of integration.

A sufficiently good approximation to the position of the saddle can be found by including the first exponential in (A 1) and the term of first order in z in the second exponential, and then using Stirling's formula (24). With (5) we find

$$z_{\text{saddle}} \sim -\mathrm{i}\,(t/K^2)\ln(n/N(t)). \tag{A 2}$$

Expanding to second order about this saddle and evaluating the resulting gaussian integral we obtain

$$|T_n|_{\text{saddle}} \approx \frac{K}{\ln(n/N)\sqrt{(2\pi nt)}}\exp\left\{-\frac{t}{2K^2}\ln^2\left(\frac{n}{N}\right)\right\} \times \exp\{\tfrac{1}{2}\operatorname{Re}[(t+z_{\text{saddle}})\ln(1+z_{\text{saddle}}/t)-z_{\text{saddle}}]\}. \tag{A 3}$$

The first exponential dominates the second.

Ultimately, convergence of (11) is determined not by this saddle contribution but by the nearest singularity to which the integration contour C_- can be deformed. This is the square-root branch point corresponding to one of the zeros of the gamma function in (2) namely

$$z_{\text{branch}} = -t-\tfrac{1}{2}\mathrm{i}. \tag{A 4}$$

We write $z = z_{\text{branch}} - \mathrm{i}\sigma$, expand the integrand to lowest order in σ, and integrate along both sides of the cut descending from z_{branch}. Thus we find, after a little reduction,

$$\begin{aligned} T_{n\,\text{branch}} &= \frac{\mathrm{i}\exp\{-(K^2/2t)(t+\frac{1}{2}\mathrm{i})^2\}}{n(t+\frac{1}{2}\mathrm{i})\,\pi\sqrt{2}}\int_0^\infty \mathrm{d}\sigma\,\sqrt{\sigma}\exp\left\{-\sigma\left(\ln n\sqrt{\pi}+\mathrm{i}K^2\left(1+\frac{\mathrm{i}}{2t}\right)\right)\right\} \\ &\approx \frac{\mathrm{i}\exp\{-\frac{1}{2}K^2t\}}{2\sqrt{(2\pi)}\,tn\ln^{\frac{3}{2}}n}, \end{aligned} \tag{A 5}$$

where the approximation requires $t \gg 1$. This decrease is very slow, but is sufficient to make the n-sum in (11) converge, because

$$\sum_{n=M}^{\infty}\frac{1}{n\ln^{\frac{3}{2}}n}\xrightarrow[M\to\infty]{}\frac{1}{2\ln^{\frac{1}{2}}M}. \tag{A 6}$$

The crossover between the saddle and branch contributions to T_n occurs when their dominant exponentials are equal, that is when

$$(t/2K^2)\ln^2(n/N) = \tfrac{1}{2}K^2t, \quad \text{i.e. } n = N\exp(K^2) \tag{A 7}$$

as asserted earlier. Summarizing, we have, retaining only leading orders,

$$\left.\begin{aligned} |T_n| &\sim \exp\{-(t/2K^2)\ln^2(n/N)\} \quad (N \ll n \ll N\exp(K^2)), \\ |T_n| &\sim \exp\{-\tfrac{1}{2}K^2t\}/n\ln^{\frac{3}{2}}n \quad (n \gg N\exp(K^2)). \end{aligned}\right\} \tag{A 8}$$

Appendix B. Asymptotics of the expansion coefficients

We seek approximations for the quantities $b_m(t)$ defined by (15) and (16), for t large with m fixed, and also for t large with $m \gg t$. Defining

$$a_s(t) \equiv [\operatorname{Im}\mathrm{i}^s\psi^{(s-1)}(\tfrac{1}{4}+\tfrac{1}{2}\mathrm{i}t)]/2^s s! \tag{B 1}$$

our problem is to solve

$$\sum_{m=3}^{\infty} z^m b_m = \exp\left\{\mathrm{i}\sum_{s=3}^{\infty} z^s a_s\right\}-1 \tag{B 2}$$

for b_m.

For low m this can easily be achieved by direct expansion of the exponential and use of the following approximation (Abramowitz & Stegun 1964) for polygamma functions:

$$\psi^{(n)}(w) \approx (-1)^{n+1}(n-1)!/w^n \quad (|w| \text{ large, } n \text{ fixed}). \tag{B 3}$$

We find

$$b_3 = \mathrm{i}a_3 \approx -\frac{\mathrm{i}}{12t^2}, \quad b_4 = \mathrm{i}a_4 \approx \frac{\mathrm{i}}{24t^3},$$
$$b_5 = \mathrm{i}a_5 \approx -\frac{\mathrm{i}}{40t^4}, \quad b_6 = \mathrm{i}a_6 - \tfrac{1}{2}a_3^2 \approx \frac{\mathrm{i}}{60t^5} - \frac{1}{288t^4}. \tag{B 4}$$

The direct calculation is much more difficult for large m. For a start, polygamma functions of high order have an asymptotic approximation different from (B 3) (in fact (B 3) must be multiplied by nw). And each high b_m is a complicated combination of many a_m, ranging from approximately

$$\frac{(\mathrm{i}a_3)^{m/3}}{(\frac{1}{3}m)!} \approx \frac{(-\frac{1}{12}\mathrm{i})^{m/3}}{(\frac{1}{3}m)!\,t^{2m/3}} \tag{B 5}$$

to

$$\mathrm{i}a_m \sim 1/t^m \tag{B 6}$$

and neither extreme dominates. Therefore we adopt a different strategy, based on Darboux's principle of the nearest singularity (Dingle 1973).

To apply this, we first write the solution of (15) as

$$b_m = G^{(m)}(0)/m!, \tag{B 7}$$

where the superscript denotes the mth derivative, and (cf. (2))

$$G(z) = \left[\frac{\Gamma\{\frac{1}{4}+\frac{1}{2}\mathrm{i}(t+z)\}}{\Gamma\{\frac{1}{4}-\frac{1}{2}\mathrm{i}(t+z)\}}\right]^{\frac{1}{2}} \exp\{\mathrm{i}\chi(z)\} \tag{B 8}$$

with

$$\chi(z) = -[\tfrac{1}{2}(t+z)\ln\pi + \theta(t) + z\theta'(t) + \tfrac{1}{2}z^2\theta''(t)] \tag{B 9}$$

(we do not indicate the t dependence explicitly).

Darboux's principle asserts that the high derivatives $G^{(m)}(0)$ are determined by the singularities of $G(z)$ (this can be justified by writing b_m as a contour integral surrounding $z = 0$, and expanding the contour to hit the singularities). In (B 8) the singularities are square-root branch points at the poles of the gamma functions, namely

$$z = -t \pm \tfrac{1}{2}\mathrm{i}(2n+\tfrac{1}{2}), \quad n = 0, 1, \ldots. \tag{B 10}$$

Of these, the dominant contribution comes from

$$z = -t + \tfrac{1}{2}\mathrm{i} \tag{B 11}$$

(this can be confirmed by repeating for the other branch points the argument which follows).

Expanding about this point, we find

$$G(z) \xrightarrow[\text{as } z \to -t+\frac{1}{2}\mathrm{i}]{} \frac{\sqrt{2}\exp\{-\frac{1}{4}\mathrm{i}\pi\}}{\pi^{\frac{1}{4}}[z-(-t+\frac{1}{2}\mathrm{i})]^{\frac{1}{2}}} \exp\{\mathrm{i}\chi(-t+\tfrac{1}{2}\mathrm{i})\}. \tag{B 12}$$

Differentiating m times, where m is large, swells the domain of applicability of this formula. Setting $z = 0$, using (B 7), and approximating the function χ for large t with the aid of (B 9) and (24), then leads to

$$b_m(t) \underset{m\to\infty}{\longrightarrow} \left(\frac{2}{\pi^3 e m^2 t}\right)^{\frac{1}{4}} \frac{(-1)^m}{(t+\frac{1}{2}\mathrm{i})^m} \exp\{\tfrac{1}{8}\mathrm{i}(2t-\pi)\}. \tag{B 13}$$

We need to know how large m must be in order for this limiting form to be a good approximation. The answer requires knowledge of any competing contributions to b_m that are eventually dominated by (B 13). The relevant contribution comes from a saddle of the integrand (cf. Appendix A) of the loop integral representing b_m, namely

$$b_m = \frac{1}{2\pi\mathrm{i}} \oint \mathrm{d}z \frac{G(z)}{z^{m+1}}, \tag{B 14}$$

where the loop encircles the origin. The relevant saddle is close to $z \approx (t^2 m)^{\frac{1}{3}}$; it gives a contribution whose dominant factors differ from those in (B 13) by the replacement of t^{-m} by $(t^2m)^{-m/3}$ (the contribution is the same as that given by (B 5)). The crossover, beyond which (B 13) dominates this saddle contribution, is therefore $m \sim t$. This is consistent with Darboux's principle: for $m < t$ the saddle is closer to $z = 0$ than the branch point $z = -t+\frac{1}{2}\mathrm{i}$ which generates (B 13), and for $m > t$ the branch point is nearer.

References

Abramowitz, M. & Stegun, I. A. 1964 *Handbook of mathematical functions*. Washington, D.C.: National Bureau of Standards.

Aurich, R. & Steiner, F. 1992 From classical periodic orbits to the quantization of chaos. *Proc. R. Soc. Lond.* A. (Submitted.)

Aurich, R., Sieber, M. & Steiner, F. 1988 *Phys. Rev. Lett.* **61**, 483–487.

Balian, R. & Bloch, C. 1974 *Ann. Phys.* (*N.Y.*) **85**, 514–545.

Balazs, N. L. & Voros, A. 1986 *Phys. Rep.* **143**, 109–240.

Berry, M. V. 1983 Semiclassical Mechanics of regular and irregular motion. In *Les Houches Lecture Series Session XXXVI* (ed. G. Iooss, R. H. G. Helleman & R. Stora), pp. 171–271. Amsterdam: North Holland.

Berry, M. V. 1985 *Proc. R. Soc. Lond.* A **400**, 229–251.

Berry, M. V. 1986 Riemann's zeta function: a model for quantum chaos? In *Quantum chaos and statistical nuclear physics* (ed. T. H. Seligman & H. Nishioka), Springer Lecture Notes in Physics no. 263, pp. 1–17.

Berry, M. V. 1987 *Proc. R. Soc. Lond.* A **413**, 183–198.

Berry, M. V. 1988 *Nonlinearity* **1**, 399–407.

Berry, M. V. 1991 Some quantum-to-classical asymptotics. In *Les Houches Lecture Series 52* (ed. M. J. Giannoni, A. Voros & J. Zinn-Justin), pp. 251–303. Amsterdam: North-Holland.

Berry, M. V. & Keating, J. P. 1990 *J. Phys.* A **23**, 4839–4849.

Berry, M. V. & Tabor, M. 1977 *Proc. R. Soc. Lond.* A **356**, 375–394.

Bohigas, O. & Giannoni, M. J. 1984 Chaotic motion and random-matrix theories. In *Mathematical and computational methods in nuclear physics* (ed. J. S. Dehesa, J. M. G. Gomez & A. Polls). Springer Lecture Notes in Physics no. 209, pp. 1–99.

Brent, R. P. 1979 *Math. Comp.* **33**, 1361–1372.

Bogomolny, E. B. 1992 Semiclassical quantisation of multidimensional systems. *Nonlinearity*. (Submitted.)

Cvitanovic, P. & Eckhardt, B. 1989 *Phys. Rev. Lett.* **63**, 823–826.

Delsarte, J. 1966 *J. Anal. Math.* (*Jerusalem*) **17**, 419–431.

Dingle, R. B. 1973 *Asymptotic expansions: their derivation and interpretation*. New York: Academic Press.

Eckhardt, B. 1988 *Phys. Rep.* **163**, 205–297.

Eckhardt, B. & Aurell, E. 1989 *Europhys. Lett.* **9**, 509–512.

Edwards, H. M. 1974 *Riemann's zeta function*. New York and London: Academic Press.

Gabcke, W. 1979 Neue Herleitung und Explizite Restabschätzung der Riemann–Siegel–Formel. Ph.D. thesis, Göttingen.

Gradshteyn, I. S. & Ryzhik, I. N. 1980 *Table of integrals, series and products*. New York: Academic Press.

Gutzwiller, M. C. 1971 *J. math. Phys.* **12**, 343–358.

Gutzwiller, M. C. 1990 *Chaos in quantum mechanics*. Berlin: Springer.

Hannay, J. H. & Ozorio de Almeida, A. M. 1984 *J. Phys.* A **17**, 3429–3440.

Haselgrove, C. B. 1963 *Tables of the Riemann zeta function*. Cambridge University Press.

Keating, J. P. 1992*a* *Proc. R. Soc. Lond.* A **436**, 99–108.

Keating, J. P. 1992*b* The Riemann zeta-function and quantum chaology. In *Proceedings of Enrico Fermi International School of Physics*, Course CXIX. North-Holland.

Maslov, V. P. & Fedoriuk, M. V. 1981 *Semiclassical approximation in quantum mechanics*. Boston: Reidel.

Odlyzko, A. M. 1987 *Math. Comp.* **48**, 273–308.

Robbins, J. M. 1991 *Nonlinearity* **4**, 343–363.

Sieber, M. & Steiner, F. 1991 On the quantization of chaos. Preprint, University of Hamburg.

Titchmarsh, E. C. 1986 *The theory of the Riemann zeta-function*, 2nd edn. Oxford: Clarendon Press.

Voros, A. 1987 *Communs math. Phys.* **110**, 439–465.

Received 16 October 1991; accepted 21 November 1991

Chapter 5
Superoscillations

In 1990, 30 years after he and David Bohm had discovered the AB effect in Bristol (Chapter 1), Yakir Aharonov visited again. He likes to frame his discoveries as paradoxes, and during the visit, he told me: "I can imagine opening a window in a box containing only red light, and out would come a gamma ray". Paradoxical indeed! It was several years before I understood what he meant: a band-limited function ('red light') can oscillate arbitrarily faster ('a gamma ray') than its fastest Fourier component, over arbitrarily long intervals.

Such 'faster than Fourier' behaviour [5.1], which I called 'superoscillation', is possible because functions are exponentially small where they superoscillate. The phenomenon does not contradict the uncertainty principle, because this concerns variances of functions, which are insensitive to very small values. Only 10 years later, following a question from Miles Padgett, did I realise [B404, B457] that in a sense I had known about superoscillations since 1974, when Nye and I had emphasised the sub-wavelength behaviour of waves near phase singularities (Chapter 1).

Superoscillations originated in the context of Aharonov's and his colleagues' scheme of 'weak measurement', in which an operator is measured in a preselected state, and the only outcomes recorded are those compatible with a post-selected state. This quantum variant of large deviation theory results in a 'weak value', that can lie far outside the spectrum of the operator. In superoscillations, the pre-selected state is a function of position, the operator is momentum, the post-selected state is a chosen value of position, and the weak value is the local wave vector. In a general statistical analysis of the ensemble of all pre- and post-selected states [5.2], Pragya Shukla and I showed that for operators with many eigenvalues, the distribution of weak values is universal, and 'superweak' (i.e. outside the spectrum) values are unexpectedly common. For operators with just two eigenvalues, Mark Dennis, Shukla, Ben McRoberts and I showed [B437] that the superweak probability is one-third — the same value that Dennis, Alasdair Hamilton and Johannes Courtial had shown, in a pioneering calculation, to apply to superoscillations of monochromatic random waves in the plane (speckle patterns).

If the system being measured is coupled to a pointer, weak values can be measured as shifts of the pointer, and superweak values correspond to 'supershifts', outside the shifts lying within the spectrum of the operator. Shukla and I showed [5.3] that such supershifts depend on particular analytic properties of the initial pointer wavefunction. Stephen Barnett and I predicted [B468] that superoscillations near an optical vortex could be detected as a momentum 'superkick' on an atom acting as a 'pointer'.

Sandu Popescu and I suggested [B388] that superoscillation could provide the basis for sub-wavelength imaging. This was an argument based on the paraxial approximation to the Helmholtz equation, but later analysis [B461] showed that a non-paraxial scheme is in principle possible. It turned out that similar schemes, defying the conventional diffraction limit, were already known in antenna theory and had been adapted to microscopy [B467]; superoscillation theory provides a unifying description. Although sub-wavelength resolution based on superoscillations has been successfully achieved in practice, this approach is limited because superoscillations are associated with near-perfect destructive interference, rendering them intrinsically vulnerable to noise [B494].

In an extreme application, superoscillations can be employed to represent fractals. We demonstrated ([B499], with Sam Morley-Short) that the celebrated Weierstrass non-differentiable function, including frequencies up to 2^{16}, can be reproduced to visual accuracy by a superoscillatory (band-limited) function with highest frequency 1.

FASTER THAN FOURIER

Michael Berry
H.H. Wills Physics Laboratory, Tyndall Avenue, Bristol BS8 1TL, U.K.

Written to celebrate the 60th Birthday of Yakir Aharonov: deep, quick, subtle.

ABSTRACT

Band-limited functions $f(x)$ can oscillate for arbitrarily long intervals arbitrarily faster than the highest frequency they contain. A class of integral representations exhibiting these 'superoscillations' is described, and by asymptotic analysis the origin of the phenomenon is shown to be complex saddles in frequency space. Computations confirm the existence of superoscillations. The price paid for superoscillations is that in the infinitely longer range where $f(x)$ oscillates conventionally its value is exponentially larger. For example, to reproduce Beethoveen's ninth symphony as superoscillations with a 1Hz bandwidth requires a signal $\exp\{10^{19}\}$ times stronger than with conventional oscillations.

1. Model for superoscillations

My purpose is to decribe some mathematics inspired by Yakir Aharonov during a visit to Bristol several years ago. He told me that it is possible for functions to oscillate faster than any of their Fourier components. This seemed unbelievable, even paradoxical; I had heard nothing like it before, and learned only recently of just one related paper[1] in the literature on Fourier analysis (see §4). Nevertheless, Aharonov and his colleagues had constructed such 'superoscillations' using quantum-mechanical arguments[2]. Here I will exhibit a large class of them, and use asymptotics and numerics to study their strange properties in detail.

Consider functions $f(x)$ whose spectrum of frequencies k is band-limited, say by $|k|\leq 1$, so that on a conventional view f should oscillate no faster than $\cos(x)$. But we wish f to be superoscillatory, that is to vary as $\cos(Kx)$, where K can be arbitrarily large, for an arbitrarily long interval in x. A representation that achieves this is

$$f(x,A,\delta)=\frac{1}{\delta\sqrt{2\pi}}\int_{-\infty}^{\infty}\mathrm{d}u\exp\{\mathrm{i}xk(u)\}\exp\left\{-\frac{1}{2\delta^2}(u-\mathrm{i}A)^2\right\} \tag{1}$$

where the wavenumber function $k(u)$ is even, with $k(0)=1$ and $|k|\leq 1$ for real u, A is real and positive, and δ is small. Examples are

$$k_1(u)=\frac{1}{1+\frac{1}{2}u^2},\quad k_2(u)=\operatorname{sech}u,\quad k_3(u)=\exp\left\{-\tfrac{1}{2}u^2\right\},\quad k_4(u)=\cos u \tag{2}$$

56

Aharonov's reasoning (he suggested Eq.(1) with k_4) was that when δ is small the second exponential would act like a 'complex delta-function' and so project out the value of the first exponential at $u=\mathrm{i}A$. Thus f should vary as

$$f \approx \exp\{\mathrm{i}Kx\} \quad \text{where } K = k(\mathrm{i}A) \tag{3}$$

Under the conditions above Eq.(2), k increases from $u=0$ along the imaginary axis, so that $K>1$, (and for the given examples can be arbitrarily large), and so corresponds to superoscillations. What follows is a study of the small-δ asymptotics of the integral representing f. As well as justifying Aharonov's argument, this will dissolve the paradox posed by superoscillations, by showing that when $x>\mathrm{O}(1/\delta^2)$ they get replaced by the expected $\cos(x)$, and f gets exponentially large.

2. Asymptotics

The aim is to get an asymptotic approximation for small δ to the integral defining f, Eq.(1), which is valid uniformly in x. To achieve this, it is convenient to define

$$\xi \equiv x\delta^2 \tag{4}$$

so that Eq.(1) can be written

$$f\left(\xi/\delta^2, A, \delta\right) = \frac{1}{\delta\sqrt{2\pi}} \int_{-\infty}^{\infty} \mathrm{d}u \, \exp\left\{-\frac{1}{\delta^2}\Phi(u,\xi,A)\right\} \quad \text{where } \Phi \equiv \tfrac{1}{2}(u-\mathrm{i}A)^2 - \mathrm{i}\xi\, k(u) \tag{5}$$

For small δ, f can now be approximated by the saddle-point method, that is by deforming the path of integration through saddles u_s of the exponent and replacing Φ by its quadratic approximation near u_s. f is dominated by the saddle with smallest $\mathrm{Re}\Phi$. Saddles, whose location depends on ξ (and also A) are defined by

$$\frac{\mathrm{d}\Phi}{\mathrm{d}u} = 0, \quad \text{i.e. } u_\mathrm{s} = \mathrm{i}\left[\xi k'(u_\mathrm{s}) + A\right] \tag{6}$$

Application of the saddle-point method now gives the main result:

$$f \approx \frac{\exp\left\{\mathrm{i}x\,k(u_\mathrm{s}) - \frac{1}{2\delta^2}(u_\mathrm{s} - \mathrm{i}A)^2\right\}}{\sqrt{1 - \mathrm{i}x\delta^2 k''(u_\mathrm{s})}} \tag{7}$$

To interpret this formula, it is necessary to understand the behaviour of the dominant saddle as ξ varies.

When $\xi<<1$, that is $x<<\delta^{-2}$, Eq.(6) gives $u_s \approx iA$, and (7) reduces to Eq.(3); this is the regime of superoscillations. When $\xi>>1$, that is $x>>\delta^{-2}$, the saddles are the zeros of $k'(u)$; assuming for simplicity that k has a single maximum at u=0 (as in the first three functions in Eq.(2)), this is the only real saddle, and (7) reduces to

$$f \approx \frac{1}{\delta\sqrt{x|k''(0)|}}\exp\left\{ix-\tfrac{1}{4}\pi\right\}\exp\left\{\frac{A^2}{2\delta^2}\right\} \tag{8}$$

This is the behaviour to be expected conventionally, that is on the basis of the frequency content of f; in the infinite range of validity of Eq.(8), f is O(exp$\{A^2/2\delta^2\}$ and so exponentially amplified relative to the superoscillation regime.

As x increases, the saddle moves from iA to 0 along a curved track, illustrated in figure 1. This is the dominant saddle u_s; its track resembles figure 1 for all $k(u)$ of this type that I have studied. There are other solutions of Eq.(6), whose arrangement and motion are complicated and depend on the details of $k(u)$, but they are not dominant and so do not compromise the validity of Eq.(7) as the leading-order approximation to the integral defining f, Eq.(1).

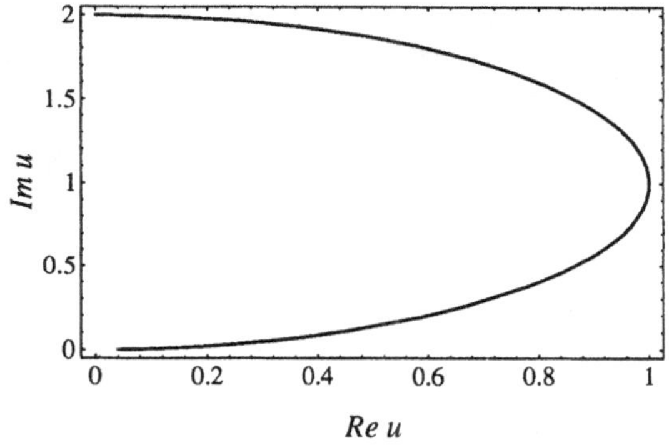

Figure 1. Track of leading saddle u_s as ξ increases from 0 to ∞, for the wavenumber function $k_5(u)$ in Eq.(10), for A=2 (the track is similar for any $k(u)$ with a single maximum)

In understanding the oscillations, it is helpful to study the local wavenumber, defined as

$$q(\xi) \equiv -\mathrm{Im}\frac{\partial \Phi\{u_s(\xi),\xi,A\}}{\partial \xi} = \mathrm{Re}\,k(u_s(\xi)) \tag{9}$$

As illustrated in figure 2, $q(\xi)$ decreases smoothly from $k(iA)$ (which is real) to 1 as ξ increases. Note that the decrease is rapid (this is true for all $k(u)$ that I have studied). This has the important implication that to observe superoscillations it is necessary to keep ξ

58

much smaller than unity, and if we want to allow x to be large, in order to observe *many* superoscillations, δ must be correspondingly smaller, Eq.(4), and the exponential amplification in the regime of conventional oscillation, Eq.(8), will be correspondingly larger.

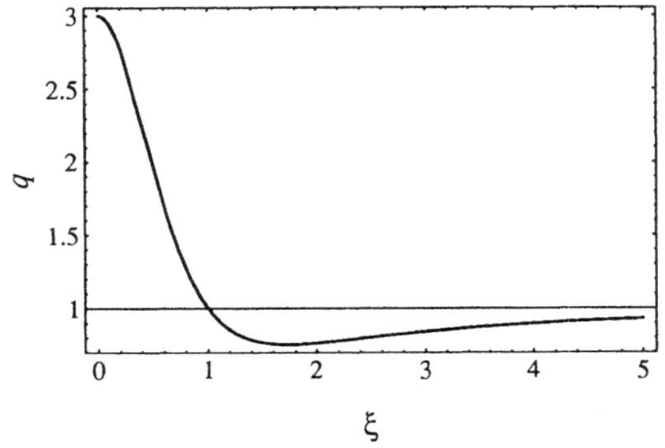

Figure 2. Local wavenumber $q(\xi)$, Eq.(9), for the $k_5(u)$ in Eq.(10), for A=2

None of the wavenumber functions in Eq.(2) gives an f whose integral representation can be evaluated exactly in terms of special functions. However, if we choose the wavenumber function

$$k_5(u) = 1 - \tfrac{1}{2}u^2 \tag{10}$$

we can ensure that it is band-limited ($|k|<1$) by restricting the range of integration in Eq.(1) to $|u|\leq 2$. The resulting truncated integral is

$$f(x,A,\delta) = \frac{1}{\delta\sqrt{2\pi}} \int_{-2}^{2} \mathrm{d}u \exp\left\{\mathrm{i}x\left(1-\tfrac{1}{2}u^2\right)\right\} \exp\left\{-\frac{1}{2\delta^2}(u-\mathrm{i}A)^2\right\} \tag{11}$$

which be expressed in terms of error functions:

$$\begin{aligned} f(x,A,\delta) = & \frac{1}{2\sqrt{1+\mathrm{i}x\delta^2}} \exp\left\{\frac{\mathrm{i}x\left(2+A^2+2\mathrm{i}x\delta^2\right)}{2\left(1+\mathrm{i}x\delta^2\right)}\right\} \times \\ & \times \left[\operatorname{erf}\left\{\frac{2+\mathrm{i}A+2\mathrm{i}x\delta^2}{\delta\sqrt{2+2\mathrm{i}x\delta^2}}\right\} + \operatorname{erf}\left\{\frac{2-\mathrm{i}A+2\mathrm{i}x\delta^2}{\delta\sqrt{2+2\mathrm{i}x\delta^2}}\right\}\right] \end{aligned} \tag{12}$$

It is instructive to examine this in detail. The superoscillation wavenumber, Eq.(3), is

$$K = k_5(\mathrm{i}A) = 1 + \tfrac{1}{2}A^2 \tag{13}$$

There is a single saddle, at (figure 1)

$$u_s(\xi) = \frac{\mathrm{i}A}{1 + \mathrm{i}\xi} \tag{14}$$

and the local wavenumber is (figure 2)

$$q(\xi) = 1 + \frac{A^2(1 - \xi^2)}{2(1 + \xi^2)^2} \tag{15}$$

For this case, the saddle-point approximation, Eq.(7) gives

$$f(x, A, \delta) \approx \frac{1}{\sqrt{1 + \mathrm{i}x\delta^2}} \exp\left\{ \mathrm{i}x \left[1 + \frac{A^2}{2(1 + x^2\delta^4)} \right] \right\} \exp\left\{ \frac{A^2\delta^2 x^2}{2(1 + x^2\delta^4)} \right\} \tag{16}$$

However, the asymptotics of (11) includes contributions from the end-points $u=\pm 2$ as well as the saddle u_S. This can be seen by realising that the steepest path between -2 and +2 runs from -2 to infinity in the negative half-plane, through u_S to infinity in the positive half-plane, and back to +2. The end-point contributions oscillate conventionally, with the wavenumber -1, so we must be sure that they do not mask the superoscillations that exist for small ξ. The condition for this is that the absolute value of the Gaussian in (11) must not exceed unity at the end-points. Thus

$$\exp\left\{ \frac{A^2 - 4}{2\delta^2} \right\} \leq 1, \qquad \text{i.e.} \qquad A \leq 2 \tag{17}$$

(we include the equality because the end-point contribution is smaller than that from the saddle by a factor δ). Eq.(13) now implies that the maximum rate of superoscillation obtainable with this model is K=3. (It is worth remarking that x=0, A=2 lies on the anti-Stokes line for the error functions in Eq.(12), that is, where the exponential contribution from the saddle exchanges dominance with those from the end-points.)

The representation Eq.(1) does not have the form of a Fourier transform, namely (for a band-limited function)

$$f(x, A, \delta) = \int_{-1}^{1} \mathrm{d}q \exp\{\mathrm{i}xq\} \bar{f}(q) \tag{18}$$

It is however easy to cast it into this form. The transform $\bar{f}(q)$ depends on the inverse function of $k(u)$; this is multivalued, and the path of integration can be deformed into a loop around a cut extending along the real axis negatively from the branch point at $q=1$ (the ends of the loop are pinned to the cut, at $q=-1$ for k_5 and at the essential singularity $q=0$ for k_1, k_2, and k_3). Again there is a dominant saddle, which for small ξ lies at $q=K$, and the loop can be expanded to pass through this. All previous results can be reproduced in this way.

3. Numerics

The aim here is twofold: to compare the saddle-point approximation Eq.(7) with the exact integral (1), and to exhibit the superoscillations. I carried out computations of f for the wavenumber functions k_1, k_2, and k_3 (Eq.(2)), but will display results only for $\mathrm{Re}\, f$ ($\mathrm{Im}\, f$ is similar) for k_5 (Eq.(10)), with the truncated integral of Eq.(11), for which the results are very similar. The computations will be exhibited for the fastest superoscillations, namely $K=3$, that is $A=2$ (Eq.(17)), choosing $\delta=0.2$.

Figure 3 shows the results. The superoscillations for small x, with period $2\pi/3$, are shown on figure 3a, and figure 3b shows a range of x where there are conventional oscillations, with period more than 3 times greater (actually about 8.4 - cf. figure 2, where $\xi \sim 1.6$ corresponds to $x \sim 40$). In both cases, the approximation (in this case Eq.(16)) agrees well with the exact expression, Eq.(12). For example, the fractional error is 0.18 for $x=2$, and 2.8×10^{-18} for $x=42$. Note the enormous ratio of the sizes of f for large and small x; from Eq.(16), this can be estimated as $\exp(36)\sim10^{16}$ (the asymptotic ratio of Eq.(8) is not attained in figure 3b). The transition between the superoscillation and conventional regimes is clearly shown in figure 3c.

In these computations, the value $A=2$ is the largest for which the saddle dominates the end-points. The competition between contributions shows up most clearly at $x=0$, for which (12) gives

$$f(0,A,\delta) = \mathrm{Re}\,\mathrm{erf}\left\{\frac{1}{\delta}\left(\sqrt{2}+\mathrm{i}\frac{A}{\sqrt{2}}\right)\right\} \tag{19}$$

For $A<2$, f is well approximated by the saddle contribution of unity, for $A>2$, the end-points dominate and f increases exponentially, Eq.(17), masking the superoscillations for small x. This is illustrated in figure 4. Even at the critical value $A=2$, that is, on the anti-Stokes line for the function (19), the exact value $f=0.945$ is close to the saddle-point value $f\sim1$.

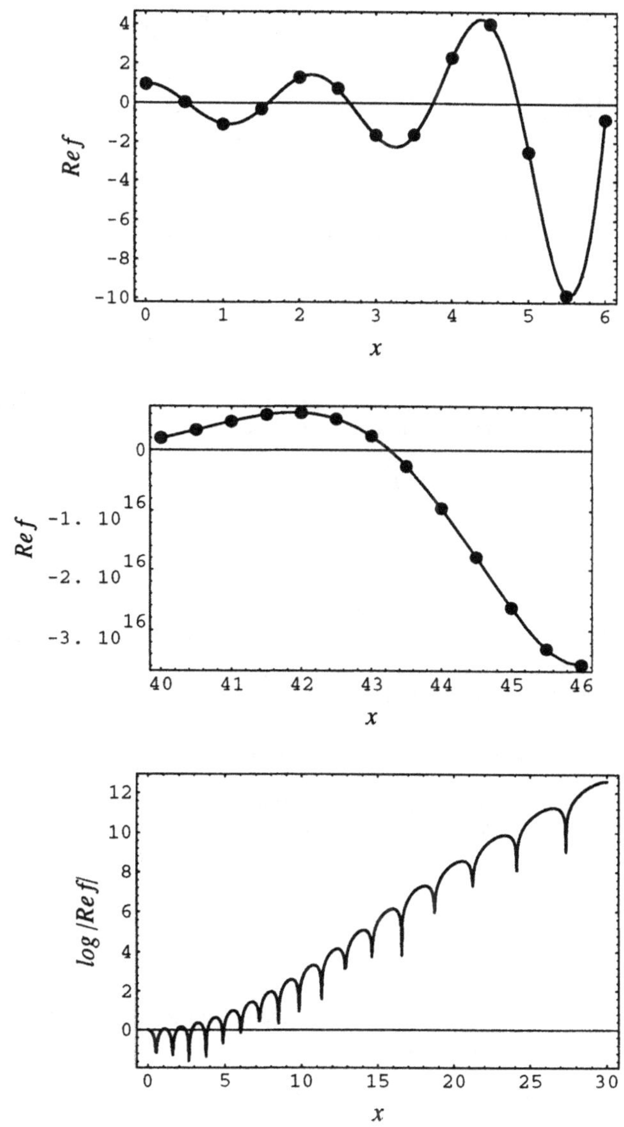

(3a)

(3b)

(3c)

Figure 3. Computations of $f(x,2,0.2)$ for the truncated integral, Eq.(11), showing (a), superoscillations, and (b) conventional oscillations. Circles: exact expression, Eq.(12); full lines: saddle-point approximation, Eq.(16). In (c) the logarithms are base 10

62

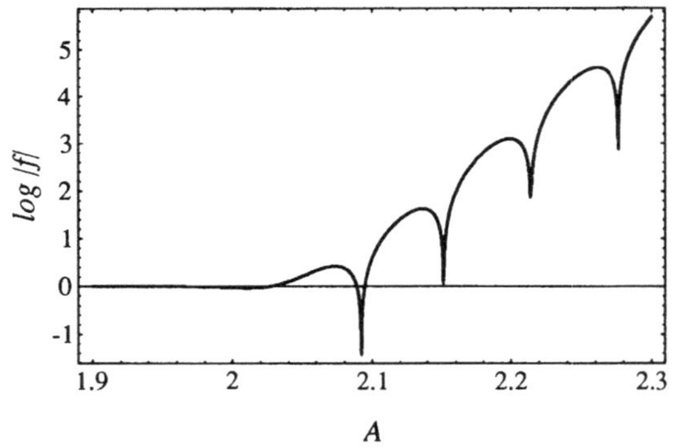

Figure 4. Computations of log $|f(0, A, 0.2)|$, Eq.(19), for the truncated integral Eq.(11); logarithms are base 10. Note the exponential growth after crossing the anti-Stokes line at A=2

4. Beethoven at 1Hz

Professor I. Daubechies has informed me that superoscillations are known in signal processing, in the context of oversampling. This is sampling a function faster than the Nyquist rate, i.e. at points $x=n\pi$ where the function is band-limited by $|k|\leq 1$. If a function is oversampled in a finite range, extrapolation outside this range is exponentially unstable[2]. She quotes B. Logan as saying that it is possible in principle to design a bandlimited signal with a bandwidth of 1Hz that would reproduce Beethoven's ninth symphony exactly. With the superoscillatory functions described in this paper it is possible to give an explicit recipe for constructing this signal, as I now explain.

We require superoscillations for the duration T (~4000s) of the symphony. Therefore the desired signal $B(t)$ can be represented as periodic outside this interval, namely

$$B(t) = \sum_{-N}^{N} B_n \exp\left\{ i \frac{2\pi n t}{T} \right\} \tag{20}$$

Here N is the order of the Fourier component corresponding to the highest frequency $\nu_{max} \equiv N/T$ (~20kHz) it is desired to reproduce.

To approximate this with a signal band-limited by frequency ν_0 (=1Hz) we make the replacement

$$\exp\left\{ i \frac{2\pi n t}{T} \right\} \to \Phi_n(t) \tag{21}$$

where (cf.Eq.(1)) Φ_n is the superoscillatory function

$$\Phi_n(t) \equiv \frac{1}{\delta_n\sqrt{2\pi}} \int_{-\infty}^{\infty} du \exp\{i2\pi t\nu(u)\}\exp\left\{-\frac{1}{2\delta_n^2}(u - iA_n)^2\right\} \tag{22}$$

Here the frequency function $\nu(u)$ never exceeds (for real u) its band-limited value $\nu(0) \equiv \nu_0$, and A_n and δ_n will now be determined by the requirement that Φ_n superoscillates with frequency n/T for time T.

The superoscillation frequency of $\Phi_n(t)$ is $\nu(iA_n)$ (cf. Eq.(3)). Thus from Eq.(21) A_n must satisfy

$$\nu(iA_n) = \frac{n}{T} \tag{23}$$

We fix δ_n by requiring that the superoscillations are maintained for time T, in the sense that the replacement of Eq.(21) remains a good approximation. For this we require the next correction to the superoscillatory exponential that $\Phi_n(t)$ represents. Expanding the saddle-point approximation to Eq.(22) (analogous to Eq.(7)) for small t, we find

$$\Phi_n(t) \approx \exp\left\{i\frac{2\pi nt}{T}\right\}\exp\left\{2\pi^2\delta_n^2\left[-\nu'^2(iA_n)\right]t^2\right\} \tag{24}$$

The second factor is an increasing exponential, because $\nu'(iA_n)$ is imaginary, and must remain close to unity for $0<t<T$. Thus

$$\delta_n << \left[2\pi|\nu'(iA_n)|T\right]^{-1} \tag{25}$$

Choosing A_n and δ_n as in Eqs.(23) and (25) guarantees that the signal $B_n(t)$, with its frequencies up to ν_{max}, will be imitated for time T. When $t>T$ the imitation will grow rapidly in strength, and eventually, that is when it is oscillating at the frequency ν_0 corresponding to its Fourier content, it will acquire an amplification factor corresponding to its largest Fourier component $n=N$. An argument analogous to that leading to Eq.(8) gives this factor as

$$F = \exp\left\{\frac{A_N^2}{2\delta_N^2}\right\} >> \exp\left\{A_N^2\pi^2T^2|\nu_N'(iA_N)|^2\right\} \tag{26}$$

with A_N determined by Eq.(23) with the right-hand side set equal to ν_{max}.

Let us calculate this amplification for the model frequency function

$$\nu(u) = \nu_0 \exp\{-u^2\} \tag{27}$$

(cf. $k_3(u)$ in Eq.(2)). We find

$$A_N^2 = \log\left\{\frac{\nu_{\max}}{\nu_0}\right\} \tag{28}$$

and hence, from Eq.(26),

$$F >> \exp\left\{4\pi^2 \log^2\left(\frac{\nu_{\max}}{\nu_0}\right)\nu_{\max}^2 T^2\right\} \tag{29}$$

For Beethoven's ninth symphony this gives

$$F >> \exp\left\{10^{19}\right\} \tag{30}$$

This amplification will not be achieved until a time t_F, which can be estimated by the argument preceding Eq.(8) as

$$t_F \sim \left[\nu_0 \delta_N^2\right]^{-1} \sim \frac{\nu_{\max}^2 T^2}{\nu_0} \sim 10^8 \text{years} \tag{31}$$

Other choices for $\nu(u)$ give similar expressions and numerical estimates.

The estimate of Eq.(30) indicates that to reproduce music as superoscillations requires a signal with so much energy as to be hopelessly impractable, but more modest bandwidth compression might be feasible.

5. Concluding remarks

Aharonov's discovery, elaborated here, could have applications in several branches of physics. One possibility is the use of superoscillations for bandwidth compression as discussed in §4. Another example, also in signal processing, concerns the observation of oscillations faster than those expected on the basis of applied or inferred filters. These would conventionally be interpreted as high frequencies leaking through imperfect filters, but the arguments presented here show that the phenomenon could have a quite different origin, namely superoscillations compatible with perfect filtering.

Perhaps more interesting are the possible applications of superoscillatory functions of two variables, representing images. One envisages new forms of microscopy, in which structures much smaller than the wavelength λ would be resolved by representing them as superoscillations. (This is different from conventional superresolution, which is based on the fact that Fourier components larger than $2\pi/\lambda$ can be present in the field near the surface of an object, but decay exponentially away from the object because the wavenumber in the perpendicular direction is imaginary. With superoscillations, the larger Fourier components are not present.)

Superoscillations can probably exist in random functions $f(x)$: arbitrarily long intervals, in which f is exponentially small relative to elsewhere, could superoscillate. Consider how this might be achieved. If f is Gauss-distributed, its statistics are completely described by its autocorrelation function, which by the Wiener-Khinchin theorem is the Fourier transform of the power spectrum $S(q)$ of f. Even if f is band-limited, it ought to be possible to choose $S(q)$ with analytic structure (saddles with Re $q > 1$, etc.) such that the autocorrelation superoscillates as it falls from its initial value. This idea is worth pursuing.

On the purely mathematical side, it is clear that superoscillations carry a price: the function is exponentially smaller than in the regime of conventional oscillations, with the exponent increasing with the size of the interval of superoscillations. We have seen examples of this, but there ought to be a general theorem (perhaps based on a version of the uncertainty principle).

6. Acknowledgments

I thank Professors Jeeva Anandan and John Safko for arranging, and generously supporting my participation in, Yakir Aharonov's birthday meeting, and inviting me to write this paper, and Professor Ingrid Daubechies for suggestions leading to the calculation of §4.

7. Reference

1. H.J. Landau, IEEE Trans. Inf. Theory. **IT-32** (1986) 464
2. Y. Aharonov, J. Anandan, S. Popescu and L. Vaidman, *Phys.Rev.Lett.* **64** (1990) 2965.

IOP PUBLISHING JOURNAL OF PHYSICS A: MATHEMATICAL AND THEORETICAL

J. Phys. A: Math. Theor. **43** (2010) 354024 (9pp) doi:10.1088/1751-8113/43/35/354024

Typical weak and superweak values

M V Berry[1] **and P Shukla**[2]

[1] H H Wills Physics Laboratory, Tyndall Avenue, Bristol BS8 1TL, UK
[2] Department of Physics, Indian Institute of Technology, Kharagpur, India

Received 1 February 2010, in final form 11 March 2010
Published 12 August 2010
Online at stacks.iop.org/JPhysA/43/354024

Abstract

Weak values, resulting from the action of an operator on a preselected state when measured after postselection by a different state, can lie outside the spectrum of eigenvalues of the operator: they can be 'superweak'. This phenomenon can be quantified by averaging over an ensemble of the two states, and calculating the probability distribution of the weak values. If there are many eigenvalues, distributed within a finite range, this distribution takes a simple universal generalized lorentzian form, and the 'superweak probablility', of weak values outside the spectrum, can be as large as $1-1/\sqrt{2} = 0.293\ldots$. By contrast, the familiar expectation values always lie within the spectral range, and their distribution, although approximately gaussian for many eigenvalues, is not universal.

PACS numbers: 02.50.Ey, 03.65.Ta, 03.67.Lx

1. Introduction

In the familiar type of quantum measurement, the expectation value of an operator $\hat{A}$ in a normalized preselected state $|\psi\rangle$ is

$$A_{\rm exp} = \langle\psi|\hat{A}|\psi\rangle. \tag{1.1}$$

If the spectrum of $\hat{A}$ is bounded, with all eigenvalues A_n lying in the range $A_{\rm min} \leqslant A_n \leqslant A_{\rm max}$, $A_{\rm exp}$ can never lie outside this range. The situation is different for the 'weak measurements' introduced by Aharonov and his colleagues [1], in which the effect of $\hat{A}$ on $|\psi\rangle$ is detected, by coupling to a system sensitive to $\hat{A}$, when a different state $|\phi\rangle$ is postselected. There, what is detected is the 'weak value'

$$A_{\rm weak} = \frac{\langle\phi|\hat{A}|\psi\rangle}{\langle\phi|\psi\rangle} \equiv A + {\rm i}A'. \tag{1.2}$$

This is usually a complex number, with the following significance [2–4].

If the initial state of the detector is represented by a real wavepacket (e.g. a minimal gaussian), the effect of the weak measurement is to shift the expectation value of $\hat{A}$ by a multiple of Re A_{weak}. In a weak measurement, Re A_{weak} can lie outside the interval $A_{\min} \leqslant A_n \leqslant A_{\max}$: it can be 'superweak'. (The expectation of the dynamical variable canonically conjugate to $\hat{A}$ is shifted by an amount proportional to Im A_{weak}, with the constant of proportionality depending on the width of the initial packet.)

Here we ask: How are weak values typically distributed, if the states $|\psi\rangle$ and $|\phi\rangle$ are randomly distributed? Are superweak values common or rare? To answer these questions, we calculate the probability distribution $P(A)$ of Re A_{weak} over an ensemble of states $|\psi\rangle$ and $|\phi\rangle$, and hence the superweak probability

$$P_{\text{super}} = \int_{-\infty}^{A_{\min}} \mathrm{d}A\, P(A) + \int_{A_{\max}}^{\infty} \mathrm{d}A\, P(A) \tag{1.3}$$

of the weak value lying outside the spectrum of $\hat{A}$.

The calculation (section 2) concerns the spectrum of $\hat{A}$ containing N eigenvalues distributed in the interval $A_{\min} \leqslant A_n \leqslant A_{\max}$, when $N \gg 1$, that is for high-dimensional Hilbert spaces. The result is that the distribution $P(A)$, and therefore the superweak probability, is universal, in the sense of being independent of N for $N \gg 1$, of the statistics assumed for $|\psi\rangle$ and $|\phi\rangle$, and of whether the eigenvalues A_n are randomly distributed or regularly (rigidly) arranged. The distribution is different if the states $|\psi\rangle$ and $|\phi\rangle$ are assumed real in the $\hat{A}$ basis, but is still universal for this class of states (of course Im $A_{\text{weak}} = 0$ for real states).

In equation (1.2), the matrix element is divided by the overlap $\langle\phi|\psi\rangle$ between the pre- and post-selected states, leading to the interpretation of A_{weak} as a quantum conditional probability [2]. This denominator will play an important part in our calculations – indeed, the very existence of superweak values depends on it.

The type of statistical calculation we perform here was pioneered by Botero [5], who obtained a different result by calculating the distribution of weak values for a different situation: when the preselected state $|\psi\rangle$ is fixed and only the postselected state $|\phi\rangle$ is random. In another anticipation, Dennis [6] calculated the probability distribution of local wavenumbers $|\boldsymbol{k}(\boldsymbol{r})|$ (gradient of phase of the wavefunction) for random monochromatic waves (the preselected states $|\psi\rangle$) in the plane, satisfying the Helmholtz equation with wavenumber k_0. $\boldsymbol{k}(\boldsymbol{r})$ is the real part of the weak value of $\hat{A}$ = momentum/$\hbar$, and the postselected state is a position eigenstate, that is $|\phi\rangle = |\boldsymbol{r}\rangle$. Superweak values of $\boldsymbol{k}(\boldsymbol{r})$ correspond to superoscillations [7, 8], where the local wave oscillates faster than the wavenumber k_0 of all the fourier components of the wave. The resulting superoscillation probability is 1/3. Later, this two-dimensional calculation was extended to waves in spaces of any dimension [9].

In section 3 we calculate the very different probability distribution of expectation values (1.1). This is not universal: it depends on N, and is different for randomly spaced and regularly arranged eigenvalues.

2. Universal weak value distributions

In terms of the spectrum of $\hat{A}$, defined by

$$\hat{A}|\alpha_n\rangle = A_n|\alpha_n\rangle, \quad 1 \leqslant n \leqslant N, \quad A_{\min} \leqslant A_n \leqslant A_{\max}, \tag{2.1}$$

we can expand the states $|\psi\rangle$ and $|\phi\rangle$:

$$|\psi\rangle = \sum_{n=1}^{N} \psi_n|\alpha_n\rangle, \qquad |\phi\rangle = \sum_{n=1}^{N} \phi_n|\alpha_n\rangle. \tag{2.2}$$

Thus the weak value (1.2) can be written

$$A = \mathrm{Re}\frac{\sum_{n=1}^{N}\phi_n^*\psi_n A_n}{\sum_{n=1}^{N}\phi_n^*\psi_n}. \tag{2.3}$$

We seek the distribution of values of A that are typical, in the sense that $|\psi\rangle$ and $|\phi\rangle$ are regarded as random states, and for $N \gg 1$. To implement this, we write

$$\phi_n^*\psi_n \equiv b_n = b_{n1} + \mathrm{i}b_{n2}, \tag{2.4}$$

and treat the numbers b_{n1} and b_{n2}, as well as the eigenvalues A_n, as independent random variables with zero mean. Thus the distribution of A, given by (2.3), is

$$\begin{aligned} P(A) &= \left\langle \delta\left(A - \mathrm{Re}\sum_{n=1}^{N} b_n A_n \Big/ \sum_{n=1}^{N} b_n\right)\right\rangle \\ &= \frac{1}{2\pi}\int_{-\infty}^{\infty}\mathrm{d}s\exp(-\mathrm{i}As)\left\langle\exp\left(\mathrm{i}s\,\mathrm{Re}\sum_{n=1}^{N} b_n A_n \Big/ \sum_{n=1}^{N} b_n\right)\right\rangle, \end{aligned} \tag{2.5}$$

in which $\langle\cdots\rangle$ denotes an average over the distribution of the random variables. The denominator in the exponential makes direct averaging awkward, so we treat

$$\sum_{n=1}^{N} b_n \equiv c = c_1 + \mathrm{i}c_2 \tag{2.6}$$

as a constraint, implemented by a double δ-function which can then be written as fourier integrals:

$$\begin{aligned} P(A) = {} & \frac{1}{(2\pi)^3}\int_{-\infty}^{\infty}\mathrm{d}s\int_{-\infty}^{\infty}\mathrm{d}c_1\int_{-\infty}^{\infty}\mathrm{d}c_2\int_{-\infty}^{\infty}\mathrm{d}t_1\int_{-\infty}^{\infty}\mathrm{d}t_2\exp\{-\mathrm{i}\,(As + c_1 t_1 + c_2 t_2)\} \\ & \times\left\langle\exp\left\{\mathrm{i}\sum_{n=1}^{N}\left[sA_n\,\mathrm{Re}\left(\frac{b_{n1}+\mathrm{i}b_{n2}}{c_1+\mathrm{i}c_2}\right) + t_1 b_{n1} + t_2 b_{n2}\right]\right\}\right\rangle. \end{aligned} \tag{2.7}$$

The central limit theorem implies that for $N \gg 1$ the exponent, denoted X, is a gaussian random variable with zero mean, because it is a sum over many random numbers, irrespective of the distribution of b_{n1} and b_{n2}. Thus we can use the relation

$$\langle\exp(\mathrm{i}X)\rangle = \exp\left(-\tfrac{1}{2}\langle X^2\rangle\right). \tag{2.8}$$

With no essential loss of generality, we can shift the spectral range to $-A_{\max} \leqslant A_n \leqslant A_{\max}$, and we will make the simplification that the distribution of the A_n is symmetric about $A_n = 0$, with variance $\langle A_n^2\rangle$, and define the common variance of b_{n1} and b_{n2}:

$$\langle b_{1n}^2\rangle = \langle b_{2n}^2\rangle \equiv B^2. \tag{2.9}$$

Thus the average in (2.7) becomes

$$\begin{aligned} & \left\langle\exp\left\{\mathrm{i}\sum_{n=1}^{N}\left[sA_n\,\mathrm{Re}\left(\frac{b_{n1}+\mathrm{i}b_{n2}}{c_1+\mathrm{i}c_2}\right) + t_1 b_{n1} + t_2 b_{n2}\right]\right\}\right\rangle \\ & \quad = \exp\left\{-\frac{1}{2}N\left[\frac{s^2B^2\langle A_n^2\rangle}{\left(c_1^2+c_2^2\right)} + B^2\left(t_1^2+t_2^2\right)\right]\right\}. \end{aligned} \tag{2.10}$$

In the integration (2.7), N and B can now be eliminated by the following scaling of the integration variables:

$$B\sqrt{N}\{t_1, t_2\} \to \{t_1, t_2\}, \qquad (B\sqrt{N})^{-1}\{c_1, c_2\} \to \{c_1, c_2\}, \tag{2.11}$$

leaving an expression free from parameters, i.e. universal. The integrations over s, t_1 and t_2 are gaussian, leaving

$$P(A) = \frac{1}{(2\pi)^{3/2}\sqrt{\langle A_n^2\rangle}} \int_{-\infty}^{\infty} \mathrm{d}c_1 \int_{-\infty}^{\infty} \mathrm{d}c_2 \sqrt{(c_1^2 + c_2^2)} \exp\left\{-\frac{1}{2}(c_1^2 + c_2^2)\left(1 + \frac{A^2}{\langle A_n^2\rangle}\right)\right\}. \quad (2.12)$$

This is also an elementary integral, giving our main result

$$P(A) = \frac{\langle A_n^2\rangle}{2\left(\langle A_n^2\rangle + A^2\right)^{3/2}}. \quad (2.13)$$

Note that this does not involve the spectrum's boundary value $A_{\max}$. The cauchy (lorentz)-like, rather than gaussian, form (2.13) has been generated by the integrations over the denominator variables c_1 and c_2, reflecting the essential role of the overlap $\langle\phi|\psi\rangle$ in the weak value formula (1.3).

For the probability of superweak values, (1.3) gives

$$P_{\text{super}} = 2\int_{A_{\max}}^{\infty} \mathrm{d}A\, P(A) = 1 - \frac{A_{\max}}{\sqrt{\langle A_n^2\rangle + A_{\max}^2}} \quad (2.14)$$

Interesting special eigenvalue distributions are

(a) uniform:

$$\langle A^2\rangle = \frac{1}{3}A_{\max}^2 \Rightarrow P_{\text{super}} = 1 - \frac{\sqrt{3}}{2} = 0.133\,98\ldots$$

(b) random-matrix (semicircle):

$$\langle A^2\rangle = \frac{1}{4}A_{\max}^2 \Rightarrow P_{\text{super}} = 1 - \frac{2}{\sqrt{5}} = 0.105\,57\ldots \quad (2.15)$$

(c) concentrated at $\pm A_{\max}$:

$$\langle A^2\rangle = A_{\max}^2 \Rightarrow P_{\text{super}} = 1 - \frac{1}{\sqrt{2}} = 0.292\,89\ldots$$

The largest superweak probability occurs for case (c), where $N/2$ eigenvalues are concentrated near $-A_{\max}$ and $N/2$ near $+A_{\max}$.

To test the formula for $P(A)$, we simulated (2.3) by sampling A_n from a uniform distribution (case (2.15*a*)), and b_{n1} and b_{n2} from gaussian distributions, over the interval $\{-1,1\}$, for different values of N. As figure 1 illustrates, the fit is excellent, even for $N = 5$. The fit is equally good if b_{n1} and b_{n2} are sampled from a uniform distribution; only the variance B^2 is changed, but this quantity has been scaled away.

The foregoing argument survives almost unchanged if the eigenvalues are regularly arranged on $\{-A_{\max}, A_{\max}\}$, rather than randomly distributed, that is (corresponding to case (2.15*a*)

$$A_n = \left(\frac{2\,(n-1)}{N-1} - 1\right)A_{\max}, \qquad \{1 \leqslant n \leqslant N\}. \quad (2.16)$$

The only changes in $P(A)$ and P_{super} arise from the variance of the eigenvalues, which is now

$$\langle A_n^2\rangle = \frac{(N+1)}{3(N-1)}A_{\max}^2. \quad (2.17)$$

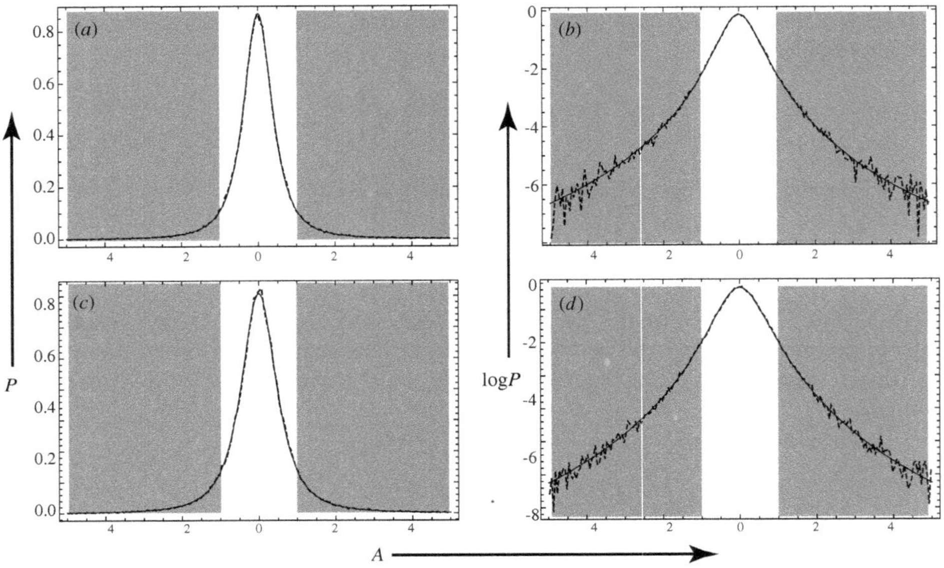

Figure 1. (*a*), (*c*) full curves: theoretical probability distribution $P(A)$ of weak values (equation (2.13)) for eigenvalues uniformly distributed (case 2.15*a*) on the range $\{-1,1\}$; dashed curves: $P(A)$ computed from (2.3) with 10^5 sample spectra with (a) 100 eigenvalues, (c) 5 eigenvalues. (b), (d) as (a), (c) for logP(A). In the shaded regions the weak values are superweak, lying outside the spectrum of $\hat{A}$.

An intriguing observation is that for $A' = \text{Im}\, A_{\text{weak}}$ the probability distribution $P(A')$ is identical to that for $A = \text{Re}\, A_{\text{weak}}$. This is surprising at first, but is obvious from the analysis leading to (2.13), and we have confirmed it by numerical simulation.

However, the distribution of weak values is different if the eigenstates $|\alpha_n\rangle$, and the states $|\psi\rangle$ and $|\phi\rangle$, can be represented as real, for example if there is time-reversal symmetry. Then b_2 and c_2 are zero and the calculation is simpler. We do not give the details but only the formulas that replace (2.13) and (2.14) for this different universality class:

$$P_{\text{real}}(A) = \frac{\sqrt{\langle A_n^2\rangle}}{\pi\left(\langle A_n^2\rangle + A^2\right)}, \qquad P_{\text{super, real}} = 1 - \frac{2}{\pi}\tan^{-1}\frac{A_{\max}}{\sqrt{\langle A_n^2\rangle}}. \tag{2.18}$$

As in the more general complex case (illustrated in figure 1), simulation gives good agreement. Of course $A' = \text{Im}\, A_{\text{weak}} = 0$ for this case.

3. Expectation value distributions

In the eigenbasis (2.1) of $\hat{A}$, the expectation value (1.1) is

$$A_{\exp} = \frac{\sum_{n=1}^{N} |\psi_n^2| A_n}{\sum_{n=1}^{N} |\psi_n^2|}. \tag{3.1}$$

This form of writing is convenient for numerical simulations, because it is not necessary to normalize the states $|\psi\rangle$. But for $N \gg 1$ the denominator is self-averaging, so it is not necessary to include it explicitly provided the probability distribution of the coefficients $|\psi_n|^2$ is chosen to satisfy

$$\langle|\psi_n|^2\rangle = \frac{1}{N}. \tag{3.2}$$

Thus the probability distribution of the expectation values is

$$\begin{aligned} P(A_{\rm exp}) &= \left\langle \delta\left(A_{\rm exp} - \sum_{n=1}^{N} |\psi_n^2| A_n\right)\right\rangle \\ &= \frac{1}{2\pi}\int_{-\infty}^{\infty} {\rm d}s \exp(-{\rm i}sA_{\rm exp})\left\langle \exp\left({\rm i}s\sum_{n=1}^{N}|\psi_n^2|A_n\right)\right\rangle. \end{aligned} \tag{3.3}$$

By the central limit theorem, the sum in the exponential is a gaussian variable, so, using (2.8),

$$\begin{aligned} P(A_{\rm exp}) &= \frac{1}{2\pi}\int_{-\infty}^{\infty} {\rm d}s \exp(-{\rm i}sA_{\rm exp}) \exp\left\{-\frac{1}{2}s^2\left\langle\left(\sum_{n=1}^{N}|\psi_n^2|A_n\right)^2\right\rangle\right\} \\ &= \frac{1}{\sqrt{2\pi\left\langle\left(\sum_{n=1}^{N}|\psi_n^2|A_n\right)^2\right\rangle}} \exp\left\{-\frac{1}{2}A_{\rm exp}^2 \bigg/ \left\langle\left(\sum_{n=1}^{N}|\psi_n^2|A_n\right)^2\right\rangle\right\}. \end{aligned} \tag{3.4}$$

The average is

$$\left\langle\left(\sum_{n=1}^{N}|\psi_n^2|A_n\right)^2\right\rangle = N\langle A_n^2\rangle\langle|\psi_n|^4\rangle. \tag{3.5}$$

For the coefficients $|\psi_n|^2$, the simplest distribution follows from choosing gauss-distributed ${\rm Re}\,\psi_n$ and ${\rm Im}\,\psi_n$:

$$P(|\psi_n|^2) = \tfrac{1}{2}\exp\left(-\tfrac{1}{2}|\psi_n|^2\right). \tag{3.6}$$

Thus in (3.5)

$$\langle|\psi_n|^4\rangle = \frac{2}{N^2}, \tag{3.7}$$

and (3.4) gives

$$P(A_{\rm exp}) = \frac{1}{2}\sqrt{\frac{N}{\pi\langle A_n^2\rangle}}\exp\left(-\frac{1}{4}N\frac{A_{\rm exp}^2}{\langle A_n^2\rangle}\right). \tag{3.8}$$

Figures 2(*b*)–(*d*), again for the uniform distribution (2.15*a*), show that this is an excellent approximation. It does not vanish outside the interval $-A_{\rm max} \leqslant A_{\rm exp} \leqslant A_{\rm max}$ as the exact $P(A_{\rm exp})$ must, but for $N \gg 1$ the outside contributions are minuscule. Unlike the distribution (2.13) of weak values, the formula (3.8) is not universal: $P(A_{\rm exp})$ gets narrower as N increases, and the width depends on the distribution of the coefficients $|\psi_n|^2$.

Although the large N approximation (3.8) is good even for rather small N, as figures 2(*b*), (*c*) illustrate, it does fail for $N = 2$. Then $P(A_{\rm exp})$ can be evaluated exactly for the uniform distribution (2.15a), as shown in the appendix and illustrated in figure 2(*a*):

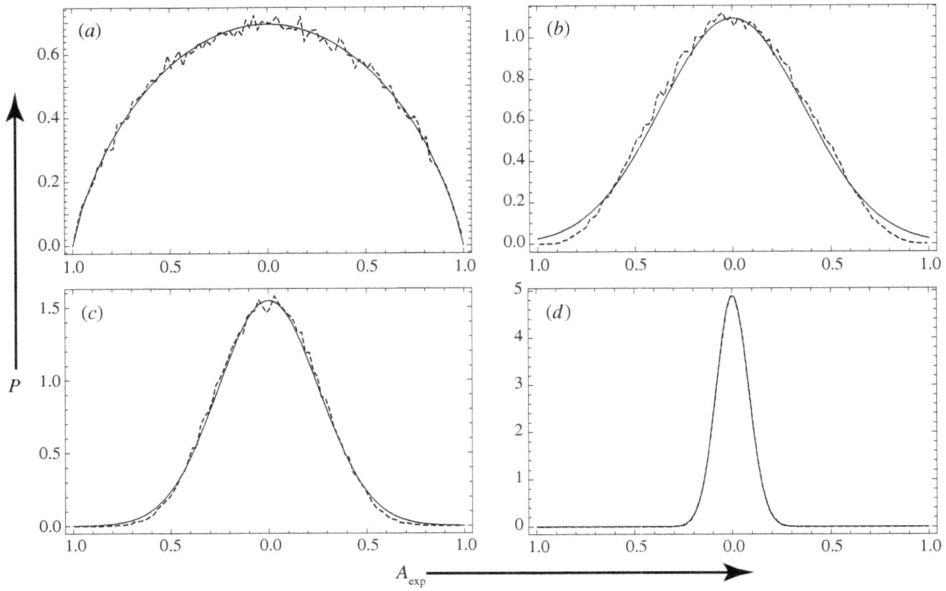

Figure 2. Probability distributions $P(A_{\text{exp}})$ of expectation values for eigenvalues uniformly distributed (case 2.15a) on the range $\{-1,1\}$, for (*a*) $N = 2$, (*b*) $N = 5$, (*c*) N = 10, (*d*) $N = 100$. Full curves: theoretical distributions, (3.8) for (*b*)–(*d*), and (3.9) for (*a*); dashed curves: $P(A_{\text{exp}})$ computed from (3.1) with 10^5 sample spectra.

$$P(A_{\text{exp}}) = \frac{1}{A_{\text{max}}}\left[\log 2 - \frac{1}{2}\left(\left(1 - \frac{A_{\text{exp}}}{A_{\text{max}}}\right)\log\left(1 - \frac{A_{\text{exp}}}{A_{\text{max}}}\right) + \left(1 + \frac{A_{\text{exp}}}{A_{\text{max}}}\right)\log\left(1 + \frac{A_{\text{exp}}}{A_{\text{max}}}\right)\right)\right]. \tag{3.9}$$

In section 2 we found that if the eigenvalues are regularly arranged, as in (2.15), the distribution of weak values is slightly modified (cf (2.17). For $P(A_{\text{exp}})$ the effect is more significant. Instead of (3.5)–(3.7), the relevant average is

$$\begin{aligned}\left\langle\left(\sum_{n=1}^{N}|\psi_n^2|A_n\right)^2\right\rangle &= \sum_{m=1}^{N}\sum_{n=1}^{N}A_mA_n(\delta_{mn}\langle|\psi_n|^4\rangle + (1-\delta_{nm})\langle|\psi_m|^2\rangle\langle|\psi_n|^2\rangle)\\ &= \frac{2}{3N}\left(\frac{1+1/N}{1-1/N}\right)A_{\text{max}}^2 - \frac{1}{3N}\left(\frac{1+1/N}{1-1/N}\right)A_{\text{max}}^2\\ &= \frac{1}{3N}\left(\frac{1+1/N}{1-1/N}\right)A_{\text{max}}^2,\end{aligned} \tag{3.10}$$

so instead of (3.8) the distribution is

$$P(A_{\text{exp}}) = \frac{1}{A_{\text{max}}}\sqrt{\frac{3N(1-1/N)}{2\pi(1+1/N)}}\exp\left(-\frac{3}{2}N\left(\frac{1-1/N}{1+1/N}\right)\frac{A_{\text{exp}}^2}{A_{\text{max}}^2}\right). \tag{3.11}$$

This shows that even for $N \gg 1$ the width of the distribution is smaller by $\sqrt{2}$ for regularly arranged eigenvalues than for randomly distributed ones.

Finally, for $N = 2$, and eigenvalues $A_1 = -A_{\max}$, $A_2 = +A_{\max}$, the argument in appendix shows that the expectation values are uniformly distributed:

$$P(A_{\exp}) = \frac{1}{2A_{\max}}\Theta(A_{\max} - |A_{\exp}|), \tag{3.12}$$

in which Θ denotes the unit step. This particularly simple form is a consequence of the choice of the distribution (3.6) for $|\psi_n|^2$; for other distributions, $P(A_{\exp})$ is not uniform.

4. Concluding remarks

The main results (2.13) and (2.14) indicate an unanticipated universality in the distribution of weak and superweak values. Superweak values – results of weak measurements outside the spectral range of the operator being measured – are quite common: in the rather general situation considered here, there can be an almost 30% chance (for the case (2.15*c*)) of a weak value being superweak. We have considered two universality classes: the generic case, in which the overlap of the pre- and post-selected states with the eigenstates of the operator being measured are complex numbers, and the case of time-reversal symmetry, in which the overlaps are real. For the more familiar expectation values, the distribution is not universal; it depends on the number of states in the spectrum, and the probability distribution of the coefficients.

The weak value probability distribution (2.13) is similar to, but in general distinct from, those found recently for superoscillations in monochromatic waves in two [6] and D [9] dimensions. In these studies, a single eigenvalue was considered (the wavenumber k_0) which was degenerate (corresponding to the different directions of plane waves, i.e. momentum eigenstates). By contrast, we have considered discrete nondegenerate eigenvalues distributed over a finite range. The two cases coincide when $D = 1$ for superoscillations (section 3 of [9]) and our eigenvalue distribution is concentrated at the extremes (case (2.15*c*)).

Extensions of our analysis can be envisaged. For example, the spectrum of eigenvalues could be continuous, or the spectral range could be infinite (in which case there would be no superweak values); or the eigenstates could be degenerate. The ensemble of pre- and post-selected states that we have considered is equivalent to a density matrix, which in some applications could represent a thermal ensemble.

Acknowledgments

We are grateful for the hospitality of the School of Physical Sciences, Jawaharlal University, New Delhi, where this work was done. MVB thanks Dr M Dennis and Professor S Popescu for helpful conversations. MVB's research is supported by the Leverhulme Trust.

Appendix

This is the derivation of (3.9) and (3.12). With the notation

$$|\psi_1|^2 \equiv x, \qquad |\psi_2|^2 \equiv y, \tag{A.1}$$

the probability distribution of expectation values when there are just two eigenvalues, uniformly distributed between $-A_{\max}$ and $+A_{\max}$, is

$$\begin{aligned} P(A_{\exp}) &= \int_{-\infty}^{\infty} \mathrm{d}x\, P(x) \int_{-\infty}^{\infty} \mathrm{d}y\, P(y) \left\langle \delta\left(A_{\exp} - \frac{A_1 x + A_2 y}{x+y}\right)\right\rangle \\ &= \int_{-\infty}^{\infty} \mathrm{d}x\, x P(x) \left\langle P\left(x \frac{(A_{\exp} - A_1)}{(A_2 - A_{\exp})}\right) \frac{(A_2 - A_1)}{(A_2 - A_{\exp})^2}\right\rangle, \end{aligned} \tag{A.2}$$

in whch the average is

$$\langle\cdots\rangle = \frac{1}{2A_{\max}}\int_{-A_{\max}}^{A_{\max}} \mathrm{d}A_1 \int_{-A_{\max}}^{A_{\max}} \mathrm{d}A_2 \cdots \tag{A.3}$$

With the distribution (3.6) for $|\psi_n|^2$, the integral over x gives

$$P(A_{\exp}) = \left\langle \frac{1}{|A_2 - A_1|}\Theta\left(1 - \left|\frac{2A_{\exp} - A_1 - A_2}{A_1 - A_2}\right|\right)\right\rangle. \tag{A.4}$$

Evaluating the average (A.3) leads to (3.9). If the eigenvalues are fixed at $A_1 = -A_{\max}$, $A_2 = +A_{\max}$, there is no need to average, and (A.4) becomes (3.12).

References

[1] Aharonov Y and Rohrlich D 2005 *Quantum Paradoxes: Quantum Theory for the Perplexed* (Weinheim: Wiley)

[2] Steinberg A M 1995 Conditional probabilities in quantum theory, and the tunneling time controversy *Phys. Rev.* A **52** 32–42

[3] Jozsa R 2007 Complex weak values in quantum measurement *Phys. Rev.* A **76** 044103

[4] Lobo A C and Ribeiro C A 2009 Weak values and the quantum phase space *Phys. Rev.* A **80** 012112

[5] Botero A 1999 Sampling weak values: a non-linear Bayesian model for non-ideal quantum measurements *PhD Thesis* Physics Texas, Austin http://arxiv.org/abs/quant-ph/0306082v1

[6] Dennis M R, Hamilton A C and Courtial J 2008 Superoscillation in speckle patterns *Opt. Lett.* **33** 2976–8

[7] Berry MV 1994 *Faster than Fourier in Quantum Coherence and Reality; in Celebration of the 60th Birthday of Yakir Aharonov* ed J S Anandan and J L Safko (Singapore: World Scientific) pp 55–65

[8] Berry M V and Popescu S 2006 Evolution of quantum superoscillations, and optical superresolution without evanescent waves *J. Phys.* A **39** 6965–77

[9] Berry M V and Dennis M R 2009 Natural superoscillations in monochromatic waves in D dimensions *J. Phys. A: Math. Theor.* **42** 022003

IOP PUBLISHING JOURNAL OF PHYSICS A: MATHEMATICAL AND THEORETICAL

J. Phys. A: Math. Theor. **45** (2012) 015301 (14pp) doi:10.1088/1751-8113/45/1/015301

Pointer supershifts and superoscillations in weak measurements

M V Berry [1,3] **and Pragya Shukla** [2]

[1] H H Wills Physics Laboratory, Tyndall Avenue, Bristol BS8 1TL, UK
[2] Department of Physics, Indian Institute of Technology, Kharagpur, India

E-mail: asymptotico@physics.bristol.ac.uk

Received 13 September 2011
Published 29 November 2011
Online at stacks.iop.org/JPhysA/45/015301

Abstract
The association between large shifts of a pointer in a weak measurement and fast oscillations in an associated function involving the pre- and post-selected states has been clarified in a recent paper (Aharonov *et al* 2011 *J. Phys. A: Math. Theor.* **44** 365304). Here we explore the association further for the case of an observable with N discrete eigenvalues, by calculating and illustrating how the supershift emerges, even for $N = 2$, as the uncertainty in the pointer position increases. This happens if the initial pointer wavefunction is Gaussian or Lorentzian but not if it is exponential.

PACS numbers: 02.50.Ey, 03.65.Ta, 03.67.Lx

1. Introduction and formulation

Quantum weak measurements rely on what appears to be a mathematical miracle. In the now-standard scenario [1–3], the value of a bounded operator $\hat{A}$ is measured not as the expectation value in a pre-selected state $|i\rangle$ but also after post-selection by a state $|f\rangle$. If the result of the measurement is registered by the shift of a pointer whose position variable is q, then this shift can correspond to values of $\hat{A}$ far outside its spectrum. Our purpose here is look in detail at the shift of the pointer, complementing and illustrating a recent mathematical study [4].

If the initial state of the pointer is $|\phi\rangle$, then the initial state of the pointer+system is $|i\rangle|\phi\rangle$. If the measurement is a brief impulsive coupling to the pointer momentum $\hat{p}$, with strength λ, and we project the state of the system onto $|f\rangle$, representing post-selection, then the final state of the pointer alone is [1]

$$|\psi\rangle = \langle f|\exp(-\mathrm{i}\lambda\hat{A}\hat{p})|i\rangle|\phi\rangle. \tag{1.1}$$

Let the eigenvalues and eigenfunctions of the bounded operator $\hat{A}$ be given by

$$\hat{A}|n\rangle = A_n|n\rangle, \qquad A_{\min} \leqslant A_n \leqslant A_{\max}. \tag{1.2}$$

[3] Author to whom any correspondence should be addressed.

Then, in position representation, the pointer wavefunction after the measurement is

$$\psi(q) = \langle f|\exp(-i\lambda\hat{A}(-i\hbar\partial_q))|i\rangle\phi(q)$$
$$= \sum_{n=-N}^{N} \langle f \mid n\rangle\langle n \mid i\rangle\phi(q-\lambda\hbar A_n). \quad (1.3)$$

Since the final wavefunction $\psi(q)$ is a superposition of shifted copies of its initial state, with each copy representing one of the eigenvalues A_n, it would seem impossible for the net shift to lie outside the spectrum of $\hat{A}$, that is, outside the range in (1.2). Nevertheless, this mathematical miracle can happen: if the initial wavefunction $\phi(q)$ is broad enough, then for certain pairs $|i\rangle$, $|f\rangle$ the series (1.3) can, by coherent interference, give rise to a reproduction of itself centred far from any of the copies, albeit greatly reduced in strength (i.e. 'weak'). We call this phenomenon 'supershift' of the pointer.

In these circumstances the shift has been identified [2] as (λ times) the 'weak value' of $\hat{A}$, given by

$$A_w = \frac{\langle f|\hat{A}|i\rangle}{\langle f \mid i\rangle} = \frac{\sum_{-N}^{N}\langle f \mid n\rangle A_n\langle n \mid i\rangle}{\sum_{-N}^{N}\langle f \mid n\rangle\langle n \mid i\rangle}. \quad (1.4)$$

(In general A_w is complex, and this has interesting implications [5, 6], but for present purposes it suffices to consider only real weak values.)

If the pre- and post-selected states are nearly orthogonal, that is if their overlap $\langle f \mid i\rangle \ll 1$, this can be arbitrarily large. In particular, A_w can lie outside the spectrum $A_{\min} \leqslant A \leqslant A_{\max}$: it can be 'superweak'. (In the opposite case, where $|i\rangle$ and $|f\rangle$ are the same, A_w is just the ordinary expectation value, not superweak, and the phenomena to be discussed here do not occur.) Several statistical studies [6–10], of different degrees of generality and in which $\hat{A}$ could represent the spins or momenta of quantum particles, have revealed that superweak values are surprisingly common in ensembles of states $|i\rangle$ and $|f\rangle$.

It is important to distinguish between superweak values of an observable and their manifestation as supershifts. From (1.4), A_w depends only on the operator being measured and the pre- and post-selected states, while from (1.3) the supershift also depends on the initial state of the pointer. Not all pointer states will produce a supershift proportional to A_w, and our focus here is on the conditions under which they do—that is, the weak value is superweak and the pointer is correspondingly supershifted.

In section 2 we recapitulate the known relation [4] between these shifts beyond the spectrum—the pointer supershifts—and the phenomenon of superoscillation, in which functions can vary arbitrarily faster than any of their Fourier components [11–13]. Section 3 considers the pointer shifts associated with a well-studied superoscillatory function, leading in section 4 to the conditions for supershifts if the initial pointer state is Gaussian. Pictures illustrate how the superposition (1.3) turns into a supershifted copy of $\phi(q)$ as its width Δ increases. The results, embodied in figures 3 and 4 to follow, spectacularly illustrate and vindicate the original insight [2, 4] underlying weak measurement and the prediction of supershifts.

The supershift is a kind of 'resurrection from the dead' involving the tail of $\phi(q)$; it may or may not happen, depending on the analytic form of the tail and on the pair $|i\rangle$, $|f\rangle$, and section 5 illustrates this with two non-Gaussian initial pointer states. As shown in section 6, supershifting can occur even when $\hat{A}$ has only two eigenvalues.

To expose the problem in its essentials, it suffices to consider a spectrum of $(2N+1)$ equally spaced eigenvalues, that is,

$$A_n = An, \qquad -N \leqslant n \leqslant N \quad (1.5)$$

(this could represent an integer spin component, for example). Then, with the scaling

$$q \to \lambda\hbar A q, \qquad \phi(q - \lambda\hbar A_n) \to \phi(q - n), \tag{1.6}$$

and defining the coefficients

$$C_n = \langle f \mid n\rangle\langle n \mid i\rangle, \tag{1.7}$$

the final pointer wavefunction is

$$\psi(q) = \sum_{n=-N}^{N} C_n \phi(q - n) \tag{1.8}$$

and the weak value of $\hat{A}$ is

$$A_w = \frac{\sum_{-N}^{N} nC_n}{\sum_{-N}^{N} C_n}. \tag{1.9}$$

We note that the set of coefficients C_n is compatible with infinitely many different pre- and post-selected states $|i\rangle$ and $|f\rangle$, for example the following choice, for any set of phases α_n:

$$\begin{aligned} |i\rangle &= \sum_{-N}^{N} \exp(\mathrm{i}\,(\alpha_n + \arg C_n))\sqrt{|C_n|}|n\rangle, \\ |f\rangle &= \sum_{-N}^{N} \exp(\mathrm{i}\,\alpha_n)\sqrt{|C_n|}|n\rangle. \end{aligned} \tag{1.10}$$

For this class of choices, there is a normalization restriction:

$$\langle i \mid i\rangle = \langle f \mid f\rangle = 1 = \sum_{-N}^{N} |C_n|. \tag{1.11}$$

Whatever the choice, (1.7) implies

$$\sum_{-N}^{N} C_n = \langle f \mid i\rangle. \tag{1.12}$$

If $|i\rangle$ and $|f\rangle$ are nearly orthogonal this sum must be small, so the C_n must have different phases or signs.

An important feature of the pointer wavefunction after the weak measurement is the mean pointer position $\langle q\rangle$. We will study this in detail later, but note the following general formula, derived from (1.8) for the common situation where $\phi(-q) = \phi^*(q)$ (i.e. $\phi(q)$ is an even function if it is real):

$$\langle q\rangle = \frac{\frac{1}{2}\sum_{-N}^{N}\sum_{-N}^{N} C_m^* C_n (m+n) F\left(\frac{1}{2}(m-n)\right)}{\sum_{-N}^{N}\sum_{-N}^{N} C_m^* C_n F\left(\frac{1}{2}(m-n)\right)}, \tag{1.13}$$

where

$$F(u) = \int_{-\infty}^{\infty} \mathrm{d}q q \phi^*(q-u)\phi(q+u). \tag{1.14}$$

For the pointer to be supershifted, $\langle q\rangle$ must lie outside the spectral range $-N \leqslant n \leqslant N$. However, as we discuss later, an unambiguous measurement of the weak value requires more: the width of the final pointer wavefunction should be small enough not to significantly overlap the spectral range. If this condition holds, the weak value can be determined from a small number of measurements [1, 3]. If it does not, many measurements are necessary to determine the supershift $\langle q\rangle$.

2. Connection with superoscillations

In terms of the Fourier transform of the initial pointer state,

$$\phi(q) = \int_{-\infty}^{\infty} \mathrm{d}p \exp(\mathrm{i}pq)\bar{\phi}(p), \tag{2.1}$$

the final pointer state can be written as

$$\psi(q) = \int_{-\infty}^{\infty} \mathrm{d}p \exp(\mathrm{i}pq)S(p)\bar{\phi}(p), \tag{2.2}$$

in which $S(p)$ is the Fourier series

$$S(p) = \sum_{-N}^{N} C_n \exp(-\mathrm{i}np). \tag{2.3}$$

Note that in these formulas $S(p)$ describes the system being measured, and $\bar{\phi}(p)$ describes the pointer registering the measurement.

From (1.12), the overlap of the pre- and post-selected states is given by the value of $S(p)$ at the origin, that is

$$\langle f \mid i \rangle = S(0). \tag{2.4}$$

More importantly, the weak value (1.9) depends on the derivative of $S(p)$ at the origin, because

$$A_w = \frac{\mathrm{i}\partial_p S(0)}{S(0)} = \mathrm{i}\partial_p \log S(0) = -\partial_p \arg S(0) + \mathrm{i}\partial_p \log|S(0)|. \tag{2.5}$$

These results exemplify the recently emphasized more general connection [4] between the Fourier and Taylor coefficients of $S(p)$.

We will see that under suitable circumstances supershifts can be regarded as a consequence of superoscillations in $S(p)$ close to the origin, corresponding, from (2.5), to large weak values. Imagine that, as will occur in the example to follow and as has been discussed in detail elsewhere [4],

$$S(p) \approx B\exp(-\mathrm{i}aNp), \qquad (p \approx 0,\ a > 1). \tag{2.6}$$

This is superoscillatory because the phase gradient aN—the weak value in this case (cf (2.5))—is larger than the maximum eigenvalue $A_N = N$ (cf (1.5)) in the Fourier series (2.3): it is superweak. Then, if the main contribution to (2.2) comes from the neighbourhood of $p = 0$, we get

$$\psi(q) \approx B\phi(q - aN), \tag{2.7}$$

that is, a supershifted pointer.

This oversimplified argument does not take account of the fact that supershifting requires not only a superweak value and associated superoscillations in the system being measured, but also a suitable initial pointer wavefunction. We will understand this in more detail in the next sections, with examples where the supershift happens (sections 4 and 5.1) and where it does not (section 5.2).

3. Model for superoscillations

The much-studied function [12]

$$S(p) = \frac{(-1)^N}{a^{2N}}(\cos(\tfrac{1}{2}p) - \mathrm{i}a\sin(\tfrac{1}{2}p))^{2N} \quad (a > 1, N \gg 1) \tag{3.1}$$

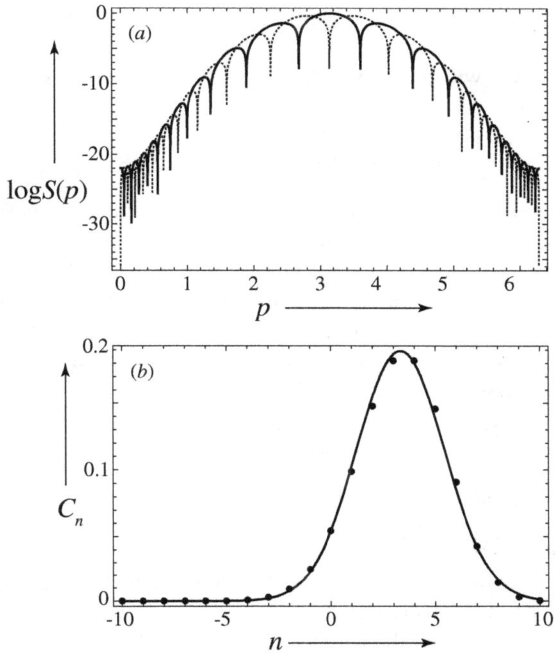

Figure 1. (*a*) Superoscillatory function $S(p)$ (3.1) underlying weak measurement, for $N = 10$, $a = 3$. Full curve: log|Re$S(p)$|, dashed curve: log|Im$S(p)$|. (*b*) Dots: Fourier coefficients $|C_n|$ (3.2) of $S(p)$; full curve: Gaussian approximation (3.3) with maximum at $n = 1/a = 1/3$.

has the form (2.3), with Fourier coefficients

$$C_n = \frac{(2N)!}{(2a)^{2N}}(a^2-1)^N\left(\frac{a+1}{a-1}\right)^n\frac{(-1)^n}{(N-n)!(N+n)!}. \tag{3.2}$$

Elementary expansion about $p = 0$ confirms the superoscillatory behaviour (2.6), and this is illustrated in figure 1(*a*). This superoscillation results from a delicate conspiracy of the phases in the sum (2.3), and cannot be detected in the strengths of the Fourier coefficients (3.2). These are not only restricted to the range $|n| \leqslant N$ but also peaked within this range at the mean value $\langle n\rangle = N/a$: Stirling's formula leads to the approximation

$$|C_n| \approx \frac{a}{\sqrt{\pi N(a^2-1)}}\exp\left\{-\left(n-\frac{N}{a}\right)^2\frac{a^2}{N(a^2-1)}\right\}, \tag{3.3}$$

illustrated in figure 1(*b*).

This set of coefficients can be interpreted as representing the class of pre- and post-selected states (1.10), automatically normalized in this case because (cf (1.11))

$$\sum_{-N}^{N}|C_n| = S(\pi) = 1. \tag{3.4}$$

From (2.4), the overlap of the states is

$$\langle f\mid i\rangle = \frac{(-1)^N}{a^{2N}}, \tag{3.5}$$

which is exponentially small under the conditions in (3.1), indicating near-orthogonality of $|i\rangle$ and $|f\rangle$.

In the following, we will need to understand $S(p)$ beyond the simplest approximation (2.6). To the next order,

$$S(p) = \frac{(-1)^N}{a^{2N}} \exp\left\{-\mathrm{i}aNp + \tfrac{1}{4}N(a^2-1)p^2 + \cdots\right\}. \tag{3.6}$$

The term in p^2 captures the rapid 'anti-Gaussian' increase of $|S(p)|$ away from $p = 0$, evident in the logarithmic plot for figure 1(a). From this it is clear that in order for the supershift (2.7) to emerge from the exact pointer wavefunction formula (2.2)—that is, for the superoscillatory behaviour (2.6) of $S(p)$ to dominate the integral over p—the anti-Gaussian increase must be dominated by the decay of the Fourier transform $\bar{\phi}(p)$ of the initial pointer wavefunction.

Globally, the increase in $|S(p)|$ from its superoscillatory value $1/a^{2N}$ at $p = 0$ to its maximum value unity at $p = \pi$ is given by

$$|S(p)| = \frac{\left(\cos^2\frac{1}{2}p + a^2\sin^2\frac{1}{2}p\right)^N}{a^{2N}}. \tag{3.7}$$

This contributes to the modulus

$$M(p) = |S(p)||\bar{\phi}(p)|, \tag{3.8}$$

of the integrand in (2.2), which (as recognized in the parallel development [4]) will be important in understanding the pointer wavefunction.

Near the maximum at $p = \pi$, the local behaviour of $S(p)$ is given by the expansion

$$\begin{aligned} S(\pi+\eta) &= \left(\cos\frac{1}{2}\eta - \frac{\mathrm{i}}{a}\sin\frac{1}{2}\eta\right)^{2N} \\ &= \exp\left(-\mathrm{i}\frac{N}{a}\eta - \frac{1}{4}N\eta^2\left(1-\frac{1}{a^2}\right) + \cdots\right). \end{aligned} \tag{3.9}$$

If the neighbourhood of the maximum would dominate the integral (2.2), as might naively be expected, the final wavefunction would be proportional to $\phi(q - N/a)$, that is, it would be concentrated at the mean of the spectrum, rather than supershifted as in (2.7). And indeed we will see that there are situations where this does happen, so there is no supershift. Note the contrast between the Gaussian decay of (3.9) and the anti-Gaussian increase (3.6) away from $p = 0$.

4. Condition for supershifts: Gaussian initial pointer state

As emphasized earlier, supershift depends on the initial pointer state $\phi(q)$ (or $\bar{\phi}(p)$) as well as on the states $|i\rangle$ and $|f\rangle$ and the spectrum of the operator being measured, i.e. on $S(p)$. In this section we consider the most familiar initial pointer wavefunction, namely the Gaussian

$$\phi(q) = \frac{1}{\Delta\sqrt{2\pi}}\exp\left(-\frac{q^2}{2\Delta^2}\right), \qquad \text{i.e.} \quad \bar{\phi}(p) = \frac{1}{2\pi}\exp\left(-\frac{1}{2}\Delta^2p^2\right), \tag{4.1}$$

with width Δ. This is a real Gaussian, representing a pointer initially at rest. It is not difficult to generalize to a complex Gaussian, for which the initial momentum of the pointer has a nonzero expectation value. But although an initially moving pointer has interesting implications for the weak value (see [5] and the development [6]), in the present context this would simply add complication and we avoid it.

Substituting into (2.2) for the final pointer state, we see the Gaussian decay of (4.1) competing with the anti-Gaussian in (3.6). The decay must dominate if the main contribution

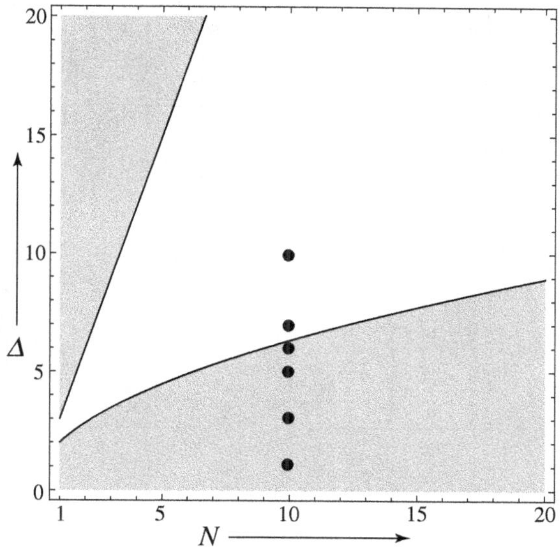

Figure 2. Permitted region of supershifts (white) in the plane Δ, N for $a = 3$, determined by the inequalities (4.3). The dots correspond to the pointer wavefunctions in figure 3.

to the integral over p is to be dominated by the neighbourhood of the origin, so that the superoscillations (2.6) in $S(p)$ give rise to the supershift (2.7). To see what this implies, we examine the modulus of the integrand, which from (3.8) and (3.9) is

$$M(p) = \frac{1}{2\pi a^{2N}} \exp\left(-\frac{1}{2}\Delta^2 p^2 + N\log\left(\cos^2\frac{1}{2}p + a^2\sin^2\frac{1}{2}p\right)\right)$$
$$= \frac{1}{2\pi a^{2N}} \exp\left(-\frac{1}{2}\left(\Delta^2 - \frac{1}{2}(a^2-1)\right)p^2 + \cdots\right). \quad (4.2)$$

For small Δ, the anti-Gaussian dominates, and the maximum of $M(p)$ is close to the maximum of $S(p)$ at $p = \pi$, and $\psi(q)$ is close to $\phi(q - N/a)$. As Δ increases, the maximum shifts to smaller p, until the coefficient of p^2 in (4.2) changes sign. Thereafter, the maximum remains at $p = 0$ and $\psi(q)$ is proportional to the supershifted function (2.7).

This shows that supershifts require the width Δ to be large enough. But if Δ is too large, the supershift Na will not be clearly separated from the spectral range $|q| \leqslant N$. The requirement for resolving the supershift is that Δ should not exceed the supershift itself. These considerations imply the inequalities

$$\Delta_c = \sqrt{\tfrac{1}{2}N(a^2-1)} < \Delta < Na, \quad (4.3)$$

so the existence of supershifts requires Δ to lie in a range between $\sqrt{N}$ and N, as illustrated in figure 2.

Assuming that these inequalities are satisfied and the integral is dominated by the region near $p = 0$, the final pointer wavefunction is

$$\psi(q) \approx \frac{(-1)^N}{a^{2N}2\pi}\int_{-\infty}^{\infty} \mathrm{d}p \exp\left\{\mathrm{i}p(q-Na) - p^2\left(\frac{1}{2}\Delta^2 - \frac{1}{4}N(a^2-1)\right)\right\}$$
$$= \frac{(-1)^N}{a^{2N}\sqrt{2\pi\left(\Delta^2 - \frac{1}{2}N(a^2-1)\right)}} \exp\left\{-\frac{(q-Na)^2}{2\left(\Delta^2 - \frac{1}{2}N(a^2-1)\right)}\right\}. \quad (4.4)$$

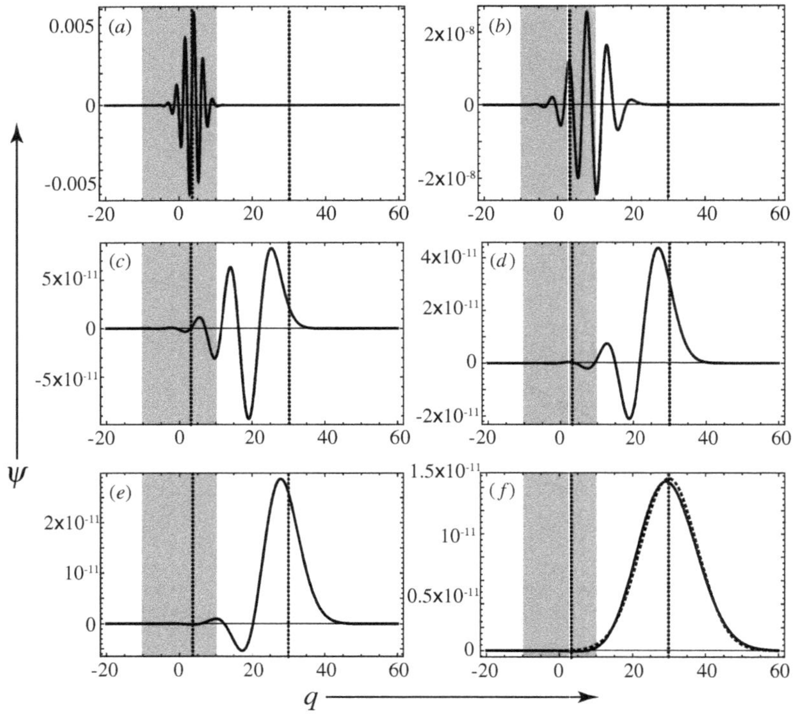

Figure 3. Pointer wavefunctions after a weak measurement, for $N = 10$ and $a = 3$, for initial Gaussian wavefunction widths (*a*) $\Delta = 1$; (*b*) $\Delta = 3$; (*c*) $\Delta = 5$; (*d*) $\Delta = 6$; (*e*) $\Delta = 7$; (*f*) $\Delta = 10$, showing how the supershift at $q = Na = 30$ emerges from the spectral range $|q| \leqslant N$ (shaded). The dashed curve in (*f*) shows the supershift approximation (4.4); the critical width (4.3) is $\Delta_c = \surd 40 = 6.32$. The vertical dotted lines outside the shaded regions indicate the weak values, at $q = Na = 30$, and those within the shaded regions indicate the spectral means, at $q = N/a = 3.33$. Note how the wavefunctions get weaker as Δ increases.

This predicts that the weak measurement indeed results in a supershifted final pointer state, with its width approaching that of the initial state (4.1) as Δ increases.

Figure 3 shows the pointer states $\psi(q)$ for different widths Δ, computed exactly from (1.8) with fixed values $N = 10$ and $a = 3$. For small Δ (figure 3(*a*)), the separate positive- and negative-weighted copies of the initial wavefunction $\phi(q)$ are clearly separated and all lie within the spectral range, here $|q| \leqslant N = 10$ (shaded). The strongest copies are close to the maximum $q = N/a = 3.33$ of the coefficients $|C_n|$ (cf (3.3)). As Δ approaches the crucial value Δ_c—the lower boundary of the permitted region (4.3) shown in figure 2—$\psi(q)$ shifts out of the spectral range and its negative excursions get weaker (figures 3(*b*) and (*c*)). Finally, as Δ passes Δ_c and has entered the permitted region (figure 3(*d*)), the supershift emerges clearly (figure 3(*f*)) and $\psi(q)$ is well approximated by the Gaussian (4.4), translated by the weak value $A_w = Na = 30$.

For the Gaussian initial pointer state, the mean final pointer position $\langle q \rangle$ can be calculated from (1.13) and (1.14) using

$$F(u) \propto \exp\left(-\frac{u^2}{\Delta^2}\right). \tag{4.5}$$

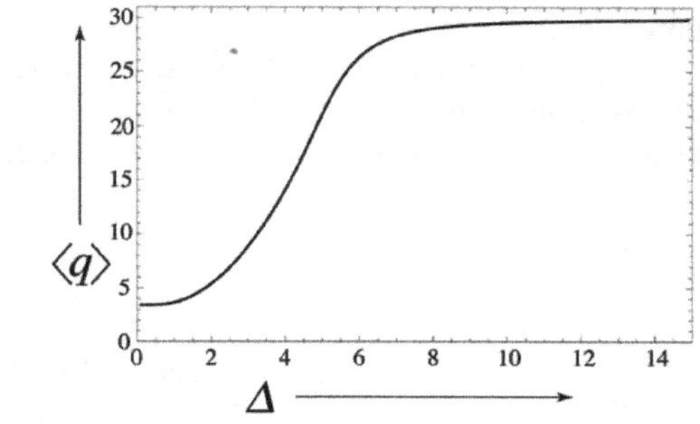

Figure 4. Pointer shift for Gaussian initial wavefunction, calculated from (1.13), (1.14) and (4.5) for $N = 10$, $a = 3$.

Figure 4 shows how as Δ increases the pointer shifts from its value N/a within the spectrum to the superweak value Na.

As can be seen from the scales in figure 3, the supershifted pointer state is enormously weaker than the initial state. This is a consequence of the dominating region of $S(p)$ near $p = 0$ being smaller by $1/a^{2N}$ than near $p = \pi$ where it attains its maximal value of unity. This attenuation can be quantified by calculating the normalization of the final state:

$$\langle \psi \mid \psi \rangle = \int_{-\infty}^{\infty} \mathrm{d}q\, \psi^*(q)\psi(q) = \sum_{-N}^{N}\sum_{-N}^{N} C_m^* C_n \int_{-\infty}^{\infty} \mathrm{d}q\, \phi^*(q-n)\phi(q-m). \tag{4.6}$$

In the limit of $\Delta \rightarrow 0$ of a narrow initial packet, the copies of $\phi(q)$ contribute separately. In the opposite limit $\Delta \gg 2N$ of supershifts, in which the width of the state exceeds that of the spectrum, the copies overlap and add coherently. Thus, with $\phi(q)$ in (4.1),

$$\begin{aligned} \langle \psi \mid \psi \rangle &\rightarrow \sum_{-N}^{N} |C_n|^2 = \frac{1}{2\Delta\sqrt{\pi}} \qquad (\Delta \ll 1) \\ \langle \psi \mid \psi \rangle &\rightarrow \left| \sum_{-N}^{N} C_n \right|^2 = |\langle f \mid i \rangle|^2 = \frac{1}{a^{4N} 2\Delta\sqrt{\pi}} \qquad (\Delta \gg N), \end{aligned} \tag{4.7}$$

indicating that the supershifted state is exponentially weakened. This weakening, and indeed the supershift itself, is the result of post-selecting with the state $|f\rangle$ that is nearly orthogonal to $|i\rangle$.

5. Non-Gaussian pointer states

The mechanism generating the supershift requires the initial pointer Fourier transform $\bar{\phi}(p)$ to possess a tail decaying fast enough to obliterate the maximum of $S(p)$ at $p = \pi$. As we have seen, the mechanism works if $\phi(q)$ is Gaussian. Here we will examine two non-Gaussian initial pointer wavefunctions: one where the mechanism operates, giving rise to supershifts, and one where it does not. We have explored the presence or absence of supershifts for several other initial pointer wavefunction, with different super-Gaussian or sub-Gaussian large $|q|$ decays, but the results are similar and we do not give the details, except for noting one point.

Since the initial pointer Fourier transform $\bar{\phi}(p)$ must decay rapidly, it might seem that the most pronounced supershift will be generated by the fastest-decaying function of all, namely the square pulse $\Theta(|p|\text{-}1/\Delta)$. And, as is easily confirmed, the pointer is indeed supershifted in this case, and the required condition is similar to (4.3) for the Gaussian (we do not show the calculation). But the corresponding pointer wavefunction $\phi(q)\sim\sin(q/\Delta)/q$ possesses such long-range oscillations that to ensure that the supershifted $\psi(q)$ does not overlap the spectral range the values of N and Δ must be much larger than in the Gaussian case.

5.1. Lorentzian initial pointer state

For this case,

$$\phi(q) = \frac{1}{1+(q/\Delta)^2}, \qquad \text{i.e. } \bar{\phi}(p) = \frac{1}{2}\Delta\exp(-|p|\Delta). \tag{5.1}$$

Instead of (4.2) the modulus of the integrand in (2.2) is

$$M(p) = \frac{1}{2\pi a^{2N}}\exp\left(-\Delta|p| + N\log\left(\cos^2\frac{1}{2}p + a^2\sin^2\frac{1}{2}p\right)\right). \tag{5.2}$$

The singular decrease represented by $|p|$ ensures that for any width Δ, however small, $M(p)$ always decays away from the origin. But for small Δ this decay stops at a shallow minimum, after which $M(p)$ rises to its large maximum near $p=\pi$. In this regime, $\psi(q)$ is concentrated near the mean of the spectrum, at $q = N/a$. As Δ increases, the maximum and minimum approach and annihilate. This occurs where the exponent in (5.2) possesses a flat inflection: at Δ_c, p_c, where

$$\Delta_c = \frac{N(a^2-1)}{2a}, \qquad p_c = \arcsin\frac{2a}{a^2+1}. \tag{5.3}$$

For $\Delta > \Delta_c$, $M(p)$ decreases monotonically away from the origin and we expect $\psi(q)$ to be supershifted to the weakened Lorentzian

$$\psi(q) = \frac{(-1)^N\Delta^2}{a^{2N}(\Delta^2+(q-aN)^2)}. \tag{5.4}$$

Figures 5(*a*)–(*d*) show that this transformation occurs for the Lorentzian as it did for the Gaussian, with the major change and supershift occurring as Δ passes Δ_c. Figure 6(*a*) shows the mean pointer position rising to its supershifted value $\langle q\rangle = Na = 30$ as Δ increases.

Note from (5.3) that $\Delta_c > N$ if $a > 1+\sqrt{2} = 2.41$, and it is clear from figure 5(*d*), for which $\Delta = N$, that the supershifted Lorentzian overlaps the spectral region substantially. This means that for Lorentzians the supershift is not perfectly resolved—in contrast to Gaussians (cf figure 3(*f*), for which $\Delta = N$).

5.2. Exponential initial pointer state

For this case, simply the reverse of (5.1),

$$\phi(q) = \exp\left(-\frac{|q|}{\Delta}\right), \qquad \bar{\phi}(p) = \frac{\Delta}{\pi(1+(p\Delta)^2)}. \tag{5.5}$$

Here the decay of $\bar{\phi}(p)$ is too slow to give rise to supershifts, even when Δ is large. In this case, the modulus $M(p)$ (equations (3.7) and (3.8)) does decay for small p, but the decay is soon followed by a shallow minimum after which $M(p)$ rises to its maximum at $p=\pi$. To see the behaviour near $p=0$, we use (5.5) for p small but $p\Delta \gg 1$. Then

$$M(p) \approx \frac{1}{\Delta^2 a^{2N}}\exp\left(-2\log p + \frac{1}{4}Np^2(a^2-1)\right) + \cdots \tag{5.6}$$

which has the minimum

$$p_{\min} = \frac{2}{\sqrt{N(a^2-1)}}, \qquad M(p_{\min}) = \frac{N(a^2-1)}{4\Delta^2 a^{2N}}, \tag{5.7}$$

a result confirmed by numerics for $\Delta \gg 1$.

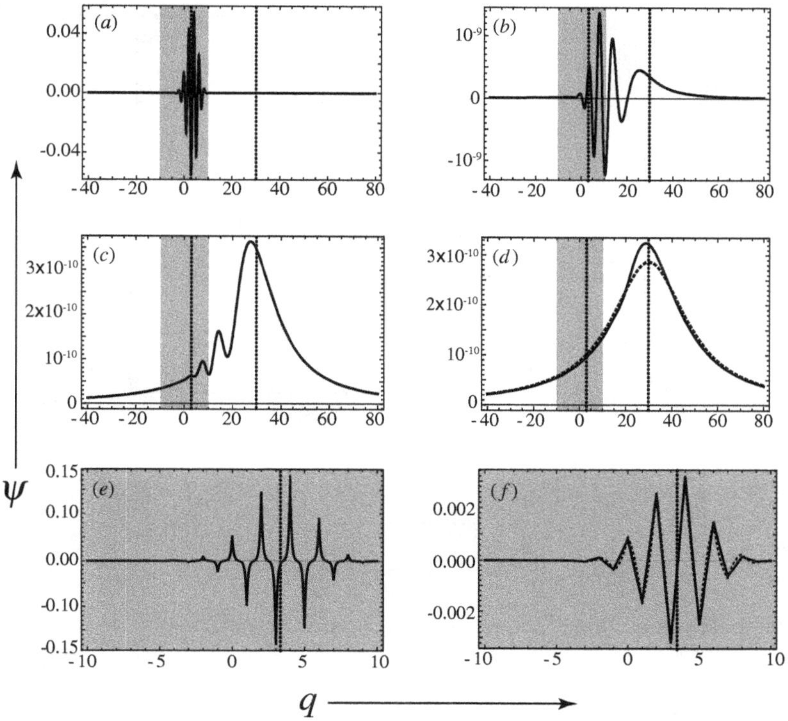

Figure 5. As figure 3, for (*a*)–(*d*) Lorentzian initial wavefunction (section 5.1), with (*a*) $\Delta = 1$, (*b*) $\Delta = 11$, (*c*) $\Delta = 15$, (*d*) $\Delta = 20$; the dashed curve in (*d*) shows the supershift approximation (5.4); the critical width (5.3) is $\Delta_c = 40/3 = 13.33$. (*e*), (*f*) Exponential initial wavefunction (section 5.2): (*e*) $\Delta = 0.1$, (*f*) $\Delta = 30$; the dashed curve in (*f*) shows the approximation (5.8).

Instead of supershifting, the pointer remains near the minimum for all initial widths. But as Δ increases, $\psi(q)$ undergoes an interesting transformation from the obvious superposition of decaying exponentials implied by (1.8) and illustrated in figure 5(*e*), to a form determined not only by the maximum of $S(p)$ at $p = \pi$ but also by the maxima at $p = (2m+1)\pi$. Use of the expansion (3.9) leads to

$$\psi(q) = \frac{4a}{\Delta\sqrt{\pi^3 N(a^2-1)}} \exp\left(-\frac{(q-N/a)^2 a^2}{N(a^2-1)}\right) \sum_{m=0}^{\infty} \frac{\cos((2m+1)\pi q)}{(2m+1)^2}. \tag{5.8}$$

This sum represents a series of saw teeth, which is modulated by a Gaussian centred on the spectral mean N/a. Figure 5(*f*) illustrates this behaviour and shows that the approximation accurately represents the exactly computed wavefunction. The absence of supershifting is further illustrated by $\langle q \rangle$ is figure 6(*b*), which remains stubbornly close to the spectral mean as Δ increases.

6. Supershifts for two-state observables

The richness that we have explored here, in which pointer wavefunctions $\psi(q)$ are supershifted towards the weak value of the observable $\hat{A}$ being measured, as the width Δ of the initial pointer

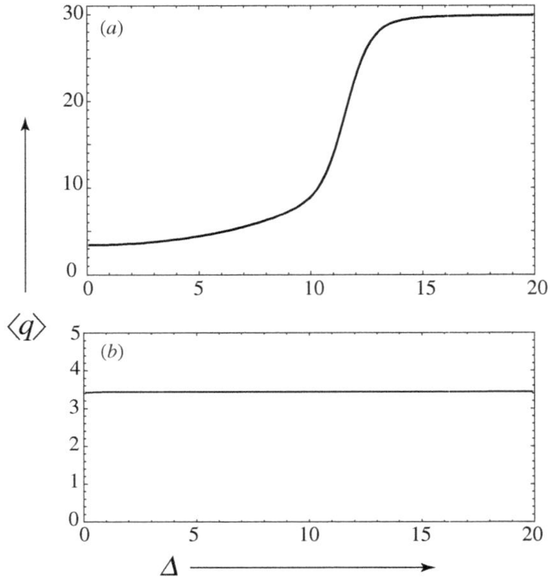

Figure 6. As figure 4, for (*a*) Lorentzian initial wavefunction (section 5.1), showing the development of supershift; (*b*) exponential initial wavefunction (section 5.2), for which no supershift develops.

wavefunction $\phi(q)$ increases, is caused by the coherent superposition (1.8) of copies of $\phi(q)$: one corresponding to each of the eigenvalues of $\hat{A}$. One might think that the mechanism depends on $\hat{A}$ having many eigenvalues, and would fail completely in the extreme of a 2-state observable (e.g. spin 1/2). But this is not the case: as is well known and as we now describe in detail (see also [6]), there can be supershifts even when $N = 2$.

Let the two eigenvalues be at +1 and −1. It will suffice to choose the corresponding coefficients (1.7) real, and in the form

$$C_{+1} = \cos\left(\theta - \tfrac{1}{4}\pi\right), \qquad C_{-1} = \sin\left(\theta - \tfrac{1}{4}\pi\right). \tag{6.1}$$

This is convenient because then $\theta = 0$ corresponds, according to (1.12), to $|i\rangle$ and $|f\rangle$ being orthogonal. From (1.8), the pointer wavefunction is

$$\psi(q) = \cos\left(\theta - \tfrac{1}{4}\pi\right)\phi(q-1) + \sin\left(\theta - \tfrac{1}{4}\pi\right)\phi(q+1), \tag{6.2}$$

and the weak value (1.9) is

$$A_w = \cot\theta. \tag{6.3}$$

A short calculation now gives the mean pointer position (1.13) after the weak measurement, for the Gaussian initial wavefunction (4.1):

$$\langle q\rangle = \frac{\sin 2\theta}{1 - \exp(-1/\Delta^2)\cos 2\theta}. \tag{6.4}$$

Supershifts correspond to $|\langle q\rangle| > 1$. Figure 7(*a*) shows the range of Δ, $|\theta| < \pi/2$ for which this occurs, namely

$$\tan\left(\frac{1}{4}\pi - |\theta|\right) < \exp\left(-\frac{1}{\Delta^2}\right). \tag{6.5}$$

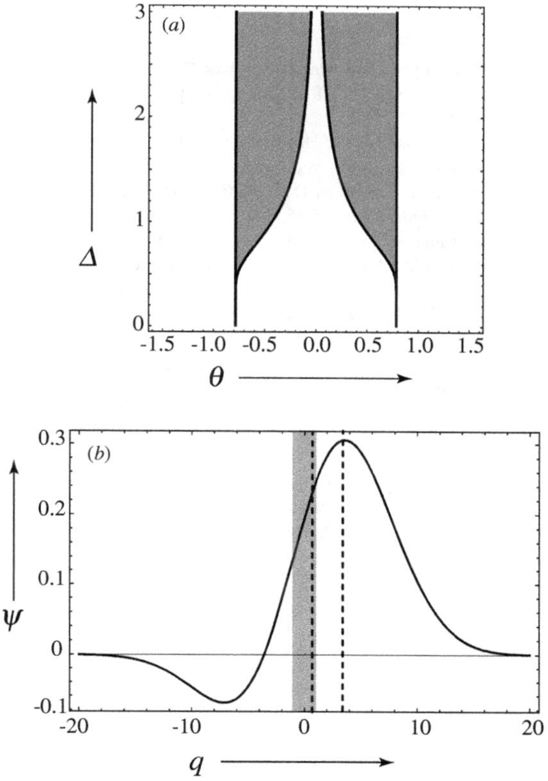

Figure 7. Two-state weak measurement. (*a*) Regions of supershifts (shaded), i.e. $|\langle q\rangle| > 1$ in (6.4). (*b*) Pointer wavefunction, as figure 3, for $\Delta = 5$, $\theta = 1/(5\sqrt{2})$.

The largest supershifts occur near $\theta = 0$, where $|i\rangle$ and $|f\rangle$are orthogonal and $A_w \gg 1$, and $\Delta \gg 1$. In this regime,

$$\langle q\rangle \approx \frac{2\theta\Delta^2}{2\theta^2\Delta^2+1}. \tag{6.6}$$

This can be arbitrarily large if $\theta = A/\Delta$, and, as can easily be confirmed, the choice $A = 1/\sqrt{2}$ gives the largest values:

$$\theta = \frac{1}{\Delta\sqrt{2}}, \qquad \langle q\rangle \approx \frac{\Delta}{\sqrt{2}}. \tag{6.7}$$

Figure 7(*b*) shows a pointer wavefunction (6.2) for a large supershift: $\Delta = 5$, $\theta = 1/(5\sqrt{2})$, corresponding to $\langle q\rangle = 3.606$. The supershift is clearly visible, but the wave overlaps considerably with the spectral range $|q| < 1$, in contrast with the separation that can be achieved for large N (cf figure 3(*f*)). The overlap is unavoidable, since, from (6.7), $\langle q\rangle/\Delta \approx 1/\sqrt{2}$; this bound is the price paid for getting supershifts for $N = 2$.

Acknowledgment

We thank Professor Sandu Popescu for several helpful suggestions.

References

[1] Aharonov Y and Rohrlich D 2005 *Quantum Paradoxes: Quantum Theory for the Perplexed* (Weinheim: Wiley)

[2] Aharonov Y, Popescu S and Tollaksen J 2010 A time-symmetric formulation of quantum mechanics *Phys. Today* **63** 27–33

[3] Aharonov Y, Albert D Z and Vaidman L 1988 How the result of a measurement of a component of the spin of a spin $1/2$ particle can turn out to be 100 *Phys. Rev. Lett.* **60** 1351–4

[4] Aharonov Y, Colombo F, Sabadini I, Struppa D C and Tollaksen J 2011 Some mathematical properties of superoscillations *J. Phys. A: Math. Theor.* **44** 365304

[5] Jozsa R 2007 Complex weak values in quantum measurement *Phys. Rev.* A **76** 044103

[6] Di Lorenzo A and Egues J C 2008 Weak measurement: effect of the detector dynamics *Phys. Rev.* A **77** 042108

[7] Dennis M R, Hamilton A C and Courtial J 2008 Superoscillation in speckle patterns *Opt. Lett.* **33** 2976–8

[8] Berry M V and Dennis M R 2009 Natural superoscillations in monochromatic waves in D dimensions *J. Phys. A: Math. Theor.* **42** 022003

[9] Berry M V and Shukla P 2010 Typical weak and superweak values *J. Phys. A: Math. Theor.* **43** 354024

[10] Berry M V, Dennis M R, McRoberts B and Shukla P 2011 Weak value distributions for spin $1/2$ *J. Phys. A: Math. Theor.* **44** 205301

[11] Berry M V 1994 *Faster than Fourier Quantum Coherence and Reality: in Celebration of the 60th Birthday of Yakir Aharonov* ed J S Anandan and J L Safko (Singapore: World Scientific) pp 55–65

[12] Berry M V and Popescu S 2006 Evolution of quantum superoscillations, and optical superresolution without evanescent waves *J. Phys. A: Math. Gen.* **39** 6965–77

[13] Kempf A and Ferreira P J S G 2004 Unusual properties of superoscillating particles *J. Phys. A: Math. Gen.* **37** 12067–76

Chapter 6
Philosophy

My research themes of asymptotics and singularities led to a contribution to the philosophical problem of theory reduction: how a theory (e.g. classical mechanics) can be a special case of a more general theory (e.g. quantum mechanics) that has a completely different conceptual basis. The new insight [6.1] stemmed from the recognition that in physics the emergence of a theory occurs as a limiting case when a parameter in the more general theory (e.g. a dimensionless combination proportional to Planck's constant) tends to zero. And in almost every case (quantum to classical, waves to rays, statistical mechanics to thermodynamics...), the reduction is both complicated and enriched by the fact that the limits are singular [6.2].

It follows that the problem of reduction must involve the mathematical theory of singular limits, namely asymptotics. This is often mathematics at the frontier of current research; examples are the renormalization group theory of critical phenomena in statistical mechanics, and applications of resurgence in non-perturbative quantum field theory. My initial attempts to present singular limits to philosophers went down like a lead balloon, but now this approach to reduction has started to be taken seriously ([6.3], [6.4]).

Singular limits are associated with singularities. In well-developed areas of physics, there can be several levels of explanations, each with its theory and associated singularities [6.1]. In optics, the singularities at the geometrical, scalar wave and vector wave levels are, respectively, caustics, phase singularities and polarization singularities. Each level's singularities are dissolved at the deeper level, an insight leading to the prediction, with Mark Dennis, that optical phase singularities must possess quantum cores (see Chapter 1 and [B364, B404]).

Logic, Methodology and Philosophy of Science IX
D. Prawitz, B. Skyrms and D. Westerståhl (Editors)

ASYMPTOTICS, SINGULARITIES AND THE REDUCTION OF THEORIES

MICHAEL BERRY

H H Wills Physics Laboratory, Tyndall Avenue, Bristol BS8 1TL, UK

1. Introduction

In science we strive to integrate our experiences, observations, and experiments into a single explanatory framework - 'a theory of everything'. Of course this goal has not been achieved, and probably never will be. What we have instead are the partial descriptions provided by biology, chemistry, physics, etc., and, within these, the various subfields such as fluid mechanics and quantum mechanics. The different areas of study do not fit tidily together. Particular difficulties arise when a more general description is supposed to encompass an older, less general, one, usually by providing a microscopic explanation of its principles. It is hoped that a less general theory can thus be 'reduced' to a more general one. But this comfortable picture is often spoilt by certain classes of higher-level, or 'emergent', phenomena which are well described by the older theory but obstinately refuse to emerge from the supposedly encompassing one.

To illustrate the point with a familiar example, consider life. Is it contained in, or implied by, Schrödinger's equation for the 10^{23} electrons and nuclei in an organism, plus rules for incorporating the environment? I suspect that most scientists, especially physicists, would, if pressed, answer yes, but be uncomfortable. The discomfort stems from a dilemma. We know that writing down the Schrödinger equation and gazing at it is not a promising strategy for finding a cure for AIDS, or learning why we do not live for ever. But we feel that invoking something else, outside physics, at a fundamental level, is mysticism. Somehow, life might emerge from physics in some limit (possibly involving increasing complexity), but we have no clear idea how to convert this dream into science.

Of course this problem of reduction has been studied a great deal by philosophers. Sometimes the discussion centres on the conflict between the

Reprinted from Proceedings of the Ninth International Congress of Logic, Methodology and Philosophy of Science, (1994).

two views summed up by the terms 'correspondence' and 'incommensurability': in brief, two theories correspond if one can be deduced as a special case of the other, and are incommensurate if their foundations are logically incompatible. My intention here is to present an idea which seems to capture an essential aspect of the problem of reduction of emergent phenomena and which goes some way towards dissolving the antinomy between incommensurability and correspondence, but which has not to my knowledge been considered by philosophers. I will confine myself to reductions of theories within physics, but of course hope that the idea could eventually prove useful in grander contexts such as the reduction of biology to physics (or chemistry).

To begin, realise that theories in physics are mathematical; they are formal systems, embodied in equations. Therefore we can expect questions of reduction to be questions of mathematics: how are the equations, or solutions of equations, of one theory, related to those of another? The less general theory must appear as a particular case of the encompassing one, as some dimensionless parameter - call it δ - takes a particular limiting value. A general way of writing this scheme is

$$\text{encompassing theory} \rightarrow \text{less general theory} \quad \text{as } \delta \rightarrow 0 \qquad (1)$$

Thus reduction must involve the study of limits, that is asymptotics. The crucial question will be: what is the nature of the limit $\delta \rightarrow 0$? We shall see that very often reduction is obstructed by the fact that the limit is *highly singular.* Moreover, the type of singularity is important, and the singularities are not only directly connected to the existence of emergent phenomena but underlie some of the most difficult and intensively-studied problems in physics today.

There is one aspect of the study of limits in physics which has attracted the attention of philosophers, beginning with Berkeley, that I will not be considering here, even though there are interesting and subtle points still to be brought out. This centres on the fact that the limit $\delta = 0$ is always an idealization; in any actual situation, δ is always finite. Instead of discussing this important matter, which involves the relation between the world and our models of it, I shall remain firmly in the realm of theory.

Before proceeding to examples, I must disambiguate an irritating terminological orthogonality. Philosophers consider the less general theory as being 'reduced by' the encompassing theory, because the latter employs principles that are more elementary to explain more phenomena [1]. Physicists, however, find it more natural to think of the reduction as occurring the other way, that is by the more general theory 'reducing to' the less general one as $\delta \rightarrow 0$ because the less general one is a special case (thus the function $\cos\theta$ 'reduces to' 1 as $\theta \rightarrow 0$).

2. Singular limits and emergent phenomena

Here are six examples of the scheme (1) in physics, together with the meaning of the dimensionless parameter δ.

- special relativity $\rightarrow$ Newtonian mechanics, $\delta = \nu/c$.
- general relativity $\rightarrow$ special relativity, $\delta = Gm/c^2a$.
- statistical mechanics $\rightarrow$ thermodynamics, $\delta = 1/N$.
- viscous (Navier-Stokes) flow $\rightarrow$ inviscid (Euler) flow, $\delta = 1/Re = \eta/\rho a\nu$.
- wave optics $\rightarrow$ ray optics, $\delta = \lambda/a$.
- quantum mechanics $\rightarrow$ classical mechanics, $\delta = \hbar/S$.

Here the meaning of the symbols is as follows. ν : speed of body; c: light speed; G: Newton's gravitational constant; m: mass of body; a: typical linear dimension of body; N: number of particles; Re: Reynolds' number; η : viscosity; ρ: density; λ :wavelength; $\hbar$: Planck's constant; S: typical classical action.

Reduction in its simplest form is well illustrated by the first example. Every physics student learns that one form of the connection between the encompassing theory of special relativity and the less general theory of Newtonian mechanics is contained in the 'low speed' series expansion

$$\sqrt{1-\delta^2} = 1 - \frac{1}{2}\delta^2 - \frac{1}{8}\delta^2 + \cdots \tag{2}$$

The left side represents special relativity, and the right side is a convergent Taylor series whose first term represents Newtonian mechanics. Mathematically, special relativity is analytic in δ at $\delta = 0$, so that the limit is unproblematic (the hyper-relativistic limit $\delta = 1$ is singular, but that is a different matter).

My main point will be that this simple state of affairs is an exceptional situation. Usually, limits of physical theories are not analytic: they are singular, and the emergent phenomena associated with reduction are contained in the singularity. Often, these emergent phenomena inhabit the borderland between theories.

To begin, consider the third example, namely the reduction of thermodynamics by statistical mechanics as the number of particles ($N = 1/\delta$) increases to infinity (the 'thermodynamic limit'). Standard arguments [2] involving large-N asymptotics show that for a fluid the thermodynamic equation of state, e.g. the pressure $P(V,T)$ as a function of volume and temperature, can (in principle and to a large extent in practice) be derived from the principles of statistical mechanics and a knowledge of the forces between the atoms. But the reduction runs into difficulty near the *critical point* P_c, V_c, T_c , where the compressibility $\kappa \equiv [-V(\partial P/\partial V)_T]^{-1}$ is infinite. The problem is to find the form of the divergence of κ as $T \to T_c$. This is a power-law, whose exponent is wrongly given by otherwise useful models such as the Van der Waals theory.

The reason for the difficulty is fundamental, and only after a decade of concentrated effort was it clarified, and techniques developed for the correct calculation of 'critical exponents'. Thermodynamics is a continuum theory, so reduction has to show that density fluctuations arising from interatomic forces have a finite (and microscopic) range. This is true everywhere except at the critical point, where there are fluctuations on all scales up to the sample size. Thus at criticality the continuum limit does not exist, corresponding to a new state of matter [3]. In terms of our general picture, the critical state is a singularity of thermodynamics, at which its smooth reduction to statistical mechanics breaks down; nevertheless, out of this singularity emerges a large class of new 'critical phenomena', which can be understood by careful study of the large-N asymptotics.

A particularly vicious example, at the cutting edge of applied mathematics nowadays, is the fourth on the above list, namely the mechanics of a fluid as its viscosity is decreased or its speed is increased (so that δ gets smaller). Exact solution of the Navier-Stokes equation for smooth flow down a pipe, driven by a pressure difference ΔP , predicts that the mass flow rate is proportional to ΔP . For small δ, however, experiment shows a rate close to $\sqrt{\Delta P}$. The reason is that the predicted flow is unstable, and the true flow is not smooth but disorderly, that is, *turbulent*. In turbulence [4-6], instead of viscous dissipation vanishing smoothly as $\delta \to 0$, the dissipation concentrates onto a set of zero measure which is fractal in form. Again the limit $\delta \to 0$ is singular, and out of the singularity emerges an important phenomenon, namely turbulence, whose mathematical nature is still far from understood.

3. Quantum and classical mechanics

Now we come to the examples I shall discuss in most detail - not because they are more fundamental than the others but because they lie closest to my own research interests [7] - namely the reduction of ray theory (e.g. geometrical optics) to wave theory, and (closely related) of classical to quantum mechanics. Here, singular limits abound, even in the simplest problems, as the following example shows.

A wave (of light, sound or water, for example) travelling along the x-axis with speed ν can be represented by

$$\psi = \cos\left\{\frac{2\pi}{\lambda}(x - \nu t)\right\} \tag{3}$$

In the ray limit (where for example geometrical optics provides a consistent and serviceable description of, for example the operation of telescopes and cameras), we have $\lambda \to 0$. But this limit is singular! ψ is non-analytic at $\lambda = 0$, so that it cannot be expanded in powers of λ ; instead, this wavefunction oscillates infinitely fast and takes all values between -1 and $+1$ infinitely often in any finite range of x or t. Only if we consider the wave *intensity*, corresponding to ψ^2, and average over a small interval corresponding to the finite resolution of a detector, do we get the finite and smooth result corresponding to the intensity of the system of parallel rays corresponding to (3); often it is convenient to average over time (reflecting the fact that for light or sound the wave frequency is too high to measure directly):

$$\left\langle\psi^2\right\rangle_t = \left\langle\cos^2\left\{\frac{2\pi}{\lambda}(x - \nu t)\right\}\right\rangle_t = \frac{1}{2} \tag{4}$$

Now consider the superposition of two such waves, with speeds ν and $-\nu$, giving

$$\psi = \cos\left\{\frac{2\pi}{\lambda}(x - \nu t)\right\} + \cos\left\{\frac{2\pi}{\lambda}(x + \nu t)\right\} = 2\cos\left\{\frac{2\pi x}{\lambda}\right\}\cos\left\{\frac{2\pi \nu}{\lambda}\right\} \tag{5}$$

and the time average

$$\left\langle\psi^2\right\rangle_t = 2\cos^2\left\{\frac{2\pi x}{\lambda}\right\} \tag{6}$$

This describes a spatially fixed interference pattern such as that produced by a double slit. Again there is a powerful singularity at $\lambda = 0$. To eliminate it requires an extra average, this time spatial, and then we obtain

$$\left\langle\psi^2\right\rangle_{t,x} = 1 \tag{7}$$

Thus to obtain from wave theory the simple fact that in ray theory two beams of intensity 1/2 add to give intensity 1, with no interference, requires a double average over a mathematically pathological function.

Having seen that interference is associated with a $\cos^2(1/\lambda)$ singularity in the ray limit, we now examine the anatomy of other sorts of wave singularity. An interesting case occurs when waves reach places that rays do not.

602

Examples are the outside of a glass-air interface within which total internal reflection occurs, the dark side of a rainbow, and the thin layer of air near a hot road in which mirage reflections are seen. In the ray limit, the wave and its intensity are zero, but there are nevertheless waves present, whose amplitude is typically

$$\psi \propto \exp\left\{\frac{-\text{function of } x}{\lambda}\right\} \tag{8}$$

Again this is singular, and cannot be expanded in a power series in λ (all terms are zero).

An important role is played by the *transition* between these two sorts of singularity ($\cos^2\{1/\lambda\}$ and $\exp\{-1/\lambda\}$). (This is somewhat analogous to the transition T through T_c in thermodynamics, for large N.) The transition happens across a *caustic*, which is an envelope of a family of rays (a generalized focal surface in space, or line in the plane), marking the boundary between regions with different numbers of rays. In the simplest case, the regions have two rays and no rays, corresponding to the 'interference' and 'penetration' regimes represented by (6) and (8). A caustic is a collective phenomenon, a property of a family of rays that is not present in any individual ray. Probably the most familiar example is the rainbow. The singularity across a caustic must interpolate between (6) and (8). How this happens was first elucidated by Airy in 1838 as part of an attempt to understand supernumerary rainbows, that is oscillations on the lit side of the bow, in the intensity of light of a given colour. It was necessary for him to invent a new function $Ai(z)$, oscillatory for $z < 0$ and decaying for $z > 0$. In terms of $Ai(z)$, the wave across a caustic has the form

$$\psi = \frac{1}{\lambda^{1/6}} Ai\left\{\frac{Kx}{\lambda^{2/3}}\right\} \tag{9}$$

In this transition, the emergent phenomenon is the fringe pattern associated with a caustic: in the ray limit $\lambda \to 0$, its intensity grows as $\lambda^{-1/3}$, and the spacing of the fringes shrinks as $\lambda^{2/3}$.

Caustics can themselves have singularities, whose classification is the province of catastrophe theory [8]. At such places, the envelope of rays is itself singular. These singular envelopes are decorated with wave patterns ψ whose $\lambda \to 0$ singularities (shrinking fringe spacings and diverging intensities) depend on the geometry of the catastrophe. Such 'diffraction catastrophes' have intricate and beautiful structures [9,10], and constitute a hierarchy of nonanalyticities, of emergent phenomena par excellence. The patterns inhabit the borderland between the wave and ray theories, because when λ is zero the fringes are too small to see, whereas when λ is too large the overall structure of the pattern cannot be discerned: they are *wave* fringes decorating *ray* singularities.

Quantum mechanics is a particular wave theory, whose corresponding ray theory is classical mechanics, and where Planck's constant $\hbar$ plays the role of wavelength λ (through De Broglie's relation $\lambda = 2\pi\hbar/p$ where p is momentum). Its relation to classical mechanics should be through the *semiclassical limit* $\hbar \to 0$. When the limit is not singular, we have the correspondence principle: quantum observables tend to their classical counterparts as $\hbar \to 0$. Usually, though, the limit is singular, and then the correspondence principle, while often a useful guide [7], is too crude to be a substitute for mathematical asymptotics. From the analogy with other sorts of waves we expect that the nonanalyticities and emergent caustic phenomena described above will occur in quantum mechanics, and these have indeed been seen in the scattering of electrons, nuclei and atoms. In addition, the $\hbar \to 0$ limit is enriched by another limit, which is fundamental, namely the *long-time limit* $t \to \infty$.

There are several reasons to study the long-time limit in conjunction with the semiclassical limit:

■ Spectra of atoms and molecules involve the quantized energies of these systems when in stationary states. These are states that persist over infinite time, so their semiclassical study – spectra near the classical limit – inescapably involves $t \to \infty$ too.

■ Experiments on atoms traversing strong oscillating fields begin to probe the combined $\hbar \to 0, t \to \infty$ limit.

■ It is only after infinite time that chaos may occur in the classical orbits. Chaos [11, 12] is unpredictability arising from exponential sensitivity to initial conditions in a bounded region. Therefore any attempt to study how classical chaos is reflected in the semiclassical limit of quantum mechanics ('quantum chaology' [13, 14]) must evidently involve $t \to \infty$ as well.

The essential point is that *the two limits do not commute*: taking the classical limit first, and the long-time limit second, leads to a different result from taking the limits in the reverse order. Such a clash of limits implies a singularity at the origin of the plane with coordinates $\hbar, 1/t$. One way to try to resolve the clash is to take both limits at once, in a controlled way, i.e.

$$\hbar \to 0, \ t \to \infty, \ \hbar t \equiv \tau = \text{constant} \tag{10}$$

In the one case where it has been possible to take the combined limit explicitly [15], for a system whose classical dynamics is trivial, analysis shows that the point $\hbar = 1/t = 0$ is truly a 'dragon's lair' , so singular that the behaviour exhibits a fantastic complexity which depends on the *arithmetic nature* of τ.

When the classical orbits are chaotic, the clash of limits generates some remarkable emergent phenomena. I will briefly describe just one: *the statistics of spectral fluctuations.* Consider a bound quantum system, that is one with a discrete spectrum of energy levels, and ask about the distribution of these levels in the semiclassical limit. The simplest fact about the levels is that as $\hbar \to 0$ they get closer together – their mean spacing is proportional to $\hbar^N$, where N is the number of freedoms. This must happen, because in the classical limit the levels form a continuum. (It is worth pausing to remark that this particular passage to the limit provides a nice illustration of the 'incommensurability' and 'correspondence' approaches to reduction. In the first, it is emphasized that for any finite $\hbar$, however small, the spectrum is always discrete: the classical continuum is never reached, and so cannot be said to be logically contained in the semiclassical limit. On the other hand, when $\hbar$ is sufficiently small the inevitably finite resolution of any spectroscopic measuring device means that the results of all observations will be the same as if the spectrum were continuous, and the correspondence principle can be said to apply.)

Now imagine looking at the set of levels with a microscope [14] whose power is proportional to the mean level density, thus generating a rescaled spectrum consisting of a set of numbers whose mean density remains constant as $\hbar \to 0$. What is the statistical nature of the fluctuations of this set of numbers about its (unit) mean density? The answer is remarkable: apart from trivial exceptions, the fluctuations are *universal* [14, 16], that is, independent of the details of the system and dependent only on whether the orbits of its classical counterpart are regular or chaotic. Paradoxically, the spectral fluctuations are those of a sequence of random numbers (Poisson distribution) when the classical motion is regular, and are more regularly distributed (exhibiting the level repulsion characteristic of the eigenvalues of random matrices) when the classical motion is chaotic. We are beginning to understand this quantum universality [7] in terms of semiclassical asymptotics: it arises from a similar universality in the distribution of long-period classical orbits.

Universality of the spectral fluctuations is a novel qualitative phenomenon emerging from quantum mechanics in the combined semiclassical long-time limit. It was not predicted by analysis of the Schrödinger equation, but was discovered in numerical experiments (and later seen in real experiments) motivated by some physical arguments. Nevertheless Schrödinger's equation does contain it, albeit well concealed behind some very tricky (and incompletely explored) asymptotics.

4. Divergent series

So far, we have considered only the leading-order behaviour in the parameter δ whose vanishing describes how the encompassing theory reduces to the less general one. In those cases where any sort of mathematical treatment was possible, the leading-order behaviour was quite complicated (cf. equation (9)), and this of course reflects the singular nature of the limit. But determination of the leading order is only the first step: a complete treatment requires understanding the series consisting of all the correction terms – usually involving powers of δ. The determination of such series is still in its infancy, but it has been carried out for certain of the simpler problems of wave physics described in §3.

The most important characteristic of such series, and one which almost certainly extends to all series associated with singular reductions, is that they *diverge*. This was one of the factors prompting a re-examination [17] of the mathematics and physics of divergent series. The main results reinforce earlier indications [18] that the divergent tail, conventionally discarded as mathematically meaningless, contains important information in coded form. When decoded, these tails not only enable the function being expanded to be approximated to previously unequalled levels of accuracy but also describe physical effects, associated with the reduction that the asymptotics is attempting to describe, which are qualitatively different from those contained in the leading terms. Examples are the exponentially weak births of rays beyond caustics [19] and the generation of transitions between quantum states [20] in the adiabatic limit of slow driving.

It seems clear that these ideas, and further developments of them, must be involved in any complete description of how the less general theory is embedded in the structure of the encompassing theory.

5. Concluding remarks

Even in what philosophers might regard as the simplest reductions, between different areas within physics, the detailed working-out of how one theory can contain another has been achieved in only a few cases and involves sophisticated ideas on the forefront of physics and mathematics today. This is because in all nontrivial reductions the encompassing theory is a singular perturbation (parameterised by δ) of the less general one. The singularities are reflected in the quantities of the encompassing theory being nonanalytic at $\delta = 0$, and the nonanalyticities describe emergent phenomena in the borderland between the theories. As examples of these phenomena I described thermodynamic critical behaviour in fluids, fluid turbulence, interference patterns decorating optical caustics, and the chaology-dependent statistics of energy-level fluctuations in quantum mechanics.

It should be clear from the foregoing that a subtle and sophisticated understanding of the relation between theories within physics requires real mathematics, and not only verbal, conceptual and logical analysis as currently employed by philosophers. One can hope that these ideas generalize beyond physics (for example to the reduction of biology or chemistry). This would mean that the problem of theory reduction would itself have been 're-duced', to the mathematical asymptotics of singularities. From the evidence so far, the task will be far from easy, and will require the development of new physical ideas and new mathematical concepts and techniques.

Finally, I would be the first to admit that the ideas explored here lack precision in several respects, and have not been presented in their final form. I hope they will benefit from the attention of philosophers.

References

[1] BULLOCK, A. and STALLYBRASS, O. 1977, *The Fontana Dictionary of Modern Thought.* (Collins, London). "Reduction: ... In philosophy ... the process whereby concepts ... that apply to one type of entity are redefined in terms of concepts ... of another kind, normally one regarded as more elementary ...".

[2] TOLMAN, R.C. 1938, *The Principles of Statistical Mechanics* (Oxford, University Press).

[3] WILSON, K.G. 1975, Rev. Mod. Phys. 47, 773-840.

[4] MANDELBROT, B.B. 1982, *The Fractal Geometry of Nature* (San Francisco: Freeman).

[5] FRISCH, U, 1983, in *Chaotic Behaviour in Deterministic Systems*, eds G. Iooss, R.H.G. Helleman and R. Stora, Les Houches Lecture Series XXXVI (Amsterdam: North-Holland) pp665-704.

[6] SREENIVASAN, K.R. and PRASAD, R. 1989, Physica D38, 332-339.

[7] BERRY, M.V. 1991, in *Chaos and Quantum Physics*, eds M-J Giannoni, A. Voros and J. Zinn-Justin, Les Houches Lecture Series LII (Amsterdam: North-Holland).

[8] POSTON, T. and STEWART, I.N. 1978, *Catastrophe Theory and its Applications* (London: Pitman).

[9] BERRY, M.V. and UPSTILL,C.1980, *Progress in Optics* 18, 257-346.

[10] BERRY, M.V. 1990, *Current Science* (India), 59, 1175-1191.

[11] STEWART, I.N. 1989 *Does God Play Dice? The Mathematics of Chaos* (Oxford: Blackwell).

[12] BERRY, M.V. 1990, Proc.Roy. Institution of Gt. Britain, 61, 189-204.

[13] BERRY,M.V., 1989, Physica Scripta 40 335-336.

[14] BERRY, M.V., 1987, Proc.Roy.Soc. A 413, 183-198.

[15] BERRY, M.V., 1988, Physica D, 33, 26-33.

[16] BOHIGAS, O. and GIANNONI, M-J.1984 in *Mathematical and Computational Methods in Nuclear Physics*, eds., J.S.Dehesa, J.M.G.Gomez and A.Polls, Lecture Notes in Physics 209 (N.Y.:Springer-Verlag) pp1-99.

[17] BERRY, M.V. 1991 *Asymptotics, superasymptotics, hyperasymptotics...* in Asymptotics beyond all orders, ed. S. Tanveer (New York: Plenum) in press.

[18] DINGLE, R.B. 1973, *Asymptotic Expansions: their Derivation and Interpretation* (London: Academic Press).
[19] BERRY, M.V., 1989, Publ. Math.of the Institut des Hautes Études scientifique, 68 211-221.
[20] BERRY, M.V. 1990 Proc.Roy.Soc.Lond, A429, 61-72.

Singular Limits

Michael Berry

Biting into an apple and finding a maggot is unpleasant enough, but finding half a maggot is worse. Discovering one-third of a maggot would be more distressing still: The less you find, the more you might have eaten. Extrapolating to the limit, an encounter with no maggot at all should be the ultimate bad-apple experience. This remorseless logic fails, however, because the limit is singular: A very small maggot fraction ($f \ll 1$) is qualitatively different from no maggot ($f = 0$). Limits in physics can be singular too—indeed they usually are—reflecting deep aspects of our scientific description of the world.

In physics, limits abound and are fundamental in the passage between descriptions of nature at different levels. The classical world is the limit of the quantum world when Planck's constant h is inappreciable; geometrical optics is the limit of wave optics when the wavelength λ is insignificant; thermodynamics is the limit of statistical mechanics when the number of particles N is so large that $1/N$ is negligible; mechanics of a slippery fluid is the limit of mechanics of a viscous fluid when the inverse Reynolds number $1/R$ can be disregarded. These limits have a common feature: They are all singular—they must be, because the theories they connect involve concepts that are qualitatively very different. As I explain here, there are both reassuring and creative aspects to singular limits. And by regarding them as a general feature of physical science, we get insight into two related philosophical problems: how a more general theory can reduce to a less general theory and how higher-level phenomena can emerge from lower-level ones.

The coherence of our physical worldview requires the reassurance that, singularities notwithstanding, quantum mechanics does reduce to classical mechanics, statistical mechanics does reduce to thermodynamics, and so on, in the appropriate limits. We know that when calculating the orbit of a spacecraft (and indeed knowing that it has an orbit) we can safely use classical mechanics, rather than having to solve the Schrödinger equation. An engineer designing a bridge can rely on continuum elasticity theory, without needing to know the atomic arrangements underlying the equation of state of the materials used in the construction. However, getting these reassurances from fundamental theory can involve subtle and unexpected concepts.

MICHAEL BERRY *(http://www.phy.bris.ac.uk/uk/staff/berry_mv.html) is Royal Society Research Professor in the physics department of Bristol University, in the UK.*

Perhaps the simplest example is two flashlights shining on a wall. Their combined light is twice as bright as when each shines separately: This is the optical embodiment of the equation $1 + 1 = 2$. But we learned from Thomas Young almost exactly two centuries ago that this mathematics does not describe the intensity of superposed light beams: To account for wave interference, *amplitudes* must be added, and the sum then squared to give the intensity. This involves the phases of the two waves, $\pm\phi$ say, and gives the intensity as $|\exp(i\phi) + \exp(-i\phi)|^2 = 2 + 2\cos 2\phi$, which can take any value between 0 and 4. So, what becomes of $1 + 1 = 2$? Young himself, responding to a critic who claimed that the wall should be covered with interference fringes, agreed, but pointed out that "the fringes will demonstrably be invisible . . . a hundred . . . would not cover the point of a needle." Underlying this explanation is a singular limit: The unwanted $\cos 2\phi$ does not vanish but oscillates rapidly. If the beams make an angle θ, the fringe spacing is $\lambda/2\theta$, vanishing in the geometrical limit of small λ. The limit is singular because the cosine oscillates infinitely fast as λ vanishes. Mathematically, this is an essential singularity of a type dismissed as pathological to students learning mathematics, yet here it appears naturally in the geometrical limit of the simplest wave pattern.

Young's "demonstrable" invisibility requires an additional concept, later made precise by Augustin Jean Fresnel and Lord Rayleigh: The rapidly varying $\cos 2\phi$ must be replaced by its average value, namely zero, reflecting the finite resolution of the detectors, the fact that the light beam is not monochromatic, and the rapid phase variations in the uncoordinated light from the two flashlights. Only then does $1 + 1 = 2$ apply—a relation thus reinterpreted as a singular limit.

Nowadays this application of the idea that the average of a cosine is zero, elaborated and reincarnated, is called decoherence. This might seem a bombastic redescription of the commonplace, but the applications of decoherence are far from trivial. Decoherence quantifies the uncontrolled extraneous influences that could upset the delicate superpositions in quantum computers. And, as we have learned from the work of Wojciech Zurek and others, the same concept governs the emergence of the classical from the quantum world in situations more sophisticated than Young's, where chaos is involved. For example, the chaotic tumbling of Saturn's satellite Hyperion, regarded as a quantum rotator with about 10^{60} quanta of angular momentum, would, according to an unpublished calculation by Ronald Fox, be suppressed in a few decades by the discrete nature of the energy spectrum. However, nobody expects to witness this suppression, because Hyperion is not isolated: Just one photon arriving from the Sun (whose reemission enables our observations) destroys the coherence responsible for quantization in a time of the order of 10^{-50} seconds, and reinstates classicality.[1] Alternatively stated, decoherence suppresses the quantum suppression of chaos.

Other reassurances are equally hard to come by. For example, formally obtaining thermodynamics from statistical mechanics involves applying the mathematical saddle-point method to an infinite-dimensional integral. But although such reassurances about the appropriate application of earlier, less general theories are welcome, they look backward rather than forward. However, there is a creative side to singular limits: They lead to new physics. For large N, where a central idea is symmetry-breaking, this creative side is concisely expressed in Philip Anderson's celebrated phrase: More is different.[2] The vast literature on critical phenomena reflects the fact that the large-N limit of statistical mechanics is singular at a critical point because there the continuum postulated in the thermodynamic limit is never reached, even when averaging over distances far exceeding the spacing between atoms. Correlations span arbitrarily large distances, and the critical state—the new physics—is a fractal. The zero-viscosity limit of fluid mechanics is singular because of the still-mysterious phenomenon of turbulence, whose definitive understanding would earn one of the Clay Foundation's $1 million prizes.

In quantum mechanics (and indeed the physics of waves of all kinds), a range of new phenomena lurk in the borderland with classical mechanics. High-lying energy levels display remarkable universality: Their statistics depend only on whether the corresponding classical orbits are regular or chaotic, and on certain global symmetries. In the chaotic case (see the column "Quantum Chaos and the Bow–Stern Enigma" by Daniel Kleppner, PHYSICS TODAY, August 1991, page 9), the associated wavefunctions resemble random functions of position decorated by "scars" along classical periodic orbits. (See the article "Postmodern Quantum Mechanics," by Eric J. Heller and Steven Tomsovic, PHYSICS TODAY, July 1993, page 38.) For regular motion, the dominant feature is focusing, and the classical paths are singular on caustics; the caustics are decorated with striking and characteristic interference patterns (see my column "Why Are Special Functions Special?" in PHYSICS TODAY, April 2001, page 11). Such postmodern quantum effects are emergent phenomena par excellence: The discrete states they describe are essentially nonclassical, but can be unambiguously identified only for highly excited states, that is, under near-classical conditions.

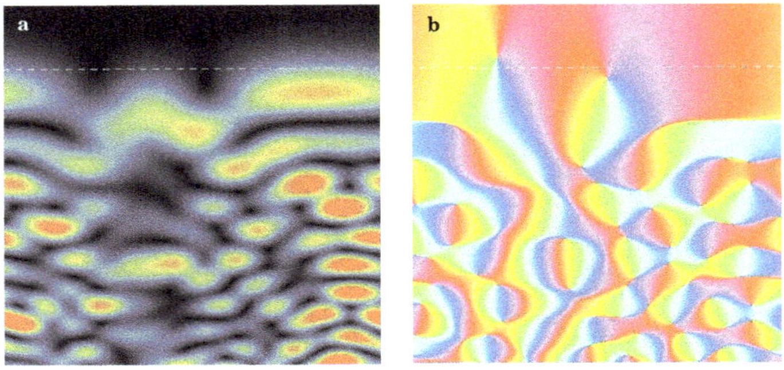

ACROSS THE BOUNDARY between classically allowed and forbidden regions in a two-dimensional chaotic system, the density of trajectories falls discontinuously to zero. This classical limit is singular because, in the corresponding semiclassical quantum wavefunctions shown here as simulations (with the classical boundary indicated by dashed lines), (**a**) the probability density (color-coded from red at maxima to black at zeros) fluctuates smoothly, and (**b**) the phase (color-coded by hue) varies smoothly except at points where all colors meet (points that are themselves singularities).

New ideas in physics often inspire, or are inspired by, new ideas in mathematics, and singular limits are no exception. Underlying critical phenomena is the renormalization group, which determines how systems transform, or remain invariant, under changes of scale—a fertile idea that is essentially mathematical but whose foundations have not been rigorously established. The quantum–classical connection involves divergent infinite series (for example, in powers of h), and the divergence can be traced precisely to the singularity of the limit. Some quantum phenomena involving divergent series are nonclassical reflection above a smooth potential barrier, weak quantum transitions caused by slowly varying external forces, and the representation of spectra in terms of classical periodic orbits. Mathematicians long regarded divergent series with suspicion; in 1828, Niels Henrik Abel wrote that they "are an invention of the devil, and it is shameful to base on them any demonstration whatsoever." But such series are often the best (even the only) way to calculate physical quantities, and applied mathematicians, disregarding Abel's censure, have freely developed sophisticated manipulations and regularizations of the divergences. An elementary example of a divergent series is $1 + 2 + 3 + 4 + \ldots$, which can be resummed to give the value $-1/12$; this looks like a joke, but is the unambiguous result of zeta function regularization, widely used in quantum field theory. Much more violent divergences arise from singular limits associated with the integrals and differential equations of physics, and have been tamed using more sophisticated methods. This is the domain of mathematical asymptotics and singular perturbation theory. In the 1990s, the long overdue beginnings of a rigorous mathematical theory were established.

Singular limits carry a clear message, which philosophers are beginning to hear:[3] The physics of singular limits is the natural philosophy of renormalization and divergent series. Perhaps they are recognizing that some problems of theory reduction can themselves be reduced to tricky questions in mathematical asymptotics—an extension of the traditional philosophical method, of argumentation based on words. Usually, we think of "applications" of science going from the more general to the more specific—physics to widgets—but this is an application that goes the other way: from physics to philosophy. One wonders if it counts with those journalists or administrators who like to question whether our research has applications. Probably not.

References

1. M. V. Berry, in *Quantum Mechanics: Scientific Perspectives on Divine Action*, R. J. Russell, P. Clayton, K. Wegter-McNelly, J. Polkinghorne, eds., Vatican Observatory Publications, Vatican City State, and The Center for Theology and the Natural Sciences, Berkeley, Calif., (2001), p. 41.
2. P. W. Anderson, *Science* **177**, 393 (1972).
3. R. W. Batterman, *The Devil in the Details: Asymptotic Reasoning in Explanation, Reduction, and Emergence*, Oxford U. Press, New York (2002). ■

Foreword

Newton's third law does not apply to the interaction between philosophers ('them') and physicists ('us'). It has usually been asymmetrical, with 'us' influencing 'them', without 'them' acting on 'us'. In a way this is natural, because the raw material that philosophers study are the discoveries and theories of science and the interactions between scientists, while the primary preoccupation of physicists is not the study of philosophy or philosophers. I do not deny that there have been eminent scientists (Einstein, Poincaré, Bohr...) who have pondered on the philosophical significance of the scientific picture of the world, and much of what they said has been immediately appreciated by practicing scientists. But their wise intellectual interventions have usually been outside the philosophical mainstream.

This book by Sergio Chibbaro, Lamberto Rondoni and Angelo Vulpiani (CRV) is an exception. Although their day job is the practice of theoretical physics, they have something genuinely new to say about the physicist's picture of the world, that should be of interest to philosophers. Their focus is on what has long been studied by philosophers as 'the problem of reduction'. This concerns the relations between different levels of description of physical phenomena.

Optics is a good example. Light can be described in terms of the rays of geometrical optics, as interfering waves, as electromagnetic fields, or as the photons of quantum field theory. These are levels of increasing generality; each encompasses phenomena described at the previous levels and also includes new phenomena that were unexplainable earlier. But the concepts and mathematical expressions of these levels are very different, and moving between them is almost always challenging. It is far from straightforward to derive the formula relating the object and image of a simple lens by starting from the field operators of quantum optics supposedly the deepest of our current pictures of light.

This illustrates a wider difficulty. One can have a general theory—or even, as some envisage, the theory of everything—which stubbornly resists attempts to employ it to explain phenomena that were well understood at more elementary levels: at the more sophisticated level, they are emergent. This was cleverly caricatured in Ian McEwen's novel *Solar*: a string theorist, caught by his wife in a compromising situation with another woman, tries to reassure her: "Darling, I can explain everything". It often happens that theories claiming great explanatory

Reprinted from 'Reductionism, Emergence and Levels of Reality: The Importance of Being Borderline', Springer, (2014).

reach are in fact powerless to explain many particular phenomena. A political analogy comes to mind: the ideologist who loves all humanity but behaves badly to every individual person he encounters.

The resolution of these difficulties starts from the observation that the theories of physics are mathematical, and relations between them involve limits as some parameter vanishes: wave optics 'reduces to' geometrical optics when the wavelength is negligibly small, quantum physics 'reduces' to classical physics when Planck's constant can be neglected, etc. Therefore understanding relations between levels must involve the study of limits, that is, mathematical asymptotics. And the central reason why 'reduces to' is so problematic is the fact that the limits involved are usually singular. These singular limits should be not regarded as a nuisance, and certainly not as deficiencies of the more general theories. On the contrary, they should be embraced with enthusiasm, because they are responsible for fundamental phenomena inhabiting the borderlands between theories—phenomena at the forefront of physics research, such as critical phenomena in statistical mechanics, fluid turbulence and the universal statistics of the energy levels of highly excited quantum systems.

CRV fully appreciate these ideas; hence their subtitle *The Importance of Being Borderline*. And they explore them in depth and in detail. There are some philosophers who have grasped the significance of singularities in mathematical asymptotics for the understanding of theory reduction—Robert Batterman and Alisa Bokulich come to mind—but they remain a minority. The value of CRV's account is that it is the first full-length and wide-ranging exposition of this point of view by physicists who are sensitive to the concerns of philosophers.

Bristol, UK, December 2013 Michael Berry

Brit. J. Phil. Sci. **61** (2010), 889–895

REVIEW

ALISA BOKULICH
Reexamining the Quantum-Classical Relation: Beyond Reductionism and Pluralism

Cambridge, Cambridge University Press 2008, 208 pp.
£42.00 (Hardback), ISBN: 978-0-521-85720-8

Michael Berry
H H Wills Physics Laboratory, Tyndall Avenue,
Bristol BS8 1TL, UK
tracie.anderson@bristol.ac.uk

Quantum mechanics and classical mechanics are magnificent structures, each with vast explanatory reach. In their usual formulations, the two theories look very different. Classical physics describes the motion of particles in terms of their positions and velocities, influenced by forces acting between them and from outside. In quantum physics, dynamical variables are operators, acting on states in a Hilbert space of vectors, with evolution determined by a Hamiltonian operator. Nevertheless, there is considerable overlap in the phenomena they describe. Although nobody planning an extraterrestrial mission would use Schrödinger's equation as a starting point for programming the trajectories of rockets, few physicists doubt that quantum mechanics applies to planets and spacecraft as well as atoms. The subtle and intricate relations between the classical and quantum worlds are the subject of this very welcome book by Alisa Bokulich.

The quantum-classical connection is a special case of the philosophical problem of theory reduction. As Bokulich explains, conventional approaches to reduction are centred on two contrasting views. In the first ('imperialism' or 'theoretical serial monogamy'), a more general theory, once conceived and validated by experiment or observation, immediately supersedes its less general predecessor theory. Thus, chemistry is regarded as a branch of quantum physics, notwithstanding difficulties in calculating molecular structure and spectra or reaction rates. In the second approach ('isolationism' or 'promiscuous realism'), we live in a 'dappled world', in which each science maintains its separate domain of applicability, with chemistry, biology, geology, etc., retaining their repertoires of concepts and techniques. Bokulich rejects

doi:10.1093/bjps/axq022

Reprinted from The British Journal for the Philosophy of Science, Volume 61, (2010).

this polarization, and most physicists who reflect on their craft would agree with her.

She begins her analysis of the relations between quantum and classical physics by a sensitive and detailed exposition of the views of three quantum pioneers. Werner Heisenberg regarded classical and quantum mechanics as perfect, complete, and separate structures that will never be changed, each describing phenomena in their separate domains of applicability. Nowadays this seems a strangely restricted view; as Bokulich points out, it not only fails to delineate what the 'separate domains of applicability' are, but is also discordant with the manner in which classical concepts played an essential role in Heisenberg's creation of quantum mechanics. Niels Bohr's view was more nuanced. For him, quantum mechanics is a 'rational generalization' of classical mechanics, involving correspondences between the two theories at every level. Usually, Bohr's 'correspondence principle' is interpreted narrowly, as relating frequencies of light involved in transitions between states with large quantum numbers to the frequencies of associated classical trajectories (Bokulich does not emphasize that the quantum numbers must be close as well as large). A more general interpretation of correspondence is that classical and quantum predictions must agree in the limit when Planck's constant h can be neglected. Even this fails to capture what Bohr intended, which is that classical concepts permeate quantum physics not just in the 'classical limit' but at every level. Paul Dirac's view was very different, perhaps reflecting his education in electrical engineering and applied mathematics. For him, both classical and quantum mechanics are 'open theories': approximate descriptions of phenomena, evolving and changing in response to new experiments and theoretical insights.

Of the three approaches, Dirac's is closest to that advocated by Bokulich, but there are two important features ignored by the pioneers and also by almost all philosophers of science with the notable exception of Robert Batterman. The first, that she discusses but in my view does not emphasize enough, is that the limit $h \to 0$ is mathematically singular. This was missed by the pioneers because of their emphasis on the connections between the two theories at the level of formalism (e.g., relating classical Poisson brackets to the commutators of quantum operators), rather than the much richer and more subtle connections between the solutions of the formalism (e.g., wavefunctions and energy levels). Bokulich illustrates the singular limit with the example of particles encountering a potential barrier with sufficient energy to cross it: classically, they slow down but eventually all are transmitted; quantum mechanically some are reflected, the fraction being exponentially small in h and therefore mathematically singular. This is an unnecessarily sophisticated example; the singular limit arises even in the most elementary situation where two equally intense beams of particles overlap. Classically, the intensity is the sum of those in the separate beams $(1 + 1 = 2)$. Quantum mechanically, the beams interfere, and the spacing of the fringes separating regions of constructive and destructive interference is proportional to h; across the fringes the intensity varies from zero and four times that of each beam $(1 + 1 \neq 2)$ and the classical limit is achieved only after spatial averaging ('decoherence'), reflecting the inability of experiment to resolve fringes on infinitely fine scales. A more far-reaching manifestation of the singularity is that the classical limit $h \to 0$ and the long-time limit $t \to \infty$ cannot be interchanged. This apparently arcane observation resolves much of the confusion about the quantal implications of classical chaos, i.e., persistent instability, which emerges in the long-time limit: confined quantum systems can mimic chaotically evolving classical ones, but ultimately (after times that get larger as h gets smaller) the chaos is suppressed.

The second important previously neglected feature is Bokulich's central and original contribution: classical structures, far from being redundant anachronisms superseded by concepts from quantum mechanics, are playing an important role in explaining quantum phenomena. The explanations are based on mathematical approximations to the full quantum theory that have come to be known collectively as 'semiclassical mechanics'. As vividly expressed in an adaptation of a quotation attributed to Boris Kinber in the related context of the ray limit of wave optics, this is 'sewing the quantum flesh on the classical bones'. Bokulich elaborates this point of view with three case studies, which form the intellectual core of her book.

The first is the ground state of the helium atom: the lowest energy level of two electrons repelling each other and attracted by the nucleus. This played an important historical role in the development of quantum mechanics. When Bohr created what has come to be called 'old quantum theory', he explained the energy levels of the single electron in the hydrogen atom by postulating quantum rules that selected particular classical orbits. This spectacular success stimulated others to try to generalize the rules to determine the energy levels of more complicated systems. Of these attempts, the most far-reaching was Einstein's, but even his generalization could be applied only to classical trajectories that were multiply periodic; in modern terms, non-chaotic trajectories. In particular, nobody succeeded in quantizing helium. This failure led to old quantum theory being regarded as merely a way-station leading to the full quantum theory, within which accurate numerical schemes for solving the Schrödinger equation enabled the ground state of helium to be determined in agreement with experiments. These computations established that quantum mechanics gives a correct description, but did not provide insight into the nature of the quantum state of the electrons. Only in the early 1990s was it realized, by the late Dieter Wintgen and his colleagues, that the failure of early attempts to quantize helium was not a failure of old quantum theory but arose from inability to identify the correct classical structure in this example of the three-body problem (two electrons and the nucleus). When the relevant classical orbits were identified, and incorporated into a modern semiclassical approximation to quantum mechanics, the energy thus calculated agreed closely with the 'exact' energy that had been computed from Schrödinger's equation. This study broke new ground in the classical three-body problem, an example (in the spirit of Dirac's 'open theory' approach to science) of quantum physics leading to deeper understanding of classical physics.

Why bother—why make an approximate calculation when an exact formalism is available? Bokulich offers three reasons, which also apply to her two other examples:

> First, the semiclassical treatments provide an *investigative tool* [. . .] to investigate physical domains that might not yet be accessible either experimentally or with a fully quantum calculation. Second, they provide a *calculational tool*: semiclassical calculations [. . .] can be less cumbersome than full quantum calculations. Finally, they provide an *interpretive tool*: [. . .] physical insight into the structure of a problem, in the way that a fully quantum-mechanical approach might not.

The second case study is the spectrum of high excited energy levels of atoms in strong magnetic fields; these are called Rydberg atoms. Bokulich writes that

> These atoms call to mind Tom Stoppard's play *Hapgood*, in which he writes 'there is a straight ladder from the atom to the grain of sand, and the only real mystery in physics is the missing rung. Below it, [quantum] particle physics; above it, classical physics; but in between, metaphysics' [. . .]. As an atom that *is* the size of a grain of sand, Rydberg atoms are ideal tools for studying the 'metaphysics' of the relation between classical and quantum mechanics.

In the 1960s, experiment had revealed unexpected resonant structure in the spectrum, associated with the outermost electron, for energy ranges in which this electron was expected to be torn off, leading to an ionized atom with a much simpler spectrum. It took twenty years for the explanation to emerge, and as with helium the concepts were semiclassical. The first step was to understand that motion of the electron is largely chaotic, as the result of conflict between the elliptical 'Kepler' orbits the electron would have under the action of the nucleus alone, without the magnetic field, and the helical paths it would have in the magnetic field alone. The second step was to modify a semiclassical theory that had been developed by Martin Gutzwiller, in which the quantum spectrum—the collective of energy levels—is related to the spectrum of classical periodic orbits, that is, the set of those orbits that repeatedly traverse the same path. One modification was to smooth the spectrum, reflecting the finite resolution of the experiments; this had been anticipated by Roger Balian and Claude Bloch, and led to a representation in terms of a small subset of the infinity of classical orbits. The more fundamental modification, by John Delos and collaborators, was to realize that the force the electron experiences has a singularity at the position of the atomic nucleus, so the relevant orbits are those that start and end at the position of the nucleus: closed, but not periodic. Then a semiclassical calculation, exquisitely blending classical concepts with the quantum notion of wave interference, succeeded in reproducing the full variety of the observed spectrum.

The third case study concerns calculations for 'quantum billiards': wavefunctions representing quantum particles confined in planar enclosures where the corresponding classical trajectories would be chaotic. Standard quantum algorithms enable the states to be computed. In 1984, Eric Heller made the important observation that some of the states exhibit regions of high intensity, centred on classical periodic orbits, and gave a semiclassical argument indicating why such 'scars' should exist, even though the orbits are all unstable—yet another example of classical reasoning giving insight into a quantum phenomenon.

Bokulich seems to imply that scars conflict with my 1977 prediction that the wavefunctions of highly excited states can be modelled by random functions represented by many interfering waves, based on the semiclassical argument that typical chaotic trajectories pass many times through the same region. In fact there is no conflict, and the reason is an instructive illustration of the subtlety of the semiclassical quantum domain. Heller's scars are an asymptotic phenomenon, in the sense that the scars get more distinct for high excited states where they are less obscured by the finite quantum wavelength. But they are transitory asymptotic phenomena: although there are always some scarred states, they get rarer higher in the spectrum, and are of measure zero in the extreme asymptotic regime. There, almost all states are not concentrated on periodic orbits; indeed, they explore the enclosure uniformly on average (according to an earlier theorem by Alexander Shnirelman), and randomly on finer scales, both features reflecting the behaviour of the classical billiard motion.

The co-existence of scars and random waves without contradiction provides a fine illustration of another distinction that Bokulich mentions but does not explore, referring instead to discussions of it by Robert Batterman: between 'universal' and 'particular' phenomena. This entered physics during the 1960s in the attempt to understand how the thermodynamic properties of materials emerged from the statistical mechanics of their microscopic constituents. The particular phenomena are system specific and distinguish one material from another; they include the boiling point of water and the temperature at which iron loses its magnetism. More interesting and fundamental are universal phenomena, such as the manner in which the compressibility becomes infinite as critical points are approached, according to scaling laws that are quantitatively identical for a vast range of materials. The understanding of this universality generated deep and unanticipated insights, including the emergence of fractal structures on all scales, obstructing the smooth reduction of statistical physics to thermodynamics: the many-particle limit is singular, as is the classical limit $h \to 0$. In quantum billiards, scars are particular phenomena, because the individual periodic orbits on which the wave intensity is concentrated depend on the shape of the billiard boundary. The random-wave regime is universal: quantitatively the same for all boundary shapes provided only that the corresponding classical motion is chaotic.

The distinction between the universal and the particular appears not only in the morphology of quantum wavefunctions of classically chaotic systems, but also in the arrangement of their high-lying energy levels. The universal features are the statistics of correlations between close-lying levels, for example, the probability distribution of the spacings between nearest neighbours. In a footnote, Bokulich speculates '[...] that one could cook up a [...] statistical asymptotic agreement between classical and quantum mechanics [...] without having [...] law-like correspondence between the classical and quantum structures.' Exactly this 'statistical asymptotic agreement' is provided by random-matrix theory, a branch of mathematics that is logically independent of quantum theory and has many applications in other areas of science but which reproduces with high accuracy the statistics of energy levels. The explanation of random-matrix universality is an application of Gutzwiller's relation between spectra and periodic orbits. According to this, short-range structure in the energy spectrum is associated with very long classical periodic orbits—a connection that is a consequence of the uncertainty principle relating time and energy. But long classical orbits display a beautiful universality of their own, discovered in the early 1980s by John Hannay and Alfredo Ozorio de Almeida, and it soon became clear that the quantum universality follows from this, though the details are subtle and still being elaborated. An unanticipated implication of the connection between spectra and periodic orbits was that correlations between distant levels are associated with short orbits, and because these are not universal the level correlations will not be universal either; these are the particular phenomena, depending on the details of the classical system, just like wavefunction scars in quantum billiards.

Bokulich's detailed case studies raise an important question, to which she devotes the final chapters. The discovery of quantum mechanics as a deeper theory revealed classical trajectories as temporary structures that dissolve under close scrutiny and can now be discarded. How can nonexistent objects form the basis of explanation and give insight, as semiclassical physics clearly indicates? The same problem arose in optics, where the explanatory structures are the caustics: singularities on which ray families are focused, clearly visible as the brightest features in optical images (e.g., the dancing lines of focused sunlight on the bottoms of swimming pools). In the 1960s, it became clear, through the mathematical discoveries of René Thom and Vladimir Arnold, that caustics are restricted to certain universal forms. It was soon realized that this universality extends to the wave patterns that decorate caustics in the limiting regime of small wavelengths and smooth away their singularities. Some people wondered why we were using properties of nonexistent singularities, when the same phenomena could be 'explained' without them, by numerically solving the fundamental wave equations.

I confess to being puzzled that people find these questions puzzling. The explanatory structures (periodic orbits, ray caustics) are models, constructs that help our understanding. We do not find it problematic that the answer to the question 'Where is my dinner?' is 'On the table', even though a table is our convenient name for a model, assigning significance to what we now know to be an assembly of molecules consisting of atoms whose electrons and nuclei move in largely empty space. Bokulich surveys several philosophical theories of how models can explain, and concludes that none of them can give a satisfactory account of current understanding based on semiclassical analysis as in her three case studies. She ends by sketching her own 'interstructuralism': an attempt to replace the contrasting ideas of the imperialist and isolationist theories with an approach that combines aspects of both, closer to Bohr and Dirac than to Heisenberg. Her conclusions justify the explanatory power of concepts from a superseded theory when used in the theory that replaces it. In brief, 'fictions can explain'.

Bokulich fully appreciates many subtleties that practicing physicists occasionally understand intuitively, but are rarely explicit about. Her ideas are refreshing and original and presented with clarity and erudition. I unreservedly recommend her book to anyone wanting to understand the intricate connections between the classical and quantum worlds.

Chapter 7
Tributes

These thirteen contributions record my appreciation of some scientists I greatly admire. With the exception of Dirac [7.5], they are not on everyone's all-time favourites list (Galileo, Newton, Maxwell, Einstein...), but they have influenced the content and style of my science. My debt to my supervisor Dingle [7.8] should be clear from Chapter 4. John Ziman ([7.7], see also [B394]) was the intellectual attractor who brought me to Bristol.

The untimely death of S. Pancharatnam [7.2] meant that I never met him, and learned only later ([B167, B212], and [1.7]) about his seminal geometric phase research in polarization optics. His other writings, and unrelated chance comments by Anton Zeilinger and Nimrod Moiseyev, led me to a variety of physical applications [B293, B350, B355, B372] of the mathematics (previously unfamiliar to me) of degeneracies of non-Hermitian operators.

I also never met Pancharatnam's uncle, the optical scientist C. V. Raman [7.1]. But I became personally friendly with Raman's son, the radio astronomer and intrepid lone sailor and flyer V. Radhakrishnan. And I had the pleasure of knowing his other nephews: the astrophysicist S. Chandrasekhar, the liquid crystal scientist S. Chandrasekhar, and the crystallographer S. Ramaseshan (to whom Mark Dennis and I dedicated [B355] on his and John Nye's 80th birthdays). A brilliant scientific dynasty, now gone.

Space prevents me describing my researches inspired by the fractals of Benoit Mandelbrot ([7.6], see also [B112, B80, B82, B149, B274, B275, B283, B334, B336]). He too became a personal friend — see [B453].

Independence of mind

Michael Berry

Journey into Light: Life and Science of C.V. Raman. By G. Venkataraman. *Indian Academy of Sciences: 1988. Pp. 570. Distributed in Britain and the United States by Oxford University Press, £22.50.*

THE past two years have seen centenaries of the birth of several of India's most brilliant citizens: Nehru, founder of the modern state; Ramanujan, diviner of amazing mathematical formulae; and Raman, the prolific and original physicist whom this book celebrates. Raman's life and work were complex and many-sided, causing great difficulties of judgement and emphasis for any biographer. Venkataraman (who in spite of his name does not share with several other talented Indian scientists the distinction of being a relative of Raman) solves these problems by anchoring his story firmly in the science. He has produced a model of scientific biography, written with respect and affection for its subject but with a clear-eyed perception of his faults, in a relaxed and gently witty style.

Raman's work was on the physics of waves and overwhelmingly centred on optics. He is best known for the effect that bears his name, but that was not discovered until he was 40 years old, when he was already established internationally as an authority on a variety of subtle, classical interference and diffraction phenomena such as the colours of heated metals and layers of bubbles. He had also studied the production of musical tones. An interesting discovery was that Indian drums have skins whose thickness varies radially in such a way that the frequencies of the overtones are integer multiples of the fundamental, so that they sound more harmonious than their Western counterparts, whose skins are radially uniform and generate irrational overtones.

The Raman effect is the scattering of light with a change of frequency (that is, inelastic scattering); it occurs in liquids, solids and isolated molecules. The change in frequency occurs when the light exchanges energy with the internal vibration or rotation states of whatever scatters the light. Compared with the elastic, or Rayleigh, scattering, for which the incident and scattered beams have the same frequency, the Raman effect is weak, which is why it was not discovered earlier. Raman (with Krishnan) found it by deploying simple apparatus with great experimental skill. The discovery was announced in *Nature*. Venkataraman mentions anecdotal evidence that the paper was first rejected. New evidence confirms this: apparently there were two unfavourable referees, both Fellows of the Royal Society, who provoked in the editor (Sir Richard Gregory) the opinion that any paper inspiring such vehement opposition must have something good in it and so should be published! The paper stimulated a great deal of immediate and continuing interest, not only by demonstrating a fundamental optical effect fully concordant with the (then) new quantum mechanics but also by providing a sensitive probe of the vibrational and (for molecules) rotational structure of matter.

Raman's creative work did not stop

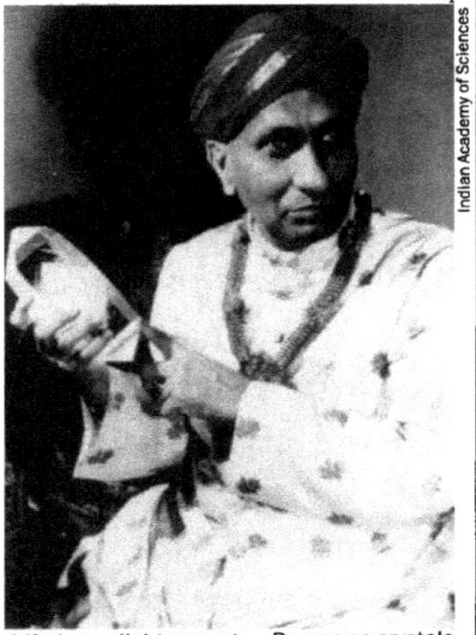
Indian Academy of Sciences

Offering enlightenment — Raman on crystals.

with the Raman effect. I was surprised to learn that he discovered or anticipated several phenomena usually regarded as having been found decades later by others. One of these is what is now called 'speckle'. This is the mottled appearance of coherent light scattered by static randomness such as grains of powder on a screen or irregularities on a rough painted wall. Nowadays speckle is a familiar accompaniment of images produced by lasers, but Raman (with Ramachandran) saw it in light from a mercury lamp filtered by a pinhole. He had a complete understanding of the phenomenon, 20 years before lasers were invented.

Another anticipation is of what are now called 'soft modes'. A soft mode in a solid is a vibration whose frequency vanishes as a critical temperature is approached. The vanishing indicates the weakening of the structure, which changes abruptly at the critical temperature. Raman (with Nedungadi), using Raman spectroscopy, observed the decrease in a lattice vibration frequency of quartz and correctly understood its association with a structural phase transition.

A high point of Raman's work after the discovery of the Raman effect was his theory (with Nath) of the diffraction patterns produced by light traversing refractive-index corrugations made by irradiating a liquid with ultrasound. As Venkataraman puts it, in a chapter charmingly titled "Son et lumière",

> Raman loved waves, and this problem had light waves as well as sound waves. What more could he ask for? . . . The full-fledged Raman–Nath theory was put to a stringent test only in 1963 . . . [and] is as good as one would want it to be, provided one takes the trouble of carefully solving the equations.

Recently I learned that this theory too is an anticipation: exactly the same equations have been rediscovered in the theory of free-electron lasers.

The failures of great scientists are often as interesting as their successes, and Raman had his share of failures. Perhaps the best known is his controversy with Born about the vibration spectrum of diamond. Because of a persistent mathematical misunderstanding of the dynamics of crystal lattices, he never accepted the fact that the vibration frequencies form a continuum. He predicted a discrete spectrum, on the basis of a 'quasi-molecular' theory of his own, and he carried out experiments that seemed to confirm it. As Venkataraman explains, Raman's theory, although wrong, picked out a particular subset of the vibrational modes, at which the spectrum has singularities that Born did not know about but whose existence was later demonstrated by Van Hove. His experiments had fairly low resolution and so he saw only the singularities, masquerading as sharp lines. In other words,

> Raman was locally correct but globally wrong. The part he got right is an important part and reflects his native brilliance. He would not have erred the way he did had he been trained properly in quantum physics.

I long suspected that Raman never properly understood the connection between wave optics and geometrical optics. That suspicion is confirmed by the account here of his ideas about the mirage. By imagining the air above a hot surface as a stack of slabs of slightly different refractive indices, he convinced himself that refraction could never make a downward-sloping ray turn upwards, because a ray once horizontal would remain so. In other words, refraction could never simulate reflection. This is a misunderstanding of the law of refraction in a continuously varying medium; the same argument, applied to an obliquely fired projectile, would predict that it would never fall but would continue horizontally on reaching its greatest height. The mistake led him to think the mirage could be explained only

Reprinted from Nature, Volume 338, (1989).

SPRING BOOKS

with a wave theory. He constructed such a theory, and it was correct, but rather than being an anticipation of later work, it was a rediscovery of a result published by Airy in 1838.

Raman's collaborator on the theory of the mirage was his talented nephew Pancharatnam, fated soon to die at a tragically early age. Venkataraman gives a detailed account of Pancharatnam's ideas on the interference of polarized light. Only now are we beginning to appreciate their originality and depth — for example his anticipation of the geometric phases now in fashion.

These and many other areas of physics are explained with elegance and clarity. The author is a theoretical physicist with a flair for simple and direct exposition. Sometimes more sophisticated mathematical treatments are merited, and these are given in separate sections so as not to interrupt the main narrative.

Raman seems to have been a difficult and obstinate man. His dealings with administrators and bureaucrats, men of narrow vision compared to him, were stormy, and he made enemies among his colleagues. He became embroiled in controversy and resigned in rage, first from his chair in Calcutta and then from the directorship of the Indian Institute of Science in Bangalore. Finally he founded his own Raman Research Institute in Bangalore. That Institute continues today as a jewel in the crown of Indian science, a tribute to Raman and his successor Radhakrishnan. Venkataraman discusses these matters in great detail and with sensitivity. His book makes us appreciate how unusual Raman must have been to produce such an abundance of creative science, first in the stifling colonial atmosphere of British India, and then in the turbulent struggle for independence. □

Michael Berry is a Professor in the H.H. Wills Physics Laboratory, University of Bristol, Tyndall Avenue, Bristol BS8 1TL, UK.

• Oxford University Press will also be the distributor of Raman's collected works, in six volumes and under the title *Scientific Papers of C.V. Raman*. The books will become available in July–August, price £25, $49.95 per volume.

Pancharatnam, virtuoso of the Poincaré sphere: an appreciation

Michael Berry

H. H. Wills Physics Laboratory, Tyndall Avenue, Bristol BS8 1TL, UK.

SIVARAJ Ramaseshan graciously invited me to write an essay review of the collected works of S. Pancharatnam. As a partial response to this invitation, I am happy to show my admiration of Pancharatnam by providing the following comments on three of his papers.

Geometric phases in polarization optics

Early in 1987, I received from Rajaram Nityananda a reprint of his paper[1] with Ramaseshan, drawing attention to Pancharatnam's anticipation, in 1956, of the geometric phases that had been in fashion since 1983 (and to some extent still are). Regrettably, I did not read their paper properly, and missed its main point. But I was fortunate to visit Bangalore in July 1987, and Ramaseshan met me and I came to appreciate the relevance of Pancharatnam's paper[2]. Ramaseshan honoured me with what he thought was the last copy of Pancharatnam's collected works[3] (now, more have come to light). Slowly, I realized that Pancharatnam's phase was something I had to understand. This was not easy because his arguments made heavy use of the geometry of the Poincaré sphere, which I knew about but had little facility with. For the long flight home, I set myself the task of interpreting Pancharatnam's discovery in more modern and general terms.

That flight was a revelation. I learned that not only had this young fellow of twenty-two created the simplest example of the geometric phase, but that he had also pointed out a feature (the definition of phase difference described below) that had not, by 1987, been perceived in any of the many papers developing my work[4] of 1983. There and then I decided to follow the example of Nityananda and Ramaseshan, and write an exposition[5] of Pancharatnam's phase that would bring the full originality of its conception to a larger readership.

Pancharatnam was inspired by his mentor C. V. Raman to study the complicated interference figures produced by light beams traversing crystal plates (see page 232, box). For this he needed to compare the phases of waves in different states of polarization. A given state of polarization (given for example by the eccentricity, axes and sense of traversal of the ellipse described by the tip of the electric $\boldsymbol{E}$ or $\boldsymbol{D}$ vector) is represented by a point on the Poincaré sphere. However, this does not specify the phase of the vibration: all states related by a phase factor correspond to the same point on the sphere. Therefore phase is an additional quantity, to be attached to points on the sphere if light beams are to be described completely.

To understand interference, it is necessary to know not the absolute phase of a wave but the phase difference between light beams in different states of polarization. Apparently, nobody had asked this simple question. Pancharatnam did ask it, and gave a simple answer: the phase difference between two beams is that phase change which when applied to one of them maximizes the intensity of their superposition. It was instructive[5] to express this in terms of more general mathematics (as used for example in quantum mechanics). Each beam is represented by a spinor with two complex components (corresponding for example to the amplitudes and phases of the components of the $\boldsymbol{D}$ vector perpendicular to the propagation direction). For two beams $|A\rangle$ and $|B\rangle$, Pancharatnam's rule implies that the phase difference is the phase of their complex scalar product:

$$\text{phase difference between } |A\rangle \text{ and } |B\rangle = \text{phase of } \langle A|B\rangle. \tag{1}$$

In particular, the beams are in phase (intensity of overlap a maximum) if the scalar product is real and positive. Nowadays this is known as 'Pancharatnam's connection'. It can be derived from a more general rule, in which $|A\rangle$ is transported to a neighbouring state $|A + \mathrm{d}A\rangle$ (for example by passing the beam through an appropriate polarizing crystal), with the phase defined by the requirement

$$\langle A|\,\mathrm{d}\,|A\rangle = 0, \text{ where } \mathrm{d}|A\rangle \equiv |A + \mathrm{d}A\rangle - |A\rangle. \tag{2}$$

It can be shown[5] that if this is integrated from $|A\rangle$ to $|B\rangle$ along the shorter arc of the great circle connecting their representative points on the sphere, then $|B\rangle$ is in phase with $|A\rangle$. The rule (2) is called parallel transport because it corresponds to moving $|A\rangle$, regarded as a unit vector perpendicular to the radius of the sphere and attached to its representative point, without turning it about the radius.

Pancharatnam showed[2] that this natural stipulation of phase has a remarkable property. It is non-transitive:

Reprinted from Current Science, Volume 67, Issue 4, (1994).

if $|A\rangle$ is in phase with $|B\rangle$, and $|B\rangle$ is in phase with a third state $|C\rangle$, then $|C\rangle$ need not be in phase with $|A\rangle$. Indeed, if $|C\rangle$ is in phase with a state $|A'\rangle$ represented by the same point on the sphere as $|A\rangle$, then the phase factor accumulated after this cycle of polarization states, embodying the phase difference between $|A\rangle$ and $|A'\rangle$, is

$$\langle A|A'\rangle = \exp\{-\tfrac{1}{2}\,\mathrm{i}\,\Omega_{ABC}\}, \qquad (3)$$

where Ω_{ABC} is the solid angle of the spherical triangle ABC. It can be shown that the same result holds for more general cycles (e.g. smooth loops) when the state is continued according to the connection (2). Then the solid angle is that of the loop, and the fact that the vector does not return to its original direction reflects the nonintegrability of this connection. What Pancharatnam actually stated was not exactly the result (3), but an unsymmetrical version equivalent to it: if two beams $|A\rangle$ and $|B\rangle$, which are in phase, are passed through an analyser bringing them to the state $|C\rangle$, their phase difference is $-\frac{1}{2}\Omega_{ABC}$.

In its original form, the geometric phase I found in 1983 concerned not polarization states of light beams, but quantum states of particles, that is, Hilbert-space vectors satisfying the Schrödinger equation. In the adiabatic limit where the environment of the system (that is, its Hamiltonian) changes slowly, the parallel transport law (2) applies, *mutatis mutandis*, and, being nonintegrable, generates a geometric phase factor when parameters on which the system depends are taken round a cycle. The generalization of the solid angle is the flux through the cycle (in parameter space) of a certain 'phase field' (mathematically, a 2-form)[4].

In special cases where the phase involves an actual solid angle, this is because the phase field is that of a monopole in parameter space. For spinning particles with angular momentum $n\hbar$, the strength of the monopole is $-n$, so the phase is $-n\Omega$. Therefore the Pancharatnam situation is analogous to the further specialization to spin 1/2, which, as is well known, is a model for any 2-state system. There was an apparent discordance here, between the fact that photons have spin 1, leading to geometric phases of $\pm\Omega$ for smooth cycles of their spin direction (e.g. in coiled optical fibres[6]), and Pancharatnam's $-1/2\Omega$. But it was clear[5] that there is no contradiction, because the two solid angles are different: the first is in the space of propagation directions $\boldsymbol{k}$ for light with a fixed state of polarization, whereas the second is on the Poincaré sphere of polarization states for light travelling in a fixed direction $\boldsymbol{k}$, so that, because of transversality, there are only two states (e.g. components of $\boldsymbol{D}$ perpendicular to $\boldsymbol{k}$).

Ignorant of Pancharatnam's work, I had also calculated geometric phases for the case he studied, that is in a light beam whose state of polarization is changed. I envisaged[5,7] a medium whose birefringence and gyrotropy were changed smoothly along the beam path – for example by a varying electric field (Cotton–Mouton effect) and a varying magnetic field (Faraday effect). I was astonished (and not a little humbled) to learn that Pancharatnam had had essentially the same idea (with discrete, rather than continuous polarization changes) thirty years earlier. This was one of several areas in which the geometric phase had been anticipated[8].

In spite of its startling originality, Pancharatnam's paper[2] was completely ignored until Ramaseshan and Nityananda[1] made us aware of it. Now his contribution is properly recognized, and his paper has been cited many times. I will not review subsequent work (Bhandari does so in his accompanying article), but will simply draw attention to two very different applications of the idea. In one, Schmitzer, Klein and Dultz[9] remark that the solid angle Ω_{ABC} can change much faster than the angle of rotation of an analyser (representing C with A and B fixed), and propose this as the basis of a new type of optical switch. In the other, Nye[10] employs Pancharatnam's connection in a study of phase and polarization singularities in electromagnetic waves that vary in space.

Mirages

Although the young Pancharatnam and the elderly Raman were intellectually (as well as consanguinously) close, they wrote only one paper[11] together, about the mirage. When I read it (on the 1987 flight) I realized that it was based on a misunderstanding of the connection between wave optics and geometrical optics. I had suspected confusion on this point since a conversation in 1976 in which Ramaseshan described the Raman–Pancharatnam ideas to me.

Their claim was that the mirage cannot be explained in terms of rays but requires the wave theory. By considering light propagating in the air above a hot surface, considered as a stack of slabs with slightly different refractive indices, they convinced themselves that refraction could never make a downward-sloping ray turn upwards, because a ray once horizontal would remain so. In other words, refraction could never simulate reflection. This was a misunderstanding of the law of refraction in a continuously-varying medium, based on an incorrect limiting process. When applied to mechanics, the same argument (upside-down) would predict that an obliquely-fired projectile would never fall but would continue horizontally on reaching its greatest height. The correct slab limit shows that transverse gradients of refractive index cause rays to curve, just as transverse forces cause particle paths to curve.

Why dwell on a mistake, in what is supposed to be an appreciation? To make the point that the errors of first-rate scientists can be both instructive and productive. After concluding, wrongly, that the mirage cannot be a refraction effect, they set about constructing a wave theory for the neighbourhood of the layer where, according to them, the rays ought to turn but do not. This is the level of the caustic, where the air acts like a mirror. By using the mathematical analogy between the light wave equation with linearly varying refractive index and Schrödinger's equation for quantum particles in a region of constant force, they expressed the wave in terms of Bessel functions of order ± 1/3. Although this wave theory was constructed on the basis of a confusion, there is no doubt that it is correct. Indeed they successfully carried out experiments with heated plates to test several of its consequences.

One irony remains. In contrast to the startling originality and prescience of the polarization phase, the Raman–Pancharatnam theory of waves near a caustic turned out to be not a discovery but a rediscovery. Airy[12] had formulated essentially the same theory in 1838.

Propagation along singular crystal axes

A treasure among Pancharatnam's papers, that, if anything, pleased me more than his phase, is a surprising special case[13] of wave propagation in absorbing anisotropic crystals. In transparent crystals, the two polarization states that travel unchanged through the crystal in any direction are orthogonal (and so represented by antipodes on the Poincaré sphere). With absorption, the eigenpolarizations are no longer orthogonal, and there are even particular crystals and propagation directions (along the 'singular axes') for which they coincide, so that only one polarization can propagate.

Pancharatnam concentrated on this case, and studied what happens when the orthogonal polarization, which cannot propagate, is introduced into the crystal. Earlier authors were, he claimed, wrong: they had thought such a wave must be reflected. He argued that, on the contrary, the polarization would change gradually into the state that does propagate, and moreover would grow stronger (by a factor increasing linearly with distance) than a wave of the same intensity introduced with the correct polarization. I found this conclusion hard to believe, and his arguments (again based on Poincaré sphere geometry, see page 233, Ranganath, G. S., this issue) difficult to follow, and embarked on the task of reconstructing the theory in my own way (as I have already mentioned, it was a long flight). The results vindicated Pancharatnam completely, and moreover produced an analytical expression for the wave which appears nowhere in his papers (although I believe he must have known it). The expression is worth presenting.

Let the beam travel in the z direction. A model crystal exhibiting Pancharatnam's phenomenon has dielectric tensor with transverse components

$$\mathbf{e} = \begin{pmatrix} \varepsilon_{xx} & \varepsilon_{xy} \\ \varepsilon_{yx} & \varepsilon_{yy} \end{pmatrix} = \begin{pmatrix} ia & \frac{1}{2}(b-a) \\ \frac{1}{2}(b-a) & ib \end{pmatrix}. \quad (4)$$

This matrix has a single degenerate eigenvalue and a single eigenvector col(1, i)/√2, representing circularly polarized light. For the medium to be anisotropic, we must have $a \neq b$. For the medium to be absorbing, the rate of energy dissipation must be positive, which implies that the Hermitian matrix $\mathrm{i}(\mathbf{e}^{\dagger} - \mathbf{e})$ must have positive eigenvalues; when applied to (4) this gives the requirement

$$\mathrm{Re}\,(a+b) > |a-b|, \quad (5)$$

which is satisfied by a and b real. If the only other non-zero dielectric tensor component is ε_{zz}, then not only the electric $\boldsymbol{D}$ but also the $\boldsymbol{E}$ vector is transverse, and Maxwell's equations reduce to

$$\partial_z^2 \boldsymbol{D} + k^2 \mathbf{e} \cdot \boldsymbol{D} = 0, \quad (6)$$

where k is the free-space wave number.

The solution of (6) given (4) with the initial polarization

$$\boldsymbol{D}(0) = \begin{pmatrix} u \\ v \end{pmatrix} \quad (7)$$

is

$$\boldsymbol{D}(z) = \exp\{-\tfrac{1}{2}kz(1-\mathrm{i})\sqrt{a+b}\} \times \left[\begin{pmatrix} u \\ v \end{pmatrix} - \frac{1}{4}kz\frac{(a-b)}{\sqrt{a+b}}(1-\mathrm{i})(u+\mathrm{i}v)\begin{pmatrix} 1 \\ \mathrm{i} \end{pmatrix}\right], \quad (8)$$

as can be confirmed by substitution. For the eigenpolarization, $u = -\mathrm{i}v = 1/\sqrt{2}$, and the term involving z is zero. For any other initial polarization – in particular the orthogonal one $u = +\mathrm{i}v$ – the term involving z grows relative to the term not involving z, just as Pancharatnam discovered. (In spite of the increasing z factor, the energy flow into the medium must decay, because the medium is absorbing, and this is ensured by the exponential prefactor. A direct proof from (8) is lengthy, but the result follows from the fact that the dissipation rate (positive) is minus the divergence of the Poynting vector.)

Pancharatnam's law of singular axis propagation is an example of a general mathematical phenomenon: evolution driven by a non-Hermitian matrix **M** with a degenerate eigenvalue. The lack of Hermeticity is fundamental, because it drastically alters the nature of the degeneracy. If **M** is Hermitian (as in the quantum mechanics of spin), there are (in the simplest case) two orthogonal eigenvectors corresponding to a degenerate eigenvalue, and in the neighbourhood of the degeneracy (in the space of parameters on which **M** depends) the eigenvalues form a double cone, each sheet of which corresponds to an eigenvalue. If **M** is not Hermitian (as with (4)), there is only one eigenvector at the degeneracy, which is a branch point for the two eigenvalue sheets in its neighbourhood, and around which each eigenvector turns into the other. While writing this I found that the mirror property of a stack of transparent plates (such as glass microscope-slide cover slips or acetate overhead-projector sheets) depends on precisely this phenomenon of a non-Hermitian matrix with a degenerate eigenvalue (I thank S. Klein for posing this problem); so does the matrix governing waves incident at the critical angle on a slab of lower refractive index (I thank G. N. Borzdov for telling me this).

The phenomenon of propagation along singular axes (known as Voigt waves) is now well understood, as a result of extensive and definitive theoretical work by Fedorov and his colleagues[14]. In a crystal slab, the two waves travelling in each direction can be made to degenerate in several ways, leading to coordinate dependence that can be not only linear but quadratic[15] (for three-wave degeneracy) or cubic[16] (for four-wave degeneracy). However, Pancharatnam's pioneering paper is little known and infrequently cited.

Now, as we remember Pancharatnam's untimely death in his creative prime, and celebrate his youthful achievements, it is time to look again through all his work. Who knows what further delicious physics this will reveal?

1. Ramaseshan, S. and Nityananda, R., *Curr. Sci.*, 1986, **55**, 1225–1226.
2. Pancharatnam, S., *Proc. Indian Acad. Sci.*, 1956, **44**, 247–262 (reprinted in ref. 3, pp. 77–92).
3. Pancharatnam, S., *Collected Works*, Oxford University Press, 1975.
4. Berry, M. V., *Proc. R. Soc. London*, 1984, **A392**, 45–57.
5. Berry, M. V., *J. Mod. Opt.*, 1987, **34**, 1401–1407.
6. Tomita, A. and Chiao, R. Y., *Phys. Rev. Lett.*, 1986, **57**, 937–940.
7. Berry, M. V., in *Fundamental Aspects of Quantum Theory* (eds. Gorini, V. and Frigerio, A.), Plenum, 1986, NATO ASI series vol. **144**, pp. 267–278.
8. Berry, M. V., *Phys. Today*, 1990, **43** (12), 34–40.
9. Schmitzer, H., Klein, S. and Dultz, W., *Phys. Rev. Lett.*, 1993, **71**, 1530–1533.
10. Nye, J. F., in *Sir Charles Frank, OBE, FRS: An Eightieth Birthday Tribute* (eds. Chambers, R. G., Enderby, J. E., Keller, A., Lang, A. R. and Steeds, J. W.), Adam Hilger, Bristol, 1991, pp. 220–231.
11. Raman, C. V. and Pancharatnam, S., *Proc. Indian Acad. Sci.*, 1959, **A49**, 251–261 (reprinted in ref. 3, pp. 211–221).
12. Airy, G. B., *Trans. Camb. Phil. Soc.*, 1838, **6**, 379–403.
13. Pancharatnam, S., *Proc. Indian Acad. Sci.*, 1955, **A42**, 86–109 (reprinted in ref. 3, pp. 32–55).
14. Barkovskii, L. M., Borzdov, G. N. and Fedorov, F. I., *J. Mod. Opt.*, 1990, **37**, 85–97.
15. Borzdov, G. N., *J. Mod. Opt.*, 1990, **37**, 281–284.
16. Borzdov, G. N., *Opt. Commun.*, 1990, **75**, 205–207.

Oration delivered by Michael Berry for

Yakir Aharonov

Doctor of Science *honoris causa*, 16 May 1997

Mr Vice-Chancellor:

When Yakir Aharonov began studying physics in the 1950s at the Technion in Haifa, Israel, one of his teachers was Nathan Rosen. Twenty years earlier, Rosen had made an important contribution to the interpretation of the (then rather new) quantum mechanics. But by the time the young Aharonov became entranced by the fundamentals of that theory, its study was unfashionable, and Rosen advised him to concentrate on applications.

This fatherly advice was sabotaged by the arrival in Haifa of David Bohm. At that time Bohm was on an odyssey forced by political persecution in the USA, which was then in the grip of anticommunist hysteria. With Bohm, he began to delve into the forbidden subject, and they wrote several papers together. One was about quantum uncertainty; another was about the meaning of the equation for the electron devised by that great Bristolian physicist Paul Dirac. It was natural that when Cecil Powell offered Bohm intellectual refuge and a lectureship in Bristol, Aharonov would accompany him, and register here for a Ph.D as Bohm's student.

Imagine Aharonov arriving in the Bristol of 1957 — not the most exciting city at that time, I am informed — on his first trip abroad. According to rumour, he was embarrassed by his Israeli accent, and thought it would be a handicap in pursuing the natural enthusiasms of a young man, so he took elocution lessons in an attempt to acquire not a Bristol accent but an Oxford one. He abandoned these efforts when, on the very first occasion he ventured onto the street to try out his new voice, the person he addressed replied in Hebrew!

Back to physics, then, and a spectacular discovery with Bohm, in 1959, that now bears their name. It concerns an aspect of the microscopic behaviour of quantum particles that contrasts dramatically with that of the same particles in the 'classical' mechanics of Newton (which applies on larger scales). The Aharonov-Bohm effect is the ability of charged quantum particles (for example, electrons) to respond to electric and magnetic fields remote from them — classical particles can respond only to fields where they are. Dr (later Professor) Bob Chambers, then himself recently arrived in Bristol, carried out an elegant experiment, helped by a suggestion by Professor (now Sir) Charles Frank, and the Aharonov-Bohm effect was observed, as a shift in the pattern of fringes of interfering electron waves.

Their discovery caused an immediate sensation in the world of physics, as an apparently paradoxical but experimentally confirmed prediction of quantum theory. Over the years its importance has grown as its ramifications have emerged in one area of physics after another: molecules, atoms, solids and, most fundamentally, as a cornerstone of the 'gauge theories' that now dominate the physics of elementary particles.

After leaving Bristol with his Ph.D, Aharonov held several temporary appointments in the USA before accepting a chair at the University of South Carolina in 1966 and a chair at the University of Tel Aviv in 1967. Two chairs? Only a quantum mechanic could occupy two chairs simultaneously, and persist happily in this unstable and potentially uncomfortable arrangement for more than thirty years.

Aharonov's contributions to physics did not end with the Aharonov-Bohm effect. That was followed by many further brilliant discoveries, of which I will describe two. An object in space, unconnected to any other, looks the same after one complete turn. If the object is tethered by a rope, its turn is registered as a twist in the rope. The surprise comes after two turns: then the rope can be untwisted, and again the tethered object looks the same as before. You can perform this trick with your own hands (I learned recently that it is now accepted as a tai chi exercise). In 1967, Aharonov (with Leonard Susskind) took seriously a fact that had been noted before but dismissed as insignificant, that the distinction between tethered and untethered rotations persists into the microscopic world. In particular, some quantum particles behave as though they are linked to the rest of the world by ghostly strings, and look different after one complete turn: their waves change sign, and return to their original state only after two turns. They suggested an experiment with neutrons to detect this bizarre effect, and in 1975 it was seen by two groups of investigators.

The second discovery concerned the 'geometric phase'. This occurs in quantum systems that undergo a sequence of changes in a cycle, so as to return to their original state. After the cycle, the waves that describe the system exhibit a shift that depends on the geometry of the cycle. This shift is the geometric phase. Its existence as a general phenomenon — with the Aharonov-Bohm and Aharonov-Susskind effects as special cases — had been demonstrated already, but only under the restriction that the cycle be performed slowly. In 1987, Aharonov (working with Jeeva Anandan) removed this restriction, and gave a more general, and rather simple, reformulation of the geometric phase.

I mention a curious historical fact. In 1959, when Aharonov and Bohm were discovering their effect here, the head of department was Professor Maurice Pryce. At that time, he was studying the quantum mechanics of molecules, with Longuet-Higgins, Öpik and Sack: specifically, the interaction between the motion of the nuclei and that of the electrons. They found a strange effect: for some cyclic deformations of the molecule (a squirming combination of rotation and vibration) the wave describing the electron changes sign, and this gets reflected in the spectrum of the molecule. Now, it turns out (with the benefit of hindsight) that this too is a geometric phase, mathematically similar to the Aharonov-Susskind effect for neutrons. So, two geometric phases were discovered in Bristol at the same time (two out of seven over the years, actually). There is no evidence that the connection was appreciated then.

Theoretical physics is a strange calling. You dream and scribble and, if you are lucky, the world conforms. Popperists would argue that if the world does not conform you are luckier, because then you are surprised and so learn more. I am sure, however, that most of us would be more than happy to be lucky as often as Aharonov has been, rather than luckier.

Mr Vice-Chancellor, I present to you Yakir Aharonov as eminently worthy of the degree of Doctor of Science *honoris causa*.

Charles Frank funeral eulogy

(Canford crematorium, Bristol, 16 April 1998)

We mourn the death of Charles Frank, but that unsentimental man would have wished us rather to celebrate his life — a life that spanned most of this century, was shaped by its main events, and, at a crucial moment, helped to determine those events, to the benefit of all of us.

He was, from surface to core, a scientific man. All his passion went into endless and active curiosity about the workings of the physical world. This is not the place to enter into details of his discoveries, but I must give a glimpse of their enormous scope.

In nuclear physics, it was Charles who proposed that fusion power — that is, the power of sunshine, could be generated using muons. That would be the real cold fusion, not the fake version that made such news a few years ago. In the physics of solid crystals, it was he who explained how the manner in which they grow depends on the defects that mar their perfection. In the physics of liquid crystals, it was his paper of 1958, one of whose main purposes was — and I quote — "to urge the revival of experimental interest in its subject", that achieved just that, and incidentally led to the technology of liquid-crystal displays, for example those we wear on our wrists. In geophysics, he explained the shapes of island arcs. In biochemistry, he showed how the mathematics of competition can explain why the molecules of life are overwhelmingly of one hand (I mean left-handed or right-handed, but not both).

The distillation of these achievements into his scientific papers was sparing. If he had something to communicate, he wrote one paper, and that was it — like the mathematician Gauss, whose writings were "few, but ripe".

He had a geometric personality, by which I mean that most of his thinking was in pictures. Nobody could forget his literally hand-waving explanations of constructions in three (or more) dimensions — while driving. He had a talent for inventing brilliant images to explain difficult ideas. For example, this, to illustrate a type of stress in a sheet: "Imagine four old men, facing the same way round a square, each sucking the beard of his neighbour". Or, to illustrate those paradoxical materials that expand when they are pushed, a balloon filled with wet sand and squeezed — try it: it's very peculiar.

I remember the first time I saw him, more than thirty years ago, holding forth in the physics tea room, in long, looping Frankish sentences, dense with nested parentheses but always eventually returning to the point, retailing facts — he loved facts, the particularity of real things, although he emphasised that "Physics is not just Concerning the Nature of Things, but Concerning the Interconnectedness of all the Natures of Things" — recalling facts from his prodigious memory (it has been said of him that he never forgot anything he ever heard, read or experienced), puffing sagely on the pipe he smoked then (actually he smoked matches rather than tobacco — the evidence was the pile on the floor, whose size accurately indicated the length of his discourse) — somehow, with his powerful head and the authority in his voice, he reminded me of a lion (I know the same thought came to others) ; perhaps his hair and beard contributed to that impression.

Only once do I remember him being angry; that was in 1968, when I dared to enter the Senate House, then occupied by students, to discuss what they were revolting about. Otherwise, he was always calm, even when, on a physics department boat trip, when we were standing back-to-back, my one-year-old daughter leaned over my shoulder, and over his shoulder, grabbed his beard, and tugged at it.

We physicists, in the department — and also the university and the city — are fortunate that he chose Bristol as his home. He was fortunate to have, for more than half a century, the support of Maita: loyal, loving, formidable, fiery support. We all miss him, but remember the way our lives were enriched by this essentially scientific man.

Paul Dirac published the first of his papers on "The Quantum Theory of the Electron" seventy years ago this month. The Dirac equation, derived in those papers, is one of the most important equations in physics

Paul Dirac: the purest soul in physics

Michael Berry

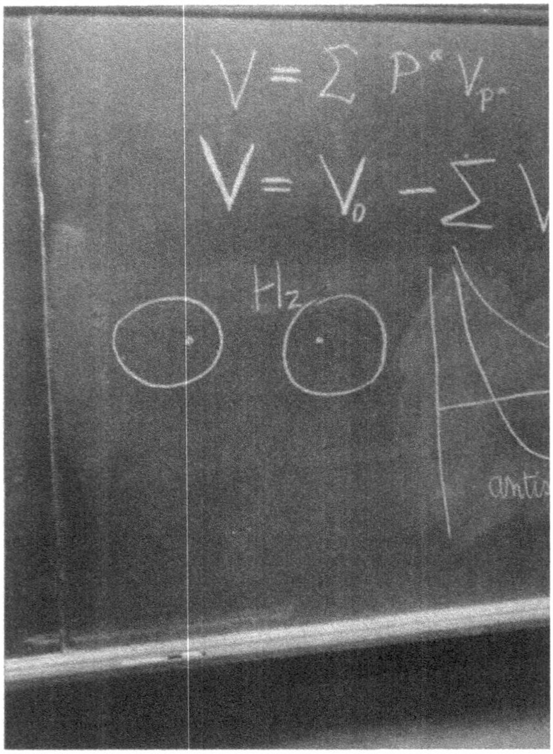

EACH day, I walk past the road where Paul Adrien Maurice Dirac lived as a child. It is pleasant to have even this tenuous association with one of the greatest intellects of the 20th century. Paul Dirac was born at 15 Monk Road in Bishopston, Bristol, on 8 August 1902, and educated at the nearby Bishop Road Primary School. The family later moved to Cotham Road, near the University of Bristol, and in 1914 the young Dirac joined Cotham Grammar School, formerly the Merchant Venturers.

Dirac was a student at Bristol University between 1918 and 1923, first in electrical engineering and then in applied mathematics. Much later, he said: "I owe a lot to my engineering training because it [taught] me to tolerate approximations. Previously to that I thought...one should just concentrate on exact equations all the time. Then I got the idea that in the actual world all our equations are only approximate. We must just tend to greater and greater accuracy. In spite of the equations being approximate, they can be beautiful."

Because Dirac was a quiet man – famously quiet, indeed – he is not well known outside physics, although this is slowly changing. In 1995 a plaque to Dirac was unveiled at Westminster Abbey in London and last year Institute of Physics Publishing, which is based in Bristol, named its new building Dirac House.

It is hard to give the flavour of Dirac's achievements in a non-technical article, because his work was so mathematical. He once said: "A great deal of my work is just playing with equations and seeing what they give."

Early days

When Dirac went to Cambridge in 1923, the physics of matter on the smallest scales – in those days this was the physics of the atom – was in ferment. It had been known for more than a decade that the old mechanics of Newton – "classical" mechanics, as it came to be called – does not apply in the microscopic world. In particular, evidence from the light coming out of atoms seemed to indicate that some quantities that in classical mechanics can take any values are actually restricted to a set of particular values: they are "quantized". One of these quantities is the energy of the electrons in an atom. This was strange and shocking. Imagine being told that when your car accelerates from 0 to 70 miles per hour it does so in a series of jumps from one speed to another (say in steps of one thousandth of a mph), with the intermediate speeds simply not existing. It did not make sense, and yet observations seemed to demand such an interpretation.

In the first attempts at a theoretical understanding, physicists tried to find the general rules for imposing these restrictions on classical mechanics – that is rules for quantization. It seemed that in order to quantize, it was necessary first to identify those quantities that do not change when their environment is slowly altered. If a pendulum is slowly shortened, for example, it swings farther and also faster, in such a way that its energy divided by its frequency stays constant. These rules worked for simple atoms and molecules but failed for complicated ones.

Dirac entered physics at the end of this baroque period. One of his first papers was an attempt at a general theory of

'Paul Dirac: The purest soul in physics' by Berry, M. V., 1998, Physics World 11 (February), 36-40.

these unchanging quantities. This is a delicate problem in classical mechanics, not solved even now. It is amazing today to read that paper. In its mathematics it is quite unlike any of Dirac's later works (for example, he brings in fine differences between rational and irrational numbers), and "pre-invents" techniques developed by other people only decades later. (I say pre-invents because the paper was forgotten until recently.)

At this time the situation in atomic physics resembled that at the end of the 16th century, when the old Earth-centred astronomy had to be made ever more elaborate in the face of more accurate observations. The difficulties of the 16th and 20th centuries were resolved in the same way: by a complete shift of thought. In atomic physics this happened suddenly, in 1925, with the discovery by Heisenberg of quantum mechanics. This seemed to throw out classical mechanics completely, though it was built in as a limiting case to ensure that, on larger scales, the new mechanics agreed with more familiar experience. The quantum rules emerged automatically, but from a mathematical framework that was peculiar. For example, it involved multiplication where the result depends on the order in which the multiplication is done. It is as though 2 multiplied by 3 is different from 3 multiplied by 2. Heisenberg found this ugly and unsatisfactory. Dirac disagreed, and just a few months after Heisenberg he published the first of a series of papers in which quantum mechanics took the definitive form we still use today.

The main idea is that the multiplied objects – objects that represent variables we can measure in experiments – should be thought of as operations. An experiment is an operation, of course, even though its result is a number. With this interpretation, it is not surprising that the order matters: we all know that putting on our socks and then our shoes gives a result different from putting on our shoes and then our socks. Dirac found the one simple rule by which *a* multiplied by *b* differed from *b* multiplied by *a*, and from which the whole of quantum mechanics follows.

The same unification was soon found to include Schrödinger's way of doing quantum mechanics, where the state of a system is represented by a wave whose strength gives the probabilities of the different possible results of measurements on it. For a while this seemed completely different from the framework that Heisenberg had used, but it quickly emerged that in fact each represents Dirac's operators in a different way. It seemed miraculous.

The Dirac equation

Although brilliant – in Einstein's words, "the most logically perfect presentation of quantum mechanics" – this was a reformulation of physics that had, admittedly only just, been discovered. Dirac's main contribution came several years later, when (still in his mid-twenties) he made his most spectacular discovery.

Before quantum mechanics, there had been another revolution in physics, with Einstein's discovery in 1905 that Newton's mechanics fails for matter moving at speeds approaching that of light. To get things right, time had to be regarded as no longer absolute: before-and-after had to be incorporated as a fourth co-ordinate like the familiar three spatial co-ordinates that describe side-to-side, forward-and-backward and up-and-down. Just as what is side-to-side and what is forward-and-backward change when you turn, so time gets mixed in with the other three co-ordinates when you move fast. Now, in the 1920s, came quantum mechanics, showing how Newton's mechanics failed in a different way: on microscopic scales. The question arose: what is the physics of particles that are at the same time small *and* moving fast?

This was a practical question: the electrons in atoms are small, and they move fast enough for the new quantum mechanics to be slightly inaccurate, since it had been constructed to have as its large-scale limit Newton's mechanics rather than Einstein's. From the start people tried to construct a quantum theory concordant with relativity, but failed to overcome technical obstructions: in particular, their attempts gave probabilities that were negative numbers – something that is nonsense, at least in the usual meaning of probability. The question boiled down to this: what are the right sort of quantum waves describing electrons? And what is the wave equation that governs the dynamics of these waves, while satisfying the requirements of relativity and giving sensible physical predictions?

Dirac's construction of his wave equation for the electron – published in two papers in the *Proceedings of the Royal Society (London)* in February and March 1928 – contained one of those outrageous leaps of imagination shared by all great advances in thought. He showed that the simplest wave satisfying the requirements was not a simple number but had four components (see box overleaf). This seemed like a complication, especially to minds still reeling from the unfamiliarity of the "ordinary" quantum mechanics. Four components! Why should anybody take Dirac's theory seriously?

Dirac with Werner Heisenberg in Chicago in 1929.

The Dirac equation

The Dirac equation for an electron moving in an arbitrary electromagnetic field can be written in many ways. In Dirac's original papers it is written as

$$\left[p_0+\frac{e}{c}A_0+\alpha_1\left(p_1+\frac{e}{c}A_1\right)+\alpha_2\left(p_2+\frac{e}{c}A_2\right)+\alpha_3\left(p_3+\frac{e}{c}A_3\right)+\alpha_4 mc\right]\psi=0$$

where $p_0=ih\partial/c\partial t$ (the energy operator), e is the charge on the electron, A_0 is the scalar potential associated with the electromagnetic field, c is the speed of light, α_i are 4×4 matrices derived from the Pauli matrices, $p_1=-ih\partial/\partial x$ is a momentum operator ($p_2=-ih\partial/\partial y$, $p_3=-ih\partial/\partial z$), A_i are the three components of the electromagnetic vector potential, m is the mass of the electron and ψ is the wavefunction of the electron.

The wavefunction ψ is a 4×1 column vector (also known as a spinor) and each element is a function of space and time, representing the spin state (up or down) of the electron and the associated positron solution. As explained in the main text, the equation was able to explain the results of all of the experiments at the time, to explain the origin of electron spin and to predict the existence of antimatter.

The equation can be written in more compact form. In §67 of *The Principles of Quantum Mechanics* (4th edn, Oxford University Press) it is written as

$$\left\{p_0+\frac{e}{c}A_0-\rho_1\left(\sigma.\mathbf{p}+\frac{e}{c}\mathbf{A}\right)-\rho_3 mc\right\}\psi=0$$

where ρ_1 and ρ_3 are 4×4 matrices (related to α_i and the Pauli matrices), σ is a three-component vector of 4×4 matrices, and $\mathbf{p}$ is a three-component vector of momentum operators. The version of the equation in Westminster Abbey is even more compact and reads $i\gamma\cdot\partial\psi=m\psi$ where γ is a 4×4 matrix and ∂ is a 4-vector.

First, and above all for Dirac, the logic that led to the theory was, although deeply sophisticated, in a sense beautifully simple. Much later, when someone asked him (as many must have done before) "How did you find the Dirac equation?" he is said to have replied: "I found it beautiful." Second, it agreed with precise measurements of the energies of light emitted from atoms, in particularly where these differed from ordinary (non-relativistic) quantum mechanics.

There are two more reasons why the Dirac equation was compelling as the correct description of electrons. To understand them, you should realize that any great physical theory gives back more than is put into it, in the sense that as well as solving the problem that inspired its construction, it explains more and predicts new things. Before the Dirac equation, it was known that the electron spins. The spin is tiny on the scale of everyday but is always the same and plays a central part in the explanation through quantum mechanics of the rules of chemistry and the structure of matter. This spin was a property of the electron, like its mass and its electric charge, whose existence simply had to be assumed before quantum mechanics could be applied. In Dirac's equation, spin did not have to be imported: it emerged – along with the magnetism of the electron – as an inevitable property of an electron that was both a quantum particle and a relativistic one.

So, electron spin was the third reason for believing Dirac's mathematically inspired equation. The fourth came from a consequence of the equation that was puzzling for a few years at first. Related to its four components was the fact that any solution of the equation where the electron had a positive energy had a counterpart where the energy was negative. It gradually became clear that these counterpart solutions could be interpreted as representing a new particle, similar to the electron but with positive rather than negative charge; Dirac called it an "anti-electron", but it soon came to be known as the positron. If an electron encounters a positron, Dirac predicted, the two charges cancel and the pair annihilates, with the combined mass transforming into radiation in the most dramatic expression of Einstein's celebrated equation $E=mc^2$. Thus was antimatter predicted. When the positron was discovered by Anderson in 1932, Dirac's immortality was assured. Dirac and Schrödinger shared the Nobel Prize for Physics in 1933.

Nowadays, positrons are used every day in medicine, in PET (positron emission tomography) scanners that pinpoint interesting places in the brain (e.g. places where drugs are chemically active). These work by detecting the radiation as the positrons emitted from radioactive nuclei annihilate with ordinary electrons nearby.

Other achievements

Having explained spin, it was natural for Dirac to try to explain electric charge, and in particular the mysterious fact that it is quantized: all charges found in nature are multiples of the charge on the electron. In classical electricity, there is no basis for this: charges can have any value.

In 1931 Dirac gave a solution of this problem in an application of quantum mechanics so original that it still astounds us to read it today. He combined electricity with magnetism, in a return to the 18th-century notion of a magnet being a combination of north and south magnetic poles (magnetic

The 1927 Solvay Congress in Brussels was attended by most of the leading physicists of the time. Dirac is in the second row, on Einstein's right. The other delegates are (left to right): front row; I Langmuir, M Planck, Madame Curie, H A Lorentz, A Einstein, P Langevin, Ch E Guye, C T R Wilson, O W Richardson; second row; P Debye, M Knudsen, W L Bragg, H A Kramers, P A M Dirac, A H Compton, L V de Broglie, M Born, N Bohr; back row; A Piccard, E Henriot, P Ehrenfest, E D Herzen, T H de Donder, E Schrödinger, E Verschaffelt, W Pauli, W Heisenberg, R H Fowler, L Brillouin.

charges), in the same way that a charged body contains positive and negative electric charges. That symmetry was lost in the 19th century with the discoveries of Oersted, Ampère and Faraday, culminating in Maxwell's synthesis of all electromagnetic and – in another example of getting out more than you put in – optical phenomena. In its place came a greater simplicity: there are only electric charges, whose movement generates magnetism (and now the motive power for much of our civilisation). The absence of isolated magnetic poles – magnetic monopoles – was built into classical electromagnetism, and also the quantum mechanics that grew out of it.

Dirac wondered if there was any way that magnetic monopoles could be brought into quantum physics without spoiling everything that had grown out of assuming that they did not exist. He found that this could be done, but only if the strength of the monopole (the "magnetic charge") was linked to that of the electric charge, and if both were quantized. This solved the original problem: for consistency with quantum mechanics, the existence of even one monopole anywhere in the universe would suffice to ensure that electric charge must be quantized. The implication is compelling: to account for the quantization of electricity, magnetic poles must exist. After this, Pauli referred to Dirac as "Monopoleon".

Alas, no magnetic monopole has ever been found. Perhaps they do not exist, or perhaps (and there are hints of this in the theory) positive and negative monopoles are so tightly bound together that they have not been separated. Much later, Dirac referred to this theory as "just a disappointment". However, the mathematics he invented to study the monopole – combining geometry with analysis – now forms the basis of the modern theories of fundamental particles.

There were two other seminal contributions to physics in those early years. I have space only to mention them. Dirac applied quantum mechanics to the way light and matter interact. This made him realize that it was necessary to quantize not only particles but the electromagnetic field itself, and led him to the first consistent theory of photons (which had been discovered several decades previously in the beginnings of quantum mechanics). This led to the elaborate and thriving quantum field theories of today.

Dirac also showed how quantum waves for many electrons had to be constructed, incorporating the philosophically intriguing fact that any two of these particles are absolutely identical and so cannot be distinguished in any way. This produced the definitive understanding of earlier rules about how quantum mechanics explains the periodic table of the elements, and provided the basis for the theory of metals and the interior of stars.

Like all scientists at the highest level, Dirac was not afraid to descend from the pinnacle and discuss more down-to-earth matters. Here are two examples. Much of our knowledge comes from light scattered by matter; in particular, that is how we see. In a clever stroke of lateral thinking, Dirac realized that the quantum symmetry between waves of light and waves of matter implied that it is also possible for material particles to be scattered by light, a ghostly possibility that could be observed, as he showed in 1933 in a paper with Peter Kapitza. This was observed for the first time about ten years ago and the manipulation of atoms by laser beams is now a thriving area of applied quantum mechanics – a fact recognized with a Nobel prize last year (*Physics World* November 1997 p51).

The second example is his Second World War work. In the Manhattan Project to develop the first nuclear bombs, it was necessary to separate isotopes of uranium. One class of methods involved the centrifugal effects of fluid streams that were made to bend. Dirac put the theory of these techniques on a firm basis, and indeed his work in this field has been described as seminal.

Dirac stories

It is not my intention to write about what sort of person Dirac was. But I must mention the genre of "Dirac stories". He was so unusual in the logic and precision of his interaction with

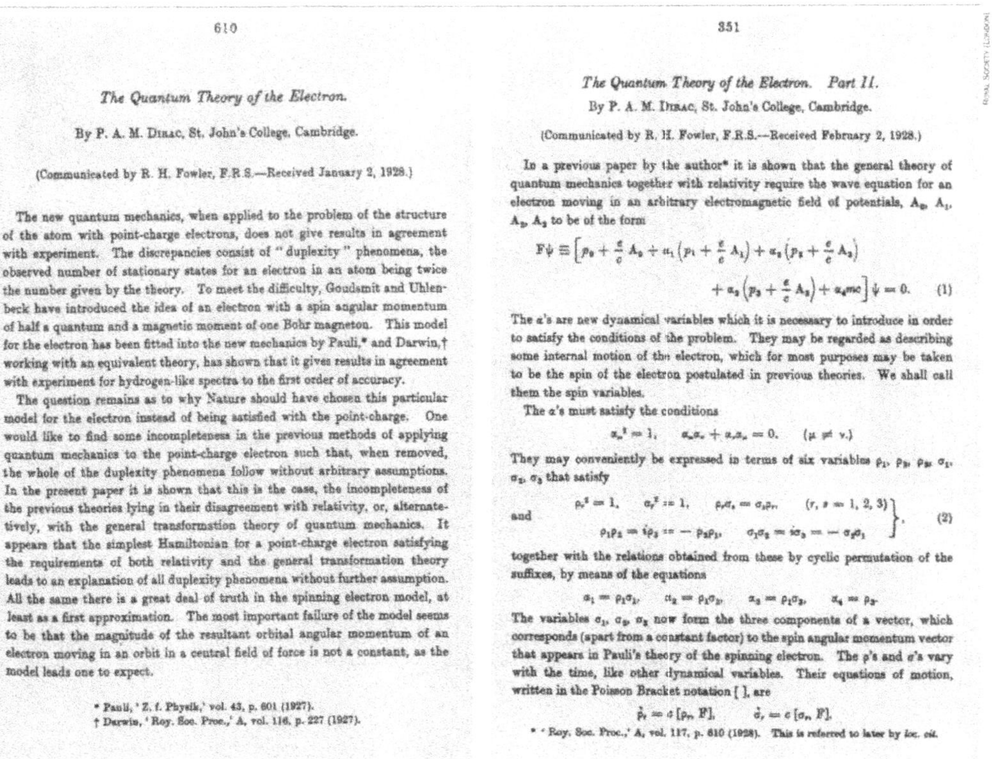

610

The Quantum Theory of the Electron.

By P. A. M. Dirac, St. John's College, Cambridge.

(Communicated by R. H. Fowler, F.R.S.—Received January 2, 1928.)

The new quantum mechanics, when applied to the problem of the structure of the atom with point-charge electrons, does not give results in agreement with experiment. The discrepancies consist of "duplexity" phenomena, the observed number of stationary states for an electron in an atom being twice the number given by the theory. To meet the difficulty, Goudsmit and Uhlenbeck have introduced the idea of an electron with a spin angular momentum of half a quantum and a magnetic moment of one Bohr magneton. This model for the electron has been fitted into the new mechanics by Pauli,* and Darwin,† working with an equivalent theory, has shown that it gives results in agreement with experiment for hydrogen-like spectra to the first order of accuracy.

The question remains as to why Nature should have chosen this particular model for the electron instead of being satisfied with the point-charge. One would like to find some incompleteness in the previous methods of applying quantum mechanics to the point-charge electron such that, when removed, the whole of the duplexity phenomena follow without arbitrary assumptions. In the present paper it is shown that this is the case, the incompleteness of the previous theories lying in their disagreement with relativity, or, alternatetively, with the general transformation theory of quantum mechanics. It appears that the simplest Hamiltonian for a point-charge electron satisfying the requirements of both relativity and the general transformation theory leads to an explanation of all duplexity phenomena without further assumption. All the same there is a great deal of truth in the spinning electron model, at least as a first approximation. The most important failure of the model seems to be that the magnitude of the resultant orbital angular momentum of an electron moving in an orbit in a central field of force is not a constant, as the model leads one to expect.

* Pauli, 'Z. f. Physik,' vol. 43, p. 601 (1927).
† Darwin, 'Roy. Soc. Proc.,' A, vol. 116, p. 227 (1927).

351

The Quantum Theory of the Electron. Part II.

By P. A. M. Dirac, St. John's College, Cambridge.

(Communicated by R. H. Fowler, F.R.S.—Received February 2, 1928.)

In a previous paper by the author* it is shown that the general theory of quantum mechanics together with relativity require the wave equation for an electron moving in an arbitrary electromagnetic field of potentials, A_0, A_1, A_2, A_3 to be of the form

$$F\psi \equiv \Big[p_0 + \frac{e}{c}A_0 + \alpha_1\Big(p_1 + \frac{e}{c}A_1\Big) + \alpha_2\Big(p_2 + \frac{e}{c}A_2\Big) + \alpha_3\Big(p_3 + \frac{e}{c}A_3\Big) + \alpha_4 mc\Big]\psi = 0. \quad (1)$$

The α's are new dynamical variables which it is necessary to introduce in order to satisfy the conditions of the problem. They may be regarded as describing some internal motion of the electron, which for most purposes may be taken to be the spin of the electron postulated in previous theories. We shall call them the spin variables.

The α's must satisfy the conditions

$$\alpha_\mu^2 = 1, \quad \alpha_\mu\alpha_\nu + \alpha_\nu\alpha_\mu = 0. \quad (\mu \neq \nu.)$$

They may conveniently be expressed in terms of six variables ρ_1, ρ_2, ρ_3, σ_1, σ_2, σ_3 that satisfy

$$\left.\begin{array}{l}\rho_r^2 = 1, \quad \sigma_r^2 = 1, \quad \rho_r\sigma_s = \sigma_s\rho_r, \quad (r, s = 1, 2, 3) \\ \text{and} \quad \rho_1\rho_2 = i\rho_3 = -\rho_2\rho_1, \quad \sigma_1\sigma_2 = i\sigma_3 = -\sigma_2\sigma_1\end{array}\right\}, \quad (2)$$

together with the relations obtained from these by cyclic permutation of the suffixes, by means of the equations

$$\alpha_1 = \rho_1\sigma_1, \quad \alpha_2 = \rho_1\sigma_2, \quad \alpha_3 = \rho_1\sigma_3, \quad \alpha_4 = \rho_3.$$

The variables σ_1, σ_2, σ_3 now form the three components of a vector, which corresponds (apart from a constant factor) to the spin angular momentum vector that appears in Pauli's theory of the spinning electron. The ρ's and σ's vary with the time, like other dynamical variables. Their equations of motion, written in the Poisson Bracket notation [ ], are

$$\dot{\rho}_r = c\,[\rho_r, F], \qquad \dot{\sigma}_r = c\,[\sigma_r, F].$$

* 'Roy. Soc. Proc.,' A, vol. 117, p. 610 (1928). This is referred to later by *loc. cit.*

Dirac's papers on the quantum theory of the electron were published in the *Proceedings of the Royal Society (London) A* in 1928 (see further reading).

the world, both in and out of physics, that tales have become attached to him and have acquired a life of their own. I suppose it matters to a historian whether they are true or apocryphal (or as Norman Mailer says, "factoids"), but to us they have a deeper resonance that transcends fact. Resisting temptation, I retell just two less well known ones.

Like many scientists, Dirac was known to sleep during (other people's) lectures, and then wake and suddenly make a penetrating remark. Once, a speaker stopped, scratched his head and declared: "Here is a minus where there should be a plus. I seem to have made an error of sign." Dirac opened one eye and said: "Or an odd number of them." Another time, Dirac was at a meeting in a castle, when another guest remarked that a certain room was haunted: at midnight, a ghost appeared. In his only reported utterance on matters paranormal, Dirac asked: "Is that midnight Greenwich time, or daylight saving time?"

Dirac's writing was famous for its clarity and simplicity. Every physicist knows his *Principles of Quantum Mechanics* – such a perfect and complete summary of his views that in later years his lectures consisted of readings from it. There is the story that he was once present when Niels Bohr was writing a scientific paper – with many hesitations and redraftings, as was his custom. Bohr stopped: "I do not know how to finish this sentence." Dirac replied: "I was taught at school that you should never start a sentence without knowing the end of it."

Many physicists have spoken of Dirac with awe. John Wheeler, referring to the sharp light of his intelligence, said "Dirac casts no penumbra." Niels Bohr said: "Of all physicists, Dirac has the purest soul." He is also reported as saying (I cannot now find this quotation): "Dirac did not have a trivial bone in his body."

The mathematician Mark Kac divided geniuses into two classes. There are the ordinary geniuses, whose achievements one imagines other people might emulate, with enormous hard work and a bit of luck. Then there are the magicians, whose inventions are so astounding, so counter to all the intuitions of their colleagues, that it is hard to see how any human could have imagined them. Dirac was a magician.

Further reading

P A M Dirac 1928 The quantum theory of the electron *Proc. R. Soc. (London)* **117** 610–612

P A M Dirac 1928 The quantum theory of the electron. Part II *Proc. R. Soc. (London)* **118** 351–361

R H Dalitz (ed) 1995 *The Collected Works of P A M Dirac 1924–1948* (Cambridge University Press)

Sir Michael Berry is Royal Society Research Professor at the H H Wills Physics Laboratory, University of Bristol, Tyndall Avenue, Bristol BS8 1TL, UK. This is a version of a talk delivered in September 1997 at the official opening of Dirac House, the headquarters of Institute of Physics Publishing

Proceedings of Symposia in pure Mathematics, volume **72.1**, 2004, 31-33

Benefiting from fractals

Michael Berry
H H Wills Physics Laboratory, Tyndall Avenue, Bristol BS81TL, United Kingdom
(http://www.phy.bris.ac.uk/staff/berry_mv.html)

Philip Morrison's review of the first English Edition [22] of Benoit Mandelbrot's book hit a precise resonance with me. For the previous few years I had been studying waves reflected from irregular surfaces, motivated by an application to geophysics. All existing theories assumed random surfaces where asperities with a single length scale perturbed a plane; this disturbed me, because it created an artificial distinction between 'roughness' and 'geography' (in this case the flat earth). Before fractals, I had no idea how to convert this unease into physics.

After fractals, the way was clear: assume a rough surface with fractal dimension *D*, and see how waves reflected from the surface carry an imprint of *D*. I called such waves 'Diffractals' [1]. An exact analytical solution of the wave equation with such a boundary condition was (and remains) unavailable, even for the statistical quantities I was interested in. Existing approximation methods failed too: ray optics (short-wave asymptotics), because the transverse length scales include the wavelength; perturbation theory, because the asperities are high compared to the wavelength; and variational methods, because there is no 'nearby' exactly solvable model. Nevertheless, I made a little progress with a Kirchhoff integral approximation, that at least showed how *D* gets imprinted on the second moment of the wave as it propagates away from the surface. Later, this monochromatic analysis was extended to pulses (echoes) [8].

Those were the early days of quantum chaology, where a useful class of models for studying high energy levels is 'quantum billiards': waves confined within boundaries of different shapes. I wondered how a fractal boundary (or even a fractal domain) might affect the asymptotic distribution of eigenvalues, and dared to publish a speculative answer [2, 7]. In a misguided attempt to be more precise than my mathematical knowledge warranted, I guessed that it would be the Hausdorff dimension that influences the asymptotics. I should have referred to *D* simply as the fractal dimension, because it was soon shown that the Minkowski dimension is more appropriate [13]. However, the essence of the conjecture has survived and has spawned a small literature [17, 20, 19, 18], extending to number theory [21]. Nevertheless, an important problem has hardly been addressed: for a billiard with fractal boundary, what is the geometrical origin of the fluctuations of eigenvalue density? For smooth boundaries, the fluctuations depend on the

Reprinted from Proceedings of Symposia in Pure Mathematics, Volume 72.1, (2004).

periodic geodesics [14, 6] bouncing inside the billiard, but reflection and hence geodesics are not defined for fractal boundaries. This question goes beyond the averages described by the Weyl formula and its extensions.

In the mid 1980s, talk of 'The Evil Empire' revived fears of nuclear war, and raised the possibility that smoke from the resulting fires would absorb incident sunlight but transmit radiated heat in a 'nuclear winter'. Ian Percival pointed out that the estimates of this cooling were based on models of the smoke particles as spheres, whereas it was already known that they aggregate into fractal clusters as the smoke ages. In another application of diffractals, we gave a mean-field theory [11] of the absorption of electromagnetic waves by fractal clusters; again D was implicated in a nontrivial way, which survives in more accurate computations [24]. As intuition might suggest, the absorption is greater for a fractals than for spheres of the same mass, so fractality makes the nuclear winter worse [23] (see also [25]). The effect is made even worse by the fact that fractal clusters fall to earth more slowly than spheres, implying a modification of the hydrodynamic Stokes law, that I was able to estimate [3].

In diffractals, waves get imprinted with the D of objects they encounter, but they are not themselves fractal, because the wavelength provides a natural scale. It was therefore a surprise to discover that there are circumstances in which waves themselves can be fractal, in the sense of possessing self-similar structures on scales between the wavelength and the size of scattering objects. Moreover, this occurs in one of the most familiar waves, namely that diffracted by a grating with sharp-edged slits [9]. This 'Talbot effect' [26] fractal is richly anisotropic, with different D lengthwise, crosswise, and diagonally. Transferring the analysis from the paraxial wave equation to the time-dependent Schrödinger equation shows that very simple nonstationary quantum waves can be fractals too [5, 10]. Another surprise was finding [16, 15] and understanding [12, 4, 27] fractal waves in the modes of unstable lasers: simply reversing one of the mirrors in the familiar stable arrangement changes the mode from a narrow Gaussian beam to a fractal filling the laser cavity.

I would never have recognised and explored these hidden territories in my intellectual habitat of wave physics without Benoit Mandelbrot's great discovery that self-similarity is commonplace rather than pathological.

References

[1] M. V. Berry, *Diffractals*, J. Phys. A, 12 (1979), pp. 781-97.

[2] M. V. Berry, *Distribution of modes in fractal resonators*, in W.Güttinger and H. Eikemeier, eds., *Structural stability in physics*, Springer, 1979, pp. 51-3.

[3] M. V. Berry, *Falling fractal flakes*, Physica, D 38 (1989), pp. 29-31.

[4] M. V. Berry, *Fractal modes of unstable lasers with polygonal and circular mirrors*, Optics. Communs, 200 (2001), pp. 321-330.

[5] M. V. Berry, *Quantum fractals in boxes*, J. Phys. A, 26 (1996), pp. 6617-6629.

[6] M. V. Berry, *Semiclassical theory of spectral rigidity*, Proc. Roy. Soc. Lond., A400 (1985), pp. 229-251.

[7] M. V. Berry, *Some geometric aspects of wave motion: wavefront dislocations, diffraction catastrophes, diffractals*, Proc. Symp. App. Maths., 36 (1980), pp. 13-28.

[8] M. V. Berry and T. M. Blackwell, *Diffractal echoes*, J. Phys. A, 14 (1981), pp. 3101-3110.

[9] M. V. Berry and S. Klein, *Integer, fractional and fractal Talbot effects*, J. Mod. Optics, 43 (1996), pp. 2139-2164.

[10] M. V. Berry, I. Marzoli and W. P. Schleich, *Quantum Carpets, carpets of light*, Physics World (2001), pp. 39-44.

[11] M. V. Berry and I. C. Percival, *Optics of fractal clusters such as smoke*, Optica Acta, 33 (1986), pp. 577-591.

[12] M. V. Berry, C. Storm and W. van Saarloos, *Theory of unstable laser modes: edge waves and fractality*, Optics Communications, 197 (2001), pp. 393-402.

[13] J. Brossard and R. Carmona, *Can one hear the dimension of a fractal?*, Commun. Math. Phys., 104 (1986), pp. 103-122.

[14] M. C. Gutzwiller, *Periodic orbits and classical quantization conditions*, J. Math. Phys., 12 (1971), pp. 343-358.

[15] G. P. Karman, G. S. McDonald, G. H. C. New and J. P. Woerdman, *Fractal modes in unstable resonators*, Nature, 402 (1999), pp. 138.

[16] G. P. Karman and J. P. Woerdman, *Fractal structure of eigenmodes of unstable-cavity lasers*, Opt. Lett., 23 (1998), pp. 1909-1911.

[17] M. L. Lapidus, *Fractal drum, inverse spectral problems for elliptic operators and a partial resolution of the Weyl-Berry conjcture*, Trans. Amer. Math. Soc., 325 (1991), pp. 465-529.

[18] M. L. Lapidus, J. W. Neuberger, R. J. Renka and C. A. Griffith, *Snowflake harmonics and computer graphics: Numerical computation of spectra on fractal domains*, Intern. J. Bifurcation & Chaos, 6 (1996), pp. 1185-1210.

[19] M. L. Lapidus and M. M. H. Pang, *Eigenfunctions of the Koch snowflake drum*, Commun. Math. Phys, 172 (1995), pp. 359-376.

[20] M. L. Lapidus and C. Pomerance, *Counterexamples to the modified Weyl-Berry conjecture on Fractal drums*, Math. Proc. Camb. Phil. Soc., 119 (1993), pp. 167-178.

[21] M. L. Lapidus and M. van Frankenhuysen, *Fractal geometry and number theory: complex dimensions of fractal strings and zeros of zeta function*, Birkhäuser, Boston, 1999.

[22] B. B. Mandelbrot, *Fractals: Form, Chance and Dimension*, W. H. Freeman and Company, San Francisco, 1977.

[23] J. Nelson, *Fractality of sooty smoke: implications for the severity of the nuclear winter*, Nature, 339 (1989), pp. 611-613.

[24] J. Nelson, *Test of a mean field theory for the optics of fractal clusters*, J. Mod. Opt., 36 (1989), pp. 1031-1057.

[25] J. A. Nelson, R. J. Crookes and S. Simons, *On obtaining the fractal dimension of a 3D cluster from its projection on a plane — application to smoke agglomerates*, J. Phys. D: Appl. Phys, 23 (1990), pp. 465-468.

[26] H. F. Talbot, *Facts relating to optical science. No IV*, Phil. Mag., 9 (1836), pp. 401-407.

[27] M. A. Yates and G. H. C. New, *Fractal dimension of unstable resonator modes*, Opt. Commun., 208 (2002), pp. 377-380.

Remembering John Ziman

Michael Berry, Bristol University

John Ziman Memorial Day, The Royal Society, 20 May 2005

To illuminate some aspects of John Ziman's intellectual character and his years in Bristol, I have to be a little personal.

In early 1965, I was finishing my Ph.D in St Andrews, and had just been awarded a fellowship for postdoctoral research (Bob Chambers, a member of the committee that interviewed me is here today). But where to go for that research, in a subject not seriously studied in any British university? My supervisor, Bob Dingle, told me that "A bright young scientist has just taken the Chair of Theoretical Physics at Bristol. His name is John Ziman. You might enjoy a book he has just written with his friend Jasper Rose, about Oxford and Cambridge universities". I wasn't particularly interested in the old English universities, but *Camford Observed* captivated me. What the authors described in Oxbridge was exactly the superiority, conservatism and complacency that had disappointed me about St Andrews, and I loved the elegant and witty style in which they wrote. I responded to John Ziman as a kindred spirit.

So, I wrote raising the possibility of holding my fellowship in Bristol, and John at once invited me to meet him. The meeting was memorable. He took me to lunch, where I met two of the friends he had made in Bristol: the late Stephan Körner, the distinguished philosopher, and the late Howard Hinton, the distinguished entomologist. In the conversation that day, ranging over science, the philosophy of science, politics, literature... I enjoyed an intellectuality I had dreamed of for years but been starved of in the two universities I had attended previously. Bristol seemed the very model of what a university should be. It was clearly the place for me, so I was delighted when John welcomed me into the department.

So I arrived in Bristol, forty years and five days ago today, and joined the embryonic theoretical physics group. I was somewhat intimidated by my two colleagues. Derek Greenwood, who had been appointed before John arrived, impressed me by having a formula (or half of one) named after him. Peter Lloyd was an inspired catch, an autodidact John had brought back from his sabbatical in Australia, and utterly brilliant — if he had not diminished his career by going back to Australia for family reasons, it could well have been him standing here today instead of me.

Over the following months and years, I saw John's style of day-to-day intellectual leadership. It was highly interactive, though not with me because my research was on different topics and we never worked directly together. He began each academic year with a meeting where he outlined the contribution of each member of the group, who became a node in a complicated diagram, with many cross-links that John emphasized, and which gradually took shape during the presentation.

In those days I was an algebraic kind of scientist; theoretical physics was a collection of formulas, spare and pregnant encodings of reality. John gained his insights and understandings in a different way: by drawing pictures, simple diagrams that grew before our eyes each morning over coffee and each afternoon over tea. Several other senior scientists in Bristol also had geometric imaginations, and this style soon infected me.

One little vignette will give the flavour of John's cultural style. We often discussed the books we had been reading. I described with enthusiasm Thomas Mann's *The Magic Mountain* — with such enthusiasm, indeed, that John bought it and read it himself. Several weeks later, he told me: "You were right: it is a wonderful novel. But you're too young to understand it!" Perhaps he was right; I ought to read it again and see.

Another vignette will give the flavour of the free-and-easy academic career progression of the 1960s. About 18 months after I arrived, John reminded me that my fellowship would soon expire — something that in spite of having a young family and a vast mortgage (all of three thousand pounds!) I had not been fully aware of. "You seem be able do research on your own ("self-winding" was a favourite metaphor of his), and you seem to be able to teach (I had given one undergraduate lecture course), so would you like a permanent job: a lectureship?" "Oh, I hadn't though about it, Ok, I suppose — thanks very much". "Wait a minute...there will have to be a formal university interview. Do you agree to that?" And so I stayed in Bristol.

Over the next decade, John continued to build up the theory group and lead the condensed-matter part of it. But his interests were shifting. Following *Camford Observed*, which was about college and university life, he became fascinated by the wider international social culture of science. His first book devoted to this subject, *Public Knowledge*, was written soon after he arrived in Bristol, and more followed. The intellectual break came in 1979, with his book *Models of Disorder*, summarising the many years he had spent distilling (deconstructing, we might say today) the essence of a highly mathematical subject: how to describe irregularities and waves scattered by them; in its quantum incarnation this describes the important physics of the electronic properties of matter — electrical resistance and magnetic susceptibility, for example. I quote the last sentence of that book:

"... I take the opportunity ... to announce that [as far as physics is concerned] this is, for me, THE END"

That period — I refer to the late 1970s and early 1980s — was a time of considerable personal upheaval in the physics department. How can I put this delicately? In a way we were the scandal centre of Bristol University; four professors in the department, including John and two others here today, were (for different reasons) in irregular situations, having separated from our wives and embarked on new private lives. John was the Head of our Department then, and when that came to an end in 1982, he decided that as he was no longer a practising physicist it was wrong to remain in Bristol as a professor of the subject, so he left to be with Joan and concentrate on his interests in science and society.

Speaking personally — no, not just personally, for the opinion was shared by others in the department — I was disappointed when John abandoned theoretical physics. There was a poignant moment when he decided to sell his collections of physics books. I thought he had a lot more physics in him, and although I read and enjoyed, and largely agreed with, all that he wrote about the structure of science as a social phenomenon and as a profession, I thought his contributions to physics were more valuable, and he could have carried on making them. But he was a wise man who knew his own mind and heart, so who am I to think he went in the wrong direction?

For more detail, see:

Berry, M V and Nye, J F, 'John Michael Ziman' *Biographical Memoirs of the Royal Society*, 52, 479–491.

Published online by the Royal Society of Edinburgh, 2011:

http://www.royalsoced.org.uk/cms/files/fellows/obits_alpha/dingle_robert.pdf

Robert Balson Dingle

Robert ('Bob') Dingle was born on March 26, 1926 in Manchester. He studied at Cambridge University (Tripos Part I 1945, Part II 1946) and began research in theoretical physics under the supervision of D R Hartree, earning a Ph.D from Cambridge in 1952 after spending the year 1947-1948 visiting Bristol under the supervision of Professors Mott and Fröhlich. Following research positions in Delft in the Netherlands and Ottawa in Canada, he was appointed to a Readership at the University of Western Australia. In June 1960 he arrived in St Andrews as the first occupant of the Chair of Theoretical Physics. He was elected to the Royal Society of Edinburgh in 1961. After a sabbatical period in Canada, California and Western Australia, he remained in St Andrews until his early retirement through ill-health in 1987.

His original field of research was theoretical condensed-matter physics, in which he made major original contributions in several areas, described in nearly forty scientific papers. The topics included quantum and statistical physics, magnetic properties and surface reflectivity of metals, anomalous skin effect, scattering theory in semiconductors, the conductivity of thin wires, and liquid helium II. Some of this work is remembered eponymously: the Dingle temperature, Dingle-Holstein resonance, and the Dingle factor.

During this work, he encountered deep mathematical difficulties, associated with the approximate evaluation of integrals and the solution of differential equations, leading to infinite series ('asymptotic expansions') that were usually divergent. Dingle realised that existing techniques for making sense of such series, and getting useful results from them, were often crude and ill-founded, and he devoted the remainder of his research to mathematical asymptotics. It was in this area that he made his most profound and lasting contributions, described in twenty papers and culminating in his definitive and magisterial exposition 'Asymptotic expansions: their derivation and interpretation' (Academic Press 1973).

Before Dingle, almost every scientist who encountered a divergent series regarded it as meaningful only up to an inherent vagueness, usually associated with the remainder after discarding the divergent tail of the series. Much effort by mathematicians was devoted to establishing precise limits ('error bounds') on this vagueness. Dingle's approach was startlingly different: building on nineteenth-century insights by Stokes, and avoiding what he regarded as a too-limited approach by Poincaré, he regarded a divergent series as an exact coding of the function it represents. Decoding ('interpreting') such series is exact in principle, and in practice can lead to vastly improved approximations. By identifying common patterns in the divergent series commonly arising in physics and applied mathematics, he was able to establish systematic interpretive rules, now recognised as providing a solid foundation for asymptotics and the first fundamental advance in the subject for nearly a century.

In subsequent decades, several other scientists arrived independently at similar concepts, but priority was undoubtedly Dingle's, his methods were more effective, and he developed the techniques in much greater detail.

Recognition was not immediate. In large measure this was the result of Dingle's style as a scientist. He did not rush to publish each incremental advance as a separate paper, breathlessly announced at conference after conference. Rather, he was oblivious to what Ramón y Cajal called 'the sour flattery of celebrity'. And although he enjoyed several collaborations, he worked mostly alone, rarely travelled to conferences, and by modern standards of physical science his papers were 'few, but ripe'.

He was a committed and sometimes provocative teacher, remembered for his dry and often mischievous wit and the generous hospitality provided by him and his wife Helen. As a research supervisor, his advice was economical, but always helpful and perfectly to the point.

On the administrative side, he chaired the Governing Committee of the NATO Scottish Summer Schools in Physics for several years, as well as being the Director of the 1962 and 1967 Schools. He was responsible for remodelling the first-year mathematics teaching at St Andrews; he was convenor of the Project Committee for the construction of the Student Union building, overseeing the project from start to finish; and he represented the university Senate on the Union Governing Board.

Outside science, he enjoyed music (his family was musically accomplished). And his enjoyment of fine wine and good food was perfectly complemented by Helen's legendary skills as a cook. His keen interest in local history and architecture led to painstaking research and an unusual and detailed map of old St Andrews.

He died on March 2, 2010, in St Andrews, and is survived by his wife and their daughters Judith and Susie.

Michael Berry,
John Cornwell

Robert Balson Dingle. Tripos I & II, PhD (Cantab), Born 26 March 1926, elected FRSE 6 March 1961, died 2 March, 2010.

Remarks at Balazs Györffy memorial event

Michael Berry, Bristol 17 November 2012
(slightly edited)

It's an honour to be asked to speak at this afternoon of remembering, this celebration for our dear colleague and friend Balazs Györffy. But it's an honour I never wanted. By rights — and I mean statistically — it should have been the other way round: the men in my family die young. But Balazs was so strong, so superfit, that, as one of our former colleagues wrote, if anyone would be immortal it would be Balazs.

He was born in Hungary in 1938. When the revolution in 1956 was brutally suppressed, Balazs took a chance and defected to America, where he had the typical immigrant experience: taking menial jobs to support himself (he often reminisced about working in a Jewish funeral home). His prowess as a swimmer got him a scholarship to Yale: for a while he held the world record for the butterfly stroke. This was a lifelong enthusiasm. We admired at the way he charged through the water, a human speedboat.

But physics — theoretical physics — soon became a competing obsession, and after his first graduation he was accepted to study for a Ph.D with one of the world's leading optical scientists, the Nobelist Willis Lamb. At that time — the mid-1960s — lasers were new and strange. Balazs created the theory of an unfamiliar type: the ring laser. But soon his interests shifted from photons en masse to electrons in crowds, in condensed matter: crystals, disordered alloys, liquids, superconductors... This subject requires fiendishly difficult calculations in quantum physics, and Balazs pioneered a powerful technique — the coherent potential approximation — that soon got him noticed.

He moved to England. As one of his referees later wrote in support of his appointment here: "He finds the political and social systems of Eastern Europe and the USA both distasteful, and thinks Britain, by comparison, has unique virtues". After a few years in London and Sheffield, he came to Bristol, as a lecturer, in 1970. John Ziman had arrived a few years before and was building up the theoretical physics group. One of our lecturers left, leaving a vacancy that was advertised. There were several excellent candidates but after the interviews and presentations there was no doubt Balazs would get the job. I remember Ziman telling me: "This Györffy is not only a good physicist; he is a mensch" That Yiddish-German term of praise describes someone who is strong, manly in a sense with no hint of machismso — a person of integrity, full of life, warm, generous, positive, comfortable in his skin: Balazs to a T. Another colleague, John Nye, recently wrote to me: "I remember well when we welcomed him to Bristol, expecting him to stir us all up, as he did."

What a handsome man he was! A beautiful man! Everyone immediately loved him — Carole too: what a glamourous and popular couple they were! They arrived in Bristol at the same time as I was leaving for three months abroad, so it was natural to lend them my flat during that time. When I came back, we shared the place for a while and became the best of friends — lifelong friends. Having Balazs as a friend meant having someone you could trust completely.

He was a man of wide culture, gregarious and congenial, and deeply intellectual. According to the same referee for his job here: "[Györffy] is interested in intellectual matters generally, and one tends to find on his desk books...such...as 'The Influence of Fifteenth Century Florentine Art on Bagpipe Folk Music in Scotland'". He and Carole soon established regular Saturday-morning gatherings, first in a local café and later in their home. All sorts of people — writers, philosophers, scientists, political activists — discussed pretty well everything. No subject was off limits in salon Györffy. Related to this, and fast-forwarding to now, there has been a very welcome resurgence of intellectuality in our university's relations with the City of Bristol; Balazs's involvement was as an active member of the committee arranging Art Lectures.

One aspect of Balazs's character that we soon discovered was his keen sense of social justice. For years he resented an experience in Sheffield: it rankled with him that, in court for a motoring offence, the magistrate ordered him to take his hands out of his pockets. He never understood how his infringement of the traffic laws was connected with where he parked his hands. More seriously, he became a lifelong active supporter of the Labour party, involved in local and national elections, and was a committed school governor for a number of years.

Driving was a temporary phase in Balazs's life. He would have been fine, apart from sharing the road with other motorists. I remember being a passenger while we were on the road discussing physics. A tricky point came up, that needed his full attention; he stopped abruptly in the middle of a busy road, oblivious to traffic forced to manoeuvre around both sides of the obstruction he had created, until the question was resolved. And there was a famous occasion when Balazs was driving a colleague up the M6 to a conference in Manchester, again while animatedly discussing physics. After several hours, they wondered why the journey was so long, until they saw a sign: welcome to Scotland.

The absent-minded professor is a familiar cliché. Balazs was an effortless exponent. Once, arriving back from a trip to the USA, he left his bag in a restaurant in London, with $1000 in it (he got it back). His most spectacular accomplishment of this type was to forget his passport and then talk his way, without documents, out of the UK, into Germany, and then through the old Soviet checkpoint into East Berlin (and back).

Physicists have different ways of working. Balazs's was to be strongly interactive: the most strongly collaborative physicist I have known — I would say, relentlessly collaborative. On every scientific trip abroad he accumulated several new colleagues and new research projects. Often, these collaborators would eventually arrive in Bristol, to continue working with him as post-docs. They came from all over the world. An interesting consequence was that sometimes there were several researchers from the same country, so we heard a variety of languages in animated conversation in the theoretical physics corridor: over the years, Spanish, Italian, and of course Hungarian. Now, Bristol has an admirable policy across the university of welcoming new colleagues regardless of where they come from. But Balazs did it decades ago: the internationality of science personified.

He was a much-loved teacher, especially in the way he inspired generations of research students with his graduate lectures. And he certainly inspired me. I'll be technical for a moment. In 1976, when I was starting to think about what became quantum chaology, he drew my attention to a then-obscure branch of mathematical physics called random-matrix theory. This observation turned out to be seminal (though still not widely appreciated). It stimulated a major theme of my own scientific life, and led, through a chain of connections I won't elaborate, to a worldwide explosion of research activity that continues still, with Bristol physics, and Bristol mathematics, playing a major part. From little acorns, mighty oak trees grow.

He was unswervingly loyal to the physics department. For him, the highest form of loyalty was not compliance or a comfortable conformity; no, it was robust criticism of any initiative or policy that he regarded as misguided. He knew in his bones what successive Heads of Physics have been wise enough to understand but university management occasionally needs to be reminded of: that universities thrive on disagreement: it is their fuel. They are models for the wider world, of how we can disagree without being disagreeable.

[to Balazs's relatives who were present] In this your hardest time, I hope you can take some comfort from what you surely know already, but I'll say it again: in this building, in this department, in Bristol university, among all the people here today and many others worldwide, Balazs was not only respected: he was very much loved.

Tribute to Vladimir Arnold

Boris Khesin and Serge Tabachnikov, Coordinating Editors

Vladimir Arnold, an eminent mathematician of our time, passed away on June 3, 2010, nine days before his seventy-third birthday. This article, along with one in the next issue of the *Notices*, touches on his outstanding personality and his great contribution to mathematics.

A word about spelling: we use "Arnold", as opposed to "Arnol'd"; the latter is closer to the Russian pronunciation, but Vladimir Arnold preferred the former (it is used in numerous translations of his books into English), and we use it throughout.

Arnold in His Own Words

In 1990 the second author interviewed V. Arnold for a Russian magazine *Kvant* (Quantum). The readership of this monthly magazine for physics and mathematics consisted mostly of high school students, high school teachers, and undergraduate students; the magazine had a circulation of about 200,000. As far as we know, the interview was never translated into English. We translate excerpts from this interview;[1] the footnotes are ours.

Q: *How did you become a mathematician? What was the role played by your family, school, mathematical circles, Olympiads? Please tell us about your teachers.*

A: I always hated learning by rote. For that reason, my elementary school teacher told my parents that a moron, like myself, would never manage to master the multiplication table.

My first mathematical revelation was when I met my first real teacher of mathematics, Ivan Vassilievich Morozkin. I remember the problem about two old ladies who started simultaneously from two towns toward each other, met at noon, and who reached the opposite towns at 4 p.m. and 9 p.m., respectively. The question was when they started their trip.

Vladimir Igorevich Arnold

We didn't have algebra yet. I invented an "arithmetic" solution (based on a scaling—or similarity—argument) and experienced a joy of discovery; the desire to experience this joy again was what made me a mathematician. A. A. Lyapunov organized at his home "Children Learned Society". The curriculum included mathematics and physics, along with chemistry and biology, including genetics that was just recently banned[2] (a son of one of our best geneticists was my classmate; in a questionnaire, he wrote: "my mother is a stay-at-home mom; my father is a stay-at-home dad").

Q: *You have been actively working in mathematics for over thirty years. Has the attitude of society towards mathematics and mathematicians changed?*

A: The attitude of society (not only in the USSR) to fundamental science in general, and to mathematics in particular, is well described by I. A. Krylov

Boris Khesin is professor of mathematics at the University of Toronto. His email address is `khesin@math.toronto.edu`.

Serge Tabachnikov is professor of mathematics at Pennsylvania State University. His email address is `tabachni@math.psu.edu`.

[1] *Full text is available in Russian on the website of* Kvant *magazine (July 1990),* `http://kvant.mirror1.mccme.ru/`.

DOI: http://dx.doi.org/10.1090/noti810

[2] *In 1948 genetics was officially declared "a bourgeois pseudoscience" in the former Soviet Union.*

Reprinted from Notices of the American Mathematical Society, Volume 59, Number 3, (March 2012).

Michael Berry

Memories of Vladimir Arnold

My first interaction with Vladimir Arnold was receiving one of his notoriously caustic letters. In 1976 I had sent him my paper (about caustics, indeed) applying the classification of singularities of gradient maps to a variety of phenomena in optics and quantum mechanics. In my innocence, I had called the paper "Waves and Thom's theorem". His reply began bluntly:

> Thank you for your paper. References:...

Michael Berry is professor of physics at the University of Bristol. His email address is Tracie.Anderson@bristol.ac.uk.

There followed a long list of his papers he thought I should have referred to. After declaring that in his view René Thom (whom he admired) never proved or even announced the theorems underlying his catastrophe theory, he continued:

> I can't approve your system of referring to English translations where Russian papers exist. This has led to wrong attributions of results, the difference of 1 year being important—a translation delay is sometimes of 7 years...

and

> ...theorems and publications are very important in our science (...at present one considers as a publication rather 2-3 words at Bures or Fine Hall tea, than a paper with proofs in a Russian periodical)

and (in 1981)

> I hope you'll not attribute these result [sic] to epigons.

He liked to quote Isaac Newton, often in scribbled marginal afterthoughts in his letters:

> A man must either resolve to put out nothing new, or to become a slave to defend it

and (probably referring to Hooke)

> Mathematicians that find out, settle and do all the business must content themselves with being nothing but dry calculators and drudges and another that does nothing but pretend and grasp at all things must carry away all the invention as well of those that were to follow him as of those that went before.

(I would not accuse Vladimir Arnold of comparing himself with Newton, but was flattered to be associated with Hooke, even by implication.)

I was not his only target. To my colleague John Nye, who had politely written "I have much admired your work...," he responded:

> I understand well your letter, your admiration have not led neither to read the [reference to a paper] nor to send reprints....

This abrasive tone obviously reflected a tough and uncompromising character, but I was never offended by it. From the beginning, I recognized an underlying warm and generous personality, and this was confirmed when I finally met him in the late 1980s. His robust correspondence arose from what he regarded as systematic neglect by Western scientists of Russian papers in which their results had been anticipated. In this he was sometimes right and sometimes not. And he was unconvinced by my response that scientific papers can legitimately be cited to direct readers to the most accessible and readable source of a result rather than to recognize priority with the hard-to-find original publication.

He never lost his ironic edge. In Bristol, when asked his opinion of perestroika, he declared: "Maybe the fourth derivative is positive." And at a meeting in Paris in 1992, when I found, in my conference mailbox, a reprint on which he had written: "to Michael Berry, admiringly," I swelled with pride—until I noticed, a moment later, that every other participant's mailbox contained the same reprint, with its analogous dedication!

In 1999, when I wrote to him after his accident, he replied (I preserve his inimitable style):

> ...from the POINCARÉ hospital...the French doctors insisted that I shall recover for the following arguments: 1) Russians are 2 times stronger and any French would already die. 2) This particular person has a special optimism and 3) his humour sense is specially a positive thing: even unable to recognize you, he is laughing.... I do not believe this story, because it would imply a slaughtering of her husband for Elia, while I am still alive.

(Elia is Arnold's widow.)

There are mathematicians whose work has greatly influenced physics but whose writings are hard to understand; for example, I find Hamilton's papers unreadable. Not so with Arnold's: through his pellucid expositions, several generations of physicists came to appreciate the significance of pure mathematical notions that we previously regarded as irrelevant. "Arnold's cat" made us aware of the importance of mappings as models for dynamical chaos. And the exceptional tori that do not persist under perturbation (as Kolmogorov, Arnold, and Moser showed that most do) made us aware of Diophantine approximation in number theory: "resonant torus" to a physicist = "rational number" to a mathematician.

Most importantly, Arnold's writings were one of the two routes by which, in the 1970s, the notion of genericity slipped quietly into physics (the other route was critical phenomena in statistical mechanics, where it was called universality). Genericity emphasizes phenomena that are typical rather than the special cases (often with high symmetry) corresponding to exact solutions of the governing equations in terms of special functions. (And I distinguish genericity from abstract generality, which can often degenerate into what Michael Atiyah has called "general nonsense".)

This resulted in a shift in our thinking whose significance cannot be overemphasized.

It suddenly occurs to me that in at least four respects Arnold was the mathematical counterpart of Richard Feynman. Like Feynman, Arnold made massive original contributions in his field, with enormous influence outside it; he was a master expositor, an inspiring teacher bringing new ideas to new and wide audiences; he was uncompromisingly direct and utterly honest; and he was a colorful character, bubbling with mischief, endlessly surprising.

Acknowledgments

The photographs are courtesy of the Arnold family archive, F. Aicardi, Ya. Eliashberg, E. Ferrand, B. and M. Khesin, J. Pöschel, M. Ratner, S. Tretyakova, and I. Zakharevich.

References

[1] V. B. ALEKSEEV, *Abel's Theorem in Problems and Solutions*, based on the lectures of Professor V. I. Arnold, with a preface and an appendix by Arnold and an appendix by A. Khovanskii, Kluwer Academic Publishers, Dordrecht, 2004.

[2] D. ANOSOV, Geodesic flows on closed Riemannian manifolds of negative curvature, *Trudy Mat. Inst. Steklov* **90** (1967).

[3] V. I. ARNOLD, On functions of three variables, *Dokl. Akad. Nauk SSSR* **114** (1957), 679-681.

[4] V. ARNOLD, Small denominators. I. Mappings of a circle onto itself, *Izvestiya AN SSSR*, Ser. Mat. **25** (1961), 21-86.

[5] ———, Small denominators and problems of stability of motion in classical and celestial mechanics, *Uspekhi Mat. Nauk* **18** (1963), no. 6, 91-192.

[6] ———, Instability of dynamical systems with many degrees of freedom, *Dokl. Akad. Nauk SSSR* **156** (1964), 9-12.

[7] ———, The cohomology ring of the group of dyed braids, *Mat. Zametki* **5** (1969), 227-231.

[8] V. I. ARNOLD, The cohomology classes of algebraic functions that are preserved under Tschirnhausen transformations, *Funkt. Anal. Prilozhen* **4** (1970), no. 1, 84-85.

[9] V. ARNOLD, The situation of ovals of real algebraic plane curves, the involutions of four-dimensional smooth manifolds, and the arithmetic of integral quadratic forms, *Funkt. Anal. Prilozhen* **5** (1971), no. 3, 1-9.

[10] ———, The index of a singular point of a vector field, the Petrovsky-Oleinik inequalities, and mixed Hodge structures, *Funct. Anal. Appl.* **12** (1978), no. 1, 1-14.

[11] V. I. ARNOLD, A. B. GIVENTAL, A. G. KHOVANSKII, A. N. VARCHENKO, *Singularities of functions, wave fronts, caustics and multidimensional integrals*, Mathematical Physics Reviews, Vol. 4, 1-92, Harwood Acad. Publ., Chur, 1984.

[12] V. ARNOLD, S. GUSEIN-ZADE, A. VARCHENKO, *Singularities of Differentiable Maps, Vol. I. The Classification of Critical Points, Caustics and Wave Fronts*, Birkhäuser, Boston, MA, 1985.

[13] V. ARNOLD, M. SEVRYUK, Oscillations and bifurcations in reversible systems, in *Nonlinear Phenomena in Plasma Physics and Hydrodynamics*, Mir, Moscow, 1986, 31-64.

[14] V. I. ARNOLD, V. A. VASSILIEV, Newton's "Principia" read 300 years later, *Notices Amer. Math. Soc.* **36** (1989), no. 9, 1148-1154; **37** (1990), no. 2, 144.

[15] V. ARNOLD, From superpositions to KAM theory, in *Vladimir Igorevich Arnold, Selected-60*, PHASIS, Moscow, 1997, 727-740 (in Russian).

[16] V. ARNOLD, B. KHESIN, *Topological Methods in Hydrodynamics*, Springer-Verlag, New York, 1998.

[17] V. ARNOLD, From Hilbert's superposition problem to dynamical systems, in *The Arnoldfest*, Amer. Math. Soc., Providence, RI, 1999, 1-18.

[18] V. I. ARNOLD, I. G. PETROVSKII, Hilbert's topological problems, and modern mathematics, *Russian Math. Surveys* **57** (2002), no. 4, 833-845.

[19] M. ATIYAH, J. BERNDT, *Projective planes, Severi varieties and spheres, Surveys in Differential Geometry*, Vol. VIII (Boston, MA, 2002), 1-27, Int. Press, Somerville, MA, 2003.

[20] M. AUDIN, *Cobordismes d'immersions lagrangiennes et legendriennes*, Travaux en Cours, 20, Hermann, Paris, 1987.

[21] D. BERNSTEIN, On the number of roots of a system of equations, *Funkt. Anal. i Prilozhen.* **9** (1975), no. 3, 1-4.

[22] V. I. DANILOV, A. G. KHOVANSKII, Newton polyhedra and an algorithm for calculating Hodge-Deligne numbers, *Math. USSR—Izv.* **29** (1987), 279-298.

[23] I. DOLGACHEV, Conic quotient singularities of complex surfaces, *Funkt. Anal. i Prilozhen.* **8** (1974), no. 2, 75-76.

[24] A. ESKIN, A. OKOUNKOV, Asymptotics of numbers of branched coverings of a torus and volumes of moduli spaces of holomorphic differentials, *Invent. Math.* **145** (2001), 59-103.

[25] A. GABRIELOV, Dynkin diagrams of unimodal singularities, *Funkt. Anal. i Prilozhen.* **8** (1974), no. 3, 1-6.

[26] D. GUDKOV, Topology of real projective algebraic varieties, *Uspekhi Mat. Nauk* **29** (1974), no. 4, 3-79.

[27] M. KHOVANOV, P. SEIDEL, Quivers, Floer cohomology, and braid group actions, *J. Amer. Math. Soc.* **15** (2002), 203-271.

[28] A. G. KHOVANSKII, *Topological Galois Theory*, MTSNMO, Moscow, 2008.

[29] M. KONTSEVICH, A. ZORICH, Connected components of the moduli spaces of Abelian differentials with prescribed singularities, *Invent. Math.* **153** (2003), 631-678.

[30] A. KOUCHNIRENKO, Polyèdres de Newton et nombres de Milnor, *Invent. Math.* **32** (1976), 1-31.

[31] N. KUIPER, The quotient space of $\mathbf{C}P^2$ by complex conjugation is the 4-sphere, *Math. Ann.* **208** (1974), 175-177.

[32] J. MARSDEN, *Steve Smale and Geometric Mechanics*, The Collected Papers of Stephen Smale, vol. 2, 871-888, World Scientific Publ., River Edge, NJ, 2000.

[33] I. G. PETROVSKII, On the topology of real plane algebraic curves, *Ann. of Math.* (2) **39** (1938), 187-209.

[34] M. SEVRYUK, My scientific advisor V. I. Arnold, *Matem. Prosveshchenie, Ser. 3* **2** (1998), 13-18 (in Russian).

Martin Gutzwiller and his periodic orbits

Michael Berry, H H Wills Physics Laboratory, Tyndall Avenue, Bristol BS8 1TL, UK

In the 1970s, physicists were made aware, largely through the efforts of the late Joseph Ford, that classical hamiltonian mechanics was enjoying a quiet revolution. The traditional emphasis had been on exactly solvable models, with as many conserved quantities as degrees of freedom, in which the motion was integrable and predictable. Examples are the Kepler ellipses of planetary motion, and the simple pendulum: 'as regular as clockwork'. The new research, incorporating Russian analytical mechanics and computer simulations inspired by statistical mechanics, revealed that most (technically, 'almost all') dynamical systems behave very differently. There are few conserved quantities, and motion, in part or all of the phase space, is nonseparable and unpredictable, that is, unstable: initially neighbouring orbits diverge exponentially. This is classical chaos.

It was quickly realised that this classical behaviour must have implications for quantum physics, especially semiclassical physics, e.g. for the arrangement of high-lying energy levels and the morphology of eigenfunctions. The study of these implications became what is now called quantum chaos (though I prefer the term quantum chaology). This is an area of research in which Martin Gutzwiller made a seminal contribution, described in the following, which I have adapted from a speech honouring his 70th birthday. Since a substantial part of my own scientific life has been devoted to the development and application to Martin's ideas, I won't attempt to be detached.

Martin published the last of his series of four papers [1-4] on periodic orbits exactly forty years ago. I encountered them at that time, while Kate Mount and I were writing our review of semiclassical mechanics. That was prehistoric semiclassical mechanics: before catastrophe theory demystified caustics, before asymptotics beyond all orders lifted divergent series to new levels of precision, and above all before we knew about classical chaos.

Reprinted from SPS Communications, Volume 37, (2012).

Of Martin's series of papers, the most influential was the last one [4], containing the celebrated 'Gutzwiller trace formula'. That was a tricky calculation, based on the Van Vleck formula for the semiclassical propagator, giving the density of quantum states (actually the trace of the resolvent operator) as a sum over classical periodic orbits. In particular, Martin calculated the contribution from an individual unstable periodic orbit. Nowadays we can see this as one of the 'atomic concepts' of quantum chaology, but in those days chaos was not appreciated. But he emphasized the essential novelty of his calculation in a similar way: it applies even when the classical dynamics is nonseparable. I'm rather proud of what we wrote at the beginning of 1972, as the last sentence of our review:

"Finally, the difficulties raised by Gutzwiller's (1971) theory of quantization, which is perhaps the most exciting recent development in semiclassical mechanics, should be studied deeply in order to provide insight into the properties of quantum states in those systems, previously almost intractable, where no separation of variables is possible."

The trace formula could be approximated by taking just one periodic orbit and its repetitions. This led to an approximate 'quantization formula' that gave good results when applied to the lowest states of an electron in a semiconductor, whose mass depended on direction. I am referring to the birth of Martin's treatment of the anisotropic Kepler problem [5].

For a few years, his calculation was widely misinterpreted (among the ignorant it is misinterpreted even today) as implying a relation between the individual energy levels and individual periodic orbits of chaotic systems. One might call this the 'De Broglie interpretation' of the trace formula: that there is a level at each energy for which the action of a periodic orbit is a multiple of Planck. This is nonsense: the simplest calculation shows that the number of levels is hopelessly overestimated – in a billiard, for example, there is an 'infra-red catastrophe', that is, the prediction of levels at arbitrarily low energies.

Martin's papers quickly inspired others. In 1974, Jacques Chazarain showed that the trace formula could be operated 'in reverse', so that a sum over energy levels generated a function whose singularities were the actions of periodic orbits. This was exact, not semiclassical, and led (often unacknowledged) to what later came to be called 'inverse quantum chaology' and 'quantum recurrence spectroscopy'. In 1975 Michael Tabor and I generalized some of the results in the first of Martin's semiclassical papers [1] to get the general trace formula for integrable systems, where the periodic orbits are not isolated but fill tori. In nuclear physics, similar formulas had been obtained by Strutinsky in the context of the shell model. Tabor and I used our result to show that the level statistics in integrable systems are Poissonian - more about that later. William Miller and André Voros resolved a puzzle about the application of the trace formula for a stable orbit: by properly quantizing transverse to the orbit, they restored the missing quantum numbers; then Martin's single-orbit quantization rule makes sense, as the 'thin-torus' limit of Bohr-Sommerfeld quantization.

Probably Martin didn't realize that his formula was so fashionable at that time that it induced a certain hysteria. Michael Tabor and I were quietly finishing the work I just described when we learned that William Miller wanted to visit us in Bristol, to talk about his new work on periodic orbits. We convinced ourselves that this must be the same as ours, and laboured day and night (up a ladder, actually, because Michael was helping me paint my new house) to get our paper written and submitted before he arrived. We were foolish to panic, because William's work was completely different.

An awkward feature of stable orbits, recognized clearly by Martin in those early days, was that focusing occurs along them, leading for certain repetition numbers and stability indices to divergences of the contributions he calculated, associated with bifurcations. That awkwardness was removed in 1985 by Alfredo Ozorio de Almeida and John Hannay, who applied ideas from catastrophe theory that had come into semiclassical mechanics in the 1970s. Their development of Martin's formula became popular much later, when the features they predicted could be detected numerically.

In the early 1970s, Ian Percival made us aware of the amazing developments in classical mechanics by Arnold and Sinai, before chaos became popular. Percival insisted that semiclassical mechanics must take account of chaos. Later, we learned more about chaos from Joseph Ford. Of course Martin had paved the way with his trace formula for unstable orbits.

A persistent question was whether the formula could generate asymptotically high levels for a chaotic system. My opinions fluctuated. In 1976 I thought it could not, arguing that long orbits - required to generate the high levels - were so unstable that the Van Vleck propagator would not be valid for them. Instead, I thought (using ideas developed by Balian and Bloch) that periodic orbits could at best describe spectra smoothed on scales that were large compared with the mean spacing – but still classically small, so that some detail beyond the Weyl rule was accessible, though still not individual levels. This question is still not settled definitively, but my pessimistic opinion was changed by two developments.

The first was energy level statistics. In the 1970s, following a suggestion from Balazs Gyorffy, I imported from nuclear physics the idea that random matrices could be relevant in the quantum mechanics of chaos. The first application of this suggestion was not to chaotic systems at all, but to integrable systems, where it was shown – as I just mentioned – that the levels are not distributed according to random-matrix theory. That work inspired Allan Kaufman and Steven McDonald to the first calculation of level spacings for a chaotic system: the stadium. Then I did the same for Sinai's billiard. In those days we were fixated on the spacings distribution. My way of deriving level repulsion was a generalization of Wigner's: through the codimension of degeneracies. This gave the same result as random-matrix theory for small spacings, and explained the differences between the different ensembles, but gave no clue as to why random-matrix theory worked for all spacings, and why it was connected with classical chaos.

Then came Oriol Bohigas and Marie-Joya Giannoni and Charles Schmit. What they did, in the early 1980s, was simple but very important. They repeated the calculations that Kaufman and McDonald and I had done, for the same systems and using the same numerical methods, but instead of focusing on the one statistic of the level spacing they appreciated that the random-matrix analogy is much broader: it predicts all the spectral statistics, in particular long-range ones. They calculated one of these: the spectral rigidity (equivalent to the number variance).

Their observation was enormously influential. In particular, it was central to my construction in 1985 of the beginnings of the semiclassical theory of spectral statistics from Martin's atoms: the periodic orbits. Another crucial ingredient in this was also a development of periodic-orbit theory: the inspired realization by John Hannay and Alfredo Ozorio de Almeida that the Gutzwiller contributions of long orbits obey a sum rule whose origin is classical and whose structure is universal - that is, independent of details. Pure mathematicians (Margulis, Parry, Pollicott) had found similar rules - more general in that they applied to dissipative as well as hamiltonian systems, but also more restricted in that Hannay and Ozorio's theory applied also to integrable systems (where Tabor and I had found their particular result in 1977 but failed to appreciate its general significance). Thus periodic orbits were able to reproduce key formulas from random-matrix theory, and random-matrix universality found a natural explanation as the inheritance by quantum mechanics of the classical universality of long orbits. There was more: the periodic orbit theory of spectral statistics showed clearly and simply why and how random-matrix theory must break down for correlations involving sufficiently many levels. There were misty mathematical aspects – now being clarified – of those arguments, but the formulas were not misty, and were the first step in convincing me that long orbits in Martin's trace formula were meaningful.

The second step sprang from the realization - increasingly urgent in the early 1980s - that the series of periodic orbits in the trace formula does not converge. The cause was realized by Martin in 1971 [4]:

"Even more serious is the fact that there is usually more than a countable number of orbits in a mechanical system, whereas the bound states of a Hamiltonian are countable."

The failure of the trace formula to converge was emphasized especially by André Voros, who pointed out that this defect is shared by the formally exact counterpart of the formula for billiards with constant negative curvature, namely the Selberg trace formula. And later Frank Steiner taught us that trace formulas can sometimes converge conditionally, in ways depending delicately on the topology of the orbits (expressed as Maslov phases). Eventually these concerns about convergence led naturally to the study of zeta functions. The idea there is to find a function where the energy levels are zeros, rather than steps or spikes as in the density of states. The grandparent of all these objects is Riemann's zeta function of number theory. I learned its possible relevance to quantum chaology from Oriol Bohigas, and also from Martin's semiclassical interpretation of the Faddeev-Pavlov scattering billiard, where Riemann's zeta function gives the phase shifts [6, 7]. It is amazing that Martin had already realized the connection with zeta functions in his 1971 paper. He wrote:

"This response function is remarkably similar to the so-called zeta functions which mathematicians have invented in order to survey and classify the periodic orbits of abstract mechanical systems."

(He cited Smale). And in 1982 Martin explicitly wrote a semiclassical zeta function of the kind we consider today, and used it in conjunction with some tricks from statistical mechanics to sum the periodic orbits for the anisotropic Kepler system [7, 8].

A crucial ingredient turned out to be the Riemann-Siegel formula, that makes the sum over integers for the Riemann zeta function converge. I realized this in 1986, and later developed the idea with Jon Keating [9]; we were helped by André Voros's precise definitions of the regularized products in these zeta functions. The result was an adaptation of the trace formula to give a convergent sum over periodic orbits, soon employed to good effect by Keating and Martin Sieber [10] (see the figure). A related idea was the invention of cycle expansions by Predrag Cvitanovic and Bruno Eckhardt; in these, essential use is made of symbolic dynamics to speed the convergence of the sum over orbits. This application of coding to semiclassical mechanics was also originally Martin's idea: he used it in the 1970s and early 1980s to classify and then estimate the sum over the orbits, again for the anisotropic Kepler problem [7, 8].

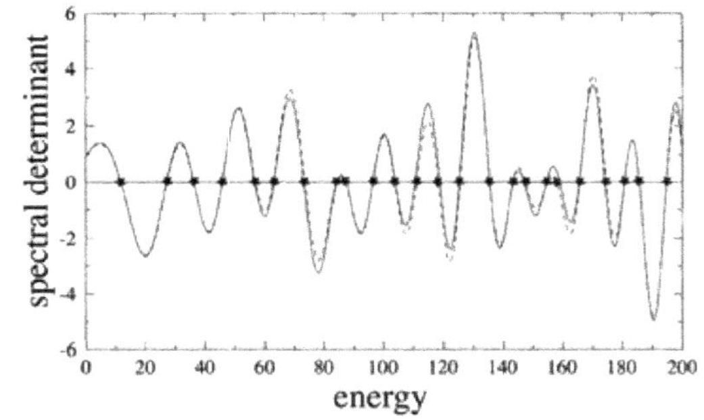

Quantum spectral determinant (zeta function) for a particle confined between branches of a hyperbola, calculated exactly (dashed curve) and from a renormalized version [9,10] of Gutzwiller's sum over the unstable classical periodic orbits (full curve); the energy levels are the zeros, indicated by stars. Reproduced from [10], with permission.

The two applications of Martin's periodic-orbit ideas that I have just described, to spectral statistics and to zeta functions, were combined by Eugene Bogomolny and Jonathan Keating. This development, and more recent insights from Martin Sieber, Fritz Haake and Sebastian Müller, are taking the derivation of random-matrix formulas from quantum chaology to new levels of sophistication and refinement.

In the mid-1980s, Eric Heller discovered that for some chaotic systems the wavefunctions of individual states are scarred by individual short periodic orbits, in ways that depend on how unstable these are. From this came further extensions of Martin's ideas, to new sorts of spectral series of periodic orbits, not involving traces, and for Wigner functions as well as wavefunctions.

In spite of all this progress, we are still unable to answer definitively and rigorously the central question Martin posed in 1971 [4]:

"What is the relation between the periodic orbits in the classical system and the energy levels of the corresponding quantum system?"

Of course the trace formula itself is one such relation, but I am sure that what Martin meant is: how can periodic orbits be used for *effective* calculations of *individual* levels. For the lowest levels there is no problem, but – and again I quote from Martin's 1971 paper -

"the semiclassical approach to quantum mechanics is supposed to be better the larger the quantum number"

and to reproduce the spectrum for high levels, using even the convergent versions of the trace formula that are now available, requires an exponentially large number of periodic orbits. This is a gross degree of redundancy unacceptable to anybody who appreciates the spectacular power of asymptotics elsewhere. Martin's old ideas continue to challenge us.

A few years ago, I refereed an application for research funding for a German-British collaboration. This required me to comment on the applicants' "timetable for research" and their "list of deliverables". I wrote "In science there are no deliverables; researches are not potatoes". Martin Gutzwiller ignored these toxic fashions. What makes him so attractive as a scientist is that he refuses to follow *any* fashion; instead, he generates ideas that *become* the fashion.

References

[1] Gutzwiller, M. C.,1967, The Phase Integral Approximation in Momentum Space and the Bound States of an Atom *J. Math. Phys.* **8**, 1979-2000

[2] Gutzwiller, M. C.,1969, The Phase Integral Approximation in Momentum Space and the Bound States of an Atom II *J. Math. Phys.* **10**, 1004-1020

[3] Gutzwiller, M. C.,1970, The Energy Spectrum According to Classical Mechanics *J. Math. Phys.* **11**, 1791-1806

[4] Gutzwiller, M. C.,1971, Periodic orbits and classical quantization conditions *J. Math. Phys.* **12**, 343-358

[5] Gutzwiller, M. C.,1973, The Anistropic Kepler Problem in Two Dimensions *J. Math. Phys.* **14**, 139-152

[6] Gutzwiller, M. C.,1983, Stochastic behavior in quantum scattering *Physica D* **7**, 341-355

[7] Gutzwiller, M. C.,1982, The Quantization of a Classically Ergodic System *Physica D* **5**, 183-207

[8] Gutzwiller, M. C.,1977, Bernoulli Sequences and Trajectories in the Anisotropic Kepler Problem *J. Math. Phys.* **18**, 806-823

[9] Berry, M. V. & Keating, J. P.,1992, A new approximation for zeta(1/2 +it) and quantum spectral determinants *Proc. Roy. Soc. Lond.* **A437**, 151-173

[10] Keating, J. P. & Sieber, M.,1994, Calculation of spectral determinants *Proc. Roy. Soc. Lond.* **A447**, 413-437

After graduating from Exeter and St Andrews, **Michael Berry** entered Bristol University, where he has been for considerably longer than he has not. He is a physicist, focusing on the physics of the mathematics...of the physics. Applications include the geometry of singularities (caustics on large scales, vortices on fine scales) in optics and other waves, the connection between classical and quantum physics, and the physical asymptotics of divergent series. He delights in finding the arcane in the mundane – abstract and subtle concepts in familiar or dramatic phenomena:

- Singularities of smooth gradient maps in rainbows and tsunamis;
- The Laplace operator in oriental magic mirrors;
- Elliptic integrals in the polarization pattern of the clear blue sky;
- Geometry of twists and turns in quantum indistinguishability;
- Matrix degeneracies in overhead-projector transparencies;
- Gauss sums in the light beyond a humble diffraction grating.

Analysis and Applications, Vol. 12, No. 4 (2014) ix–x

DOI: 10.1142/S0219530514020011

A tribute to Frank Olver (1924–2013)

Michael Berry
H. H. Wills Physics Laboratory
University of Bristol, UK
asymptotico@bristol.ac.uk

My parents came from the same part of London as Frank Olver, so I recognized his London ways: smart dry wit, and of course the accent, unchanged after more than sixty years in the USA.

Frank's life's work was devoted to the asymptotics of special functions, that is, understanding familiar mathematical objects — familiar because we encounter them often enough to give them names (Bessel, hypergeometric, ...) — in extreme situations. It is interesting how this subject has waned and waxed: waned in the 1970s, when some people of limited cultural perspective thought that computers would render analytical mathematics redundant; and waxed, as now, when special functions and their asymptotics are recognised as more important than ever. Frank deserves a large part of the credit for this "climate change" of opinion among those who apply mathematics.

People are not isolated, and we can understand them by comparing and contrasting them with others. Frank happened to be one vertex of an equilateral triangle of asymptotists of his generation, dominating their subject during the second half of the last century: three unique and complementary personalities — and I mean human personalities as well as scientific personalities. Each had many disciples, now cooperating in the further development of asymptotics.

First was Robert ("Bob") Dingle (1926–2010). He had the broad vision that asymptotic series are exact coded representation of functions, that can be decoded with a multi-stage resummation scheme of remarkable universality (a modern term that precisely fits). His book *Asymptotics: Their Derivation and Interpretation* envisaged this general scheme, with insight but without rigor, and illustrated its first stage with an enormous number of examples. A second volume was promised but never written. When Dingle died, he left all his papers to me. These include a great deal that was never published, about expansions of functions of asymptotic series, including implicit functions such as zeros. The papers include many letters, including some to and from Frank; it is clear that there was mutual respect but not complete agreement. Dingle was an interesting character: a passionate lover of classical music, and a connoisseur of fine food and scotch whiskies — a bon viveur, despite a life blighted by chronic sickness.

Martin Kruskal (1925–2006). He too had a vision, of asymptotics as possibly underpinned or legitimised by the generalization of arithmetic that involved surreal numbers — a surreal vision indeed. He and Frank were the two senior asymptotists in the 1995 Newton Institute program on exponential asymptotics. It is difficult to imagine two more different characters. In seminars, Kruskal was bimodal, the two modes being: hyperactive — unable to let the speaker utter a word without interrupting and questioning or disagreeing; or slumped in sleep, to a degree verging on the narcoleptic. Although he did not invent exponential asymptotics, he did much to popularize the subject with the phrase "asymptotics beyond all orders". His important contributions were the applications to nonlinear equations, including some describing widely-studied phenomena.

And now Frank Olver: the third vertex of the asymptotics triangle. His vision was to make asymptotics useful, by combining rigour with practicality. With down-to-earth persistence, he examined almost every special function, in almost every real or complex domain of its argument and order, resulting in a cornucopia of formulas, each with its validity precisely delineated. In recent years, during what in the UK would have been his retirement, he developed the modern resummation schemes, again in detail with a view to making them useful in practical calculations.

And, of course, there is the online Digital Library of Mathematical Functions (DLMF). This is a magnificent achievement. Frank dominated its conception and execution — execution in painstaking detail, to such a degree that those of us who were also involved in its development sometimes wondered whether the DLMF would ever see daylight: whether Frank's insistence that every detail had to be correct was an instance of the perfect being the enemy of the good. Not so: everything about the DLMF justifies Frank's careful stewardship of the project from beginning to end. I write this not only as an editor, but also as a user who has made the decision that when writing a paper where I need to refer to a formula in the DLMF, that is the only source I will cite, on the grounds that it is not only the most comprehensive but also the most readily accessible.

Frank's departure marks the end of an era for asymptotics. His modest and unpretentious manner concealed enormous accomplishments and influence. We miss him.

Remembering Akira Tonomura

Michael Berry

H H Wills Physics Laboratory,
Tyndall Avenue, Bristol BS8 1TL, United Kingdom
E-mail: asymptotico@physics.bristol.ac.uk

Three of Tonomura's fundamental quantum physics experiments are discussed from a personal perspective.

My memories of Akira Tonomura are of a gentle and quiet man, always courteous in his dealings with colleagues, a virtuoso experimenter who transformed the electron microscope into the Stradivarius of scientific instruments, on which he played beautiful physics music. As a theorist I cannot comment technically on his many contributions. Instead I will make brief remarks about three of them.

The first is his demonstration[1] of the Aharonov-Bohm (AB) effect, intended to settle controversies associated with the inevitable failure of the idealization, assumed in elementary presentations of the effect, that the electrons are completely isolated from the magnetic flux. To eliminate leakage from the ends of a conventional finite solenoid, he confined the flux within a toroidal magnet, and to eliminate almost all penetration by the electrons he coated the toroid with a superconductor. The principle had been proposed by Kuper[2] (with the inessential difference that the magnetic flux would be confined in a hollow torus rather than a solid one). Because the flux in a superconductor is quantized (in units of $h/2e$), the experiment did not test the general AB effect, for which the flux is arbitrary; but it did demonstrate the important special case where the AB phase shift is π – as well as providing direct evidence of the value of the flux quantum (if this had been h/e there would have been no effect).

Even with a superconductor there is always some penetration of the electrons, so the flux cannot "completely shielded" as claimed in the title of Tonomura's paper. This is important, because as had been proved by Roy[3], if there is any penetration, however small, the AB phase shift can be interpreted in terms of fields rather than potentials. Nevertheless, the fact that the phase shift remains finite as the limit of zero penetration is approached supports the usual interpretation in terms of potentials. I commented on this[4] (in the same year – 1986 – as Tonomura's paper appeared) as an example of the need to be careful when considering idealizations in physics. Tonomura clearly appreciated the same point, commenting eloquently and wisely[1] "Since experimental realization of absolutely zero field is impossible, the continuity of physical phenomena in the transition from negligibly small field should be accepted instead of perpetual demands for the ideal; if a discontinuity is asserted, only a futile agnosticism results".

The second is his demonstration[5] in 1989 of electron two-slit interference, with the pattern developing gradually by the detection of individual electron impacts. This has been voted the most beautiful experiment in physics[6]. It illustrates convincingly and with the utmost simplicity the wave-particle duality that is fundamental to quantum physics. Its priority has been the subject of some controversy[6], because the buildup of the pattern by individual electrons had already been observed in a pioneering experiment by Merli et al and published in 1976[7, 8], together with an award-winning movie. However, as Tonomura points out[6], his experiment improved on Merli's in several respects: (a) it had lower electron intensity (so the possibility of there being two or more electrons in the apparatus at any time is negligible), (b) it was sufficiently stable for the buildup to take place very slowly (during 20 minutes), and (c) it was sufficiently sensitive to detect the electrons with almost 100% efficiency. As with the AB torus experiment, Tonomura's demonstration was definitive.

The third is his creation of vortices (= phase singularities = nodal lines = wavefront dislocations) in an electron beam[9]. This was particularly gratifying to us in Bristol, where we have emphasized vortices[10] as generic singularities of waves of all types and have explored these topological features in detail theoretically[11-14] – including vortices generated by transmission through spiral phase plates[15], exactly as employed in the experiment by Tonomura. His emphasis was on the orbital angular momentum carried by the vortex beam – an aspect much studied in recent years[16]. Almost all earlier experiments were carried out with classical light; the novelty of Tonomura's[9] was that it demonstrated vortices in the much more challenging quantum physics of electrons.

The Japanese government has agreed that the direction of research pursued by Tonomura will continue, and that is a welcome decision. Nevertheless the premature passing of this supremely talented experimenter leaves a sadness that is hard to overstate.

References

1. Tonomura, A., Okasabe, N., Matsude, T., Kawasaki, T. & Endo, J.,1986, Evidence for Aharonov-Bohm Effect with Magnetic Field Completely Shielded from Electron Wave *Phys. Rev. Lett* **56**, 792-795
2. Kuper, C. G.,1980, Electromagnetic potentials in quantum mechanics: a proposed test of the Aharonov-Bohm effect *Physics Letters A* **79**, 413-416
3. Roy, S. M.,1980, Condition for nonexistence of Aharonov-Bohm effect *Phys. Rev. Lett* **44**, 111-114
4. Berry, M. V.,1986, *The Aharonov-Bohm effect is real physics not ideal physics in Fundamental aspects of quantum theory* eds. Gorini, V. & Frigerio, A. (Plenum, Vol. 144, pp. 319-320.

5. Tonomura, A., Endo, J., Matsuda, T., Kawasaki, T. & Ezawa, H.,1989, Demonstration of single-electron buidup of an interference pattern *Am. J. Phys.* **57**, 117-120
6. Crease, R.,2002, The double-slit experiment *Physics World (online edition)* September 2002.http://physicsworld.com/cws/article/print/2002/sep/01/the-double-slit-experiment
7. Merli, P. G., Missiroli, G. F. & Pozzi, G.,1976, On the statistical aspect of electron interference phenomena *Am. J. Phys.* **44**, 306-307
8. Rosa, R.,2012, The Merli-Missiroli-Pozzi Two-Slit Electron-Interference Experiment *Phys. Perpect.* **14**, 178-195
9. Tonomura, T. & Uchida, M.,2010, Generation of electron beams carrying orbital angular momentum *Nature* **464**, 737-739
10. Nye, J. F. & Berry, M. V.,1974, Dislocations in wave trains *Proc. Roy. Soc. Lond.* **A336**, 165-90
11. Nye, J. F.,1999, *Natural focusing and fine structure of light: Caustics and wave dislocations* (Institute of Physics Publishing, Bristol)
12. Berry, M. V.,1998, Much ado about nothing: optical dislocation lines (phase singularities, zeros, vortices...) *SPIE* **3487**, 1-5
13. Berry, M. V.,2001, in Singular Optics 2000, eds. Soskin, M. (SPIE, Alushta, Crimea), SPIE **4403**, pp. 1-12.
14. Dennis, M. R., O'Holleran, K. & Padgett, M. J.,2009, Singular Optics: Optical Vortices and Polarization Singularities *Progress in Optics* **53**, 293-363
15. Berry, M. V.,2004, Optical vortices evolving from helicoidal integer and fractional phase steps *J.Optics. A* **6**, 259-268
16. Allen, L., Barnett, S. M. & Padgett, M. J.,2003, *Optical Angular Momentum* (IoP Publishing, Bristol)

Tribute to Richard Gregory

(psychologist, Bristol University)
Remarks at what he called his fun(n)eral, 30 May 2010

Michael Berry

What a mind! — a delicious mind, animated by wisdom, wit, and whimsy, unable to have an unoriginal thought, casting light on everything it lighted on.

An example from my own trade. In physics, our cherished second law of thermodynamics — disorder inevitably increases — declares:

"You can't unscramble an egg". "Yes you can," said Richard,

"Feed scrambled egg to a chicken".

Deconstructing those six words takes you deep into understanding what the second law means.

A big man, with generosity to match. Soon after he arrived in Bristol, I sent him a paper about distorted images in perfectly reflecting spheres: "Reflections on a Christmas-tree bauble". He responded by presenting me with a beautiful old witch ball that he had inherited — his generosity was physical as well as intellectual.

We miss you Richard, all the more for the sad irony that you — an unsentimental man — would surely have appreciated: the software needs the hardware and, in the end, your material brain destroyed your beautiful mind.

Chapter 8
Travels

In my family, nobody was scientifically-educated. My father was a taxi-driver (see [9.13]) and my mother a dressmaker; both left school at 14. When the romance of scientific discovery captivated me as a youngster, I had no idea that the life scientific would involve travelling, to a degree then enjoyed only by royalty, politicians, business people and sports teams. What attracts me about such travels — so far, to more than 60 countries — are the contrasts between the universality of the science as practised in different parts of the world (albeit with minor differences of style), and the diversity of cultures, scenery and (especially!) food.

Occasionally, I travel to places regarded as dangerous: Belfast during the troubles, Israel during the intifadas... But I am a cautious person, and estimate the risks as considerably smaller than being involved in a traffic accident. And I don't regard the purpose of life as simply to remain alive for the longest possible time (this perspective failed to reassure my late mother).

I haven't systematically kept travel diaries, but hope these four pieces capture some of my varied experiences.

NEWS Bristol University NEWSLETTER 21 MARCH 1991

Heisenberg's sófa

Not many people go, or, these days, are sent to Siberia. Professor Berry went there on a scientific visit. This is his report.

Novosibirsk is in Southern Siberia, at the same latitude as Newcastle and close to the centre of Asia. It is seven hours' flying time and seven time zones from here. The journey is awkward because it is necessary to change in Moscow between two airports on opposite sides of the city. The inconvenience was more than compensated for by the company of Andrei, one of my hosts, who come 1,500 miles from Novosibirsk to meet me, and Yuri, a scientist from Moscow who made the opposite journey to spend the week listening to my lectures.

When we finally arrived, it was minus 23° Centigrade–far colder than it has been here, and below the zero of Fahrenheit, but not too cold by Siberian standards. At such temperatures one's beard hairs prickle, and I wonder whether the cause is increased rigidity or a coating of ice. The snow is dry and powdery, and blows between the bleak apartments in whirling 'snow devils' that dramatise the air's turbulence. On some days the snowflakes are the classic hexagonal needles which we do not often see here; on others they are prismatic, with facets that glitter in the sun.

I had been invited by the Siberian branch of the Soviet Academy of Science for a heavy programme of lectures and discussions in the 'science city' of Akademgorodok, about 20 miles from Novosibirsk. This work was a success, after an initial battle in which I successfully resisted an attempt to introduce intolerable friction into my lectures by having them translated sentence by sentence (they eventually agreed that English English is much easier to understand than American English, even though most scientists there are more familiar with the latter). Much of the research is of world-class, in spite of harrassment by the 1980s dogma, still dismally familiar here, about research needing to be made 'self-supporting' or even 'profitable', through industrial sponsorship. It seems that their industrialists are no more interested in supporting fundamental research than ours are. One difference from here, which I hope Senate House caterers will note, is that the coffee served during committee meetings is laced with brandy. The institutes are distributed in a forest of buildings which are handsome outside but (like our Physics Department here) drab and old-fashioned inside. In one office, I was invited to sit on a tattered and uncomfortable ancient hard leather sofa, which, it was claimed, had been 'liberated' from Heisenberg's office in Germany by Russian troops at the end of the war. A large number of molecules have thus been transferred from Heisenberg's trousers to mine.

Everything I had read about the meagre stocks of food in the shops was true. However, Novosibirsk is relatively well supplied, and in the shops I saw at least milk, bread, a little meat and fish, and enormous jars of pickles and jams. But I heard about a food shop near Moscow University which contained nothing at all. In spite of this and other difficulties, the flame of hospitality burns very bright. I was invited every day to people's homes, and feasted on *pelmeny* (small pastries stuffed with meat), some of the best potato salads I have ever tasted, several sorts of forest mushrooms, exotic tarragon-flavoured lemonade from Georgia, and of course vodka. I was disappointed not to meet Andrei's little children, but was told that they had been dispatched to granny, 'because we know that in England children are seen and not heard' (evidently they decided to play safe and not let them be seen either). Partly through fear of crime, many people have big dogs in their small flats. Once I was welcomed by a 'friendly puppy', well disguised as 100 pounds of solid Rottweiler.

There is an upsurge in religion of all sorts, from Greek Orthodox Christianity to UFOlogy. Even after 70 years of official atheism, all the buildings my hosts (who are not religious) were proud to show me were churches, mostly recently reopened. Incredulity met my suggestion that one day they will show visitors round Stalin's vast dams and power stations, as happens now with the monuments of our own heroic industrial age. The churches were elaborately decorated, smelt heavily of incense, and were occupied mainly by old women, who frequently kissed the floor, door and walls. My friend Boris compared the services favourably with a Protestant one he had attended in America, 'which reminded me of a Soviet trade union meeting'.

On the last day, Andrei and I went fishing on the ice, escorted by my other host Sergei, who is proud to declare himself a hunter. We drove along the road towards Mongolia and turned right onto Lake Ob. The ice is a metre thick, and perfectly clear apart from diaphanous cracks and occasional bubbles of air. I was surprised by frequent echoing bangs and rumbles, mighty borborygms as the lake digested the stresses in its ice. Driving is safe but requires a special technique, in which the two available dimensions are fully exploited in the search for a route where the blown snow is neither too thin (because the ice is too slippery), nor too thick (because you get stuck and have to push, as we did several times). To find the fish you must drill a hole. This is about six inches wide, and made (quite easily, as I found) with an enormous left-handed drill. To Sergei's chagrin, we failed to catch any fish.

In the popular image, Soviet bureaucracy is inflexible, graceless and inefficient, and I am sure that there is truth in that. But my own experience (no doubt reflecting the privileges accorded to an 'English specialist') was pleasantly different. Everybody was helpful and smiling, even passport controllers, customs officials, air hostesses, and (after some persuasion) restaurant staff. So there is hope that after the lifting of the Iron Curtain we are now seeing the opening of the *nyet* curtains.

Michael Berry is a Professor in the Department of Physics.

Visiting Nablus

Michael Berry, Physics Department occasional newsletter, June 1993,University of Bristol

Some of my Israeli colleagues urged me not to visit occupied Palestine, or, if I insisted on going, to take a gun. They cited examples of well-meaning visitors who had been attacked or even killed. I took comfort from the fact that these were always the same few examples, and decided to go anyway (unarmed), trusting the hospitality of my hosts to protect me from possible assailants. However, prudence did suggest that driving through Palestine in an Israeli rental car might be risky, so my wife dropped me by the checkpoint at Tulkaram. My host was waiting in a Palestinian-registered car on the other side, to drive me to Nablus, a few miles away in the West Bank. It was unnerving to stop in a back street just moments later and be told "Now we have to change to another car", but this turned out to be connected more with the mysteries of insurance than the beginning of an abduction.

"Here, everything is dominated by politics", my host declared as we drove past bleak-looking fields whose farmers had been forced to change their pattern of cultivation because they were denied water for irrigation. Just outside Tulkaram was a camp inhabited by refugees from the founding of Israel in 1948. I was told that the high wire fences were not to keep people in but to stop them stoning passing Israeli vehicles. When I was driven through another such camp later in the day, I was surprised to find not the ragged tents I had expected but permanent houses and shops, looking rather like villages I have seen in India and Africa (and not the poorest). Apparently the UN, recognising that there is no realistic hope of the occupants ever returning to their original homes in Israel, is trying to drop the designation 'refugee' in the hope that gradually the villages will become integrated into the surrounding community.

The Al-Najah National University houses 4000 students crowded onto a small campus (they have more land but for the moment cannot develop it because it adjoins an Israeli jail). About half the students were women, a few partially veiled but most in rather glamorous Western clothes. I was pleased to see some accompanied by their small children (as I was in Bristol for several years). Pictures of Yasser Arafat were everywhere, and student politics was being conducted vociferously at full volume through loudhailers (on that day, the shouters were calling for a committee to be set up to investigate the mysterious appearance and disappearance of unauthorised tents on the campus).

I had been invited by somebody whose research was related to mine, but we decided that a general talk about physics would be more appropriate than a research seminar. Fortunately the electricity failure occurred just after I had shown the critical slide. After initial shyness, there were a number of questions from students, and I was amused that most were completely unrelated to the subject of my talk ("How do you explain the twin paradox in relativity?; What is quantum tunnelling?"). During the day there was almost no scientific discussion with members of the physics department. All had obtained doctorates in the USA or Canada, but were too occupied with heavy teaching loads, and too hampered by poor library and almost nonexistent laboratory facilities, to conduct research in Nablus.

After lunch, we went for a tour of the area. The stony slopes peppered with olive trees, and round-topped Ottoman lookout stations on the hilltops, form a distinctive and strangely beautiful landscape. We drove past many old Arab villages and new Jewish settlements, and I heard about life's daily irritations and humiliations under the occupation, and more serious oppressions, all delivered in a quiet monotone. "Soon after the start of the intifada, a local commander decided this village needed to be taught a lesson. He ordered all that all the men between twelve and fifty be taken away and their hands and legs broken, except for one boy whose legs were spared so that he could walk home and report what had happened. There was no publicity about this event until an army officer, sickened by what he saw, spoke out; as a result, the military governor of Nablus was replaced." I have no reason to doubt the good faith of my informants, but of course cannot know whether these stories are accurate reports, or exaggerations, or inventions.

At this point I perceived that although my hosts knew that my wife is Israeli and therefore almost certainly Jewish, they did not realise that I am Jewish too. This became apparent to me when they asked whether I had suffered by what they imagined would be the Israeli government's disapproval of mixed marriages. Probably they were misled by my name and my (to them) unfamiliar non-American English accent. More from mischief than prudence, I decided not to enlighten them immediately. It pleased me that although I heard many criticisms of the policies of Israel, not a single anti-Semitic remark was uttered or implied. And when I finally 'came out', they did not turn a hair.

The high point of the tour, literally as well as metaphorically, was a visit to the family village of a university official. We had tea in the hilltop mansion of one of his relatives who although only in his thirties, was apparently a millionaire (from processing turkey meat for the Israeli market), with (at the latest count) seventeen children.

We returned to Nablus through a checkpoint which was "one of the worst" — although apparently the soldiers now have instructions to be polite in their questioning. My driver was taken to the back of the vehicle and asked about me. Although English and Arabic are official languages in Israel, the questioning was in Hebrew for a long time, until the soldier came to realise he was not being underestood. And although the soldier (who looked like a raw recruit) had seen my British passport, he repeatedly asked when I would be returning to America. Perhaps he was simply stupid.

I stayed the night in Nablus, and was slightly embarrassed to be accommodated in a university flat which was the home of a British woman teacher of English, who in spite of my hosts' assurances had not been informed of my arrival. She told me that since the closure of the border (to Palestinians) several months ago, support for the intifada was petering out. Several organisations, vying for support in the population, frequently call strikes on different days, causing confusion among shopkeepers who have to telephone each other each morning and ask: "Should we be on strike today?" Shops near military centres tended to remain open, because those who would intimidate the owners were themselves intimidated from taking reprisals by the proximity of Israeli soldiers. (So far, nothing seems to intimidate the intimidators' intimidators.)

The next morning, before returning me to Tulkaram, my host took me to buy a characteristic and delicious Nablus confection — syrupy cereal threads over a base of cheese — "as a present for your wife". At the border, I was the only person crossing on foot. The guard inspecting the documents of the drivers ignored me as I walked into Israel, en route back to the tranquil and companionable security and intellectual intensity of the Weizmann Institute, where I was enjoying my annual visit.

A week in Beirut

Michael Berry
Bristol University Newsletter, 28 January 1999 (vol. 29, no. 8)

"This is Hizbollah country", our guide reassured us as we approached the Roman temple complex of Baalbeck (Heliopolis), with the world's largest standing columns back-lit by snow-covered mountains. Indeed there were checkpoints every few miles — Syrian and Lebanese, carefully separated — and posters of President Assad and Ayatollah Khomeini (also, incongruously, the Restaurant Lady Diana). By two days we missed a huge rally in Baalbeck, celebrating 'Jerusalem day'.

Thence to the ruins of the early Moslem town of Anjar, whose isolated pairs of arches resembled nothing so much as the sign of an ur-McDonald's. Lunch that day was near Lebanon's disused railway line. I was prevented from inspecting it by a lone Syrian soldier. According to our guide, keeping Lebanon stable is the main reason for the Syrian army's presence there. It felt odd to be so powerful that with one glance at the rusty grassy rails I would risk destabilising the entire country.

We were in Lebanon for a conference to inaugurate the Center for Advanced Mathematical Sciences (CAMS) at the American University of Beirut. The meeting was both a declaration and a hope that normality is returning to the country after so many years of war. The purpose of CAMS is to support research at the highest level and attract expatriate scientists back to Lebanon, thereby stimulating science throughout the country. Even before CAMS moves into new premises in the rebuilt main university building (replacing the previous one, destroyed by a bomb in 1991), a nucleus of several highly accomplished mathematical physicists is working there.

For most of the invited foreign speakers, it was our first visit to Lebanon. Many of us had to acquire a second, 'Israel-free', passport, so as not to be turned back at airport immigration control. But the 'inspection' was a cursory flip through the pages while the 'inspector' looked elsewhere. The meeting combined abundant and full-hearted hospitality with intellectual intensity covering a wide range of topics.

At the opening ceremony, soldiers were everywhere, bayonets gleaming. I was intimidated by the thought of being protected by them, then reassured to discover that their purpose was rather to guard the prime minister, and they left when he did. The president of the university surprised us with his frankness in quoting the mathematician André Weil's opinion that mathematics is better than sex because its pleasure persists undiminished for hours; "If only Bill Clinton had studied mathematics…", he mused.

The meeting, and CAMS, was generously supported by the owner of Beirut's most luxurious hotel, on the mountain above the city, with splendid views over the city and the Mediterranean. Academics are not accustomed to sleeping in beds wider than their length (for a moment, I wondered if I was sleeping sideways). Each day, we were driven ten miles down to the University through the dense and reckless traffic. It was surprising (and, in the circumstances, ironic) to see how much Lebanon resembles Israel (especially Haifa), mainly in the uncontrolled proliferation of concrete brutality alongside elegant and expensive modern shops. Because of the war, though, downtown Beirut is more dilapidated than Israeli cities, with few buildings unmarked by bullets and many completely destroyed; but the suburbs are more elegant, with many splendid old houses preserved.

I reveal here my main nonscientific reason for travelling to exotic locations: food. In this, Lebanon surpassed all my expectations. As well as the familiar hummous and smoked aubergines (baba ganoush), I enjoyed kebabs with unexpected hot yogurt (whose stability seemed to defy gastrophysical laws until I learned that it was kept from separating by eggwhite), a tongue and brain sandwich, and kibbeh naye, the exquisite Lebanese equivalent of steak tartare: raw meat ground with spices and served with a fluffy garlic sauce. It was sad to reflect how, through a nervousness and exaggerated response to tiny risks, we rarely enjoy such delights in this country. The delicious sweetness of jallab, a raisin syrup drizzled with nuts, was soured when I noticed that the café's most elaborate confection was called Hitler. The reason (which to my hosts' embarassment I insisted they translate) was "to show how much we appreciate what he did". I am sure the semitic boys in that café did not realise how prominent they themselves would be on Hitler's list of despised peoples.

I walked along the Corniche near the university to see the spectacular Pigeon Rock in the sea. Suddenly it began to rain. Mediterranean rain seems splashier than ours (because it hits dustier streets?). The shower turned thundery, then torrential, then to hail, then back to rain. I nursed like a baby the lecture I was carrying, that I had spent two days writing on transparencies with water-soluble pens. Desperate, I waded through the mud-flowing street into a taxi, discovering that I was sharing it with a young woman, glamorous but sinister because her full red lips were outlined in deepest black.

On the last day, a trip to Byblos (Jbaile), advertised as the world's oldest continuously inhabited city. I was proud to see hoardings advertising the BBC, before realising that this was the Byblos Beauty Centre. I preferred the unearthly beauty of the vast Jeite caverns, the upper dry and the lower flooded, to be visited by boat in eerie silence, and with a greater variety of types of stalactites and stalagmites than I have seen (ears, leaves, sheets, spikes, knobbles...). We wondered how many parameters a mathematical model would need to generate all these forms (the square root of cauliflowers times mushrooms, perhaps).

Odessa, little and large

Michael Berry
Bristol University Newsletter, 6 December 2000

I was curious to see Odessa — Ukraine's principal port on the northern shore of the Black Sea — because that was the city from where my grandparents emigrated to London in 1906. A few weeks earlier, I had been in Little Odessa, otherwise known as Brighton Beach, Brooklyn, New York City (and reputed money-laundering centre of the USA). Little Odessa is populated almost entirely with emigrants from Ukraine, and the ambience is markedly unamerican: shop signs in Cyrillic, bad-tempered waiters, no credit cards accepted...

On international business flights, most passengers are men, but on the short journey from Istanbul to the real Odessa nearly all my fellow travellers were women — looking prosperous and glamorous in a slightly old-fashioned way, and laden with shopping they were bringing home to Odessa.

I had been invited by the Institute of Optics of Odessa University. They specialise in photographic science, and were proud that Lippmann, one of the pioneers of colour photography, was a native of their city. However, they reserved particular reverence for Sir Nevill Mott, who elucidated the physics of the photographic process in the 1940s, and they were surprised to learn that at that time he was head of our Bristol physics department. Against my wishes, my hosts insisted on prolonging my lecture and hobbling my delivery by providing simultaneous translation, even though almost everybody in the audience could understand my spoken English.

Odessa is an architecturally homogeneous nineteenth-century city, not unlike Paris or Bucharest, albeit a little tattier though being sensitively restored. I was taken to the Potemkin Steps, and proudly informed that the baby in the famous pram scene in Eisenstein's film 'Battleship Potemkin' later became a director of the physics department. Viewed from the top, the steps are invisible, and only the horizontal pavements separating the several flights can be seen, whereas from the bottom only the steps can be seen; my hosts seemed unaware that this locally celebrated geometrical property is shared by almost all steps consisting of several flights. Another tourist sight is the English Club. When I asked why this was so named, I was told that it had been a club for Victorian sea captains: "No women; no conversation" (no comment).

Since my hotel was the second best in Odessa, I expected it to be good, but my hosts warned me that I should expect only the *second* best. And indeed, the door to the refrigerator in my room fell off when I tried to open it, and the shower head suddenly flew off the wall and drenched my clothes, towels, toilet paper ... I was disturbed by a sequence of mysterious phone calls, in which a young woman first asked questions in Russian and, when I didn't understand, began screaming as though she were being tortured; I suspected a trick, and indeed was told that this is a common prostitute's ruse for gaining entry into hotel rooms.

Odessans like to be first and best, especially with jokes. For example, common throughout eastern Europe is the explanation of why policemen always go in threes: one to read, one to write, and one to guard the two intellectuals. But in the Odessa version, one student is telling this story to another, when he feels a tap on the shoulder, and is terrified to discover that he had been overheard by three policemen, one of whom demands his documents, then turns to his companions and asks: "Now, which of you two can read?"... People were generally friendly in this jocular way, but my hosts were evidently embarrassed when we were accosted on the street by one person who made what was obviously an antisemitic gesture of extreme crudity.

I found an internet cafe. Most of the other customers were boys exploring violent and pornographic sites, but there was another foreigner, an American evidently hotmailing his travel diary home, who asked: "Excuse me, what city is this?" I could not believe this stereotype of the ignorant tourist, but he told me that he had travelled overnight on a cruise from Yalta, and had forgotten today's port of call. Later, in the harbour, I saw his brand-new eleven-storey cruise ship.

On the road outside the city, we encountered a surreal convoy of poor people celebrating a wedding. With the bride and groom in pushchairs, the men blowing raucous horns and dressed in clowns' clothes, pressing vodka and sausages on every motorist who slowed to observe them, and the thickly rouged women reeking of perfume (which they transferred to me with abundant and insistent kisses), it resembled a scene from a film by Fellini.

I needed to get to Crimea to attend an optics meeting. The only way was by overnight train, and my hosts advised me to buy a return ticket for a companion to ensure my safety. But the 'companion' was a physicist who wanted to attend the same meeting, so the safety scare could have been a scam to get his fare paid (it was only £10, but not a trivial sum for a university scientist in Ukraine, so I paid up). Ukraine is a country of unreformed smokers, and I wondered whether my night on the sleeper would be tormented by their fumes. With unintentional ambiguity, I was firmly informed that in train compartments “anybody smokes”. Fortunately they meant nobody, and the night passed comfortably. We arrived in Simferopol, capital of Crimea, fresh for the conference on the coast in Alushta; but that is another story.

Chapter 9
Miscellaneous

This ragbag of a chapter starts [9.1] with what is becoming a new research theme, being developed with Pragya Shukla, whose significance has yet to emerge. Curl forces are classical forces depending only on position and not derivable from a scalar potential: their curl is non-zero. Although non-conservative, these forces are not dissipative, and, because there is (usually — but see [B475]) no underlying Hamiltonian or Lagrangian, Noether's theorem does not apply: a curl force can possess a symmetry without a corresponding conservation law, and there can be a conservation law without a corresponding symmetry. Although largely unfamiliar to physicists, curl forces describe [B466] the effective dynamics of polarizable particles in fields of light.

My explanation [B271] of a tantalising scientific toy led to a somewhat surreal episode: a collaboration [9.2] with Andre Geim explaining his observation of a levitating frog, and leading to the 2000 Ig Nobel Prize. (The prize was initially offered to him, but he insisted on sharing it with his collaborator, i.e. me.)

Three reviews [9.3–9.5] give my opinions on a range of scientific topics, and they are followed by two editorials [9.6 and 9.7] from my seven years as editor-in-chief of *Proceedings A of the Royal Society of London* (for more on a current publications issue, see [B459]).

The pieces [9.8–9.15] are speeches and reminiscences musing on a variety of personal and scientific matters.

The final item [9.16] comes close to summarising my scientific credo in a few sentences.

IOP Publishing New J. Phys. 18 (2016) 063018 doi:10.1088/1367-2630/18/6/063018

New Journal of Physics

The open access journal at the forefront of physics

Deutsche Physikalische Gesellschaft DPG
IOP Institute of Physics

Published in partnership with: Deutsche Physikalische Gesellschaft and the Institute of Physics

CrossMark

OPEN ACCESS

RECEIVED
18 March 2016

REVISED
19 May 2016

ACCEPTED FOR PUBLICATION
25 May 2016

PUBLISHED
13 June 2016

PAPER

Curl force dynamics: symmetries, chaos and constants of motion

M V Berry[1,3] and Pragya Shukla[2]

[1] H H Wills Physics Laboratory, Tyndall Avenue, Bristol BS8 1TL, UK
[2] Department of Physics, Indian Institute of Science, Kharagpur, India
[3] Author to whom any correspondence should be addressed.

E-mail: asymptotico@bristol.ac.uk **and** shukla@phy.iitkgp.ernet.in

Keywords: Newtonian, follower force, circulatory force, nonhamiltonian, vortices

Supplementary material for this article is available online

Abstract

This is a theoretical study of Newtonian trajectories governed by curl forces, i.e. position-dependent but not derivable from a potential, investigating in particular the possible existence of conserved quantities. Although nonconservative and nonhamiltonian, curl forces are not dissipative because volume in the position–velocity state space is preserved. A physical example is the effective forces exerted on small particles by light. When the force has rotational symmetry, for example when generated by an isolated optical vortex, particles spiral outwards and escape, even with an attractive gradient force, however strong. Without rotational symmetry, and for dynamics in the plane, the state space is four-dimensional, and to search for possible constants of motion we introduce the *Volume of section*: a numerical procedure, in which orbits are plotted as dots in a three-dimensional subspace. For some curl forces, e.g. optical fields with two opposite-strength vortices, the dots lie on a surface, indicating a hidden constant of motion. For other curl forces, e.g. those from four vortices, the dots explore clouds, in an unfamiliar kind of chaos, suggesting that no constant of motion exists. The curl force dynamics generated by optical vortices could be studied experimentally.

1. Introduction

This concerns Newtonian dynamics driven by forces $\boldsymbol{F}(\boldsymbol{r})$ depending on position $\boldsymbol{r}$ (but not velocity), whose curl is not zero so they are not derivable from a scalar potential [1]. Thus the acceleration (assuming unit mass for convenience) is

$$\ddot{\boldsymbol{r}} = \boldsymbol{F}(\boldsymbol{r}), \quad \nabla \times \boldsymbol{F} \neq 0. \tag{1.1}$$

Motion governed by curl forces is nonconservative: the work done by $\boldsymbol{F}(\boldsymbol{r})$ depends on the path. But, in contrast with other nonconservative contexts such as velocity-dependent frictional forces, the force (1,1) is not dissipative. This is because [1] the flow preserves volume in the position–velocity state space $(\boldsymbol{r},\ \boldsymbol{v} = \dot{\boldsymbol{r}})$: there are no attractors. In the absence of a potential, there is usually no underlying hamiltonian or lagrangian structure, so Noether's theorem does not apply: the link between symmetries and conservation laws is broken, as elementary examples [1] demonstrate.

Most curl forces are nonhamiltonian (a special class [2], which is hamiltonian, will play no role in this paper). Without a hamiltonian, there is no conserved energy. Our aim here is to investigate whether there are other conserved functions of the variables $(\boldsymbol{r}, \boldsymbol{v})$. Alternatively stated, we ask about the dimensionality of regions in state space explored by typical orbits. The extreme case, of most interest, would be where there are no conserved quantities at all, and motions explore regions of full dimensionality densely. We will not be able to answer these questions definitively using analytical arguments, but will present numerics indicating rich structures that deserve to be explored further. In particular, some curl forces $\boldsymbol{F}(\boldsymbol{r})$, apparently not special, indicate the existence of (so far unidentified) constants of motion; and, more interesting other $\boldsymbol{F}(\boldsymbol{r})$, also not apparently not special, indeed seem to explore regions of full dimensionality in the $(\boldsymbol{r}, \boldsymbol{v})$ state space, suggesting a type of chaos not previously encountered.

Although the usefulness of curl forces as physical models has been the subject of dispute in engineering mathematics [3], their applicability in optics is not in doubt and their nonconservative nature has been recognized [4–9]. In this paper we will use optical curl forces as examples, to illustrate the more general curl force dynamics which is our main interest—even though it is known that not all curl forces can be realised optically (see appendix C of [2]). We consider the force on a small polarizable particle in a monochromatic light field $\psi(\boldsymbol{r})$; if a is the ratio of imaginary and real parts of the polarizability [9], the optical force is proportional to

$$\boldsymbol{F} = -\nabla\,|\psi|^2 + a\,\mathrm{Im}[\psi^*\nabla\psi]. \tag{1.2}$$

The second term is a curl force if

$$\nabla\times\mathrm{Im}[\psi^*\nabla\psi] = \mathrm{Im}[\nabla\psi^*\times\nabla\psi] \neq 0. \tag{1.3}$$

We should dispel a possible confusion. In optics, the term 'curl force' has sometimes [10, 11] been used in a different sense from (1.1), to denote forces that are the curl of a vector potential $\boldsymbol{A}$. To relate the two terminologies, we first note that any $\boldsymbol{F}$ can be separated into its curl-free and divergence-free parts, given respectively by a scalar and a vector potential:

$$\begin{aligned}&\boldsymbol{F} = \boldsymbol{F}_{\mathrm{grad}} + \boldsymbol{F}_{\mathrm{curl}},\ \text{where}\ \boldsymbol{F}_{\mathrm{grad}} = -\nabla\phi, \quad \boldsymbol{F}_{\mathrm{curl}} = \nabla\times\boldsymbol{A},\\ &\Rightarrow \nabla\times\boldsymbol{F}_{\mathrm{grad}} = 0, \quad \nabla\cdot\boldsymbol{F}_{\mathrm{curl}} = 0.\end{aligned} \tag{1.4}$$

In this representation, the curl condition (1.1) is

$$\nabla\times\boldsymbol{F} = \nabla\times\boldsymbol{F}_{\mathrm{curl}} = \nabla\nabla\cdot\boldsymbol{A} - \nabla^2\boldsymbol{A} \neq 0. \tag{1.5}$$

When this holds, we call $\boldsymbol{F}_{\mathrm{curl}}$ a *pure curl* force.

The separation (1.4) is not unique, because a gradient $\nabla\phi_1$ can be added and subtracted from each part:

$$\boldsymbol{F}_{\mathrm{grad1}} = \boldsymbol{F}_{\mathrm{grad}} - \nabla\phi_1, \quad \boldsymbol{F}_{\mathrm{curl1}} = \boldsymbol{F}_{\mathrm{curl}} + \nabla\phi_1. \tag{1.6}$$

To maintain the separation, ϕ_1 must be related to a new vector potential $\boldsymbol{A}_1$ by

$$\nabla\times\boldsymbol{A}_1 = \nabla\phi_1,\ \text{i.e.}\ \nabla^2\phi_1 = 0, \quad \nabla\times\nabla\times\boldsymbol{A}_1 = \nabla\nabla\cdot\boldsymbol{A}_1 - \nabla^2\boldsymbol{A}_1 = 0. \tag{1.7}$$

For motion in the plane $\boldsymbol{r} = (x, y)$, $\boldsymbol{A}_1$ can be chosen to lie in the perpendicular direction $\boldsymbol{e}_z$, and its magnitude—the 'stream function' A_1—satisfies

$$\nabla\times A_1\boldsymbol{e}_z = (\partial_y A_1, -\partial_x A_1) = \nabla\phi_1 = (\partial_x\phi_1, \partial_y\phi_1), \tag{1.8}$$

whose solution (see e.g. [12]) is

$$A_1(x, y) = C - \int_0^x \mathrm{d}x'\partial_y\phi_1(x', 0) + \int_0^y \mathrm{d}y'\partial_x\phi_1(x, y'). \tag{1.9}$$

An optical curl force, defined by (1.2) and (1.3), is also a pure curl force, in the sense defined by (1.4) and (1.5), if the following condition holds (equivalent to the stationary continuity equation for the ψ current):

$$\nabla\cdot\mathrm{Im}[\psi^*\nabla\psi] = \mathrm{Im}[\psi^*\nabla^2\psi] = 0,\ \text{e.g. if}\ \nabla^2\psi = A\psi\,(A\ \text{real}). \tag{1.10}$$

So, optical curl forces generated by fields satisfying the Helmholtz or Laplace equations are also pure curl forces.

Our examples, illustrating what we think are general features of curl force dynamics, will be concerned with curl forces in the plane, usually generated by optical fields with one or more vortices. The conservative (i.e. gradient-force) counterpart of this dynamics has been extensively studied (also in three dimensions) [13–17]; in addition, aspects of the curl force motion have been studied, emphasising its azimuthal form near vortices [18, 19].

We emphasise that although curl forces can be regarded as mathematically fundamental (the most general case of position-dependent Newtonian forces), they are not fundamental physically. They are effective forces—here, as elsewhere in physics, the inevitable result of idealisations. Our focus here is on the unexpected features of the dynamics generated by the simplest Newtonian force (1.1). So, rather than considering these idealisations in detail, we now comment briefly on several of them.

In (1.2), the curl force, giving rise to nonhamiltonian particle motion, is associated with internal energy dissipation within the particle (specified by the parameter a); part of this energy goes into radiation, but this energy, and its replenishment by the sources of the field, are not considered explicitly. Internal dissipation has a curious implication, discussed elsewhere [9]: in contrast with statistical mechanics, where microscopic elastic dynamics generates macroscopic friction, in the present case microscopic dissipation is associated with macroscopic motion that is not dissipative.

In a quantum treatment of optical forces, on atoms rather than classical polarizable particles [18, 19], the nonhamiltonian effective forces governing the centre-of-mass motion emerge from a more fundamental hamiltonian formulation of the total system, based on the optical Bloch equations.

Another idealisation, that we discussed in detail [20] and whose violation involves additional forces, is the adiabatic separation between the slow translation and the fast internal dynamics of the particle. Further additional forces arise for particles whose size, relative to the optical wavelength, is not small [21, 22].

The structure of the paper is as follows. Section 2 concerns forces with rotational symmetry. We show the unexpected result that in a large class of cases the existence of a curl force means that the particle always escapes to infinity; even an arbitrarily strong accompanying attractive gradient force fails to restrain it.

Section 3 concerns forces that do not possess rotational symmetry. The state space $(\boldsymbol{r}, \boldsymbol{v})$ is four-dimensional. As a numerical tool for displaying possible constants of motion, we introduce (section 3.1) the *Volume of section* (VoS), by analogy with the Poincaré surface of section familiar in hamiltonian dynamics. The VoS is a numerical procedure, in which the state variables at times along an orbit which satisfy a particular condition (e.g. crossing the x axis), are plotted as dots in the three-dimensional space of the remaining state variables. We give two examples of curl force dynamics associated with several optical vortices of zero total strength. In one (section 3.2) the dots lie on a surface, indicating the existence of a hidden constant of motion (hidden, in the sense that we do not know its functional form). In the other (section 3.3), the dots form a cloud, indicating no constants at all, i.e. orbits exploring a 4D region of the state space. This seems to be an unfamiliar kind of chaos. Readers interested only in this latter case should look at the 'dust clouds' in figure 9.

Questions raised by this study, and the possibility of experimentally detecting the new features we have identified, are discussed in the concluding section 4.

2. Rotational symmetry: separability, and inevitable escape from a single optical vortex

It is convenient to write the most general rotationally symmetric dynamics in the plane $\boldsymbol{r} = (x, y) = r(\cos\theta, \sin\theta)$, of the type (1.1), in the form

$$\ddot{\boldsymbol{r}} = -g(r)\boldsymbol{e}_r + \frac{h(r)}{r}\boldsymbol{e}_\theta. \tag{2.1}$$

The first term is a radial gradient force (with potential given by the integral of $g(r)$), attractive if $g(r) > 0$, and the second is an azimuthal curl force. In terms of the angular momentum

$$J = r^2\dot{\theta}, \tag{2.2}$$

the dynamical equations for the evolution $r(t)$ and $J(t)$ are

$$\ddot{r} = \frac{J^2}{r^3} - g(r), \quad \dot{J} = h(r). \tag{2.3}$$

This is already a reduction from four freedoms in (1.1) (i.e. x, y, v_x, v_y) to three (i.e. $r, J, \dot{r}$), reflecting the fact that changing the starting azimuth $\theta(0)$ simply rotates the trajectory. We assume that $h(r)$ has the same sign for all r—positive, say. Then (2.3) shows that the angular momentum always increases—an obvious consequence of the torque associated with the azimuthal force, as well as illustrating the non-applicability of Noether's theorem (angular momentum not conserved, although there is rotational symmetry).

A further reduction is possible, to two freedoms. Generalizing our earlier analysis [1], to include the radial force in (2.1), we transform the independent time variable t to J, so we now write $r(J)$ instead of $r(t)$. We have, using (2.3)

$$\frac{\mathrm{d}r}{\mathrm{d}J} \equiv r' = \frac{\dot{r}}{\dot{J}} = \frac{\dot{r}}{h(r)}. \tag{2.4}$$

Thus we get the $r(J)$ dynamics

$$r'' + (r')^2\frac{\partial}{\partial r}(\log[h(r)]) = \frac{(J^2/r^3 - g(r))}{h(r)^2}. \tag{2.5}$$

Thus rotational symmetry has separated the dynamics, with reduction to two freedoms: r, r': the θ coordinate has been eliminated. We can make the r equation simpler by defining the new radial variable

$$R \equiv \int_{r_0}^{r} \mathrm{d}r_1 h(r_1). \tag{2.6}$$

Thus we seek the dynamics $R(J)$ rather than $r(J)$ or the original $r(t)$. The relevant derivatives are

$$R' = r'h(r), \quad \text{and } R'' = r''h(r) + (r')^2\frac{\partial}{\partial r}h(r), \tag{2.7}$$

IOP Publishing New J. Phys. 18 (2016) 063018 M V Berry and P Shukla

leading to the final equation of radial motion

$$R'' = \frac{(J^2/r^3 - g(r))}{h(r)}. \tag{2.8}$$

Explicitly

$$R = R(J), \quad r = r(R(J)), \quad R'' = \frac{d^2R(J)}{dJ^2}. \tag{2.9}$$

Once (2.8) has been solved for $R(J)$, $r(J)$ can be determined by inverting (2.6). Then the time variable can be reinstated from the second equation in (2.3):

$$t(J) = \int_{J_0}^{J} \frac{dJ_1}{h(r(J_1))}. \tag{2.10}$$

Thus we have $r(t)$ and $J(t)$. Finally, the azimuth can be found from (2.2):

$$\theta(t) = \int_{t0}^{t} \frac{dt_1 J(t_1)}{r^2(t_1)}. \tag{2.11}$$

This completes the separation of the radial and angular dynamics, reducing the original four-freedom system to two, i.e. R, R'. The reduced dynamical equation (2.8) is of hamiltonian form:

$$H(R, P, J) = E(J) = \frac{1}{2}P^2 + U(R, J), \text{ where}$$
$$U(R, J) = \frac{J^2}{2r(R)^2} + \int_{\text{const.}}^{r(R)} dr' g(r'). \tag{2.12}$$

We note that this is J dependent, and since J corresponds to time, the energy E is not conserved; in fact it always increases:

$$\frac{dE(J)}{dJ} = \frac{\partial U}{\partial J} = \frac{J}{r} \geqslant 0. \tag{2.13}$$

The repulsive part $J^2/2r(R)^2$ of the potential $U(R)$ in (2.12) always increases. Although this argument is formulated in the (R, J) plane, in the original (r, t) plane it carries the unexpected consequence that the particle always recedes from the origin, however strongly attractive the radial force $g(r)$ is (except when it is a hard wall, in which case the particle spirals ever closer to the wall, ever faster).

An important special case of rotational dynamics is generated by optical forces from a single isotropic vortex of order m, whose light wave is

$$\psi(\mathbf{r}) = (x + iy)^m = r^m \exp(im\theta). \tag{2.14}$$

According to (1.2), the radial and azimuthal forces in (2.1) are

$$g(r) = r^{2m-1}, \quad h(r) = ar^{2m}, \tag{2.15}$$

and the curl force is a pure curl force according to (1.10). Thus (2.8) simplifies to

$$\rho'' = B[J^2\rho^{-(2m-3)/(2m+1)} - \rho^{-1/(2m+1)}],$$
$$\text{where } \rho = \frac{2m+1}{a}R, \ B = \frac{2m+1}{a^2}. \tag{2.16}$$

This is a variant of the Emden-Fowler equation [23, 24], with two source terms instead of one; a general solution in closed form seems unavailable. The variables are related by

$$\rho = r^{2m+1}, \quad \dot{J} = a\rho^{2m/(2m+1)}, \tag{2.17}$$

and the hamiltonian (2.12) simplifies to

$$H(\rho, p_\rho, J) = E(J) = \frac{1}{2}p_\rho^2 + U(\rho, J), \text{ where}$$
$$U(\rho, J) = \frac{1}{2}B(2m+1)\left(J^2\rho^{-2/(2m+1)} + \frac{\rho^{2m/(2m+1)}}{m}\right). \tag{2.18}$$

The potential, whose perpetual increase causes the particle to escape, is illustrated in figure 1.

In an alternative expression of the dynamics (2.1), position is denoted by the complex variable

$$z = x + iy. \tag{2.19}$$

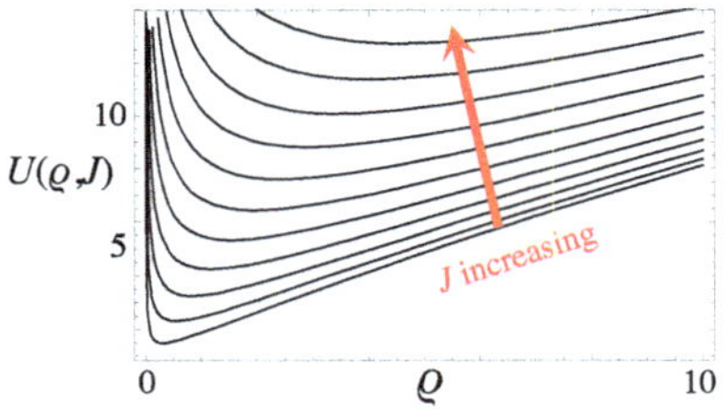

Figure 1. Potential in (2.18) for $m = 2$ and different J.

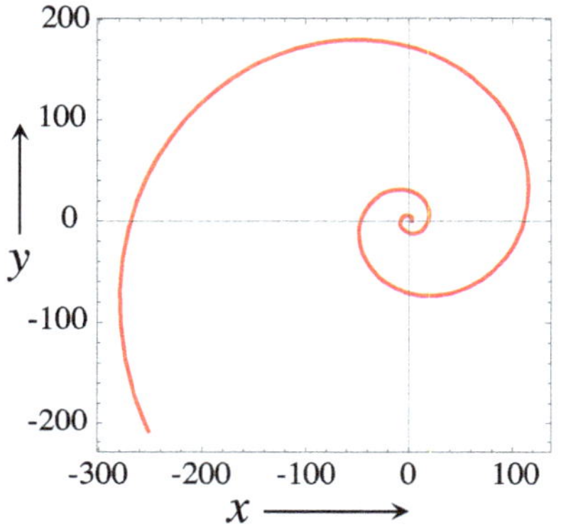

Figure 2. One-vortex track (2.22), for $(0 \leqslant t \leqslant 0.99t_0)$, $t_0 = 1, m = 2, a = 1$, corresponding to initial conditions $(x_0, y_0, v_{x0}, v_{y0}) = (3.269, 0, 3.269, 11.642)$.

Now the evolution (1.1) is

$$\ddot{z} = \frac{(-g(|z|) + ih(|z|)/|z|)}{|z|}z. \tag{2.20}$$

In this representation, the one-vortex dynamics takes the form of a stationary nonlinear Schrödinger equation (with t analogous to a coordinate):

$$\ddot{z} = |z|^{2m-2}(-1 + ia)z. \tag{2.21}$$

Although a general solution seems unavailable, a particular solution can be found, representing a particle escaping to infinity at time $t = t_0$ while spiralling logarithmically (figure 2). For $m > 1$, this is

$$z(t) = \frac{C\exp(iq\log(t_0/(t_0 - t)))}{(t_0 - t)^{1/(m-1)}}, \tag{2.22}$$

in which

$$q = \frac{m + 1 + \sqrt{(m + 1)^2 + 4ma^2}}{2a(m - 1)}, \quad C = \left(\frac{q(m + 1)}{a(m - 1)}\right)^{1/(2m-2)}. \tag{2.24}$$

This solution corresponds to the initial conditions

$$x(0) = \frac{C}{(t_0)^{1/(m-1)}}, \quad y(0) = 0, \quad v_x(0) + iv_y(0) = \frac{C(1 + i(m - 1)q)}{(m - 1)(t_0)^{m/(m-1)+iq}}. \tag{2.25}$$

For the excluded case $m = 1$, (2.21) is a linear equation, whose spiralling solutions are

$$z(t) = c_+\exp(it\sqrt{1 - ia}) + c_-\exp(-it\sqrt{1 - ia}). \tag{2.26}$$

IOP Publishing New J. Phys. 18 (2016) 063018 M V Berry and P Shukla

3. More optical vortices: seeking chaos

3.1. The VoS

What distinguishes dynamics under curl forces from hamiltonian dynamics? A fundamental difference would be motion exploring a full-dimension region of the $(\boldsymbol{r}, \boldsymbol{v})$ state space. For motion in the $\boldsymbol{r} = (x, y)$ plane, this would be exploration of a four-dimensional region. As a numerical tool for investigating this possibility, we introduce the *Volume of section* VoS, defined as follows. We select times t_n along an orbit $\boldsymbol{r}(t)$ satisfying some condition, for example $x(t_n) = 0$. At such times (identified numerically), we plot the other three variables, for example $(y(t_n), v_x(t_n), v_y(t_n))$, and examine the dot patterns in this three-dimensional space—the VoS—after long times. There are four natural choices for the VoS, corresponding to $x(t_n) = 0, y(t_n) = 0, v_x(t_n) = 0, v_y(t_n) = 0$; we will denote these by VoSx, VoSy, VoSv_x, VoSv_y. Although the full dynamics in 4D state space is volume-preserving, the absence of underlying symplectic structure means that the 3D map between successive dots on each VoS does not preserve volume. This contrasts with the area-preserving 2D Poincaré map on a 3D constant-energy hypersurface in hamiltonian dynamics.

If there is one conserved quantity, as in a hamiltonian system, the dots in each VoS will lie on a surface, and if the dynamics is chaotic the surface will be partly or wholly filled. If there is an additional constant, as in an integrable system, the dots will lie on a curve. And in the situation we are contemplating, in which there is no conserved quantity, the dots will fill a volume.

3.2. Two vortices

To explore the possibilities, we need to choose suitable forces $\boldsymbol{F}(\boldsymbol{r})$. As we have seen, the optical curl force from a single vortex always leads to escape, even in the presence of an attractive gradient force. The escape is associated with the continuous increase of angular momentum caused by the torque from the curl force. It is natural to try to avoid this by exploring the curl force from two vortices of opposite strength, so at large distances the net torque tends to zero. The simplest such optical field, representing two vortices on the x axis, at $x = +1$ with strength $+1$ and at $x = -1$ with strength -1, is

$$\psi_1(\boldsymbol{r}) = (x - 1 + \mathrm{i}y)(x + 1 - \mathrm{i}y). \tag{3.1}$$

This generates a curl force. Although it is not a pure curl force, because (see (1.10)) $\nabla^2\psi_1 = 4$, it can easily be made so, for example by adding $-2y^2$.

We note in passing that the associated gradient force

$$\boldsymbol{F}_{1\mathrm{grad}}(\boldsymbol{r}) = -\nabla\,|\psi_1(\boldsymbol{r})|^2 = -4\{x(r^2 - 1), y(r^2 + 1)\}, \tag{3.2}$$

is integrable as well as hamiltonian. Of course the energy

$$E_1 = \frac{1}{2}(v_x^2 + v_x^2) + |\psi_1(\boldsymbol{r})|^2 = \frac{1}{2}(v_x^2 + v_x^2) + (r^2 + 1)^2 - 4x^2 \tag{3.3}$$

is conserved. And, as can easily be confirmed, the following quantity is also conserved:

$$K_1 = (\boldsymbol{r} \times \boldsymbol{v}.\, \boldsymbol{e}_z)^2 - 4v_y^2 - 8y^2(r^2 + 2). \tag{3.4}$$

Figure 3 shows the track of an orbit in the (x, y) plane, with a pattern clearly illustrating the integrability.

These conserved quantities are destroyed when the gradient force is combined with the corresponding curl force according to (1.2):

$$\boldsymbol{F}_{1\,\mathrm{curl+grad}}(\boldsymbol{r}, a) = -4\{x(r^2 - 1), y(r^2 + 1)\} + 2a\{-2xy, x^2 - y^2 - 1\}. \tag{3.5}$$

Figure 4 shows the pattern of directions of this force, asymptotically attractive and inspiralling to the two vortices (circles), with a stagnation point (square) on the negative y axis.

This is a promising choice for a curl force, because it fails the 'anisotropic hamiltonian' test derived elsewhere (equation (2.6)) in [2]. Figure 5 shows an orbit generated by this force, in the (x, y) and (v_x, v_y) (hodograph) planes. It appears irregular, indicating that the system is not integrable: there is at most one constant of motion. Of course this cannot be energy because the force is non-conservative.

To investigate whether in fact there is such a constant of motion, we show in figure 6 the corresponding VoS patterns. The dots clearly lie on surfaces, indicating that a constant of motion, that is, a conserved function of $\boldsymbol{r}$ and $\boldsymbol{v}$, associated with the force (3.5), does exist. We do not know what this constant is: the complicated form of the surfaces (including the holes, which do not fill up in simulations for longer times) suggests that it is not a simple function. Its existence raises the possibility that with a suitable change of variables the constant could represent a hamiltonian H that generates the motion, as in (2.12) for the rotationally symmetric case—though in this two-vortex case any such H would be time-independent.

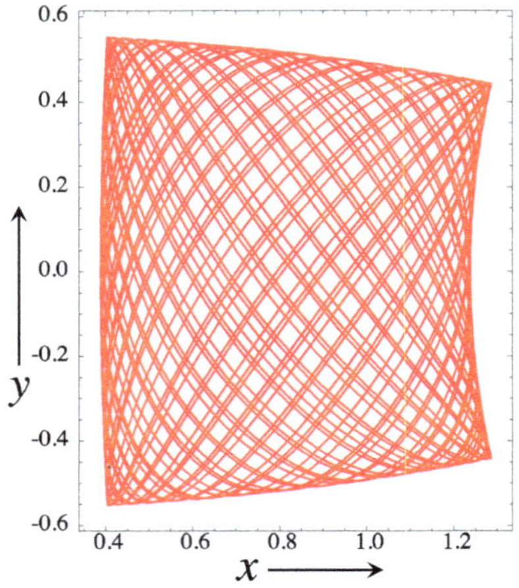

Figure 3. Orbit in (x, y) plane for the gradient (i.e. not curl) force $\boldsymbol{F}_{1\text{grad}}$ (equation (3.2)) from two vortices, for $0 \leqslant t \leqslant 100$ and with $(x_0, y_0, v_{x0}, v_{y0}) = (1/2, 1/2, 1/2, 1/2)$.

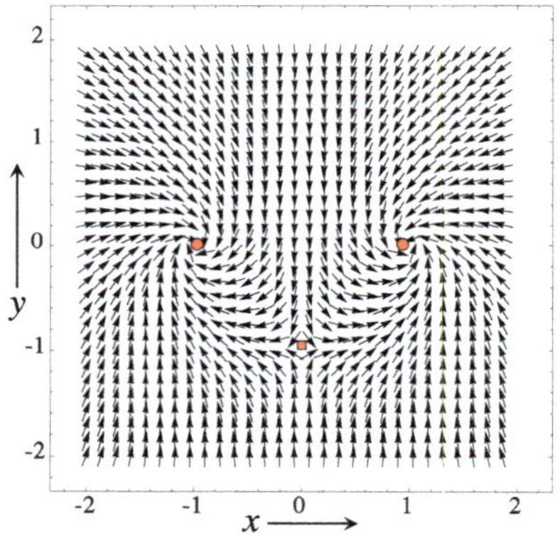

Figure 4. Two-vortex force $\boldsymbol{F}_{1\text{curl+grad}}$ (equation (3.5)) for $a = 2$.

3.3. Four vortices

We have investigated two-vortex fields more general than (3.5), in which the gradient force is not of optical type, suggesting curl force dynamics with no constants of motion. But forces in which both the curl and gradient parts are optical are not only conceptually simple but could also be explored experimentally. For this reason, we now consider the optical force from four alternating-sign vortices arranged on a square:

$$\psi_2 = (x + 1 + \mathrm{i}(y + 1))(x + 1 - \mathrm{i}(y - 1))(x - 1 + \mathrm{i}(y - 1))(x - 1 - \mathrm{i}(y + 1)). \tag{3.6}$$

Again we create the corresponding curl + gradient force according to (1.2):

$$\begin{aligned} \boldsymbol{F}_{2\,\text{curl+grad}} = &-8(x(r^6 + 4x^2 - 12y^2),\ x(r^6 + 4y^2 - 12x^2)) \\ &+ 8a(x(r^2(x^2 - 3y^2) + 4),\ -y(r^2(y^2 - 3x^2) + 4)). \end{aligned} \tag{3.7}$$

Figure 7 shows the pattern of directions of this force, inspiralling to the four vortices (circles), with three stagnation points (squares) on the x axis, and asymptotically attractive.

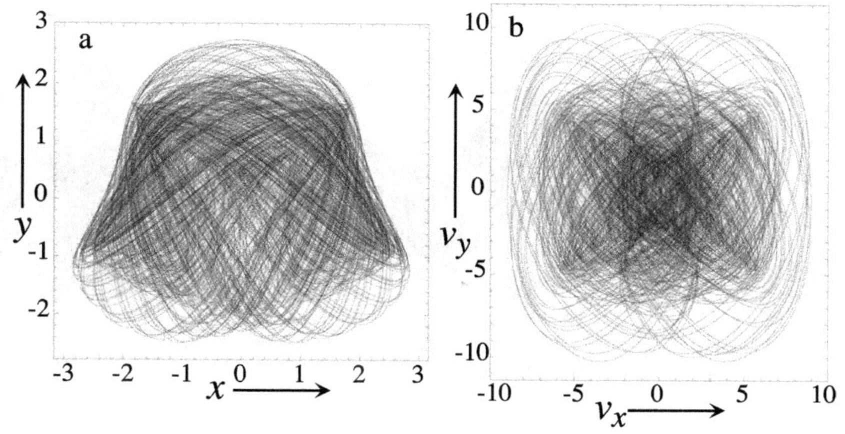

Figure 5. Orbit in the (x, y) plane (left) and the (v_x, v_y) (hodograph) plane (right) for the two-vortex force $\boldsymbol{F}_{1\text{curl}+\text{gtad}}$ (equation (3.5)), for $a = 2$, $(x_0, y_0, v_{x0}, v_{y0}) = (1/2, 0, -1, 2)$, and $0 \leqslant t \leqslant 1000$.

Figure 8 shows an orbit generated by this force, in the (x, y) and in the (v_x, v_y) (hodograph) planes. As with the two-vortex orbit in figure 5, this looks irregular, indicating that the dynamics is not integrable, and again raising the question of whether any constant of motion exists.

Figure 9 shows the corresponding VoS dot patterns for this four-vortex dynamics. They are very different from the two-vortex VoS patterns in figure 6. It is hard to interpret these 'dust cloud' patterns as anything other than irregularly exploring a four-dimensional region in the state space. If correct, this means that curl forces can generate an unfamiliar kind of chaos—contrasting with hamiltonian chaos, where because of energy conservation a hypersurface is explored. Preliminary computations support the conjecture that orbits with nearby initial conditions separate exponentially. The dust cloud patterns appear to possess a complicated structure; there are 'holes', almost or totally devoid of dots, which do not fill up in simulations for longer times; and some regions appear denser than others, almost hinting that the orbit would condense onto a chaotic attractor—which of course it cannot do because the full dynamics is 4D volume-preserving.

Irregular trajectories generated by curl forces, have been demonstrated in simulations before [16, 19] though not emphasised as nonhamiltonian and for times too short to identify constants of motion.

4. Concluding remarks

This study reveals a variety of structures in orbits governed by rather simple curl forces, including those exerted on small particles from optical waves with vortices. For a single optical vortex, the orbits spiral outwards and always escape. When there are two vortices, numerics indicates a hidden constant of motion, even though there is no conserved energy. Most interesting are other cases, for example four vortices, where it seems there are no conserved quantities at all. Further computations, not reported here, show that these different behaviours are not exceptional.

We regard this as an exploratory study, raising several questions:

- Are the dust cloud patterns such as those in figure 9 typical in situations where orbits are bounded?
- How can their structures be characterised?
- For optical fields with an infinite periodic array of vortices, with total strength zero in each unit cell, can motion under the associated curl plus gradient forces be chaotic and explore the full state space dimensionality—and, if so, is this typical or exceptional? (This would extend previous studies [14, 15] of orbits in optical lattices under conservative forces.)

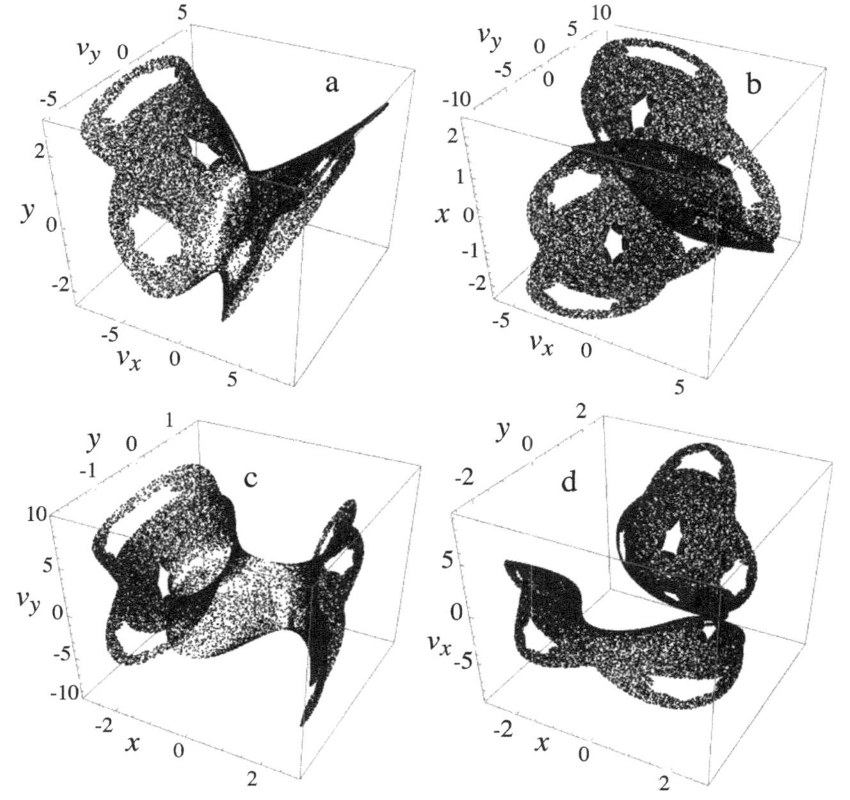

Figure 6. Two-vortex volume of surface patterns corresponding to figure 5, for ($0 \leqslant t \leqslant 50\,000$): (a) VoS$x$; (b) VoS$y$; (c) VoS$v_x$; (d) VoS$v_y$. To enable the surface to be visualised more clearly, we include separately in the supplementary material figure 6e.wrl, by which the VoSx pattern can be manually rotated and magnified.

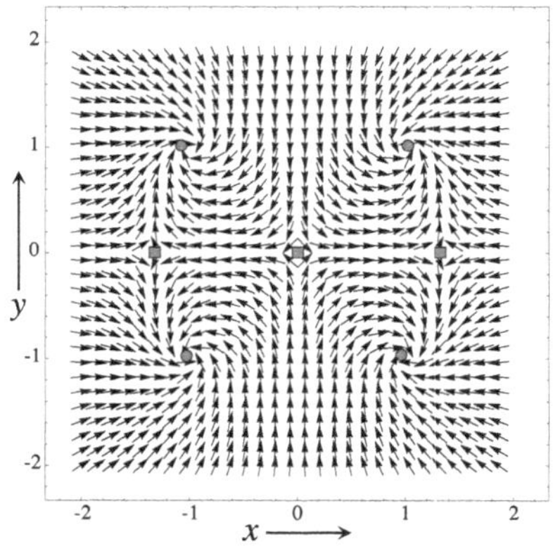

Figure 7. Four-vortex force $\boldsymbol{F}_{2\mathrm{curl+grad}}$ (equation (3.7)) for $a = 1.8$.

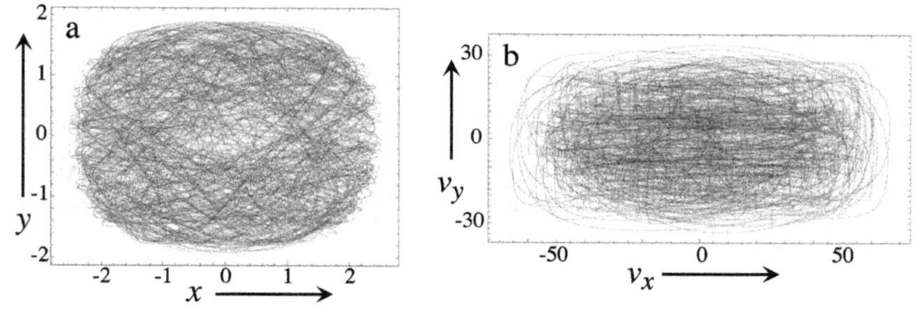

Figure 8. Orbit in the (x, y) plane (left) and the (v_x, v_y) (hodograph) plane (right) for the four-vortex force $\boldsymbol{F}_{2\mathrm{curl+gtad}}$ (equation (3.7)), for $a = 1.8$, $(x_0, y_0, v_{x0}, v_{y0}) = (0,0,-0.1,-0.1)$, and $(0 \leqslant t \leqslant 75)$.

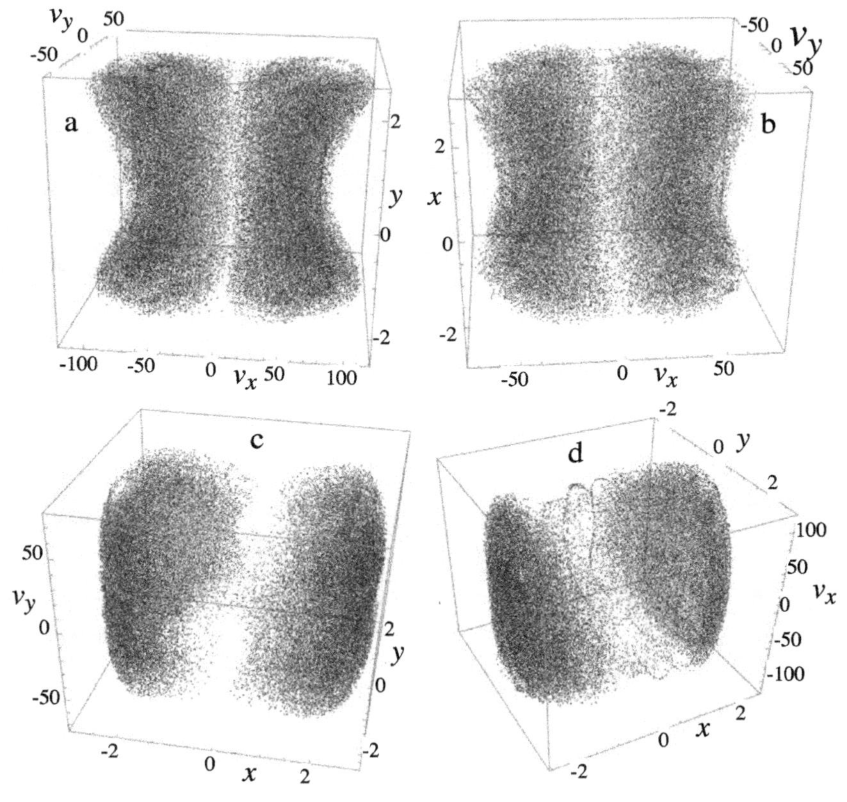

Figure 9. Four-vortex volume of surface 'dust cloud' patterns corresponding to figure 8, for $(0 \leqslant t \leqslant 10\ 000)$: (a) VoS$x$; (b) VoS$y$; (c) VoS$v_x$; (d) VoS$v_y$. To enable the dust cloud to be visualised more clearly, we also include separately in the supplementary material figure 9e.wrl, by which the VoSx pattern can be manually rotated and magnified.

- Are periodic orbits dense, as in hamiltonian systems? This is not a trivial question because, as we discussed elsewhere (section 4 of [2]), the nonconservative nature of curl forces imposes strong restrictions on the forms of periodic orbits.
- Where numerics strongly suggests a constant of motion, as in the case considered in section 3.2 and illustrated in figure 6, is there any analytical way, general for curl forces (1.1), to establish its existence and characterise it?
- Can the curl force dynamics we have identified theoretically be seen experimentally, in the motion of small polarizable particles, governed by forces from optical fields with several vortices? This is not straightforward: to see the Newtonian particle motion generated by (1.1), the particles would need to be trapped in a vacuum, unlike the viscosity-dominated motion in many experiments [25–28]—which also emphasise effects of the angular momentum, for example particle rotation, rather than the centre of mass motion considered here.

Acknowledgments

We thank three referees for their detailed reading of the paper, and their very helpful criticisms and suggestions. MVB's research is supported by the Leverhulme Trust.

References

[1] Berry M V and Shukla P 2012 Classical dynamics with curl forces, and motion driven by time-dependent flux *J. Phys. A: Math. Theor.* 45 305201
[2] Berry M V and Shukla P 2015 Hamiltonian curl forces *Proc. R. Soc.* A 471 20150002
[3] Elishakoff I 2005 Controversy associated with the so-called 'follower forces': critical overview *Appl. Math. Rev.* 58 117–42
[4] Ashkin A and Gordon J P 1983 Stability of radiation-pressure particle traps: an optical Earnshaw theorem *Opt. Lett.* 8 511–3
[5] Nieto-Vesperinas M, Sáenz J J, Gómez-Medina R and Chantada L 2010 Optical forces on small magnetodielectric particles *Opt. Express* 18 11430–43
[6] Shimizu Y and Sasada H 1998 Mechanical force in laser cooling and trapping *Am. J. Phys.* 66 960–7
[7] Wu P, Huang R, Tischer C, Jonas A and Florin E-L 2009 Direct measurement of the nonconservative force field generated by optical tweezers *Phys. Rev. Lett.* 103 108101
[8] Pesce G, Volpe G, De Luca A C, Rusciano G and Volpe G 2009 Quantitative assessment of non-conservative radiation forces in an optical trap *Eur. Phys. Lett.* 86 38002
[9] Berry M V and Shukla P 2013 Physical curl forces: dipole dynamics near optical vortices *J. Phys. A: Math. Theor.* 46 422001
[10] Albaladejo S, Marqués M I, Laroche M and Sáenz J J 2009 Scattering forces from the curl of the spin angular momentum *Phys. Rev. Lett.* 102 113602
[11] Gómez-Medina R, Nieto-Vesperinas M and Sáenz J J 2011 Nonconservative electric and magnetic optical forces on submicron dielectric particles *Phys. Rev.* A 83 033825
[12] Berry M V and Dennis M R 2011 Stream function for optical energy flow *J. Opt.* 13 064004
[13] Volke-Sepúlveda K and Jáuregui R 2009 All-optical 3D atomic loops generated with Bessel light fields *J. Phys. B: At. Mol. Opt.* 42 085303
[14] Pérez-Pascual R, Rodríguez-Lara R and Jáuregui R 2011 Chaotic dynamics of thermal atoms in labyrinths created by optical lattices *J. Phys. B: At. Mol. Opt. Phys.* 44 035303
[15] Castaneda J A, Pérez-Pascual R and Jáuregui R 2013 Chaotic dynamics of dilute thermal atom clouds on stationary optical Bessel beams *J. Phys. B: At. Mol. Opt.* 46 145306
[16] Carter A R, Babiker M, Al-Amri M and Andrews D L 2006 Generation of microscale current loops, atom rings, and cubic clusters using twisted optical molasses *Phys. Rev.* A 73 021401
[17] Lloyd S M, Babiker M and Yuan J 2012 Interaction of electron vortices and optical vortices with matter and processes of orbital angular momentum exchange *Phys. Rev.* A 86 023816
[18] Babiker M, Power W L and Allen L 1994 Light-induced torque on moving atoms *Phys. Rev. Lett.* 73 1239–42
[19] Allen L, Babiker M, Lai W K and Lembessis V E 1996 Atom dynamics in multiple Laguerre–Gaussian beams *Phys. Rev.* A 54 4259
[20] Berry M V and Shukla P 2014 Superadiabatic forces on a dipole: exactly solvable model for a vortex field *J. Phys. A: Math. Theor.* 47 125201
[21] Bliokh K Y, Bekshaev A and Nori F 2014 Extraordinary momentum and spin in evanescent waves *Nat. Commun.* 5 3300
[22] Antognozzi M *et al* 2016 Direct measurements of the extraordinary optical momentum and transverse spin-dependent force using a nano-cantilever *Nat. Phys.* in press (doi:10.1038/nphys3732)
[23] Polyanin A D and Zaitsev V F 2003 *Handbook of Exact Solutions for Ordinary Differential Equations* (Boca Raton: Chapman and Hall/CRC)
[24] Polyanin A D 2004 Emden-Fowler equation (http://eqworld.ipmnet.ru/en/solutions/ode/ode0302.pdf)
[25] Volke-Sepulveda K and Garcés-Chávez V 2002 Orbital angular momentum of a high-order Bessel light beam *J. Opt. B: Quantum Semiclass. Opt.* 4 S82–9
[26] O'Neill A T, MacVicar I, Allen L and Padgett M J 2002 Intrinsic and extrinsic nature of the orbital angular momentum of a light beam *Phys. Rev. Lett.* 88 053601
[27] Curtis J E and Grier D G 2003 Structure of optical vortices *Phys. Rev. Lett.* 90 133901
[28] Garcés-Chávez V, McGloin D, Padgett M J, Dultz W, Schmitzer H and Dholakia K 2003 Observation of the transfer of the local angular momentum density of a multiringed light beam to an optically trapped particle *Phys. Rev. Lett.* 91 093602

Eur. J. Phys. 18 (1997) 307–313. Printed in the UK PII: S0143-0807(97)84689-2

Of flying frogs and levitrons

M V Berry† and A K Geim‡

† H H Wills Physics Laboratory, Tyndall Avenue, Bristol BS8 1TL, UK
‡ High Field Magnet Laboratory, Department of Physics, University of Nijmegen, Toernooiveld, 6525 ED Nijmegen, The Netherlands

Received 4 June 1997

Abstract. Diamagnetic objects are repelled by magnetic fields. If the fields are strong enough, this repulsion can balance gravity, and objects levitated in this way can be held in stable equilibrium, apparently violating Earnshaw's theorem. In fact Earnshaw's theorem does not apply to induced magnetism, and it is possible for the total energy (gravitational + magnetic) to possess a minimum. General stability conditions are derived, and it is shown that stable zones always exist on the axis of a field with rotational symmetry, and include the inflection point of the magnitude of the field. For the field inside a solenoid, the zone is calculated in detail; if the solenoid is long, the zone is centred on the top end, and its vertical extent is about half the radius of the solenoid. The theory explains recent experiments by Geim *et al*, in which a variety of objects (one of which was a living frog) was levitated in a field of about 16 T. Similar ideas explain the stability of a spinning magnet (Levitron^{TM}) above a magnetized base plate. Stable levitation of paramagnets is impossible.

Samenvatting. Magnetische velden stoten diamagnetische voorwerpen af. Zulke velden kunnen zo sterk zijn dat zij de zwaartekracht opheffen. Het is op deze wijze mogelijk zulke voorwerpen te laten zweven. Dit vormt een stabiel evenwicht, wat in tegenspraak schijnt te zijn met Earnshaw's Theorema. Echter Earnshaw's Theorema is niet langer geldig als het magnetisme veld geinduceerd is. De totale energie (bevattende bijdragen van het magnetisme en de zwaartekracht) kan toch een lokaal minimum vertonen. Algemene criteria voor zo'n minimum zullen worden opgesteld. Verder zal worden aangetoond dat voor een cilindrisch symmetrisch veld, langs zijn symmetrie as altijd een zone gevonden kan worden waarin een stabiel evenwicht bestaat. Voor het veld binnen een solenoïde zal deze zone in detail bepaald worden. Als deze spoel voldoende land is bevindt deze zone zich aan het uiteinde van de spoel. De lengte van deze zone langs de symmetrie as van het veld is ongeveer de helft van de straal van de spoel. Deze theorie geeft een goede verklaring voor de experimenten van Geim *et al*. In deze experimenten werden een grote verscheidenheid aan verschillende voorwerpen (waaronder een levende kikker) tot zweven gebracht in velden van ongeveer 16 T. Analoge theoriën verklaren de stabiliteit van een roterend permanent magneetje (Levitron^{TM}) boven een magneetische grondplaat. Het is onmogelijk om paramagnetische voorwerpen stabiel te doen zweven.

1. Introduction

It is fascinating to see objects floating without material support or suspension. In the 1980s, this became a familiar sight when pellets of the new high-temperature type II superconductors were levitated above permanent magnets, and vice versa (Brandt 1989) (levitation of type I superconductors had been achieved much earlier (Arkadiev 1947, Shoenberg 1952)). Recently, two other kinds of magnetic levitation have captured the attention of physicists and the general public. In the Levitron^{TM} (Berry 1996, Simon *et al* 1997, Jones *et al* 1997), a permanent magnet in the form of a spinning top floats above a fixed base that is also permanently magnetized. In diamagnetic levitation, recently achieved by A K Geim with J C Maan, H Carmona and P Main (Rodgers 1997), small objects (live frogs and grasshoppers, waterdrops, flowers, hazelnuts ...) float in the large (16 T) magnetic field inside a solenoid.

As well as being striking to the eye, magnetic levitation is particularly surprising to physicists because of the obstruction presented by Earnshaw's theorem (Earnshaw 1842, Page and Adams 1958, Scott 1959). This states that no stationary object made of charges, magnets and masses in a fixed configuration can be held in stable equilibrium by any combination of static electric, magnetic or gravitational forces, that is, by any forces derivable from a potential satisfying Laplace's equation. The proof is simple: the stable equilibrium of such an object would require its energy to possess a minimum, which is impossible because the energy must satisfy Laplace's equation, whose solutions have no isolated minima (or maxima), only saddles.

Our purpose here is to explain how stable magnetic levitation of diamagnets can occur despite Earnshaw's theorem. To do this, we obtain formulas for the

0143-0807/97/040307+07$19.50

energy and equilibrium of a diamagnet in magnetic and gravitational fields (section 2), and then derive the general conditions for the stability of the equilibrium (section 3). Stability is restricted to certain small zones, which we calculate in detail (section 4) for the field inside a solenoid. Finally, we describe (section 5) the diamagnetic levitation experiments carried out by Geim *et al*.

The explanation of the stability of the diamagnets is mathematically related to that of the Levitron™, but since the Levitron™ has already been treated in several papers we will restrict ourselves here to mentioning the similarities and differences between the two cases. We do not consider the levitation of high-temperature superconductors; this is stabilized by a different mechanism, involving dissipation (dry friction) caused by flux lines jumping between defects that pin them (Brandt 1990, Davis *et al* 1988). Nor do we discuss traps for microscopic particles, some of which are similar to the Levitron™ (Berry 1996) and some of which evade Earnshaw's theorem through time-dependent fields (Paul 1990).

2. Energy and equilibrium

Let the magnetic field inside a vertical solenoid at position $\boldsymbol{r} = \{x, y, z\}$ be $\boldsymbol{B}(\boldsymbol{r})$ (figure 1), with strength $B(\boldsymbol{r}) = |\boldsymbol{B}(\boldsymbol{r})|$, and let the gravitational field have acceleration g. The object that will be levitated in these fields has mass M, volume V (and density $\rho = M/V$), and magnetic susceptibility χ. For diamagnetic materials, $\chi < 0$ (the special case $\chi = -1$ corresponds to superconductors, i.e. perfect diamagnets), so we write $\chi = -|\chi|$. For paramagnets $\chi > 0$, but as we will show in section 3 levitation is impossible for these materials. We will be interested in substances for which $|\chi| \ll 1$. Then, to a close approximation, the induced magnetic moment $\boldsymbol{m}(\boldsymbol{r})$ is

$$\boldsymbol{m}(\boldsymbol{r}) = -\frac{|\chi| V \boldsymbol{B}(\boldsymbol{r})}{\mu_0}. \tag{1}$$

(In a more accurate treatment (Landau *et al* 1984), incorporating the distortion of the ambient field by the object, there is a shape-dependent correction to (1); for a sphere, the r.h.s. is divided by $(1 - |\chi|/3)$. In general, the relation between $\boldsymbol{B}$ and $\boldsymbol{M}$ is tensorial.)

By integrating the work $-\mathrm{d}\boldsymbol{m} \cdot \boldsymbol{B}$ as the field is increased from zero to $\boldsymbol{B}(\boldsymbol{r})$, we can obtain the total magnetic energy of the object and, adding this to the gravitational energy, the total energy:

$$E(\boldsymbol{r}) = mgz + \frac{|\chi| V}{2\mu_0} B^2(\boldsymbol{r}). \tag{2}$$

For the object to be floating in equilibrium, the total force $\boldsymbol{F}(\boldsymbol{r})$ must vanish. Thus

$$\boldsymbol{F}(\boldsymbol{r}) = -\nabla E(\boldsymbol{r}) = -mg\boldsymbol{e}_z - \frac{|\chi| V}{\mu_0} B(\boldsymbol{r}) \nabla B(\boldsymbol{r}) = 0 \tag{3}$$

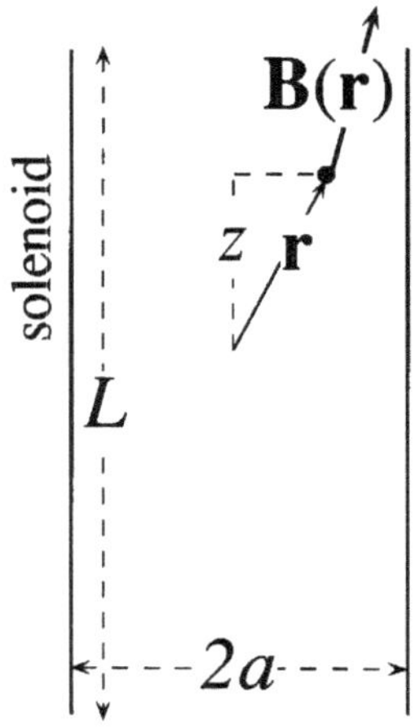

Figure 1. Geometry and notation for field in a solenoid.

where $\boldsymbol{e}_z$ is the upwards unit vector. All the fields we are interested in will have rotation symmetry about $\boldsymbol{e}_z$ (continuous for a solenoid, discrete for the Levitron™ whose base is square). So, considering equilibria on the axis and denoting the field strength by $B(z)$, the equilibrium condition becomes

$$B(z)B'(z) = -\frac{\mu_0 \rho g}{|\chi|}. \tag{4}$$

Note that this involves only the density of the levitated object, not its mass.

For the Levitron™, the spinning-top is magnetized with magnetic moment $\boldsymbol{m}$ directed along the symmetry axis of the top. The purpose of the spin is to keep $\boldsymbol{m}$ gyroscopically oriented in the direction for which the force $\nabla \boldsymbol{m} \cdot \boldsymbol{B}(\boldsymbol{r})$ from the base is upwards, that is, with $\boldsymbol{m}$ antiparallel to the effective dipole representing the base, since unlike dipoles repel (unlike unlike poles). Thus magnetic repulsion can balance gravity. (Without spin, the magnet orients itself parallel to the dipole representing the base, and is therefore attracted to the base, and falls.) The magnetic torque causes $\boldsymbol{m}$ to precess about the local direction of $\boldsymbol{B}(\boldsymbol{r})$. If this precession is fast enough (in comparison with the rate at which the direction of $\boldsymbol{B}(\boldsymbol{r})$ changes as the top bobs and weaves during its oscillations about equilibrium), a dynamical adiabatic theorem (Berry 1996) ensures that the angle between $\boldsymbol{m}$ and $\boldsymbol{B}(\boldsymbol{r})$ is preserved. For the Levitron™, $\boldsymbol{m}$ is approximately antiparallel to $\boldsymbol{B}(\boldsymbol{r})$, so this angle is close to 180°, and the energy is

$$E(\boldsymbol{r}) = mgz - \boldsymbol{m} \cdot \boldsymbol{B}(\boldsymbol{r}) \approx mgz + |\boldsymbol{m}| B(\boldsymbol{r}). \tag{5}$$

Comparing (2) and (5), we see that the energy, and therefore the equilibrium, of both a diamagnetically levitated object and the Levitron™, depends on the magnitude $B(\boldsymbol{r})$ of the field; at the end of section 3 we will see that this dependence is crucial to stability in both cases.

3. Stability

For levitation, the equilibrium must be stable, so that the energy must be a minimum, that is, the force $\boldsymbol{F}(\boldsymbol{r})$ must be restoring. We begin by showing that this excludes the levitation of paramagnetic objects. A necessary condition for stability is

$$\oiint \boldsymbol{F}(\boldsymbol{r}) \cdot \mathrm{d}\boldsymbol{S} < 0 \qquad (6)$$

where the integral is over any small closed surface surrounding the equilibrium point. From the divergence theorem, this implies $\nabla \cdot \boldsymbol{F}(\boldsymbol{r}) < 0$, and hence, from (2) written for paramagnets, that is with $|\chi|$ replaced by $-\chi$, that

$$\nabla^2 B^2(\boldsymbol{r}) < 0. \qquad (7)$$

But

$$\begin{aligned}\nabla^2 B^2(\boldsymbol{r}) &= \nabla^2 \left(B_x^2 + B_y^2 + B_z^2\right) \\ &= 2\Big[|\nabla B_x|^2 + \left|\nabla B_y\right|^2 + |\nabla B_z|^2 \\ &\quad + B_x \nabla^2 B_x + B_y \nabla^2 B_y + B_z \nabla^2 B_z\Big] \\ &= 2\Big[|\nabla B_x|^2 + \left|\nabla B_y\right|^2 + |\nabla B_z|^2\Big] \geq 0 \qquad (8)\end{aligned}$$

where the last equality follows from the fact that the components of $\boldsymbol{B}$ satisfy Laplace's equation (because there are no magnetic monopoles, so that $\nabla \cdot \boldsymbol{B} = 0$, and no currents within the solenoid, so that $\nabla \times \boldsymbol{B} = 0$). Therefore the necessary condition (6) for stability is violated, and stable levitation of paramagnets is impossible. That is why the equations in section 2 were written in the form appropriate for diamagnets.

Equation (8) is the essential step in the proof that the magnitude $B(\boldsymbol{r})$ of a magnetic field in free space can possess a minimum but not a maximum. This theorem is 'well known to those who know well' (and particularly by physicists who construct traps for microscopic particles) but we do not know who first proved it. It applies to any field that is divergenceless and irrotational. To a good approximation, it applies to velocity fields in the ocean, with the surprising consequence that there is no point within the Pacific Ocean where the water is flowing faster than at all neighbouring points; therefore places where the current has maximum speed lie on the surface.

The sufficient conditions for stability (as opposed to (6), which is merely necessary) are that the energy must increase in all directions from an equilibrium point satisfying (3), that is

$$\partial_x^2 E(\boldsymbol{r}) > 0 \qquad \partial_y^2 E(\boldsymbol{r}) > 0 \qquad \partial_z^2 E(\boldsymbol{r}) > 0. \qquad (9)$$

For diamagnets, it now follows from (2) that

$$\begin{aligned}&\partial_z^2 B^2(\boldsymbol{r}) > 0 \quad \text{(vertical stability)} \\ &\partial_x^2 B^2(\boldsymbol{r}) > 0 \quad \partial_y^2 B^2(\boldsymbol{r}) > 0 \quad \text{(horizontal stability)}.\end{aligned} \qquad (10)$$

Because of the rotational symmetry, the last two conditions are equivalent. Now we show that the conditions can be conveniently expressed in terms of the magnetic field on the axis, $B(z)$, and its derivatives $B'(z)$ and $B''(z)$.

We begin by introducing the magnetic potential $\Phi(\boldsymbol{r})$, satisfying

$$\boldsymbol{B}(\boldsymbol{r}) = \nabla\Phi(\boldsymbol{r}) \qquad (11)$$

and its derivatives on the axis

$$\phi_n(z) \equiv \partial_z^n \Phi(0, 0, z). \qquad (12)$$

From the fact that Φ satisfies Laplace's equation, and rotational symmetry, there follows

$$\partial_x^2 \Phi(0, 0, z) = \partial_y^2 \Phi(0, 0, z) = -\tfrac{1}{2}\phi_2(z). \qquad (13)$$

Therefore the potential close to the axis can be written

$$\begin{aligned}\Phi(\boldsymbol{r}) &= \phi_0(z) + \tfrac{1}{2}\left(x^2 \partial_x^2 \Phi(0, 0, z)\right. \\ &\quad \left. + y^2 \partial_y^2 \Phi(0, 0, z)\right) + \ldots \\ &= \phi_0(z) - \tfrac{1}{4}(x^2 + y^2)\phi_2(z) + \ldots\ . \qquad (14)\end{aligned}$$

From (11), the field strength can now be written

$$\begin{aligned}B^2(\boldsymbol{r}) &= \phi_1^2(z) + \tfrac{1}{4}(x^2 + y^2) \\ &\quad \times (\phi_2^2(z) - 2\phi_1(z)\phi_3(z)) + \ldots\ . \qquad (15)\end{aligned}$$

The stability conditions (10) can now be expressed in terms of $\phi_n(z)$, and thence in terms of the field on the axis:

$$\begin{aligned}&D_1(z) \equiv B'(z)^2 + B(z)B''(z) > 0 \\ &\qquad \text{(vertical stability)} \\ &D_2(z) \equiv B'(z)^2 - 2B(z)B''(z) > 0 \\ &\qquad \text{(horizontal stability)}.\end{aligned} \qquad (16)$$

For the Levitron™, where the magnetic energy (5) depends on $B(\boldsymbol{r})$ rather than $B^2(\boldsymbol{r})$, a similar analysis leads to the same horizontal stability condition, and the simpler vertical stability condition $B''(\boldsymbol{r}) > 0$.

Mathematically, the reason why diamagnets and the Levitron™ can be levitated in spite of Earnshaw's theorem is that the energy depends on the field strength $B(\boldsymbol{r})$, which unlike any of its components does not satisfy Laplace's equation and so can possess a minimum. Physically, the diamagnet violates the conditions of the theorem because its magnetization $\boldsymbol{m}$ is not fixed but depends on the field it is in, via (1). Microscopically, this is because diamagnetism originates in the orbital motion of electrons and so is dynamical. In the Levitron™, the magnitude of $\boldsymbol{m}$ is fixed but its direction is slaved to the direction of $\boldsymbol{B}(\boldsymbol{r})$ by an adiabatic mechanism that is also dynamical (at the macroscopic level) because it relies on the fast precession of the top.

The (non-dissipative) stability of permanent magnets levitated above a (concave upwards) bowl-shaped base of type I superconductor (e.g. lead) (Arkadiev 1947) is similar to that of the diamagnets we have been considering. The superconductor is a perfect diamagnet ($\chi = -1$), and so the permanent magnet above it is repelled by the field of the image it induces (Saslow 1991). If the magnet moves sideways, the image gets closer, so that the energy increases.

4. Stable zones

On the axis of a solenoid, or above the base of a Levitron^{TM}, the field $B(z)$ decreases monotonically as z increases from 0 to ∞, and there is an inflection point at some height z_i, that is $B''(z_i) = 0$. At z_i, both discriminants D_1 and D_2 in (16) are obviously positive, so the equilibrium is stable at z_i. Simple geometrical arguments show that D_1 has a zero at a point $z_1 < z_i$, and vertical stability requires $z > z_1$; similarly, D_2 has a zero at a point $z_2 > z_i$, and horizontal stability requires $z < z_2$. This establishes the existence of a stable zone on the axis, namely $z_1 < z < z_2$, within which diamagnetic objects can be levitated.

It is necessary for the equilibrium position satisfying (4) to lie in the stable zone. This can be achieved by changing the current in the solenoid, which scales the magnetic field strength $B(\boldsymbol{r})$ while preserving the geometry of the field lines and therefore the stable zone determined by (16).

In the Levitron^{TM}, the stable zone is $z_i < z < z_2$, and, since the base is a permanent magnet whose field cannot easily be altered, the equilibrium height of the floating top can be brought into this interval by adding or removing small washers to change the weight Mg.

As a model to study in detail, we consider the field inside a long solenoid of length L and radius a (figure 1). Then, defining the scaled variables

$$\xi \equiv x/a, \quad \eta \equiv y/a, \quad \zeta \equiv z/L \quad \text{and} \quad \delta \equiv 2a/L \tag{17}$$

and the field B_0 at the centre of the solenoid $z = 0$, we have, introducing obvious notations,

$$\frac{B(\zeta,\delta)}{B_0} \equiv \mathrm{B}(\zeta,\delta) = \tfrac{1}{2}\sqrt{1+\delta^2} \times \left(\frac{1-2\zeta}{\sqrt{(1-2\zeta)^2+\delta^2}} + \frac{1+2\zeta}{\sqrt{(1+2\zeta)^2+\delta^2}} \right). \tag{18}$$

There are inflections close to the ends $\zeta = \pm 1/2$ of the solenoid; levitation occurs near the top end, that is $\zeta = +1/2$, where the field gradient is negative as required by (4). Figure 2 illustrates this field, and the corresponding discriminants (16), for $\delta = 0.1$. The stable zone is $\zeta_1 = 0.487083 < \zeta < \zeta_2 = 0.510223$.

For thin solenoids ($\delta \ll 1$), some simplification is possible, since then the second term in (18) can be approximated by unity near $\zeta = 1/2$. A short analysis shows that in this limit the inflection and stable zone are, when expressed in the original z coordinate,

$$\left. \begin{aligned} & z_i = \tfrac{1}{2}L \\ & z_1 = \tfrac{1}{2}L - 0.258199a < z < z_2 = \tfrac{1}{2}L + 0.204124a \\ & (L \gg a). \end{aligned} \right\} \tag{19}$$

For fat solenoids ($\delta \gg 1$), simplification is again possible, because then the field is that on the axis of a current loop, namely

$$B(z) = \frac{B_0}{(1+(z/a)^2)^{3/2}} \qquad (a \gg L). \tag{20}$$

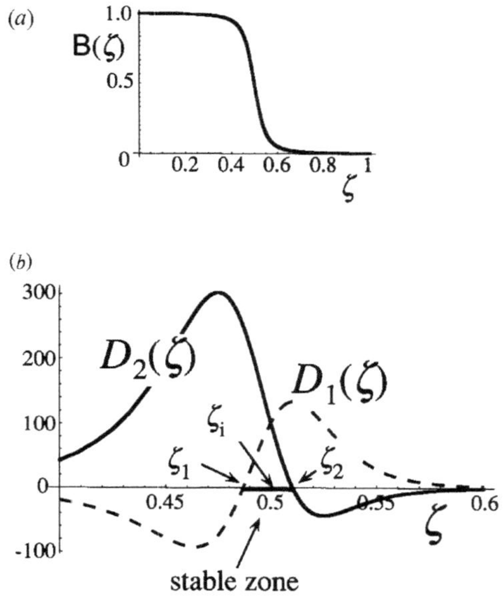

Figure 2. (a) Field on the axis inside a solenoid with $\delta = 2a/L = 0.1$; (b) the discriminants $D_1(\zeta)$ and $D_2(\zeta)$ defined by (16), and the stable zone where both are positive

From (16), the inflection and stable zone are

$$\left. \begin{aligned} & z_i = \tfrac{1}{2}a \\ & z_1 = \tfrac{1}{\sqrt{7}}a = 0.378a < z < z_2 = \sqrt{\tfrac{2}{5}}a = 0.6325a \\ & (a \gg L). \end{aligned} \right\} \tag{21}$$

By Ampère's equivalence between distributions of magnetization and current loops, the field (20) is the same as that on the axis of a uniformly magnetized disc. Therefore, with the vertical stability condition $B''(\boldsymbol{r}) > 0$ (see the remark following equation (16)), (21) leads to the stable zone previously calculated (Berry 1996) for a Levitron^{TM} with a circular disc base, namely $a/2 < z < a\sqrt{(2/5)}$. (If the base of the Levitron^{TM} is a ring, rather than a disc, the stable region is much higher, namely $1.6939a < z < 1.8253a$, and this explains the operation of the recently developed 'superlevitron'.)

It is instructive to display spatial contour maps of the energy (2) as the field B_0 at the centre of the solenoid is varied, showing the appearance and disappearance of the minimum as the equilibrium enters and leaves the stable zone. We employ the dimensionless field β and energy E defined by

$$\begin{aligned} & B_0^2 \equiv \beta^2 \frac{\rho g L \mu_0}{|\chi|} \\ & E(\boldsymbol{r}) \equiv \frac{|\chi| V B_0^2}{2\mu_0} \mathrm{E}(\xi,\eta,\zeta;\beta,\delta) \end{aligned} \tag{22}$$

where, in terms of (15) and the field profile (18),

$$\mathsf{E}(\xi,\eta,\zeta;\beta,\delta) \equiv \frac{2}{\beta^2}\zeta + \tfrac{1}{4}\big[\mathsf{B}(\zeta,\delta)^2 + \tfrac{1}{4}(\xi^2+\eta^2) \times \{\mathsf{B}'(\zeta,\delta)^2 - 2\mathsf{B}(\zeta,\delta)\mathsf{B}''(\zeta,\delta)\}\big] \tag{23}$$

(the primes denote $\partial/\partial\zeta$). From the equilibrium condition (4), the field $\beta(z)$ for which the diamagnet floats at height ζ is

$$\beta(z)^2 = -\big[\mathsf{B}(\zeta,\delta)\mathsf{B}'(\zeta,\delta)\big]^{-1}. \tag{24}$$

Figure 3 shows the E landscape as the field β is decreased through the stable range, for a solenoid with $\delta = 0.1$. At the top of the range (figure 3(*b*) $\beta = \beta_2 = 0.513563$, corresponding to equilibrium at the upper limit $z = z_2$ of the stable zone, and at the bottom of the range (figure 3(*d*)) $\beta = \beta_1 = 0.417998$, corresponding to equilibrium at the lower limit $z = z_1$ of the stable zone. At β_2 the minimum is born (along with two off-axis saddles) from the splitting of an axial saddle; at β_1, the minimum dies as it annihilates with another axial saddle. We caution against quantitative reliance on the details of these landscapes near the wall of the solenoid (e.g. near $\xi = 0.05$ in figure 3), because they are based on the quadratic approximation (23), which is strictly valid only close to the axis.

Stably levitated diamagnets can make small, approximately harmonic, oscillations near the energy minimum, and these are observed as the gentle bobbing and weaving of the objects. Larger oscillations will be anharmonic. The region they explore has the form of a conical pocket (figure 3(*c*)), in which motion is almost certainly nonintegrable and probably chaotic. We think this would repay further study, but here confine ourselves to estimating the greatest lateral extent of the region in which the oscillations occur. From figure 3, it is reasonable to define this as the distance $R = \sqrt{(x^2+y^2)}$ from the axis to the off-axis saddles for the field that corresponds to equilibrium at z_i, namely $\beta = 0.445301$. It follows from (23) that these saddles lie at $z = z_2$, and use of (4) then leads to

$$R^2 = 4L^2 \frac{\big[\mathsf{B}(\zeta_i,\delta)\mathsf{B}'(\zeta_i,\delta) - \mathsf{B}(\zeta_2,\delta)\mathsf{B}'(\zeta_2,\delta)\big]}{\mathsf{B}(\zeta_2,\delta)\mathsf{B}'''(\zeta_2,\delta)}. \tag{25}$$

For thin solenoids, this can be evaluated as

$$R = 0.75569a \qquad (L \gg a). \tag{26}$$

When $\delta = 0.1$ this gives $R/L = 0.0377$, in agreement with figure 3(*c*) (which was calculated without the thin-solenoid approximation).

5. Experiment

Most diamagnetic materials have susceptibilities of order $\chi \approx -10^{-5}$. For water, $\chi = -8.8 \times 10^{-6}$ (Kaye and Laby 1973), and using $\rho = 1000\ \mathrm{kg\,m^{-3}}$

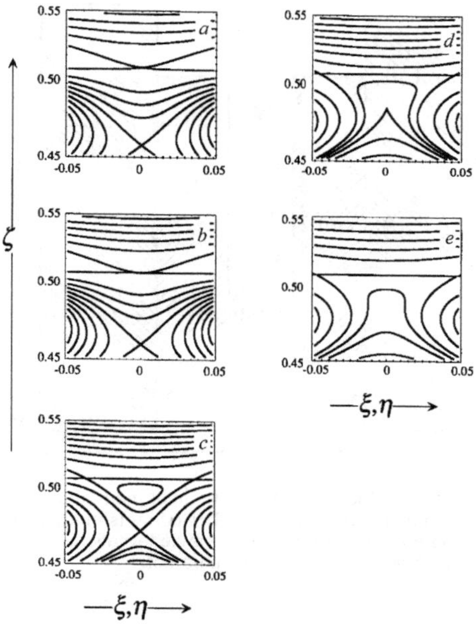

Figure 3. Contours of the scaled energy $\mathsf{E}(\xi,\eta,\zeta;\beta,\delta)$ (gravitational + magnetic) for a diamagnet, for different values of the dimensionless field β (defined by (22)) at the centre of a solenoid with $\delta = 0.1$.
(*a*) $\beta = 0.527046$;
(*b*) $\beta = \beta_2 = 0.513563$, i.e. levitation at z_2;
(*c*) $\beta = 0.445301$, i.e. levitation at z_i;
(*d*) $\beta = \beta_1 = 0.417998$;
(*e*) $\beta = 0.411693$, i.e. levitation at z_1.

the equilibrium condition (4) gives the required product of field and field gradient as

$$B(z)B'(z) = -1400.9\ \mathrm{T^2\ m^{-1}}. \tag{27}$$

This has been achieved in experiments involving one of us (Geim *et al*) with a Bitter magnet whose geometry is shown in figure 4(*a*). The operation of this electromagnet consumed 4 MW, but we emphasize that this is power dissipated in the coils, not power required for levitation—indeed, with the field of a persistent current in a superconducting magnet levitation can be maintained without supplying any energy.

The measured field profile is shown in figure 5. The inflection point is at $z_i = 78$ mm, where the field is $B(z_i) = 0.63B_0$ and the gradient of the field at z_i is $-8.15B_0\ \mathrm{T\ m^{-1}}$, from which the required central field is predicted via (27) to be

$$B_0 = 16.5\ \mathrm{T}. \tag{28}$$

From the measured data we have calculated the discriminants D_1 and D_2 defined by (16), and thence the stable zone, which is predicted to be $z_1 = 67.5\ \mathrm{mm} < z < z_2 = 87.5$ mm.

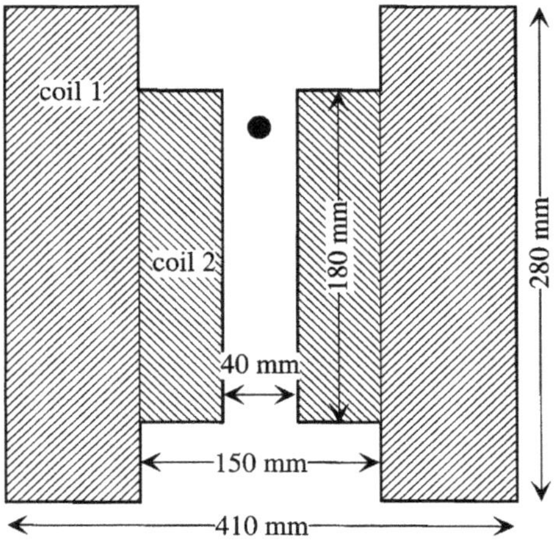

Figure 4(*a*). Geometry of coils in Bitter magnet used for levitating diamagnetic objects. The currents in the two coils were equal. The region of stable levitation is near the top of coil 1, and marked with a dot.

Figure 4(*b*). Frog levitated in the stable region.

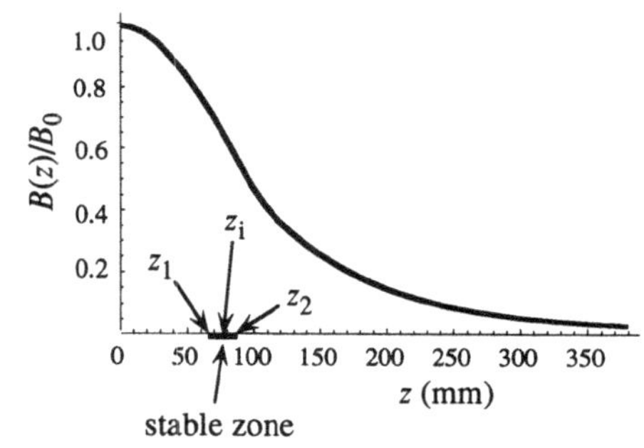

Figure 5. Profile of field on axis of Bitter magnet in figure 4, measured at intervals of 10 mm, showing the stable zone near the top of coil 1.

A variety of diamagnetic objects was inserted into the magnet, and the current through the coils adjusted until stable levitation occurred (figure 4(*b*)). The corresponding fields B_0 were all close to the calculated 16 T, and the objects always floated near the top of the inner coil, as predicted. Careful observation of a (3 mm diameter) plastic sphere showed that it could be held stably in the range (69 ± 1) mm$< z <$ (86 ± 1) mm, in very good agreement with theory.

The induced dipole m (equation (1)) responsible for the levitation of a diamagnet can be regarded as equivalent to a current $I = |m|/A$ circulating in a loop of area A embracing it. For an object of radius 10 mm, such as the very young frog that was levitated (figure 4(*b*)), this current is about 1.5 A (corresponding to a field $B \approx 10^{-5} B_0 \approx 1.5$ Gauss induced inside the frog). Of course this represents the summation of microscopic currents localized in atoms, not the bulk transport of charge, so the living creatures were not electrocuted. Indeed, they emerged from their ordeal in the solenoid without suffering any noticeable biological effects—see also Schenck (1992) and Kanal (1996).

As we showed earlier, it is impossible to levitate paramagnets stably. Balance of forces can however be achieved, and from (4) with the sign reversed it is clear that this occurs for $z < 0$, and close to the centre of the solenoid—rather than near the bottom—because $\chi_{\text{paramagnetic}} \approx 10^{-3} \approx 100\chi_{\text{diamagnetic}}$; this position is vertically stable but laterally unstable. Nevertheless, some paramagnetic objects (Al, several types of brass, stainless steel, paramagnetic salts with Mn and Cu) were suspended in this way, but not levitated: they were held against the side wall of the inner coil. On a few occasions, paramagnets floated without apparent contact, but were found to be buoyed up by a rising current of paramagnetic air; when this was inhibited, for example by covering the ends of the solenoid with gauze, the objects slipped sideways and were again held against the wall.

6. Discussion

Our treatment of diamagnetic levitation has neglected at least three small effects that could have interesting consequences. The first arises from the shape-dependence of the induced magnetic moment. For living organisms (e.g. frogs) trapped in the energy minimum this could be exploited to provide an escape mechanism. If the frog is initially in equilibrium, there are no forces on it. By changing shape (e.g. from a sphere to an ellipsoid) the induced moment will change (Landau *et al* 1984), and the force will no longer be zero, so the frog will start to oscillate about a slightly different point. By repeating this manoeuvre at the frequency of oscillations in the minimum, the oscillations will be amplified by parametric resonance until the frog leaves the stable zone. This is a tiny effect, because the shape-dependence of m is of the order $|\chi| \approx 10^{-5}$, so escape would require 10^5 such 'swimming strokes'; therefore the frog would have to be persistent as well as highly coordinated. (In practice, the frog does try to swim—but in the ordinary way, by paddling the air in the solenoid—but nevertheless remains held in the energy minimum, for the entire observation—up to 30 minutes.)

The second effect arises from the finite extent of any real levitated object. Its equilibrium depends on the total magnetic force, which must balance the weight. The local force balance (4) will occur only at one height z_b in the body. For $z < z_b$, the net force on each element will be upwards, and for $z > z_b$ the net force will be downwards. Therefore the object will be compressed to an extent that depends on how much BB' varies across it, that is on the curvature of $B^2(z)$ at z_b. A land-based living creature would be unlikely to feel this effect, since it is already accustomed to a much greater inhomogeneity: the external upward force that balances gravity is concentrated in a molecular layer in the soles of its feet.

The third effect occurs for objects that are diamagnetically inhomogeneous, so that their different parts (e.g. flesh and bone for a living organism) have different χs. Then, as just described for an extended object, the force balance will be different at different points. This could cause strange sensations; for example, if $|\chi|_{\text{flesh}} > |\chi|_{\text{bone}}$ the creature would be suspended by its flesh with its bones hanging down inside, in a bizarre reversal of the usual situation that could inspire a new (and expensive) type of face-lift (since $|\chi|_{\text{bone}} \approx |\chi|_{\text{water}}$ (Schenck 1992) this would require $|\chi|_{\text{flesh}} > |\chi|_{\text{water}}$).

Acknowledgement

AKG thanks the staff at the High Field Magnet Laboratory (University of Nijmegen) for technical assistance, and the European Community Program 'Access to Large Scale Facilities' for financial support.

References

Arkadiev V 1947 A floating magnet *Nature* **160** 330

Berry M V 1996 The Levitron^{TM}: an adiabatic trap for spins *Proc. R. Soc.* A **452** 1207–20

Brandt E H 1989 Levitation in Physics *Science* **243** 349–355

—— 1990 Rigid levitation and suspension of high-temperature superconductors by magnets *Am. J. Phys.* **58** 43–9

Davis L C, Logothetis E M and Soltis R E 1988 Stability of magnets levitated above superconductors *J. Appl. Phys.* **64** 4212–8

Earnshaw S 1842 On the nature of the molecular forces which regulate the constitution of the luminiferous ether *Trans. Camb. Phil. Soc.* **7** 97–112

Jones T B, Washizu M and Gans R 1997 Simple theory for the Levitron^{TM} *J. Appl. Phys.* in press

Kanal E 1996 International MR Safety Central Web Site (`http://kanal.arad.upmc.edu/mrsafety.html`)

Kaye G W C and Laby T H 1973 *Tables of Physical and Chemical Constants* (London: Longman)

Landau L D, Lifshitz E M and Pitaevskii L P 1984 *Electrodynamics of Continuous Media* (Oxford: Pergamon)

Page L and Adams N I Jr 1958 *Principles of Electricity* (New York: Van Nostrand)

Paul W 1990 Electromagnetic traps for charged and neutral particles *Rev. Mod. Phys.* **62** 531–40

Rodgers P 1997 *Physics World* **10** 28

Saslow W M 1991 How a superconductor supports a magnet, how magnetically 'soft' iron attracts a magnet, and eddy currents for the uninitiated *Am. J. Phys.* **59** 16–25

Schenck J F 1992 Health and physiological effects of human exposure to whole-body four-tesla magnetic fields during MRI *Ann. Acad. Sci. NY* **649** 285–301

Scott W T 1959 Who was Earnshaw? *Am. J. Phys.* **27** 418–9

Shoenberg D 1952 *Superconductivity* (Cambridge: Cambridge University Press)

Simon M D, Heflinger L O and Ridgway S L 1997 Spin stabilized magnetic levitation *Am. J. Phys.* **65** 286–92

62 Physics World June 1992

reviews

Michael Berry

We are not aliens

Understanding the Present: Science and the Soul of Modern Man Bryan Appleyard 1992 Picador 283pp £14.99hb

WHEN he was a child, Appleyard tells us, he was astonished by the ability of his father (an engineer) to calculate the volume of water in a water tower, but "sensed something dangerous and ominous in this wisdom". During a later career as a journalist, in which he was often concerned with scientists and technological issues related to science, this unease matured and has now become the mainspring of an intricately connected, wide-ranging and passionate attack on science. I will try to summarise it.

Appleyard argues that science is astonishingly effective, but "effectiveness is not truth", and science is not 'true' because what it tells us changes with each revolution (mechanics changing from Newtonian to quantum, etc), and because its apparent triumphs are achieved only after grotesque oversimplifications which select "only those problems that can be solved by the known method"; moreover, its essentially mathematical content is inaccessible to ordinary people, which real truth ought not to be. Science has caused "appalling spiritual damage" and made us abandon "our true selves" by exposing us as always "destined to be the wrong size" to comprehend the scientists' cosmos of general relativity or the subatomic quantum world. This devaluation of the human is largely to blame for the indefensible relativism and fundamental instability of our liberal society. Then there is the atom bomb, which "suddenly revealed science itself as an uncontrollable extension of the human will to destruction", and our unpredicted effects on the environment, which show that "our arrogant simplicity has been humbled by . . . awesome . . . complexity". And now, with the theory of algorithms and the foundations of logic, applied to artifical intelligence, science is preparing "to cross the inner frontier of the self", a development that could yet be resisted in what would be "the humbling of science".

One of Appleyard's aims is to make scientists "as morally and philosophically answerable as the rest of us". Therefore it might please him to learn that I found his book painful to read. But the pain stems not from a guilty recognition of our shortcomings but rather from seeing our work so inadequately and ineptly dismissed. For a start, there is nothing essentially new here: Appleyard is putting a modern gloss on the eighteenth century romantic rejection of Newtonianism, well summarised by Hamann's "The Tree of Knowledge has robbed us of the Tree of Life".

Appleyard's dismissal of the truths of science as evanescent is doubly faulty. First, he underplays the role of experiment and observation, which not only generate simple facts (such as that matter is made of atoms) that change our view of the world and are not subsequently overturned, but repeatedly surprise us and challenge our simplifications. (Anyone who doubts this should reflect on the discoveries in the last decade of quasicrystals, high-temperature superconductivity and the quantum Hall effect – all fundamental and unexpected even in such a theory-driven area as condensed-matter physics.) And second, he forgets that a truth does not disappear when it is subsequently discovered to be a small part of a larger truth.

SHUT UP, CLEVERCLOGS -
I'M NOT LISTENING!
LA LA LA!

He might be less apprehensive if he realised the enormous incompleteness of science even in its own terms. While he rightly condemns glimpses of the end of physics as "crude historical arrogance", he seems not to appreciate that (for example) our understanding of the implications of quantum mechanics is still so rudimentary that, given the Schrödinger equation applied to a collection of nuclei and electrons, we would probably not yet have predicted even the existence of molecules. We are able to do quantum chemistry only because we know beforehand that molecules exist.

There is a contradiction in his treatment of the inaccessibility of the insights of science. He rejects the definition of our subject as "organised common sense", and finds it "absurd, almost sentimental, to claim, as would Bronowski, that a plough is like a CD-player". This is because the way the CD-player works is incomprehensible from the outside, unlike a plough. But there is a continuity between the two, and to the engineer the CD-player is a mechanism as transparent in its operation as a plough is to a peasant. To see Appleyard's contradiction, return to that water tower. I am sure he soon learned enough mathematics to demystify every step of his father's calculation. He graduated from the plough to the CD-player. Yet he still finds the mathematics sinister. This can only be because he wants to. I cannot think why.

Nor do I know anybody who feels diminished by learning that they occupy a mesoscale, between the quantum and the cosmos. On the contrary, most non-scientists I encounter feel exhilarated by being thus connected with realms that cannot be perceived by the unaided senses. Moreover, this connection provides a healthy antidote to the often-criticised reductionist disarticulation of the world into components, in a vision of the Universe as a unity (of course incompletely glimpsed – inevitably in my view). Appleyard dismisses this enlargement with contempt: "The stars . . . are no more wondrous than an . . . electric kettle . . . we are supposed to be impressed . . . to be *grateful*."

To 'blame' science for its evil misapplications such as nuclear weapons is like 'blaming' literature for the inflammatory evil of political propaganda (or, for that matter, the corrosive triviality of soap powder advertising). Among scientists and technologists there is roughly the same distribution of high-mindedness, confusion, cowardice, tolerance, greed, etc, as in any other subset of the population. It is naive to think otherwise.

And I find unconvincing his equation of science with the less appealing aspects of modern liberalism, where there are no certainties and "the possibility of choosing one point of view as more valuable than another" is denied. In my experience (especially in American universities) it is the physicists who see most clearly through the "perverse phenomenon of liberal authoritarianism" and the idiotic excesses of 'political correctness'. Why is this? It is because we have in science (and apparently not recognised by Appleyard) a strong culture of consensuality, where a community works together to winkle out the truth in a sort of cumulative (and generally good-humoured) error-correcting which is, I suspect, unique among human activities; in many cases, the rights and wrongs of a scientific issue really do get settled, even if there is confusion along the way.

After all his raging against science, Appleyard has a responsibility to articulate an alternative vision of society. Instead of "the enforced neutrality of scientific liberalism", he wants us to move towards "realisation of specific excellence within a social context", or (quoting Updike), the recognition that "Existence itself . . . feels like an ecstasy, rather, which we only have to be still to experience". Not much for a scientist to disagree with here, but what about the need for "irreducible affections, values and convictions . . . these need not be defended further because they will express only our kinship with our culture and that kinship will be beyond appeal"? Well, which culture? Is it to be Judeo-Christian culture, English culture, North European culture, Western culture, Thatcherite mercantilism, Medieval Catholicism? Whichever he favours – he does not say – how does he seek to inculcate its myths? Would he eliminate science, for example, by making it illegal to continue to solve the Schrödinger equation, or publish the solutions? And what about those in our troubled but gloriously multicultural society who might have different visions?

Appleyard disappoints us. For him, the answer is beyond words; he provides a quotation from the philosopher Cavell, interpreting this as "saying that it is the philosophers who are wrong rather than the people who simply get on with their lives" – or, as Duke Ellington said: "Too much talk stinks up the place." Such anti-intellectual populism is not only pitifully inadequate, but also a bit rich, coming from one who has just spent 200 pages telling us how science has destroyed people's vision of their place in the world.

Nevertheless, I do recommend scientists to take a little time away from real work at the bench, or dreaming in the bath or at the computer, to learn one man's systematic misunderstandings and distortions of what we do. It might seem a waste of time, but we ought to be aware of a current of opposition that could be on the rise in these mean and shabby times.

Moreover, Appleyard is articulate and well informed; he knows the history of science, and has read his Bohm, his chaology, his Gödel. Unfortunately, though, the vision presented here, however intense and deeply felt, is no more than an elaborate rationalisation of that infantile revulsion at the water tower. As such it is, ultimately, deeply superficial.

Michael Berry is in the Department of Physics, University of Bristol, UK

book reviews

Enigma variations on the nuclear stage

Copenhagen
A play by Michael Frayn
Performed at the National Theatre, London
Methuen: 1998. 128pp. £6.99 (pbk)

Michael Berry

Meeting of minds: Burke (left), Kestelman and Marsh succeed in the fusion of science and drama.

In *Copenhagen*, Michael Frayn re-creates one of the more mysterious episodes of the Second World War, when Werner Heisenberg travelled to German-occupied Denmark in 1941 to visit his fellow physicist Niels Bohr. Why did he go there?

Historians — as well as the protagonists themselves, with their faulty and shifting memories — have long puzzled over Heisenberg's motives. Did he seek to enlist Bohr's assistance in developing a German nuclear bomb? Was he pumping Bohr for information about a possible parallel programme by the Allies? Was he alerting Bohr to the existence of the German programme, in the hope that Bohr would provoke the Allies to accelerate their own programme? Was he seeking a reason to delay, or even sabotage, the German nuclear effort? And why did the German programme fail? Was it through rivalry and division of resources between competing teams, or because a crucial neutron diffusion rate was wrongly assumed instead of being calculated?

In the play, these possibilities are explored in detail and with sensitivity, concentrating on the contrapuntal relationship between the two physicists' personalities, their careers, and their scientific styles. Bohr, 16 years Heisenberg's senior, initiated the modern theory of the atom in 1913, with his proposal that the energies of electrons in atoms are restricted by quantum rules similar to those that Planck and Einstein had applied to light. This was not only bold but outrageous, because it had no basis in theoretical physics — indeed, it flatly contradicted the established 'classical' physics of the day.

Twelve astonishing years followed, in which the most intense concentration of scientific effort was devoted to the search for the fundamental quantum theory underlying Bohr's atom. An important centre of that research was Bohr's newly established institute in Copenhagen. He imported a stream of brilliant young physicists, and sought unremittingly to uncover the physical significance of their theories and formalisms. This led him into philosophy, and to his principle of complementarity. His style was slow, tentative, almost inarticulate, repeatedly redrafting his papers in the hope of reaching a version that matched what he was groping for.

Heisenberg, competitive, insecure, barely into his twenties when he went to Copenhagen, was instantaneous in his responses, diamond-hard in the precision of his thought. He occupied himself with mathematical formalism rather than wordy interpretations. It was he who reached the first consistent quantum mechanics: spare, algebraic, a landscape alien to theoretical physicists yet capable of reproducing all the hitherto baffling experimental observations on atoms. Complementing Bohr's complementarity was Heisenberg's uncertainty principle, a precise statement of the limits within which a particle's position and motion can be known.

The structure of the play mirrors these contrasts. Bohr appears mumbling, reflective, but with a simple goodness ("I don't think anyone has yet discovered a way you can use theoretical physics to kill people") and clear moral insight; Heisenberg, swift-tongued but morally ambiguous. The 1941 meeting is repeatedly re-enacted and revisited, echoing Bohr's repeated redraftings, in attempts to get at what really happened. As each dilemma gets focused on, another recedes, vague, into a mist.

Underlying the momentous events that followed the Copenhagen encounter are facts about atomic nuclei, understandable only in terms of quantum physics. Frayn does a splendid job of explaining these subtle and tricky matters, in some detail yet without technicalities. Through the exchanges of the protagonists, we get clear accounts of fission, the production of nuclei, chain reactions (where the play as staged corrects an arithmetic error in the published version), the important distinction between slow and fast neutrons, diffusion rates, and quantum uncertainty and interference.

On a stage bare but for three chairs, our attention is gripped and held by three actors: Bohr, played by David Burke, Heisenberg by Matthew Marsh, and Bohr's wife Margrethe by Sara Kestelman. Margrethe is the chorus, mercilessly exposing Heisenberg's evasions and Bohr's good-natured, even fatherly, indulgence towards his former colleague. Marsh and Burke bring out beautifully the obvious affection between the two very different men, strained to the utmost by the excruciatingly hard times — and not helped by Heisenberg's clumsy invitation to the Bohrs to make use of his ski-hut in Bavaria, apparently forgetting that Bohr was half-Jewish.

With *Copenhagen*, Frayn helps to create a genre, alongside Tom Stoppard, with *Hapgood*, inspired by quantum mechanics, and *Arcadia*, based on chaology, and Mike Maran, with *Surely You're Joking, Mr Feynman*, about that unique and colourful physicist. This acceptance of science as a legitimate subject for dramatization, rather than something separate and technical, is both welcome and overdue. In human culture, as with nuclei, fusion is more powerful than fission but harder to achieve. □

Michael Berry is in the Physics Department, University of Bristol, Bristol BS8 1TL, UK.

Reprinted from Nature, Volume 394, (1998).

Endless progress within our reach

Michael Berry appreciates the relentless scientific optimism of one of the UK's most original thinkers

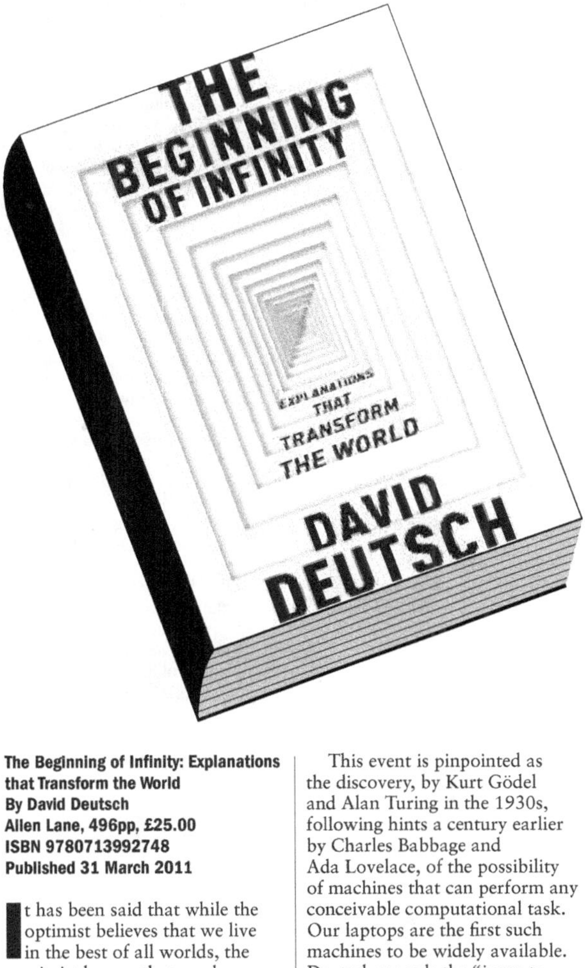

The Beginning of Infinity: Explanations that Transform the World
By David Deutsch
Allen Lane, 496pp, £25.00
ISBN 9780713992748
Published 31 March 2011

It has been said that while the optimist believes that we live in the best of all worlds, the pessimist knows that we do. David Deutsch's book is profoundly optimistic, but in a way that carefully avoids these extremes. His optimism is based not on Panglossian reassurance, but on a principled and passionate confidence in people's constructive inventiveness. He thinks that the potential for unlimited creativity has arisen only once, as the culmination of a process that began in the Enlightenment.

This event is pinpointed as the discovery, by Kurt Gödel and Alan Turing in the 1930s, following hints a century earlier by Charles Babbage and Ada Lovelace, of the possibility of machines that can perform any conceivable computational task. Our laptops are the first such machines to be widely available. Deutsch regards the "jump to universality" as a seminal moment in human culture, representing an idea with enormous reach. "Reach" is one of Deutsch's favourite words (with almost the largest number of entries in the book's index); it refers to a concept with implications far beyond its original context.

He rejects techno-pessimism: our imagined inability to cope with the pace and consequences of the inventions that stem from science. To illustrate this, he reports that when colour television became popular in the 1970s, a friend argued that its dependence on the rare element europium for the red phosphor would lead to society splitting into those who could afford it and those who could not – a dangerous situation originating in a technology that nobody needed.

Now we see this argument as absurd and based on a failure to anticipate the development of other kinds of colour display, such as those used in our computer screens and based on liquid crystals. The story illustrates Deutsch's distinction between prediction, rationally based on present knowledge, and prophecy, based on the inability to imagine future knowledge.

Why did the jump to universality occur in our Western society and not elsewhere? Deutsch rejects the explanations of Karl Marx, Friedrich Engels and Jared Diamond that the dominance of the West is a consequence of geography and climate, emphasising instead the distinction between open societies, where criticism and innovation flourish – Athens, for example – and closed societies such as Sparta where change is suppressed. In stressing the importance of learning through mistakes ("error correction is the beginning of infinity"), he is a follower and admirer of Karl Popper ("science as misconception").

Another example of a closed society that Deutsch cites is Easter Island's. He contrasts David Attenborough's interpretation of it – a magnificent civilisation whose failure resulted from the destruction of the environment that sustained it – with what he regards as Jacob Bronowski's more fundamental assessment of it as a society that was unable to change, with its statues like "frozen frames in a film that is running down, mark[ing] a civilization which failed to take the first step on the ascent of rational knowledge".

> **Deutsch rejects techno-pessimism: our 'inability' to cope with the consequences of the inventions that stem from science**

Reprinted from Times Higher Education (31 March), (2011).

Like Popper, Deutsch emphasises the open-ended nature of knowledge creation. We are now just scratching the surface of understanding the world and our place in it, and we always will be ("the beginning of infinity" again). Likewise, our evolution, now proceeding at a greatly accelerated pace by cultural transmission between generations – by memes rather than genes – is only in its initial phase.

I wonder how Deutsch would classify the medieval Islamic societies, and the earlier Indian cultures, where seminal discoveries were made (in optics and mathematics, for example) centuries before they were rediscovered in the West. Perhaps their flair for innovation vanished because the brilliant individual researchers in these societies were not embedded in a critical mass of scientists, so that the culture of collaboration, so important in the later lift-off of European science, never developed.

Deutsch has made fundamental contributions to the developing technologies based on quantum information, so it is not surprising that the heart of his analysis of explanations (and the book's most difficult chapter) concerns quantum mechanics – which is, after all, currently our deepest explanation of physical reality. He favours the "many-worlds" interpretation. His "multiverse" incorporates many universes, in which every one of the possibilities in every quantum process is actualised in an unimaginably vast sequence of splittings; of these, we have direct experience only of ours.

His argument hinges on a careful deconstruction of the concept of identicalness: an analysis of classical and quantum doppelgängers, heavily dependent on the subtle and slippery notion of fungibility. Something is fungible if the identity of an individual instance of it is irrelevant: when I lend you a pound and then you pay me back, it is irrelevant that you return the same coin, and even if you do, it is meaningless to claim that it is the same pound.

Quantum particles such as electrons are sometimes fungible and sometimes not, depending on how they interact with the rest of the world. Deutsch claims that observations of interference between quantum waves demonstrate the existence of the multiverse, but admits that this is a minority view. Most physicists are unconvinced; in my opinion, postulating this infinity of entities with which we cannot communicate is the most extravagant violation of Occam's razor.

Given the emphasis on identicalness and quantum physics, I was disappointed to see no discussion of the most important manifestation of the combination of the two. The rule specifying how the waves that represent indistinguishable quantum particles such as electrons register an interchange of two of them is a mighty fact about our world. It explains the structure of the atoms in the periodic table of the chemical elements, why these are stable, how lasers work, why wires conduct electricity, and much else.

Deutsch criticises the opinion that quantum theory needs no interpretation – disparagingly described as the "shut up and calculate" school – and calls it "bad philosophy". I disagree. The principles of any currently fundamental theory cannot be explained: they are simply postulated. That is what "fundamental" means. Asking for an "interpretation" amounts to explaining the theory in terms of deeper concepts. But that would be new physics; it would be more fundamental.

I do not claim that quantum mechanics is the ultimate microscopic science; one day, it will surely be superseded, probably as the result of experiments that are incompatible with it. There are currently no such experiments, so quantum physics is the best theory we have: the fundamental one. The alternative mathematical formulations of quantum mechanics, all yielding the same predictions for experimental results, may inspire different pictures that assist our intuition in applying the theory to different phenomena, but no one of them should be regarded as privileged.

This is Deutsch at his most ambitious, seeking to understand the implications of our scientific explanations of the world. At first sight, his book looks like an intellectual smorgasbord, with short chapters – some of which could stand as separate essays – dealing with beauty in art and nature (he thinks this is absolute; I do not), the origins of morality, memes, quantum physics, computational complexity, electoral voting systems (he argues that the British first-past-the-post system is the least worst of many flawed alternatives) and artificial intelligence. But they are all linked (albeit sometimes tenuously) with his theme: the possibility of endless transformation of the world by the creation of new knowledge.

I do not agree with everything that Deutsch writes, but I enthusiastically recommend this rich, wide-ranging and elegantly written exposition of the unique insights of one of our most original intellectuals.

Michael Berry is a physicist at the University of Bristol, studying optics and quantum physics and the mathematical relationships between different levels of explanation in physics.

PROCEEDINGS OF THE ROYAL SOCIETY A

Proc. R. Soc. A (2012) **468**, 1
doi:10.1098/rspa.2011.0564

EDITORIAL

Papers we reject

We are publishing many fine papers across the whole of the physical sciences, but here I want to focus on those we do not publish. Our current rejection rate is approaching 80 per cent. This is partly a result of the journal's popularity: we receive many more submissions than our page budget allows, so there are some good papers we simply do not have space for. But our main reasons for rejection are different: papers are too specialized for our readers, or report advances that are incremental rather than fundamental.

For a paper to be publishable it must report work that is correct and new. But these obvious conditions are not sufficient, as this absurd example demonstrates: $3921 \times 4723 = 18\,518\,883$. This result is surely correct, and has almost certainly never been published: it is 'new'. To be publishable in *Proc. R. Soc. A*, much more is required. In addition to the criteria listed on the journal's home page ('high-quality, original, fundamental articles of interest to a wide range of scientists'), I like to see an element of surprise. And in connection with 'new', I ask authors to delete the strident phrase 'for the first time': would they be submitting a paper if it were 'for the second time'?

Perhaps unsurprisingly for such a prestigious journal, we receive a steady stream of papers proposing a new theory of the universe, or a replacement for Einstein's relativity, or an 'obvious' proof of the Riemann hypothesis. Such papers usually fail several tests that any new idea must satisfy: demonstrable consistency with everything existing theory can explain, quantitative explanations of what existing theory fails to explain and/or unambiguous prediction of something not yet observed. Rejecting them takes much of our Editorial Board members' time, because, although we are under no obligation to publish any article that we receive, we recognize the seriousness of all authors, including those who are mistaken. Authors of rejected papers sometimes complain that their discoveries are being suppressed, but they are now free to publish on the Internet (if they do, it could constitute prior publication, disqualifying their papers from *Proc. R. Soc. A*).

To avoid misunderstanding, I declare that we welcome speculative papers that meet our quality standards. My own view is that all our current scientific concepts are probably wrong, but in a very specific sense: they will eventually be discovered to be special cases of much more general theoretical schemes (just as quantum mechanics superseded classical, by revealing it as a limiting case). I would be delighted if the paper that sparks the next scientific revolution were published in *Proc. R. Soc. A*, but such an advance will have to surmount a very high barrier indeed.

Michael Berry

Reprinted from Proceedings of the Royal Society A, Volume 468, (2012).

PROCEEDINGS OF THE ROYAL SOCIETY A

rspa.royalsocietypublishing.org

Editorial

CrossMark click for updates

Cite this article: Berry M. 2013 Impact and influence: valedictory editorial. Proc R Soc A 469: 20120698.
http://dx.doi.org/10.1098/rspa.2012.0698

Impact and influence: valedictory editorial

For the past seven years, I have been the steward of the journal where some of the world's best scientists published their seminal work: Maxwell, Kelvin, Marconi, the Braggs, Ramanujan, Dirac, Raman, Crick and Watson—to name a few. As Editor-in-Chief, I have felt the weight of this tradition as a responsibility to be taken very seriously. But it has also been an agreeable responsibility; the staff at the Royal Society, and the members of the Editorial Board, have been unfailingly helpful. And I have enjoyed negotiating the challenges and contrasts that come with the job: trying to keep the quality high, dealing with authors and referees who are brilliant, conscientious, argumentative, irascible, careless, eccentric, etc.

Proceedings of the Royal Society A emphasizes quantitatively rigorous approaches to the physical sciences. The success of this method has revolutionized technology and transformed everyday life. But it might also underlie current enthusiasm for using catchwords or numbers to assess achievements that are too subtle to be captured so simplistically. The extreme of this tendency is to sum up scientific accomplishments with the single word 'impact'. This carries connotations of hardness, of a sudden and immediate force, and misrepresents the delicate and complicated ways in which a scientific publication can impinge on, infiltrate, insinuate, and otherwise influence its own and other sciences and the world outside science, over timescales that can vary from days to years to decades. For *Proceedings of the Royal Society A*, the associated number that popularly represents citations—the 'impact factor' (itself much cited)—is currently close to 2: low by comparison to those in the life sciences but respectable for physical sciences. But, as I never tire of pointing out, impact factor is a short-term measure, representing citations over two years. On the longer-term criterion of citation half-life (a rough indicator of long-lasting influence), this journal stands highest in the Royal Society's list—indeed off the ten-year scale on which this quantity is calculated.

I arrived as Editor after publishing many papers in *Proceedings of the Royal Society A*: nearly 50 (possibly more than any other author in this journal), over 40 years—though I am awkwardly aware that, as implied above, numbers do not indicate quality. One consequence of this long prior involvement has been a reluctance to change much; in common with some of my predecessors, I have been a rather conservative editor. But it is time for change. I welcome my successor, Sir Mark Welland, who comes fresh to the position and may be more willing to innovate.

Michael Berry
Editor

Royal Society Publishing
Informing the science of the future

Reprinted from Proceedings of the Royal Society A, Volume 469, (2013).

Getting a knighthood

(Bristol, June 1996)

It's unreal, isn't it? I've always had a strong sense of the ridiculous, of the absurd, and this is working overtime now. Many of you must have been thinking, 'Why Berry?' Well, since I am the least knightly person I know, I have been asking the same question, and although I haven't come up with an answer I can offer a few thoughts.

How does a scientist get this sort of national recognition? One way is to do something useful to the nation, that everyone agrees is important to the national wellbeing or even survival. Our home-grown example is of course Sir Charles Frank's war work. Or, one can sacrifice several years of one's life doing high-level scientific administration, not always received with gratitude by the scientific hoi polloi but important to the smooth running of the enterprise. Sir John Kingman is in this category. Another way — at least according to vulgar mythology — is to have the right family background. Well, my father drove a cab in London and my mother ruined her eyes as a dressmaker. Another way is to win the Nobel Prize or one of the other huge awards like the Wolf or King Faisal prizes, like Sir George Porter as he then was, or Sir Michael Atiyah. The principle here I suppose is, to those that have, more shall be given.

I don't qualify on any of these grounds. The nearest I come is to have have won an fairly large number of smaller awards. I got one last Friday: the Dirac medal of the International Centre for Theoretical Physics at Trieste. Perhaps the principle is that a lot of small prizes equals one big one. I doubt it. That would be silly. Probably you know the story about the journalist who went to visit a senior scientist who had just won a big prize, and, seeing an enormous number of certificates and citations proudly hanging on his walls, asked: "How did you get all these?" The scientist replied by pointing to an insignificant little certificate: "Once I did a really fine piece of research, and was awarded this one. Then I got this other prize because I had received the first one, the next one because I had received the second one, and so on". So I really don't know why I find myself in the position I'm in now. The citation reads: for services to physics. Well, I don't know about services, but although I have done some physics over the years, we all know that doing science is its own reward — I often marvel that we even get paid for having such a rich and satisfying life. I must have friends somewhere, I suppose.

So ignorant am I of this strange matter of honours that I don't even know what a Knight Bachelor is. I thought I knew what bachelor meant, having been in that state an indecent number of times, although not recently, but when I looked it up I found:

A young knight, not old enough, or having too few vassals, to display his own banner, and who therefore followed the banner of another.

Hmm...

A knighthood is an national award, but although I am often grateful for the freedoms and opportunities I have had through having been born and grown up here in this particular historical period, I am not a particularly patriotic person. I travel too much, and see science as too thoroughly and essentially international, for that. The loyalty I do have, and where I think today's honour is appropriately directed, is to this department. Over the years I have had such an inspiring and brilliant series of colleagues: John Nye, John Hannay — they all seem to have been called John, don't they? — Jon Keating, Jon Robbins, many non-Johns, of course, and in the early years not a collaborator but a mentor who protected me when I crawled out on various unfashionable scientific limbs, namely John Ziman.

But the department is much more than one's research collaborators. This Royal Fort — or, on letters I have received, the Royal Fork, Royal Foot, Royal Font, or Royal Fart, or Professor M V H H Wills, or the H H Wills Physics La Oratory, or, my alltime favourite, that gave me delusions of grandeur for ages: Professor M V Berry, H H..... But I digress. Over the many years that I have been here — more years than not, actually — I have enjoyed the most supportive and tolerant working environment that could be imagined. I visit many departments, in this country and abroad, and I can tell you that there isn't a place like it for helpfulness, pulling together, absence of feuding and bitterness, and plain friendliness. The spirit of this place has been sustained by sensitive leadership, but it exists because eveybody shares it. I mean everybody: not just our academic colleagues and students but cleaners, porters, secretaries, the workshop.

Now I'm getting emotional, so I'd better stop. What I'm getting towards is this: this knighthood has my name on it, but I think of it as ours.

Wolf Prize

(Shared with Yakir Ahoronov)

Knesset, Israel 10 May 1998

I am a quantum mechanic. So is Yakir Aharonov. A technical term in our subject is *the entangled state*. Anyone who has a conversation with Yakir gets into an entangled state, with contradictions, digressions and interruptions all mixed up — Talmudic, I suppose. But now, here, Yakir can't interrupt me, as I declare what an honour and delight it is to share this occasion with such a quick, deep and subtle man. There is only one sadness that I'm sure he shares with me: that David Bohm, with whom he did some of his seminal research, is no longer living; if he were, he would surely be here tonight.

Although we've never collaborated directly, Yakir's life and mine have been entangled too. We both worked in Bristol, though not at the same time — in other words, we underlapped there: I'm still in Bristol, and he came and went in the 1950s, while I was at school, though I should point out that in 1959, when he was discovering the Aharonov-Bohm effect, I applied to enter Bristol University to study physics, but they turned me down without even an interview (now, they can't get rid of me).

In all honesty, I must confess to certain uneasinesses about recognition of the kind we are getting today. In fair trade, you can bake a loaf of bread and sell it. If you want to earn more money, you can't (or shouldn't) sell the same loaf again; you must bake another one, and sell that. But in science such is the concentration on particular achievements that you get rewarded again and again for the same thing, and today is the culmination of those rewards. I'm not complaining, but it is peculiar.

In the experimental sciences, it is often fairly clear who discovered what. But in the theoretical and mathematical sciences, discoveries have a more ambiguous quality. Credit for what is sometimes a complicated evolution of ideas, involving many people, tends to crystallize about particular individuals. Tonight it's Yakir and I who are benefiting from this crystallization of recognition, and we're certainly not complaining, as I said before. But one ought to get things in perspective, and so I offer (not entirely seriously) three laws. First, there is the fact that it is hard to achieve justice in the attribution of scientific ideas, and this leads to

Arnold's Law (after the famous Russian mathematician): No discovery is credited to the right person. (Of course this applies to Arnold's Law too.)

The second law says that it is not only hard but *impossible* to achieve justice, because it seems that one can always find a precursor to any thought. So we have

Berry's Law: Nothing is ever discovered for the first time. (This too is self-referential.)

Here now is the third law, which I offer not because it applies to us (nor do the others, really) but because it is deep. It is a quotation from a philosopher:

Whitehead's Law: To come close to a true theory and to realise its precise application are two very different things, as the history of science teaches us. Everything of importance has been said before by someone who did not discover it.

Now, and finally, prizes. It feels very good to have one's work recognised, and to get a prize. Part of the pleasure is the unexpectedness — almost none of us works to get prizes — doing science has its own delights, and nobody who has not experienced the thrill of discovery can truly understand it. Moreover, getting recognised and getting prizes need not be the same thing, I will illustrate with a story. It is told that in a certain mountain village in Spain there takes place, every year, a poetry competition. From near and far, contestants come to declaim their verses, hoping to win. There are three prizes:

Third prize: a silver rose.

Second prize: a gold rose.

First prize: a real rose.

Well, tonight, it seems, we have something better: a Zeroth prize: the real rose of recognition as well as a — rather substantial — gold rose.

On behalf of Yakir as well as myself, thank you to the Wolf Foundation, for this honour and for your generosity, and for these wonderful and memorable days in Jerusalem, and thanks to you, President Weizmann, for so graciously presenting this prize on behalf of the State of Israel.

Convocation Address, 19 April 2000

Cornell Undergraduate Research Board

This morning I had the pleasure of listening to several of your talks, and I've seen the programme of your activities today. Very impressive it is too. In my department — physics — at the University of Bristol, England, we have encouraged undergraduate research for many years, but there's not a university-wide organization like yours.

In England, we're a little shy of talking about ourselves. Here you're less burdened by such inhibitions — I know, I've seen Mr Jerry Springer. So I'll try to adapt to the local convention (but don't expect me to go as far as he does) and tell you something about what a life in science, and research in particular, is like. I'll talk about science, knowing that many of you are doing research in the humanities rather than science, but most of what I'll say applies if you replace 'science' by 'scholarship'.

My family wasn't academic, or even educated beyond the bare minimum of literacy, and my childhood wasn't the happiest, though my mother did succeed in protecting me from most of the brutalities she had to suffer. Perhaps science was an escape into what I saw as a purer world — I don't know, and it doesn't really matter. But rather early something happened to make me know what I didn't want to do, which was to go into the kind of business some of my relatives engaged in.

I had a Saturday job in my uncle's store, selling trousers. He used to ask why I wasted my time reading books, when I could leave school and join his business. One day, a customer asked for trousers of a particular sort, which I showed him. 'What do these cost?' 'Three pounds' (That was a long time ago.) 'No', he said, 'I wanted something a bit better' At that moment my uncle walked by, took the trousers, and said: 'We do have something better — please wait' and then disappeared into the back of the store. About five minutes later, he came back, with the same pair of trousers, saying, 'Look at these; but they're more expensive: five pounds'. Of course the customer bought them. This made me feel bad. Not for me, such a life.

Then came something else. A cartoon in a magazine, that influenced me in a negative way. It showed a weedy pale little fellow lying on a beach — of course I identified with him, apart from the glasses, that I didn't need in those days. He was surrounded by adoring girls — gorgeous in the obnoxious stereotype of those days, that is, boobs rather than brains. They were ignoring a bronzed hunk, bursting with unemployed testosterone, because he couldn't compete with my lookalike, who was reading a textbook on nuclear physics. Nowadays I could spend an hour deconstructing the negative images and presuppositions that cartoon implied, but even then I was embarrassed by it. It didn't fit the image of science I had then, even in my glimmering innocence.

Don't worry: this isn't going to be my life story. But that early idealistic belief in the honesty and decency of scientists was pretty well been borne out by my later experience. This is something people have different opinions about. There are some big guns on the other side. When Michael Faraday started working at the Royal Institution as a laboratory assistant, in the early 1800s, he told his mentor, the great chemist Humphrey Davy, that he wanted to work in science, because scientists were free of the jealousies and petty motivations that afflict other professions. Humphrey Davy replied, 'Just you wait, young man, you'll soon learn what the world is like, that scientists are just as mean and self-centred as anyone else.'

Maybe he was right. Maybe I'm lacking the gene for competitiveness — or perhaps it's the meme (that is, it's culturally constructed and propagated). We have something called *Who's Who* — I think there's a similar book in this country. In my entry, you can read my list of interests: just 'anything but sport'. Not that I'm against exercise and physical activity — indeed I approve of it — though mostly in theory, as you can probably see. But I could never understand why anyone would want to win. And I'm not blind — I see, especially in areas of science where serious money is involved, squabbles for priority, funding, etc. But my own experience has been that overall people behave pretty decently, sharing ideas, data, etc. The image is of colleagues rather than competitors. Those few who are obviously driven by a desire to beat their friends are usually regarded as having a mild personality disorder; they are not admired. If that seems old-fashioned, embarrassingly naïve, even, I'm unapologetic: that's how I've found it to be, and the way I encourage my students to keep it.

Research can get you into some quite absurd situations. Soon after I arrived in Bristol, I had a visit from a graduate student from the veterinary department. He was studying horses, and wanted me to help with a paper he was trying to understand. 'But I'm a physicist; why have you come to me?' 'Because I heard you're a mathematical type'. He was studying horses' hearts. They are like little batteries, pulsing tiny currents round the body once per heartbeat. If you can measure the electric potential at points on the surface of the horse, you can infer the polarity and strength of the little battery, and get something called the heart vector, and how this changes during each heartbeat — the heart loop — which is a powerful non-invasive way to diagnose heart diseases.

The mathematical theory underlying this inference was in the paper he was trying to understand, written by Denis Gabor (he who invented holography). The problem was that my visitor knew very little mathematics, in particular not calculus, which was a big drawback since the paper was based on integrals. His question was: Does Gabor's theory apply to a real horse, or only to an ideal cylindrical horse? The situation was desperate, because the poor fellow had spent three years making his horse a coat of several hundred potentiometers, each measuring the electric potential fifteen times during a heartbeat; he had a mass of data, and three weeks to make sense of it. An emergency requires quick action, and so it was that I found myself helping him build a little device to measure the direction of the perpendicular at each point on the surface of the horse.

A year later, when I had forgotten all about it, the guy sent me the paper he'd written about his work. In the Acknowledgments, he thanked me for helping with the mathematics, and, for funding his research, the Horserace Betting Levy Board — that's our government agency that imposes taxes on gamblers who place bets at racetracks. I'm sure those gamblers didn't realize that their addiction supported my Monty Pythonesque activity of taking a surface integral over a horse.

Mostly, day-to-day research is pretty technical, but we should never lose sight of the perspective that what we're doing is contributing to building a patchwork of enormous scope, whose magic is that we are connecting very different areas of experience. Nowadays we are all encouraged to tell the nonspecialist public about what we do — outreach, you call it; for us, it's Public Understanding of Science (notwithstanding the unfortunate acronym). Sometimes, what we do has obvious practical application, and then it's easy to communicate.

But very often our motives are more intellectual, and then, rather than pretend applications that aren't there (it might help us find a cure for cancer, as people used to say in the 1970s), it's better to stress this aspect of unity, of connections. I find that, contrary to what media people tell us — they insist on dumbing things down — non-scientific people appreciate that: the magic of abstract ideas — even mathematical ones — that apply to widely differing situations. I realize that not all areas of research lend themselves to this style of exposition. It is very much to my own taste to discover 'the arcane in the mundane'. I'll show you an example — but again, don't worry: you'll not be getting a physics lecture today.

In the 1970s, mathematicians devised a branch of geometry called catastrophe theory, based on a remarkable theorem classifying certain shapes. These describe ways in which smooth changes in something can generate sudden changes in something else. If the first something is weight on a bridge, and the second something is collapse of the bridge, that's a catastrophe in the everyday sense; hence the name. Some of the applications were controversial — invoking the mathematics to describe the dynamics of prison riots, or anorexia nervosa, or the switch from anger to fear — and this led to misguided attacks on the mathematics itself. A genuine application, within physics, was to give a proper understanding of the *focusing of light* in natural situations, where there isn't symmetry — that's in contrast to the perfect lenses we try to build into our cameras and binoculars. The resulting theory gets deep into the physics of light, and especially the relation between light rays and light waves — that's a case where reductionism — so controversial elsewhere — works. Here now is a piece of bathroom-window glass, wiggly on a scale of millimetres, and here is a laser, whose beam of pure light is also about a millimetre across. Shine it through the glass, whose wiggles bring the light to a focus and also give rise to interference between different rays striking the same region of the screen, and the resulting patterns exhibit the whole zoology described by the mathematicians' catastrophes. Similar patterns — with the interference detail blurred away — are generated on vast scales by the light from distant galaxies focused by the gravity of intervening stars, and more familiarly, on the bottom of swimming pools, by sunlight focused by the water waves on the surface.

It isn't all fun, of course. Research can be frustrating in several ways. Most of the time, it doesn't work. My son did a Ph.D in mathematics, but left afterwards because he couldn't bear the uncertainty of research — sometimes weeks getting nowhere. Now he makes a pretty good living as an inspector with our income tax service — what you call the IRS — a job with security, and perfectly suited to his mean-minded intelligence. When he started, I made the usual weak joke about having to be careful about my tax return. 'Oh, don't worry, we don't care about professors — their income is insignificant'.

There is not security in research. I tell my graduate students that if they study with me they shouldn't expect to get a job afterwards. Their only reasonable motivation should be to spend several exhilarating years in total freedom, creating at the forefront of knowledge. In fact, all my students did get jobs — academic positions for those who deserved them (about half), and respectable other work for the rest. (In case you're wondering, I don't have tenure — I gave it up about ten years ago when I applied for the research professorship I hold now. Every five years, I have to reapply for my job.)

Here's a useful concept, to ponder when your research gets stuck. What is the elementary particle of sudden understanding? You can call it the *clariton*. You all know immediately what I mean by a clariton: that wonderful 'Aha!' moment, when all is revealed, and the way ahead opens up. But unfortunately there are also the unwelcome *anticlaritons*, that arrive and annihilate the insights of yesterday.

A different source of frustration is bureaucracy. In Britain there's an ignorant opinion that all bad trends in our society originate over here. And some do — for example advertisements on television (I still find them vulgar and intrusive), and commercial medicine, that our governments have allowed to threaten the kinder system we used to have. But in research we have our home-grown nonsense, in the form of endless official demands for falsely-quantified information about our researches, justified in the name of public accountability. Public accountability has become a mantra that's hard to argue against, even though it leads quickly to something like Heisenberg's uncertanty principle: by asking for continual self-justification, you inhibit the very freedom to create and speculate that keeps research alive.

This is not the place for a a systematic treatment of this dismal subject, but let me share a delicious example. Our National Physical Laboratory — the counterpart of your National Institute of Standards and Technology — puts out annual reports. In one, from a section called 'Performance targets' we read 'Target number three: Research Milestones'. Now, we live in a scientific age, so everything must be defined, and indeed we have: 'Detailed definition of measure: Research milestones are those which are agreed with the customer'. I am not joking. That year, the goal was 0.49 research milestones per scientist per year, to be increased by 3% per year over a four-year period. In the first year, an average of 0.48 research milestones were achieved. A near miss. Must do better. In the following year, though, the score was 0.79. But that's too good! In that annual report, we read: 'The comfortable achievement of this target, while pleasing, probably indicates the unsatisfactory nature of this measure. A revised target will be set....' In the next few years, the number of milestones increased, until: 'The National Physical Laboratory is making increasing use of project management techniques which encourage the breakingdown of programmes into modules, and in which the measurement of success by milestones is *even more significant* than it was at the time when the target was set.'

At about that time, I met our minister of science, and told him this ridiculous story. He didn't say much, and looked a bit embarrassed. A week later, I understood why, when I read that he had just presented our National Physical Laboratory with an award for that year's best annual report.

I don't know how it is here, as far as research being contaminated by ignorant bureaucracy is concerned, but let me offer a one-word self-therapy that sometimes helps. It comes from the physicist Richard Feynman, who used it on himself whenever he was tormented by people at meetings who were forever telling him trivial things he already knew. It can work with bureaucracy too. The word is: Disregard. Try it.

Research can involve difficult issues of time management. (I've caught myself using this word 'issues', that I don't like — a few years ago we hardly heard it, now everything is an issue — I suspect it's a Bill Gatesism, a Microsoftheadedness.) I can't help you with time management, because I don't do it. Several years ago, I gave a series of lectures in Chicago, on several different subjects. Afterwards, my host paid me a compliment: 'It was very good for our students to encounter someone with such a well-conceived and well-implemented research plan'. I told him. 'Alas, I have to disappoint you. There's no research plan. Most days, I have no idea what I'm going to think about.' I take no responsibility for any consequences that might result if any of you follow my example (accept no liability, as I think I have to say here.)

What else? Yes. Travel. Growing up in a poor family, it never occurred to me that as a scientist I would travel beyond the dreams of everybody except the most wealthy. Recently my work has taken me to Germany, Palestine, Israel, Lebanon, Romania, New Zealand, Ukraine, France, India, here of course…I forget all the places. As well as being fun, it's also inspiring and beautiful, meeting my colleagues there and finding we communicate perfectly, and do science in the same way. This universality of the free play of the intellect is something worth pointing out, especially now when there is a different emphasis, on the separate identities of the cultures of different peoples. I have no problem with that; I see no contradiction, but rather complementarity — indeed, it could be that only when people appreciate and repect their separateness can they get truly together (on the microlevel of pairwise relationships, we know those couples who are proud to share every little thing; usually, they divorce).

One more thing. In academic life there are little turf wars, games people play with the self-image of their discipline and those of others. You know what I mean: physicists versus chemists, mathematicians versus physicists, engineers versus scientists, sociologists versus everybody…Even within disciplines there are silly little image wars. Sometimes, fashionable, often highly-technical, subjects can exert an intimidating influence on people not involved with them. Well, don't be intimidated by what other people do. Here's another story.

Back in 1985, visiting CalTech, I had just started studying some quantum physics related to some mathematics called the zeta function, itself related to a famously untransparent mathematical speculation called the Riemann hypothesis. It doesn't matter what all that means. At the same time, CalTech was a centre of superstring theory, a speculative development that some people thought might give the 'theory of everything'. The people who did (and do) that research were (and still are) fearsomely clever. I met one of them, who asked what I was working on. When I told him, he fixed me with a pitying stare. 'Yes, we have zeta functions throughout string theory. I expect the Riemann hypothesis will be proved in a few months, as a baby example of string theory.' That certainly intimidated me, but I take some pleasure in reporting that now, fifteen years later, we're still waiting.

At the same time, the president of CalTech, a physicist, apologized that he couldn't come to my colloquium, because he had to go to Palm Springs to explain string theory to Ronald Reagan. How I wish I could have been a fly on the wall...

Sometimes, the special subjects we choose to study can sound bizarre to outsiders — and it's not only science that can seem strange. Several years ago, I visited the Institute for Advanced Study attached to the University of Jerusalem. In that place, two programs run at the same time. On the first day, I went to coffee and there were all these people I didn't know, obviously from the other program. We exchanged greetings, and one of them asked what my program was. I told him: quantum mechanics. 'And what's yours?' I asked. 'The historicity of emotions'. I wondered what was the difference between historicity and history, but didn't get a clear answer. So, to lighten the conversation, I told him I always wanted to know what made the Romans laugh. Actually I do know — at least a little — because I have a friend who studies Roman satire, but I thought this person might give an interesting slant on it. 'Roman humour, huh! you asked the wrong person. My emotion is hate.' If you know a better conversation-stopper, please tell me.

Oh, help! It's Wednesday, ...I shouldn't be here. There's another silly little turf war — more teasing, really — between theorists, like me, and experimentalists, who think we're lazy. One of them (Leon Lederman) said: 'You'll never get a theoretical physicist to give a talk on a Wednesday, because it spoils two weekends.' So I'd better stop.

Michael Berry
H H Wills Physics Laboratory, Tyndall Avenue
Bristol BS8 1TL, United Kingdom

Response at Honorary Degree Ceremony

Weizmann Institute, 10 November 2003

Passionate rationality

I've known this abode of passionate rationality for nearly twenty years. Passionate rationality? One example: the inspiration that comes from startling connections — like last week, in the journal *Nature*, a picture from the neurophysiology department here, representing what a cat sees with its eyes shut — identical to pictures we have been making in the physics of random waves. The sciences seem very different, but a mathematical theme links them. Another: when the colour of gold is calculated from the quantum mechanics of electrons, the predicted colour is silver! But gold is heavy, so its electrons move fast, and to get the colour right it's necessary to incorporate Einstein's relativity: gold is relativistic silver. Such delights, such passionate rationality.

I said *abode* of passionate rationality. Perhaps *enclave* is more correct. Outside this campus, there are passions aplenty, but *rationality*? Not always — even here in Israel — and still less just across your borders (wherever they are). But as scientists, sharing perhaps the only truly transhuman culture, it's our duty, if we must build walls, to build bridges too (forgive the cliché). No contradiction in doing both at once — certainly for those of us who are saturated with quantum mechanics.

Bridge-building could be: quiet encouragement of a growing discussion among Islamic scholars, about how passionate rationality got lost in their world. Scientific discovery has been described as "a chariot ... driven by Greeks, Romans, people of all kinds, lately Moslems, now Jews and Christians." A thousand years ago, Islamic scholars drove science forward. I know this from my own work in optics.

But — and now I quote Ziauddin Sardar — "Around the fourteenth century ... Muslims, consciously and deliberately, abandoned scientific enquiry in favour of the path of ignorance and blind imitation ... The intellectual devastation we see in the Muslim world today is ... a product of this mentality. The once great tradition of Islamic science has degenerated into a few research programmes on nuclear weapons and chemical and biological warfare ... The Muslim world does not contain a single world class university."

Their new debate is an attempt to reclaim an earlier tradition within Islam, of systematic original thinking. According to the prophet Muhammed:

"Seeking knowledge for one hour is better than praying for seventy years"

As an atheist Jew, I couldn't put it better myself — though commenting on a religious debate within Islam does seem absurd — with the dominant religion in this country, I can at least consult a rabbi (my brother, actually).

So, how to give quiet encouragement? I am aware of the difficulties and frustrations — both ways — in reaching out to your scientist neighbours. I know of some attempts, but few in my own area of theoretical physics. Several years ago, in Beirut, I helped launch the Center for Advanced Mathematical Studies. Some of their research is closely related to research here; what a pity there is no direct interaction. Now there's a glimmer of a plan to create a Center of Excellence in Mathematics and Physics in Bir Zeit University — your neighbour. Here's a wonderfully Middle-Eastern touch from the message I received: "... we may be able to get the land donated ... by the President of Bir Zeit University ... who is also the land owner". Dream on, Ilan [President of Weizmann Institute in 2003].

Procrastination is tempting and easy. I speak from an academic culture where delaying was an art form. We used to complain that 'in Bristol nothing can be done for the first time'; in Oxbridge colleges, any initiative could be destroyed by invoking the 'principle of unripe time'. Not a good model here, now. Better, let the Weizmann Institute be a centre for the export of passionate rationality, a quest I'm proud and — especially today — honoured to share.

Published in '100. A collection of words and images to mark the Centenary of the University of Bristol', (Ed: Barry Taylor, University of Bristol 2008)

My (nearly) half-century in Bristol

by Michael Berry

The H H Wills Physics Laboratory is the highest point in the centre of Bristol, so it can be seen from many places across the city and is easy to find ('Keep going up'). When I arrived in 1965, damp behind the ears with a new PhD from St Andrews, I was impressed not only by the location of the building but also by its imposing 1927 design as a mock castle. Then, the physics department was clamorous with builders constructing the 'new wing'. Four decades later, as my retirement looms, the builders are back, renovating the whole laboratory.

With a young family, it seemed sensible to consider buying a house. One that we viewed was, in the estate agent's clichés, 'In need of some redecoration' (we glimpsed a dead cat on the rotting staircase leading to the basement), and 'in the up-and-coming area of Kingsdown'. He was right on both counts, but we decided we could never afford the princely asking price of £1,200.

With a post-doctoral fellowship, and a physics project different from what my colleagues in the department were studying, it was easy to settle into research with few distractions. But I soon discovered that physics has a surreal side. A mathematical enquiry from a research student in the veterinary department was channelled to me. He had spent three years measuring electric signals on the surface of a horse that he had covered with detectors, with the aim of deducing the electrical properties of the heart, which acts as a weak battery whose functioning gives a good indication of health. The mathematics he did not understand deduced what was happening inside the horse from measurements on its outside. His first question was 'Do these formulae apply to a real horse, or only an ideal cylindrical horse?' Thus I found myself applying calculus to horses. A year later, I received the student's published paper, in which he expressed gratitude: for help from me, and funding from the Horserace Betting Levy Board.

An early responsibility was as co-ordinator of staff meetings, that is, custodian of departmental democracy. We introduced practices that were unfamiliar in the university then but are commonplace now, such as having an agenda circulated beforehand, and inviting student representatives with equal rights to speak and vote — though votes were (and still are) rare in a department with decisions made largely by consensus. We agreed to the unheard-of practice of allowing students to bring notes into examinations, so questions could probe understanding as well as memory. In an extreme application of this principle, still remembered by some of the students, I held an 'infinite examination', in which there was no time limit. This was a failure: the best results were obtained by those students who had finished within the usual three hours; and I had forgotten to bring anything to eat and so was starving, unlike the weaker students who had taken my advice and brought sandwiches to sustain them during their largely futile scribblings, which for the stragglers lasted eight hours.

In those days, academic appointments were made differently. After I had been in Bristol for nearly two years, John Ziman, who had arrived in 1964 as the new professor of theoretical physics, called me into his office and pointed out a fact that I was dimly aware of but whose significance had not sunk in, namely that my research funding would soon come to an end and I would be out of a job. 'You seem to be able to teach, so would you like a lectureship?' I mumbled that it seemed a good idea. 'OK, but you will have to go through the formality of an interview by an appointments committee'. And so, with no advertisement, no references taken up, and no citations scrutinised, I found myself with a permanent appointment in the same university that in 1959 had rejected my application to enter as an undergraduate.

The 1960s was the decade of the counterculture in America and riots in Paris. In December 1968, the turbulence reached Bristol University, in the form of a student occupation of Senate House. The 'great cause' that raised passions and inspired the sit-in was not opposition to the war in Vietnam or apartheid in South Africa, or the wish for a root-and-branch restructuring of society; rather it was outrage at students from other colleges in Bristol not being allowed to use the facilities of the university Students' Union building (I joke not). But passions there were, on both sides. Professors who had fled from the Nazis and fought with the British Army, but happened to disagree with the aims or tactics of the students, were denounced as fascists. And I, as almost the only lecturer who entered Senate House to listen to the interminable spontaneous seminars that were (un)organised by the students, was denounced by some of my colleagues as a traitor to academic freedom by giving legitimacy to the students who were threatening it. But the protest fizzled out after a week, as the occupiers gave higher priority to going home for Christmas.

A good result of the occupation was the setting up of a university Committee for Communications and Relationships, including students and lecturers (I was the youngest), and chaired with wisdom and sensitivity by our Nobel prize-winning recent head of physics, Professor Cecil Powell. Following our recommendations, many changes were made in the organisation of the university. Some had been pioneered in the physics department. Others, including the setting up of a University Newsletter, were new. This was inspired by the university's response to the student occupiers' very effective distribution of information about the progress of the sit-in. In recent years, the Newsletter has mutated into several university publications and our website — sources of news that in this information age we take for granted.

Some of my colleagues complain about the bureaucracy, largely imposed from outside the university, associated with research (too much time applying for grants) and teaching (too much time spent documenting). As a research professor for the past 20 years, I have not suffered in this way, but I think my colleagues are right. One can call it the Heisenberg principle of accountability: time spent measuring a creative activity can soon inhibit it.

It is easy for criticism of government attempts to micromanage intellectual activity to spill over into more local criticism, of the university administration. This is a negative view that I do not share. It is true that I am largely ignorant of the administration and how it works — indeed, for my first few years in Bristol I did not even know the name of the Vice-Chancellor. But this is meant as a compliment: the best administrations are largely invisible. On the few occasions when I have needed specific help from Senate House, this has been effective, pain-free and good-natured.

There are many ways of doing theoretical physics. Over the years, mine has changed. When I arrived, my style was algebraic: a typical day would be spent writing pages of equations. But soon, under the influence of senior Bristol physicists, I realised that richer understanding of concepts and phenomena could be achieved by supplementing the algebra with pictures. A massive boost to this emphasis on the visual came in the late 1980s, with the arrival of small computers, powerfully equipped with graphics capabilities. Now, images have become a valuable way of exploring the mathematical content of physical theories (and mathematics itself), and have led to discoveries that were only later verified by more traditional methods.

I was always concerned with the beauty of equations, and now this aesthetic has been transferred to the pictures that represent them. In this I am not alone: 'science and art' and 'the art of science' are terms increasingly being used to describe appealing images originating in scientific research. I prefer the more accurate descriptions in the title of a research project on this theme: 'Envisioning Science'; and the name of a series of conferences on the topic: 'Image and Meaning'. We increasingly recognise images as powerful tools, not only for furthering our own understanding but for communicating between scientists and to people who are not scientists.

I consider myself fortunate to have spent all these years in Bristol University. Occasionally, possibilities arose to move to other institutions, in the UK and abroad, including some that are superficially more prestigious, but I always concluded that the working environment in the physics department was as close to perfect as I could imagine, so I stayed. In large measure, these conditions were created by my colleagues. I am referring not only to my research students, from whose direct collaboration I have benefited enormously, even though my approach to physics has sometimes appeared solitary. In addition, my fellow professors and lecturers have displayed generosity and congeniality, perfectly exemplifying the co-operative spirit that animates science, contrasting sharply with the competitiveness that the media loves to emphasise but which is in fact relatively rare.

Physicists, especially theoretical physicists, have the reputation, perhaps not entirely unjustified, of adopting an attitude of intellectual superiority. I have already mentioned the fact that our department is at the top of Bristol. An even more appropriate location occurred to me during one of the balloon rides for which our city is celebrated: ballooning is the perfect occupation for a theoretical physicist: looking down on the world, supported by hot air.

Physics for taxi-drivers

With the International Year of Physics due to start next month, physicists need to be able to explain how their subject relates to other people. **Michael Berry** gives two examples and calls for more from the physics community

Most people enjoy conversations with taxi-drivers. I occasionally fantasize the driver beside me morphing into my father, who was a London cabbie and never shy about sharing his opinions with his fares. A recent journey began with a question: "I often drive between the airport and the university physics department, but I don't understand what you do there. It seems very clever, but does it have anything to do with the real world?" My response was to tell the driver two stories.

The first began with a factory producing consumer goods – surely as connected with "the real world" as anyone who uses that dismal expression would wish. This particular factory makes CD players, which enable anyone to hear music, reproduced almost perfectly, anywhere in the world: in the desert, up mountains, in forests, on the seas and so on. This is a new development in human history. Previously, people who wanted to hear music had to be physically present when and where it was performed (or, more recently, within range of radio reception and tuning in at the right time). In a sense, the CD player represents the ultimate cultural democracy: making available to many what could previously be enjoyed only by a few.

But the CD player is also a "quantum physics machine", in several senses. First, it contains a laser, the beam of which reflects from the pits on the disk that encode the music. This was an application by engineers of a device invented 30 years earlier, and, moreover, an unanticipated application – recall the cliché that the laser was an invention looking for an application. The laser itself was an application by physicists of ideas about the quantum physics of light, initiated by Einstein four decades earlier still – an application that was surely not anticipated by Einstein himself. And, of course, each of the millions of transistors that guide the flow of electrical information is itself a direct application of the quantum mechanics of electrons in the periodic environment of a crystal.

From this chain of associations comes an unexpected conclusion: quantum mechanics has democratized music.

Note the connection between three abstractions: quantum mechanics, which raises doubts about previous conceptions of physical reality; democracy, a relatively recent ideal in the organization of human affairs; and music, where sounds unrelated to any in the natural world, and continually changing in ways incomprehensible to each previous generation, exert strange powers over our deepest emotions.

Tell me more – what do physicists do?

Where and when

For my second story, I pointed to the taxi's GPS navigation device. To decode the signals from the GPS satellites to discover our location, it is necessary to incorporate various relativistic effects that influence the time it takes for the signals to travel to and from the satellites. If these effects were ignored, disastrous navigation errors would soon result. So the GPS is a "relativity physics machine" – to my knowledge, the first such consumer device. It enables the solution, for all everyday purposes, of the ancient problem of knowing where we are, much as the vibrations of little quartz crystals in wristwatches recently solved the analogous problem of knowing when we are.

I have deceived you. The driver's question caught me off balance, and what he got from me was not what you have just read but ill-articulated and probably not very convincing versions of these two stories. But they did seem to give him an unfamiliar perspective on what we do, because he went on to ask me about my own research in theoretical physics, which has no applications as far as I know. (Actually it does have applications in philosophy, where it contributes to the long-standing problem of how our different levels of descriptions of the physical world – e.g. classical and quantum mechanics – are related.)

However, physicists should avoid the serious misjudgement of claiming these revolutionary devices for ourselves, as though we alone created them. No physicist, or group of physicists, would ever have produced a CD player. As I have already mentioned, the invention of the CD player also involved engineers, while mathematicians developed and optimized the codes that transform the music into light and electricity. Moreover, to get any invention into people's hands requires factories to produce it, businesses to finance and sell it, advertisers to tell people about it and so forth. The world is strangely and wonderfully connected.

Nevertheless, it is good for people to see that our subject affects them in ways they usually have no conception of. Not just taxi-drivers, of course: I made the CD player story the centrepiece of a recent prize-giving speech at a secondary school, and it seemed to go down well with the 11–18-year-old students in the audience.

Be prepared

Now I come to the point. 2005 is the International Year of Physics, when we can hope for a higher profile for our subject. We should anticipate more questions like that of my taxi-driver, and be ready with a good supply of convincing stories. As it happens, both of mine involve Einstein, which is appropriate for the year that also celebrates the centenary of his monumental early contributions. Every physicist knows dozens more.

My reason for writing this article is to encourage readers to share their favourite stories by sending them to this magazine, which will publish a selection of the best. The type of stories I have in mind are characterized by connections between the intellectual and the everyday, where an abstract concept in physics leads, perhaps after many decades, to a direct effect on people's lives. Stories should be not more than 500 words and should be sent to *Physics World* before 31 January 2005.

• Stories should be marked "Physics for taxi-drivers" and can be sent by post to *Physics World*, Dirac House, Temple Back, Bristol BS1 6BE, UK, or by e-mail to pwld@iop.org

Michael Berry is Royal Society research professor at the H H Wills Physics Laboratory, Bristol University, UK, Web www.physics.bristol.ac.uk/staff/berry_mv.html

Retirement speech

Quartier Vert restaurant, Bristol, 25 September 2008

If I'm run over crossing the road tomorrow, I'll be stereotyped in the newspapers as "Pensioner and grandfather". Until the last moment, you think it will never happen. I'm still in denial.

When people occasionally ask "Where did you grow up?", I tell them that I never did. But the closest to growing up has been my forty-three years in which our beautiful department has been my home. I arrived in 1965. There were builders all over the place, and the 'new wing' was half-made. Our address was still the Royal Fort — or, as in letters I received over the years, the Royal Foot, the Royal Fork, the Royal Fart (from Japan), my American Express bills were addressed to H H Wills la Oratory, and my favourite: M V Berry, H H. I remember my first meeting with Cecil Powell, standing side-by-side in the third-floor men's toilet. "The plumbing in this room has been leaking since they built the place in 1927", he said. That was in 1965. As half of us know, it still leaks after eighty years.

And now, when (as Bob Evans [head of Bristol physics] delicately puts it) I make the "smooth transition to University Research Fellow and Emeritus Professor", it's once again amid the bustle of builders. My grand challenge for him is: get that plumbing sorted.

I have been enormously fortunate. First, to have enjoyed free education. Without that, I would never have considered going to university (only one of my twenty-five cousins had a university education), and might have ended up in my uncle's shop selling trousers ("Why waste time reading all those books?"). And fortunate in coming to Bristol, almost by accident. I had been awarded a Fellowship to move from my Ph.D, in St Andrews, to Sussex (Bob Chambers [a Bristol professor] was on the interview panel in London). My supervisor, Bob Dingle, told me about a young professor, newly arrived in Bristol, who I might enjoy meeting: one John Ziman. I read the first of John's science and society books, 'Camford Observed'. This identified in Oxford and Cambridge exactly the stifling traditionalist mindset I had disliked in St Andrews, so although I had no interest in the condensed-matter physics he was doing I thought he might be an interesting person, and came here to visit him. What an inspiring day! Lunch with him and the entomologist Howard Hinton and the philosopher Stefan Körner — all of them have died — this was the intellectual life I had dreamed of at university but had never experienced. With John's encouragement, I decided that day to transfer my Fellowship here. It was a good move.

Two years later, John pointed out that my Fellowship was coming to an end. "Oh is it?" — employment was something I hadn't thought much about, despite having a young family and a huge mortgage (£3000). "Your teaching seems satisfactory, so would you like a permanent job?" "Ok, why not? Thank you." So easy — and a sweet irony, since Bristol had rejected me as an undergraduate eight years before.

My intellectual trajectory was unusual. I hadn't worked, and didn't want to work, on mainstream physics — condensed matter or high-energy or particle physics. In my Ph.D, Bob Dingle gave me an optics problem that he hoped could be solved with the mathematical methods he had developed. When these didn't work, he left me alone to sort it out myself. I needed his help occasionally, and in those few conversations (I could count them on the fingers of one hand) his advice was perfectly focused, and indispensable. In Bristol, I pursued weird visions about connecting quantum and classical physics using that same mathematics I had learned from Dingle in St Andrews. I didn't fly abroad until I was nearly thirty — hard to imagine now, when my carbon footprint rivals Tony Blair's, or Naomi Campbell's. I didn't attend a physics conference until I was thirty, and didn't collaborate with a scientist senior to me until even later. John Ziman somehow appreciated my physics enthusiasms, and protected me when I was publishing papers on unfashionable topics. A mentor, in a way, though we never collaborated scientifically.

The senior scientist I collaborated with was John Nye. From him I experienced first-hand what I was already absorbing from Charles Frank: thinking about physics geometrically and simply (as Einstein is reputed to have said: "as simple as possible, but not simpler"). From John Nye I learned what it means to be a scientific gentleman — how to treat other scientists respectfully, how to collaborate decently, how to disagree robustly without being disagreeable — a role model hard to live up to.

Actually I do have a contribution to particle physics. I announce it now. A new particle: the elementary particle of sudden understanding — the *clariton*. Any scientist will recognise the "Aha!" moment when this particle is created. But there is a problem: all too frequently, today's clariton is annihilated by tomorrow's *anticlariton*. So many of our scribblings disappear beneath a rubble of anticlaritons.

I have been very lucky with wonderful colleagues — not only John Nye as mentioned, but my students and post-docs. Their arrival has often coincided with new phases (no pun intended) of my scientific life, which I've been able to develop in collaboration with them. These have been the happiest interactions — continuing now with those who have remained in Bristol or returned to Bristol: the Johns — Hannay, Keating, Robbins — and Mark Dennis — not forgetting those who have made successful careers elsewhere: Alfredo Ozorio de Almeida, Bernard Buxton, Francis Wright, Colin Upstill, Michael Wilkinson, Chris Howls (who ran off with my secretary), Duncan O'Dell, Mike Jeffrey, and many others. I insist on acknowledging their contributions to any success I have achieved. In some ways I'm an intellectual loner, avoiding fashionable or over-populated subjects and large collaborations — not because I disparage them (far from it) but because I'm not very competitive and also because I need quiet mental space to get deeply into problems. But even for a loner, science is still an intensely collaborative activity (only nutters think otherwise), and having day-to-day interactions with colleagues who have learned how I think has been indispensable.

This is a rather large department, and I have benefited from that. Even before my Research Professorship began in 1988, my administrative and teaching loads were light. The wisdom of successive heads of the department has protected us from tortures from the managerialists and the accountabilitarians. My secretaries have been quietly supportive: Lilian for many years, Dianne (the lean years, for those who remember her), Vicki, Abla, Jenny, Maggie, and now Tracie. And also indispensable have been the other support staff: from our administrators to porters and cleaners, all working in the background to help us do our work. I'm grateful to to all of them.

Domestically I've been fortunate too — though perhaps not as well off financially as if I'd been a binary professor. 'Binary professor'? — that's a professor with zero or one wife. In this subject area lurk multiple mischiefs, so I must mind what I say. I've enjoyed the love of several women who have to their credit tolerated and supported my obsessive absorption in this weird activity. You scribble for months, write papers incomprehensible to most human beings, and magically acquire air tickets to the world's exotic places (I just returned from Crimea and Istanbul; next week it's Uzbekistan), dinners in the best restaurants — including this evening — altogether the most delightful life. But this can leave your loved ones feeling (if not being) neglected. For more than a quarter-century, this support has been from Monica, who has her own scientific life as many in this department know, and who in spite of not being the world's most patient person has tolerated me. At the risk of embarrassing you, I thank you publicly now.

You haven't seen the last of me. I'll still work full-time in the department when I'm not flying here and there singing my songs while people still want to hear them and while the juices still flow. I dreamed, and still dream, of making one piece of physics ('making' is the right word — we aim to discover things about the world, but our theories are human creations nevertheless) — making one piece of physics as beautiful as a single note from Louis Armstrong's trumpet. But I recognise the dangers of self-delusion: we all know, but don't always admit, this logical truth: after a certain age, every day that passes increases the probability that our best work is behind us. Balazs Gyorffy might recall that many years ago he and I made a pact: to each tell the other if we detect signs of flakiness or intellectual senility.

My genes are not auspicious. I've outlived most of the men in my family by many years. So even though I'm nowhere near 100, as the jazz musician Eubie Blake claimed to be (mistakenly, as it turned out), I echo what he said: "If I'd known I would live this long, I'd have taken better care of myself." So far so good. Thank you all.

Night thoughts of a theoretical physicist

Michael Berry
H H Wills Physics Laboratory, Tyndall Avenue, Bristol BS8 1TL

Originally written in 2000 to celebrate the 75th birthday of John Ziman (1925–2005)[13] but not published. This version has been very slightly updated.

For more than forty years, I have been able to make physics with delight and enthusiasm. In large measure, I owe this privilege to John Ziman's protection and mentoring during my early years in Bristol. We never worked on the same physics, and what I did then was unappreciated for a long time, so it remains a mystery how he could divine that I might have a spark of intellect, and shield this from potentially hostile winds so that it had the chance to grow into a flame.

Mindful of John's transmutation from a physicist to one who seeks to elucidate the complex web of social interactions associated with science, I note another transmutation. Those who administer science in practical ways have changed: from people who saw it as their duty to serve our interests, into members of a coherent profession, not always sympathetic to what we do. This development has ominous aspects, and my aim here is to draw attention to a few of them.

In the 1950s, there was a cartoon showing a collection of girls on a beach. Ignoring a hunk with bulging muscles, the busty bimbos in bikinis clustered admiringly round a weedy little fellow in glasses reading a book on nuclear physics. How embarrasing are the stereotypes of this cameo from those distant days: women as decorations, men with muscles as stupid, glasses as an indicator of nerdy intellectuality, nuclear physics as a route to sexual gratification... I draw particular attention to the picture of science as an activity to be admired for the power it represents and gives. The promise of unlimited cheap energy, and the actuality of the bomb, made it seem that the physics of the nucleus would transform our lives just as the physics of electromagnetism had done (and is still doing). As we all know, neither the fears nor the hopes were realised.

Modern reactions are more complicated, but include a current of instinctive revulsion against science as something that threatens us. This particular current is not new. Isaiah Berlin[1] has documented a similar one in the late eighteenth-century romantic reaction against science, summed up in Hamann's phrase "The tree of knowledge has destroyed the tree of life". From nowadays, I quote Bryan Appleyard[2], who when a child was astonished by his father's ability to calculate the volume of water in a water-tower, but "sensed something ominous in this wisdom", and Fay Weldon[3] "We all did science at school. We all know that when our experiments came up 'wrong' they were simply overlooked, ignored. The scientists just can't face the notion of a variable universe. We can." This is silly, of course, but Appleyard and Weldon are not silly people and in part they were motivated by an understandable reaction to a type of science popularisation that Appleyard calls the 'crude ahistorical arrogance' of claims for theories of everything.

It ought not to be necessary any longer to argue for what is good about science, what makes it worth doing. But in these postmodern (or is it postpostmodern?) days it is still true that:

• Science is one of the few activities connecting nations, cultures, religions — somewhat like sport, but with the crucial difference that co-operation, rather than competition, is strongly built in. In science we usually speak about our foreign colleagues, rather than our foreign competitors. National boundaries are irrelevant. I am indifferent to, and often ignorant of, the nationality of those whose work connects with mine. This is a healthy antidote to newly erupting nationalisms. (To avoid misunderstanding, I emphasize that I am referring to the attitudes of most scientists, and not to journalists and research councils, who from ignorance or mischief — put excessive stress on the competitive and national aspects of science.)

• Science is a model for the rational and civilised exploration of disagreements, largely without rancour and in a way that leads to progress. This is a healthy antidote to fundamentalisms, now on the rise worldwide.

• From science come inspiring and magical connections between very different things. This observation counters one of our commonest criticisms: that by the reductionist disarticulation of the world into its parts, which are then studied separately, we lose the sense of the whole. One of my favourite connections starts with the question: Why is matter hard? Atoms consist mostly of empty space, so why doesn't matter squash down, with all the electrons collapsing into their lowest quantum energy states near the nuclei? Because this is prevented by the Pauli exclusion principle: no two electrons can be in the same state. And where does that come from? It could well originate in a property of rotation in three-dimensional space[4]: holding a glass of wine, you can turn it completely twice (that is, through 720°) and find at the end of this contortion that your arm is untwisted (this does not work for a single turn). I find that 'two into none' connection, that totally unexpected association of microscopic hardness with geometry[5], miraculous.

• And now descending from the sublime to the mundane, science brings economic benefits. Some people, narrow of vision, call this 'the real world'. There have been reports[6] on physics-based industries showing how profitable they are, relative to others. I will not describe that in detail, but instead will give a small example from my own experience, to show the surprising ways such benefits might accrue. In one of his lovely films, David Attenborough showed insects floating on a sunlit pond, and pointed out their curious shadows: unlike more familiar shadows, these underwater ones have bright edges. The reason is that surface tension bends the water near where the insects float, and the light is sharply focused by these curved surfaces. This prompted a systematic study[7] of bright shadows, including the similar ones cast on the bottoms of rivers by little whirlpools on the surface. In these shadows, the light focuses onto a ring. That was in 1983. The paper was noticed by Michael Gorman, a physicist at Houston. It inspired him to make a plastic lens of unusual shape, whose function is to mimic whirlpools with their ring focusing. He surprised me by announcing that he had patented this construction with the hope of profiting from it. For example, angioplastic surgeons were interested in shining a powerful laser through a tiny version of the lens, at the end of an optical fibre, and with the ring focus bore holes through blocked arteries. (At my time of life, this application is close to my heart.)

Notwithstanding the economic transformations wrought by applied science, it remains true that none of us does fundamental science for the money. (By 'fundamental' I mean what is — somewhat disparagingly — called 'curiosity-driven' research — to be contrasted with the greed-driven variety, I suppose.) Some scientists may be more business minded, but this is always a small part of what drives them. I have always been surprised and grateful to live in a society so civilised as to pay me anything at all to pursue these obsessions in complete freedom. Support for science is a fragile miracle; we cannot expect non-scientists will automatically understand why it should continue. Science is not a natural activity for most people. I would not find it easy to argue the case for supporting me with an unemployed lone parent, or with someone from our recreated beggar class.

Over the decades during which I have been in science, my union, the Association of University Teachers (now the Universty and College Union), has always pressed vigorously for higher salaries, and of course I have benefited from their successes. But it is hard for me to support such wage claims with an easy conscience, because they are based on the biased application of relativities. The arguments often run like this: my salary has slipped relative to that of workers in profession A, or is much less than I would get in country B, so I ought get more. But nobody argues that this means the A workers get too much, or points out that there is a country C where scientists get much less than we do, and takes this to imply we are paid too much. There seems no way to establish the absolute value of dreamers (or anybody else for that matter).

Let us assume, however, that it is greater than zero — that is, that fundamental science, pursued without thought of applications, is regarded as worthwhile. Even small science costs money — not much, but the amounts are large in comparison to what individual scientists can afford from their own pockets. In the traditional university system (of course I mean the one I grew up in) small projects were often paid for out of departmental funds. This was possible because we had what came to be called 'the well-found laboratory'. That no longer exists. Departmental funds for research have been cut and cut and cut. There never was much flesh — let alone fat — and now we are well into the bone. So somebody who has a bright or wild idea, and wants to abandon everything else to pursue it, must either be a theorist (we can change course quickly) or else get a grant. This takes time, which destroys a flexibility that the old system allowed.

In spite of these negative indications, the situation is not completely bleak. It is sometimes asserted (for example by the organisation Save British Science, with whose views I usually agree) that modern granting agencies have a built in bias towards 'safe' projects with a well-defined timetable and assured outcome, and against chancy proposals. Certainly the application forms for grants, with their demands for 'timelines' and 'deliverables' reinforce this view. But I must say that my (admittedly limited) experience is different. I have never used 'chanciness' as an argument against a project and have never heard, on any funding committee, anybody else do so. Nor have my own more speculative projects been rejected for this reason. People can be more sensible than the systems they have to work with.

There is an unfortunate trend in our universities (and indeed throughout the professions) towards greater bureaucracy. Hardly a week passes without my head of department getting a demand for elaborately quantified 'information'; it is not clear who is being informed or to what worthwhile use the data will be put. He tries hard to protect us — the rank and file — from these demands, but inevitably some of it filters down. How many papers did we publish last year? Itemise the list into twelve categories — yes, twelve! Which are the best ones? How could we contract our group into 25% less space, because the Bursar's formula proves that we have too much. How much time do we spend on research? On teaching? Administering? A senior scientist in what used to be the Royal Signals and Radar establishment at Malvern told me that their management required his day to be broken down into units of six minutes. On the first try, 25% of the laboriously collected entries were found to have been wrongly typed in and after two months there was still no reliable printout of the results. Do the questioners not know that many of our best ideas come at random times: standing in front of students, under the shower, filling in idiotic questionnaires...?

In recent years, universities have acquiesced in having several new layers of bureaucracy imposed on them. One is the Research Assessment (I resist the half-apologetic and self-mocking modifier 'Exercise', carrying the suggestion that the activity is merely a practice run for the real thing). Another is the recent Teaching Quality Assessment (TQA), where 'swat teams' of judges require the preparation of several filing-cabinets-worth of paperwork in case it might be required during the few days of their visit. Those who conduct these investigations are our colleagues, polite, well-meaning, respectable, occasionally distinguished. But I see little if any good in their activities, and much harm. The deputy head of one mathematics department had to retire early as the result of a nervous breakdown caused by the demands of the TQA. In many departments, research planning has degenerated into short-term strategies aimed at 'doing well' in the next assessment. Unforgivably, this has infected the highest levels of academe with the perception that the purpose of intellectual activity is to get high scores in tests — a vulgar quiz-show measure of mental achievement, the very same 'examinitis', that every serious thinker regards with derision when it breaks out among students. Fortunately, sensible voices[8] begin to advocate radical downsizings of the two assessments.

When William Waldegrave was our UK minister of science, I grumbled to him about the growing bureaucracy of checking-up. I think he appreciated the point, but his response was a bit lame: "It's the spirit of the age: everything must be measured." They — the tormentors, I mean — think this kind of assessment is both democratic — because it is a way of reporting back to the taxpayer — and scientific — because it is done with numbers. I was able to give him a delicious example — a masterpiece of false quantification — of how ridiculous it can be in practice. This is from an annual report of the National Physical Laboratory[9, 10], from the section called "Performance Targets". Target number three was "Research Milestones". Everything must be defined, and indeed we have: "Detailed definition of measure: Research milestones are those which are agreed with the customer". I am not joking. The goal was 0.49 research milestones per scientist per year, to be increased by 3% for each scientist over a four-year period. In the first year, 0.48 research milestones were achieved. A near miss; must do better. In the second year, though, it had improved to 0.79. Now comes the giveaway: "The comfortable achievement of this target, while pleasing, probably indicates the unsatisfactory nature of this measure. A revised target will be set...". In successive years, the number of milestones increased, until[10]: "... NPL is making increasing use of project management techniques which encourage the breaking down of programmes into modules, and in which the measurement of success by milestones is *even more significant* than it was at the time when the target was set" [my italics].

I suppose a salary was wasted employing somebody to produce that. It is not only rubbish: it is dishonest. If you doubt this, try to imagine an organisation reporting: "Last year we failed to meet any of our targets." Of course that would never happen. It is all uncomfortably reminiscent of the old Soviet five-year plans, glossed up by a modern advertising agency: nobody can ever fail and, as a very British touch, nobody should succeed too well either.

The minister listened quietly, but looked a bit uncomfortable. A week later, I discovered why: he had just presented the National Physical Laboratory with that year's Price Waterhouse award for the best annual report.

For many of my colleagues, this amounts to harassment. It could be fatal. Successful basic research involves intense concentration, undistracted, over long periods. Newton was asked how he managed to penetrate so deeply into fundamental problems, and replied "by thinking continuously about them". Even those of us with full-time research positions know how easy it is for prime thinking time to become marginalized. There are so many other legitimate demands refereeing papers and people, writing talks, etc. Of course, we all develop defensive strategies — into the bin, unread, goes anything from the University Industrial Liaison Officer, all Eurobumf etc.

This threatened and actual degradation of many scientists' working lives seems the unintended outcome of an ideological cocktail whose ingredients are *accountability*, *measurability* and a *business-based model* of efficiency.

It is hard to quarrel with the view that because we are spending public money we ought to be able to justify what we do with it. And 'accountability' is a mantra, spreading through our national life. Simply to utter it is to make any case appear unanswerable. But it cannot be right that by a sort of Heisenberg uncertainty principle the justification interferes with the research process itself — and the 'Planck constant' for science (as for creative activity of almost any kind) is pretty large. Yet there is an answer to the accountabilitarians, a single unfashionable word: trust.

The new 'science' of objective measurement of research output seems based on the hope that this can be as precise as crystallography or nuclear magnetic resonance. I doubt this, and in any case do not see how it can replace intuition about the promise of a young scientist or a new idea. Quantitative research assessment is a new profession in a rapid growth phase; it will probably stabilise, but at a size large enough to be a nuisance. On the scale of intellectual respectability, it probably beats astrology, but graphology and mechanised lie-detection spring uncomfortably to mind. Along with the measurables (citations, impact factors, H-indices...), comes the pretence of precision, with a vocabulary of empty evocations: the different universities have almost identical 'mission statements' declaring their 'commitment' to 'excellence', to 'quality'.

Worst of all is the business efficiency paradigm. I am sure many academics regard this with particular hatred. One manifestation is that we are judged by how much money we bring into the University in the way of grants, etc. This distorted image of universities as businesses is poisoning relations between administrations and academics. In former times, administrations were small, for the most part invisible, and, in my experience at least, unfailingly obliging when we needed their help. As individuals, university administrators still do try to be helpful. But they are beginning to regard us differently. Academics who do not attract — or even seek — grants are increasingly regarded as a charge on the others — even though, as we all know, such individuals can play a very valuable part in the life of a department. In an unguarded moment, a former registrar at my university referred to those she served as her 'customers'. To administrators, we no longer belong to departments; physics is a 'cost centre'. Am I alone in finding this a repulsive image, reflecting a grotesque misunderstanding of the nature and meaning of our activities? Universities are not businesses. When they try to act as businesses they are usually not very successful; in my own university there have been several costly failures at profit-making. In fact the proper business of a university, or a department, should not be business at all. What we are paid to 'produce' in a physics department is not profits (in the form of grants or sponsorship) but physics.

To avoid confusion, I should declare that I have no objection to scientists or anyone else getting rich from their intellectual or entrepreneurial efforts. Closer links between universities and businesses are welcome too; Bristol University's collaboration with Hewlett-Packard's Basic Research Institute for Mathematical Sciences (sadly short-lived) was admired worldwide, and bore splendid intellectual fruit. Commercialization is obnoxious only when forced on us, when we are told that "Scientific excellence in its own right is not enough", that we should see "the spread of the enterprise culture across the university", and that, even if we are dismal failures at moneymaking, we can at least pretend: "if the university could present what it already does in the language of enterprise, it will benefit"[11].

I do not wish to exaggerate. It is still a wonderful privilege to do science in Britain. It is easy to find the grass greener elsewhere, especially across the Atlantic, but my experience of the hard to summarise (because much more varied) American situation is that they are suffering many of the same problems (it seems that management fads are adopted here just when they have been shown to fail there). There is a trickle of distinguished reverse brain drainers, which could one day exceed the conventional one. An optimistic sign here is the appointment (albeit half-hearted and occasional) of ministers responsible for science. Usually, they have little if any first hand knowledge of what we do, and in any case this is the British way: appoint an amateur. But at least one minister did write some remarkably good sense about the fundamental aims of science. I hope some politician will have the courage and vision to reverse the futilities being imposed on us — futile, because repeated reorganizations that follow each other faster than the response time of the academic system can result only in innovation exhaustion.

Finally, here is an unfashionable thought, to annoy those who would continue to 'reform' our institutions of higher learning. It was inspired by a biography[12] of the mathematician Ramanujan. While in his creative prime, Ramanujan lacked employment. According to a friend who helped him get work in the port office in Madras, what he needed was a job without too many duties, that would give him leisure. The biographer points out that he was using that word in a different sense from today's. To us, it implies something trivial: 'freedom from' work. But for Ramanujan it carried the positive connotation of 'freedom to' — in his case, create mathematics. It is leisure in this sense that we need to reclaim, and then preserve, in our universities.

References

1. Berlin, I. The crooked timber of humanity: chapters in the history of ideas (John Murray, London, 1990).
2. Appleyard, B. Understanding the present: science and the soul of modern man (Picador, London, 1992).
3. Weldon, F. Thoughts we dare not speak aloud *Daily Telegraph*, London, 2 December 1991, p. 14.
4. Hartung, RW. Pauli principle in Euclidean geometry. *Amer. J. Phys.* 1979, **47**: 900–910.
5. Berry, MV, Robbins, JM. Indistinguishability for quantum particles: spin, statistics and the geometric phase. *Proc. R. Soc. A.* 1997, **453**: 1771–1790.
6. Campbell, P. PBIs Lauded, *Physics World* 1991, **5**, p. 7.
7. Berry, MV, Hajnal, JV. The shadows of floating objects and dissipating vortices. *Optica Acta* 1983, **30**: 23–40.
8. Cox, D, Halsey, AH. Simple but radical, we think these changes would help, *The Guardian*, London 16 July 1999.
9. NPL. National Physical Laboratory Annual Report and Accounts 1991–1992.
10. NPL. National Physical Laboratory Annual Report and Accounts 1994–1995.
11. Bristol University *Newsletter*, 22 April 1999, p. 2.
12. Kanigel, R. The man who knew infinity: a life of the genius Ramanujan (Scribner's, London and Sydney, 1991).
13. Berry, M V and Nye, J F, John Michael Ziman *Biographical Memoirs of the Royal Society*, **52**, 479–491.

Les Déchiffreurs: Voyage en mathématiques

eds: Jean-François Dars, Annick Lesne & Anne Papillault
published in French by Éditions Belin, 2008. pp. 134-5

Michael Berry

H H Wills Physics Laboratory, Tyndall Avenue, Bristol BS8 1TL, United Kingdom
http://www.phy.bris.ac.uk/people/berry_mv/index.html

The arcane in the mundane

How delightful to discover our abstractions clothing Nature's realities:

Singularities of smooth gradient maps in rainbows and tsunamis
The Laplace operator in oriental magic mirrors
Elliptic integrals in the polarization pattern of the clear blue sky
Geometry of twists and turns in quantum indistinguishability
Matrix degeneracies in overhead-projector transparencies
Gauss sums in the light beyond a humble diffraction grating

More fundamentally, we are repeatedly astonished to find, recently-developed and quietly waiting, exactly the mathematics we need for physics: Riemannian geometry awaiting general relativity, matrices awaiting quantum physics, fibre bundles awaiting gauge theories of fundamental forces... Should we be astonished? I think not. We are beings of finite intelligence in an infinite inscrutable universe. In science, our individual intelligences cooperate, and we can understand more. But still, we are able to comprehend only those structures in the natural world that mirror our mental constructs. And at any stage of humanity's development, the most sophisticated constructs are those of our mathematics. Therefore our deepest penetration into the natural world is limited by our latest mathematics. As mathematics develops, more subtle features of the universe become accessible to our understanding. While our species survives, I see no end to this process — no 'Theory of everything'.

So, "The unreasonable effectiveness of mathematics in the natural sciences" is not unreasonable at all; on the contrary, it is inevitable. Not unreasonable, but wonderful!

Reprinted from Les Déchiffreurs: Voyage en mathématiques, Éditions Belin (2008).

Chapter 10

Awards, Honorary Degrees, etc.

1978 Maxwell Medal and Prize of the Institute of Physics
1982 Elected to the Royal Society of London
1983 Elected Fellow of the Royal Society of Arts
1983 Elected Fellow of the Royal Institution
1986 Elected Member of the Royal Society of Sciences, Uppsala
1989 Elected member of the European Academy
1990 Julius Edgar Lilienfeld Prize of the American Physical Society
1990 Paul Dirac Medal and Prize of the Institute of Physics
1990 Elected to Indian Academy of Sciences
1990 Royal Medal of the Royal Society
1991 Doctor of Science degree from Exeter University
1993 Naylor Prize, London Mathematical Society
1994 Moët-Hennessey Louis-Vuitton Science for Art Science pour l'Art Prize (Paris)
1994 Honorary Professorship, Wuhan University
1995 Hewlett-Packard Europhysics Prize
1995 Elected Foreign Member of the National Academy of Science of the USA
1995 Elected Member of the London Mathematical Society
1996 Dirac Medal and Prize of the International Centre for Theoretical Physics, Trieste
1996 Honorary Doctorate, Trinity College Dublin
1996 Knight Bachelor, Queen's Birthday Honours, 15 June
1997 Kapitsa Medal of the Russian Academy of Sciences
1997 Lego Master Builder Badge **(see end picture)**
1997 Honorary Doctorate, Open University
1998 Wolf Prize in Physics
1998 Elected a Governor of the Weizmann Institute
1998 Honorary Doctorate, St Andrews University
1998 Honorary Doctorate, University of Warwick
1999 Honorary Fellow, Institute of Physics
2000 Elected Foreign Member of Royal Netherlands Academy of Arts and Sciences
2000 Ig Nobel Prize in Physics
2001 Onsager Medal (Norwegian Technical University, Trondheim)
2001 Honorary Doctorate, University of Ulm
2002 Novartis/Daily Telegraph 'Visions of Science' competition, 1st prize (Science as Art), 3rd prize (Science Concepts)
2003 Honorary Doctorate, Weizmann Institute, Rehovot, Israel
2005 Elected to Royal Society of Edinburgh
2005 Pólya Prize, London Mathematical Society

2006	Chancellor's Medal, University of Bristol
2006	Honorary Doctorate, Technion, Haifa, Israel
2007	Honorary Doctorate, University of Glasgow
2009	Honorary Doctorate, Universities of Metz and Nancy
2010	Elected first Honorary Member of the Mexican Mathematical Society
2012	Distinguished Visiting Professor, Marymount University, Virginia
2012	Honorary Doctorate, Russian-Armenian (Slavonic) University, Yerevan
2013	Honorary Doctorate, University of Hyderabad, India
2014	Richtmyer Memorial Lecture Award, American Association of Physics Teachers
2015	Lorentz Medal, Royal Netherlands Academy of Arts and Sciences
2016	Moyal Medal, Macquarie University, Australia
2017	Honorary Fellowship, University of Bristol

Chapter 11
Publications

When citing these papers, please give the full original publication details as listed below.

1965

1 Berry, M. V., Clark, R. C., & Rijnierse, P. J., 1965 'Note on the invariance of the phase difference between two waves', *Proc. Phys. Soc.* **86**, 242–244.

2 Berry, M. V., 1965 'The diffraction of light by ultrasound', Ph.D. Thesis, St Andrews University.

1966

3 Berry, M. V., 1966 *The diffraction of light by ultrasound* (Academic Press).

4 Berry, M. V., 1966 'Solution of the Raman-Nath Equation for light diffracted by ultrasound at normal incidence', *Physica* **32**, 1582–1590.

5 Berry, M. V., 1966 'Semiclassical scattering phase shifts in the presence of metastable states', *Proc. Phys. Soc.* **88**, 285–292.

6 Berry, M. V., 1966 'Uniform approximation for potential scattering involving a rainbow', *Proc. Phys. Soc.* **89**, 479–490.

1967

7 Lloyd, P., & Berry, M. V., 1967 'Wave propagation through an assembly of spheres IV. Relations between different multiple scattering theories', *Proc. Phys. Soc.* **91**, 678–688.

8 Berry, M. V., 1967 *Uniformly approximate solutions for short-wave problems.* (SERC research report).

1968

9 Berry, M. V., 1968 'Liquid Structure Theory', *Engineering Outline* 152, 779–782.

1969

10 Berry, M. V., 1969 ‘Uniform approximation: a new concept in wave theory’, *Science Progress* (Oxford) **57**, 43–64.

11 Berry, M. V., 1969 ‘Uniform approximations for glory scattering and diffraction peaks’, *J. Phys. B* **2**, 381–392.

12 Berry, M. V., 1969 ‘Rheology’, *Engineering Outline* **187**, 121–124.

1970

13 Berry, M. V., 1970 ‘Liquid surface physics’, *Engineering Outline* **230**, 581–584.

14 Berry, M. V., & Gibbs, D. F., 1970 ‘The interpretation of optical projections’, *Proc. R. Soc. A* **314**, 143–152.

1971

15 Berry, M. V., & Reznek, S. R., 1971 ‘A simple theory for the densities of coexistent liquid and vapour through the transition region’, *J. Phys. A* **4**, 77–84.

16 Berry, M. V., 1971 ‘Diffraction in crystals at high energies’, *J. Phys. C* **4**, 697–722.

17 Berry, M. V., 1971 ‘Transition from quantum to classical theory for HEED’, *Proc. 25th Anniv. meeting Emag Inst. Phys.* (UK), 122–124.

18 Berry, M. V., 1971 ‘The molecular mechanism of surface tension’, *Physics Education* **6**, 79–84.

1972

19 Berry, M. V., 1972 Review of *Introduction to meterological optics* by R. A. R. Tricker, *Science Progress* **60**, 125–128.

20 Berry, M. V., Durrans, R. F., & Evans, R., 1972 ‘The calculation of surface tension for simple liquids’, *J. Phys. A.* **5**, 166–170.

21 Berry, M. V., 1972 'On deducing the form of surfaces from their diffracted echoes', *J. Phys. A* **5**, 272–291.

22 Berry, M. V., 1972 'Reflections on a Christmas-tree bauble', *Physics Education* **7**, 1–6.

23 Berry, M. V., & Mount, K. E., 1972 'Semiclassical approximations in wave mechanics', *Reps. Prog. Phys.* **35**, 315–397.

24 Doyle, P. A., & Berry, M. V., 1972 'Semiclassical prediction of increased penetration near certain voltages', *Proc. EMCON* **72**, 452–453.

25 Berry, M. V., & Buxton, B. F., 1972 'A new interpretation of bend contours in terms of semiclassical mechanics', *Proc. EMCON* **72**, 454–455.

26 Berry, M. V., & Ozorio de Almeda, A., 1972 'Towards a manageable theory for cross-grating HEED', *Proc. EMCON* **72**, 456–457.

27 Nye, J. F., Berry, M. V., & Walford, M. E. R., 1972 'Measuring the change in thickness of the Antarctic ice sheet', *Nature* **240**, No. 97, 7–9.

28 Berry, M. V., 1972 Review of *Relativistic Quantum Theory*, by Lifshitz *et al.*, *Times Higher Education Supplement* (24.11.1972).

1973

29 Berry, M. V., 1973 'The statistical properties of echoes diffracted from rough surfaces', *Phil. Trans. R. Soc. A* **273**, 611–654.

30 Berry, M. V., & Doyle, P. A., 1973 'Interpreting electron micrographs of amorphous solids', *J. Phys. C* **6**, L6–9.

31 Doyle, P. A., & Berry, M. V., 1973 'Absorption and penetration in the semiclassical theory of high-energy electron diffraction', *Z. für Naturforsch* **28a**, 571–576.

32 Berry, M. V., & Ozorio de Almeida, A. M., 1973 'Semiclassical approximation of the radial equation with two-dimensional potentials', *J. Phys. A* **6**, 1451–1460.

33 Berry, M. V., Buxton, B. F., & Ozorio de Almeida, A. M., 1973 'Between wave and particle — the semiclassical method for interpreting high-energy electron micrographs in crystals', *Radiation effects* **20**, 1–24.

1974

34 Nye, J. F., & Berry, M. V., 1974, *Proc. R. Soc. A* **336**, 165–190, 'Dislocations in wave trains'.

35 Berry, M. V., 1974 'Simple fluids near rigid solids — statistical mechanics of density and contact angle', *J. Phys. A* **7**, 231–245.

36 Atkinson, P., & Berry, M. V., 1974 'Random noise in ultrasonic echoes diffracted by blood', *J. Phys. A* **7**, 1293–1302.

37 Buxton, B. F., & Berry, M. V., 1974 'A general theory of the critical voltage effect' in *High-voltage electron microscopy* (eds: Swann *et al.*), Academic Press, 60–63.

38 Berry, M. V., 1974 Review of *The optics of rays, wavefronts and caustic*, by O. N. Stavroudis, *Science Progress* **61**, 595–597.

1975

39 Berry, M. V., & Greenwood, D. A., 1975 'On the ubiquity of the sine wave', *Am. J. Phys.* **43**, 91.

40 Berry, M. V., 1975 Review of *Introductory eigenphysics* by C A Croxton, *Nature* **254**, 465.

41 Berry, M. V., 1975 'Cusped rainbows and incoherence effects in the rippling-mirror model for particle scattering from surfaces', *J. Phys. A* **8**, 566–584.

42 Berry, M. V., 1975 Review of *Gravitation*, by G W Misner, K Thome and J A Wheeler, *Science Progress* **62**, 356–360.

43 Berry, M. V., 1975 Review of 'Quantum Physics and Ordinary Language' by T Bergstein, *Times Higher Educational Supplement*.

44 Berry, M. V., 1975 'Theory of radio echoes from glacier beds', *Journal of Glaciology* **15**, 65–74.

45 Berry, M. V., 1975 'Liquid Surfaces', IAEA-SMR-15/9, *Surface Science* **Vol. 1**, 291–327.

46 Berry, M. V., 1975 Review of *System Identification: method and applications* by H H Kagiwada, *Science Progress* **62**, 638–640.

47 Berry, M. V., 1975 'Attenuation and focussing of electromagnetic surface waves rounding gentle bends', *J. Phys. A* **8**, 1952–1971.

1976

48 Berry, M. V., 1976 'Waves as catastrophes', *Physics Bulletin* (March), (107–108).

49 Berry, M. V., 1976 'Waves and Thom's theorem', *Advances in Physics* **25**, 1–26.

50 Berry, M. V., 1976 Review of *Twentieth Century Physics* by Joseph Norwood, *Nature* **261**, 174.

51 Berry, M. V., & Tabor, M., 1976 'Closed orbits and the regular bound spectrum', *Proc. R. Soc. A* **349**, 101–123.

52 Buxton, B. F., & Berry, M. V., 1976 'Bloch wave degeneracies in systematic high energy electron diffraction', *Phil. Trans. R. Soc. A* **282** (No. 1308), 485–525.

53 Berry, M. V., 1976 Essays on Maxwell and Einstein, in *The greatest thinkers* (ed: E. de Bono), Weidenfeld and Nicolson.

54 Berry, M. V., 1976 *Principles of cosmology and gravitation*, Cambridge University Press.

1977

55 Berry, M. V., & Tabor, M., 1977 'Calculating the bound spectrum by path summation in action-angle variables', *J. Phys. A* **10**, 371–379.

56 Berry, M. V., & Nye, J. F., 1977 'Fine structure in caustic junctions', *Nature* **267**, 34–36.

57 Berry, M. V., & Mackley, M. R., 1977 'The six roll mill: unfolding an unstable persistently extensional flow', *Phil. Trans. R. Soc. A* **287** (No. 1337), 1–16.

58 Berry, M. V., 1977 'Focusing and twinkling: critical exponents from catastrophes in non-Gaussian random short waves', *J. Phys. A* **10**, 2061–2081.

59 Navascues, G., & Berry, M. V., 1977 'The statistical mechanics of wetting', *Mol. Phys.* **34**, 649–664.

60 Berry, M. V., & Hannay, J. H., 1977 'Umbilic points on Gaussian random surfaces', *J. Phys. A* **10**, 1809–1821.

61 Berry, M. V., & Tabor, M., 1977 'Level clustering in the regular spectrum', *Proc. R. Soc. A* **356**, 375–394.

62 Berry, M. V., 1977 'Semi-classical mechanics in phase space: a study of Wigner's function', *Phil. Trans. R. Soc. A* **287**, 237–271.

63 Berry, M. V., 1977 'Remarks on degeneracies of semiclassical energy levels', *J. Phys. A* **10**, L193–194.

64 Berry, M. V., 1977 'Regular and irregular semiclassical wavefunctions', *J. Phys. A* **10**, 2083–2091.

65 Berry, M. V., 1977 about 200 entries on physics, in *The Fontana dictionary of modern thought*, (eds: A. Bullock and O. Stallybrass) Collins/Fontana.

66 Berry, M. V., 1977 'Catastrophe theory: a new mathematical tool for scientists', *J. Sci. Ind. Res.* (New Delhi) **36** No. 3, 103–105.

66a Berry, M. V., 1977 Letter in *Nature* **270**, 382–383.

1978

67 Berry, M. V., 1978 'Disruption of wavefronts: statistics of dislocations in incoherent Gaussian random waves', *J. Phys. A* **11**, 27–37.

68 Berry, M. V., 1978 Review of 'Catastrophe theory: selected papers 1972–1977' by E C Zeeman, *Nature* **271**, 486.

69 Berry, M. V., 1978 Review of Elements of wave propagation in random media by B J Uscinski, *Physics Bulletin* **29**, 177.

70 Berry, M. V., & Hannay, J. H., 1978, *Nature* **273**, 573 Comment on 'Topography of Random Surfaces' by R S Sayles and T R Thomas, *Nature* **271**, 431–434, 1978.

71 Navascues, G., & Berry, M. V., 1978 'A statistical-mechanical theory for the solid-liquid interface', in *Wetting, spreading and adhesion* (ed: Padday), Academic Press, 83–92.

72 Berry, M. V., 1978 Review of *Catastrophe theory: the revolutionary new way of understanding how things change* by A Woodcock and M Davis, *Nature* **274**, 930.

73 Berry, M. V., 1978 Review of *Catastrophe theory and its applications* by T Poston and I Stewart, *Nature* **275**, 75–76.

74 Berry, M. V., 1978 'Les Jeux de lumière dans l'eau', *La Recherche* **92**, 760–768.

75 Berry, M. V., 1978 'Catastrophes in semiclassical mechanics', in *Rencontre de CARGESE sur les singularité es et leurs applications*' (ed: F Pham), 1975, 133–135.

76 Berry, M. V., 1978 'Regular and Irregular Motion', in *Topics in Nonlinear Mechanics* (ed: S Jorna), *Am. Inst. Ph. Conf. Proc.* No. **46,** 16–120.

1979

77 Berry, M. V., & Balazs, N. L., 1979 'Evolution of semiclassical quantum states in phase space', *J. Phys. A* **12**, 625–642.

78 Berry, M. V., & Balazs, N. L., 1979 'Nonspreading wave packets', *Am. J. Phys.* **4**, 264–267.

79 Berry, M. V., Nye, J. F., & Wright, F. J., 1979 'The elliptic umbilic diffraction catastrophe', *Phil. Trans. R. Soc. A* **291**, 453–484.

80 Berry, M. V., 1979 'Diffractals', *J. Phys. A* **12**, 781–797.

81 Berry, M. V., 1979 'Catastrophe and fractal regimes in random waves', in *Structural stability in physics* (eds: W. Güttinger and H. Eikemeier), Springer, 43–50.

82 Berry, M. V., 1979 'Distribution of modes in fractal resonators', in *Structural stability in physics* (eds: W. Güttinger and H. Eikemeier), Springer, 51–53.

83 Berry, M. V., 1979 'Catastrophe and stochasticity in semiclassical quantum mechanics', in *Structural stability in physics* (eds: W. Güttinger and H. Eikemeier), Springer, 122–125.

84 Berry, M. V., Balazs, N. L., Tabor, M., & Voros, A., 1979 'Quantum maps', *Ann. Phys. N.Y.* **122**, 26–63.

85 Berry, M. V., 1979 'Forms of Light', *The Sciences* **19**, No. 8, 18–20.

1980

86 Berry, M. V., & Wright, F. J., 1980 'Phase-space projection identities for diffraction catastrophes', *J. Phys. A* **13**, 149–160.

87 Berry, M. V., 1980 'Quantization of mappings and other simple classical models', in *Nonlinear Dynamics* (ed: R. H. G. Helleman), *Ann. N.Y. Acad. Sci.* Vol. **357**, 183–202.

88 Berry, M. V., 1980 'Some geometric aspects of wave motion: wavefront dislocations, diffraction catastrophes, diffractals', in *Geometry of the Laplace operator* (eds: R Osserman and A Weinstein), *Proc. Symp. App. Maths* **36**, AMS, 13–28.

89 Berry, M. V., & Upstill, C., 1980 'Catastrophe optics: morphologies of caustics and their diffraction patterns', *Progress in Optics XVIII*, 257–346.

90 Berry, M. V., & Lewis, Z. V., 1980 'On the Weierstrass-Mandedlbrot fractal function', *Proc. R. Soc. A* **370**, 459–484.

91 Berry, M. V., 1980 'L' Atmosfera come laboratorio di ottica', *L' Astronomia* **5**, 35–38.

92 Berry, M. V., 1980 Review of *Lattice Path Counting and Applications* by Sri Gopal Mohanty, *Nature* **285**, 597.

93 Berry, M. V., 1980 Review of *Bifurcation Theory and Applications in Scientific Disciplines* (eds: O. Gürel and O. E. Rössler), *Nature* **286**, 191.

94 Berry, M. V., 1980 Review of *Works on the Foundations of Statistical Physics*, by N S Krylov, *Nature* **286**, 542.

95 Hannay, J. H., & Berry, M. V., 1980 'Quantization of linear maps on a torus — Fresnel diffraction by a periodic grating', *Physica* **1D**, 267–290.

96 Berry, M. V., Chambers, R. G., Large, M. D., Upstill, C., & Walmsley, J. C., 1980 'Wavefront dislocations in the Aharonov-Bohm effect and its water wave analogue', *Eur. J. Phys.* **1**, 154–162.

97 Berry, M. V., 1980 'Exact Aharonov-Bohm wave function obtained by applying Dirac's magnetic phase factor', *Eur. J. Phys.* **1**, 240–244.

1981

98 Berry, M. V., 1981 'Quantizing a classically ergodic system: Sinai's billiard and the KKR method', *Ann. Phys.* **131**, 163–216.

99 Richens, P. J., & Berry, M. V., 1981 'Pseudo-integrable systems in classical and quantum mechanics', *Physica* **1D**, 495–512.

100 Berry, M. V., & Shepherd, P. J., 1981 'Physics and Weapons', *Physics Bulletin*, 238.

101 Berry, M. V., 1981 'A curious multifoliate caustic in the magnetic Green function', *Eur. J. Phys.* **2**, 22–28.

102 Berry, M. V., 1981 'Regularity and chaos in classical mechanics, illustrated by three deformations of a circular billiard', *Eur. J. Phys.* **2**, 91–102.

103 Berry, M. V., & Blackwell, T. M., 1981 'Diffractal echoes', *J. Phys. A* **14**, 3101–3110.

104 Korsch, H. J., & Berry, M. V., 1981 'Evolution of Wigner's phase-space density under a non-integrable quantum map', *Physica* **3D**, 627–636.

105 Berry, M. V., 1981 'Singularities in Waves', in *Les Houches Lecture Series Session XXXV* (eds: R. Balian, M. Kléman and J-P Poirier), North-Holland: Amsterdam, 453–543.

1982

106 Berry, M. V., 1982 'Wavelength-independent fringe spacing in rainbows from falling neutrons', *J. Phys. A* **15**, L385–388.

107 Berry, M. V., 1982 Review of *Semiclassical approximation in quantum mechanics* by V. P. Maslov and M. V. Fedoriuk, *Phys. Bull* **33**, 241.

108 Berry, M. V., 1982 'Universal power-law tails for singularity-dominated strong fluctuations', *J. Phys. A* **15**, 2735–2749.

109 Berry, M. V., 1982 Review of '*The Accidental Universe*' by Paul Davies, *Nature* **300**, 133–134.

110 Berry, M. V., 1982 'Semiclassical weak reflections above analytic and nonanalytic potential barriers', *J. Phys. A* **15**, 3693–3704.

1983

111 Berry, M. V., & Hajnal, J. V., 1983 'The shadows of floating objects and dissipating vortices', *Optica Acta* **30**, 23–40.

112 Berry, M. V., 1983 Review of *The fractal geometry of nature* by B. B. Mandelbrot, *New Scientist* **97** No. 1342, 250.

113 Berry, M. V., 1983 Review of *Image analysis and mathematical morphology* by J. Serra, *Physics Bulletin* **34**, 252.

114 Walker, J. G., Berry, M. V., & Upstill, C., 1983 'Measurement of twinkling exponents of light focused by randomly rippling water', *Optica Acta* **30**, 1001–1010.

115 Berry, M. V., 1983 'Semiclassical Mechanics of Regular and Irregular Motion', in *Les Houches Lecture Series Session XXXVI* (eds: G. Iooss, R. H. G. Helleman and R. Stora), North Holland: Amsterdam, 171–271.

116 Berry, M. V., 1983 Review of *Regular and stochastic motion* by A. J. Lichtenberg and M. A. Lieberman, *Nature* **305**, 456.

117 Berry, M. V., Hannay, J. H., & Ozorio de Almeida, A. M., 1983 'Intensity moments of semiclassical wavefunctions', *Physica* **8D**, 229–242.

1984

118 Berry, M. V., 1984 'Structures in semiclassical spectra: a question of scale', in *The Wave-Particle Dualism* (eds: S. Diner, D. Fargue, G. Lochak, F. Selleri), D. Reidel, 231–252.

119 Berry, M. V., & Wilkinson, M., 1984 'Diabolical points in the spectra of triangles', *Proc. R. Soc. A* **392**, 15–43.

120 Berry, M. V., 1984 'Quantal phase factors accompanying adiabatic changes', *Proc. R. Soc. A* **392**, 45–57.

121 Wright, F. J., & Berry, M. V., 1984 'Wave-front dislocations in the sound-field of a pulsed circular piston radiator', *J. Acoust. Soc. Amer.* **75**, 733–748.

122 Berry, M. V., 1984 'The adiabatic limit and the semiclassical limit', *J. Phys. A* **17**, 1225–1233.

123 Berry, M. V., & Klein, G., 1984 'Newtonian trajectories and quantum waves in expanding force fields', *J. Phys. A* **17**, 1805–1815.

124 Berry, M. V., 1984 'Incommensurability in an exactly-soluble quantal and classical model for a kicked rotator', *Physica* **10D**, 369–378.

125 Berry, M. V., Indekeu, J. O., Tabor, M., & Balazs, N. L., 1984 'Nonlocal maps', *Physica* **11D**, 1–24.

126 Berry, M. V., & Robnik, M., 1984 'Semiclassical level spacings when regular and chaotic orbits coexist', *J. Phys. A* **17**, 2413–2421.

127 Berry, M. V., 1984 'Comment on "New Representation of Quantum Chaos"', *Physics Letters* **104A**, 306–309.

128 Berry, M. V., 1984 Review of *Sunsets, twilights and evening skies*, by Aden and Marjorie Meinel, *Phys. Bull.* **35**, 338.

129 Berry, M. V., 1984 Review of *Universality in chaos* ed. by Predrag Cvitanovic, *Phys. Bull.* **35**, 437.

130 Berry, M. V., 1984 Obituary of P A M Dirac, *Bristol University Newsletter* 15 November, 2.

131 Berry, M. V., 1984 'Patterns from nature', *Input* **37**, 1164–1171.

1985

132 Berry, M. V., 1985 'Classical adiabatic angles and quantal adiabatic phase', *J. Phys. A* **18**, 15–27.

133 Berry, M. V., 1985 Review of *Symplectic techniques in physics* by Victor Guillemin and Shlomo Sternberg, *Phys. Bull.* **36**, 177.

134 Berry, M. V., 1985 'Aspects of Degeneracy', in *Chaotic behavior in quantum systems,* (ed: Giulio Casatil), Plenum: New York, 123–140.

135 Berry, M. V., 1985 'Quantum, classical and semiclassical adiabaticity', in *Theoretical and Applied Mechanics* (eds: F I Niordson and N Olhoff), Elsevier North Holland: Amsterdam, 83–96.

136 Robnik, M., & Berry, M. V., 1985 'Classical billiards in magnetic fields', *J. Phys. A* **18**, 1361–1378.

137 Berry, M. V., 1985 Review of *An idiot's fugitive essays on science: methods, criticism, training, circumstances* by C Truesdell, *Nature* **315**, 779.

138 Tanner, L. H., & Berry, M. V., 1985 'Dynamics and optics of oil hills and oilscapes', *J. Phys. D* **18**, 1037–1061.

139 Berry, M. V., 1985, 'A problem in semiclassical adiabatic theory', in *Méthodes Semiclassiques en Méchanique Quantique* (eds: B. Helffer et D. Robert), Publications de L'Université de Nantes, 23–27.

140 Berry, M. V., 1985 'Semiclassical theory of spectral rigidity', *Proc. R. Soc. A* **400**, 229–251.

141 Berry, M. V., 1985 Interview in *Corriere della Provincia* (Como), September 16.

142 Berry, M. V., 1985 Review of *Deterministic chaos* by H. G. Schuster, *Nature* **318**, 241.

143 Berry, M. V., 1985 'Ipotesi di scala e fluttuazioni non gaussiane nella teoria catastrofica della onde', (eds: Paolo Bisogno, Augusto Forti), *Prometheus* **1**, 41–79 (Italian translation of 'Scaling and non-Gaussian fluctuations in the catastrophe theory of waves').

1986

144 Berry, M. V., 1986 'Twinkling exponents in the catastrophe theory of random short waves', in *Wave propagation and scattering* (ed: B. J. Uscinsci), Oxford Clarendon Press, 11–35.

145 Berry, M. V., & Robnik, M., 1986 'Statistics of energy levels without time-reversal symmetry: Aharonov-Bohm chaotic billiards', *J. Phys. A* **19**, 649–668.

146 Robnik, M., & Berry, M. V., 1986 'False time-reversal violation and energy level statistics: the role of anti-unitary symmetry', *J. Phys. A* **19**, 669–682.

147 Berry, M. V., & Mondragon, R. J., 1986 'Diabolical points in one-dimensional Hamiltonians quartic in the momentum', *J. Phys. A* **19**, 873–885.

148 Berry, M. V., & Robnik, M., 1986 'Quantum states without time-reversal symmetry: wavefront dislocations in a non-integrable Aharonov-Bohm billiard', *J. Phys. A* **19**, 1365–1372.

149 Berry, M. V., & Percival, I. C., 1986 'Optics of fractal clusters such as smoke', *Optica Acta* **33**, 577–591.

150 Berry, M. V., 1986 'Spectral zeta functions for Aharonov-Bohm quantum billiards', *J. Phys. A* **19**, 2281–2296.

151 Berry, M. V., 1986 Review of *The beauty of fractals: images of complex dynamical systems* (by H-O Peitgen and P. H. Richter) Springer, *Nature* **323**, 590.

152 Berry, M. V., 1986 Review of *Symmetry: Unifying Human Understanding*, ed. by Istvan Hargittai, *New Scientist* 16 October, 60.

153 Berry, M. V., 1986 'The unpredictable bouncing rotator: a chaology tutorial machine', in *Dynamical systems: a renewal of mechanism* (eds: S. Diner, D. Fargue, G. Lochak), World Scientific, 3–12.

154 Berry, M. V., 1986 'Riemann's zeta function: a model for quantum chaos?', in *Quantum chaos and statistical nuclear physics* (eds: T. H. Seligman and H. Nishioka), Springer Lecture Notes in Physics No. 263, 1–17.

155 Berry, M. V., 1986 'Fluctuations in numbers of energy levels', in *Stochastic processes in classical and quantum systems* (eds: S. Albeverio, G. Casati and D. Merlini), Springer Lecture Notes in Physics No. 262, 47–53.

156 Berry, M. V., 1986 'Adiabatic phase shifts for neutrons and photons', in *Fundamental aspects of quantum theory* (eds: V. Gorini and A. Frigerio), Plenum, NATO ASI series vol. 144, 267–278.

157 Berry, M. V., 1986 'The Aharonov-Bohm effect is real physics not ideal physics', in *Fundamental aspects of quantum theory* (eds: V. Gorini and A. Frigerio), Plenum, NATO ASI series vol. 144, 319–320.

1987

158 Berry, M. V., 1987 'Disruption of images: the caustic-touching theorem', *J. Opt. Soc. Amer.* **A4**, 561–569.

159 Berry, M. V., 1987 'Interpreting the anholonomy of coiled light', *Nature* **326**, 277–278.

160 Berry, Michael, & Berry, Monica, 1987 Review of T*he problems of mathematics* by Ian Stewart, Oxford Univ. Press, *New Scientist* 2 April, 56.

161 Berry, M. V., & Mondragon, R. J., 1987 'Neutrino billiards: time-reversal symmetry-breaking without magnetic fields', *Proc. R. Soc. A* **412**, 53–74.

162 Berry, M. V., 1987 'Improved eigenvalue sums for inferring quantum billiard geometry', *J. Phys. A* **20**, 2389–2403.

163 Berry, M. V., 1987 'Quantum chaology' (The Bakerian Lecture), *Proc. R. Soc. A* **413**, 183–198.

164 Berry, M. V., 1987 'Quantum phase corrections from adiabatic iteration', *Proc. R. Soc. A* **414**, 31–46.

165 Berry, M. V., 1987 'Quantum physics on the edge of chaos', *New Scientist* 19 November, 44–47.

166 Berry, M. V., 1987 Review of *Chaos: making a new science* by James Gleick, *Nature* **330**, 293–294.

167 Berry, M. V., 1987 'The adiabatic phase and Pancharatnam's phase for polarized light', *J. Mod. Optics* **34**, 1401–1407.

168 Keating, J. P., & Berry, M. V., 1987 'False singularities in partial sums over closed orbits', *J. Phys. A* **20**, L 1139–1141.

169 Berry, M. V., 1987 *Semiclassical chaology, in Quantum measurement and chaos* (eds: E. R. Pike and Sarben Sarkar), Plenum, NATO ASI series B vol. 161, 81–87.

1988

170 Berry, M. V., 1988 Interview 'Chaos and Order', *Cogito* Vol. 2 No. 1, 1–5.

171 Berry, M. V., & Goldberg, J., 1988 'Renormalization of curlicues', *Nonlinearity* **1**, 1–26.

172 Berry, M. V., & Hannay, J., 1988 'Classical non-adiabatic angles', *J. Phys. A* **21**, L325–331.

173 Scharf, R., Dietz, B., Kus, M., Haake, F., & Berry, M. V., 1988 'Kramers' degeneracy and quartic level repulsion', *Europhys. Lett.* **5**, 383–389.

174 Berry, M. V., 1988, 'The electron at the end of the universe', in *A passion for science* (eds: L. Wolpert and A. Richards), Oxford University Press, 38–51.

175 Berry, M. V., 1988 'Semiclassical formula for the number variance of the Riemann zeros', *Nonlinearity* **1**, 399–407.

176 Berry, M. V., 1988 Review of *Mathematics and the unexpected* by Ivar Ekeland, *Nature* **335**, 22.

177 Berry, M. V., 1988 'Classical chaos and quantum eigenvalues', in *Order and Chaos in nonlinear physical systems* (eds: Stig Lundqvist, Norman H. March and Mario P. Tosi), Plenum Press: New York and London, 341–348.

178 Berry, M. V., 1988 'The geometric phase', *Scientific American* **259** (6), 26–34.

179 Berry, M. V., 1988 'Random renormalization in the semiclassical long-time limit of a precessing spin', *Physica D* **33**, 26–33.

180 Berry, M. V., 1988, 'Le dé passement interne des paradignes de la physique classique' (translation — 'Breaking the paradigms of classical physics from within'), in 'Logos et Théorie des catastrophes' (ed: J. Petitot), Editions Patino, Geneva, 106–117.

1989

181 Berry, M. V., 1989 'Uniform asymptotic smoothing of Stokes's discontinuities', *Proc. R. Soc. A* **422**, 7–21.

182 Berry, M. V., 1989 Review of 'The Science of Fractal Images' (eds: H-O Peitgen and D. Saupe), *New Scientist* 11 March, 66.

183 Berry, M. V., 1989 Review of *Journey into Light: Life and Science of C. V. Raman* by G. Venkataraman, *Nature* **338**, 685–686.

184 Berry, M. V., 1989 'Quantum scars of classical closed orbits in phase space', *Proc. R. Soc. A* **423**, 219–231.

185 Berry, M. V., 1989 *Principles of Cosmology and Gravitation* (Adam Hilger) (corrected reprint of item 54).

186 Berry, M. V., 1989 'Studies of Semiclassical Spectra: where next?', in *Atomic Physics II* (eds: S. Haroche, J. C. Gay, S. Grynberg), World Scientific, 277–279.

187 Berry, M. V., 1989 'The quantum phase, five years after', in 'Geometric Phases in Physics' (eds: A. Shapere, F. Wilczek), World Scientific, 7–28.

188 Mondragon, R. J., & Berry, M. V., 1989 'The quantum phase 2-form near degeneracies: two numerical studies', *Proc. R. Soc. A* **424**, 263–278.

189 Berry, M. V., 1989 'Fringes decorating anticaustics in ergodic wavefunctions', *Proc. R. Soc. A* **424**, 279–288.

190 Berry, M. V., 1989 'Stokes' phenomenon; smoothing a Victorian discontinuity', *Publ. Math. of the Institut des Hautes Études Scientifique* **68**, 211–221.

191 Berry, M. V., 1989 'Quantum chaology, not quantum chaos', *Physica Scripta* **40**, 335–336.

192 Berry, M. V., 1989 'Falling fractal flakes', *Physica D* **38**, 29–31.

193 Berry, M. V., 1989 Review of 'Does God play dice: the mathematics of chaos' by Ian Stewart, *Times Higher Ed. suppl.*

194 Berry, M. V., 1989 Review of 'Multiphase Averaging for classical systems' by P. Lochak and C. Meunier, *Bull. Lond. Math. Soc.*

195 Berry, M. V., 1989 Review of 'Buckminster Fuller's Universe: an appreciation', by Lloyd Steven Seiden, *Physics World* (November), 45.

196 Berry, M. V., 1989 'Chaology; the emerging science of unpredictability', *Proc. Roy. Institution of Gt Britain* **61**, 189–204.

1990

197 Berry, M. V., 1990 'Waves near Stokes lines', *Proc. R. Soc. A* **427**, 265–280.

198 Berry, M. V., 1990 'Quantum adiabatic anholonomy', in *Anomalies, phases, defects* (eds: U. M. Bregola, G. Marmo and G. Morandi), Naples: Bibliopolis, 125–181.

199 Berry, M. V., 1990 'Generalized rainbows in wave physics: how catastrophe theory has helped', in *Rainbows and catastrophes* (ed: N. Neskovic), Boris Kidric Institute, Belgrade, 19–23.

200 Berry, M. V., & Howls, C. J., 1990 'Fake Airy functions and the asymptotics of reflectionlessness', *J. Phys. A* **23**, L243–L246.

201 Berry, M. V., 1990 'Histories of adiabatic quantum transitions', *Proc. R. Soc. A* **429**, 61–72.

202 Berry, M. V., 1990 'Catastrophes and waves' (account of [49] as 'Citation Classic'), *Current Contents* **21** (no. 19), 14.

203 Berry, M. V., 1990 'Quantum theory near the classical limit', *Highlights in Physics* (SERC), 50–51.

204 Berry, M. V., & Howls, C. J., 1990 'Stokes surfaces of diffraction catastrophes with codimension three', *Nonlinearity* **3**, 281–291.

205 Berry, M. V., & Lim, R., 1990 'The Born-Oppenheimer electric gauge force is repulsive near degeneracies', *J. Phys. A* **23**, L655–L657.

206 Berry, M. V., 1990 'Geometric amplitude factors in adiabatic quantum transitions', *Proc. R. Soc. A* **430**, 405–411.

207 Berry, M. V., 1990 'Quantum asymptotics of rainbows' (account of [6] as 'Citation Classic'), *Current Contents* **21** (no. 33), 16.

208 Berry, M. V., & Howls, C. J., 1990 'Hyperasymptotics', *Proc. R. Soc. A* **430**, 653–668.

209 Berry, M. V., 1990 'Minds, quantum measurement, and gravity', *Physics World* (October), 21–22.

210 Berry, M. V., & Keating, J. P., 1990 'A rule for quantizing chaos?', *J. Phys. A* **23**, 4839–4849.

211 Berry, M. V., 1990 'Budden and Smith's 'additional memory' and the geometric phase', *Proc. R. Soc. A* **431**, 531–537.

212 Berry, M. V., 1990 'Anticipations of the geometric phase', *Physics Today* **43** (no. 12), 34–40.

213 Berry, M. V., 1990 'Beyond Rainbows', *Current Science* **59**, 1175–1191.

1991

214 Goldberg, J., Smilansky, U., Berry, M. V., Schweitzer, W., Wunner, G. & Zeller, G., 1991 'The parametric number variance', *Nonlinearity* **4**, 1–14.

215 Berry, M. V., 1991 Review of 'The correspondence between Sir George Gabriel Stokes and Sir William Thomson, Baron Kelvin of Largs' ed. David B Wilson, *Physics World* (March), 52.

216 Berry, M. V., 1991 'Heisenberg's Sofa', *Bristol University Newsletter* (21 March), 8.

217 Berry, M. V., 1991 'Bristol Anholonomy Calendar', in 'Sir Charles Frank OBE FRS, an eightieth birthday tribute' (eds: R. G. Chambers, J. E. Enderby, A. Keller, A. R. Lang and J. W. Steeds), Adam Hilger, Bristol, 207–219.

218 Berry, M. V., 1991 'Wave Geometry: A plurality of singularities', in 'Quantum Coherence' (ed: Jeeva S Anandan), World Scientific, 92–98.

219 Berry, M. V., 1991 'What's wrong with these conference proceedings?', *Physics World* (July), 12–13.

220 Berry, M. V., 1991 'Introductory remarks', in 'Quantum Chaos' (eds: H. A. Cerdeira, R. Ramaswamy, M. C. Gutzwiller and G. Casati), World Scientific, vii–viii.

221 Lim, R., & Berry, M. V., 1991 'Superadiabatic Tracking of Quantum Evolution', *J. Phys. A* **24**, 3255–3264.

222 Berry, M. V., 1991 'Infinitely many Stokes smoothings in the Gamma function', *Proc. R. Soc. A* **434**, 465–472.

223 Berry, M. V., & Howls, C. J., 1991 'Hyperasymptotics for integrals with saddles', *Proc. R. Soc. A* **434**, 657–675.

224 Berry, M. V., 1991 'Quantal reflections of classical chaos', in 'Nonlinear and chaotic Phenomena in Plasmas, Solids and Fluids' (eds: W. Rozmus and J. A. Tuszynski), World Scientific, (Abstract) 2.

225 Berry, M. V., 1991 'Stokes phenomenon for superfactorial asymptotic series', *Proc. R. Soc. A* **435**, 437–444.

226 Berry, M. V., 1991 Letter about the green flash, *New Scientist* 30 November.

227 Berry, M. V., 1991 'Some quantum-to-classical asymptotics', in *Les Houches Lecture Series* LII (1989) (eds: M-J Giannoni, A. Voros and J. Zinn-Justin), North-Holland: Amsterdam, 251–304.

1992

228 Berry, M. V., 1992 'Rays, wavefronts and phase: a picture book of cusps', in 'Huygens' principle 1690–1990; theory and applications', (eds: H. Bok, H. A. Ferwerda), North-Holland: Amsterdam, 97–111.

229 Berry, M. V., 1992 'Quantum physics on the edge of chaos', (reprint of item 165), in 'The New Scientist Guide to Chaos' (ed: Nina Hall), Penguin Books: London, 184–195.

230 Berry, M. V., 1992 'Quantum chaology: our knowledge and ignorance', in 'New Trends in Nuclear Collective Dynamics', (eds: Y. Abe, H. Horiuchi and K. Matsuyanagi), *Springer proceedings in Physics* **58**, 177–181.

231 Berry, M. V., 1992 'True Quantum Chaos?' An Instructive Example', in 'New Trends in Nuclear Collective Dynamics' (eds: Y. Abe, H. Horiuchi and K. Matsuyanagi), *Springer proceedings in Physics* **58**, 183–186.

232 Robbins, J. M., & Berry, M. V., 1992 'The geometric phase for chaotic systems', *Proc. R. Soc. A* **436**, 631–661.

233 Berry, M. V., & Keating, J. P., 1992 'A new asymptotic representation for $\zeta(\frac{1}{2}+it)$ and quantum spectral determinants', *Proc. R. Soc. A* **437**, 151–173.

234 Berry, M. V., 1992 'Asymptotics, superasymptotics, hyperasymptotics', in 'Asymptotics beyond all orders' (eds: H. Segur and S. Tanveer), Plenum: New York 1991, 1–14.

235 Berry, M. V., 1992 Review of 'Understanding the Present: Science and the Soul of Modern Man' by Bryan Appleyard, *Physics World* (June), 62–63.

236 Berry, M. V., 1992 'Catastrophe Theory' (original considerably edited), *Encyclopedia Britannica* (15th ed.) Vol. 2, 948.

237 Berry, M. V., 1992 'Chaos' (original considerably edited), *Encyclopedia Britannica* (15th ed.) Vol. 3, 92.

238 Robbins, J. M., & Berry, M. V., 1992 'Discordance between quantum and classical correlation moments for chaotic systems', *J. Phys. A* **25**, L961–L965.

239 Berry, M. V., 1992 Review of 'Genius: Richard Feynman and Modern Physics' by James Gleick, *Physics World* (December), 40.

240 Berry, M. V., 1992 Review of 'Pi in the sky: Counting, Thinking, and Being' by John Barrow, *Nature* **360**, 376–377.

1993

241 Berry, M. V., & Howls, C. J., 1993 'Infinity interpreted', *Physics World* (June), 35–39.

242 Berry, M. V., & Robbins, J. M., 1993 'Classical geometric forces of reaction: an exactly solvable model', *Proc. R. Soc. A* **442**, 641–658.

243 Berry, M. V., & Robbins, J. M., 1993 'Chaotic classical and half-classical adiabatic reactions: geometric magnetism and deterministic friction', *Proc. R. Soc. A* **442**, 659–672.

244 Berry, M. V., & Howls, C. J., 1993 'Unfolding the high orders of asymptotic expansions with coalescing saddles: singularity theory, crossover and duality', *Proc. R. Soc. A* **443**, 107–126.

245 Berry, M. V., & Lim, R., 1993 'Universal transition prefactors derived by superadiabatic renormalisation', *J. Phys. A* **26**, 4737–4747.

246 Berry, M. V., 1993 'Quantum chaology, prime numbers and Riemann's zeta function' (abstract and bibliography), *Inst. Phys. Conf. Ser.* **133**, 133–134.

247 Berry, M. V., 1993 Review of 'Light and color in the Outdoors' by M. G. J. Minnaert, *Physics World* (December), 46–47.

248 Berry, M. V., 1993 Review of 'Lectures on Mechanics' by J. E. Marsden, *Bull. Lond. Math. Soc.* **25**, 411–412.

248a Berry, M. V., 1993 'Visiting Nablus', Bristol University Physics Department occasional newsletter.

1994

249 Berry, M. V., & Howls, C. J., 1994 'Overlapping Stokes smoothings: survival of the error function and canonical catastrophe integrals', *Proc. R. Soc. A* **444**, 201–216.

250 Berry, M. V., 1994 Reprints of papers 202, 49, 79, 114, in 'Selected papers on geometrical aspects of scattering' (ed: P. L. Marston), SPIE Optical Engineering Press: Washington.

251 Berry, M. V., 1994 Review of 'The quark and the jaguar' by Murray Gell-Mann, *Nature* **364**, 529.

252 Berry, M. V., 1994 'Evanescent and real waves in quantum billiards, and Gaussian Beams', *J. Phys. A* **27**, L391–L398.

253 Robbins, J. M., & Berry, M. V., 1994 'A geometric phase for $m = 0$ spins', *J. Phys. A* **27**, L435–L438.

254 Berry, M. V., 1994 Review of 'The beat of a different drum: the life and science of Richard Feynman' by Jagdish Mehra, *Physics World* (August), 49.

255 Berry, M. V., 1994 'Supernumerary ice-crystal halos?', *Applied Optics* **33**, 4563–4568.

256 Berry, M. V., & Wilson, A. N., 1994 'Black-and-white fringes and the colours of caustics', *Applied Optics* **33**, 4714–4718.

257 Berry, M. V., 1994 'Pancharatnam, virtuoso of the Poincaré sphere: an appreciation', *Current Science* **67**, 220–223.

258 Berry, M. V., & Keating, J. P., 1994 'Persistent current flux correlations calculated by quantum chaology', *J. Phys. A* **27**, 6167–6176.

259 Berry, M. V., 1994 'Michael Berry, un géomètre des ondes' (interview), *La Recherche* **269**, 1066–1067.

260 Berry, M. V., 1994 'Asymptotics, singularities and the reduction of theories', *Proc. 9th Int. Cong. Logic, Method., and Phil. of Sci. IX* (eds: D. Prawitz, B. Skyrms and D. Westerståhl), 597–607.

261 Berry, M. V., & Howls, C. J., 1994 'High orders of the Weyl expansion for quantum billiards: resurgence of periodic orbits, and the Stokes phenomenon', *Proc. R. Soc. A* **447**, 527–555.

262 Berry, M. V., 1994 'Faster than Fourier', in *Quantum Coherence and Reality; in celebration of the 60th Birthday of Yakir Aharonov* (eds: J. S. Anandan and J. L. Safko), World Scientific: Singapore, 55–65.

1995

263 Berry, M. V., 1995 'Some two-state quantum asymptotics', in *Fundamental Problems of Quantum Theory* (eds: D. M. Greenberger and A. Zeilinger), *Ann. N.Y. Acad. Sci.* **755**, 303–317.

264 Berry, M. V., 1995 'Quantum mechanics, chaos and the Riemann zeros', in *Quantum systems: new trends and methods* (eds: A O Barut, I D Feranchuk, Ya M Shnir and L M Tomil'chik), World Scientific, 387–392.

265 Berry, M. V., 1995 'The Riemann-Siegel expansion for the zeta function: high orders and remainders', *Proc. R. Soc. A* **450**, 439–462.

266 Berry, M. V., 1995, 'Natural Focusing', in 'The Artful Eye' (eds: Richard Gregory, John Harris, Priscilla Heard and David Rose), Oxford University Press, 311–323.

267 Berry, M. V., 1995 Review of 'Nature's Numbers: Discovering Order and Pattern in the Universe' by Ian Stewart, *Physics World* (December), 52–53.

268 Berry, M. V., 1995 'Three comments on the Aharonov-Bohm effect', in 'Advances in Quantum Phenomena' (eds: E. S. Beltrametti and J. M. Levy-Leblond), Plenum: New York and London, NATO ASI Series B: Physics Vol. 347, 353–354.

1996

269 Berry, M. V., & Klein, S., 1996 'Geometric phases from stacks of crystal plates', *J. Mod. Opt.* **43**, 165–180.

270 Berry, M. V., & Klein, S., 1996 'Colored diffraction catastrophes', *Proc. Natl. Acad. Sci. USA* **93**, 2614–2619.

271 Berry, M. V., 1996 'The Levitron^{TM}: an adiabatic trap for spins', *Proc. R. Soc. A* **452**, 1207–1220.

272 Berry, M. V., 1996 'Levitron Physics', explanatory leaflet distributed with the Levitron by *Fascinations Toys and Gifts* (Seattle).

273 Berry, M. V., & Morgan, M. A., 1996 'Geometric angle for rotated rotators, and the Hannay angle of the world', *Nonlinearity* **9**, 787–799.

274 Berry, M. V., & Klein, S., 1996 'Integer, fractional and fractal Talbot effects', *J. Mod. Opt.* **43**, 2139–2164.

275 Berry, M. V., 1996 'Quantum fractals in boxes', *J. Phys. A* **29**, 6617–6629.

1997

276 Berry, M. V., 1997 Review of 'Would-be worlds: how simulation is changing the frontiers of science', by John Casti, *Nature* **385**, 33.

277 Berry, M. V., 1997 Interview by Nina Hall: 'Caustics, catastrophes' and quantum chaos', *Nexus News* February, 4–5.

278 Berry, M. V., & Sinclair, E. C., 1997 'Geometric magnetism in massive chaotic billiards', *J. Phys. A* **30**, 2853–2861.

279 Berry, M. V., 1997 Interview with M Postma and M Sallé, *Afleidung* (Amsterdam) 3 (April), 11–13.

280 Berry, M. V., 1997 Report 'A professorial riddle' on Möbius spring ring, *Lego Review* January, 10.

281 Berry, M. V., & Klein, S., 1997 'Transparent mirrors: rays, waves and localization', *Eur. J. Phys.* **18**, 222–228.

282 Berry, M. V., 1997 Oration for Professor Yakir Aharonov, *Bristol University Newsletter* **27** (22 May), 3.

283 Berry, M. V., 1997 'Quantum and optical arithmetic and fractals', in *The Mathematical Beauty of Physics* (eds: J-M Drouffe and J-B Zuber), World Scientific: Singapore, 281–294.

284 Berry, M. V., & Klein, S., 1997 'Diffraction near fake caustics', *Eur. J. Phys.* **18**, 303–306.

285 Berry, M. V., & Geim, A. K., 1997 'Of flying frogs and levitrons', *Eur. J. Phys.* **18**, 307–313.

286 Berry, M. V., & Robbins, J. M., 1997 'Indistinguishability for quantum particles: spin, statistics and the geometric phase', *Proc. R. Soc. A* **453**, 1771–1790.

287 Berry, M. V., & Klein, S., 1997 'Die Farben von Kaustiken: Katastrophen in Regentropfen und Strukturglas', *Phys. Blätt* **53**, 1095–1098.

288 Berry, M. V., 1997 Desert island book review (M Abramowitz and I A Stegun: Handbook of Mathematical Functions), *New Scientist* **2109** (22 November), 50.

289 Berry, M. V., 1997 'Slippery as an eel', Review of 'The Fire Within the Eye' by David Park, *Physics World* 10 (December), 41–42.

290 Berry, M. V., 1997 'Aharonov-Bohm geometric phases for rotated rotators', *J. Phys. A* **30**, 8355–8362.

1998

291 Berry, M. V., 1998 'Paul Dirac: The purest soul in physics', *Physics World* **11** (February), 36–40.

292 Berry, M. V., Foley, J. T., Gbur, G., & Wolf, E., 1998 'Non-propagating string excitations', *Am. J. Phys.* **66**, 121–123.

293 Berry, M. V., & O'Dell, D. H. J., 1998 'Diffraction by volume gratings with imaginary potentials', *J. Phys. A* **31**, 2093–2101.

294 Berry, M. V., Keating, J. P. & Prado, S., 1998 'Orbit bifurcations and spectral statistics', *J. Phys. A* **31**, L245–L254.

295 Berry, M. V., 1998 'Lop-sided diffraction by absorbing crystals', *J. Phys. A* **31**, 3493–3502.

296 Berry, M. V., 1998 'Much ado about nothing: optical dislocation lines (phase singularities, zeros, vortices...)', *Singular optics* (ed: Soskin, M. S.) Frunzenskoe: Crimea, *SPIE* **3487**, 1–5.

297 Berry, M. V., 1998 'Paraxial beams of spinning light', *Singular optics* (ed: Soskin, M. S.) Frunzenskoe: Crimea, *SPIE* **3487**, 6–11.

298 Berry, M. V., 1998 Review of 'Copenhagen' (A play by Michael Frayn), *Nature* **394** (20 August), 735.

299 Berry, M. V., 1998 'Wave dislocations in non-paraxial Gaussian beams', *J. Mod. Optics* **45**, 1845–1858.

300 Berry, M. V., 1998 'Extreme twinkling, and its opposite' (summary), in 'Gravitation and Relativity: at the turn of the millennium' (Proc GR15) (eds: N. Dadhich and J. Narlikar), IUCAA: Pune, 23–24.

301 Berry, M. V., 1998 Foreword to 'Global properties of simple quantum systems — Berry's phase and others' by Hua-Zhong Li (Shanghai Scientific and Technical Publishers).

302 Berry, M. V., 1998 Foreword to 'Introduction to quantum computation' (eds: Hoi Kwong Lo, Sandu Popescu and Tim Spiller), World Scientific.

1999

303 Berry, M. V., Bhandari, R., & Klein, S., 1999 'Black plastic sandwiches demonstrating biaxial optical anisotropy', *Eur. J. Phys.* **20**, 1–14.

303a Berry, M. V., 1999 'A week in Beirut', *Bristol University Newsletter*, Vol. 29 No. 8 (28 January).

304 Berry, M. V., & Bodenschatz, E., 1999 'Caustics, multiply-reconstructed by Talbot interference', *J. Mod. Opt.* **46**, 349–365.

305 Berry, M. V., & O'Dell, D., 1999 'Ergodicity in wave-wave diffraction', *J. Phys. A* **32**, 3571–3582.

306 Berry, M. V., & Keating, J. P., 1999 '$H = xp$ and the Riemann zeros', in 'Supersymmetry and trace formulae: chaos and disorder' (eds: I. V. Lerner and J. P. Keating), Plenum: New York, 355–367.

307 Berry, M. V., & Keating, J. P., 1999 'The Riemann Zeros and Eigenvalue Asymptotics', *SIAM Review* **41**, 236–266.

308 Berry, M. V., 1999 'A theta-like sum from diffraction physics', *J. Phys. A* **32**, L329–L336.

309 Berry, M. V., 1999 'Aharonov-Bohm beam deflection: Shelankov's formula, exact solution, asymptotics and an optical analogue', *J. Phys. A* **32**, 5627–5641.

310 Berry, M. V., & Shelankov, A., 1999 'The Aharonov-Bohm wave and the Cornu spiral', *J. Phys. A* **32**, L447–L455.

311 Berry, M. V., 1999 'Darkness behind the curtain', *Bristol University Newsletter* Vol. 30 No. 1, 12.

312 Berry, M. V., 1999 Foreword to 'Optical Vortices' (eds: M. Vasnetsov and K. Staliunas), New York: Nova Science Publishers.

313 Berry, M. V., 1999 Reprints of papers 158, 255 and 256, in 'On Minnaert's Shoulders: Twenty years of the "Light and Color" Conferences', Optical Society of America CD-ROM (ed: C. L. Adler).

2000

314 Berry, M. V., 2000 'Millennium essay: Making waves in physics. Three wave singularities from the miraculous 1830s', *Nature* **403** (6 January), 21.

315 Berry, M. V., 2000 'Connections', *Impact* (Leiden University Physics Student Magazine) **11**, 12.

316 Berry, M. V., 2000 Review of 'Natural focusing and fine structure of light' by J. F. Nye, *Contemp. Phys.* **41**, 118–119.

317 Berry, M. V., 2000 Review of 'The Genius of Science: A Portrait Gallery of Twentieth-Century Physicists' by A. Pais, *Physics World* June, 56.

318 Berry, M. V., 2000 'Spectral twinkling', in *Proc. Int. School of Physics "Enrico Fermi"*, course **CXLIII** (eds: G. Casati, I. Guarneri and U. Smilansky), IOS Press: Amsterdam, 45–63.

319 Berry, M. V. & Robbins, J. M., 2000 'Quantum Indistinguishability: alternative constructions of the transported basis', *J. Phys. A* **33**, L207–L214.

320 Berry, M. V., Keating, J. P., & Schomerus, H., 2000 'Universal twinkling exponents for spectral fluctuations associated with mixed chaology', *Proc. R. Soc. A* **456**, 1659–1668.

321 Berry, M. V., & Dennis, M. R., 2000 'Phase singularities in isotropic random waves', *Proc. R. Soc. A* **456**, 2059–2079.

322 Berry, M. V., & Robbins, J. M., 2000 'Quantum Indistinguishability: Spin-statistics without Relativity or Field Theory?', in *Spin-Statistics Connection and Commutation Relations* (eds: R. C. Hilborn & G. M. Tino), American Institute of Physics **CP545**, 3–15.

323 Berry, M.V., 2000 'Odessa, little and large', *Bristol University Newsletter* (6 December), 8.

2001

324 Berry, M. V., & Dennis, M. R., 2001 'Polarization singularities in isotropic random waves', *Proc. R. Soc. A* **457**,141–155.

325 Bender, C. M., Berry, M. V., Meisinger, P. M., Savage, V. M., & Simsek, M., 2001 'Complex WKB analysis of energy-level degeneracies of non-Hermitian Hamiltonians', *J. Phys. A* **34**, L31–L36.

326 Berry, M. V., 2001 'Why are special functions special?', *Physics Today* (April), 11–12.

327 Berry, M. V., 2001 'Spectral twinkling: a new example of singularity-dominated strong fluctuations (summary)', *Physica Scripta* **T90**, 15.

328 Berry, M. V., 2001 'Knotted zeros in the quantum states of hydrogen', *Found. Phys.* **31**, 659–667.

329 Berry, M. V., Marzoli, I., & Schleich, W., 2001 'Quantum carpets, carpets of light', *Physics World* (June), 39–44.

330 Berry, M. V., 2001 'Geometry of phase and polarization singularities, illustrated by edge diffraction and the tides', in 'Second International Conference on Singular Optics (Optical Vortices): Fundamentals and Applications', Bellingham: Washington, *SPIE* **4403**, 1–12.

331 Berry, M. V., 2001 'Asymptotics of Evanescence', *J. Modern Optics* **48**, 1535–1541.

332 Berry, M. V., & Dennis, M. R., 2001 'Knotted and linked phase singularities in monochromatic waves', *Proc. R. Soc. A* **457**, 2251–2263.

333 Berry, M. V., & Dennis, M. R., 2001 'Knotting and unknotting of phase singularities: Helmholtz waves, paraxial waves and waves in 2+1 spacetime', *J. Phys. A* **34**, 8877–8888.

334 Berry, M. V., Storm, C., & van Saarloos, W., 2001 'Theory of unstable laser modes: edge waves and fractality', *Optics Commun.* **197**, 393–402.

335 Berry, M. V., 2001 'Spectral twinkling: a new example of singularity-dominated strong fluctuations (summary)', *Physica Scripta* **T90**, 15.

336 Berry, M. V., 2001 'Fractal modes of unstable lasers with polygonal and circular mirrors', *Optics Communications* **200**, 321–330.

337 Berry, M. V., 2001 'Chaos and the semiclassical limit of quantum mechanics (is the moon there when somebody looks?)', in *Quantum Mechanics: Scientific perspectives on divine action* (eds: Robert John Russell, Philip Clayton, Kirk Wegter-McNelly and John Polkinghorne), Vatican Observatory CTNS publications, 41–54.

338 Berry, M. V., 2001 'Indistinguishable spinning particles', in *XIIIth International Congress of Mathematical Physics* (eds: A. Fokas, A. Grigoryan, T. Kibble and B. Zegarlinski), International Press of Boston, 29–30.

2002

339 Berry, M. V., & Keating, J. P., 2002 'Clusters of near-degenerate levels dominate negative moments of spectral determinants', *J. Phys. A* **35**, L1–L6.

340 Berry, M. V., 2002 'Statistics of nodal lines and points in chaotic quantum billiards: perimeter corrections, fluctuations, curvature', *J. Phys. A* **35**, 3025–3038.

341 Berry, M. V., 2002 'Singular Limits', *Physics Today* (May), 10–11.

342 Berry, M. V., 2002 'Exuberant interference: rainbows, tides, edges, (de) coherence ...', *Phil. Trans. R. Soc. A* **360**, 1023–1037.

343 Berry, M. V., 2002 Comments on Stephen Wolfram's 'A new kind of science', *The Daily Telegraph* 15 May, 22.

344 Berry, M. V., & Ishio, H., 2002 'Nodal densities of Gaussian random waves satisfying mixed boundary conditions', *J. Phys. A* **35**, 5961–5972.

345 Bender, C. M., Berry, M. V., & Mandilara, A., 2002 'Generalized PT symmetry and real spectra', *J. Phys. A* **35**, L467–L471.

346 Berry, M. V., 2002 'Coloured phase singularities', *New Journal of Physics* **4**, 66.1–66.14.

347 Berry, M. V., 2002 'Exploring the colours of dark light', *New Journal of Physics* **4**, 74.1–74.14.

348 Berry, M. V., 2002 'Paul Dirac: the purest soul in physics', *Nonesuch* (University of Bristol) Autumn, 22–25 (shortened version of paper 291).

349 Berry, M. V., 2002 'Deconstructing rainbows', Review of 'The Rainbow Bridge: Rainbows in Art, Myth and Science' by Raymond L Lee and Alastair B Fraser, *Physics World* **15** (No. 11), 49.

2003

350 Berry, M. V., 2003 'Mode degeneracies and the Peterman excess-noise factor for unstable lasers', *Journal of Modern Optics* **50** No. 1, 63–81.

351 Berry, M. V., 2003 'Outstanding visions of science', *Excellence in Science* (The Royal Society of London), February, 9.

352 Berry, M. V., 2003 'The Art of Physics: Snapshots from Recent Research', for *re*:search (Bristol University Magazine) March, 3, 5, 7, 9, 11, 13 & 15.

353 Berry, M. V., 2003 'Paraxial beams of spinning light', in *Optical Angular Momentum* (eds: L. Alten, Stephen M. Barnett and Miles J. Padgett), Institute of Physics Publishing, 65–74 (reprint of paper 297).

354 Berry, M. V., 2003 'Making Light of Mathematics', *Bull. Amer. Math. Soc.* **40**, 229–237.

355 Berry, M. V., & Dennis, M. R., 2003 'The optical singularities of birefringent dichroic chiral crystals', *Proc. R. Soc. A* **459**, 1261–1292.

356 Berry, M. V., & Dennis, M. R., 2003 'The singularities of crystal optics', in *Proceedings of ICO conference on Polarization Optics*, University of Joensuu press (Physics), 18–19 (summary of paper 355).

357 Dennis, M. R., & Berry, M. V., 2003 'Polarization singularities in paraxial and nonparaxial fields', in *Proceedings of ICO conference on Polarization Optics*, University of Joensuu Press (Physics), 20.

358 Berry, M. V., 2003 'Quantum Chaology', in *Quantum: a guide for the perplexed* by Jim Al-Khalili, Weidenfeld and Nicolson, 104–105.

2004

359 Berry, M. V., 2004 'Optical vortices evolving from helicoidal integer and fractional phase steps', *J. Optics. A* **6**, 259–268.

360 Berry, M. V., 2004 'Conical diffraction asymptotics: fine structure of Poggendorff rings and axial spike', *J. Optics. A* **6**, 289–300.

361 Berry, M. V., & Dennis, M. R., 2004 'Black polarization sandwiches are square roots of zero', *J. Optics. A* **6**, S24–S25.

362 Berry, M. V., 2004 'The electric and magnetic polarization singularities of paraxial waves', *J. Optics. A* **6**, 475–481.

363 Berry, M. V., 2004 'Sightings' (interview with Felice Frankel), *Amer. Sci.* **92** (no. 3), 268–269.

364 Berry, M. V., & Dennis, M. R., 2004 'Quantum cores of optical phase singularities', *J. Optics. A* **6**, S178–S180.

365 Berry, M. V., 2004 'Riemann-Silberstein vortices for paraxial waves', *J. Optics. A* **6**, S175–S177.

366 Berry, M. V., Dennis, M. R., & Soskin, M. S., 2004 'The plurality of optical singularities', *J. Optics. A* **6**, (Editorial Introduction to special issue).

367 Zafra, R., Bergeman, T., Berry, M. V., Balian, R., & Voros, A., 2004 'Nandor Balazs (obituary)', *Physics Today* (May), 74 and Michael Berry's extended version.

368 Berry, M. V., 2004 'Index formulae for singular lines of polarization', *J. Optics. A* **6**, 675–678.

369 Berry, M. V., 2004 'The study of empirical laws that determine unpredictable events' (reprint of paper 196), in *History and Philosophy of Science for African Undergraduates* (ed: Helen Lauer), Hope Publications: Ibadan Nigeria, Chapter 26, 383–395.

370 Berry, M. V., 2004 'Asymptotic dominance by subdominant exponentials', *Proc. R. Soc. A* **460**, 2629–2636.

371 Berry, M. V., 2004 Transcript of TV interview, in 'Talking Science' by Adam Hart-Davis, John Wiley & Sons: Chichester 2004, 24–40.

372 Berry, M. V., 2004 'Physics of non-Hermitian degeneracies', *Czech. J. Phys.* **54**, 1039–1047.

373 Berry, M. V., Dennis, M. R. & Lee, R. L. Jr., 2004 'Polarization singularities in the clear sky', *New Journal of Physics* **6**, 162 (doi: 10.1099/1367-2630/1/162) includes press release from the journal.

374 Berry M. V., 2004 'Physics for taxi-drivers', *Physics World* (December), 15.

375 Berry, M. V., 2004 'Benefiting from fractals' (A tribute to Benoit Mandelbrot), *Proc. Symp. Pure Mathematics* **72.1**, 31–33.

375a Berry, M. V., 2004 'Living with Physics', in 'One Hundred Reasons to be a Scientist' (ed: K. Sreenivasan), Abdus Salam Centre for Theoretical Physics: Trieste, 47–49.

2005

376 Berry, M. V., 2005 'Tsunami asymptotics', *New Journal of Physics* **7**, 129.

377 Berry, M. V., 2005 'Universal oscillations of high derivatives', *Proc. R. Soc. A* **461**, 1735–1751.

378 Berry, M. V., & Ishio, H., 2005 'Nodal-line densities of chaotic quantum billiard modes satisfying mixed boundary conditions', *J. Phys. A* **38**, L513–L518.

379 Berry, M. V., 2005 'The optical singularities of bianisotropic crystals', *Proc. R. Soc. A* **461**, 2071–2098.

380 Ahmed, Zafar, Bender, Carl M., & Berry, M. V., 2005 'Reflectionless potentials and PT Symmetry', *J. Phys. A* **38**, L627–L630.

381 Berry, M. V., Jeffrey, M., & Mansuripur, M. R., 2005 'Orbital and spin angular momentum in conical diffraction', *J. Optics. A: Pure Appl. Opt.* **7**, 685–690.

382 Berry, M. V., 2005 'Phase vortex spirals', *J. Phys. A* **38**, L745–L751.

2006

383 Berry, M. V., 2006 'Oriental magic mirrors and the Laplacian image', *Eur. J. Phys.* **27**, 109–118.

384 Berry, M. V., 2006 Review of 'The equations' by Sandor Bais, *Nature* **2**, 65.

385 Berry, M. V., 2006 'Inaugural editorial', *Proc. R. Soc. A* **462**, 1.

386 Berry, M. V., & Jeffrey, M. R., 2006 'Chiral conical diffraction', *J. Opt. A: Pure Appl. Opt.* **8**, 363–372.

387 Berry, M. V., Jeffrey, M. R., & Lunney, J. G., 2006 'Conical diffraction: observations and theory', *Proc. R. Soc. A* **462**, 1629–1642.

388 Berry, M. V. & Popescu, S., 2006 'Evolution of quantum superoscillations, and optical superresolution without evanescent waves', *J. Phys. A* **39**, 6965–6977.

389 Berry M. V., 'Proximity of degeneracies and chiral points', *J. Phys. A* **39**, 10013–10018.

390 Berry, M. V., 2006 Review of 'Fantastic realities' by Frank Wilczek, *Nature* **442**, 870.

391 Berry, M. V., 2006 Inaugural podcast for Institute of Physics, posted 1 August 2006 http://podcasts.iop.org/index.php?post_id=114098%22.

392 Berry, M. V., & Jeffrey, M. R., 2006 'Conical diffraction complexified: dichroism and the transition to double refraction', *J. Optics A* **8**, 1043–1051.

393 Berry, M. V., 2006 'Optical vorticulture', in 'Topology in Ordered Phases' (eds: Satoshi Tanda, Toyoki Matsuyama, Migaku Oda, Yasuhiro Asano & Kousuke Yakubo), World Scientific, 3–4. Book includes a cd-rom containing all Bristol papers on phase and polarization singularities up to 2005.

394 Berry, M. V., & Nye, J. F., 'John Michael Ziman', in *Biographical Memoirs of the Royal Society* **52**, 479–491.

394a Berry, M. V., 2006 'The interface between mathematics and physics' (Panel discussion), *Irish. Math. Soc. Bulletin* **58**, 33–54.

2007

395 Berry, M. V., & Dennis, M. R., 2007 'Topological events on wave dislocation lines: birth and death of loops, and reconnection', *J. Phys. A* **40**, 65–74.

396 Berry, M. V., 2007 'Vortex-free complex landscapes and umbilic-free real landscapes', *J. Phys. A* **40**, F185–F192.

397 Berry, M. V., 2007 'Wave dislocations threading interferometers', *Proc. R. Soc. A* (online reference doi:10.1098/rspa.2007.1842).

398 Berry, M. V., 2007 'Looking at coalescing images and poorly resolved caustics', *J. Optics A* **9**, 649–657.

399 Berry, M. V., 2007, 'Focused tsunami waves', *Proc. R. Soc. A* **463**, 3055–3071.

400 Berry, M. V., & Jeffrey, M. R., 2007 'Conical diffraction: Hamilton's diabolical point at the heart of crystal optics', *Progress in Optics* **50**, 13–50.

2008

401 Berry, M. V., 2008 'Three quantum obsessions', *Nonlinearity* **21**, T19–T26.

402 Berry, M. V., & McDonald, K. T., 2008 'Exact and geometrical-optics energy trajectories in twisted beams', *J. Optics A* **10**, 035005.

403 Berry, M. V., & Dennis, M. R., 2008 'Boundary-condition-varying circle billiards and gratings: the Dirichlet singularity', *J. Phys. A* **41**, 135203.

404 Berry, M. V., 'Waves near zeros', in *Coherence and Quantum Optics IX* (The Optical Society of America, Washington DC), (eds: N. P. Bigelow, J. H. Eberly & C. R. Stroud Jr.), 37–41.

405 Berry, M. V., 2008 'The arcane in the mundane', in *Les Déchiffreurs: Voyage en mathématiques* (eds: Jean-François Dars, Annick Lesne & Anne Papillault), Éditions Belin: France), 134–135.

406 Berry, M. V., 2008 'Optical lattices with PT symmetry are not transparent', *J. Phys. A* **41**, 244007.

407 Berry, M. V., 2008, 'Divagações nocturnas de um físico teórico', Gazeta de Fisica, May 2008, http://tektix.serveftp.com:8080/gfisica/index.jsp?page=articles&id=52&lang=pt (translation of unpublished article U10).

408 Berry, M. V., & Shukla, P., 2008, 'Tuck's incompressibility function: statistics of zeta zeros and eigenvalues', *J. Phys. A* **41**, 385202.

409 Berry, M. V., 2008 Report and speech at John Ziman plaque unveiling, Physics South-West, November, 5–6.

410 Berry, M. V., & Pollard, B., 2008 'Physics in Bristol', *Phys. Perspect.* **10**, 468–480.

411 Berry, M. V., 2008 'My (nearly) half-century in Bristol', in *100. A collection of words and images to mark the Centenary of the University of Bristol* (ed: Barry Taylor, University of Bristol).

2009

412 Berry, M. V., & Dennis, M. R., 2009 'Natural superoscillations in monochromatic waves in D dimension', *J. Phys. A* **42**, 022003.

413 Berry, M. V., 2009 'Hermitian boundary conditions at a Dirichlet singularity: the Marletta-Rozenblum model', *J. Phys. A* **42**, 165208.

414 Berry, M. V., 2009 'Optical currents', *J. Optics. A* **11**, 004001.

415 Berry, M. V., 2009 'Transitionless quantum driving', *J. Phys. A* **42**, 365303 (9pp).

416 Berry, M. V., 2009 'John Michael Ziman', in the *Oxford Dictionary of National Biography*.

417 Berry, M. V., & Shukla, P., 2009 'Spacings distributions for real symmetric 2x2 generalized Gaussian ensembles', *J. Phys. A* **42**, 485102 (13pp).

2010

418 Berry, M. V., 2010 'Editorial', *Proc R. Soc. A* **466**, 1–2.

419 Berry, M. V., & Shukla, P., 2010 'High-order classical adiabatic reaction forces: slow manifold for a spin model', *J. Phys. A* **43**, 045102 (27pp).

420 Berry, M. V., 2010 'Geometric phase memories', *Nature Physics* **6**, 148–150.

421 Berry, M. V., & Howls, C. J., 2010 'Integrals with coalescing saddles', Chapter 36 of the NIST Digital Library of Mathematical Functions (eds: Frank W. J. Olver, Daniel W. Lozier, Ronald F. Boisvert & Charles W. Clark), Cambridge University Press. Available online at http://dlmf.nist.gov/

422 Berry, M. V., 2010 "Some reflections after the meeting" in 'Introductory section of Special Issue on Spin-Statistics' *Foundations of Physics* **40**, 681–683.

423 Berry, M. V., 2010 'Conical diffraction from an N-Crystal cascade', *J. Opt.* **12**, 075704 (8pp).

424 Berry, M. V., 2010 'Horse calculus', *Annals of improbable research* **16** no. 4, 10–11) (corrigendum *J. Opt.* **12** (2010) 089801).

425 Berry, M. V., & Howls, C. J., 2010 'Axial and focal-plane diffraction catastrophe integrals', *J. Phys. A* **43**, 375206 (13pp).

426 Berry, M. V., 2010, 'Aptly named Aharonov-Bohm effect has classical analogue, long history', Letter in *Physics Today* (August), 8.

427 Berry, M. V., & Popescu, S., 2010 'Semifluxon degeneracy choreography in Aharonov-Bohm billiards', *J. Phys. A* **43**, 354005 (11pp).

428 Berry, M. V., 2010 'Asymptotics of the many-whirls representation for Aharonov-Bohm scattering', *J. Phys. A* **43**, 354002 (9pp).

429 Berry, M. V., & Shukla, P., 2010 'Typical weak and superweak values', *J. Phys. A* **43**, 354024 (9pp).

430 Berry, M. V., 2010 'After-dinner remarks at the 60th birthday celebration for Celso Grebogi', in *Nonlinear Dynamics and Chaos: Advances and Perspectives* (eds: Marco Thiel, Jürgen Kurths, Carmen Romero, Alessandro Moura & György Karoly), Springer 2010, 7–9.

431 Berry, M. V., 2010 'Quantum backflow, negative kinetic energy, and optical retro-propagation', *J. Phys. A* **43**, 415302 (15pp).

432 Berry, M. V., 2010 Foreword to 'New Directions in Linear Acoustics and Vibration: Quantum Chaos, Random matrix Theory, and Complexity' (eds: Matthew Wright & Richard Weaver) Cambridge: University Press.

433 Berry, M. V., 2010 Review of 'Reexamining the Quantum-Classical Relation: Beyond Reductionism and Pluralism', by Alisa Bokulich, Brit. *J. Phil. Sci.* **61**, 889–895, doi: 10.1093/bjps/axq022.

2011

434 Berry, M. V., & Shukla, P., 2011 'Slow manifold and Hannay angle for the spinning top', *Eur. J. Phys.* **32**, 115–127.

435 Berry, M. V., 2011 'Lateral and transverse shifts in reflected dipole radiation', *Proc. R. Soc. A* **467**, doi:10.1098/rspa.2011.0081.

436 Berry, M. V., 2011 Review of 'The Beginning of Infinity: Explanations that Transform the World' by David Deutsch, *Times Higher Education*, (31 March), 50–51.

437 Berry, M. V., Dennis, M. R., McRoberts, B., & Shukla, P., 2011 'Weak value distributions for spin ½', *J. Phys. A* **44**, 205301.

438 Berry, M. V., & Dennis, M. R., 2011 'Stream function for optical energy flow', *J. Opt.* **13**, 064004.

439 Berry, M. V., & Cornwell, J., 2011 Obituary of Robert Balson Dingle, published online by the Royal Society of Edinburgh: http://www.royalsoced.org.uk/cms/files/fellows/obits_alpha/dingle_robert.pdf.

440 Berry, M. V., & Keating, J. P., 2011 'A compact Hamiltonian with the same asymptotic mean spectral density as the Riemann zeros', *J. Phys. A* **44**, 285203 (14pp).

441 Berry, M. V., & Uzdin, R., 2011 'Slow non-Hermitian cycling: exact solutions and the Stokes phenomenon', *J. Phys. A* **44**, 435303 (26pp).

442 Berry, M. V., 2011 'Optical polarization evolution near a non-Hermitian degeneracy', *J. Optics* **13**, 115701 (15pp).

443 Berry, M. V., Brunner, N., Popescu, S., & Shukla, P., 2011 'Can apparent neutrino superluminal speeds be explained as a quantum weak measurement?", *J. Phys. A* **44**, 492001 (5pp) (http://arxiv.org/abs/1110.2832).

2012

444 Berry, M. V., 2012 'Editorial: Papers we reject', *Proc. R. Soc. A* **468**, 1 (doi:10.1098/rspa.2011.0564).

445 Berry, M. V., & Shukla, Pragya, 2012 'Pointer supershifts and superoscillations in weak measurements', *J. Phys. A* **45**, 015301 (14pp).

446 Berry, M. V., 2012 'Causal wave propagation for relativistic massive particles: physical asymptotics in action', *Eur. J. Phys.* **33**, 279–294.

447 Berry, M. V., 2012 'Martin Gutzwiller and his periodic orbits', in *The legacy of Martin Gutzwiller*, *Communs. Swiss Phys. Soc.* **37**, 27–30.

448 Berry, M. V., & Dennis, M. R., 2012 'Reconnections of wave vortex lines', *Eur. J. Phys.* **33**, 723–731.

449 Berry, M. V., 2012 'Superluminal speeds for relativistic random waves', *J. Phys. A* **45**, 185308 (14pp).

450 Berry, M. V., & Shukla, Pragya, 2012 'Classical dynamics with curl forces, and motion driven by time-dependent flux', *J. Phys. A* **45**, 305201 (18pp).

451 Berry, M. V., 2012 'Riemann zeros in radiation patterns', *J. Phys. A* **45**, 302001 (9pp).

452 Berry, M. V., 2012 Contribution to 'Tribute to Vladimir Arnold' (eds: B. Khesin and S. Tabachnikov), *Not. AMS.* **59**, 396–398.

453 Berry, M. V., 2012 Contribution to 'Glimpses of Beniot B Mandelbrot (1924–2010)' (eds: M. Barnsley & M. Frame), *Not. AMS.* **59**, 1056–1063.

454 Berry, M. V., 2012 'Hearing the music of the primes: auditory complementarity and the siren song of zeta', *J. Phys. A* **45**, 382001 (7pp).

454a Berry, M. V., 2012 'Beware the double colon', *Ann. Improb. Res.* **18** (no. 6), 2.

2013

455 Berry, M. V., 2013 'Impact and influence: valedictory editorial', *Proc. R. Soc. A* **469**, 20120698.

456 Berry, M. V., & Shukla, Pragya, 2013 'Hearing random matrices and random waves', *New. J. Phys.* **15**, 013026 (11pp).

457 Berry, M. V., 2013 'A note on superoscillations associated with Bessel beams', *J. Opt.* **15**, 044006 (5pp).

458 Berry, M. V., 2013 'Circular lines of circular polarization in three dimensions, and their transverse-field counterparts', *J. Opt.* **15**, 044024 (5pp).

459 Berry, M. V., 2013 'Much ado about rather little', *Learned Publishing* **26**, 77.

460 Berry, M. V., 2013 'Classical limits', in *The Theory of the Quantum World (Proceedings of the 25th Solvay Physics Conference on Physics)* (eds: David Gross, Marc Henneaux & Alexander Sevrin), World Scientific: Singapore, 52–54.

461 Berry, M. V., 2013 'Exact nonparaxial transmission of subwavelength detail using superoscillations', *J. Phys. A* **46**, 205203 (15pp).

462 Berry, M. V., 2013 Review of 'Time Reborn: From the Crisis of Physics to the Future of the Universe' by Lee Smolin, in *Times Higher Education* (27 June), 50.

463 Berry, M. V., 2013 'Five momenta', *Eur. J. Phys.* **34**, 1337–1348.

464 Berry, M. V., 2013 'Curvature of wave streamlines', *J. Phys. A* **46**, 395202 (6pp).

465 Berry, M. V., 2013 'Raman and the mirage revisited: confusions and a rediscovery', *Eur. J. Phys.* **34**, 1423–1437.

466 Berry, M. V., & Shukla, Pragya, 2013 'Physical curl forces: dipole dynamics near optical vortices', *J. Phys. A* **46**, 422001 (9pp).

467 Berry, M. V., 2013 'Superoscillations, endfire and supergain', in *Quantum Theory: a Two-Time Success Story; Yakir Aharonov Festschrift* (eds: D. Struppa & J. Tollaksen), Springer, 327–336.

468 Barnett, S. M., & Berry, M. V., 2013 'Superweak momentum transfer near optical vortices', *J. Opt.* **15**, 125701 (6pp).

2014

469 Berry, M. V., & Shukla, Pragya, 2014 'Superadiabatic forces on a dipole: exactly soluble model for a vortex field', *J. Phys. A* **47**, 125201 (16pp).

470 Berry, M. V., 2014 'Remembering Akira Tonomura', in *In Memory of Akira Tonomura: Physicist and Electron Microscopist*, (eds: K. Fujikawa and Y. A. Ono), World Scientific: Singapore, 30–32.

471 Baeriswyl, D., Berry, M. V., & Vollhardt, D., 2014 'Martin Charles Gutzwiller', *Physics Today* (June) 60.

472 Berry, M. V., 2014 'A tribute to Frank Olver (1924-2013)', in *Analysis and Applications* **12** No. 4, ix–x.

473 Berry, M. V., & Moiseyev, N., 2014 'Superoscillations and supershifts in phase space: Wigner and Husimi function interpretations', *J. Phys. A* **47**, 315203 (14pp).

474 Berry, M. V., 2014 Foreword to *Reductionism, Emergence and Levels of Reality: The Importance of being Borderline* by Sergio Chibbaro, Lamberto Rondoni & Angelo Vulpiani, Springer: Heidelberg, New York, Dordrecht & London, vii–viii.

2015

475 Berry, M. V., & Shukla, Pragya, 2015 'Hamiltonian curl forces', *Proc. R. Soc. A* **471**, 20150002 (13pp).

476 Berry, M. V., 2015 'Nature's optics and our understanding of light', *Contemporary Physics* (celebrating the International Year of Light) **56**, 2–16.

477 Aiello, Andrea, & Berry, M. V., 2015 'Note on the helicity decomposition of spin and orbital optical currents', *J. Optics* **17**, 062001 (4pp).

478 Berry, Michael, 2015 'Chasing the Silver Dragon', *Physics World* (July), 45–47.

479 Berry, M. V., 2015 'The squint Moon and the witch ball', *New J. Phys.* **17**, 060201 (11pp).

480 Berry, M. V., 2015 Review of 'Einstein, his space and times' by Steven Gimbel (Yale University Press), in *Jewish Renaissance* (July 2015), 60.

481 Berry, M. V., 2015 'Riemann zeros in radiation patterns: II. Fourier transforms of zeta', *J. Phys. A* **48**, 385203 (8pp).

482 Berry, M. V., & Howls, C. J., 2015 'Divergent series: taming the tails', in *The Princeton Companion to Applied Mathematics* (ed: N. J. Higham) Princeton, NJ: Princeton University Press, 634–640.

2016

483 Berry, M. V., 2016 'Riemann's saddle-point method and the Riemann-Siegel formula', for: The Legacy of Bernhard Riemann After One Hundred and Fifty Years, International Press (USA) and Higher Education Press (China), eds: Shing-Tung Yau, Frans Oort and Lizhen Ji, 69–78.

484 Berry, M. V., 2016 Contribution to 'Our hero: how Einstein shaped our lives', *Jewish Renaissance* (April 2016), 15.

485 Berry, M. V., 2016 'The elliptic integral in the sky', in 'Findings on Light' (eds: Hester Aardse & Astrid Alben), Lars Müller Publishers, 78–81.

486 Berry, M. V., & Shukla, Pragya, 2016 'Curl force dynamics: symmetries, chaos and constants of motion', *New. J. Phys.* **18**, 063018.

487 Berry, M. V., 2016 'Forward-smooth high-order uniform Aharonov-Bohm asymptotics', *J. Phys. A* **49**, 305204 (16pp).

488 Berry, M. V., & Baeriswyl, D., 2016 'Martin C. Gutzwiller, 1925–2014' submitted to *Biographical Memoirs of the National Academy of Sciences of the USA*, 18pp. http://www.nasonline.org/publications/biographical-memoirs/memoir-pdfs/gutzwiller-martin.pdf.

489 Berry, M. V., 2016 Oration for Professor Tim Palmer CBE FRS, University of Bristol.

490 Berry, M. V., 2016 'Representing superoscillations and narrow Gaussians with elementary functions', *Milan Journal of Mathematics* **84**, 217–230.

491 Berry, M. V., 2016 Reprints of papers 120 and 181, in *Wolf Prize in Physics* (ed: Tsvi Piran), World Scientific: Singapore, 474–536.

492 Berry, M. V., 2016 Interview about Paul Dirac, for Tom Brothwell's Bristol History Podcast, episode 7: https://soundcloud.com/bristolhistorypodcast.

2017

493 Berry, M. V., & Dennis, M. R., 2017 'Vortices, natural and deliberate' in 'Roadmap on structured light' (eds: Halina Rubinsztein-Dunlop & Andrew Forbes), *J. Optics* **18**, 013001.

494 Berry, M. V., 2017 'Suppression of superoscillations by noise', *J. Phys. A* **50**, 025003 (9pp).

495 Berry, M. V., 2017 'Approaches to our history', letter to *Physics Today* (March), 11–12.

496 Berry, M. V., 2017 'Stable and unstable Airy-related caustics and beams', *J. Optics* **19**, 055601 (7pp).

497 Berry, M. V., 2017 'Laplacian magic windows', *J. Optics* **19**, 06LT01.

498 Berry, M. V., 2017 'Dingle's self-resurgence formula', *Nonlinearity* **30**, R25–R31.

499 Berry, M. V., & Morley-Short, S., 2017 'Representing fractals by superoscillations', *J. Phys. A* **50**, 22LT01.

Unpublications

U1 Oration for Yakir Aharonov on receiving a Doctor of Science *honoris causa*, 1997.

U2 Funeral eulogy for Sir Charles Frank, 1998.

U3 Address at Balazs Györffy memorial event, 2012.

U4 Getting a knighthood, 1996.

U5 Wolf Prize acceptance speech, 1998.

U6 Honorary degree acceptance response (Weizmann Institute), 2003.

U7 Convocation address (Cornell University), 2000.

U8 Retirement speech, 2008.

U9 'Emotional asymptotics' (poem), 2000.

U10 'Night thoughts of a theoretical physicist', 1999.

U11 Address on John Ziman Memorial Day, 2005.

U12 Tribute to Richard Gregory, 2010.

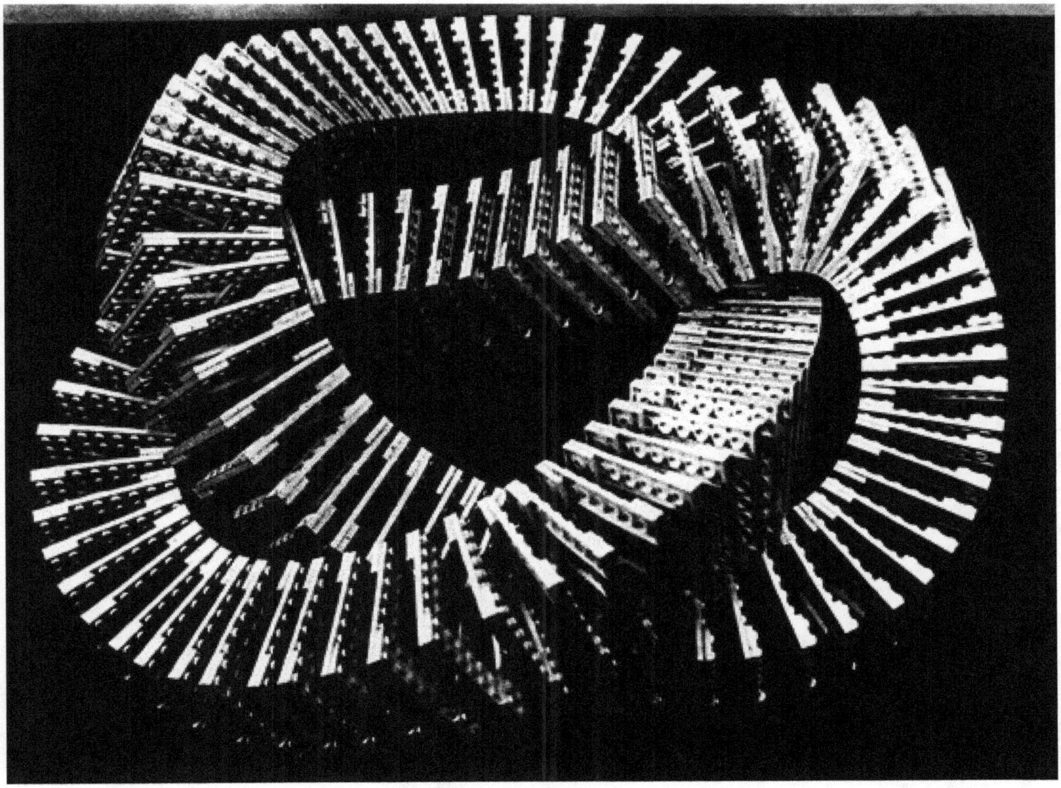

This LEGO creation [B280] demonstrates that left- or right-handedness is not intrinsic but depends on the scale on which an object is viewed — here the basic square spiral, the Möbius quarter-twist, or the trefoil knot.

www.ingramcontent.com/pod-product-compliance
Lightning Source LLC
Chambersburg PA
CBHW081209220326
41598CB00037B/6721
* 9 7 8 9 8 1 3 2 2 1 2 0 8 *